GATE PREP SERIES

GATE

Graduate Aptitude Test in Engineering

2022

General Aptitude & Engineering Mathematics

G. K. PUBLICATIONS (P) LTD.

Title : GATE 2022 : General Aptitude & Engineering Mathematics - Guide

Language : English

Editor's Name : Vinit Garg

Typeset & Published by :

Career Launcher Infrastructure (P) Ltd.

A-45, Mohan Cooperative Industrial Area, Near Mohan Estate Metro Station, New Delhi - 110044

Marketed by :

G.K. Publications (P) Ltd.

Plot No. 9A, Sector-27A, Mathura Road, Faridabad, Haryana-121003

ISBN : **978-93-90820-78-8**

Printer's Details : Made in India, New Delhi.

For product information :

Visit ***www.gkpublications.com*** or email to ***gkp@gkpublications.com***

GATE
General Aptitude
&
Engineering Mathematics

Contents

Analytical Aptitude

Spatial Aptitude

Quantitative Aptitude

Engineering Mathematics

GATE
Graduate Aptitude Test in Engineering

IIT Institutes

GATE 2021 conducted by
Indian Institute of Technology, Mumbai

GATE 2022 will be conducted by
Indian Institute of Technology, Kharagpur

Indian Institute of Technology, Delhi

Indian Institute of Technology, Chennai

Indian Institute of Technology, Guwahati

Indian Institute of Technology, Roorkee

Indian Institute of Technology, Kanpur

Preface

The Graduate Aptitude Test in Engineering (GATE) is the largest online Engineering Entrance exam in India for the PG courses in Engineering. Admissions to the IITs, IISc Bangalore, NITs and many state run universities as well as private universities and institutes in the country are based on the GATE scores. There are more than 37 PSUs that use GATE score for recruitments. A large number of corporates are also using GATE score as a tool to screen students for placements.

GK Publications is well known as the "publisher of choice" to students preparing for GATE and other technical test prep examinations in the country. We published first set of books in 1994 when GATE exam, both objective and conventional, was conducted in the paper and pencil environment, and used as a check point for entry to post graduate courses in IITs and IISCs. At that time, students had little access to technology and relied mainly on instructor led learning followed by practice with books available for these examinations.

A lot has changed since then!

Today, GATE is conducted in an online only mode with multiple choice and numerical based questions. The score is valid for three years and is used not only for post graduate courses but is also used by major PSUs for recruitment. Today's students have easy access to technology and the concept of a monologue within the classroom has changed to dialogue where students come prepared with concepts and then discuss topics. They learn a lot of things on the go with their mobile devices and practice for mock tests online.

We, as a leading publisher of GATE books, have also embraced change. Today, our books are no more guides only but come with a fully supported mobile app and a web portal. The mobile App provides access to video lectures, short tests and regular updates about the exam. The web portal in additional to what is available on the App provides full length mock tests to mimic the actual exam and help you gauge your level of preparedness. The combination of practice content in print, video lectures, and short and full length tests on mobile and web makes this product a complete courseware for GATE preparation.

We also know that improvement is a never ending process and hence we welcome your suggestions and feedback or spelling technical errors if any. Please write to us at gkp@gkpublications.com

We hope that our small effort will help you prepare well for the examination.

We wish you all the best!

GK Publications Pvt. Ltd.

About GATE

The Graduate Aptitude Test in Engineering (GATE) conducted by IISc and IITs has emerged as one of the bench mark tests for engineering and science aptitude in facilitating admissions for higher education (M.Tech./Ph.D.) in IITs, IISc and various other Institutes/Universities/Laboratories in India. With the standard and high quality of the GATE examination in 27 disciplines of engineering and science, Humanities and Social Sciences subjects, it identifies the candidate's understanding of a subject and aptitude and eligibility for higher studies. During the last few years, GATE score is also being used as one of the criteria for recruitment in Government Organizations such as Cabinet Secretariat, and National/State Public Sector Undertakings in India. Because of the importance of the GATE examination, the number of candidates taking up GATE exams has increased tremendously. GATE exams are conducted by the IITs and IISc as a computer based test having multiple choice questions and numerical answer type questions. The questions are mostly fundamental, concept based and thought provoking. From 2017 onwards GATE Exam is being held in Bangladesh, Ethiopia, Nepal, Singapore, Sri Lanka and United Arab Emirates. An Institute with various nationalities in its campus widens the horizons of an academic environment. A foreign student brings with him/her a great diversity, culture and wisdom to share. Many GATE qualified candidates are paid scholarships/assistantship, especially funded by Ministry of Human Resources Development, Government of India and by other Ministries. IIT, Kharagpur is the Organizing Institute for GATE 2022.

Why GATE?

Admission to Post Graduate and Doctoral Programmes

Admission to postgraduate programmes with MHRD and some other government scholarships/ assistantships in engineering colleges/institutes is open to those who qualify through GATE. GATE qualified candidates with Bachelor's degree in Engineering/Technology/Architecture or Master's degree in any branch of Science/Mathematics/Statistics/Computer Applications are eligible for admission to Master/Doctoral programmes in Engineering/Technology/Architecture as well as for Doctoral programmes in relevant branches of Science with MHRD or other government scholarships/ assistantships. Candidates with Master's degree in Engineering/Technology/Architecture may seek admission to relevant Ph.D programmes with scholarship/assistantship without appearing in the GATE examination.

Financial Assistance

A valid GATE score is essential for obtaining financial assistance during Master's programs and direct Doctoral programs in Engineering/Technology/Architecture, and Doctoral programs in relevant branches of Science in Institutes supported by the MHRD or other Government agencies. As per the directives of the MHRD, the following procedure is to be adopted for admission to the post-graduate programs (Master's and Doctoral) with MHRD scholarship/assistantship. Depending upon the norms adopted by a specific institute or department of the Institute, a candidate may be admitted directly into a course based on his/her performance in GATE only **or** based on his/her performance in GATE **and** an admission test/interview conducted by the department to which he/she has applied **and/or** the candidate's academic record. If the candidate is to be selected through test/interview for post-graduate programs, a minimum of 70% weightage will be given to the performance in GATE and the remaining 30% weightage will be given to the candidate's performance in test/interview and/or academic record, as per MHRD guidelines. The admitting institutes could however prescribe a minimum passing percentage of marks in the test/interview. Some colleges/institutes specify GATE qualification as the mandatory requirement even for admission without MHRD scholarship/assistantship.

To avail of the financial assistance (scholarship), the candidate must first secure admission to a program in these Institutes, by a procedure that could vary from institute to institute. Qualification in GATE is also a minimum requirement to apply for various fellowships awarded by many Government organizations. Candidates are advised to seek complete details of admission procedures and availability of MHRD scholarship/assistantship from the concerned admitting institution. The criteria for postgraduate admission with scholarship/assistantship could be different for different institutions. The management of the post-graduate scholarship/assistantship is also the responsibility of the admitting institution. Similarly, reservation of seats under different categories is as per the policies and norms prevailing at the admitting institution and Government of India rules. *GATE offices will usually not entertain any enquiry about admission, reservation of seats and / or award of scholarship / assistantship.*

PSU Recruitments

As many as 37 PSUs are using GATE score for recruitment. It is likely that more number of PSUs may start doing so by next year. Below is the list of PSUs:

MDL, BPCL, GAIL, NLC LTD, CEL, Indian Oil, HPCL, NBPC, NECC, BHEL, WBSEDCL, NTPC, ONGC, Oil India, Power Grid, Cabinet Secretariat, Govt. of India, BAARC, NFL, IPR, PSPCL, PSTCL, DRDO, OPGC Ltd., THDC India Ltd., BBNL, RITES, IRCON, GHECL, NHAI, KRIBHCO, Mumbai Railway Vikas Corporation Ltd. (MRVC Ltd.), National Textile Corporation, Coal India Ltd., BNPM, AAI, NALCO, EdCIL India.

Important :

1. Admissions in IITs/IISc or other Institutes for M.Tech./Ph.D. through GATE scores shall be advertised separately by the Institutes and GATE does not take the responsibility of admissions.
2. Cabinet Secretariat has decided to recruit officers for the post of Senior Field Officer (Tele) (From GATE papers of EC, CS, PH), Senior Research Officer (Crypto) (From GATE papers of EC, CS, MA), Senior Research Officer (S&T) (From GATE papers EC, CS, CY, PH, AE, BT) in the Telecommunication Cadre, Cryptographic Cadre and Science & Technology Unit respectively of Cabinet Secretariat. The details of the scheme of recruitment shall be published in National Newspaper/Employment News by the concerned authority.
3. Some PSUs in India have expressed their interest to utilize GATE scores for their recruitment purpose. The Organizations who intend to utilize GATE scores shall make separate advertisement for this purpose in Newspapers etc.

Who Can Appear for GATE?

Eligibility for GATE

Before starting the application process, the candidate must ensure that he/she meets the eligibility criteria of GATE given in Table.

Eligibility Criteria for GATE 2022

Degree/Program	Qualifying Degree/Examination	Description of Eligible Candidates	Expected Year of Completion
B.E. / B.Tech. / B.Pharm.	Bachelor's degree in Engineering / Technology (4 years after 10+2 or 3 years after B.Sc. / Diploma in Engineering / Technology)	Currently in the 3rd year or higher grade or already completed	2022
B. Arch.	Bachelor's degree of Architecture (5- year course) / Naval Architecture (4- year course) / Planning (4- year course)	Currently in the 3rd year or higher grade or already completed	2023 (for 5-year program), 2022 (for 4-year program)
B.Sc. (Research) / B.S.	Bachelor's degree in Science (Post-Diploma/4 years after 10+2)	Currently in the 3rd year or higher grade or already completed	2022
Pharm. D. (after 10+2)	6 years degree program, consisting of internship or residency training, during third year onwards	Currently in the 3rd/4th/5th/6th year or already completed	2024
M.B.B.S.	Degree holders of M.B.B.S. and those who are in the 5th/6th/7th semester or higher semester of such programme.	5th, 6th, 7th or higher semester or already completed	2022
M. Sc. / M.A. / MCA or equivalent	Master's degree in any branch of Arts/Science/Mathematics/Statistics/ Computer Applications or equivalent	Currently in the first year or higher or already Completed	2022
Int. M.E./ M.Tech. (Post-B.Sc.)	Post-B.Sc Integrated Master's degree programs in Engineering/ Technology (4-year program)	Currently in the 1st / 2nd/3rd/4th year or already completed	2024
Int. M.E./ M.Tech. or Dual Degree (after Diploma or 10+2)	Integrated Master's degree program or Dual Degree program in Engineering/Technology (5-year program)	Currently in the 3rd /4th/5th year or alreadycompleted	2023
B.Sc. / B.A. / B.Com.	Bachelor degree in any branch of Science / Arts / Commerce (3 years program)	Currently in the 3rd year or already completed	2021
Int. M.Sc. / Int. B.S. / M.S.	Integrated M.Sc. or 5-year integrated B.S.-M.S. program	Currently in the 3rd year or higher or already completed	2022
Professional Society Examinations (equivalent to B.E. / B.Tech. / B.Arch.)	B.E./B.Tech./B.Arch. equivalent examinations of Professional Societies, recognized by MoE/UPSC/AICTE (e.g. AMIE by Institution of Engineers-India, AMICE by the Institute of Civil Engineers-India and so on)	Completed Section A or equivalent of such professional courses	NA

In case a candidate has passed one of the qualifying examinations as mentioned above in 2020 or earlier, the candidate has to submit the degree certificate / provisional certificate / course completion certificate / professional certificate / membership certificate issued by the society or institute. In case, the candidate is expected to complete one of the qualifying criteria in 2022 or later as mentioned above, he/she has to submit a certificate from Principal or a copy of marks card for section A of AMIE.

Certificate From Principal

Candidates who have to submit a certificate from their college Principal have to obtain one from his/her institution beforehand and upload the same during the online submission of the application form.

Candidates With Backlogs

Candidates, who have appeared in the final semester/year exam in 2022, but with a backlog (arrears/failed subjects) in any of the papers in their qualifying degree should upload a copy of any one of the mark sheets of the final year,

OR

obtain a declaration from their Principal along with the signature and seal beforehand and upload the same during the online submission of the application form.

GATE Structure

Structure of GATE

GATE 2022 will be conducted on 27 subjects (papers). Table below shows the list of papers and paper codes for GATE 2022. A candidate is allowed to appear in ANY ONE or UP TO TWO papers of the GATE examination. However, note that the combination of TWO papers in which a candidate can appear MUST be from the pre-defined list as given in Table. Also note that for a paper running in multiple sessions, a candidate will be mapped to appear for the examination in one of the sessions ONLY.

List of GATE Papers and Corresponding Codes

GATE Paper	Code	GATE Paper	Code
Aerospace Engineering	AE	Instrumentation Engineering	IN
Agricultural Engineering	AG	Mathematics	MA
Architecture and Planning	AR	Mechanical Engineering	ME
Bio-medical Engineering	BM	Mining Engineering	MN
Biotechnology	BT	Metallurgical Engineering	MT
Civil Engineering	CE	Petroleum Engineering	PE
Chemical Engineering	CH	Physics	PH
Computer Science and Information Technology	CS	Production and Industrial Engineering	PI
Chemistry	CY	Statistics	ST
Electronics and Communication Engineering	EC	Textile Engineering and Fibre Science	TF
Electrical Engineering	EE	Engineering Sciences	XE*
Environmental Science & Engineering	ES	Humanities & Social Sciences	XH**
Ecology and Evolution	EY	Life Sciences	XL***
Geology and Geophysics	GG		

Note: *Environmental Science and Engineering (ES) and Humanities and Social Sciences (XH) are two new papers introduced in GATE 2022.*

*XE Paper Sections	Code	**XH Paper Sections	Code	***XL Paper Sections	Code
Engineering Mathematics (Compulsory) (15 marks)	A	Reasoning and Comprehension (Compulsory) (25 marks)	B1	Chemistry (Compulsory) (25 marks)	P
Any TWO optional Sections (2x35 = 70 marks)		Any ONE optional Section (60 marks)		Any TWO optional Sections (2x30 = 60 marks)	
Fluid Mechanics	B	Economics	C1	Biochemistry	Q
Materials Science	C	English	C2	Botany	R
Solid Mechanics	D	Linguistics	C3	Microbiology	S
Thermodynamics	E	Philosophy	C4	Zoology	T
Polymer Science and Engineering	F	Psychology	C5	Food Technology	U
Food Technology	G	Sociology	C6		
Atmospheric and Oceanic Sciences	H				

*XE (Engineering Sciences), **XH (Humanities & Social Sciences), ***XL (Life Sciences), papers are of general nature and will be comprised of Sections listed in the above table

Note: *Each subject/paper is of total 100 marks. General Aptitude (GA) section of 15 marks is common for all papers. Hence remaining 85 marks are for the respective subject/paper code.*

Combination of Two Papers Allowed to Appear in GATE 2022 (subject to availability of infrastructure and schedule)

Code of The First (Primary) Paper	Codes of Papers Allowed as The Second Paper
AE	XE
AG	ES
AR	CE
BM	BT / XL
BT	BM / XL
CE	AR / ES
CH	CY / PE / XE
CS	MA
CY	CH / XL
EC	IN / PH
EE	IN
ES	AG / CE
EY	XL
GG	MN / PE / PH
IN	EC / EE / PH
MA	CS / PH / ST
ME	XE
MN	GG / XE
MT	PH / XE
PE	CH / GG / XE
PH	EC / GG / IN / MA / MT / ST
PI	XE
ST	MA / PH
TF	XE
XE	AE / CH / ME / MN / MT / PE / PI / TF
XH	----
XL	BM / BT / CY / EY

General Aptitude Questions

All the papers will have a few questions that test the General Aptitude (Language and Analytical Skills), apart from the core subject of the paper.

Duration and Examination Type

All the papers of the GATE 2022 examination will be for 3 hours duration and they consist of 65 questions for a total of 100 marks. Since the examination is an ONLINE computer based test (CBT), at the end of the stipulated time (3-hours), the computer screen will automatically close the examination inhibiting any further action.

Candidates will be permitted to occupy their allotted seats 40 minutes before the scheduled start of the examination. Candidates can login and start reading the instructions 20 minutes before the start of examination. The late login time (if any) recorded by the computer system MUST NOT be beyond 30 minutes from the actual starting time of the examination. Under NO circumstances, will a candidate be permitted to login after 30 minutes from the actual examination starting time. Candidates will NOT be permitted to leave the examination hall before the end of the examination.

Pattern of Question Papers

GATE 2022 may contain questions of THREE different types in all the papers:

(i) **Multiple Choice Questions (MCQ)** carrying 1 or 2 marks each, in all the papers and sections. These questions are objective in nature, and each will have choice of four answers, out of which ONLY ONE choice is correct.

Negative Marking for Wrong Answers: For a wrong answer chosen in a MCQ, there will be negative marking. For 1-mark MCQ, 1/3 mark will be deducted for a wrong answer. Likewise, for 2-mark MCQ, 2/3 mark will be deducted for a wrong answer.

(ii) **Multiple Select Questions (MSQ)** carrying 1 or 2 marks each in all the papers and sections. These questions are objective in nature, and each will have choice of four answers, out of which ONE or MORE than ONE choice(s) is / are correct.

Note: There is NO negative marking for a wrong answer in MSQ questions. However, there is NO partial credit for choosing partially correct combinations of choices or any single wrong choice.

(iii) **Numerical Answer Type (NAT)** Questions carrying 1 or 2 marks each in most of the papers and sections. For these questions, the answer is a signed real number, which needs to be entered by the candidate using the virtual numeric keypad on the monitor (keyboard of the computer will be disabled). No choices will be shown for these types of questions. The answer can be a number such as 10 or –10 (an integer only). The answer may be in decimals as well, for example, 10.1 (one decimal) or 10.01 (two decimals) or –10.001 (three decimals). These questions will be mentioned with, up to which decimal places, the candidates need to present the answer. Also, for some NAT type problems an appropriate range will be considered while evaluating these questions so that the candidate is not unduly penalized due to the usual round-off errors. Candidates are advised to do the rounding off at the end of the calculation (not in between steps). Wherever required and possible, it is better to give NAT answer up to a maximum of three decimal places.

Note: *There is NO negative marking for a wrong answer in NAT questions.*
Also, there is NO partial credit in NAT questions.

Marking Scheme – Distribution of Marks and Questions

General Aptitude (GA) Questions

In all papers, GA questions carry a total of 15 marks. The GA section includes 5 questions carrying 1-mark each (sub-total 5 marks) and 5 questions carrying 2-marks each (sub-total 10 marks).

Question Papers other than GG, XE, XH and XL

These papers would contain 25 questions carrying 1-mark each (sub-total 25 marks) and 30 questions carrying 2-marks each (sub-total 60 marks) consisting of some MCQ type questions, while the remaining may be MSQ and / or NAT questions.

GG (Geology and Geophysics) Paper

Apart from the General Aptitude (GA) section, the GG question paper consists of two parts: Part A and Part B. Part A is compulsory for all the candidates. Part B contains two sections: Section 1 (Geology) and Section 2 (Geophysics). Candidates will have to attempt questions in Part A and questions in either Section 1 or Section 2 of Part B. The choice of Section 1 OR Section 2 of Part B has to be made at the time of online registration in GOAPS. At the examination hall, candidate cannot request for change of section.

Part A consists of 25 questions carrying 1-mark each (sub-total 25 marks and some of these will be MSQ and/or numerical answer type questions while remaining questions will be MCQ type). Either section of Part B (Section 1 and Section 2) consists of 30 questions carrying 2-marks each (sub-total 60 marks and some of these will be MSQ and/or numerical answer type questions while remaining questions will be MCQ type).

XE Paper (Engineering Sciences)

A candidate appearing in the XE paper has to answer the following:

- GA – General Aptitude carrying a total of 15 marks.
- Section A – Engineering Mathematics (Compulsory): This section contains 11 questions carrying a total of 15 marks: 7 questions carrying 1-mark each (sub-total 7 marks), and 4 questions carrying 2-marks each (sub-total 8 marks). Some questions will be of numerical answer type while remaining questions will be MCQ type.
- Any two of XE Sections B to H: The choice of two sections from B to H can be made during the examination after viewing the questions. Only TWO optional sections can be answered at a time. A candidate wishing to change midway of the examination to another optional section must first choose to deselect one of the previously chosen optional sections (B to H). Each of the optional sections of the XE paper (Sections B through H) contains 22 questions carrying a total of 35 marks: 9 questions carrying 1-mark each (sub-total 9 marks) and 13 questions carrying 2-marks each (sub-total 26 marks). Some questions will be of MSQ and/or numerical answer type while remaining questions will be MCQ type.

XH Paper (Humanities and Social Sciences)

A candidate appearing in the XH paper has to answer the following:

- **GA - General Aptitude** carrying a total of 15 marks.
- **Section B1 – Reasoning and Comprehension (Compulsory):** This section contains 15 questions carrying a total of 25 marks: 5 questions carrying 1-mark each (sub-total 5 marks) and 10 questions carrying 2-marks each (sub-total 20 marks). Some questions will be of MSQ and/or numerical answer type while remaining questions will be MCQ type.
- **Any ONE of XH Sections C1 to C6:** The ONE choice of section from C1 to C6 has to be made at the time of online registration in GOAPS. At the examination hall, candidate cannot request for change of section. Each of the optional sections of the XH paper (Sections C1 through C6) contains 40 questions carrying a total of 60 marks: 20 questions carrying 1-mark each (sub-total 20 marks) and 20 questions carrying 2-marks each (sub-total 40 marks). Some questions will be of MSQ and/or numerical answer type while remaining questions will be MCQ type.

XL Paper (Life Sciences)

A candidate appearing in the XL paper has to answer the following:

- **GA – General Aptitude** carrying a total of 15 marks.
- **Section P–Chemistry (Compulsory):** This section contains 15 questions carrying a total of 25 marks: 5 questions carrying 1-mark each (sub-total 5 marks) and 10 questions carrying 2-marks each (sub-total 20 marks). Some questions will be of MSQ and/or numerical answer type while remaining questions will be MCQ type.
- **Any two of XL Sections Q to U:** The choice of two sections from Q to U can be made during the examination after viewing the questions. Only TWO optional sections can be answered at a time. A candidate wishing to change midway of the examination to another optional section must first choose to deselect one of the previously chosen optional sections (Q to U). Each of the optional sections of the XL paper (Sections Q through U) contains 20 questions carrying a total of 30 marks: 10 questions carrying 1-mark each (sub-total 10 marks) and 10 questions carrying 2-marks each (sub-total 20 marks). Some questions will be of MSQ and/or numerical answer type while remaining questions will be MCQ type.

GATE Score

After the evaluation of the answers, the actual (raw) marks obtained by a candidate will be considered for computing the GATE Score. For multi-session papers (subjects), raw marks obtained by the candidates in different sessions will be converted to Normalized marks for that particular subject. Thus, raw marks (for single session papers) or normalized marks (for multi-session papers) will be used for computing the GATE Score, based on the qualifying marks.

Calculation of Normalized Marks for Multi-Session Papers

In GATE 2022 examination, some papers may be conducted in multi-sessions. Hence, for these papers, a suitable normalization is applied to take into account any variation in the difficulty levels of the question papers across different sessions. The normalization is done based on the fundamental assumption that "in all multi-session GATE papers, the distribution of abilities of candidates is the same across all the sessions". This assumption is justified since the number of candidates appearing in multi-session papers in GATE 2022 is large and the procedure for allocation of session to candidates is random. Further, it is also ensured that for the same multi-session paper, the number of candidates allotted in each session is of the same order of magnitude.

Based on the above, and considering various normalization methods, the committee arrived at the following formula for calculating the normalized marks for the multi-session papers.

Nsormalization mark of jth candidate in the ith session $\widehat{M_{ij}}$ is given by

$$\widehat{M_{ij}} = \frac{\bar{M}_t^g - M_q^g}{\bar{M}_{ti} - M_{iq}} (M_{ij} - M_{iq}) + M_q^g$$

where

M_{ij} : is the actual marks obtained by the jth candidate in i^{th} session

$\bar{M}_t^g$: is the average marks of the top 0.1% of the candidates considering all sessions

M_q^g : is the sum of mean and standard deviation marks of the candidates in the paper considering all sessions

$\bar{M}_{ti}$: is the average marks of the top 0.1% of the candidates in the ith session

M_{iq} : is the sum of the mean marks and standard deviation of the ith session

Calculation of GATE Score for All Papers

For all papers for which there is only one session, actual marks obtained by the candidates will be used for calculating the GATE 2022 Score. For papers in multi-sessions, normalized marks will be calculated corresponding to the raw marks obtained by a candidate and the GATE 2022 Score will be calculated based on the normalized marks.

The GATE 2022 score will be computed using the formula given below.

$$\text{GATE Score} = S_q + (S_t - S_q)\frac{(M - M_q)}{(\bar{M}_t - M_q)}$$

where

- M : marks obtained by the candidate (actual marks for single session papers and normalized marks for multi-session papers)
- M_q : is the qualifying marks for general category candidate in the paper
- $\bar{M}_t$: is the mean of marks of top 0.1% or top 10 (whichever is larger) of the candidates who appeared in the paper (in case of multi-session papers including all sessions)
- S_q : 350, is the score assigned to M_q
- S_t : 900, is the score assigned to $\bar{M}_t$

In the GATE 2022 the qualifying marks (M_q) for general category student in each subject will be 25 marks (out of 100) or $\mu + \sigma$, whichever is larger. Here μ is the mean and σ is the standard deviation of marks of all the candidates who appeared in the paper.

After the declaration of results, GATE Scorecards can be downloaded by the GATE qualified candidates ONLY.

The GATE 2022 Committee has the authority to decide the qualifying mark/score for each GATE paper. In case of any claim or dispute with respect to GATE 2022 examination or score, the Courts and Tribunals in Mumbai alone will have the exclusive jurisdiction to entertain and settle them.

GATE Syllabus

General Aptitude

Verbal Aptitude

Basic English Grammar: tenses, articles, adjectives, prepositions, conjunctions, verb-noun agreement, and other parts of speech.

Basic Vocabulary: words, idioms, and phrases in context, Reading and comprehension, Narrative sequencing.

Quantitative Aptitude

Data Interpretation: data graphs (bar graphs, pie charts, and other graphs representing data), 2-and 3-dimensional plots, maps, and tables.

Numerical Computation and Estimation: ratios, percentages, powers, exponents and logarithms, permutations and combinations, and series, Mensuration and geometry, Elementary statistics and probability.

Analytical Aptitude

Logic: Deduction and induction, Analogy, Numerical relations and reasoning.

Spatial Aptitude

Transformation of shapes: translation, rotation, scaling, mirroring, assembling, and grouping Paper folding, cutting, and patterns in 2 and 3 dimensions.

Engineering Mathematics

Linear Algebra: Matrix algebra, Systems of linear equations, Eigen values and Eigen vectors.

Calculus: Functions of single variable, Limit, continuity and differentiability, Mean value theorems, Evaluation of definite and improper integrals, Partial derivatives, Total derivative, Maxima and minima, Gradient, Divergence and Curl, Vector identities, Directional derivatives, Line, Surface and Volume integrals, Stokes, Gauss and Green's theorems.

Differential Equations: First order equations (linear and nonlinear), Higher order linear Differential Equations with constant coefficients, Cauchy's and Euler's equations, Initial and boundary value problems, Laplace transforms, Solutions of one dimensional heat and wave equations and Laplace equation.

Ordinary Differential Equation (ODE): First order (linear and non-linear) equations; higher order linear equations with constant coefficients; Euler-Cauchy equations; Laplace transform and its application in solving linear ODEs; initial and boundary value problems.

Partial Differential Equation (PDE): Fourier series; separation of variables; solutions of one-dimensional diffusion equation; first and second order one-dimensional wave equation and two-dimensional Laplace equation.

Complex Analysis: Analytic functions, Cauchy's integral theorem, Cauchy's integral formula; Taylor's and Laurent's series, residue theorem, Sequence series and Conversion costs.

Probability and Statistics: Definitions of probability and sampling theorems, Conditional probability, Mean, median, mode and standard deviation, Random variables, Poisson, Normal and Binomial distributions.

Numerical Methods: Numerical solutions of linear and non-Linear Algebraic equations Integration by trapezoidal and Simpson's rule, single and multi-step methods for Differential Equations.

Discrete Mathematics: Propositional and first order logic. Sets, relations, functions, partial orders and lattices. Groups. Graphs: connectivity, matching, coloring. Combinatorics: counting, recurrence relations, generating functions, moniod.

Transform Theory: Fourier Transform, Laplace Transform, z-Transform.

Vector Analysis: Vectors in plane and space, vector operations, gradient, divergence and curl, Gauss's, Green's and Stoke's theorems.

GENERAL APTITUDE
Verbal Aptitude

1 CHAPTER English Grammar

ERRORS IN USE OF ARTICLES

ARTICLES

- Indefinite (A, An)
- Definite (The)

Use of '*An*'

1. Before words beginning with vowel sounds [a, e, i, o, u are called vowels, others are consonants].
   e.g. an apple, an egg, an owl.
2. Before words beginning with silent 'h' but sounds as vowel.
   e.g. an hour, an honourable man, an heir, an honest man.
3. F, H, L, M, N, R, S, X are letters that are not vowels but begin with vowel sound 'M' has the sound of 'em'. So, 'an' is used before abbreviations beginning with vowels of these letters.
   e.g. an M.L.A., an R.A.F., an N.C.C. officer, an F.I.R., an X-ray, an H.E. school, an S.P.

Use of '*A*'

1. In the sense of one.
   e.g. He couldn't speak a word to save himself.
   With 'one' (since 'one' begins with sound of 'w')
   e.g. a one-man show, a one-rupee note.
2. Before words beginning with consonant sound
   e.g. a boy, a box, a dog.
3. With Vowel letters having consonant value.
   e.g. a university, a unique article, a euphenism, a unit, a European language
4. With units and rate (per).
   e.g. He earns rupees five hundred a month.
5. In exclamatory expressions before singular countable nouns.
   *e.g. What **a** pretty girl !*
6. When two subjects or articles are thought of as a single unit.
   *e.g. He was ready with **a** cup and saucer.*
7. With certain expressions of quantity.
   *e.g. **a** lot of , **a** dozen, **a** great deal of, **a** couple.*
8. With a person's name to indicate that the person is perhaps unknown to the person addressed.
   e.g. A Mr. roy is at the door.
9. With a special meal (to celebrate something or in someone's honour).
   *e.g. I called my friends to **a** lunch to celebrate my success.*
10. To make a common noun of a proper noun.
    *e.g. This man is '**a** second Newton'.*
    (This phrase means 'a philosopher as great as Newton')

Use of '*The*'

1. When we speak of a particular person or thing already referred to.
   *e.g. I dislike **the** follow.*
2. When a singular noun represents a whole class.
   e.g. The mango is considered the king among fruits.
3. With name of
   (*i*) gulfs, rivers, oceans, islands and mountains e.g. the Himalayas, the Indian ocean, the Persian Gulf, the Red sea, the Andaman islands, the Brahmaputra river.
   (*ii*) Certain books
   e.g. the Vedas, the Puranas, the Bible, the Ramacharitmanas.
   (*iii*) Musical instruments
   e.g. the flute, the violin, the tabla, the trumpet.
   (*iv*) The inventions
   e.g. I hate the telephone for its constant ringing.
   (*v*) Parts of body
   *e.g. He was wounded in **the** leg.*
   (*vi*) Religious groups
   e.g. the Sikhs, the Hindus, the Parsees.
   (*vii*) Names enforcing law
   e.g. the Police, the Navy, the Air Force.
   (*viii*) Political parties
   e.g. the Congress, the Janata Pary, the B.J.P.
   (*ix*) Aeroplanes, ships, trains etc
   e.g. the Makalu (aeroplane), the Vikrant (Ship), the Rajdhani express (train).
   (*x*) Before names of an empire, dynasty or historical event
   e.g. the Gupta dynastry, the Old Stone Age, the First World War, the American Revolution.
   (*xi*) Clubs, foundations etc.
   e.g. the Lion's Club, the Ford Foundation.
   (*xii*) Before common nouns denoting unique things
   e.g. the sun, the sky, the earth, the worl.
   (*xiii*) With superlatives
   e.g. *He is **the** best boy in the class.*
   (*xiv*) With ordinals
   *e.g. He took **the** first taxi that came his way.*
   (*xv*) Before the comparative degree
   e.g. ***The more** they get, **the** more they want.*
   (*xvi*) Before an adjective when the noun is understood
   *e.g. **The** poor would favour him.*

ERRORS IN USE OF NOUNS

1. Nouns which are used in singular form

(i) Scenery, information, furniture, advice, machinery, stationery, news, poetry, business, mischief, fuel, issue, repair, bedding.

e.g. *The **scenery** of this place is worth seeing.*

(ii) Physics, Mathematics, Economics, Classics, Ethics, Athletics, innings, gallows.

e.g. ***Economics** is a very interesting subject.*

(iii) Brick, bread, fruit, word (as 'promise')

e.g. *Let me buy some **fruit**.*

(iv) Words like dozen, score, hundred, thousand, million when preceded by a numeral.

e.g. *He bought ten **dozen** oranges.*

(v) Expressions as a ten-rupee note, a two-hour journey, a four-mile walk, a Five-year plan, a six-man committee etc.

e.g. *A ten-rupee note is lying there.*

2. Nouns used only in plural form

(i) Cattle, police, poultry, people, gentry, peasantry, artillery.

e.g. *The **police have** caught the thief.*

(ii) Scissors, trousers, stockings, spectacles, shorts, alms, remains, riches, goods, measles.

e.g. *My scissors are very sharp.*

3. Nouns used both as singular and plural in the same form

(i) Dear, sheep, fish, apparatus, wages.

e.g. *The **wages** of sin **is** death.*

*The **wages** of the workers **have** been raised.*

*I saw **a sheep** grazing in the field.*

***Sheep are** sold cheaper than goat.*

(ii) Collective nouns as jury, public, team, audience, committee, government, audience, congregation, orchestra.

e.g. *The **team are** looking quite fit.*

*The **team has** not turned up yet.*

4. Use of Collective Nouns

- **crowd** of people
- **herd** of cattle/cows
- **team** of players
- **flight** of birds
- **bouquet** of flowers
- **Shoal/school** of fish
- **bundle** of sticks
- **army** of soldiers
- **flock** of sheep
- **crew** of sailors
- **swarm** of bees
- **garland** of beads
- **gang** of thieves
- **library** of books
- **mob** of angry people
- **fleet** of ships/cruisers
- **pack** of wolves
- **pack** of cards
- **loaf** of bread
- **heap** of corn
- **stock** of grain
- **class** of pupils
- **bunch** of grapes
- **herd** of lions
- **pile** of books
- **hoarde** of nomads
- **block** of flats
- **file** of papers

5. **One of** or **any of** is followed by plural words.

e.g. *I want **one of** the **books** kept on the table.*

***Any of** these **tools** may serve the purpose.*

6. Plural nouns are used with fractions and decimal over 1.

e.g. *It took us one and a half **hours**.*

ERRORS IN USE OF PRONOUNS

Pronouns are words used to replace nouns or noun groups already mentioned.

Nominative/ Vocative case (comes before verb)	***Accurative/ Objective case (comes after verb)***	***Possessive case***	***Reflexive Pronoun***
I	Me	My	Myself
We	Us	Our	Ourselves
He	Him	His	Himself
She	Her	Her	Herself
They	Them	Their	Themselves
Who	Whom	Whose	—

1. The pronoun '*One*' must be followed by '*one's*'.

e.g. *One must do one's duty to one's country.*

2. When 'one' means 'one in number', the pronoun for it is third person singular pronoun (he, she, it).

e.g. *One of them has given up one's studies.* (×)

One of them has given up his studies. (✓)

3. 'Everyone' or 'Everybody' must be followed by 'his'.

e.g. *Everyone should love his country.*

4. Each, every, anyone, anybody must be followed by the singular pronoun of their person.

e.g. *Anyone can do this if he tries.*

5. 'Let' is followed by pronoun in the objective case.

e.g. *Let him go.*

6. 'But' and 'except' are followed by pronoun in the objective case.

e.g. *Everyone attended the party except him.*

'such as' is followed by pronoun in the subjective case.

e.g. *I have no liking for such a man as he.*

7. Verbs like enjoy, avail, pride, resign, apply, acquit, assert, absent are followed by reflexive pronouns.

e.g. *He absented himself from the class.*

We enjoyed ourselves at the party.

8. Reflexive pronouns are never used with verbs keep, conceal, quality, spread, rest, stay.
   e.g. I stayed away from my class.
   He qualified in the test.
9. When first, second and third person singular pronouns (I, you and He) are used together, they are placed in the order : You, he and I.
   e.g. You he and I are neighbours.
   In case of plural pronouns, 'we' comes first, then 'you' and then 'they'.
   e.g. We, you and they must work together.
   But if we have only two persons including first, then first person pronoun is written first.
   e.g. I and Sanjeev have done this job.
10. 'Who' denotes subject and 'whom' denotes object.
    e.g. Who do you think did the job ?
11. 'Whose' is used for persons and 'which' for lifeless objects.
    e.g. This is the table which I was talking about.
12. 'Which' conveys additional information and 'that' explains a certain thing.
    e.g. *I will tell you the first thing which I remember.*
13. The following expressions usually take 'that' in place of 'who' or 'which'.
    Only, Any, It is, All, Superlatives.
    e.g. He is the only man that can do it.
    Any man that listens to you is a fool.
14. *'Each other'* is used for two; *'one another'* for more than two.
    e.g. Rahul and Renu love each other.
15. The complement of the verb *to be*, when it is expressed by a pronoun, should be in Nominative case.
    e.g. It was he who did it.
16. When the same person is the subject and object, it is necessary to use reflexive pronouns.
    e.g. I cut me shaving this morning. (×)
    I cut myself shaving this morning. (✓)
17. When a pronoun is the object of a verb or preposition it should be in objective case.
    e.g. These books are for you and I. (×)
    These books are for you and me. (✓)
    Between him and me there is an understanding.
18. The Relative Pronoun should be placed as near as possible to the antecedent.
    e.g. I have read Shakespeare's works who was a great dramatist. (×)
    I have read the works of Shakespeare who was a great dramatist.(✓)
19. The case of the pronoun following 'than' and 'as' is decided by mentally supplying the verb and completing the sentence.
    e.g. She is taller than I (am)
    I love more than (I love) him.

PARTS OF SPEECH

The **parts of speech** explain how a word is used in a sentence.

There are eight main parts of speech (also known as word classes): **nouns, pronouns, adjectives, verbs, adverbs, prepositions, conjunctions** and **interjections.**

Most **parts of speech** can be divided into sub-classes. Prepositions can be divided into prepositions of time, prepositions of place etc. Nouns can be divided into proper nouns, common nouns, concrete nouns etc.

It is important to know that a word can sometimes be in more than one part of speech. For example with the word *increase*.

Increase can be a verb e.g. Prices *increased*

and *increase* can also be a noun e.g. There was an *increase* in the number of followers.

The eight main parts of speech in English are:

1. NOUN

A noun is the name of a person, place, thing, or idea.

man... Butte College... house... happiness

A noun is a word for a person, place, thing, or idea. Nouns are often used with an article (*the*, *a*, *an*), but not always. Proper nouns always start with a capital letter; common nouns do not. Nouns can be singular or plural, concrete or abstract. Nouns show possession by adding *'s*. Nouns can function in different roles within a sentence; for example, a noun can be a subject, direct object, indirect object, subject complement, or object of a preposition.

Examples of nouns: Daniel, London, table, dog, teacher, pen, city, happiness, hope

Example sentences:

- **Steve** lives in **Sydney**. **Mary** uses **pen** and **paper** to write **letters**.
- The young **girl** brought me a very long **letter** from the **teacher**, and then she quickly disappeared. Oh my!

2. PRONOUN

A pronoun is a word used in place of a noun.

She... we... they... it

A pronoun is a word used in place of a noun. A pronoun is usually substituted for a specific noun, which is called its antecedent. In the sentence above, the antecedent for the pronoun *she* is the girl. Pronouns are further defined by type: personal pronouns refer to specific persons or things; possessive pronouns indicate ownership; reflexive pronouns are used to emphasize another noun or pronoun; relative pronouns introduce a subordinate clause; and demonstrative pronouns identify, point to, or refer to nouns.

Examples of pronouns: I, you, we, they, he, she, it, me, us, them, him, her, this, those

Example sentences:

- Mary is tired. **She** wants to sleep. **I** want **her** to dance with **me**.
- The young girl brought **me** a very long letter from the teacher, and then **she** quickly disappeared. Oh my!

3. VERB

A verb expresses action or being.

jump... is... write... become

The verb in a sentence expresses action or being. There is a main verb and sometimes one or more helping verbs. (*"She can sing." Sing* is the main verb; *can* is the helping verb.) A verb must agree with its subject in number (both are singular or both are plural). Verbs also take different forms to express tense.

Examples: go, speak, run, eat, play, live, walk, have, like, are, is

Example sentences:

- I **like** Woodward English. I **study** their charts and **play** their games.
- The young girl **brought** me a very long letter from the teacher, and then she quickly **disappeared**. Oh my!

4. ADJECTIVE

An adjective modifies or describes a noun or pronoun.

pretty... old... blue... smart

An adjective is a word used to modify or describe a noun or a pronoun. It usually answers the question of which one, what kind, or how many. (Articles [a, an, the] are usually classified as adjectives.)

Examples: big, happy, green, young, fun, crazy, three

Example sentences:

- The **little** girl had a **pink** hat.
- The **young** girl brought me a very **long** letter from the teacher, and then she quickly disappeared. Oh my!

5. ADVERB

An adverb modifies or describes a verb, an adjective, or another adverb.

gently... extremely... carefully... well

An adverb describes or modifies a verb, an adjective, or another adverb, but never a noun. It usually answers the questions of when, where, how, why, under what conditions, or to what degree. Adverbs often end in -ly.

Examples: slowly, quietly, very, always, never, too, well, tomorrow, here

Example sentences:

- I am **usually** busy. **Yesterday**, I ate my lunch **quickly**.
- The young girl brought me a **very** long letter from the teacher, and **then** she **quickly** disappeared. Oh my!

6. PREPOSITION

A preposition is a word placed before a noun or pronoun to form a phrase modifying another word in the sentence.

by... with.... about... until

(by the tree, with our friends, about the book, until tomorrow)

A preposition is a word placed before a noun or pronoun to form a phrase modifying another word in the sentence. Therefore a preposition is always part of a prepositional phrase. The prepositional phrase almost always functions as an adjective or as an adverb. The following list includes the most common prepositions:

Examples: at, on, in, from, with, near, between, about, under

Example sentences:

- I left my keys **on** the table **for** you.
- The young girl brought me a **very** long letter from the teacher, and **then** she **quickly** disappeared. Oh my!

7. CONJUNCTION

A conjunction joins words, phrases, or clauses.

and... but... or... while... because

A conjunction joins words, phrases, or clauses, and indicates the relationship between the elements joined. Coordinating conjunctions connect grammatically equal elements: and, but, or, nor, for, so, yet. Subordinating conjunctions connect clauses that are not equal: because, although, while, since, etc. There are other types of conjunctions as well.

Examples: and, or, but, because, so, yet, unless, since, if.

Example sentences:

- I was hot **and** exhausted **but** I still finished the marathon.
- The young girl brought me a very long letter from the teacher, **and** then she quickly disappeared. Oh my!

8. INTERJECTION

An interjection is a word used to express emotion.

Oh!... Wow!... Oops!

An interjection is a word used to express emotion. It is often followed by an exclamation point.

Examples: Ouch! Wow! Great! Help! Oh! Hey! Hi!

Example sentences:

- Wow! I passed my English test. Great! – Ouch! That hurt.
- The young girl brought me a very long letter from the teacher, and then she quickly disappeared. **Oh my**!

ERRORS IN USE OF PREPOSITIONS

A preposition is a word used with a noun or pronoun to show its relation to some other word in a sentence.

PREPOSITIONS OF TIME

1. **At** is used
   (*a*) with a definite point of time
   e.g. I usually get up at 5 o'clock.
   (*b*) with festivals
   e.g. He will come at Holi.
2. **In** is used :
   (*a*) with the parts of the day, (with **noon**, use **at**), months, seasons and years
   e.g. He takes a walk in the afternoon.
   (*b*) with the future tense referring to the period in which action may take place
   e.g. You must be careful in future.
3. **On** is used : with days and dates
   e.g. My brother will arrive on Monday.
4. **By** refers to the latest time at which an action will be over
   e.g. The examination will be over by 5 p.m.
5. **For** is used with perfect continuous tense showing the duration of action
   e.g. I have been here for three years.
6. **Since** is used with the point of time when action begins and continues :
   e.g. He has been ill since last Tuesday.
7. **From** refers to the starting point of action.
   e.g. This water came from a spring.

PREPOSITIONS OF POSITION

1. **At** refers to an exact point.
   e.g. The tourist stayed at the Tourist Hotel.
   He studied at Oxford.
2. **In** refers to larger areas.
   e.g. He lives in Bombay.
3. **Between** is used for two persons or two things.
   e.g. Share these sweets between him and me.
4. **Among** is used with more than two persons or things but before the word which starts with a consonant letter.
   e.g. Divide the sweets among the three boys.
5. **Amongst** is also used with more than two persons or things but before the word which starts with a vowel letter.
   e.g. Divide the sweets amongst us.
6. **Above** is used for **higher than.**
   e.g. The sun rose above the horizon.
7. **Under** is used for **vertically below.**
   e.g. It is shady under the trees.
8. **Below** is used for **lower than.**
   e.g. When the sun sets it goes below the horizon.
9. **Over** is used for **vertically above.**
   e.g. There is an aircraft coming over.
10. **Beneath** means a lower position.
   e.g. the ground was slippery beneath her.

PREPOSITIONS OF DIRECTION

1. **To** is used to express motion from one place to another.
   e.g. We walked to the river and back.
2. **Towards** refers to direction.
   e.g. He saw me running towards him.
3. **Into** denotes motion towards the inside of something.
   e.g. She fell into a ditch.
4. **At** refers to aim.
   e.g. He aimed at the bird.
5. **For** denotes direction.
   e.g. I will leave for Pune today.
6. **Against** shows pressure.
   e.g. She cleaned the edge of her knife against the plate.
7. **Off** refers to separation.
   e.g. He was wiping sweat off his face.
8. **From** refers to the point of departure.
   e.g. The man parted from his friends.

OTHER USES OF PREPOSITION

1. **About** shows nearness.
   e.g. His father is about to retire.
2. **Along** stands for **in the same line.**
   e.g. She led them along the corridor.
3. **After** refers to sequence.
   e.g. She came after me.
4. **Across** means from one side.
   e.g. He blew a cloud of smoke across the table.
5. **Before** stands for **in front of.**
   e.g. A thief was brought before the judge.
6. **Behind** means **at the the back of.**
   e.g. She sat down behind the hedge.
7. **Beyond** means **on the farther side of.**
   e.g. This is beyond his power.
8. **Beside** means **by the side of.**
   e.g. *I sat down beside my wife.*
9. **Besides** means **in addition to.**
   e.g. He is guilty of five killings and more besides.

WORDS FOLLOWED BY PREPOSITIONS

A

- abound with
- abstain from
- according to
- accordance with
- accuse of
- accustomed to
- acquaint with
- acquit of
- adverse to
- affiliate to/with
- afflict with
- afraid of
- agree to/on
- aim at
- allegation
- about
- allude to
- alternate with
- alternative to
- angry with
- apply to (body)
- appreciation of
- apprise of
- approximate of
- arraign against
- arrest in
- arrive at/in
- ashamed of
- aspire to/after/at
- associate with
- assure of
- attribute to

B, C

- beset with
- beware of
- blink at
- boast of
- break off
- bungle over
- careful of
- cause of
- centre on
- charge with
- clamour for
- clamp on
- coerce into
- collide with
- comment on
- compatible with
- complain of
- comply with
- compliment on
- composed of
- concur in
- condole with
- conducive to
- confer on
- confirmation of
- conform to
- congratulate on
- conscious of
- contiguity with
- converge on
- convict of
- co-opt to
- cope with
- correspond with
- converge on
- convict of
- co-opt to
- cope with
- correspond with
- cost of
- credit with
- criticism of
- cure of

D

- decide on
- debar from
- delight in
- deliberate on
- denude of
- depart to
- departure for
- depend on/upon
- deprive of
- derive from
- derogate from
- derogatory to
- desist from
- despair of
- detrimental to
- devoid of
- difficulty in
- diffident of
- disabuse of
- disagree with
- disgusted with
- dispose of
- dissociate from
- divest of
- divide into
- dressed in

E

- effort in (—ing)
- election to
- embroil in
- emphasis on
- employ in
- enamoured of
- encroach on
- endow with
- enlarge on
- engage in
- essential to
- expostulate with
- extend to

F, G, I

- fed up with
- focus on
- forbid to
- fraught with
- fritter in
- full of
- give to
- grapple with
- good at
- identical with
- impose on
- incur on
- indict for
- indifferent to
- indulge in
- infatuate with
- infect with
- infest with
- information on
- insensible to
- insist on
- interest in
- intimate to
- intoxicate with
- intrigue with
- intrude on
- investigation of
- involve in

J, K, L

- jealous of
- jeer at
- jump at
- just to
- knock at
- lash at
- lay siege to
- leave for
- live on (income)
- long for

O, P

- oblivious of
- operate on
- opinion on
- overcome with
- pay a call on
- persevere in
- persist in
- plan for
- plead for
- pleased with
- prefer to
- preference for
- present with
- prevent from
- priority to
- probe into
- prohibit from
- proud of

R, S

- recoil from
- reconcile to
- remand to
- remonstrate with
- repent of
- replete with
- reply to
- research on
- resolve on/to
- sail for
- satiate with
- satisfied with
- sentence to
- sick of
- side with
- sparing
- square with
- stickler for
- subversive of
- succeed in
- sue for
- superior to
- supply with
- sure of
- surprised at
- susceptible to
- suspect of
- sympathy for

T, U, V, W

- testify to
- thank for
- throw at
- tinker at
- tire of
- trace to
- trample on
- translate into
- try to (do)
- unequal to (the task)
- vie with
- vote for (a candidate)
- wait for (or await)
- ways of
- wish for

SOME SPECIAL CASES

- absolve from (blame) / of (sin)
- admit to (hospital, place, class)
- appear for (examination)/at
- begin on (not from Sunday)
- blow down (the roof)/off (steam)/over (pass off)
- buy in (market, store)/at (shop)
- come/go by (bus, train)
- come into (conflicit with)
- compare with (one man height with another's)/to (one man's height to a building)
- consist of (to denote the substance)/in (define the subject)
- deal with (somebody)/ in (goods)
- die of (disease)/from (other causes)
- differ from (to be different from)/with (to disagree)
- due to (used after a noun, not after a verb)
- entrust to (somebody a job)
- entrust (somebody) with a job)
- fill with (enthusiasm)/ in (a form)
- fire at (when not hit)/on (when injured to killed)
- glad of (the news)/at (having arrived safe)
- hostility to/in hand (being attended to)/ on hand (in possession of)
- ingratiate (oneself) with
- inquire into (matter) / of (person)/about. often (thing)
- invest with (power/in (blank)
- key of (the door)/to (a problem)
- level (charge) against.
- look at (consider evidence)/to (somebody for aid)
- meeting (called) for (Sunday)
- motive in (doing)/for (an act)
- neglect of (duty)/in (doing a thing)
- nominated to (board)
- noted for (good things)
- owing to (used after a verb, not after a noun)
- part from (people)/with (things)
- preventive for priority
- put out (light)/off (trip)
- Shoot at (when injured)/in (leg, head)
- take up (job)/to (a hobby)
- taste of (sensation)/for (liking)
- tear off (remove)/up (to pieces)
- vest with (The P.M. is vested with power to ...)/ in (Powers are vested in the P.M. to ...)

ERRORS IN USE OF CONJUNCTIONS

1. **Scarcely or hardly** is followed by **when**.
   e.g. Hardly had I slept when the telephone rang.
2. **Though** is followed by **yet**.
   e.g. Though he worked hard yet he failed.
3. **No sooner** is followed by **than.**
   e.g. No sooner did we reach there than it began to rain.
4. **Not only** is followed by **but also.**
   e.g. Not only did he help her, but also dropped her home safely.
5. **Lest** is negative and so should not be followed by **not**; it is followed by **should**.
   e.g. Work hard lest you should fail.
   If **would** or **may** is used in place of **should**, then **else** should be used in place of **lest.**
   e.g. Give him water, else he may die.
6. **Both** is complemented by **and**, not by **as well as.**
   e.g. Both Amit and Satish are good at Science.
7. **So.....as** is used in negative sentences, whereas **as .....as** is used in affirmative sentences.
   e.g. He is not so tall as his brother.
8. **Other** is followed by **than**.
   e.g. He has no other claim than his wealth.
9. The word **reason** is not followed by **because,** but by **that.**
   e.g. The reason why he didn't go was that his mother was ill.
10. **Because** denotes reason.
    in order that denotes purpose.
    e.g. He went to the doctor because he was ill.
11. Words such as regard, describe, define, treat, mention, depict, portray are followed by **as**.
    e.g. I regard her as my sister.
12. **As** and **since** are also used to express reason.
    e.g. As he was not there, I spoke to his brother.
13. **Neither** is followed by **nor** and both are followed by same auxiliary verb.
    e.g. Neither Ravi nor Raju was seen.
14. **Either** is followed by **or**.
    e.g. Either you are or he is to blame.
15. **Unless, until, if not, so that** should not be followed by **not.**
    e.g. Wait here until I come.
16. **If** is used in conditional series
    Whether is used in uncertainty.
    e.g. I don't know whether he was present.
17. To express **time before** use **until** or **till** and to express **how long** use **as long as.**
    *e.g. Work **as long as** you live.*
18. **Such** is followed by **as.**
    e.g. We talked about such subjects as the weather.
    Such is followed by **that** if we emphasize degree of something by mentioning the result.
    e.g. The extent of the disaster was such that not a single man could survive.
19. **Like** is followed by pronoun; **as** is followed by a clause.
    e.g. He runs like me.

ERRORS IN SUBJECT–VERB AGREEMENT

1. Singular subject must have singular verb.
   e.g. He writes; I write.
2. Plural subject must have plural verb.
   e.g. They write; We write.
3. Two subjects joined by **and** will always take a plural verb.
   e.g. The doctor and nurse work together.
   The doctor and nurses work together.
   The doctors and nurse work together.
4. Two singular subjects joined by **or** or **nor** will take a singular verb.
   e.g. A doctor or a nurse is working in the hospital.
5. A singular subject and a plural subject joined by **or** or **nor** will take a singular or plural verb depending on which subject is nearer the verb.
   e.g. Neither Deepak nor his friends are joining the tour.
   Neither his friends nor Deepak is joining the tour.
6. If the subject is singular and the predicate is plural, the verb must agree with its subject and not its predicate.
   e.g. Physical conditioning and mental attitude are winning combination.
   The winning combination is physical conditioning and mental attitude.
7. Indefinite pronouns such as
   someone, somebody, each, nobody, anyone, anybody, one, no one, everyone, everybody, either, neither, etc. always take a singular verb.
   e.g. Each of my friends calls me once a month.
8. Indefinite pronouns which indicate more than one (several, few, both, many) always take plural verbs
   e.g. Both of the books required careful reading.
9. Collective nouns (fleet, army, committee, crowd) are singular when the group works together as a unit and hence take singular verbs.
   e.g. The jury has reached its verdict.
10. Collective nouns are plural when the members of the group are acting individually and hence take plural verbs.
    e.g. The jury have argued for five hours.
11. Some words (such as news, measles, mumps etc.) end in —*s* but represent a single thing. These words need singular verbs.
    e.g. The 7 o'clock news is about to begin.
12. Some words (such as scissors, trousers, spectacles, shorts etc.) end in —*s* and seem to represent a single thing, but they are two parts to that single thing. These words take plural verbs.
    e.g. The scissors are on the table.
13. Words (such as politics, ethics, athletics etc.) that end in —ies are usually singular and hence take singular verbs.
    e.g. Mathematics is his favourite subject.
    But these words are singular when they refer to a study, science or practice. If these words have modifiers with them, they become plural and hence take plural verbs.
    e.g. His politics are somewhat divided.
14. Title of the books need singular verbs.
    e.g. Great Expectations is a good book.
15. Some nouns in the plural form represent an amount, a fraction, or an element of time. These nouns are considered singular and hence take singular verbs.
    e.g. Sixty minutes is enough to finish this work.
16. If two subjects are joined together by **as well as** the verb will act according to the first subject.
    e.g. Students as well as the teacher are playing.
17. The subject **Many a .....** is always followed by the singular verb.
    e.g. Many a man was drowned in the sea.
18. If two subjects are joined together by **with** the verb will act according to the subject.
    e.g. The principal together with his students was seeing the final match.
19. If subject is **The number of ...** use a singular verb.
    e.g. The number of books is very small.
20. If the subject begins with **A number of ....** (= many), use a plural verb.
    e.g. A number of books are missing.
21. If two subjects express one idea, use a singular verb.
    e.g. Bread and butter is wholesome food.
22. When adjectives such as **much, less, little** and **more** are used as nouns, they must have a singular verb.
    e.g. A little of good habits makes our life happy.

ERRORS IN THE USE OF TENSES

The changed forms of a verb that indicate time of the action are called **tenses** of the verb.

1. When the verb in the Principal Clause is in the Past tense, the verbs of the Subordinate Clauses should be in the Past tense.
   e.g. He said that he had finished his work.
2. But a past tense in the Principal Clause may or may not be follwoed by the Past tense in the subordinate clause if the latter expresses unviersal or habitual truth.
   e.g. The teacher said that the earth revolves round the sun.
3. Any tense may be used in the sub-ordinate clause if it gives a comparison by using the word **than.**
   e.g. The teacher liked Anil better than he liked me.
4. Any tense can be used when the subordinate clause is in a quotation.
   e.g. I said, "I am going to Delhi today".
5. The Present Perfect Tense (subject + has/have + V_3) cannot be used when an expression of Past time (yesterday, last night, ago etc.) is used.
   e.g. We wrote to you yesterday about his mother.

6. In conditional sentences the subordinate clause, beginning with if, is generally in Present or Past tense but not in future tense.
   e.g. If I go to Delhi, I shall see the Red Fort.
7. With the phrases **as if** and **as though**, Past tense and plural form of the verb should be used.
   e.g. He behaves as if he were a millionaire.
8. Past Perfect tense (subject + had + V_3) is used when a sentence refers to two past actions and one of them occurs earlier than the other.
   e.g. When my uncle came to India I had passed M.A.
9. Words like **usually, generally, often, whenever** etc. are used in Present Indefinite tense.
   e.g. I usually go to play cricket.
10. If the action began in the past and is still continuing in the present, use Present Perfect tense (Subject + has/have + been V_1 + ing).
   e.g. I have been reading in this class for two months.
11. Do not use the Future or the Present tense after such expressions as **suppose that, it is high time, it is time, as if** etc.
   e.g. It is high time that you went home.

GERUND & INFINITIVE

Use of Gerund (V_1 + ing)

1. When an action is being considered in general sense, gerund is used as subject.
   e.g. Swimming is his favourite pastime.
2. Gerund is used as subject in short prohibitions
   e.g. Smoking is prohibited.
3. Verbs such as help, stop, detest, avoid, finish, dread, mind, prevent, dislike, risk, deny, recollect, no good, no use, resent, delay, postpone, defer, enjoy, forgive, pardon, excuse, suggest etc. are followed by the gerund.
   e.g. It is no use saying.
4. A gerund is placed after a preposition :
   e.g. She is tired of walking.
   But the prepositions **except** and **but** are followed by an infinitive.
5. A gerund, showing physical activity, is put after **come** or **go**
   e.g. He is to go shooting.
6. Nouns, followed by a gerund, are put in the possessive case.
   e.g. He hates his brother's drinking everyday.
7. If there is a sense of dislike, hesitation, risk etc. in a sentence, use gerund.
   e.g. I dislike reading cheap novels.
8. Adjective before gerund are put in the possessive case.
   e.g. I do not mind your going away.
9. **With a view to** is always followed by a gerund and not an adjective.
   e.g. We go to school with a view to studying.

INFINITIVE (TO + V_1)

1. Verbs such as learn, remember, promise, swear, consent, neglect, refuse, purpose, try, endeavour, attempt, fail, care, hope, decide, hesitate, prepare, determine, undertake, manage, arrange, seem, forget, agree, regret etc. are followed by infinitive:
   e.g. We endeavour to invite such persons.
   We promise to do this work soon.
2. Verbs such as order, tell, invite, oblige, allow, permit, compel, teach, instruct, advise, tempt, encourage, urge, request, show, remind, forbid, warm etc. are followed by object and infinitive
   e.g. The teacher instructed the students to go.
3. Verbs or expressions like—will, can, do, must, may, let are followed by infinitive without **to**
   e.g. Let him do this work.
4. Expressions like would rather, would sooner, rather than, sooner than, had better are followed by infinitive without **to**
   e.g. You had better tell him.
5. **See** can be used with infinitive or without it
   e.g. Go to see him.
6. The infinitive is used after adjectives like delight, angry, glad, astonished
   e.g. I was astonished to find him.
7. The verb know is never directly followed by the infinitive. It is followed by a conjunction and then the infinitive
   e.g. Do you know to play the harmonium ? (×)
   Do you know how to play the harmonium ?(✓)
8. **...had better** is always followed by the infinitive without to
   e.g. You had better to stop taking the medicine which has harmful side-effects. (×)
   You had better stop taking the medicine which has harmful side-effects.(✓)
9. Infinitive is not used with **to** after verbs of perception like-to see, to hear, to feel, to watch etc.
   e.g. I saw him cross the street.
10. Avoid using split-infinitive :
   e.g. He planned to not go on a vacation this year.(×)
   He planned not to go on a vacation this year.(✓)

AUXILIARY VERBS

These are helping verbs.

1. **May** implies permission, doubt or possibility.
   e.g. It may rain tonight.
2. **Might** is the past form of **may** but it does not necessarily represent past time.
   Often it implies more doubt than **may.**
   e.g. If the clouds are salted, the rains might come.
3. **Can** is used to express ability.
   e.g. He can do this work.
4. **Could** is the past form of **can** but it does not necessarily represent past time. Often it implies a more uncertain condition than **can.**
   e.g. She could refuse, but she never does.

5. **Should** is the past form of **shall.** It is used in sub-ordinate clauses after **in case** and sometimes after **if.**

   e.g. I shall get some money in case brother comes.

   It is also used in past sentences with **so that** and **in order that :**

   e.g. He turned the stereo down very low so that he should not disturb him.

- **Should** can also be used in sub-ordinate clauses when we are expressing the idea that something must be done or is important. The fact is that this happens after verbs like command, order, request, insist, suggest, advise etc. & after adjectives like important, vital, essential, necessary, eager, anxious, concerned etc.

  e.g. I am anxious that nobody should be hurt.

- **Should** is also used in sub-ordinate clauses in sentences where we express personal reactions to events. We express our reactions with words like—amazing, interesting, shocked, sorry, normal, natural, it's a shame etc.

  e.g. I am sorry you should think I did it on purpose.

6. **Must** is followed by the infinitive without **to.**

   e.g. I must get up at five tomorrow.

   Must is used to give strong advice or orders :

   e.g. I really must stop drinking.

   Must is used to say that we are sure about something :

   e.g. I am in love—that must be nice.

   Must is used after a past reporting verb.

   e.g. I felt there must be something wrong.

7. **Should** and **Ought** have similar meanings, but **ought** is followed by **to. Ought to** has a more objective force and is used when we are talking about laws, duties and regulations.

   e.g. We ought to see her tomorrow.

   Should and **ought to** can also be used to talk about strong probability.

   e.g. He has bought thirty pints of whisky—that ought to be enough.

   To talk about things which did not happen, although they were supposed to, we use **should** and **ought to** with the perfect infinitive :

   e.g. The taxi should have arrived at 8.30.

8. **Do** is used to make question and negative forms of ordinary verbs.

   e.g. Do you know Rajat ?

   Do is used in question-tags and short answers

   e.g. Does he know I am here ? —Yes, he does.

   Do is used before an imperative to make the request more persuasive

   e.g. Do accompany us.

   Do is not used in questions which have **who, what or which** as their subject

   e.g. Who said that ?
   What happened ?
   Which one's that ?

   But if **who, what** or **which** is the object of the sentence, **do** is used

   e.g. Who did you see ?
   Which department do you want ?

   Do is also used with an **—ing** form when we want to talk of activity that takes a certain time or that is repeated. In this case we find determiners like —the, my, some, much, etc.

   e.g. I usually do most of my washing on Sundays.

ERRORS IN USE OF ADJECTIVES

Word qualifying a noun or pronoun is called an **adjective.**

1. The adjectives ending in —ior (prior, junior, senior, superior, inferior, posterior) take 'to' and not 'than' after them.

   e.g. He is senior to me.

2. Some adjectives like unique, ideal, perfect, extreme, complete, universal, infinite, perpetual, chief, entire, round, impossible are not compared.

   e.g. It is the most unique book. (×)
   It is a unique book (✓)

3. Comparative degree is used in comparing two things or persons.

   e.g. It is the better of the two books.

   Superlative degree is used in comparing more than two things or persons.

   e.g. He *is the best* of the three boys.

4. Double comparatives and double superlatives must not be used.

   e.g. He is more wiser than his brother. (×)
   He is wiser than his brother. (✓)

5. When we compare two qualities in the same person or thing, the comparative ending —er is not used.

   e.g. You are wiser than old. (×)
   You are more wiser than old. (✓)

6. When two adjectives in superlative or comparative degree are used together, the one formed by adding 'more' or 'most' must follow the other adjective.

   e.g. He is more intelligent and wiser than his brother. (×)
   He is *wiser* and *more intelligent* than his brother. (✓)

7. When two adjectives with differing degrees of comparison are used they should be complete in themselves.

   e.g. He is as wise, if not wiser than his brother. (×)
   He is as wise as, if not wiser than his brother. (✓)

8. When two changes happen together, comparative degree is used in both.

   e.g. The higher you go, the cooler you feel.

9. When comparative degee is used in superlative sense, it is followed by '*any other*'.

e.g. Kapil is better than any bowler. (×)

Kapil is better than any other bowler. (✓)

10. Compound adjective formed by adding 'worth' is placed after the noun it qualifies.

e.g. This is a worth seeing sight. (×)

This is a sight worth seeing. (✓)

11. When two or more comparatives are joined by **'and'**, they must be in the same degree.

e.g. Russel was one of the wisest and most learned men of the world.

12. When there are two objects of comparison, then to avoid repetition of noun, 'that' is used for singular noun and 'those' for plural noun.

e.g. The climate of Ranchi is better than Gaya. (×)

The climate of Ranchi is better than that of Gaya. (✓)

13. If comparison is made by using 'other', 'than' is used instead of 'but'.

e.g. He turned out to be no other than my old friend.

14. **Likely, certain** and **sure** are followed by **'to'**.

e.g. He is likely to win.

CONFUSED ADJECTIVES

1. **Beautiful** is used for woman; **handsome** for man.

   e.g. He is a handsome youth.

   She is a beautiful girl.

2. **Less** refers to quantity, **fewer** denotes number.

   e.g. He takes no less than a litre of milk.

   They have fewer books than I have.

3. **Last** is the final one; **Latest** is last upto the present.

   e.g. Z is the last letter of the alphabet.

   This is the latest edition of the book.

4. **Each** is used for one of two or more things; **every** is used for more than two things, taken as a group.

   e.g. Each of the two boys was wrong.

   He read every book I gave him.

5. **Older** refers to persons or things.

   e.g. This tree is older than that.

   Elder refers to persons only.

   e.g. He is my elder brother.

6. **Little** means 'not much'.

   A little means 'at least some'.

   e.g. He slept little.

   He slept a little.

7. **Farther** means 'more distant'.

   Further means 'additional'.

   e.g. Mumbai is farther than Delhi.

   I shall get further information.

8. **Latter** means the second of two things.

   e.g. *Keats and Byron are romantic poets, but I prefer the latter.*

   Later refers to time.

   e.g. She came to school later than I.

ERRORS IN USE OF ADVERBS

Adverbs are words that add information about the verb.

1. Adverbs of manner, place and time are usually placed after the verb or object.

   e.g. He was running slowly.

   Reena does her work carefully.

2. Adverbs of frequency (e.g. never, often, usually, always, rarely, etc.) and other adverbs (like already, almost, just, quite, nearly, hardly) are normally put between subject and verb. If there is more than one word in the verb, they are put after first word.

   e.g. He often goes to Delhi.

   I quite agree with you.

   But if verb is 'am', 'is' and 'are', adverb is placed after the verb.

   e.g. I am *never* late for school.

3. The adverb *enough* is placed after the adjective.

   e.g. She is cunning enough to tackle him.

4. 'Ever' is sometimes incorrectly used for 'never'.

   e.g. He seldom or ever tells a lie. (×)

   He seldom or never tells a lie. (✓)

5. Adverb 'not' shouldn't be used with words having negative meaning.

   e.g. The teacher forbade me not to go. (×)

   The teacher forbade me to go. (✓)

6. The word 'only' should be placed immediately before the word it modifies.

   e.g. Hari answered only two questions.

7. An adverb should not be used before an infinitive.

   e.g. He quickly did the job. (×)

   He did the job quickly. (✓)

8. The auxiliaries *have to* and *used to* come after the adverb.

   e.g. He often used to go to cinema.

SOME MORE TIPS OF USAGE

1. Avoid the use of **lots, a lot** and **a whole lot** in the sense of **much** or **a great deal.**

   e.g. He expects to earn a lot of money on his sale of farm products. (×)

   He expects to earn a great deal of money on his sale of farm products. (✓)

2. **Plenty** is a noun and is always followed by **of.**

   e.g. He has plenty of room in his old house.

3. **Due to** always modifies a noun and not a verb. Hence, no sentence should begin with **due to,** it must be used only after some form of the verb **to be**.

   e.g. *His death was due to natural causes.*

4. **Alright** is incorrect, use **all right.**

   e.g. *I think it is a quite alright if you stay.* (×)
   I think it is quite all right if you stay. (✓)

5. If the gender is not determined, use the pronoun of the masculine gender.

   e.g. If anybody has got the book let her return it (×)
   If anybody has got the book let him return it (✓)

6. When two nouns are closely connected, apostrophe —'s will be written after second only.

   e.g. *Ravi's and Shikha's mother is ill.* (×)
   Ravi and Shikha's mother is ill. (✓)

7. **Both** and **as well as** cannot be used together in the same sentence because both convey the same sense.

   e.g. *Both Pinki as well as Pooja are beautiful* (×)
   Both Pinki and Pooja are beautiful. (✓)

8. **Supposing** and **if** cannot be used together in the same sentence.

   e.g. *Supposing if he fails, what will he do ?*(×)
   If he fails, what will he do ? (✓)

9. **Endure** suggests **suffering,** usually in silence and **tolerate,** means to allow it with some degree of approval.

   e.g. *He endured the pain without complaint.*
   He will not tolerate laziness.

10. Avoid the use of **had** with **ought.**

    e.g. *You had not ougth to do it.* (×)
    You ought not do it. (✓)

11. **As** is used when one compares things or persons of equal or about equal size or quality, **so** is used when one compares things or persons which are unequal.

    e.g. *He weighs as much as his father.* (×)
    He does not weigh so much as his father. (✓)

12. Avoid using the double comparative and double superlative.

    e.g. *This pen is the most costliest in this store.* (×)
    This pen is the costliest in this store. (✓)

13. **Cent per cent** and **word by word** are wrong uses. The real terms are **hundred percent** and **word for word.**

14. When two comparatives are used in a sentence for proportion, **the** is used before both of them.

    e.g. *The higher we go, the cooler it is.*

15. **Do the needful** is incorrect, write **do what is necessary.**

16. Never say **family members,** say **members of the family.**

    e.g. *His family members have gone to Mumbai.* (×)
    The members of his family have gone to Mumbai (✓)

17. **Else** is always followed by **but** and never by **than.**

    e.g. *It is nothing else but your pride which makes you say such a thing.*

18. **All of** is used in conversation, not in standard written English.

    e.g. *She gave all of her antiques to the museum* (×)
    She gave all her antiques to the museum (✓)

19. **Talking terms** is wrong, use **speaking terms.**

    e.g. *He is not on talking terms with his brother* (×)
    He is not on speaking terms with his brother (✓)

20. Verbs like resemble, recommend, comprise, order, accompany, reach, join, shirk, attack, emphasise, discuss, fear, succeed, resist, request, assist, benefit, afford etc. are not followed by any preposition when used in the active voice.

    e.g. *He emphasised on the need for discipline.* (×)
    He emphasised the need for discipline. (✓)
    She will accompany with you to Chennai. (×)
    She will accompany you to Chennai. (✓)

21. The phrase **type of, sort of** and **kind of** do not follow **a** or **an.**

    e.g. *What kind of a friend are you ?* (×)
    What kind of friend are you ? (✓)

22. A full hour number follows **o'clock,** but in fraction it does not follow.

    e.g. *He left this place by 10.40 o'clock.* (×)
    He left this place by 10.40. (✓)

23. **Yet** means up to the present time something that has not happened.

    e.g. *His brothers are not married even yet.*

EXERCISE

MCQ TYPE QUESTIONS

Directions (Q. 1 – 20)

In this section you are required to spot errors in sentences. Read each sentence to find out whether there is any error in any of the parts. Errors, if any, are only in one of the parts. No sentence has more than one error. When you detect an error, in any of the parts of the sentence, choose the corresponding alphabet.

1. (*a*) The scenery here is not good
(*b*) I have lost my furnitures
(*c*) We have received no information
(*d*) He told his mother this news
(*e*) No error.

2. (*a*) Please excuse the trouble
(*b*) He took pains over his work
(*c*) In India there are many poor
(*d*) Scouts wear shorts
(*e*) No error.

3. (*a*) I spent the holidays with my family
(*b*) Gone him some blotting paper
(*c*) Write this in your note book
(*d*) Gone my kind regards to all
(*e*) No error.

4. (*a*) We had a good play of football
(*b*) We saw a play
(*c*) We like acting
(*d*) He is a tall man
(*e*) No error.

5. (*a*) The boy was wearing a new suit
(*b*) He took offence at this
(*c*) Please put your sign here
(*d*) Please put your signature here
(*e*) No error.

6. (*a*) He is my cousin brother
(*b*) We are all brothers
(*c*) I spent the middle of the day working
(*d*) He had a large amount of money
(*e*) No error.

7. (*a*) He got into bad company
(*b*) One of my servant tells me
(*c*) Some of my servants tell me
(*d*) The front of the house
(*e*) No error.

8. (*a*) I had a bathe in the sea
(*b*) I have hurt a toe
(*c*) Each of these days play games
(*d*) None of us went
(*e*) No error.

9. (*a*) A man should work hard
(*b*) A man should not waste his time
(*c*) A boy should not waste his time
(*d*) Here is my cup; please fill it
(*e*) No error.

10. (*a*) Have you a pencil ? I have not got one
(*b*) Is he coming ? Yes I think so
(*c*) He enjoyed during the holidays
(*d*) I asked for my pencil but he did not give it to me
(*e*) No error.

11. (*a*) The boy who does best he will get a prize
(*b*) Whoever does best will get a prize
(*c*) Who did this ? I
(*d*) He and I are brothers
(*e*) No error.

12. (*a*) I went for a walk with some friends
(*b*) He is wiser than I
(*c*) The master tested the boy if he could read English
(*d*) I shall see whether the brakes work well
(*e*) No error.

13. (*a*) Everyone is frightened when he sees a tiger
(*b*) None of us has seen him
(*c*) People strave when they have no money
(*d*) The size of the shoe should be the same as this shoe
(*e*) No error.

14. (*a*) Everyone knows this
(*b*) Every man knows this
(*c*) These all mangoes are ripe
(*d*) He hold the book in both hands
(*e*) No error.

15. (*a*) I have no any friends
(*b*) Neither man has come
(*c*) All idle man should do some work or other
(*d*) Shakespeare is greater than any other poet
(*e*) No error.

16. (*a*) Open your book at page six
(*b*) He is in class ninth
(*c*) This article costs ten rupees
(*d*) He came a second time
(*e*) No error.

17. (*a*) King George VI
(*b*) He is older than I
(*c*) He is better than I
(*d*) Raipur is Hotter thanSimla
(*e*) No error.

18. (*a*) He is worse than I
(*b*) A horse is more useful than a car
(*c*) He gets a small salary
(*d*) I our library the number of books is less
(*e*) No error.

19. (*a*) He is the more clever of the two
(*b*) He is the cleverer of the two
(*c*) From the three he is more clever
(*d*) Of the two plans this is the better
(*e*) No error.

20. (*a*) He is becoming strong
(*b*) There is a very good teacher in that class
(*c*) He will spend the rest of his life here
(*d*) This is a sight worth seeing
(*e*) No error.

Directions (Q. 21 – 30)

Each question below has a sentence, from the choices provided, identify the one which best restates the given sentence and mark its number as the answer.

21. Uneasy lies the head that wears a crown.
(*a*) Uneasily lies the head that wears a crown.
(*b*) Uneasy is the head wearing a crown.
(*c*) The head which wears the crown lies uneasy.
(*d*) The head which is wearing the crown lies uneasy.

22. Had I been a little early, I would not have missed the train.
(*a*) Had I been a little early, I would not have missed the train.
(*b*) Had I been a little early than now, I would not have missed the train.
(*c*) Had I been a little earlier, I would not have missed the train.
(*d*) Had I only been a little early than now, I would not have missed the train.

23. She is as intelligent as her sister if not more.
(*a*) She is more intelligent than her sister.
(*b*) She is as intelligent as her sister if not more intelligent.
(*c*) She and her sister are equally intelligent.
(*d*) She is less intelligent than her sister.

24. You will not succeed unless you are optimistic.
(*a*) You will not be successfull unless you are optimistic.
(*b*) You will not succeed until you are an optimist.
(*c*) You will not succeed unless you don't be optimistic.
(*d*) You will not succeed unless you don't be an optimistic.

25. The students were advised to follow the instructions of the examiner.
(*a*) The instructions of the examiner were followed by the students.
(*b*) The students followed the examiner's instructions.
(*c*) The examiner's instructions were advised to be followed by the students..
(*d*) It has been advised to the students that they should follow the examiner's instructions.

26. Despite being ill, he attended the classes.
(*a*) Despite of being ill, he attended the classes.
(*b*) Despite his illness, he attended the classes.
(*c*) Inspite of his being ill, he attended classes.
(*d*) The classes were attended by him inspite of being ill.

27. That man is aggressive by nature is a hard fact of life and no one can deny it.
(*a*) That man is aggressive by nature is a hard fact of life which none can deny.
(*b*) That man is aggressive by nature is a hard fact of life and no one can deny it.
(*c*) That man is aggressive by nature is a hard fact of life and not one can deny.
(*d*) That man is aggressive by nature is hard for anyone to deny.

28. He is as old as I.
(*a*) He is as old as me.
(*b*) He and I are equally old.
(*c*) He and I are equally older.
(*d*) Both he and I are of the same age.

29. A man becomes wiser with age and experience.
(*a*) A man as he is aged and experienced becomes wiser.
(*b*) A man with age and experience becomes wiser.
(*c*) A man as he is aged and experienced becomes the wiser.
(*d*) With age and experience, a man becomes wiser.

30. Hazards of life cannot be negated but they can be quite effortlessly evaded.
(*a*) Hazards in life can be quite effortlessly evaded and completely nullified.
(*b*) Hazards of lilfe cannot be avoided but they can be made ineffective.
(*c*) To be made ineffective, hazards of life must be avoided.
(*d*) It may not be possible to nullify the hazards of life but they can quite easily be avoided.

Directions(Q. 31 –40)

Each sentence is broken into four parts a, b, c, d. Mark the part which has an error. Ignore errors of punctuation.

31. (*a*) Every man, woman
(*b*) and child
(*c*) in the house on fire
(*d*) have been saved.

32. (*a*) One of the
(*b*) best lawyers in town
(*c*) have been
(*d*) hired.

33. (*a*) I request you (*b*) kindly to
(*c*) come to me (*d*) immediately.

34. (*a*) My friend's mother
(*b*) is the principal
(*c*) of a
(*d*) girl's college.

35. (*a*) To succeed in these tests
(*b*) it is absolutely necessary
(*c*) for us
(*d*) to aim for speed and accuracy.

36. (*a*) The aeroplane is a
(*b*) powerful instrument of war
(*c*) and their peacetime role
(*d*) is just as important.

37. (*a*) The depletion of ozone in the atmosphere
(*b*) is a reminder to all of us
(*c*) both in the developed and developing world
(*d*) that they cannot continue to use contraptions which give out noxious gases.

38. (*a*) Scotsmen are notorious
(*b*) for their thrift
(*c*) and he is the butt
(*d*) of many exaggerated jokes.

39. (*a*) A mixed economy
(*b*) is it in which
(*c*) the public sector
(*d*) and the private sector co - exist.

40. (*a*) Although the beaver's hind feet
(*b*) are webbed for swimming
(*c*) their front feet
(*d*) are small and handlike.

Directions(Q. 41 – 45)

Given below are five sentences, each of which may or may not have errors.
Mark (a), if there is only one error.
Mark (b), if there are two errors.
Mark(c), if there are more than two errors.
Mark (d), if there is no error.

41. "This feels like one big dream," says a weary American woman to her husband, as they sit in a plane flying home.

42. The worst mistake a non - profit organization can make is to take all the money it is given and be beholden to doing things on someone else's terms.

43. Although most hotels in Bhutan has a distinctly local flavour, some pander with the perceived tastes of Western tour groups - and can be rather bland.

44. The crustaceans find sanctuary in the jords, often clinging to the walls in large, gregarious groups.

45. In these tropical lowlands the Maya built an agrarian civilization that eventually supported one of the highest population densities in the pre - industrial America.

Directions (Q. 46–50)

Each of the following questions has two sentences A and B.
Mark (a), if you think sentence A has an error.
Mark (b), if you think sentence B has an error.
Mark (c), if you think both sentences A and B have errors.
Mark (d), if you think neither sentence has an error.

46. A. The boss himself or his secretary answer the phone on Saturday.
B. Neither the quality nor the prices have changed.

47. A. The members re-commended that all delinquents be fined.
B. A stone lying in one position for a long time may gather moss.

48. A. Until I received that letter, I was hoping to have had a visit from Krishnan.
B. Follow the main road for a mile; then you need to take the next road at the left.

49. A. The people to watch closely are the ones ruling behind the political scene.
B. Give the tools to whoever can use them best.

50. A. The magazine has been published continuous since 1951, and it has the funniest cartoons you can possibly imagine.
B. I feel glumly when I read his nightmarish tales.

ANSWERS

MCQ Type Questions

1. (*b*)	**2.** (*a*)	**3.** (*e*)	**4.** (*a*)	**5.** (*c*)	**6.** (*a*)	**7.** (*b*)	**8.** (*c*)	**9.** (*e*)	**10.** (*c*)
11. (*a*)	**12.** (*c*)	**13.** (*d*)	**14.** (*c*)	**15.** (*a*)	**16.** (*b*)	**17.** (*d*)	**18.** (*d*)	**19.** (*c*)	**20.** (*a*)
21. (*c*)	**22.** (*a*)	**23.** (*c*)	**24.** (*a*)	**25.** (*d*)	**26.** (*b*)	**27.** (*a*)	**28.** (*d*)	**29.** (*d*)	**30.** (*d*)
31. (*d*)	**32.** (*c*)	**33.** (*b*)	**34.** (*d*)	**35.** (*d*)	**36.** (*c*)	**37.** (*d*)	**38.** (*c*)	**39.** (*b*)	**40.** (*c*)
41. (*a*)	**42.** (*d*)	**43.** (*b*)	**44.** (*d*)	**45.** (*d*)	**46.** (*a*)	**47.** (*d*)	**48.** (*c*)	**49.** (*d*)	**50.** (*c*)

EXPLANATIONS

MCQ TYPE QUESTIONS

21. Uneasily does not denote a state, which is required in the context. 'Wearing a crown' denotes that the head is doing it, 'is wearing' denotes a continuous action, which is impossible, as the head cannot do it. Hence choice (*c*) restates the given sentence correctly.

22. All the statements have the correct combination of tenses. So, we need not check it. Let us find the errors in other statements. 'Early than now' and 'earlier' suggest that I reached early; in that case, I did not miss the train. So, choices (*b*), (*c*), and (*d*) do not state the same idea as in the question statement. Hence correct choice is (*a*).

23. The given statement states that both she and her sister are equally intelligent. This is given in statement. Hence correct choice is (*c*).

24. 'Unless' denotes condition, 'until' denotes time. Choice (*b*) has adverb clause of time.With unless, another negative is not used. 'Unless ..... don't' is wrong. So, Choices (*c*) and (*d*) are incorrect. Hence correct choice is (*a*).

25. Choices (*a*) and (*b*) state the same idea *[*(*a*) – passive voice; (*b*) Active voice] that the students followed the instructions, which differs from the question statement. Instructions cannot be advised. So, choice (*c*) is meaningless. Choice (*d*) states the context of the given statement aptly.

26. Despite means inspite of. Despite of is a wrong usage. 'His' need not be used in choice (*c*), as the sentence is about one person. Choice (*d*) means that the classes were ill. Hence correct choice is (*b*).

27. Choice (*a*) best restates the given sentence, though choice (*b*) is the repetition of the given statement. 'Not one can deny' in choice (*c*) and for any one to deny' in choice (*d*) make them incorrect. Hence correct choice is (*a*).

28. In this type of comparison, the pronouns are in subjective case 'Me', which is in objective case is wrongly used in choice (*a*). Equally is not used with 'old'. Of the same age is the correct expression. Hence correct choice is (*d*).

29. Choice (*a*) and (*b*) mean that the prerequisite for wisdom is age and experience, which is incorrect. Just by growing old and working for a long period, a person cannot improve his wisdom. 'The wiser' is used only when we compare wisdom with some other quality. Choice (*d*) is the correct restatement of the given one. Hence correct choice is (*d*).

30. The given statement says that one cannot completely nullify the hazards of life but they can be evaded quite easily. Choice (*a*) which states that hazards cannot be avoided is wrong (*b*) states that hazards can be made ineffective and hence this is also wrong (*c*) does not convey the meaning as stated and hence it too is incorrect. (*d*) correctly restates the idea and is our answer.

31. Every man refers to a singular noun, therefore.... has been saved.

32. has

33. drop 'kindly'

34. Girl's implies possessive noun, whereas the college is meant for girls.

35. aim at, aim for is wrong usage.

36. delete their - add : its

37. delete they - add : we

38. delete he is - add : they are.

39. delete it - add : that

40. delete their - add : its.

41. One sits 'on' a plane.

42. There are no errors.

43. Hotels 'have' and some pander 'to....'

44. There are no errors in this sentence.

45. There are no errors.

48. (A) hoping to have a visit;

(B) .... then take the next road to

50. (A) continuously; (B) glum.

2 CHAPTER Sentence Completion

A Sentence Completion question consists of an incomplete sentence, which has one or two blanks. From the answer choices given, you are asked to choose the appropriate word / words that can take the place of the blanks in order to best complete the meaning of the given sentence.

Directions:

Each of the sentences below has one or two blanks; each blank indicates omission of a word or set of words. Beneath the sentence are given four answer choices, each containing a word or a set of words. Choose the word or set of words for each blank that gives the best meaning to the sentence as a whole.

A thinking man resists the vulgar prosperity that ______ ever to ______ by preserving and communicating heroic sentiments.

A. disintegrates ... banality

B. transforms... neutrality . .

C. decays... luxury

D. retrogrades ... barbarism

E. level ...hurtly

The theory underlying a sentence completion stimulus is that the meaning and the tone of a sentence can be gathered without the missing word/ words. If so, it is possible for you to supply the appropriate word(s) to complete the sentence.

Invariably, in every sentence completion question, there will be a clue or two that could be used as lead(s) toward identifying the correct word(s).

These clues could be structural (syntax/grammar clues) or semantic (meaning clues). The word(s) that you choose should agree and merge with the sentence structurally and functionally or semantically.

Every sentence completion question can be analyzed for the negative or positive nature or tone of words to be filled in.

What is measured by a Sentence completion question?

i. Skill of vocabulary

Sentence completion question is a test of your word power and diction. It is only if you know the meaning of word(s) that you will be able to choose confidently, the appropriate one(s) to fill the incomplete sentence. Moreover the question sentence also may have difficult words in itself, knowing the meaning of which is essential for understanding the sentence holistically and thus to choose the appropriate word(s).

Look at the following question:

The department can use information received from dissatisfied employees and __ to launch an investigation against the company.

A. pettifoggers

B. buccaneers

C. ambassadors

D. whistleblowers

The knowledge of the meaning of the words in choices is essential for you to select the answer to this question with confidence. Let's analyze meaning of these words.

A Pettifogger is one who practices chicanery or unlawful business. Can this word fit the sentence? No.

A Buccaneer is a pirate; again not a word that would go with dissatisfied employees to supply information about a company.

A whistleblower is one who informs about corruption or wrongdoing. Yes, this fits the sentence.

An Ambassador is not one to do anything with a company. In fact this is the first choice you could eliminate.

ii. Skill of semantics and syntax

Every sentence has a semantic dimension (meaning) and a syntactical dimension (word arrangement). Knowledge of these two is a mandatory condition for you to be able to choose the appropriate word. You must understand the meaning of a sentence in order to identify the appropriate choice word that fills the blank, and completes the sentence coherently.

You should have a good understanding of the functions of structural words in a sentence such as unless, rather, yet, while unfortunately, in contrast, despite.

Look at the following question:

The meaning of words is destroyed if you change them, whereas the meaning of a thought______ whatever word you think proper to use.

A. alters B. transforms

C. remains D. settles

E. Ruins

When you read the sentence you must recognize the function of'whereas' in the sentence. It signals a shift in the meaning of the sentence.

'One thing is destroyed whereas the other is not' the answer thus is (C) remains.

The semantic feature of a sentence refers to its meaning, and how the meaning of the given words in a sentence helps us to identify the word to complete the sentence.

Study the following sentence:

A cult brand _____ the imagination of a small group who spreads the word, makes converts, helps turn a fringe product into a______ name.

A. ceases... redoubtable

B. elevates ... dynamic

C. fires... dream

D. seizes... mainstream

The first clue is the word Imagination; hence options (A) and (D) it, ceases and seizes can easily be eliminated. You have to work for fires and elevates. To fire One's Imagination is a standard idiomatic usage. Moreover the meaning of the sentence implies turning a fringe (peripheral) product into (the opposite of it) a dream product. Hence, answer is (C).

iii. Skill of identifying clues

A sentence completion question is not merely a test of your vocabulary. It also tests your ability to analyze the given sentence and to identify the structural and semantic clues that decide on the meaning and nature of the word(s) to be chosen. You will find one or two clues in a sentence completion question. Common key words such as and or but are all more valuable for us in sentence completion. These are structural clues. (These are dealt with in detail later)

Sentence Completion Strategies

In order to answer sentence completion question correctly, you should first read the entire sentence as you see it, without trying to fill the word(s). Desist from working on each blank separately. The second part of the sentence may have a bearing on its first part. And both words must fit into the meaning of the sentence; therefore read your choice into the sentence supplying and evaluating both words.

Each sentence contains a few crucial clues that determine the answers, and finding these clues will guide you to the correct answer. Look for what is directly stated. In sentence completion question, you may also find poetic and literary language, which tests your understanding of figurative usage. The correct answer is the one that is directly implied in the sentence.

- Determine where the sentence is heading, using structural clues.
- Anticipate answers in your words.
- Look for answers that are similar to yours.
- Fix the words in the sentence to see if they get with the whole sentence.

Sentence completion strategy in a nutshell.

- Strategy 1

  Read the entire sentence.

  In both one- and two-blank sentences read the entire sentence before you inspect the choices. An overall understanding of the sentence is essential.

- **Strategy 2**

  Identify structural clues to determine the nature of the word(s) required.

  You can determine the positive or negative tone of the expected word using structural clues. And identify the words whose synonym, antonym, parallel word or cause/effect you have to supply.

- **Strategy 3**

  Substitute your own words.

  Place your own words in the blanks assuming that answer choices are not given, going by the clues you have identified.

- Strategy 4

  Identify the words similar to the ones you substituted.

  If the word you have substituted goes with the requirements of the sentence, you are sure to find, a word(s) similar in meaning and tone to the ones you have substituted.

- **Strategy 5**

  Read the sentence again with word(s) filled in.

  Now read the entire sentence with the words filled in to see if the word(s) fit literally and figuratively.

A. THOUGHT CONTINUITY CLUES

In some sentences, the first part of the sentence and the second part are parallel ideas. There is no shift in meaning. The missing word in the sentence is parallel to, or has a meaning similar to a word already given in the sentence. One part of the sentence may be a support to another. Some other times, though, one part of the sentence may be complementary to the other in meaning. These clues continue the sentence in the same direction. The positive or negative tone of what follows is not changed by these clue words.

The words that indicate such continuity are:

- Similarly
- Because
- And
- Since
- Thus
- In fact
- Moreover
- Indeed
- In addition
- Consequently

Study the following sentences:

1. The film script showed depth and maturity; likewise the treatment showed profundity.
2. His argument was convincing and persuasive.
3. Shakespeare lives forever; in fact he is immortal.

B. PARALLEL IDEA CLUES

These clues are similar to the thought continuity clues. The word that completes the sentence is in the same tone, as in the word already given. The word(s) that complete the sentence complements a word or phrase already given.

i. All that the young ask of the political class is compassion and concern/ sympathy/ care.

ii. On question about political and electoral reforms, the respondents showed great forethought and vision/rationality.

In the above sentences you may observe that the words given and the words that have to be filled in are parallel/ complementary or in the same tone.

C. THOUGHT REVERSAL CLUES

These terms change the direction of the sentence. They make one part of the sentence contradictory to the other part. The word or phrase that completes the sentence is opposite in meaning to a word or phrase already given. The positive or negative charge of an answer is changed by these clues.

Such sign – word include:

- But
- On the contrary
- In spite of
- Yet
- However
- Unless
- Rather
- Although
- While
- Unfortunately
- Nonetheless
- Ironically
- Paradoxically
- On the other hand
- Contradictorily
- Contrastively
- Despite
- Even though
- Even if

Illustrations:

His argument was flawed but persuasive.

There is no cynicism in the way the young deal with issues; on the contrary they are optimistic toward life.

The emphasis will not be on physical assets but on intangibles.

In the sentences above the word in italics are structural clues, which indicate that the words to be filled in are opposite in meaning to the words, that are already given in the sentences.

D. CAUSE EFFECT CLUES

While studying a sentence completion question, you should be able to gauge as to which direction the sentence is heading. If the sentence has an idea, it is possible to infer what would logically follow the causal phenomenon. In other words, we can anticipate the type of effect that the given cause would lead to. Similarly by analyzing the nature of an effect (sentences in which effect is given), we can also infer the plausible causal factor that would've contributed to the type of effect given. The easiest way is to identify the tone of the cause or effect to be supplied into the sentence-negative or positive

Let's work on sentences with cause effect relationship.

Torture is so prevalent and ________of the helpless so common that these helpless beings are certainly ______

When they pair out their story of agony.

(1) indemnifications... unhinged

(2) felicitation... believable

(3) disparaging ...credulous

(4) victimization ... credible

In the above sentence you observe that there are two clues: parallel and causal. Signaled by the word and the first blank needs a word parallel to torture i.e. victimization. The intensity of these two results in the possibility of the stories believable or credible.

Look at another example:

Higher education in India can no longer be allowed to remain as a/ an _____ instrument that crushes the spirit of the individual while goading him into a ________ and consumerist mode.

(1) innerving... boisterous

(2) emasculating... materialist

(3) invigorating ... pragmatic

(4) relegating ... humanistic

An instrument that crushes the spirit of the individual cannot be innerving or invigorating (both means energizing) nor can it be riveting; since rivet means to hold firmly and relegate means to lower in rank is closer but the latter word cannot be humanistic (3) The word that is parallel to consumerist is materialist.

In a similar mode identify the cause effect ideas in the incomplete sentences in the following drill and choose appropriate words to complete them.

E. WORD DEFINATION CLUES

Yet another common clue found in sentence completion stimuli is the meaning or definition of the word to be filled given as part of the incomplete sentence itself. It is similar to the thought continuity clue, except that in case the meaning of the word or definition of the word to be selected is already given.

Let us study an example :

The great irony of Angle Computers is that very posture that won Steve jobs' a cult following creating an exclusive group of ______ technicians who ______ refused to bow to IBM, hobbled his company.

iconoclastic......stubbornly

conservative......invariably

rebellious.......timidly

traditional........out rightly

In the sentence above you would've already observed the words 'refused to bow', which signals the word to be placed in the second blank. This phrase clearly defines the word stubbornly.

Another example :

The crisis in modern education is not merely a constraint of ____ of funds but of _______ practices, unsuited for contemporary system.

scarcity......doctrinaire

fulsomeness......authoritative

restrictiveness.......autocratic

paucity.......anachronistic

In the above sentence the key to the letter word is 'unsuited for contemporary education', you can now easily eliminate unsuitable choices and arrive at choice D, i.e., anachronist.

EXERCISE

MCQ TYPE QUESTIONS

Directions for Q. 1 to 41 : *In each of the following sentences, a part/parts of the sentence is left unfinished. Beneath each sentence, four different ways of completing the sentence are indicated. Choose the best alternative among the four.*

1. For some of our ruling politicians, social justice has apparently come to mean that __________.
(*a*) they have a right to plunder public money from the treasury
(*b*) they have a right to the plunder of public money from the treasury
(*c*) theirs is the right to plunder public money from the treasury
(*d*) the right is theirs to plunder public money from the treasury

2. Some critics believe that Satyajit Ray never quite came back to the great beginning he made in his path-breaking film Pather Panchali. __________ have endured decades of well-travelled bad prints to become a signpost in cinematic history.
(*a*) The bizarre history of its misty origins
(*b*) Its haunting images
(*c*) Its compelling munificence
(*d*) The breathtaking awe it inspires

3. __________, the more they remain the same.
(*a*) The less the dynamism
(*b*) The more things change
(*c*) The more pronounced the transformation
(*d*) The more the merrier

4. Revenues are likely to register a dramatic increase if __________ on the hundred million people who are said to comprise the risingIndian middle class.
(*a*) a flat tax is collected
(*b*) a flat tax is placed
(*c*) taxes at a flat rate are charged
(*d*) tax is imposed at a flat rate

5. I am an entertainer. __________ I have to keep smiling because, deep in my heart, laughter and sorrow have an affinity.
(*a*) Even if I have tears in me
(*b*) Despite conditions of extreme adversity
(*c*) Inspite of misery around me
(*d*) Although I have yet to make it big

6. The stock market is probably __________ And the way the markets are plunging says a lot about investor confidence.
(*a*) an ideal indication of the health of public sentiment
(*b*) the least imperfect mechanism for judging the quantity of the sentiment of the public.
(*c*) the best indicator of public sentiment
(*d*) the best barometer to assess the sentiment of the public

7. Home is __________ one has yet to improve.
(*a*) a discovery over which
(*b*) a certain over which
(*c*) an invention on which
(*d*) an institution upon which

8. The highest reward for a man's toil is not what he gets for it but what __________.
(*a*) he makes out of it.
(*b*) he gets for others.
(*c*) he has overcome through it.
(*d*) he becomes by it.

9. An act of justice closes the book on a misdeed; an act of vengeance __________.
(*a*) reopens the first chapter
(*b*) writes an epoilogue
(*c*) writes one of its own
(*d*) opens new books

10. The trail is the thing , not the end of the trail. Travel too fast and you miss __________.
(*a*) all you are travelling for.
(*b*) all the sights you are supposed to see.
(*c*) the very excitement of your travel.
(*d*) all the enjoyments of travel.

11. __________ that in this apparent mess, two things not be interfered with.
(*a*) It is important.
(*b*) it is of cardinal important.
(*c*) It should be urgently under-stood.
(*d*) It cannot be emphasised.

12. A country's wealth is its people. But instead of drawing out the strengths of the people, instead of drawing out their talent, this use of religion debases, degrades and depresses than __________.
(*a*) in greater and greater measure
(*b*) further
(*c*) beyond reasonable limit
(*d*) more and more

13. Wines that yielded a good commercial profit __________ in the same limited areas of France as of now.
(*a*) seem to have been produced
(*b*) appear to have a remarkable semblance
(*c*) bear a significant similarity in terms of production to those grown
(*d*) appear to have been similarly produced

14. This is about __________ a sociological analysis can penetrate.
(*a*) the other limits that
(*b*) just how far into the subject that
(*c*) the relative distance that
(*d*) as far as

15. The safest general characterisation of the European philosophical tradition as it has developed up to now, with all its diverse proponents, is that it consists of a________Plato.
(*a*) series of footnotes to
(*b*) set of prologues to
(*c*) collection of chapters on
(*d*) string of commentatries to

16. The ideas that these companies used seem so simple with __________ that their competitors will now __________ themselves for not thinking of them first.
(*a*) advanced technology, hit
(*b*) analysis, abuse
(*c*) passage of time, curse
(*d*) hindsight, kick

17. No doubt, it was our own government but it was being run on borrowed ideas, using ___________, solutions.
(*a*) worn out (*b*) second hand
(*c*) impractical (*d*) appropriate

18. Overall, all the recent policy changes by the government only amount to a _________ in the sugar industry.
(*a*) superficial attemp at liberalisation
(*b*) drop in the ocean
(*c*) large change
(*d*) small regulating authority

19. In pusuance of their decision to resist what they saw as anti-labour policies, the company employees, union launched agitation to __________.
(*a*) show their virility
(*b*) reaffirm their commitment to the company.
(*c*) bring down the government
(*d*) demonstrate their strength

20. Scientists have found _______ evidence that Neanderthals, now an extinct species of humans __________ Europe, were cannibals.
(*a*) grisly, living in
(*b*) incontrovertible, a cross
(*c*) chilling inhabiting
(*d*) proper, in

21. The telephone symbolises that awkard _________ in all communication technologies ; while it _________ to bring us together, it keeps us apart.
(*a*) paradox, needs (*b*) irony, intends
(*c*) paradox , tries (*d*) irony wishes

22. The world is fast moving into a period of 'civilisational clash' in which the primary identification of people and countries will not be ____________ as during the Cold War, but ____________.
(*a*) ideological, technological
(*b*) cultural, ethnic
(*c*) political, economic
(*d*) ideological, cultural

23. In our country, the challenges are to raise _______ incomes to reduce poverty, and to _________inefficient public sector enterprises.
(*a*) farm, liberalise
(*b*) middle-class, privatise
(*c*) workers, take over
(*d*) rural, restructure

24. The interest generated by the Soccer World Cup is __________ compared to the way cricket __________ the nation.
(*a*) milder, fascinates
(*b*) lukewarm, electrifies
(*c*) tepid, inspires
(*d*) unusual, grips

25. Though one eye is kept firmly on the _________, the company now also promotes ______contemporary art.
(*a*) present, experimental
(*b*) future, popular
(*c*) present, popular
(*d*) market, popular

26. The law prohibits a person from felling a sandalwood tree, even if it grows on one's own land, without prior permission from the government. As poor people cannot deal with the government, this legal provision leads to a rip-roaring business for _________, who care neither for the _________, nor for the trees.
(*a*) middlemen, rich (*b*) the government, poor
(*c*) touts, rich (*d*) touts, poor

27. It will take some time for many South Koreans to _____________ the conflicting images of North Korea, let alone to _____________ what to make of their northern cousins.
(*a*) reconcile, decide (*b*) understand, clarify
(*c*) make out, decide (*d*) reconcile, understand

28. In these bleak and depressing times of ______ prices, non-performing governments and ___________ crime rates, Sourav Ganguly has given us, Indians, a lot to cheer about.
(*a*) escalating, increasing
(*b*) spiralling, booming
(*c*) spiralling, soaring
(*d*) ascending, debilitating

29. The manners and _________ of the nouveau riche is a recurrent _____________ in the literature.
(*a*) style, motif
(*b*) morals, story
(*c*) wealth, theme
(*d*) morals, theme

30. Many people suggest _________and still others would like to convince people not to buy pirated cassettes.

(*a*) to bring down audio-cassette prices to reduce the incidence of music piracy, others advocate strong legal action against the offenders,

(*b*) bringing down audio-cassette prices to reduce the incidents of music piracy, others are advocating strong legal action against offenders,

(*c*) bringing down audio-cassette prices to reduce the incidence of music piracy, others advocate strong legal action against offenders,

(*d*) audio-cassette prices to be brought down to reduce incidence of music piracy, others advocate that strong legal action must be taken against offenders,

31. The ancient Egyptians believed_________so that when these objects were magically reanimated through the correct rituals, they would be able to function effectively.

(*a*) that it was essential that things they portrayed must have every relevant feature shown as clearly as possible

(*b*) it was essential for things they portray to have had every relevant feature shown as clearly as possible,

(*c*) it was essential that the things they portrayed had every relevant feature shown as clearly as possible,

(*d*) that when they portrayed things, it should have every relevant feature shown as clearly as possible

32. Archaeologists believe that the pieces of red-ware pottery excavated recently near Bhavnagar and _________shed light on a hither to dark 600-year period in the Harappan history of Gujarat.

(*a*) estimated with a reasonable certainty as being about 3400 years old,

(*b*) are estimated reasonably certain to be about 3400 years old

(*c*) estimated at about 3400 years old with reasonable certainty,

(*d*) estimated with reasonable certainty to be about 3400 years old,

33. The genocides in Bosnia and Rwanda, apart from being mis-described in the most sinister and _______manner as 'ethnic cleansing', were also blamed, in further hand-washing rhetoric, on something dark and interior to ________ and perpetrators alike.

(*a*) innovative; communicator

(*b*) enchanting; leaders

(*c*) disingenuous; victims

(*d*) exigent; exploiters

34. As navigators, calendar makers, and other ____of the night sky accumulated evidence to the contrary, ancient astronomers were forced to______that certain bodies might move in circles about points, which in turn moved in circles about the earth.

(*a*) scrutinizers; believe

(*b*) observers; agree

(*c*) scrutinizers; suggest

(*d*) observers; concede

35. Every human being, after the first few days of his life, is a product of two factors: on the one hand, there is his _______ endowment; and on the other hand, there is the effect of environment, including _______.

(*a*) constitutional; weather

(*b*) congenital; education

(*c*) personal; climate

(*d*) genetic; pedagogy

36. Exhaustion of natural resources, destruction of individual initiative by governments, control over man's minds by central ______ of education and propaganda are some of the major evils which appear to be on the increase as a result of the impact of science upon minds suited by _______ to an earlier kind of world.

(*a*) tenets; fixation

(*b*) aspects; inhibitions

(*c*) institutions; inhibitions

(*d*) organs; tradition

37. But ________________ are now regularly written not just for tools, but well-established practices, organisations and institutions, not all of which seem to be ________________ away.

(*a*) reports, withering (*b*) stories, trading

(*c*) books, dying (*d*) obituaries, fading

38. The Darwin who _______________ is most remarkable for the way in which he ________________ the attributes of the world class thinker and head of the household.

(*a*) comes, figures

(*b*) arises, adds

(*c*) emerges, combines

(*d*) appeared, combines

39. Since her face was free of ______________ there was no way to ___________ if she appreciated what had happened.

(*a*) make-up, realise

(*b*) expression, ascertain

(*c*) emotion, diagnose

(*d*) scars, understand

40. In this context, the _________ of the British labour movement is particularly __________.
(*a*) affiar, weird
(*b*) activity, moving
(*c*) experience, significant
(*d*) atmosphere, gloomy

41. Indian intellectuals may boast, if they are so inclined, of being ______________ to the most elitist among the intellectual ______________ of the world.
(*a*) subordinate, traditions
(*b*) heirs, cliques
(*c*) ancestors, societies
(*d*) heir, traditions

Directions for Q. 42 to 47.

Fill the gaps in the passages below with the most appropriate word from the options given for each gap. The right words are the ones used by the author. Be guided by the author's overall style and meaning when you choose the answers.

Von Nuemann and Morgenstern assume a decision framework in which all options are thoroughly considered, each option being independent of others, with a numerical value derived for the utility of each possible outcome (these outcomes reflecting, in turn, all possible combinations of choices). The decision is then made to maximize the expected utility.

(42) _____, such a model reflects major simplifications of the way decisions are made in the real world. Humans are not able to process information as quickly and effectively as the model assumes; they tend not to think (43) _____ as easily as the model calls for; they often deal with a particular option without really assessing its (44) _____, and when they do assess alternatives, they may be extremely nebulous about their criteria of evaluation.

42. (*a*) Regrettably (*b*) Firstly
(*c*) Obviously (*d*) Apparently

43. (*a*) quantitatively (*b*) systematically
(*c*) scientifically (*d*) analytically

44. (*a*) implications (*b*) disadvantages
(*c*) utility (*d*) alternatives

In a large company, (45) _____ people is about as common as using a gun or a switch-blade to (46) _____ an argument. As a result, most managers have little or no experience of firing people, and they find it emotionally traumatic; as result, they often delay the act interminably, much as an unhappy spouse will prolong a bad marriage. And when the firing is done, it's often done clumsily, with far worse side effects than are necessary.

Do the world-class softward organizations have a different way of firing people ? No, but they do the deed swiftly, humanely, and professionally.

The key point here is to view the fired employees as a "failed product" and to ask how the process (47) _____ such a phenomenon in the first place.

45. (*a*) dismissing (*b*) punishing
(*c*) firing (*d*) admonishing

46. (*a*) resolve (*b*) thwart
(*c*) defeat (*d*) close

47. (*a*) derived (*b*) engineered
(*c*) produced (*d*) allowed

Directions for Q. 48 to 57 :

Fill up the blanks, numbered [48], [49]....up to [57], in the two passages below with the most appropriate word from the options given for each blank.

"Between the year 1946 and the year 1955, I did not file any income tax returns." With that [48] statement, Ramesh embarked on an account of his encounter with the Income Tax Department. "I originally owed Rs. 20,000 in unpaid taxes. With [49] and [50], the 20,000 became 60,000. The Income Tax Department then went into action, and I learned first hand just how much power the Tax Department wields. Royalties and trust funds can be [51]; automobiles may be [52], and auctioned off. Nothing belongs to the [53] until the case is settled."

48. (*a*) devious (*b*) blunt
(*c*) tactful (*d*) pretentious

49. (*a*) interest (*b*) taxes
(*c*) principal (*d*) returns

50. (*a*) sanctions (*b*) refunds
(*c*) fees (*d*) fines

51. (*a*) closed (*b*) detached
(*c*) attached (*d*) impounded

52. (*a*) smashed (*b*) seized
(*c*) dismantled (*d*) frozen

53. (*a*) purchaser (*b*) victim
(*c*) investor (*d*) offender

At that time the White House was as serene as a resort hotel out of season. The corridors were [54]. In the various offices, [55] gray men in waistcoats talked to one another in low-pitched voices. The only colour, or choler, curiously enough, was provided by President Eisenhower himself. Apparently, his [56] was easily set off; he scowled when he [57] the corridors.

54. (*a*) striking (*b*) hollow
(*c*) empty (*d*) white

55. (*a*) quiet (*b*) faded
(*c*) loud (*d*) stentorian

56. (*a*) laughter (*b*) curiosity
(*c*) humour (*d*) temper

57. (*a*) paced (*b*) strolled
(*c*) stormed (*d*) prowled.

ANSWERS

MCQ Type Questions

1. (*a*)	**2.** (*b*)	**3.** (*b*)	**4.** (*d*)	**5.** (*b*)	**6.** (*c*)	**7.** (*d*)	**8.** (*d*)	**9.** (*c*)	**10.** (*c*)
11. (*a*)	**12.** (*d*)	**13.** (*a*)	**14.** (*d*)	**15.** (*a*)	**16.** (*d*)	**17.** (*b*)	**18.** (*a*)	**19.** (*d*)	**20.** (*c*)
21. (3)	**22.** (*c*)	**23.** (*d*)	**24.** (*b*)	**25.** (*b*)	**26.** (*d*)	**27.** (*a*)	**28.** (*c*)	**29.** (*d*)	**30.** (*c*)
31. (*c*)	**32.** (*d*)	**33.** (*c*)	**34.** (*d*)	**35.** (*d*)	**36.** (*d*)	**37.** (*d*)	**38.** (*c*)	**39.** (*b*)	**40.** (*c*)
41. (*d*)	**42.** (*d*)	**43.** (*d*)	**44.** (*d*)	**45.** (*c*)	**46.** (*a*)	**47.** (*d*)	**48.** (*b*)	**49.** (*a*)	**50.** (*d*)
51. (*c*)	**52.** (*b*)	**53.** (*d*)	**54.** (*c*)	**55.** (*a*)	**56.** (*d*)	**57.** (*d*)			

EXPLANATIONS

MCQ TYPE QUESTIONS

1. They in choice (*a*) means 'some of our ruling politicians' (the subject). Also grammatically 'they have a right to plunder ....' is correct wheras'...... right to the punder ......' is incorrect.

2. 'Bizarre' means 'weird'. Hence, choice (*a*) is ruled out. "Its haunting images have endured ' fits syntactically. choice (*c*) can also be ruled out as 'munificence' means generosity and this is not intended in the sentence. Also, the word 'prints' in the given sentence gets with the word 'images' of choice (*b*).

3. 'They' in the part after the comma must refer to somebody or some-thing. Hence, only choice (*b*) will fit as 'they' in that case refers to 'things'.

4. The right word to precede the proposition 'on' is imposed. Tax is not collected on .... people. Tax is not placed on people, Tax is not charged on people. Hence, choices (*a*), (*b*) and (*c*) are not the answers. Tax is imposed ..... on ..... .

5. Choice (*d*) is ruled out as it does not logically fit. Choice (*a*) is also ruled out because of the expression 'tears in me'.

6. The key words are 'plunging' and 'investor confidence'. If markets are plunging, investor confidence is low. This aspect is brought out best by choice (*c*).

7. (*a*) 'Discovery'refers to something that existed previously and came to knowledge later. Home is not a discovery.

(*b*) Home is not a certain. It is a set up which comes into being as a result of a certain composition of various factors.

(*c*) Home is not a new invention of anyone at any stage of development.

So choice (*d*) is the best answer. Home is an institution just like 'marriage. Home is a reality only when its members accept certain principles.

8. (*a*) A man's toil may help to make out wealth, but it is not the ultimate aim.

(*b*) A man's toil may secure a lot of things for his children or friends . That too is not something great.

(*c*) A man's toil may help him to overcome many difficulties, but that does not give him much pleasure.

So, choice (*d*) is the best answer. What one becomes by one's toil is greater than what one gets.

9. (*a*) 'Reopens the first chapter' means begins from the very start.

(*b*) 'Writes an epilogue' means ends the book.

(*c*) 'Opens new books' means creates new problems.

The above three are only partly correct. An act of justice pardons for ever an act of vengeance creates new problems. It is a saying. "An act of justice closes the book on a misdeed; an act of vengeance writes one of its own."

10. Here, the trail is the important thing and hence the excitement of the trail is what you will be looking for.

11. (*b*) The expression must be "it is of cardinal importance".

(*c*) There is no urgency in understanding, "It should be understood" is enough.

(*d*) Instead of"it cannot be,' the expression should be it must be'.

So choice (*a*) is the best, It is important that the two things are not to be interfered with.

13. (*b*) It is not mentioned with what do wines bear semblance.

(*c*) The 'similarity' is only in terms of production. The similarity in production does not achieve commercial profit.

(*d*) For the same commercial profit wines need not be produced in the same of similar manner.

So, choice (*a*) is the best. It does not mention that the wines of similar profit are similarily produced or have semblance with one another.

14. (*a*) A sociological analysis does not stop at the outer limits of the subjects.

(*b*) No one can indicate how far into the subject can a sociological analysis go.

(*c*) 'Relative distance' refers to the distance of something else which is not mentioned.

So, choice (*d*) is the best. This indicates the probable distance to which the sociological analysis can go to.

30. Choices (*a*) and (*d*) are ruled out because 'suggest' does not go with 'to+ verb' form. Choice (2) cannot fit in the blank because of the error in parallelism. The appropriate structure is "Many people suggest.....and others advocate ....".

31. The tense in choices (*a*) and (*b*) is not consistent. Moreover, choice (*a*) uses both 'essential' and 'must'. thus rendering either one of these words redundant. In choice (*d*) 'things' the plural form cannot take it'.

32. Choice (*d*) fits into the blank in choice (*a*) the usage of the indefinite article is erroneous. Choices (*b*) and (*c*) are erroneous due to the incorrect ordering of words.

33. 'Disingenuous' means 'insincere' and this best fits the context as the phrase 'ethnic cleansing' is marked by apostrophes indicating a misrepresentation. Also, the word 'victims' is best in contrast to 'perpetrators'. Hence, option (*c*) is correct.

36. 'Navigators, calendar makers' are best clubbed under the category of 'observers' rather than 'scrutinizers', as the latter term is more used to refer to specialists in a particular field. 'Students' can also fit to a certain extent and this leaves us with options (*b*), (*d*). 'Concede' has more to do with defeat or revision which is more apt in this context than 'agree' or 'conclude'. Hence, option (*c*) is correct.

42. Going by options, 'regrettably' doesn't fit as there is nothing to regret. 'Firstly' is no choice and nothing 'secondly' is mentioned. Paragraph has nothing aggresive tone so 'obviously' does not suit. 'Apparently' is only and most suitable option.

43. All options are suitable for the space but to choose one, the best, is required here 'Analytically' appears to be best. It expresses to analyse something using statistically proven methods to reach a decision.

44. It is a very obvious choice determined from the paragraph giving it an access.

45. 'Firing' comes as best suited option as paragraph explains it as a subject of discussion.

46. An argument is always 'resolved', it is never thwart, defeat or closed.

47. Going by options makes choice easier. Only "how the process allowed......." suits logically.

48. 'Blunt' means `very direct' with respect to a remark. 'Devious' in choice (*d*) means 'dishonest' and is incorrect as the statement is a frank one. For the same reason 'tactful' in choice (*c*) is also not appropriate.

49. 'Interest' is the most appropriate work because the original amout he owed to the department accumulated to the said figure. 'Taxes is redundant while. 'principal' usually refers to a loan (in this case it isn't). 'Returns' is what one gets rather than pays.

50. 'Sanctions is used for an official permission or order that limits trade and is rarely used for an amout. 'Fees' are paid for a service and 'refunds' are what one gets back. 'Fines is the best choice because of the penalty component.

51. 'Attach' is the word used for officially taking something away. 'Impound' means 'confiscate' and is generally used in the context of illegal goods or contraband. Other choices are in no way comparable.

52. 'Smashed in choice (*a*) and 'dismantled' in choice (*c*) are far from being logical. 'Frozen is a wrong choice. 'Frozen is used in the context of making bank accounts inoperable etc. Automobiles are seized.

53. Clearly the topic is about a person who has violated incom tax regulations and he should be called an 'offender' of the law.

55. Choices (*c*) and (*d*) have similar meaning and are misnomers in the given context, as the keywords 'lowpitched voices' require a word that is synonymous with 'quiet' or 'calm'. The usage of the adjective 'faded' in the sentence to describe men is untenable.

56. The given sentence depicts the President as a very angry person as suggested by the keyword 'scowled' Choices (*a*), (*b*) and (*c*) are either positive or neutral and cannot fit in the sentence.

57. 'Paced' in choice (*a*) is not consistent with the mood of the President as depicted in the first part of the sentence. It means 'to walk fast'. 'Strolled' in (*b*) means 'Walk casually' and hence is discarded. 'Stormed' in choice (*c*) refers to 'attack a place suddenly' or 'say something in a loud, angry way'. But here, neither of them fits in. 'Prowled' refers to moving in a restless manner and hence fits.

3 CHAPTER Synonyms

INTRODUCTION

A Synonym may be defined as a word that has the same meaning or the same general meaning as a particular word in the same Language or is in some applications a more or less satisfactory application of it. Thus the synonym of a word can be an exact equivalent in meaning or should atleast convey its meaning in one way or the other or should show in what sense it can best be used.

SOME IMPORTANT WORDS WITH THEIR SYNONYMS

A

abandon : desert, for sake, jilt, leave, evacuate, quilt, vacate, abdicate, renounce.

abduct : kidnap, run away with, seize, snatch.

abet : aid, assist, back, condone, help, promote, sanction, support, sustain, uphold.

abeyance : adjournment,deferral, discontinuation, inactivity, intermission, postponement, recess.

abide : accept, bear, brook, endure, dwell, linger, live, lodge, reside, sojourn, persist.

ability : adeptness, aptitude, capability, capacity, competence, dexterity, endowment, expertise.

ablaze : afire, aflame, alight, blazing, burning, fiery, flaming, aglow, brilliant, flashing, gleaming, illuminated, radiant, sparkling.

abstain : avoid, decline, deny, desist, renounce, refrain, withhold, forgo.

abundance : affluence, ampleness, bounty, copiousness, exuberance, pro-fusion, fortune, wealth.

abysmal : bottomless, boundless, complete, deep, extreme, incalculable, profound, vast.

accede : accept, admit, agree, comply, concede, concur, consent, endorse, assume, attain.

access : admission, admittance, approach, avenue, course, entering, outburst, paroxysm.

acclaim : applaud, approve, celebrate, eulogize, exalt, hail, praise, salute, clapping.

accompany : attend, chaperon, conduct, convoy, squire, usher, supplement.

accord : agree, assent, concur, conform, harmonize, bestow, concede, endow, render, vouchsafe.

accuse : allege, blame, cite, denounce, impench, incriminate, indict, recriminate.

accustom : acclimatize, acquaint, adapt, descipline, familiarize, habituate, inure.

acquaint : advice, announce, apprise, disclose, enlighten, notify, reveal, tell.

acrid : caustic, harsh, irritating, pungent, stinging, acrimonious, mordant, nasty, sarcastic, trenchant, vitriolic.

acrimony : bitterness, harshness, churlishness, peevishness, rancour, spleen, tartness, virulence.

adamant : determined, firm, immovable, inexorable, intransigent, obdurate, resolute, stubborn, unrelenting, flinty, steely, stony.

adept : adroit, dexterous, masterful, proficient, versed, expert, genius.

adequate : capable, commensurate, competent, requisite, tolerable, sufficient.

adhere : cleave, cling, faster, obey, observe, respect.

adieu : conge, goodbye, valediction.

adjourne : defer, delay, discontinue, prorogue.

admirable : commendable, excellent, exquisite, wonderful, worthy.

admirer : bean, lover, wooer, devotee, enthusiast, votary.

adrift : drifting, unanchored, aimless, amiss, astray.

adroit : clever, cunning, deft, dexterous, ingeneous, nimble, proficient, quick-witted, able.

audacious : courageous, daring, dauntless, intrepid, rash.

austere : exacting, forbidding, grave, grim, harsh, rigorous, severe, solemn, sterr, stringent.

averse : antipathetic, disinclined, hostile, inimical, loath, reluctant, unwilling.

avid : ardent, eager, fanatical, ferrent, passionate, zealous, avaricious,

B

baptism : immersion, purification, sprinkling, debut.

bastion : citadel, defence, fortress, mainstay.

blaspheme : abuse, anathematize, curse, desecrate, execrate, profane, revile, swear.

breach : break, aperture, charm, cleft, fissure, infraction, infringement, estrangement.

broaden : augment, develop, fatten, spread, stretch.

brood : agonize, dwell, mope, ponder, repine, ruminate.

C

catastrophe: affliction, adversity, cataclysm, devastation, disaster, fiasco, culmination.

chaos : anarchy, bedlam, confusion, disorganization. entropy, tumult, lawlessness,

chastity : celibacy, continence, innocence, modesty, purity.

clan : band, brotherhood, fraternity, gens.

clinch : assure, conclude, confirm, determine, secure.

colloquial : conversational, demotic, everyday, familiar, idiomatic, informal, vernacular.

commotion : ado, agitation, bustle, disorder, disturbance, excitement.

complicate : confuse, entangle, involve, muddle.

contour : curve, figure, lines, profile, relief.

conviction : assurance, certainty, confidence, ferrour.

creditable : admirable, commendable, deserving, estimable, honourable, loudable.

crest : apex, zenith, pinnacle, summit, ridge.

culminate : climax, close, conclude, finish.

cynical : sarcastic, scornful, sneering, distrustful.

D

daunt : dismay, alarm, appal, frighten, intimidate, scare, subdue, terrify.

deed : achievement, act, action, exploit, feat.

deflate : collapse, confract, empty, exhaust, shrink, chasten.

deliberate : cogitate, consult, debate, mediate, ponder, reflect.

diligent : active, attentive, busy, careful, constant.

dire : alarming, appalling, awful, catastrophic, cruel, disastrous.

discreet : careful, cautious, circumspect, diplomatic, judicious, politic, prudent.

dissent : decline, differ, disagree, protest, refuse, disagreement, discord.

ditch : channel, drain, furrow, trench, discord.

dodge : dant, duck, shift, sidestep, avoid, deceive.

dubious : doubtful, hesitant, uncertain, unconvinced, undecided, unsure, ambiguous

dwell : abide, establish, inhabit, settle, sojourn.

E

edgy : anxious, irritable, nervous, touchy.

elusive : difficult to catch, shifty, slippery, tricky, baffling, fleeting, puzzling, subtle.

emblem : badge, gest, device, figure, image, representation.

enchant : beguile, bewitch, captivate, charm, delight, enamour, enrapture, fascinate, hypnotize.

enrage : aggrarate, anger incense, incite, inflame, irritate, provoke.

exalt : advance, dignify, elevate, honour, promote, acclaim, applaud, extol.

exhilarating: breathtaking, cheering, exalting, exciting, vigorating, stimulating.

expedite : accelerate, advance, assist, dispatch, facilitate, hasten, precipitate.

F

fame : celebrity, eminence, glory, honour, prominence, renown, repute, stardom.

famine : scarcity, dearth, destitution, hunger, starvation.

feeble : weak, delicate, etiolated, exhausted, infirm, languid, puny, shilpit.

fierce : barbarous, brutal, cruel, dan-gerous, ferocious, murderous, passionate, savage.

fishy : doubtful, dubious, funny, improbable, suspect, suspicious, deadpan, dull.

flawless : faultless, impeccable, perfect, unblemished.

foresee : anticipate, divine, envisage, forebode, portend, predict, prophesy, presage.

fortress : castle, citadel, fastness, redoubt, stronghold.

fraternity : brotherhood, association, company, comradship.

fretful : captious, complaining, fractions, irritable, peevish, splenetic, touchy.

G

gaiety : animation, blitheness, cheerfulness, elation, exhilaration, glee, humour.

gaudy : bright, brilliant, flashy, florid, garish, glaring, meretricious, showy, vulgar.

gestation : development, evolution, incubation, pregnancy, repening.

gracious : affable, amiable, beneficent, benevolent, cheritable, chivalrous, cordial.

grimace : face, frown, mouth, scowl, sneer.

grubby : dinty, filthy, frowzy, manky, shabby.

H

haggle : bargain, barter, beat down, dicker, higgle, dispute, quarrel, wrangle.

haphazard : accidental, arbitrary, chance, fluky, random, aimless, careless.

hatred : abomination, animosity, animus, antagonism, antipathy, enmity.

hectic : animated, boisterous, chaotic, frantic, furious, turbulent, wild.

hilarious : amusing, comical, convivial, exhilarated, humorous, jolly, jovial, joyful, joyous.

holocaust : annihilation, carnage, destruction, devastation, massacre.

hurricane : cyclone, gale, storm, tempest, tornado, twister.

I

idle : dead, empty, gathering, dust, inactive, jobless, redundant.

immense : enormous, extensive, giant, gigantic, huge, illimitable,

: interminable, jumbo.

implicated : incriminated, involved, suspected, under suspicion.

impudence : assurance, audacity, boldness, bumptiousness, insolence.

incense : anger, enrage, excite, aroma, balm, fragrance, redolence, scent.

indent : order, request, dint, nick, notch.

inept : awkward, bumbling, bungling, clumsy, gauche, inexpert, maladroit, absurd.

insolent : abusive, bold, contemptuous, impertinent, impudent, insulting, rude, saucy.

instigate : actuate, encourage, forment, impel, incite, influence, provoke, persuade.

J, K, L, M

jaunty : airy, breezy, buoyant, carefree, dapper, gay, lively, perky, showy, smart.

jurisdiction : authority, command, control, influence, prerogative, sway.

keepsake : emblem, favour, memento, relic, reminder, souvenir, symbol, token.

latch : bar, bolt, catch, fastening, lock, sneck.

link : component, constituent, division, association, attachment.

meander : ramble, snake, straving, stroll, wander, bend, curve, loop, twist.

melancholy : dejection, depression, despondency, gloom.

N, O, P, Q

nostalgic : homesick, longing, regretful, sentimental, wistful.

obnoxious : abhorrent, abominable, detestable, disagreeable, disgusting, hateable, horrid.

onerous : burdensome, crushing, demanding, difficult, exciting, exhausting, exigent.

paw : grab, handle, roughly, manhandle, maul, molest.

percolate : drain, drip, exude, filter, leache, ooze, penetrate, permeate, pervade.

piety : devotion, dutifulness, faith, grace, piousness, religion, sanctify, veneration.

pudding : dessert, sweet.

quandary : bewilderment, stick, delicate, situation, difficult, dilemma, embarrassment.

quiver : agitate, oscillate, palpitate, pulsate, shudder, tremble, vibrate.

R, S, T

rapt : absorbed, carried away, enthralled, entranced, fascinated, gripped, intent.

reveal : announce, betray, broadcast, communicate, disclose, divulge, impart, proclaim.

rowdy : bristerous, disordely, loutish, noisy, unruly, uproarious, brawler.

sage : acute, canny, intelligent, judicious, polite, prudent, sagacious.

scorn : contempt, despite, disdain, disparagement, slight, sneer, curl, condemn.

shake : fluctuate, bump, joggle, jounce, quake, shiver, shudder, totter, tremble.

sinner : offender, reprobate, trespasser.

snigger : giggle, laugh, smirk, sneer, titter.

stimulus : encouragement, incentive, incitement, inducement, provocation.

tepid : slightly warm, luke warm.

timid : afraid, bashful, cowardly, diffident.

U, V, W, Y, Z

unmask : disclose, discover, expose, reveal, unveil.

urbane : civil, civilized, cosmopolitan, courteous, cultivated, debonair.

vehement : ardent, eager, earnest, emphatic, enthusiastic, fervent, fervid, fierce, forcible.

weired : bizarre, creepy, cerie, freakish, ghostly, mysterious, queer, spooky.

writhe : contont, distant, jerk, struggle, thrash, wiggle, twist, wriggle.

yearn : ache, covet, crave, desire, hanker, hunger, itch, languish.

zenith : acme, apex, apogee, climax, height, meridian, peak, pinnacle, summit, top.

EXERCISE

MCQ TYPE QUESTIONS

Directions: Each of the questions below consists of a word followed by four lettered words or phrases. Choose the lettered word or phrase that is most nearly same in meaning to the word.

1. Abase
(*a*) incur (*b*) tax
(*c*) estimate (*d*) humiliate

2. Aberration
(*a*) deviation (*b*) abhorrence
(*c*) dislike (*d*) absence

3. Abet
(*a*) conceive (*b*) wager
(*c*) encourage (*d*) evade

4. Aabeyance
(*a*) obedience (*b*) discussion
(*c*) excitement (*d*) suspended action

5. Abjure
(*a*) discuss (*b*) renounce
(*c*) run off secretly (*d*) perjure

6. Ablution
(*a*) censure (*b*) forgiveness
(*c*) mutiny (*d*) washing

7. Abnegation
(*a*) blackness (*b*) self-denial
(*c*) selfishness (*d*) cause

8. Aborigine
(*a*) first design
(*b*) absolution
(*c*) finale
(*d*) primitive inhabitant

9. Abortive
(*a*) unsuccessful
(*b*) consuming
(*c*) financing
(*d*) familiar

10. Abstinence
(*a*) restrained (*b*) vulgar display
(*c*) deportment (*d*) reluctance

11. Abstruse
(*a*) profound (*b*) irrespective
(*c*) suspended (*d*) protesting

12. Abut
(*a*) stimulate (*b*) grasp
(*c*) oppose (*d*) adjoin

13. Abysmal
(*a*) bottomless (*b*) eternal
(*c*) meteric (*d*) diabolic

14. Accede
(*a*) fail (*b*) compromise
(*c*) correct (*d*) consent

15. Acclivity
(*a*) index
(*b*) report
(*c*) upslope of a hill
(*d*) negotiator

16. Astute
(*a*) sheer (*b*) noisy
(*c*) astral (*d*) clever

17. Atrocity
(*a*) endurance (*b*) fortitude
(*c*) session (*d*) heinous act

18. Atrophy
(*a*) capture (*b*) waste away
(*c*) govern (*d*) award prize

19. Attenuate
(*a*) appear (*b*) be absent
(*c*) weaken (*d*) testify

20. Atypical
(*a*) superfluous (*b*) fortitude
(*c*) unusual (*d*) clashing

21. Audacity
(*a*) boldness (*b*) asperity
(*c*) strength (*d*) stature

22. Augment
(*a*) make noble (*b*) anoint
(*c*) increase (*d*) harvest

23. Auxiliary
(*a*) righteous (*b*) prospective
(*c*) assistant (*d*) archaic

24. Avarice
(*a*) easiness (*b*) greed
(*c*) statement (*d*) invoice

25. Avert
(*a*) entertain (*b*) transform
(*c*) turn away (*d*) lead toward

26. Awry
(*a*) recommended
(*b*) commiserating
(*c*) started
(*d*) crooked

27. Baleful
(*a*) doubtful (*b*) virtual
(*c*) deadly (*d*) conventional

28. Balmy
(*a*) venturesome (*b*) dedicated
(*c*) mild (*d*) fanatic

29. Banal
(*a*) philosophical (*b*) trite
(*c*) dramatic (*d*) heedless

30. Baneful
(*a*) intellectual (*b*) thankful
(*c*) decisive (*d*) poisonous

31. Boisterous
(*a*) conflicting (*b*) noisy
(*c*) testimonial (*d*) grateful

32. Bombastic
(*a*) sensitive (*b*) pompous
(*c*) rapid (*d*) sufficient

33. Bouillon
(*a*) insight
(*b*) chowder
(*c*) gold
(*d*) clear soup

34. Brackish
(*a*) careful (*b*) salty
(*c*) chosen (*d*) tough

35. Braggadocio
(*a*) weaponry (*b*) boasting
(*c*) skirmish (*d*) encounter

36. Brazen
(*a*) shameless (*b*) quick
(*c*) modest (*d*) pleasant

37. Brindled
(*a*) equine (*b*) pathetic
(*c*) hasty (*d*) spotted

38. Brochure
(*a*) opening (*b*) pamphlet
(*c*) censor (*d*) bureau

39. bucolic
(*a*) diseased (*b*) replusive
(*c*) rustic (*d*) twinkling

40. Buxom
(*a*) voluminous (*b*) indecisive
(*c*) convincing (*d*) plump

41. Cache
(*a*) lock
(*b*) hiding place
(*c*) tide
(*d*) automobile

42. Cacophony
(*a*) discord
(*b*) dance
(*c*) applause
(*d*) type of telephone

43. Callow
(*a*) youthful (*b*) holy
(*c*) mild (*d*) colored

44. Candid
(*a*) vague (*b*) outspoken
(*c*) experienced (*d*) anxious

45. Chaste
(*a*) loyal (*b*) timid
(*c*) curt (*d*) pure

46. Chide
(*a*) unite (*b*) fear
(*c*) record (*d*) scold

47. Chimerical
(*a*) developing (*b*) brief
(*c*) distant (*d*) fantastic

48. Choleric
(*a*) musical (*b*) episodic
(*c*) hot headed (*d*) global

49. Churlish
(*a*) marine (b) economical
(*c*) impolite (*d*) compact

50. Ciliated
(*a*) foolish (*b*) swift
(*c*) early (*d*) hairy

51. Circuitous
(*a*) indirect (*b*) complete
(*c*) obvious (*d*) aware

52. City
(*a*) galvanize (*b*) visualize
(*c*) locate (*d*) quote

53. Clandestine
(*a*) abortive (*b*) secret
(*c*) tangible (*d*) doomed

54. Claustrophobia
(*a*) lack of confidence
(*b*) fear of spiders
(*c*) love of books
(*d*) fear of closed places

55. Cleft
(*a*) split (*b*) waterfall
(*c*) assembly (*d*) adherence

56. Cliche
(*a*) increase (*b*) vehicle
(*c*) morale (*d*) platitude

57. Coerce
(*a*) recover (*b*) begin
(*c*) force (*d*) license

58. Cognizance
(*a*) policy (*b*) knowledge
(*c*) advance (*d*) omission

59. Cohere
(*a*) hold together
(*b*) occur simultaneously
(*c*) recollect
(*d*) materialize

60. Conjecture
(*a*) magic (*b*) guess
(*c*) position (*d*) form

61. Connoisseur
(*a*) gourmand (*b*) lover of art
(*c*) humidor (*d*) delinquent

62. Consanguinity
(*a*) kinship (*b*) friendship
(*c*) bloodletting (*d*) relief

63. Consensus
(*a*) general agreement (*b*) project
(*c*) insignificance (*d*) sheaf

64. Construe
(*a*) explain (*b*) promote
(*c*) reserve (*d*) erect

65. Contaminate
(*a*) arrest (*b*) prepare
(*c*) pollute (*d*) beam

66. Contentious
(*a*) squealing (*b*) surprising
(*c*) quarrelsome (*d*) smug

67. Continence
(*a*) humanity (*b*) research
(*c*) embryology (*d*) self-restraint

68. Contraband
(*a*) purpose (*b*) rogue
(*c*) rascality (*d*) difficulty

69. Contrite
(*a*) smart
(*b*) penitent
(*c*) restful
(*d*) recognized

70. Controvert
(*a*) turnover (*b*) contradict
(*c*) mind (*d*) expolain

71. Convene
(*a*) propose (*b*) restore
(*c*) question (*d*) gather

72. Conversant
(*a*) ignorant
(*b*) speaking
(*c*) incorporated
(*d*) familiar

73. Copious
(*a*) plentiful (*b*) cheating
(*c*) dishonorable (*d*) adventurous

74. Corpulent
(*a*) regenerate (*b*) obese
(*c*) different (*d*) hungry

75. Decimate
(*a*) kill (*b*) disgrace
(*c*) search (*d*) collide

76. Declivite
(*a*) trap (*b*) quadrangle
(*c*) quarter (*d*) downward slope

77. Decollete
(a) flavored (b) demure
(c) flowery (d) low-necked

78. Decorous
(a) momentary (b) emotional
(c) suppressed (d) proper

79. Decrepitude
(a) feebleness (b) disease
(c) coolness (d) melee

80. Default
(a) failure to act
(b) tendency to err
(c) desire to remedy
(d) debt

81. Defection
(a) determination (b) desertion
(c) invitation (d) affection

82. Defile
(a) maincure (b) ride
(c) pollute (d) assemble

83. Degraded
(a) surpprised (b) lowered
(c) ascended (d) learned

84. Deleterious
(a) delaying
(b) experimental
(c) harmful
(d) graduating

85. Deluge
(a) confusion (b) deception
(c) flood (d) mountain

86. Denigrate
(a) refuse (b) blacken
(c) terrify (d) admit

87. Denouement
(a) action (b) scenery
(c) resort (d) solution

88. Depravity
(a) wickedness (b) sadness
(c) heaviness (d) tidiness

89. Deranged
(a) insane (b) systematic
(c) neighborly (d) alphabetical

90. Dwindle
(a) blow (b) inhabit
(c) spin (d) lessen

91. Ecstasy
(a) joy (b) speed
(c) treasure (d) warmth

92. Edify
(a) mystify (b) suffice
(c) improve (d) erect

93. Efface
(a) countenance
(b) encourage
(c) recognize
(d) rub out

94. Effigy
(a) requisition (b) organ
(c) charge (d) dummy

95. Egregious
(a) pious (b) shocking
(c) anxious (d) sociable

96. Egress
(a) entrence (b) bird
(c) exit (d) double

97. Elated
(a) debased
(b) respectful
(c) drooping
(d) gay

98. Elusive
(a) deadly (b) eloping
(c) evasive (d) simple

99. Emaciated
(a) garrulous (b) primeval
(c) vigorous (d) thin

100. Emancipate
(a) set free (b) take back
(c) make worse (d) embolden

101. Embellish
(a) doff (b) don
(c) abscond (d) adon

102. Embroil
(a) cherish (b) overtheat
(c) entangle (d) assure

103. **Emendation**
(a) correction
(b) interpretation
(c) exhumation
(d) inquiry

104. **Eminent**
(a) purposeful (b) high
(c) delectable (d) currious

105. **Flored**
(a) ruddy (b) rusty
(c) ruined (d) patient

106. **Foil**
(a) bury (b) frustrate
(c) shield (d) desire

107. **Foment**
(a) spoil (b) instigate
(c) interrogate (d) spray

108. **Foolhardy**
(a) strong (b) unwise
(c) brave (d) futile

109. **Foppish**
(a) scanty (b) radical
(c) orthodox (d) dandyish

110. **Foray**
(a) exursion (b) contest
(c) ranger (d) intuition

111. **Formidable**
(a) dangerous
(b) outlandish
(c) grandiloquent
(d) impenetrable

112. **Foster**
(a) accelerate (b) fondle
(c) become infected (d) raise

113. **Franchise**
(a) subway (b) discount
(c) license (d) reason

114. **Fritter**
(a) sour (b) chafe
(c) dissipate (d) cancel

115. **Frugality**
(a) foolishness (b) extremity
(c) indifference (d) economy

116. **Fulminate**
(a) fulfill (b) contemplate
(c) talk nonsense (d) protest

117. **Furor**
(a) excitement (b) worry
(c) flux (d) anteroom

118. **Furtive**
(a) underhanded (b) coy
(c) brilliant (d) quick

119. **Gadfly**
(a) humorist
(b) nuisance
(c) scholar
(d) bum

120. **Galleon**
(a) liquid measure (b) ship
(c) armada (d) company

121. **Garisth**
(a) sordid (b) flashy
(c) prominent (d) lusty

122. **Garner**
(a) prevent (b) assist
(c) collect (d) compute

123. **Garnish**
(a) paint (b) gamer
(c) adorn (d) abuse

124. **Garrulity**
(a) credulity
(b) senility
(c) loquaciousness
(d) speciousness

125. **Garrulous**
(a) arid (b) hasty
(c) sociable (d) talkative

126. **Gauche**
(a) rigid (b) swift
(c) awkward (d) taciturn

127. **Gaudy**
(a) holy (b) showy
(c) sentimental (d) mild

128. **Gaunt**
(a) victorious (b) tiny
(c) stylish (d) haggard

129. Genuflect
(*a*) falsify (*b*) trick
(*c*) project (*d*) bend the knee

130. Germane
(*a*) bacteriological
(*b*) middle european
(*c*) prominent
(*d*) relevant

131. Germtinal
(*a*) creative (*b*) excused
(*c*) sterilized (*d*) primitive

132. Gist
(*a*) chaff (*b*) summary
(*c*) expostulation (*d*) expiation

133. Glib
(*a*) slippery (*b*) fashionable
(*c*) antiquated (*d*) articulate

134. Gnome
(*a*) fury (*b*) giant
(*c*) dwarf (*d*) native

135. Inclement
(*a*) unfavorable (*b*) abandoned
(*c*) kindly (*d*) selfish

136. Incompatible
(*a*) capable
(*b*) reasonable
(*c*) faulty
(*d*) alienated

137. Inconsequential
(*a*) disorderly (*b*) insignificant
(*c*) subsequent (*d*) insufficient

138. Incontinent
(*a*) insular (*b*) complaisant
(*c*) crass (*d*) wanton

139. Incorrigible
(*a*) narrow (*b*) straight
(*c*) inconceivable (*d*) unreformable

140. Incriminate
(*a*) exacerbate (*b*) involve
(*c*) intimidate (*d*) lacerate

141. Inculcate
(*a*) exculpate (*b*) educate
(*c*) exonerate (*d*) prepare

142. Indigent
(*a*) lazy (*b*) pusillanimous
(*c*) penurious (*d*) affluent

143. Indigniy
(*a*) pomposity (*b*) bombast
(*c*) obeisance (*d*) insult

144. Indolence
(*a*) sloth (*b*) poverty
(*c*) latitude (*d*) aptitude

145. Indubitably
(*a*) flagrantly (*b*) doubtfully
(*c*) carefully (*d*) cortainly

146. Inebriety
(*a*) revelation (*b*) drunkenness
(*c*) felony (*d*) starvation

147. Inept
(*a*) outward (*b*) spiritual
(*c*) foolish (*d*) clumsy

148. Infallible
(*a*) final
(*b*) unbelievable
(*c*) perfect
(*d*) inaccurate

149. Infirmity
(*a*) disability (*b*) age
(*c*) inoculation (*d*) hospital

150. Ingenuous
(*a*) clever (*b*) stimulating
(*c*) naive (*d*) worried

151. Inimical
(*a*) antagonistic
(*b*) anonymous
(*c*) fanciful
(*d*) accurate

152. Innocuous
(*a*) not capable (*b*) not dangerous
(*c*) not eager (*d*) not frank

153. Insinuate
(*a*) resist (*b*) suggest
(*c*) report (*d*) rectify

154. Insipid
(*a*) witty (*b*) flat
(*c*) wily (*d*) talkative

155. Integrate
(*a*) tolerate (*b*) unite
(*c*) flow (*d*) copy

156. Inter
(*a*) bury (*b*) amuse
(*c*) relate (*d*) frequent

157. Interdict
(*a*) acclain (*b*) dispute
(*c*) prohibit (*d*) decide

158. Intermittent
(*a*) heavy (*b*) fleet
(*c*) occasional (*d*) fearless

159. Intractable
(*a*) culpable (*b*) flexible
(*c*) unruly (*d*) efficient

160. Intransigence
(*a*) lack of training (*b*) stubbornness
(*c*) novelty (*d*) timidity

161. Intrepid
(*a*) cold (*b*) hot
(*c*) understood (*d*) courageous

162. Intrinsic
(*a*) extrinsic (*b*) abnormal
(*c*) above (*d*) basic

163. Inundate
(*a*) abuse (*b*) deny
(*c*) swallow (*d*) flood

164. Inveigh
(*a*) speak violently (*b*) orate
(*c*) disturb (*d*) apply

165. Lethargy
(*a*) dynamism (*b*) orate
(*c*) disturb (*d*) apply

166. Loquacious
(*a*) verbose (*b*) tacturn
(*c*) rational (*d*) alluring

167. Loath
(*a*) tired (*b*) unwilling
(*c*) sickly (*d*) spirited

168. Mawkish
(*a*) sentimental (*b*) true
(*c*) certain (*d*) devious

169. Mediocre
(*a*) average (*b*) bitter
(*c*) medieval (*d*) industrial

170. Melee
(*a*) heat (*b*) brawl
(*c*) attempt (*d*) weapon

171. Mellifluous
(*a*) porous (*b*) honeycombed
(*c*) strong (*d*) smooth

172. Menial
(*a*) intellectual (*b*) clairvoyant
(*c*) servile (*d*) arrogant

173. Mentor
(*a*) guide (*b*) genius
(*c*) talker (*d*) philosopher

174. Mesmerize
(*a*) remember (*b*) hypnotize
(*c*) delay (*d*) bore

175. Meticulous
(*a*) steadfast (*b*) recent
(*c*) quaint (*d*) painstaking

176. Miasma
(*a*) dream (*b*) noxious fumes
(*c*) scenario (*d*) quantity

177. Militant
(*a*) combative (*b*) dramatic
(*c*) religious (*d*) quaintity

178. Minon
(*a*) monster (*b*) quorum
(*c*) majority (*d*) dependent

179. Mirage
(*a*) dessert (*b*) illusion
(*c*) water (*d*) mirror

180. Misanthrope
(*a*) benefactor (*b*) philanderer
(*c*) man-hater (*d*) aesthete

181. Mischance
(*a*) gamble (*b*) ordinance
(*c*) aperture (*d*) anecdote

182. Misdemeanor
(*a*) felony (*b*) peccadillo
(*c*) indignity (*d*) flat

183. Modish
(*a*) sentimental (*b*) stylish
(*c*) vacillating (*d*) contrary

184. Mollify
(*a*) avenge (*b*) attenuate
(*c*) attibute (*d*) appease

185. Monetary
(*a*) boring (*b*) fascinating
(*c*) fiscal (*d*) stationary

186. Moratorium
(*a*) burial (*b*) gathering
(*c*) delay (*d*) refusal

187. Mordant
(*a*) dying (*b*) trenchant
(*c*) fabricating (*d*) controlling

188. Moribund
(*a*) dying (*b*) trenchant
(*c*) fabricating (*d*) controlling

189. Motley
(*a*) active (*b*) disguised
(*c*) variegated (*d*) somber

190. Muggy
(*a*) attacking (*b*) fascinating
(*c*) humid (*d*) characteristic

191. Mulct
(*a*) swindle (*b*) hold
(*c*) record (*d*) pring

192. Multilingual
(*a*) variegated (*b*) polyglot
(*c*) multilateral (*d*) polyandrous

193. Mundane
(*a*) global (*b*) futile
(*c*) spatial (*d*) worldly

194. Munificent
(*a*) grandiose (*b*) puny
(*c*) philanthropic (*d*) poor

195. Musty
(*a*) flat (*b*) necessary
(*c*) indifferent (*d*) nonchalant

196. Myopic
(*a*) visionary (*b*) nearsighted
(*c*) moral (*d*) glassy

197. Nascent
(*a*) incipient (*b*) ignorant
(*c*) loyal (*d*) treacherous

198. Phlegmatic
(*a*) calm (*b*) cryptic
(*c*) practical (*d*) salivary

199. Physiognomy
(*a*) posture (*b*) head
(*c*) physique (*d*) face

200. Piebald
(*a*) motley (*b*) coltish
(*c*) hairless (*d*) thoroughbred

201. Pillage
(*a*) hoard (*b*) plunder
(*c*) versify (*d*) denigrate

202. Pinion
(*a*) express (*b*) report
(*c*) reveal (*d*) restrain

203. Pinnacle
(*a*) foothills (*b*) card game
(*c*) pass (*d*) peak

204. Pious
(*a*) historic (*b*) devout
(*c*) multiple (*d*) fortunate

205. Pique
(*a*) pyramid (*b*) revolt
(*c*) resentment (*d*) struggle

206. Placate
(*a*) determine (*b*) transmit
(*c*) pacify (*d*) allow

207. Plagiarism
(*a*) theft of funds (*b*) theft of ideas
(*c*) belief in god (*d*) arson

208. Plaintive
(*a*) moumful (*b*) senseless
(*c*) persistent (*d*) rural

209. Platitude
(*a*) fatness (*b*) bravery
(*c*) dimension (*d*) trite remark

210. Polemic
(*a*) blackness (*b*) lighting
(*c*) magnetism (*d*) controversy

ANSWERS

MCQ Type Questions

1. (*d*)	**2.** (*a*)	**3.** (*c*)	**4.** (*d*)	**5.** (*b*)	**6.** (*d*)	**7.** (*b*)	**8.** (*d*)	**9.** (*a*)	**10.** (*a*)
11. (*a*)	**12.** (*d*)	**13.** (*a*)	**14.** (*d*)	**15.** (*d*)	**16.** (*d*)	**17.** (*d*)	**18.** (*b*)	**19.** (*c*)	**20.** (*c*)
21. (*a*)	**22.** (*c*)	**23.** (*c*)	**24.** (*b*)	**25.** (*c*)	**26.** (*d*)	**27.** (*c*)	**28.** (*c*)	**29.** (*b*)	**30.** (*d*)
31. (*b*)	**32.** (*b*)	**33.** (*d*)	**34.** (*b*)	**35.** (*b*)	**36.** (*a*)	**37.** (*d*)	**38.** (*b*)	**39.** (*c*)	**40.** (*d*)
41. (*b*)	**42.** (*a*)	**43.** (*a*)	**44.** (*c*)	**45.** (*d*)	**46.** (*d*)	**47.** (*d*)	**48.** (*c*)	**49.** (*c*)	**50.** (*d*)
51. (*a*)	**52.** (*d*)	**53.** (*b*)	**54.** (*d*)	**55.** (*a*)	**56.** (*d*)	**57.** (*c*)	**58.** (*b*)	**59.** (*a*)	**60.** (*b*)
61. (*b*)	**62.** (*a*)	**63.** (*a*)	**64.** (*a*)	**65.** (*c*)	**66.** (*c*)	**67.** (*d*)	**68.** (*d*)	**69.** (*b*)	**70.** (*b*)
71. (*d*)	**72.** (*d*)	**73.** (*a*)	**74.** (*b*)	**75.** (*a*)	**76.** (*d*)	**77.** (*d*)	**78.** (*d*)	**79.** (*a*)	**80.** (*a*)
81. (*b*)	**82.** (*c*)	**83.** (*b*)	**84.** (*c*)	**85.** (*c*)	**86.** (*b*)	**87.** (*d*)	**88.** (*a*)	**89.** (*a*)	**90.** (*d*)
91. (*a*)	**92.** (*c*)	**93.** (*d*)	**94.** (*d*)	**95.** (*b*)	**96.** (*c*)	**97.** (*d*)	**98.** (*c*)	**99.** (*d*)	**100.** (*a*)
101. (*d*)	**102.** (*c*)	**103.** (*a*)	**104.** (*b*)	**105.** (*a*)	**106.** (*b*)	**107.** (*b*)	**108.** (*b*)	**109.** (*d*)	**110.** (*a*)
111. (*a*)	**112.** (*c*)	**113.** (*c*)	**114.** (*c*)	**115.** (*d*)	**116.** (*d*)	**117.** (*a*)	**118.** (*a*)	**119.** (*b*)	**120.** (*b*)
121. (*b*)	**122.** (*c*)	**123.** (*c*)	**124.** (*c*)	**125.** (*d*)	**126.** (*c*)	**127.** (*b*)	**128.** (*d*)	**129.** (*d*)	**130.** (*d*)
131. (*a*)	**132.** (*b*)	**133.** (*d*)	**134.** (*c*)	**135.** (*a*)	**136.** (*d*)	**137.** (*b*)	**138.** (*d*)	**139.** (*d*)	**140.** (*b*)
141. (*b*)	**142.** (*c*)	**143.** (*d*)	**144.** (*a*)	**145.** (*d*)	**146.** (*b*)	**147.** (*d*)	**148.** (*c*)	**149.** (*a*)	**150.** (*c*)
151. (*a*)	**152.** (*b*)	**153.** (*b*)	**154.** (*b*)	**155.** (*a*)	**156.** (*a*)	**157.** (*c*)	**158.** (*c*)	**159.** (*c*)	**160.** (*b*)
161. (*d*)	**162.** (*d*)	**163.** (*d*)	**164.** (*a*)	**165.** (*b*)	**166.** (*a*)	**167.** (*b*)	**168.** (*d*)	**169.** (*a*)	**170.** (*b*)
171. (*d*)	**172.** (*c*)	**173.** (*a*)	**174.** (*b*)	**175.** (*d*)	**176.** (*b*)	**177.** (*a*)	**178.** (*d*)	**179.** (*b*)	**180.** (*c*)
181. (*d*)	**182.** (*b*)	**183.** (*b*)	**184.** (*d*)	**185.** (*c*)	**186.** (*c*)	**187.** (*b*)	**188.** (*a*)	**189.** (*c*)	**190.** (*c*)
191. (*a*)	**192.** (*b*)	**193.** (*d*)	**194.** (*c*)	**195.** (*a*)	**196.** (*b*)	**197.** (*a*)	**198.** (*a*)	**199.** (*d*)	**200.** (*a*)
201. (*b*)	**202.** (*d*)	**203.** (*d*)	**204.** (*b*)	**205.** (*c*)	**206.** (*c*)	**207.** (*b*)	**208.** (*a*)	**209.** (*d*)	**210.** (*d*)

4 CHAPTER

Antonyms

The basic idea of a antonym question is to find an opposite for a word. Antonym items are first and foremost a test of vocabulary.

If you have no idea of the meaning of the given word, there's not much you can do. When you don't know the meaning of an antonym, don't waste a lot of time trying to fiture it out. In other words, once you recognize that you are out of ammunition, just do a random guess, confirm it, and move on the next item.

Each antonym starts with a capitalized word, which is followed by four words of phrases in lowercase letters. The task is to find the word or phrase that is most nearly opposite in meaning to the bold word.

Plan to Answer Antonym Questions

1. Define the capitalized word.
2. Think of a meaning that is opposite to this word.
3. Read all the answer choices. Eliminate those that do not relate to the meaning you thought of. If only one choice remains, mord it and go on.
4. If more than one choice remains, go back and refine your thinking about the bold word.
5. Pick the answer choice that is most nearly opposite to the bold word.

Tedious

(*a*) unlimited (*b*) confined
(*c*) enthralling (*d*) appetizing

(*i*) The capitalized word means "boring" or "tiresome"

(*ii*) The opposite of "boring" would be something like *exciting or interesting*.

(*iii*) Looking at all the answer choices, you can immediately eliminate (*a*), and (*b*) as having nothing to do with excitement. But that leaves (*c*) *enthralling and* (*d*) *apperizing*.

(*iv*) Of the two remaining choices, *enthralling, which means "captivating"* or fascinating," is more nearly opposite in meaning to *tedious than appetizing*, which means "savory," or "delicious".

(*v*) Mark (*c*), enthralling, as the correct choice.

Note :

(*i*) **Makes sure to read all the answers :** To answer antonym questions you'll sometimes have to distinguish fine shades of meaning. That's why you have to consider very answer choice.

(*ii*) **Verb or Noun?** If the part of speech of the capitalized word is ambiguous, check the answer choices.

6. Sometimes the test writer selects a word you are likely to be familiar with, but sets up the question to test a meaning you do not ordinarily associate with the word. Here is an example:

e.g. precipitous

(*a*) pleasantly sweet
(*b*) overly ambitious
(*c*) agreeably situated
(*d*) gently sloping

The best choice is (*d*). We most often use the word precipitous to mean rash or foothardy, but its central meaning is related to precipice, a sharp drop-off.

7. **Eleminate non-answers**
The simple way to eleminate answer choices is to toss out any words that don't have opposite meanings.

8. **Concentrate on answer choices**
Words can have different meanings based upon their part of speech. If you are uncertain about the part of speech of the bold word, just check the naswer choices. If they're all verbs, for example, so is the bold word.

9. **Play around with the part of speech of a word**
You can alter the part of speech of the bold word and answer choices in your own mind. Sometimes an antonym will use a word you know but as a part of speech that is unfamiliar to you.

***e.g.* sublimity**

(*a*) erosion (*b*) baseness
(*c*) conciseness (*d*) insistence

The best choice is (*b*). You may know the word sublimity better as the adjective sublime, meaning "lofty, high, or noble". So you may find it easire to think about the antonym by changing *sublimity* to the more familiar form *sublime*". As you think about each answer choice, you would then change it in your mind to an adjective *Baseness*, therefore, would become *base*; and *base* is an opposite of *sublime*.

10. **Use word connotations to eliminate answer choices**
Even if you don't know the exact meaning of a word, you may have a vague recollection of the context in which you first encountered it. So you may know whether the word has positive overtones or negative ones. This recollection may be sufficient to get a correct answer.

***e.g.* raffish**

(*a*) grotesque (*b*) delinquent
(*c*) uncaring (*d*) noble

The correct choice is (*d*). Let's assume that you do not know that is means "low, vulgar, the base. 'Let's further assume, however, that you have a vague knowledge of the word. You've seen it used to describe a character who is dishonest and not trustworthy. So even though you don't know the exact meaning of the word, you know that it has negative overtones. Sine you an looking for the opposite of a word with negative overtones, you would eliminate every answer choice with negative overtones. As it turns out, this strategy workd perfectly with this antonym; only one word is left. *Noble* is the only word with positive overtones.

11. Unlock word meanings by taking the word apart

Even when you encounter a word for the first time, you may be able to ascertain its meaning from its parts.

***e.g.* congnoscitive**

(*a*) courageous (*b*) expensive
(*c*) unconscious (*d*) redundant

The best answer is (*c*). This is very unusual word, but you can probably figure out its meaning by looking at its root cog-. This is the same root found in words such as *cognition and recognize*, and it has to do with knowledge. So we infer that *cognoscitive* has something to do with awareness, and (c) looks like a good opposite. A word of caution, however : Don't spend too much time trying to decipher the meaning of a word.

EXERCISE

MCQ TYPE QUESTIONS

Directions : *Each question below consists of a word printed in capital lettes, followed by four lettered words of phrases. Choosed the lettered word or phrase that is most nearly opposite in meaning to the word in bold letters.*

1. Abrogate
(*a*) transgress (*b*) signify
(*c*) allevate (*d*) ratify

2. Admonition
(*a*) premonit ion (*b*) hallucination
(*c*) escape (*d*) commendation

3. Adulation
(*a*) youth (*b*) purity
(*c*) brightness (*d*) criticism

4. Advocate
(*a*) define (*b*) oppose
(*c*) remove (*d*) inspect

5. Affable
(*a*) rude (*b*) ruddy
(*c*) needy (*d*) useless

6. Affecfted
(*a*) weary (*b*) unfriendly
(*c*) divine (*d*) unfeigned

7. Affluence
(*a*) poverty (*b*) fear
(*c*) persuasion (*d*) consideration

8. Agility
(*a*) awkwardness (*b*) solidity
(*c*) temper (*d*) harmony

9. Alacrity
(*a*) slowness (*b*) plenty
(*c*) fifth (*d*) courtesy

10. Alleviate
(*a*) endure (*b*) worsen
(*c*) enlighten (*d*) maneuver

11. Allure
(*a*) hinder (*b*) repel
(*c*) ignore (*d*) leave

12. Aloof
(*a*) triangular (*b*) gregarious
(*c*) comparable (*d*) honorable

13. Amalgamate
(*a*) equip (*b*) separate
(*c*) generate (*d*) materialize

14. Ambiguous
(*a*) salvageable (*b*) corresponding
(*c*) responsible (*d*) clear

15. Amble
(*a*) befriend (*b*) hasten
(*c*) steal (*d*) browse

16. Ambulatory
(*a*) convalescent (*b*) valedictory
(*c*) bedridden (*d*) emergency

17. Ameliorate
(*a*) make slow (*b*) make sure
(*c*) make young (*d*) make worse

18. Amicable
(*a*) penetrating (*b*) compensating
(*c*) unfriendly (*d*) zigzag

19. Amoral
(*a*) unusal (*b*) unfriendly
(*c*) ethical (*d*) suave

20. Amorphous
(*a*) nauseous (*b*) obscene
(*c*) providential (*d*) definite

21. Amplify
(*a*) distract (*b*) infer
(*c*) publicize (*d*) decrease

22. Analogous
(*a*) not comparable (*b*) not capable
(*c*) not culpable (*d*) not corporeal

23. Anthematize
(a) locate (b) deceive
(c) regulate (d) bless

24. Anemic
(a) pailid (b) cruel
(c) red-blooded (d) ventilating

25. Animated
(a) worthy (b) dull
(c) humorous (d) lengthy

26. Animus
(a) pterodactyl (b) bastion
(c) giraffe (d) favor

27. Anomaly
(a) desperation (b) requisition
(c) registry (d) regularity

28. Anonymous
(a) desperate (b) signed
(c) defined (d) expert

29. Antediluvian
(a) transported (b) subtle
(c) isolated (d) modern

30. Antipathy
(a) profundity (b) objection
(c) willingness (d) foundness

31. Antithesis
(a) velocity (b) maxim
(c) similarity (d) acceleration

32. Aphasia
(a) volubility (b) necessity
(c) pain (d) crack

33. Baroque
(a) polished (b) constant
(c) transformed (d) simple

34. Beatific
(a) glorius (b) dreadful
(c) theatrical (d) crooked

35. Belittle
(a) disobey (b) forget
(c) magnify (d) extol

36. Bellicose
(a) peaceful (b) naval
(c) amusing (d) piecemeal

37. Benign
(a) tenfold (b) peaceful
(c) blessed (d) malignant

38. Benison
(a) curse (b) bachelor
(c) wedding (d) orgy

39. Berate
(a) grant (b) praise
(c) refer (d) purchase

40. Bestial
(a) animated (b) noble
(c) zoological (d) clear

41. Bigotry
(a) arrogance (b) approval
(c) mourning (d) tolerance

42. Bizarre
(a) roomy (b) veiled
(c) subduced (d) normal

43. Blanch
(a) bleach (b) scatter
(c) darken (d) analyze

44. Bland
(a) caustic (b) meager
(c) soft (d) uncooked

45. Blase
(a) fiery (b) clever
(c) intriguing (d) ardent

46. Bleak
(a) pale (b) sudden
(c) dry (d) cheerful

47. Blithe
(a) spiritual (b) profuse
(c) cheerless (d) hybrid

48. Capacious
(a) warlike (b) cordial
(c) curious (d) not spacious

49. Capricious
(a) satisfied (b) insured
(c) photographic (d) steadfast

50. Captious
(a) tolerant (b) capable
(c) frivolous (d) winning

51. Carnal
(a) impressive (b) minute
(c) spiritual (d) actual

52. Carnivorous
(a) gloomy (b) tangential
(c) productive (d) vegetarian

53. Carping
(a) rapid (b) uncritical
(c) unintellectual (d) illegal

54. Castigation
(a) commendation (b) patience
(c) generosity (d) understatement

55. Categorical
(a) negative (b) ironic
(c) impartial (d) qualified

56. Catholic
(a) religious (b) pacific
(c) narrow (d) weighty

57. Celerity
(a) assurance (b) state
(c) acerbity (d) delay

58. Celibate
(a) investing (b) married
(c) retired (d) commodious

59. Censure
(*a*) process (*b*) enclose
(*c*) interest (*d*) praise

60. Centrifugal
(*a*) centripetal (*b*) ephemeral
(*c*) lasting (*d*) barometric

61. Cessation
(*a*) premium (*b*) gravity
(*c*) beginning (*d*) composition

62. Chaffing
(*a*) achieving (*b*) serious
(*c*) caitalistic (*d*) sneezing

63. Coy
(*a*) weak (*b*) airy
(*c*) brazen (*d*) old

64. Cozen
(*a*) amuse (*b*) treat honestly
(*c*) prate (*d*) shackle

65. Craven
(*a*) desirous (*b*) direct
(*c*) bold (*d*) civilized

66. Crux
(*a*) affiliction (*b*) spark
(*c*) events (*d*) trivial point

67. Cryptic
(*a*) tomblike (*b*) futile
(*c*) famous (*d*) candid

68. Cupidity
(*a*) anxiety (*b*) tragedy
(*c*) generosity (*d*) entertainment

69. Curtail
(*a*) mutter (*b*) lengthen
(*c*) express (*d*) burden

70. Cynical
(*a*) trusting (*b*) effortless
(*c*) conclusive (*d*) gallant

71. Dank
(*a*) dry (*b*) guiltless
(*c*) warm (*d*) babbling

72. Dapper
(*a*) unintelligent (*b*) untidy
(*c*) uncertain (*d*) ungrateful

73. Dauntless
(*a*) stolid (*b*) cowardly
(*c*) irrelevant (*d*) peculiar

74. Dearth
(*a*) life (*b*) abundance
(*c*) brightness (*d*) terror

75. Debacle
(*a*) progress (*b*) refusal
(*c*) masque (*d*) cowardice

76. Debilitate
(*a*) bedevil (*b*) repress
(*c*) strengthen (*d*) animate

77. Debonair
(*a*) awkward (*b*) windy
(*c*) balmy (*d*) strong

78. Derogatory
(*a*) roguish (*b*) immediate
(*c*) openionated (*d*) praising

79. Desecrate
(*a*) desist (*b*) integrate
(*c*) confuse (*d*) consecrate

80. Despicable
(*a*) streeing (*b*) worthy of esteem
(*c*) inevitable (*d*) featureless

81. Destitute
(*a*) affluent (*b*) dazzling
(*c*) stationary (*d*) characteristic

82. Devoid
(*a*) latent (*b*) eschewed
(*c*) full of (*d*) suspecting

83. Devout
(*a*) quiet (*b*) dual
(*c*) impious (*d*) straightforward

84. Diabolical
(*a*) mischievous (*b*) lavish
(*c*) seraphic (*d*) redolent

85. Diatribe
(*a*) mass (*b*) range
(*c*) eulogy (*d*) elegy

86. Diffidence
(*a*) sharapness (*b*) boldness
(*c*) malcontent (*d*) dialogue

87. Dilate
(*a*) procrastinate (*b*) contract
(*c*) conclude (*d*) participate

88. Dilatory
(*a*) narrowing (*b*) prompt
(*c*) enlarging (*d*) portentous

89. Diminution
(*a*) expectation (*b*) context
(*c*) validity (*d*) appreciation

90. Din
(*a*) lightness (*b*) safety
(*c*) silence (*d*) hunger

91. Disabuse
(*a*) crash (*b*) violate
(*c*) renege (*d*) deceive

92. Disconsolate
(*a*) unprejudiced (*b*) thankful
(*c*) threatrical (*d*) joyous

93. Enervate
(*a*) strengthen (*b*) sputter
(*c*) arrange (*d*) scrutinize

94. Enhance
(*a*) degrade (*b*) doubt
(*c*) scuff (*d*) gasp

95. Ennui
(*a*) hate (*b*) excitement
(*c*) seriousness (*d*) humility

96. Enunciate
(*a*) pray (*b*) request
(*c*) deliver (*d*) mumble

97. Ephemeral
(*a*) sensuous (*b*) passing
(*c*) popular (*d*) eternal

98. Equable
(*a*) flat (*b*) decisive
(*c*) stormy (*d*) dishonest

99. Equanimity
(*a*) agitation (*b*) stirring
(*c*) volume (*d*) identity

100. Equilibrium
(*a*) imbalance (*b*) peace
(*c*) inequity (*d*) directness

101. Equitable
(*a*) able to leave (*b*) able to learn
(*c*) unfair (*d*) preferable

102. Equivocal
(*a*) mistaken (*b*) quaint
(*c*) azure (*d*) clear

103. Erratic
(*a*) unromantic (*b*) free
(*c*) popular (*d*) steady

104. Erroneous
(*a*) accurate (*b*) dignified
(*c*) curious (*d*) abrupt

105. Erudite
(*a*) professorial (*b*) stately
(*c*) short (*d*) ignorant

106. Ethereal
(*a*) long-lasting (*b*) earthy
(*c*) ill (*d*) critical

107. Eulogistic
(*a*) pretty (*b*) critical
(*c*) brief (*d*) stern

108. Euphonious
(*a*) strident (*b*) lethargic
(*c*) literary (*d*) significant

109. Evasive
(*a*) frank (*b*) correct
(*c*) empty (*d*) fertile

110. Exasperate
(*a*) confide (*b*) formalize
(*c*) placate (*d*) betray

111. Excoriate
(*a*) scandalize (*b*) encourage
(*c*) avoid (*d*) praise

112. Exculpate
(*a*) blame (*b*) prevail
(*c*) acquire (*d*) ravish

113. Execrable
(*a*) innumerable (*b*) philosophic
(*c*) physical (*d*) excellent

114. Excrate
(*a*) decipher (*b*) sadden
(*c*) integrate (*d*) admit

115. Exhume
(*a*) decipher (*b*) sadden
(*c*) integrate (*d*) inter

116. Exodus
(*a*) neglect (*b*) consent
(*c*) entry (*d*) gain

117. Exonerate
(*a*) forge (*b*) accuse
(*c*) record (*d*) doctor

118. Exorbitant
(*a*) moderate (*b*) partisan
(*c*) military (*d*) barbaric

119. Extemporaneous
(*a*) rehearsed (*b*) hybrid
(*c*) humilating (*d*) satetesmanlike

120. Extraneous
(*a*) modern (*b*) decisive
(*c*) essential (*d*) effective

121. Extrinsic
(*a*) reputable (*b*) inherent
(*c*) swift (*d*) ambitious

122. Extrovert
(*a*) clown (*b*) hero
(*c*) ectomorph (*d*) introvert

123. Grandiose
(*a*) false (*b*) ideal
(*c*) proud (*d*) simple

124. Gratuitous
(*a*) warranted (*b*) frank
(*c*) ingenuous (*d*) frugal

125. Gregarious
(*a*) antisocial (*b*) anticipatory
(*c*) glorious (*d*) horrendous

126. Grisly
(*a*) suggestive (*b*) doubtful
(*c*) untidy (*d*) pleasant

127. Gullible
(*a*) incredulous (*b*) fickle
(*c*) tantamount (*d*) easy

128. Gusto
(*a*) noise (*b*) panic
(*c*) atmosphere (*d*) distaste

129. Ghusty
(*a*) clam (*b*) noisy
(*c*) fragrant (*d*) routine

130. Hackneyed
(*a*) carried (*b*) original
(*c*) banned (*d*) timely

131. Haggard
(*a*) shrewish (*b*) inspired
(*c*) plump (*d*) maidenly

132. Halcyon
(*a*) wasteful (*b*) prior
(*c*) subsequent (*d*) martial

133. Haphazard
(*a*) safe (*b*) indifferent
(*c*) deliberate (*d*) tense

134. Hapless
(*a*) cheerful (*b*) consistent
(*c*) fortunate (*d*) considerate

135. Heed
(*a*) ignore (*b*) hope
(*c*) overtake (*d*) nurture

136. Heretic
(*a*) sophist (*b*) believer
(*c*) interpreter (*d*) pacifist

137. Heterogeneous
(*a*) orthodox (*b*) pagan
(*c*) unlikely (*d*) similar

138. Hibernal
(*a*) musical (*b*) summerlike
(*c*) local (*d*) seasonal

139. Hilarity
(*a*) gloom (*b*) heartiness
(*c*) weakness (*d*) casualty

140. Hirsute
(*a*) scaly (*b*) bald
(*c*) erudite (*d*) quiet

141. Hortatory
(*a*) inquiring (*b*) denying
(*c*) killing (*d*) dissuading

142. Hovr
(*a*) commence (*b*) soothe
(*c*) leave (*d*) transform

143. Hunbbub
(*a*) calm (*b*) fury
(*c*) capital (*d*) axle

144. Hummock
(*a*) ummusical (*b*) scorn
(*c*) wakefulness (*d*) vale

145. Husbandry
(*a*) sportsmanship (*b*) dishonesty
(*c*) wastefulness (*d*) friction

146. Hybrid
(*a*) productive (*b*) special
(*c*) purebred (*d*) oafish

147. Hyperbole
(*a*) velocity (*b*) climax
(*c*) curve (*d*) understatement

148. Hypercritical
(*a*) tolerant (*b*) false
(*c*) extreme (*d*) inarticulate

149. Hypocritical
(*a*) sincere (*b*) narrow-minded
(*c*) shameful (*d*) amiable

150. Hypothetial
(*a*) rational (*b*) fantastic
(*c*) wizened (*d*) axiomatic

151. Ignoble
(*a*) produced by fire (*b*) worthy
(*c*) given to questioning (*d*) huge

152. Illusive
(*a*) not deceptive (*b*) not certain
(*c*) not obvious (*d*) not choherent

153. Irksome
(*a*) interesting (*b*) lazy
(*c*) tireless (*d*) devious

154. Irrelevant
(*a*) lacking piety (*b*) fragile
(*c*) congruent (*d*) pertinent

155. Irreparable
(*a*) legible (*b*) correctable
(*c*) proverbial (*d*) concise

156. Irtreverent
(*a*) related (*b*) mischievous
(*c*) respective (*d*) pious

157. Jaded
(*a*) upright (*b*) stimulated
(*c*) aspiring (*d*) applied

158. Jaundiced
(*a*) whitened (*b*) inflamed
(*c*) quickened (*d*) unbiased

159. Jaunty
(*a*) youthful (*b*) ruddy
(*c*) strong (*d*) sedate

160. Jeopardy
(*a*) patience (*b*) courage
(*c*) safety (*d*) willingness

161. Jettison
(*a*) salvage (*b*) submerge
(*c*) descend (*d*) decelerate

162. Jocular
(*a*) arterial (*b*) bloodless
(*c*) verbose (*d*) serious

163. Judicious
(*a*) punitive (*b*) unwise
(*c*) criminal (*d*) licit

164. Kindle
(*a*) dislike (*b*) quench
(*c*) gather (*d*) sparkle

165. Lachrymose
(*a*) cheering (*b*) smooth
(*c*) passionate (*d*) curt

166. Lackadasical
(*a*) monthly (*b*) possessing time
(*c*) ambitious (*d*) pusillanimous

167. Laconic
(a) milky (b) verbose
(c) wicked (d) flagrant

168. Lampoon
(a) darken (b) praise
(c) abandon (d) sail

169. Languor
(a) vitality (b) length
(c) embarrassment (d) wine

170. Latent
(a) trim (b) forbidding
(c) execrable (d) obvious

171. Lavish
(a) hostile (b) unwashed
(c) timely (d) frugal

172. Laudatory
(a) dritying (b) disclaiming
(c) defamatory (d) inflammatory

173. Lax
(a) salty (b) strick
(c) shrill (d) boring

174. Lechery
(a) trust (b) compulsion
(c) zeal (d) purity

175. Lethargic
(a) convalescent (b) beautiful
(c) enervating (d) invigorating

176. Levite
(a) bridge (b) dam
(c) praise (d) solemnity

177. Lilliputian
(a) destructive (b) proper
(c) gigantic (d) elegant

178. Limpid
(a) erect (b) turbid
(c) tangential (d) timid

179. Lithe
(a) stiff (b) limpid
(c) facetious (d) insipid

180. Livid
(a) alive (b) mundane
(c) positive (d) undiscolored

181. Loath
(a) loose (b) evident
(c) deliberate (d) eager

182. Loquacious
(a) taciturn (b) sentimental
(c) soporific (d) soothing

183. Moist
(a) crisp (b) hard
(c) dry (d) parched

184. Mask
(a) hit (b) expose
(c) benign (d) pompous

185. Moisture
(a) dryness (b) delegate
(c) dampness (d) fdefect

186. Magnify
(a) produce (b) support
(c) reduce (d) destroy

187. Mendacious
(a) broken (b) efficacious
(c) truthful (d) destructive

188. Manumit
(a) throw (b) lock
(c) promise (d) enslave

189. Nebulous
(a) starry (b) clear
(c) cold (d) fundamental

190. Nefarious
(a) various (b) lacking
(c) benign (d) pompous

191. Negation
(a) postulation (b) hypothecation
(c) affirmation (d) violation

192. Neophyte
(a) veteran (b) satellite
(c) desperado (d) handwriting

193. Niggardly
(a) protected (b) biased
(c) prodigal (d) bankrupt

194. Nocturnal
(a) harsh (b) marauding
(c) patrolling (d) daily

195. Noisome
(a) quiet (b) dismayed
(c) fragrant (d) sleepy

196. Notorious
(a) fashionable (b) renowned
(c) inactive (d) interepid

197. Obdurate
(a) yielding (b) fleeting
(c) finite (d) fascinating

198. Obese
(a) skillful (b) cadaverous
(c) clever (d) unpredictable

199. Objective
(a) indecisive (b) apathetic
(c) markedly inferior (d) emotionally involved

200. Obligatory
(a) demanding (b) optional
(c) facile (d) friendly

201. Obloquy
(a) praise (b) rectangle
(c) circle (d) dialogue

202. Obsequious
(a) successful (b) democratic
(c) supercilious (d) ambitious

203. Obsession
(*a*) whim (*b*) loss
(*c*) phobia (*d*) delusion

204. Obsolete
(*a*) heated (*b*) desolate
(*c*) renovated (*d*) frightful

205. Obstreperous
(*a*) turbid (*b*) quiet
(*c*) remote (*d*) lucid

206. Obtuse
(*a*) sheer (*b*) transparent
(*c*) tranquil (*d*) shrewd

207. Odious
(*a*) fragrant (*b*) redolent
(*c*) fetid (*d*) delightful

208. Odium
(*a*) noise (*b*) liking
(*c*) dominant (*d*) hasty

209. Omnipotent
(*a*) weak (*b*) democratic
(*c*) despotic (*d*) passionate

210. Omniscient
(*a*) sophisticated (*b*) ignorant
(*c*) essential (*d*) trivial

211. Opiat
(*a*) distress (*b*) sleep
(*c*) stimulant (*d*) laziness

212. Opportune
(*a*) occasional (*b*) fragrant
(*c*) fragile (*d*) awkward

213. Opportunist
(*a*) man of destiny (*b*) man of principle
(*c*) changeling (*d*) adversary

214. Opprobrium
(*a*) delineation (*b*) aptitude
(*c*) majesty (*d*) praise

215. Optimum
(*a*) pessimistic (*b*) knowledgeable
(*c*) worst (*d*) minimum

216. Opulence
(*a*) pessimism (*b*) patrionsm
(*c*) potency (*d*) poverty

217. Ostentatious
(*a*) inactive (*b*) unassuming
(*c*) impolite (*d*) illicit

218. Overweening
(*a*) humble (*b*) impotent
(*c*) avid (*d*) acrimonious

219. Pejorative
(*a*) positive (*b*) legal
(*c*) determining (*d*) delighting

220. Pellucid
(*a*) logistical (*b*) philandering
(*c*) incomperhensible (*d*) vagrant

221. Penchant
(*a*) distance (*b*) imminence
(*c*) dislike (*d*) attitude

222. Penurious
(*a*) imprisoned (*b*) captivated
(*c*) generous (*d*) vacant

223. Perfunctory
(*a*) official (*b*) through
(*c*) insipid (*d*) vicarious

224. Perigee
(*a*) eclipse (*b*) planet
(*c*) apogee (*d*) refugee

225. Peripatetic
(*a*) wordly (*b*) stationary
(*c*) disarming (*d*) seeking

226. Permeable
(*a*) perishable (*b*) effective
(*c*) plodding (*d*) impenetrable

227. Pernicious
(*a*) practical (*b*) comparative
(*c*) harmless (*d*) tangible

228. Perpetual
(*a*) momentary (*b*) standard
(*c*) serious (*d*) industrial

229. Perspicuity
(*a*) grace (*b*) feature
(*c*) review (*d*) vagueness

230. Pert
(*a*) polite (*b*) perishable
(*c*) moral (*d*) deliberate

231. Pertinacious
(*a*) vengeful (*b*) consumptive
(*c*) superficial (*d*) skilled

232. Pertient
(*a*) understood (*b*) living
(*c*) discontented (*d*) irrelevant

233. Petulant
(*a*) angry (*b*) moral
(*c*) declining (*d*) uncomplaining

234. Precipitate
(*a*) dull (*b*) anticipatory
(*c*) cautious (*d*) considerate

235. Prefatory
(*a*) outstanding (*b*) magnificent
(*c*) conclusive (*d*) intelligent

236. Prelude
(*a*) intermezzo (*b*) diva
(*c*) aria (*d*) aftermath

237. Presumption
(*a*) assertion (*b*) activation
(*c*) motivation (*d*) humility

238. Pretentious
(*a*) ominous (*b*) calm
(*c*) unassuming (*d*) futile

239. Prim
(*a*) informal (*b*) prior
(*c*) exterior (*d*) private

240. Pristine
(*a*) cultivated (*b*) condemned
(*c*) irreligious (*d*) cautious

241. Probity
(*a*) aregert (*b*) assumption
(*c*) corruptibility (*d*) extent

242. Prodigal
(*a*) large (*b*) thrifty
(*c*) consistent (*d*) compatible

243. Prodigious
(*a*) infinitesimal (*b*) indignant
(*c*) indifferent (*d*) indisposed

244. Profane
(*a*) sanctify (*b*) desecrate
(*c*) define (*d*) manifest

245. Prolific
(*a*) unworkable (*b*) backward
(*c*) barren (*d*) controversial

246. Prolix
(*a*) stupid (*b*) indifferent
(*c*) redundant (*d*) pithy

247. Prophylactic
(*a*) causing growth (*b*) causing disease
(*c*) antagonistic (*d*) brushing

248. Propinquity
(*a*) remoteness (*b*) uniqueness
(*c*) health (*d*) virtue

249. Propitious
(*a*) rich (*b*) induced
(*c*) promoted (*d*) unfavorable

250. Prosaic
(*a*) pacified (*b*) reprieved
(*c*) pensive (*d*) imaginative

251. Protean
(*a*) amateur (*b*) catholic
(*c*) unchanging (*d*) rapid

252. Protract
(*a*) make circular (*b*) shorten
(*c*) further (*d*) rectrace

253. Provident
(*a*) unholy (*b*) rash
(*c*) miserable (*d*) disabled

254. Provincial
(*a*) wealthy (*b*) crass
(*c*) literary (*d*) sophisticated

255. Psychotic
(*a*) dangerous (*b*) clairvoyant
(*c*) criminal (*d*) sane

256. Puerile
(*a*) fragrant (*b*) adult
(*c*) lonely (*d*) feminine

257. Pugnacious
(*a*) pacific (*b*) feline
(*c*) mature (*d*) angular

258. Puissant
(*a*) pouring (*b*) fashionable
(*c*) articulate (*d*) weak

259. Pulchritude
(*a*) ugliness (*b*) notoriety
(*c*) bestiality (*d*) masculinity

260. Punctilious
(*a*) happy (*b*) active
(*c*) vivid (*d*) careless

261. Punitive
(*a*) large (*b*) humorous
(*c*) rewarding (*d*) restive

262. Pusillanimous
(*a*) poverty-stricken (*b*) chained
(*c*) posthumous (*d*) courageous

263. Putative
(*a*) colonial (*b*) quarrelsome
(*c*) undisputed (*d*) powerful

264. Resilient
(*a*) pungent (*b*) foolish
(*c*) worthy (*d*) unyielding

265. Restive
(*a*) buoyant (*b*) placid
(*c*) remorseful (*d*) resistant

266. Retentive
(*a*) forgetful (*b*) accepting
(*c*) repetitive (*d*) avoiding

267. Reticence
(*a*) fatigue (*b*) fashion
(*c*) treachery (*d*) loquaciousness

268. Retrograde
(*a*) progressing (*b*) inclining
(*c*) evaluating (*d*) concentrating

269. Revere
(*a*) advance (*b*) dishonor
(*c*) age (*d*) precede

270. Rife
(*a*) direct (*b*) scant
(*c*) peaceful (*d*) grim

271. Robust
(*a*) weak (*b*) violent
(*c*) vicious (*d*) villainous

272. Rotundity
(*a*) promenade (*b*) nave
(*c*) grotesqueness (*d*) slimness

273. Rubble
(*a*) artificial facade (*b*) unbroken stone
(*c*) pale complexion (*d*) strong defense

274. Ruddy
(*a*) robust (*b*) witty
(*c*) wan (*d*) exotic

275. Rudimentary
(*a*) pale (*b*) polite
(*c*) asinine (*d*) developed

276. Rueful
(*a*) trite (*b*) content
(*c*) capable (*d*) capital

277. Rustic
(*a*) urban (*b*) slow
(*c*) corroded (*d*) mercenary

278. Ruthless
(*a*) merciful (*b*) majestic
(*c*) mighty (*d*) militant

279. Sadistic
(*a*) happy (*b*) quaint
(*c*) kindhearted (*d*) vacant

280. Sagacious
(*a*) foolish (*b*) bitter
(*c*) voracious (*d*) veracious

281. Sallow
(*a*) salacious (*b*) ruddy
(*c*) colorless (*d*) permitted

282. Salubrious
(*a*) salty (*b*) bloody
(*c*) miasmic (*d*) maudlin

283. Salvage
(*a*) remove (*b*) outfit
(*c*) burn (*d*) lose

284. Sanctimonious
(*a*) hypothetical (*b*) paltry
(*c*) mercenary (*d*) pious

285. Sanguine
(*a*) choleric (*b*) sickening
(*c*) warranted (*d*) pessimistic

286. Satiety
(*a*) emptiness (*b*) warmth
(*c*) erectness (*d*) straightness

287. Scanty
(*a*) collected (*b*) remote
(*c*) invisible (*d*) plentiful

288. Scurrilous
(*a*) savage (*b*) scabby
(*c*) decent (*d*) volatile

289. Secular
(*a*) vivid (*b*) clerical
(*c*) punitive (*d*) positive

290. Sedentary
(*a*) vicarious (*b*) loyal
(*c*) accidental (*d*) active

291. Sedulous
(*a*) indolent (*b*) guileless
(*c*) vindictive (*d*) upright

292. Senlity
(*a*) verility (*b*) loquaciousness
(*c*) forgetfulness (*d*) youth

293. Sententious
(*a*) paragraphed (*b*) positive
(*c*) posthumous (*d*) wordy

294. Tractable
(*a*) unmanageable (*b*) irreligious
(*c*) mortal (*d*) incapable

295. Traduce
(*a*) exhume (*b*) increase
(*c*) purchase (*d*) extol

296. Tranquillity
(*a*) lack of sleep (*b*) lack of calm
(*c*) emptiness (*d*) renewal

297. Transient
(*a*) carried (*b*) close
(*c*) permanent (*d*) removed

298. Tremulous
(*a*) steady (*b*) obese
(*c*) young (*d*) healthy

299. Trenchant
(*a*) lacking bite (*b*) imperious
(*c*) inessential (*d*) unafraid

300. Trepidation
(*a*) slowness (*b*) amputation
(*c*) fearlessness (*d*) adroitness

301. Trite
(*a*) correct (*b*) original
(*c*) distinguished (*d*) premature

302. Truculent
(*a*) juicy (*b*) overflowing
(*c*) peaceful (*d*) determined

303. Tumult
(*a*) scarcity (*b*) defeat
(*c*) cooleness (*d*) secrenity

304. Turbid
(*a*) clear (*b*) improbable
(*c*) invariable (*d*) honest

305. Turbulence
(*a*) reaction (*b*) approach
(*c*) impropriety (*d*) calm

306. Turgid
(*a*) rancid (*b*) shrunken
(*c*) cool (*d*) explosive

307. Turpitude
(*a*) amplitude (*b*) heat
(*c*) wealth (*d*) virtue

308. Tyro
(*a*) infant (*b*) rubber
(*c*) personnel (*d*) expert

309. Unearth
(*a*) conceal (*b*) gnaw
(*c*) clean (*d*) fling

310. Unfeigned
(*a*) pretended (*b*) fashionable
(*c*) wary (*d*) switched

311. Ungainly
(*a*) ignorant
(*b*) graceful
(*c*) detailed
(*d*) dancing

312. Unimpeachable
(*a*) fruitful
(*b*) rampaging
(*c*) faulty
(*d*) pensive

313. Unkempt
(*a*) bombed (*b*) washed
(*c*) neat (*d*) tawdry

314. Unruly
(*a*) chatting (*b*) obedient
(*c*) definite (*d*) lined

315. Unseemly
(*a*) effortless (*b*) proper
(*c*) conducive (*d*) pointed

316. Unsullied
(*a*) tranished (*b*) countless
(*c*) soggy (*d*) papered

317. Untenable
(*a*) supportable (*b*) tender
(*c*) sheepish (*d*) tremulous

318. Unwitting
(*a*) clever (*b*) intense
(*c*) sensitive (*d*) intentional

319. Vacillation
(*a*) remorse (*b*) relief
(*c*) respect (*d*) steadfastness

320. Valedictory
(*a*) sad (*b*) collegiate
(*c*) derivative (*d*) salutatory

321. Valor
(*a*) admonition
(*b*) injustice
(*c*) cowardice
(*d*) generosity

322. Vanguard
(*a*) regiment
(*b*) rear
(*c*) echelon
(*d*) protection

323. Vaunted
(*a*) unvanquished
(*b*) fell
(*c*) belittled
(*d*) believed

ANSWERS

MCQ Type Questions

1. (*d*) **2.** (*b*) **3.** (*d*) **4.** (*b*) **5.** (*a*) **6.** (*d*) **7.** (*a*) **8.** (*a*) **9.** (*a*) **10.** (*b*)
11. (*b*) **12.** (*b*) **13.** (*b*) **14.** (*d*) **15.** (*b*) **16.** (*c*) **17.** (*d*) **18.** (*c*) **19.** (*c*) **20.** (*d*)
21. (*d*) **22.** (*a*) **23.** (*d*) **24.** (*c*) **25.** (*b*) **26.** (*d*) **27.** (*b*) **28.** (*b*) **29.** (*d*) **30.** (*d*)
31. (*c*) **32.** (*a*) **33.** (*d*) **34.** (*b*) **35.** (*d*) **36.** (*a*) **37.** (*d*) **38.** (*a*) **39.** (*b*) **40.** (*b*)
41. (*d*) **42.** (*d*) **43.** (*c*) **44.** (*a*) **45.** (*d*) **46.** (*d*) **47.** (*c*) **48.** (*d*) **49.** (*d*) **50.** (*a*)
51. (*c*) **52.** (*d*) **53.** (*b*) **54.** (*a*) **55.** (*d*) **56.** (*c*) **57.** (*b*) **58.** (*a*) **59.** (*b*) **60.** (*a*)
61. (*c*) **62.** (*b*) **63.** (*c*) **64.** (*b*) **65.** (*c*) **66.** (*d*) **67.** (*d*) **68.** (*c*) **69.** (*b*) **70.** (*a*)
71. (*a*) **72.** (*b*) **73.** (*b*) **74.** (*b*) **75.** (*a*) **76.** (*c*) **77.** (*a*) **78.** (*d*) **79.** (*d*) **80.** (*b*)
81. (*a*) **82.** (*c*) **83.** (*c*) **84.** (*d*) **85.** (*c*) **86.** (*b*) **87.** (*b*) **88.** (*b*) **89.** (*d*) **90.** (*c*)
91. (*d*) **92.** (*d*) **93.** (*a*) **94.** (*a*) **95.** (*b*) **96.** (*d*) **97.** (*d*) **98.** (*c*) **99.** (*a*) **100.** (*a*)
101. (*c*) **102.** (*b*) **103.** (*d*) **104.** (*a*) **105.** (*d*) **106.** (*b*) **107.** (*b*) **108.** (*a*) **109.** (*a*) **110.** (*c*)
111. (*d*) **112.** (*a*) **113.** (*d*) **114.** (*b*) **115.** (*d*) **116.** (*c*) **117.** (*b*) **118.** (*a*) **119.** (*b*) **120.** (*c*)
121. (*b*) **122.** (*d*) **123.** (*d*) **124.** (*a*) **125.** (*a*) **126.** (*d*) **127.** (*a*) **128.** (*d*) **129.** (*d*) **130.** (*b*)
131. (*c*) **132.** (*d*) **133.** (*c*) **134.** (*c*) **135.** (*a*) **136.** (*b*) **137.** (*d*) **138.** (*b*) **139.** (*a*) **140.** (*b*)
141. (*d*) **142.** (*c*) **143.** (*a*) **144.** (*d*) **145.** (*c*) **146.** (*c*) **147.** (*d*) **148.** (*a*) **149.** (*a*) **150.** (*d*)
151. (*b*) **152.** (*a*) **153.** (*a*) **154.** (*d*) **155.** (*b*) **156.** (*d*) **157.** (*b*) **158.** (*d*) **159.** (*d*) **160.** (*c*)
161. (*a*) **162.** (*d*) **163.** (*a*) **164.** (*b*) **165.** (*a*) **166.** (*c*) **167.** (*d*) **168.** (*b*) **169.** (*a*) **170.** (*d*)
171. (*d*) **172.** (*c*) **173.** (*b*) **174.** (*d*) **175.** (*d*) **176.** (*d*) **177.** (*c*) **178.** (*b*) **179.** (*a*) **180.** (*d*)
181. (*d*) **182.** (*a*) **183.** (*c*) **184.** (*b*) **185.** (*a*) **186.** (*c*) **187.** (*c*) **188.** (*d*) **189.** (*b*) **190.** (*c*)
191. (*c*) **192.** (*a*) **193.** (*c*) **194.** (*d*) **195.** (*c*) **196.** (*b*) **197.** (*d*) **198.** (*d*) **199.** (*d*) **200.** (*b*)
201. (*a*) **202.** (*c*) **203.** (*a*) **204.** (*c*) **205.** (*b*) **206.** (*d*) **207.** (*d*) **208.** (*b*) **209.** (*a*) **210.** (*a*)
211. (*c*) **212.** (*d*) **213.** (*b*) **214.** (*d*) **215.** (*c*) **216.** (*d*) **217.** (*b*) **218.** (*a*) **219.** (*a*) **220.** (*c*)
221. (*c*) **222.** (*c*) **223.** (*b*) **224.** (*c*) **225.** (*b*) **226.** (*d*) **227.** (*c*) **228.** (*a*) **229.** (*d*) **230.** (*a*)
231. (*c*) **232.** (*d*) **233.** (*d*) **234.** (*c*) **235.** (*c*) **236.** (*d*) **237.** (*d*) **238.** (*c*) **239.** (*a*) **240.** (*a*)
241. (*c*) **242.** (*b*) **243.** (*a*) **244.** (*a*) **245.** (*c*) **246.** (*d*) **247.** (*b*) **248.** (*a*) **249.** (*d*) **250.** (*d*)
251. (*c*) **252.** (*b*) **253.** (*b*) **254.** (*d*) **255.** (*d*) **256.** (*b*) **257.** (*a*) **258.** (*d*) **259.** (*a*) **260.** (*b*)
261. (*c*) **262.** (*d*) **263.** (*c*) **264.** (*b*) **265.** (*a*) **266.** (*d*) **267.** (*a*) **268.** (*b*) **269.** (*b*) **270.** (*b*)
271. (*a*) **272.** (*d*) **273.** (*b*) **274.** (*c*) **275.** (*d*) **276.** (*b*) **277.** (*a*) **278.** (*a*) **279.** (*c*) **280.** (*a*)
281. (*b*) **282.** (*c*) **283.** (*d*) **284.** (*d*) **285.** (*d*) **286.** (*d*) **287.** (*d*) **288.** (*c*) **289.** (*b*) **290.** (*d*)
291. (*a*) **292.** (*d*) **293.** (*d*) **294.** (*a*) **295.** (*d*) **296.** (*b*) **297.** (*c*) **298.** (*a*) **299.** (*a*) **300.** (*c*)
301. (*b*) **302.** (*c*) **303.** (*d*) **304.** (*a*) **305.** (*d*) **306.** (*b*) **307.** (*d*) **308.** (*d*) **309.** (*a*) **310.** (*a*)
311. (*b*) **312.** (*c*) **313.** (*d*) **314.** (*b*) **315.** (*b*) **316.** (*a*) **317.** (*a*) **318.** (*d*) **319.** (*d*) **320.** (*d*)
321. (*c*) **322.** (*b*) **323.** (*d*)

5 CHAPTER Miscellaneous

VOCABULARY FOR READING COMPREHENSION

Not appreciating the subtle differences between words like - sarcastic and sardonic, ironical, cynical,... (the list goes on), many threaten your chances of a high score in R.C. (especially in inferential and true tone passages).

The beauty of English gets more enhanced if understood in the right perspective. On the other hand, one word misunderstood many stand you in for a comical interpretation of a serious passage. The 'Vocabulary for R.C.' section becomes a unique section, as it tries to encompass a kaleidoscope of tones, moods, opinions, characteristics, analogies which have statistically found a higher representation in the R.C. passages of management entrance exams.

Anarchist : A person who does not believe in any kind of system, law and order, governance and goes an extra mile to prove his view on political disorder.

Agnostic : A person who is unsure of the existence of God. Unlike atheists he is a little flexible with respect to his belief about God, but still unsure.

Atheism/atheist : A belief in the non-existence of God.

Authoritarian : A person following complete obedience to authority. Hitler is a good example of the kind.

Example : The authoritarian regime of Taliban finally met its nemesis.

Analytical : Characterized by logical and reasonable basis. Analytically sound managers - need of the hour.

Critic : A person who shows faults in something and usually expresses a base/low opinion.

Example : Saurav Ganguly's critics have suddenly found a new voice, after his litany of failures with shortpitch ball. A person who judges the quality of something especially works of art, literature, music, etc.

Example : He is a music critic, i.e. he judges

Callous : Having or showing no sympathy for other people's feelings or suffering.

Example : A callous disregard for the feeling of others.

Bacchanalian : Orgy; referred to a festival which was celebrated with dancing, song and revelry.

Blasphemy : The act of insulting or showing contempt or lack of reverence towards God.

Example : Blasphemy and independent thinking are very often confused.

Empiricism : The practice of relying on observation and experiment especially in the natural sciences.

A theory that all knowledge originates in experience.

Example : An empirical analysis of the situation yields different results.

Critical (for somebody/something) : It means indicating faults in someone.

The jehadis were quite critical of Pakistan supporting US against Afghanistan.

Critique : Critical analysis

Example: Presented in this column is a critique of the govt's economic policies.

Cynic : A person who believe that people do not do things for good, sincere or noble reasons, but only for their own advantage: concerned only with one's own interests.

Example : Stop giving these cynical smiles .

Example : Women in Afghanistan had over a period time started viewing the cynical Taliban regime with much apprehension.

Example : His adventurism was very close to being termed as a cynical disregard for the safety of others.

Contemplative : Pondering over or considering thoughtfully.

Example : Contemplate changing careers.

Commendation : Praise or approval.

Castigating : To chastise, punish or to criticize severely.

Example : Hansie Cronje's castigation by the sports media.

Disparagement : The art or process of speaking in a belittling way; decrying, or reducing someone in rank or esteem.

Example : Disparaging someone's character or achievements.

Disinterested (noun) : Freedom from bias or self-interest, impartiality

Example : A disinterested act of kindness. Different from uninterested being without an interest.

Disillusion (ed) : To destroy the but mistaken beliefs or ideals.

Example : Disillusioned voters want an alternative to the two parties.

Disingenuous : Not straightforward; crafty.

Exploratory : To investigate systematically; to examine.

Example : Explore every suggestion given.

Ecclesiastical : Of or relating to a church, as an organized institution.

Example : The period witnessed strong anti-ecclesiastical view from the masses.

Eclectic : Selecting what appears to be the best form diverse sources systems, or styles. An eclectic approach.

Example : Different from the non-holistic: emphasizing the importance or the whole and the interdependence of its parts.

Frivolous : Something which demands lesser important or someone who is lacking in seriousness.

Example : The Zimbabweans are not frivolous when it comes to playing them in their own soil.

Indignation : Anger aroused by something unjust, unworthy, or mean.

Example : Kapil Dev's indignation is just.

Inquisitive : Someone who has the propensity to ask questions.

Example : My first class in the college saw many inquisitive students trying to unnerve me.

Ironic : Something which happens in the opposite way to what expected and often causes amusement because of this.

Example : It is ironic that he died of AIDS, though he was the one who discovered the first AIDS vaccine.

Myth : It is something untrue.

Mystify : To puzzle someone/something.

Morose : Unhappy and bad tempered; but not saying much.

Example: Morose mannerisms.

Moron : A very stupid person.

Misnomer : A wrong use of a name, word or description.

Example : To call CAT exam, a catwalk would be a misnomer.

Mystic : Something/someone which has a hidden meaning or spiritual power and arouses feeling of deep respect; and wonder.

Example : Mystical writings.

Mysticism : To believe that knowledge of God and a real truth may be reached by directing one's mind through spiritual insight.

NARRATIVE SEQUENCE

Narrative sequence is used to identify what happens at the beginning, middle and end of a sample paragraph and also identify the transition word in a given paragraph. Places of these identify words in given sentences used for to create a story.

When we read, it is important to be able to retell the events in a story. Those events need to be retold in the correct order for the story to make sense. Here, we will learn how to sequence events in a narrative.

Notes:

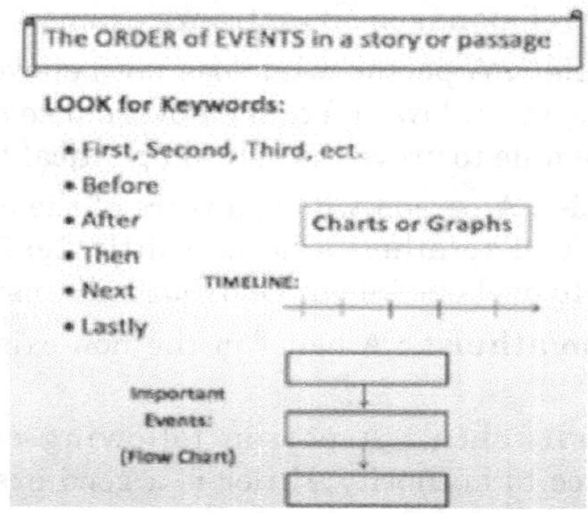

Sequence of Events

Have you ever seen a really good movie? Maybe it was an action-packed superhero movie or a fantasy adventure to a far-off land. When a director creates a new film, he or she has to make sure the scenes help the movie make sense from the opening scene until the credits start rolling.

Likewise, as a reader, it's important for you to identify the sequence of events in a story. The **sequence of events** is the order of events as they occur in a story. Just like when the heroes can't stop the world-ending doomsday machine before it's been built in a superhero movie, the sequence of events in story occurs in a precise order so that it makes sense to the person reading it.

So how do you identify the sequence of events? Get your popcorn because it's showtime!

Organize the Story

Example Sequence of Events Chart

The Boy Who Cried Wolf

Beginning	Middle	End
A wolf was eating the sheep of a town	The towns people came to help and found out the boy was lying.	A wolf really did approach the boy
The boy was bored of watching the sheep	The boy did this a couple more time.	The boy cried for help, but no one came because they thought he was lying again
He decided to pretend the wolf was attacking the sheep.	The town's people were upset.	The wolf attacked the boy and the sheep

Movie directors often use storyboards to help them visualize what will happen in each scene. Storyboards show what will happen in each camera shot just like the panels of a comic book tell a story. As a reader, you can also think of the sequence of events like a storyboard.

After you've read a story, you can organize the story into three sections and ask yourself the following questions:

a) What happens at the beginning of the story?
b) What happens in the middle of the story?
c) What happens at the end of the story?

The beginning of the story usually introduces the main characters, setting, and problem faced by or purpose of the characters. The middle part of the story usually focuses on the characters as they try and solve the problem. The end of the story focuses on how the problem is resolved, or the end of the characters' journey.

It may help to create a chart, like this one, to help you organize the different events that make up the beginning, middle, and end of a story.

Sequence of Events:

Here we taught the importance of understanding the events of a story in order.

Structure of a Narrative: Easy Level

As this lesson explained, there are certain things that usually happen at the **beginning, middle,** and **end** of a story. Think of some stories that you are very familiar with. Nursery rhymes and fairy tales are a great place to start because they tend to be simple and easy to remember. Choose at least three stories and divide them into their beginning, middle, and end sections. Check back on this lesson if you get stuck. When you feel confident that you've divided the stories properly, get a family member or teacher to check your work.

Story Scramble: Medium Level

There is more to a story than just beginning, middle, and end. Many events can happen in a single story, just like in the example of the tortoise and the hare in this story. Think of a book that you've recently read. On one piece of paper, write a clear list of events in the story. Now, cut them up into separate sentences and mix them all up. Can you put them back together again in the right order? Make sure that the end result of your activity makes logical sense.

Write Your Own: Hard Level

If you feel really confident in your understanding of narrative structure, try writing your own short story. Think carefully about the order of events as you write them. What does the audience have to be aware of for the events to make sense? What is your story's beginning, middle, and end? Consider these questions as you write.

Narrative Sequencers:

Narrative Sequencer is nothing but a word which tells you where in the paragraph transition words will take place.

Steps of sequencers:

1. First
2. Next
3. Then
4. After that
5. Finally

Example 1:

Add sequencers in the correct spot in the following sentence:

I'll buy coffee and go to the movie.

(a) I'll buy first coffee and go then to the movie
(b) Then, I'll buy coffee and first go to the movie
(c) First, I'll buy coffee and then go to the movie
(d) I'll buy coffee and first go to the movie

Anser : (c)

Example 2:

Andrea loves to play soccer, but she broke her arm and cannot play for a while. Since breaking her arm, Andrea has been very sad. A friend of hers suggested that Andrea find another hobby to take her mind off of her injury. Andrea took her friend's advice and tried knitting. She loved it! Now, Andrea spends her time knitting while waiting for her arm to heal, and she isn't sad anymore.

(a) Andrea broke her arm and cannot play soccer for a while.

(b) Andrea's friend suggested that she find another hobby, so she tried knitting

(c) Andrea isn't sad anymore.

(d) Andrea's arm healed and she started playing soccer again.

Answer (b)

SPEAKERS

Banal : A person whose speech is dull, flat, boring, unimaginative

Cogent : A person who can express his views brilliantly & convincingly

Garrulous : A person who goes on speaking a lot, all meaningless

Grandiloquent : A person who always uses heavy words

Inarticulate : A person who is very awkward while speaking

Laconic : A person saying few words full of deep meaning

Loquacious : A person who speaks just too much, quantity wise

Taciturn : An extremely quite person speaking very little Syn : Reticent/ Reserved/Close- lipped

Verbose : A person who repeats the same idea again and again

Ventriloquist : A person who can throw his voice creating an illusion that it is coming from another source

Voluble : A very fluent and rapid speaker

Vociferous : A very noisy speaker

Magniloquent : A person who talks big

Eloquent : A person who uses great, artistic or emotional language

Soporific : A person whose speech puts the audience to sleep.

Egotist : A person who constantly boasts of his own achievements.

Soliloquist : A person talking to oneself.

JARGONS

These lists will help you remember groups of words and their interconnections. This will help you solve questions based on verbal analogies etc.

(1) Words Related to law:

admonition, evidence, prosecution, advocate, forgery, reprieve, allege, gibbet, scourage, attorney, incarceration, scoundrel, barrister, indictment, sentences, borstal, judicial, surety, circumstantial, culpable, perjury, vault, culprit

(2) Words related to death:

bier, decease, requiem, burial, epitaph, rites, cenotaph, obituary

(3) Words related to Human Nature:

arrogant, foppish, obsequious, beneficients, garrulous, peculiar, benefactor, haughty, propensity, benevolent, impudent, reckless, courteous, insolent, rogue, cringing, knave, sheer, dauntless, malevolent, sober, deficient, malicious, suave, enthusiastic.

Mendacious, taunt, finical, obstinate, tantalise

(4) Words related to Energy and Electricity:

aerial, crystal, galena, atmospheric, detector, insulate, battery, distortion, leads, bias, ebonite, neutralise, capacity, emitter, oscillate, phenomenon, circuit, energy, condenser, ether, radiating, current, frequency, radio, receiver. Telephone, sensitiveness,

(5) Words related to Automobile:

accelerator, differential, piston, armature, dynamo, radiator, axle, exhaust, silencer, sparking plug, bonnet, gear, bearings, generator, sump, carburetor, ignition, switch, clutch, induction, tyres, cylinder, magneto, valves, decarbonise, petrol, windscreen

(6) Words Related to Games and Entertainment:

aquatic, cycling, rehearsal, arena, hockey, sculling, athletics, opera, tennis, chess, pantomime, theatre, croquet, regatta, whistle

(7) Words Related to Geography:

agriculture, frigid, plantation, arable, frontier, sedimentary, atoll, furrow, solar, barrier, glade, survey, cascade, insular, swamp, climate, inundation, temperature, confluence, irrigation, tornado, continent, knoll, torrential, contour, mound, tillage, crater, meteor, typhoon, configuration, monsoon, vegetation, cyclone, oasis, verdure, eclipse, peninsular, zephyr, expanse, planetary, zone

(8) Words related to Aviation:

accessory, aviation, monoplane, aerodrome, aviator, propeller, aeroplane, hydroplane, pancake, rigid, skid, volplane, rudder, stunt

(9) Words related to Metallurgy:

amalgam, filament, platinum, brazen, flux, radium, copra, galvanise, shellac, concrete, kiln, solder, ductile, mercury, tenacious, ebonite, nickel

(10) Words Related to Age and History :

antecedent, octogenarian, epoch, anterior, era, perpetual, anniversary, evanescent, prior, antique, jubilee, senile, centenary, simultaneous, matinee, chronometer, nonturnal, sycnhronise, contemporary, oblivion, ultimate, ephemeral

(11) Words Related to Shipping :

angler, freight, scupper, auxiliary, funnel, scuttle, berth, gear, sloop, bilge, keel, splice, bulkhead, mariner, startboard, bulwark, motor, stowaway, bunkers, naval, tackle, buoy, navigable, tarpaulin in, butt, nautical, tempest, canvas, pilot, towboat, chronometer, pirate, turbine, damage, quarantine, ventilator, derelict, rocket, veer, draught, saloon, wreckage, eddy, salvage, wrack, flare

(12) Words Related to Arms and Ammunition:

ammunition, knobkerrie, scimitar, bludgeon, knout, sheath, carbine, missile, sheathe, corslet, munitions, shrapnel, dynamite, poniard, visor, gully, quarrel

(13)Words related to Battle and Warfare:

adversary, dispatch, organise, armistice, enrolment, parley, assault, equipment, preliminary, belligerent, reinforcement, escort, casualty, reverberating, havoc, censor, hazard, revolution, cipher, indecisive, scheme, code, convoy, massacre, sniper, conscription, mutineer, strategic, decipher, mutual, strategic, declaration, neutral, violation, despatch, opponent

(14) Words Related to Navy:

admiralty, explosive, squadron, arsenal, magazine, submarine, artificer, monitor, submerge, boastswain, naval, submersion, camouflage, pinnacle, torpedo, cruiser, privateer, turret

(15) Words Related to Food and Cookery:

abattoir, essence, quail, albumen, farina, raisin, appetiser, frugal, relish, banquet, gammon, refrigerator, beverage, giblet, salad, caterer, gluttonous, sauce, cheddar, griskin, sausage, cocoa, intestine, spice, cinnamon, kidney, sponge, crumpet, knead, sumptuous, condensed, mace, steak, chutney, maize, slaughter, cutlet, mustard, tallow, chopper, nutriment, teal, cleaver, nitrogenous, tongue, cruet, offal, unleavened, culinary, palatable, vanilla, curry, prince, Vienna, dough, porridge, vegetables, dessert, pudding, viands, diet, preserve, victuals, digestible, provisions, vinegar

PHOBIAS, MANIAS AND COMPLEXES

FEAR	PHOBIA
Air	Aerophobia
Animals	Zoophobia
Beauty	Gallophobia
Birth	Genophobia
Blood	Hematophobia
Burglars	Scelerophobia
Burial alive	Taphephobia
Change	Neophobia
Childbirth	Maleusiophobia
Children	Pedophobia
Colours	Chromophobia
Crowd	Ochlophobia
Darkness	Nyctophobia
Death	Thanatophobia
Depth	Bathophobia
Disease	Pathophobia
Doctors	Latrophobia
Ghost	Phasmophobia
Hand	Chirophobia
Hair	Trichophobia
Heat	Thermophobia
Hell	Stygiophobia
Horses	Hippophobia
Knives	Aichmophobia
Large things	Megalophobia
Light	Photophobia
Lightning	Astrophobia
Males	Androphobia
Medicine	Pharmacophobia
Motherhood	Metrophobia
Motion	Kinesophobia
Pain	Agophobia; odynophobia
People	Demophobia
Poison	Toxicophobia
Poverty	Peniophobia
Red	Erythrophobia
Sexual intercourse	Coitophobia
Sinning	Peccatophobia
Sleep	Hypnophobia
Small things	Microphobia
Snake	Ophidiophobia

Solitude	Autophobia; monophobia
Sound	Acousticophobia
Stairs	Climacophobia
Stealing	Kleptophobia
Study	Logophobia
Sunlight	Heliophobia
Thirteen (The number)	Triksiaidekaphobia
Water	Hydrophobia
Wealth	Plutophobia
Women	Gynaephobia
Work	Ergophobia
Writing	Graphophobia

Mania

Pyromaniac	Has a passion for fire
Dipsomaniac	Has an uncontrollable compulsion to drink
Monomaniac	Has an obsession with one thing
Kleptomaniac	Has a compulsion to steal
Manic depressive	Has alternating moods of extreme depression and uncontrollable exaltation
Schizophrenia	A split personality
Melancholic	Has a fixed condition of despondency
Paranoic	Has delusions of persecutions
Oedipus complex	A problem in "mama's boys" hostile to father and excessively devoted to mother
Electra complex	Female version of Oedipus complex, girl is hostile to mother and devoted to father

Plagued Pairs

These generally consist of Homonyms: words that have different meanings and different spellings, but sound the same (two, too)

Spellonyms: (a made - up term are words that have different meanings but are spelled so nearly the same (torturous and tortuous)

Study these words carefully

To abjure	To reject, Renounce
To adjure	To charge or command solemnly
To accept	Receive
Except (prep)	But
To except(v)	To exclude, Leave out (I will expect Peter from this exam)
To affect	To make a difference
To effect (v)	To bring about, cause to happen
Effect (n)	The result or outcome
To adapt	To modify
To adopt	To take as one's own
Averse	Strongly against
Adverse	Unfavourable, hostile
Capitol	Government building
Capital	First rate, a money stake, net worth
Censor (n)	Person judging offensive matter
To censor (v)	To delete questionable matter
Censure (n)	Hostile criticism
To Censure (v)	To criticize harshly
Complacent	Self-satisfied
Complaisant	Amiable, Docile
Complement (n)	Something that completes, goes well with
To Complement (v)	To supply a lack, to complete
Compliment	Approving remark
To compliment	To speak approvingly
Contemptible	Hateful, deserving of contempt or scorn
Contemptuous	Hateful, deserving of contempt or scorn
Contemptuous	Showing contempt
Credible	Plausible
Creditable	Worthy or credit
Credulous	Gullible, naive
To demur	To object
Demure	Well-behaved
To deprecate	To disapprove of (with regret)
To depreciate	To go down in value
Discrete	Separate, Distinct
Discreet	Careful
Disinterested	Without prejudice or bias
Uninterested	Not interested
Diverse	Different
Divers	Plural of diver
To elicit	To draw forth
Illicit	Illegal
To elude	To escape
To allude	To refer to indirectly
To emend	To correct by editing
To amend	Improve
Equable	Uniform, steady
Equitable	Fair, Just

Factious	Inclined to dispute
Factitious	Artificial, Sham
Gourmet	Lover of fine food
Gourmand	Indiscriminate lover of food
To hoard	To save for yourself
Horde	Swarm, Crowd
To immigrate	To move to another country
Emigrate	To leave one's country and move elsewhere
Ingenious	Inventive, clever
Ingenuous	Innocent, Naive
Interment	Burial
Internment	Imprisonment
To lie	To recline as on a bed
To lay	To put or place
Loath	Reluctant
Loathe	Detest
Marital	Of marriage
Martial	Warlike
Nauseated	Sick to the stomach
Nauseous	Revolting or Disgusting
Personable	Pleasant
Personal	Private
Perspicacious	Perceptive
Perspicuous	Clear
To prophesy	To foretell
Prophecy	That which is foretold
To prosecute	To bring legal action
To persecute	To harass
To regale	To entertain
Regal	Referring to royalty
Social	Referring to human society
Sociable	Friendly
Tortuous	Winding, twisted
Torturous	Cruelly painful
Urbane	Sophisticated
Urban	Referring to cities
Venal	Corrupt
Venial	Excusable
Vocation	Career
Avocation	Hobby

WORDS FROM MYTHS & LEGENDS

Achilles' heel	Week spot; place of vulnerability
Aeolian	Referring to winds. Aeolus was God of the winds
Apollo and Adonis	A handsome desirable male Apollo was the Sun God of the Greeks and Romans. Adonis was a mortal man, lover of Venn
Atlantean, Atlantic	From Atlas; Indicates strength
Aurora	The dawn - Aurora was Goddess of the dawn
Bacchanal	An orgy; a bacchanl is a reveler or carouser at parties
Bacchanalian	Drunken revelry or wild party
Callope	Stream whistle or anything making a similar shrill sound
Cassandra	One who prophesies trouble.
Centaur	Having the head and arms of a man and body of a horse.
Chimera	Imaginative creature; wild and scary
Cupidity	Greed or avarice. Cupid was the young god who embodied sexual desire
Erotic	Of or pertaining to sexual love. Eros (cupid) was the God of passionate love
Furor (e)	Frenzy, excitement, Rage
Fury	An avenging spirit, Virago.
Gorgon	Ugly, frightening person repulsive creature
Halcyon	Peaceful, tranquil. Halcyon was a widow who was metamorphosed by Gods into a kingfisher
Harpy	Grasping, greedy, scary female. Harpies were part women, part bird
Hector	To bully or annoy
Helen (of Troy)	A woman of incomparable beauty (a face that launched a thousand ships)
Herculean	Extra ordinarily strong, Difficult
Hydra-headed	An evil with many sources; Difficult to control
Iridescent	Having a rainbow like play of colours
Jovial	Good-natured, Jolly
Junoesque	Queenly, Statuesque (said of women) Juno wife of Jupiter was queenly
Lethargy	Sluggishness. The river Lethe ran through Hades (the underworld) and whoever drank the water forgot his past

Martial	Warlike, Suited for army or military life (from Mars, God of war)
Mentor	Respected guide, Teacher
Mercurial	Volatile in mood, Easily changeable
Mnemonics	The art of improving memory
Morphine	An addictive narcotic. Morpheus was God of dreams
Muse	Source of inspiration, a poet
Narcissism	Self-love; extreme egotism
Nemesis	Relentless pursuer of evil doers. Nemesis was Goddess of retribution
Neptune	Figuratively, the ocean. A Neptunist thinks the world emerged from water
Odyssey	Long wandering or voyage, the hero of Homer's Odyssey.
Olympian	Majestic, superior to others
Paean	Hymn or song of praise. Paean was the physician of Gods
Pandora's box	Source of evil. Malign influence needing to be kept under control
Parnassus	Having to do with poems; to climb Mount Parnassus is to write poetry
Pegasus	Extremely swift horse; Poetic inspiration
Phoenix	Symbol of immortality
Plutocracy	Government of the wealthy. Pluto was the god of the underworld
Promethean	Unusually original and creative, life-giving; Prometheus formed man from clay
Protean	Assuring many shapes or forms. Proteus, the sea god, kept changing shape
Psyche	The soul, Mind. Psyche was the mortal girl loved by Eros
Saturnalia	Wild party or orgy. Saturn, God of sowing
Sibyl	Prophetess of Ancient Greeks and Romans
Stentorian	Extremely loud(Stentor was the greek herald)
Stygian	Gloomy, Dark, Frightening
Tantalise	To torment or tease.
Terspsichorean	Referring to dance. Terpsichorean was the Muse of the Dance
Titanic	Gigantic in power
Zeus	Ruler, king Zeus was king of Greek gods. A Zeus has absolute power.
Cyclopean	Huge, Massive
Erotic	Pertainig to sexual desire
Gorgon	An ugly person, A horrifying face
Harpy	A grasping, ravenous person
Janus - Faced	Two faced, having two distinct faces
Jovial	Merry, joyous, Happy
Mentor	A teacher or a wise counselor/ coach
Mercurial	Swift, Active
Narcissism	Self love & admiration, Narcissistic feelings
Nemesis	An agent of punishment, Upholder of moral code
Odyssey	Long dangerous journey
Olympian	Majestic, Awe-inspiring, datached and aloof
Paean	A song of praise, Triumph or thanks giving
Palladium	A safeguard
Phoenix	A symbol of immortality
Protean	Changing in shape
Saturnine	Heavy, Dull, Gloomy, Serious, Morose
Stentorian	Very loud voiced; Bellowing

WORDS DEPICTING SUBJECTS OF STUDY

TERM	SUBJECT/ STUDY OF
Aetiology	Cause of diseases
Anthropology	Mankind
Astrology	Heavenly bodies
Cardiology	Heart function and diseases
Carpology	Fruits and seeds
Cetology	Whales
Cartography	Maps
Choreography	Dance
Chronology	Dates
Craniology	Skulls
Criminology	Crimes and criminals
Cryptology	Codes

Dactylology	Finger prints
Demography	Population statistics
Dermatology	Human skin
Endocrinology	Glands
Entomology	Insects
Epigraphy	Ancient inscriptions
Etymology	Word origins
Genealogy	Ancestry
Gerontology	Old age
Gynaecology	Women's disorders
Haematology	Blood
Harpetology	Reptiles and amphibians
Histology	Plant and animal tissue
Horology	Time, Timepieces
Hypnology	Sleep
Ichthyology	Marine life (fish)
Lexicography	Writing of dictionary
Lexicology	Dictionary
Lithology	Rocks
Metaphysics	Nature of existence, Truth and knowledge
Meteorology	Weather
Metrology	Measurement
Myology	Muscles
Obstetrics	Birth of children
Odontology	Teeth
Oenology	Wines
Oncology	Cancer, Tumors
Ontology	Nature of existence
Ophiology	Snakes
Opthalmology	Eyes
Orology	Mountains
Osteology	Bones
Osteopathy	Manipulating bones and muscles
Otology	Ears
Paedology	Children
Palaeography	Old manuscripts
Palaeontology	Fossils
Pathology	Diseases
Pedagogy	Teaching
Pedology	Soil
Penology	Prisons and treatment of criminals
Pharmacology	Drugs
Philology	Languages
Physiology	Life process
Phytology	Plants
Podiatry	Feet
Polemology	Wars
Potamology	River
Psephology	Elections
Psychiatry	Treatment of mental illness
Psychology	Mind
Radiology	Radiation
Seismology	Earth quakes
Selenology	The moon
Speleology	Caves
Topography	Surface features of a region
Topology	Shapes and surfaces
Toxicology	Poison

WORDS FROM CHARACTERS OF LITERATURE

Babbit	An American business man, vastly pleased with himself; of Lewis' novel Babbit
Brobdingnagian	From Swift's Gulliver's Travels, synonym for gigantic
Bumble	Mr. Bumble was the pompous Orphanage official in Oliver Twist, anyone like him.
Cinderella	Like the fairy tale girl who goes from rags to riches.
DonJuan	An unsavoury lover, the original character killed men and seduced women, any fellow who is interesting to know.
Falstaffian	Shakespeare's bawdy, good humoured fat man in his historical plays, resemblance to this character.
Frankenstein	Name of the scientist in Mary Shelley's book by the same name, who created the monster; now the monster that destroys its creator.
Friday	Robinson Crusoe's original, a helper who is dependable (man Friday)
Galahad	Noble of all knights, both good looking and gallant
Jekyll&Hyde	One who alternates between good and evil; from Stevenson'sstory, Hyde ends with Jekyll trapped in the murderous Hyde personality
Lilliputian	Opposite of Brobdingnagians; anything miniature or tiny
Lothario	Well known lover, character in the play 'The Fair Penitent'

Malapropism	Mrs Malaprop from Sheridan's play 'theRivals' gained immortality by her verbal mistakes, a hilarious misuse of words
Pickwickian	Dicken's character of' Tickwick Papers": Pickwickian words are used in an esoteric special way (not in accordance with their usual meaning)
Pooh-bah	In Gilbert and Sullivan's opera The Mikado' was the Lord', any politician who holds several offices is a pooh-bah
Quixotic	Don Quixote was the lovable impractical old man in Cervante's novel with the same title. Quixotic schemes are unrealistic
Robot	Czech playwright karel Capek's play titled R. U. R (Rossum's universal Robots); a robot is anyone who works without thought
Scrooge	The penny-pinching miser of Christmas Carol (Dicken's); now any grasping, covetous person without kindness
Simon Legree	Stowe's character in 'Uncle Tom's Cabin', the character was a slaver; anyone who works others mercilessly
Uncle Tom	In uncle Tom's Cabin, Uncle Tom was saintly old black. Today a black who is seen as quick to co-operate with whites
Uriah Heep	From Dicklen's David Copper field, a term to describe someone hypocritical.
Walter Mitty	Thurber' s lovable character from 'The secret life of Walter Mitty; the embodiment of fantasies.

WORDS FROM PEOPLE'S & PLACE NAMES

People's Name

America	(from the sailor Amerigo Vespucci) The two great continents
Ampere	(from Andre Ampere) Unit of electric current
Atlas	(from Greek demigod Atlas) A man holding the globe on his back; a collection of maps ;
Bowdlerize	(from Dr Bowdler, an art critic) To make a book readable by deleting offensive portions
Boycott	(from Captain Boycott) Ostracize, break connections with
Braille	(from Louis Braille) A system of printing for the blind
Chauvinism	(from Nicholas Chauvin) Exaggerated loyalty or patriotism
Chimerical	(from Greek monster Chimaera) Ideal but impractical
Cologne	(from German city Cologne) Famous as "Eau de cologne
Derrick	(from a hangman Derrick) A hoisting apparatus
Fahrenheit	(from D.G. Fahrenheit) Unit of temperature measurement
Frankenstein	(from the person Frankenstein) A horrible monster
Galvanize	(from Dr Galvani) To electrify; to stir into action; Galvanization = electrolytic technique
Gargantua	(from Gargantua) Anything on a large scale; massive
Gordian Knot	(from king Gordian) A very tough problem
Guy	(from Guy Fawkes, a drama character) Ridiculous name for a man
Hector	(from Greek hero Hector) To quarrel, tease, bully someone
Herculean	(from Roman Hercules) Very heavy, very difficult
Hobson's Choice	(from Thomas Hobson) Really no choice at all
Jingoism	(from Jingo) Feeling of belligerence towards foreigners
Lilliputian	(from Lilliput) Very small, Miniature, Diminutive
Lothario	(from Lothario, a character) A flirt; seducer of women
Macadam	(from John Macadam) Roads made of broken stones
Malapropism	(from Mrs. Malaprop) A contusion of words; verbal blunder
Martinet	(from Gen. Martinet) A person very strict about discipline

Masochism	(from Leopoid Masoch) Enjoyment through self torture
Maverick	(from Samuel Maverick) An unpredictable person
Nemesis	(from Greek Gold Nemesis) An agent of restribution
Nicotine	(from Jean Nicot) Poisonous drug in Tobacco
Ohm	(from GS. Ohm) Unit of electrical resistance
Procrustean	(from Greek robber Procrustes) To be able to mould
Quixotic	(from Don Quixote) Unrealistic visionary, Impractical idea
Robot	(from Robota) a machine or human totally mechanized (no soul)
Rodomontade	(from king Rodomonte) Very boasting, Blustering, Bragging
Solecism	(from Greek province Soloi) Blunder in speech or writing
Spoonerism	(from Rev. W.A. Spooner) Habit of using first letters of words in phrases
Thespian	(from Thespis, the father of Greek tragedy) An actor
Utopian	Ideal but impractical, Unfeasible, Impossible, Chimerical
Volt	(from A. Volta) Unit of electric potential
Yahoo	(from a tribe called Yahoos) A degraded human specimen

Place Name

Bohemian	Bohemia was believed to be the home of all the gypsy tribes that roamed through Europe
Canopy	Greek word konops means mosquito; canopy, a cloth covering for protection
Donny brook	The Irish fair at Donny brook, a scene of cheerful violence , a club
Dungaree	Jeans dating back to the cotton cloth woven in Dungaree
EL Dorado	"Indians in South America used to cover their king with gold dust. “He was Eldorado, the Gold-covered One; a place of incredible wealth."
Hackney	Now means a horse of English breed, a poor servant who does tiring work
Laconic	Laconic was the home of the Greek city Sparta. To be laconic is to be brief and succinct of speech, to be terse.
Limerick	The tiny town of Limerick, Ireland, gave its name to a most enjoyable verse form, the five-line limerick. Modern limeriks are often suggestive
Marathon	Anything that goes on forever. The original long-distance runner fell dead at the end of his run, at Marathon, Greece.
Meander	The Asian river meanders, winds and twists its way to the sea. If a speech or movement of winds and twists it is said to meander
Sardonic	A herb form Sardinia is so bitter that if people eat it they go insane. A sardonic expression is scornful
Spartan	To live without luxuries is to follow a Spartan life. Spartans were macho men.
Utopia	Title of many books and imaginary places; unattainable
Waterloo	"To meet one’s Waterloo" is to repeat Napolean’s dismal performance in his last battle and lose.

YOUNG ONES & CRISES OF ANIMALS

Young Ones of Animals

Bear	Cub
Buffalo	Calf
Cat	Kitten
Cow	Calf
Dog	Puppy
Duck	Duckling
Frog	Tadpole
Goat	Kid
Hare	Leveret
Hen	Chicken
Horse	Colt

Lion	Cub
Sheep	Lamb
Swan	Cygnet
Tiger	Cub
Eagle	Eaglet

Words Indicative of Sound

Eagle	Eaglet
Ape	Gibber
Ass	Bray
Bear	Growl
Bird	Twitter, Warble
Bull	Bellow
Camel	Grunt
Cat	Mew, Purr
Cattle	Low
Crow	Crow
Dog	Bark, Snarl, Howl, Yelp, Whine
Dove	Coo
Duck	Quack
Eagle	Scream
Elephant	Trumpet
Fly	Buzz
Frog	Croak
Fox	Yelp
Goat	Bleat
Geese	Cackle, Gaggle
Hawk	Scream
Hen	Cluck
Hog	Grunt
Horse	Neigh
Monkey	Chatter
Mouse	Squeak
Nightingale	Sing
Owl	Hoot, Screech
Parrot	Talk
Pig	Grunt
Pigeon	Coo
Puppy	Yelp
Raven	Croak
Rook	Caw
Serpent	Hiss
Tiger	Growl
Vulture	Scream
Wolf	Howl

GROUP TERMS

Deer	Herd
Fish	Shoal
Geese	Flock
Corn	Stack
Wood	Stack
People	Tribe
Birds	Flight
Wheat	Sheaf
Wolves	Pack
Hay	Bundle
Grain	Sheaf
Sticks	Bundle
Soldiers	Regiment
Plantains	Bunch
Partridges	Convoy
Ants	Nest
Bees	Hive, Swarm
Cattle	Herd
Stars	Cluster, Galaxy
Arrows	Sheaf, Quiver
Locusts	Swarm, Flight
Chickens	Brood
Pigeons	Brace
Ladies	Bevy
Sailors	Crew
Camels	String
Musicians	Band
Pistols	Brace
Cigar	Box
Arms	Pile
Swine	Herd
Sheep	Flock
Steps	Flight
Flowers	Posy
Horses	Stud
Oxen	Yoke
Hens	Brood
Birds	Flock
Savages	Hoard
Hounds	Leash
Men	Host

Puppies	Litter
Peacocks	Muster
Plants	Nursery
Oxen	Team
Fruits	Basket
People	Throng, Colony, Clique
Sticks	Faggot
Sardines	Family
Dogs	Kennel
Jurymen	Panel
Constables	Posse
Magistrates	Bench
Actors	Troupe
Guns	Battery
Actors	Company
Advisers	Council
Pictures	Gallery
Flowers	Nosegay
Boats	flotilla
Cavalry	Squadron
Shrubs	Shrubbery
Cars, ships	Fleet
Animals, Plants	Genus
Worshippers	Congregation
Merchants	Syndicate
Stars	Constellation
Fruit trees	Orchard

WORDS OF FOREIGN ORIGIN

French Words

Aplomb	Complete confidence
Au Courant	Well informed
Avant Garde	Describes artist and writers who are unorthodox and untraditional
Bete Noire	Person or subject hated
Brouhaha	Furor, Uproar
Carte Blanche	Complete authority
Cause Celebre	Situation/case arousing attention and discussion
Coup D' Etat	An unexpected action (of government) usually implying the use of force,
C'estLaVie	That's life
DejaVu	Sense of having undergone
De Mode	Out of fashion (antonym: A La Mode)
De Rigueur	Indispensable; Something absolutely required
Enfant Terrible	Unconventional person
En Passant	By the way
Espirit De Corps	Morale
Fait Accompli	An irreversible act done without warning
Faute de Mieux	For want of something better
Faux Pas	A social error
Fin De Siecle	End of century (specifically 19th century)
Haute Couture	High fashion
Haute Cuisine	Excellent cooking
Haut Monde	The elite
Laissez Faire	Policy of non-interference
Mot Juste	Perfect expression
Manque	Failing to achieve a desired goal
Noblesse Oblige	Obligation imposed on high positions
Noveau Riche	Ostentatious person
Par Excellence	Highest degree
Parti Pris	Bias
Raison D' Etre	Reason for existence
Roman A Clef	Novel in which real characters and events figure under disguise Sang Froid Self control
Savoir Faire	Instinctive ability to act appropriately
Soi Disant	So called
Soupcon	Suggestion, Inkling
Tete-a-tete	Close conversation between two persons
Vis-a-vis	In regard to
Volte Face	Complete reversal

Latin Words

Apriori	Self evident, known independently
Ad Hominem	Directed at someone personally
Ad Libitum	Improvise (V), Off-the-cuff

AdLib	Remark(N)
Ad Nauseam	To the point of disgust
Alma Matter	One's old school, College
Alter Ego	One's other self
Alumnus	Former student ,
Amicus Curiae	Disinterested advisor
Anno Domini (Ad)	In the year of the lord
Ante Meridiem (Am)	Between midnight and noon
Annus Mirabilis	Year of wonders, Disasters
Bonafide	Genuine
Bis	Twice
Caput Mortuum	Worthless residue
Caries	Decay (of bone or teeth)
Caveat	Warning
Ceteris Paribus	Other things being equal
CuiBono	For whose good?
De Facto	Functioning in reality
De Jure	Functioning in accordance
Deus Ex Machina	Person or thing that suddenly resolves a problem
Ex Gratia	Referring to payment made as a favour
Ex Post Facto	After the event '
Ignis Fatuus	A false hope or illusion
InToto	Completely
In Extremis	In extreme circumstances
In Loco Parentis	In Place of a parent, acting as a guardian
Inter Alia	Among other things
Ipso Facto	As an immediate consequence of a fact
Magnum Opus	A great work; Crowning achievement
Mea Culpa	An acknowledgement of guilt
Modus Vivendi	A way of getting along with another person
Non Sequitir	An illogical remark
Obiter Dictum	An incidental remark
Per Se	By itself
Persona Grata	An acceptable person
Persona Non Grata	An unacceptable person
Prima Facie	On first appearance 3'
ProRata	Proportionately

German Words

Angst	A feeling of anxiety
Blitzkrieg	Sudden attack
Ersatz	Artificial
Fraulein	Lady
Haus Frau	House wife
Kaffee	Coffee
Kaiser	Emperor
Kaput	Totally ruined
Kitsch	Worthless
Lebensraum	Living space
Putsch	Coup
Quid Pro Quo	Something given for something received
Q.E.D	That the point has been made
Sine Die	With no fixed date, Indefinitely
Sinc Quanon	An indispensable condition
Status Quo	Present condition
Sui Generis	Unique; In a class by himself herself
Vade Mecum	A reference or hand book
Vox Populi	Public opinion

Spanish Words

Bonanza	A huge profit
Caballero	Gentleman
Cantina	A bar
Cortes	Parliament
Fiesta	Festival
Hidalgo	Minor nobleman
Incommunicado	Without contact with outside world
Junta	Secret Council
Manana	Soon, Tomorrow
Peccadillo	Small error
Siesta	Afternoon nap

Japanese Words

Banzai	Cheer of triumph
Bonsai	Growing miniature plants
Bushido	Samual code
Geisha	Young woman who accompanies and entertains men

Haiku	Delicate form of Japanese poetry
Hara Kiri	Suicide
Ikebana	Flower arrangement
Kamikaze	Suicide pilot (World War II)
Kimono	Long, Loose robe tied with sash
Mikado	Japanese emperor
Nisei	Children of Japanese immigrants born in the USA
Sake	Rice wine
Samurai	Warrior
Sayonara.	Good Bye
Shinto	Japanese religion
Sumo	Form of wrestling
Zaibatsu	Small group (of families) Controlling Japanese Industries
Reich	Republic
Schadenfreude	Feeling of enjoyment from misfortune of others
Strafe	Punish
Wunderkind	Prodigy
Arrivederci	Good Bye
Bravura	A show of daring
Brio	Vivacity
Carabiniere	Policeman
Chiaroscuro	Sharp contrast
Cognoscenti	Person with expert knowledge
Ciao	Used as a greeting at meeting or parting
Dilettante	Amateur or Dabbler
Dolce Far Niente	Carefree existence
Dolce Vita	The good life
Diminuendo	With gradually decreasing intensity or volume
Embroglio	Violent, Embarrassing quarrel
Lingua Franca	A common language
Palazzo	Place
Prima Donna	Leading female singer at an opera
Presto	Quickly
Verismo	Realism .
Virtuoso	An expert performer

WORD FREQUENCY

Word Frequency questions have a set of four sentences, each having a blank. Each of these sets is succeeded by four words. You have to find out the maximum number of sentences in which any one of the four given words fits. Use this number to mark your answer. If any of the four words fits into all four sentences, your answer shall be (4).

If a word fits only into three sentences (and if that is the highest frequency in comparison to all the other words) your answer shall be (3) and so on.

Let us consider an example first. This is a reproduced CAT question.

(I) The mediator helped reach a/an _____ agreement.

(II) The structure of most animals and insects exhibits_____symmetry.

(III) The two countries signed a_____cultural exchange pact. (TV) Due to the disqualification of one of the three candidates, the election was reduced to a contest.

(l) Amicable (2) Bilateral

(3) Historical (4) Multilateral

Here, your answer would be (2). This is because the word 'bilateral' fits into all the four blanks. The word 'amicable' fits the blank in sentences (I) and (II) only, while the word 'multilateral* can fit only sentence (I).

WORD ANALOGY

Questions on Word Analogy appear recurrently in aptitude tests. In these, you have to select from a number of choices, the pair of words that relates together in the same fashion as the given pair of words in the question is. Since such question demands that you should be able to analyze and identify basic relationships (apart from having a good vocabulary in English language), it is widely believed to provide a good assessment of mental ability. Every analogy gives you a pair of words represented as below:

The symbol (:) stands for 'is related to' or simply 'is to'. The base pair is followed by several choices, and the whole question is written as follows:

ORCHESTRA: INSTRUMENTAL:

a. Trio: harmonic

b. Choir: vocal

c. Band: martial

d. pianist: solo

The symbol (::) stands for 'in the same way as' or simply 'as'.

In working out an analogy, you must proceed by first determining the nature of relationship that exists between the two words in the base pair. In this example, analysis indicates that orchestra is a large group of people playing various musical instruments together.

The next step is to study the relationship between the words in the pairs of answer choices. In our example, a study of the first pair indicates that a trio is not a large group performing harmonic music; in choice 2, a 'choir' is a large group performing vocal music. In choice (3) Band is a group but is not known to perform martial music.

Similarly, in choice (4) pianist is not a group. Of all the options given, the words in the second pair match most closely with the relationship expressed in the base pair. Consequently, option (b) provides the best answer.

IMPORTANT: You must remember that the pair of words of the answer choice must match the base pair in 'form' as well as 'relationship'. If, for example, the base pair depicts a relationship between two nouns, the correct answer will also consist of two nouns. If, moreover, the base pair is made up of two plurals, the correct answer will be made up of two plurals. Sometimes a quick scrutiny of options will help you immediately discard the choices that do not have the same form as the base pair.

Strategy For Analogy

(i) Identify the relationship between the words in base pair by linking them using a sentence,

(ii) Look out for contextual meaning of words,

(iii) The relationship between the words could be vertical or horizontal.

(iv) The sides should match; if the 'parent' is on the left side and the 'child' on the right in the question pair, then the correct answer choice should also follow the same order in sense.

CAUTION: Do read the instruction given in a test closely; sometimes the direction can be to select the option that LEAST similar to the question words.

Word analogies frequently illustrate common relationships. The following are worth noting.

COMMON TYPES OF ANALOGY

	Analogy Types	Example
1	Action and its typical quality/adjective	spark plug-Car
2	Cause - effect	Clouds - rain
3	Degree / intensity	Tepid - not
4	Object / substance-natural component	Pearl – oyster
5	Object related to natural location	Camel - desert
6	Word-synonym	Suave - sophisticated
7	Word-antonym	Fertile - barren
8	Person related to tool	Carpenter - saw
9	Person related to major trait	Philosopher - wisdom
10	Person related to skill/interest	Scientists - research
11	Person related to least desirable characteristic	Stickler - approximation
12	One of a kind /class / type	Daisy-flower
13	Substance / raw material related to end product	Sheep - wool
14	Thing related to logical action.	Pestle - pound
15	ear clichéd expression	Friend - congeniality
16	Implied comparison	Clouds - sun
17	Relationship of description	Speech – articulate
18	Phenomenon/entity-measuring device/unit	Decibel - sound
19	Object is to thing with which it is made	Sweater - acrylic
20	Branch of study and subject	Ornithology-bird
21	Person is to area of specialization	Ichthyologist - marine life

22	Extract is to source	Gold-ore
23	Symbol and purpose	Lighthouse - danger
24	Legendary term - symbolic meaning	Malapropism - verbal
25	Object is to case in which it is kept	Quiver - arrows
26	Plan is to action	Agenda - meeting
27	Member is to particular group	Captain - team
28	Relationship of size	Boulder - pebble
29	Action and its typical quality/adjective	Movement - graceful
30	Relationship of structural/functional similarity	Drawbridge – castle
31	Idiomatic expression	On-off

ODD WORD OUT

"Odd word Out" is a vocabulary-based questions that test your ability to classify words based on their common attributes. As this test requires you to know the meanings of many words, your performance in such questions to a great extent depends upon your word power. These questions are better answered after strengthening your vocabulary.

IDIOMS & PHRASES

In an idiom and phrases, questions are asked to identify its literal meaning from the choices. These idioms may have their origins in other languages like Latin, French or Greek. A list of the most commonly used idioms are given in the Word Bank issued to you as part of module 1. Learn them well and look up to the dictionary for more of them. It is also interesting as well as important to collect and make a list of those idioms that you come across in your daily reading.

EXERCISE

MCQ TYPE QUESTIONS

Word Frequency

Directions for (Que. 1 - 30): Read the information carefully and answer the questions given below it.

Each of the following questions has a set of four sentences, each containing a blank.

Each set is followed by four words. Find out the maximum number of times that any of the words fits each set.

Use this number to indicate you answer.

For example, if any one of the words fits into all four sentences, your answer will be (*d*) However, if it fits only into three sentences (and that is also the highest frequency for any of the four words) your answer will be 3 and so on.

1. (I) This pole, erected for its aesthetic appeal, has turned into a popular ...for birds.

(II) I did not expect that old cupboard to... such a fancy price at the auction.

(III) The... was rusty and creaked whenever the door was closed or opened.

(IV) The dog was trained to... the newspaper every morning.

(*a*) hinge (*b*) perch
(*c*) bring (*d*) fetch

2. (I) The ... adjourned all business for the day as a mark of respect for the departed leader.

(II) For several centuries before it was proved that the earth was round, scientists believed that the earth was ...

(III) What is referred to as an apartment in America, is called a... in England.

(IV) Driving on a(n)... road is easier than driving on an uneven one.

(*a*) even (*b*) flat
(*c*) house (*d*) oral

3. (I) The troops were ordered to... the forests to locate the deserter.

(II) Since the heavy duty equipment failed, the stones had to be moved by..... .means

(III) The PM had to... to popular pressure in the face of documented evidence.

(IV) This... explains how to assemble, use and maintain the system.

(*a*) physical (*b*) comb
(*c*) bow (*d*) manual

4. (I) It is important that your concentration should not... particularly in the reading comprehension section.

(II) We tried t... down several passing cars in the hope of getting a lift.

(III) Showing disrespect to the national... is an act to treason.

(IV) The hurricane was on the... when we entered the city limits.

(*a*) wane (*b*) flag
(*c*) wave (*d*) emblem

5. (I) The leaning tower or Pisa is a(n)... structure.

(II) Of the two living dodos, one died and made the other...

(III) Tigers are protected in order to prevent them from becoming...

(IV) The... details of his escapade were very interesting.

(*a*) extinct (*b*) unique
(*c*) rare (*d*) astounding

6. (I) The first day's play was washed out due to inclement...

(II) After the accident, his thinking has become very...

(III) His success has not come easily, he has had to... many a storm in his career.

(IV) Because of the incessant... we could not go out.

(*a*) rain (*b*) cloudy
(*c*) weather (*d*) turmoil

7. (I) When the effect of the painkiller wore off, he was racked by... pain.

(II) In any right angled triangle, at least two of the angles are...

(III) The philanthropist was moved by the ... poverty he saw around him.

(IV) The lawyer's commitment was... and unqualified.

(*a*) total (*b*) prevalent
(*c*) complete (*d*) acute

8. (I) His... in the subject stood him in good stead.

(II) A mere... to do well will not do; it must backed by will power and hard work.

(III) In driving a car, you have to change gears to negotiate a/an ...

(IV) The RBI has decreased lending ... rates by 2%,

(*a*) interest (*b*) inclination
(*c*) support (*d*) curve

9. (I) Why should he apologise? He has not... any mistake.

(II) He has risen to the top only because he has been ... to his profession.

(III) Since the supply situation is unclear, I hope you have not... any dates to the client.

(IV) Sharma's... was there for all to see; she worked beyond the call of duty.

(*a*) diligence (*b*) capacity

(*c*) committed (*d*) confirmed

10. (I) Because of the transporter's strike and the consequent delay in material supply, Harsha's work schedule wasway off the

(II) ... my words; this boy is destined for greatness.

(III) This single failure continues to be the only... on his impeccable resume.

(IV) Manners are the... of a gentleman.

(*a*) 37mark (*b*) stain

(*c*) blot (*d*) impression

11. (I) When the machinery in the... broke down, the production manager was summoned immediately.

(II) This... is a rare one because it does not produce either flowers or seeds.

(III) This puzzle requires more than three hours to put the ... together.

(IV) The teacher must... the idea of lateral thinking in her student

(*a*) parts (*b*) unit

(*c*) encourage (*d*) plant

12. (I) Because of the wide spread protests the government was forced to ... the law.

(II) In the bilateral contest, one contestant polled 29% of the votes, giving the other an ... majority.

(III) This countryside... with natural minerals.

(IV) In relative terms, your salary may not be much, but in... terms, Rs.2,00,000 per month is a sizeable amount.

(*a*) rescind (*b*) abounds

(*c*) absolute (*d*) financial

13. (I) When Rip Van Winkle woke up, the world around him ... no resemblance to the one he had known.

(II) This gun is a Smith and Wesson with a 0.38 ...

(III) We avoided Ram in the office because he was such a...

(IV) The tidal ware... down on the ship with relentless fury.

(*a*) barrel (*b*) bore

(*c*) crashed (*d*) tear

14. (I) His insightful... on juvenile rehabilitation has become the foundation for several laws.

(II) No sooner did he hear the gun's ... than he collapsed.

(III) The value of this... is inestimable.

(IV) The company's ... into software exports has proved profitable.

(*a*) essay (*b*) report

(*c*) article (*d*) study

15. (I) In this era of liberation, one popular, request from local entrepreneurs is for a level playing... .

(II) Some of the greatest scientist have had little formal training in their...

(III) The scandal-ridden Pakistan team had to... one of its weakest teams in recent years.

(IV) In the absence of insightful reporters, the senators had a... day at the press conference.

(*a*) ground (*b*) field

(*c*) area (*d*) enter

16. (I) Without the... that normally surrounded him, the musician was simple human being with his own share of weaknesses and eccentricities. :

(II) It is amazing that the... on the vase remained even after centuries.

(III) This text is tautological and you can grasp its meaning even if you... over it.

(IV) The assassin was hard to trace because of the... that surrounded him.

(*a*) gloss (*b*) glaze

(*c*) web (*d*) skim

17. (I) The miner used his... to open up the tunnel.

(II) Prince William had to ... at the vines to get to the castle.

(III) This horse is a ... and cannot be used for racing purposes.

(IV) The brick maker took the dried bricks out of the... and put them on the truck.

(*a*) pick (*b*) hack

(*c*) thoroughbred (*d*) chop

18. (I) The... that we normally talk about in a pencil is actually graphite.

(II) Sherlock Homes is known for his ability to follow even a simple... to unravel acomplex problem.

(III) General Mac Arthur is a man who... by example.

(IV) It is always a difficult task for the... speaker in a debate.

(*a*) clue (*b*) first

(*c*) lead (s) (*d*) principle

19. (I) As the best lawyer specializing in air disaster cases, he was without ...

(II) During adolescence... pressure plays a major role in molding the value system.

(III) A... resembles an apple, though it is different in taste.

(IV) Sam asked Anitha to... out of the window and see if the guests had arrived.

(*a*) social (*b*) peer

(*c*) look (*d*) equal

20. (I) This is not a matrix because it has several columns but only one...

(II) The sergeant ordered the cadets to stand in single... at shoulder lengths.

(III) The... containing the papers on the merger was lost in transit.

(IV) The mechanic is looking for a... to even out the rough surface.

(*a*) row (*b*) order

(*c*) file (*d*) hammer

21. (I) Because of the strong winds, the cargo in the ship moved unevenly and caused the ship to ...

(II) This Ayurvedic drug is popular for its ability to... asthmatic disorders. ;

(III) Despite the trainer's best efforts, the dog would not come to... ;

(IV) The new pair of socks was small and would not fit ,at the... .

(*a*) tilt (*b*) heel

(*c*) foot (*d*) alleviate

22. (I) The annual boat race is in the month after April, that is in.....

(II) The rate of interests works out to a/an ... 8% of given the highest rate of 10% and the lowest pf 6%.

(III) His academic record is no ... of his intellectual caliber.

(IV) Not helping his friend who was indisposed, was a ... act.

(*a*) average (*b*) may

(*c*) measure (*d*) mean

23. (I) The ... on the newly asphalted road had not dried and stuck to the wheels of our car.

(II) The diva's voice surprisingly cracked when she moved into the song at a high ...

(III) The cricket match had to be called off because the vandals had damaged the ...

(IV) The advertising agency spent many days in developing its ... for the new business.

(*a*) pitch (*b*) proposal

(*c*) level (*d*) tar

24. (I) The ... has come to be synonymous with the psychiatrist's profession.

(II) Despite his age, Raman had a(n)... posture.

(III) During the last year, the government has sanctioned Rs. 1,000 to... new structures.

(IV) The committee has taken steps to... charges of misconduct and bias.

(*a*) avoid (*b*) erect

(*c*) renovate (*d*) couch

25. (I) Frazier was out for the... and lost the match by K.O. to Muhammed Ali.

(II) While coarse yams are not widely exported, yams of finer... are very much in demand.

(III) Dracula was, in terms of social hierarchy, a ...

(IV) To pass the bill in the assembly, the ruling party had to ,.. on the support of at least 10 opposition members.

(*a*) depend (*b*) vampire

(*c*) count (*d*) grade

26. (I) Trade in waste metal or... is carried out through a well organized network,

(II) Following wide spread copying and paper leakage, the board decided to ... the test.

(III) Tilak is a belligerent person and picks up a ... with almost everyone he meets.

(IV) The child cut out the picture of the dinosaur for the use in his ... book.

(*a*) scrap (*b*) perch

(*c*) bring (*d*) fetch

27. (I) Dr. Shekar is our... representative in Bangalore.

(II) Most dental treatment calls for only... anesthesia.

(III) Once the foreigners left, only the... people remained.

(IV) The invaders wrested...

(*a*) area (*b*) local

(*c*) primary (*d*) control

28. (I) She was a key... for the prosecution.
(II) You should not... a book by its cover.
(III) Thousand of people gathered to... fireworks display.
(IV) Because he was personally involved, the lawyer could not take up the...
(*a*) case (*b*) judge
(*c*) witness (*d*) court

29. (I) You must ...more time if you want to make a considered decision.
(II) Batsmen wearing rubber soles get little ...when running on the pitch.
(III) She want to the market to... some grocery.
(IV) The.., department in an organization is normally responsible for vendor development.
(*a*) buy (*b*) purchase
(*c*) finance (*d*) grip

30. (I) A...is an accepted of male office wear.
(II) Of the two living dodos, one died and made the other...
(III) A sailor's knot is one of the most difficult knots to...
(IV) "... your gun from its holster", Adams said.
(*a*) draw (*b*) bow
(*c*) shirt (*d*) tie

Word Analogy

Directions for (Que. 31-40): Read the information carefully and answer the questions given below it.

Each question below consists of a pair of words that are related. Complete the blank with suitable word from the answer choices given to form an analogous pair.

31. Pediatrician : _____ :: Dermatologist: skin
(*a*) Children (*b*) Feet
(*c*) Plant (*d*) Hair

32. Blandish : Coax : : Asseverate :_________
(*a*) Affirm (*b*) Repeat
(*c*) Complain (*d*) Twist

33. Exigency: Emergency:: Expansive :______
(*a*) Hostile (*b*) Friendly
(*c*) Proactive (*d*) Clumsy

34. Motif : decoration : : Eulogy : __________
(*a*) Diatribe (*b*) Lament
(*c*) Praise (*d*) Satire

35. Illustration : Clarity :: Constriction : _____
(*a*) Wide (*b*) Upright
(*c*) Swiftness (*d*) Narrow

36. Cosmology: Universe :: Morphology:____
(*a*) Structure (*b*) Function
(*c*) Radio (*d*) Size

37. Answer : Test :: Denouement : _______
(*a*) Symphony
(*b*) Mystery story
(*c*) Horse race
(*d*) Complete understanding

38. Astronauts: Space :: Argonauts :_________
(*a*) Fire (*b*) Ship
(*c*) Birds (*d*) Sea

39. Snake : Reptilian : : Fox : ____________
(*a*) Leotard (*b*) Vulpine
(*c*) Tiger (*d*) Feline

40. Genuine: Simulated:: Unaffected: _______
(*a*) Elevated (*b*) Bombastic
(*c*) Dynamic (*d*) Emulated

Directions for(Que) 41 - 70): Read the information carefully and answer the questions given below it

In each of the following questions identify the pair of words, which has a relationship similar to the pair in question

41. **Wool : Acrylic**
(*a*) Rayon : Silk (*b*) Winter : Spring
(*c*) Cotton : Terylene (*d*) Rubber : Plastic

42. **Bouquet : Flower**
(*a*) Skin : Body (*b*) Chain : Link
(*c*) Page : Book (*d*) Product : Factory

43. **Germ : Disease**
(*a*) Man : Woman (*b*) War : Destruction
(*c*) Doctor : Medicine (*d*) Owner : Shop

44. **Ignominy : Disloyalty**
(*a*) Fame: Heroism (*b*) Derelict : Fool
(*c*) Death : Victory (*d*) Martyr : Man

45. **Gazelle : Swift**
(*a*) Horse : Slow (*b*) Swan : Graceful
(*c*) Lion : Roar (*d*) Lamb : Bleat

46. **Ascetic : Luxury**
(*a*) Capitalist : Communist
(*b*) Teacher : Classroom
(*c*) Misogynist : Women
(*d*) Scholar : Knowledge

47. **Augur : Future**
(*a*) Knight : Medieval (*b*) Post : Century
(*c*) Historian : Past (*d*)Commuter : Present

48. Spear : Dart
(*a*) Knife : Sword
(*b*) Door : Window
(*c*) Mountain : Molehill
(*d*) Cannon : Gun

49. Buzz : Hum
(*a*) Whisper: Noise (*b*) Echo : Sound
(*c*) Crack : Whip (*d*) Tinkle : Clang

50. Argument : Debate
(*a*) Violence : Peace
(*b*) Fight : Contest
(*c*) Opponent : Challenge
(*d*) Abuse : Scold

51. Cow : Herd
(*a*) Child : family (*b*) Peacock : Muster
(*c*) Bald : Hair (*d*) Horse : Jockey

52. Bird : Ornithology
(*a*) Zoology : Animal
(*b*) Stars : Galaxy
(*c*) Archaeology : Antiquity
(*d*) Plant : Botany

53. Wood : Carve
(*a*) Clay : Mould (*b*) Pipe : Blow
(*c*) Paper : Burn (*d*) Trees : Sway

54. Open : Secretive
(*a*) Mystery : Detective
(*b*) Honest : Truth
(*c*) forthright : Snide
(*d*) Better : Best

55. Wave : Crest
(*a*) Pinnacle : Nadir (*b*) Mountain : Peak
(*c*) Sea : Ocean (*d*) High : Low

56. Letter : Word
(*a*) Club : People (*b*) Page : Book
(*c*) Product : Factory (*d*) Picture : Crayon

57. Spasm : Pain
(*a*) Sound : Noise (*b*) Flash : Light
(*c*) Touch : Delicate (*d*) Love : Hate

58. Myth : Story
(*a*) Fiction : Story (*b*) Bonnet : Hat
(*c*) Pencil : Paper (*d*) Novel : Tale

59. Saturnine : Mercurial
(*a*) Redundant : Concise
(*b*) Saturn : Venus
(*c*) Heavenly : Starry
(*d*) Wolf : Sly

60. Wheat : Chaff
(*a*) Wine : Drugs (*b*) Crop : Corn
(*c*) Laughter : Revelry (*d*) Cloud : Rain

61. Intimidate: Fear
(*a*) Maintain : Satisfaction
(*b*) Astonish : Wonder
(*c*) Soothe : Concern
(*d*) Tame: Lion

62. Handcuffs : Robber
(*a*) Leash : Dog (*b*) Rope : Tie
(*c*) Shoes: Feet (*d*) Paper : Kite

63. Celebrate : Marriage
(*a*) Window : Bedroom
(*b*) Lament : Bereavement
(*c*) Pot : Pan
(*d*) Crime : Penalty

64. Banish : Apostate
(*a*) Reward : Traitor (*b*) Request : Assistant
(*c*) Remove : Officer (*d*) Welcome : Ally

65. Affirm : Hint
(*a*) Say : Deny (*b*) Assert : Convince
(*c*) Confirm : Reject (*d*) Charge : Insinuate

66. Boiling : Lukewarm
(*a*) Cold : Hot (*b*) Cold : Freezing
(*c*) Cloudy : Evening (*d*) Freezing : Cold

67. Dishonesty : Distrust
(*a*) Hand : Paper
(*b*) Money : Pickpocket
(*c*) Carelessness : Accident
(*d*) Murder : Jail

68. Immigration : Entrance
(*a*) Native : Foreigner
(*b*) Emigration : Departure
(*c*) File : Knife
(*d*) Migration : Bird

69. Emulate : Mimic
(*a*) Slander : Defame
(*b*) Praise : Flatter
(*c*) Aggravate : Promote
(*d*) Express : Imply

70. Impeach : Dismiss
(*a*) Arraign : Convict
(*b*) Exonerate : Charge
(*c*) Plant : Reap
(*d*) Imprison : Police

Odd Word Out

Directions for (Que. 71 - 110): Read the information carefully and answer the questions given below it. Each Question has a set of four words. Choose the word that does not belong to the group.

71. (*a*) Incapacitate (*b*) Debauch (*c*) Weaken (*d*) Debilitate

72. (*a*) Scepter (*b*) Throne (*c*) Crown (*d*) Staff

73. (*a*) Minaret (*b*) Dome (*c*) Porch (*d*) Steeple

74. (l) Moat (*b*) Tower (*c*) Drawbridge (*d*) Piazza

75. (*a*) Portrait (*b*) Caricature (*c*) Cartoon (*d*) Colossus

76. (*a*) Decipher (*b*) Destroy (*c*) Exterminate (*d*) Decimate

77. (*a*) Collie (*b*) Pekinese (*c*) Mongrel (*d*) Alsatian

78. (*a*) Cantata (*b*) Oratorio (*c*) Opera (*d*) Sonata

79. (*a*) Lobby (*b*) Portico (*c*) Reception (*d*) Corridor

80. (*a*) Beans (*b*) Peas (*c*) Barley (*d*) Lentils

81. (*a*) Succinct (*b*) Terse (*c*) Circumlocutory (*d*) Brief

82. (*a*) Pink (*b*) Red (*c*) Indigo (*d*) Green

83. (*a*) Bugle (*b*) Trumpet (*c*) Flute (*d*) Violin

84. (l) Tennis (*b*) Rugby (*c*) Football (*d*) Hockey

85. (*a*) Deleterious (*b*) Pernicious (*c*) Surreptitious (*d*) Harmful

86. (l) Arrow (*b*) Javelin (*c*) Dart (*d*) Spear

87. (l) Soft (*b*) Strident (*c*) Shrill (*d*) Grating

88. (*a*) Corporeal (*b*) Somatic (*c*) Physical (*d*) Affective

89. (*a*) Snide (*b*) Sneer (*c*) Smugness (*d*) Slyness

90. (*a*) Audacious (*b*) Skittish (*c*) Fidgety (*d*) Nervous

91. (*a*) Cygnet (*b*) Ewe (*c*) Kitten (*d*) Fawn

92. (*a*) Nylon (*b*) Silk (*c*) Jute (*d*) Cotton

93. (*a*) Coral (*b*) Amethyst (*c*) Diamond (*d*) Ruby

94. (*a*) Pedagogue (*b*) Scientist (*c*) Scholar (*d*) Philanthropist

95. (*a*) Febrile (*b*) Fecund (*c*) Prolific (*d*) Fertile

96. (*a*) Pulpit (*b*) Pew (*c*) Chalice (*d*) Chancel

87. (*a*) Virus (*b*) Bacteria (*c*) Algae (*d*) Ant

98. (*a*) Debutante (*b*) Decent (*c*) Debonair (*d*) Decadent

99. (*a*) Galley (*b*) Deck (*c*) Caboose (*d*) Ship

100. (*a*) Editor (*b*) Author (*c*) Publisher (*d*) Jockey

101. (*a*) Poise (*b*) Polish (*c*) Sophistication (*d*) Shoe

102. (*a*) Attire (*b*) Dress (*c*) Kimono (*d*) Monogram

103. (*a*) Poetry (*b*) Haiku (*c*) Sonnet (*d*) Elegy

104. (l) Dine (*b*) Dime (*c*) Fine (*d*) Line

105. (*a*) Triangle (*b*) Rectangle (*c*) Square (*d*) Parallelogram

106. (*a*) Botany (*b*) Sericulture (*c*) Horticulture (*d*) Culture

107. (*a*) Paradigm (*b*) Carpet (*c*) Pattern (*d*) Framework

108. (*a*) Cow (*b*) Mare (*c*) Lion (*d*) Goose

109. (*a*) Poise (*b*) Balance (*c*) Stability (*d*) Motility

110. (*a*) Keen (*b*) Plexus (*c*) Poignant (*d*) Biting

Idioms & Phrases

Directions for (Que. 111 - 140): Read the information carefully and answer the questions given below it.

Study the following idioms and choose their meanings from the answer choices given below them.

111. By the seat of one's pants
(*a*) Through the tough method
(*b*) Using guess work or instinct
(*c*) Through maintaining a secret
(*d*) Using one's strength

112. In good season
(*a*) Perfect timing
(*b*) Having a pleasant temper
(*c*) Sufficiently early
(*d*) Safe from danger

113. In a sense
(*a*) To be sane (*b*) To some extent
(*c*) To be awake (*d*) In a moment

114. At second hand
(*a*) In a sale
(*b*) Through something that is already used
(*c*) By the minute
(*d*) Through an intermediate source

115. Set one's face against
(*a*) To confront (*b*) To infuriate
(*c*) To contradict (*d*) To be opposed to

116. Dictum
(*a*) Value (*b*) Theory
(*c*) Order (*d*) Thesis

117. Dojo
(*a*) Extinct (*b*) Distinct
(*c*) Idiot (*d*) Seminary

118. Fio-ru-it
(*a*) He/she flourished
(*b*) It flowered
(*c*) She gave birth
(*d*) It bore fruit

119. Fracas I
(*a*) Mess
(*b*) Disorderly mob
(*c*) Disorderly light
(*d*) Ire

120. Ruckus
(*a*) Brain-washing (*b*) Dirt
(*c*) Wild (*d*) Uproar

121. Savant
(*a*) Elderly Person
(*b*) Learned Person
(*c*) Person with Good Taste
(*d*) A ruler

122. Grantee
(*a*) One who gives a grant
(*b*) One who receives a grant
(*c*) Someone who assures one of something
(*d*) Someone who assures an assurance

123. Mayhem
(*a*) Deliberate violence
(*b*) Confusion
(*c*) Self-sacrifice
(*d*) Aggression

124. Megalomania
(*a*) Exaggerated self-importance
(*b*) Fear of food
(*c*) Love for food
(*d*) Love for Insanity

125. In Toto
(*a*) In accordance to (*b*) In existence
(*c*) In fear of (*d*) Completely

126. Xeno
(*a*) Warrior (*b*) Foreign
(*c*) Angry (*d*) Feminine

127. Kaput
(*a*) Immobile (*b*) Worthless
(*c*) Totally ruined (*d*) Royal

128. Kitsch
(*a*) Vague (*b*) Immaterial
(*c*) Worthless (*d*) Destroyed

129. Persona Grata
(*a*) Grateful person
(*b*) Acceptable person
(*c*) Huge personality
(*d*) Person with a pure character

130. Diaspora
(*a*) Dispersed Group (*b*) Collected mob
(*c*) Varied interests (*d*) Colorfully magnificent

131. Let sleeping dogs lie
(*a*) Leave an existing situation as it is
(*b*) To encourage someone to tell a lie
(*c*) To shirk responsibility
(*d*) Let somebody sleep

132. To go to the dogs
(*a*) To be destroyed
(*b*) To degenerate
(*c*) To be eaten up
(*d*) To have a bad life

133. To pump up
(*a*) To flatter
(*b*) To disintegrate
(*c*) To infuse with enthusiasm
(*d*) To improve something

134. To put on the ritz
(*a*) To put on an accent
(*b*) To dress very well
(*c*) To cheat somebody
(*d*) To assume an attitude of wealth

135. To put oneself out
(*a*) To commit suicide
(*b*) To leave the country
(*c*) To take pains & do something
(*d*) To fall sick intentionally

136. To say put
(*a*) To remain in a single position
(*b*) To laze around
(*c*) To be in captivity
(*d*) To remain in order

137. In the saddle
(*a*) In deep trouble
(*b*) To travel constantly
(*c*) In control
(*d*) Be in an embarrassing situation

138. From scratch
(*a*) From absolutely nothing
(*b*) From waste material
(*c*) From strong evidence
(*d*) From the living

139. Up to scratch
(*a*) Till Death
(*b*) Up to the standard
(*c*) Till one becomes useless
(*d*) Till one becomes irritable

140. Go to sea
(*a*) To leave one's children
(*b*) To realize the 'truth' of life
(*c*) To lose a lot of money
(*d*) Set out on a voyage

ANSWERS

MCQ Type Questions

1. (*b*) **2.** (*b*) **3.** (*b*) **4.** (*c*) **5.** (*b*) **6.** (*b*) **7.** (*c*) **8.** (*c*) **9.** (*c*) **10.** (*c*)

11. (*c*) **12.** (*b*) **13.** (*d*) **14.** (*c*) **15.** (*d*) **16.** (*b*) **17.** (*d*) **18.** (*d*) **19.** (*b*) **20.** (*b*)

21. (*b*) **22.** (*b*) **23.** (*d*) **24.** (*b*) **25.** (*d*) **26.** (*d*) **27.** (*c*) **28.** (*b*) **29.** (*c*) **30.** (*c*)

31. (*a*) **32.** (*a*) **33.** (*b*) **34.** (*c*) **35.** (*d*) **36.** (*a*) **37.** (*b*) **38.** (*d*) **39.** (*d*) **40.** (*a*)

41. (*c*) **42.** (*b*) **43.** (*b*) **44.** (*a*) **45.** (*b*) **46.** (*c*) **47.** (*c*) **48.** (*d*) **49.** (*d*) **50.** (*b*)

51. (*a*) **52.** (*d*) **53.** (*a*) **54.** (*c*) **55.** (*b*) **56.** (*b*) **57.** (*b*) **58.** (*b*) **59.** (*a*) **60.** (*a*)

61. (*c*) **62.** (*b*) **63.** (*b*) **64.** (*a*) **65.** (*b*) **66.** (*d*) **67.** (*c*) **68.** (*b*) **69.** (*b*) **70.** (*a*)

71. (*b*) **72.** (*d*) **73.** (*c*) **74.** (*d*) **75.** (*d*) **76.** (*b*) **77.** (*a*) **78.** (*c*) **79.** (*d*) **80.** (*c*)

81. (*c*) **82.** (*a*) **83.** (*d*) **84.** (*a*) **85.** (*c*) **86.** (*a*) **87.** (*a*) **88.** (*d*) **89.** (*c*) **90.** (*a*)

91. (*b*) **92.** (*a*) **93.** (*a*) **94.** (*d*) **95.** (*a*) **96.** (*d*) **97.** (*d*) **98.** (*a*) **99.** (*d*) **100.** (*d*)

101. (*d*) **102.** (*d*) **103.** (*a*) **104.** (*b*) **105.** (*a*) **106.** (*d*) **107.** (*b*) **108.** (*c*) **109.** (*d*) **110.** (*b*)

111. (*c*) **112.** (*d*) **113.** (*a*) **114.** (*c*) **115.** (*d*) **116.** (*b*) **117.** (*d*) **118.** (*a*) **119.** (*a*) **120.** (*c*)

121. (*b*) **122.** (*c*) **123.** (*b*) **124.** (*a*) **125.** (*d*) **126.** (*b*) **127.** (*c*) **128.** (*c*) **129.** (*b*) **130.** (*a*)

131. (*a*) **132.** (*b*) **133.** (*c*) **134.** (*d*) **135.** (*c*) **136.** (*a*) **137.** (*c*) **138.** (*a*) **139.** (*b*) **140.** (*d*)

Analytical Aptitude

1 CHAPTER Logical Reasoning Ability

REASONING

Reasoning is the process of drawing conclusions from facts. These conclusions must follow inevitably from the facts from which they are drawn. Reasoning is not concerned with some conclusion that has a good chance of being true when facts are true. Indeed, reasoning as used here refers to logical reasoning, not of commonsense reasoning or probabilistic reasoning. The only conclusions that are acceptable are those that follow logically from the supplied facts.

Types of Reasoning

Reasoning can be classified in two major parts :

(i) Verbal Reasoning

(ii) Non-verbal Reasoning

Non-verbal reasoning involves the ability to understand and analyze visual information and solve problems using visual reasoning. This reasoning test will test your non-verbal reasoning ability as the questions appear in diagrammatic and pictorial form. Such test is also called diagrammatic or abstract reasoning tests, this test discussed in next chapter of spatial aptitude. For example: identifying relationships, similarities and differences between shapes and patterns, recognizing visual sequences and relationships between objects, and remembering these.

VERBAL REASONING

Verbal reasoning tests words, letters and numbers, and require logical reasoning and a reasonable knowledge of the English Language. It is also necessary to be familiar with simple manipulations with figures, like addition, subtractions division and multiplication. The problems of number in *Test of Reasoning* test how logical we are i.e., how well we reason and think while carrying out simple arithmetic manipulations.

CLASSIFICATION OF VERBAL REASONING

1. SERIES COMPLETION

In verbal series, words, letters or digits are given in a specific sequence/order and we have to find out next word, letter or digit to complete the given series.

Numbers or alphabetical letters, are generally called *terms of the series*. These terms follow a certain pattern throughout. In the questions we have to identify last one or two terms to continue the series or to find a missing term in between given terms to continue the sequence followed in the question. There is no set pattern and each question may follow a different pattern or sequential arrangement of letters or digits, which have to detect using common sense and reasoning ability.

Types of series completion Questions

Mainly following five types of question are asked.

I. Alphabet series

In this series, given alphabets follow a particular sequence or order. We have to detect the pattern from the given alphabets and find missing alphabet or the next alphabet to continue the pattern.

- There are no set rules.
- There can be omission of alphabets in an order.
- Alphabets may also be omitted in an increasing/decreasing order, which may be direct increase or decrease.
- There can also be alternate order.
- There may also be alternate sequences
- There may be several other patterns in the letter series.
- To tackle letter series questions, vary position of the alphabet and its position number in both forward and backward sequences.
- To continue the series after Z, we again begin with A. In other words, the sequence is kept in a circular order.
- In solving these questions pattern of the alphabet series should be noted.

Some Skipping Patterns.

(*i*) ***Regular Order :*** Number of alphabets skipped remains the same.

Example. A, D, G, J, ...?

Ans. M

(*ii*) ***Increasing Order :*** Each time the number of alphabets skipped increases in a given pattern.

Example. A C F J O ?

Ans. U

Here, each time number of letters skipped increases by one.

(*iii*) ***Decreasing Order :*** Each time the number of letters skipped decreases in a given pattern.

Example. A G L P S ?

Ans. U

Here number of letters skipped decreases by one each time, *i.e.,* first 5, then 4, then 3, and so on.

(*iv*) ***Interlinked Series :*** In this two or more different series are attached together. These different series follow their own different rules.

Example. A D F J M R ?

Ans. V

Here, there are two interlinked series.

II. Letter Series

This type of questions usually consist of a series of small letters which follow a certain pattern. However, some letters are missing from the series. Then these missing letters are given in a proper sequence as one of the alternatives.

In these type of questions we have letters of the alphabet given here. We have to first identify the patterns that the series of letters follow. Then we have to find the missing letter based on the pattern already identified. In number series we saw different patterns that the numbers in the series can follow- like squares, cubes. In letter series, obviously, patterns like squares, cubes will not be possible. In letter series, in increasing or decreasing difference. The position of the letters in the English alphabet is considered to be the value of the alphabet in questions on letter series. Also when we are counting after we count from A to Z we again start with A, i.e. we treat the letters as being cyclic in nature like in Number Series. In this type of Letter Series also, we can have a "combination " of series , i.e. two series are combined and given . We need to identify the patterns in the two series to find out the missing letter. Sometimes, there will be some special types of series also Let us look at a few examples to understand question on Letter Series.

Example. aaa... b...a...

(*a*) baa (*b*) abb (*c*) bab (*d*) aab (*e*) bbb

Ans. (*a*)

Method :

- First blank space should be filled in by 'b', so that we have two a's followed by two b's.
- Second blank space should be filled in either by 'a', so that we have four a's followed by two b's, or by 'b' so that we have three a's followed by three b's.
- Last space must be filled in by 'a'.
- *Thus, we have two possible answers :* 'baa' and 'bba'.

But, only 'baa' appears in the alternatives. So, the answer is (*a*).

Note : In case, we had both the possible answers in the alternatives, we would have chosen the one that forms a more prominent pattern, which is aabb/ aaabbbaa. Thus, our answer would have been "*bba*".

Example 1 : Find the next letter in the series

D, G, J, M, P, ______

(*a*) Q (*b*) R

(c) S (*d*) T

Solution (c)

Three letters are added to each letter to get the next letter in the series

i.e. D + 3, G + 3, J + 3, M + 3, P + 3, S and P = 16 + 3 = 19 and the 19th letter in the alphabet is S.

Example 2 : Find the next letter in the series

A, B, D, H, ______

(*a*) L (*b*) N

(c) R (*d*) P

Solution (d)

Each letter in the given series is multiplied with 2 to get the next letter in the series. $A \times 2 \Rightarrow 1 \times 2 = 1$ And the 2nd letter is B, $B \times 2 \Rightarrow 2 \times 2 = 4$ and the 4th letter is D.

Similarly $H \times 2 \Rightarrow 8 \times 2 = 16$ and the 16th letter is P.

Example 3 : What is the next letter in the series?

B, D, G, K, P, ________

(*a*) S (*b*) V

(c) W (*d*) X

Solution (b)

B + 2, D + 3, G + 4, K + 5, P + 6,________

P + 6 = 16 + 6 = 22 and the 22nd letter is V

Example 4 : I, X, J, W, K, V, L, ______

(*a*) M (*b*) U

(c) S (*d*) T

Solution (b)

The given series is an alternate series, I + 1, J + 1, K + 1, L is one series and X – 1, W – 1, V – 1, ______is the other series .

X – 1 = W, W – 1 = V and V – 1 = 22 – 1 = 21 and the 21st letter is U

III. Number Series

For better understanding, we will classify this into the following broad categories.

(i) Difference series

(ii) Product series

(iii) Squares /Cubes series

(iv) Miscellaneous series

(v) Combination series

(a) Difference Series

The difference series can be further classified as follows.

(a) Number series with a constant difference.

(b) Number series with an increasing or decreasing difference.

In the number series with a **Constant difference,** there is always a constant difference between two consecutive numbers. For example, the numbers of the series1, 4, 7, 10, 13... are such that any number is obtained by adding a constant figure of 3 to preceding term of the series

If we have to find the next number in the above series, we need to add a 3 to the last term 13. Thus, 16 is the next term of the series.

Under the series with constant difference, we can have series of odd numbers or series of even numbers also.

In the series with **increasing /decreasing difference,** the difference between consecutive terms keeps increasing (or decreasing as the case may be). For example let us try to find out the next number in the series 2, 3, 5, 8, 12, 17, 23,.....

Here, the difference between the first two terms of the series is 1; the difference between the third and the fourth terms is 3 and so on. That is the difference between any pair of consecutive terms is one more that the difference between the first number of this pair and the number immediately preceding this number. Here, since the difference between 17 and 23 is 6, the next difference should be 7. So the number that comes after 23 should be (23 + 7) = 30

We can also have a number series where the difference is in decreasing order (unlike in the previous example where the difference is increasing). For example, let us find out the next term of the series 10, 15, 19, 22, 24,.....

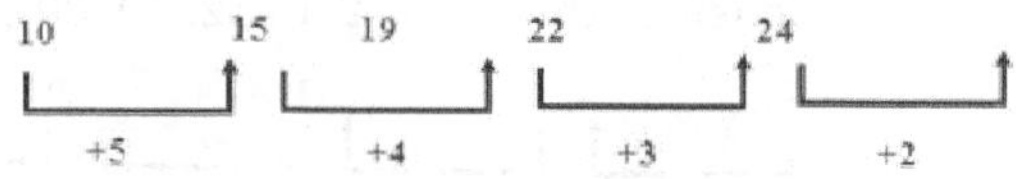

Here the difference between 1st & 2nd, 2nd & 3rd, 3rd & 4th numbers, etc are 5, 4, 3, 2 and so on since the difference between 22 and 24 is 2, the next difference should be 1. So, the number that comes after 24 should be 25.

(b) Product Series

A product series is usually a number series where the terms are obtained by a process of multiplic-ation. Here also, there can be different types of series. We will look at these through examples.

Consider the series 2, 4, 8, 16, 32, 64.....

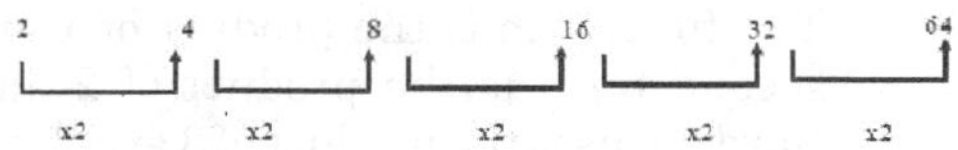

Here, each number in the series is multiplied by 2 to get the next term. So the term that comes after 64 is 128. So, each term is multiplied by a fixed number to get the next term. Similarly we can have a series where we have numbers obtained by dividing the previous term with a constant number. For example, in the series 64, 32, 16, 8.... Each number is obtained by dividing the previous number by 2 (or in other words, by multiplying the previous term by ½). So here, the next term will be 4 (obtained 8 with 2).

Consider the series 4, 20, 80, 240....

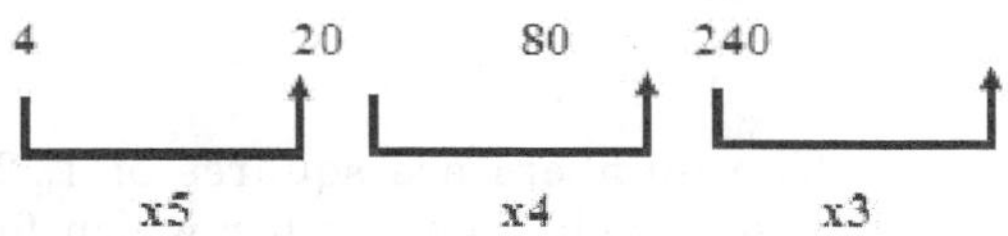

Here, the first term is multiplied by 5 to get the second term, the second term is multiplied by 4 to get the third term; the third term is multiplied by 3 to get the fourth term now, we have to multiply the fourth term by 2 to get the fifth term i.e. 480. So each term is multiplied by a decreasing factor (or it could also be an increasing factor) to get the next term. That is, with whatever number a particular term is multiplied to get the next term, this latest term is multiplied by a number different from the previous multiplying factor to get the next term of the series. All the multiplying factor follow a certain pattern (normally of increasing or decreasing order).

Consider the series 2, 6, 12, 20, 30....

2 → 6 → 12 → 20 → 30
+4 +6 +8 +10

This can be looked at a series of increasing differences. The difference of consecutive pairs of terms are 4 (between 2 and 6), 6 (between 6 and 12), 8 (between 12 and 20), 10 (between 12 and 20), 10 (between 20 and 30) and so on. Hence, the difference between 30 and the next term should be 12 and so the next terms will be 42. But this series can also be looked at as a product series

2	6	12	20	30
↓	↓	↓	↓	↓
1×2	3×3	3×4	4×5	5×6

The first term is the product of 1 and 2; the second term is the product of 2 and 3; the third terms is the product of 3 and 4; the fourth term is the product of 4 and 5 and the fifth term is the product of 5 and 6. Hence the next term will be the product of 6 and that is 42.

(c) Squares/Cubes Series

There can be series where all the terms are related to the squares of numbers or cubes of numbers. With squares/ cubes of numbers as the basis, there can be many variations in the patterns of the series. Let us look at various possibilities of series based on squares/ cubes

Each term of the series may be the square of a natural number such as 1, 4, 9, 16....

1	4	9	6
↓	↓	↓	↓
1^2	2^2	3^2	4^2

The numbers are squares of 1, 2, 3, 4... respectively. The number which follows 16 (which is the square of 4) will be 25 (which is the square of 5).

The terms of the series may be the squares of odd numbers (for example, 1, 9, 25, 49....) or even numbers (for example, 4, 16, 36, 64...)

The terms of the series could be such that a number and its square are both given one after the other and such pairs are given in some specific patterns. For example, take the series 2, 4, 3, 9, 4, 16 ...

2 4 → 3 9 → 4 16
+1 +1

Here, 2 is followed by its square 4; then comes the number 3 (which is one more than 2) followed by its square 9 and so on. Hence the next number in the series is 5 and the one after that is square i.e. 25.

Similarly each term could be the square root of its predecessor, For example, in the series 81, 9, 64, 8, 49, 7, 36, ... 81 is the square root of 9, 64 is the square root of 8, and so on. Therefore the next number which follow in the series should be the square root of 36 ie. 6.

The terms of the series could be the squares of natural numbers increased or reduced by certain number. For example, in the series 3, 8, 15, 24....

We have {Squares of natural numbers-1} as the terms. The first term is $2^2 - 1$; the second term is $3^2 - 1$, the third term is $4^2 - 1$ and so on. Hence, the next term will be $6^2 - 1$ i.e. 35 [Please note that the above series can also be looked at as a series with increasing differences.]

The difference between the 1st & 2nd term, the 2nd & 3rd terms, and on are 5, 7, 9, and so on. Hence the next difference should be11 giving us the next term as 35. There could also be series with {squares of natural numbers + some constant}.

Like we have seen with squares of numbers. We can have similar series with cubes of numbers. For example take series 1, 8, 27, 64...

1	8	27	64
↓	↓	↓	↓
1^3	2^3	3^3	4^3

Here, all the terms are cubes of natural numbers So, the next terms will be 5^3, i.e. 125.

Consider the series 2, 9, 28, 65....

2	9	28	68
↓	↓	↓	↓
1^3+1	2^3+1	3^3+1	4^3+1

Here, the terms are {Cubes of natural numbers +1}. The first terms is $1^3 + 1$, the second term is $2^3 + 1$, the third term is $3^3 + 1$ and so on. Hence the next term will be $5^3 + 1$ i.e. 126.

(d) Combination Series

A number series which has more than one type of (arithmetic) operation performed or more that one series combined together is combination series. The series that are combined can be two series of the same type or could be different type of series as described above. Let us look at some example.

First let us look at those series which are formed by more than one arithmetic operation performed on the terms to get the subsequent terms.

Consider the series: 2, 6, 10, 3, 9, 13, 4, 12, ...

Here the first term 2 is multiplied by 3 to get the second term, and 4 is added to get the third term. The next term is 3 (one more than the first term 2) and it is multiplied by 3 to get 9 (which is the next terms) and then 4 is added to get the next term 13. The next term 4 (which is one more than 3) which is multiplied by 2 to get 12. Then 4 is added to this to get the next number 16.

Consider the series: 1, 2, 6, 21, 88..... Here we can observe that 88 is close to 4 time s 21. It is in fact $21 \times 4 + 4$. So, if we now look at the previous term 21, it is related to the previous term 6 as $6 \times 3 + 3$. Now we get the general patterns: to get any term, multiply the previous term with k and then add k where k is a natural number with values in increasing order from 1. So, to get the second term, the first term has to be multiplied with 1 and then 1 is added. To get the third term, the second term is multiplied with 2 and then 2 is added and so, on Hence after 88 the next term is $88 \times 5 + 5$ i.e. 445.

Now, let us look at a series that is formed by combining two (or more) series can be of the same type or of different types described above.

Consider the series 8, 12, 9, 13, 10, 14..... Here the 1st, 3rd, 5th... terms which are 8, 9, 10... form one series whereas the 2nd ,4th ,6th etc. terms which are 12, 13,14 forms another series. Here both series that are being combined are two simple constant difference series. Therefore the missing number will be the next term of the first series 8, 9, 10 ... which is equal to 11.

Consider the series 0, 7, 2, 17, 6, 31, 12, 49, 20.... Here the series consisting of 1st, 3rd, 5th .... terms (i.e., the series consisting of the odd terms) which is 0, 2, 6, 12, 20,... is combined with another series consisting of 2nd 4th 6th .... terms(i.e., the series consisting of the even terms) which is 7, 17, 31, 49...The first series has the differences in increasing order 2, 4, 6, 8, 10 and so on. The second series also has the difference in increasing order 10, 14, 18... Since the last term 20 belongs to the first series a number from the second series will be obtained by adding 22 to 49 that is 71.

Consider the series 1, 1, 2, 4, 3, 9, 4, 16,... Here one series consisting of odd terms, which is 1, 2, 3, 4....is combined with the series of even terms which is 1, 4, 9, 16..... The first series is a series of natural numbers. The second series is the squares of natural numbers. Hence, the next term is 5.

Consider the series: 1, 1, 4, 8, 9, 27 ...... Here we have to find out the term that should be in place of the question mark. The odd terms forms one series 2,5,9,14,20 .... Where the difference is increasing The difference are 3,4,5,6.... This series is combined with the series of even terms 4,9,16... where the terms are squares of numbers 2,3,4.... Hence, the term that should come in place of the question mark is the next term of the second series which is 5^2, ie., 25.

(e) Miscellaneous Series

There are series that do not come under the other patterns and are of general nature but are important and are fairly common. Even here, sometimes there can be specific patterns in some cases.

Take the series 3, 5, 7, 11, 13 ..... This is a series of consecutive **PRIME NUMBERS.** It is an important series and the student should look out for this as one of the patterns. The next term in this series is 17.

There can also be variations using prime numbers. Take the series 9, 25, 49, 121... In this series, the terms are squares of prime numbers. Hence the next term is 13^2 i.e. 169.

Take the series 15, 35, 77.... The first term is 3×5; the second term is 5×7; the third term is 7×11; here the terms are the product of two consecutive product of 11 and 13 i.e. 143.

Take the series 8, 24, 48, 120, 168.... Here the 2nd term is 3 times the first term and the 3rd term is 2 times the 2nd term, but after that it does not follow this pattern any more, if you look at the terms carefully ,you will find that the terms are {one less that squares of prime numbers}. Hence the next term will be $17^2 - 1$ i.e. 288.

Consider the series 1, 4, 9, 1, 6, 2, 5, 3.... At first sight there is nothing we can say about the series. This is actually a series formed by a squares is in two or more digits, each of the digits is written as a separate term of the series. Thus the first terms are 1, 4 and 9 the squares of 1, 2 and 3 respectively. After this we should get 16 (which is the square of 4) since this different terms 1 and 6 in the series. Similarly, the next square 25 is written as two different terms 2 and 5 in the series. So, the next square 36 should be written as two terms 3 and 6. Of these 3 is already given, so the next terms of the series is 6.

Consider the series 1, 1, 2, 3, 5, 8...

1 1 2 3 5 8

↓ ↓ ↓ ↓

1+1 1+2 2+3 3+5

Here, each term starting with the third number, is the sum of the two preceding terms, After the first two terms as given (1 and 1), then onwards, to get any term, we need to add the two terms that come immediately before that position . Hence to get the next term of the series, we should take the two preceding terms 5 and 8 and add them up to get 13. So the next term of the series is 13. The term after this will be 21(= 8 + 13)

A General Approach to Number Series

The best way of approaching the number series questions is to first observe the difference between terms. If the difference is constant, it is a constant difference series. If the difference is increasing or decreasing by a constant number, then it is series with a constant increasing or decreasing difference, then try out the product series approach .For this first divided the second term with the first term, third with the second, and so on. If the numbers obtained are the same then it is a product series. Alternatively try writing each term of the series as a product of two factors and see if there is any pattern that can be observed, if still there is no inference, but the difference is increasing or decreasing in a rapid manner, then check out the square series. If the increase is very high and it is not a square series, then try out the cube series. If the difference is alternatively decreasing and increasing (or increasing for some time and alternately decreasing), then it should most probably be a mixed series. Therefore test out the series with alternate numbers. If still the series is not solved, try out the general numbers.

IV. Mixed Series

In this type of number series, numbers obeying various orders of two or more different types of series are arranged alternately in a single number series.

Rules:

- Difference between consecutive numbers is same.
- Differences between consecutive numbers are in Arithmetic Progression (A.P.).
- Difference between consecutive numbers is a perfect square.
- Differences between consecutive numbers are multiples of a number.
- Differences between consecutive numbers are prime numbers.
- Difference between consecutive numbers is a perfect cube.
- Difference between consecutive numbers are in Geometric Progression (G.P.)
- Ratio between each consecutive number is the same.
- Ratio between each consecutive numbers is in Arithmetic Progression (A.P)
- Ratio between consecutive number is perfect square number.
- Ratio between consecutive number is the multiple of a number.
- Ratio between consecutive numbers is a prime number.
- Ratio between consecutive numbers is a perfect cube number.
- Ratios between consecutive number are in Geometric Progression (G.P.)

V. Correspondence Series

This type of series consists of three sequences with three different elements (usually capital letters, digits and small letters). On the basis of the similarity in positions in the three sequences, a capital letter is found to correspond with a unique digit and a unique small letter, whenever it occurs. We are required to trace out this correspondence and accordingly choose the elements to be filled in at the desired places.

2. ANALOGY

'*Analogy*' means 'correspondence'.

In these type of questions based on analogy, a particular relat-ionship is given and another similar relationship has to be identified from the given alternatives.

Analogy means similarity or some kind of relationship. Analogy test, therefore tests the ability to see a relationship between two words and to recognize a similar relationship between two other words. It is a test of verbal reasoning ability rather than vocabulary. The key to analogy success is being able to express the relationship between the pair of words.

Kinds of Relationships with Examples

1. Instrument and Measurement

- Thermometer : Temperature (thermometer is an instrument used to measure temperature).
- Barometer : Pressure
- Anemometer : Wind vane
- Odometer : Speed
- Scale : Length
- Balance : Mass
- Sphygmomanometer: Blood Pressure
- Rain Gauge : Rain
- Hygrometer : Humidity
- Ammeter : Current
- Screw Gauge : Thickness
- Seismograph : Earthquakes
- Taseometer : Strains

2. Quantity and Unit

- Mass : Kilogram (kilogram is the unit of mass)
- Length : Metre
- Force : Newton
- Energy : Joule
- Resistance : Ohm
- Volume : Litre
- Angle : Radians
- Time : Seconds
- Potential : Volt
- Work : Joule
- Current : Ampere
- Luminosity : Candela
- Pressure : Pascal
- Area : Hectare
- Temperature : Degrees
- Power : Watt
- Conductivity : Mho
- Magnetic field : Oersted/Tesla

3. Individual and Group

- Soldiers : Army (group of soldiers is called an army).
- Cattle : Herd
- Flowers : Bouquet
- Grapes : Bunch
- Singer : Chorus
- Artist : Troupe
- Fish : Shoal
- Sheep : Flock
- Riders : Cavalcade
- Bees : Swarm
- Man : Crowd
- Sailors : Crew
- Nomads : Horde

4. Animal and Young one

- Cow : Calf (calf is the young one of cow)
- Horse : Pony
- Cat : Kitten
- Sheep : Lamb
- Butterfly : Caterpillar
- Insect : Larva
- Dog : Puppy
- Cock/Hen : Chicken
- Lion : Cub
- Duck : Duckling
- Man : Child

5. Male and Female

- Horse : Mare (mare is the female horse)
- Dog : Bitch
- Stag : Doe
- Son : Daughter
- Lion : Lioness
- Sorcerer : Sorceress
- Drone : Bee
- Gentleman : Lady
- Nephew : Niece
- Tiger : Tigress

6. Individual and Class

- Monkey : Mammal (monkey belongs to the class of mammal).
- Man : Mammal
- Ostrich : Bird
- Snake : Reptile
- Butterfly : Insect
- Whale : Mammal
- Rat : Rodent
- Lizard : Reptile

7. Individual and Dwelling Place

- Dog : Kennel (dog lives in a kennel)
- Bee : Apiary
- Cattle : Shed
- Lion : Den
- Poultry : Farm
- Monk : Monastery
- Fish : Aquarium
- Birds : Aviary
- Horse : Stable

8. Study and Topic

- Ornithology : Birds (ornithology is the study of birds).
- Seismology : Earthquakes
- Botany : Plants
- Onomatology : Names
- Ethnology : Human Races
- Ontology : Reality
- Herpetology : Amphibians
- Pathology : Diseases
- Astrology : Future
- Anthropology : Man
- Palaeography : Writings
- Ichthyology : Fishes
- Semantics : Language
- Nephrology : Kidney
- Concology : Shells
- Haematology : Blood
- Craniology : Skull
- Mycology : Fungi
- Entomology : Insects
- Zoology : Animals
- Occultism : Supernatural

9. Worker and Tool

- Warrior : Sword (sword is the tool used by a warrior).
- Carpenter : Saw
- Woodcutter : Axe
- Tailor : Needle
- Labourer : Spade
- Soldier : Gun
- Sculptor : Chisel
- Chef : Knife
- Doctor : Stethoscope
- Farmer : Plough
- Author : Pen

10. Tool and Action

- Needle : Sew (needle is used for sewing).
- Knife : Cut
- Sword : Slaughter
- Mattock : Dig

- Filter : Purify
- Steering : Drive
- Pen : Write
- Spanner : Grip
- Spoon : Feed
- Microscope : Magnify
- Gun : Shoot
- Shovel : Scoop
- Chisel : Carve
- Oar : Row
- Axe : Grind
- Auger : Bore
- Spade : Dig

11. Worker and Working Place

- Chef : Kitchen (chef works in a kitchen)
- Farmer : Field
- Warrior : Battlefield
- Engineer : Site
- Sailor : Ship
- Pilot : Cockpit
- Beautician : Parlour
- Artist : Theatre
- Actor : Stage
- Mechanic : Garage
- Lawyer : Court
- Scientist : Laboratory
- Teacher : School
- Doctor : Hospital
- Clerk : Office
- Servant : House
- Driver : Cabin
- Grocer : Shop
- Painter : Gallery
- Waiter : Restaurant
- Worker : Factory
- Umpire : Pitch
- Gambler : Casino

12. Worker and Product

- Mason : Wall (mason builds a wall)
- Farmer : Crop
- Teacher : Education
- Hunter : Prey
- Chef : Food
- Carpenter : Furniture
- Judge : Justice
- Author : Book
- Choreographer : Ballet
- Goldsmith : Ornaments
- Producer : Film
- Butcher : Meat
- Architect : Design
- Cobbler : Shoes
- Tailor : Clothes
- Poet : Poem
- Dramatist : Play

13. Product and Raw Material

- Prism : Glass (prism is made of glass).
- Butter : Milk
- Cloth : Fibre
- Wall : Brick
- Road : Asphalt
- Furniture : Wood
- Book : Paper
- Shoes : Leather
- Sack : Jute
- Pullover : Wool
- Omelette : Egg
- Metal : Ore
- Jewellery : Gold
- Rubber : Latex
- Linen : Flax
- Jaggery : Sugarcane
- Oil : Seed
- Wine : Grapes
- Paper : Pulp
- Fabric : Yarn

14. Part and Whole Relationship

- Pen : Nib (nib is a part of a pen).
- Pencil : Lead
- House : Kitchen
- Fan : Blade
- Class : Student
- Room : Window
- Aeroplane : Cockpit
- Book : Chapter

15. Word and Intensity

- Anger : Rage (rage is of higher intensity than Anger)
- Wish : Desire
- Kindle : Burn
- Sink : Drown
- Quarrel : War
- Error : Blunder
- Famous : Renowned
- Unhappy : Sad
- Refuse : Deny

16. Word and Synonym

- Abode : Dwelling (abode means almost the same as Dwelling)
- Blend : Mix
- Ban : Prohibition
- Assign : Allot
- Vacant : Empty
- Abduct : Kidnap
- Dearth : Scarcity
- Dissipate : Squander
- Sedate : Calm
- Brim : Edge
- House : Home
- Solicit : Request
- Presage : Predict
- Haughty : Proud
- Flaw : Defect
- Fierce : Violent
- Fallacy : Illusion
- Substitute : Replace
- Mend : Repair
- Alight : Descend
- Presume : Assume

17. Word and Antonym

- Attack : Defend (defend means opposite of Attack)
- Advance : Retreat
- Cruel : Kind
- Best : Worst
- Fresh : Stale
- Ignore : Notice
- Initial : Final
- Condense : Expand
- Chaos : Peace
- Create : Destroy
- Gradual : Abrupt
- Sink : Float

- Robust : Weak
- Gentle : Harsh
- Deep : Shallow
- Cordial : Hostile
- Affirm : Deny
- Mourn : Rejoice
- Lethargy : Alertness

18. Part

Gill : Fin (Gill and fin are two parts of a fish)

19. Type

Sword : Weapon(Sword is a type of weapon)

20. Tool & Worker

Paint Brush : Artist (Paintbrush is a tool to artist)

21. Degree

Warm : Hot (Warm is a greater degree of hot)

Fond : Doting (Fond is less extreme than doting)

22. Sign of

Grimace : Pain(Grimace is a sign of pain)

Brush : Discomfiture (A blush signifies discomfiture)

23. A place for

Miner : Quarry (A miner works in a quarry)

24. Sequence relationship

Foreward : Appendix (In a book a foreword precedes the Appendix)

25. Worker and Work

Poet : Sonnet (A sonnet is written by a poet)

Mason : Wall (A mason builds a wall)

26. Cause and Effect

Soporofic : Sleepiness (Soporofic causes sleepiness)

Common Types of Analogy Questions

(i) Completing Analogous Pair

In this type of questions, given two words are, which related to each other in some way. Another word is also given and we required to find out the relationship between first two words and choose the word from the given alternatives, which bears the same relationship to the third word, as the first two bear.

Example. Doctor : Nurse : : ? : Follower

(*a*) Employer (*b*) Leader (*c*) Worker
(*d*) Manager (*e*) Union

Solution : (*b*)

Just as a nurse follows the Doctor's instructions, so also a follower works as directed by the leader.

(ii) Choosing Analogous Pair

In this type of questions, a pair of words is given, followed by four pairs of words as alternatives and we are required to choose the pair in which words bear the same relationship to each other as the words of the given pair bear.

Example. Sink : Float : : ?

(*a*) Brim : Edge (*b*) Mend : Repair
(*c*) Gentle : Harsh (*d*) Flow : defect
(*e*) Solicit : Request

Solution : (*b*)

Just as sink is antonym of float, so also gentle is opposite of harsh.

(iii) Choosing Similar word

In this type of questions, a group of three words is given, followed by four other words as alternatives. The candidate is required to choose the alternative, which is similar to the given three words.

Example. Sitar : Guitar : Tanpura

(*a*) Trumpet (*b*) Violin (*c*) Harmonium
(*d*) Mridanga (*e*) Fluet

Solution : (*b*)

Sitar, Guitar and Tanpura are all string instruments. Violin is also a string instrument.

(iv) Detecting Analogies

In this type of questions, we are required to trace out hidden analogy or common characteristic among the given words or to choose the word which possesses the same characteristic as the given word.

Example. Judo : Karate : Taekwando

(*a*) They are names of martial arts.
(*b*) They can be performed by obese persons.
(*c*) They are performed on stage.
(*d*) They are important items of Asian Games.
(*e*) They have origin in Kerala

Solution : (*a*)

Judo, Karate and Taekwando are martial arts and (*a*) is most suitable description for all the three.

(v) Three word Analogy

In this type of questions, a group of three inter-related words is given and we are required to trace out the relationship among these three words and choose another group with similar analogy, from among the alternatives provided.

Example. Pen : Pencil : Ink

(*a*) Orange : Banana : Juice
(*b*) Table : Chair : Wood
(*c*) Cow : Milk : Curd

(*d*) Fish : Shark : Water

(*e*) Car : Engine : Circle

Solution : (*a*)

Pen contains ink and pencil belongs to the same category as pen i.e. stationery. Similarly, orange contains juice and banana belongs to the same category as orange, *i.e.* fruits.

(vi) Number Analogy

It includes deals two types of questions :

(*i*) Choosing a similarly related pair as the given number pair on the basis of relation between the numbers in each pair.

Example. 582 : 194 :: 258 : ?

(*a*) 82 (*b*) 86

(*c*) 92 (*d*) 58

Solution : (*b*)

Just as 582 is related to 194 as 582 is 3 times 194, in the same way 258 is 3 times the 86.

(*ii*) Choosing a number similar to a group of numbers on the basis of certain common properties that they possess.

Example. Which number belongs to given set of numbers?

2, 3, 5, 7, 11, 13, 17, 19, 23......

(*a*) 15 (*b*) 9 (*c*) 21 (*d*) 29 (*e*) 27

Solution : (*d*)

Given set of numbers belong to the prime numbers. In the given option, only prime number is 29.

(vii) Alphabet Analogy

In this type of questions, two groups of alphabets related to each other in same way, are given and we are required to find out this relationship and choose a group of alphabets which is related in the same way to a third group provided in the question.

Example. B C D : P Q R : : X Y Z : ?

(*a*) FGH (*b*) LMN (*c*) RST (*d*) TUV (*e*) STU

Solution : (*b*)

Each letter of the first group is moved fourteen steps forward to obtain the corresponding letter of the second group. A similar relationship will exist between the third and fourth groups.

How to Answer Analogy Questions

- Figure out how the capitalized words are related.
- Create a sentence that expresses that connection.
- Test the choices with your sentence and eliminate the ones that don't work.
- If we are left with more than one answer—or no answer at all—go back and refine your sentence.
- Choose the best answer. If none of the choices fits exactly, choose the one that works best.

Plan to Solve Analogy Questions

(i) Make your Sentence More Precise

The analogies get more difficult as we work our way through each group. Use the common categories as a starting point, but be prepared to refine the relationship by making your sentence more precise.

Example. Grain : Silo ::

(*a*) Pilot : plane (*b*) judge : courtroom

(*c*) water : reservoir (*d*) clock : time

If we apply the "place where" idea without thinking, here is what happens.

A silo is a place where we would find grain.

(*a*) A *plane* is a place where you would find a *pilot*.

(*b*) A *courtroom* is a place where you would find a *judge*.

(*c*) A *reservoir* is a place where you would find *water*.

(*d*) A *clock* is a place where you would find *time*.

We can eliminate (d), but that still leaves you with four possible answers. Now is express the relationship between silo and grain more precisely?

A silo is a place where grain is stored.

(*a*) A plane is a place where a pilot is stored.

(*b*) A courtroom is a place where a judge is stored.

(*c*) A reservoir is a place where water is stored.

(*d*) A highway is a place where automobiles are stored.

(ii) Analogies Work only if there's a clear cut connection

An analogy depends upon a necessary connection between pairs of words based on the meaning of the words. This clear connection must exist for both the original capitalized word pair and the correct answer choice, i.e. we can eliminate any answer choice for which we cannot describe a necessary relationship between the words. Use the "clear connection" test to rule out answer choices even when we don't know one of the capitalized words.

Example. Letter : Alphabet : :

(*a*) note : scale

(*b*) ocean : merchandise

(c) expert : automobile

(d) victory : farmland

We dont't know what the capitalized words are in this analogy, but you can still eliminate choices (b), (c) and (d). There is no clear-cut connection between the words of those choices. The actual analogy above is LETTER : ALPHABET :: note : scale.

(iii) Some Analogies Work Better by Working Backward

Sometimes the capitalized words fall easily into a sentence that expresses their relationship and sometimes they don't. If we are having trouble making up a sentence that relates the two words, be prepared to shift gears. Try reversing the order of the original word pair.

Example. *Ice : Glacier ::*

(a) train : trestle (b) sand : dune

(c) path : forest (d) feather : bird

If we can't come up with a sentence relating ICE to GLACIER, try relating GLACIER to ICE :

A *glacier* is made of *ice*.

If we reverse the order of the capitalized words, we must also reverse the order of the words in each answer choice.

(a) A *trestle* is made up of a *train*.

(b) A *dune* is made up of *sand*.

(c) A *forest* is made up of a *path*.

(d) A *bird* is made up of a *feather*.

Clearly (b) exhibits the same relationship as the original pair.

(iv) Check from the Answer Choices

Many words have different meanings depending upon whether they are used as nouns or verbs or adjectives. If we are not sure how one of the capitalized words is being used, just check the answer choices. In analogies, all the answer choices will have the same grammatical structure as the capitalized words pair. That means if the answer choices are noun : noun, the capitalized pair will be NOUN : NOUN. On the other hand, if the answer choices are adjective : noun, then the capitalized pair will be ADJECTIVE : NOUN.

If we have eliminated all the answer choices with words that we know or sort of know, and we are left with a couple of choices containing words that we have never seen before, just guess and move on.

3. CLASSIFICATION

'Classification' means arrangement of given items on the basis of some common character. In this test, a group of certain items are given, out of which some are similar to in some manner and one is different from the rest. We are required to choose this one item which does not fit into the given group.

Types of Classification Questions

(i) Choosing Odd word

In this type of questions, four (or five) words are given out of which one is dissimilar to the others and students have to find that odd one from the given set.

Example. *Choose the word which is least like the other words in the group.*

(a) chair (b) cupboard (c) Table

(d) paper weight (e) Bed

Ans. (d)

Here, all except paper weight are furnitures.

(ii) Choosing Odd pair of words

In this type of questions, certain pairs of words are given out of which words in all the pairs except one, bear a certain common relationship. We are required to decipher this relationship and choose the pair in which words are differently related, as the answer.

Example. *Choose the odd pair of words.*

(a) Loom : Cloth (b) Table : Drawer

(c) Book : Page (d) Car : Wheel

Ans. (a)

In all other pairs, second is part of the first.

(iii) Choosing Odd numeral

In this type of questions, certain numbers are given, out of which all except one are alike in some manner while one is different and this number is to be chosen as the answer.

Example. *Choose the number which is different from others in the group.*

(a) 49 (b) 64

(c) 121 (d) 144

(e) 156

Ans. (e)

Each of the numbers except 156 is complete square.

(iv) Choosing the Odd numeral pair/group

In this type of questions, certain pairs/groups of numbers are given out of which all except one are similar in some manner while one is different. The numbers in these similar pairs may have the same property or may be related to each other according to the same rule. We are is required to choose the odd pair/group.

Example. *Choose the numeral pair / group which is different from others.*

(*a*) 83 – 75 (*b*) 58 – 50 (*c*) 49 – 42
(*d*) 25 – 17 (*e*) 131– 123

Ans. (*c*)

In each of the pairs, first number is eight more than the second.

(v) Choosing the Odd letter group

In this type of questions, usually four groups of letters are given. Three of them are similar to each other in some manner while one is different and we have to the answer.

Example. *Choose the group of letters which is different from others.*

(*a*) SU (*b*) PN
(*c*) IK (*d*) BD

Ans. (*b*)

All other are two alternate letters

4. CODING-DECODING

A CODE is a '*system of signals*'. Therefore, coding is a method of transmitting a message between the sender and the receiver without a third person knowing it.

Coding and Decoding Test is set up to judge the candidate's ability to decipher the rule that codes a particular word/ message and break the code to decipher the message.

Types of Coding Decoding Questions

(i) Letter Coding

In these questions, real alphabets in a word are replaced by certain other alphabets according to a specific rule to form its code and we are required to detect the common rule and answer the questions accordingly.

Example. If TAP is coded as SZO, then how is FREEZE coded?

(*a*) EQDFYG (*b*) ESDFYF (*c*) GQFDYF
(*d*) EQDDYD (*e*) EPDEZD

Solution : (*d*)

Each letter in the word TAP is moved one step backward to obtain the corresponding letter of the code.

	S	Z	O
–1	↑	↑	↑
	T	A	P

Thus, in FREEZE, F will be coded as E, R as Q, E as D and Z as Y. So, the code becomes EQDDYD.

(ii) Number Coding

In these questions, either numerical code values are assigned to a word or alphabetical code letters are assigned to the numbers and we are is required to analyse the code as per the directions.

Example. *If CHINA is written as 38126 and NEPAL is 25769, how is PLAINE is coded ?*

Solution : Clearly Alphabets are coded as shown:

C	*H*	*I*	*N*	*A*	*E*	*P*	*L*
3	8	1	2	6	5	7	9

Hence, PLAINE is coded as 796125.

Example. In a certain code '37' means 'which class' and '583' means 'caste and class'. What is the code for 'caste' ?

(*a*) 3 (*b*) 7 (*c*) 8
(*d*) Either 5 or 8

Solution : (*d*)

Given, which class' ≡ '37' ...(*i*)

and'caste and class' ≡ '583' ...(*ii*)

From (*i*) and (*ii*), the common word 'class' is coded as '3'.

From (*ii*), 'caste' is coded as '5' or '8'.

(iii) Substitution

In this type of questions, some particular objects are assigned code names. Then a question is asked that is to be answered in the code language.

Example. If cook is called *butler*, butler is called manager, manager is called teacher, teacher is called clerk and clerk is called principal, who will teach in a class ?

(*a*) Cook (*b*) Butler (*c*) Manager
(*d*) Teacher (*e*) Clerk

Solution : (*e*)

A 'teacher' teaches in a class and as given 'teacher' is called 'clerk'. So, a 'clerk' will teach in the class.

(iv) Mixed letter Coding

In this type of questions, three or four complete messages are given in the coded language and the code for a particular word is asked. To analyse such codes, any two messages bearing a common word are picked up. The common code word will mean that word. Proceeding similarly by picking up all possible combinations of two, the entire message can be analysed.

Example. If 'tee see pee' means 'Drink fruit juice'; 'see kee lee'means 'Juice is sweet' and 'fee ree mee' means. "He is intelligent', which word in that language means 'sweet' ?

(*a*) see (*b*) kee (*c*) lee

(*d*) pee (*e*) None of these

Solution : (*b*)

In first and second statements, common word is 'Juice' and common code word is 'see'. So, 'see' means 'Juice'. In second and third statements, common word is 'is' and common code is lee'. So, 'lee' means 'is'. Thus, in second statement, remaining word 'sweet' is coded as 'kee'.

(v) Mixed number Coding

In this type of questions, a few groups of numbers each coding a certain short message, are given. Through a comparison of the given coded messages, taking two at a time, the candidate is required to find the number code for each word and then formulate the code for the message given.

Example. In a certain code, '786' means 'study very hard', '958' means 'hard work pays' and '645' means 'study and work'. Which of the following is the code for 'very' ?

(*a*) 8 (*b*) 6 (*c*) 7 (*d*) 5

Solution : (*c*)

In first and second statements, common word is '*hard*'and common code digit is '8'. So, '8' means 'hard'.

In the first and third statements, common word is 'study' and the common code digit is '6'. So, '6' means 'study'.

Thus, in the first statement, '7' means 'very'.

(vi) Deciphering individual letter codes by analysis

In this type of questions, certain sample words are given along with their codes. The candidate is required to decipher individual codes for different letters by comparing, taking two words at a time, and then answer the given questions accordingly.

5. BLOOD RELATIONS

In these tests, the success of a candidate depends upon the knowledge of the blood relations.

Following Blood relations help solve these Questions :

Mother's or father's son — Brother
Mother's or father's daughter — Sister
Mother's or father's brother — Uncle
Mother's or father's sister — Aunt
Mother's or father's father — Grand father
Mother's or father's mother — Grand mother
Son's wife — Daughter-in-Law
Daughter's husband — Son-in-Law
Husband's or wife's sister — Sister-in-Law
Husband's or wife's brother — Brother-in-Law
Brother's son — Nephew
Brother's daughter — Niece
Uncle or aunt's son or daughter— Cousin
Sister's husband — Brother-in-Law
Brother's wife — Sister-in-Law
Grandson's or Grand daughter's daughter — Great grand daughter

Types of Blood Relation Questions

(*i*) Deciphering Jumbled up descriptions

In this type of questions, a round-about description is given in the form of certain small relationships and direct relationship between the persons concerned is to be deciphered.

Example. Pointing to a photograph, a man said, *"I have no brother or sister but that man's father is my father's son."* Whose photograph was it ?

(*a*) His own (*b*) His son's

(*c*) His father's (*d*) His nephew's

Solution : (*a*)

Since man has no brother, his father's son is he himself.

(*ii*) Relation Puzzle

In this type, mutual blood relations or other informations of more than two persons are mentioned and information about any two is mentioned.

Example. A and B are brothers. C and D are sisters. A's son is D's brother. How is B related to C.

(*a*) Father (*b*) Brother (*c*) Grandfather

(*d*) Uncle (*e*) None of these

Solution : (*d*)

B is the brother of A; A's son is D's brother. This means D is the daughter of A. Since C and D are sisters, C is also the daughter of A. So, B is the uncle of C.

(*iii*) Coded Relations

In such questions, relationships are represented by certain codes or symbols such as +, –, ×, ÷,*, Δ etc. Then relationships between certain persons, given in the form of these codes, are to be analysed.

Example. If A + B means A is the sister of B; A – B means A is the brother of B; A × B means A is the daughter of B, which of the following shows the relation that E is the maternal uncle of D?

(*a*) D + F × E (*b*) D – F × E (*c*) D × F + E

(*d*) D × F × E (*e*) None of these.

Solution : (*c*)

E is the maternal uncle of D means D is the daughter of the sister (say F) of E i.e. D × F + E.

The easiest and non-confusing way to solve these types of problems would be to draw a family tree diagram and increase the levels in the hierarchy as shown below:

1st	GRAND PARENTS	Grandfather, Grandmother, Granduncle, Grandaunt
2nd	PARENTS & IN-LAWS	Father, Mother, Uncle, Aunt, Father- in-low, Mother in law
3rd	SIBLINGS, SPOUSE & IN-LAWS	Brother, Sister, Cousin, Wife, Husband, Brother in law, Sister in law
4th	CHILDREN & IN-LAWS	Son, Daughter, Niece, Nephew, Son in law, Daughter in law
5th	GRAND CHILDREN	Grandson, Granddaughter

Mother's or Father's son	Brother
Mother's or Father's daughter	Sister
Mother's or Father's brother	Uncle
Mother's or Father's sister	Aunt
Mother's or Father's mother	Grandmother
Mother's or Father's father	Grandfather
Grandmother's brother	Grand uncle
Grandfather's sister	Grand aunt
Grandfather's brother	Grand uncle

Grandmother's sister	Grand aunt
Sister's or Brother's son	Nephew
Sister's or Brother's Daughter	Niece
Uncle or Aunt's son or daughter	Cousin
Son's wife	Daughter in law
Daughter's husband	Son-in-law

Husband's or Wife's sister	Sister- in- law
Husband's or Wife's brother	Brother- in- law
Sister's husband	Brother- in law
Brother's wife	Sister- in law
Children of same parents	Siblings (could be all brothers, all sisters or some brothers and some sisters)
Children	Son, Daughter
Children's Children	Grandson, Granddaughter (Grandchildren)

In addition, remember the word spouse which means either husband or wife.

Grandfather and grandmother will come in the first stage. Mother, father uncle and aunt will come in the second stage: sister brother and cousin will come at the third stage: son daughter, niece and nephew will come in the fourth stage and finally, granddaughters and grandsons will come. The above stage are made from the point of view of an individual.

In type I questions, the relationship between two people is given through a roundabout way of relating them through other people. We have to go through the series of relationship can be given I in the questions. The relationship can be given as a simple statements or as a statement made by a person. In the first example given below, a person is involved in making a statement whereas in the second question, there is no person involved in making a statement.

Example 1 : A's father's mother-in law only daughter's son is b. How is A related to B?

(*a*) Brother (*b*) Sister

(*c*) Nephew (*d*) Cannot be determined

Solution (d)

A's father's mother- in –law's only daughter is A's mother. A's mother's son is A's brother. But A can either brother or sister to B.

Example 2 : If A's father is B, C is the father of B and D is A's mother, then how is C related to D?

(*a*) Father (*b*) Grandfather

(*c*) Father- in law (*d*) Uncle

Solution (c)

A's father is B and mother is D. Therefore D is B's wife and C is the father of B. Hence C is D's father-in law.

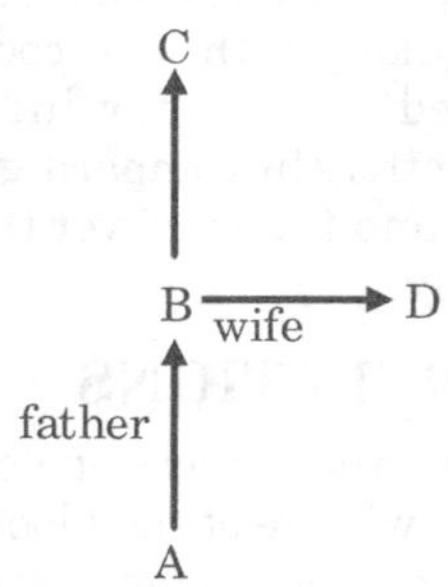

Example 3 : A + B means A is the son of B

A – B means A is the daughter of B.

A × B means A is the father of b.

A ÷ B means A is the mother of B.

If M × N + O – P ÷ Q, then how is M related to?

(*a*) Husband (*b*) Cousin

(*c*) Brother- in law (*d*) Uncle

Solution (c)

M × N + O – P ÷ Q Means M is the father of N, N is the son of O, O is the daughter of P, P is the mother of Q.

M is the father of N and N is the son of O means m is the husband of O.O is the daughter of P and P is the mother of Q means O is the sister of Q.

M is the husband of O and O is the sister of Q means M is the brother- in law of Q.

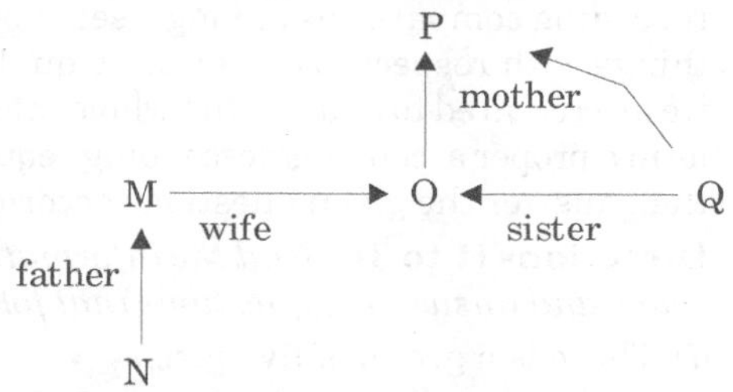

Example 4 : A + B means A is the son of B

A – B means A is the daughter of B.

A × B means A is the father of B.

A ÷ B means A is the mother of B.

Which of the following means S is the son- in law of P?

(*a*) P + Q ÷ R × S – T (*b*) P × Q ÷ R – S + T

(*c*) P + Q × R – S ÷ T (*d*) P × Q – R ÷ S × T

Solution (b)

P + Q ÷ R × S – T Means P is the son of Q. Q is the daughter of R, R is the father of S and S is the daughter of T. Hence S is the nephew of P.

P × Q ÷ R – S + T Means P is the father of Q, Q is the mother of R, R is the daughter of S and S is the son of T. Hence S is the son-in-law of P.

R + Q × R – S ÷ TMeans P is the son of Q, Q is the father of R, R is the daughter of S and S is the mother of T. Hence S is the mother of P.

P × Q – R ÷ S × T Means P is the father of Q, Q is the daughter of R, R is the mother of S and S is the father of T. Hence S is the son of P.

Example 5 : Pointing to a person, Raju said, "He is the only brother of my father's mother's daughter". How is the person related to Raju?

Solution (b)

Raju's father's mother's daughter is Raju's father's sister. Raju's father's sister's only brother is Raju's father. Hence the person is Raju's father.

6. PUZZLE TEST

This type of questions put in the form of puzzles involving certain number or items, be it persons or things and we are required to analyse the given information, condense it in a suitable form and answer the questions asked.

Type of Puzzle Tests

(i) Classification type Questions

This type consists of questions in which certain items belonging to different groups or possessing different qualities are given along with some clues with the help of which the candidate is required to group and analyse the given items and answer the questions accordingly.

Directions (1 to 5) : *Read the following information carefully and answer the questions that follow*:

(*i*) Five friends P, Q, R, S and T travelled to five different cities of Chennai, Calcutta, Delhi, Bangalore and Hyderabad by five different modes of transport of Bus, Train, Aeroplane, Car and Boat from Mumbai.

(*ii*) The person who travelled to Delhi did not travel by boat.

(*iii*) R went to Bangalore by car and Q went to Calcutta by aeroplane.

(*iv*) S travelled by boat whereas T travelled by train.

(*v*) Mumbai is not connected by bus to Delhi and Chennai.

Example 1 : Which of the following combinations of person and mode is not correct ?

(*a*) P — Bus (*b*) Q — Aeroplane

(*c*) R — Car (*d*) S — Boat

(*e*) T — Aeroplane

Example 2 : Which of the following combinations is true for S ?

(*a*) Delhi — Bus (*b*) Chennai — Bus

(*c*) Chennai — Boat (*d*) Data inadequate

(*e*) None of these

Example 3 : Which of the following combinations of place and mode is not correct ?

(*a*) Delhi — Bus

(*b*) Calcutta — Aeroplane

(*c*) Bangalore — Car

(*d*) Chennai — Boat

(*e*) Hyderabad — Bus

Example 4 : The person travelling to Delhi went by which of the following modes ?

(*a*) Bus (*b*) Train

(*c*) Aeroplane (*d*) Car

(*e*) Boat

Example 5 : Who among the following travelled to Delhi ?

(*a*) R (*b*) S

(*c*) T (*d*) Data inadequate

(*e*) None of these

Solutions : *Given information can be analysed as follows :*

Mode of Transport :

R travels by Car, Q by Aeroplane, S by Boat and T by Train.

Now, only P remains. So, P travels by Bus.

Place of Travel :

R goes to Bangalore, Q to Calcutta.

Now, bus transport is not available for Delhi or Chennai. So, P who travels by Bus goes to Hyderabad.

S travels by boat and hence, S did not go to Delhi. So, S goes to Chennai.

Now, only T remains. So, T goes to Delhi.

Person	Place	Mode
P	Hyderabad	Bus
Q	Calcutta	Aeroplane
R	Bangalore	Car
S	Chennai	Boat
T	Delhi	Train

1. **(*e*).** Incorrect combination is T — Aeroplane
2. **(*c*).** Correct combination for S is Chennai — Boat
3. **(*a*).** Incorrect combination is Delhi — Bus
4. **(*b*).** T travelled to Delhi by Train
5. **(*c*).** T travelled to Delhi

(ii) Seating / Placing Arrangements

In this type of questions, some clues regarding seating or placing sequence (linear or circular) of some persons or items is given. The candidate is required to form the proper sequence using these clues and answer the questions accordingly.

Example. Four girls are sitting on a bench to be photographed. Shikha is to the left ofReena. Manju is to the right of Reena. Rita is between Reena and Manju. Who would be second from the left in the photograph ?

(*a*) Reena (*b*) Shikha (*c*) Manju

(*d*) Rita (*e*) Either Shikha or Reena

Solution : (*a*)

Shikha is to the left of Reena and Manju is to her right.

Rita is between Reena and Manju.

Hence order is : Shikha, Reena, Rita, Manju.

In the photograph, Reena will be second from left.

(iii) Comparison Type Questions

In this type of questions, clues are given regarding comparisons among a set of persons or things with respect to one or more qualities and we are required to analyse the whole information, form a proper ascending/descending sequence and then answer the given questions accordingly.

Directions (1 to 3) : *Read the information given below and answer the questions that follow :*

(*i*) There is a group of five girls.

(*ii*) Kamini is second in height but younger than Rupa.

(*iii*) Pooja is taller than Monika but younger in age.

(*iv*) Rupa and Monika are of the same age but Rupa is tallest between them,

(*v*) Neelam is taller than Pooja and elder to Rupa.

Example 1 : If they are arranged in the ascending order of height, who will be in third position ?

(*a*) Monika (*b*) Rupa

(*c*) Monika or Rupa (*d*) Data inadequate

(*e*) None of these

Example 2 : If they are arranged in the descending order of their ages, who will be in fourth position ?

(*a*) Monika or Rupa (*b*) Kamini

(*c*) Monika (*d*) Data inadequate

(*e*) None of these

Example 3 : To answer the question "who is the youngest person in the group", which of the given statements is superfluous ?

(*a*) Only (*i*) (*b*) Only (*ii*)

(*c*) Only (*v*) (*d*) Either (*i*) or (*iv*)

(*e*) None of these

Solution : First find sequence of heights

By (*iii*), we have $M < P$

By (*v*), we have $P < N$.

Now, Rupa is tallest and Kamini is second in height, hence sequence of heights is :

$$M < P < N < K < R.$$

Now, find age sequence :

By (*ii*), we have K < R.

By (*ii*), we have P < M.

By (*iv*), we have R = M.

By (*v*), we have R < N.

Hence sequence of ages is:

P < K < R = M < N or K < P < R = M < N

1. (*e*) In increasing order of height, Neelam is in third position.

2. (*e*) In descending order of ages, Neelam will be in fourth position (because Monika and Rupa both lie at third position).

3. (*a*) Only statement (*i*) is not necessary.

(iv) Sequential Order of Things

In this type of questions, some clues are given regarding order of occurrence of certain eventsand we are required to analyse the given information, frame the right sequence and then answer the questions accordingly.

Directions (1 to 4) : *Read the following information carefully and answer the questions given below it :*

(*i*) Eight doctors P, Q, R, S, T, U, V and W visit a charitable dispensary every day except on a holiday i.e. Monday.

(*ii*) Each doctor visits for one hour from Tuesday to Sunday except Saturday. The timings are 9 a.m. to 1 p.m. and 2 p.m. to 6 p.m.; 1 p.m. to 2 p.m. is lunch break.

(*iii*) On Saturday, it is open only in the morning i.e. 9 a.m. to 1 p.m. and each doctor visits for only half an hour.

(*iv*) No other doctor visits the dispensary before doctor Q and after doctor U.

(*v*) Doctor-W comes immediately after lunch break and is followed by R.

(*vi*) S comes in the same order as P in the afternoon session.

Example 1 : Doctor P visits in between which of the following pairs of doctors ?

(*a*) S and V (*b*) U and W

(*c*) R and W (*d*) R and U

(*e*) None of these

Example 2. At what time the visit of doctor R is over on Sunday?

(*a*) 1 p.m. (*b*) 3 p.m.

(*c*) 4p.m. (*d*) 5p.m.

(*e*) None of these

Example 3 : At what time the visit of Doctor T would be over on Saturday?

(*a*) 10 a.m.

(*b*) 11 a.m.

(*c*) Either 10 a.m. or 11 a.m.

(*d*) Data inadequate

(*e*) None of these

Example 4 : If the lunch break and subsequent visiting hours are reduced by 15 minutes, at what time Doctor U is expected to attend the dispensary ?

(*a*) 3.15 p.m. (*b*) 4 p.m.

(*c*) 4.15 p.m. (*d*) 4.45 p.m.

(*e*) None of these

Solutions : First form sequence of visit using (*iv*), (*v*) and (*vi*).

From (*iv*), Q visits first and U visits last.

From (*v*), W visits first after break and is followed by R.

From (*vi*), P visits after break.

Hence sequence of visit after break becomes W R P U

Also, S has the same position in morning session as P in afternoon session. Thus, sequence of visit before break is Q, T/V, S, V/T.

1. (*d*). P visits between R and U.

2. (*c*). Time of visit of W is 2 p.m. to 3 p.m., Time of visit of R is 3 p.m. to 4 p.m. So, the visit of doctor R is over at 4 p.m.

3. (*c*). T visits either second or fourth. So, the time of visit on Saturday will be either 9.30 a.m. or 10.30 a.m. Thus, T's visit will be over at either 10 a.m. or 11 a.m.

4. (*b*). As mentioned, lunch break will be over and doctor W will visit at 1.45 p.m., doctor R will visit at 2.30 p.m., doctor P will visit at 3.15 p.m. and U will visit at 4 p.m.

(v) Selection based on given conditions

In this type of questions, a few essential criteria for selection of a group of items are given and we have to keep these conditions in mind and make the required selection as per the directions given in each question.

(vi) Jumbled problems

In this type of questions, some mixed clues regarding three or more qualities of given items or persons is given and we are required to analyse this mixed information with respect to different qualities and classify the items accordingly.

7. DIRECTION SENSE

In this type of test, the questions consist of a sort of direction puzzle. A successive follow-up of directions is formulated and we are required to ascertain final direction or the distance between two points. The test is meant to judge ability to trace and follow correctly and sense the direction correctly.

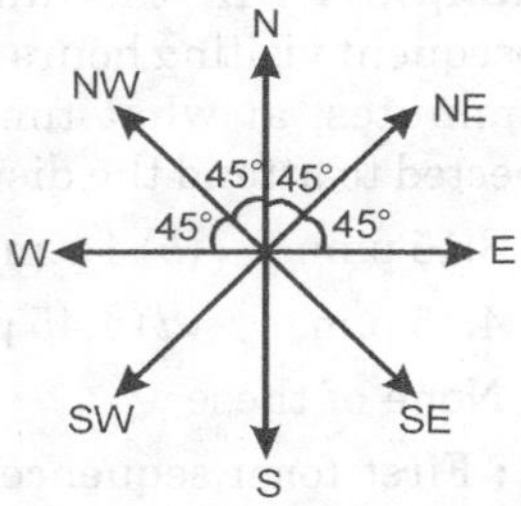

Above figure shows four main directions : North N, South S, East E, West W

Four cardinal directions : North East NE, North West NW, South East SE, South West SW

Pythagoras Theorem

The theorem is applicable only in right angle triangles, in which from theorem one angle is of 90°

$$AC^2 = AB^2 + BC^2$$

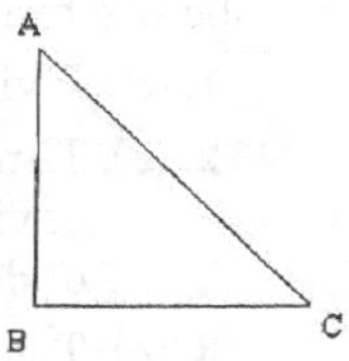

where, AC is Hypotense
AB is Perpendicular (Height)
BC is Base

Shortcut method

1. If 2 opposite turns are coming together, they will cancel each other
   Example. left, right and right, left.
2. If similar turns are coming together they will opposite the earlier direction.
   Example. if a person is going in North and taking two left turns it means now person is moving in opposite to North i.e. South.
3. Start from either left side or right side

NOTE

The distance from a particular point after travelling a distance of x meters in the horizontal direction and a distance of y meters in the vertical direction is equal to $\sqrt{x^2 + y^2}$ (please note that in common usage, North South direction is referred to as "vertical " direction and the East West direction is referred to as the "horizontal" direction).

Example 1 : A person travels a distance of 7 km towards east from his house, then travels 7 km towards north and then a distance 7 km towards east and finally 7 km towards north. What is the vertical distance traveled by him?

Solution :

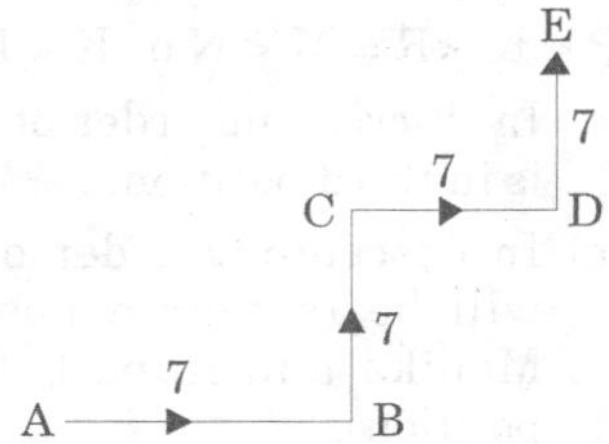

Let A and E be the initial and the final positions.
The vertical distance travelled

$= BC + ED\ (7 + 7)$ km $= 14$ km

Example 2 : A person starts from his house and travels 8m towards west then he travels 6 m towards his left then 8m towards west ad then 4m towards south. Finally he turns right and travels 5m. What is the horizontal distance travelled by him?

Solution :

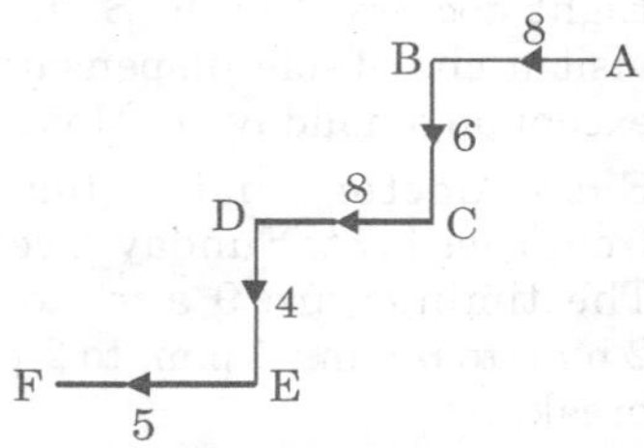

Let A and F be the initial and the final positions.

∴ Horizontal distance traveled

$= FE + DC + BA = 5 + 8 + 8 = 21$ m

Example 3 : Surya travels 3 m towards east and then turns right and travels 4 m. What is the distance between the initial & the final position of Surya?

Solution :

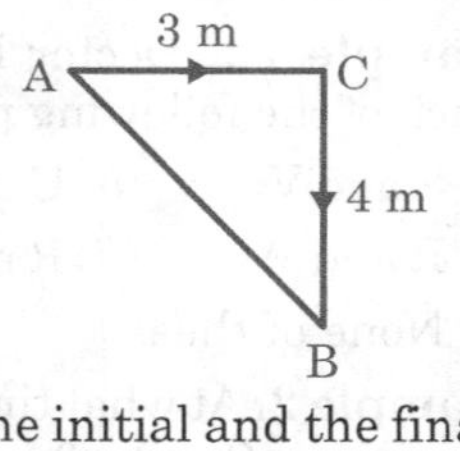

Let A and B the initial and the final positions of Surya.

$$AB = \sqrt{AC^2 + BC^2}$$

$$= \sqrt{3^2 + 4^2} = 5m$$

Example 4 : Starting from his house, Sachin walks a distance of 8 m towards north, then he turns left and walks 6m, then walks 3m towards south an finally travels 6m towards west to reach his office. What is the distance between his house and office and also find in which direction is his office situated with respect to his house.

Solution :

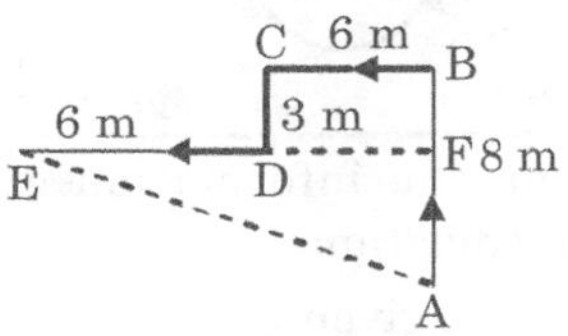

Let A be the Sachin's house and E his office. The distance between A and E

$$= \sqrt{(EF)^2 + (AF)^2}$$

$$= \sqrt{(ED + CB)^2 + (AB - FB)^2}$$

$$= \sqrt{(ED + CB)^2 + (AB - CD)^2}$$

$$= \sqrt{12^2 + 5^2} = 13m$$

Example 5 : Starting from her house, Nisha traveled 8 m towards west then turned right and travelled 15 m. She then traveled 20 m towards east, followed by 20 m to south to reach a hostel. How far is her house from the hostel and in which direction?

Solution :

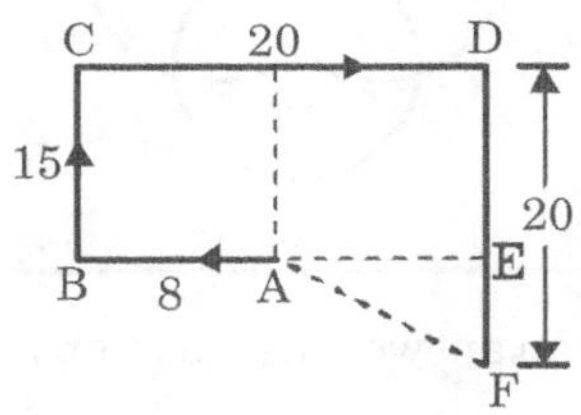

Let A and f be the initial and final positions.

Now $AF = \sqrt{AE^2 + EF^2}$

$AE = CD - BA = 20 - 8 = 12m$

$EF = DF - DE - CB = 20 - 15 = 5m$

$\therefore \quad AF = \sqrt{144 + 25} = \sqrt{169} = 13m$

Her house is towards north- west from the hostel

8. LOGICAL VENN DIAGRAMS

This type of questions aim at analysing ability to relate a certain given group of items and illustrate it diagrammatically.

Venn Diagrams are diagrammatic representation of sets, using geometrical figures like circles, triangles rectangles etc. Each geometrical figure represents a group as shown in the examples. The area common to two or more figure represent those elements which are common to two or more groups. There are various models in Venn Diagrams which we will discuss with examples.

Venn Diagram Type I

In these kind of questions. There are many geometrical figures representing different groups. Let's discuss this type with the help of the following example.

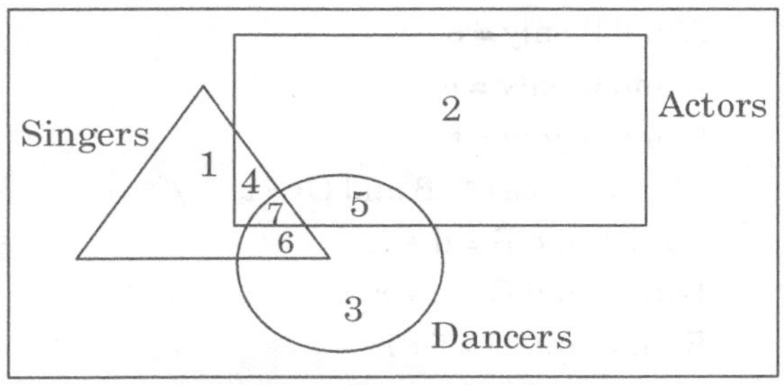

This diagram consists of three groups –Singers, Actors and Dancers represented by a triangle, a rectangle and a circle respectively. There are seven regions represented by numbers from 1 to 7 where each region represents a different group

Region 1 → Represents only Singers

Region 2 → Represents only Actors

Region 3 → Represents only Dancers

Region 4 → Represents only Singers & Actors

Region5 → Represents only Actors & Dancers

Region6 → Represents only Dancers & Singers

Region7 → Represents Singers, Actors & dancers.

Hence the various areas as shown represent different groups i.e. region 6 represents those Singers who are only Dancers as region 6 is the intersection of thee triangle and the circle, but not the square. You may be triangle and the circle, but not the square. You may be asked questions like "Which region represents

Dancers who are only Singers?" The answer to this question, as explained above will be region 6.

Venn Diagram Type II

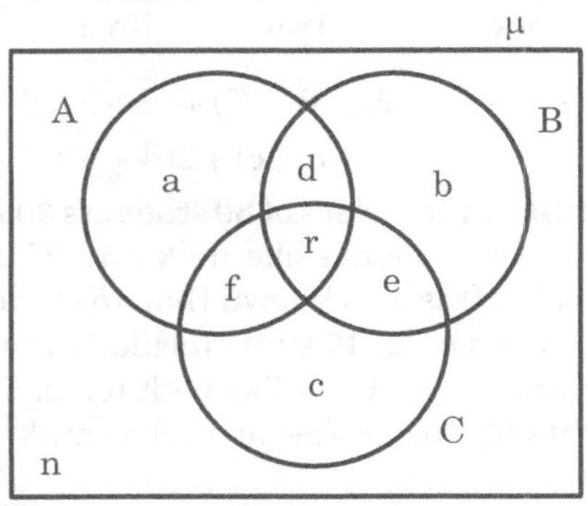

In this type, two three or four different groups could be given with some elements common to two or more groups. Let us observe the diagram given below.

Here A, B and C are three different groups and the various regions can be explained as given below.

Only A = a

Only B = b

Only C = c

A and B only = d

B and C only = e

C and A only = f

All the three (A, B and C) = r

Both A and B = d + r

Both B and C = e + r

Both C and D = f + r

Neither A nor B nor C = n

A, B or C and none = μ Also μ = (A ∩ B ∩ C) + n

He, the rectangle represents the sample space, which consists of three groups A, B and C and also n, which is the number of people belonging to neither A, nor B nor C. Some more formula are as given under.

(i) $A' = (b + e + c) + n$ where A′ = A complement (not in A)

(ii) $B' = (a + f + c) + n$; where B′ = B complement (not in B)

(iii) $C' = (a + d + b) + n$ where C′ = C complement (not in C)

(iv) $A - B = A - (A \cap B)$

(v) $A \Delta B = (A - B) \cup (B - A)$

(vi) Number of people (or things) belonging to at least one out of the three groups

$= (A \cup B \cup C)$

$= (a + b + c) + (d + e + f) + r$

↑ Exactly one ↑ Exactly two ↑ Exactly three

(vii) $A + B + C = (A \cup B \cup C) + (d + e + f) + 2r$

$= (a + b + c) + 2(d + e + f) + 3r$

Example 1 : In a class of 80 students 30 students like tea, 35 students like milk and 25 students like coffee. It is also known that 15 students like both tea and milk, 17 students like both milk and coffee and 14 students like both tea and coffee 9 students like tea, coffee as well as milk.

Solution :

μ = 80

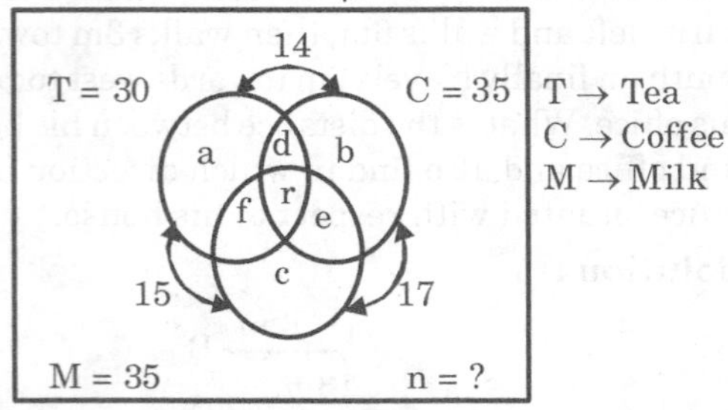

Based on the information given, we get the following diagram

Now r = 9 is given

⇒ d = 14 − 9 = 5; e = 15 − 9 = 6; f = 17 − 9 = 8

Also a = 30 − (d + r + e) = 30 − (5 + 9 + 6) = 10

b = 25 − (d + r + f) = 25 − (5 + 9 + 8) = 3

c = 35 − (e + r + f) = 35 − (6 + 9 + 8) = 12

Now, $T \cup C \cup M = a + b + c + d + e + f + r$

$= 10 + 13 + 12 + 6 + 8 + 5 + 9$

$= 53$

⇒ $n = \mu - (T \cup C \cup M) = 80 - 53 = 27$

Hence, we can now get the following diagram from the above data

μ = 80

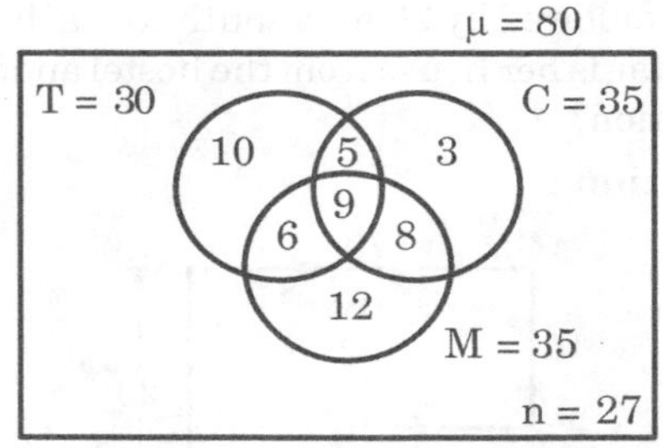

Now, we can answer the questions of the following nature.

Example 2 : What is the total number of students who like neither tea nor coffee nor milk?

Solution :

$n = \mu - (T \cup C \cup M) = 80 - 53 = 27$

Example 3 : How many students do not like Tea?

Solution :

$T' = b + z + c + n = 3 + 8 + 12 + 27 = 50$

(or) $T' = \mu - T = 80 - 30 = 50$

Example 4 : What is the value of T-C?

Solution :

$(T - C) = a + f$

(i.e. the regions of A excluding B)

= 10 + 6 = 16

Example 5 : What is the value of C∆M ?

Solution :

$$C\Delta M = (C - M) \cup (M - C)$$
$$= (b + d) + (d + c)$$
$$= (3 + 5) + 6 + 12 = 26$$

Example 6 : What is the total number of students who like both tea and coffee but not milk?

Solution :

Number of students who like both tea and coffee but not milk = d = 5

Venn Diagram Type III

In this type Venn diagrams are used to established relationship between the given groups. In other words two or more groups are given and the Venn diagram, which most correctly establishes a relation between them has to be chosen out of the various Venn diagrams given in the choices. Let us look at some of the examples given below.

Examples 1: Animals, Cat, Dog

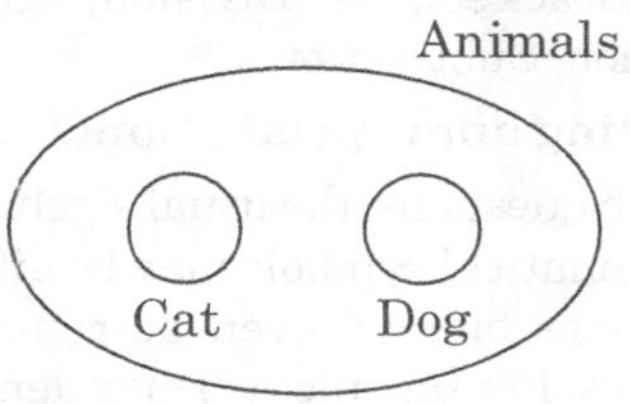

Solution:

Here, in animals we have many species of which cat and dog are two different kinds of species, having nothing in common. So the above diagram is the most appropriate representation of the given groups.

Examples 2 : Month, week, day

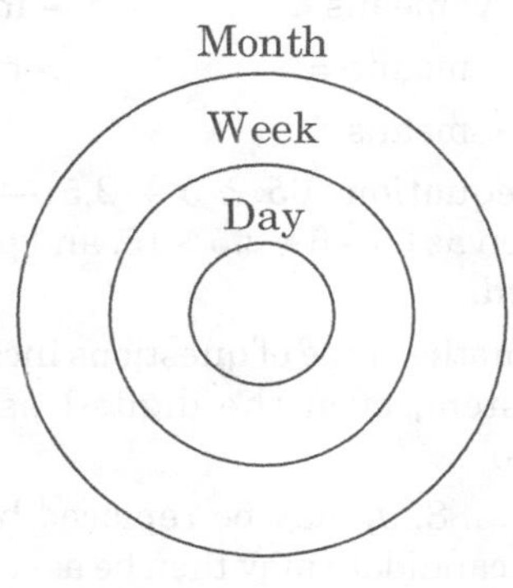

Solution:

We know that day is a part of the week and week is a part of the month. So the above diagram is the most appropriate representation of the given groups.

Examples 3 : Mars Earth, Jupiter

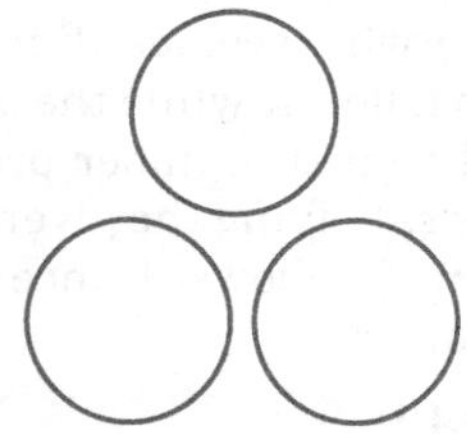

Solutions:

We know that Mars, Earth and Jupiter are there independent entities having nothing in common. So the above diagram is the most appropriate representation of the given groups.

9. ALPHABETICAL QUIBBLE

In this type of questions, we are given alphabets from A to Z. The position of a letter is given in the form of a puzzle and we are required to find this letter. However, sometimes a random letter series is given and we are required to find out how many times a letter satisfying the conditions specified in the question occurs.

Example. *Following question is based on the following alphabet series.*

ABCDEFGHIJKLMNOPQRSTUVWXYZ

Which letter is exactly midway between H and S in the given alphabet ?

(*a*) L (*b*) M

(*c*) N (*d*) O

(*e*) No such letter

Ans. (*e*)

From given series we get that M & N are in midway between H and S. So, exactly midway is no such letter.

10. NUMBER, RANKING & TIME SEQUENCE

(i) Number Test

In this type of questions, generally you are given a long series of numbers and we are required to find out how many times a number satisfying the conditions, specified in the question, occurs.

Example : How many 5's are there in the following sequence which are immediately followed by 3 but not immediately preceded by 7 ?

8953253855687335775365335738

(*a*) One (*b*) Two

(*c*) Three (*d*) Four

(*e*) More than four

Solution : (*c*)

A number which comes after a given number is said to follow it while the one which comes before the given number precedes it. Thus, numbers satisfying the given conditions, can be shown as follows. Hence there are three such numbers.

(ii) Ranking Test

In this type of test, generally ranks of a person both from top and bottom are mentioned and total number of persons is asked. However, sometimes this question is put in the form of a puzzle of interchanging seats by two persons.

Example : Anil ranked ninth from the top and thirty eighth from the bottom in a class. How many students are there in the class?

(*a*) 45 (*b*) 46 (*c*) 47
(*d*) 48 (*e*) 49

Solution : (*b*)

Whole class consists of :

(*i*) 8 students who have a rank higher than Anil

(*ii*) Anil; and

(*iii*) 37 students who have rank lower than Anil i.e.,(8 + 1 + 37) = 46 students.

(iii) Time Sequence Test

Example : Raman remembers that his brother's birthday is after fifteenth but before eighteenth of February whereas his sister Kamla remembers that her brother's birthday is after sixteenth but before nineteenth of February. On which day in February is Raman's brother's birthday ?

(*a*) 16th (*b*) 17th (*c*) 18th
(*d*) 19th (*e*) None of these

Solution : (*b*)

According to Raman, brother's birthday is on one of the days among 16th and 17th February According to Kamla, the brother's birthday is on one of the days among 17th and 18th February. Hence Raman's brother's birthday is on the day common to both the above groups i.e., 17th February.

11. MATHEMATICAL OPERATIONS

This type questions are based on simple mathematical operations.

Here, four fundamental operations :

Addition, subtraction, multiplication and division

and statements such as

'less than', 'greater than', 'equal to', 'not equal to', etc.

are represented by symbols, different from the usual ones.

The questions involving these operations are set using artificial symbols. The candidate has to substitute the real signs and solve the questions accordingly, to get the answer.

Type of Problems in Mathematical Operation

(i) Solving by Substitution

In this type of questions, we are provided with substitutes for various mathematical symbols, followed by a question involving calculation of an expression or choosing the correct/ incorrect equation. We are required to put in the real signs in the given equation and then solve the questions as required.

While solving a mathematical expression, proceed according to the rule BODMAS

i.e., Brackets, Of, Division, Multiplication, Addition, Subtraction.

(ii) Deriving appropriate Conclusions

In such questions, the usual algebraic signs and mathematical symbols may be given different meanings or may even be replaced by new symbols. For instance (÷) may denote addition, (×) may denote subtraction (<) may denote division and so on. The simplest way to solve such an equation is to rewrite the problem replacing the given symbols with the signs they are supposed to stand for. For instance, if :

+ means – ÷ means +
'V' means < – means >
× means = > means ×
&, < means ÷

Then equation, $65 < 5 \div 2.5 — 15$ will be rewritten as $65 \div 5 + 2.5 > 15$, and problem may be solved.

In yet another type of questions included under this system, even the digits (instead of the symbols)

0, 1, 2, ... 8, 9, may be replaced by alphabets and the candidate may then be asked to perform calculations and solve the given problem. For instance, codes may be used as (a = 1, b = 2, c = 3 etc.) or in reverse (i.e., z = 1, y = 2, x = 3 and so on).

Following question employs such a code.

Example 1 : If y × w = s and u — v = z, what is r/x ?

(*a*) y (*b*) z (*c*) s (*d*) x

Solution : If we count backwards from z, we get y as the 2nd letter, w as the 4th letter, and s as the 8th letter. In this order u is the 6th, v is the 5th and z the 1st letter of the code.

So, $2 \times 4 = 8$ and $6 - 5 = 1$

Hence, r = 9 and x = 3.

So $\frac{r}{x} = \frac{9}{3} = 3$. So the answer is (*d*) x.

In certain cases, arithmetical operations may not be defined. Consider the following question.

Example 2 : If 8 + 12 = 2, and 7 + 13 = 3, then what is the value of 10 + 18 ?

(*a*) 9 (*b*) 6 (*c*) 4 (*d*) 7

Solution : $\frac{12-8}{2} = 2$, and $\frac{13-7}{2} = 3$. So, code employed here is special.

In x + y, '+' stands for $\frac{y-x}{2}$

$\therefore$ $10 + 18 = \frac{18-10}{2} = 4$, so the answer is (c).

12. LOGICAL SEQUENCE OF WORDS

In this type of questions, a group of words is given. The candidate is required to arrange these words in a meaningful order such as the sequence of occurrence of events, sequence from a part to the whole, sequence of increasing/decreasing size, value, intensity etc., and then choose the correct sequence accordingly.

Example 1 : *Arrange the following in a meaningful sequence :*

1. Consultation 2. Illness
3. Doctor 4. Treatment
5. Recovery

(*a*) 2, 3, 1, 4, 5 (*b*) 2, 3, 4, 1, 5
(*c*) 4, 3, 1, 2, 5 (*d*) 5, 1, 4, 3, 2
(*e*) 2, 3, 4, 5, 1

Solution : (*a*)

Illness occurs first. One then goes to the doctor and after consultation with him, undergoes treatment to finally attain recovery. Thus, the correct order is 2, 3, 1, 4, 5.

Example 2 : *Arrange the following in a logical order :*

1. Euphoria 2. Happiness
3. Ambivalence 4. Ecstasy
5. Pleasure

(*a*) 4, 1, 3, 2, 5 (*b*) 3, 2, 5, 1, 4
(*c*) 2, 1, 3, 4, 5 (*d*) 1, 4, 2, 5, 3
(*e*) 5, 4, 3, 2, 1

Solution : (*b*)

All the given words stand for 'Joy', but the intensity increases in the order — Ambivalence, Happiness, Pleasure, Euphoria, Ecstasy. Thus, the correct order is 3, 2, 5, 1, 4.

13. DECISION MAKING

It with questions in which we have to decide the course of action to be taken upon a candidate who has applied for a vacancy or allotment or membership to an institution, keeping in mind the essential requisites and the data given for the candidate.

Type I Decision Test

The decision about each candidate has to be made from amongst the five choices named (*a*), (*b*), (*c*), (*d*), (*e*) which state the courses of action to be taken as per the candidate's potentials.

Direction (1 to 5) : *Read the following information to answer the given questions*

Following are the criteria for selecting & marketing officer by a company. The candidate must :

(1) be a graduate with at least 50% marks.

(2) have secured at least 40% marks in the written test.

(3) not be less than 24 years and more than 29 years as on 10th October, 2007.

(4) should have work experience of at least two years as an officer. However, if a candidate:

(5) fulfils all other criteria except at (4) above but has a diploma in Marketing Management, his/her case is to be referred to General Manager, Marketing.

(6) fulfils all other criteria except at (3) above but has worked as Marketing Officer at least for three years, his/her case is to be referred to Director, Marketing.

Based on the above criteria and information given in each of the following questions, we have to take the decision in regard to each case. We are not to assume anything. These cases are given as on 10th October, 2007.

Mark answer (*a*) if candidate is to be appointed;

Mark answer (*b*) if the candidate is not to be appointed;

Mark answer (*c*) if the data given are not sufficient to take decision;

Mark answer (*d*) if to be referred to General Manager - Marketing; and

Mark answer (*e*) if to be referred to Director – Marketing.

Example 1 : Amit, born on 5th June, 1983, has done his post-graduation in Marketing Management with first class. He has secured 50% marks in the written test. He has been working in an organisation as a Marketing Officer for the last four years.

Example 2 : Rohit has been working in an organisation as Officer for the last ten years. His date of birth is 17th February, 1964. He has secured 60% marks in the degree examination and 40% marks in the written test.

Example 3 : Manju is a first class graduate and has done a diploma in Marketing Management. She has secured 50% marks in the written test. She was 23 years old as on 5th September, 2006.

Example 4 : Nitin was born on 25th August, 1985. He has secured 60% and 50% marks in graduation and in the written test, respectively. He has been working in an organisation as Officer for the last four years.

Example 5 : Suman is a graduate with first class and has secured 60% marks in the written test. She has been working as an Officer for the last three years. She was born on 20th May, 1982.

Solutions

1. (*a*). All conditions of eligibility are satisfied.
2. (*e*). Condition (3) is violated but it can be waived by condition (6).
3. (*d*). The candidate satisfies all conditions except (4). But he fulfils condition (5) so that (4) is waived.
4. (*e*). The candidate satisfies condition (2) instead of condition (3).
5. (*a*). All conditions of eligibility are satisfied.

Type II Assertion and Reason

This test is meant to judge the candidate's knowledge and his ability to reason out correctly. In this test, two statements referred to as the Assertion (A) and Reason (R) respectively are provided. Five alternative comments on these are given and the correct one is to be chosen.

Direction (1 and 2) : *For the Assertions (A) and Reasons (R) below, choose the correct alternative from the following*:

(*a*) Both A and R are true and R is the correct explanation of A.

(*b*) Both A and R are true but R is NOT the correct explanation of A.

(*c*) A is true but R is false.

(*d*) A is false but R is true.

(*e*) Both A and R are false.

Example 1 : Assertion (A) : Moon cannot be used as a satellite for communication.

Reason (R) : Moon does not move in the equatorial plane of the earth.

Solution (*a*) :

Since R explains A.

Example 2 : Assertion (A) : Salt is added to cook food at higher altitudes.

Reason (R): Temperature is lower at higher altitudes.

Solution (*b*) :

Since both statements A and R are separately true but R does not explain A.

Type III Situation Reaction Test

In this test, certain situation is described and the candidate is required to choose the most suitable reaction to the given situation from amongst the alternatives provided. This test judges reasoning power of the candidate and his ability to act correctly and promptly to a situation that may arise in emergency.

Direction (1 to 3) : While travelling in your car, certain persons stop you on the way asking you to take an injured child to the hospital. You would :

(*a*) ask them to leave your way and then drive away.

(*b*) ask them to first call the police.

(*c*) immediately take the child to hospital.

(*d*) get out of the car and ask some other person to help them.

(*e*) None of these

Solution (*c*)

Situation described above demands that the person should immediately render the help asked for and take the child to the hospital.

Example 1 : While playing cricket in the school, suddenly when you hit the ball, it strikes your classmate on the forehead and blood starts oozing out. You would :

(*a*) run away from the field.

(*b*) start fighting with the boy why he came in the way.

(*c*) blame somebody else for the accident.

(*d*) take the boy to the first aid room.

(*e*) None of these

Solution (*d*)

In the above situation, urgent need is to provide first aid to the boy so that the bleeding may stop.

Example 2 : You are visiting a place for the first time and are travelling in a bus. Suddenly you realise that the driver is taking the bus to a lonely place with no right intentions. You would :

(*a*) with the help of some other passengers, try to baffle the driver and take over the bus.

(*b*) sit and wait to face the repercussions.

(*c*) jump out of the running bus.

(*d*) console the worried passengers.

Solution. (*a*)

When a wrong doing is expected, immediate action to prevent it is the need.

Example 3 : You have gone to enjoy a Diwali Mela organised by a club. Suddenly you come across a lost child crying desperately.

You would :

(*a*) neglect and walk away.

(*b*) ask the child to find his parents.

(*c*) ask him to stop crying and wait patiently for his parents.

(*d*) contact with the club authorities and make an announcement for the parents.

Solution (*d*)

Immediate need is to find the child's parents and for this, best way is to announce the child's name and appearance so that his parents might know where the child is.

14. CUBES

A cube is a solid which has 6 faces, 8 corners and 12 edges.

Following are some of the important types based on which questions have been asked in various aptitude tests.

(i) Number of cuts given to a cube is given, find the maximum number of identical pieces that can be produced.

(ii) Number of pieces a cube is cut into is given, find the least number of cuts required to produce these many pieces.

(iii) Miscellaneous types.

It must be remembered that in order to obtain the maximum number of pieces, the cuts given to a large cube must be divided as equally as possible in three different directions.

Let us discuss each of these types with the help of examples.

(i) Number of cuts = given;

Maximum number of pieces =?

(a) Whenever the number of cuts made to a cube is a multiple of 3.

If 'n' is the number of pieces along each edge, and

$$n = \left(\frac{\text{Total cuts}}{3}\right) + 1, \text{ then}$$

Maximum number of identical pieces = n^3

Number of pieces with 3 faces visible = 8

Number of pieces with 2 faces visible

$= 12(n-2)$

Number of pieces with 1 face visible

$= 6(n-2)^2$

Number of pieces with no face visible

$= (n-2)^3$

Let us take an example where 30 cuts are made to a large cube to get the maximum number of pieces.

Here, the value of 'n' will be $\frac{30}{3}+1 = 11$

Then,

- Maximum number of pieces

  $= n^3 = 11^3 = 1331$
- Number of 3 – face visible pieces = 8.
- Number of 2 – faces visible pieces

  $= 12(n-2) - 12(11-2)$

  $= 108.$
- Number of 1 – face visible pieces

  $- 6(n-2)^2 = 6(11-2)^2 = 486.$
- Number of no face visible pieces

  $= (n-2)^3$

  $= (11-2)^3 = 729.$

(b) Whenever the number of cuts made to the large cube is NOT a multiple of 3.

In such a case, there is no direct formula for finding out the 2 - faces visible, 1 - face visible or no face visible pieces, but the maximum number of pieces produced can be found out. Let us consider that a large cube is given 4 cuts. Now 4 cuts are divided into three directions as 1 cut, 1 cut, 2 cuts. We know that the number of pieces along each edge is one more than the number of cuts along that edge (when all cuts are made in the same direction - parallel to the same pair of faces). We can represent the above as shown below:

$$\begin{array}{ccc} 1 & 1 & 2 \\ +1 & +1 & +1 \\ \hline 2 \times & 2 \times & 3 = 12 \text{ pieces.} \end{array}$$

Similarly, 11 cuts will give us maximum 100 pieces, as shown below:

$$\begin{array}{ccc} 3 & 4 & 4 \\ +1 & +1 & +1 \\ \hline 4 \times & 5 \times & 5 = 100 \text{ pieces.} \end{array}$$

(ii) Number of pieces = given,

Least number of cuts =?

We use the reverse process. To get the least number of cuts to produce 100 pieces, factorise 100 into 3 factors with least difference between them, and then subtract one from each of them, as shown below:

$$\begin{array}{cccc} 100 = & 4 \times & 5 \times & 5 \\ & -1 & -1 & -1 \\ \hline & 3 + & 4 + & 4 = 11 \text{ cuts} \end{array}$$

(iii) Miscellaneous types:

Apart from the above two types, a cube can be coloured with one, two or more colours, and then is given some number of cuts.

Let us study and practice each of the above types with the questions given in the exercise.

15. ARRANGEMENTS

Linear Sequencing

Linear sequencing is essentially arranging the items in a sequence (in a single line.) The questions of this type are also referred to as "Seating Arrangement". The word "Seating Arrangement" should not be misconstrued- it should not be treated as questions involving only persons sitting a specified conditions. Essentially, these questions involve arranging subjects (people or things) according to the given conditions. The arrangement is done only on one "axis" and hence, the positions of the subjects assume importance in terms of order like first position second position, etc.

Directions (1 to 5) : *Read the data given below carefully and answer the questions.*

Seven persons Paul, Queen, Rax, Sam, Tom, Unif and Vali are sitting in a row facing us. Rax and Sam sit next to each other. There must be exactly four persons between Queen and Vali. Sam sits to the immediate right of Queen.

Example 1 : If Paul and Tom are separated exactly by two persons, then who sits to the immediate left of Vali?

(*a*) Paul (*b*) Tom

(*c*) Unif (*d*) Rax

Example 2 : If Queen is not sitting at either extremes of the row, then who among the following has as many persons on his left as on his right ?

(*a*) Sam (*b*) Unif

(*c*) Rax (*d*) Vali

Example 3 : If Queen sits at one extreme, then who is at the other extreme?

(*a*) Paul (*b*) Tom

(*c*) Vali (*d*) Cannot determined

Example 4 : Tom sits to the right of Queen, and Paul is separated from Tom by exactly three persons. Then, who is sitting to the immediate left of Vali?

(*a*) Sam (*b*) Unif

(*c*) Tom (*d*) Rex

Example 5 : In how many different ways can the seven persons sit in a row?

(*a*) 3 (*b*) 2

(*c*) 10 (*d*) 12

Solution:

Let us write down the conditions given in short form and then represent them pictorially Also, let us treat the left of the persons sitting as "left" and their right as "right" for interpreting the conditions

Rax and Sam sit next to each other

→ RS or SR.

There are exactly 4 persons between Queen and Vali

Q ___ ___ ___ ___V or V ___ ___ ___ ___ Q

Sam sits to the immediate right of Queen → SQ.

Now let us analysis the data /conditions that we are given and then put the three conditions together. Let us number the steps from our left to right as Seat 1 to Seat 7.

Since S is to the right of Q and since R and SA have to be next to each other, R can come only to the immediate right of S. Thus, R, S and Q will be in the order RSQ. Since there are four persons between Q and V, Q can be placed in seats for R and S. Hence, there are only two seats available for Q .Let us fix the positions of R, S and V in each of these two positions of Q and write them down. The directions Left and Right are as shown below.

→R L→

Arrangement I :

1234567y

V R S Q

ArrangementI II :

1234567

V R S Q

These are the only two arrangements possible for the four persons V,R,S and Q .The other three persons Paul ,Tom and Unif can sit in the three vacant seats in any order, as no inforamtion is given about them. Now let us look at each of the questions.

1.Solution (c)

Paul and Tom are separated by exactly two persons. Arrangement I is the only one possible as in Arrangement II, Paul and Tom cannot have exactly two persons between them . So, we have the arrangement as follows.

T/P, V,U,P/T, R,S,Q

So, Unif must be sitting to theimmediate left of Vali.

2.Solution (c)

If Queen is not at the extreme right, then only arrangement II mentioned above is possible . The peson who has as many persons on his left as on his right can only be the person who is sitting in the middle seat i.e. seat 4. In this arrangement Rex is sitting in seat 4.

3. Solution (d)

"Queen sits at one extreme" means that we should look at arrangement I. In this arrangement I. In this arrangement, any one out of the three persons Paul, Tom and Unif can be in seat 1, i.e. extreme right

4. Solution (b)

If Tom and Paul are separted by exactly three pesons, then only arrnagement II is possible. So Tom and Paul have to be in seats 3 and 7, Since it is also given that Tom is to the right of Queen, Tom has to be in seat 3 and Paul, in seat 7. So the arrangement must be follows:

V, U, T, R, S, Q, P

The person sitting to the immediate left of Vali is Unif.

5. Solution (d)

We have two possible arrangements – "Arrangemnt I and Arrangement II" that we have already looked at In each arrangement. The remaining three people can sit in the remaining three seats in 6 ways. Thus, a total of 12 ways of seating the seven persons is possible.

Directions (6 to 10):

Read the data given below carefully and answer the questions that follow.

Seven boys, Rajan, Shyam, Vardhan , Mithra, Vimal, Raj and Kishan – are sitting in a row Shyam sits to the immediate left of Vardhan and third to the right of Rajan, wheras Mithra, who sits at the extreme left is next ot Kishan.

Example 6 : Who is sitting to the immediate right of Shyam?

(*a*) Mithra (*b*) Kishan

(*c*) Vamal (*d*) Vardhan

Example 7 : If Vardhan and Kishan exchange place with each other without changing the rest of the arrangement that is already done, then who will sit to the immmedaite left of Rajan ?

(*a*) Kishan (*b*) Raj

(*c*) Vimal (*d*) Vardhan

Example 8 : If only Shyam sits betweeen Raj and Vardhan , who is exactly in the middle of the row?

(*a*) Raj (*b*) Vardhan

(*c*) Vimal (*d*) Rajan

Example 9 : Which of the following cannot confirm the seating arrangement of all the boys?

(*a*) Raj is to the immediate right of Rajan , wheres Vimal is to the left of Shyam.

(*b*) Mithra and Raj have two persons between them.

(*c*) Raj and Kishan have two persons between them.

(*d*) Rajan and Shyam have two person in betweeen them.

Example 10 : After arranging all the boys as per the conditions given in the data, if Rajan now exchange his place with Mithra, and Vardhan exchange his place with Vimal then how many persons will be there between Vimal and Rajan ?

(*a*) Three (*b*) Two

(*c*) Five (*d*) One

Solution:

Let us denote Left and Right as shown below:

→L R→

Now let us represent the data given in pictroial form (We use R for Raj and Rn for Rajan; Va for Vardhan; Vi for vimal; S for Shyam; M for Mithra and K for Kishan.)

Mithra sits at the extreme left –next to Kishan →MK ___ ___ ___ ___

Shyam sits to the immediate left of Vardhan and third to the right of Rajan → Rn ___ ___ S Va

Putting both the above condition together Va can go only to the extreme right postion. Thus, We have the arrrangement as M K Rn ___ ___ S Va.

Raj and Vimal occupy the two vacant seats between Rajan aand Shyam.

6. Solution (d)

From the seating arrangement figure above Vardhan is to the immediate right of Shyam

7. Solution (d)

If Kishan and Vardhan exchange places as can be seen from the arrangement, the person to the immediate left of Rajan will be Vardhan

8. Solution (c)

If Shyam sits between Raj and Vardhan then the seating arrangement is as follows. Mithra, Kishan, Rajan, Vimal, Raj, Shyam, Vardhan. Then, Vimal will be exactly in the middle of the row.

9. Solution (d)

Statement (a) gives the arrangement as Mithra, Kishan, Rajan, Raj, Vimal, Shyam, Vardhan

Statement (b) gives the seating arrangement as Mithra, Kishan, Rajan, Raj, Vimal Shyam, Vardhan

Statement (c) gives the seating arrangement as: Mithra, Kishan, Rajan, Vimal, Raj, Shyam, Vardhan

So only statement (d) cannot makes the seating arrangement unique while others can.

10. Solution (c)

The arrangement is M K Rn R/Vi Vi/R S Va

Rajan exchange his place with Mithra and Vimal with Vardhan then we have the following arrangement Rn K M R/Va Va/R S Vi

While we still do not know the exact position of Vardhan (or which place Vimal sits), we can see that there are five persons between Rajan and Vimal

Circular Arrangement

Questions on circular arrangement involve seating people around a table or arrangement of things in a circular manner (for example, different colors beads string to form a necklace). In case of people sitting around a table, the table could be of any shape – rectangular, square, circular or any other.

The data given in such sets of questions specify the positions of some or all of the individuals (or things) in the arrangement. The positions are specified through conditions involving specified persons sitting (or not sitting) opposite each other or a paritcular person sitting to the right or left of another person etc.

Once you read the data, first draw the shape specified in the data and then draw the slots in the seating arrangement.

Six people around a circular table Eight people around a cirucular table

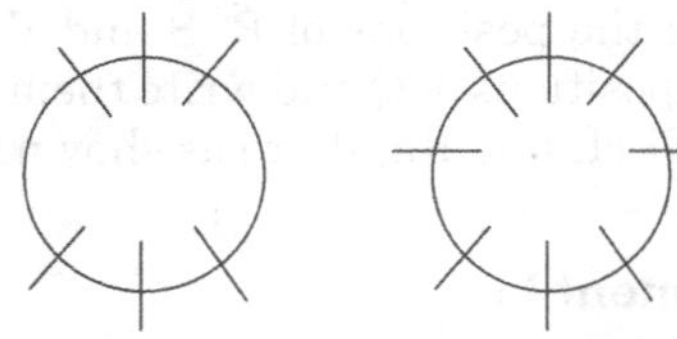

Statement like "A and B are sitting farthest from each other" or "A and B sit across the table" imply that A and B sit opposite each other.

On the other hand, you should remember that unlike in straight –line arrangement, the words "immediately" and "directly" do not play any role in circular arrangement. In general, there is no left side or right side (unless we are talking of 'immediate right' or 'immediate left').

So, if it is given that C sits to right of B, then it is clear that C must be to the immediate right of B. Go "anti- clockwise" if anybody's right has to be located and go "clockwise" if somebody left has to be located.

16. CLOCKS

The hour hand and the minute hand of a clock move in relation to each other continuously and at any given point of time, they make an angle between 0° and 180° with each other.

If the time shown by the clock is known, the angle between the hands can be calculated. Similarly, if the angle between two hands is known, the time shown by the clock can be found out.

When we say angle between the hands, we normally refer to the acute/obtuse angles (upto 180°) between the two hands and not the reflex angles (> 180°).

For solving the problems on clocks, the following points will be helpful.

(1) Minute hand covers 360° in 1 hour, ie. in 60° minutes. Hence **Minute Hand Covers 6° Per Minute.**

(2) Hour hand covers 360° in 12 hours. Hence hour hand covers 30° per hour. Hence, **Hour Hand Covers 1/2° Per Minute**. The following additional points also should be remembered, in a period of 12 hours, the hands make an angle of

(a) 0° with each other (ie. they coincide with each other) 11 times.

(b) 180° with each other (i.e. they point exactly in opposite directions) 11times.

(c) 90° or any other angle with each other 22 times.

NOTE

(1) We can also solve the problems on clocks using the method of "Relative Velocity".

(2) In 1 minute Hand covers 6° and Hour hand covers1/2°

(3) Therefore, Relative Velocity = 6 – 1/2 = 5½° per minute. Alternately, in 1 hour hand covers 5 minute divisions.

∴ Relative Speed = 60 – 5 = 55 minutes per hor. However, adopting the approach of actual angles covered is far simplest and does not create any confusion.

(4) Any angle other than (0° and 180°) is made 22 times in a period of 12 hours.

(5) In a period of 12 hours there are 11 coincidences of the two hands, when the two hands are in a straight line facing opposite directions.

(6) The time gap between any two coincidences is 12/11 hours or $65^{5/11}$ minutes

(7) If the hands of a clocks (which do not show the correct time) coincide every 'p' minutes. Then, If $65^{5/11}$ then the watch is going slow or losing time.

If then the watch is going fast or gaining time.

(8) To calculate the angle 'θ' between the hands of a clocks, we use the following formula (where m = minutes and h = hours)

(i) $\theta = \frac{11}{2}m - 30h\left(\text{when}\frac{11}{2}m > 30h\right)$

(ii) $\theta = 30h - \frac{11}{2}m\left(\text{when}30h > \frac{11}{2}m\right)$

Example 1 : What is the angle between the minute hand and the hour hand of a clock at 3 hour 40 minutes?

(*a*) 20° (*b*) 70°

(*c*) 90° (*d*) 130°

Solution (d)

The angle between the hands can be calculated by $\theta = \left|\frac{11}{2}m - 30h\right|$ where m is minutes and h is hours.

Here, m = 40 and h = 3

$$\therefore \quad \theta = \left|\frac{11}{2} \times 40 - 30 \times 3\right| = |220 - 90| = 130°$$

The angle between the two hands is 130°

Example 2 : Find the time between 2 and 3 O'clock at which the minute hand and the hour hand

(a) Make an angle of 60° with each other

(b) Overlap.

(c) are perpendicular to each other.

(d) are on the same straight line but are facing opposite directions.

Solution

(1) In the formula

$$\theta = \left|\frac{11}{2}m - 30h\right|$$

$$\theta = 60° \text{ and } h = 2$$

$$\therefore \quad 60 = \frac{11}{2}m - 30 \times 2$$

$$\frac{11}{2}m = 120$$

$$m = \frac{240}{11} = 21\frac{9}{11} \text{min past2}$$

$$\text{(or)} \quad 60 = 30 \times 2 - \frac{11}{2}m$$

$$\therefore \quad \frac{11}{2}m = 0$$

$$m = 0$$

Therefore, the angle between the hour hand and the hour hand and the minute hand is at 2 O'clock and at $21\frac{9}{11}$ minutes past 2O'clock.

(2) When the two hands overlap, the angle between them is

$$\theta = \left|\frac{11}{2}m - 30h\right|$$

$$\therefore \quad \theta = 0° \text{ and } h = 2$$

$$0 = \frac{11}{2}m - 30 \times 2$$

$$m = \frac{120}{11} = 10\frac{10}{11} \text{min past2.}$$

(3) When two hands are perpendicular θ = 90° and h = 2

$$\therefore \quad \theta = \left(\frac{11}{2}m - 30h\right) \text{or} \left(30h - \frac{11}{2}m\right)$$

$$90 = \frac{11}{2}m - 30 \times 2$$

$$\frac{11}{2}m = 150$$

$$m = \frac{300}{11} = 27\frac{3}{11} \text{minutes past2}$$

$$\text{(Or)} \quad 90 = 30 \times 2 - \frac{11}{2}m$$

$$\frac{11}{2}m = 30$$

As m cannot be negative, this case is not possible. So, the hands are perpendicular to each other only once i.e. at $27\frac{3}{11}$ minutes past 2 O'clock.

(4) When two hands are pointing opposite directions and are on a straight line the angle between them would be 180° ie. $\theta = 180°$ and $h = 2$

$$180° = \frac{11}{2}m - 3h$$

$$\frac{11}{2}m = 180 + 60 = 240$$

$$m = \frac{480}{11} = 43\frac{7}{11}$$

So, at $43\frac{7}{11}$ minutes past 2 O'clock the hands will be at 180°.

17. CALENDARS

Suppose you are asked to fine the day of the week on 30th June 1974 it would be a tough job to find because you do not know the method. The method of finding the day of the week lies in the number of "odd day"

NOTE

Every 7th day will be the same day count wise, i.e. if today is Monday then 7th day counting from Tuesday onwards will once again be Monday. Odd days are the days remaining after completion of an exact number of weeks. An Odd day is the reminder obtained on dividing the total number of days with seven.

Example: 52 days ÷ 7 = 3 odd days

Leap and Non-Leap Year

A non –leap year has 365 days whereas a leap year has one extra day because of 29 days in the month of February .Every year which is divisible by 4 is called leap year. Leap year consists of 366 days, (52 complete weeks +2 days). The extra two days are the odd days. So, a leap year has two odd days.

A non-leap year consists of 365 days (52 complete weeks +1 day). The extra one day is the odd day.

NOTE

Every century, year which is a multiple of 400, is a leap year. A century year which is not divisible by 400 is a non-leap year

Example:

400,800, 1200, 1600...are leap years

500,700, 900, 1900.. are non-leap years

Counting the Number of Odd Days

100 years consist of 24 leap years +76 ordinary years. (100 years when divided by 4 we get 25. But at the 100th year is not a leap year, hence only 24 leap year).

= 2 × 24 odd days + 1 × 76 odd days
= 124 days
= 17 weeks + 5 days

The extra 5 days are the odd days.

So, 100 years contains 5 odd days.

Similarly, for 200 years we have10 extra days (1 week +3 day)

∴ 200 years contains 3 odd days.

Similarly 300 years contains 1 odd day and 400 years contains 0 odd days.

Counting of Number of Odd Days, When only one Date is Given

Here we take January 1st AD as the earlier date and we assume that this day is a Monday. We take is its previous day i.e. Sunday as the reference day. After this the above mentioned method is applied to count the number of odd days and find the day of the week for the week for the given date.

Counting Number of Odd Days, When Two Dates Are Given

Any month which has 31 days has 3 odd days (∵ 31 ÷ 7 leaves 3 as reminder) and any month which has 30 days has 2 odd days (30 ÷ 7 leaves 2 as reminder).

Then, the total numbers of odd days are calculated by adding the odd days for each month. The value so obtained is again divided by 7 to get the final number of odd days. The day of the week of the second date is obtained by adding the off days the day of the week of the earlier date.

Example 1 : If you were born on 14th April 1992 which was a Sunday then on which day of the week does your birthday fall in 1993?

(*a*) Monday (*b*) Tuesday
(*c*) Wednesday (*d*) cannot be determined

Solution (a)

14th April 1992 to 14 the April 1993 is a complete year, which has 365 days.

Hence the number of odd days from 14th April 1992 to 14th April 1993 is 1.

Hence 14th April 1993 is one day after Sunday ie. Monday

Example 2 : If 1st Jan 1992 is a Tuesday then on which day of the week will 1st Jan 1993 fall?

(*a*) Wednesday (*b*) Thursday
(*c*) Friday (*d*) Saturday

Solution (b)

Since 1992 is a leap year there are 2 odd days. Hence 1st January 1992 is two days after Tuesday i.e. Thursday.

Example 3 : If 1st April 2003 was Monday, then which day of the week will 25th December of the same year be?

(*a*) Tuesday (*b*) Wednesday

(*c*) Thursday (*d*) Friday

Solution (b)

The number of days from 1st April to 25th December

$(29+31+30+31+31+30+31+30+25)$ days $= 268$ days.

$$= \frac{268}{7} = 38 + 2\text{odd days.}$$

Hence 25th December is two days after Monday ie. Wednesday

Example 4 : On which day of the week does 4th June 2001 fall?

(*a*) Monday (*b*) Tuesday

(*c*) Wednesday (*d*) Thursday

Solution (a)

4th June 2001 ⇒ (2000) year+1st January to 4th June 2001

We know that 2000 years have zero odd days. The number of odd days from 1st January to 4th June 2001

Month: Jan + Feb + Mar + Apr + May + June

Odd day: 3 + 0 + 3 + 2 + 3 + 4

$$\frac{15}{7} = 1\text{odd day.}$$

Hence 4th June 2001 was a Monday.

Example 5 : Which year will have the same calendar as that of 2005?

(*a*) 2006 (*b*) 2007

(*c*) 2008 (*d*) 2011

Solution (d)

Year: 2005 + 2006 + 2007 + 2008 + 2009 + 2010

Odd days: 1 + 1 + 1 + 2 + 1 + 2 + 1

Total numbers of odd days from 2005 to 2010 are 7 = 0 odd days.

Hence 2011 will have the same calendar as that of 2005

Example 6 : What day of the week was 18th April 1901?

(*a*) Monday (*b*) Tuesday

(*c*) Wednesday (*d*) Thursday

Solution (d)

18th April 1901 ⇒ (1600 + 300) years + 1st January to 18th April 1901

1600 years have = 0 odd days

300 years have – 1 odd day. The number of days from 1st January 1901 to 108 days ≅ 3 odd days

∴ Total number of odd days = 3 + 1 = 4

Hence 18th April 1901 is Thursday

18. DEDUCTIONS

In number of competitive exams there are a few questions on "deductions". Typically here, each question consists of two statements–on the basis of which a deduction has to be made. The answer has to be chosen from the given four (or five) choices and that will be the deduction made.

(If no conclusion can be drawn or the answer is not obtained then the choice which is normally worded as "none of the above" has to be marked).

These questions can be answered by representing the given statements by Venn Diagrams. However, here we will look at arriving at the deduction by using some simple rules.

First let us look at some basic terms used in the rules and understand what they mean. The two statements given in the question are called 'premises' & the answers are conclusion.

Eg. All dogs are cats—(i)

All cats are pigs —(ii)

The premises normally start with the words All, No Some and Some –not

The word "All" has its synonyms as –Every, Any, Each, whereas the word "Some " can also be replaced by Many, Few, A little, Most of, Much of, More etc.

These words are referred to as qualifiers (also termed as quantifiers).

A premise consists of a subject and a predicate wherein the first term [e.g. "dogs" in statement (i)] is the subject and the second term [e.g. "cat" in statement (i)] the predicate. Similarly, in statement (ii),"cats" is called the subject and "pigs" is the predicate.

The word that occurs in both the premises is known as the 'middle term' ("cat" in the example, give above). The answer or "conclusion" should consists of the other two words ("dog and "pig's in the above example) and the middle term should not appear in the answer.

The premises can be divided into

(a) Universal statements and

(b) Particular statements.

This classification of the premises into the above categories is dependent on the qualifier used in the premise. For example, the statements where "All" is used are called Universal statements and the statements where "Some" is used are called Particular statements where "Some" is used are called Particular statements. Premises can also be divided into

(i) Positive (affirmative) statements and

(ii) Negative statements)

If there is a negative term like "not" or "no" in the statement it is called a negative premise. Otherwise it is called a positive premise or an affirmative statement. The combination of the two different categories of classification leads to four different premises as given in Table I below.

Table I

	Affirmative	Negative
Universal	All A	No E
Particular	Some many l	Some not many O

The subject and the predicate are either distributed (√) Or not distributed (×) depending on what kind of a statement it is (particular affirmative etc.)

Table II shows the distribution Pattern of the subject and the predicate.

Table II

	Subject	Predicate
Universal affirmative	√	x
Universal negative	√	√
Particular affirmative	x	x
Particular negative	x	√

Note: (√) indicate distributed (×) indicates undistributed.

Rules for Deductions

(1) Every deduction should contain three and only three distinct terms.

(2) The middle term must be distributed at least once in the premises.

(3) If one premise is negative, then the conclusion must be negative.

(4) If one premise is particular then the conclusion must be particular

(5) If both the premises are negative, no conclusion can be drawn

(6) If both the premises are particular, no conclusion can be drawn

(7) No term can be distributed in the conclusion, if it is not distributed in the premises.

We take example of each type and look at them in details.

Example 1 :

All dogs are cats — (i)

All cats are pig — (ii)

As the first statement is a Universal affirmative statement. The subject (dog) has to be distributed (√) and the predicate (cats) is not distributed (×). As the second statement is also Universal affirmative, the subject cat is distributed (√) and predicate pigs is not distributed (×). The above answer/ logic is arrived at on the basis of Table II.

The middle term ("cat is the middle term as it occurs in both the premises) is distributed once in the premises. Hence it satisfies Rule[2] As "dogs" is distributed in the premise and "pigs" is undistributed in the deduction also, they should appear accordingly. The type of statement that satisfies both of them is Universal affirmative statement i.e. a statement with "All" Hence the answer will be

"All dogs are pigs"

The answer cannot be "All pigs are dogs", because Rule [7] states that no term can be distributed in the conclusion if it is not distributed in the premises. As "pigs" is not distributed in the premise. It cannot be distributed in the conclusion (because if we take "All pigs are dogs", then the subject "pigs" will be distributed). Hence the conclusion "All pigs will dog" is wrong.

Example 2 :

All cats are dogs — (i)

All cats are pigs - - (ii)

Statement (i) is Universal affirmative and hence the subject "cats" is distributed and the predicate "dogs "is not distributed as per Table II.

Statement (ii) is also Universal affirmative and hence the subject "cats" is distributed and thee predicate "pigs" is not distributed as per Table II.

Here, the middle term "cats" ("cats" is the middle term as it is occurring in both the premises) is distributed; hence we can draw a conclusion.

The answer should contain the terms "dogs" and "pigs" and both the terms are not distributed Referring to Table II ,we find that this is possible only in Particular affirmative [the conclusion cannot start with qualifier 'All' as the subject in "All" should be distributed]. According to Rule 7 a term cannot be distributed in the conclusion if it is not distributed in the premises. So the answer will be

"Some dogs are pigs" Or "Some pigs are dogs"

Example 3 :

All dogs are cats — (i)

All pig are cats — (ii)

Statements (i) is a Universal affirmative and hence the subject "dogs" is distributed and the predicate "cats" is not distributed. In statement (ii) which is also a Universal affirmative the subject "Pigs" is distributed and the predicate "cats" is not distributed this is arrived at one the basis of Table II.

The middle term "cats" ["cats" is the middle term as it occurs in both the statements]is not distributed in either of the two statements From Rule [2] which states that the middle term should be distributed at least once in the premises for drawing a conclusion we cannot draw any conclusion in this case.

Example 4 :

All cats are dogs. — (i)

Some cats are pigs — (ii)

The first statement is a Universal affirmative premise and hence the subject "cats" is distributed and the predicate "dogs" is not distributed(x). The second statement is Particular affirmative and hence both the subject "cats" and the predicate "pigs" are not distributed (x)as per Table II. As we have a particular premise, the conclusion. Should also be a particular one as per Rule [4]. The middle term is distributed hence we can draw a conclusion .So the answer will be

"Some dogs are pigs" Or "Some pigs are dogs"

Example 5 :

All dogs are cats — (i)

No cats are pigs — (ii)

As the first premise is a Universal affirmative, the subject (dogs) is distributed and the predicate (cats) is not distributed. In the second premise which is a Universal negative, the first term (cats) and the second term (pigs) are both distributed (as per Table II) As the middle term is distributed at least once in the premises. Rule [2] is satisfied and hence we can draw a conclusion.

From Rule 3] which states that if one of the premises is negative the conclusion should be negative, the conclusion should be negative and as both the terms "dogs" and "pigs" are distributed, the conclusion should be a Universal negative statement. Hence the answer will be

"No dogs are pigs" Or "No pigs are dogs"

Example 6 :

All dogs are cats — (i)

Some cats are not pigs — (ii)

Since the first statement is a Universal affirmative "dogs" is distributed and "cats" is not distributed. Since the second statement is a Particular negative, "cats" is not distributed and "pigs" is distributed (as per Table II)

In the above given example no conclusion can be drawn , as rule [2] states that the middle term("cats" in the above example as it occurs in both the premises) should be distributed at least once in the premises. This is not satisfied.

Example 7 :

All cats are dogs — (i)

Some cats are not pigs — (ii)

The first statement is a universal affirmative and hence "cats" is distributed and "dogs" is not distributed The second statement is a Particular negative and hence "cats" is not distributed and "pigs" is distributed (as per Table II)

Here, the middle term (cats) is distributed and hence we can draw a conclusion.

The conclusion should be particular negative as Rule [3] states that if a premise is negative. Also Rule[4] state that if a premise is Particular, the conclusion should also be Particular. Hence the conclusion should be a particular negative.

In particular negative, we know that the subject is not distributed and the predicate is distributed.

The terms "dogs" and "pigs" should come in the conclusion. Also since "dogs" is not distributed in the premise, it cannot be distributed in the conclusion, as per Rule [7]

As per the above reasoning only "pigs" can be the predicate in the conclusion and hence "dogs" will be the subject

Thus the answer will be "Some dogs are not pigs"

Example 8 :

No dogs are cats — (i)

No cats are pigs — (ii)

We cannot draw any conclusion, as Rule [5] states that if both premises are negative, we cannot draw any conclusion.

Example 9 :

No dogs are cats

Some cats are not pigs.

As both the premises are negative hence as per Rule [5] we cannot draw any conclusion (Please note that the first premise is a Universal negative and hence the subject (dogs) is distributed and the predicate (cats) is also distributed as per Table II.

The second statement is a Particular negative and hence the subject (cats) is not distributed and the predicate (pigs) is distributed a per Table II)

Example 10 :

Some cats are not pigs — (i)
Some cats are dogs — (ii)

As the first premise is a Particular negative the subject (cats) is not distributed and the predicate (pigs) is distributed. In the second premise both the subject and the predicate (cats and dogs respectively) are not distributed since the premise is a particular affirmative (as per Table II).

No conclusion can be drawn as both the premises are particular as per Rule [6]

Example 11 :

Some cats are not dogs — (i)
Some cats are not pigs — (ii)

We cannot get an answer from the two premises as Rule [5] states that from two negative premises, no conclusion can be drawn. Also Rule [6] states that from two particular premises no conclusion can be drawn.

19. DATA SUFFICIENCY

Question on Data Sufficiency may consist of two statements or three statements. The questions asked on Data Sufficiency are either based on Mathematical or Reasoning. A firm grasp over the basis of all topics being covered under Quantitative Ability and Reasoning is the prerequisite for doing well in Data Sufficiency.

Data Sufficiency, as the name suggests, tests the ability of the student to identify whether the data given are sufficient to answer a question or not.

Hence, in the question given on Data Sufficiency, the student must stop at the stage of determining the sufficiency of the data and he/she is not expected to solve the problem completely. For that reason most of the question given on data sufficiency will be such that the student is not required to do any calculation but can answer the question by observation. However, in some cases, it is difficult to say whether the data are sufficient or not unless complete calculations are done.

In Data Sufficiency questions, typically, there will be a question followed by some data. The data are given in the form of two statements (statement 1 and 2 or A and B) or three statements. (Statements 1, 2 and 3 or statements A,B and C).

The student has to decide the answer choice for the question depending on whether the data given in the statements are sufficient to answer the questions or not. Direction is provided to the students as to how to arrive at the answer choice. While, as mentioned above, though there can be three statements, giving data, it is question with two statements that are very common.

Before we understand the solving part of questions on data sufficiency, let us look at the direction given on the Data Sufficiency.

Each question is followed by two statements giving certain data. You have to decide whether the information provided in the statements is sufficient for answering the question.

Choose A:

If the question can be answered by using one of the statements alone, but cannot not be answered by using the other statement alone.

Choose B:

if the question can be answered by using either statement alone.

Choose C:

if the question can be answered by using both statements together, but cannot be answered by using either statement alone.

Choose D:

If the question cannot be answered even by using both the statements together.

For solving any Data Sufficiency with two statements, one has to follow four steps in sequential order.

The steps involves are as follows.

Step1 :

Look at the sufficiency of the statement 1 alone.

Step2 :

Look at the sufficiency of statement 2 alone.

Step3 :

If anyone or either of the statements sufficient, then stop.

Step4 :

If each of the statement is independently insufficient, then combine them.

Step5 :

See whether the combination of the statements is sufficient or not.

The following flow chart explains pictorially this set of directions and shows how one can go about deciding the answer choice for any question.

Flow Chart

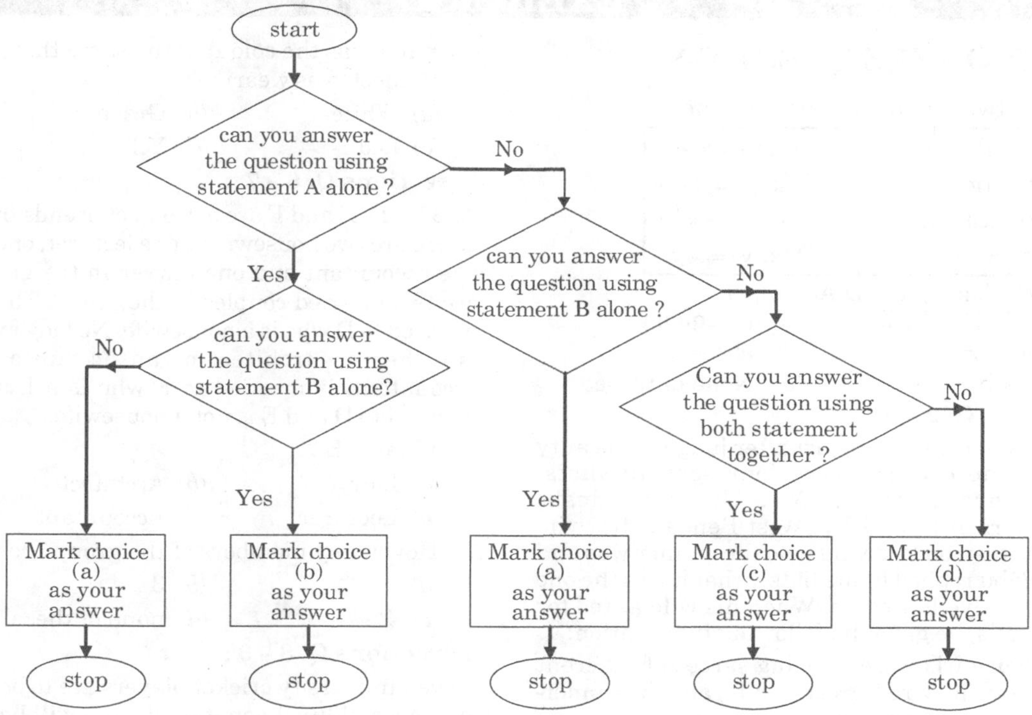

Example 1 : what is the age of mother?

1. Two years ago, the father's age was six times the son's age.
2. After 18 years, the ratio of the ages of son, mother and father will be 5:9:10.

Solution:

From statement 1, let the present age of the father and the son be f and s respectively. Then we have

$(f-2) = 6(s-2) \rightarrow (1)$

Using statement2, after 18 yearsfather's age is twice son's age. $(f+18) = 2(s+18)$ (2)

Each statement is independently not sufficient.

Combining both the statements, father's age and son's age can be obtained by simplifying (1) and (2). Using the ratio of their ages, the mother's age also can be obtained. Choice (A)

Example 2 : what is the two digit number xy?

1. The sum of the units digit and the tens digit of xy is 15.
2. Difference between the digits of xy is 1.

Solution:

Given two digit number is xy.

From statement 1, we have $x + y = 15$ from which we cannot find x, y uniquely.

$\therefore$ Statement 1 is not sufficient to answer.

From statement 2, we have $x - y = 1$ or $y - x = 1$.

This alone is not sufficient to answer.

Combining both the statements, we have

$x + y = 15$ ——(1)

$x - y = 1$ ——(2)

or $y - x = 1$ ——(3)

From (1) and (2), we have $x=8$ &

$y=7$.

From (1) and (3), we have $x = 7$ and $y = 8$

Both the statements together also are not sufficient Choice (D)

Example 3 : what is the measure of angle C in triangle ABC?

1. Angle A, B and C is in the ratio 2 : 1 : 3.
2. $A + B = 130°$

Solution:

Statement 1 alone is sufficient to answer as we know the ratio of the angle.

From statement 2, as $A + B = 130°$ and as sum of the angles of any triangle is 18, we have

$= 180° - 130° = 50°$

Statement 2 alone is sufficient to answer the question. Choice (B)

EXERCISE

MCQ TYPE QUESTIONS

1. Given two numbers x and y, define

Addition	A(x, y) = x + y
Subtraction	S (x, y) = x – y
Multiplication	M (x, y) = x, y
Division	D (x, y) = x/y

S[M(D(A(a, b), 2), D(A(a, b), 2)), M(D(S(a, b), 2), D(S(a, b),2))] is equal to

(*a*) $(a - b)^2$ (*b*) ab
(*c*) $(a + b)^2$ (*d*) None of these

Directions Q. 2 – 5 :

Kamal Babu came home just after judging a beauty contest where there were four semi-finalists, Ms. Andhra Pradesh, Ms. Uttar Pradesh, Ms. Maharashtra and Ms. West Bengal. His wife was very keen on knowing who the winner was and Kamal Babu replied immediately that it was the one wearing the yellow saree. When his wife asked for more details, he gave the following information:

* The four girls were wearing sarees of different colours (yellow, red, green, white) and the runner-up was wearing green.
* The four girls were sitting in a row, and Ms. West Bengal was not sitting at either end.
* There was only one runner-up and she was sitting next to Ms. Maharashtra.
* The girls wearing yellow and white sarees occupied the seats at either end.
* Ms. West Bengal was nither the winner nor the runner-up.
* Ms. Maharashtra was wearing white.
* The winner and the runner-up were not sitting next to each other.
* The girl wearing the green saree was not Ms. Andhra Pradesh.

Answer the following questions based on the above information:

2. What was the colour of the saree that Ms. Andhra Pradesh was wearing?
(*a*) White (*b*) Yellow
(*c*) Red (*d*) Cannot be determined

3. Between which two was Ms.West Bengal sitting?
(*a*) Ms. Andhra Pradesh and Ms. Maharashtra
(*b*) Ms. Andhra Pradesh and Ms. Uttar Pradesh
(*c*) Ms. Uttar Pradesh and Maharashtra
(*d*) Cannot be deternmined

4. Who was wearing the red saree?
(*a*) Ms. Andhra Pradesh
(*b*) Ms. Uttar Pardesh
(*c*) Ms. West Bengal
(*d*) Cannot be determined

5. What was the colour of the saree that Ms. Uttar Pradesh was wearing?
(*a*) White (*b*) Green
(*c*) Red (*d*) Yellow

Directions Q. 6 – 7 :

A, B, C, D, E and F are a group of friends from a club. There are two housewives, one lecturer, one architect, one accountant and one lawyer in the group. There are two married couples in the group. The lawyer is married to D who is a housewife. No lady in the group is either an architect or an accountant C. The accountant, is married to F who is a lecturer. A is married to D and E is not a housewife.

6. What is E ?
(*a*) Lawyer (*b*) Architect
(*c*) Lecturer (*d*) Accountant

7. How many members of the group are males?
(*a*) 2 (*b*) 3
(*c*) 4 (*d*) none of these

Directions Q. 8 – 9 :

Seven university cricket players are to be honoured at a special luncheon. the players will be seated on the dais along one side of a single rectangular table.

A and G have to leave for the lunch early and must be seated at the extreme right end of the table which is closest to the exit.

B will receive the Man of the Match award and must be in the center chair.

C and D, who are bitter rivals for the position of wicket keeper, dislike each other and should be seated as far apart as possible.

E and F are best friends and want to sit together.

8. Which of the following may not be seated at either end of the table ?
(*a*) C (*b*) D
(*c*) G (*d*) F

9. Which of the following pairs may not be seated together ?
(*a*) E & A (*b*) B & D
(*c*) C & F (*d*) G & D

Directions Q. 10 –13 :

A study was conducted to ascertain the relative importance that employees in five different countries assigned to five different traits in their Chief Executive Officers. The traits were compassion (C), decisiveness (D), negotiation skills (N), public visibility (P), and vision (V). The level of dissimilarity between two countries is the maximum difference in the ranks allotted by the two countries to any of the five traits. The following table indicates the rank order of the five traits for each country.

Country					
RANK	INDIA	CHINA	JAPAN	MALAYSIA	THAILAND
1	C	N	D	V	V
2	P	C	N	D	C
3	N	P	C	P	N
4	V	D	V	C	P
5	D	V	P	N	D

10. Which of the following pairs of countries are most dissimilar?

(*a*) China and Japan

(*b*) India and China

(*c*) Malaysia and Japan

(*d*) Thailand and Japan

11. Which of the following countries is least dissimilar to India?

(*a*) China (*b*) Japan

(*c*) Malaysia (*d*) Thailand

12. Which amongst the following countries is most dissimilar to India?

(*a*) China (*b*) Japan

(*c*) Malaysia (*d*) Thailand

13. Three of the following four pairs of countries have identical levels of dissimilarity. Which pair is the odd one out?

(*a*) Malaysia and China

(*b*) China and Thailand

(*c*) Thailand and Japan

(*d*) Japan and Malaysia

Directions Q. 14 – 17 :

Four families decided to attend the marriage ceremony of one of their colleagues. One family has no kids, while others have at least one kid each. Each family with kids has at least one kid attending the marriage.

Given below is some information about the families, and who reached when to attend the marriage.

The family with 2 kids came just before the family with no kids.

Shanthi who does not have any kids reached just before Sridevi's family.

Sunil and his wife reached last with their only kid.

Anil is not the husband of Joya.

Anil and Raj are fathers.

Sridevi's and Anita's daughters go to the same school.

Joya came before Shanthi and met Anita when she reached the venue.

Raman stays the farthest from the venue.

Raj said his son could not come because of his exams.

14. Which woman arrived third ?

(*a*) Shanthi (*b*) Sridevi

(*c*) Anita (*d*) Joya

15. Name the correct pair of husband and wife ?

(*a*) Raj and Shanthi (*b*) Sunil and Sridevi

(*c*) Anil and Sridevi (*d*) Raj and Anita

16. Of the following pairs, whose daughters go to the same school ?

(*a*) Anil and Raman (*b*) Sunil and Raman

(*c*) Sunil and Anil (*d*) Raj and Anil

17. Whose family is known to have more than one kid for certain ?

(*a*) Raman's (*b*) Raj's

(*c*) Anil's (*d*) Sunil's

Directions Q. 18 – 20 :

Answer the following questions based on the statements given below:

(*i*) There are three houses on each side of the road.

(*ii*) These six houses are labeled as P, Q, R, S, T and U.

(*iii*) The houses are of different colours, namely, Red, Blue, Green, Orange, Yellow and White.

(*iv*) The houses are of different heights.

(*v*) T, the tallest house, is exactly opposite to the Red coloured house.

(*vi*) The shortest house is exactly opposite to the Green coloured house.

(*vii*) U, the Orange coloured house, is located between P and S.

(*viii*) R, the Yellow coloured house, is exactly opposite to P.

(*ix*) Q, the Green coloured house, is exactly opposite to U.

(*x*) P, the White coloured house, is taller than R, but shorter than S and Q.

18. What is the colour of the house diagonally opposite to the Yellow coloured house?

(*a*) White (*b*) Blue

(*c*) Green (*d*) Red

19. Which is the second tallest house?

(*a*) P (*b*) S

(*c*) Q (*d*) cannot be determined

20. What is the colour of the tallest house?

(*a*) Red (*b*) Blue

(*c*) Green (*d*) Yellow

Directions for questions 21 to 25:

Select the correct alternative from the given choices

21. If five cuts are made on a cube, what is the minimum number of pieces obtained?

(*a*) 18 (*b*) 6
(*c*) 16 (*d*) 25

22. If six cuts are made on a cube, what is the maximum number of identical pieces obtained?

(*a*) 16 (*b*) 18
(*c*) 36 (*d*) 27

23. If two, three and four cuts are made parallel to different faces of a cube, then what is the number of identical pieces obtained?

(*a*) 60 (*b*) 30
(*c*) 48 (*d*) 24

24. What is the minimum number of cuts required to cut a cube into 216 identical pieces obtained?

(*a*) 36 (*b*) 18
(*c*) 15 (*d*) 12

25. What is the minimum number of cuts required to cut a cube into 24 identical pieces?

(*a*) 6 (*b*) 8
(*c*) 9 (*d*) 12

Directions for questions 26 to 28:

These questions are based on the following information A cube, painted on all its faces, is cut into 125 identical smaller cubes.

26. How many smaller cubes have no face painted?

(*a*) 100 (*b*) 81
(*c*) 64 (*d*) 27

27. How many smaller cubes have only one faced painted?

(*a*) 36 (*b*) 54
(*c*) 64 (*d*) 108

28. How many smaller cubes have exactly two faces painted?

(*a*) 36 (*b*) 54
(*c*) 48 (*d*) 60

Directions for questions 29 to 31:

These questions are based on the following information Each of 216 small identical cubes are painted blue on all faces and all these cubes are arranged to form a large cube. Now all the faces of the large cube are painted pink.

29. How many small cubes have only one color on them?

(*a*) 96 (*b*) 125
(*c*) 64 (*d*) 48

30. How many small cubes have exactly two faces painted pink?

(*a*) 36 (*b*) 48
(*c*) 64 (*d*) 80

31. How many small cubes have exactly three faces painted blue?

(*a*) 8 (*b*) 4
(*c*) 2 (*d*) 6

Directions for questions 32 to 34:

These questions are based on the following information A cube is painted in black and green, each on three faces such that any two faces with same color are adjacent to each other. Now this cube is cut into 60 identical pieces using 2, 3, and 4 cuts parallel to different faces.

32. How many smaller pieces have exactly two faces painted in black color?

(*a*) 5 (*b*) 9
(*c*) 18 (*d*) 27

33. How many smaller pieces have both the colors on them?

(*a*) 9 (*b*) 18
(*c*) 6 (*d*) 24

34. How many smaller pieces have no face painted?

(*a*) 6 (*b*) 9
(*c*) 11 (*d*) 1

Directions: Q. No 35 to 39: These questions are based on the information given below:

Four ladies A, B, C and D and four gentlemen E, F, G and H are sitting in a circle round a table facing each other.

(*a*) No two ladies or two gentlemen are sitting side by side.
(*b*) C who is sitting between G & E is facing D.
(*c*) F is between D and A is facing G.
(*d*) H is to the right of B

35. Who is sitting to the left of A?

(*a*) E (*b*) F
(*c*) G (*d*) h

36. E is facing whom?

(*a*) F (*b*) B
(*c*) G (*d*) None of these

37. Who are immediate neighbor of B?

(*a*) G and H (*b*) E and F
(*c*) E and H (*d*) F and H

38. Who is sitting between E and G?

(*a*) A (*b*) B
(*c*) C (*d*) D

39. Who is sitting to the right of D?

(a) F (b) A
(c) E (d) C

Directions Q. No 40 to 44

Study the following information carefully and answer the question given below: A, B, C, D, E, F, G and H are sitting around a circle facing the centre. B is second to the right of D who is third to the right of F, C is second to the left of A who is second to the left of F, G is third to the right of E.

40. In which of the following combinations is the first person sitting between the second and the third persons?

(a) GCD (b) FGH
(c) EFH (d) ABE

41. Who is third to the right of H?

(a) Data inadequate
(b) D
(c) C
(d) G

42. Who is the immediate right of A?

(a) Data inadequate
(b) E
(c) F
(d) B

43. What is H's position with respect to B?

(a) Fifth to the right
(b) Third to the left
(c) Third to the right
(d) None of these

44. Who is the immediate left of G?

(a) H
(b) F
(c) Data inadequate
(d) D

Directions for questions 45 to 49 :

Read the following information and answer the questions that follow.

P,Q,R,S and T sit around a table. P sits two seats to the left of R and Q sits two seats to the right of R.

45. If S sits in between Q and R, who sits to the immediate right of P?

(a) T (b) S
(c) Q (d) R

46. Which of the following cannot be the correct seating arrangement of the five persons in either the clockwise direction or the anti-clockwise directon ?

(a) P,Q,R,S,T (b) PS,R,T,Q
(c) P,Q,S,R,T (d) P,T,R,S,Q

47. If S is not sitting next to Q, who sitting between Q and S?

(a) R (b) P
(c) T (d) Both R and P

48. If a new person U joins the group such that the intial conditions for the seating arrangement should be observed and also a new condition that U does not sit next to R be satisfied then which of the following statements is true?

(a) U sits to the immediate right of S
(b) U sits to the immediate left of T
(c) U sits to the immediate left of P.
(d) P is to the right of Q

49. If a new person U joins the group such that the initial condtions for the seating arrangement should be observed and also a new condition that U does not sit next to P,S or T be satisfied then who will be the neighbours of P (one either side)?

(a) S and T (b) S and Q
(c) T and R (d) R and Q

Directions for questions 50 to 59:

Select the correct alternative from the given choices

50. What is the angle covered by the minute hand in 22 minutes?

(a) 66° (b) 110°
(c) 132° (d) 220°

51. By how many degrees does an hour hand move in one quarter of an hour?

(a) 5° (b) 7.5°
(c) 10° (d) 12.5°

52. By how many degrees will the minute hand move, in the same time, in which the hour hand moves ?

(a) 54° (b) 84°
(c) 72° (d) 60°

53. What is the angle between the hands of the clocks, when it shows 40 Minutes past 6?

(a) 40° (b) 70°
(c) 80° (d) 90°

54. When the clocks shows 3 hours 14 minutes what is the angle between the hands of the clocks?

(a) 10° (b) 12°
(c) 13° (d) 14°

55. What is the angle between the two hands of a clock when the time is 25 minutes past 7 O'clock?

(*a*) 66½° (*b*) 66½°

(*c*) 72½° (*d*) 69½°

56. When the clock shows 20 minutes past 11 O'clock, what is the angle between the two hands of the clocks?

(*a*) 220° (*b*) 120°

(*c*) 230° (*d*) 140°

57. At the time between 9 and 10 O'clock, will both the hands of the clocks coincide?

(*a*) $43\frac{3}{11}$ Minutes past 9 O'clock

(*b*) $45\frac{6}{11}$ Minutes past 9 O'clock

(*c*) $45\frac{6}{11}$ Minutes past 9 O'clock

(*d*) $49\frac{6}{11}$ Minutes past 9 O'clock

58. At what time between 4 and 5 O'clock are the hands of a clock in the opposite directions?

(*a*) $52\frac{3}{11}$ minutes past 4 O'clock

(*b*) $54\frac{6}{11}$ minutes past 4 O'clock

(*c*) $51\frac{6}{11}$ minutes past 4 O'clock

(*d*) $53\frac{9}{11}$ minutes past 4 O'clock

59. The angle between the hands of a clock is 20° and the hour hand is in between 2 and 3. What is the time shown by the clock?

(*a*) 2 hours $7\frac{3}{11}$ minutes

(*b*) 2 hours $14\frac{6}{11}$ minutes

(*c*) 2 hours $15\frac{5}{11}$ minutes

(*d*) Both (a) and (b)

Directions for questions 60 to 72:

Select the correct alternative from the given choices.

60. The number of odd days in non-leap year is

(*a*) 0 (*b*) 1

(*c*) 2 (*d*) 3

61. How many odd days are there in 352 days?

(*a*) 1 (*b*) 2

(*c*) 3 (*d*) zero

62. The number of odd days for the year 1996 is

(*a*) 0 (*b*) 1

(*c*) 2 (*d*) 3

63. Which among the following years is a leap year?

(*a*) 3000 (*b*) 3100

(*c*) 3200 (*d*) 3300

64. What will be next leap year after 2096?

(*a*) 2100 (*b*) 2101

(*c*) 2104 (*d*) 2108

65. If 5th January 2001 was a Friday then 25th December 2001 was a

(*a*) Monday (*b*) Tuesday

(*c*) Wednesday (*d*) Thursday

66. If 21st march 2000 was on Monday, what day of the week will be 21st March 2003?

(*a*) Tuesday (*b*) Friday

(*c*) Thursday (*d*) Wednesday

67. If 14th February 2001 was a Wednesday then what day of the week will be 14th February 2101 (i.e. after a century)?

(*a*) Friday (*b*) Saturday

(*c*) Sunday (*d*) Monday

68. If in a calendar year, there are 541 days and 10 days a week then how many odd days will be there in that year?

(*a*) 1 (*b*) 2

(*c*) 3 (*d*) 4

69. If 8th February 1995 was a Wednesday then 8th February 1994 was on which day?

(*a*) Wednesday (*b*) Thursday

(*c*) Tuesday (*d*) Monday

70. If 11th August 1985 was a Sunday then which day of the week was 13th August 1986?

(*a*) Tuesday (*b*) Wednesday

(*c*) Thursday (*d*) Friday

71. If 1st January 2012 is a Sunday then which days of the week will the New Year is celebrated in 2016?

(*a*) Friday

(*b*) Sunday

(*c*) Wednesday

(*d*) Saturday

72. If holidays are declared only on Sundays and in a particular year 12th March is a Sunday, is 23rd September in that year a holiday?

(*a*) Yes

(*b*) No

(*c*) Yes, if it is a leap yea

(*d*) No, if it is a leap year

Direction for Q. 73 to 75:

Read the following information and answer the questions that follow.

Abhishek, Bharat, Chetan, Dinesh Eklvya and Fraklin are friends married to Richa, Stuti, Urmila, Varsha, Trupti and Wanda not necessarily is same order.

Following facts are known about them-

(1) Richa and stuti are Abhishek's sisters

(2) Neither Richa nor trupti are wives of Chetan

(3) Wanda is wife of Eklavya and Varsh is wife of Bharat

(4) Dinesh is not married to Richa, stuti or trupti

73. Who is Abhishek's wife?

(*a*) Richa (*b*) Urmila

(*c*) Trupti (*d*) cannot be determined

74. Richa is wife of

(*a*) Chetan (*b*) Franklin

(*c*) Bharat (*d*) Dinesh

75. Dinesh is married to

(*a*) Urmila (*b*) Barsha

(*c*) Trupti (*d*) Richa

Directions for Qu. 76 to 79

Read the following information and answer the questions that follow.

XYZ, an institute involved in training the students for MBA, MCA, MMS and other entrance exams runs six batches A, B, C, D, E and F, each of which is scheduled of at least once a week.

(1) Sunday is a holiday.

(2) The B batch meets all days except Friday and Saturday.

(3) The C batch meets 4 days in succession.

(4) The F batch meets only from Monday to Thursday.

(5) The E batch is scheduled every day, but not on Thursday and Saturday.

(6) On alternate days, the A batch is scheduled.

(7) The A and D batches never meet on the same days.

(8) The C batch does not meet on Mondays and Tuesday.

(9) The D batch is scheduled only once a week on either Wednesday or Friday.

76. Which batches meet for the same number of classes during the week?

(*a*) B, A, F only (*b*) E, B, C only

(*c*) E, F, A only (*d*) None of these

77. If a certain class of D is scheduled on the same today as that of B batch, then how many batches meet on Friday?

(*a*) 1 (*b*) 2

(*c*) 3 (*d*) 4

78. If there is one such day that all batches are to meet, then which batch would gain an additional day?

(*a*) A (*b*) B

(*c*) D (*d*) E

79. Which two batches can never be conducted on the same day?

(*a*) C and D (*b*) C and E

(*c*) A and D (*d*) None of these

Directions for Qu. 80 to 82

Read the following information and answer the questions that follow.

Abhita is bright girl who has a CD box on her study table. The box has three horizontal row each can be filed with 4 CD cases. She has also put her "Jewellery box" in the CD box which consume the space of two CD ceases.

The following information is known

(1) The jewellery box is in the bottom row.

(2) The CD cases with "Window XP" and "Windows 2000" are horizontally adjacent to each other.

(3) The CD case with Linux is in the bottom row.

(4) "Utility tools" is placed horizontally adjacent to "Game1" CD.

(5) A vertical columns containing "Linux" and "music" also contains "Movies"

(6) "Net tools" is just below the "Windows XP" and "CS II" is just above "Game 2"

(7) "Net tools" is not at the corner.

80. Which CD is just below "Net tools"?

(*a*) Utility tools (*b*) Games 1

(*c*) CS II (*d*) None of these

81. If "Movie" is in second row, then how many CDs are between "Movies" and "Game2"?

(*a*) 0 (*b*) 1

(*c*) 2 (*d*) Cannot be determined

82. If "Music is in top row, how many CDs are between "Music" and "Utlity tools".

(*a*) 0 (*b*) 1

(*c*) 2 (*d*) Cannot be determined

Directions:

In each of the question below are given two statements followed by two conclusions numbered I and II. You have to take the two given statements to be true even if they seem to be at variance from commonly known facts and decide which of the given conclusion (s) logically follow (s) from the two given statements, disregarding commonly known facts. Given answer as

(*a*) If only conclusion I follows

(*b*) If only conclusion II follows

(*c*) If either I or II follows

(*d*) If neither I nor II follows

(*e*) If both I and II follow

83. Statements:

(1) Some players are singers.

(2) All singers are tall.

Conclusions:

I. Some players are tall.

II. All players are tall.

84. Statements:

(1) Some vegetables are fruits.

(2) No fruit is black.

Conclusions:

I. Some fruits are vegetables.

II. No vegetable is black.

85. Statements:

(1) Some stone are bricks.

(2) Some bricks are trees.

Conclusions:

I. Some stones are trees.

II. Some trees are bricks.

86. Statements:

(1) All tigers are ships.

(2) Some ships are cupboards.

Conclusions:

I. Some tigers are cupboards.

II. Some cupboards are tigers.

87. Statements:

(1) Some stones are cups.

(2) Some cups are black.

Conclusions:

I. Some blacks are not cups.

II. Some cups are stones.

Directions: Each question is followed by two statements giving certain data. You have to Decide whether the information provided in the statements is sufficient for answering the question.

Choose A : if the question can be answered by using one of the statements alone, but cannot not be answered by using the other statement alone.

Choose B : if the question can be answered by using either statement alone.

Choose C : if the question can be answered by using both statements together, but cannot be answered by using either statement alone.

Choose D : if the question cannot be answered even by using both the statements together.

88. What is the age of Ramesh?

1. The average age of Ramesh, Ramu and Ravi is 76 years.
2. Ramesh is three years younger to both, Ramu and Ravi.

89. Is (n – 1) even, given that n is a prime number?

1. (n – 2) is a whole number.
2. (n + 1) is an odd number.

90. How many coins are there in the purse that has coins in the denominations of 25 paisa, 50 paisa and one rupee?

1. The value of the coins in the respective denominations is in the ratio 1:2:5.
2. The total amount in the three denominations is 64 Rs.

91. What is the monthly salary of Q?

1. Q earns 6,000 Rs more than R who earns 3,000Rs less than P.
2. The total monthly salary of P and Q is 27,000.

92. Does Raju makes Profit on selling two articles?

1. On one article Raju makes a profit of 10% and on the other Raju incurs a loss of 10%.
2. Both the articles are sold at the same price.

NUMERICAL TYPE QUESTIONS

1. Given two numbers x and y, define

Addition	A(x, y) = x + y
Subtraction	S (x, y) = x – y
Multiplication	M (x, y) = x, y
Division	D (x, y) = x/y

If x = 2, y = 3 and z = 5, then value of M[A(S (x, y), M (y, z)), S (A(x, z), S (z, y))] is ____

Directions Q. 2 – 3 :

These questions are based on the following data

A, B, C, D are standing in a queue. It is known that C and D are not standing adjacent to each other and that B is not in the third place.

2. If A is not in the second place and B is in the first place, then which is the place for D is ____

3. If A is not standing adjacent to B and D is the last in the queue, then the place for C is ____

Directions Q. 4 – 5 :

A boy is asked to put in a basket one mango when ordered 'One', one orange when ordered 'Two', one apple when ordered 'Three' and is asked to take out from the basket one mango and an orange when ordered 'Four'. A sequence of orders is given as 1 2 3 3 2 1 4 2 3 1 4 2 2 3 3 1 4 1 1 3 2 3 4

4. The ______total organges were in the basket at the end of the above sequence.

5. The ______total fruits will be in the basket at the end of the above order sequence.

Directions Q. 6 – 7

Each of the 11 letters A, H, I, M, O, T, U, V, W, X and Z appears same when looked at in a mirror. They are called symmetric letters. Other letters in the alphabet are asymmetric letters.

6. The______ number of four-letter computer passwords can be formed using the symmetric letters (no repetition allowed)

7. The ______ number of three-letter computer passwords can be formed (no repetition allowed) with at least one symmetric letter.

Directions Q. 8 – 9 :

A string of three English letters is formed as per the following rules:

(*a*) The first letter is any vowel

(*b*) The second letter is m, n or p.

(*c*) If the second letter is m then the third letter is any vowel which is different from the first letter

(*d*) If the second letter is n then the third letter is *e* or *u*

(*e*) If the second letter is p then the third letter is the same as the first letter.

8. The ______ strings of letters can possible be formed using the above rules.

9. The ______ strings of letters can possibly be formed using the above rules such that the third letter of the string is e.

ANSWERS

MCQ Type Questions

1. (*b*)	**2.** (*b*)	**3.** (*b*)	**4.** (*c*)	**5.** (*b*)	**6.** (*b*)	**7.** (*b*)	**8.** (*d*)	**9.** (*a*)	**10.** (*a*)
11. (*b*)	**12.** (*d*)	**13.** (*d*)	**14.** (*a*)	**15.** (*b*)	**16.** (*c*)	**17.** (*b*)	**18.** (*d*)	**19.** (*d*)	**20.** (*b*)
21. (*b*)	**22.** (*d*)	**23.** (*a*)	**24.** (*c*)	**25.** (*a*)	**26.** (*d*)	**27.** (*b*)	**28.** (*a*)	**29.** (*c*)	**30.** (*b*)
31. (*a*)	**32.** (*b*)	**33.** (*b*)	**34.** (*a*)	**35.** (*b*)	**36.** (*d*)	**37.** (*a*)	**38.** (*c*)	**39.** (*a*)	**40.** (*d*)
41. (*a*)	**42.** (*a*)	**43.** (*d*)	**44.** (*c*)	**45.** (*a*)	**46.** (*a*)	**47.** (*b*)	**48.** (*c*)	**49.** (*a*)	**50.** (*c*)
51. (*b*)	**52.** (*c*)	**53.** (*a*)	**54.** (*c*)	**55.** (*c*)	**56.** (*d*)	**57.** (*c*)	**58.** (*b*)	**59.** (*d*)	**60.** (*b*)
61. (*b*)	**62.** (*c*)	**63.** (*c*)	**64.** (*c*)	**65.** (*b*)	**66.** (*c*)	**67.** (*d*)	**68.** (*a*)	**69.** (*c*)	**70.** (*b*)
71. (*a*)	**72.** (*b*)	**73.** (*c*)	**74.** (*b*)	**75.** (*a*)	**76.** (*d*)	**77.** (*b*)	**78.** (*a*)	**79.** (*c*)	**80.** (*d*)
81. (*c*)	**82.** (*b*)	**83.** (*a*)	**84.** (*a*)	**85.** (*b*)	**86.** (*d*)	**87.** (*b*)	**88.** (*c*)	**89.** (*a*)	**90.** (*c*)
91. (*c*)	**92.** (*b*)								

Numerical Type Questions

1. 70	**2.** 105	**3.** 2	**4.** 2	**5.** 11	**6.** 7920	**7.** 12870	**8.** 35	**9.** 10

EXPLANATIONS

MCQ TYPE QUESTIONS

1. S[M(D(A(a, b)2, D(A(a, b)2)), M(D(S(a,b)2)) D (s(a,b)2)]

$$= S\left[\left|M\frac{a+b}{2},\frac{a+b}{2}\right|\left|M\frac{a-b}{2};\frac{a-b}{2}\right|\right]$$

$$= \left(\frac{a+b}{2}\right)^2 - \left(\frac{a-b}{2}\right)^2 = ab$$

6 – 7.

A → Lawyer → Male
B → Housewife → Female
C → Accountant → Male
D → Housewife → Female
E → Architect → Male
F → Lecturer → Female
Married Couple : AD & CF

8—9. The seating arrangement is shown below:

Left 7	6	5	4	3	2	Right 1
D/C	E/F	F/E	B	C/D	G/A	A/G

10 – 13.

Level of Dissimilarity is equal to Maximum difference in ranks alloted to any two countries on any of the five traits.

For example, dissimilarity between India and China.

Trait	India'sRank	China's Rank	Difference
C	1	2	1
P	2	3	1
N	3	1	2
V	4	5	1
D	5	4	1

From the above table, maximum difference is 2.
So dissimilarity level between India and China is 2.

10. Calculating the level of dissimilarity for India Vs remaining countries.
China : For the parameter N (3 – 1) = 2
Japan : For the parameter D (5 – 1) = 4
Malaysia : For the parameter C (4 – 1) = 3
Thailand : For the parameter V (4 – 1) = 3
Hence least level of dissimilarity is 2 which is for China.

11. From above calculations, highest level of dissimilarity is 4 which is for Japan.

12. *The level of dissimilarity between*
China and Japan is for the parameter D (4 – 1) = 3
India and China is for the parameter N (3 – 1) = 2
Malaysia and Japan is for the parameter V (4 – 1) = 3
Thailand and Japan is for the parameter D (5 – 1) = 4

13. *The level of dissimilarity between*
Malaysia and China is for the parameter V (5 – 1) = 4
China and Thailand is for the parameter V (5 – 1) = 4
Thailand and Japan is for the parameter D (5 – 1) = 4
Japan and Malaysia is for the parameter V (4 – 1) = 3

14 to 17 :

The given set of questions is based on distribution and sequencing.

Firstly, let us put the names of women and men,
Women : Shanthi, Sridevi, Joya, Anita
Men : Sunil, Anil, Raj, Raman

It is given that(2 kids family) ←(came before)— (No kid family)

Also Shanti (0 kids) —(came before)→ Sridevi

As Sunil and his wife came with their only kid, hence Sunil cannot be the husband of Shanthi.

As Anil and Raj are fathers, hence neither of them is the husband of Shanthi. Hence <u>Raman is the husband of Shanthi,</u>

Also, Joya came before Shanthi and met Anita on the venue, this gives the following order of their reaching the venue. [Anita, Joya, Shanthi, Sridevi]

As Sunil reached last, <u>Sunil is the husband of Sridevi</u>. This means that <u>Raj is Joya's husband and Anil is Anita's husband.</u>

Following table gives the desired details

Order	Women	Men	No. of Kids
1	Anita	Anil	at least 1
2	Joya	Raj	2
3	Shanthi	Raman	0
4	Sridevi	Sunil	1

14. Shanthi arrived third.

15. Out of the given choices. Sunil and Sridevi is the correct pair.

16. The daughters of Anita and Sridevi go to the same school. Anita's husband is Anil and Sridevi's husband is Sunil. Hence, Sunil and Anil is the correct answer.

17. Raj's family consists of two kids for certain (Anil has at least one kid, hence could be having exactly one kid also).

18 to 20 :

From statements (vii), (viii) and (ix), we get

P U S
Orange
R Q
Yellow Green

Then, T is opposite S and the colour of S is red.

From (x), the colour of house P is white, hence the colour of house T is blue.

P	U	S
White	Orange	Red
R	Q	T
Yellow	Green	Blue

Now, from (v) and (x) , we get

T > S, Q > P > R

From (vi), U is the shortest.

i.e., the order of houses in the descending order of their heights is T, S/Q, Q/S, P, R, U.

18. R is yellow coloured house and the house diagonally opposite R is S. And the colour of house S is Red.

19. The second tallest house is either S or Q.

20. The tallest house is T and its colour is blue.

21. If all the five cuts are made parallel to one face, the number of pieces obtained is 5 + 1 = 6

22. If two cuts are made parallel to each perpendicular plane, then the number of pieces obtained is $3^3 = 27$

23. $(2 + 1) \times (3 + 1) \times (4 + 1) = 3 \times 4 \times 5 = 60$

Identical pieces are formed

24. $(n + 1)^3 = 216$ (n = number of cuts parallel to each face) $\Rightarrow n + 1 = 6$

$n = 5$

$\therefore$ 3n = 15 cuts are required

25. $24 = 2 \times 3 \times 4 = (1 + 1)(2 + 1)(3 + 1)$

$\therefore$ 1 + 2 + 3 = 6 cuts are required

Solutions for questions 26 to 28:

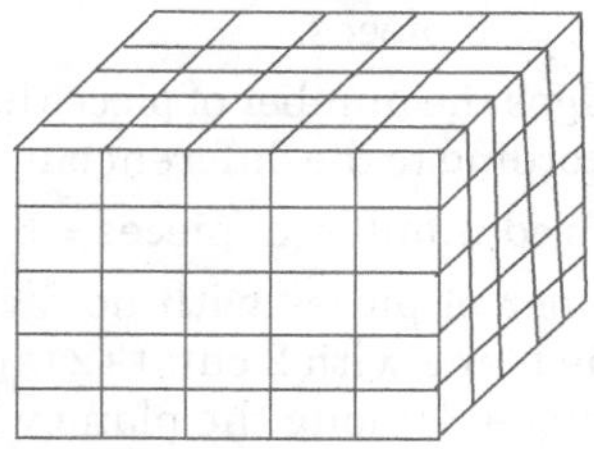

Here $n^3 = 125$

$\Rightarrow n = 5$

26. Cubes with no face painted:

$(n - 2)^3 = (5 - 2)^3 = 27$

27. Cubes with one face painted:

$6 \times (n - 2)^3 = 6(3)^3 = 54$

28. Cubes with two face painted:

$12(n - 2) = 12(3) = 36$

Solutions for questions 29 to 31:

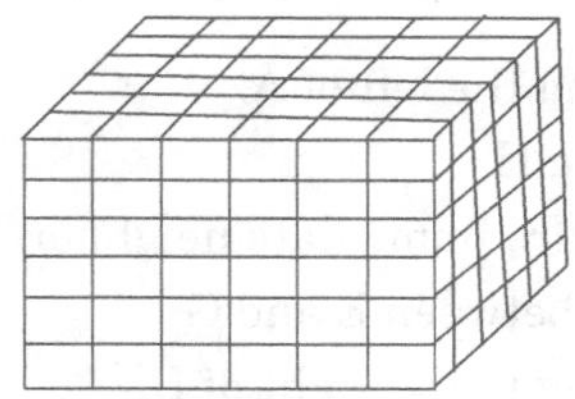

Here $n^3 = 216$

$\Rightarrow n = 6$

29. Cubes with exactly one color

$= (n - 2)^3 = 4^3 = 64$

30. Cubes with exactly two faces painted pink

$= 12(n - 2) = 12 \times 4 = 48$

31. The cubes at the corners have exactly three faces painted blue i.e., 8 cubes

Solutions for questions 32 and 34:

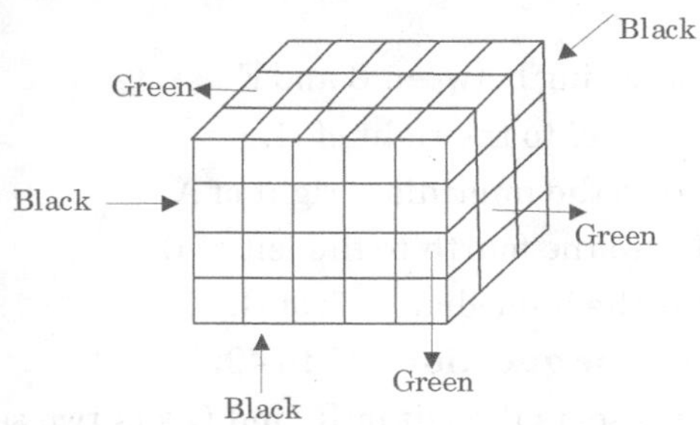

Comers (8)	**Edges(12)**	**Faces (6)**
BBB-1	BB-3	B-3
GGG-1	GG-3	G-3
GGB-3	BG-6	
BBG-3		

32. At the corners, three such pieces are there.

On the edges, the number of such pieces = 1 (along the plane with 2 cuts) + (along the plane with 3 cuts) + 3 (along the plane with 4 cuts).

$\therefore$ Required number of pieces = 3 + 6 = 9

33. Except two corners, all the corners have both the colors $\Rightarrow$ 6 pieces.

On the edges the number of pieces having both the colors according to the different cuts in each plane.

$\therefore$ Required number of pieces = 6 + 12 = 18

34. The number of pieces with no face painted = 1 (along the plane with 2 cuts) +2 (along the plane with 3 cuts) + 3 (along the plane with 4 cuts)

Solution for questions 35 to 39:

All information can be drawn in a circle

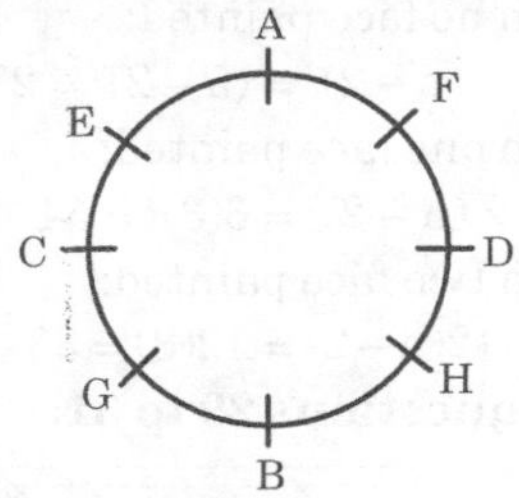

35. F is sitting the left of A.

36. E is facing H

37. G and H are immediate neighbors' of B

38. C sitting between E and G

39. F is sitting to the right of D.

Solution for questions 40 to 44:

All information can be drawn in a circle

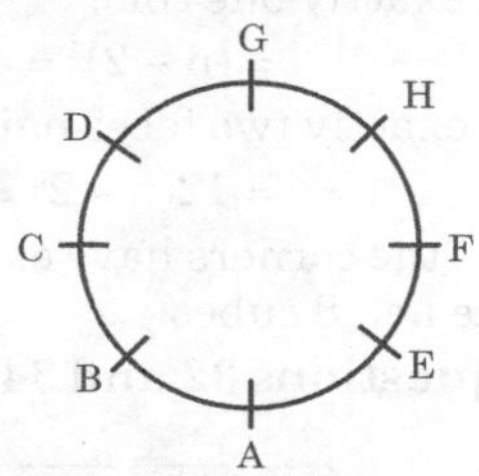

40. A is sitting between B and E.

41. C is thief to the right of H.

42. E is to the immediate right of A.

43. H is to the fourth to the left of B.

44. H is the immediate left of G.

Solution for questions 45 to 49:

P sits two seats the left of R, and Q sits two seats to the right of R. We can represent this information in the diagram below.

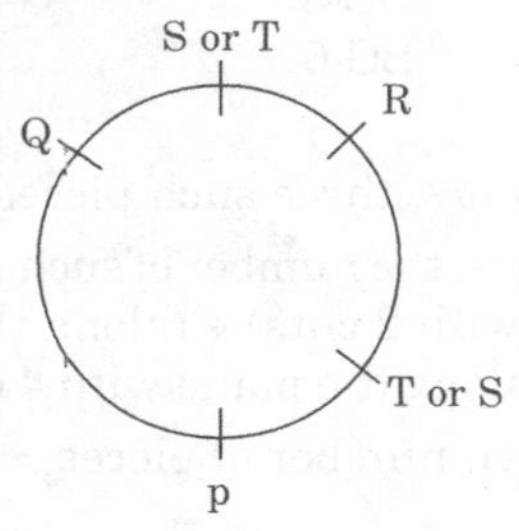

45. If S sits between Q and R, then the arrantgements is as follows

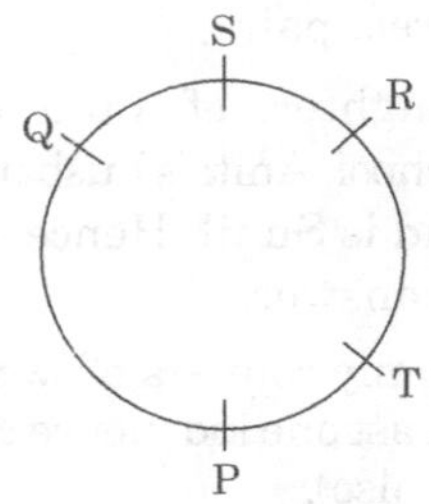

As can be seen from the diagram, T is to the immediate right of P.

46. We will take each choice and see whether it fits in the arrangement that we represented through a diagram in the analysis of the data (the same diagram is reproduced below).

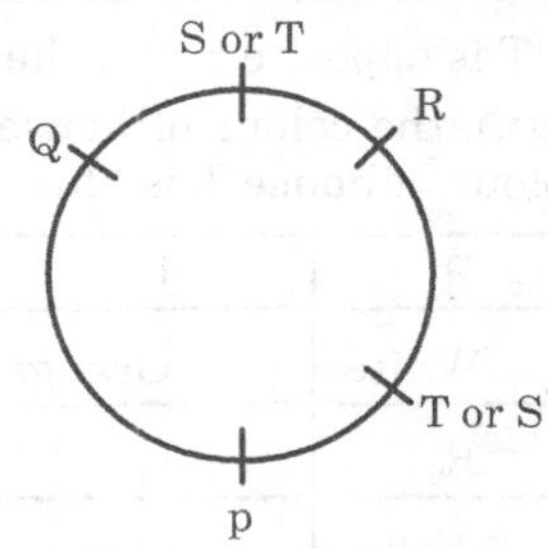

We can see that the arrangement given in choice (a) is not possible and hence the answer choice is (a)

47. If S is not next to Q, then the seating arrangement is fixed as follows

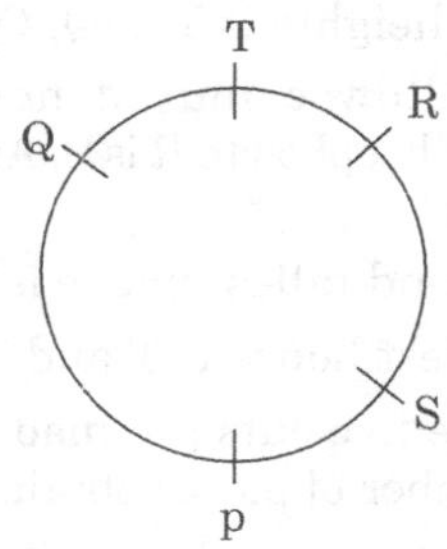

Now, P between Q and S.

48. On the basis of the diagram that we drew , we find that to accommodate U we have to create a new space between P and Q

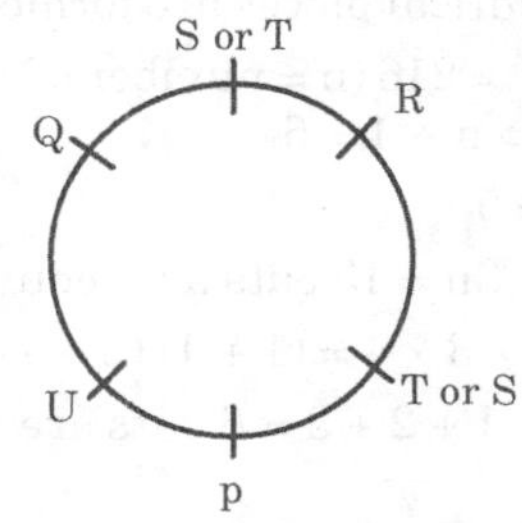

Hence, Choice (c) is the correct answer.

49. We create a new slot for the sixth person B. But since U will not sit next to P, S or T, he will have to sit between R and Q. The arrangement will then look as follows:

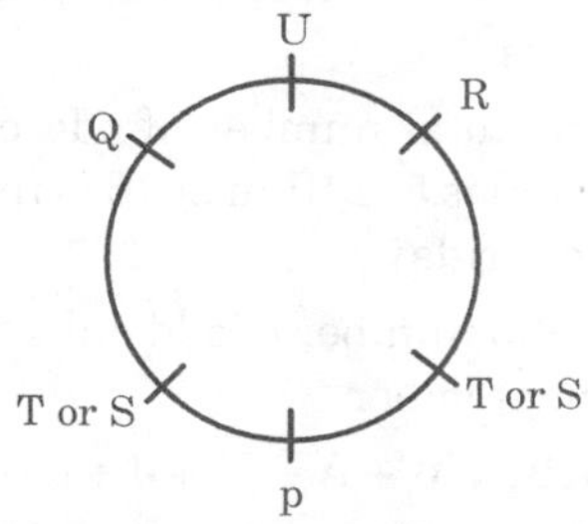

As we can see from the diagram, the neighbors of P will be T and S.

50. The angle covered by the minute hand in 22 minutes is $22 \times 6° = 132°$

51. The hour hand covers 360° in 12 hours.

$\therefore$ It covers $\frac{1°}{2}$ in one minute. In quarter of an hour i.e., in 15 minutes the hour hand will move $15 \times \frac{1°}{2} = 7.5°$

52. The hour hand will move by 6° in 12 minutes, so minute hand will move $12 \times 6° = 72°$ in 12 minutes, as the minute had moves by 6° in one minute

53. Angle will be $\theta = \left(\frac{11}{2}m - 30\,h\right)$

$$\left(\frac{11}{2} \times 20 - 30 \times 6\right) = 40°$$

54. The angle between the hands will be

$\theta = \left|\frac{11}{2}m - 30\,h\right|$ Here h = 3 and m = 14

$$\theta = \frac{11}{2} \times 14 - 30 \times 3$$

$$\theta = |77 - 90| = 13°$$

55. Angle between two hands is given by

$\theta = \left|\frac{11}{2}m - 30h\right|$ Here h = 7 and m = 25

$$\theta = \left|\frac{11}{2} \times 25 - 30 \times 7\right| = \frac{275 - 210}{2} = \frac{145°}{2} = 72\frac{1°}{2}$$

56. Angle between the two hands is given by

$\theta = \left|\frac{11}{2}m - 30h\right|$ here m = 20 and h = 11

$$\Rightarrow \theta = \left|\frac{11}{2} \times 20 - 30 \times 11\right| = 220$$

As angle is more than 180° the angle must be $360° - 220° = 140°$

57. When hands coincide with each other the angle between them is 0. Therefore angle between hands is given by

$$\theta = 30 - \frac{11}{2}m\left(Q30h > \frac{11}{2}m\right) \text{here } h = 9$$

$$270 \times \frac{2}{11} = m$$

$$\therefore \quad m = 49\frac{1}{11} = 12 \text{ minutes past 9 hours.}$$

58. When hands of a clock are in opposite direction the angle between them is 180°

Therefore $\theta = \left|\frac{11}{2}m - 30h\right|$ where θ

= 180° and h = 4

$$180 = \frac{11}{2}m - 120$$

$$\frac{11}{2}m = 300$$

$$m = \frac{600}{11}m = 54\frac{6}{11} \text{minutes}$$

So, at 54 minutes past 4 the hands are in opposite direction.

59. Given θ = 20° and h = 2

$$\theta = \frac{11}{2}m - 30h \text{ or } 30h - \frac{11}{2}m$$

$$20 = \frac{11}{2}m - 30 \times 2; \frac{11}{2}m = 80$$

$$m = \frac{160}{11} = 14\frac{6}{11} \text{minutes}$$

$$\text{or} \quad 20 = 30 \times 2 - \frac{11}{2}m; \frac{11}{2}m = 40$$

$$m = \frac{80}{11} = 7\frac{3}{11} \text{minutes}$$

Therefore the angle between the hands will be 20° at hors $14\frac{6}{11}$ minutes past 2 and $7\frac{3}{11}$ minutes past 2.

60. Total numbers of days in an ordinary year are 365. Number of odd days

$$\frac{365}{7} = 52 \text{ weeks} + 1 \text{ odd day}$$

Therefore number of odd days is 1.

61. The number of days in 352 days

$= \frac{352}{7}$

= 50 weeks + 2 odd days.

Hence, the total number of odd days is 2

62. The year 1996 is a leap year. Total number of days in a leap year are 366 days. Number of odd

days $= \frac{366}{7} = 52$ weeks + 2 odd days.

63. Century years which are divisible by 400 are leap years. As 300 and 3300 are not divisible by 400 they are not leap years, but 3200 is a leap year.

64. For a century year to be a leap year it should be divisible by 400. As 2100 is not divided by 400 it is not a leap year. The next leap year is 2104

65. Total numbers of odd days from 5th January 2001 to 25th Dec 2001 are

Months: Jan + Feb + Mar + Apr + May + June + July + Aug + Sep + Oct + Nov + Dec

Odd days: 26 + 0 + 3 + 2 + 3 + 2 + 3 + 3 + 2 + 3 + 2 + 25 =74

$\frac{74}{7} = 10$ Weeks + 4 odd days

4 days from Friday is Tuesday.

66. 21st March, 2000 is Monday and the year is a leap year. So, none of the next 3 year is the leap year. So the day of the week will be 3 days beyond Monday i.e. Thursday

67. A century has 5 odd days

$\therefore$ 5 day beyond Wednesday, Thursday, Friday, Saturday, Sunday, Monday.

Hence 14th February 2101 will be on Monday

68. To find the number of odd days, we have to find remainder of $\frac{541}{10}$ the reminder is one. Hence there is one odd day.

69. 1994 is not a leap year.

$\therefore$ It has only 1 odd day.

$\therefore$ 8th Feb 1995 is one day before Wednesday.

Hence 8th Feb 1994 is a Tuesday

70. It is given that 11th August 1985 was Sunday then 13th August 1985 is Tuesday. 1985 is not a leap year hence it has only one odd day. So 13th August 1986 is one day after Tuesday ie. Wednesday

71. The total number of years from 2012 to 2016 is four out of which 2013, 2014 and 2015 are non leap years. Hence there is 1 odd day in each of these years 2012 is a leap year , hence it has 2 odd days.

$\therefore$ The total number of odd days in these four years is 5. 1st Jan 2016 is five days to Sunday i.e. Friday

72. The total number of odd days from 12th March to 23rd September

Months : M + A + M + J + J + A + S

Odd day : 5 + 2 + 3 + 2 + 3 + 3 + 2 = 20days.

$\frac{20}{7} = 6$ odd day

Hence 23rd September is 6 days to Sunday ie. Saturday, so 23rd September is not a holidays.

Solution for 73 to 75:

From the given information we can conclude Richa and Stuti are not Abhishek's wives Further Wanda and Varsha are wives of Eklavya and Bharat respectively. In this manner we analyses the information.

Wife / Husband	Richa	Stuti	Trupti	Urmila	Varsha	Wanda
Abhishek	×	×				
Bharat					✓	
Chetan	×		×			
Dinesh	×	×	×			
Eklavya						
Franklin						

Now as Varsha and Wanda are married to Bharat and Eklavya they cannot be wives of anyone else Further, as every husband has one wife.

Dinesh is married to Urmila, Chetan, is married to stuti, Abhishek is married to Trupti and Franklin is married to Richa.

Wife / Husband	Richa	Stuti	Trupti	Urmila	Varsha	Wanda
Abhishek	×	×	✓	×	×	×
Bharat	×	×	×	×	✓	×
Chetan	×	✓	×	×	×	×
Dinesh	×	×	×	✓	×	×
Eklavya	×	×	×	×	×	✓
Franklin	✓	×	×	×	×	×

Now all the question can be answered

Solutions for 76 to 79:

On the basis of the given information we get the following schedule:

Batch	Mon	Tue	Wed	Thu	Fri	Sat
A	×	✓	×	✓	×	✓
B	✓	✓	✓	✓	×	×
C	×	×	✓	✓	✓	✓
D	×	×		×		×
E	✓	✓	✓	×	✓	×
F	✓	✓	✓	✓	×	×

Here Batch D meets either on Wednesday or on Friday

Solution for questions 80 to 82:

CS II	UT or Game 1	UT or Game 1	Music or Movies
Game 2	Win 2000	Win XP	Music or Movies
Jewellery Box		Net tools	Linux

Case 2

Music or movies	UT or Game 1	UT or Game 1	CS II
Music or Movies	Win XP	Win 2000	Game 2
Linux	Net Tools	Jewellery Box	

Case 3

83. Statements:

1. Some players are singers
2. All singers are tall

Conclusion: Some players are tall.

84. Statements:

1. Some vegetables are fruit
2. No fruit is black

Conclusion: Some fruits are vegetable.

85. Statements:

1. Some stones are bricks
2. Some bricks are trees

Conclusion: Some trees are bricks

86. Statements:

1. All tigers are ship
2. Some ships are cupboards

Conclusion: No conclusion

87. Statements:

1. Some stones are cups
2. Some cups are black

Conclusion: Some cups are stones

88. From statement I, we have

Ramesh + Ramu + Ravi = 228 years

But from this we cannot find any one of their ages.

From statement II, we have

Ramesh = Ramu-3 and Ramesh = Ravi-3

Again to find Ramesh's age,we should know either Ramu's or Ravi's age.

Using both the statements, we have

Ramesh + Ramesh + 3 + Ramesh + 3 = 228

∴ Ramesh age is 74years

89. If n is a prime number, then n is either even prime (2) or odd prime (in all other cases). So n-1 is either even prime or odd. Using statement 2 alone, as n-1 is odd so n-1 is not an even number.

90. Using statement 1 alone let the total value of the each variety of coin be Rs. x, Rs. 2x and Rs. 5x.

Using statement 2 alone is not sufficient.

Combining both the statements,

$$x + 2x + 5x = 64 \Rightarrow x = 8$$

The value of each variety of coins is Rs. 8, Rs. 16 and Rs. 40

Since the denominations and the total value of the coins I each denomination is known, the number of each dominations can be found out. Then, we can find the total numbers of coins.

91. Statement 1 alone is not sufficient as the relation between the CP of the two articles or the SP of the two articles is not known. Statement 2 alone is not sufficient as the relation between the CP of the two articles or the SP of the two articles is not known. Using both the statements, as both the article are sold at the same price and at equal percent of profit and loss on the two respectively, loss is made on selling an article. Hence question can be answered using both the statements.

92. From statement 1, ABCD may be square or rhombus, but since all the squares are rhombus, therefore statement 1 alone is sufficient to answer the question.

From statement 2, as $AC \perp BD$ and $AC^2 \neq BD^2$ and opposite sides are equal.

∴ The quadrilateral is a rhombus.

∴ Statement 2 alone is sufficient to answer the question.

NUMERICAL TYPE QUESTIONS

1. M[A(S(x, y), M(y, z)), S(A(x, z), S (z, y))]
= M[A((2 – 3), (3, × 5)), S((2 + 5), (5 – 3))]
= M[A(–1, 15), S(7, 2)]
= M[14, 5] = 70.

2. Since B is in 1, and C and D are not adjacent, the only places possible for them are 2 and 4.

3. Since B is not in the third place and A is not adjacent to B, and D is last in the queue. The position of A will be beside D and B will be at the other corner. C will be between B and A. Hence the positions will be BCAD. Place of C is 2, hence choice (b)

The sitting position is as follows.

White	Green	Red	Yellow
Maharashtra	U.P.	West Bengal	A.P.

From the above table,
Ms. Andhra Pradesh ⟶ Yellow

4. (1) (2) 3 3 2 1 4 (2) 3 (1) 4 (2) (2) 3 3 (1) 4 1 (1) (3) (2) 3 4
Total no. oranges were in basket = 2

5. Total fruits in the basket = 11

6. Total number of four letter password formed
$= {}^{11}C_4 \times \angle 4 = 11 \times 10 \times 9 \times 8 = 7920$

7. Required number of passwords
$= {}^{26}C_3 \times \angle 3 - {}^{15}C_3 \times \angle 3$
$= 26 \times 25 \times 24 - 15 \times 14 \times 13 = 12870$

8. [ ] [ ] [ ]
1st Letter 2nd letter 3rd letter

There are 5 vowels, a, e, i, o,, u,.
hence 1st letter can be any of the 5 vowels ;
∴ 5 different methods. ...(i)
The second letter can be any of the m, n, or p. But each has its own rule for the third letter. Hence, each case to be considered separately.

[1 Vowel] [m]

In the above case; third letter is one of the remaining 4 vowels
Hence there are 5 × 4 = 20 ways of forming such a string. ...(ii)
If the second letter is n, the third is either e or u, $i.e.$ 2 different ways.
∴ total of 5 × 2 ways = 10 ways ...(iii)
When 2nd position is filled by P, the third letter is same as the first letter. As the 1st letter is already selected, third can be filled only one way.
Hence, there are 5× 1 = 5 ways ....(iv)
From equations (ii), (iii) and (iv),
Total number of strings that can be formed
= 20 + 10 + 5 = 35

9. Third letter find is 'e'.
Once again considering each case of the second position being filled by m, n or p, we have

[1st position] [n] [e]
2nd position 3rd position

The 3rd position shall be different form vowel in first position, $i.e.$ vowel for 1st position can be selected out of the vowels a, i, o, u only.
Hence 4 different ways ....(i)
When 2nd possible is filled by n, occupying 2nd position, does not create any restriction on the vowel is position 1.

[e] [n] [ ]
3rd 2nd 1st

Hence, 1st position can be occupied by any of the 5 vowels.
Hence, 5 different ways ....(ii)
When 2nd position is filled by p,

[e] [p] [ ]
3rd 2nd 1st

In this case, letters in 3rd position out 1st position shall be the same.
Hence, 1st position shall be filled by 'e'.
$i.e.$ only one way of filling ....(iii)
Total number of possible strings = 4 + 5 + 1 = 10

Spatial Aptitude

1

CHAPTER

Spatial Aptitude

TRANSFORMATIONS

In a plane, you can slide, flip, turn, enlarge, or reduce figures to create new figures. These corresponding figures are frequently designed into wallpaper borders, mosaics, and artwork. Each figure that you see will corresopnd to another figure. These corresponding figures are formed using transformations.

A **transformation** maps an initial image, called a preimage, onto a final image, called an image.

The geometric **transformation** is a bijection of a set that has a geometric structure by itself or another set. If a shape is transformed, its appearance is changed. After that, the shape could be congruent or similar to its preimage. The actual meaning of transformations is a change of appearance of something.

Types of Transformations

When the size of a shape is increased or reduced then the image of the shape will be similar to the pre-image. The **similar figures** have dimensions equal in proportion.

But in the case of congruent, the transformation of objects is done by using rotation, reflection or translation. The shape is turned or flipped to transform into another shape.

There are four major types of transformations namely:

- Rotation
- Translation
- Dilation
- Reflection

Rotation

The *rotation* transformation leaves the transformed object the same size, but turned relative to the original position.

The rotation is controlled by a *base point* (which does not move) and an angle. As illustrated in the figure, if the base point is away from the object being transformed, the effect of the rotation may well include "movement" of the original shape.

In three dimensional rotations, the angle is often (but not always) measured in one of the primary coordinate system planes.

Remember, when performing multiple rotations, that Rotation "A" followed by Rotation "B" does not produce the same result as Rotation "B" followed by Rotation "A".

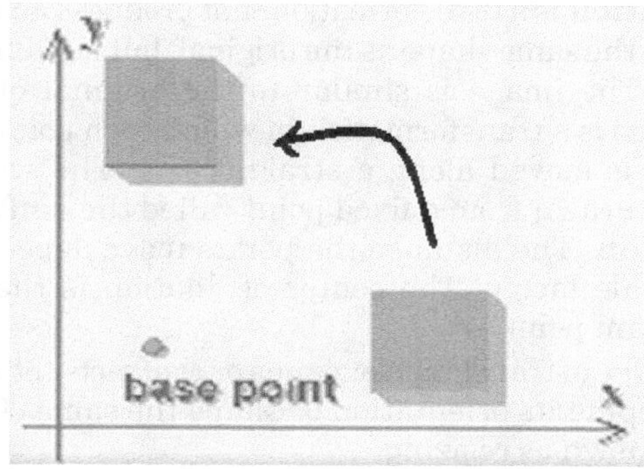

This type of transformation has an object about a fixed point without changing its size or shape.

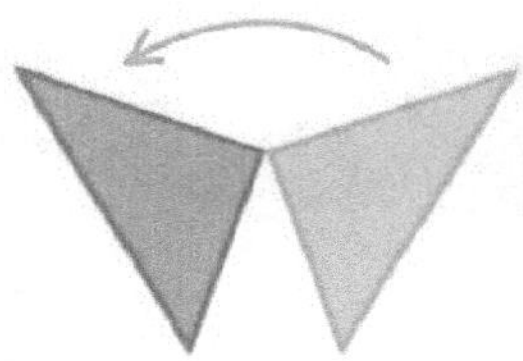

In the above figure, you can see, that the shape is rotated to form its image.

To describe you need:

- **direction** (clockwise or counterclockwise)
- **degree**
- **center point of rotation** (this is where compass point goes)

Translation

This type of translation is defined as moving the object in space by keeping its size, shape or orientation constant. In a translation, each point of the shape must be moved in the same direction and for the same distance. When you are doing a translation, the primary object is called the pre-image, and the object after the translation is called the image.

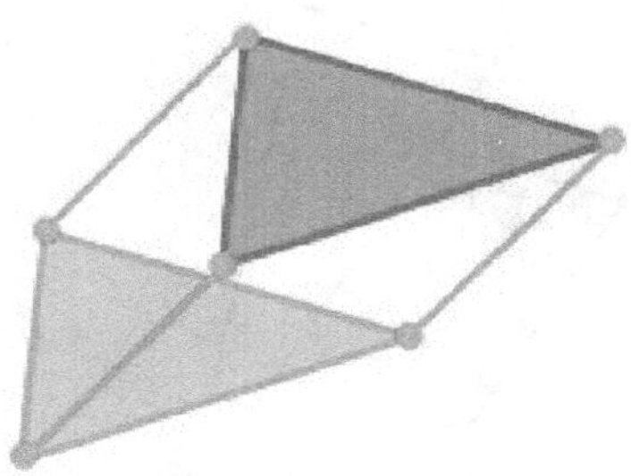

To describe you need:

- **direction** (left/right/up/down)
- **magnitude** (number of units)

Coordinate Notation:

$$(x, y) \rightarrow (x \pm a, y \pm b)$$

Dilation

A **dilation** is a transformation that produces an image that is the same shape as the original, but is a different size. (The image is similar to the original object). Dilation is a transformation in which each point of an object is moved along a straight line. The straight line is drawn from a fixed point called the **center of dilation**. The distance the points move depends on the scale factor. The center of dilation is the only invariant point.

This type of translation expands or contracts the object by keeping its orientation or shape the same. This is also known as resizing.

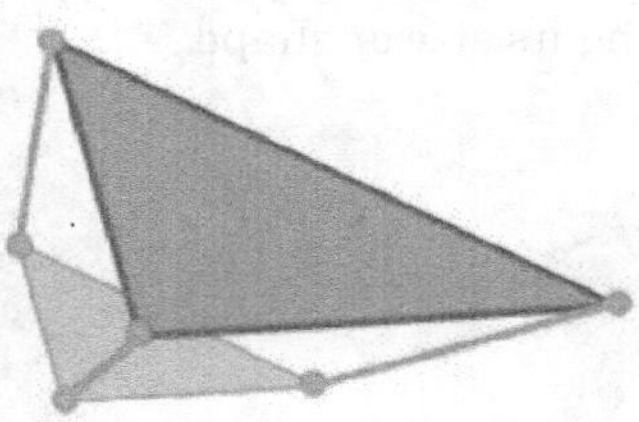

$$\text{Scale factor} = \frac{\text{image length}}{\text{original length}}$$

$$= \frac{\text{distance of image from center of dilation}}{\text{distance of object from center of dilation}}$$

If the scale factor is greater than 1, the image is an **enlargement**.

If the scale factor is between 0 and 1, the image is a **reduction**.

Example:

The figure shows two similar triangles PQR and P'Q'R'. Triangle P'Q'R' is a dilation of triangle PQR. We say that triangle PQR is transformed onto triangle P'Q'R' by a dilation with center at O and scale factor $\frac{OP'}{OP}$

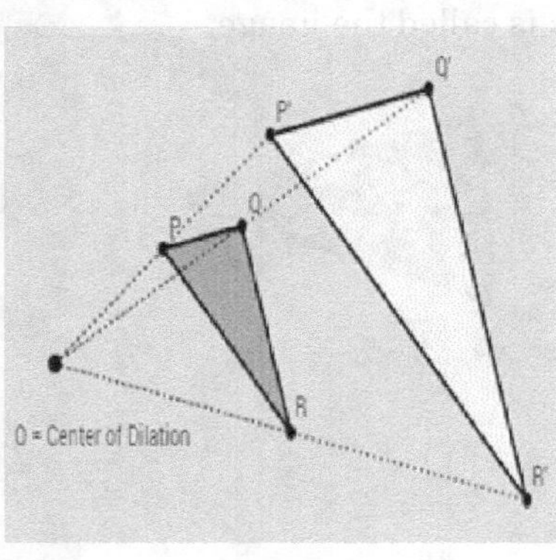

The following diagrams show the triangle ABC dilated with different scale factors. Scroll down the page for more examples and explanations of dilations.

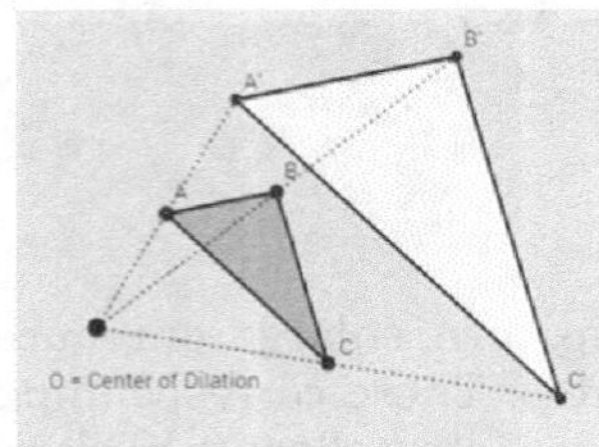

Scale Factor of 2

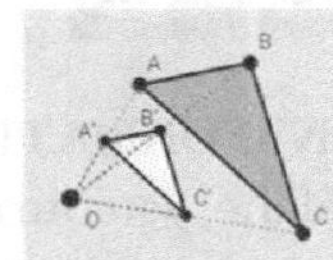

Scale Factor of 0.5

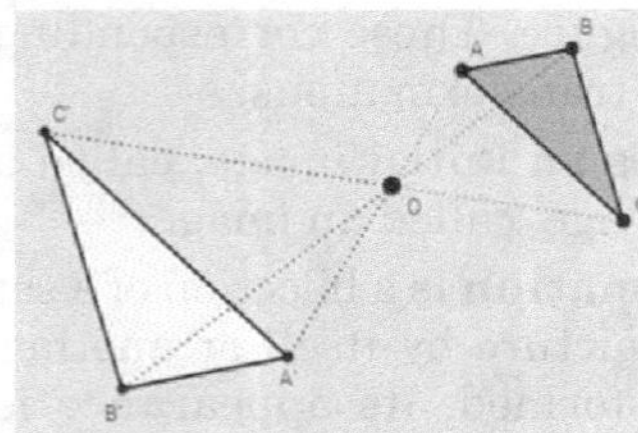

Scale Factor of -1.5

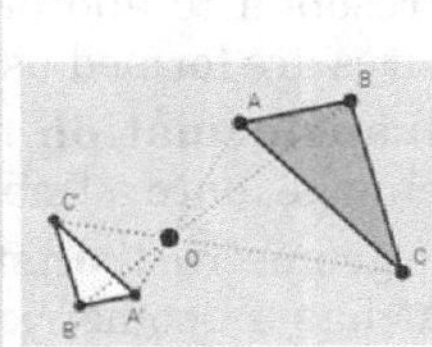

Scale Factor of -0.5

Dilation with scale factor > 1

We will first look at enlargements which are dilations with scale factors greater than 1

Example:

Enlarge triangle PQR with O as the center of dilation and a scale factor of 2.

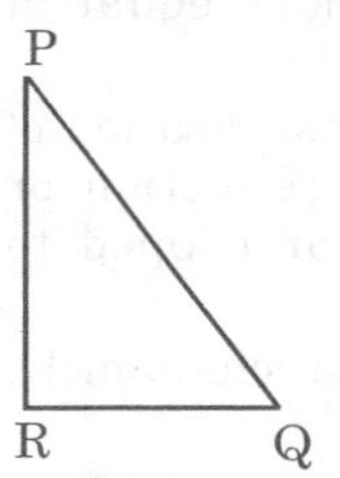

Solution:

Step 1: Measure OP.

Step 2: Extend the line OP to the point P' such that OP' = 2OP.

Step 3: Repeat the steps for all the vertices: point Q to get Q' and point R to get R'.

Step 4: Join the points P'Q'R' to form the image.

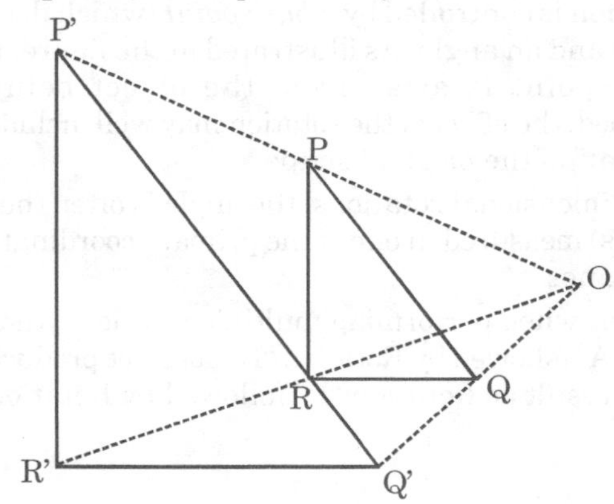

To describe you need:

- **Center point of the dilation**
- **Scale factor**

Coordinate Notation (if centered at the origin):

$(x, y) \rightarrow (ax, by)$

We find the scale factor by the help of scaling.

Scaling

An object which is **scaled** remains the same shape, but is reduced or enlarged in size in a proportionate fashion. The scaling operation shown in the figure involves enlarging the black square. Notice how the point **p1** is transformed into **p2**. Notice also that there is an anchor point (**a**) in the figure below, which remains in the same position.

This point is often one point of the object being scaled, but need not be.

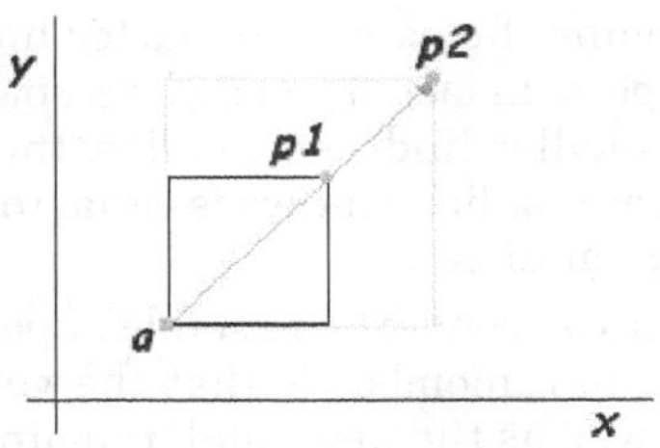

The simplest scaling operation simply multiplies all coordinates of the scaled object by a uniform amount. Since this changes all non-zero coordinates, it also effectively **moves** the object, an unintended side-effect.

The "anchor point" for such a scaling operation is the origin itself, which remains unmoved.

Scaling process:

In the scaling process, we either compress or expand the dimension of the object.

Scaling operation can be achieved by multiplying each vertex coordinate (x, y) of the polygon by scaling factor s_x and s_y to produce the transformed coordinates as (x', y').

So, $x' = x * s_x$ and $y' = y * s_y$.

The scaling factor s_x, s_y scales the object in X and Y direction respectively. So, the above equation can be represented in matrix form:

$$\begin{bmatrix} X' \\ Y' \end{bmatrix} = \begin{bmatrix} Sx & 0 \\ 0 & Sy \end{bmatrix} \begin{bmatrix} X \\ Y \end{bmatrix}$$

Or P' = S.P

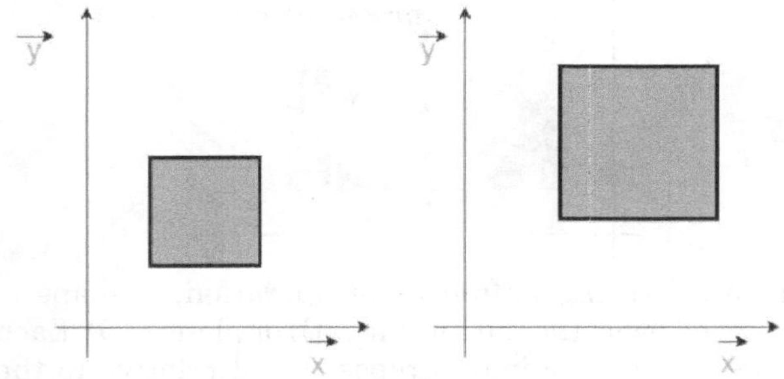

If the scaling factor S is less than 1, then we reduce the size of the object. If the scaling factor S is greater than 1, then we increase size of the object.

Algorithm:

1. Make a 2x2 scaling matrix S as:

   S_x 0

   0 S_y

2. For each point of the polygon.

   (i) Make a 2x1 matrix P, where P[0][0] equals to x coordinate of the point and P[1][0] equals to y coordinate of the point.

   (ii) Multiply scaling matrix S with point matrix P to get the new coordinate.

3. Draw the polygon using new coordinates.

Reflection

The image of an **object** as seen in a mirror is its **mirror reflection** or mirror image. In such an image, the right side of the object appears on the left side and vice versa. A mirror-image is therefore said to be laterally inverted and the **phenomenon** is called the lateral inversion.

Some letters don't change upon reflection. In other words, they are the same as that of their Mirror Images. The letters having identical mirror images are A, H, I, M, O, T, U, V, W, X, Y. Similarly, in small letters we have i, l, o, v, w, and x that have the same Mirror Image as that of their original images. The number 1 and 8 are the only two numbers with their identical mirror images.

Any object that has a bilateral **symmetry** will have its mirror image the same as that of the object. By that, we mean that if you were to draw a line from top to bottom that splits the object into two parts, such that both the parts can be superimposed onto each other, the object's Mirror Image will be same as that of the object's image.

This type of translation is called reflection or mirroring because it flips the object across a line by keeping its shape or size constant.

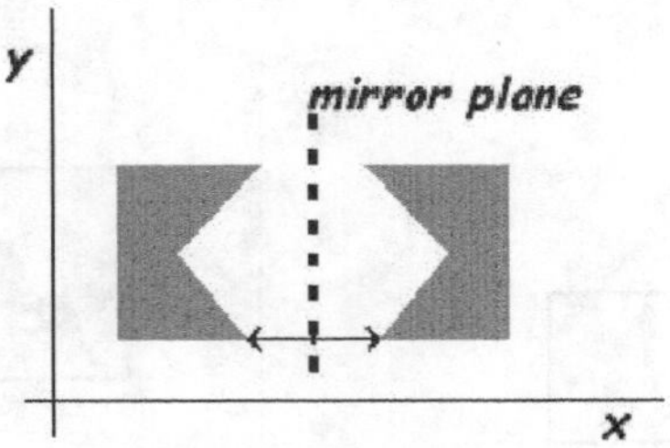

In a *mirroring* or *reflection* operation, a shape is "flipped over" the mirror line (2D) or plane (3D). Each point in the original is repositioned relative to the mirror line/plane, ending up an equal distance away, but on the opposite side.

The easiest mirror operation uses one of the primary coordinate system axes as the plane and simply multiplies all x coordinates (or all y-coordinates, or z-coordinates) by –1.

More general mirroring operations usually rely on the user selecting or indicating a line (or plane) in space. Because a line is somewhat easier to establish than a plane, a line is sometimes used, assuming that the mirror plane is perpendicular to the line, passing through its midpoint.

Topologically, mirroring turns objects inside-out, so programs maintaining consistently ordered surface orientations must compensate for the inversion by reversing the ordering of the faces which have been mirrored.

To describe you need:

- **the equation of a line**

Example 1:

Directions: In each of the following questions, there are four alternatives, which follow the **word** or number marked as a question. You have to select one alternative, which exactly matches with the mirror of the word/ number in the question.

Which among the following illustrations specifies the correct mirror image of P R E C A R I O U S?

(a) ƧUOIЯAƆƎЯꟼ (b) SUOIRACERP

(c) SUOPRECARI (d) SPRECARIOU

Ans.: (a)

Example 2:

Directions: In each of the following questions, there are four alternatives, which follow the **word** or number marked as a question. You have to select one alternative, which exactly matches with the mirror of the word/ number in the question.

Which among the following illustrations specifies the correct mirror image of P E R F E C T I O N?

(a) NOITCEFERP (b) RPEFECTION

(c) ИOITƆƎꟻЯƎꟼ (d) ERPFECTION

Ans.: (c)

NOTES:

Rigid transformations preserve lens measures, perimeter, and area. The image and preimage are CO. These transformations include:

- Rotations
- Reflections
- Translations

Non-Rigid transformations preserve angle measure only. The side lengths and perimeter are not equal, but are in proportion. The image and preimage are similar. This transformation is a:

- Dilation

GEOMETRICAL TRANSFORMATIONS

- Conversion of One Shape to Another.
- Assembling
- Grouping

Conversion from one Shape to Another

Each and every solid that exists occupies some volume. When you convert one solid shape to another, its volume remains the same, no matter how different the new shape is. In fact, if you melt one big cylindrical candle to 5 small cylindrical candles, the sum of the volumes of the smaller candles is equal to the volume of the bigger candle.

Hence, when you convert one solid shape to another, all you need to remember is that the volume of the original, as well as the new solid, remains the same. Let us discuss some examples to understand this better.

Conversion of Shapes Examples

Example 1:

An iron ball of radius 21 cm is melted and recast into 27 spherical balls of the same radius. Find the radius of each spherical ball.

Solution:

Volume of the iron ball = $\left(\frac{4}{3}\right)\pi r^3$

$= \left(\frac{4}{3}\right) \times \left(\frac{22}{7}\right) \times 21 \times 21 \times 21$

$= 38{,}808$ cm^3

Let us assume that the radius of the smaller balls is r.

So volume of 27 smaller balls = $27 \times \left(\frac{4}{3}\right)\pi r^3$

$= 36\pi r^3$

Thus we have, the volume of the big iron ball = volume of 27 smaller balls

$\Rightarrow 38{,}808 \text{ cm}^3 = 36\pi r^3$

$\Rightarrow r^3 = 343\ cm^3$

So r = 7 cm

Therefore, the radius of each spherical ball is 7 cm.

Example 2:

How many cylindrical candles, 1.75 cm in diameter and of thickness 2 mm, must be melted to form a cuboid candle of dimensions 5.5 cm × 10 cm × 3.5 cm?

Solution:

Radius of cylindrical candle = $\frac{1.75}{2}$ = 0.875 cm

Volume of one cylindrical candle = $\pi r^2 h$

$= \pi \times (0.875)^2 \times (0.02)\ cm^3$

$= \pi \times 0.0153125$

$= 0.048125\ cm^3$

Volume of cuboid candle = 5.5 × 10 × 3.5

$= 192.5\ cm^3$

Thus, number of cylindrical candles

$$= \frac{\text{Volume of cuboid candle}}{\text{Volume of one cylindrical candle}}$$

$$= \frac{192.5}{0.048125} = 4000$$

Example 3:

A copper rod with a diameter 1 cm and length 8 cm is drawn into a wire of length 18 m of uniform thickness. Determine the thickness of the wire.

Solution:

Given that, Copper rod diameter = 1 cm.

Length of the copper rod = 8 cm

Length of new wire = 18 m = 18 × 100

= 1800 cm.

We know that the rod should be in the cylindrical shape.

Hence, the volume of the rod = $\pi \times (½)^2 \times 8$

$= 2\pi\ cm^3$

Therefore, the volume of the copper rod = $2\pi\ cm^3$

If "r" is the radius of cross-section of the wire, then the volume of the wire is given as:

The volume of the wire = $\pi \times r^2 \times 1800$

Since the volume of the copper rod and the volume of the new wire should be equal, then we can write

$\Rightarrow \pi \times r^2 \times 1800 = 2\pi$

$$\Rightarrow r^2 = \frac{2\pi}{1800\pi}$$

$$\Rightarrow r^2 = \frac{1}{900}$$

$$\Rightarrow r = \frac{1}{30}$$

Hence, the thickness of the wire should be the diameter of the cross-section of the new wire.

$$\text{Thickness} = \left(\frac{1}{30}\right) \times 2 = \frac{1}{15}\ \text{cm.}$$

Thus, the thickness of wire is approximately equal to 0.067 cm.

Assembling:

Assembling is a process of finding the combined image with the help of visualization of it's different-different shapes.

Example 1:

(a) (b) (c) (d)

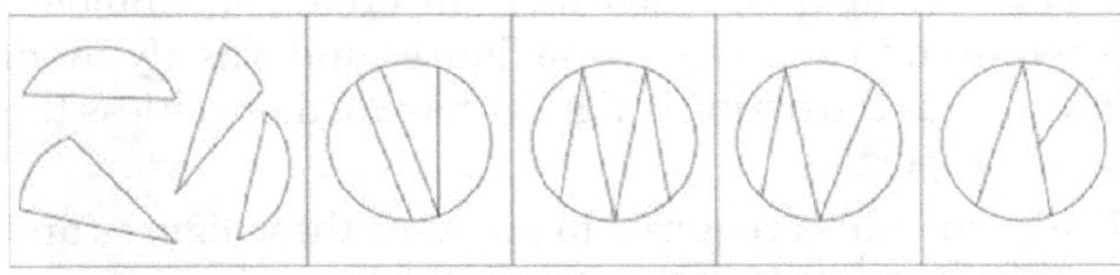

Answer. (c)

Example 2:

(a) (b) (c) (d)

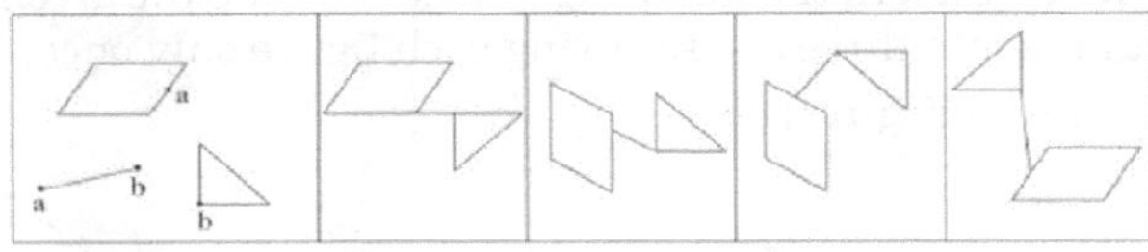

Answer. (c & d)

Example 3:

(a) (b) (c) (d)

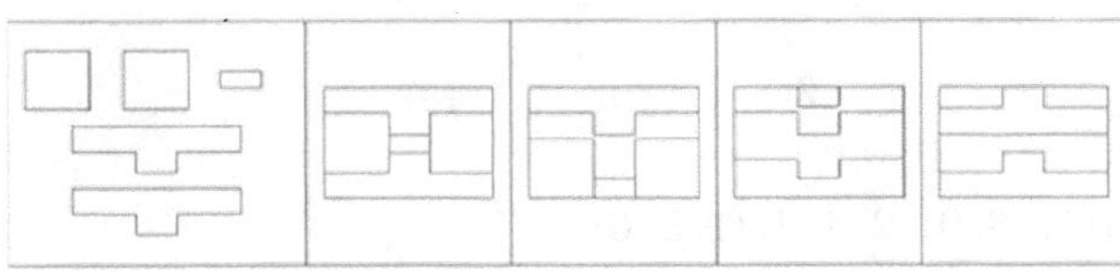

Answer. (a)

Example 4:

(a) (b) (c) (d)

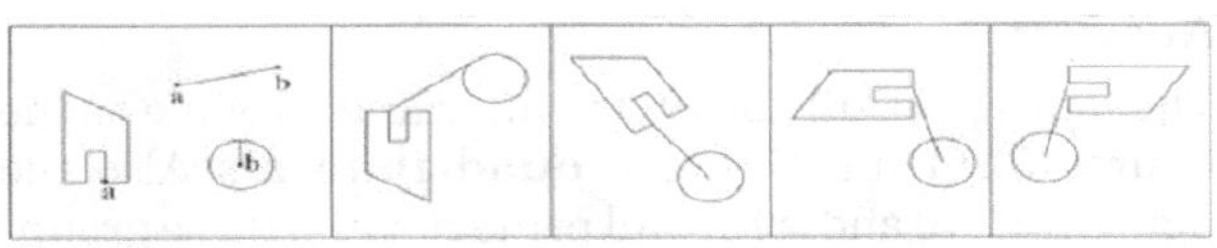

Answer. (b, c & d)

Example 5:

(a) (b) (c) (d)

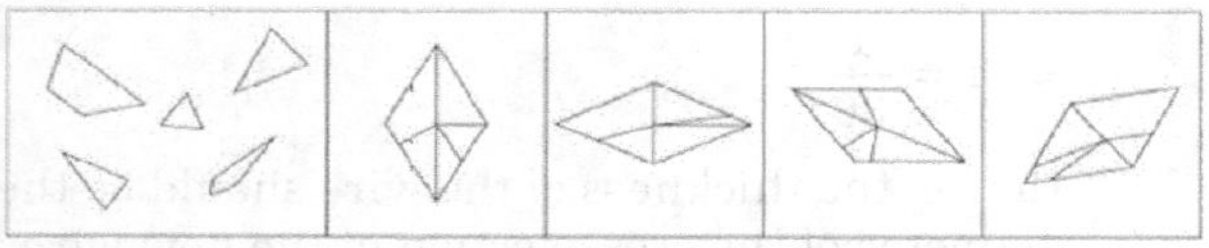

Answer. (d)

Grouping of Figures

In Grouping of Figures, different figures are put together on the basis of a certain rule, pattern or property. Here in this section on the grouping of figures, we will solve many questions that are similar to the previous year questions and concepts. In these questions, a given set of figures will be featured. Some of these figures can be grouped together while as others can't be. The task is to find out the rules and/or similarities between each individual set.

In this type of questions, you are given a set of usually 6, 7 or 9 etc. figures, which are numbered. The candidate is required to analyses these figures and classify them into groups consisting of figures having more or less the same properties.

The candidate is required to analyses these figures and classify them into groups consisting of figures having more or less than the same properties. The best answer is to be selected from a given set of fairly close alternatives.

Example 1:

Directions: In the following question, group the given figures into three classes using each figure only once.

Problem Figure:

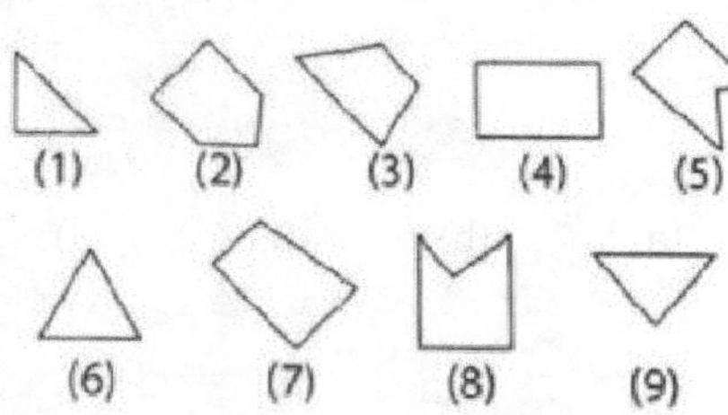

(a) (7, 8, 9) (2, 4, 3) (1, 5, 6)

(b) (1, 3, 3) (4, 5, 7) (6, 8, 9)

(c) (1, 6, 8) (3, 4, 7) (2, 5, 9)

(d) (1, 6, 9) (3, 4, 7) (2, 5, 8)

(e) none of the above

Answer:

The figures (1), (6) and (9) are all triangles while as the figures (3), (4) and (7) are all **quadrilaterals**. Also, the figures (2), (5) and (8) are all pentagons. So these groups are identical and the correct answer is thus D.

Example 2:

Directions: There are two classes of three figures each. Class 'A' figures differ in a certain way from the figures in class 'B'. Which of the four answer **figures** belong to class 'A'?

Problem Figure:

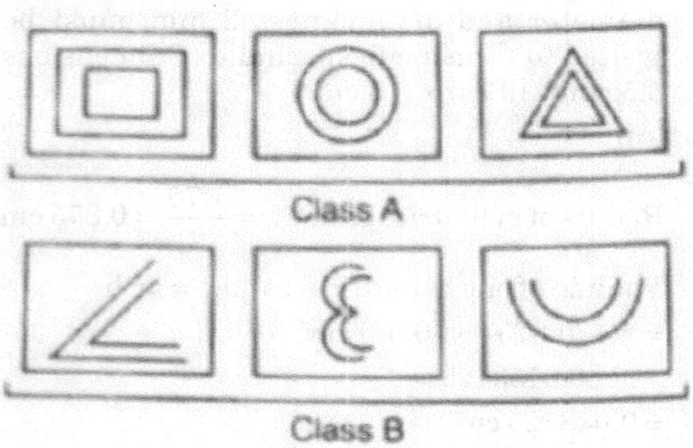

Answer Figures:

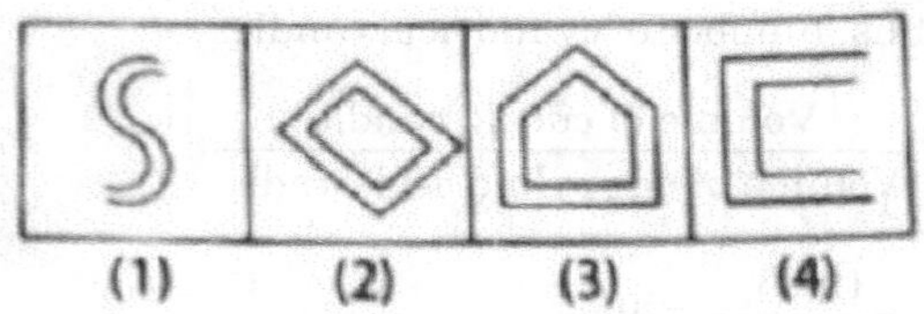

(a) Both (1) and (3)

(b) (1) and (2) both

(c) Both (2) and (4)

(d) (2) and (3) only

(e) None of the above

Answer:

The answer here is very simple. Each figure in class A consists of two similar closed figures, which are placed one inside the other. Therefore the answer is that (2) and (3) both are correct and the correct option is (d). (2) and (3) only.

Example 3:

In the following question, group the given figures into three classes using each figure only once.

Problem Figure:

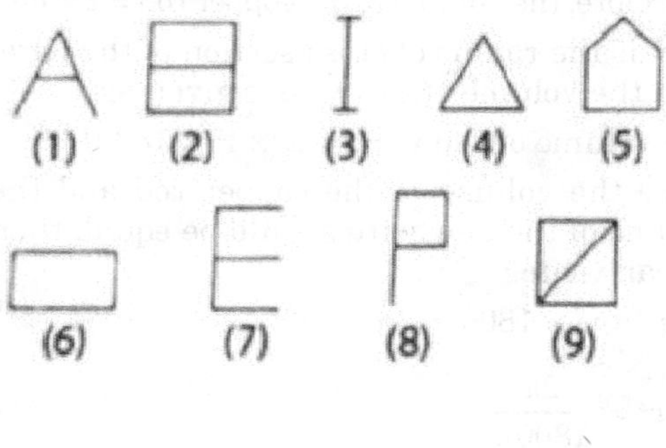

(a) (1, 3, 4) (2, 5, 9) (6, 7, 8)
(b) (1, 2, 3) (4, 5, 6) (7, 8, 9)
(c) (1, 5, 9) (2, 4, 7) (3, 6, 8)
(d) (3, 7, 8) (1, 6, 5) (4, 2, 9)
(e) none of the above

Answer:

If we start grouping the figures on the basis of sides, we have: The figures that are made by three lines are (1, 3, 4) Also, the figures which are made by four **lines** are (6, 7, 8) Similarly, the figures made by the five lines are (2, 5, 9). These three groups are present in the option A. Thus the answer is (a).

Example 4:

In the following question, group the given figures into three classes using each figure only once.

Problem Figure:

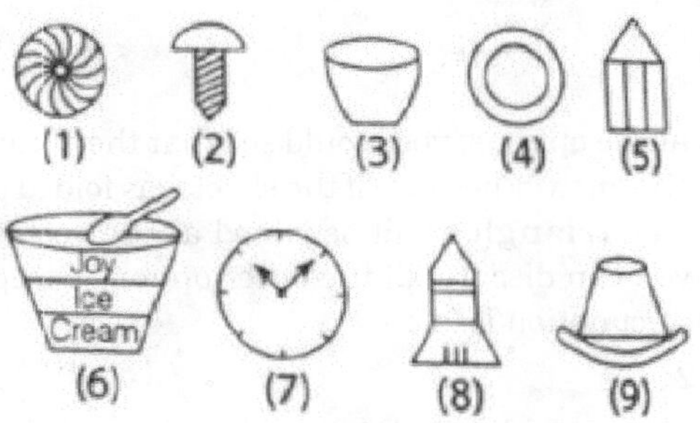

(a) (1, 4, 7) (2, 5, 8) (3, 6, 9)
(b) (1, 3, 6) (2, 5, 8) (4, 7, 9)
(c) (1, 2, 4) (3, 5, 8) (6, 7, 9)
(d) (1, 4, 9) (2, 5, 8) (3, 6, 7)
(e) none of the above

Answer:

There are figures that have one small and one big circle are figure 1, figure 4, and figure 7. Figures which have one sharp point are the figure2, figure 5, and figure 8. Also, the figures that have **shapes** like buckets are the figure 3, figure 6, and figure 9.

Example 5:

In the following question, group the given figures into three classes using each figure only once.

Problem Figure:

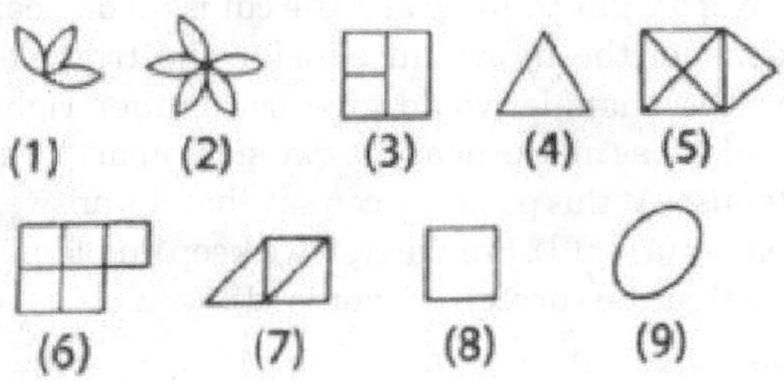

(a) (1, 2, 3) (4, 8, 9) (5, 7, 6)
(b)(4, 5, 7) (3, 1, 2) (7, 8, 9)
(c) (1, 2, 7) (8, 9, 4) (2, 3, 6)
(d) (3, 5, 6) (8, 7, 4) (9, 1, 2)
(e) none of the above

Answer:

The figures 1, 3 and 7 have three petals.three rectangles and three triangles respectively. So, they should be grouped in one group figures 4, 8 and 9 have one triangle, one **rectangle** and one circle respectively. So, they should be grouped into one group. On the other hand, the figures between 2, 5 and 6 have five petals, five triangles and five **squares**, respectively. So, they should be grouped into one group. Therefore the correct answer is (c). (1, 2, 7) (8, 9, 4) (2, 3, 6).

Example 6:

In the following question, group the given figures into three classes using each figure only once.

Problem Figure:

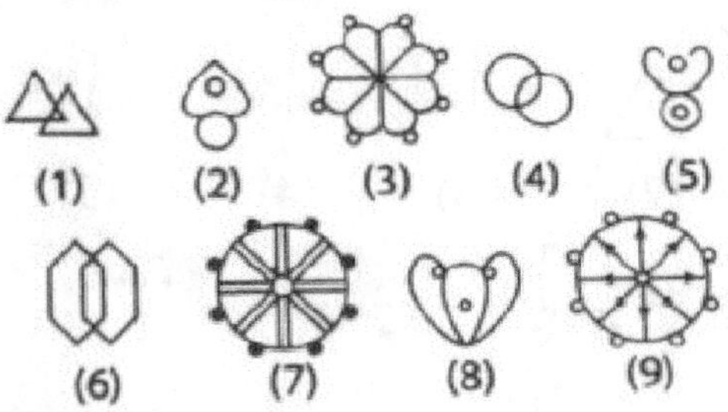

(a) (1, 4, 8) (2, 5, 7) (3, 9, 6)
(b) (1, 4, 6) (2, 5, 8) (3, 7, 9)
(c) (1, 4, 6) (2, 5, 7) (3, 8, 9)
(d) (1, 2, 3) (4, 5, 6) (7, 8, 9)

Answer:

The figures 1, 4 and 6 have two **triangles**, two circles and two **hexagons**, respectively. Figures 2, 5 and 8 are similar in shape. In figures 3, 7 and 9 there are eight small **circles** around a big circle. Now let us move on to more solved examples on the Grouping of Figures.

Example 7:

In the following question, group the given figures into three classes using each figure only once.

Problem Figure:

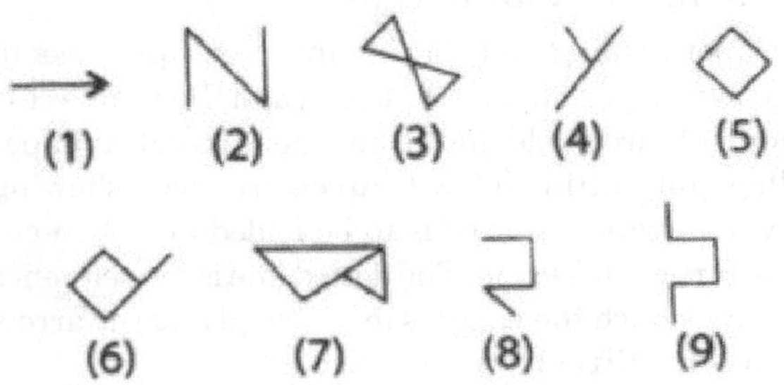

(a) (1, 5, 6) (2, 3, 4) (7, 8, 9)

(b) (1, 2, 4) (3, 5, 8) (6, 7, 9)

(c) (5, 6, 7) (1, 2, 4) (3, 8, 9)

(d) (1, 2, 4) (3, 5, 7) (6, 8, 9)

(e) none of the above

Answer:

In the question figure we see that the figures that we label 1, 2 and 4 all consist of three lines. Also, the figures 3, 5 and 8 all consist of four lines each. Similarly, we can see that the figures 6, 7 and 9 all consist of five lines. Therefore this gives us the grouping of the figures. Thus the correct option is (b).

Example 8:

In the following question, group the given figures into three classes using each figure only once.

Problem Figure:

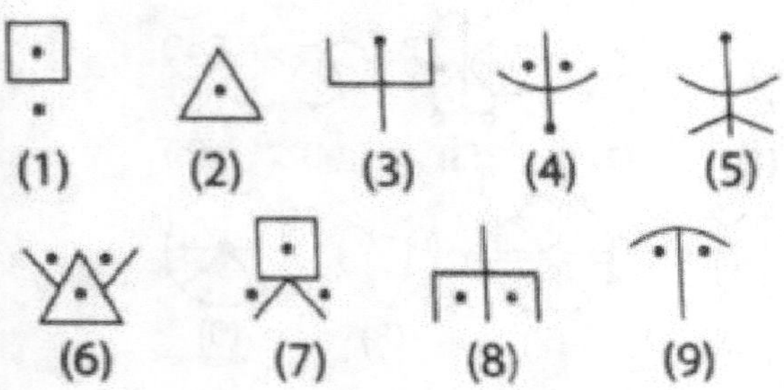

(a) (1, 7, 8) (2, 6, 5) (3, 4, 9)

(b) (1, 8, 9) (2, 3, 5) (4, 6, 7)

(c) (2, 3, 5) (1, 7, 8) (4, 6, 9)

(d) (2, 6, 7) (1, 3, 4) (5, 8, 9)

(e) none of the above

Answer:

As is clear from the question figure, the figures 2, 3 and 5 have one point or in other words, we can say that they are a one-pointed system. Also, we can see that the figures 1, 8 and 9 are made up of four lines. Below we have some practice problems on the Grouping of Figures.

PAPER FOLDING AND CUTTING

The Paper Folding and Cutting involves a **process** in which a transparent sheet is folded and then some cuts and/or folds are made. In the questions based on Paper Folding and Cutting a few **figures** are given showing the way in which a piece is to be folded and then cut from a particular section. The dotted line is the reference line along which the paper is to be folded and the arrow indicates the **direction** of the fold.

Thus, these figures indicate the **sequence** in which the paper is to be folded. The designs from the cut will appear on each one of the folds made on the paper. In questions based on paper folding and cutting it can be asked to find either the folded or unfolded pattern of the sheet. Let us see some examples below. We also have some really tricky problems that have been put under the higher **order** thinking section.

Example 1:

A square transparent sheet with a pattern is given. Figure out from among the four alternatives as to how the pattern would appear when the transparent sheet is folded at the middle line shown?

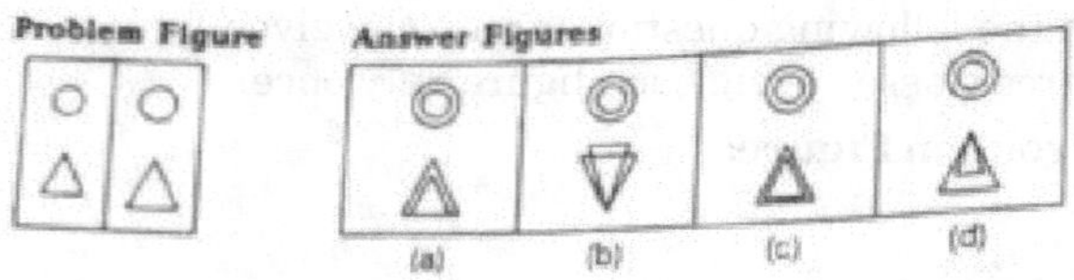

Answer:

If you look at the options, you would see that the option (a) can't be the answer because if the sheet was folded at the middle the **triangles** will be joined at the vertex. Similarly, you can discard all the other options except (d). The correct option is (d).

Example 2:

A piece of paper is folded and punched as shown below in the question figures. from the given answer figures, indicate how it will appear when opened?

Problem Figures

Answer Figures

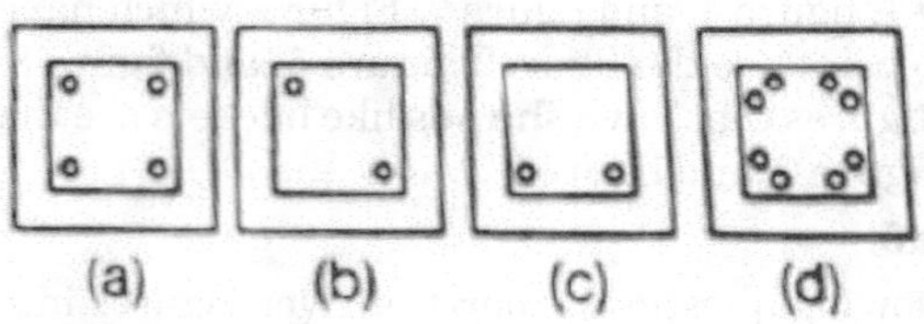

Answer:

In the last problem figure, we see that the paper is folded in the form of a right triangle and the cut is made near the midpoint of the hypotenuse inside the triangle. Unfolding the triangle would give us another right triangle which has marks near the two sides apart from the hypotenuse. At this point, we can say that the answer could be either (a) or (d). Similarly, if we keep unfolding, we will see that the correct answer is (d).

Example 3:

Directions: In each of the following problems, a **square** transparent sheet with a pattern is given. Figure out from amongst the four alternatives as to how the pattern would appear when the transparent sheet is folded along the dotted line

Transparent Sheet **Answer Figures**

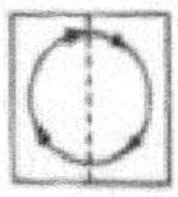

(a) (b) (c) (d)

Answer:

The right half of the transparent sheet is being folded along the dotted line and placed on the left half of the sheet. The figure thus obtained resembles the answer figure (c).

Example 4:

Directions: In the following question, a piece of paper is folded and cut as shown below in the question figures. From the given answer figures, indicate how it will appear when opened?

Problem Figures

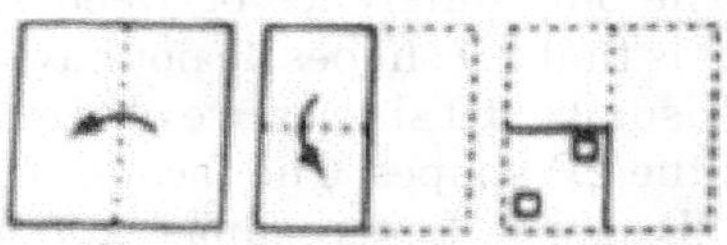

x y z

Answer Figures

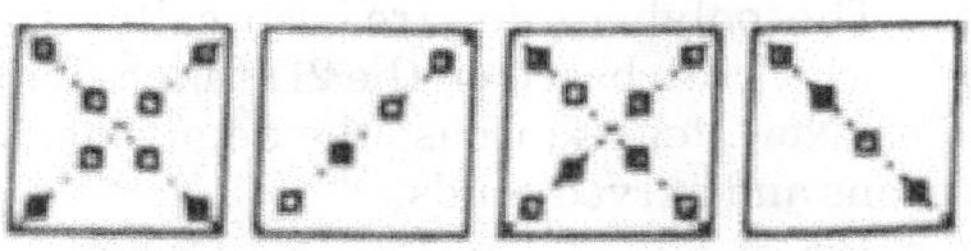

(a) (b) (c) (d)

Answer: In these types of questions, the easiest thing to do is to start with the last picture and undo the steps taken in the question. Let us see, if we start from the figure z, and move to y, we will get a figure that has square cuts as half diagonals in the left half only. From the figure y we have to now move to the figure x, that would give a figure similar to the one in (a). Hence the correct answer here is (a).

Example 5:

A piece of paper is folded and punched as shown below in the question figures. From the given answer figures, indicate how it will appear when opened?

Problem Figures

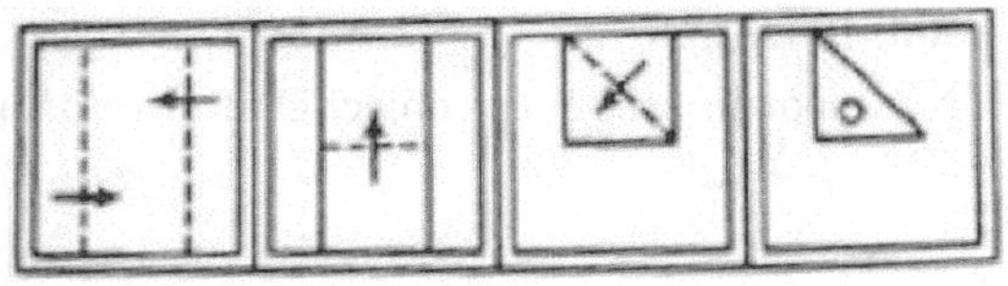

Answer Figures

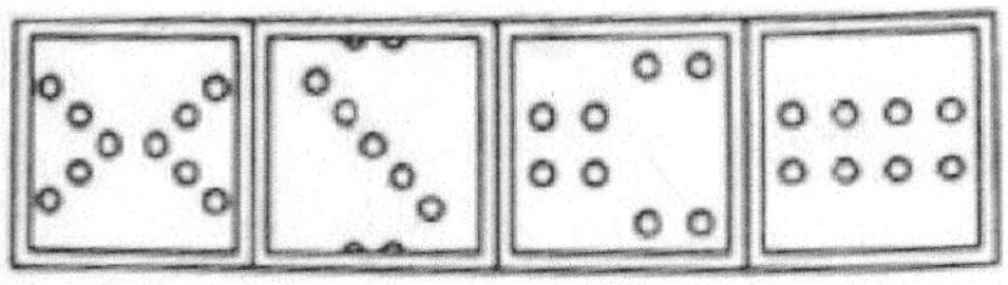

(a) (b) (c) (d)

Answer:

As we have done with the other questions, we will begin with the last problem figure and move on to the first one, while tracing out all the spots on each step. Whence we do that we will see that out of the four options, the only one that represents the answer figure is the one in option (a).

PATTERNS IN 2 AND 3 DIMENSIONS

2D Shape's Pattern:

In geometry, a shape or a figure that has a length and a breadth is a 2D shape pattern. In other words, a plane object that has only length and breadth is 2 dimensional. Straight or **curved lines** make up the sides of this shape. Also, these figures can have any number of sides. In general, plane figures made of lines are known as polygons. For example, triangle and square are **polygons**.

Examples of 2D shapes

Rectangle, circle, square, triangle, quadrilateral and pentagon are some examples of 2D shapes.

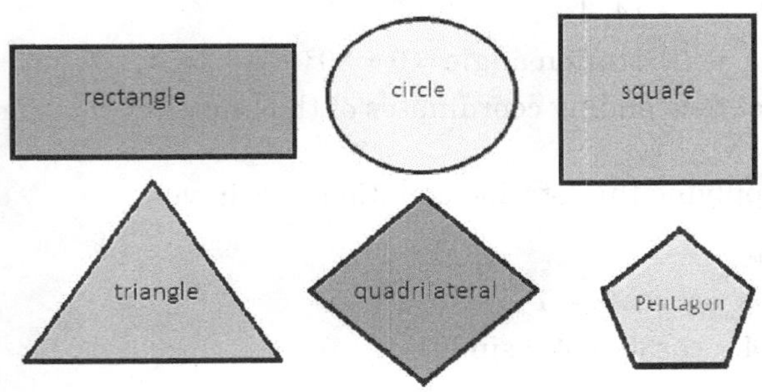

2D Rotation:

2D Rotation is a process of rotating an object with respect to an angle in a two dimensional plane.

Consider a point object O has to be rotated from one angle to another in a 2D plane.

Let-

- Initial coordinates of the object O = (X_{old}, Y_{old})

- Initial angle of the object O with respect to origin = ϕ
- Rotation angle = θ
- New coordinates of the object O after rotation = (X_{new}, Y_{new})

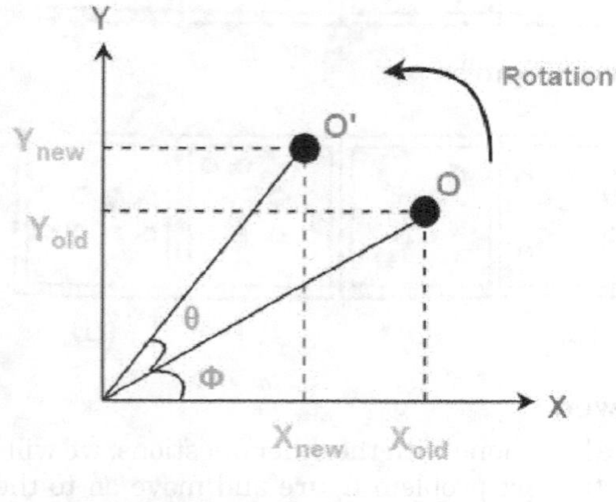

2D Rotation

This rotation is achieved by using the following rotation equations-

- $X_{new} = X_{old} \times \cos\theta - Y_{old} \times \sin\theta$
- $Y_{new} = X_{old} \times \sin\theta + Y_{old} \times \cos\theta$

Example:

Given a line segment with starting point as (0, 0) and ending point as (4, 4). Apply 30-degree rotation anticlockwise direction on the line segment and find out the new coordinates of the line.

Solution:

We rotate a straight line by its end points with the same angle. Then, we re-draw a line between the new end points.

Given-

- Old ending coordinates of the line = (X_{old}, Y_{old}) = (4, 4)
- Rotation angle = θ = 30°

Let new ending coordinates of the line after rotation = (X_{new}, Y_{new}).

Applying the rotation equations, we have-

X_{new}

$= X_{old} \times \cos\theta - Y_{old} \times \sin\theta$

$= 4 \times \cos 30° - 4 \times \sin 30°$

$$= 4 \times \left(\frac{\sqrt{3}}{2}\right) - 4 \times \left(\frac{1}{2}\right) = 2\sqrt{3} - 2$$

$= 2(\sqrt{3} - 1) = 2(1.73 - 1)$

$= 1.46$

Y_{new}

$= X_{old} \times \sin\theta + Y_{old} \times \cos\theta$

$= 4 \times \sin 30° + 4 \times \cos 30°$

$$= 4 \times \left(\frac{1}{2}\right) + 4 \times \left(\frac{\sqrt{3}}{2}\right) = 2 + 2\sqrt{3}$$

$= 2(1 + \sqrt{3}) = 2(1 + 1.73) = 5.46$

Hence, New ending coordinates of the line after rotation = (1.46, 5.46).

3d Shapes Pattern:

We come across a number of objects of different shapes and sizes in our day-to-day life. There are golf balls, doormats, ice-cream cones, coke cans, and so on. These objects have different characteristic properties such as length, breadth, diameter, etc. which set them apart from one another. But no matter how different their dimensions are, all of them occupy space and have three dimensions. So, they are referred to as three-dimensional or solid shapes. In this article, we will discuss the various 3D shapes, surface area and volumes, and the process of making 3D shapes using nets with the help of 2D Shapes.

In Geometry, 3D shapes are known as three-dimensional shapes or solids. 3D shapes have three different measures such as length, width, and height as its dimensions. The only difference between 2D shape and 3D shapes is that 2D shapes do not have a thickness or depth. Usually, 3D shapes are obtained from the rotation of the 2D shapes. The faces of the solid shapes are the 2D shapes. Some examples of the 3D shapes are a cube, cuboid, cone, cylinder, sphere, prism and so on. The 3D shapes consist of both curved shaped solid and the straight-sided polygon called the polyhedron. The polyhedrons are also called the polyhedrons, which are based on the 2D shapes with straight sides. Now, let us discuss the details about the polyhedrons and curved solids.

Examples of 3D Shapes

Cuboid, cube, cylinder, sphere, pyramid and cone are a few examples of 3D shapes

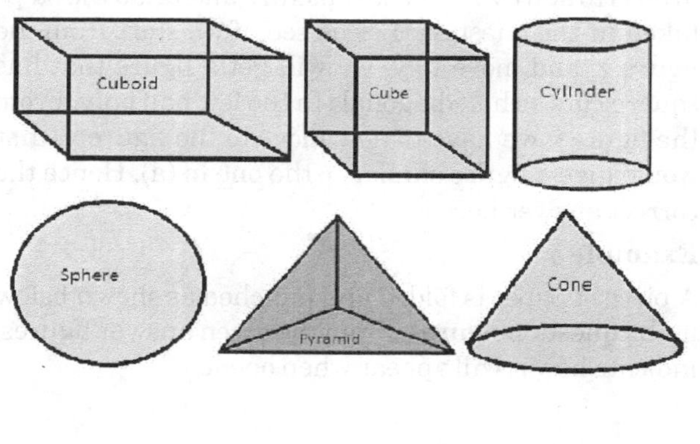

3D Rotation:

3D Rotation is a process of rotating an object with respect to an angle in a three-dimensional plane.

Consider a point object O has to be rotated from one angle to another in a 3D plane.

Let-

- Initial coordinates of the object O = (X_{old}, Y_{old}, Z_{old})
- Initial angle of the object O with respect to origin = ϕ
- Rotation angle = θ
- New coordinates of the object O after rotation = (X_{new}, Y_{new}, Z_{new})

In 3 dimensions, there are 3 possible types of rotation-

- X-axis Rotation
- Y-axis Rotation
- Z-axis Rotation

For X-Axis Rotation-

This rotation is achieved by using the following rotation equations-

- $X_{new} = X_{old}$
- $Y_{new} = Y_{old} \times \cos\theta - Z_{old} \times \sin\theta$
- $Z_{new} = Y_{old} \times \sin\theta + Z_{old} \times \cos\theta$

For Y-Axis Rotation-

This rotation is achieved by using the following rotation equations-

- $X_{new} = Z_{old} \times \sin\theta + X_{old} \times \cos\theta$
- $Y_{new} = Y_{old}$
- $Z_{new} = Y_{old} \times \cos\theta - X_{old} \times \sin\theta$

For Z-Axis Rotation-

This rotation is achieved by using the following rotation equations-

- $X_{new} = X_{old} \times \cos\theta - Y_{old} \times \sin\theta$
- $Y_{new} = X_{old} \times \sin\theta + Y_{old} \times \cos\theta$
- $Z_{new} = Z_{old}$

Example:

Given a homogeneous point (1, 2, 3). Apply rotation 90 degree towards X, Y and Z axis and find out the new coordinate points.

Solution:

Given-

- Old coordinates = (X_{old}, Y_{old}, Z_{old}) = (1, 2, 3)
- Rotation angle = θ = 90°

For X-Axis Rotation-

Let the new coordinates after rotation = (X_{new}, Y_{new}, Z_{new}).

Applying the rotation equations, we have-

- $X_{new} = X_{old} = 1$
- $Y_{new} = Y_{old} \times \cos\theta - Z_{old} \times \sin\theta$
  $= 2 \times \cos 90° - 3 \times \sin 90°$
  $= 2 \times 0 - 3 \times 1 = -3$
- $Z_{new} = Y_{old} \times \sin\theta + Z_{old} \times \cos\theta$
  $= 2 \times \sin 90° + 3 \times \cos 90°$
  $= 2 \times 1 + 3 \times 0 = 2$

Thus, New coordinates after rotation = (1, –3, 2).

For Y-Axis Rotation-

Let the new coordinates after rotation = (X_{new}, Y_{new}, Z_{new}).

Applying the rotation equations, we have-

- $X_{new} = Z_{old} \times \sin\theta + X_{old} \times \cos\theta$
  $= 3 \times \sin 90° + 1 \times \cos 90°$
  $= 3 \times 1 + 1 \times 0 = 3$
- $Y_{new} = Y_{old} = 2$
- $Z_{new} = Y_{old} \times \cos\theta - X_{old} \times \sin\theta$
  $= 2 \times \cos 90° - 1 \times \sin 90°$
  $= 2 \times 0 - 1 \times 1 = -1$

Thus, New coordinates after rotation = (3, 2, –1).

For Z-Axis Rotation-

Let the new coordinates after rotation = (X_{new}, Y_{new}, Z_{new}).

Applying the rotation equations, we have-

- $X_{new} = X_{old} \times \cos\theta - Y_{old} \times \sin\theta$
  $= 1 \times \cos 90° - 2 \times \sin 90°$
  $= 1 \times 0 - 2 \times 1 = -2$
- $Y_{new} = X_{old} \times \sin\theta + Y_{old} \times \cos\theta$
  $= 1 \times \sin 90° + 2 \times \cos 90°$
  $= 1 \times 1 + 2 \times 0 = 1$
- $Z_{new} = Z_{old} = 3$

Hence, New coordinates after rotation = (–2, 1, 3).

Nets for Building 3-D Shape Pattern:

A geometric net is a two-dimensional 'pattern' for a three-dimensional object that is unfolded along its edges. It represents each face of the figure in two dimensions. In other words, a net is a pattern made when the surface of a three-dimensional figure is laid out. Thus, showing each face of the figure. A solid may have different nets.

For example, a box is solid. It's a 3D object with the shape of a cuboid. Below is a net pattern for a box. Copy an enlarged version of the net and try to make the box by folding and gluing the faces together. You may use suitable units.

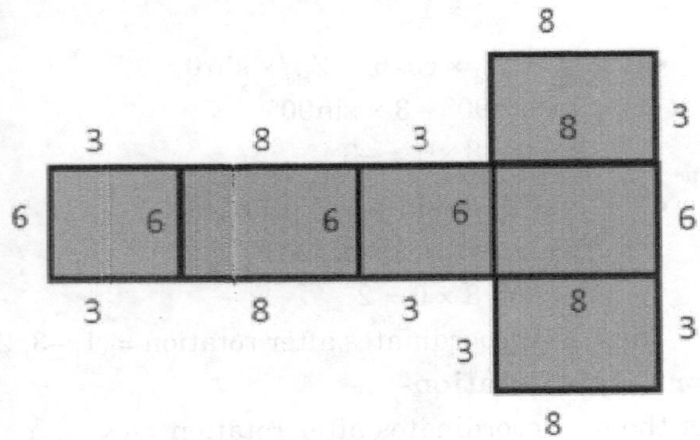

Furthermore, different shapes have different nets.

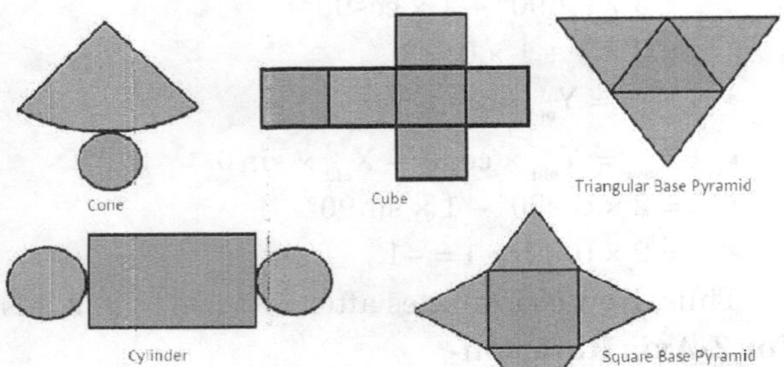

Example of 3D Shape Pattern:

Polyhedron

A solid shape bound by polygons forms a **polyhedron**. The word polyhedral is the plural of word polyhedron.

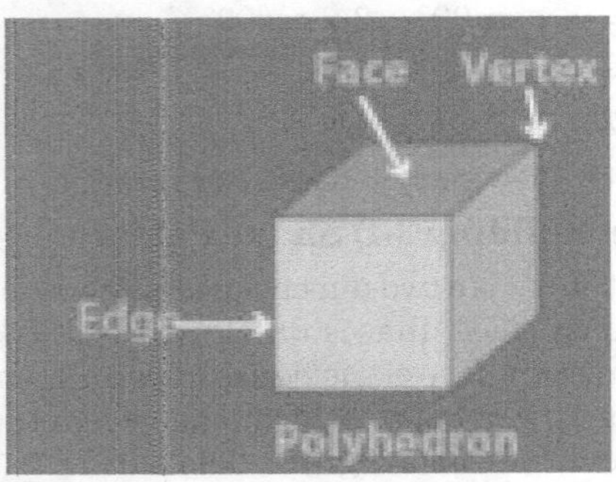

- **Faces:** Polygons forming a polyhedron are its faces.
- **Edges:** Line segments common to intersecting faces of a polyhedron are its edges.
- **Vertices:** Points of intersection of edges of a polyhedron are its vertices.

By Euler's Formula:

For Polyhedron

$$F + V = E + 2$$

Where,

F = Number of faces of polyhedron

V = Number of vertices of polyhedron

E = Number of Edges of polyhedron

A net is a 2- dimensional representation of a 3-dimensional figure that is unfolded along its edges so that each face is in 2- dimension. So, a given polyhedron can have different nets.

Polyhedrons are straight-sided solids, which has the following properties:

- It should have flat sides are called the faces
- It must have the corners, called vertices
- Polyhedrons should have straight edges.

Like polygons in two-dimensional shapes, polyhedrons are also classified into regular and irregular polyhedrons and convex and concave polyhedrons. The most common examples of polyhedral are cube, cuboid, pyramid, prism. Some other examples of regular polyhedron are tetrahedron, octahedron, dodecahedron, icosahedron, and so on. These regular polyhedrons are also known as the platonic solids, whose faces are identical to each face.

For example, the most commonly used example of a polyhedron is a cube, which has 6 faces, 8 vertices, and 12 edges.

In a polyhedron, three or more edges meet at a point to form a vertex. Some examples of polyhedrons are cuboid, cube, pyramid, and triangular pyramid.

Curved Solids

Apart from polyhedrons, there are 3D shapes with curved shapes such as the sphere, cone, cylinder, etc. For example, cones have a circular base that narrows smoothly from the circular base to the point called the apex. All these shapes have curved faces, and hence they are called curved solids.

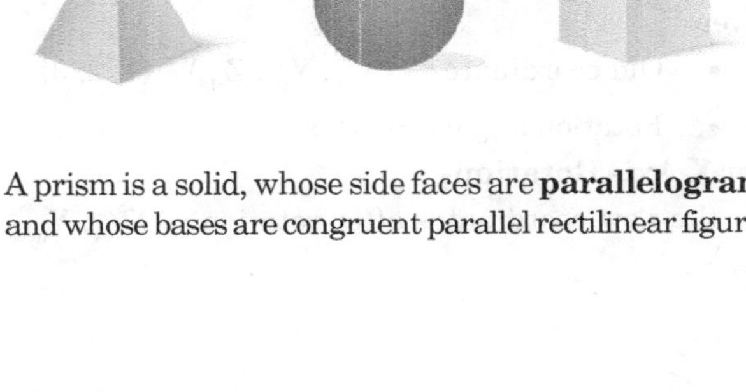

A prism is a solid, whose side faces are **parallelograms** and whose bases are congruent parallel rectilinear figures.

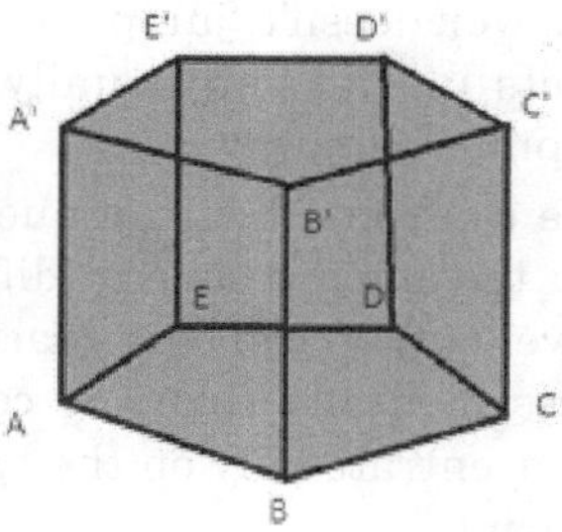

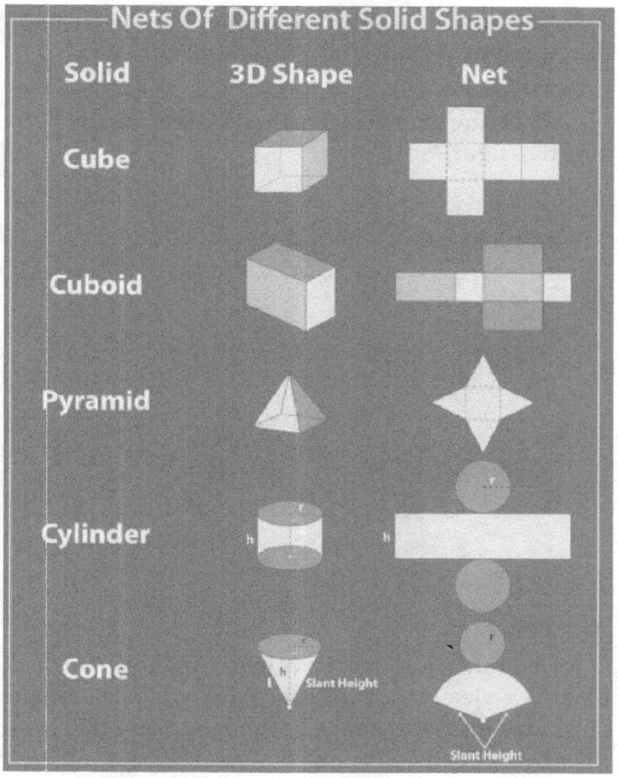

Shape Matching in 2D & 3D Pattern:

1. Shape Matching – Two-Dimensional pattern:

Let take an example, you are asked to look at two groups of simple, flat objects and find pairs that are exactly the same size and shape.

Each group has 25 small drawings of these two-dimensional objects. The objects in the first group are labelled with numbers and are in **numerical** order. The objects in the second group are labelled with letters in alphabetical order.

Each drawing in the first group is exactly the same as a drawing in the second group. The objects in the second group have been **moved** and some have been **rotated**.

Example:

Which shape in Group 2 corresponds to a shape in Group 1?

Group 1

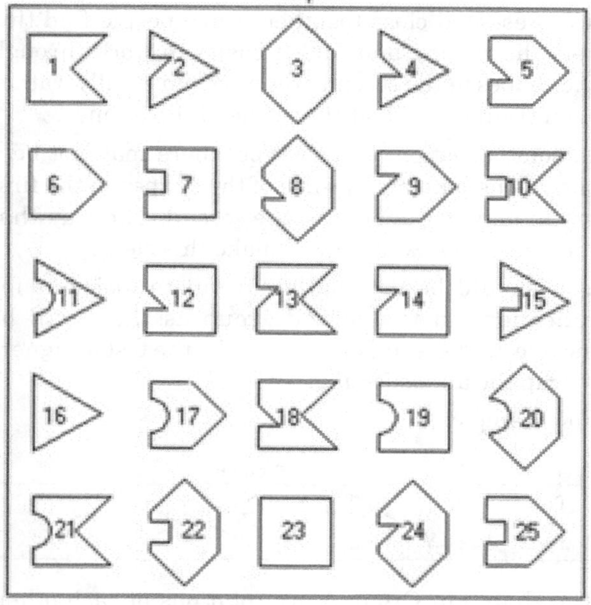

Group 2

Answers:

1. X;	2. P;	3. M;	4. V;	5. G;
6. A;	7. D;	8. T;	9. C;	10. B;
11. W;	12. E;	13. U;	14. Y;	15. F;
16. S;	17. H;	18. K;	19. J;	20. L;
21. O;	22. N;	23. Q;	24. R;	25. I.

This is very much a speed test, as you will be given far more of these questions than you can comfortably answer in the given time.

These questions use a large number of shapes that are presented close together. Some people find this very distracting and find it easier to work through the shapes in the second group systematically, rather than trying to look at the whole group at once.

In some questions of this type, there may not be a one-to-one match and some of the shapes in the first group may not appear in the second. The way that the question is worded will make this clear.

You should be especially careful to look out for reflections in the second group as these will be incorrect. These are often put in by the test designers to trap the unsuspecting.

For example:

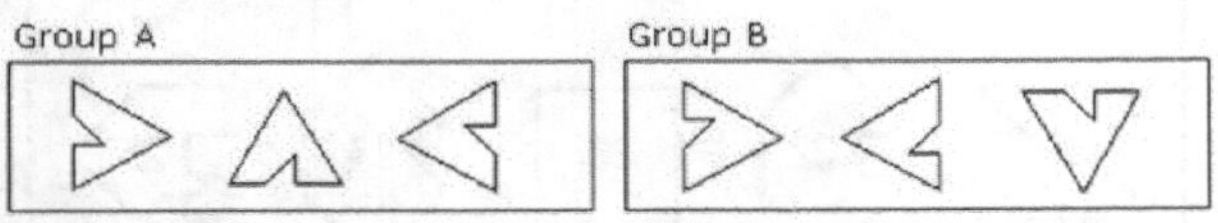

The shapes in Group A are rotations of each other. The shapes in Group B are also rotations of each other. However, none of the shapes in Group A are rotations of those in Group B.

2. Visual Comparison – Two Dimensional pattern:

In these types of question, you will be presented with several objects, only two of which are identical. Once again, these are speed questions and you will need to work quickly and attempt to answer as many as possible in the given time.

Example:

Which two pictures are identical?

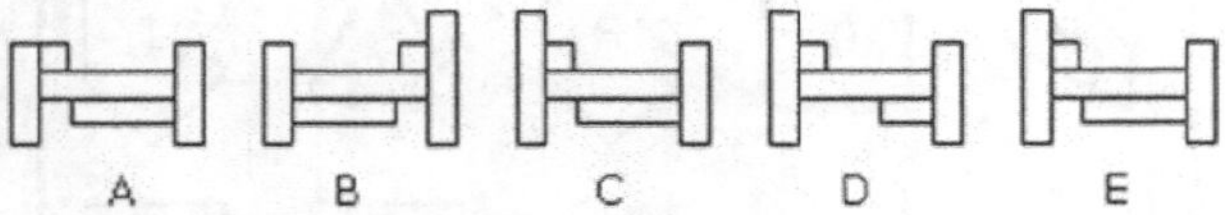

Answer:

C and E are the only two pictures that are identical.

The best strategy for these types of questions is to begin with the shape on the left and work through the shapes to the right of it, systematically looking for an exact match. If there isn't one then move on to the second shape and repeat the process.

It can be quite difficult to discipline yourself to adopt this systematic approach, as you may think that it is quicker just to look at all of the shapes until the answer 'jumps out' at you. The problem with this is that if the answer doesn't 'jump out' fairly quickly then panic sets in and you usually resort to the systematic approach anyway.

When you are confronted with a question like this that asks you to compare several different shapes, the other answer options can be distracting. It is much easier and less mentally tiring to cover the other shapes and concentrate only on the two shapes that you are comparing.

3. Group Rotation – Two Dimensional pattern:

This is slightly more complex than the rotations in the two-dimensional shape matching question.

Example:

Which of the answer figures is a rotation of the question figure?

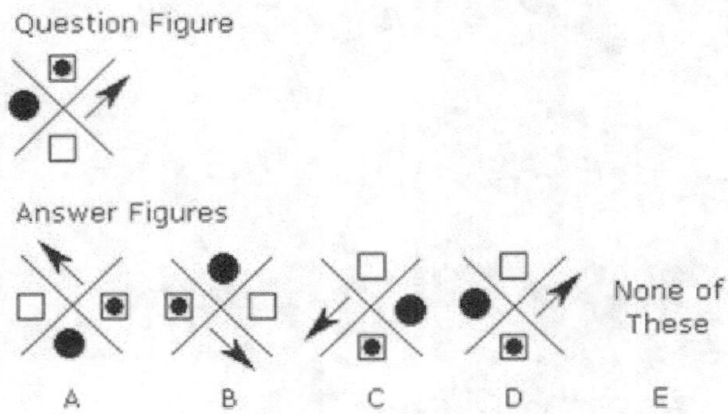

Answer: C

Once again, you need to be careful that you don't identify reflections. The best strategy is to choose the most asymmetrical shape in the group – in this case, the arrow. Then determine how the shape would look going clockwise, anticlockwise and opposite.

In the example above, the white square is clockwise from the arrow. This means that A, B and D cannot be rotations of the question figure. This leaves only C as a possibility, which can quickly be checked element by element.

4. Combining Two–Dimensional Pattern:

These questions show you a series of two-dimensional pattern. One of the shapes has been cut up into pieces. The questions present you with the pieces and you are asked to work out which of the shapes has been cut-up.

Example:

Which of the complete shapes can be made from the components shown?

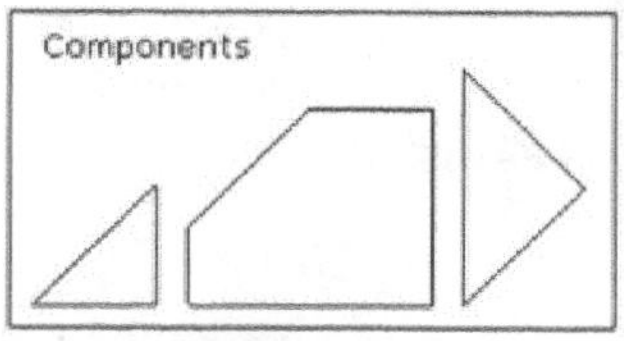

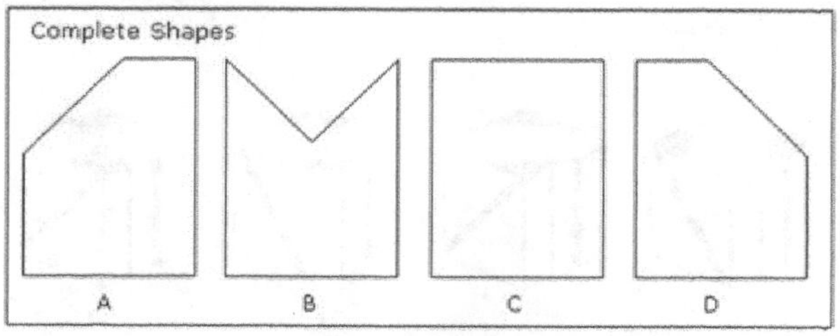

Spatial Ability

Answer: B

The best strategy for answering these questions is to look at the complete shapes and see if there are any distinct features that would make it impossible to construct such a shape from the components. Sometimes there are one or two shapes that can be immediately discounted based on size alone.

One thing to remember is that if the complete shapes don't have any bits sticking out (they usually don't) then the components must fit together so that sides of the same length are together. This reduces the number of combinations considerably.

5. Two-Dimensional Mirror Reflections:

These questions will assess how you view two-dimensional objects from a different perspective. You will be presented with an image and will be asked to identify its mirror image.

Example:

Which answer shows a reflection of the image below?

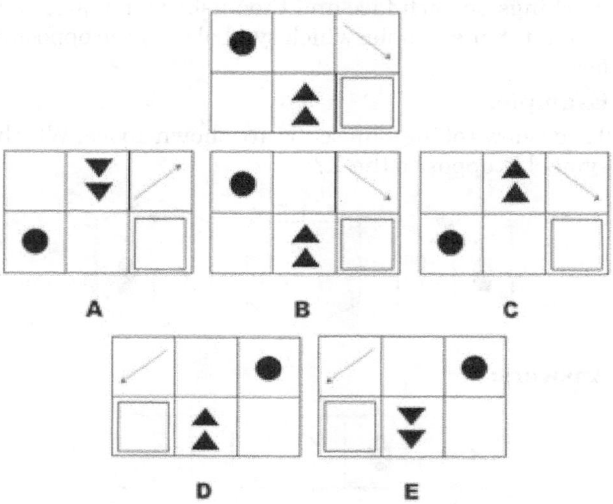

Answer: D

First, check the answer options to see if any elements have changed position. If so, you can eliminate these answers. This will eliminate answer A straight away as the square and arrow have swapped places.

Then look for any elements which have been manipulated differently. Answer E appears to be a mirror image, but the triangles have been incorrectly inverted.

Answer C is an example where the square and arrow have changed position, but the triangles are also incorrectly inverted. Therefore, this can not be a reflection and can be eliminated.

Answer B is exactly the same as the question image, it hasn't been reflected at all. This answer can also be eliminated.

6. Two-Dimensional Maps

Map questions will assess your ability to follow basic instructions and visualise a route. Normally, you will be given a two-dimensional plan or map and a series of directions via two or three questions that all relate to the same map.

You just need to use the instructions to navigate the map. As long as you have a basic sense of direction, this will seem relatively straightforward, but you will need to do so quickly and accurately.

Example:

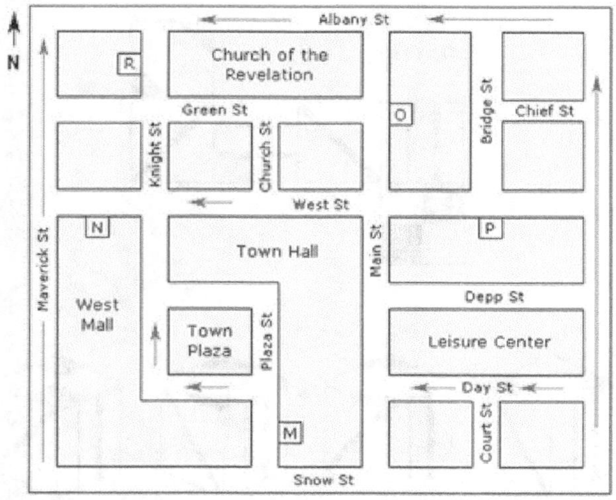

Officer Wilkinson is in Depp St and can see the Town Hall to her right. What direction is she facing?

A	B	C	D
North	South	East	West

She turns and walks to the junction with Main St. She turns left and proceeds two blocks before turning right, then taking the next right, and walking half a block. Which location is nearest to her current position?

A	B	C	D
M	N	R	P

Answers: B and A

The secret to getting the answers correct is imagining yourself walking down the street, making sure you are facing the correct way when deciding whether to turn left or right. You should also revise compass directions and make sure you are confident with each one, as well as the intermediate points such as north-west and south-east.

The first part of this question is pretty straightforward if you know your compass directions. You should immediately spot that to have the Town Hall to her right she must be facing south (answer B).

The next part of the question is a simple case of following the directions given. Remember that you have to see the map from the walker's perspective when choosing left or right.

7. Cubes in Two and Three-Dimension's Pattern:

These questions show a flat (two-dimensional) pattern that can be folded to make a cube. The answer options will usually be four three-dimensional cubes.

The pattern and the cubes have symbols or markings on each face. You need to look at the pattern and decide which of the cubes, if any, could be made from it.

Example:

Which of the cubes shown could be made from the pattern?

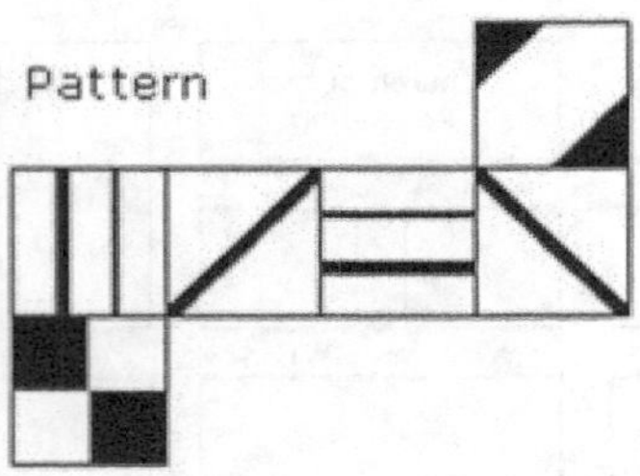

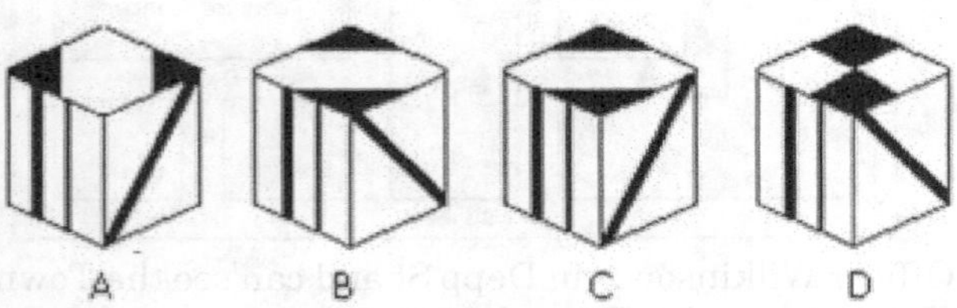

Answer: A

The key to these questions is to remember that only three faces of the cube can be shown in the illustration, this means that you need only consider the relationship between the **three visible elements** on each cube and see if the same relationship exists in the pattern.

The best strategy for this type of question is to call one face of the cube the 'front' and then name the other faces of the cube in relation to it.

For example:

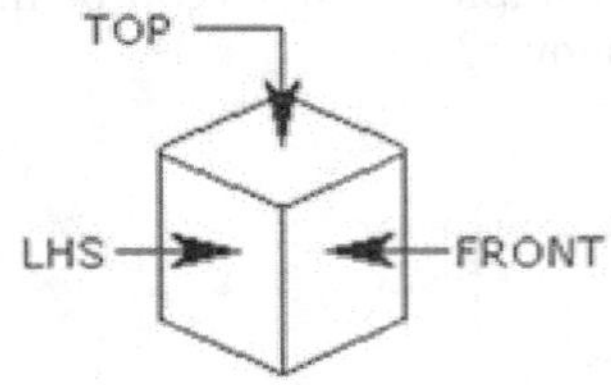

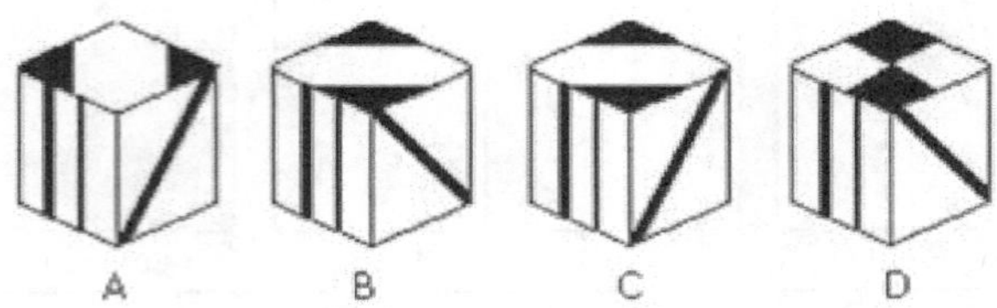

Obviously, this is an arbitrary decision as you could look at a cube from any side. However, thinking of the problem this way makes it much easier to see the relationship between the faces of the cube.

You can then:

- Look at the front of the cube
- Find that face on the pattern
- Use the pattern to identify the 'top' and eliminate any options that do not match
- Use the pattern to identify the other face that touches the 'front' (in this example LHS) and eliminate any options that do not match

The problem may also be posed the other way around. Here you would have a single three-dimensional cube and several two-dimensional patterns, only one of which, when folded, will make the cube.

8. Cube Views in Three Dimensions pattern:

These questions show you several (usually three) views of a three-dimensional cube with unique symbols or markings on each face and then asks you a question about it. For example, which symbol is on the opposite face?

Example:

Three views of the same cube are shown above. Which symbol is opposite the X?

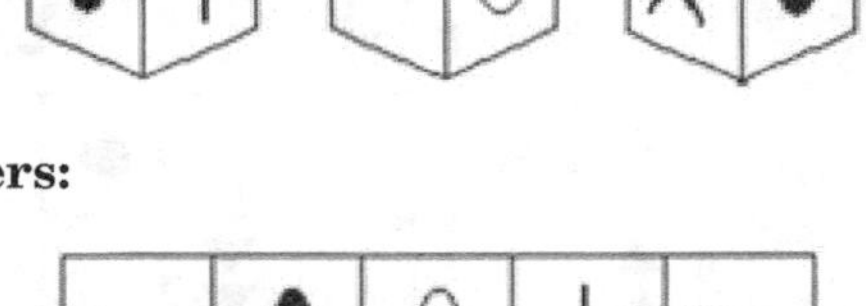

Answers:

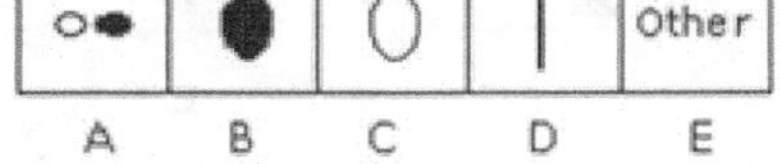

Answer: D

Some people seem to have a natural talent for imagining objects in three dimensions and find these questions straightforward. However, if you're not one of them and you find thinking in three dimensions difficult, there are other ways to get to the answer.

In the question above, for example, you can simply use a process of elimination. If you can see a symbol next to the 'X' on the cube, then it cannot be opposite.

The second and third cubes eliminate A, B and C. This leaves only D and 'other' as possibilities. D has edges shared with A and B, which would be consistent with the third cube illustrated. Therefore, D is correct.

Although it is not usually specified in the instructions, it is almost always true that in these questions each symbol is used only once. This means that even in cases where elimination is not possible, it is sometimes quite easy to see the solution without mentally manipulating the cube too much.

In the example above, you can simply compare the first and third illustrations. The third illustration shows a 90-degree anticlockwise rotation (looking at the cube from above) of the first illustration. Therefore, D must be opposite the 'X'.

9. Other Solids in Two- and Three-Dimension's Pattern:

These questions are similar to the cube questions above but rather than cubes, they use other solid shapes which may be irregular.

In some respects, these questions are easier than the cube questions as there are more relationships to work with. In other words, each face of the solid shape has a shape of its own rather than just being square.

Example:

Which of the solid shapes shown could be made from the pattern?

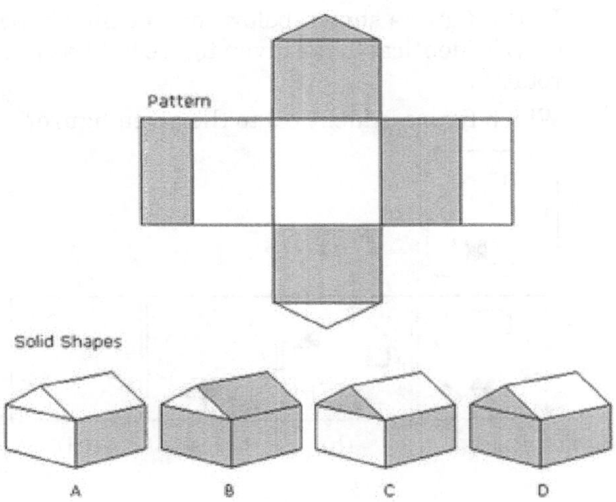

Answer: D

Shape A can be eliminated because it shows an unshaded face below a triangular face, both of these faces (below the triangular face) are shaded on the pattern. Shape C can be eliminated for the same reason.

Shape B can be eliminated as it has a shaded roof above the shaded side, which does not appear on the pattern.

Once again, you can reduce these problems to the relationship between the visible elements on the three-dimensional picture. This makes things easier because even though the solid shape may have more faces than a cube, it is unusual for more than four faces to be shown. This means that you need to consider the relationship between the four visible faces, paying particular attention to shading or other patterns on them.

10. Block Counting in Three Dimensions Pattern:

In these questions, you will be given a three-dimensional drawing of a shape formed from a group of blocks, some of which may be hidden.

You will be asked to identify how many blocks have been used to make up the shape. You will need to visualize the blocks you cannot see.

Example:

How many blocks make up the shape below?

28	18	25	27	26
A	B	C	D	E

Answer: E

Try to break up the blocks in your mind. For example, in the exercise above, there is a block of three 5-cube pillars (which together makes 15 blocks), a group of three 3-cube pillars (which together makes 9 blocks), and two single blocks.

Adding together 15 + 9 + 2 gives 26

This is quicker and produces easier sums, leading to more robust answers.

EXERCISE

MCQ TYPE QUESTIONS

ROTATION

1. Below are two separate objects, each has a dot placed in one corner, and each has five possible rotations. Choose the answer option in which the dots are placed in the same corners of the objects as in the model.

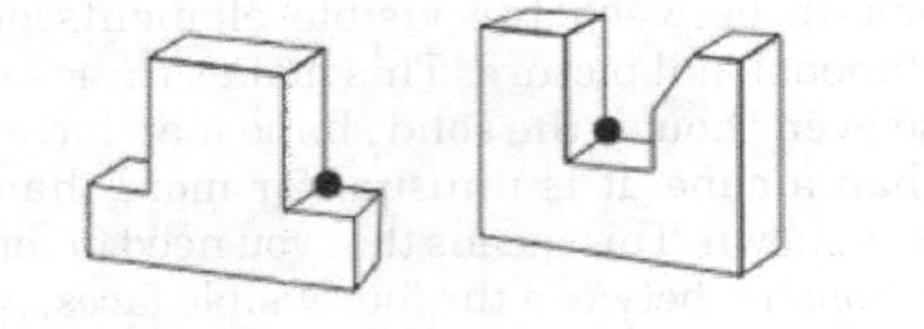

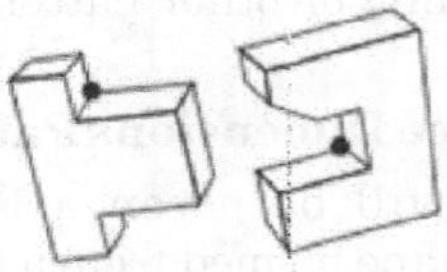

(a)

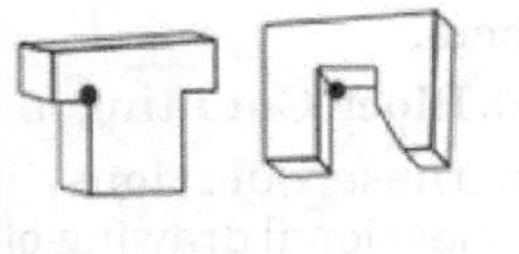

(b)

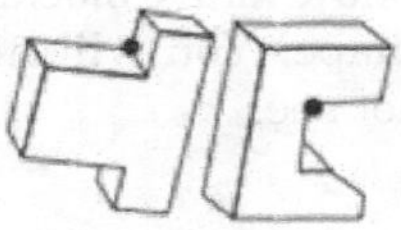

(c)

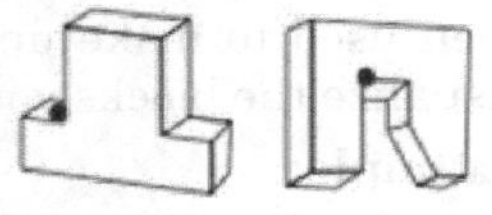

(d)

2. Below are two separate objects, each has a dot placed in one corner, and each has five possible rotations. Choose the answer option in which the dots are placed in the same corners of the objects as in the model.

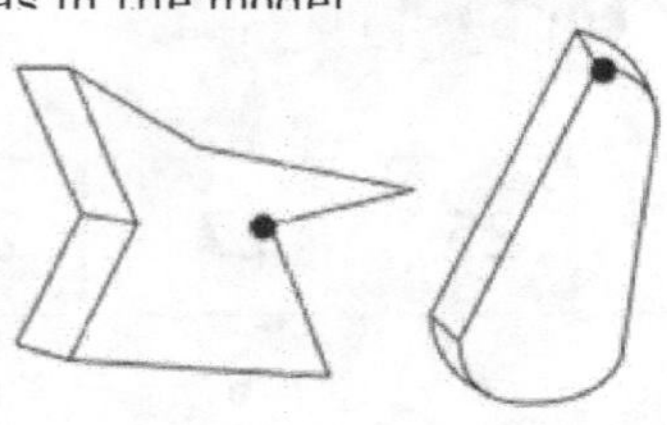

(a)

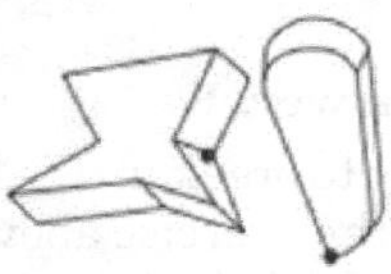

(b)

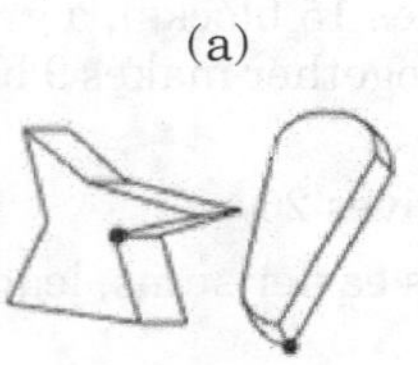

(c)

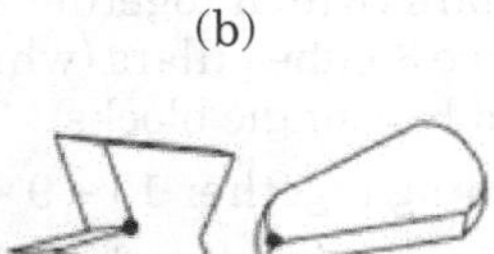

(d)

3. Below are two separate objects, each has a dot placed in one corner, and each has five possible rotations. Choose the answer option in which the dots are placed in the same corners of the objects as in the model.

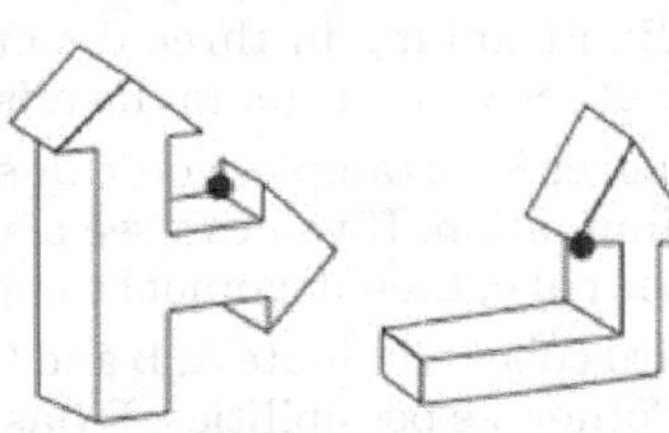

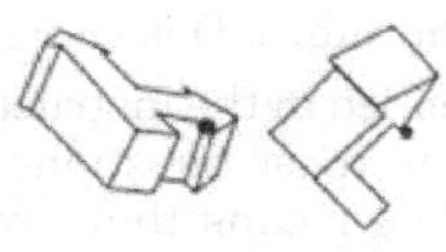

(a)

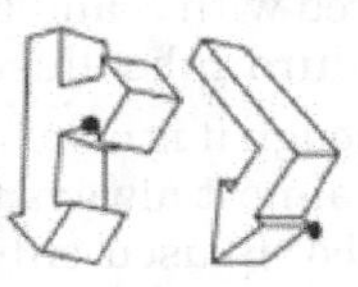

(b)

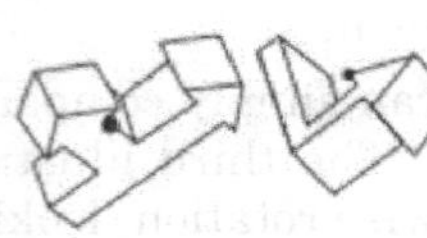

(c)

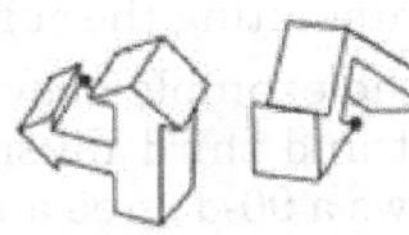

(d)

4. In the figures shown below, one of the shapes (a-d) is identical to the given figure but has been rotated.

Which figure is identical to the given figure?

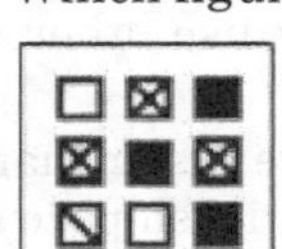

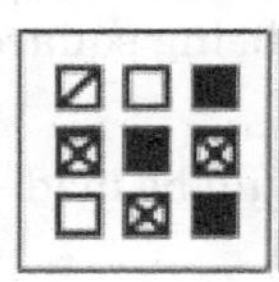

(a)

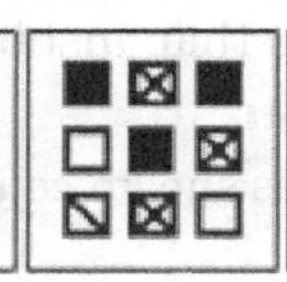

(b)

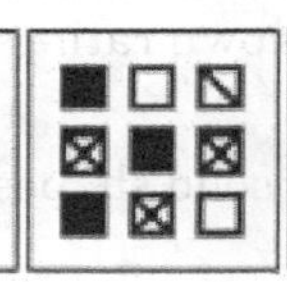

(c)

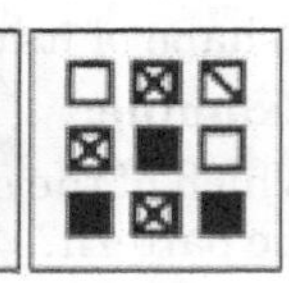

(d)

5. In the figures shown below, one of the shapes (a-d) is identical to the given figure but has been rotated.

Which figure is identical to the given figure?

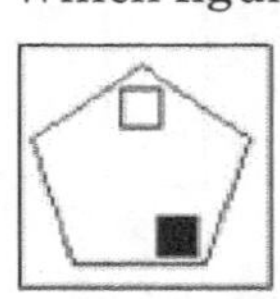

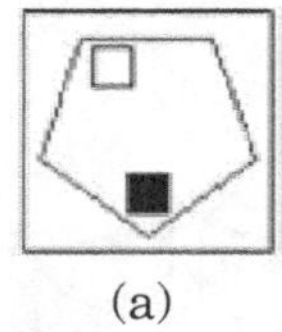

(a)

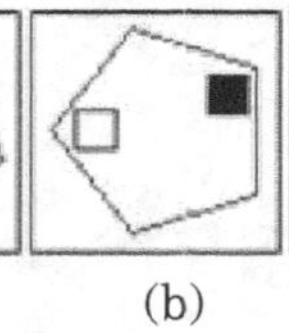

(b)

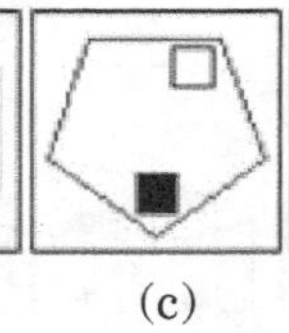

(c)

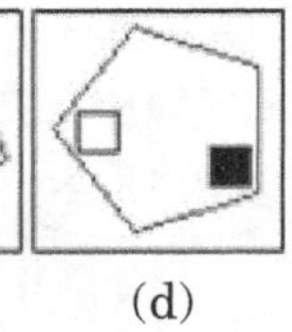

(d)

6. In the figures shown below, one of the shapes (a-d) is identical to the given figure but has been rotated.
Which figure is identical to the given figure?

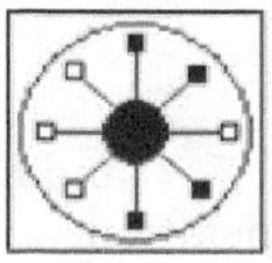

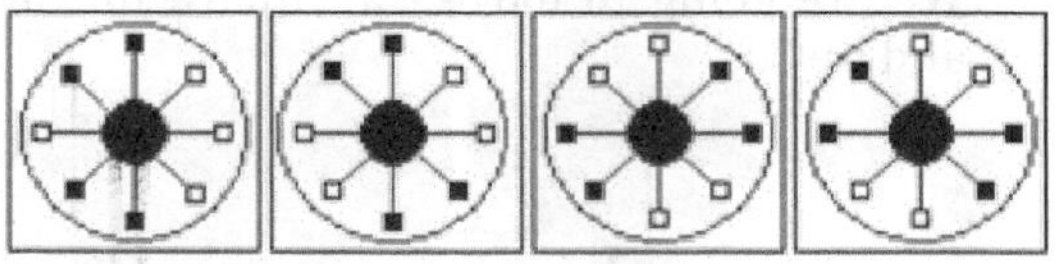

(a) (b) (c) (d)

7. In the figures shown below, one of the shapes (a-d) is identical to the given figure but has been rotated.
Which figure is identical to the given figure?

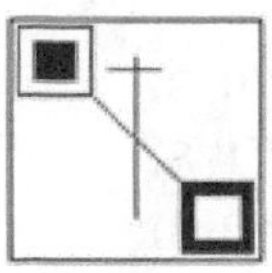

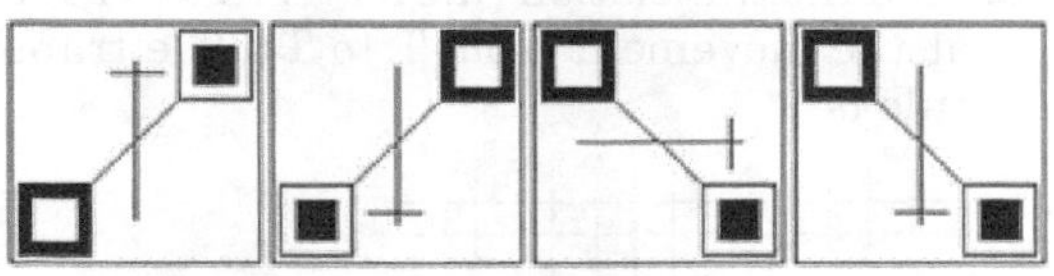

(a) (b) (c) (d)

8. In the figures shown below, one of the shapes (a-d) is identical to the given figure but has been rotated.
Which figure is identical to the given figure?

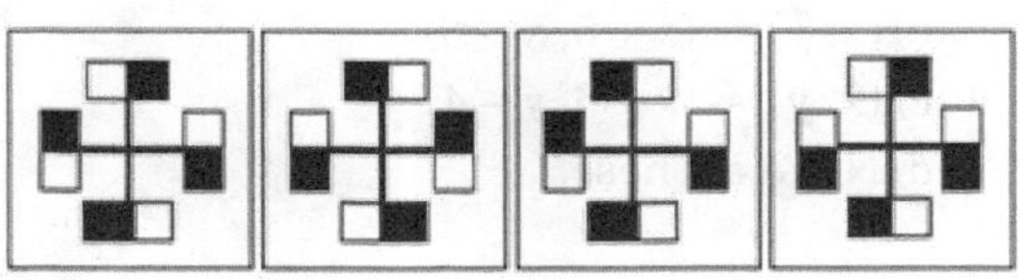

(a) (b) (c) (d)

9. Which figure is a rotation of the object?

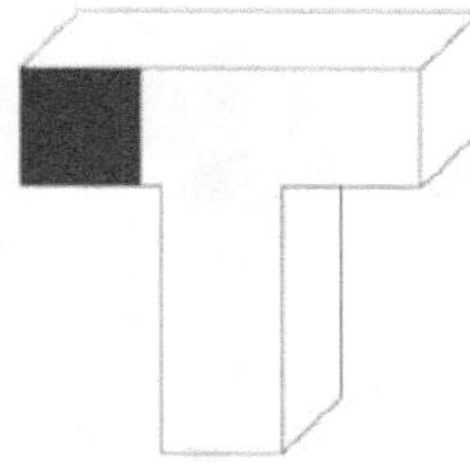

(a) 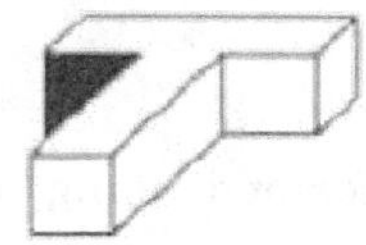 (b) 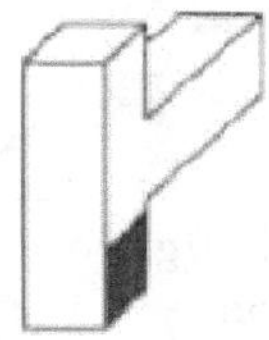

(c) 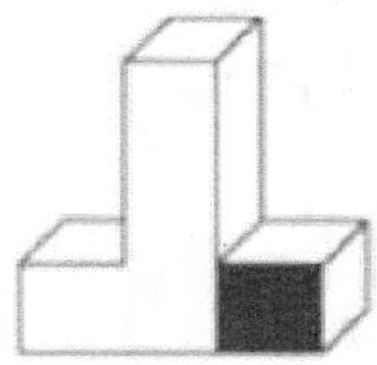 (d) 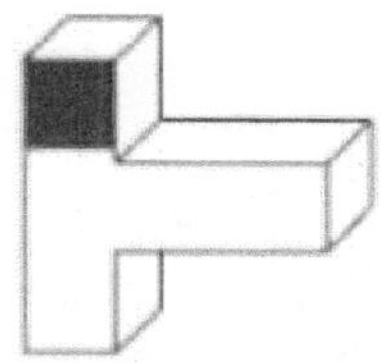

10. Which figure is a rotation of the object?

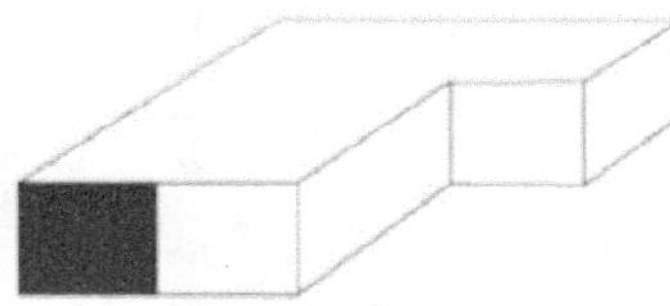

(a) 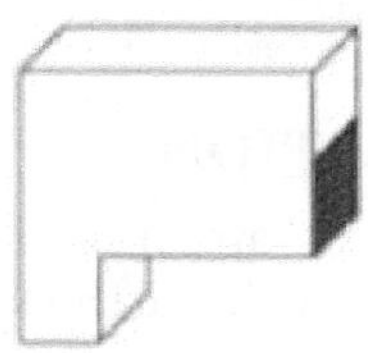 (b) 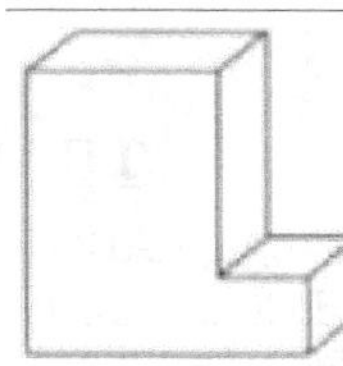

(c) 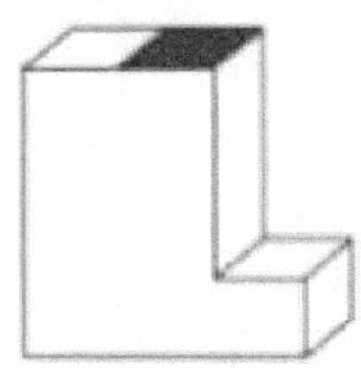 (d) 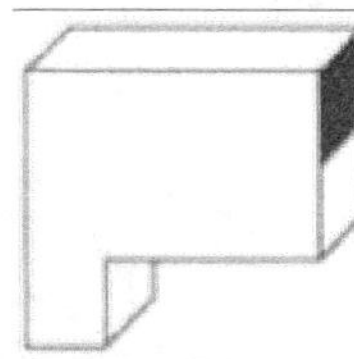

11. Which of the 4 figures presented (A, B, C, D) is a rotation of the first?

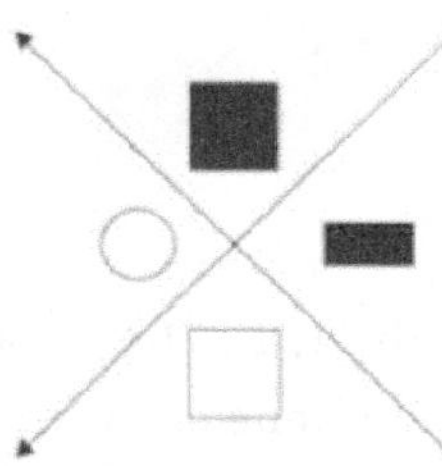

(a) 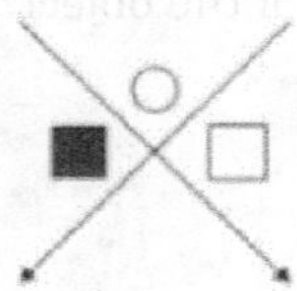 (b) 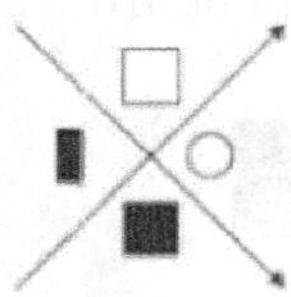

(c) 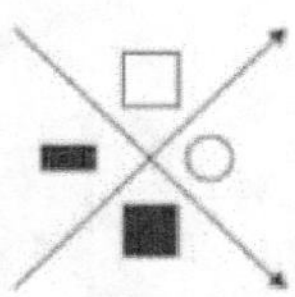 (d) 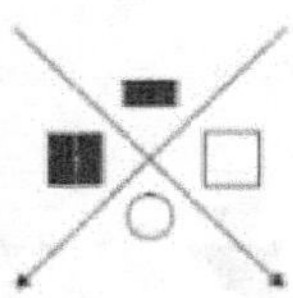

12. Which figure can be formed using only 2D rotation?

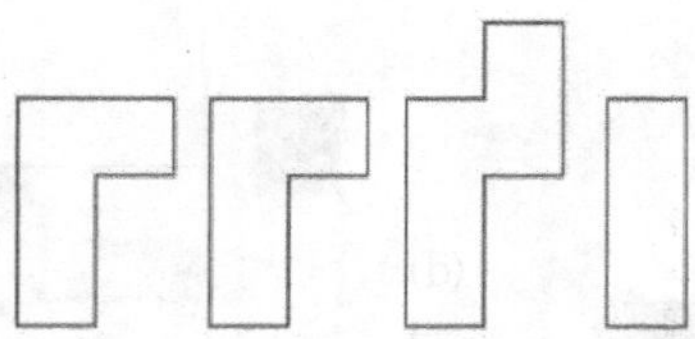

(a) 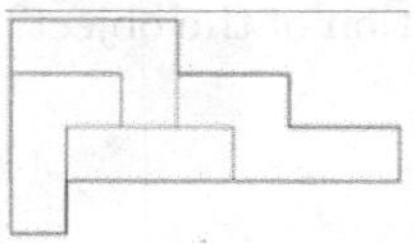 (b) 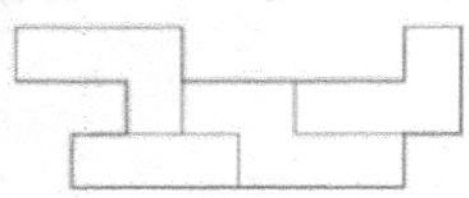

(c) 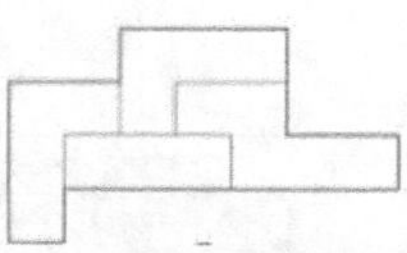 (d) 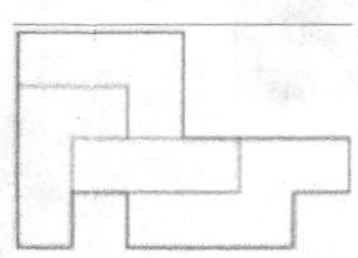

TRANSLATION

13. Triangle ΔABC has coordinates A(3, –1), B(7, –5) and C(–2, –2). Translate ΔABC to the left 4 units and up 5 units. Determine the coordinates of ΔA'B'C'.

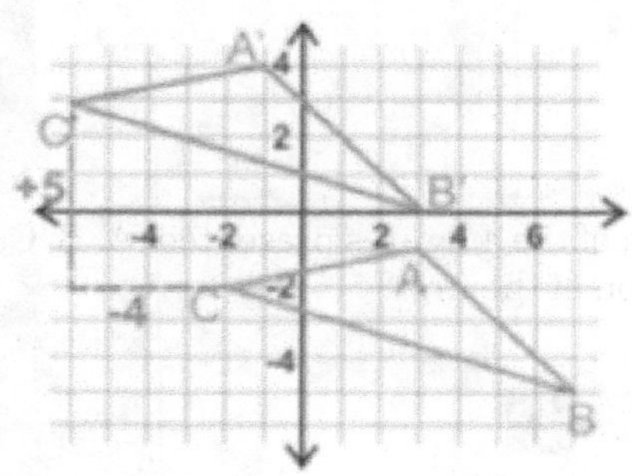

(a) (–1, 4), (3, 0), (–6, 3) (b) (1, 4), (0, 3), (–4, 3)

(c) (–1, 5), (3, 0), (–6, 3) (d) None of these.

14. Using the translation (x, y) → (x + 2, y – 5), what is the image of A(–6, 3), B(2, 5), C(3, 1)?

(a) (–4, –2), (4, 0), (5, –4)

(b) (4, 2), (0, 4), (–5, 4)

(c) (–4, –2), (4, 0), (–5, –4)

(d) None of these.

15. Graph square S(1, 2), Q(4, 1), R(5, 4) and E(2, 5). Find the image after the translation (x, y) → (x – 2, y + 3). Then graph and label the image.

We are going to move the square to the left 2 and up 3.

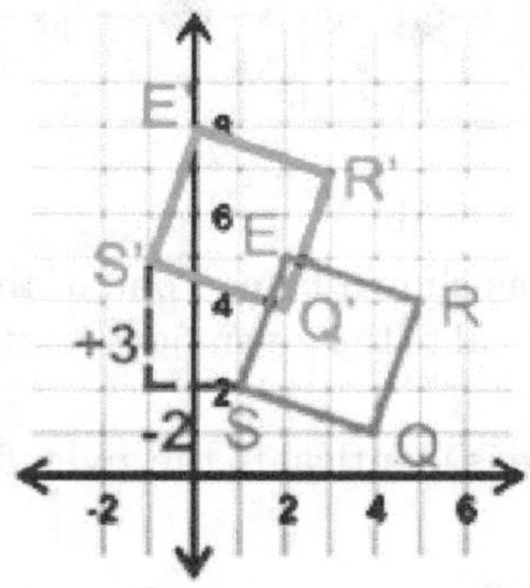

(a) S'(–1, 5), Q'(2, 4), R'(3, 7), E'(0, 8)

(b) S'(–2, 5), Q'(2, 4), R'(3, 7), E'(1, 8)

(c) S'(–1, 5), Q'(2, 4), R'(1, 7), E'(2, 8)

(d) None of these.

16. Find the translation rule for ΔTRI to ΔT'R'I'. Look at the movement from T to T'. The translation rule is

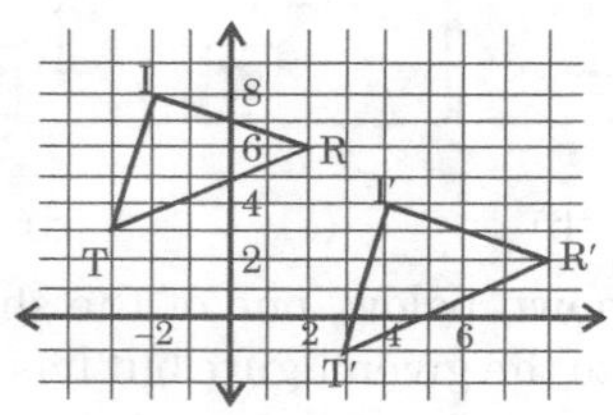

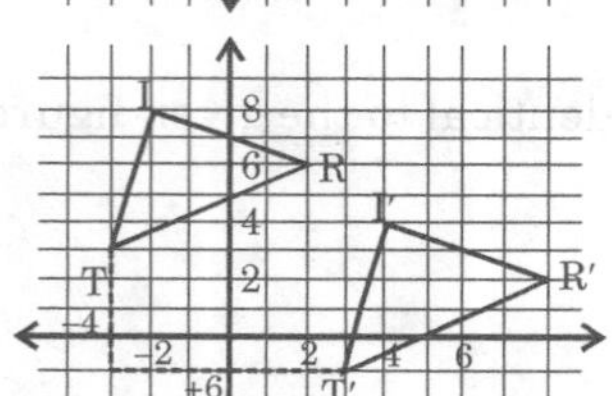

(a) (x, y) → (x + 6, y – 5).

(b) (x, y) → (x + 5, y – 3).

(c) (x, y) → (x + 6, y – 4)

(d) None of these.

17. Find the translation rule for:

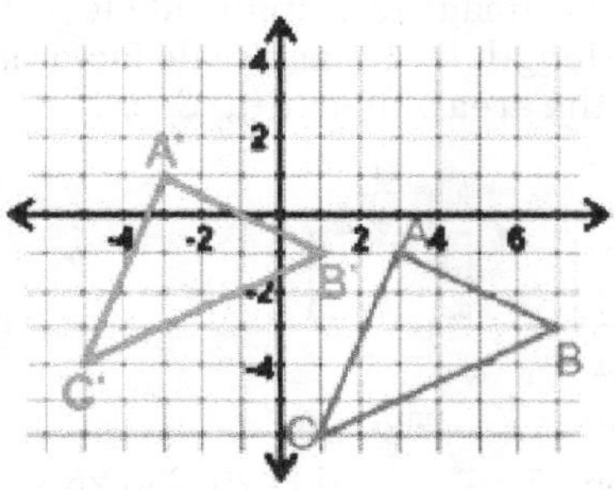

(a) (x, y) → (x – 7, y + 4).
(b) (x, y) → (x – 6, y + 2).
(c) (x, y) → (x – 5, y + 1).
(d) None of these.

18. Find the translation rule for:

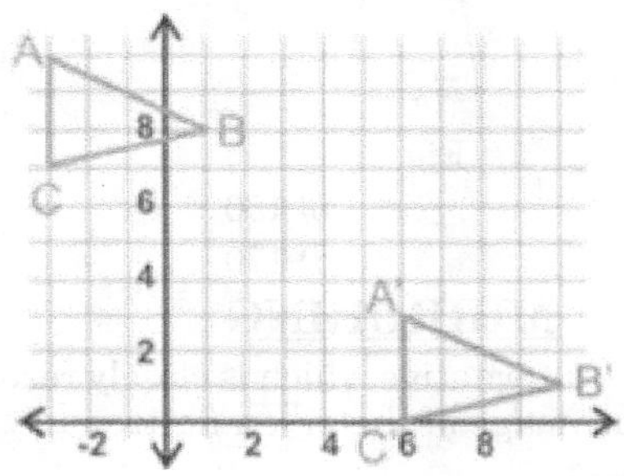

(a) (x, y) → (x – 7, y + 7).
(b) (x, y) → (x + 7, y + 9).
(c) (x, y) → (x + 9, y – 7).
(d) None of these.

19. Find the translation rule for:

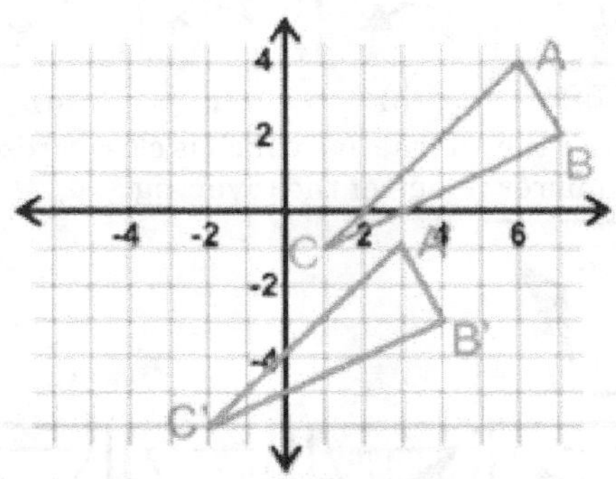

(a) (x, y) → (x – 3, y – 5).
(b) (x, y) → (x + 7, y + 9).
(c) (x, y) → (x + 9, y – 7).
(d) None of these.

20. Find the translation rule for:

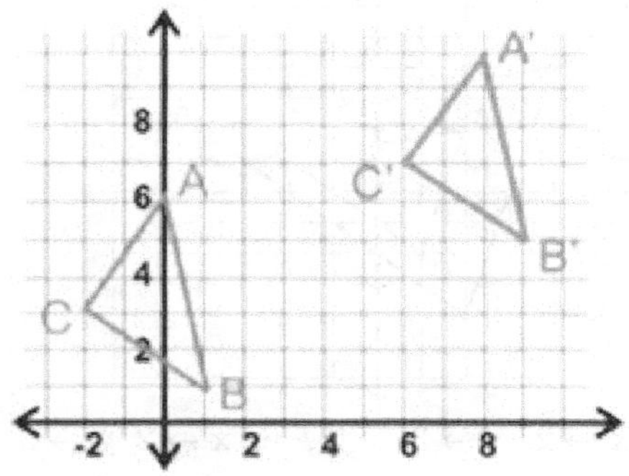

(a) (x, y) → (x + 6, y – 5).
(b) (x, y) → (x + 7, y + 9).
(c) (x, y) → (x + 8, y + 4).
(d) None of these.

DILATION OR SCALING

21. An art supply store sells several sizes of drawing triangles. All are dilations of a single basic triangle. The basic triangle and one of its dilations are shown on the grid. Find the scale factor of the dilation.

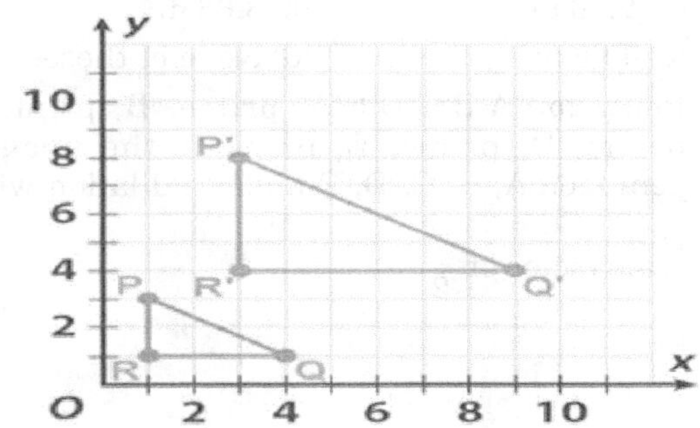

(a) x → 2, y → 2
(b) x → 1.5, y → 2
(c) x → 2, y → 1.5
(d) x → 2, y → 1.5

22. Find the scale factor of the dilation.

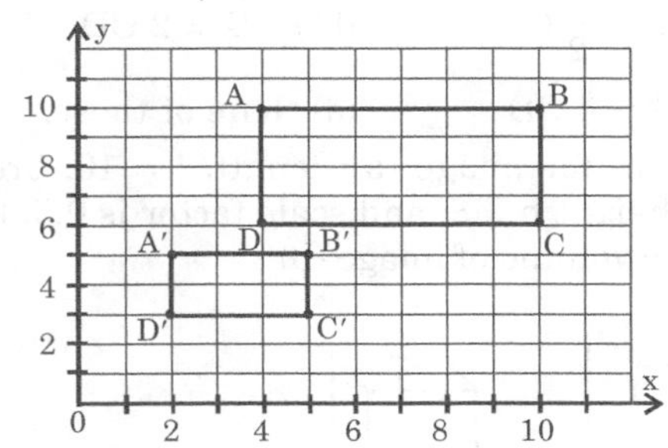

(a) x → 2, y → 2 b) x → 2, y → 3
(c) x → 3, y → 2 d) x → 3, y → 3

23. Find the scale factor of the dilation.

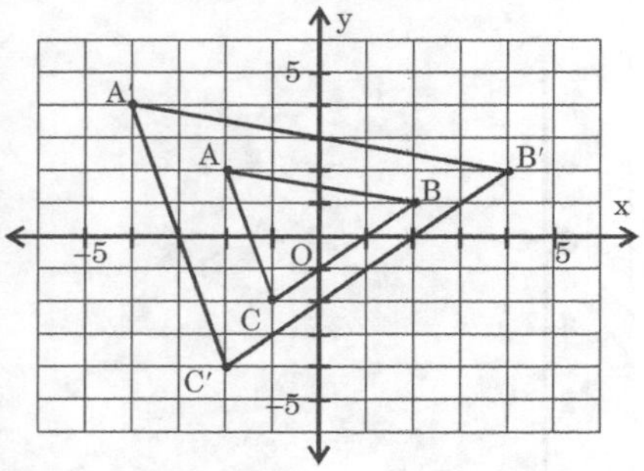

(a) $x \rightarrow 2,\ y \rightarrow 3$ (b) $x \rightarrow 2,\ y \rightarrow 2$
(c) $x \rightarrow 3,\ y \rightarrow 2$ (d) None of these

24. As given, the area of shaded portion of image (a) is $24m^2$ and scale factor is 3, then find the area of image (b).

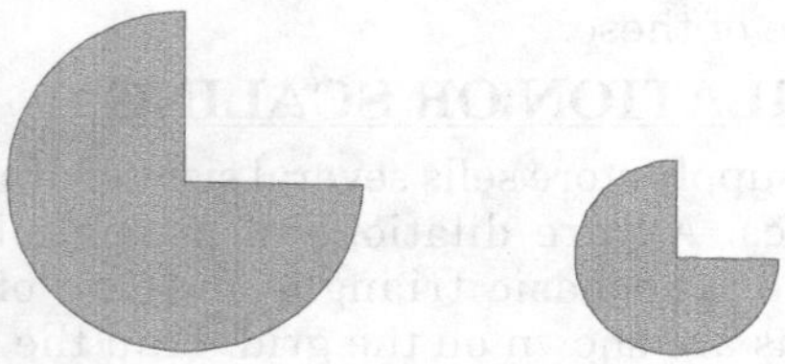

Fig. (a) Fig. (b)

(a) $2.66\ m^2$ (b) $3.66\ m^2$
(c) $4.66\ m^2$ (d) None of these.

25. Pentagon A'B'C'D'E'A, prime, B, prime, C, prime, D, prime, E, prime is the image of pentagon A, B, C, D, E under a dilation with a scale factor of $\frac{1}{2}$.

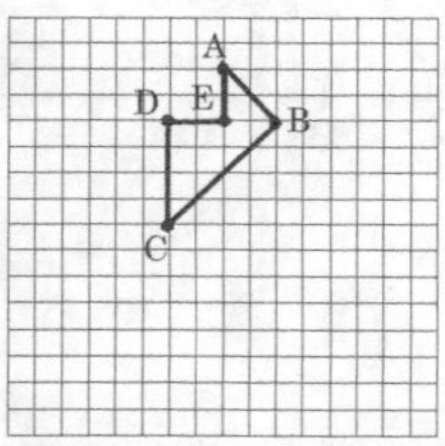

What is the relation between $C'D'$ and CD?

(a) $C'D' = \frac{1}{2}$ CD (b) $C'D' = 2$ CD
(c) $C'D' = 4$ CD (d) None of these.

26. As given, for image (a); length l = 10, breath b = 6, height h = 8, and scale factor is 0.5, then find the volume of image (b)

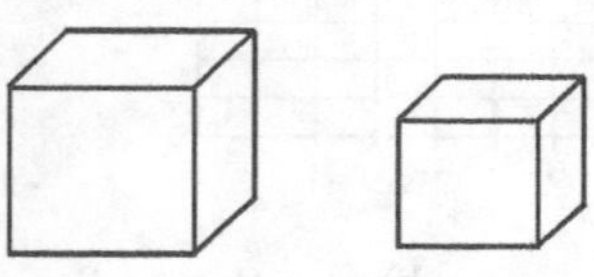

Fig. (a) Fig. (b)

(a) 40 (b) 50
(c) 60 (d) 70

27. As given, for image (a); small side length l = 8, large side length l = 10, and scale factor is 0.75 , then find the area of image (a) & (b).

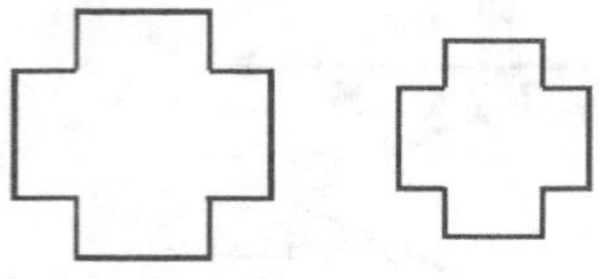

Fig. (a) Fig. (b)

(a) 320, 136.25 (b) 420, 236.25
(c) 520, 336.25 (d) None of these

28. As given, for image (a); small side length l = 8, large side length l = 10, and scale factor is 0.75 , then find the area of image (b).

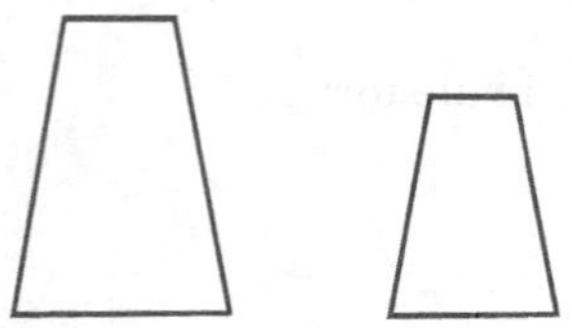

Fig. (a) Fig. (b)

(a) 140 (b) 150
(c) 160 (d) 170

MIRRORING

29. Choose the alternative which is closely resembles the mirror image of the given figure.

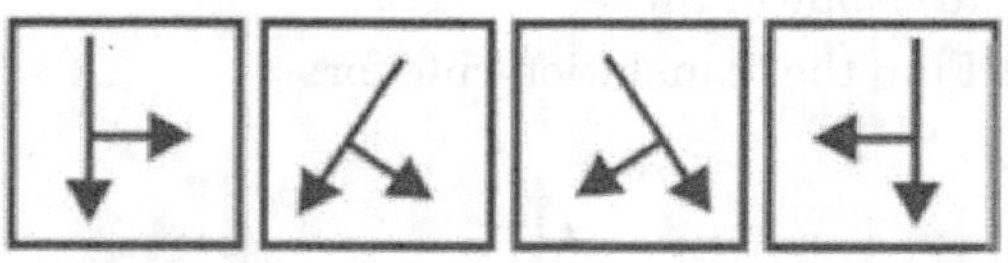

(a) (b) (c) (d)

30. Choose the alternative which is closely resembles the mirror image of the given fig

(a) (b) (c) (d)

31. Choose the alternative which is closely resembles the Mirror image of the given figure.

(a) (b) (c) (d)

32. Choose the alternative which is closely resembles the Mirror image of the given figure.

(a) (b) (c) (d)

33. Choose the alternative which is closely resembles the mirror image of the given figure.

(a) (b) (c) (d)

34. Choose the alternative which is closely resembles the Mirror image of the given figure.

(a) (b) (c) (d)

35. Choose the alternative which is closely resembles the Mirror image of the given figure.

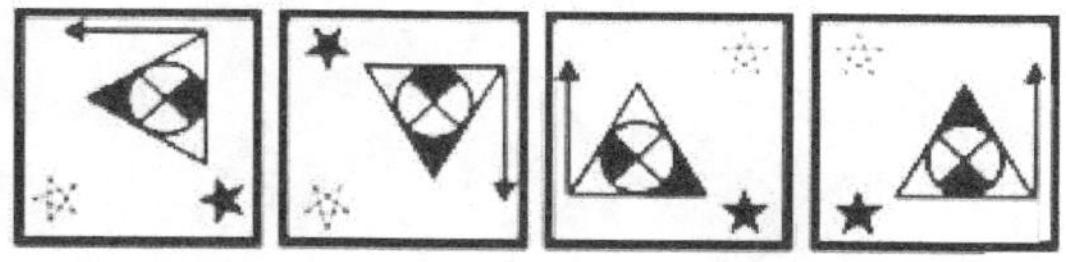

(a) (b) (c) (d)

36. Choose the alternative which is closely resembles the Mirror image of the given figure.

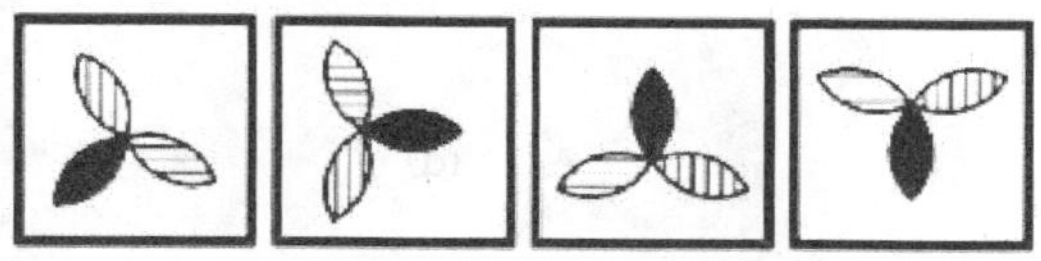

(a) (b) (c) (d)

37. Choose the alternative which is closely resembles the mirror-image of the given combination.

LETTERS

(a) LƎTTƎRS (b) LETTEЯS

(c) ƧЯƎTTƎ⅃ (d) ƧЯƎTTEL

38. Choose the alternative which is closely resembles the mirror-image of the given combination.

NiCaRaGuT

(a) TuGaRaCiN (b) TuƊaЯaƆiИ

(c) TuƊaЯaƆiИ (d) TuƊɒЯɒƆiИ

39. Choose the alternative which is closely resembles the mirror-image of the given combination.

INFORMATIONS

(a) ƧNOITAMЯOꟻNI (b) ƧИOITAMЯOꟻИI

(c) IИԀOᴚWⱯꓕIONS (d) ƧИOITAMЯOꟻИI

40. Directons: Find the mirror image of following words/numbers from amongst the given alternatives.

x (a) (b) (c) (d)

41. Choose the alternative which is closely resembles the mirror image of the given figure.

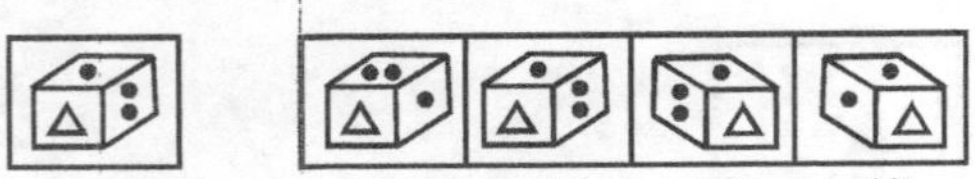

x (a) (b) (c) (d)

42. Which figure is a vertical reflection across the x axis?

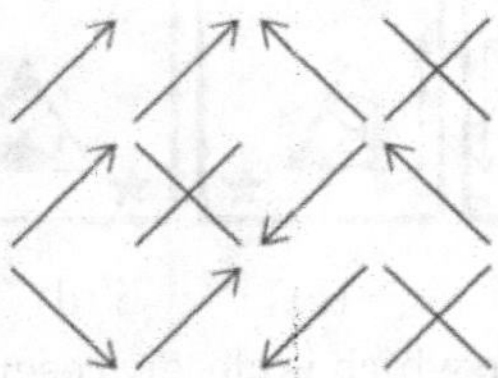

(a)

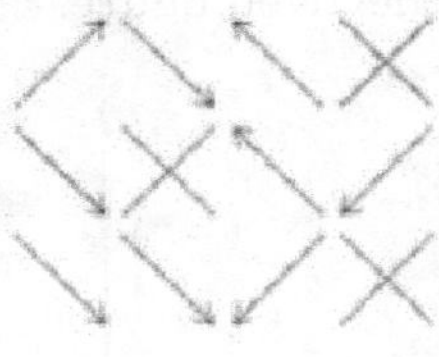

(b)

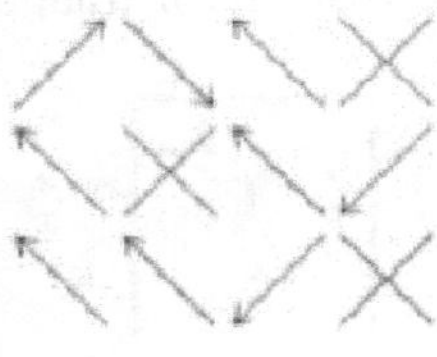

(c)

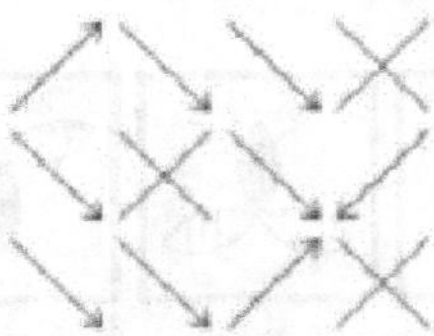

(d)

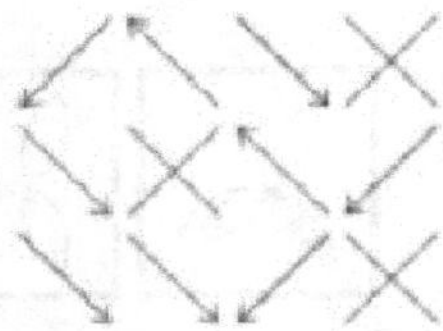

Directions for questions 43 to 57: If a mirror is placed, then which of the answer figures is the mirror image of the given figure?

43. R E S T

(a) T Ƨ E Я (b) T Ƨ Ǝ Я
(c) ⊥ Ƨ Ǝ Я (d) T S E R

44. 1 5 8 2 4

(a) ɭ 5 8 2 4 (b) ᔭ 2 8 5 1
(c) ᔭ 2 8 5 ɭ (d) 4 2 8 5 1

45. M 3 5 K G 1

(a) M Ɛ 5 ꓘ Ә ɭ (b) 1 G K 5 3 M
(c) ɭ Ә K 5 Ɛ M (d) ɭ Ә ꓘ 5 Ɛ M

46.

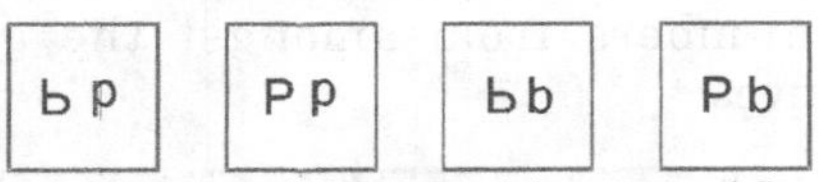

(a) (b) (c) (d)

47.

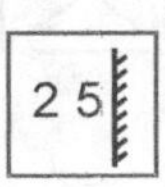

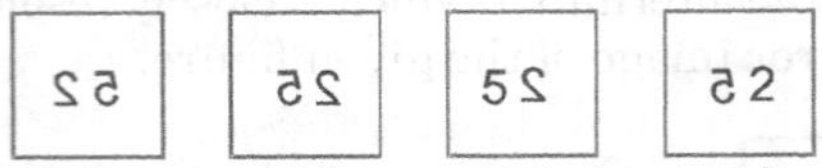

(a) (b) (c) (d)

48.

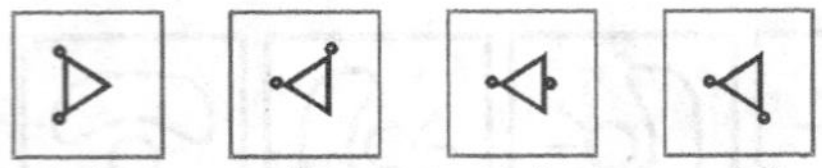

(a) (b) (c) (d)

49.

(a) (b) (c) (d)

50.

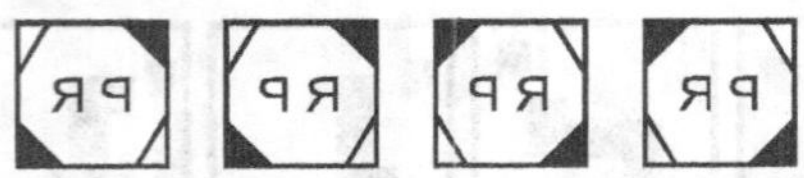

(a) (b) (c) (d)

51.

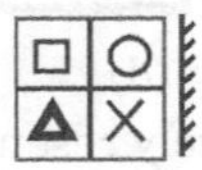

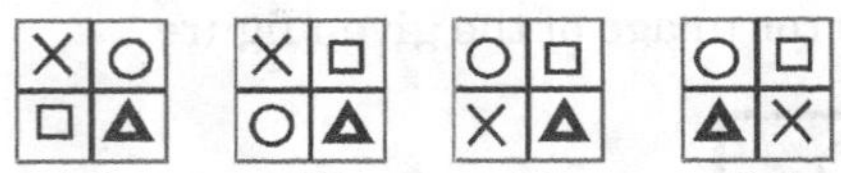

(a) (b) (c) (d)

52.

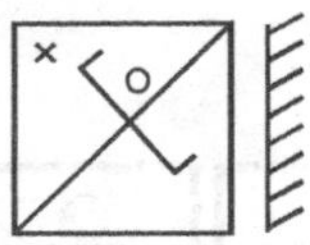

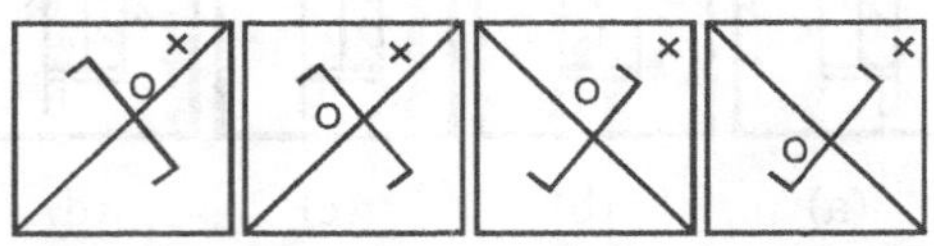

(a) (b) (c) (d)

53.

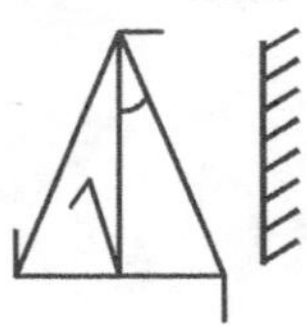

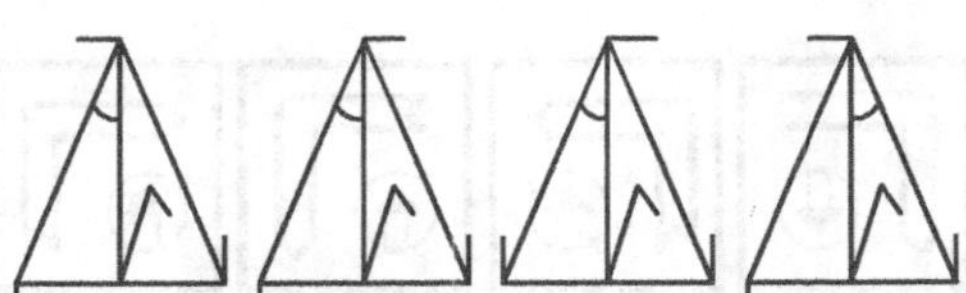

(a) (b) (c) (d)

54.

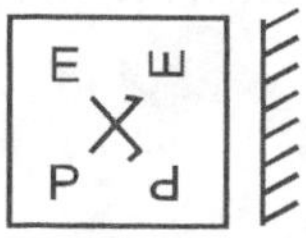

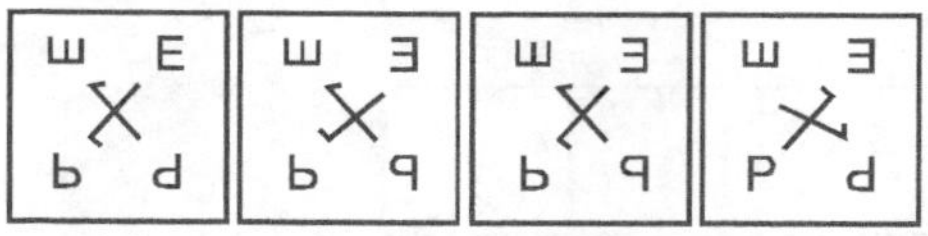

(a) (b) (c) (d)

55.

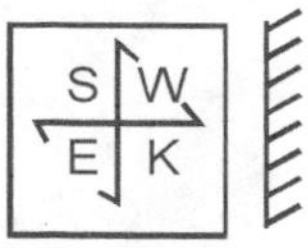

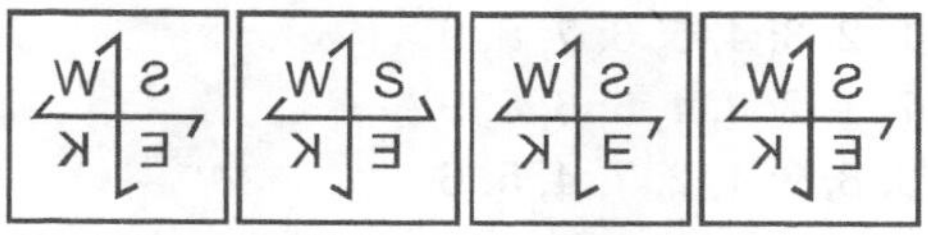

(a) (b) (c) (d)

56.

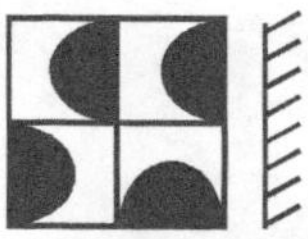

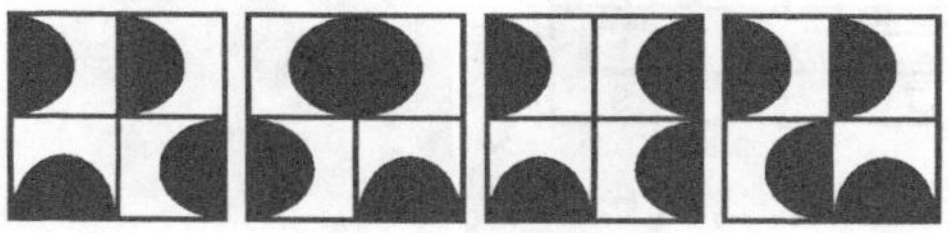

(a) (b) (c) (d)

57.

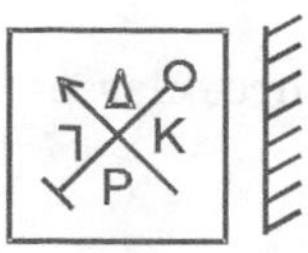

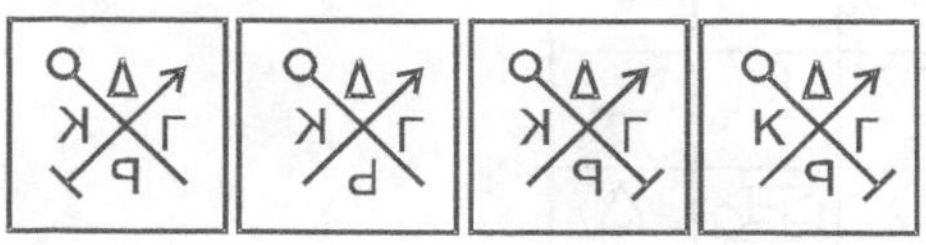

(a) (b) (c) (d)

GROUPING

58. Group the following figures into three classes with identical properties and select the correct alternative

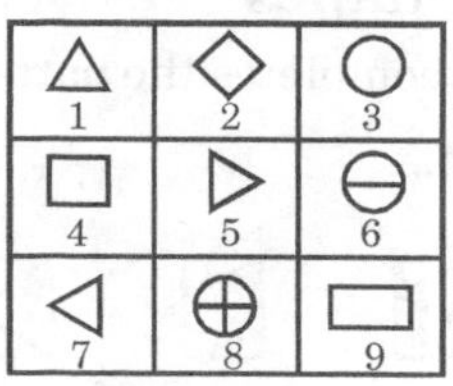

(a) 1, 5, 8; 2, 4, 7; 3, 6, 9
(b) 1, 4, 7; 2, 5, 8; 3, 6, 9
(c) 1, 5, 7; 2, 4, 9; 3, 6, 8
(d) 1, 5, 7; 3, 4, 9; 2, 6, 8

59. Group the following figures into three classes with identical perpoerties and select the correct alternative.

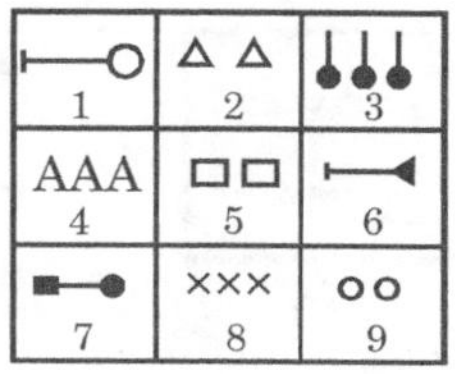

(a) 1, 2, 3; 6, 5, 4; 7, 9, 8
(b) 1, 6, 7; 2, 5, 9; 3, 4, 8
(c) 1, 6, 7; 2, 5, 8; 3, 4, 9
(d) 1, 3, 7; 2, 5, 9; 4, 6, 8

60. Choose the correct sequence which are the group of identical image.

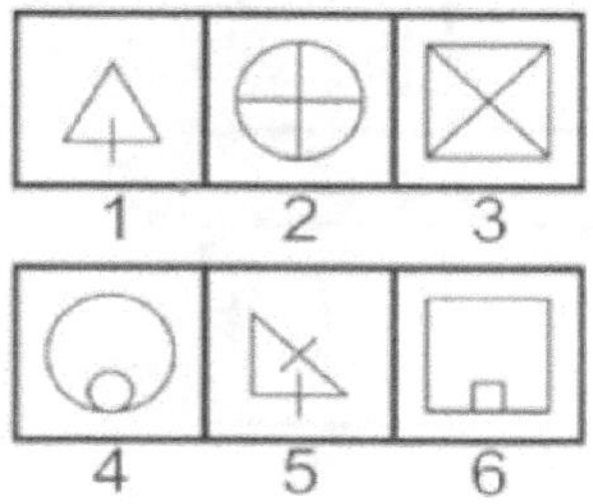

Options

(a) 2, 3; 4, 6; 5, 1 (b) 1, 5; 2, 3; 4, 6
(c) 1, 4; 3, 2; 5, 6 (d) 1, 3; 4, 2; 5, 6

61. Choose the correct sequence which are the group of identical images.

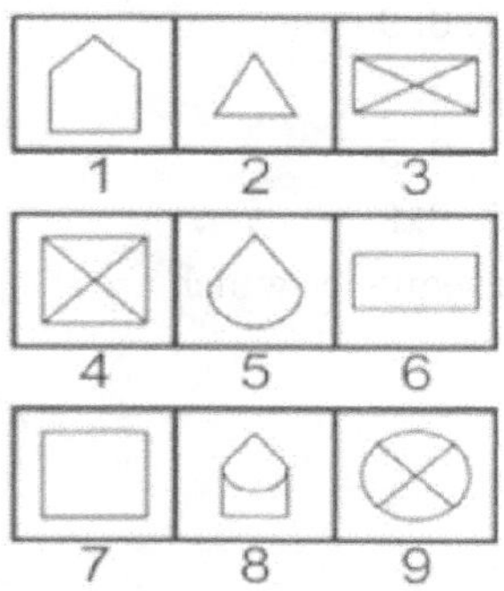

Options

(a) 1, 5, 8; 2, 6, 7; 3, 4, 9
(b) 1, 2, 3; 4, 5, 6; 7, 8, 9
(c) 4, 5, 6; 7, 8, 9; 1, 2, 3
(d) 1, 4, 7; 2, 5, 8; 3, 6, 9

62. Choose the correct sequence which are the group of identical images.

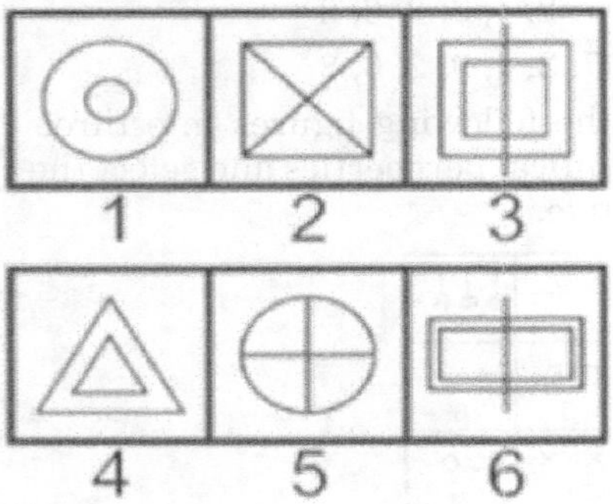

Options

(a) 1, 2; 3, 4; 5, 6 (b) 1, 4, 6, 3; 2, 5

(c) 1, 3; 4, 5; 2, 6 (d) 1, 5; 2, 3; 4, 6

63. Choose the correct sequence which are the group of identical images.

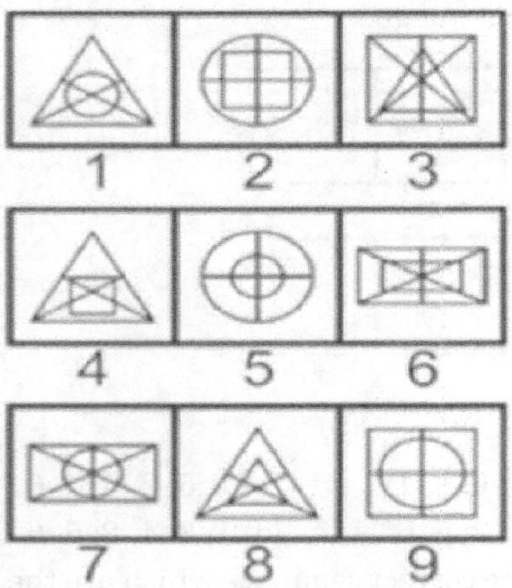

Options

(a) 1, 3, 5; 2, 4, 6; 7, 8, 9

(b) 1, 4, 8; 2, 5, 9; 3, 6, 7

(c) 1, 2, 3; 4, 6, 7; 8, 9, 5

(d) 1, 6, 7; 2, 3, 4; 5, 8, 9

64. Choose the correct sequence which are the group of identical images.

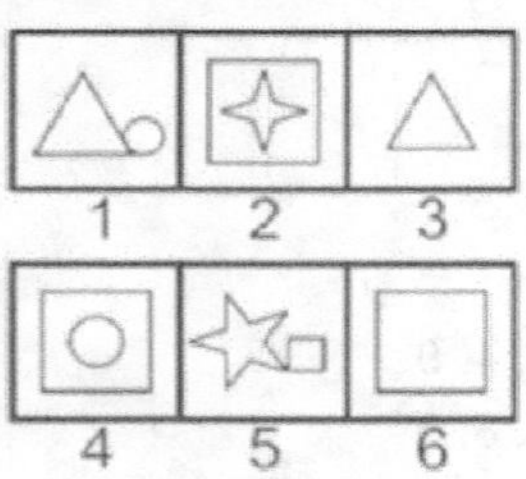

(a) 1, 3; 2, 4; 5, 6 (b) 1, 4; 3, 5; 2, 6

(c) 1, 6; 3, 4, 2, 5 (d) 1, 5; 2, 4; 3, 6

65. Choose the correct sequence which are the group of identical images.

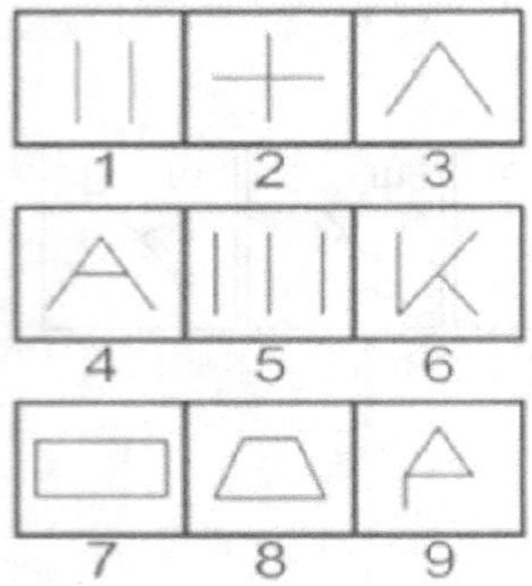

Options

(a) 3, 2, 4; 1, 5, 6; 7, 8, 9

(b) 1, 2, 3; 4, 5, 6; 7, 8, 9

(c) 7, 1, 2; 3, 4, 5; 6, 8, 9

(d) 3, 8, 9; 1, 2, 7; 4, 5, 6

66. Group the given figures into three classes using each figure only once.

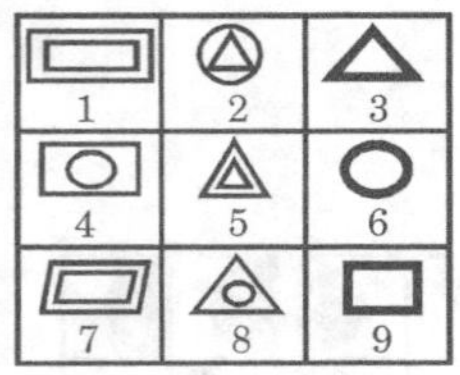

(a) 1, 5, 7; 2, 4, 6; 3, 9, 8

(b) 1, 5, 7; 2, 4, 8; 3, 6, 9

(c) 1, 4, 7; 2, 5, 8; 3, 6, 9

(d) 1, 7, 9; 3, 5, 8; 2, 4, 6

67. Group the given figures into three classes using each figure only once.

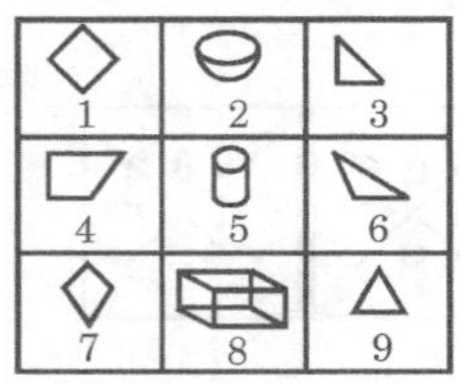

(a) 1, 4, 7; 2, 5, 8; 3, 6, 9

(b) 1, 4, 7; 2, 5, 9; 3, 6, 7

(c) 1, 3, 4; 2, 5, 8; 6, 7, 9

(d) 1, 2, 3; 4, 5, 6; 7, 8, 9

2D & 3D Shapes

68. Identify the figure that completes the pattern.

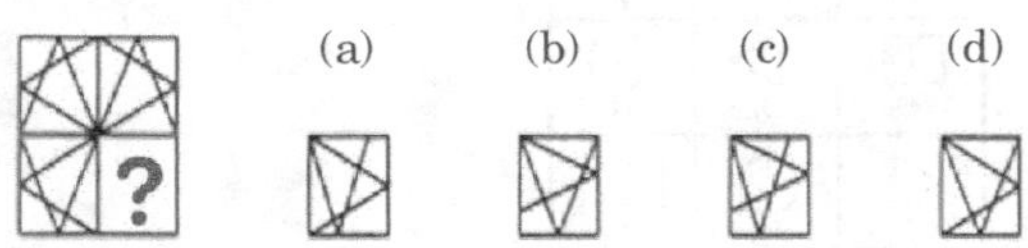

69. Identify the figure that completes the pattern.

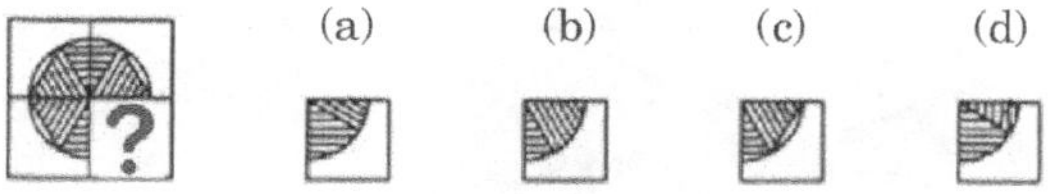

70. Which of the shadows on the right can result from casting light on one of the 3D-shape's sides?

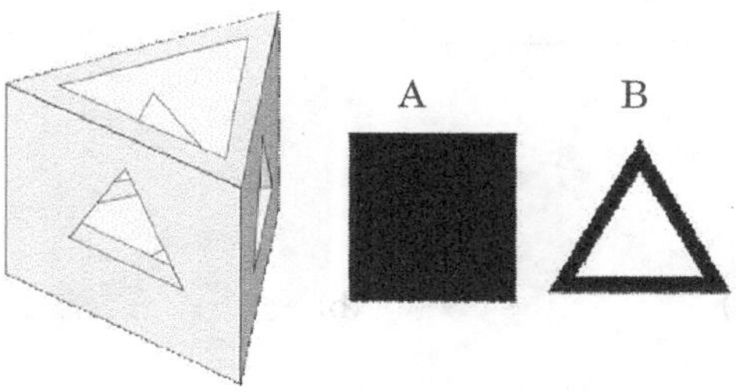

(a) A (b) B
(c) Both A & B (d) None of these.

71. Choose the image that completes the pattern:

Which of the shape below continues the sequence:

(a) 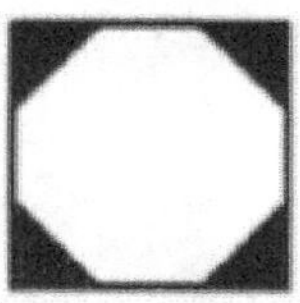 (b) 

(c)  (d) 

72. Choose the image that completes the pattern:

Which of the shapes below continues the sequence:

(a)  (b) 

(c)  (d) 

73. Choose the image that completes the pattern:

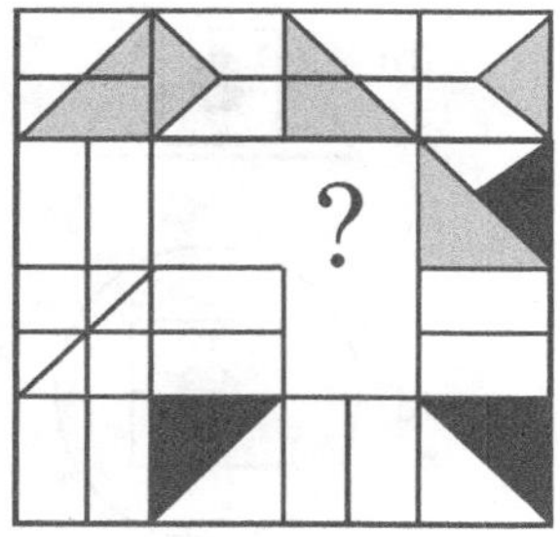

Which of the shapes below replaces the missing box above?

(a) 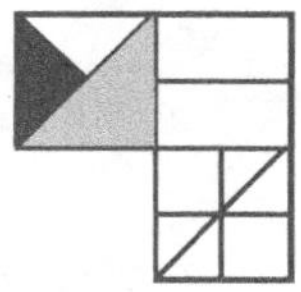 (b) 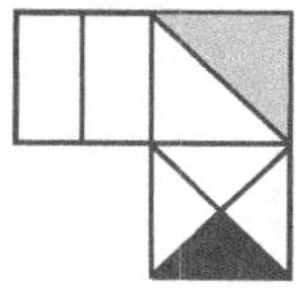

(c) 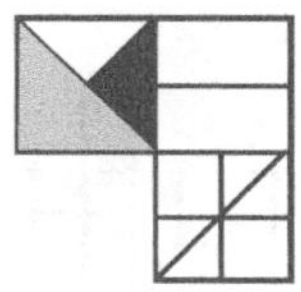 (d) 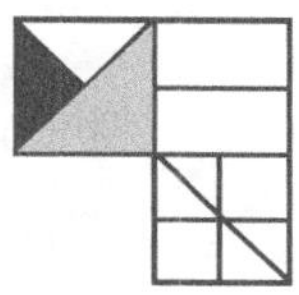

74. Choose the image that completes the pattern:

Which of the shapes below continues the sequence:

(a)  (b) 

(c)  (d) 

75. Choose the image that completes the pattern:

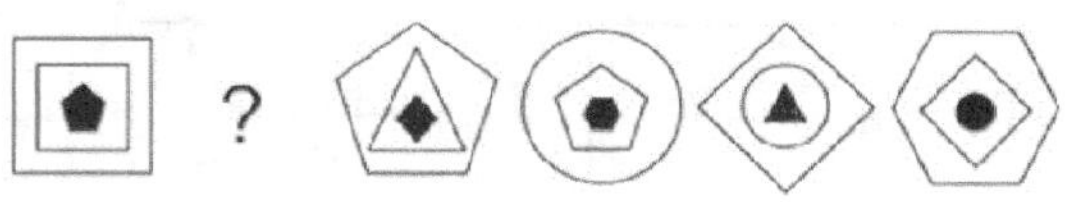

Which of the shapes below replaces the missing box above?

(a) 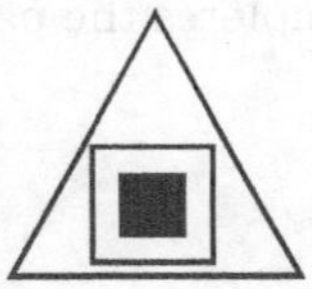 (b) 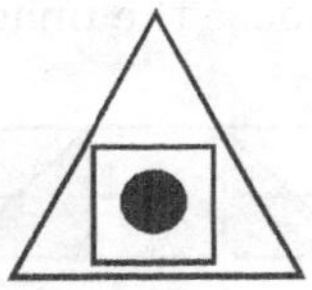

(c) 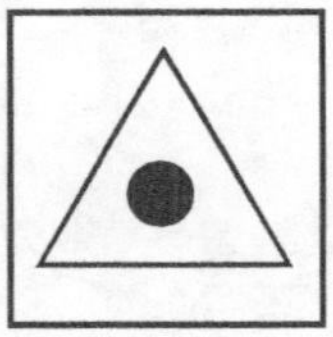 (d) 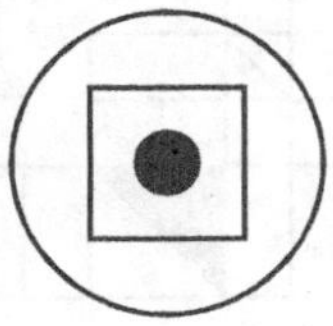

76. Which piece corresponds to the given solid?

(a) 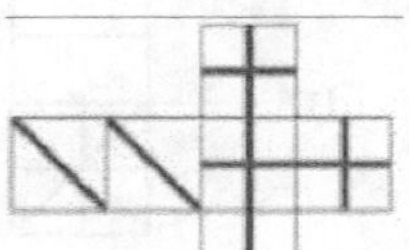 (b) 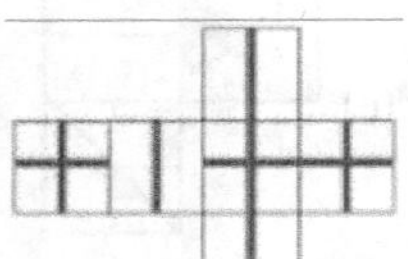

(c) 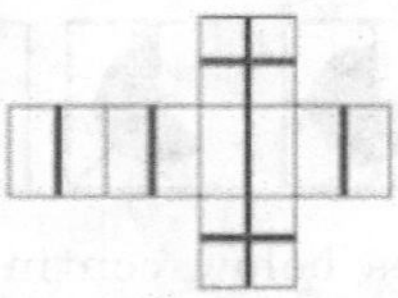 (d) 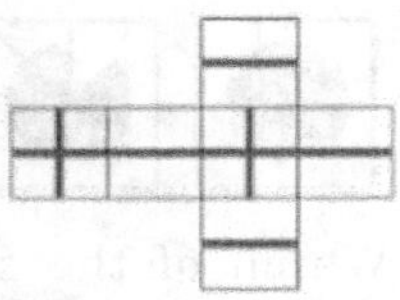

77. Which piece corresponds to the given solid?

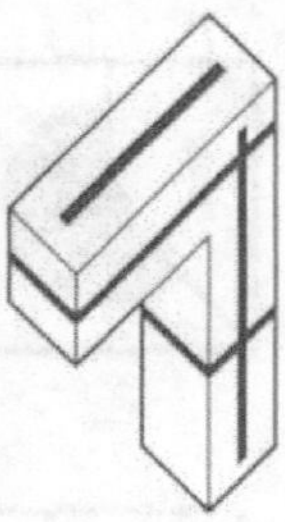

(a) 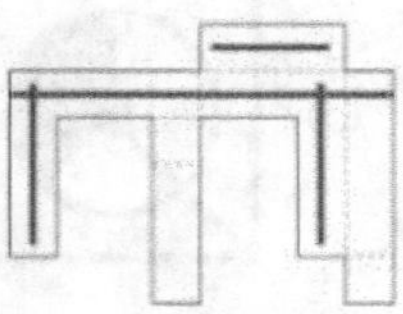 (b) 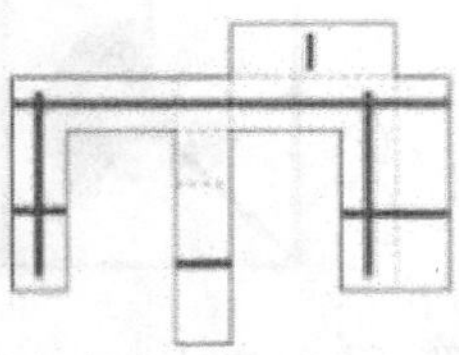

(c) 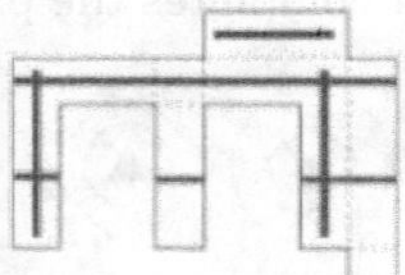 (d) 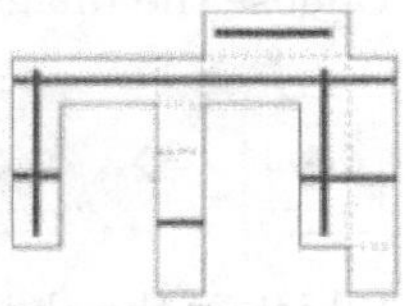

78. To which object does the given top view correspond?

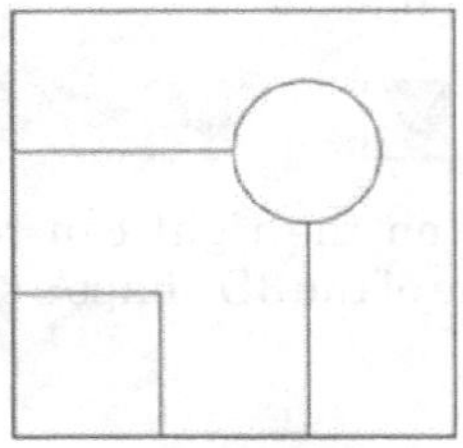

(a)  (b) 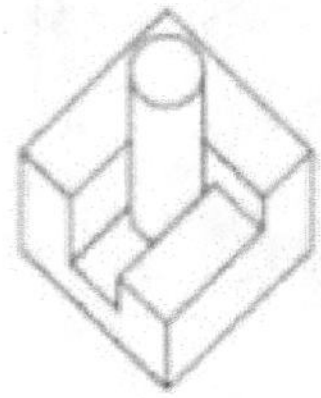

(c)  (d) 

79. Which of these images is the top view of the given object?

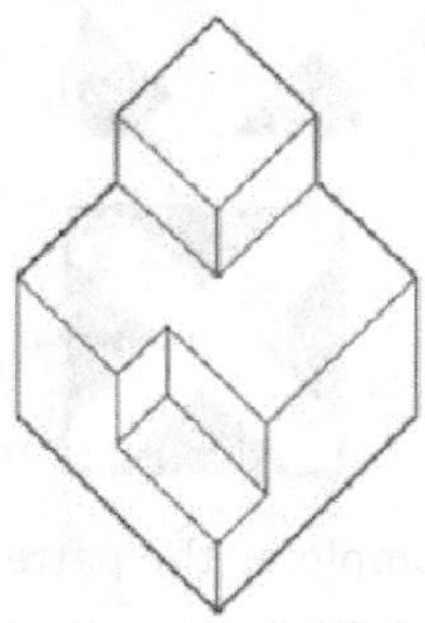

(a) 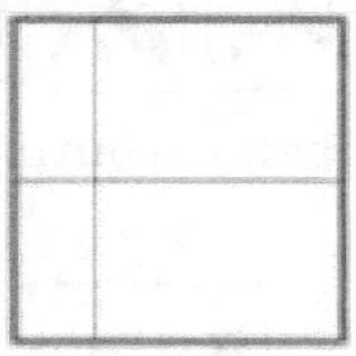 (b) 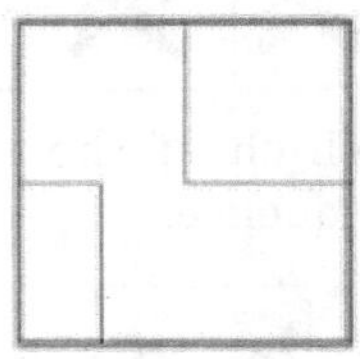

(c) 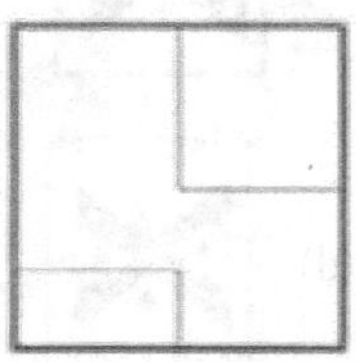 (d) 

PAPER FOLDING AND CUTTING

Directions for questions 80 to 104: In each of the following questions a paper is folded and cut as shown below in the question figure. From the given answer figures, indicate how it will appear when opened?

80. Question Figures

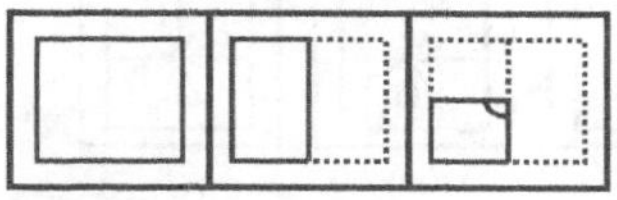

Answer Figures

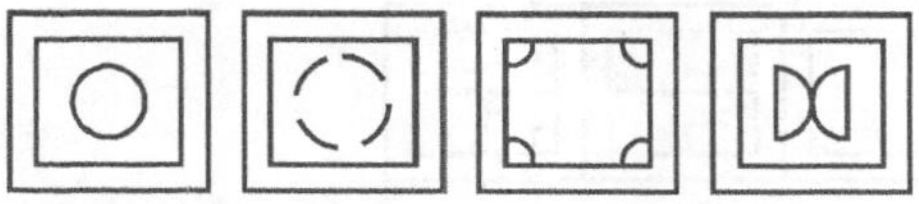

(a) (b) (c) (d)

81. Question Figures

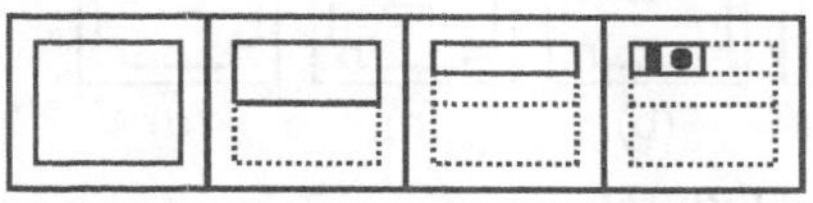

Answer Figures

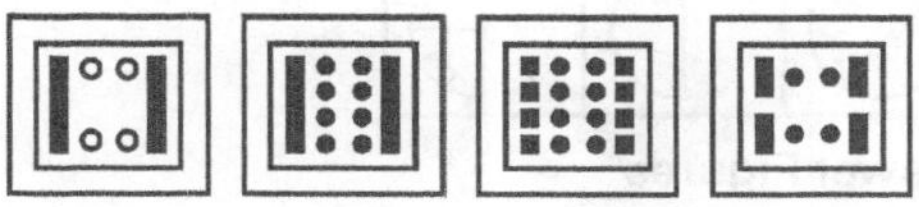

(a) (b) (c) (d)

82. Question Figures

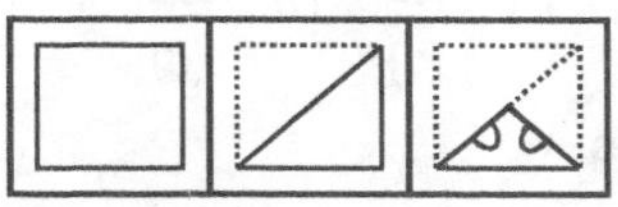

Answer Figures

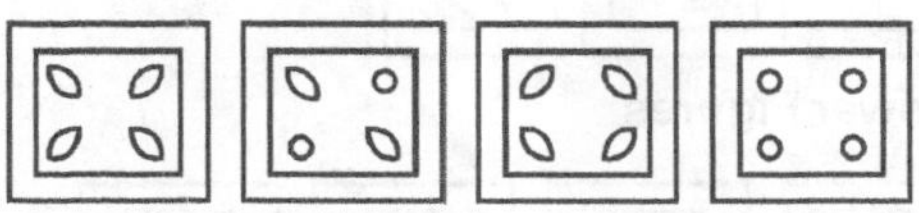

(a) (b) (c) (d)

83. Question Figures

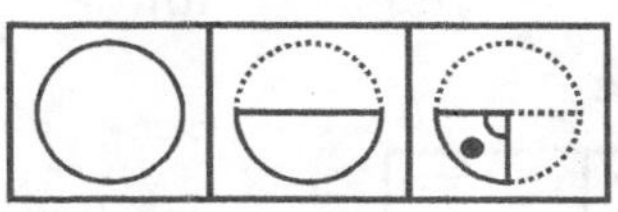

Answer Figures

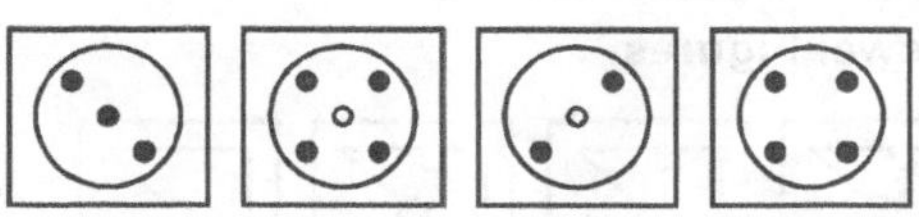

(a) (b) (c) (d)

84. Question Figures

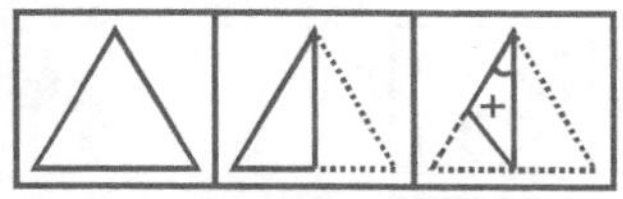

Answer Figures

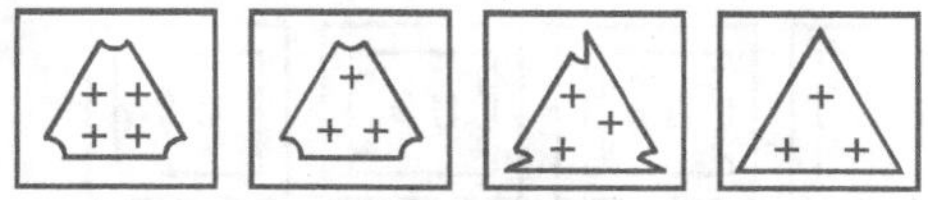

(a) (b) (c) (d)

85. Question Figures

Answer Figures

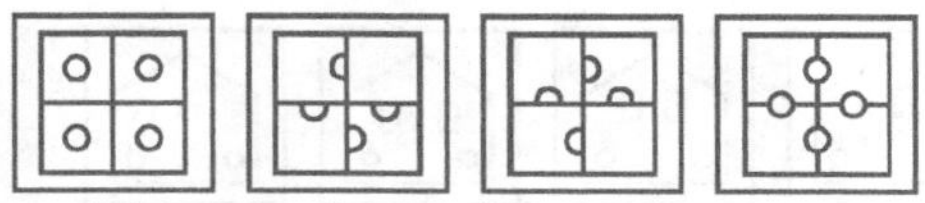

(a) (b) (c) (d)

86. Question Figures

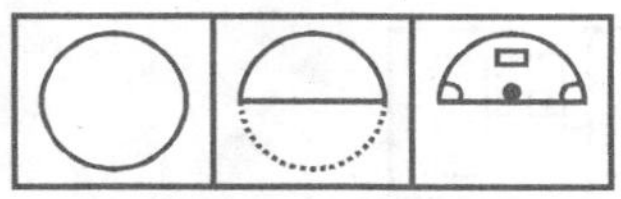

Answer Figures

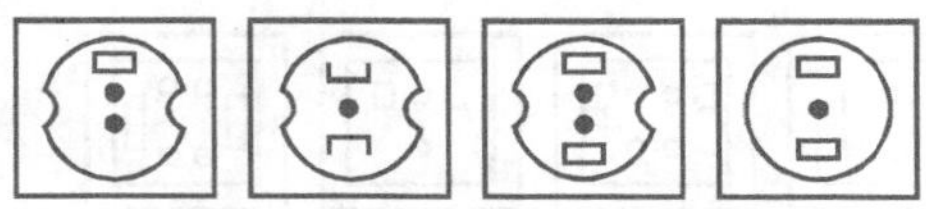

(a) (b) (c) (d)

87. Question Figures

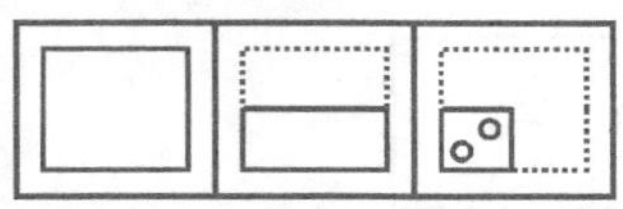

Answer Figures

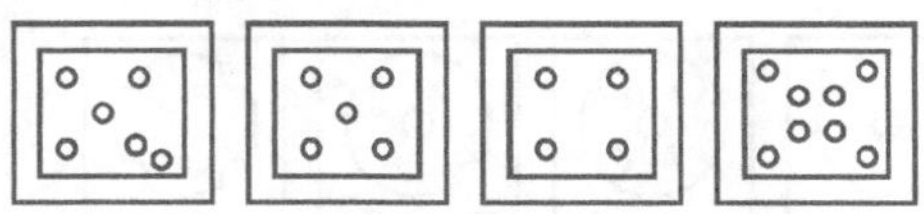

(a) (b) (c) (d)

88. Question Figures

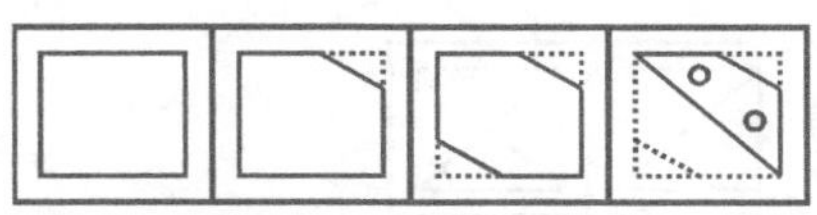

Answer Figures

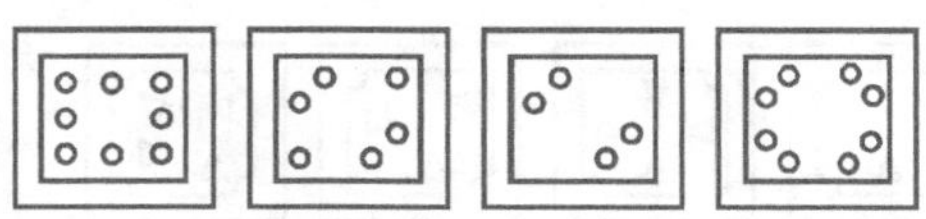

(a) (b) (c) (d)

89. Question Figures

Answer Figures

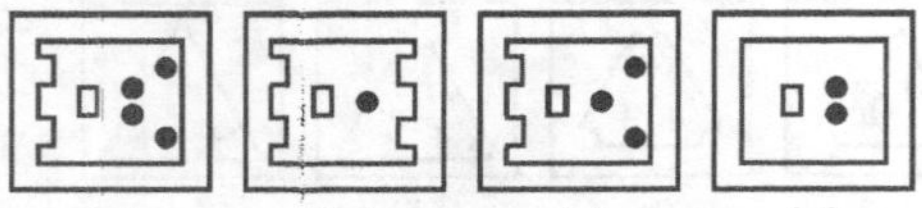

(a) (b) (c) (d)

90. Question Figures

Answer Figures

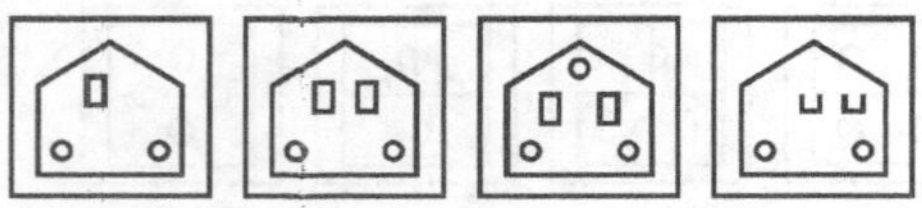

(a) (b) (c) (d)

91. Question Figures

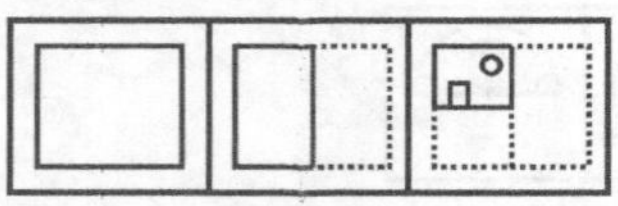

Answer Figures

(a) (b) (c) (d)

92. Question Figures

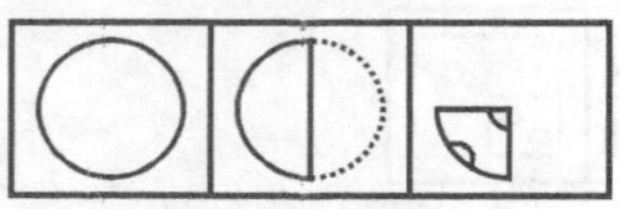

Answer Figures

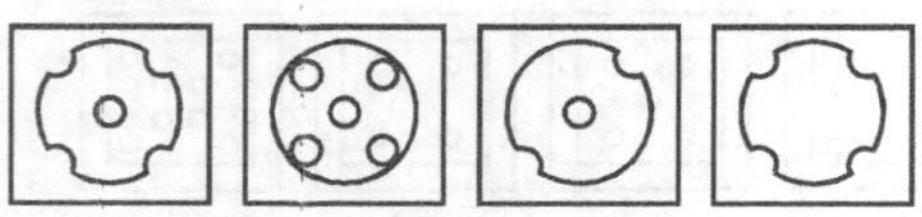

(a) (b) (c) (d)

93. Question Figures

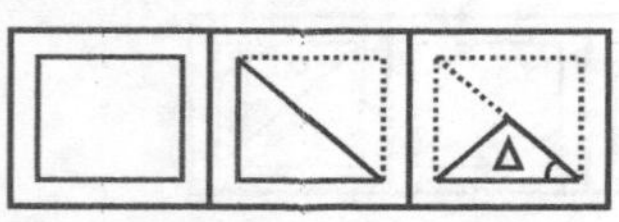

Answer Figures

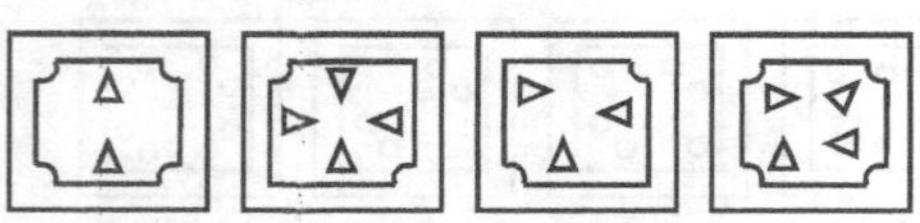

(a) (b) (c) (d)

94. Question Figures

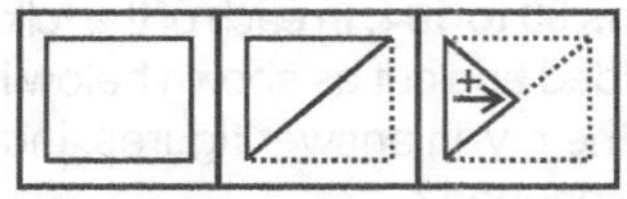

Answer Figures

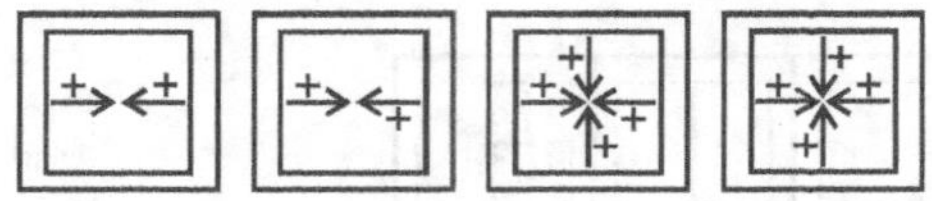

(a) (b) (c) (d)

95. Question Figures

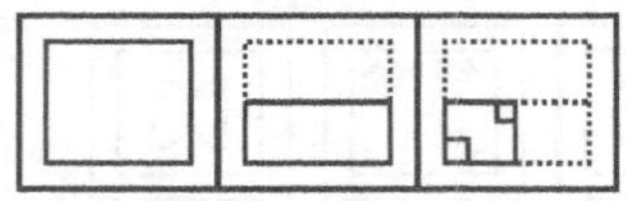

Answer Figures

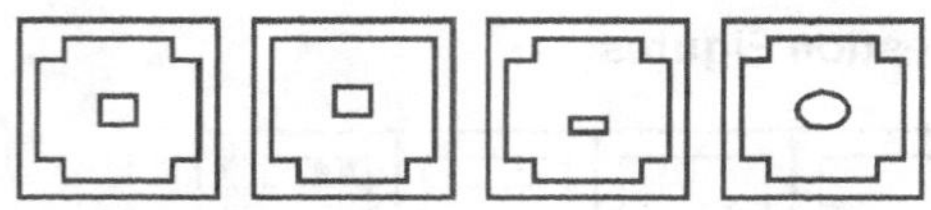

(a) (b) (c) (d)

96. Question Figures

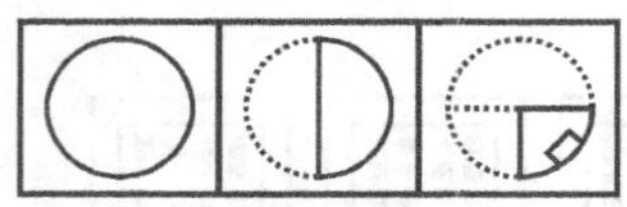

Answer Figures

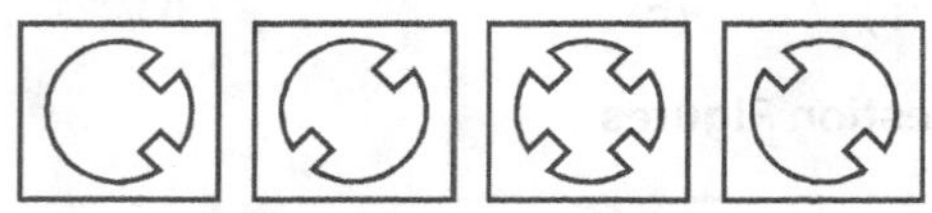

(a) (b) (c) (d)

97. Question Figures

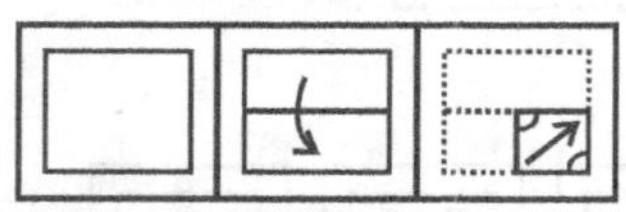

Answer Figures

(a) (b) (c) (d)

98. Question Figures

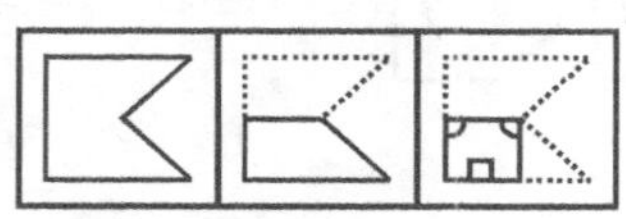

Answer Figures

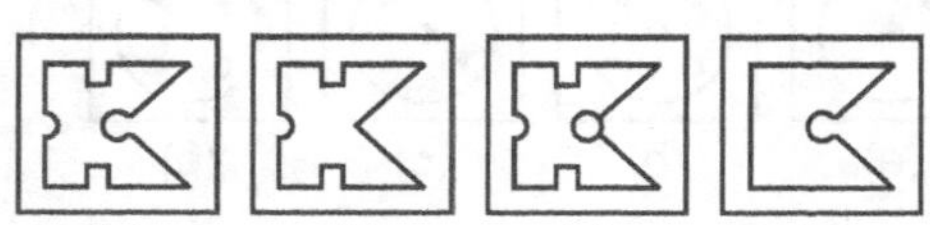

(a) (b) (c) (d)

99. Question Figures

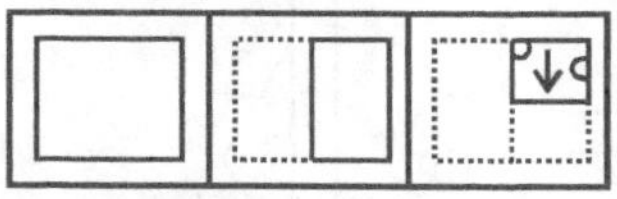

Answer Figures

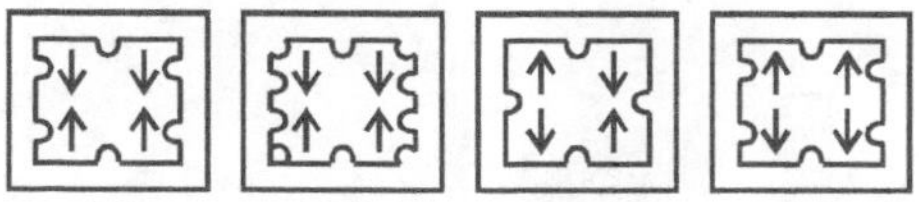

(a) (b) (c) (d)

100. Question Figures

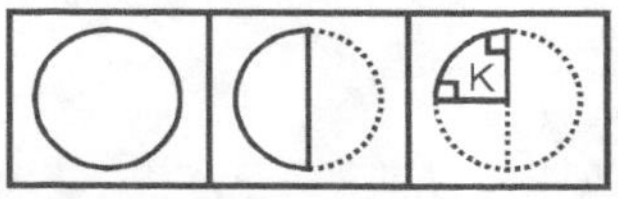

Answer Figures

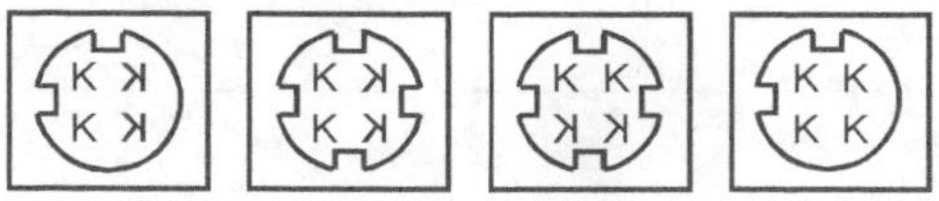

(a) (b) (c) (d)

101. Question Figures

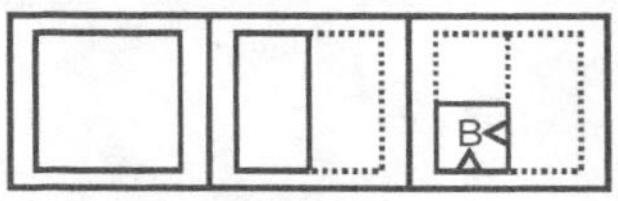

Answer Figures

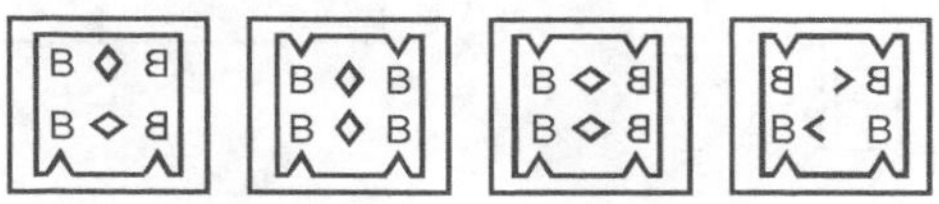

(a) (b) (c) (d)

102. Question Figures

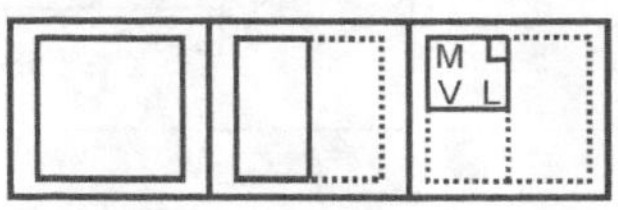

Answer Figures

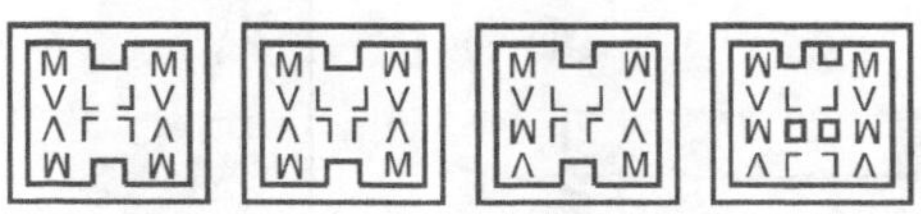

(a) (b) (c) (d)

103. Question Figures

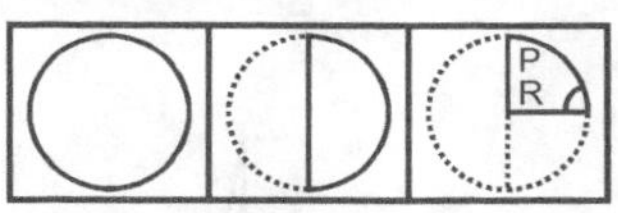

Answer Figures

(a) (b) (c) (d)

104. Question Figures

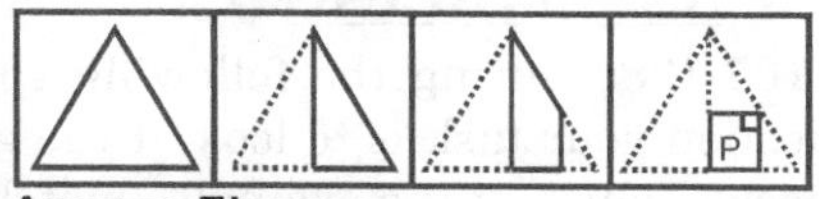

Answer Figures

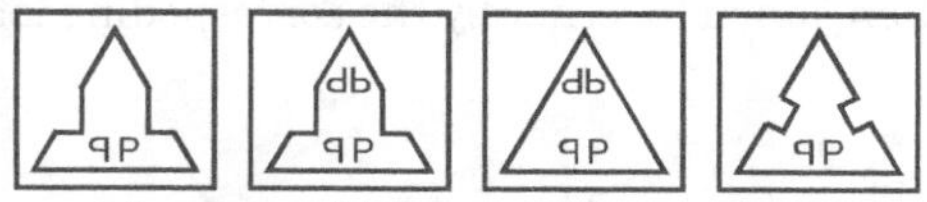

(a) (b) (c) (d)

105. Which object can be made by folding the given shape?

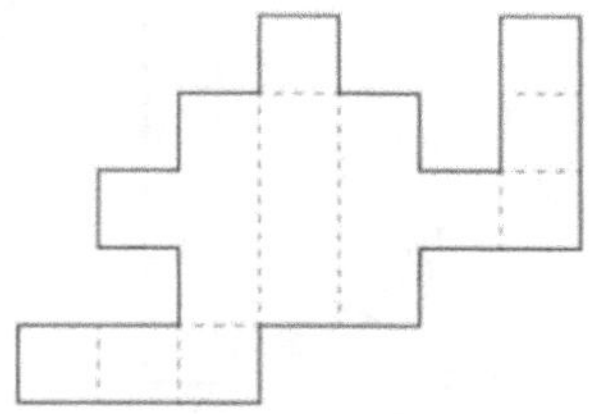

(a) 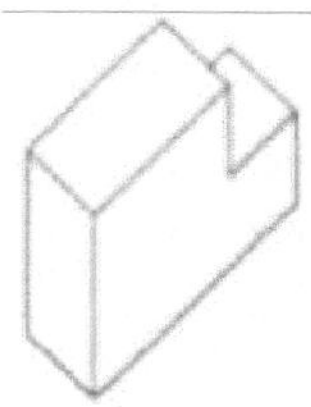

(b) 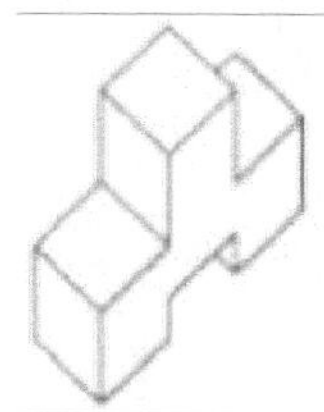

(c) 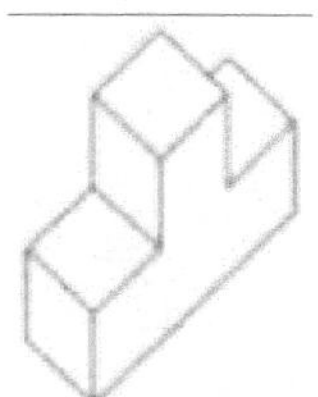

(d) 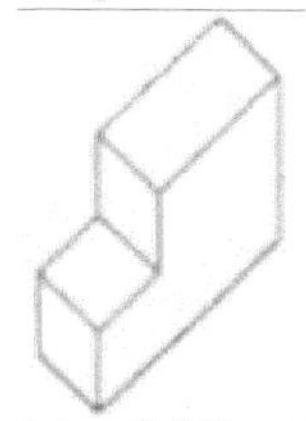

106. Which cube can be formed by folding the given shape?

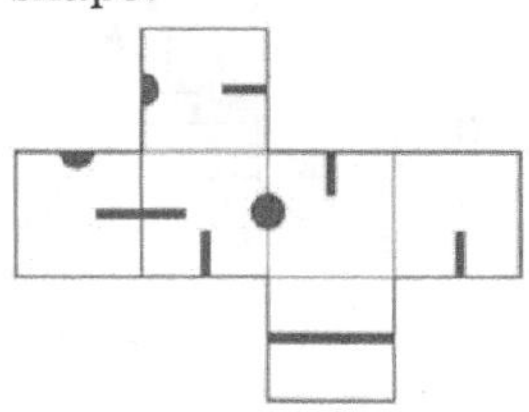

(a) 

(b) 

(c) 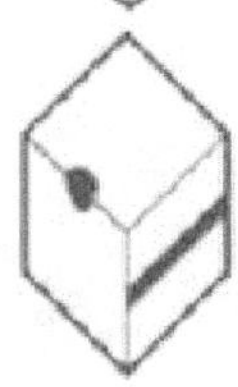

(d) 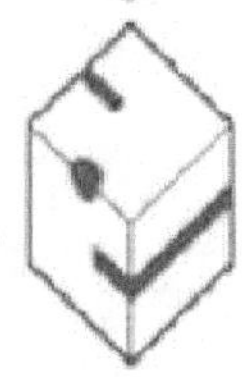

ASSEMBLING

Directions 107-138: During the following spatial reasoning question your task is to look at the given shapes and decide which of the examples match the shape when joined together by the correspoding letters.

107. 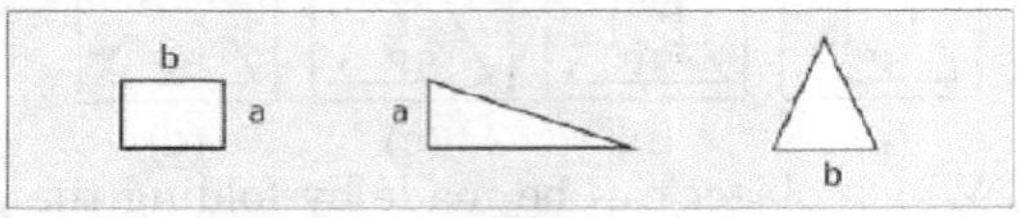

(a) 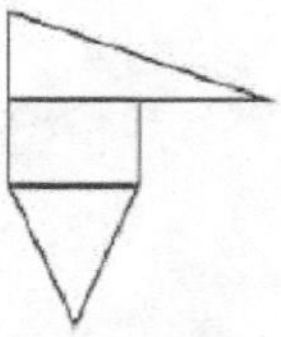

(b) 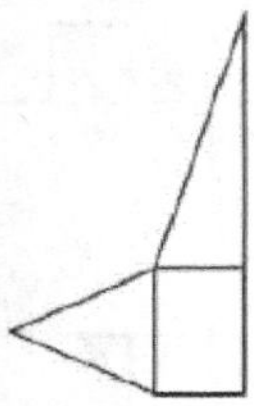

(c) 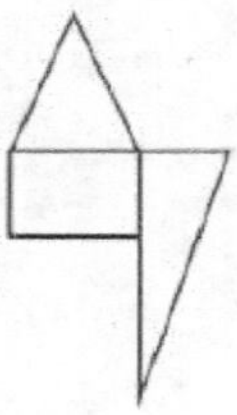

(d) 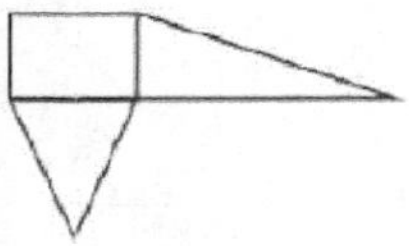

108. 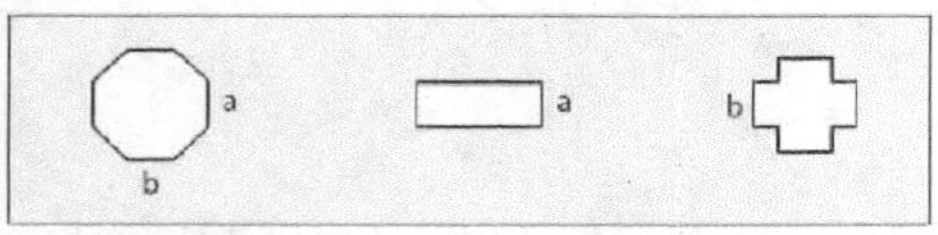

(a) 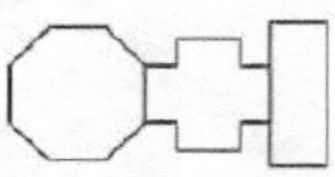

(b) 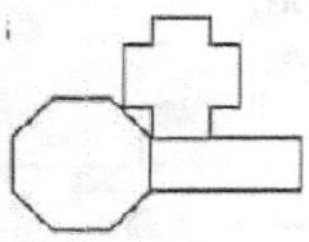

(c) 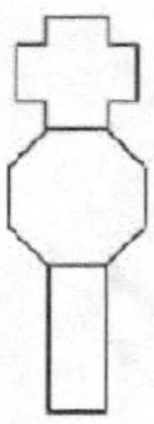

(d) 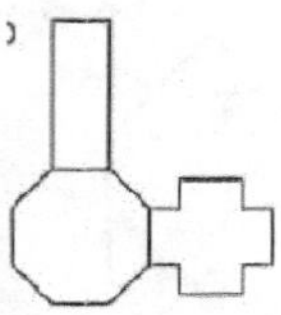

109. 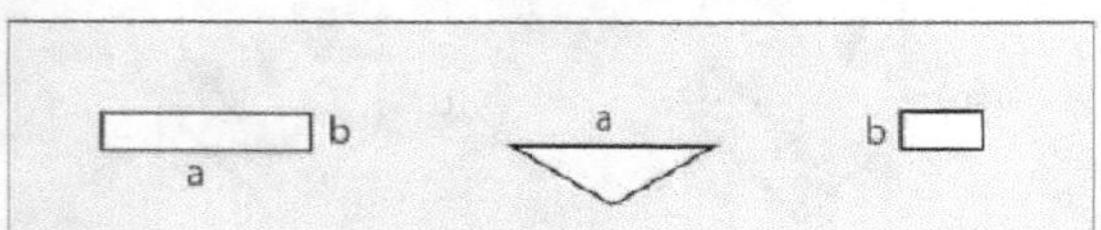

(a) 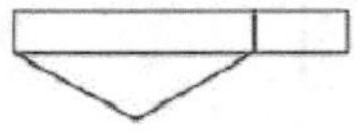

(b) 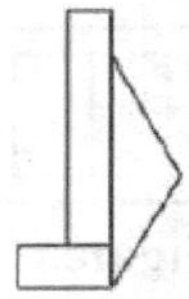

(c) 

(d) 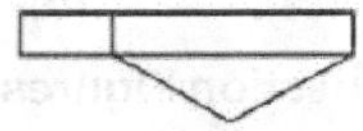

110. 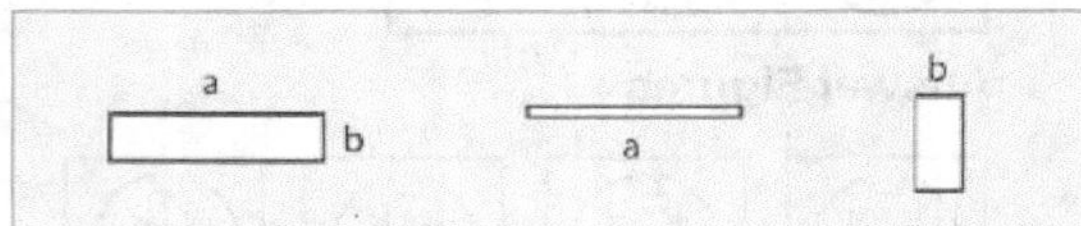

(a) 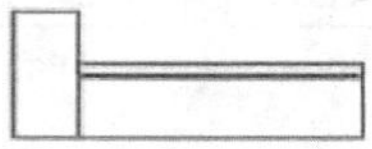

(b) 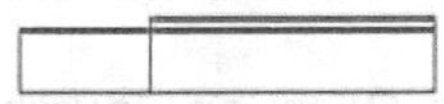

(c) 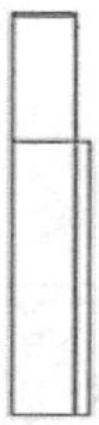

(d) 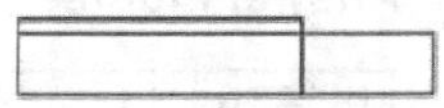

111. 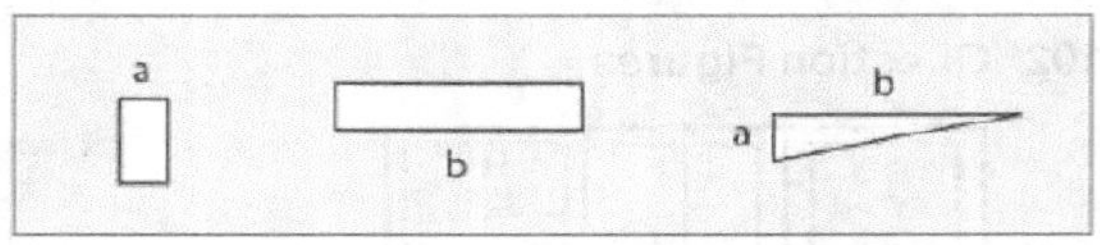

(a) 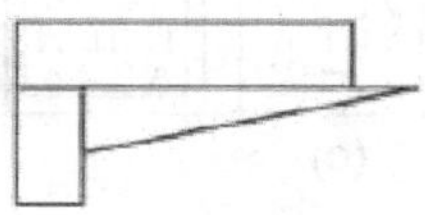

(b) 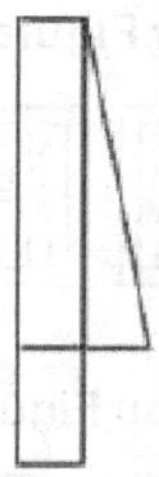

(c) 

(d) 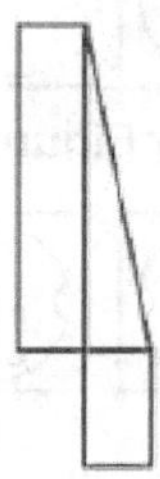

112.

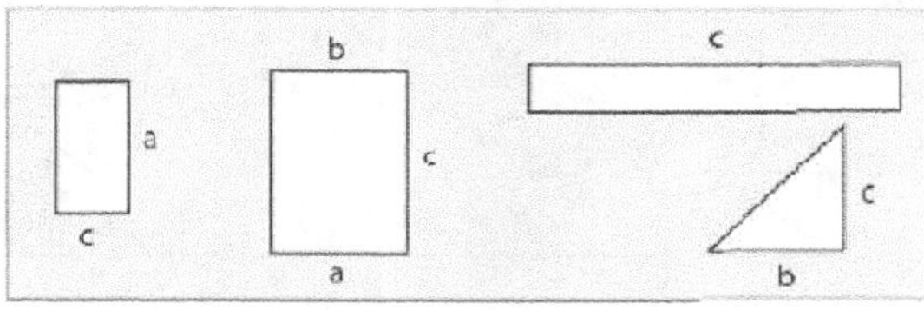

(a) 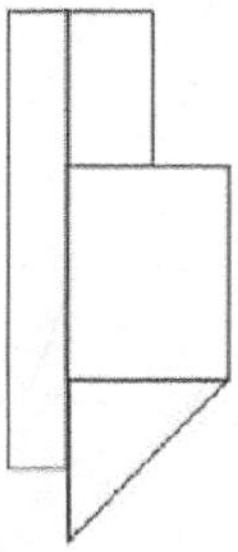 (b) 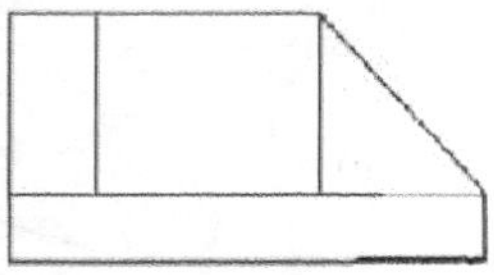

(c) 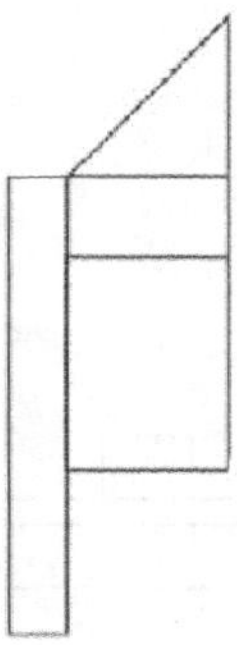 (d) 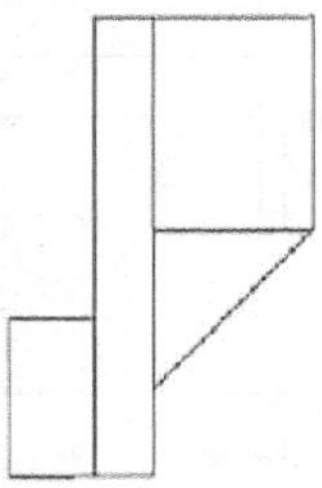

113.

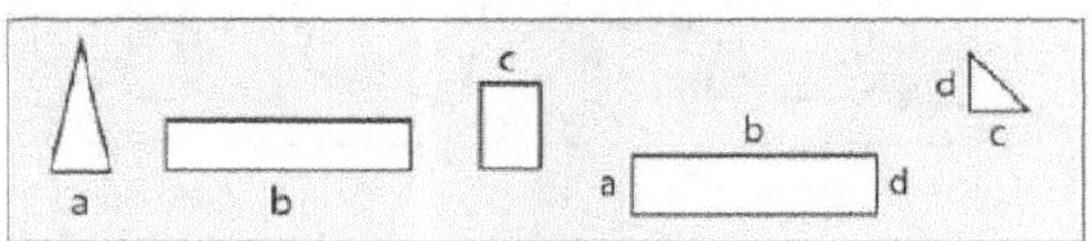

(a) 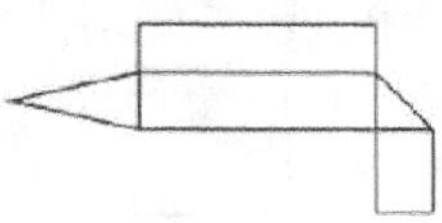 (b) 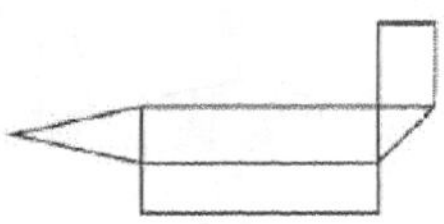

(c) 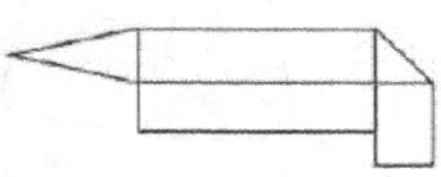 (d) 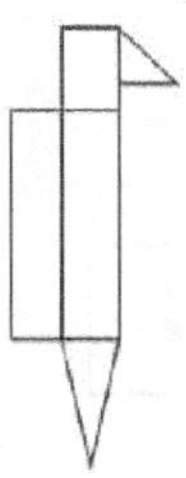

114.

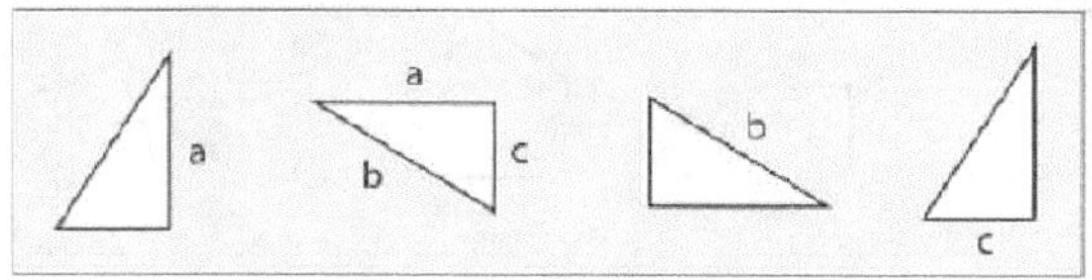

(a) 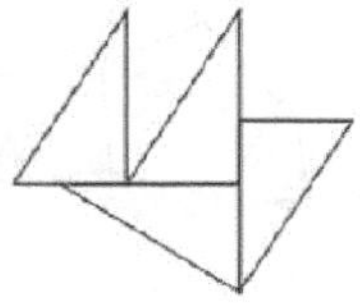 (b) 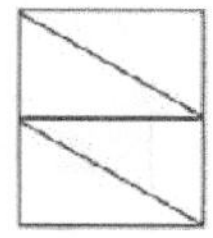

(c) 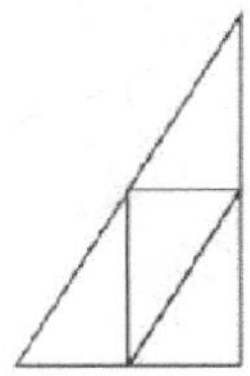 (d) 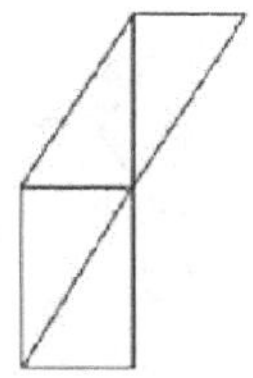

115.

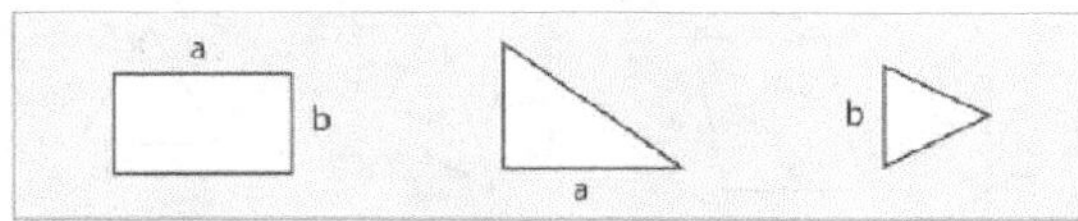

(a) 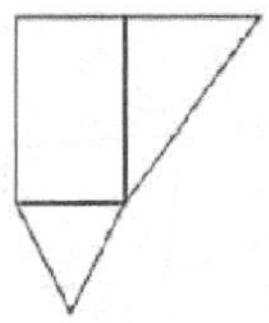 (b) 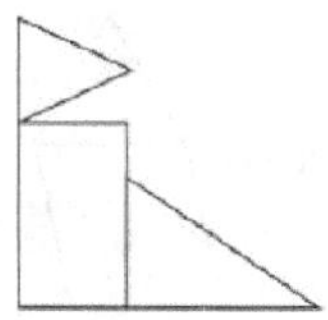

(c) 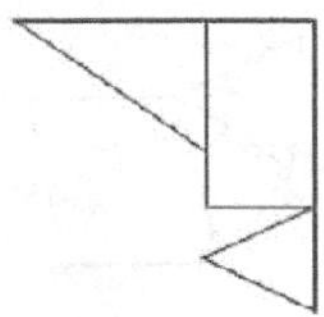 (d) 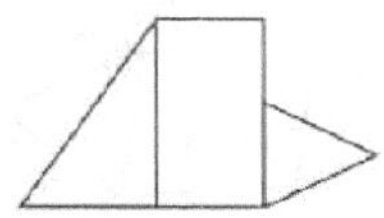

116.

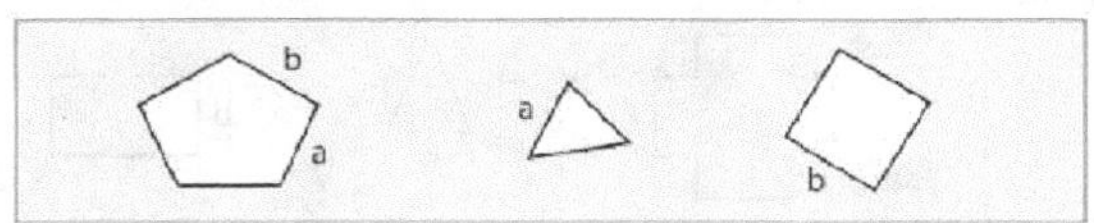

(a) 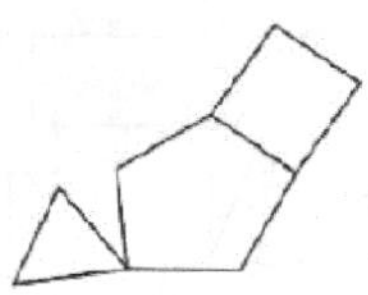 (b) 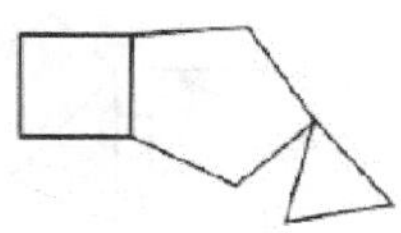

(c) 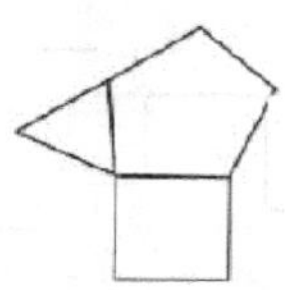 (d) 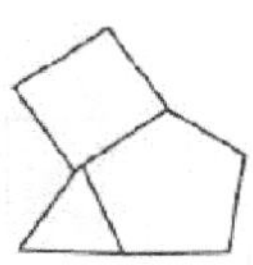

117.

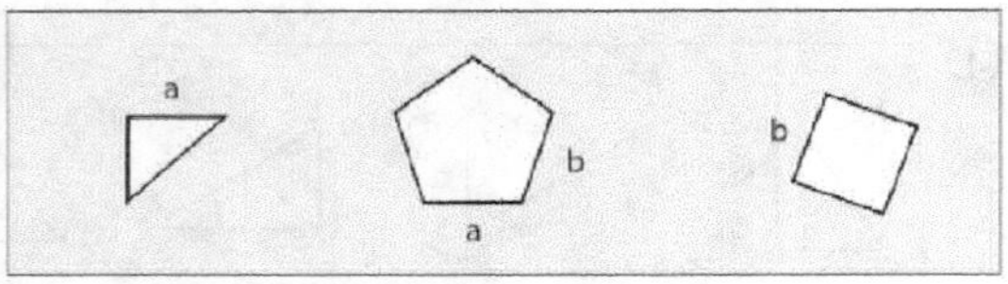

(a) 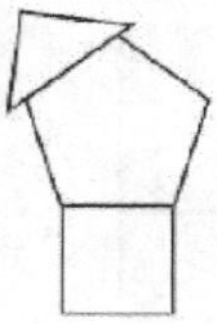

(b) 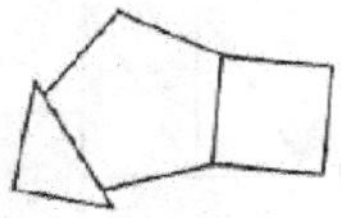

(c) 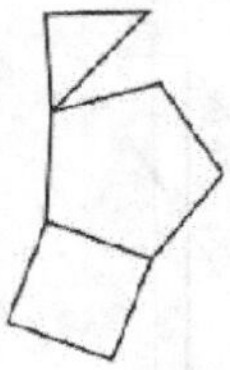

(d) 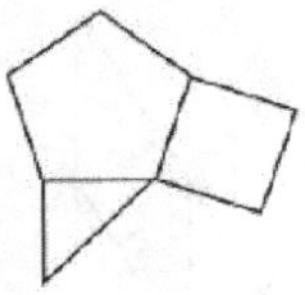

118.

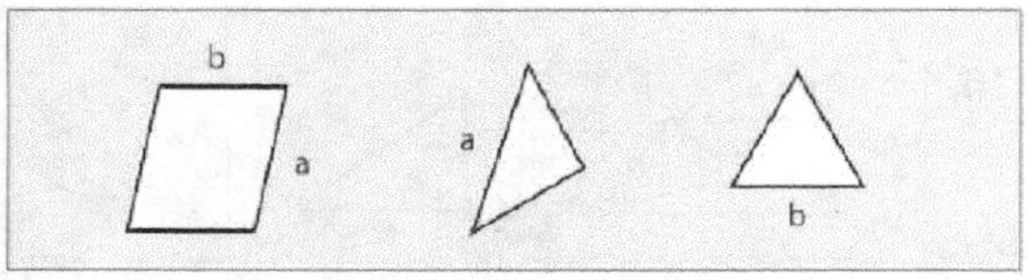

(a) 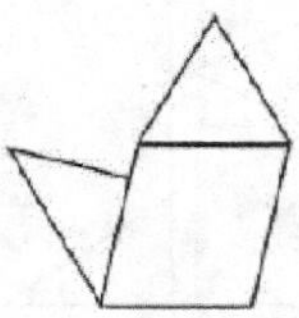

(b) 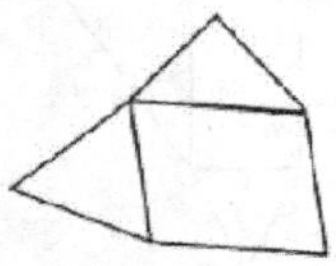

(c) 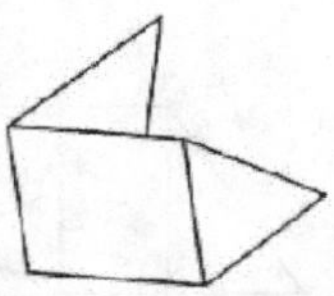

(d) 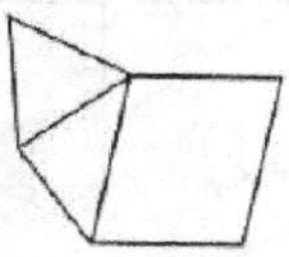

119.

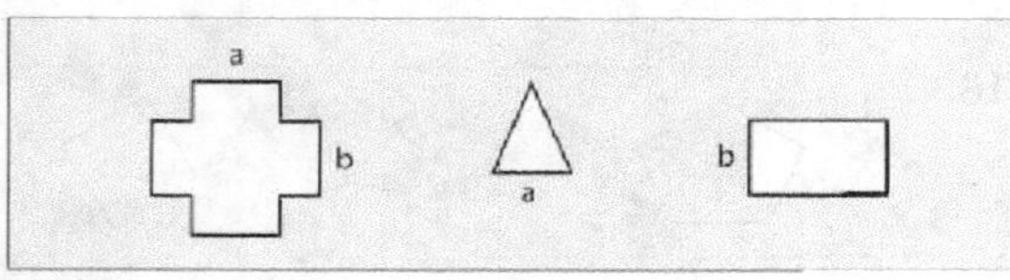

(a) 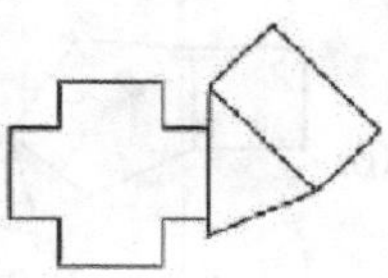

(b) 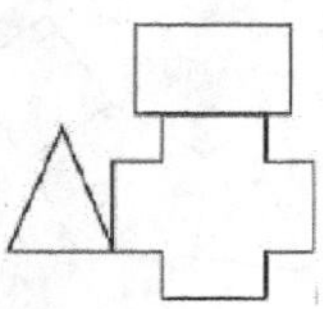

(c) 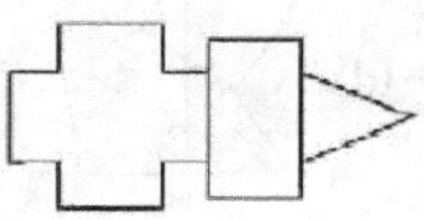

(d) 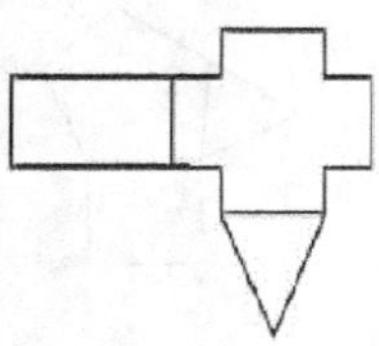

120.

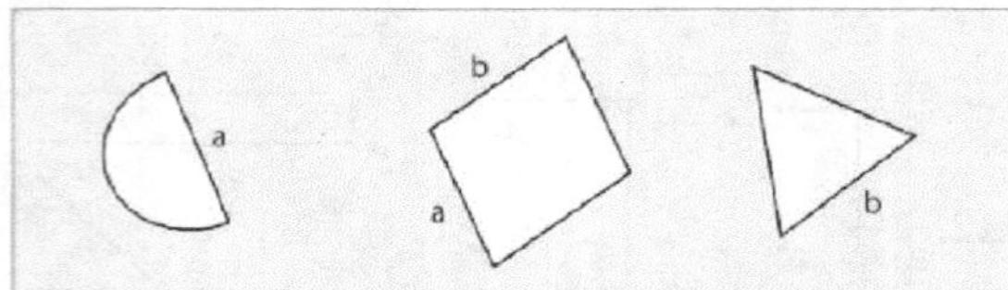

(a) 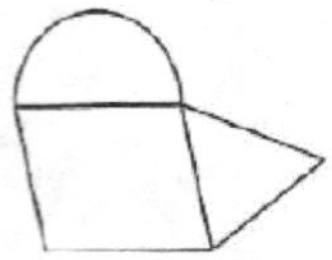

(b) 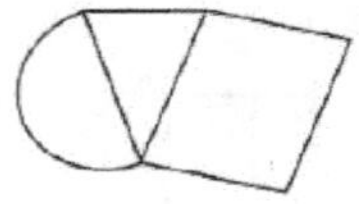

(c) 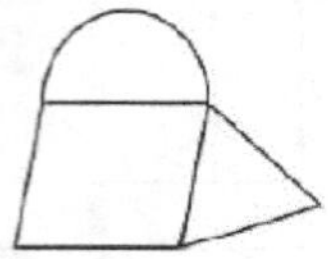

(d) 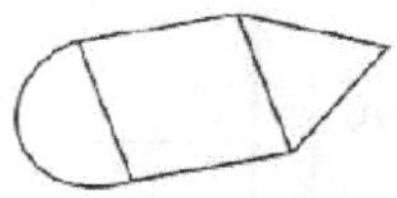

121.

(a) 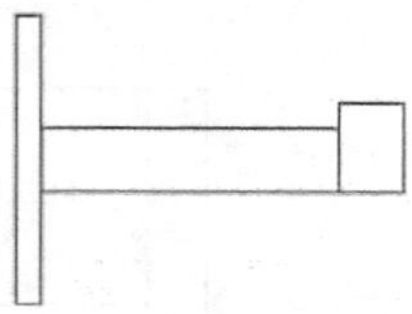

(b) 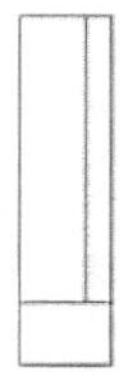

(c) 

(d) 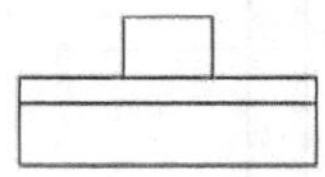

122.

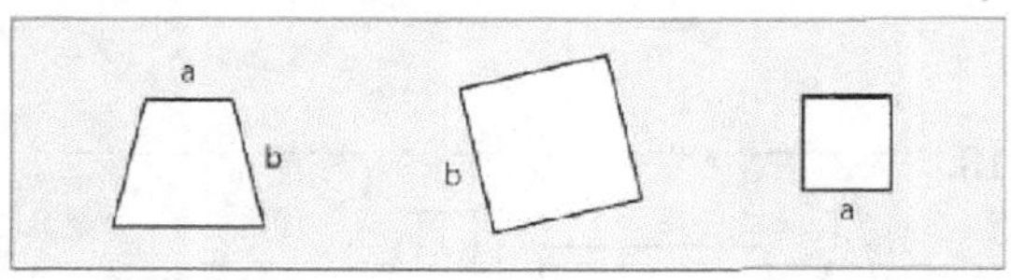

(a) 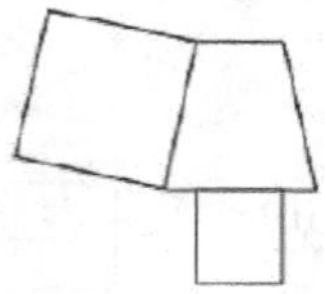

(b) 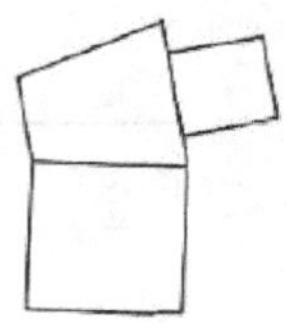

(c) 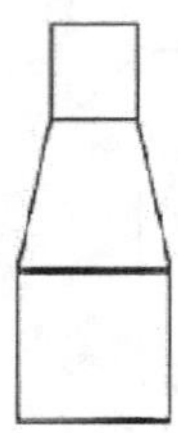

(d) 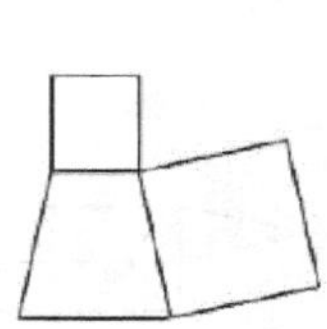

123.

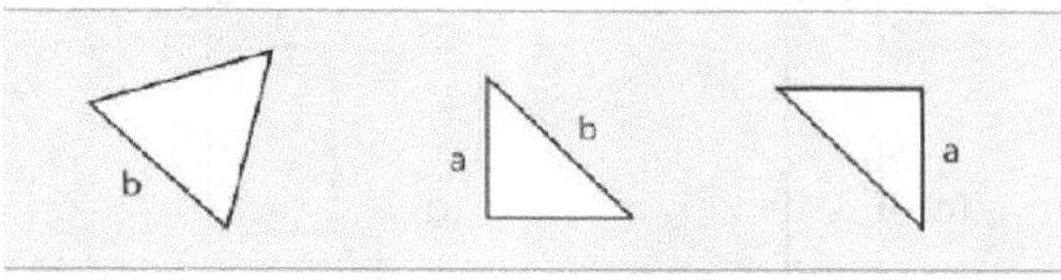

(a) 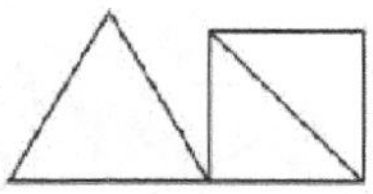 (b) 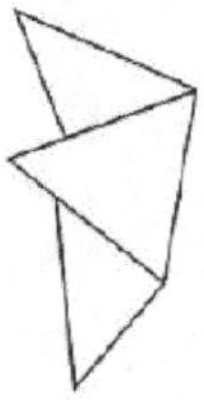

(c) 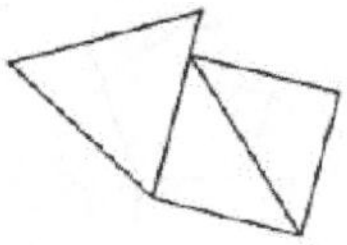 (d) 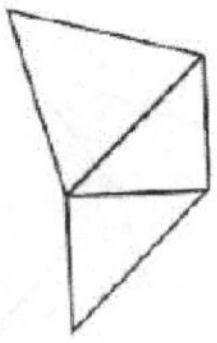

124.

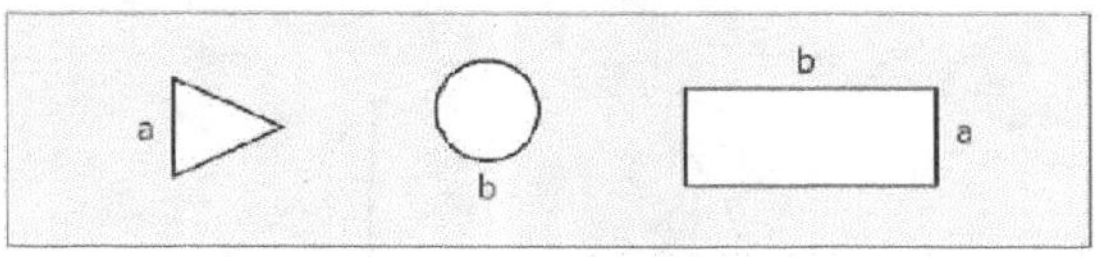

(a) 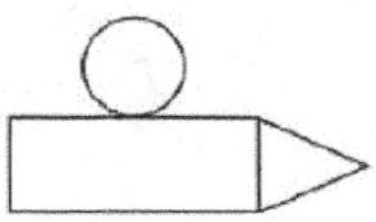 (b) 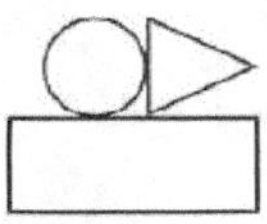

(c) 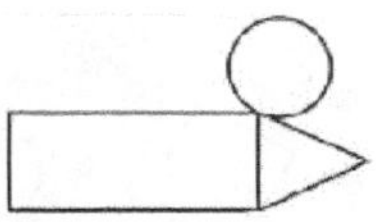 (d) 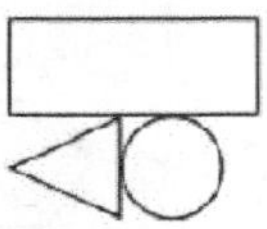

125.

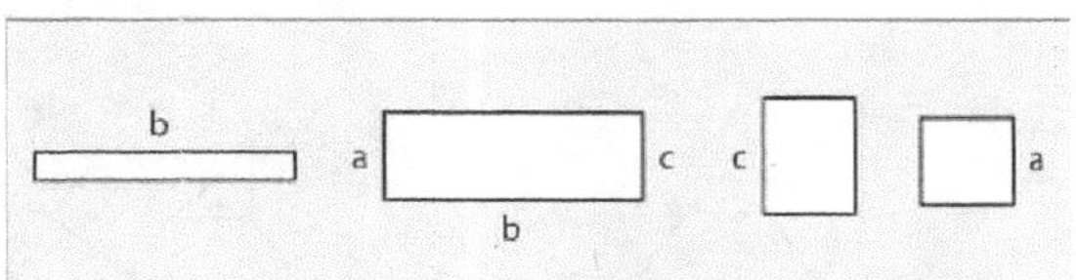

(a) 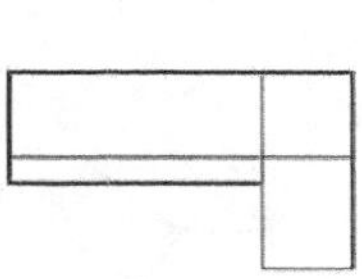 (b) 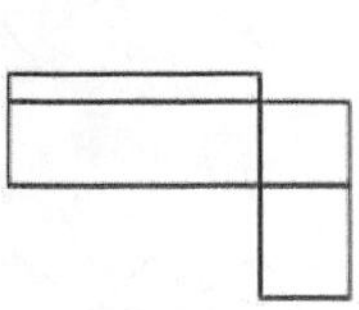

(c) 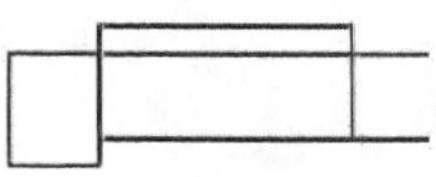 (d) 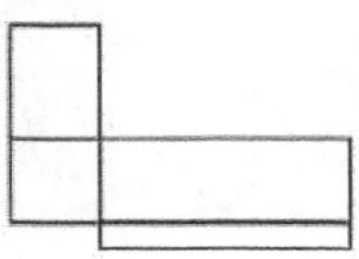

126.

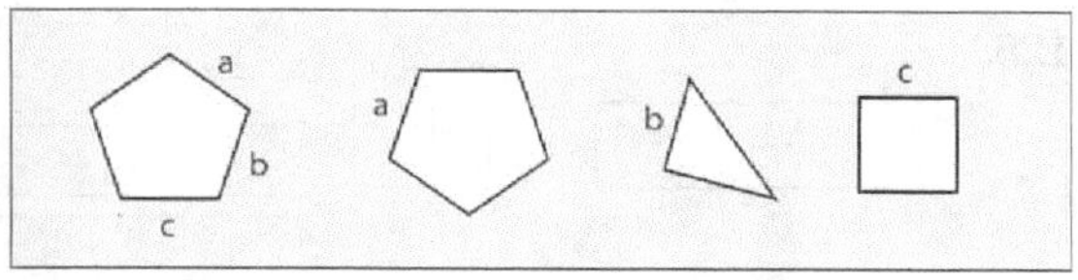

(a) 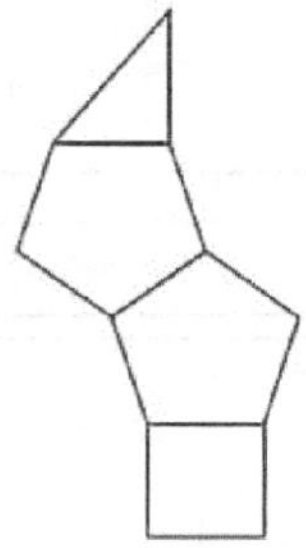 (b) 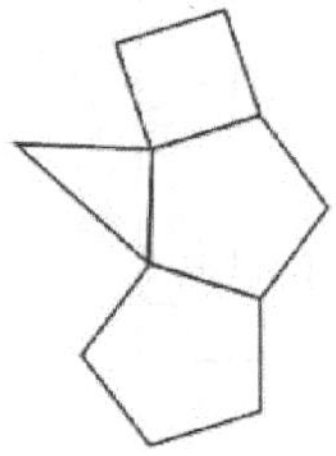

(c) 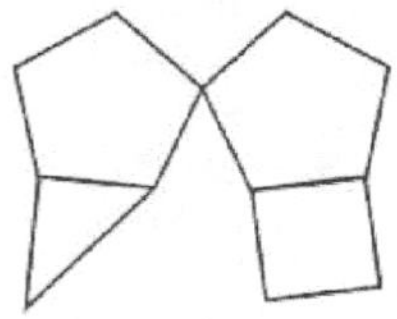 (d) 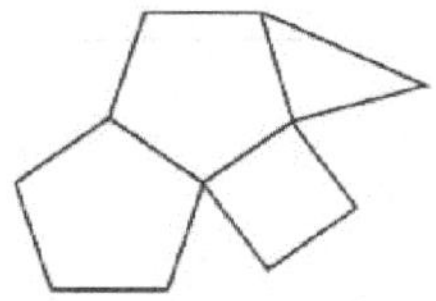

127.

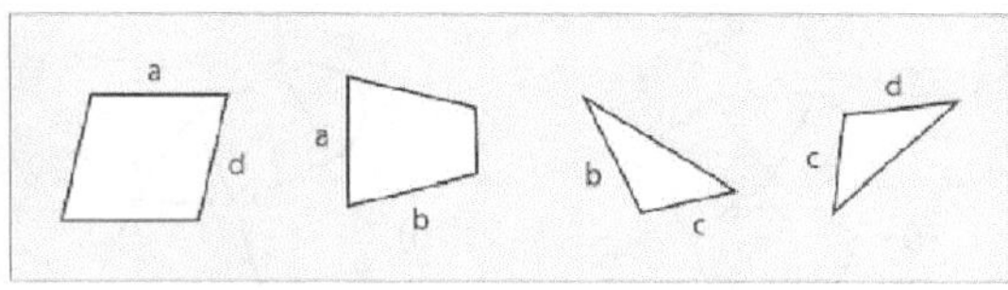

(a) 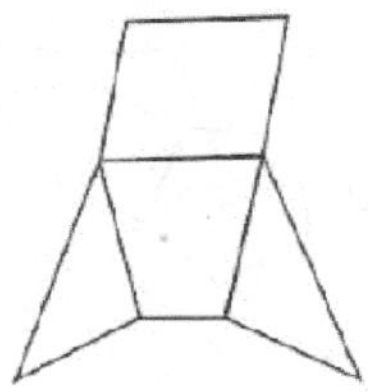 (b) 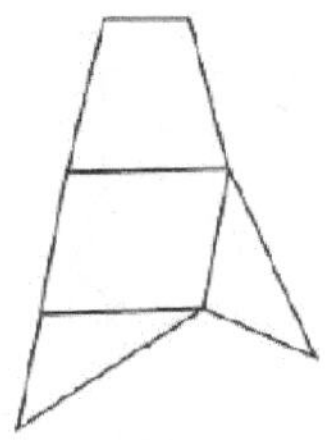

(c) 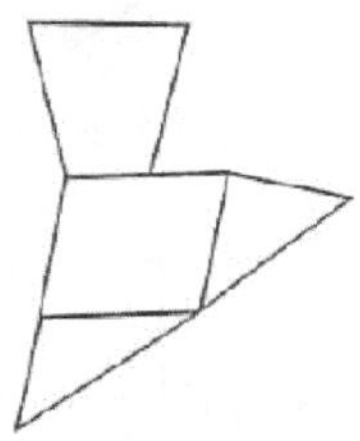 (d) NONE OF THESE

128.

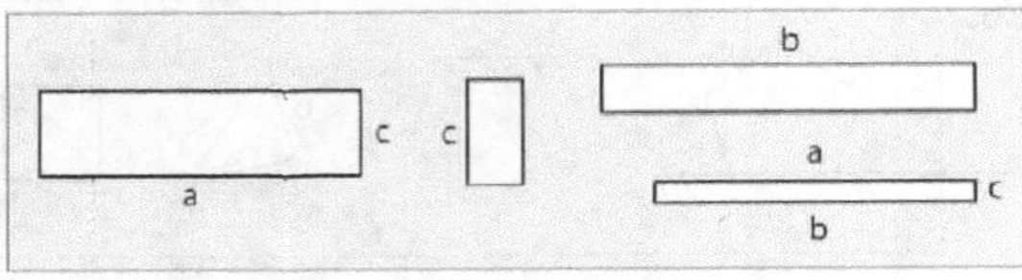

(a) 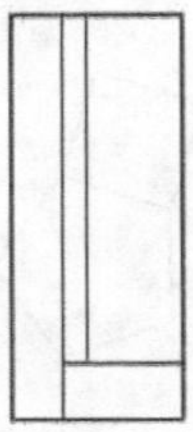

(b) 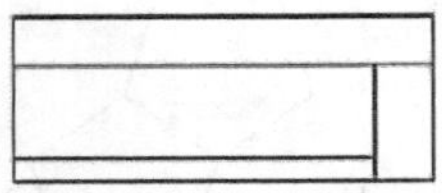

(c) 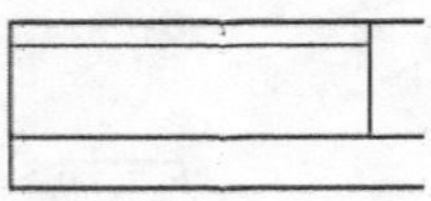

(d) NONE OF THESE

129.

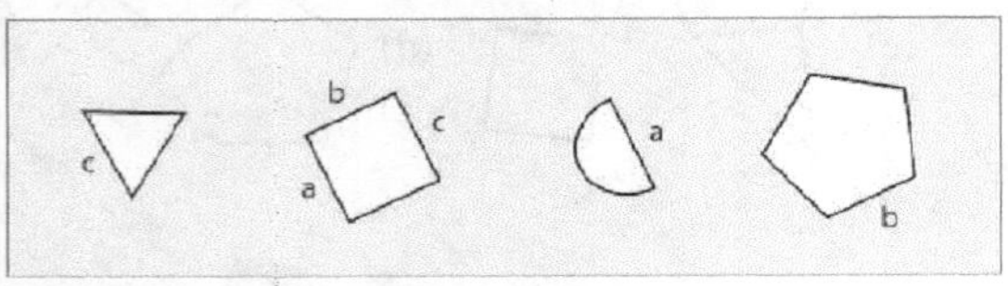

(a) 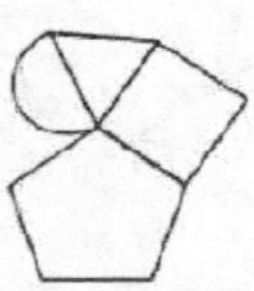

(b) 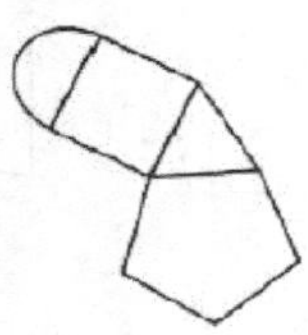

(c) 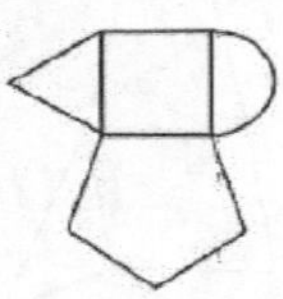

(d) 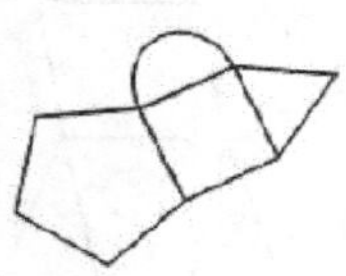

130.

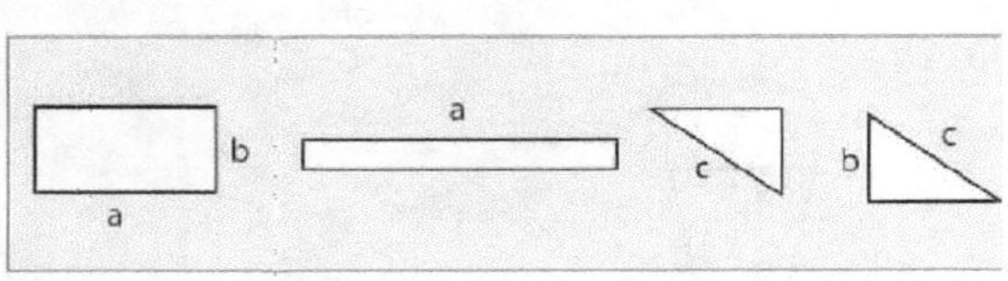

(a) 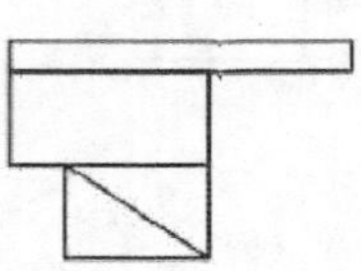

(b) 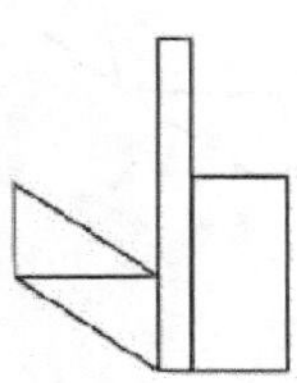

(c) 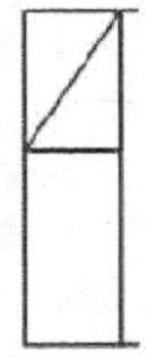

(d) 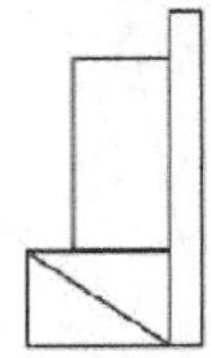

131.

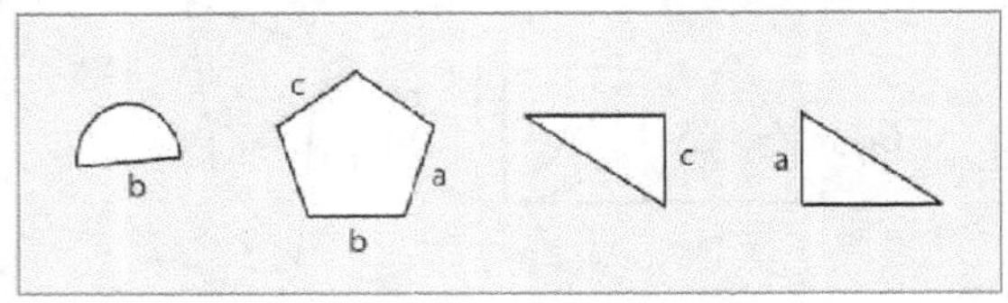

(a) 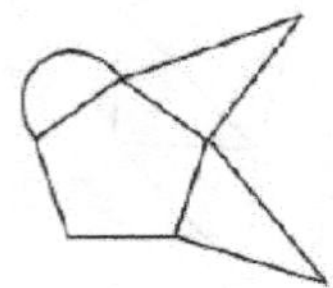

(b) 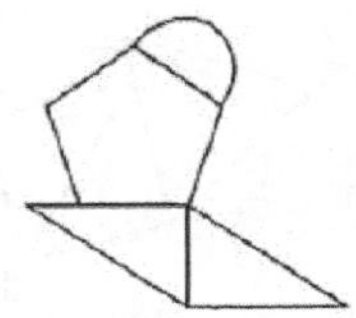

(c) 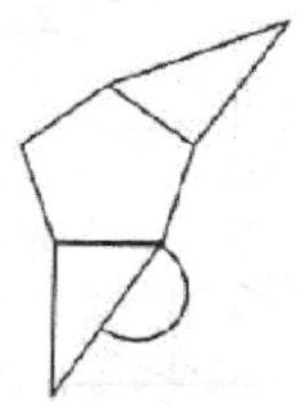

(d) 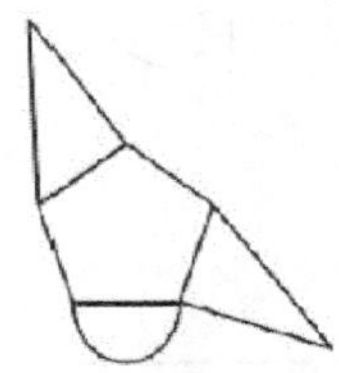

132.

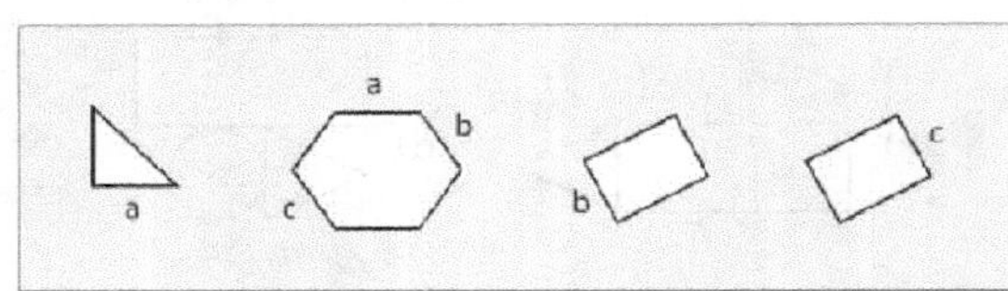

(a) 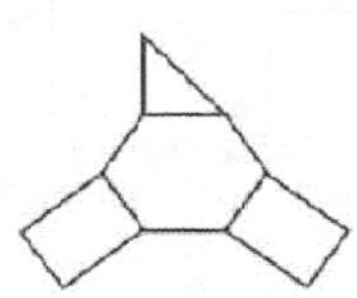

(b) 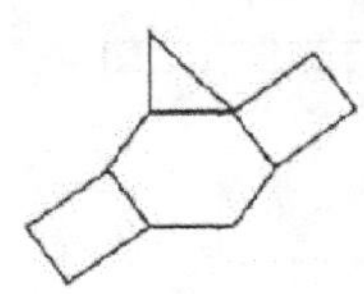

(c) 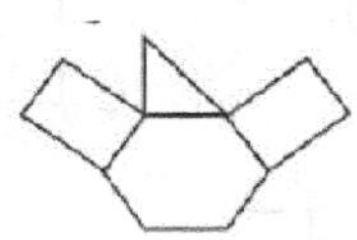

(d) 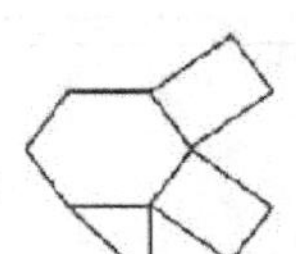

133. 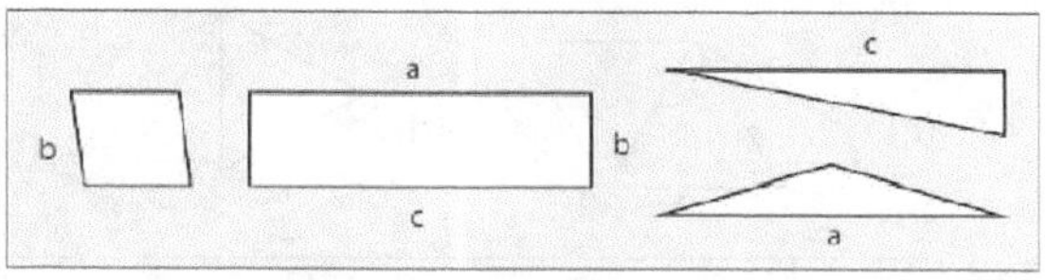

(a) 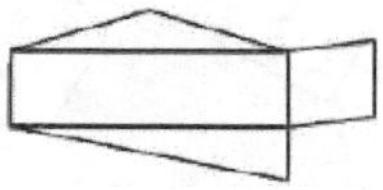 (b) 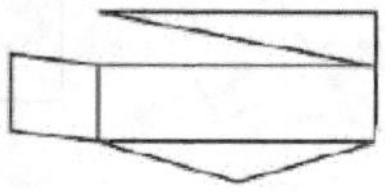

(c) 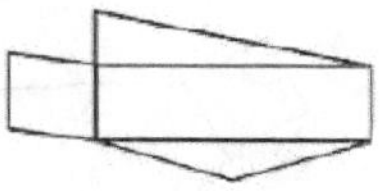 (d) NONE OF THESE

134. 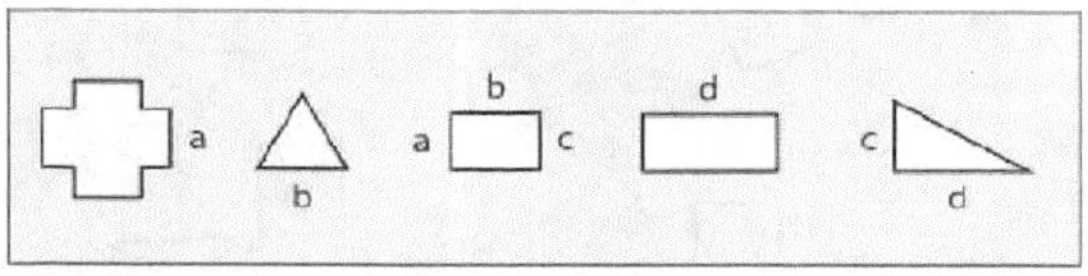

(a) 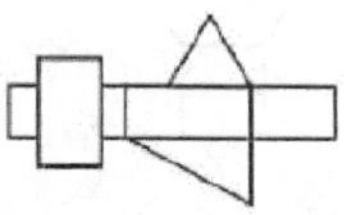 (b) 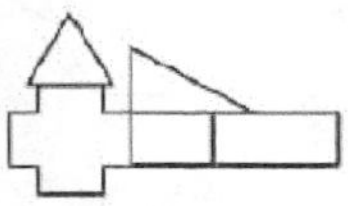

(c) 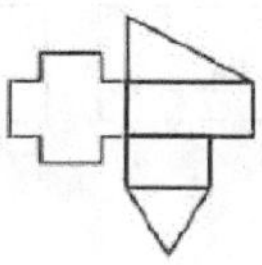 (d) 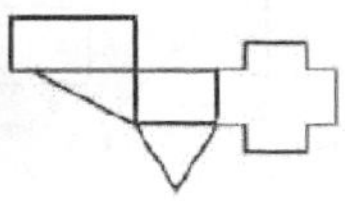

135. 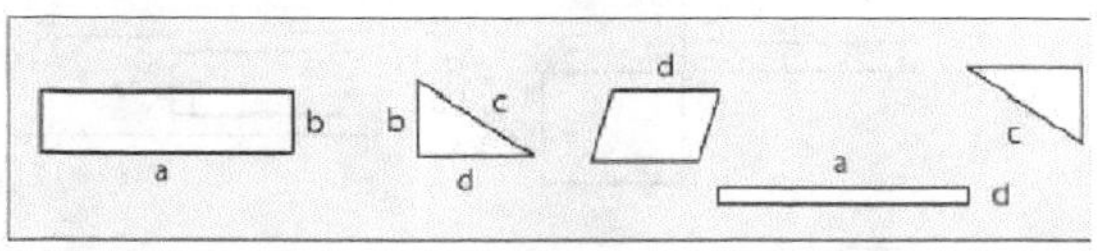

(a) 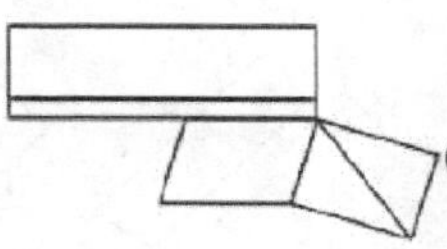 (b) 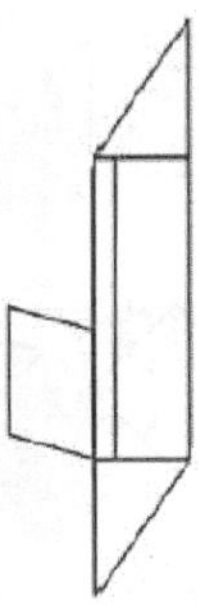

(c) 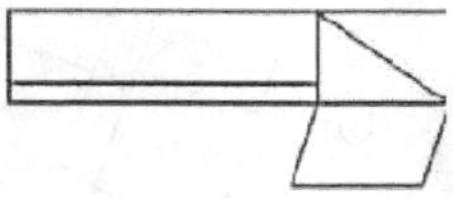 (d) NONE OF THESE

136. 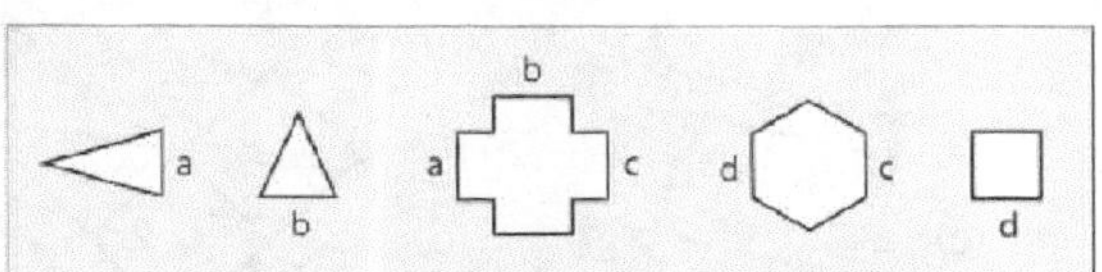

(a) 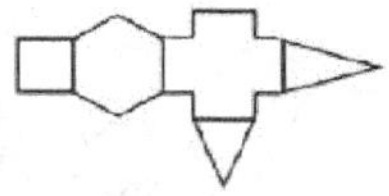 (b) 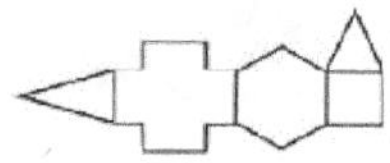

(c) 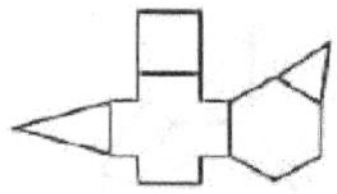 (d) 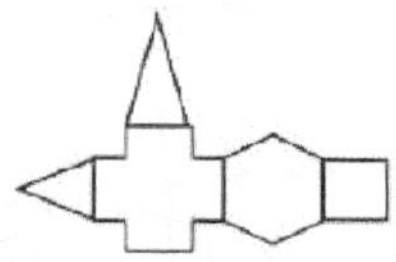

137. 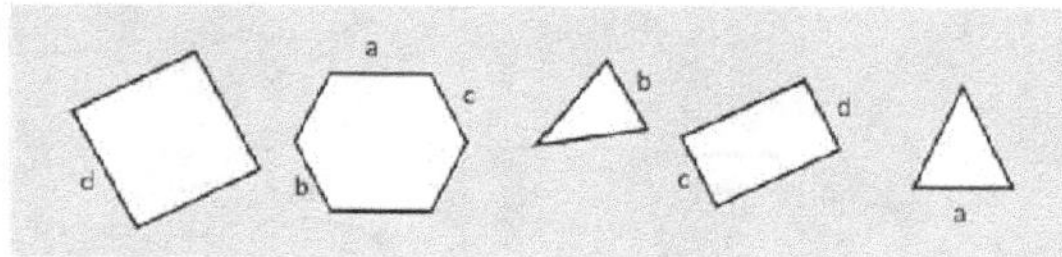

(a) 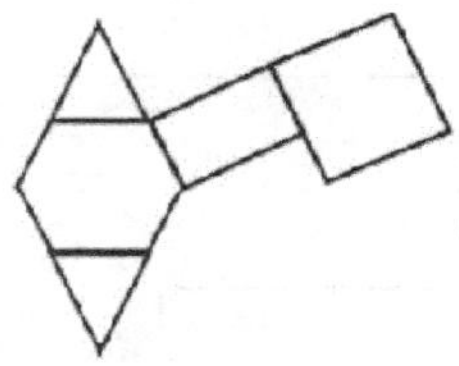 (b) 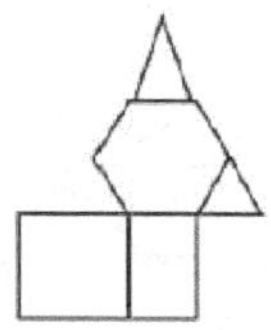

(c) 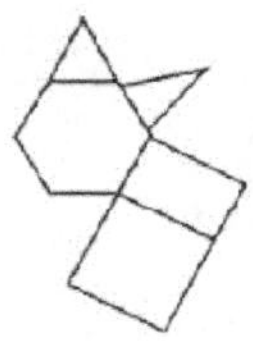 (d) NONE OF THESE

138. 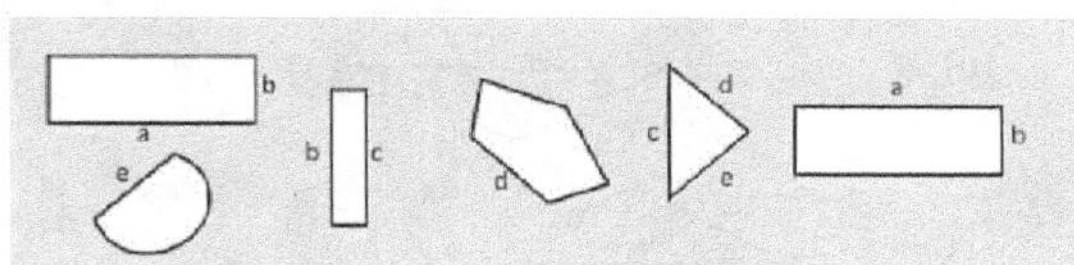

(a) 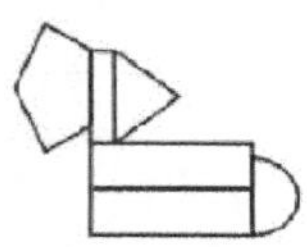 (b) 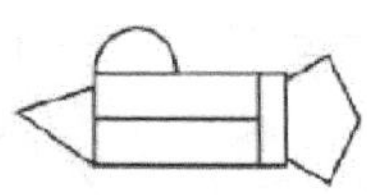

(c)  (d) 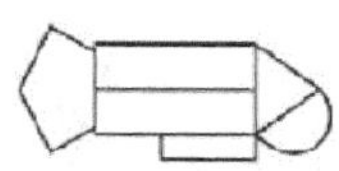

139.

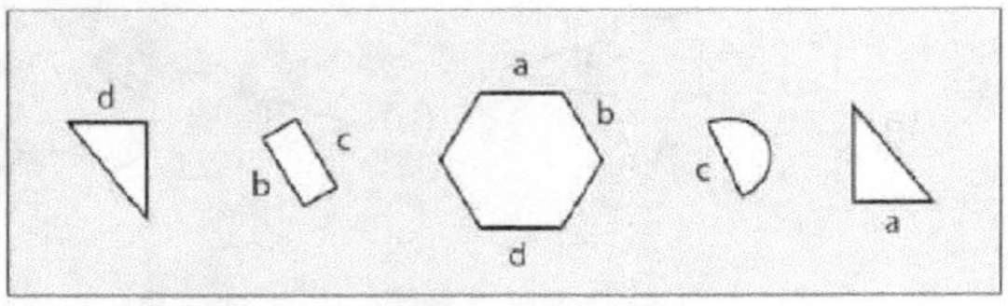

(a) 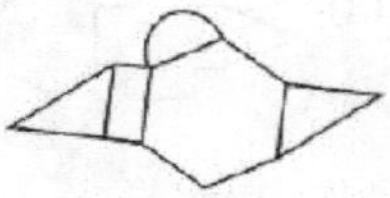
(b) 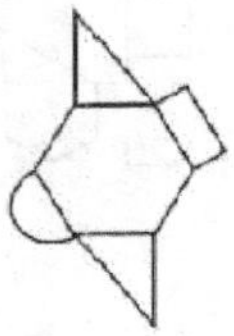
(c) 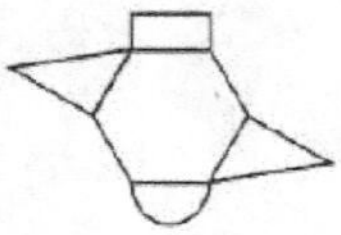
(d) 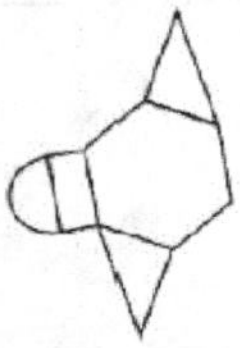

140.

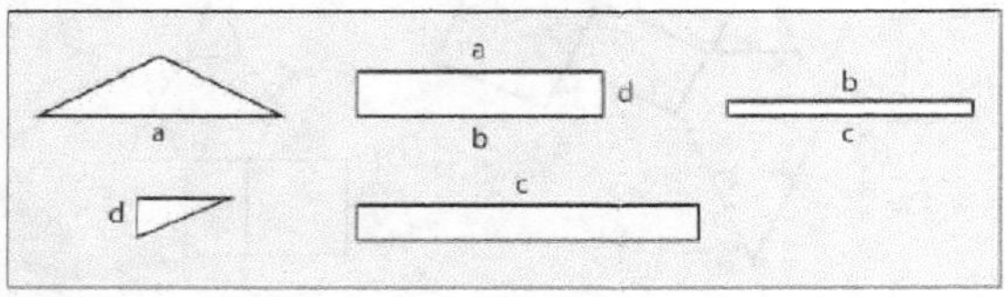

(a) 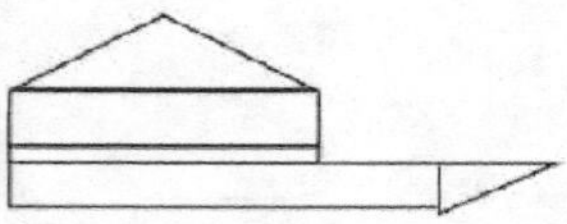
(b) 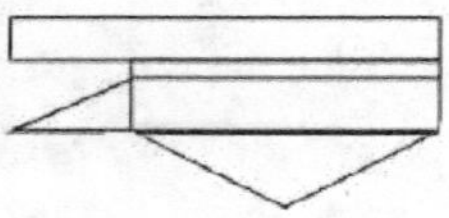
(c) 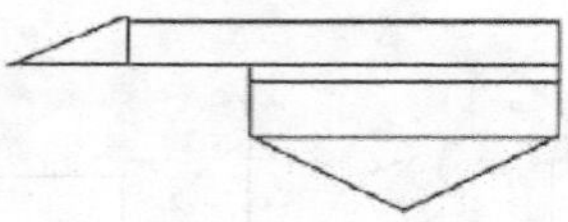
(d) 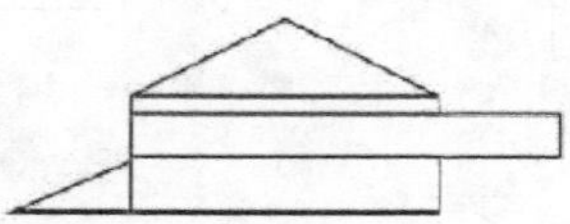

141.

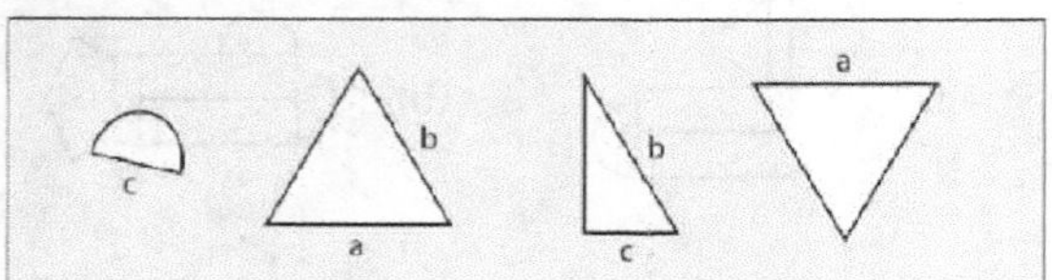

(a) 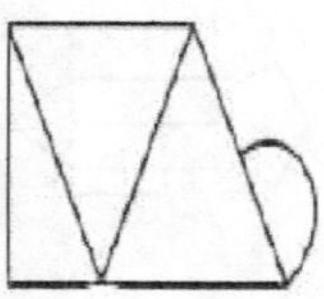
(b) 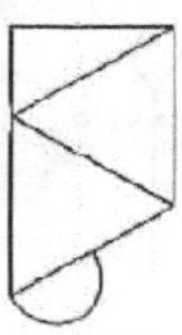
(c) 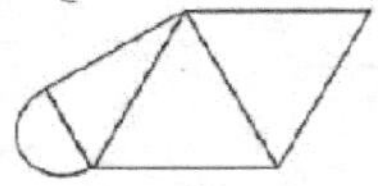
(d) 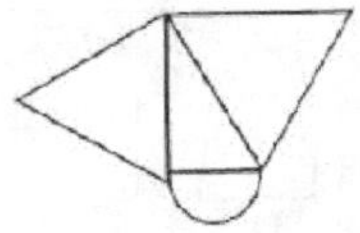

142.

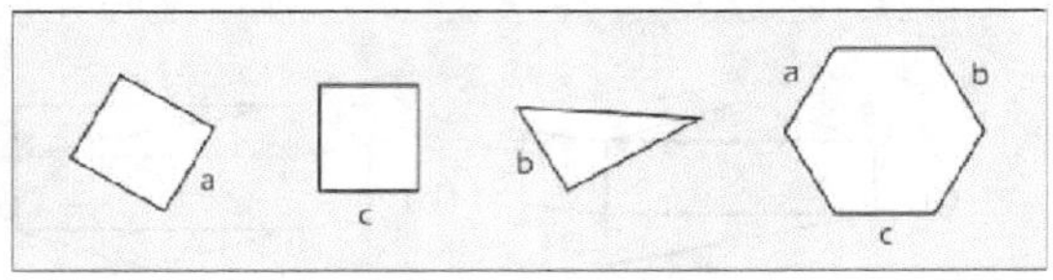

(a) 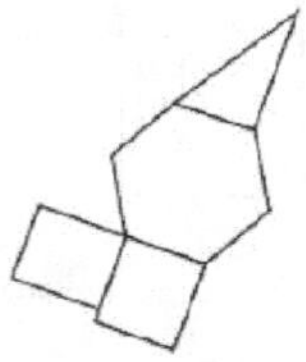
(b) 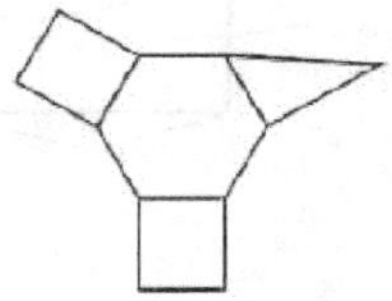
(c) 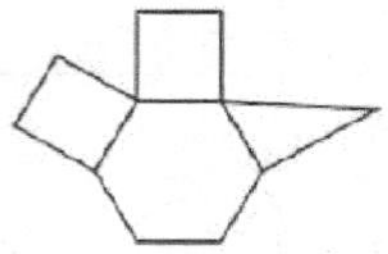
(d) 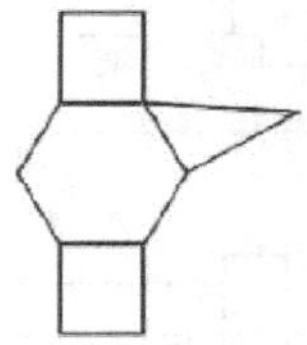

143.

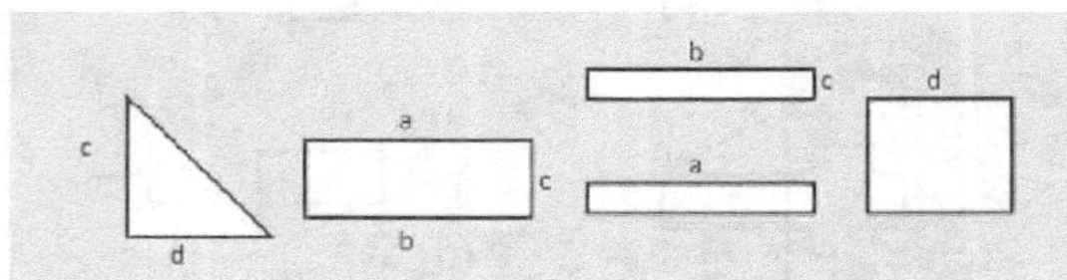

(a) 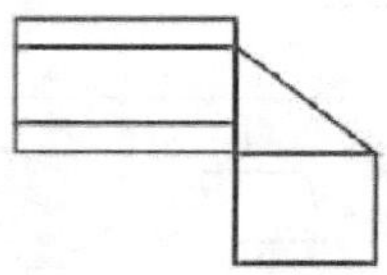
(b) 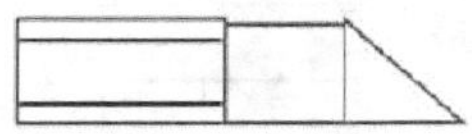
(c) 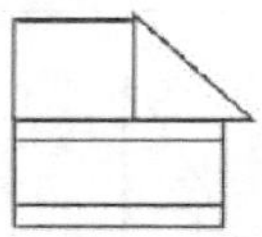
(d) 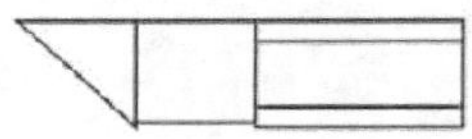

144.

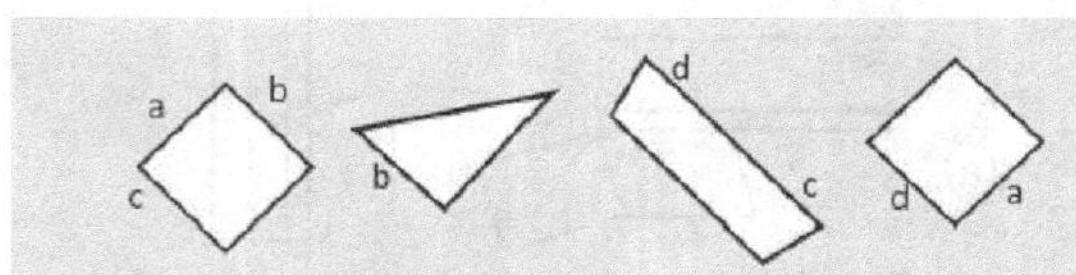

(a) 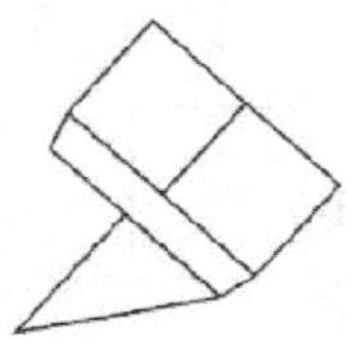
(b) 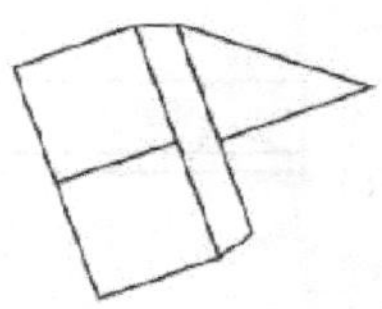
(c) 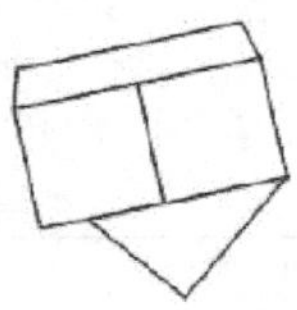
(d) 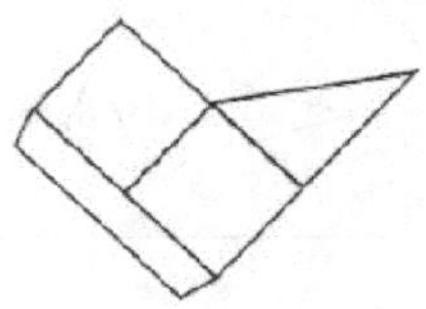

145.

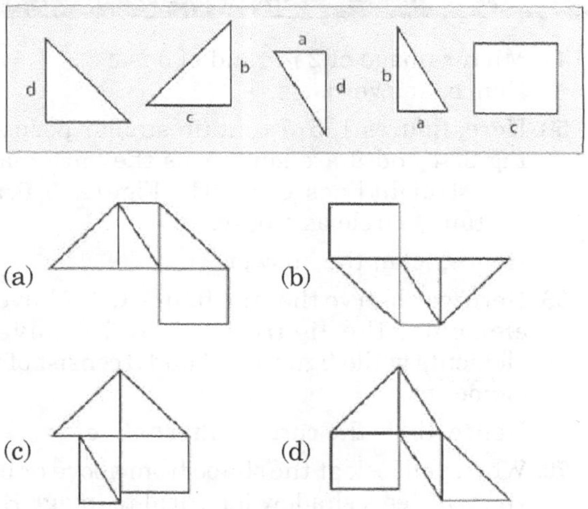

146.

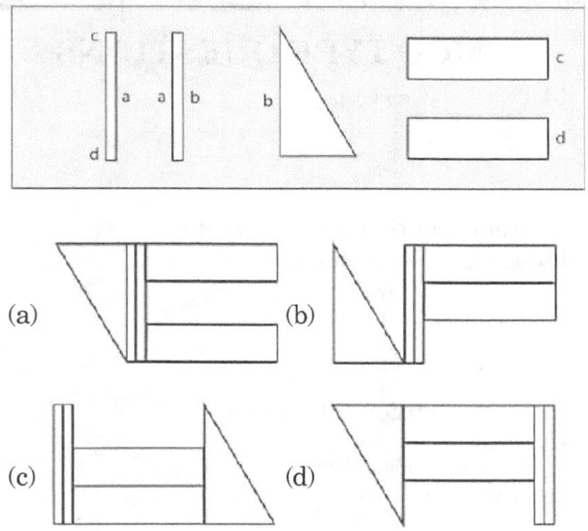

ANSWERS

MCQ Type Questions

1. (d)	**2.** (d)	**3.** (c)	**4.** (c)	**5.** (b)	**6.** (a)	**7.** (d)	**8.** (b)	**9.** (b)	**10.** (d)
11. (c)	**12.** (a)	**13.** (a)	**14.** (a)	**15.** (a)	**16.** (c)	**17.** (b)	**18.** (c)	**19.** (a)	**20.** (c)
21. (a)	**22.** (a)	**23.** (b)	**24.** (a)	**2.** (a)	**26.** (c)	**27.** (b)	**28.** (c)	**29.** (d)	**30.** (b)
31. (b&c)	**32.** (c&d)	**33.** (b)	**34.** (c)	**35.** (b&d)	**36.** (c&d)	**37.** (c)	**38.** (b)	**39.** (c&d)	**40.** (c&d)
41. (c)	**42.** (a)	**43.** (b)	**44.** (c)	**45.** (d)	**46.** (a)	**47.** (b)	**48.** (c)	**49.** (a)	**50.** (d)
51. (c)	**52.** (c)	**53.** (a)	**54.** (c)	**55.** (a)	**56.** (a)	**57.** (c)	**58.** (c)	**59.** (b)	**60.** (a&b)
61. (a)	**62.** (b)	**63.** (b)	**64.** (d)	**65.** (b)	**66.** (b)	**67.** (a)	**68.** (d)	**69.** (c)	**70.** (b)
71. (d)	**72.** (b)	**73.** (d)	**74.** (c)	**75.** (b)	**76.** (c)	**77.** (d)	**78.** (a)	**79.** (b)	**80.** (a)
81. (b)	**82.** (c)	**83.** (b)	**84.** (a)	**85.** (d)	**86.** (c)	**87.** (d)	**88.** (c)	**89.** (a)	**90.** (b)
91. (d)	**92.** (a)	**93.** (b)	**94.** (c)	**95.** (a)	**96.** (c)	**97.** (d)	**98.** (a)	**99.** (a)	**100.** (b)
101. (c)	**102.** (a)	**103.** (a)	**104.** (b)	**105.** (c)	**106.** (d)	**107.** (b)	**108.** (d)	**109.** (a)	**110.** (d)
111. (d)	**102.** (b)	**113.** (a)	**114.** (c)	**115.** (a)	**116.** (c)	**117.** (d)	**118.** (b)	**119.** (d)	**120.** (a)
121. (c)	**122.** (d)	**123.** (d)	**124.** (a)	**125.** (c)	**126.** (b)	**127.** (d)	**128.** (d)	**129.** (c)	**130.** (c)
131. (d)	**132.** (b)	**133.** (d)	**134.** (d)	**135.** (d)	**136.** (a)	**137.** (d)	**138.** (c)	**139.** (d)	**140.** (b)
141. (c)	**142.** (b)	**143.** (a)	**144.** (d)	**145.** (d)	**146.** (a)				

EXPLANATIONS

MCQ TYPE QUESTIONS

43. Letter – Mirror image

Letter		Mirror image
R	–	Я
E	–	Ǝ
S	–	Ƨ
T	–	T

Hence, option (b) is correct.

44. Number – Mirror image

Number		Mirror image
1	–	ɿ
5	–	ꞔ
8	–	8
2	–	ς
4	–	ᔨ

Hence, option (c) is correct.

45. Letter/Number Mirror image

Letter/Number		Mirror image
M	–	M
3	–	Ɛ
5	–	ꞔ
K	–	ꓘ
G	–	Ә
1	–	ɿ

Hence, option (d) is correct.

46. This case is like water-image. Hence, option (a) is correct.

47. Mirror image of 2 is ςand of 5 is ꞔ
Hence, answer is ꞔς.

58. Here, figures 1, 5, 7 contain similar porperties. Fig.2, 4 and 9 are similar as they are made of four straight lines. Similarly, Figures 3, 6, and 9 contain a circle as a base.

Hence, (c) in the answer.

59. Here we observe that the figure 1, 6, 7 have only elements, the figures 2, 5 and 9 have two elements while figures 3, 4 and 8 consist of three elements.

Hence, (b) is the correct alternative.

70. When you look at the shape from above or below, you will see a shadow identical to image B.

When you look at the shape from the side, you will see a shadow in the form of a dark square with lit triangles in it (B.N. the lit triangle is not identical to the one shown in shape itself!).

Quantitative Aptitude

1

CHAPTER

Numbers, Algebra and Data Interpretation

NUMBERS

Natural Numbers

All positive integers are natural numbers.

e.g. 1, 2, 3, 4, 5,......

There are infinite natural numbers and number 1 is the *least natural number*.

Types of Natural Numbers : *Based on divisibility there are following two types of natural numbers :*

1. *Prime Numbers:*

A natural number larger than unity is a prime number if it does not have other divisors except itself and unity.

Note: Unity *i.e* 1 is not a prime number.

To Check A Number is Prime or not:

(*i*) Take square root of the number.

(*ii*) Round of the square root to the next highest integer and call this number as Z.

(*iii*) Check for divisibility of the number N by all prime numbers below Z. If there is no number below the value of Z which divides N then the number will be prime.

e.g. 241 is prime or not?

$\sqrt{241}$ lies between 15 or 16.Hence take the value of Z =16.

Prime numbers less than 16 are 2, 3, 5, 7, 11 and 13.

241 is not divisible by any of these. Hence we can conclude that 241 is a *prime number.*

2. *Composite Numbers:*

The numbers which are not prime are known as composite numbers.

Note: 1 is neither prime not composite

Co-Primes: Two numbers a and b are said to be co-primes, if their H.C.F is 1.

e.g. (2, 3), (4, 5), (7, 9), (8, 11).....

Tests for Divisibility

A number is divisible by

(1) 2, when its unit's digit is even or 0.

(2) 3, when the sum of its digits is divisible by 3.

(3) 4 when number formed by last two right hand digits is divisible by 4, or if last two digits are 0's.

(4) 5, when its unit's digit is 5 or 0.

(5) 6, when it is divisible by 2 and 3 both.

(6) 8 when number formed by last three right hand digits is divisible by 8, or when last three digit are 0's.

(7) 9, when sum of its digits is divisible by 9.

(8) 10, when its unit's digit is 0.

(9) 11, when difference between sum of the digits in the odd and sum of the digits in the even places is 0 or a multiple of 11.

Note: When any number with an even number of digits is added to its reverse, the sum is always a multiple of 11.

(10) 12, when it is divisible by 3 and 4 both.

NUMERICAL RELATION

Let **A** and **B** be two non-empty sets. Then the relation which is usually denoted by **R** from **A to B** is defined as the set of ordered pairs (in each of which the first element is from **A** and the second element is from **B)** satisfying a certain condition.

The set of all first elements of ordered pairs of R is called "Domain" of **R.**

The set of all second elements of ordered pairs is called "range" of **R.**

Example:

If **A = {1,4,6}, B = {1, 2, 3}** then construct a relation **R** from **A to B** describing "is a multiple of".

Since **1** "is amultiple of" **1, (1, 1)** is in **R.**

Since **4** "is a multiple of" **2 (4, 2)** is in **R.**

Similarly, we can find all the ordered pairs of **R.**

Hence **R = {(1, 1) (4, 1) (4, 2) (6, 1) (6, 2) (6, 3)}**

Here, Domain of **R = {1, 4, 6}** and

Range of **R = {1, 2, 3}**

Note : Though **(4, 3)** is an ordered pair, it won't be there in **R** because **4** is not a multiple of **3.**

Graphical Rrepresentation of a Relation

A relation is graphically represented as follows.

If **A = {1, 4, 6}, B = {1, 2, 3}** then a relation **R = {(1, 1)(4,2) (6, 2)(6, 3)}** from **A** to **B** is represented through a graph as follows.

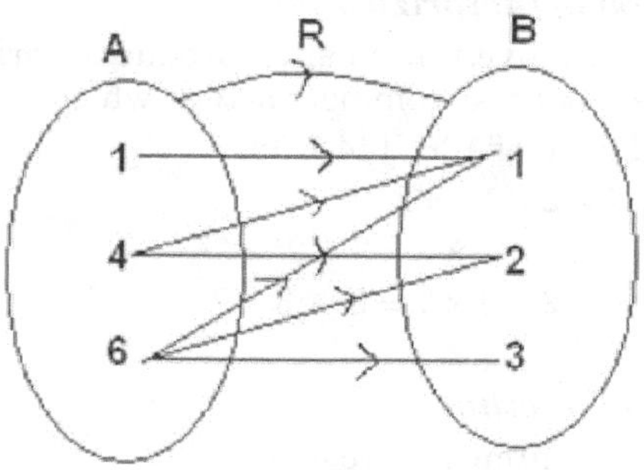

Types of Relations

The types of relations are nothing but their properties. There are different types of relations namely **reflexive, symmetric, transitive and anti symmetric** which are defined and explained as follows through real life examples.

Reflexive Relation:

A relation R is said to be reflexive over a set A if (a,a) € R for every a € R.

Example:

If A is the set of all males in a family, then the relation "is brother of" is not reflexive over A. Because any person from the set A cannot be brother of himself.

Symmetric Relation:

A relation R is said to be symmetric if (a, b) € R ⇒ (b, a) € R

Example:

If A is the set of all males in a family, then the relation "is brother of" is symmetric over A.

Because if a is brother of b then b is brother of a.

Transitive Relation:

A relation R is said to be symmetric if (a, b) € R, (b, c) € R ⇒ (a, c) € R.

Example:

If A is the set of all males in a family, then the relation "is brother of" is transitive over A.

Because if a is brother of b and b is brother of c then a is brother of c.

Anti Symmetric Relation:

A relation R is said to be anti-symmetric if (a,b) € R and (b, a) € R ⇒ a = b.

Example:

The relation "≤" on the set of real numbers is anti-symmetric because if a ≤ b and b ≤ a then a = b.

HCF AND LCM

Greatest Common Divisor (GCD)

It is the greatest divisor common to two or more given numbers. It is also called HCF (Highest Common Factor).

Method to find HCF of given Numbers.

1. ***By method of factorization:***

   Express each given number as product of primes. Now take product of common factors, which is HCF.

   e.g., Find HCF of 136, 144, 168.

   $136 = 2 \times 2 \times 2 \times 17 = 2^3 \times 17$

   $144 = 2 \times 2 \times 2 \times 2 \times 3 \times 3 = 2^4 \times 3^2$

   $168 = 2 \times 2 \times 2 \times 3 \times 7 = 2^3 \times 3 \times 7$

   $HCF = 2^3 = 8$

2. ***By division method:***

   Suppose two numbers are given.

   Divide greater number by the lesser one; divide lesser by the remainder; divide first remainder by the new remainder, and so on till there is no remainder. Last divisor is the required H.C.F.

   e.g., Find H.C.F. of 12 and 15.

```
12)15(1
   -12
    3)12(4
      -12
       00
```

∴ HCF = 3

In case more than two numbers are given, then choose any two of them and find their HCF. HCF of these two and third gives HCF of three numbers and so on.

Note: If 'Z' is HCF of 'X' and 'Y', then Z is also a factor of aX + bY where a, b are integers.

Least Common Multiple (LCM)

LCM of two or more given number is the least number which is exactly divisible by each of the given numbers.

Methods to find LCM of given Numbers.

1. **By method of factorization:**

   Resolve each one of the given numbers into prime factors, then their LCM is product of highest powers of all factors, that occur in these numbers.

   e.g., Find LCM of 136, 144, 168

   $136 = 2^3 \times 17$

   $144 = 2^4 \times 3^2$

   $168 = 2^3 \times 3 \times 7$

   $LCM = 2^4 \times 3^2 \times 17 \times 7 = 17136$

2. **Using the formula:**

   Product of numbers = HCF × LCM

   If two numbers are given, find their HCF.

$$LCM = \frac{\text{Product of two numbers}}{HCF}$$

   e.g., Find LCM of 136, 144.

$$LCM = \frac{136 \times 144}{8} = 2448$$

   If more than two numbers are given, find LCM of any two of them. Then LCM of this LCM and third number gives LCM of these three numbers and so on.

3. **Short cut method using factorization:**

   e.g. Find LCM of 18, 28, 108, 105.

2	18,	28,	108,	105
2	9,	14,	54,	105
3	9,	7,	27,	105
3	3,	7,	9,	35
3	1,	7,	3,	35
5	1,	7,	1,	35
7	1,	7,	1,	7
	1,	1,	1,	1

$LCM = 2 \times 2 \times 3 \times 3 \times 3 \times 5 \times 7 = 3780$

Alternatively : We can strike out 18 at once, since any multiple of 108 must contain 18 as a factor.

2	28,	108,	105
2	14,	54,	105
3	7,	27,	105
7	7,	9,	35
	1,	9,	5

LCM = $2 \times 2 \times 3 \times 7 \times 9 \times 5 = 3780$

Note: For any given numbers, HCF is necessarily a factor of their L.C.M.

HCF and LCM of Fractions.

$$\text{HCF of Fractions} = \frac{\text{HCF of Numerators}}{\text{LCM of Denominators}}$$

$$\text{LCM of Fractions} = \frac{\text{LCM of Numerators}}{\text{HCF of Denominators}}$$

e.g. Find HCF and LCM of $\frac{4}{5}, \frac{8}{35}, \frac{2}{15}, \frac{6}{25}$

$$\therefore \quad \text{HCF} = \frac{\text{HCF of 4, 8, 2, 6}}{\text{LCM of 5, 35, 15, 25}}$$

$$= \frac{2}{5 \times 7 \times 3 \times 5} = \frac{2}{525}$$

$$\text{LCM} = \frac{\text{LCM of 2, 8, 4, 6}}{\text{HCF of 5, 35, 15, 25}} = \frac{24}{5}$$

PROGRESSIONS

Arithmetic Progression (A.P.)

Let 'a' be the first term of an arithmetic progression; 'd' the common difference and 'n' the number of terms in the progression.

$$T_n = n^{th} \text{ term} = a + (n - 1)d$$

$$S_n = \text{Sum of n terms} = \frac{n}{2} \times [2a + (n - 1)d]$$

Progression can be represented as a, a + d, a + 2d, ..., [a + (n – 1) d]. Here, quantity d is to be added to any chosen term to get next term of the progression.

When there are n terms in an arithmetic progression, a is first term and {a + (n – 1) d} is last term.

Then $\quad S_n = \frac{n}{2} \times [\text{First Term} + \text{Last Term}]$

Geometric Progression (G.P.)

Geometric progression can be represented as a, ar, ar^2, ... where a is first term and r is common ratio of the geometric progression.

n^{th} term of the geometric progression = ar^{n-1}.

$$\text{Sum to n terms} = \frac{a(1 - r^n)}{1 - r} \text{ or } \frac{a(r^n - 1)}{r - 1}$$

$$= \frac{r \times \text{last term} - \text{First term}}{r - 1}$$

Infinite Geometric Progression.

If – 1 < r < + 1 or |r| < 1, then sum of a geometric progression does not increase infinitely; it "converges" to a particular value. Such a G.P. is called infinite geometric progression.

Sum of an infinite geometric progression, $S_\infty = \frac{a}{1 - r}$

> **Remember**
>
> Sum of first n natural numbers, $= S_n = \sum n = \frac{n(n+1)}{2}$
>
> Sum of squares of first n natural numbers,
>
> $$\sum n^2 = \frac{n(n+1)(2n+1)}{6}$$
>
> Sum of cubes of first n natural numbers,
>
> $$\sum n^3 = \left[\frac{n(n+1)}{6}\right]^2 \frac{n^2(n+1)^2}{4} = \left[\sum n\right]^2$$

ALGEBRAIC FORMULAE AND THEIR APPLICATION

An equation is a statement that *two algebraic expressions are equal*. If an equation is satisfied by any value of the variable, then equation is said to be an *identity*.

1. $(a + b)^2 = a^2 + 2ab + b^2 = (a - b)^2 + 4ab$
2. $(a - b)^2 = a^2 - 2ab + b^2 = (a + b)^2 - 4ab$
3. $(a + b)^2 + (a - b)^2 = 2(a^2 + b^2)$
4. $(a + b)^2 - (a - b)^2 = 4ab$
5. $(a + b + c)^2 = a^2 + b^2 + c^2 + 2(ab + bc + ca)$
6. $(a + b + c + d)^2 = a^2 + b^2 + c^2 + d^2 + 2a(b + c + d) + 2b(c + d) + 2cd$
7. $(a + b)(a - b) = a^2 - b^2$
8. $(x + a)(x + b) = x^2 + (a + b)x + ab$
9. $(x + a)(x + b)(x + c) = x^3 + (a + b + c)x^2 + (ab + bc + ca)x + abc$
10. $(a + b)^3 = a^3 + 3ab(a + b) + b^3$
11. $(a - b)^3 = a^3 - 3ab(a - b) - b^3$
12. $a^3 + b^3 = (a + b)^3 - 3ab(a + b) = (a + b)(a^2 - ab + b^2)$
13. $a^3 - b^3 = (a - b)^3 + 3ab(a - b) = (a - b)(a^2 + ab + b^2)$
14. $a^3 + b^3 + c^3 - 3abc = (a + b + c)(a^2 + b^2 + c^2 - ab - ac - bc)$
    If $a + b + c = 0$ then $a^3 + b^3 + c^3 = 3abc$
15. $a^n - b^n = (a - b)(a^{n-1} + a^{n-2} b + a^{n-3} b^2 + ... + b^{n-1})$ for all n.
16. $a^n + b^n = (a + b)(a^{n-1} - a^{n-2} b + a^{n-3} b^2 - ... - b^{n-1})$
    if n is even
17. $a^n + b^n = (a + b)(a^{n-1} - a^{n-2} b + a^{n-3} b^2 + ... + b^{n-1})$
    if n is odd.
18. $a^4 + a^2b^2 + b^4 = (a^2 + ab + b^2)(a^2 - ab + b^2)$

POLYNOMIALS

Factors of polynomials: An expression is said to be resolved into factors when expressions of its product are found.

HCF of polynomials: When two or more polynomials are factorized, then product of all the common factors is HCF of the polynomials.

LCM of polynomials: When two or more polynomials are factorized, then product of all the factors with highest power is LCM of the polynomials.

INEQUATIONS

Properties

1. An inequality will still hold after each side has been increased, diminished, multiplied or divided by the same positive quantity.
   i.e., if $a > b$ and $c > 0$
   $a + c > b + c$
   $ac > bc$
   $\frac{a}{c} > \frac{b}{c}$
2. In an inequality, any term may be transposed from one side to the other if its sign is changed.
   i.e., if $a - c > b$, then $a > b + c$ or $c < a - b$
3. If sides of an inequality be multiplied by same negative quantity, then sign of inequality must be reversed.
   i.e., if $a > b$ and $c < 0$, then $ac < bc$
4. If $a > b$, $a, b \geq 0$ then $a^n > b^n$ and $\frac{1}{a^n} < \frac{1}{b^n}$, or $a^{-n} < b^{-n}$; if n is a positive quantity.
5. Square of every real quantity is positive and therefore must be greater than zero
   i.e., for $a \neq b$, $(a - b)^2 > 0$; $a^2 + b^2 > 2ab$;
   Similarly, if $x > 0$, $y > 0$, then
   $$\frac{x+y}{2} \geq \sqrt{xy}$$
   Hence, Arithmetic mean of two positive quantities is greater than or equal to their Geometric mean.
6. If sum of two positive quantities is given, their product is greatest when they are equal; and if product of two positive quantities is given, their sum is least when they are equal.
7. If a, b, c.... k are n unequal quantities, then
   $$\left(\frac{a+b+c+...+k}{n}\right)^n > a \times b \times c \times d ... \times k$$
   i.e., $\frac{a+b+c+...+k}{n} > (a \times b \times c \times ... \times k)^{1/n}$

   Note: Arithmetic mean of any number of positive quantities is greater than their Geometric mean.
8. If a and b are positive and unequal, then
   $\frac{a^m b^m}{2} > \left(\frac{a+b}{2}\right)^m$ except when m is a positiveproper fraction.
   If m is a positive integer or any negative quantity, then $\frac{a^m b^m}{2} > \left(\frac{a+b}{2}\right)^m$
   If m is positive and less than 1, then $\frac{a^m b^m}{2} < \left(\frac{a+b}{2}\right)^m$
   If there are n positive quantities a, b, c ... k, then
   $$\left(\frac{a^m + b^m + c^m + ... + k^m}{n}\right) > \left(\frac{a+b+c+...+k}{n}\right)^m$$
   unless m is a positive proper fraction.
9. If a, b, c are positive and not all equal, then
   $(a+b+c)(ab+bc+ca) > 9abc$
   and, $(b + c)(c + a)(a + b) > 8abc$.
10. If x is positive and $a < b$, then $\frac{a+x}{b+x} > \frac{a}{b}$
    If x is positive and $a > b$, then $\frac{a+x}{b+x} < \frac{a}{b}$
11. If x is positive and $a > b > x$, then $\frac{a-x}{b-x} > \frac{a}{b}$
    If x is positive and $x < a < b$, then $\frac{a-x}{b-x} < \frac{a}{b}$
12. $\frac{a+c+e+...}{b+d+f+...}$ is less than greatest and greater than least of the fractions $\frac{a}{b}, \frac{c}{d}, \frac{e}{f}, ...$
13. For positive numbers, if $a > x$, $b > y$, $c > z$, then $a + b + c + ... > x + y + z + ...$ and $abc ... > xyz ...$
14. $a^2 + b^2 + c^2 \geq bc + ca + ab$
15. $(n!)^2 > n^n$, for $n > 2$.
16. For any positive integer n, $2 \leq \left(1+\frac{1}{n}\right)^n \leq 3$
17. $a^2b + b^2c + c^2a \geq 3abc$
18. $\frac{a}{b}+\frac{b}{c}+\frac{c}{d}+\frac{d}{a} \geq 4$
19. $a^4 + b^4 + c^4 + d^4 \geq 4abcd$

QUADRATIC EQUATION WITH APPLICATION TO INEQUALITIES

In a quadratic equation $ax^2 + bx + c = 0$, if $b^2 - 4ac > 0$, then

$ax^2 + bx + c = a(x - \alpha)(x - \beta)$, where α and β are two roots of the equation with $\alpha < \beta$.

$(x - \alpha)(x - \beta)$ is always positive when $x < \alpha$ or $x > \beta$ i.e., x does not lie between α and β.

$(x - \alpha)(x - \beta)$ is always negative when $\alpha < x$ or $x > \beta$ i.e., x lies between α and β.

DATA INTERPRETATION

Data can be organized in a number of ways so that larger volume of data can be presented in a more compact and precise form. Data thus presented has to be deciphered correctly by the user of the data. The deciphering of the data from its compactly presented form is called Data INTERPRETATION.

Representation of Data

Numerical data can be presented in one or more of the following ways

(1) Data Table
(2) Pie Chart
(3) Line Graph
(4) Bar Chart
(5) Others

The "Others" category covers miscellaneous form like descriptive case format etc. customized for the situation. Data can also be presented by using a combination of two or more of the above forms.

While some data can be presented in many different forms, some other may be amenable to be presented only in few ways. In real life situation, the style of data presentation is based on the end-objective. In certain situation data has to be presented as a combination of two or more forms of data presentation.

Let us understand each of the above form of data presentation with an example.

Data Table

Here data is presented in the form of table. Where any type of data can be presented in tabular form, that too is a very accurate manner, interpreting the data in table form becomes more difficult and time consuming than the other modes, all of which are basically pictorial or graphical in presentation.

Year	A	
	Appeared	Qualified
2002	250	720
2003	2750	810
2004	3000	890
2005	3250	910
2006	3720	1050

Year	B	
	Appeared	Qualified
2002	1750	460
2003	1860	490
2004	2000	520
2005	2100	640
2006	2400	830

Year	C	
	Appeared	Qualified
2002	100	120
2003	1120	200
2004	1250	300
2005	1500	600
2006	1650	780

Year	D	
	Appeared	Qualified
2002	800	120
2003	1000	220
2004	1200	300
2005	1210	340
2006	1440	480

Year	E	
	Appeared	Qualified
2002	2000	370
2003	2200	420
2004	2500	510
2005	2750	680
2006	3440	960

Data tables can be of a number of types. They can be of a single –table variety or combination of tables. One such example of table is given above.

The above table shows the number of students appeared and qualified in an entrance test from five districts A, B, C, D and E of a state.

From the above table, we can obtain the following data;

(1) Total number of students appeared and qualified from the given districts in each of the years.

(2) Percentage increase in the number of students appeared or qualified in the district over the years.

(3) Average number of students appeared or qualified

Pie-Charts

This is probably the simplest of all pictorial forms of data presentation. Here, total quantity to be shown is distributed over one complete circle or 360 degree. In pie-charts, data is essentially presented shares of various elements as proportion and percentage of the total quantity. Each element or group in a pie-chart is represented in the terms of quantity (or value, as the case may be) or as the angle made by the sector representing the elements or as a proportion of the total or as a percentage of the total.

Chart 1 gives distribution of sales of different companies.

CHART - 1

Total sales Rs. 6,000 crores

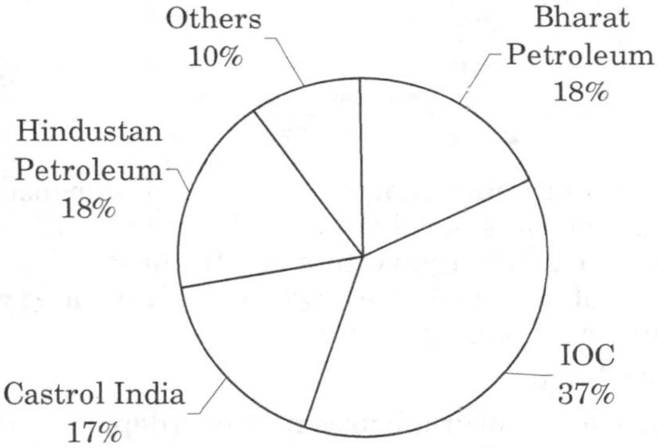

From the above pie chart, we can calculate the following;

(1) Total sales of each of the companies.

(2) Sales of a company as a percentage of the other.

(3) Conversion of these percentage values into angles for each zone.

Pie-chart are also very frequently used in combination with other forms of data or along with other pie-charts.

Two-Dimensional Graphs

This is essentially used for continuous data but can also be used for depicting discrete data provided we understand the limitation. Also known as Cartesian Graphs, they represent variation of the one parameter with respect to another parameter each shown on a different axis. These types of graphs are useful in studying the rate of change or understanding the trends through extrapolations.

These graphs can be of various types and a few of them are shown below:

CHART – 2

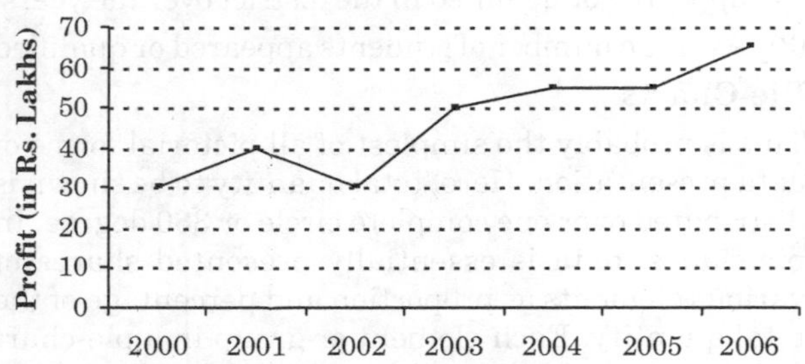

The graphs in Chart 2 show the changes in the profit of the company during a period time. One can find out trends and the growth in the profit over the years.

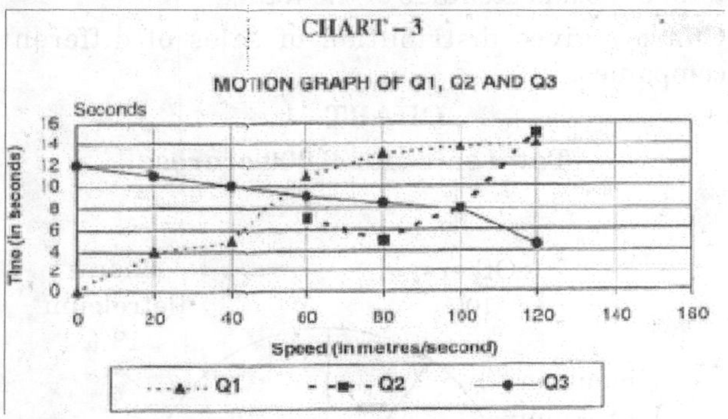

Chart 3 presents another type of two- dimensional graph which is mostly used to depict scientific data like speed, velocity, vectors etc. In the graph speed trends of three bodies Q1, Q2, Q3 is given along with their actual path of motion.

Bar Chart

This is a type of graph used mostly to depict data in a discrete way. They are accurate and comparison of variable is very convenient.

CHART-4

Import and Export of XYZ Company

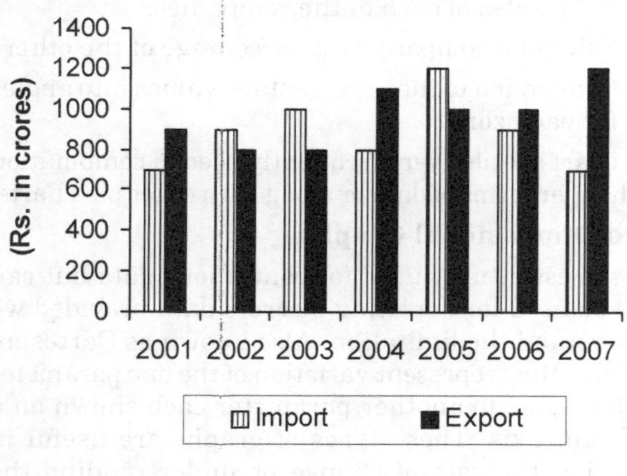

Chart 4 shows import and export of a company over different years. From this graph we can obtain the following:

(1) Percentage contribution of imports or exports to the company's total trade for different years.

(2) Relative increase or decrease in the share of imports or exports

(3) Percentage growth/average annual growth in imports or exports during given period.

Three-Dimensional Graph

The data in a triangular graph are given on each side of the triangle. Each point represents three different values, one each in each direction.

CHART 5

Students	Percentage					
A	82	66	59	76	62	65
B	76	72	65	84	74	75
C	56	78	71	66	86	70
D	64	80	68	72	66	80
E	48	68	83	88	56	60
F	60	74	79	64	80	85

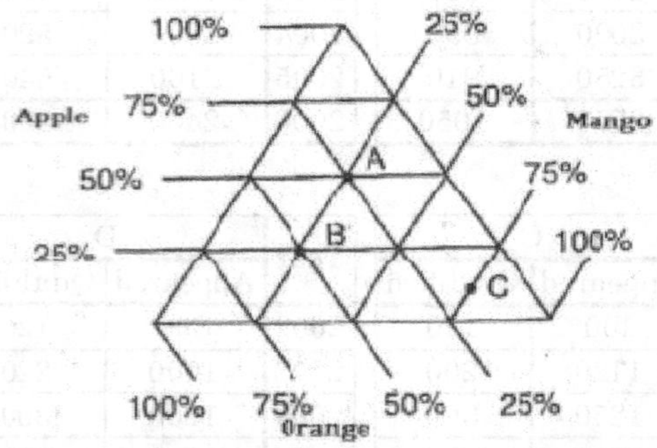

The graph in chart 5 represents the percentage of people who like the three Fruit-Apple, Mango and Orange in three cities A, B and C. For example, in city B 25% like Mango, 50% like Orange and 25% like Apple.

EXERCISE

MCQ TYPE QUESTIONS

1. If $x = b + c$, $y = c - a$, $z = a - b$, then $x^2 + y^2 + z^2 - 2xy - 2xz + 2yz$ is equal to

(*a*) $a + b + c$ (*b*) $4b^2$
(*c*) abc (*d*) $a^2 + b^2$

2. HCF of $y^3 - 4y$ and $4y(y^3 + 8)$ is

(*a*) $y(y + 2)$ (*b*) $y - 2$
(*c*) $2(y + 2)$ (*d*) $(y + 2)(y^2 - 2y + 4)$

3. A tiled floor of a room has dimensions m × m sq.m. Dimensions of the tile used are n × n sq.m. All tiles used are green tiles except diagonal tiles which are red. After some years some green tiles are replaced by red tiles to form an alternate red and green tile pattern. How many green tiles are removed?
($m \neq n$ and total number of tiles are odd).

(*a*) $\frac{m^2 - 4mn + 2n^2}{2n^2}$ (*b*) $\frac{(m-2n)^2 - n^2}{2n^2}$
(*c*) $\frac{m^2 - 4mn - n^2}{2n^2}$ (*d*) $\frac{m^2 - 4mn - 2n^2}{2n^2}$

4. If $\frac{x-4}{-x^2-6} \leq \frac{2}{-x^2-6}$ then

(*a*) $x \leq 2$ (*b*) $x \geq 2$
(*c*) $x \leq 6$ (*d*) $x \geq 6$

5. Which of the following is true?

(*a*) $\sqrt{2} + \sqrt{5} = \sqrt{7}$ (*b*) $\sqrt{2} + \sqrt{5} \leq \sqrt{7}$
(*c*) $\sqrt{2} + \sqrt{5} < \sqrt{7}$ (*d*) $\sqrt{2} + \sqrt{5} > \sqrt{7}$

6. If $6 \geq x \geq -2$ and $4 \geq y \geq -4$, then limits for $\frac{y}{x}$, where x and y are non zero integers, is

(*a*) $\frac{y}{x} \geq 2, \frac{y}{x} \leq \frac{2}{3}$ (*b*) $\frac{y}{x} \geq \frac{-2}{3}, \frac{y}{x} \leq 2$
(*c*) $\frac{y}{x} \geq \frac{-2}{3}, \frac{y}{x} \leq \frac{1}{4}$ (*d*) $\frac{y}{x} \geq -4, \frac{y}{x} \leq 4$

7. What are the limits of $\frac{2^n(n+1)^n}{n^n}$; where n is a positive integer?

(*a*) $2 \leq \frac{2^n(n+1)^n}{n^n} \leq 3$

(*b*) $2^{n+1} \leq \frac{2^n(n+1)^n}{n^n} \leq 3.2^n$

(*c*) $2^n \leq \frac{2^n(n+1)^n}{n^n} \leq 3.2^n$

(*d*) None of these

8. If $g = 10^{100}$ and $H = 10^g$, then in which interval does $g! = 1.2.3 .... 10^{100}$ lie?

(*a*) $10H < g! < H$ (*b*) $H < g! < 10^H$
(*c*) $10H < g! < 10^H$ (*d*) $10^H < g!$

9. Which of the following is true?

(*a*) Sum of four consecutive even numbers is always divisible by 8.
(*b*) Sum of four consecutive odd numbers is always divisible by 8.
(*c*) Product of any n consecutive natural numbers may not be divisible by n!.
(*d*) Product of 4 consecutive odd numbers is always divisible by 15.

10. If a and b are prime numbers, which of the following is true?

I. a^2 has three positive integer factors.
II. ab has four positive integer factors.
III. a^3 has four positive integer factors.

Codes :

(*a*) I and II only (*b*) II and III only
(*c*) All of these (*d*) None of these

11. If $7x + 6y = 420$, x and y are natural numbers, then what can be said about x?

(*a*) x is always odd
(*b*) x is always even.
(*c*) x is even only if y is odd.
(*d*) x is odd if y is even.

12. If a, b, c, d, p and q are non-zero, unequal integers and $\frac{a+bi}{c+di} = \frac{p}{q}$, then $\frac{a^2+b^2}{c^2+d^2}$ equals

(*a*) $\frac{p}{q}$ (*b*) $\frac{p^2}{q^2}$
(*c*) 1 (*d*) None of these

13. An Egyptian fraction has a numerator equal to 1, and its denominator is a positive integer. What is the maximum number of different Egyptian fraction such that their sum is equal to 1, and their denominators are equal to 10 or less?

(*a*) 3 (*b*) 5 (*c*) 7 (*d*) 9

14. What is the least number which must be subtracted from 1936 so that the remainder when divided by 9, 10, 15 will leave in each case the same remainder 7?

(*a*) 32 (*b*) 53
(*c*) 46 (*d*) 39

15. What is the greatest number consisting of six digits which on being divided by 6, 7, 8, 9, 10 leaves 4, 5, 6, 7, 8 as remainders respectively?

(*a*) 997920 (*b*) 997918
(*c*) 999999 (*d*) 997922

16. What is the least number which on being divided by 5, 6, 8, 9, 12 leaves in each case a remainder 1 but when divided by 13 leaves no remainder?

(*a*) 2987 (*b*) 3601
(*c*) 3600 (*d*) 2986

17. A group of four numbers has only one prime number amongst them. Which of the following must be true about the group?

I. HCF of the four numbers of the group is either 1 or equal to that prime number.

II. LCM of the four numbers of the group is same as product of the prime number and LCM of the remaining three numbers.

III. Product of four numbers is equal to product of the prime number × HCF of the group × LCM of the group.

(*a*) I only
(*b*) II only
(*c*) I and II only
(*d*) All of these

18. If (a, n)! is defined as product of n consecutive numbers starting from a, where a and n are both natural numbers, and if H is the HCF of (a, n)! and n!, then what can be said about H?

(*a*) H = a! (*b*) H = n!
(*c*) $H \geq n!$ (*d*) $H \geq a \times n$

19. HCF and LCM of two numbers is given. It is possible to find out the two numbers uniquely if

I. either sum or difference between the two numbers is known.

II. HCF of two numbers = LCM of two numbers.

III. $\frac{\text{LCM}}{\text{HCF}}$ = Prime number.

(*a*) I and II only
(*b*) II only
(*c*) II and III only
(*d*) I, II and III

20. A ball is dropped from a height of 12 m and it rebounds 1/2 of the distance it falls. If it continues to fall and rebound in this way, how far will it travel before coming to rest?

(*a*) 36 m (*b*) 30 m
(*c*) 48 m (*d*) 60 m

21. There are N questions in an exam.

For i = 1, 2, ..., N, there are 2^{N-1} students who answered 1 or more questions wrongly. If total number of wrong answers is 8191, then N will be

(*a*) 12 (*b*) 11 (*c*) 10 (*d*) 13

22. Two new charity organizations C_1 and C_2 were formed, with x members each, on January 1, 2003. On first day of each, subsequent month, in C_1, number of members increases by a certain number a, while in C_2, number of members increases in such a way that ratio of the number of members in a month to the preceding month bear a ratio equal to b. On May 1, 2003, both organizations had the same number of members. If a = 20x, then b will be

(*a*) 2 (*b*) 3 (*c*) 2.5 (*d*) 3.5

23. What is the ratio of common differences d_1 and d_2 of two arithmetic progressions if respective n^{th} terms are in the ratio of 2n + 3 : n – 11?

(*a*) 1 : 2 (*b*) 2 : 3
(*c*) 2 : 1 (*d*) 1 : 3

24. What is the sum upto 20 terms of the series $\frac{2}{3}+\frac{8}{9}+\frac{26}{27}+\frac{80}{81}+......?$

(*a*) $\frac{19.3^{20}+1}{3^{20}}$ (*b*) $\frac{19.3^{20}+1}{2.3^{20}}$

(*c*) $\frac{19.3^{20}-1}{2.3^{20}}$ (*d*) $\frac{39.3^{20}+1}{2.3^{20}}$

25. What is the sum to infinity of the series, $3 + 6x^2 + 9x^4 + 12x^6 + ...$ given $|x| < 1$?

(*a*) $\frac{3}{(1+x^2)}$ (*b*) $\frac{3}{(1-x^2)^2}$

(*c*) $\frac{3}{(1+x^2)^2}$ (*d*) $\frac{3}{(1-x^2)}$

26. A school has 5 divisions in a class IX having 60, 50, 55, 62 and 58 students. Mean marks obtained in a History test were 56, 64, 72, 63 and 50 by each division respectively. What is overall average of the marks per students?

(*a*) 56.8 (*b*) 58.2
(*c*) 62.4 (*d*) 60.8

27. Average age of a committee of seven trustees is the same as it was five years ago, a younger man having been substituted for one of them. How much younger was he then trustee whose place he took?

(*a*) 32 years (*b*) 35 years
(*c*) 33 years (*d*) 34 years

28. Six men A, B, C, D, E, F agree with a seventh man G to provide a sum of money among them. A, B, C, D, E, F are to subscribe ₹10 each, and G is to pay ₹3 more than the average of the seven. What is the whole sum to be provided?

(*a*) ₹73.50 (*b*) ₹ 74
(*c*) ₹ 73 (*d*) ₹ 72.50

29. Which of the following is/are true?

I. When Arithmetic Mean of a set of values and number of values in the set are known, then Geometric Mean of the set can be found out.

II. When Arithmetic Mean of a set of values and number of values in the set are known, sum of all the elements in the set can be found out.

III. When Geometric Mean of a set of values and number of values in the set are known, then Arithmetic Mean of the set can be found out.

Codes :

(*a*) Only I (*b*) Only II
(*c*) Only I and II (*d*) All except II

30. Which of the following statement/s is/are true?

I. Average of a set of values will always lie between lowest and the largest of these values.

II. If each of the values in a set is increased by a constant k, then new average of the set is increased by k.

III. If each of the values in a set is multiplied by a constant m, then new average will also be 'm' times the old average.

Codes :

(*a*) Only I and II (*b*) Only II and III
(*c*) Only I and III (*d*) I, II and III

31. Three people A, B and C weigh themselves in a particular order. First A, B, C weigh themselves individually and then AB, BC CA and ABC together respectively. Recorded weight for the last measured is 180 kgs. Then average of the 7 measures is

(*a*) 320 kgs (*b*) $\frac{360}{7}$ kgs
(*c*) $\frac{720}{7}$ kgs (*d*) Cannot be determined

32. Average age of a team of 15 employees is 36. Youngest of them is 20 years old and eldest is 36 years old. Two of them with average age 28 leave the team. If one of the two comes back on the condition that he will be made the team leader, then which of the following can possibly be average age of the new team so formed?

(*a*) 35 (*b*) 36 (*c*) 38 (*d*) 39

Directions for Q.33 to Q.37:

Study the following table carefully and answer the questions given below:

Annual Income of Various Individuals over the Years (Income in Lakhs)

Year	Individual					
	A	B	C	D	E	F
2002	5.5	3.2	4.8	6.35	6.15	3.5
2003	5.65	3.25	4.85	6.55	6.25	3.65
2004	5.7	3.725	4.95	7.155	6.425	3.75
2005	6.2	4.25	5.1	7.35	7.15	4
2006	6.5	4.5	5.2	7.4	7.25	4.25
2007	6.75	5	5.25	7.48	7.285	4.8
2008	7	5.35	5.3	8	7.3	5.1

33. What is the approximate average monthly income of A over the years?

(*a*) Rs. 44762 (*b*) Rs.48348
(*c*) Rs. 53216 (*d*) none of these

34. What is the ratio of the annual income of B to that of C in the year 2005?

(*a*) 6 : 5 (*b*) 6 : 7
(*c*) 5 : 6 (*d*) 7 : 6

35. What is the difference between the total annual income of all the individuals in the year 2003 and the total annual income of all the individuals in the year 2007?

(*a*) Rs. 736500 (*b*) Rs.645400
(*c*) Rs. 636500 (*d*) Rs.743300

36. The annual income of E in the year 2004 is approximately what per cent of the average annual income earned by E over the years?

(*a*) 98 (*b*) 86
(*c*) 88 (*d*) 94

37. What is the average annual income of F over the years?

(*a*) Rs. 416000 (*b*) Rs.425000
(*c*) Rs.401500 (*d*) none of these

Directions for Q.38 to Q.42:

Study the pie-charts carefully to answer the questions that follow:

Chart I gives : Percentage distribution of candidates studying in various disciplines of a college.

Total Number of students = 4800

Chart II gives : Percentage of breakup of girl students in each discipline. Total number of girls = 2000

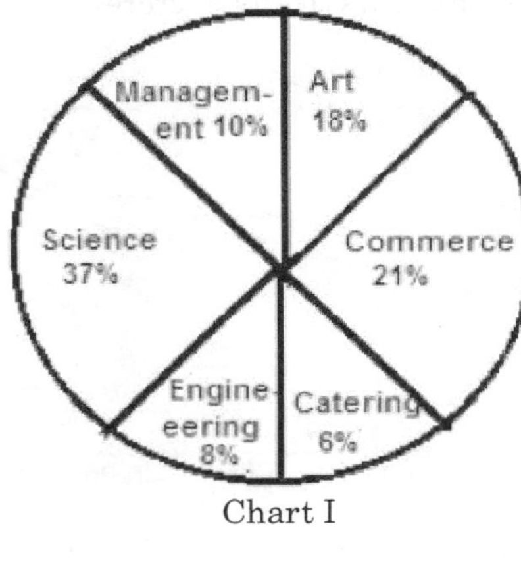

Chart I

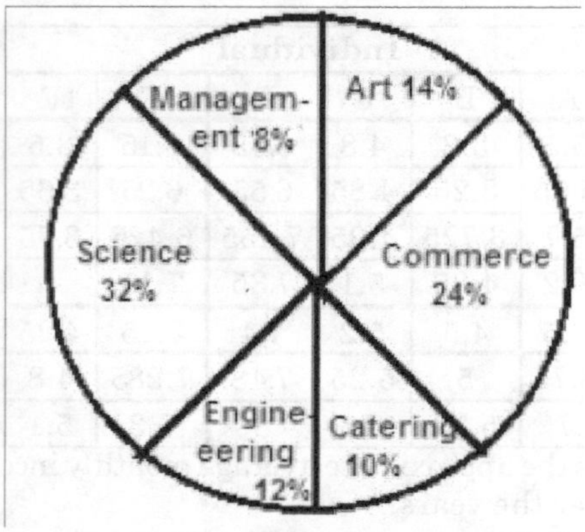

Chart II

38. What is the number of girls studying in Engineering discipline?
(*a*) 180 (*b*) 220
(*c*) 320 (*d*) none of these

39. Which discipline has the highest number of boys studying in it ?
(*a*) Science (*b*) Commerce
(*c*) Arts (*d*) None of these

40. The number of boys studying in Commerce discipline forms approximately what per cent of the total number of students studying in that discipline?
(*a*) 40 (*b*) 52
(*c*) 36 (*d*) 28

41. What is the respective ratio of number of boys studying in Mangement discipline to the boys studying in Science discipline?
(*a*) 71:20 (*b*) 18:67
(*c*) 20:71 (*d*) None of these

42. The number of girls studying in Catering discipline forms what per cent of the total number of students in the college? (rounded off two digits after decimal)
(*a*) 3.28 (*b*) 5.61
(*c*) 3.95 (*d*) 4.17

Directions for Q.43 to Q.44:

Study the following graphs carefully to answer the questions given below.

Unit of raw material manufactured and sold by a company over the years (Units in crores)

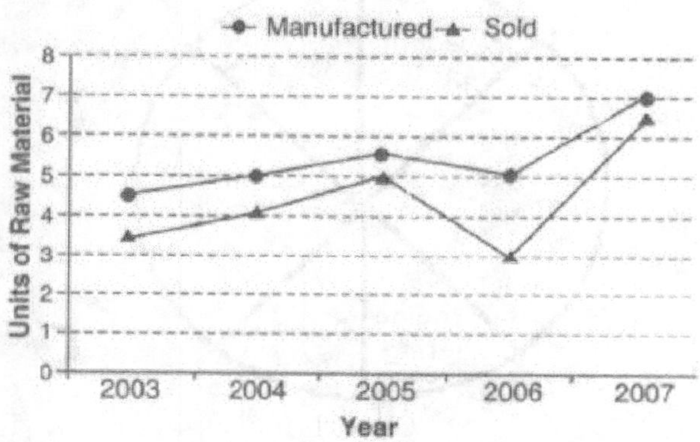

43. What is the average number of unit sold over the years?
(*a*) 440000000 (*b*) 4400000
(*c*) 440000 (*d*) 44000000

44. What is the ratio of the difference between the number of units manufactured and sold in the year 2005 to the difference between the number of units manufactured and sold in the year 2006?
(*a*) 2 : 3 (*b*) 1 : 2
(*c*) 1 : 4 (*d*) 3 : 5

Directions for Q.45 to Q.49:

Study the following graph carefully to answer the questions that follow:

Production and sale of printers of various companies in a month

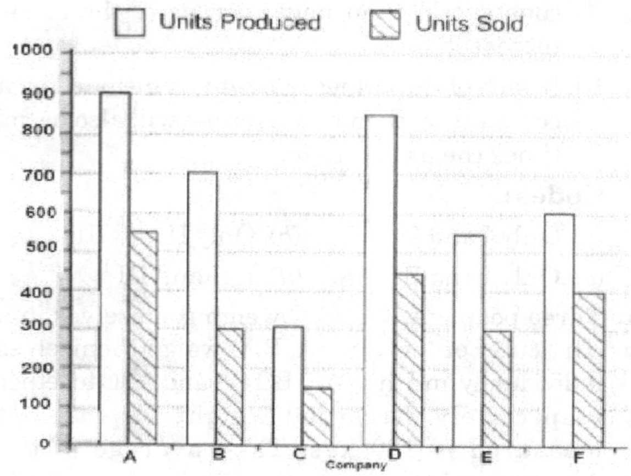

45. What is the average number of units sold by all the companies together?
(*a*) 360 (*b*) 390
(*c*) 375 (*d*) 410

46. Which company had the highest percentage of sale with respect to its production?
(*a*) D (*b*) B
(*c*) E (*d*) None of these

47. What is the average number of units produced by all the companies together?
(*a*) 675 (*b*) 650
(*c*) 625 (*d*) 600

48. The total unit sold by the companies A, B and C together is approximately what per cent of the total units produced by these companies?
(*a*) 62 (*b*) 50
(*c*) 76 (*d*) 58

49. What is the ratio of the total production of companies D and E to the total sale of the same companies?

(*a*) 28 : 15 (*b*) 9 : 5

(*c*) 15 : 11 (*d*) None of these

NUMERICAL TYPE QUESTIONS

1. If $x - y = 1$, then $x^3 - y^3 - 3xy$ equals to _______

2. When $x + y + z = 9$ and $xy + yz + zx = 11$, then $x^3 - y^3 - z^3 - 3xyz$ equals to _______

3. If $x + y + z = 0$, then $\frac{x^2}{yz} + \frac{y^2}{zx} + \frac{z^2}{xy}$ equals to _______

4. In an election for the President, if 261 valid votes are cast, for the 5 contestants then least number of votes a candidate requires to receive to win the election are is _______

5. A number when divided by sum of 555 and 445 gives two times their difference as quotient and 30 as remainder. The number is _______

6. Sum of three prime numbers is 100. If one of them exceeds another by 36, then one of the numbers is _______

7. Sum of all odd numbers up to 100 is_______

8. $A = 0.\ a_1a_1a_1 \ldots$ and $B = 0.a_2a_2a_2$, where a_1 and a_2 are multiples of 3 and also, a_1 and a_2 are distinct integers from 0 to 8. Then value of A + B is _______

9. Anil wants to divide ₹100 into a number of bags so that one can ask for any amount between ₹1 and ₹100, he can give the proper amount by giving certain number of these bags without taking out the amount from them. The minimum number of bags he will require if each bag has whole number of rupees is _______

10. The remainder when $2050 \times 2071 \times 2095$ is divided by 23 is _______

11. The smallest number which when increased by 3 is divisible by 27, 35, 25 and 21 is ______

12. _______is the least number which on being divided by 12, 21 and 35 will leave in each case the same remainder 6

13. A gardener had a number of shrubs to plant in horizontal rows. At first he tried to plant 5 shrubs in each row, then 6, then 8 and then 12, but had always 1 left. On trying 13, in one row he had none left. The smallest number of shrubs that he could have had is ______

14. Sum of an infinite geometric series is 3 and sum of an infinite geometric series formed from squares of the terms of the original series is 6. First term of the first series will be____

15. The maximum sum of the series 60, 58, 56, 54, 52, ... is_______

16. On an average my income for 15 days ₹ 7, average for the first 5 days was ₹ 6 and average for the last 9 days was ₹ 8. The income on the sixth day was ________

17. The harmonic mean of two numbers whose geometric mean and arithmetic mean is 8 and 5 respectively is ________

18. Average marks of 15 students in a class is 145, maximum marks being 150. If two lowest scores are removed, the average increases by 5. Also, two lowest scores are consecutive multiples of 9. The lowest score in the class is _______

ANSWERS

MCQ Type Questions

1. (*b*)	**2.** (*a*)	**3.** (*b*)	**4.** (*d*)	**5.** (*d*)	**6.** (*d*)	**7.** (*b*)	**8.** (*c*)	**9.** (*b*)	**10.** (*c*)
11. (*b*)	**12.** (*b*)	**13.** (*a*)	**14.** (*d*)	**15.** (*b*)	**16.** (*b*)	**17.** (*a*)	**18.** (*b*)	**19.** (*d*)	**20.** (*a*)
21. (*d*)	**22.** (*b*)	**23.** (*c*)	**24.** (*d*)	**25.** (*b*)	**26.** (*d*)	**27.** (*b*)	**28.** (*a*)	**29.** (*b*)	**30.** (*d*)
31. (*c*)	**32.** (*b*)	**33.** (*d*)	**34.** (*c*)	**35.** (*c*)	**36.** (*d*)	**37.** (*d*)	**38.** (*d*)	**39.** (*a*)	**40.** (*b*)
41. (*c*)	**42.** (*d*)	**43.** (*c*)	**44.** (*d*)	**45.** (*b*)	**46.** (*c*)	**47.** (*a*)	**48.** (*c*)	**49.** (*d*)	

Numerical Type Questions

1. 1	**2.** 432	**3.** 3	**4.** 53	**5.** 220030	**6.** 67	**7.** 2500	**8.** 1	**9.** 7	**10.** 6
11. 4722	**12.** 426	**13.** 481	**14.** 12/5	**15.** 930	**16.** 3	**17.** 12.8	**18.** 108		

EXPLANATIONS

MCQ TYPE QUESTIONS

1. $x^2 + y^2 + z^2 - 2xy - 2xz + 2yz = (x - y - z)^2$

$= (b + c - c + a - a + b)^2 = 4b^2$

2. $y^3 - 4y = y(y^2 - 4)$

$= y(y - 2)(y + 2)$

$4y(y^3 + 8) = 4y(y + 2)(y^2 - 2y + 4)$

$\therefore$ HCF $= y(y + 2)$.

3. Total red tiles initially $= 2\left(\frac{m}{n}\right) - 1$

Total number of tiles $= \frac{m^2}{n^2}$

As number of red tiles > number of green tiles (since room has odd number of tiles and corner tiles are red)

Hence total red tiles required to form an alternate pattern

$$= \frac{\left(\frac{m}{n}\right)^2 + 1}{2} = \frac{m^2 + n^2}{2n^2}$$

$\therefore$ Number of green tiles removed

$$= \frac{m^2 + n^2}{2n^2} - 2\left(\frac{m}{n}\right) + 1$$

$$= \frac{m^2 + n^2 - 4mn + 2n^2}{2n^2}$$

$$= \frac{(m - 2n)^2 - n^2}{2n^2}$$

4. $$\frac{x - 4}{-x^2 - 6} \le \frac{2}{-x^2 - 6}$$

i.e. $$\frac{x - 4}{-(x^2 + 6)} \le \frac{2}{-(x^2 + 6)}$$

i.e. $x - 4 \ge 2$ ...(cancelling a negative term reverses sign of the inequality)

i.e. $x \ge 6$.

5. $$\left(\sqrt{2} + \sqrt{5}\right)^2 = 2 + 2\sqrt{10} + 5$$

$$= 7 + 2\sqrt{10}$$

$\therefore \sqrt{2} + \sqrt{5} > \sqrt{7}$

7. $$\frac{2^n (n+1)^n}{n^n} = 2^n \left(\frac{n+1}{n}\right)^n = 2^n \left(1 + \frac{1}{n}\right)^n$$

Now, $$2 \le \left(1 + \frac{1}{n}\right)^n \le 3$$

$\therefore$ $$2.2^n \le 2^n \left(1 + \frac{1}{n}\right)^n \le 3.2^n$$

i.e., $$2^{n+1} \le 2^n \left(1 + \frac{1}{n}\right)^n \le 3.2^n$$

8. Choose k, so that

$10^H = g^\lambda = 10^{100k}$,

$$k = \frac{H}{100}$$

$\therefore$ $$10^H = g^{\frac{H}{100}} > g^g > g!$$

On the other hand,

$10H = 10 \times 10^g = 10^{g+1} < g!$

(most factors are much bigger than 10)

$10H < g! < 10^H$

9. (*a*) Four consecutive even numbers can be written as 2n, 2n + 2, 2n + 4 and 2n + 6, where n is any natural number.

$\therefore$ Sum = 2n + 2n + 2 + 2n + 4 + 2n + 6

= 8n + 12

= 4(2n + 3) not always divisible by 8.

Thus, (*a*) is not true.

(*b*) Four consecutive odd numbers can be written as 2n + 1, 2n + 3, 2n + 5, 2n + 7 where n is a natural number.

Sum = 2n + 1 + 2n + 3 + 2n + 5 + 2n + 7 = 8n + 16 divisible by 8

Thus, (*b*) is true

(*c*) In product of n consecutive natural numbers atleast one is divisible by n, atleast one by n – 1 ... till 1.

Thus product is atleast divisible by

$n \times (n-1) \times (n-2) \ldots 1 = n!$.

Thus, (*c*) is not true.

(*d*) Take four consecutive odd numbers as $7 \times 9 \times 11 \times 13$ which is not divisible by 15.

Thus, (*d*) is not true.

10. Factors of a^2 are 1, a and a^2.

Factors of ab are 1, a, b and ab.

Factors of a^3 are 1, a, a^2 and a^3.

11. $7x + 6y = 420$

Equation is of the form:

7x + even number = even number.

$\therefore$ 7x has to be even

Hence x has to be even.

12. $$\frac{a+bi}{c+di} = \frac{p}{q}$$

$\therefore$ $qa + qbi = pc + pdi$

Equating real and imaginary parts.

$qa = pc$ and $qb = pd$.

$\Rightarrow$ $a = \frac{p}{q}c$ and $b = \frac{p}{q}d$

$$\therefore \quad \frac{a^2+b^2}{c^2+d^2} = \frac{\frac{p^2}{q^2}c^2 + \frac{p^2}{q^2}d^2}{c^2+d^2} = \frac{p^2}{q^2}$$

13. We ignore $\frac{1}{7}$ and $\frac{1}{9}$ because no sum of other numbers is going to give 7th or 9th in the denominator.

Also, $\frac{1}{5}$ and $\frac{1}{10}$ are not enough to add up to anything $\left(\frac{1}{10}, \frac{2}{10}, \text{ and } \frac{3}{10}\right.$ are going to leave tenths left over no matter what else you add$\left.\right)$.

What's left is $\frac{1}{2}, \frac{1}{3}, \frac{1}{4}, \frac{1}{6}, \frac{1}{8}$.

Sum total of these is $\frac{11}{8}$. So we need all of them except $\frac{1}{8}$, which means $\frac{1}{2} + \frac{1}{3} + \frac{1}{6}$. Which is the only way to do this with egyptian fractions whose denominators are 10 or less. Hence maximum number of Egyptian fractions needed is 3.

14. L.C.M. of 9, 10 and 15 = 90

$\therefore$ $1936 = 90 \times 21 + 46$ (Remainder).

But a part of this remainder is 7.

so $46 - 7 = 39$ should be subtracted from 1936.

15. $6 - 4 = 2, 7 - 5 = 2, 8 - 6 = 2, 9 - 7 = 2, 10 - 8 = 2$

L.C.M. of 6, 7, 8, 9, 10 = 2520;

Greatest number of 6 digits = 999999

$2520 \times 396 + 2079 = 999999$

$\therefore$ Reminder = 2079.

Subtract 2079 from 999999, then we get

$999999 - 2079 = 997920$.

Subtract 2 from this number to get required number, which is 997918 and which will give the remainders 4, 5, 6, 7, 8 when divided by 5, 6, 7, 8, 9 respectively.

16. L.C.M. of 5, 6, 8, 9 and 12 is 360

Required number = 360 K + 1

$= (13 \times 27 + 9)K + 1$

$= (13 \times 27)K + (9K + 1)$

Now this number must be divisible by 13

$\therefore$ K = 10 and required number = 3601.

17. I. In case all the remaining three numbers are multiples of the prime number, HCF of the group is equal to that prime number. otherwise HCF of the group is 1. Thus, I is true.

II. **Case 1:** Let numbers be 2, 4, 6 and 8, where 2 is only prime number, then LCM = 24.

But LCM of 4, 6 and 8 = 24.

LCM = 24 & 2 = 24

Case 2 : Let numbers be 2, 9, 81 and 27 Then LCM = 162 and LCM of 9, 81 and 27 = 81.

Thus, LCM = $81 \times 2 = 162$

Hence II is not always true.

III. Let numbers be 2, 4, 6 and 8

$\therefore$ LCM = 24 and HCF = 2

Product of four numbers

$$= 2 \times 4 \times 6 \times 8 = 384$$

Also, $384 \neq 24 \times 2 \times 2$.

Thus, III is not true.

18. (a. n)! = product of n consecutive natural number starting from 'a' which is atleast divisible by n!.
(n)! = product of n consecutive natural numbers.

For n = 2 :

$$(a.\,n)! = a(a + 1) \text{ and } n! = 2$$

$a(a + 1)$ is divisible by 2!.

For n = 3 :

$$(a\,n)! = a(a + 1)(a + 2) \text{ and } n! = 6.$$

One of the factors of $a(a + 1)(a + 2)$ is divisible by 3 and other by 2.

Thus, proceeding in this manner, (a. n)! and n! have HCF = n!

$\therefore$ H = n!.

19. Let HCF be h and LCM be l.

I. Let numbers be ah and bh.

Then abh = l and (a + b)h = m

$\Rightarrow$ (a – b)h = n

Using these ah and bh can be uniquely determined. Thus, I is true

II. If HCF = LCM, then two numbers are equal and same as HCF or LCM.

Thus, II is true.

III. $\frac{\text{LCM}}{\text{HCF}}$ = a prime i.e., $\frac{\ell}{h} = P$.

Then one of the numbers is equal to h and other is equal to ℓ. Thus, III is true.

20. Total distance travelled by the ball before coming to rest

$$= 12 + 2\,(6 + 3 + 1.5 + \ldots \text{ to } ?)$$

$$= 12 + 2.\left(\frac{6}{1 - 1/2}\right)$$

$$= 12 + 2(12) = 36\text{m}$$

21. Number of students who answered 1 or more questions wrongly = 2^{N-1}.

Number of students who answered 2 or more questions wrongly = 2^{N-2}

Hence number of students who answered 1 question wrongly

$$= 2^{N-1} - 2^{N-2} = 2^{N-2}$$

Similarly it can be shown that number of students who answered 2 questions wrongly

$$= 2^{N-2} - 2^{N-3} = 2^{N-3}$$

Similarly we can find number of students who answered K questions wrongly where $K \geq 3$

Hence total number of questions attempted wrongly

$$s = 2^{N-2} + 2(2^{N-3}) + 3(2^{N-4}) + \ldots + (N-1)(2^{\circ}) + N(1) \quad \ldots(i)$$

$$\therefore \frac{s}{2} = 2^{N-3} + 2(2^{N-4}) + \ldots + (N-2)(2^{\circ}) + \frac{N-1}{2} + \frac{N}{2} \quad \ldots(ii)$$

Substracting equations (*i*) from (*ii*)

$$\frac{s}{2} = 2^{N-2} + 2^{N-3} + \ldots + 2^{\circ} + \frac{1}{2}$$

$$\Rightarrow s = 2^{N-1} + 2^{N-2} + 2^{N-3} + \ldots + 1$$

$$= 2^{N} - 1 = 8191$$

$$\Rightarrow N = 13$$

22. Number of members in C_1 on May 1, 2003 = x + 4a

Number of members in C_2 on May 1, 2003 = xb^4

$$x + 4a = xb^4 \text{ and } a = 20x$$

$$\therefore x(b^4 - 81) = 0$$

As $x \neq 0$, $b^4 - 81 = 0$

$$\therefore b = 3$$

23. n^{th} terms are in the ratio of $\frac{2n + 3}{n - 11}$.

If n = 1 in 2n + 3 series, then first term is

$T_1 = 2(1) + 3 = 5$

n = 2, then second term is 2(2) + 3 = 7

$T_2 = 7$

$d = T_2 - T_1 = 7 - 5 = 2$

$\therefore d_1 = 2$

If n = 1 in n –11series, then first term is 1 –11 = –10

If n = 2, then second term is 2 –11 = – 9

$\therefore d_2 = T_2 - T_1 = -9 - (-10) = 1$

$$\therefore \frac{d_1}{d_2} = \frac{2}{1} \text{ or } 2 : 1$$

24. $\frac{2}{3} + \frac{8}{9} + \frac{26}{27} + \frac{80}{81} + \ldots$ up to 20 terms

$$= \left(1 - \frac{1}{3}\right) + \left(1 - \frac{1}{9}\right) + \left(1 - \frac{1}{27}\right) + \left(1 - \frac{1}{81}\right) + \ldots \text{ up to 20 terms}$$

$$= 20 - \left(\frac{1}{3} + \frac{1}{3^2} + \frac{1}{3^3} + \frac{1}{3^4} + \dots + \frac{1}{3^{20}}\right)$$

$$= 20 - \frac{1}{3}\left(\frac{1-\left(\frac{1}{3}\right)^{20}}{1-\frac{1}{3}}\right)$$

$$= 20 - \frac{3^{20}-1}{2.3^{20}}$$

$$= \frac{40.3^{20} - 3^{20} + 1}{2.3^{20}} = \frac{39.3^{20}+1}{2.3^{20}}$$

25. Let $s = 3 + 6x^2 + 9x^4 + 12x^6 + \dots$

$\therefore$ $x^2s = 3x^2 + 6x^4 + 9x^6 + 12x^6 + \dots$

$\Rightarrow$ $s - x^2s = 3 + 3x^2 + 3x^4 + 3x^6 + \dots$

$\Rightarrow$ $s(1-x^2) = 3\,[1 + x^2 + x^4 + x^6 + \dots]$

$\Rightarrow$ $s(1-x^2) = 3\left[\frac{1}{1-x^2}\right]\left[\because |x| < 1\right]$

$\therefore$ $s = \frac{3}{\left(1-x^2\right)^2}$

26. Average

$$= \frac{60\times56+50\times64+55\times72+62\times63+58\times50}{60+50+55+62+58}$$

$$= \frac{3360+3200+3960+3906+2900}{285} = 60.8$$

27. Increase after 5 years should have been

$7 \times 5 = 35$ years.

But since average remains same, difference in their age is 35 years.

28. Let average of the seven men be x.

$\therefore$ $x = \frac{10\times6+(x+3)}{7}$

$\Rightarrow$ $60 + x + 3 = 7x$

$\Rightarrow$ $x = \frac{63}{6} = 10.5$

$\therefore$ Total amount $= 10.5 \times 7 = 73.5$

29. Consider a set of three numbers a, b and c.

I. $AM = \frac{a+b+c}{3}$

Thus, $a + b + c = 3AM$.

But $GM = \sqrt[3]{abc}$ which cannot be found out.

Thus, I is false,

II. $AM = \frac{a+b+c}{3}$

$\therefore$ Sum of the numbers $= a + b + c = 3AM$.

Thus, II is true.

III. $GM = \sqrt[3]{abc}$.

But $AM = \frac{a+b+c}{3}$, which cannot be found out.

Thus, III is false.

30. I. Average of a set of numbers is greater than smallest and smaller than the greatest number of the set. Thus, I is true.

II. Consider five numbers a, b, c, d and e whose average is $\frac{a+b+c+d+e}{5}$

Now if each of them is increased by k, then we have average

$$= \frac{a+k+b+k+c+k+d+k+e+k}{5}$$

$$= \frac{a+b+c+d+e}{5} + \frac{5k}{5} = \text{old average} + k.$$

Thus, II is true.

III. Consider three numbers x, y and z whose average is $\frac{x+y+z}{3}$

Now if each of them is multiplied by m then

new average $= \frac{xm+ym+zm}{3}$

$$= m\left(\frac{x+y+z}{3}\right) = m \times \text{odd average}.$$

Thus, III is true.

31. Order of measures is

A, B, C, A + B, B + C, C + A, A + B + C.

Given: A + B + C = 180

Hence, average of the 7 measures

$$= \left[\frac{[(A)+(B)+(C)+(A+B)+(B+C)+(C+A)+(A+B+C)]}{7}\right]$$

$$= \frac{4}{7}(A+B+C) = \frac{4}{7}\times180 = \frac{720}{7}\text{ kgs.}$$

32. Total age $= 15 \times 36$

After two left, to tal age $= 15 \times 36 - 2 \times 28$

$= 540 - 56 = 484$ years.

Let the age of the person who returns be x years.

Then new average = $\frac{484+x}{14}$

Now x lies between 20 and 36 (both inclusive)

$\therefore$ New average min = $\frac{484+20}{14}$ = 36 years

And New average max = $\frac{484+36}{14}$

= 37.14. years.

33. Total income of A = 43.3 lakh

Monthly income of A = $\frac{43.3}{7\times 12}$ lakh

= Rs. 51547 $\approx$ Rs. 51245

34. Required ratio = $\frac{425}{510} = \frac{5}{6}$ i.e. 5 : 6

35. Required difference

= (36.565 – 30.2) lakh – Rs. 636500

36. Required percentage

$= \frac{6425}{47810} \times 100 \times 7$

= 94%

37. Average annual income of F

$= \frac{1}{7}\begin{pmatrix}3.50+3.65+3.75+4+\\4.25+4.80+5.10\end{pmatrix}$ lakh

$= \frac{1}{7}$ (29.05) = 4.15 lakh

38. No. of girls studying in engineering discipline

$= \frac{12}{100} \times 2000 = 240$

39. It is obvious from pie-charts.

40. Number of students in Commerce discipline

$= \frac{21}{100} \times 4800 = 1008$

Number of girls in Commerce discipline

$= \frac{24}{100} \times 2000 = 480$

Number of boys = 1008 – 480 = 528

Hence, required %

$= \frac{528}{1008} \times 100 = 52\%$

41. Number of students in Management discipline

$= \frac{10}{100} \times 4800 = 1008$

Number of girls in Management discipline

$= \frac{8}{100} \times 2000 = 160$

Number of boys = 480 – 160 = 320

Number of students in Science discipline

$= \frac{37}{100} \times 4800 = 1776$

Number of girls in Science discipline

$= \frac{32}{100} \times 2000 = 640$

Number of boys 1776 – 640 = –1136

Hence, required ratio = 320 : 1136

= 20 : 71

42. Number of girls in catering discipline

$= \frac{10}{100} \times 2000 = 200$

Hence, required percentage

$= \frac{200}{4800} \times 100 = 4.1\%$

43. Required average

$= \frac{1}{5}[3.5+4+5+3+6.5]$

$= \frac{1}{5}[22] = 4.4$ crore = 44000000

44. Required ratio

$= \frac{(5.5-5)}{(5.3)} = \frac{0.5}{2}$

$= \frac{1}{4}$ ie, 1 : 4

45. Average number of units sold by all the companies together

$= \frac{1}{6}(650 + 300 + 150 + 450 + 300 + 400)$

$= \frac{1}{6} \times 2250 = 375$

46. Hence, it is maximum for company A

47. Average number of units produced by all companies together

$= \frac{1}{6}(900 + 700 + 300 + 850 + 550 + 600)$

$= \frac{1}{6} \times 3900 = 650$

48. Required percentage $= \frac{1100}{1900} \times 100 = 58\%$

49. Required ratio $= \frac{(850+550)}{(450+300)} = \frac{1400}{750}$

$= \frac{28}{15}$ ie, 28 : 15

NUMERICAL TYPE QUESTIONS

1. $x^3 - y^3 - 3xy$

$= x^3 - y^3 - 3xy(x - y)$ ...... as $(x - y) = 1$

$= (x - y)^3 = 1.$

2. $x^3 - y^3 - z^3 - 3xyz$

$= (x + y + z)(x^2 + y^2 + z^2 - xy - yz - zx)$

$= 9[(x + y + z)^2 - 3(xy + yz + zx)]$

$= 9(81 - 33) = 432.$

3. $\frac{x^2}{yz} + \frac{y^2}{zx} + \frac{z^2}{xy}$

... (since, $x + y + z = 0$; $x^3 + y^3 + z^3 = 3xyz$)

$\frac{x^3 + y^3 + z^3}{xyz} = \frac{3xyz}{xyz} = 3.$

4. Worst scenario is when other four get equal number of votes.

Let the winning candidate get x votes.

$\therefore \quad x > \frac{261}{5}$

$\Rightarrow \quad x > 52.2$

$\therefore \quad x = 53$

5. Number $= (555 + 445) \times (555 - 445) \times 2 + 30$

$= (555 + 445) \times 2 \times 110 + 30$

$= 220000 + 30 = 220030.$

6. $x + (x + 36) + y = 100$

$\Rightarrow \quad 2x + y = 64$

Therefore y must be even prime which is 2

$2x + 2 = 64 \quad \Rightarrow x = 31.$

$\therefore$ Third prime number

$= x + 36 = 31 + 36 = 67.$

7. Given numbers are 1, 3, 5.........99.

This is an A.P with a = 1, d = 2.

Let it contain n terms

$\therefore \quad 1 + (n - 1)2 = 99 \quad \Rightarrow \quad n = 50$

Hence required sum $= \frac{n}{2}$ (first term +last term)

$= \frac{50}{2}(1 + 99) = 2500.$

8. $A = 0.a_1a_1a_1 \ldots$

$\therefore \quad 10A = a_1.a_1a_1a_1.....$

$\Rightarrow \quad 9A = a_1$

$\Rightarrow \quad A = \frac{a_1}{9}$

Similarly, $\quad B = \frac{a_2}{9}.$

Now, a_1 and a_2 are multiples of 3 and are distinct. Also, these values are less than 8.

either $a_1 = 3$ and $a_2 = 6$ or $a_1 = 6$ and $a_2 = 3$

$\therefore \quad a_1 + a_2 = 9.$

$\therefore \quad A + B = \frac{a_1}{9} + \frac{a_2}{9} = \frac{a_1 + a_2}{9} = \frac{9}{9} = 1.$

9. If Anil has to give 1 rupee he needs a bag with ₹1. For 2 rupees he had two bags with ₹1 each or ₹2. bag. To have minimum bags, he has a bag with ₹ 2. Now with the two bags he can give ₹3. So next he will require a bag with ₹4. With these three he can give ₹5. ₹6 and ₹7 and next bag will be one containing ₹8 and so on. Thus he would have bags with ₹1. 2. 4. 8. 16. 32. Sum of which is 63 and remaining 37 can be put in the last bag. So total number of bags is 7.

10. $2050 \times 2071 \times 2095$

$(23 \times 89 + 3) \times (23 \times 90 + 1) \times (23 \times 91 + 2)$

$= (23[89 \times 90 + 89 + 90 \times 3] + 3) \times (23 \times 91 + 2)$

$= (23 \times K + 3) \times (23 \times 91 + 2)$

$= (23 \times [K \times 91 + K \times 2 + 91 \times 3] + 6)$

$= 23 \times L + 6$ (where K and L are constants).

Hence remainder is 6.

11. L.C.M. of 27, 35, 25 and 21 = 4725

$\therefore$ Number = 4725 – 3 = 4722.

12. L.C.M. of 12, 21 and 35 = 420

$\therefore$ Number required = 420 + 6 = 426.

13. Number is 120K + 1

$= ((13 \times 9 + 3)K + 1)$

$= (13 \times 9)K + 3K + 1,$

which is divisible by 13.

3K + 1 is divisible by 13.

$\therefore \quad K = 4.$ Number = 481.

14. $\frac{a}{1-r} = 3$

$\Rightarrow$ $3 - 3r = a$

Also, $\frac{a^2}{1-r^2} = 6$

$\Rightarrow$ $\frac{a}{1-r} \cdot \frac{a}{1+r} = 6$

$\Rightarrow$ $\frac{a}{1+r} = 2$

$\Rightarrow$ $2 + 2r = a$

$\therefore$ $3 - 3r = 2 + 2r$

$\Rightarrow$ $r = \frac{1}{5}$

$\therefore$ $a = 3 - \frac{3}{5} = \frac{12}{5}$

15. Since d = –2, maximum sum of the series will be sum of all the positive terms.

Least positive term of the series is 2 or 0 (both giving equal sum)

Let, 2 be the nth term

$2 = 60 + (n - 1)(-2)$

$\Rightarrow$ $n = 30$

$\therefore$ Required sum = 30/2 (60 + 2) = 930

16. Total income for 15 days = 15 × 7 = ₹105

Total income for first 5 days = 5 × 6 = ₹30

Total income for last 9 days = 9 × 8 = ₹72

$\therefore$ Income on sixth day

= 105 – (30 + 72)

= 105 – 102 = ₹3

17. $(GM)^2 = HM \times AM$

$HM = \frac{8 \times 8}{5} = \frac{64}{5} = 12.8$

18. Total marks of 15 students = 15 × 145 = 2175.

Average marks of 15 students (excluding two lowest scores) = 13 × (145 + 5) = 1950.

$\therefore$ Total of two lowest scores = 2175 – 1950 = 225

Given that two scores are consecutive multiples of 9

$\therefore$ $9x + 9x + 9 = 225$

$\Rightarrow$ $18x = 216$

$\Rightarrow$ $x = \frac{216}{18}$

$\therefore$ Lowest score = $9x = \frac{216}{18} \times 9 = 108$.

2 CHAPTER Percentage and Its Applications

The term percentage is quite frequently used in our day to day life especially to avoid the fractions less than 1. Instead of treating the complete entity as 1, we treat as 100, and take the ratios accordingly.

PERCENTAGE

Per Cent: This term means for every hundred. A fraction whose denominator is 100 is called *percentage* and numerator of the fraction is called *rate per cent*. It is denoted by the symbol %.

To find % equivalent of a fraction.

Express fraction with the denominator 100 and numerator is the required answer.

e.g., $\frac{11}{24} = \frac{11}{24} \times 100 = 45\frac{5}{6}\%$

To find fraction equivalent of 'a %': Divide 'a' by 100.

e.g., $21\frac{7}{8}\% = \frac{21\frac{7}{8}}{100} = \frac{175}{800} = \frac{7}{32}$

To increase a number by a given %:

Multiply the number by the factor $\frac{100 + \text{rate}}{100}$

e.g., increase 20 by 15%,

i.e $20 \times \left(\frac{100+15}{100}\right) = 20 \times \frac{115}{100} = 23$

or $20 \times 1.15 = 23$

To decrease a number by a given %:

Multiply the number by the factor $\frac{100 - \text{rate}}{100}$

e.g., decrease 30 by 20%, i.e $30 \times \left(\frac{100-20}{100}\right)$

$= 30 \times \frac{80}{100} = 24$

or $30 \times 0.8 = 24$

To find the % increase of a number:

$$\%\text{increase} = \frac{\text{total increase}}{\text{initial value}} \times 100$$

$$= \frac{\text{Final value} - \text{Initial value}}{\text{Initial value}} \times 100$$

e.g., Population of a village in 1980 was 3000 and in 1990 was 3200. Find the % increase.

$$\%\text{ increase} = \frac{3200 - 3000}{3000} \times 100$$

To find % decrease of a number:

$$\%\text{ decrease} = \frac{\text{total decrease}}{\text{Initial value}} \times 100$$

$$= \frac{\text{Initial value} - \text{Final value}}{\text{Initial value}} \times 100$$

e.g., Cost of a bike last year was ₹ 19000. Its cost this year is ₹ 17000. Find the % decrease in its cost.

$$\%\text{ decrease} = \frac{19000 - 17000}{19000} \times 100$$

$$= \frac{2000}{19000} \times 100 = 10.5\%$$

Note:

(1) If price of a commodity increases by r%, then reduction in consumption, so as not to increase expenditure is $\left(\frac{r}{100+r} \times 100\right)\%$

(2) If price of a commodity decrease by r%, then the increase in consumption so as not to decrease expenditure is $\left(\frac{r}{100-r} \times 100\right)\%$

(3) If A's income is r% more than that of B, then B's income is less than that of A by $\left(\frac{r}{100+r} \times 100\right)\%$

(4) If A's income is r% less than that of B, then B's income is more than that of A by $\left(\frac{r}{100-r} \times 100\right)\%$.

Example: If the production of rice went up from 250 MT in 2005 to 300 MT in 2006, then the percentage increase in rice production from 2005 to 2006 is calculated as follows:

Actual increase = 300 – 250 = 50MT

Percentage increase:

$$= \frac{\text{Actual increase from 2005 to 2006}}{\text{Actual production of rice in 2005}} \times 100$$

$$= \frac{50}{250} \times 100 = 20\%$$

I. If the increase on a value of 350 is 15% the new quantity is 1.15 × 350 = 402.5 (where 1.15 = 1 + 0.15, 0.15 being the decimal equivalent of 15%)

II. If the production in 2005 is given as 400 MT and the increase from 2004 to 2005 is given to be 25% then the production in 2004 will be equal to $\frac{400}{1.25}$ = 320 MT (where 1.25 = 1 + 0.25, 0.25 being the decimal equivalent of 25%)

III. Similarly, if there is a decrease of 12% on a quantity of 225, then the new quantity will be equal to 225 × 0.88 (where 0.88 = 1 – 0.12, 0.12 being the decimal equivalent of 12%). If the production in 2005 is given as 400MT and it is a decrease of 13% from 2004, then the production in 2004 will be equal to $\frac{400}{0.87}$ (where 0.87 = 1 – 0.13, 0.13 being the decimal equivalent of 13%).

On the basis of percentage increase, we can write down how many times the old value gives the new value. For example, if the percentage increase is 100%, we can conclude that the new value is 2 times the old value and if the percentage increase is 300% then the new value is 4 times the old value. If the percentage increase is 450% then the new value is 5.5 times the old value. In general if the percentage increase is p% then the new value is $\left(\frac{p}{100}+1\right)$ times the old value.

Conversely, if we know how many times the old value gives the new value, we can find out the percentage increase in the old value to get the new value. For example, if the new value is 3 times the old value, the percentage increase in the old value to get the new value is 200%. If the new value is 4.25 times the old value, then the percentage increase is 325%. In general if the new value is k times the old value, then the percentage increase is (k – 1) × 100%

Example: If 45% of a certain number is 990, then find the value of 54% of that number.

Solution:

Let the number be x

Given, $\frac{45}{100}x = 990$

$\Rightarrow x = 2200$

$\therefore$ required value = $\frac{54}{100}x = 1188$

Example: The ratio of the monthly salaries of A in December 2004 and in January 2005 was $3\frac{3}{5} : 4\frac{1}{2}$. Find the percentage increase in his salary.

Solution:

The ratio of the monthly salaries $3\frac{3}{5} : 4\frac{1}{2} = \frac{18}{5} : \frac{9}{2}$

(18)(2) : (9)(5) = 4 : 5

Let the monthly salary of A in December 2004 = 4x

His monthly salary in January 2005 would be = 5x.

Required percentage increase in his salary

$= \frac{5x - 4x}{4x}(100)\% = 25\%$

Example: There are three numbers. The first and the second numbers are 50% less and 60% less respectively than the third. What percentage of the first number is the second?

Solution:

Let the third number be 100

First number = $100\left(1 - \frac{50}{100}\right) = 50$

Second number = $100\left(1 - \frac{60}{100}\right) = 40$

$\therefore$ it is $\frac{40}{50}$ th i.e. $\frac{4}{5}$ th or 80% of the first number.

Example: Ram got 30% in a test and failed by 10 marks. If the pass marks in the test was 70, find the maximum marks in it.

Solution:

Let the maximum mark in the test be M.

Ram's mark = $\frac{30}{100}$ M

Pass mark = $\frac{30}{100}$ M + 10

$\frac{30}{100}$ M + 10 = 70 i.e., M = 200

Example: The price of an article is decreased by 20%. By what percentage must the consumption of it be increased in order to retain the expenditure on it?

Solution:

Let the initial price be Rs. 100/gm

Let the initial consumption be 1 gm

Initial expenditure = Rs. 100

New price = $100\left(1 - \frac{20}{100}\right)$

= Rs. 80/gm

New expenditure = Rs 100

$\therefore$ New consumption = $\frac{100}{80}$ gm

$= \frac{5}{4}$ gm

$\therefore$ Consumption must increase by

$\frac{\left(\frac{5}{4}-1\right)100\%}{1} = 25\%$

Example: The length of a rectangle increases by 14% and the breadth by 8%. What is the consequent percentage increase in area?

Solution:

Let length and breadth of the rectangle be l and b

Area of rectangle = lb

Length is increased by 14%

$\therefore$ New length = $1 \times 1 \times \frac{14}{100} = (1 + 0.141) = 1.141$

Breadth is increased by 8%

$\therefore$ New breadth = b + 8% b

= b + 0.08 b = (1.08)b

Area of rectangle

= 1.14l × 1.08b = (1.2312)lb

Increase in area = 1.2312lb – lb

i.e., 0.2312 lb

$\therefore$ Percentage increase = $\frac{23.12}{100}$ lb

i.e., 23.12%

SIMPLE INTEREST AND COMPOUND INTEREST

When a sum of money is lent by A to B, A is called *lender (creditor)*, B the *borrower (debtor)*.

The sum lent is called *principal* (P).

Interest (I) is the extra money paid by the borrowed to the lender for the use of the money for a specified time.

The time for which the money is borrowed is called *period* (N).

The extra amount paid per 100 rupees in a year is called *rate per cent per annum* (R).

The sum of interest and principal is called *Amount* (A).

$$A = P + I$$

Simple Interest (S.I.)

When interest is paid as it falls due, it is called simple interest i.e., throughout the load period, interest is charged on the original sum (principal) borrowed.

$$\text{S.I.} = \frac{PNR}{100}$$

e.g., Find interest to be paid on ₹ 1500 at 9% per annum for a period of 9 years.

Here P = 1500, R = 9%, N = 3 years

$$\therefore \quad \text{S.I.} = \frac{PNR}{100} = \frac{1500 \times 3 \times 9}{100} = ₹\ 405$$

Compound Interest (C.I.)

Money is said to be lent at compound interest when at the end of a year or other fixed period, the interest that has become due is not paid to the lender, but is added to the sum lent, and the amount thus obtained becomes the principal for the next year or period. The process is repeated until the amount for the last period has been found. The difference between final amount and the original principal is the *compound interest* (C.I.).

In compound interest, interest is calculated on the accrued interest also.

$$\text{Amount} = P\left[1 + \frac{R}{100}\right]^N$$

Compound Interest = Amount – Principal

Note:

(1) The time period after which the interest is added each time to form a new principal is called *conversion period*. It may be annually, semi annually or quarterly.

(2) In case, interest is paid semi annually (half yearly), N is number of half years and R is rate percent per half year *i.e.,*

Number of years × 2 = Number of half years.

Rate percent per annum ÷ 2 = Rate percent per half year.

(3) In case interest is paid quarterly, N is number of quarters and R is rate percent per quarter.

Number of years ÷ 4 = Number of quarters

Rate percent per quarter + 4 = Rate percent per quarter.

(4) When rates are different for different years, say R_1, R_2, R_3 percent of 1st, 2nd and 3rd years respectively then

$$\text{Amount} = P\left(1 + \frac{R_1}{100}\right)\left(1 + \frac{R_2}{100}\right)\left(1 + \frac{R_3}{100}\right)$$

(5) In case the time is a fraction of a year, say $y\frac{x}{z}$ years, then

$$\text{Amount} = P\left(1 + \frac{R}{100}\right)^y \times \left(1 + \frac{\frac{x}{z} \times R}{100}\right)$$

Example: A certain sum of money at C.I. amounts to Rs. 811.25 in 2 years and to Rs. 843.65 in 3 years. Find the sum of money.

Solution:

Since, $A = P\left[1 + \frac{R}{100}\right]^n$

$\Rightarrow 811.25 = P\left[1 + \frac{R}{100}\right]^2$ ... (1) and

$$\Rightarrow 843.65 = P\left[1+\frac{R}{100}\right]^3 \quad ...(2)$$

On dividing (2) by (1), we get:

$$\frac{843.65}{811.25} = 1+\frac{R}{100}$$

$$\Rightarrow 1.04 = 1+\frac{R}{100} \Rightarrow R = 4$$

Now, putting R = 4 into (1), we get

$$\Rightarrow 811.25 = P\left[1+\frac{4}{100}\right]^2 \Rightarrow P = 750$$

⇒ The sum of money is Rs.750

EQUAL ANNUAL INSTALLMENT TO PAY THE DEBT (BORROWED) AMOUNT

Let, the value of each equal annual installment = Rs. a.

Rate of interest = R% p.a.

Number of installments per year = n.

Number of years = T.

∴ Total number of installments = n × T.

Borrowed amount = B. Then,

$$a\left[\frac{100}{100+R}+\left(\frac{100}{100+R}\right)^2 ...+\left(\frac{100}{100+R}\right)^{n\times T}\right] = B$$

Example: What annual installment will be required to repay a borrowed amount of Rs.1,32,400 in 3 years at 10% per annum compounded annually?

Solution:

Let each annual installment be Rs. a.

By using the formula,

$$132400 = a\left[\frac{100}{100+10}+\left(\frac{100}{100+10}\right)^2+\left(\frac{100}{100+10}\right)^3\right]$$

$$\Rightarrow 132400 = a\left[\frac{10}{11}+\left(\frac{10}{11}\right)^2+\left(\frac{10}{11}\right)^3\right]$$

$$\Rightarrow 132400 = a\times\frac{10}{11}\left[1+\frac{10}{11}+\frac{100}{121}\right]$$

$$\Rightarrow 132400 = a\times\frac{10}{11}\times\frac{331}{121}$$

$$\Rightarrow a = 132400\times\frac{11}{10}\times\frac{121}{331}$$

⇒ a = Rs. 53,240

Example: A loan of Rs. 2000 is to be paid back in 3 equal annual installments. How much is each Installment to the nearest whole rupee, if the interest is compounded annually at $12\frac{1}{2}$% p.a.?

Solution:

$$2000 = a\left\{\frac{100}{112.5}+\left(\frac{100}{112.5}\right)^2+\left(\frac{100}{112.5}\right)^3\right\}$$

= Rs. 840 (approx.)

Population Formula: The original population of a town is P and annual increase is r%, then population in n years is $\left(1+\frac{R}{100}\right)^n$ and if annual decrease is r%, then population in n years is given by a change of sign in the formula

i.e., $\left(1-\frac{R}{100}\right)^n$

e.g., If annual increase in the population of a town is 4% and present population is 15625, what will be the population in 3 years?

Required population = 15625 $(1.04)^3$ = 17576

Hire Purchase: In a hire purchase plan, a customer can make use of the goods while paying for them. The amount paid at the time of purchase is called *down payment*. The remainder is paid in equal installments and each is the monthly installment. The difference between total amount to be paid and the cash price is called *installment charge*.

Monthly Installment

$$= \frac{\text{Amount to be paid} - \text{down payment}}{\text{Number of instalments}}$$

e.g., If a transistor is available at ₹ 400 each or ₹ 100 down payment and ₹ 70 per month for 5 months, find total amount paid for it, and the installment charge

Amount paid = 100 + 70 × 5 = ₹ 450

Installment charge = 450 – 400 = ₹ 50

PROFIT AND LOSS

Cost Price (CP): The price for which an article is bought is called its *cost price*.

Selling Price (SP): The price at which an article is sold is called its *selling price*.

Profit (Gain): The difference between selling price and cost price is called *profit*. For profit, selling price should be greater than cost price.

Loss: The difference between cost price and the selling price is called *loss*. When cost price is greater than the selling price, there is a loss.

Profit and loss is generally represented as a percent of the cost price, unless otherwise stated.

Overhead charges: If an individual has to spend some money on transportation etc., then this extra expenditure is called *overhead charges*.

Marked price (MP): The price on the label is called *marked price* or *list price*.

Discount: The reduction made on the '*marked price*' of an article is called *discount*. When no discount is given, '*selling price*' is the same as '*marked price*'.

List of Formulae

(1) Profit = SP – CP ... (SP > CP)

(2) Loss = CP – SP ... (CP > SP)

(3) $\% \text{ Profit} = \frac{\text{Profit}}{\text{CP}} \times 100 = \frac{\text{SP} - \text{CP}}{\text{CP}} \times 100$

(4) $\% \text{ Loss} = \frac{\text{Loss}}{\text{CP}} \times 100 = \frac{\text{CP} - \text{SP}}{\text{CP}} \times 100$

(5) $\text{Profit} = \frac{\text{Profit}\% \times \text{CP}}{100}$

(6) $\text{Loss} = \frac{\text{Loss}\% \times \text{CP}}{100}$

(7) $\text{SP} = \frac{100 + \text{Profit}\%}{100} \times \text{CP}$

(8) $\text{SP} = \frac{100 - \text{Loss}\%}{100} \times \text{CP}$

(9) SP = Marked Price – Discount

(10) $\text{CP} = \text{SP} \times \left(\frac{100}{100 + \text{Profit}\%}\right)$

(11) $\text{CP} = \text{SP} \times \left(\frac{100}{100 - \text{Loss}\%}\right)$

(12) $\text{Discount } \% = \frac{\text{Discount}}{\text{Marked Price}} \times 100$

Some important point:

(1) If two items are sold, each at ₹ X, one at a gain of p% and the other at a loss of p%, there is an overall loss given by $\frac{p^2}{100}\%$. The absolute value of the loss is given by $\frac{2p^2x}{100^2 - p^2}$

(2) If CP of two items is the same and % Loss and % Gain on the two items are equal, then net loss or net profit is zero.

(3) Buy x get y free *i.e.,* if x + y articles are sold at cost price of x articles, then

$$\text{percentage discount} = \frac{x}{x + y} \times 100.$$

(4) By using false weight, if a substance is sold at cost price, then

$$\text{overall gain } \% = \frac{100 + \text{Gain}\%}{100} = \frac{\text{True Scale or Weight}}{\text{False Scale or Weight}}$$

(5) In case of successive discount a% and b%, effective discount $= \left(a + b - \frac{ab}{100}\right)\%$.

Example: Two shirts were having a cost price of Rs. 200 each. One was sold at a profit of 15% and the other was sold at a loss of 15%. Find the net profit or loss.

Solution:

$$SP_1 = 200 \times \frac{(100 + 15)}{100} = \text{Rs. } 230.$$

$$SP_2 = 200 \times \frac{(100 - 15)}{100} = \text{Rs. } 170.$$

Total S.P. received = $S.P_1 + S.P_2$

= 230 + 170 = Rs. 400.

Total C.P. = 200 + 200 = 400.

Net result: No profit, No Loss.

Note: If two items are **SOLD**, each at rupees S, one at a gain of X% and other at a loss of X%, then the net result is always a loss.

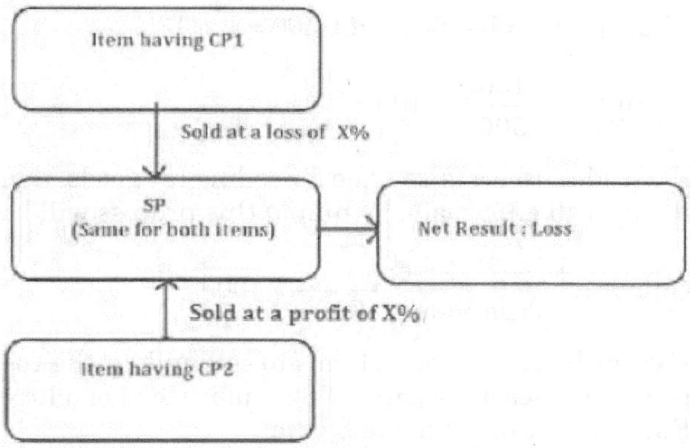

Example: Two articles were sold at Rs. 100 each. After selling it was realized that on one, a profit of 10% was made, and on the other, a loss of 10% was made. What is the net result?

Solution:

	Item 1	Item 2
Selling Price	Rs. 100	Rs. 100
Profit %	10	-
Loss %	-	10
Cost Price	$\frac{100}{1.1}$ = 90.90	$\frac{100}{0.9}$ = 111.11

Total S. P. received = 100 + 100 = Rs. 200

Total C. P. = 90.90 + 111.11 = Rs. 202.01

$$\text{Loss\%} = \frac{200 - 202.01}{202.01} \times 100$$

$$= \frac{-2.01}{202.01} \times 100 \approx -1\%$$

The same calculation can be done by a very simple formula

$$\text{Loss \%} = \frac{X^2}{100} \text{ and Value of loss} = \frac{2X^2S}{100^2 - X^2}.$$

Where X is the percentage profit and loss made on each of the items and S is the common selling price received on both. In case of discounts being offered, the price on which the discount was offered is known as the marked price and the price that is finally received is known as the selling price.

Example: A merchant purchases an item for Rs. 500. He marks the item at a price of Rs. 700 but allows a discount of 10% on cash payment. What is the total profit in terms of amount and percentage made by the merchant?

Solution:

C.P. = Rs. 500, M. P. = Rs. 700.

Hence S.P. = $700\left(1-\frac{10}{100}\right)$ = Rs. 630

Thus, profit = Rs. 630 – Rs. 500 = Rs. 130.

Profit % = $\left(\frac{130}{500}\right) \times 100 = 26\%$.

If a trader uses a false scale for selling his goods, then the overall gain made by him in this process will be

$$\text{Gain\%} = \left[\frac{\text{Error}}{\text{True Value} - \text{Error}} \times 100\right]$$

Example: A milkman claims to sell milk at the cost price but uses a measure of 800 ml instead of a litre. Find the net profit made by him.

Solution:

Using the formula given above

$$\text{Gain \%} = \left[\frac{200}{1000-200} \times 100\right] = \frac{200}{800} \times 100 = 25$$

If a tradesman defrauds (by means of a false balance or otherwise) to the tune of X% in buying and also defrauds to the tune of X% in selling, his overall percentage gain will be

$$\left[\frac{(100+\text{common gain \%})^2}{100} - 100\right]\%$$

Example: A trader defrauds the seller by 10% when he purchases goods from him, and while selling the same to a customer, he defrauds once again by 10%. Find the net gain made by the trader.

Solution:

The required answer is

$$\left[\frac{(100+10)^2}{100} - 100\right]\% = \frac{12100}{100} - 100$$

$= 121 - 100 = 21\%$

If a tradesman defrauds (by means of a false balance or otherwise) to the tune of X% in buying and also defrauds to the tune of Y% in selling, his overall percentage gain will be

$$\left[\frac{(100+X\%)(100+Y\%)}{100} - 100\right]\%$$

Example: A milkman defrauds by means of a false measure to the tune of 20% in buying and also defrauds to the tune of 25% in selling. Find his overall % gain.

Solution:

The milkman defrauds 20% in buying and also defrauds 25% in selling, so his overall % gain will be

$$\left[\frac{(100+20\%)(100+25\%)}{100} - 100\right]\% = 50\%$$

If selling price of X articles is equal to the cost price of Y articles, then the net profit percentage is given by $\frac{Y-X}{X} \times 100$

Example: The cost price of 20 pens is equal to the selling price of 25 pens. What is the net loss percentage?

Solution:

Here, cost price of 20 pens = selling price of 25 pens. So

the net loss percentage = $\frac{20-25}{25} \times 100 = -20\%$

Here minus sign indicates the loss.

PARTNERSHIP

It is an association of two or more persons who invest their money in order to carry on a certain business. A partner who manages the business is called *working partner* and the one who simply invests the money is called *sleeping partner*.

If capitals of the partners are invested for the same time, then partnership is called *simple*, and if invested for different periods, the partner is called *compound*.

If period of investment is the same for each partner, then profit or loss is divided in the ratio of their investments.

e.g., (*i*) If A and B are partners in a business, then

$$\frac{\text{Investment of A}}{\text{Investment of B}} = \frac{\text{Profit of A}}{\text{Profit of B}}$$

$$\text{or } \frac{\text{Investment of A}}{\text{Investment of B}} = \frac{\text{Loss of A}}{\text{Loss of B}}$$

(*ii*) If A, B and C are partners in a business, then Investment of A : Investment of B : Investment of C

= Profit of A : Profit of B : Profit of C

OR Loss of A : Loss of B : Loss of C

Monthly Equivalent Investment: It is the product of the capital invested and the period for which it is invested.

If period of investment is different, then profit or loss is divided in the ratio of their Monthly Equivalent Investment.

e.g., If A and B are partners in a business, then,

$$\frac{\text{Monthly Equivalent Investment of A}}{\text{Monthly Equivalent Investment of B}}$$

$$= \frac{\text{Profit of A}}{\text{Profit of B}} = \frac{\text{Loss of A}}{\text{Loss of B}}$$

i.e., $\frac{\text{Investment of A} \times \text{Period of Investment of A}}{\text{Investment of B} \times \text{Periof of Investment B}}$

$$= \frac{\text{Profit of A}}{\text{Profit of B}} = \frac{\text{Loss of A}}{\text{Loss of B}}$$

If A, B and C are partners in a business, then
Monthly Equivalent Investment of A
: Monthly Equivalent Investment of B
: Monthly Equivalent Investment of C = Profit of A
: Profit of B : Profit of C

or

Monthly Equivalent Investment of A : Monthly Equivalent Investment of B
: Monthly Equivalent Investment of C = Loss of A
: Loss of B : Loss of C

Example: A and B together invested Rs. 12000 in a business. At the end of the year, out of a total profit of Rs. 1800, A's share was Rs. 750. What was the investment of A?

Solution:

Since profits are shared in the ratio of their investments

$$\therefore \frac{\text{A's investment}}{\text{B's investment}} = \frac{\text{Profit share of A}}{\text{Profit share of B}}$$

(Money invested by A & B for the same period)

$$= \frac{750}{1800 - 750} = \frac{750}{1050} = \frac{5}{7}$$

$$\therefore \text{Investment of A} = \frac{5}{5+7} \times 12000 = \text{Rs. } 5000$$

Example: In a business A, B and C invested Rs.380, Rs. 400 & Rs. 420 respectively. Divide a new profit of Rs. 180 among the partners.

Solution:

A's Profit. B's Profit. C's Profit
= A's investment : B's investment : C's investment
= A's investment : B's investment : C's investment
= 380 : 400 : 420 = 19 : 20 : 21

Profit share of A = $\frac{19}{60} \times 180$ = Rs. 57.

Profit share of B = $\frac{20}{60} \times 180$ = Rs. 60.

Profit share of C = $\frac{21}{60} \times 180$ = Rs. 63.

Example: If the partners invest different amounts and for different period of time, then the profits at the end of the year are shared in the ratio of products of (Investment x time period) calculated for each partner.

A started a business with a capital of Rs. 10000. Four months later, B joined him with a capital of Rs. 5000. What is the share of A in a total profit of Rs. 2000 at the end of the year?

Solution:

$$\frac{\text{Profit of A}}{\text{Profit of B}} = \frac{\text{Amount} \times \text{No. of months}}{\text{Amount} \times \text{No. of months}}$$

$$= \frac{10000 \times 12}{5000 \times 8} = \frac{3}{1}.$$

$$\therefore \text{profit share of A} = \frac{3}{3+1} \times 2000 = \text{Rs. } 1500.$$

Example: A, B and C enter into a partnership. A contributes Rs. 320 for 4 months, B contributes Rs. 510 for 3 months and C contributes Rs. 270 for 5 months. If the total profit is Rs. 208, then find the profit share of each of the partners.

Solution:

A's Profit : B's Profit : C's Profit
= 320 × 4 : 510 × 3 : 270 × 5
= 1280 : 1530 : 1350 = 128 : 153 : 135

$$\therefore \text{profit of A} = \frac{128}{128+153+135} \times 208$$

$$= \frac{128}{416} \times 208 = \text{Rs. } 64$$

$$\therefore \text{profit of B} = \frac{153}{128+153+135} \times 208$$

$$= \frac{153}{416} \times 208 = \text{Rs. } 76.50$$

$$\therefore \text{profit of C} = \frac{135}{128+153+135} \times 208$$

$$= \frac{135}{416} \times 208 = \text{Rs. } 67.50$$

STOCKS AND SHARES

Face value: The price of a share is printed on the share certificate is called face value of the share. (Nominal value).

Dividend: Every shareholder of a company is entitled to a proportionate share of the profits of the company. The amount of profits on each share is called *dividend*. Dividend is expressed as a percentage of the face value.

Market Value: In the market, selling price of the share may be different from its face value. Thus, market value of a share is *fluctuable price*.

Shares:*In the market, three types of shares are available* :

(*i*) **Premium share:** If market price of a share is more than nominal value, then it is called *premium share.*

(*ii*) **Par share:** If market price of a share is equal to the nominal value, then it is called *par share.*

(*iii*) **Discount share:** If market price of a share is less than the nominal value, then it is called a *discount share.*

Types of shares: *There are two types of shares:*

(*i*) **Preference shares:** On these shares a fixed rate of dividend is paid to share holders irrespective of whether the company makes any profits or whether it runs into loss. In case the company is not able to pay the agreed dividend to preference shareholders, the dividend on cumulative preference shares goes on accumulating and is paid as and when the company can do so.

(*ii*) **Equity shares:** Profit of the company that is left over after its distribution to the preference shareholders is distributed among the equity share holders. This dividend on equity share may sometimes be more or sometimes less than the dividend on the preference share.

Stock

In order to meet expenses of a certain plan, the Government of India sometimes raises a load from the public at a certain fixed rate of interest. Bonds or promissory notes called *stocks of a fixed value* are used for sale to the public.

Note:

(1) Income per share = Rate of dividend × Nominal value of 1 share.

(2) Annual income = Income per share × Number of shares.

(3) Number of shares

$$= \frac{\text{Nominal value of all the shares}}{\text{Nominal value of one shares}}$$

$$= \frac{\text{Market value of the shares (Investment)}}{\text{Market value of one share}}$$

$$= \frac{\text{Total Dividend}}{\text{Dividend per share}}$$

(4) Rate of interest on the Invest (yield)

$$= \frac{\text{Total income} \times 100}{\text{Total investment}}$$

(5) When stoke/share is purchased, brokerage is added to cost price.

(6) When stock/share is sold, brokerage is subtracted from selling price.

(7) In most problems, face value of a share is taken to be ₹, 100 unless specified.

EXERCISE

MCQ TYPE QUESTIONS

1. The Manager (M) of the company along with his Assistant Manager (AM) and a Business Analyst (BA) invest ₹ 55000, ₹ 45000 and ₹ 60000 respectively in a partnership firm for six months. The Manager being the highest authority among the three gets one-tenth of the total profit and Assistant Manager being second highest authority gets $\frac{1}{20}$ th of the remaining and the remaining profit is divided among the three in the ratio of their investments. If the total profit was ₹ 40000, who gets the minimum share and how much?
(*a*) BA, ₹ 11400 (*b*) AM, ₹ 11419
(*c*) AM, ₹ 12500 (*d*) None of these

2. In a business partnership of three people A, B and C, B's share in total profit is ₹ 250 per ₹ 2000. At this profit his return on investment is 15%. If B earns total of ₹ 900. what is the total investment made by A and C in business? Assume that each of A, B and C have invested their money for the entire year and profits are shared in the ratio of their investments.
(*a*) ₹ 48000 (*b*) ₹ 42000
(*c*) ₹ 6000 (*d*) Data insufficient

3. Abhay and Bimal started a business in partnership. Abhay invested a capital of ₹ 50000. Bimal agrees to pay Abhay an interest on half the capital at 8% per annum. Also Bimal receives a salary of ₹ 500 per month for managing the business. After six months Abhay withdraws ₹ 20000 and Bimal pitches in an equal amount. Bimal continues to pay Abhay interest at 8% per annum for the balance amount. At the end of the year they divide the profit equally and found that total earning of Bimal in the year was exactly thrice that of Abhay. The total profit ?
(*a*) ₹ 7200 (*b*) ₹ 6800
(*c*) ₹ 5400 (*d*) ₹ 4800

4. How much stock must be bought in 3 percent $89\frac{1}{8}$ in order that by selling it at $91\frac{3}{8}$ a profit of ₹ 45 may be made?
(*a*) 20 (*b*) 30 (*c*) 40 (*d*) 50

5. Which is the better investment (the security being equal), 3 percents perpetually at par, or 5 percents at 124, the latter stock being redeemable in 20 years at par? (Simple interest to be reckoned).
(*a*) 3% at par
(*b*) 5% at 124
(*c*) Both are same
(*d*) Cannot be determined

6. What rate percent per annum does a man get by buying ₹100 share (₹ 70 paid-up) at ₹ 60 when dividend is 5% per annum?
(*a*) 4.28% (*b*) 7.14%
(*c*) $8\frac{1}{3}$ % (*d*) $5\frac{5}{6}$ %

7. A rejects 0.08% of the meters as defective .How many will he examine to reject 2?
(*a*) 2500 (*b*) 1250
(*c*) 625 (*d*) 5000

8. (a% of b) + (b% of a) is equal to
(*a*) a% of b (*b*) 2% of ab
(*c*) 20% of ab (*d*) 2% of 100 ab

9. Ryan's stock of goods is worth $9462, which is 15% more than Harland's, and 15% less than Lambert's. What is the value of stock carried by Harland and Lambert respectively?
(*a*) $ 8218, $ 10032 (*b*) $ 1419, $ 10881
(*c*) $ 8042, $ 10881 (*d*) $ 8228, $ 11132

10. Numerator of a fraction increases by 25%. What can be said about the resultant fraction?
(*a*) It is 67% less than original.
(*b*) It is equal to the original fraction.
(*c*) It is 67% more than the original fraction.
(*d*) The change cannot be determine.

11. Which of the following statement/s is/are true?
I. If two numbers are such that one is 25% more than the second, then sum of the two numbers is nine times that of the difference of the two numbers.
II. If 56% of a number A is 24% of another number B then 49% of A is 21% of B.

Codes:
(*a*) Only I (*b*) Only II
(*c*) Only I and II (*d*) None of these

12. In the Annual General Meeting of the Rotary club a chairman is to be selected for which A and B is the only two candidates. After the first round, each of them gave a speech on their policy and work plan for the coming year after which another round of election was held. The number of votes not cast for A increased by 25% in the second round over those not cast for him in the first round and A lost second round to B by twice as many votes as that by which he had won in the first round. If total 260 people voted each time what was the percent decrease in votes received by A in the second round assuming that no vote was void in any of the rounds?
(*a*) 18% (*b*) 25% (*c*) 21.43% (*d*) 30%

13. A watermelon weighs 5000 gm. 99% of its weight is water. It is kept in a drying room and after some time it turns out that it is only 98% water by weight. What is its weight now?

(*a*) 2500 gm

(*b*) 4500 gm

(*c*) 4950 gm

(*d*) None of these

14. A sum of ₹ 5000 is divided into two parts A and B such that simple interests at the rate of 20% p.a. on A and B after 2 and 3 years respectively are equal. Which of the following is false about A and B?

(*a*) Ratio of A to B is 2 : 3

(*b*) Ratio of A to B is 3 : 2

(*c*) A exceeds B by ₹ 1000

(*d*) Total interest earned on A and B is ₹ 2400

15. Which amongst the following two offers is the better one? Consider with respect to interest accrued.

I. Investing an amount compounded annually at 1% per annum for 100 years.

II. Investing the amount compounded annually at 100% per annum for 1 year.

Codes:

(*a*) First offer

(*b*) Second offer

(*c*) Both are same

(*d*) Cannot be determined

16. The population of rate in a locality x increases by 20% in one year. Observing this, the pest control committee decided to use a special kind of pesticide 'xyz' which effectively kills 160 rats in 3 months. Just after 2 years, what is the net increase or decrease in the population of rats if, initially the population of rats is 3200 and pesticide is used effectively?

(*a*) Increase of 128 rats.

(*b*) Decrease of 128 rats.

(*c*) Neither an increase nor a decrease in the population.

(*d*) None of these.

17. A man travels from A to B to buy goods which he can get 10% cheaper in B than in A. If expenses of the journey are ₹ 15 and he makes a clear saving of ₹ 10, how much does he pay for the goods?

(*a*) ₹ 225

(*b*) ₹ 200

(*c*) ₹ 150

(*d*) Cannot be determined

18. By selling 5 dozen mangoes for ₹ 156, it was found that $\frac{3}{10}$ th of the outlay was gained. What should the retail price per mango be in order to gain 60%?

(*a*) ₹ 4 (*b*) ₹ 2 (*c*) ₹ 3.2 (*d*) ₹ 4.2

19. A man sells sugar to a tradesman at a profit of 20% but the tradesman becoming bankrupt pays only 80 paise in the rupee. How much percentage does the man gain or lose by his sale?

(*a*) 2.5% (*b*) 3% (*c*) 4% (*d*) 5.2%

20. A trader allows a discount of 5 percent to his customers. What price should he mark on a article the cost price of which is ₹ 800 so as to make clear profit of 25 percent on his outlay?

(*a*) ₹ 1000 (*b*) ₹ 1053

(*c*) ₹ 1200 (*d*) ₹ 1123

21. A camera costing ₹550 is marked to be sold at a price, which gives a profit of 30%. What will be its selling price in a sale when 20% is taken off the marked price?

(*a*) ₹ 600 (*b*) ₹ 572

(*c*) ₹ 635 (*d*) ₹ 605

22. Buy two and get one free. Which of the following is true?

(*a*) The person is selling at a loss.

(*b*) The person is selling at a profit.

(*c*) The loss or gain depends on the profit from sale of two articles.

(*d*) The person is selling at the cost price.

Directions (Q.23–25): *Refer to the data below and answer the questions that follow.*

A profit-maximizing producer has the option of discriminating between markets A and B (i.e., charging different prices in markets A and B for the same product) for a product X. Demand for the product in each of the markets A and B is given by

$$Q_A = 42 - 0.2\,P_A$$

$$Q_B = 100 - 0.8\,P_B$$

Total cost in any market is given by

C = 1000 + 5Q, where Q is quantity sold in that market.

Consider two cases:

***Case 1** in which she discriminates between the markets and*

***Case 2** in which she does not discriminate.*

23. What is the approximate ratio of the price in market A to price in market B in Case 1?

(*a*) 1.65 (*b*) 2.00
(*c*) 2.25 (*d*) 2.50

24. What is the ratio of the total quantity of product X sold in Case 1 to the total quantity sold in Case 2?

(*a*) 0.50 (*b*) 1
(*c*) 1.10 (*d*) 1.05

25. What is the difference in total profit in Case 1 and the total profit in Case 2?

(*a*) ₹ 289 (*b*) ₹ 829 (*c*) ₹ – 289 (*d*) ₹ – 829

NUMERICAL TYPE QUESTIONS

1. A and B enter into a partnership. A puts in the whole capital of ₹ 45000 on the condition that the profits will be equally divided after which B will pay A interest on half the capital at 10% p.a. and receive ₹ 60 per month from A for carrying on the concern. The yearly profit in rupees, if B's income is half of A's income is ______

2. A, B and C started a business by investing ₹ 72000, ₹ 27000 and ₹ 81000. It was decided that servant's salary ₹ 2000 per month should be given from the profit. If profit is ₹ 25000 in the first month and ₹ 20000 in the second month and average profit of remaining 10 months is ₹ 21000, then share of C is ______

3. Three friends started a business of renting out air conditioners by investing ₹ 20000, ₹ 24000 and ₹ 16000, respectively. C gets 20% of total profit for repair and maintenance of the air conditioner. If in a particular year, C gets ₹ 487.50 less than total earnings of the other two, then total profit for the year in ₹ is ______

4. A person holds forty ₹ 500 shares in a concern, which pays dividend at the rate of 6 percent per annum. When the shares are at ₹ 675, he sells out and invests half the proceeds in 4 percent stock at 90. With the other half he buys a house. For which he receives an annual rent of ₹ 1440 subject to a deduction of 25P per rupees for repairs and taxes. Alteration in rupees in his annual income will be ______.

5. A person invests ₹ 34539 in three percent at 87. After receiving one year's dividend, he sells out at 89. He then invests the whole in railway stock paying 5 percent at 115.The difference in his income will be______

6. Difference of two numbers is 1660. If 7.5 % of one number is 12.5% of the other number, then two numbers are______

7. A school has only three classes which contain 40,50,60 students respectively. The pass percent of these classes are 10, 20 and 10 respectively. Then pass percent in the school will be______

8. For a sphere of radius 10cm, the numerical value of surface area is ______ percent of the numerical value of its volume.

9. At an election, there are two candidates only, a candidate who gets 43 per cent of the votes is rejected by a majority of 420 votes. Then total number of votes recorded assuming that there was no void vote are______

10. If price of wheat be raised 30%, then by _____ percent a householder must reduce his consumption of the article so as not to increase his expenditure?

11. A fraction in reduced form is such that when it is squared and then its numerator is reduced by $33\frac{1}{3}\%$ % and denominator is reduced to 20%, its result is twice the original fraction. Sum of the numerator and the denominator is _____

12. A person borrows two equal sums at the same time at 5 and 4 percent respectively and finds that if he repays the former sum with simple interest on certain date 6 months before the latter, he will have to pay in each case the same amount, viz., ₹ 1100. The amount borrowed in rupees is ______

13. Effective annual rate corres-ponding to a nominal rate of 6 percent per annum, payable half-yearly is ______

14. A tradesman puts two prices on his goods, one for ready money and the other for 6 month's credit interest being calculated at $12\frac{1}{2}$ % per annum.If credit price on an article be ₹ 26.56, then its cash price in rupees is ______

15. An article costing ₹9000 is sold at a discount which is equal to the simple interest on ₹ 3000 for N months. If rate of discount is same as the rate of interest, the N will be______months

16. A lent B ₹ 900 for a certain time at a certain rate percent per annum, which is equal to the square root of the number of the months of the time. After the time B wanted to return the money, but A, instead of taking the interest which amounted to ₹8 asked for a certain sum of money at the same rate for a certain number of years which is equal to the square root of the rate percent. Sum of money in rupees so that none of them should be the loser will be______ (Take simple interest)

17. A shopkeeper sells a microwave at a discount on its marked price of ₹ 2500. But in a clearance sale he gives an additional discount of 20% thus selling it at ₹ 1500. Then first discount was_____%

18. A sports goods shop owner has 6 packets of balls each containing the same number of balls. If he sells each ball at ₹ 8 he gains ₹ 96. If there were 1 ball less in each packet and he sells each ball at ₹ 9 then he gains ₹ 90 in the transaction. The cost price of each packet of balls in rupees is _____

19. One merchant correctly calculates his percentage profit on the cost price; another wrongly calculates it on the selling price. The difference in actual profits in rupees if both claim to make $17\frac{1}{2}$ % profit on goods sold at ₹ 3760 is_____

20. A manufacturer marks his goods in advance at 80 percent more than the cost price, but he allows 15 articles to the dozen and also 10 percent discount for cash. The profit on his outlay does he obtain from a customer who pays cash is _____%

21. A trader buys a certain amount of goods worth ₹ 22520. He decides to make a profit of 5.36% on the sale of goods worth ₹ 5000 and increase the profit percent by 3.14% for sales upto ₹ 15000 and then increase the profit percent for the sale of remaining lot such that he is able to make a profit of 25% on the sale of the full lot. Then profit in rupees that he makes on the third lot of goods is_____

22. Three shepherds A, B and C rented a pasture for a year. A grazed 22 sheeps for 4 months, B grazed 16 sheeps for 8 months and C grazed 32 sheeps for 6 months. If C's share of rent is ₹ 600, then total rent in rupees for the year was_____

23. Three bachelors, Amar, Akbar and Anthony rented a house for a year. But, Amar left after 4 months, Akbar stayed for 8 months and only Anthony stayed for the entire year. If the annual rent was ₹ 6000, then share of Akbar is_____

24. Ram, Shyam and Ghanshyam invest ₹ 2000, ₹ 10000 and ₹ 5000 to set up a stall. Ram gets 15% of the total profit for running the stall. Ram gets ₹ 500 less than Shyam and Ghanshyam together, then total profit of that year is_____

25. x, y and z start a joint venture wherein they make an annual profit of ₹ 3600. x invested one-third of the capital for one-fourth of the time, y invested one-fourth of the capital for one-half of the time, while z invested the remainder of the capital for the entire year. Then share of x in the profit is_____.

ANSWERS

MCQ Type Questions

1. (*b*)	**2.** (*b*)	**3.** (*a*)	**4.** (*a*)	**5.** (*b*)	**6.** (*d*)	**7.** (*a*)	**8.** (*b*)	**9.** (*d*)	**10.** (*c*)
11. (*c*)	**12.** (*c*)	**13.** (*a*)	**14.** (*a*)	**15.** (*a*)	**16.** (*c*)	**17.** (*a*)	**18.** (*c*)	**19.** (*c*)	**20.** (*b*)
21. (*b*)	**22.** (*c*)	**23.** (*a*)	**24.** (*b*)	**25.** (*a*)					

Numerical Type Questions

1. 9180	**2.** 103950	**3.** 2812.50	**4.** 480	**5.** 345.2	**6.** (4150, 2490)
7. $13\frac{1}{3}$	**8.** 30	**9.** 3000	**10.** $23\frac{1}{13}$	**11.** 8	**12.** 1000
13. 6.09	**14.** 25	**15.** 4	**16.** 600	**17.** 25	**18.** 48
19. 98	**20.** 29.6	**21.** 4512	**22.** 1275	**23.** 2000	**24.** 1000
25. 480					

EXPLANATIONS

MCQ TYPE QUESTIONS

1. The ratio of their investments is 11 : 9 : 12.

So, Manager gets $\frac{1}{10}(40000) = 4000 + \text{his share}$

Remaining amount = 36000.

Asstt.Manager gets $\frac{1}{20}(36000) = 1800 + \text{his share}$

Remaining amount = 36000 – 1800 = 34200

$\therefore$ M's share $= \frac{11}{32} \times 34200 + 4000$
$= 11756.25 + 4000$
$= ₹\ 15756.25 \cong ₹\ 15756$

AM's share $= \frac{9}{32} \times 34200 + 1800$
$= 9618.75 + 1800$
$= ₹\ 11418.75 \approx ₹\ 11419$

BA's share $= \frac{12}{32} \times 34200 = ₹\ 12825.$

2. B's share in the profit = 250 per 2000

If B gets ₹ 900, then

total profit $= 2000 \times \frac{18}{5} = 7200$

B's return on investment = 15%

$\therefore \frac{900}{\text{B's investment}} \times 100 = 15$

$\therefore$ B's investment = 6000

Hence profits are shared in the ratio of their investments.

Total investments $= 6000 \times \frac{7200}{900} = 48000$

$\therefore$ Investment made by A and C = 48000 – 6000 = 42000.

3. Interest paid by Bimal to Abhay for first six months

$= \frac{50000}{2} \times \frac{8}{100} \times \frac{1}{2} = ₹\ 1000$

For last six months

$= \frac{30000 - 20000}{2} \times \frac{8}{100} \times \frac{1}{2}$

$= ₹\ 200$

$\therefore$ Total salary paid to Bimal $= 12 \times 500 = ₹\ 6000$

Let total profit be ₹ x

$\therefore$ Abhay's earing $= 1200 + \frac{x - 6000}{2}$

Bimal's earning $= 6000 + \frac{x - 6000}{2} - 1200$

$\therefore \quad 3\left(1200 + \frac{x - 6000}{2}\right) = 4800 + \frac{x - 6000}{2}$

$\Rightarrow \quad 3600 + \frac{3x}{2} - 9000 = 4800 + \frac{x}{2} - 3000$

$\Rightarrow \quad \frac{3x}{2} - \frac{x}{2} = 7200;$

$\therefore \quad x = ₹\ 7200$

4. Profit per stocks $= 91\frac{3}{8} - 89\frac{1}{8} = ₹\ 2\frac{2}{8}$

$\therefore \quad \frac{45}{2\frac{1}{4}} = 20$ stocks

5. Income from a stock at 3% in 20 years at par

$= \frac{100}{100} \times 3 \times 20 = 60$

Income from a stock at 5% at 124 in 20 years

$= \frac{100}{124} \times 5 \times 20 = 80.64.$

6. He invests ₹ 60 and gets ₹ 100 share

Income = ₹ 5

Required % $= \frac{5}{60} \times 100(0.70) = 5.83\% = 5\frac{5}{6}\%$

7. Let number of meters to be examined be x.

Then 0.08% of $x = 2$.

$\Rightarrow \quad \frac{0.08}{100} \times x = 2$

$x = \frac{2 \times 100}{0.08} = \frac{2 \times 100 \times 100}{8} = 2500$

8. $\left(\frac{a}{100} \times b\right) + \left(\frac{b}{100} \times a\right) = \frac{ab}{100} + \frac{ab}{100} = \frac{2ab}{100}$

$= \frac{2}{100} \times ab = 2\%$ of ab.

9. Ryan's stock = 1.15 × Harland's stock

$\therefore$ Harland's stock $= \frac{9462}{1.15} = \$8228$

Ryan's stock = 0.85 × Lambert's stock

$\therefore$ Lambert's stock $= \frac{9462}{0.85} = \$11132.$

10. Let the fraction be $\frac{x}{y}$.

Let $x_1 = x + 0.25x = 1.25x$

$y_1 = y - 0.25x = 0.75x$

$\therefore \quad \frac{x_1}{y_1} = \frac{1.25x}{0.75x} = \frac{5}{3}\frac{x}{y}$

$$\therefore \text{ Increase} = \frac{\frac{5}{3}\frac{x}{y} - \frac{x}{y}}{\frac{x}{y}} \times 100 = \frac{2}{3} \times 100 = 66.67\%$$

Thus, resultant fraction is more than the original fraction by 67%. Thus III is true.

11. I. Let first number be x.

$\therefore$ Second number = 1.25x

Sum = x + 1.25x = 2.25x

Difference = 1.25x – x = 0.25x

Also, $9 \times 0.25x = 2.25x$

Thus I is true.

II. $\frac{56}{100} \times A = \frac{24}{100} \times B$

Multiplying by $\frac{49}{56}$ on both the sides.

$$\therefore \quad \frac{49}{56} \times \frac{56}{100} \times A = \frac{49}{56} \times \frac{24}{100} \times B$$

$$\Rightarrow \quad \frac{49}{100} A = \frac{21}{100} B$$

$\Rightarrow$ 49% of A = 21% of B.

Thus, II is true.

12. In the first round, let A get 'a' votes and B get 'b' votes.

Then, $a + b = 260$

i.e., $a = 260 - b$ ... (*i*)

In second round, B gets

$$b \times \frac{125}{100} = \frac{5b}{4}$$

and A gets $260 - \frac{5b}{4}$,

Also, $\frac{5b}{4} - \left(260 - \frac{5b}{4}\right) = 2(a - b)$

$$\frac{10b}{4} - 260 = 2(260 - b - b)$$

$$\frac{10b}{4} - 260 = 520 - 4b$$

$$4b + \frac{10b}{4} = 520 + 260$$

$$b = 120$$

Hence first round, A gets 140 and B gets 120.

In second round A gets 110 and B gets 150.

$\therefore$ % decrease in votes received by A

$$= \frac{140 - 110}{140} \times 100 = 21.43\%$$

13. Non-water matter is 1% of 5000 = 50 gm. After drying, this matter becomes 2% of the weight of the water-melon. Hence weight of the watermelon now is 2500 gm.

14. Let A = ₹ x and B = ₹ (5000 – x)

Then, $\frac{x \times 2 \times 20}{100} = \frac{(5000 - x) \times 3 \times 20}{100}$

$\Rightarrow \quad 2x = 15000 - 3x$

$\Rightarrow \quad 5x = 15000$

$\Rightarrow \quad x = 3000$

A = x = ₹ 3000

and B = 5000 – x

= 5000 – 3000

= ₹ 2000

$$\therefore \quad \frac{A}{B} = \frac{3000}{2000} = \frac{3}{2},$$

Thus, (*a*) is not true and (*b*) is true.

A – B = 3000 – 2000 = ₹ 1000.

Thus, (*c*) is true.

(*d*) The total interest

$$= \frac{3000 \times 2 \times 20}{100} + \frac{2000 \times 3 \times 20}{100}$$

= ₹ 2400.

Thus, (*d*) is true.

15. First investment gives a return

$$P\left(1 + \frac{1}{100}\right)^{100} - P$$

First term is $P\left(1 + \frac{1}{100}\right)^{100} = X$,

such that $2P < X \leq 3P$

Hence return X – P lies between P and 2P

i.e., $2P - P < X - P \leq 3P - P$.

i.e., $P < X - P \leq 2P$.

Whereas second offer.

$$= \text{Return} = P\left(1+\frac{100}{100}\right)^1 - P$$

$$= 2P - P = P.$$

Alternatively

Option of investing at 1% for 100 years is obviously better than investing at 100% for 1 year, because total compound interest for 100 years will be atleast 100% of the total amount (1% for each year), plus whatever the compounding effect will have.

16. Growth rate of rat population in 3 months

$$= 20 \times \frac{3}{12} = 5\%.$$

Increase in first 3 months = 3200 × 1.05 = 3360

Also, net decrease in 3 months = 160

∴ Rat population = 3360 – 160 = 3200

In the same way, after every 3 months, rat population remains the same.

Hence, even after 3 × 8 months i.e., 2 years, the population is maintained

17. Let CP in A be ₹ x

∴ CP in B = ₹ 0.9x

∴ 0.9x + 15 + 10 = x

⇒ 0.1x = 25

⇒ x = 250

∴ Price in B = 250 × 0.9 = ₹ 225

18. Let CP of 5 dozen mangoes be ₹ x.

SP = ₹ 156 and Gain = 0.3x

∴ 156 – x = 0.3x

⇒ x = 120

∴ SP of 60 mangoes = 120 × 1.6 = 192

∴ SP per mango = ₹ 3.2

19. Let CP be ₹ x ; SP = 1.2x

But he gets only 1.2x × 0.8 = 0.96x

∴ Loss = 0.04x

∴ Loss = 4 %

20. Selling price = 800 × 1.25 = ₹ 1000

$$\therefore \text{ Marked price} = ₹\frac{1000}{0.95} = 1053$$

21. CP = ₹ 550

∴ Marked price = 1.3 × 550 = ₹ 715

∴ Sales price = 0.8 × 715 = ₹ 572.

22. Profit or loss of the sale depends upon how much profit the person is earning on the sale of two articles.

If SP > CP of 1 article, there is a profit.

If SP < CP of 1 article, there is a loss.

If SP = CP of 1 article, there is not profit no loss.

Thus, only (*c*) is true.

23. *Following two key points need to be noted :*

(A) Producer is profit maximizing which implies she will set prices that maximize the profit function.

(B) Quantity demanded in market A is independent of quantity demanded in market B.

CASE 1:

In this case, producer charges different prices in markets A and B. This implies that she will maximize profits individually.

Equations can be re-written as

$$P_A = 210 - 5Q_A$$
$$P_B = 125 - 1.25\,Q_B$$

Profit in Market A = π_A

$$= \text{Revenue} - \text{Cost}$$
$$= Q_A \times P_A - C_A$$
$$= Q_A \times (210 - 5Q_A) - (1000 + 5Q_A)$$

At the maximum profit level, differential π_A with respect to Q_A is 0

or

$$\therefore \quad 210 - 10Q_A - 5 = 0$$

$$\Rightarrow \quad Q_A = 20.5,\ P_A = 107.5$$

and total revenue in market A

$$= 20.5 \times 107.5 = ₹\,2203.75$$

Similarly

Profit in Market B = π_B

$$= \text{Revenue Cost}$$
$$= Q_B \times P_B - C_B$$
$$= Q_B \times (125 - 1.25\,Q_B) - (1000 + 5Q_B)$$

At maximum profit level, differ-ential of π_B with respect to Q_B is 0

or $\quad 125 - 2.5Q_B - 5 = 0$

$$\Rightarrow \quad Q_B = 48,\ P_B = 65$$

and total revenue in market B

$$= 48 \times 65 = ₹\,3120$$

Using the above, we get

$$\frac{P_A}{P_B} = \frac{107.5}{65} \sim 1.65.$$

24. Data for 1 is available above. Consider case 2. In this case, prices across the two markets are the same. Hence, quantity equations can be added to give

$$Q = 142 - P$$

where Q is total quantity and P is price in both markets

$$\Rightarrow \quad P = 142 - Q$$

$\therefore \quad \pi = \text{Profit} = Q \times (142 - Q) - (1000 + 5Q)$

At the maximum profit level, differential of π with respect to Q is 0

$\therefore \quad Q = 68.5, \quad P = 73.5$

and Total Revenue = 5034.75

Given this, we note that quantities sold in both Case 1 and Case 2 are the same. Hence, the ratio is exactly 1.

25. Profit Difference between case 1 and case 2

= Revenue Difference (since quantity sold and hence, cost is the same)

$\therefore$ Profit difference = 2203.75 + 3120 – 5034.75
= 289

NUMERICAL TYPE QUESTIONS

1. Interest on ₹ 22500 = 0.1 × 22500 = ₹2250

Charges for managing the concern
= 60 × 12 = ₹ 720

If yearly profit is ₹ x, then

$$\text{B's share} = \frac{\text{A's Share}}{2}$$

$$\therefore \quad \frac{x}{2} - 2250 + 720 = \frac{1}{2}\left(\frac{x}{2} - 720 + 2250\right)$$

$\Rightarrow \quad x = ₹ 9180.$

2. Total annual profit
= 25000 + 20000 + 21000 × 10
= ₹ 255000

Actual profit
= 255000 – (2000 × 12) = ₹ 231000

Total investment
= 72000 + 27000 + 81000 = ₹180000

$$\text{A's share} = \frac{72000}{180000} \times 231000 = ₹\ 92400$$

$$\text{B's share} = \frac{27000}{180000} \times 231000 = ₹\ 34650$$

$$\text{C's share} = \frac{81000}{180000} \times 231000 = ₹\ 103950.$$

3. 80% of the total profit is divided in the ratio 20000 : 24000 : 16000 = 5 : 6 : 4

80% of total profit = 5x + 6x + 4x = 15x

$$\text{Total profit} = \frac{15x}{80\%} = 18.75x$$

$\therefore$ Share of C in profit = 4x + 20% of 18.75
= 4x + 3.75x = 7.75x

Share of A in profit = 5x

Share of B in profit = 6x

$\therefore \quad (6x + 5x) - 7.75x = 487.50$

$\Rightarrow \quad 3.25x = 487.50$

$\Rightarrow \quad x = 150$

$\therefore$ Total profit = 18.75 × 150 = ₹ 2812.50.

4. Selling price of 40 shares
= 675 ×40 = ₹ 27000

Income from half the proceeds

$$= \frac{13500}{90} \times 4 = ₹\ 600$$

Income from house = 1440 × 0.75 = 1080

Present total income = 1080 + 600 = ₹ 1680

$$\text{Earlier Income} = 40 \times 500 \times \frac{6}{100} = ₹\ 1200$$

Increase in income = ₹480

5.
$$\text{One years dividend} = \frac{34539}{87} \times 3 = ₹\ 1191$$

$$\text{Selling Price} = \frac{34539}{87} \times 89 = ₹\ 35333$$

Number of railway stocks bought

$$= \frac{35333}{115} = 307.24$$

Income from railway stocks
= 307.24 × 5 = ₹ 1536.2

Difference in income = 345.2

6. Let two numbers be x and y.

7.5% of x = 12.5% of y

$\therefore \quad 75x = 125\ y$

$\Rightarrow \quad 3x = 5y$

$\Rightarrow \quad x = \left(\frac{5}{3}\right)y.$

Now x – y = 1660

$\Rightarrow \quad \left(\frac{5}{3}\right)y - y = 1660$

$\Rightarrow \quad \left(\frac{2}{3}\right)y = 1660$

$\Rightarrow \quad y = 2490$

$\therefore$ x = 2490 +1660 = 4150.

Hence numbers are 4150, 2490.

7. Number of passed candidates

$$= \frac{10}{100} \times 40 + \frac{20}{100} \times 500 + \frac{10}{100} \times 60$$

= 4 + 10 + 6 = 20

Total students in school
= 40 + 50 + 60 =150

∴ Required percentage

$$= \frac{20}{150} \times 100 = \frac{40}{3} = 13\frac{1}{3}\%$$

8. Surface area $= 4 \times \frac{22}{7} \times r^2$

$$= \frac{3}{r}\left(\frac{4}{3} \times \frac{22}{7} \times r^3\right) = \frac{3}{r} \times \text{Volume}$$

where r = 10 cm

So we have

$$S = \frac{3}{10}\ V = \frac{3}{10} \times 100\ \%\text{ of V} = 30\ \%\text{ of V}$$

So surface area is 30 % of volume.

9. Difference in % of votes = 57 – 43 = 14%

∴ 14% is represented by 420

∴ Total number of votes $= 420 \times \frac{100}{14} = 3000.$

10. % reduction in consumption $= \left(\frac{30}{100+30} \times 100\right)\%$

$$= \left(\frac{30}{130} \times 100\right)\% = 23\frac{1}{13}\%$$

11. Let the fraction be $\frac{a}{b}$

$$\therefore \quad \frac{a^2}{b^2} \times \frac{\frac{2}{3}}{\frac{1}{5}} = 2\frac{a}{b} \quad \frac{a}{b} \neq 0 \text{ if } a \neq 0,$$

then $\frac{a}{b} \times \frac{2}{3/5} = 2$

$$\Rightarrow \quad \frac{a}{b} = \frac{3}{5}$$

∴ a + b = 8.

12.

$$1100 = P + \frac{P \times 5 \times \left(T - \frac{1}{2}\right)}{100}$$

$$1100 = P + \frac{P \times 4 \times T}{100}$$

$$\therefore \quad \frac{P \times 5 \times \left(T - \frac{1}{2}\right)}{100} = \frac{P \times 4 \times T}{100}$$

$$\Rightarrow \quad 5T - \frac{5}{2} = 4T$$

$$\therefore \quad T = \frac{5}{2}\text{ years}$$

$$\therefore \quad 1100 = P\left(1 + \frac{4 \times \frac{5}{2}}{100}\right) = P\left(\frac{110}{100}\right)$$

$$\therefore \quad P = ₹\ 1000$$

13. $\left(1 + \frac{3}{100}\right)^2 = 1.0609$

∴ Effective annual rate = 6.09%

14. Let cash price be ₹P

$$\therefore \quad 26.56 = P + \frac{P \times 12.5 \times 0.5}{100}$$

$$\Rightarrow \quad P \approx ₹\ 25.$$

15. Let R be the rate of interest.

$$\text{Discount} = \frac{9000 \times R}{100}$$

$$\text{Interest} = \frac{3000 \times R \times T}{100}$$

By the given information:

$$\frac{9000 \times R}{100} = \frac{3000 \times R \times T}{100}$$

$$\Rightarrow \quad T = \frac{1}{3}\text{ years} = 4\text{ months.}$$

16. Let N be the number of years

$$\therefore \quad R\% = \sqrt{N \times 12}$$

$$\Rightarrow \quad 48 = \frac{900 \times \sqrt{12N} \times N}{100}$$

Squaring both sides

$$N^2 = \frac{48 \times 48}{9 \times 9 \times 12} = \frac{16 \times 4}{9 \times 3} = \frac{64}{27}$$

$$\therefore \quad R = \sqrt{12 \times \frac{4}{3}} = 4\%$$

$$\therefore \quad 48 = \frac{P \times 4 \times 2}{100}$$

$$\Rightarrow \quad P = ₹\ 600$$

17. Let first discount be x%

$$\therefore \quad \frac{(100-20)}{100} \times \frac{100-x}{100} \times 2500 = 1500$$

$$\Rightarrow \quad \frac{80}{100} \times \frac{100-x}{100} \times 2500 = 1500$$

$$\Rightarrow 100 - x = \frac{1500 \times 100 \times 100}{80 \times 2500}$$

$$\Rightarrow 100 - x = 75;$$

$$\Rightarrow x = 25$$

Alternatively,

Successive discount = $\left(a + b - \frac{ab}{100}\right)\%$

$\therefore \left(\frac{2500 - 1500}{2500}\right) \times 100 = a + 20 - \frac{20a}{100}$

$\Rightarrow \frac{1000}{2500} \times 100 = a + 20 - \frac{a}{5}$

$\Rightarrow 40 - 20 = a - 0.2a$

$\Rightarrow a = \frac{20}{0.8} = 25\%$

18. Let there be n number of balls in each packet.

$\therefore \quad 6n \times 8 = CP + 96$

$\Rightarrow \quad 48n = CP + 96 \quad ...(i)$

Also, $6(n - 1) \times 9 = CP + 90$

$\Rightarrow \quad 54n - 54 = CP + 90$

$\Rightarrow \quad 54n = CP + 144 \quad ...(ii)$

Solving equations (*i*) and (*ii*), we get,

$n = 8$

$\therefore \quad CP = 48 \times 8 - 96 = 384 - 96 = ₹\,288$

Each packet cost = ₹48.

19. CP of the merchant who calculates his % profit on

$CP = \frac{3760}{1.175} = ₹\,3200$

$\therefore$ His profit $= 0.175 \times 3760 = ₹\,560$

Profit of the merchant who calculates his % profit on SP

$= 0.175 \times 3760 = ₹\,658$

$\therefore$ Difference in profit = 658 − 560 = ₹ 98.

20. Let CP be ₹ x

$\therefore$ MP = 1.8x

SP of 15 articles = 12 × 1.8x = 21.6x

$\therefore$ Discount = 0.9 × 21.6x = 19.44x

$\therefore$ Profit on 15 articles = (19.44 − 15)x = 4.44x

$\therefore$ Profit % $= \frac{4.44x}{15x} \times 100 = 29.6\%$.

21. C.P. of goods for the trader = ₹ 22520

$S.P. = \frac{125 \times 22520}{100} = ₹\,28150$

$\therefore$ Profit = ₹ 5630

Now, $5630 = 0.0536 \times 5000 + 0.085 \times 10000 + \frac{x}{100} \times 7520$

$\Rightarrow \quad 5630 = 268 + 850 + 75.2x$

$\Rightarrow \quad 75.2x = ₹\,4512,$

Which is nothing but profit from sale of third lot.

22. A's Monthly Equivalent Rent = 22 × 4

B's Monthly Equivalent Rent = 16 × 8

C's Monthly Equivalent Rent = 32 × 6

$\frac{\text{B's Monthly Equivalent Rent}}{\text{C's Monthly Equivalent Rent}} = \frac{\text{Rent paid by B}}{\text{Rent paid by C}}$

Rent paid by $B = \frac{16 \times 8}{32 \times 6} \times 600 = ₹\,400$

Similarly, rent paid by A

$= \frac{22 \times 4 \times 600}{32 \times 6} = ₹\,275$

$\therefore$ Total rent = 400 + 600 + 275 = ₹1275

23. Rate in which the rent is to be divided = 4 : 8 : 12

$\therefore$ Akbar's share of rent

$= \frac{8}{24} \times 6000 = ₹\,2000.$

24. 85% of the total profit is divided in the ratio 2 : 10 : 5.

85% of total profit = 2x + 10x + 5x = 17x

Total profit $= \frac{17x}{85\%} = 20x$

$\therefore$ Share of Ram in the profit

= 2x + 15% of 20x = 2x + 3x = 5x

Share of Shyam in the profit = 10x

Share of Ghanshyam in the profit = 5x

Now, $(10x + 5x) - 5x = 500$

$\Rightarrow \quad 10x = 500$

$\Rightarrow \quad x = 50$

$\therefore$ Total profit = 20 × 50 = ₹ 1000.

25. Let capital invested be C.

Monthly Equivalent Investment of

$x = \frac{C}{3} \times \frac{T}{4}$

Monthly Equivalent Investment of

$y = \frac{C}{4} \times \frac{T}{2}$

Monthly Equivalent Investment of

$z = \frac{5C}{12} \times T$

$\therefore$ Profit is to be shared in the ratio

$\frac{CT}{12} : \frac{CT}{8} : \frac{5CT}{12}$

i.e., 2 : 3 : 10

$\therefore$ Share of x in the profit

$= \frac{2}{15} \times 3600 = ₹\,480$

3

CHAPTER

Speed, Time and Work

SPEED

If the location of an object changes with time then it is said to be in motion. A bus running on a road, an ant crawling on a floor, etc. all are examples of objects in motion because the locations of these objects keep on changing w.r.t. their surroundings.

Let an object move from a point A to the point B through any path, then the actual length of the path followed by the object is called the distance travelled by the object.

The rate at which any moving body covers a particular distance is called its speed.

$$\text{Speed} = \frac{\text{Distance travelled}}{\text{Time taken}},$$

if the distance is constant, $\text{Speed} \propto \frac{1}{\text{Time}}$

if the time is constant, $\text{Distance} \propto \text{Speed}$

if the speed is constant, $\text{Distance} \propto \text{Time}$.

We can say that for constant distance travelled, speed is inversely proportional to the time taken. This can be explained by a simple example. To cover a distance of 100 km, if a person goes at the speed of 25 kmph, he will require 4 hours to complete the journey and travelling at a speed of 50 kmph, 2 hours will be required.

Unit of Measurement

Time is measured in seconds (s), minutes (min) or hours (hr). Distance is usually measured in metres (m), kilometres (km), miles, yards or feet.

Speed is usually measured in metres per second (mps), kilometres per hour(kmph) or miles per hour (mph).

Conversion of Units

1 hour = 60 minutes = 60 × 60 seconds.

1 kilometre = 1000 metres.

1 kilometre = 0.6214 mile.

1 mile = 1.609 kilometre.

i.e. 8 metres ≈ miles.

1 yard = 3 feet.

$$1\text{km / hr} = \frac{5}{18}\text{m / sec}$$

To convert kmph to m/s we multiply by 5/18

$$1\frac{\text{m}}{\text{sec}} = \frac{18}{5}\text{km / hr.}$$

To convert m/s to kmph we multiply by 18/5

$$1\frac{\text{km}}{\text{hr}} \approx \frac{5\,\text{miles}}{8\text{hr}}$$

$$1\frac{\text{miles}}{\text{hr}} = \frac{22}{15}\frac{\text{ft}}{\text{sec}}$$

Note: If a man changes his speed in the ratio m : n, then the ratio of time taken becomes n : m

Example: Walking at 5/6 of his usual rate, a boy reaches his school 12 min late. Then find his usual time taken to reach the school.

Solution: If the boy is walking at 5/6th of his usual speed, then the time taken should become 6/5 times the original time required. Thus, we can say that the boy is going to require 1/5th more/extra time as compared to the usual time required.

$$\left(\text{as}\frac{6}{5}\text{T} - 1\text{T} = \frac{1}{5}\text{T}\right)$$

Thus, $\frac{1}{5}\text{T} \rightarrow 12\,\text{min}$

$1\text{T} \rightarrow \text{X}$

Solve to get X = 60 min.

Example: Walking at $\frac{13}{11}$th of his usual rate, a boy takes 3 min less to reach his school. Then find his usual time taken to reach the school.

Solution: If the boy is walking $\frac{13}{11}$th of his usual speed, then the time taken should become $\frac{11}{13}$ times the original time required. Thus, we can say that the boy is going to required $\frac{2}{13}$th less time as compared to the standard time required

Thus, $\frac{2}{13}\text{T} \rightarrow 3\,\text{min}$

$1\text{T} \rightarrow \text{X}$

Solve to get X = 39/2 = 19.5 minutes.

If three men cover the same distance with speeds in the ratio a : b : c, the times taken by these three will be respectively in the ratio 1/a : 1/b : 1/c.

Note:

(1) The time taken by a train L meters long, travelling at a speed of X m/sec in passing a signal post or a telegraph pole(or a standing man) is the same as the time taken by the train to cover a distance of L meters i.e. its own length because the length of the pole is nearly zero with respect to the length of the train though its height may be significant.

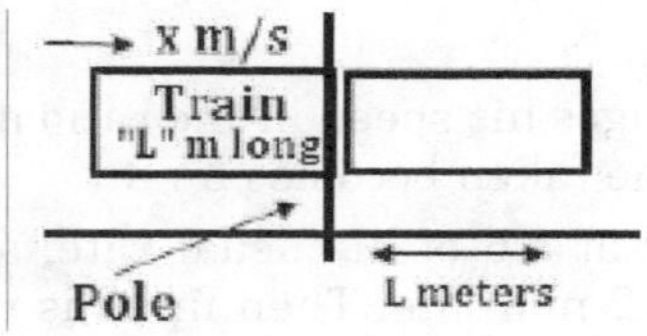

Thus, $T = \frac{L}{X}$ seconds

(2) The time taken by a train "L_1" meters long, travelling at a speed of X m/sec in passing a stationary object (a bridge ora tunnel or a train at rest or a platform) of length "L_2" meters is the same as the time taken by the train to cover a distance of $(L_1 + L_2)$ meters.

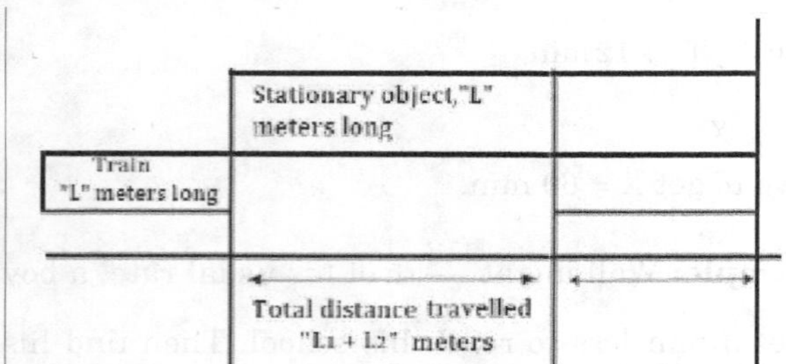

$T = \frac{L_1 + L_2}{X}$ seconds

(3) If two trains of lengths L_1 and L_2 meters respectively, are moving in the same direction with a speed of X and Y m/sec (where X > Y), then (X-Y)m/sec is called **Relative Speed** and time taken by faster train to pass the slower train

$= \frac{L_1 + L_2}{X - Y}$ seconds.

(4) If two trains of lengths L_1 and L_2 meters respectively, are moving in opposite directions (towards each other or away from each other) with a speed of X and Y m/sec, then (X + Y)m/sec is called **Relative Speed** and time taken by the trains to pass each other.

$= \frac{L_1 + L_2}{X + Y}$ seconds.

(5) If a man running at a speed X m/sec in the same direction in which a train of length L meters is running at a speed Ym/sec, then (Y - X)m/sec is called the speed of the train relative to man. Then the time taken by the train to cross the man.

$= \frac{L}{Y - X}$ seconds.

(6) If a man running at a speed X m/sec in a direction opposite to that in which a train of length L meters is running with a speed Y m/sec, then X + Y is called the speed of the train relative to man. Then the time taken by the train to cross the man.

$= \frac{L}{Y + X}$ seconds.

(7) If two trains or moving bodies are crossing each other such that

Length of the first train or object = L_1 meters,

Length of the second train or object = L_2 meters,

Time taken by the two when crossing each other in opposite direction = X sec,

Time taken by the two when crossing each other in same direction = Y sec,

Then speed of the faster train

$= \frac{L_1 + L_2}{2}\left(\frac{1}{X} + \frac{1}{Y}\right)$;

Speed of the slower train $= \frac{L_1 + L_2}{2}\left(\frac{1}{X} - \frac{1}{Y}\right)$.

Example: Two trains 110 m and 88 m long respectively are running in same direction. The first runs at the rate of 35.2 kmph and the second at the rate of 44 kmph. How long will they take to cross each other?

Solution: It is clear that the trains will cross each other when they have travelled a distance equal to the sum of their lengths = 110 + 88 = 198 m. Since they are moving in the same direction, we can find the relative speed = 44 – 35.2 = 8.8 km/hr.

⇒ Time required = distance/speed

$= \frac{198}{8.8} \times \frac{18}{5}$ second $= 81$ sec

Example: A train in motion, 66 m long overtakes a train 88 m long travelling @ 30 kmph in the opposite direction in 0.168 min. The speed of the first train is

(1) 60 kmph (2) 25 kmph

(3) 45 kmph (4) 36 kmph

Solution:

$$30 \text{ kmph} = \frac{30 \times 5}{18} \times 8.33\text{m / sec}$$

Let the speed of the first train be x m/sec.

Now total distance = 66 + 88 = 154 m.

$$\text{Time} = \frac{\text{Dis tan ce}}{\text{Speed}}$$

$$\Rightarrow 0.168 \times 60 = \frac{154}{8.33 + x}$$

$$\Rightarrow 10.08(8.33 + x) = 154$$

$\Rightarrow x = 6.95$ m/sec = 25 kmph

Ans. (2)

Example: A train travelling with constant speed crossed a 96 m long platform in 12 sec and another 141 m long platform in 15 sec. Find the length and speed of the train.

Solution: If X be the length of train in meters and Y be the speed of the train in m/s, then

(X + 96)/Y = 12, (X + 141)/Y = 15

Solve to get X = 84, Y = 15.

15 m/sec = 15×18/5 = 54 kmph.

Hence the required answer is 84m & 54kmph.

Average Speed: If a body travels $d_1, d_2, d_3 \dots d_n$ distances, with speeds $s_1, s_2, s_3 \dots s_n$ in time $t_1, t_2, t_3 \dots t_n$ respectively, then average speed of the body through the total distance is given by

$$\text{Average Speed} = \frac{\text{Total distance travelled}}{\text{Total time taken}}$$

$$= \frac{d_1 + d_2 + d_3 + \dots d_n}{t_1 + t_2 + t_3 + \dots t_n}$$

$$= \frac{s_1t_1 + s_2t_2 + s_3t_3 + \dots + s_nt_n}{t_1 + t_2 + t_3 + \dots + t_n}$$

$$= \frac{d_1 + d_2 + d_3 + \dots + d_n}{\frac{d_1}{s_1} + \frac{d_2}{s_2} + \frac{d_3}{s_3} + \dots + \frac{d_n}{s_n}}$$

Some Important Points:

(1) If a body covers the distance d_1 and d_2 at a speed of s_1 and s_2 km/hr, respectively, in time t_1 and d_2, then

Total time taken, $T = t_1 + t_2 = \frac{d_1}{s_1} + \frac{d_2}{s_2}$.

Total distance covered, $d = d_1 + d_2 = s_1t_1 + s_2t_2$.

(2) While travelling a certain distance d, if a man changes his speed in the ratio m: n, then ratio of time taken becomes n : m.

(3) If a certain distance (d), say from A to B, is covered at 'a' km/hr and same distance is covered again say from B to A in 'b' km/hr, then average speed during the whole journey is given by

$$\text{Average speed} = \left(\frac{2ab}{a+b}\right)\frac{\text{km}}{\text{hr}} \dots$$

(which is harmonic mean of a and b)

Also, if t_1 and t_2 is taken to travel from A to B and B to A respectively, then distance 'd' from A to B is given by

$$d = (t_1 + t_2)\left(\frac{ab}{a+b}\right); \; d = (t_1 - t_2)\left(\frac{ab}{a-b}\right);$$

$$d = (a-b)\left(\frac{t_1t_2}{t_1 - t_2}\right)$$

(4) If a body travels a distance 'd' from A to B with speed 'a' in time t_1 and travels back from B to A i.e., same distance with $\frac{m}{n}$ of the usual speed 'a', then change in time taken to cover the same distance is given by

$$\text{Change in time} = \left(\frac{n}{m} - 1\right) \times t_1; \text{ for } n > m$$

$$= \left(1 - \frac{n}{m}\right) \times t_1; \text{ for } m > n$$

(5) If two persons A and B start at the same time in opposite directions from two points and meet at any point in 'a' and 'b' hr, respectively after having met, then,

$$\frac{\text{A's Speed}}{\text{B's Speed}} = \frac{\sqrt{b}}{\sqrt{a}}$$

Relative speed.

(1) Time taken by a moving object 'x' meters long in passing a stationary object of negligible length from the time they meet is same as the time taken by the moving object to cover 'x' meters with its own speed.

(2) Time taken by a moving object 'x' meters long in passing a stationary object 'y' meters long from the time they meet, is same as the time taken by the moving object to cover 'x + y' meters with its own speed.

(3) If two objects of length 'x' and 'y' meters move in the same direction at 'a' and 'b' m/s, then time taken to cross each other from the time they meet

$$= \frac{\text{Sum of their length}}{\text{Relative speed}}$$

i.e., $\frac{x+y}{a-b}$ if a > b or else, $\frac{x+y}{b-a}$

(4) If two objects of length 'x' and 'y' meters, move in the opposite direction at 'a' and 'b' m/s, then time taken to cross each other from the time they meet

$$= \frac{\text{Sum of their length}}{\text{Relative speed}} = \frac{x+y}{a+b}$$

(5) If speed of a boat in still water is x km/hr and speed of the stream is y km/hr, then

Relative speed while travelling with the stream i.e., speed downstream = (x + y) km/hr.

Relative speed while travelling against the stream i.e., speed upstream = (x – y) km/hr.

(6) Also speed of the boat in still water = $\frac{1}{2}$ (Speed with stream + Speed against stream)

Speed of the river = $\frac{1}{2}$ (Speed with stream – Speed against stream)

BOATS AND STREAMS

A Few Important Terminologies

The following terms will be used often while discussing boats and streams.

Stream: It implies that the water in the river is moving or flowing.

Upstream: Going against the flow of the river.

Downstream: Going with the flow of the river.

Still Water: It implies that the speed of water is zero (generally in a lake).

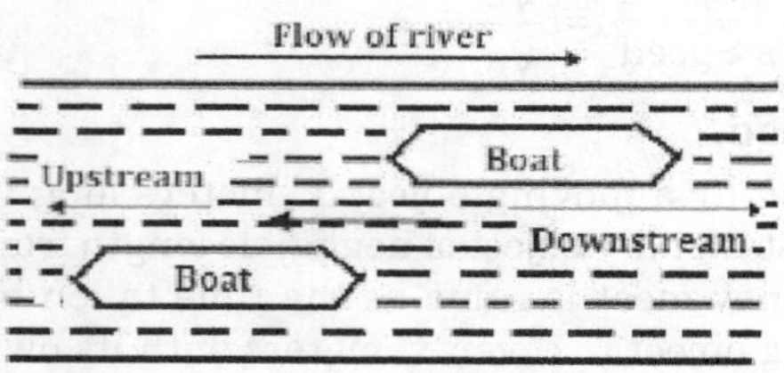

Downstream (With The Stream) Rowing

It indicates that the stream favors the man's rowing (or boating). i.e. direction of rowing and direction of flow (stream) is same.

Upstream (Against The Stream) Rowing

It indicates that the stream flows against the man's rowing (or boating) i.e. direction of rowing and direction of stream (current) are opposite.

Note: Let the speed of a boat (or man) in still water be X m/sec and the speed of the stream (or current) be Y m/sec.

Speed of boat with the stream (or Downstream of D/S) (X + Y) m/sec.

Speed of boat against the stream (or Upstream or U/S) = (X – Y) m/sec.

Speed of man/boat in still water

Speed of the stream Y = $\frac{(X+Y)-(X-Y)}{2}$

$= \frac{\text{Downstream} - \text{Upstream}}{2}$

Example: A boat is rowed down a river 28 km in 4 hours and up a river 12 km in 6 hours. Find the speed of the boat and the river.

Solution:

Downstream speed is $\frac{28}{4}$ =7 kmph,

Upstream speed is $\frac{12}{6}$ = 2 kmph,

Speed of Boat = $\frac{1}{2}$ [downstream + upstream speed]

$= \frac{1}{2}(7+2) = 4.5$ kmph.

Speed of current = $\frac{1}{2}$ [downstream-upstream]

$= \frac{1}{2}(7-2) = \frac{5}{2} = 2.5$ kmph

Example: A man rows 18 km down a river in 4 hours with the stream and returns in 12 hours. Find his speed and also the speed of the stream.

Solution:

Speed with the stream = 18/4 = 4.5 kmph.

⇒ Speed against the stream = 18/12 = 1.5 kmph.

⇒ Speed of the stream = 1/2{(4.5 – 1.5)} = 1.5 kmph.

and speed of the man = 4.5 – 1.5 = 3 kmph.

If a man capable of rowing at the speed of X m/sec in still water, rows the same distance up and down a stream flowing at a rate of Y m/sec, then his average speed throughout the journey is

$$= \frac{\text{Speed Upstream} \times \text{Speed Downstream}}{\text{Man's rate in still water}}$$

$$= \frac{(X-Y)(X+Z)}{X}$$

Note: When downstream distance = upstream distance, then

$$\frac{\text{Man's rate in still water}}{\text{Speed of stream}} = \frac{t_{up} + t_{down}}{t_{up} - t_{down}}$$

Average speed for total journey (UP + DOWN)

$$= \frac{\text{Upstream Rate} \times \text{Downstream rate}}{\text{Man's rate in still water}}$$

Total Journey time ($t_{up} + t_{down}$)

$$\frac{\text{Man's rate in still water} \times \text{Total distance}}{\text{Upstream Rate} \times \text{Downstream rate}}$$

Example: A man rows 10 km upstream and back again to the starting point in 55 min. If the speed of stream is 2 kmph, then find the speed of rowing in still water.

Solution: Let x be the speed of rowing in still water. y = speed of stream = 2 kmph.

Total time $T = \frac{55}{60}$h.

Hence,

$$\text{Total Time} = \frac{\text{Speed in still water} \times \text{Total distance}}{\text{Upstream Rate} \times \text{Downstream rate}}$$

$$\Rightarrow \frac{55}{60} = \frac{x \times 2 \times 10}{(x+2)(x-2)} \Rightarrow \frac{55}{60}(x^2 - 2^2) = 2 \times x \times 10$$

$\Rightarrow 11x^2 - 240x - 44 = 0$

$\Rightarrow (x - 22)(11x + 2) = 0$

$\therefore$ x = 22, since (–)ve value of x is not admissible.

Total Distance = Downstream distance + Upstream distance = 2 x any one side distance.

Example: A man who can swim 48 m/min in still water swims 200m against the current and 200 m with the current. If the difference between these two times is 10 min, then find the speed of the current in m per min.

Solution: Let the speed of stream be x kmph. The equation becomes

$$\frac{200}{48-x} - \frac{200}{48+x} = 10$$

$\Rightarrow 200(48 + x) - 200(48 - x) = 10[48x^2 - 2^2]$

$\Rightarrow x^2 + 40x - 2304 = 0$

On solving it we get x = 32 and x = –72 (not acceptable) i.e. speed of stream is 32 m/min.

LINEAR AND CIRCULAR RACES

The terms given below are commonly used in this topic, and a clear understanding will help the student get a good grip on the subject.

(1) **Race:** A contest of speed between participants is called a race.

(2) **Starting Point:** The point from where a race begins is called the starting point.

(3) **Race Course:** It means the path or ground on which races are run.

(4) **Finishing Point:** The point where the race finishes is called the winning post or finishing point or a goal.

(5) **Dead Heat Race:** A race in which no one is the winner because all the runners reach the winning post at the same time, is called a dead heat race.

(6) **Winner:** The person who first reaches the finishing point is called the winner.

(7) **Winner's Time:** The time taken by the winner to complete the race is called winner's time. If A and B both start from the same place, then winner's distance = Length of the race.

(8) **Loser's time:** The time taken by the loser to complete the race is called loser's time.

Linear Races

Suppose A & B are two contestants in race. If before the start of the race, A is at the starting point and B is ahead of A by 10 meters, then A is said to have given B a start of 10 meters. To cover a race of 300 meters in this case, A will have to cover a distance of 300 meters and B will have to cover (300 – 10) = 290 meters only.

Suppose A & B are two contestants in race. At the end of the race, if A is at the finishing point and B is x meters away from the finishing point, and then A is said to have beaten B by x meters in a race.

(1) **A gives B a start of x meters** implies that, if the distance between the starting point and finishing point is L meters, A covers L meters while B covers L – x meters.

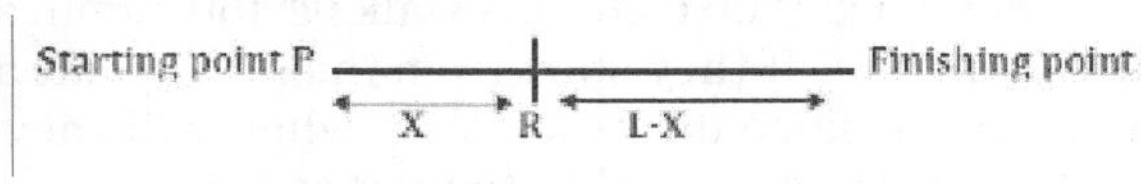

From the figure given above, it is clear that A starts at point P, but B starts at R at the same moment. For example in a 100 meters race, A gives B a start of 10 meters means, while A run 100 meters, B runs 90 meters.

(2) **A beats B by x meters** implies that, if the distance between the starting point and finishing point is L meters, A wins the race by covering L meters while B covers L – x meters only.

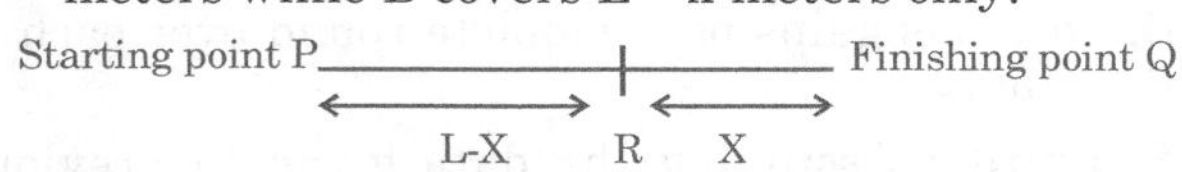

(3) **A gives B a start of t seconds**, implies that A starts the race t seconds after B starts from the starting point.

(4) **A beats B by t seconds**, implies that, A and B start together from the starting point, but A reaches the finishing point t seconds before B reaches.

(5) Start distance is the distance between the two contestants at start if they are not starting from the same position.

(6) If A & B are starting from the same point, A beats B by 'x' meters or 't' seconds means, B runs 'x' meters in 't' seconds.

Example: In a km race, A beats B by 40 meters or 7 seconds. Find A's time over the course.

Solution: Here B runs 40 meters in 7 seconds.

$\therefore$ B runs 1000 m in $\frac{1000 \times 7}{40} = 175$ seconds.

Hence, A's time over the course –(175 – 7) = 168 sec

Short-cut:

By formula,

$$\frac{\text{Winner's time}}{\text{Loser distance}} = \frac{\text{Beat time} + \text{Start time}}{\text{Beat distance} + \text{Start distance}}$$

$\therefore$ A's time $= \frac{7}{40} \times 960 = 168$ seconds.

If a race ends in a dead heat, then beat time = 0 and beat distance = 0.

Circular Races

When two or more persons start from the same place at the same time and travel round a circle in the same direction or in opposite directions, then they will be first together at the starting point again after an interval of time which is the LCM of the times in which each of them makes one complete round.

Example: A and B walk around a circular path of circumference 1200 meters. A walks @ 150 m/min and B @ 70 m/min. If they start from the same point and walk in the same direction, then when will they be first together again at the starting point?

Solution: Time for 1 round by A = 1200/150 = 8 minutes and for B = 1200/70 = 120/7 minutes.

$\Rightarrow$ Time taken to meet again at the starting point = LCM of 8 and 120/7 = 120 minutes = 2 hrs.

They will be first together again after an interval of time which is the LCM of the times in which one of the persons gains one complete round over each of the others.

Example: Assuming the data given in previous example, when will they be together again anywhere else on the course.

Solution: Now, in 1 minute, A gains over B(150 – 70) = 80 meters.

To gain 1200 meters over B, A will take 1200/80 × 1 = 15 min.

They will be first together again after an interval of time which is required to complete a round with the relative speed.

Example: Assuming the data given in previous example, when will they be together supposing they walk in opposite directions?

Solution:

Relative speed =150 + 70 = 220 m/min.

So, they will be together after 1200/220 = 60/11 minutes.

We hope that the concept of circular race is clear to you. If not, please go through the following example. It will help you to understand the concept thoroughly.

Example: Along with A and B, C is also running across the circular track at the rate of 130 m/min in the same direction. When will all the three people meet (use data of previous example)

1) at the starting point?
2) at any other point on the track?
3) At what distance (nearest) from the starting point do the three meet?

Solution:

(1) Time for 1 round by A = 8 min.
Time for 1 round by B = 120/7 min.
Time for 1 round by C = 120/13 min.
So, all the three people will meet at the starting point after the time.
= The LCM of 8, 120/7 and 120/13 i.e.120 minutes.

(2) A gains 80 meters per minute over B.
So, they will met every 1200/80 = 15min
A gains 20 meters per minute over C.
So they will meet every 1200/20 = 60min
So, all the three people will meet after the time which is equal to the LCM of 15 and 60 i.e. after 60 minutes or 1 hour.

(3) They are meeting after 60 minutes from the start.
Distance covered by A in 60 minutes
= 150 × 60 = 9000m = 600m from the starting point.
Distance covered by B in 60 minutes
= 70 × 60 = 4200 m = 600m from the starting point.
Distance covered by C in 60 minutes
= 130 × 60 = 7800m = 600m from the starting point.
So, the three people will meet at 600 m from the starting point.

Example: Now, if we assume that C starts running in the opposite direction, then the three will once again meet at the starting point after the same time interval. But can they meet anywhere else? If Yes, after what time interval and what minimum distance from the starting point?

Solution: C starts running in the opposite direction, as shown in the figure below.

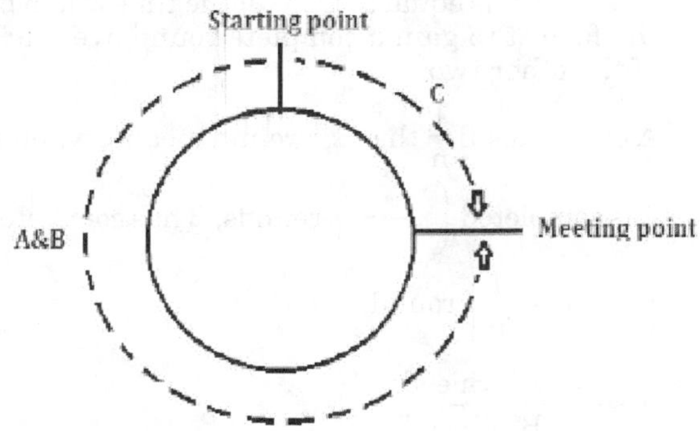

The three will once again meet at the starting point after the same time interval i.e. 2 hours.

Now A & C will meet after 1200/280 = 30/7 minutes from the starting point. B & C will meet after 1200/200 = 6 minutes from the starting point.

So, in this case the three people will meet after the time which is equal to the LCM of 30/7 and 6 i.e. after 30 minutes from the starting point.

Distance covered by A in 30 minutes

=150 × 30 = 4500m = 900m from the starting point.

Distance covered by B in 30 minutes

=70 × 30 = 2100m = 900m from the starting point.

Distance covered by C in 30 minutes

=130 × 30 = 3900m = 300m from the starting point.

The minimum distance at which they will meet = 300 m from the starting point.

Note: Throughout this chapter, units of quantities have a crucial importance. e.g. while solving a problem if we take distance in meters, we should take speed in m/sec and time in seconds. If proper units are not used and conversions not effected, then you'll find yourself in trouble.

WORK & TIME

In most of the problems based on Work and Time, either the amount of time taken to finish a given job or the amount of work done is to be calculated.

Unless otherwise specified, the amount of work done is generally taken as unity (1). Also, if it is given that a person X can finish a job in D days, then it implies that X alone can do the job in D days.

If a man can do a piece of work in N days (or hours or any other unit of time), then the work done by him in one day will be **1/N** of the total work.

Example: Anish can finish a piece of work by himself in 14 days. Then calculate the amount of work done by him in 1 day and 11 days.

Solution: The amount of work done by Anish, working alone in 14 days = 1 unit of work. So, the amount of work done by Anish, working alone in 1 day = 1/14 unit of work and in 11 days = 11/14 units of work.

Note: If A is twice as good a workman as B, then A will take half the time B takes to finish a piece of work.

Example: Suresh can finish a piece of work by himself in 42 days. Mahesh, who is 1/5 times more efficient as Suresh, requires X days to finish the work by working all by himself. Then what is the value of X?

Solution: Suresh, working alone 42 days = 1 unit of work.

Mahesh is 1/5 time more efficient than Suresh. So Mahesh is 6/5 times as efficient as Suresh. Hence Mahesh should require 5/6th of the time, the time taken by Suresh.

Therefore time taken by Mahesh = 5/6 × 42 = 35 days.

Measurement Of Work

Work can be measured by many units. Some of them are

(1) Man-Days

A certain number of men working for acertain number of days.

Example: 5 men can finish a piece of work in 13 days. This implies that the work is of 5 × 65 Man-days, i.e. if one man alone works to finish the given work then he is going to take 65 days to finish it and if 65 men work together, then the work would be finished in 1 day.

(2) Man-Hours

A certain number of men working for acertain number of hours.

Example: 3 men can finish a piece of work in 13 hours. This implies that the work is of 3 × 13 = 39 Man-hours, i.e. if one man alone works to finish the given work then he is going to take 39 hours to finish it and if 39 men work together, then the work would be finished in 1 hour.

Some others are, Man-minutes, Machine-hours etc.

Note: If A, B and C can finish a piece of work in X, Y and Z days respectively, while working alone and together they require M days to finish the work, then the amount of work done by A is M/X, B is M/Y & C is M/Z.

Concept Of Work Equivalence

In order to solve certain types of questions, there is a very logical method that can be used.

This is the WORK EQUIVALENCE method which works on a simple premise-make the LHS equal to the RHS on the basis of SAME UNITS in terms of which work can be measured.

Some Important Points:

(1) 'A gives B a start of x meters', implies that, if distance between starting point and finishing point is L meters, then A covers L meters while B covers L – x metres.

e.g., In a 100 metre race. A gives B a start of 10 metres means, while A runs 100 metres, B runs 100 –10 = 90 metres.

(2) '*A beats B by x metres*', implies that, if distance between starting point and finishing point is L meters, then A wins the race by covering L metres, while B covers L – x metres only.

(3) 'A gives B a start of t seconds', implies that A starts the race t seconds after B starts from the starting point.

(4) '*A beats B by t seconds*', implies that, A and B start together from the starting point, but A reaches the finishing point t seconds before B finishes.

Note: (*iii*) and (*iv*) both imply that B takes t seconds more than A to finish the distance.

(5) A beats B by 'x' metres or 't' seconds means, B runs 'x' metres in 't' seconds.

(6) Winner's distance = Length of the race.

(7) Distance covered by loser = Winner's distance – (Beat distance + Start distance)

(8) Time taken by winner = Time taken by loser – (Beat time + Start time)

(9) $$\frac{\text{Winner's Time}}{\text{Loser's Distance}} = \frac{\text{Loser's time}}{\text{Winner's distance}} = \frac{\text{Beat time + Start time}}{\text{Beat distance + Start distance}}$$

(10) If a race ends in a dead heat, then beat time = 0 and beat distance = 0.

(11) Two persons staring at the same time and from the same point along a circular path will be together again for the first time. When faster gains one complete round over the other, then time taken by faster person to complete one round over the other

$$= \frac{\text{Length of race course}}{\text{Relative Speed}}$$

(12) Two persons, starting at the same time from the same point along a circular path, will be together again for the first time at the same starting point, at a time which is LCM of the time taken by each to complete a round.

(13) Three persons, staring at the same time and from the same point along a circular path, will be together for the first time after the start at a time which is equal to LCM of the time taken by the fastest to gain a complete round over each of the other two.

(14) A overtakes B $\frac{1}{n}$ th of xth round means, when A has completed $\left(x - \frac{1}{n}\right)$ rounds, B has completed $\left[(x-1) - \frac{1}{n}\right]$ round.

Also, $\frac{\text{A's Speed}}{\text{B's Speed}}$

$$= \frac{\text{Rounds completed by A in a given time}}{\text{Rounds completed by B in the same time}}$$

e.g., A overtakes B in the middle of the 4th round implies, when A has completed $3\frac{1}{2}$ rounds, B has completed $2\frac{1}{2}$ rounds. 2 $\frac{\text{A's Speed}}{\text{B's Speed}} = \frac{3\frac{1}{2}}{2\frac{1}{2}}$

(15) If a can do a work in 'a' number of days, then in one day $\frac{1}{a}^{th}$ work is done.

Conversely, if a man does $\frac{1}{a}^{th}$ of a work in 1 day, then he can complete the work in $1 \div \frac{1}{a} = a$ days.

(16) If A is 'x' time as good a workman as B, then he will take $\frac{1}{x}^{th}$ of the time taken by B to do the same work.

(17) If A and B can do a piece of work in 'x' and 'y' days respectively. then working together, they will take

$\frac{xy}{x+y}$ days to finish the work and in one day.

They finish $\left(\frac{x+y}{xy}\right)^{th}$ part of the work.

(18) To compare the work done by different people, first find the amount of work each can do in the same time.

(19) If number of men to do a job is changed in the ratio a : b, then time required to do the work will be in the ratio b : a, assuming amount of work done by each of them in the given time is the same, or they are identical.

(20) If two men A and B together can finish a job in 'x' days and if A working alone takes 'a' days more than A and B working together and B working alone takes 'b' days more than A and B working together, then $x = \sqrt{ab}$.

(21) To do a piece of work, the number of men employed and the number of days required to do the work are in inverse proportion. Also, the number of men employed and the hours worked per day are in inverse proportion.

Example: A group of soldiers can completely destroy an enemy bunker in 7 days. However 12 soldiers fell ill. The remaining now can do the job in 10 days. Find the original group strength.

Solution: Here, first of all, let us see how WORK can be defined. It is obvious that work can be measured a "destruction of the enemy bunkers."

In the first case, let us say that there were S numbers of soldiers in the group. So they had to work for 7 days for the work which we call W.

$\Rightarrow S \times 7 = W \quad ...(1)$

Now 12 fell ill and the remaining did the work in 10 days. Hence the new equations

$(S - 12) \times 10 = W \quad ...(2)$

Just compare the two equations to get the answer.

$S \times 7 = (S - 12) \times 10$

$\Rightarrow 7S = 10S - 120$

$\Rightarrow 120 = 3S$

$\Rightarrow S = 40$ soldiers

Hence, there were 40 soldiers in the group initially.

Example: A group of 20 cows can graze a field of 3 acres in size in 10 days. How many cows can graze a field twice as large in 8 days?

Solution: Here, first of all, let us see how WORK can be defined. It is obvious that work can be measured as "acres grazed". In the first case, there were 20 cows in the group.

They had to work for 10 days to do the work which we call W (which = 3).

$\Rightarrow 20 \times 10 = 3 \quad ...(1)$

Do not be worried about the numerical values on either side. The point is that logically this equation is consistent as the LHS indicates "Cowdays" and the RHS indicates "Acres", both of which are correct ways of measuring work done.

Now the field is twice as large. Hence the new equation is

$\Rightarrow C \times 8 = 6 \quad ...(2)$

Just divide (2) by (1) to get the answer.

$8C/200 = 6/3$

$\Rightarrow 8C = 2 \times 200$

$\Rightarrow C = 400/8 = 50$ cows

Hence, there were 50 cows in the second group.

PIPES AND CISTERNS

A pipe connected with a cistern is called an *inlet*, if it fills the cistern.

A pipe connected with a cistern is called an *outlet*, if it empties the cistern.

filling or emptying a cistern can be considered as work done.

Some Important Points:

(1) If an inlet pipe fills a cistern in 'a' hours, then $\frac{1^{th}}{a}$ part is emptied in 1 hour.

Similary, if an outlet pipe empties a cistern in 'a' hours, then $\frac{1^{th}}{a}$ part is emptied in 1 hour.

(2) If pipe A is 'x' times bigger than pipe B, then pipe A will take $\frac{1^{th}}{x}$ of the time taken by pipe B to fill the cistern.

(3) If A and B fill a cistern in 'm' and 'n' hours, respectively, then together they will take $\frac{mn}{m+n}$ hours to fill the cistern and in one hour $\frac{m+n^{th}}{mn}$ part of the cistern will be filled.

Similarly, 'A' and 'B' empty a cistern in 'm' and 'n' hours respectively, then, together they will take $\frac{mn}{m+n}^{th}$ hours to empty the cistern and in one hour $\frac{m+n^{th}}{mn}$ part of the cistern will be empty.

(4) If an inlet pipe fills a cistern in 'm' hours and an outlet pipe empties the cistern in 'n' hours, then the net part filled in 1 hour when both the pipes are opened is $\left(\frac{1}{m}-\frac{1}{n}\right)$ i.e., $\frac{n-m}{mn}$ and the cistern will get filled in $\left(\frac{mn}{n-m}\right)$ hours.

For the cistern to get filled, m < n. If m > n, the cistern will never get filled.

In general,

Net part filled of a cistern = (Sum of work done by inlets) – (Sum of work done by outlets)

(5) If an inlet pipe fills a cistern in 'a' minutes, takes 'x' minutes longer to fill the cistern due to a leak in the cistern, then the time in which the leak will empty the cistern is given by a × $\left(1+\frac{a}{x}\right)$.

(6) If two pipes A and B can fill a cistern in 'x' minutes and if A alone can fill it in 'a' minutes more than 'x' minutes and B alone can fill it in 'b' minutes more than 'x' minutes, then $x = \sqrt{ab}$.

Example: If an inlet pipe can fill a tank in 4 hours and an outlet pipe empties the full tank in 5 hours, then what is the net part filled in 1 hour when both the pipes are opened?

Solution: The net part filled in 1 hour when both the pipes are opened = (1/4) –(1/5) = (1/20).

So, time required to fill the tank = 1/(1/20) = 20 hrs.

Example: An inlet pipe can fill the tank in 6 hours and an outlet pipe can empty the tank in 4 hours. When the tank is full and both the pipes are open, find the new part emptied in 1 hour.

Solution: The net part emptied in 1 hour

$=\frac{1}{4}-\frac{1}{6}=\frac{1}{12}$.

{∴ Time required to empty the full tank is 12 hours.}

Two pipes A and B can fill a cistern in X and Y hours respectively, while working alone. If both the pipes are opened together, then the time taken to fill the cistern will be $\frac{X \times Y}{X+Y}$.

Example: Two pipes A and B can fill a cistern in 4 and 5 hours respectively, while working alone. If both the pipes are opened together, then find the time taken to fill the cistern.

Solution: Pipe A can fill the cistern in 4 hours and Pipe B can fill the cistern in 5 hours.

When the pipes are opened together, the time taken to fill the cistern will be $\frac{4 \times 5}{4+5}=\frac{20}{9}$ hrs.

Three pipes A, B and C can fill a cistern in X, Y and Z hours respectively, while working alone. If all the three pipes are opened together, the time taken to fill the cistern will be $\frac{X \times Y \times Z}{XY+YZ+ZX}$.

Example: Three pipes A, B and C can fill a cistern in 10, 12 and 15 hours respectively, while working alone. If all the three pipes are opened together, then find the time taken to fill the cistern.

Solution: The pipes can fill the cistern in 10, 12 and 15 hours respectively.

When the three pipes are opened together, the time taken to fill the cistern

$$=\frac{10 \times 12 \times 15}{10 \times 12+12 \times 15+10 \times 15}=4\text{hrs}.$$

EXERCISE

MCQ TYPE QUESTIONS

1. A barrel full of beer has 2 taps, one midway which draws a litre in 6 minutes and the other at the bottom which draws a litre in 4 minutes. The lower tap is normally used after the level of beer in the barrel is lower than midway. The capacity of the barrel is 36 litres. A new assistant opens the lower tap when the barrel is full and draws out some beer. As a result the lower tap has been used 24 minutes before the usual time. For how long was the beer drawn out by the new assistant?

(*a*) 10 mins. (*b*) 8 mins.
(*c*) 16 mins. (*d*) 32 mins.

2. Pipe A can fill a cistern in 36 minutes and B in 48 minutes. If both the pipes are opened together, when should pipe B be closed so that the cistern may be just full in 24 minutes?

(*a*) 8 mins. (*b*) 9 mins.
(*c*) 12 mins. (*d*) 16 mins.

3. Pipes P, Q and R are attached to a tank and each can act as either an inlet or outlet pipe. Pipes P, Q and R respectively take 8, 10 and 12 hours to fill the empty tank or empty the full tank. In the first hour, pipes P and R work as inlet and Q work as outlet. In the second hour, pipes P and Q work as inlet and pipe R as outlet. In the third hour pipes Q and R work as inlet and pipe P as outlet and the process goes on like this. When will the cistern be filled?

(*a*) In the 8th hour. (*b*) In the 9th hour.
(*c*) In the 10th hour. (*d*) In the 11th hour.

4. There are 2 inlet pipes and an outlet pipe. The efficiency of one of the inlet pipes is double than that of the other. Also, the efficiency of the outlet pipe is half that of the lesser efficient inlet pipe. The empty tank gets filled in 16 hours when all the pipes are opened. How many hours will be taken to fill the empty tank when the 'lesser' efficiency inlet pipe is plugged and the rest kept opened?

(*a*) $33\frac{1}{3}$ hours (*b*) $26\frac{2}{3}$ hours
(*c*) 100 hours (*d*) 80 hours

5. A train overtakes 2 persons walking at 2 km/hr and 4 km/hr respectively in the same direction and completely passes them in 9 seconds and 10 seconds. Length of the train and its speed in km/hr respectively are

(*a*) 75 m, 18 km/hr (*b*) 80 m, 21 km/hr
(*c*) 60 m, 20 km/hr (*d*) 50 m, 22 km/hr

6. A policeman goes after a thief who has a 100 m start. The policeman runs 1 km. in 6 mins. and the thief 1 km in 10 mins. How far did the thief go before he was overtaken?

(*a*) 100 m (*b*) 125 m
(*c*) 150 m (*d*) 250 m

7. Everyday Sachin exercises for $2\frac{1}{2}$ hours. He exercises by walking, jogging and running. His speeds while walking, jogging and running in the ratio 1 : 2 : 4. Which of the following statements are true?

I. If he spends equal time on the three activities, then he runs twice the distance that he jogs.

II. If he runs at 8 km per hour and distance covered by running and jogging together is 12 km, then he walks a distance of 1 km in 1 hour.

III. If the time for which he walks, jogs and runs is in the ratio 2 : 2 : 1 and he covers a total distance of 10 km, then his speeds while walking, jogging and running are 2 km/hr, 4 km/hr and 8 km/hr res-pectively.

Codes:

(*a*) I and II (*b*) II and III
(*c*) I and III (*d*) I, II and III

8. A monkey climbing up a greased pole ascends 10 metres and slips down 2 metres in alternate minutes. If the pole is 64 metres high, how long will it take him to reach the top?

(*a*) 16 min
(*b*) 14 min and 48 sec
(*c*) 12 min
(*d*) 14 min

9. A train travelling at 10 m/sec from A to B at 7 a.m. meets a train leaving B at 7:20 a.m. and coming to A at a speed $\frac{1}{3}$ times faster than the first train. If the distance from A to B is 68 km. then, at what distance from A will the two trains meet?

(*a*) 72 km (*b*) 36 km
(*c*) 60 km (*d*) 50 km

10. Two trains A and B start from stations X and Y towards each other. B leaves station Y half an hour after train A leaves station X. Two hours after train A has started, the distance between trains A and B is $\frac{19}{30}$th of the distance between stations X and Y. How much time would it take each train (A and B) to cover the distance X to Y, if train A reaches half an hour later to its destination as compared to B?

(*a*) 8 hrs, 6 hrs (*b*) 5 hrs, 4 hrs
(*c*) 10 hrs, 9 hrs (*d*) 9 hrs, 8 hrs

11. Two men, A and B, run a 4 km race on a circular course of $\frac{1}{4}$ km. If their speeds are in the ratio of 5 : 4, how often does the winner pass the other?

(a) Once (b) Twice

(c) Thrice (d) Four times

12. A and B start from the same point to run in opposite directions round a circular path 550 yards in length, A giving B a start of 100 yards. They pass each other when A has run 250 yards. Who will come first to the starting point and at what distance will they be apart?

(a) A, 20 yards (b) B, 10 yards

(c) B, 20 yards (d) A, 10 yards

13. Two runners run in the same direction along a circular track 2 km long. The faster runner overtakes the slower one every 30 minutes. What are their speeds, if faster one completes a round one minute sooner than the other?

(a) 24 km/hr and 20 km/hr

(b) 44 km/hr and 28 km/hr

(c) 20 km/hr and 18 km/hr

(d) 20 km/hr and 16 km/hr

14. To do certain work alone 'A' takes 4 hrs, 'B' takes 5 hrs and 'C' takes 6 hrs. How long would A and B together take to do a work which C can do in $4\frac{1}{2}$ hrs.?

(a) 2 hrs 10 mins. (b) 2 hrs 30 mins.

(c) 1 hr 20 mins. (d) 1 hr 40 mins.

15. A can do piece of work in 8 days which B and C can do in 12 and 16 days respectively. If they work on alternate days, then which of the following is true?

(a) If A started the work on the first day followed by B and c, then the work will be finished on the 10th day.

(b) If B started the work on the first day followed by C and A, then the work will be finished on the 11th day.

(c) If C started the work on the first day followed by A and B, then the work will be finished on the 11th day.

(d) None of these

NUMERICAL TYPE QUESTIONS

1. Meera paints walls for a living. She sometimes has Ashok to work with her. Meera takes 2 days to paint a 10 ft by 10 ft wall. Ashok takes 3 days to do the same job. Meera has to pay Ashok for his work on a per day basis, so she'd rather do without him. Meera has recently got a contract to paint the walls and ceiling of a room of dimensions 10 ft by 15 ft and a height of 10 ft. She had promised to do the job within 9 days. The ______ number of days does she need to employ Ashok?

2. 3 pipes can fill a reservoir in 10, 15 and 20 hours respectively. If the three taps are opened one after another in the given order, with a certain fixed time gap between them, the reservoir fills in 5 hours. The time gap will be _____ hr.

3. Two pipes A and B can fill a tank in 8 hours. If only pipe A is open, then it would take 4 hours longer to fill the tank. The _______ hours it would take if only pipe B is open.

Directions (Q.4 – 5): *Refer to the data below and answer the questions that follow.*

The boiler tank in a chemical factory holds 10^5 litres. 5 tanks each having one-fifth the capacity of the boiler tank fill in 'hard water' at same rates in the boiler tank in 2 hours. The outlet of the two of smaller tanks work as inlet pipes and other two work as outlet and the fifth tank fill in the main 'boiler' at half its efficiency.

4. The number of hours taken to fill in the main boiler when all the pipes are opened at once is ______

5. Only three pipes are working, two at their full efficiency and the third one with half of its efficiency and all three are acting as an inlet. If all the three are opened alternately with the latter pipe being first to be opened to fill the boiler such that each smaller tank fills 'hard water' in the main boiler for equal time, then number of hours taken to fill the boiler to half its capacity is______

6. A contract is to be completed in 46 days and 117 men were set to work, each working 8 hours a day. After 33 days, 4/7 of the work is completed.

The _______ number of additional men may be employed so that the work may be completed in time, each man now working 9 hours a day?

7. A train travelling at 42 km/hr. passes a cyclist going in the same direction in 9 secs.; if cyclist had been going in the opposite direction, the train would have passed him in 5 secs. Length of the train is______ metres

8. A man rows upstream 13 km. and downstream 28 km. taking 5 hrs. each time. Velocity of the current in km/hr_______ is

Directions (Q. 9 – 11): *Refer to the data below and answer the questions that follow.*

The variation in the speed of a car on a particular day at the respective times is shown in the table below:

s(km/hr)	t(hr)
0	11.00 am
40	11.30 am
50	1.00 pm
85	1.30 pm
10	3.30 pm
10	4.30 pm

9. The distance travelled by the car from 11 a.m. to 1 p.m. is ______km

10. If car maintains the speed it has at 4.30 p.m. then at ______ time will the car cover the same distance as it had covered from 11.00 a.m. to 1.00 p.m.?

11. Average speed of the car from 11.00 a.m. to 4.30 p.m. is_______ kmph

12. A cheetah chase a deer which is 100 m ahead. The time in which the deer takes 10 leaps the cheetah takes only 6 leaps. In one leap, the deer covers 1 m while the cheetah covers 2 m. In ______ leaps would the cheetah catch up the deer.

13. A hare pursued by a hound who is 50 of her own leaps before him. When the hare takes 4 leaps, the hound takes 3. In one leap, the hare goes $1\frac{3}{4}$ metres and the hound $2\frac{3}{4}$ metres. In how _______ leaps will the hound overtake the hare.

Directions (Q. 14 – 15): *Refer to the graph below and answer the questions that follow.*

The engine of the new car Palio brought by Fiat in the market follows certain mileage characteristics as shown in the diagram below:

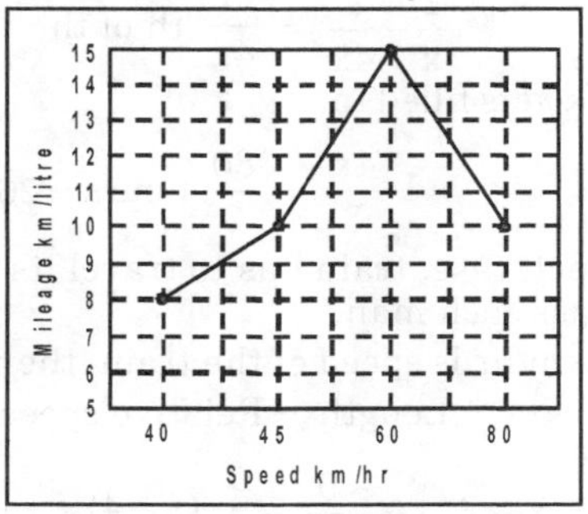

Vijay bought the new car Palio and filled exactly 10 litres in his car. He goes at a speed of 40 km/hr for first 20 km and then at a speed of 60 km/hr for next 60 km and driving at a constant speed reaches his destination which was 115 km away from his starting point.

14. The time in hours taken by Vijay if he reaches his destination in minimum possible time and finishes all the fuel.

15. In the above question, if Vijay had travelled at a constant average speed rather than changing his speed, taking the same minimum time, then the approximate saving in the fuel consumed is ______ litres

16. In a 2 km race on a circular course of $\frac{1}{4}$ of a km. A overlaps B in the middle of his 6th round. Then_____ distance (in km) will A win at the same rate of running

17. A runs $1\frac{3}{8}$ times faster than B. A gives B a start of 120 metres. _______ metre must the winning post be so that it may be a dead heat.

18. The work done by a man, a woman and a child is in the ratio of 3 : 2 : 1. There are 20 men, 30 women and 36 children in a factory. Their weekly wages amount of ₹ 780, which is divided in the ratio of work done by the men, women and children. The wages of 15 men, 21 women and 30 children for 2 weeks is _______

19. A can do a work in 6 days and B in 8 days. With the help of a boy, the three complete the work in 3 days and get ₹ 200. The boy's share in rupees, if the money is distributed in the ratio of the work done is _______

20. A can finish a job in 12 days and B when working at twice his efficiency finishes a job in 9 days. The ______ number of days will they take if they work for two days alternately working at their standard rate given that A starts first.

ANSWERS

MCQ Type Questions

1. (c) **2.** (d) **3.** (c) **4.** (b) **5.** (d) **6.** (c) **7.** (c) **8.** (b) **9.** (b) **10.** (c)

11. (c) **12.** (d) **13.** (a) **14.** (d) **15.** (c)

Numerical Type Questions

1. 6 **2.** $\frac{1}{2}$ **3.** 16 **4.** 20 **5.** 6 **6.** 81 **7.** 75 **8.** 1.5 **9.** 77.5

10. 12.15 a.m. **11.** 39.3 **12.** 300 **13.** 210 **14.** 2 **15.** $2\frac{1}{3}$ **16.** $\frac{4}{11}$

17. 440 **18.** 1170 **19.** 25 **20.** 14

EXPLANATIONS

MCQ TYPE QUESTIONS

1. Top tap is operational till 18 litres is drawn out.

$\therefore$ Time after which the lower tap is usually open $= 18 \times 6 = 108$ minutes.

$\therefore$ Time after which it is open now $= 108 - 24 = 84$ minutes.

$\therefore$ Litres drawn $= \frac{84}{6} = 14$ litres

$\therefore$ $18 - 14 = 4$ litres

were drawn by the new assistant.

$\therefore$ Time $= 4 \times 4 = 16$ minutes.

Alternatively

Upper tap was not used for 24 minutes.

$\therefore$ Litres drawn $= \frac{24}{6} = 4$ litres.

$\therefore$ Beer was drawn for $4\times 4 = 16$ minutes.

2. Let pipe B be turned-off after x minutes.

Part filled in x minutes $= x\left(\frac{1}{36}+\frac{1}{48}\right) = \frac{7x}{144}$.

Part filled in $24 - x$ minutes $= \frac{24-x}{36}$

$$\therefore \quad \frac{7x}{144}+\frac{24-x}{36} = 1$$

$$\Rightarrow \quad \frac{7x+96-4x}{144} = 1$$

$$\Rightarrow \quad x = 16 \text{ minutes}$$

3. In a cycle of 3 hours : pipes P, Q and R are working as inlet pipes for 2 hours each and they are working as outlet pipes for an hour each. So part of tank filled in 3 hours.

$$= 2x\left(\frac{1}{8}+\frac{1}{10}+\frac{1}{12}\right)-\left(\frac{1}{8}+\frac{1}{10}+\frac{1}{12}\right)$$

$$= \frac{1}{8}+\frac{1}{10}+\frac{1}{12} = \frac{15+12+10}{120} = \frac{37}{120}$$

Hence they will take $\frac{120\times 3}{37} = 9\frac{27}{37}$ hours.

Thus, tank will be filled in 10th hour.

4. Let number of hours taken by the outlet and the 2 inlet pipes be 4x, 2x and x respectively. So in 1 hour the part of empty tank filled is

$$\frac{1}{x}+\frac{1}{2x}-\frac{1}{4x} = \frac{1}{16}$$

$$\frac{5}{4x} = \frac{1}{16}$$

$$\Rightarrow \quad x = 20$$

Hence, inlet pipe with higher efficiency fills the tank in 20 hours.

So in one hour, the desired pipes fill

$$\frac{1}{x}-\frac{1}{4x} = \frac{3}{4x} \text{ th of the tank.}$$

Tank gets filled in

$$\frac{4x}{3} = \frac{80}{3} \text{ hours} = 26\frac{2}{3} \text{ hours.}$$

5. In each case, train has to travel its own length, to pass each man.

If x km/hr is speed of the train, then

Length = Relative speed $\times$ Time

$$= (x-2)\times\frac{9}{60\times 60} = (x-4)\times\frac{10}{60\times 60}$$

$$\therefore \quad 9x - 18 = 10x - 40$$

$$\Rightarrow \quad x = 22 \text{ km/hr.}$$

$$\therefore \quad \text{Length} = (22-2)\times\frac{9}{3600}$$

$$= \frac{20 \times 9}{3600} \text{ km} = 50 \text{ m.}$$

6. Policeman gains $= \frac{1}{6} - \frac{1}{10} = \frac{4}{60}$ km/min

$$= 66\frac{2}{3} \text{ m/min.}$$

Hence to gain100 m, time required

$$= \frac{100}{66\frac{2}{3}} = 1\frac{1}{2} \text{ min.}$$

Hence thief has gone ahead by

$$1\frac{1}{2} \times \frac{1}{10} \times 1000 \text{ metres} = 150 \text{ m.}$$

7. I. Ratio of speed = 1 : 2 : 4

If equal time is spent on each of these, then ratio of distance = 1 : 2 : 4

So, Sachin runs twice the distance that he jogs. Thus, I is true.

II. He runs at 8 km/hr and therefore walks at 2 km/hr.

Hence II is not true.

III. Ratio of speed= 1 : 2 : 4

Ratio of time = 2 : 2 : 1

Ratio of distance

$= 2 \times 1 : 2 \times 2 : 4 \times 1 = 2 : 4 : 4$

If Sachin covers a total distance of 10 km, then he

Walks for $\frac{2}{2+4+4} \times 10 = 2$ km,

Joggs for $\frac{4}{2+4+4} \times 10 = 4$ km.

Runs for $\frac{4}{2+4+4} \times 10 = 4$ km.

Total time taken is $2\frac{1}{2}$ hours $= \frac{5}{2}$ hours.

Thus, he walks for $\frac{2}{2+2+1} \times \frac{5}{2} = 1$ hour.

jogs for $\frac{2}{2+2+1} \times \frac{5}{2} = 1$ hour.

runs for $\frac{1}{2+2+1} \times \frac{5}{2} = \frac{1}{2}$ hour.

So speed while,

Walking $= \frac{2}{1} = 2$ km/hr

Jogging $= \frac{4}{1} = 4$ km/hr

Running $= \frac{4}{\frac{1}{2}} = 8$ km/hr.

Thus, III is true.

8. In 1 minute, monkey ascends 10 metres but he takes 1 minute to slip down 2 metres. Thus, at the end of 2 minutes, net ascending of the monkey is =10 – 2 = 8 metres.

Thus, to have a net ascending of 8 metres, process of ascending and then slipping happens once. So, to cover 64 metres, above process is repeated $\frac{64}{8}$ or 8 times.

It is clear that in 8 such happenings, the monkey will slip 7 times, because 8th time, he will ascend to the top. Thus, in climbing 7 times and slipping 7 times, he covers (7 × 8) or 56 metres.

Time taken to cover 56 metres

$$= \frac{56 \times 2}{8} = 14 \text{ minutes}$$

Remaining distance = 64 – 56 = 8 metres

Time taken to ascend 8 metres

$$= \frac{8}{10} \text{ min} = \frac{4}{5} \text{ min.}$$

Total time taken = 14 minutes $+ \frac{4}{5}$ min

= 14 mins. 48 sec.

9. If they meet t hrs after 7 a.m,

$$10 \times \frac{18}{5} \times t + 10 \times \frac{18}{5} \times \frac{4}{3}\left(t - \frac{1}{3}\right) = 68$$

$\Rightarrow$ 36t + 48t – 16 = 68

$\Rightarrow$ t = 1 hr.

They meet at a distance of

10 × 60 × 60 = 36000 m

i.e., 36 km from A.

10. Two equations are

$$T_1 - T_2 = \frac{1}{2} + \frac{1}{2} = 1 \quad ...(i)$$

where T_1 and T_2 are time taken by trains A and B to cover the whole distance

and $$\frac{2}{T_1} + \frac{1.5}{T_2} = \frac{11}{30} \quad ...(ii)$$

Solving equations (*i*) and (*ii*) ,we get

T_1 = 10 hrs. and T_2 = 9 hrs.

11. When A runs 5 rounds, B runs 4 rounds (ratio of speeds)

∴ A passes B each time A has run 5 rounds or

$5 \times \frac{1}{4} = \frac{5}{4}$ km. $= 1\frac{1}{4}$ km.

$1\frac{1}{4}$ km is contained in 4 km.

3 times.

Hence A passes B thrice.

12. When A has run 250 yards, B should have run

$$550 - 250 = 300 \text{ yards.}$$

$\therefore$ B runs 300 – 100 = 200 yards in the same time taken by A to run 250 yards.

$\therefore$ Ratio of the speeds of A and B

$$= 250 : 200 = 5 : 4$$

If speed of A is 5x yards per minute, then speed at B = 4x yards/min.

Time taken by B to reach the starting point

$$= \frac{450}{4x} = \frac{112.5}{x} \text{ min.}$$

Time taken by A to reach the starting point

$$= \frac{550}{5x} = \frac{110}{x} \text{ min.}$$

$\therefore$ A wins by $\frac{2.5}{x}$ min

$\therefore$ Distance travelled by B in $\frac{2.5}{x}$ min.

$$= \left(\frac{2.5}{x}\right) \times (4x) = 10 \text{ yards}$$

Hence B is 10 yards behind A.

13. Let n_1 and n_2 be the speeds in km/hr and let n_1 be the greater speed i.e., $n_1 > n_2$.

$$\therefore \quad \frac{2}{n_2} = \frac{2}{n_1} + \frac{1}{60}.$$

$$\Rightarrow \quad n_1 n_2 = 120(n_1 - n_2)$$

Faster runner covers $(n_1 - n_2)$ km. more than the slower runner in 1 hour.

Hence to cover one round of 2 km, he will take

$\frac{2}{n_1 - n_2}$ or $\frac{1}{2}$ hour.

$$\therefore \quad \frac{2}{n_1 - n_2} = \frac{1}{2}$$

$$\Rightarrow \quad n_1 - n_2 = 4$$

$$\Rightarrow \quad n_1 = 4 + n_2$$

$$\Rightarrow \quad n_2(4 + n_2) = 120 \times 4$$

$$\Rightarrow \quad n_2^2 + 4n_2 - 480 = 0$$

$$\Rightarrow \quad n_2 = 20 \text{ or } -24 \text{ (rejected)}$$

$$\therefore \quad n_1 = 4 + n_2 = 24 \text{ km/hr}$$

Hence speeds are 24 km/hr and 20 km/hr

Alternatively

Options may also be substituted to find the result.

14. Work done by A and B in 1 hr

$$= \frac{1}{4} + \frac{1}{5} = \frac{9}{20}^{th} \text{ part.}$$

$\therefore$ Ratio of work done by (A + B)

and $\quad C = \frac{9}{20} : \frac{1}{6}$

$\therefore$ Ratio of time taken

$$= \frac{1}{6} : \frac{9}{20} = 10 : 27$$

If C takes 27 hrs. (A + B) will take 10 hrs.

If C takes $4\frac{1}{2}$ hrs. (A + B) will take

$$\frac{10 \times 9}{2 \times 27} \cong 1 \text{ hr. } 40 \text{ mins.}$$

15. I. Work done by A, B and C together in 3 days

$$\frac{1}{8} + \frac{1}{12} + \frac{1}{16} = \frac{13}{48} \text{ th of the work.}$$

Thus, in 9 days, total work done will be

$$3 \times \frac{13}{48} = \frac{39}{48} \text{ th of the work}$$

$\therefore$ Work remaining $= 1 - \frac{39}{48} = \frac{9}{48}$ th of the work is left of which A will finish $\frac{6}{48}$ th of work on 10th day and B will finish remaining work on 11th day.

Thus, I is false

II. Work done by A, B, C together in 9 days $= \frac{39}{48}$

$$\text{Work left} = 1 - \frac{39}{48}$$

$$= \frac{9}{48} \text{ th of the whole.}$$

On the 10th day B will finish $\frac{1}{12}$ th of the whole.

$$\therefore \quad \text{Work left} = \frac{9}{48} - \frac{1}{12}$$

$$= \frac{5}{48} \text{ th of the whole.}$$

On 11th day C will finish $\frac{1}{16}$ th of the work

$$\therefore \quad \text{Work left} = \frac{5}{48} - \frac{1}{16}$$

$$= \frac{2}{48} \text{ th of the whole}$$

which will be done by A on the 12th day. Thus, work will be finished on the 12th day. Thus, II is false.

III. Work done by A, B and C in 9 days $= \frac{39}{48}$

Work left after 9 days

$$= 1 - \frac{39}{48} = \frac{9}{48} \text{ th of the whole.}$$

On the tenth day $\frac{1}{16}$ th will be finished by C.

$\therefore$ Work left $= \frac{9}{48} - \frac{1}{16}$

$= \frac{6}{48}$ th of the whole,

which will be finished by A on the eleventh day. Thus, III is true.

NUMERICAL TYPE QUESTIONS

1. Total area to be painted

$= 10 \times 15 \text{ (ceiling)} + 2 \times 10 \times 10 \text{ (wall)} + 2 \times 10 \times 15 \text{ (wall)}$

$= 650$ sq.ft

In one day Meera can paint

$10 \times \frac{10}{2} = 50$ sq.ft

Ashok can paint

$10 \times \frac{10}{3} = \frac{100}{3}$ sq.ft,

So, together they can paint $\frac{250}{3}$ sq.ft per day.

Let number of days Ashok works be x.

Work done in 9 days $= 50(9 - x) + \frac{250}{3}x = 650$

$\therefore \quad 450 - 50x + \frac{250x}{3} = 650$

$\Rightarrow \quad \frac{100}{3}x = 200$

$\Rightarrow \quad x = 6$ days.

2. Let fixed interval be x

$\therefore \quad \frac{5}{10} + \frac{5-x}{15} + \frac{5-2x}{20} = 1$

$\Rightarrow \quad (5)6 + (5-x)4 + (5-2x)3 = 60$

$\Rightarrow \quad 30 + 20 - 4x + 15 - 6x = 60$

$\Rightarrow \quad x = \frac{1}{2}$ hour.

3. Let pipe B alone take b hours longer than if both A and B were open.

$8 = \sqrt{4 \times x}$

$\Rightarrow \quad 64 = 4 \times x$

$\Rightarrow \quad x = 16$ hours.

Alternatively

Part filled if both A and B are open $= \frac{1}{8}$

Part filled if only A is open $= \frac{1}{12}$

$\therefore$ Part filled if only B is open $= \frac{1}{8} - \frac{1}{12} = \frac{1}{24}$

$\therefore$ B alone would take 24 hours, i.e., B alone would take $24 - 8 = 16$ hours longer to fill the tank.

4. Since 5 tanks of same efficiency fill the boiler in 2 hours each tank individually takes

$2 \times 5 = 10$ hours to fill it.

Tank with half the efficiency will take

$10 \times 2 = 20$ hours.

In one hour, fraction of boiler getting filled when 4 fully efficient pipes with two as inlet and two as outlet and 5th acts at 50% efficiency as an inlet is

$$\frac{1}{10} + \frac{1}{10} + \frac{1}{20} - \frac{1}{10} - \frac{1}{10} = \frac{1}{20}$$

Hence it will take 20 hours to fill the boiler tank.

5. Here fraction of tank filled in 1 hour

$$\frac{1}{10} + \frac{1}{10} + \frac{1}{20} = \frac{5}{20} = \frac{1}{4} \text{th}$$

of the tank.If they are opened alternately, then tank gets filled in $4 \times 3 = 12$ hours

Hence half of the tank gets filled in $\frac{12}{2} = 6$ hours.

6. 4/7 of work is completed.

Remaining work $= 1 - \frac{4}{7} = \frac{3}{7}$

Remaining period $= 46 - 33 = 13$ days

Less work, less men(direct proportion)

Less days, more men(Indirect proportion)

More hours/day, less men(Indirect proportion)

work 4/7 : 3/7

Days 13 : 33 :: 117 : x

hrs/day 9 : 8

$\frac{4}{7} \times 13 \times 9 \times x = \left(\frac{3}{7}\right) 33 \times 8 \times 117$

$x = (3 \times 33 \times 8 \times 117) / (4 \times 13 \times 9)$

$= 198$ men

So, additional men to be employed

$= 198 - 117 = 81$

7. Let L is length of train in km and x is speed of the cyclist, Then

(i) $\frac{3600}{9} L = 42 - x = 400L$

(ii) $\frac{3600}{5} L = 42 + x = 720 L$

On adding, we get

$1120 L = 84$

$\Rightarrow \quad L = \frac{84 \times 1000}{1120} = 75\text{m}$

8. $\frac{13}{\text{Speed upstream}} = 5$

and $\frac{28}{\text{Speed downstream}} = 5$

$\therefore$ Speed upstream $= \frac{13}{5}$ km/hr.

and Speed downstream $= \frac{28}{5}$ km/hr.

$\therefore$ Speed of river $= \frac{1}{2}\left(\frac{28}{5} - \frac{13}{5}\right)$

$= \frac{1}{2} \times \frac{15}{5} = 1.5$ km/hr

9. Graph of speed vs time is plotted as shown. Area under the graph and time axis gives the distance.

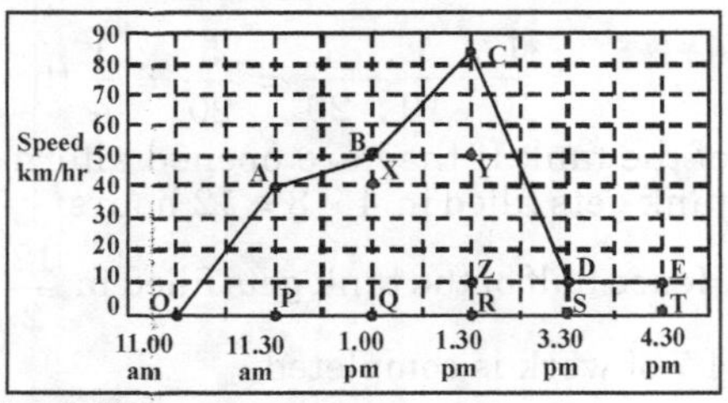

Required distance

$= A(\Delta\ OAP) + A(\Delta\ BXA) + A(\square\ APQX)$

$= \frac{1}{2} \times \left(\frac{1}{2}\right) \times 40 + \frac{1}{2} \times 10 \times \left(\frac{3}{2}\right) + \left(\frac{3}{2}\right) \times 40$

$= 10 + \frac{30}{4} + 60 = 77.5$ km.

10. Distance covered are the same.

Speed in the latter is maintained at 10 kmph.

$\therefore$ $77.5 = 10 \times x$

$\Rightarrow$ $x = 7.75$ hours.

Hence, clock time will be 4.30 + 7 hours 45 mins

= 12.15 a.m.

11. Average speed $= \frac{\text{total distance travelled}}{\text{total time taken}}$

Distance travelled from 1.00 pm to 4.30 p.m.

$= A(\square BQRY) + A(\Delta BYC) + A(\Delta CZD) + A(\square ZRTE)$

$= 50 \times \left(\frac{1}{2}\right) + \left(\frac{1}{2}\right) \times \left(\frac{1}{2}\right) \times 35 + \frac{1}{2} \times (2) \times 75 + 3 \times 10$

$= \frac{50}{2} + \frac{35}{4} + 75 + 30$

$= 130 + 8.75$

$= 138.75$ km.

Total distance travelled $= 77.5 + 138.75$

$= 216.25$ km.

$\therefore$ Average speed $= \frac{216.25}{5.5} = 39.3$ kmph.

12. While deer makes 10 leaps, cheetah makes 6 leaps.

$\therefore$ Distance covered in 10 leaps by deer

$= 10 \times 1 = 10$m

Distance covered in 6 leaps by cheetah

$= 6 \times 2 = 12$ m

Hence in six leaps, cheetah gains

$12 - 10 = 2$ m over the deer.

Cheetah has to gain 100 m over the deer.

So, total leaps required by cheetah

$= \frac{6}{2} \times 100 = 300.$

13. 50 leaps of the hare $50 \times 1\frac{3}{4} = \frac{175}{2} = 87.5$ m.

So, hound should gain $\frac{175}{2}$ m over the hare.

When hound travels $3 \times 2\frac{3}{4}$ m, then

hare travels $4 \times 1\frac{3}{4}$ m.

Hence in 3 leaps of the hound the hound gains

$\frac{33}{4} - \frac{28}{4} = \frac{5}{4}$ metres

$\therefore$ Number of leaps required $= \frac{\frac{175}{2}}{\frac{5}{4}} \times 3$

$= \frac{175}{2} \times \frac{4}{5} \times 3 = 210$ leaps.

14. Fuel required for first 20 km $= \frac{20}{8} = 2.5$ litres.

Fuel required for next 60 km $= \frac{60}{15} = 4$ litres.

$\therefore$ Petrol left $= 10 - 6.5 = 3.5$ litres.

Now, he has to travel $= 115 - (60 + 20)$

$= 35$ km in 3.5 litres.

So he can go at a speed of either 45 km/hr or 80 km/hr as mileage is 10 km/litre in both cases. But since he reaches in minimum time, he travels at 80 km/hr

$\therefore$ Time taken $= \frac{20}{40} + \frac{60}{60} + \frac{35}{80} = \frac{155}{80}$

$= 1 + \frac{75}{80}$ hours

= 1 hours 56 minutes and 25 seconds.

15. Total distance travelled = 115 km

Total time taken $= \frac{155}{80}$ hours

$$\therefore \text{ Average speed} = \frac{115}{\frac{155}{80}} = 60 \text{ km/hr}$$

$$\therefore \text{ Fuel consumption} = \frac{115}{15} = 7\frac{2}{3} \text{ litres.}$$

So, fuel saved = $10 - 7\frac{2}{3} = 2\frac{1}{3}$ litres.

16. A overlaps B in the middle of the 6th round

So, when A runs $5\frac{1}{2}$ rounds, B runs only $4\frac{1}{2}$ rounds.

$$\therefore \quad \frac{\text{A's Speed}}{\text{B's Speed}} = \frac{5\frac{1}{2}}{4\frac{1}{2}} = \frac{11}{9}$$

So when A runs 11 km B runs 9 km.

Thus, when A runs 2 km, B runs $2 \times \frac{9}{11} = \frac{18}{11}$ km.

$$\therefore \text{ A wins by } \left(2 - \frac{18}{11}\right) = \frac{4}{11} \text{ km.}$$

17. Speeds of A and B are as

$$1\frac{3}{8} : 1 = \frac{11}{8} : 1 = 11 : 8$$

In a race of 11 metres, A gain

$11 - 8 = 3$ metres over B.

To gain 120 metres, race must be of

$$120 \times \frac{11}{3} = 440 \text{ metres.}$$

18.

	Work	Number
Men	3	20
Women	2	30
Children	1	36

Ratio of wages = $(3 \times 20) : (2 \times 30) : (1 \times 36)$

$= 5 : 5 : 3$

Total wages of 20 men = $\frac{5}{13} \times 780 =$ ₹ 300

∴ Wages of a man = ₹ 15.

Similarly, wages of 30 women

$\frac{5}{13} \times 780 =$ ₹ 300

Wages of a woman = ₹ 10

Wages of 36 children $\frac{3}{13} \times 780 =$ ₹ 180

Wages of a child = ₹ 5

∴ Total wages of 15 men, 21 women and 30 children

$= 15 \times 15 + 21 \times 10 + 30 \times 5$

= ₹ 585

∴ Total wages for 2 weeks = ₹ 1170.

19. Work done by A in 3 days

$= 3 \times \frac{1}{6} = \frac{1}{2}$ of the work.

Work done by B in 3 days

$= 3 \times \frac{1}{8} = \frac{3}{8}$ of the work.

∴ Work done by the boy in 3 days

$= 1 - \left(\frac{1}{2} + \frac{3}{8}\right)$

$= \frac{1}{8}$ of the work.

∴ Ratio of work done

$= \frac{1}{2} : \frac{3}{8} : \frac{1}{8} = 4 : 3 : 1$

∴ Boy's share = $\frac{1}{8} \times$ ₹ 200 = ₹ 25.

20. A finishes the job at standard rate in 12 days,

So in 1 day, the part of job done by A = $\frac{1}{12}$

Similarly the part of job done by B in 1 day

$= \frac{1}{2 \times 9} = \frac{1}{18}$

If a works for 2 days followed by 2 days of B at standard rate.

Part of work done = $\frac{2}{12} + \frac{2}{18} = \frac{5}{18}$.

So, in (2 alternate days work of 'A' and 'B')

$3 \times 4 = 12$ days

They will finish $3 \times \frac{5}{18} = \frac{15}{18}$th of the work.

Work remaining = $1 - \frac{15}{18} = \frac{3}{18} = \frac{1}{6}$

This is finished by A alone in $\frac{\frac{1}{6}}{\frac{1}{12}} = 2$ days.

Hence, total number days = 12 + 2 = 14 days.

4

CHAPTER

Ratio, Proportion and Mixtures

RATIO

Ratio of two terms 'a' and 'b' is denoted by a : b (read as 'a is to b') and is measured by $\frac{a}{b}$.

Numerator 'a' is called *'antecedent'* and denominator 'b' is called 'consequent'.

While comparing two quantities in terms of ratio :

(*i*) Two quantities must be of the same kind.

(*ii*) Units of measurement of the two quantities must be the same.

e.g., if a = ₹ 2 and b = 50 paise,

then a : b = 200 : 50 = 4 : 1.

(*iii*) Ratio is a pure number, i.e., without any unit of measurement as it denotes how many times a given quantity is in comparison to another.

(*iv*) Ratio would stay unaltered even if both antecedent and the consequent are multiplied or divided by the same number.

e.g., $a : b = \frac{a}{b} = \frac{a \times m}{b \times m} = \frac{\frac{a}{m}}{\frac{b}{m}}$.

Properties of Ratios

1. **Invertendo :** If $\frac{a}{b} = \frac{c}{d}$, then $\frac{b}{a} = \frac{d}{c}$ i.e., inverse ratios of two equal ratios are equal.

2. **Alternendo :** If $\frac{a}{b} = \frac{c}{d}$, then $\frac{a}{c} = \frac{b}{d}$ i.e., ratios of antecedents and consequents of two equal ratios are equal.

3. **Componendo :** If $\frac{a}{b} = \frac{c}{d}$, then $\frac{a+b}{b} = \frac{c+d}{d}$ i.e., adding 1 to both sides.

4. **Dividendo :** If $\frac{a}{b} = \frac{c}{d}$, then $\frac{a-b}{b} = \frac{c-d}{d}$ i.e., subtracting 1 from both sides.

5. **Componendo-Dividendo :** If $\frac{a}{b} = \frac{c}{d}$, then $\frac{a+b}{a-b} = \frac{c+d}{c-d}$ i.e., dividing results of componendo by dividend.

6. **Equal Ratio :** If $\frac{a}{b} = \frac{c}{d} = \frac{e}{f}$; then each ratio

$$= \frac{\text{Sum of Numerators}}{\text{Sum of Denominators}} = \frac{a+c+e...}{b+d+f...}$$

The principle can also be applied after multiplying Numerator and Denominator of any fraction by the same number.

If $\frac{a}{b} = \frac{c}{d} = \frac{e}{f} = ...$, then each ratio is equal to $\left(\frac{pa^n + qc^n + re^n + ...}{pb^n + qd^n + rf^n + ...}\right)^{1/n}$, where p, q, r, n ... may have any values except that they must not all be zeros.

In the ratio a : b, a is called the antecedent (the first term) and b is called the consequent (the second term)

Note:

The ratio of two numbers a and b, written as a : b, is the fraction $\frac{a}{b}$ provided $b \neq 0$. Ratio is always reduced to its simplest form. Thus $a : b = \frac{a}{b}$, $b \neq 0$. if $a = b \neq 0$, the ratio is 1 : 1 or 1/1 = 1

1. The ratio of 4 to 6 = 4 : 6 = $\frac{4}{6} = \frac{2}{3}$

2. $\frac{2}{3} : \frac{4}{5} = \frac{2/3}{4/5} = \frac{4}{6}$

3. $5x : \frac{3y}{4} = \frac{5x}{3y/4} = \frac{20x}{3y}$

Type of Ratio

Compounded Ratio

When two or more ratios are multiplied term wise, the ratio thus obtained is called their compounded ratio. For the ratios a : b and c : d, the compounded ratio is ac : bd.

Example: What is the compounded ratio, for 2 : 3 and 4 : 5?

Solution: The compounded ratio is 2 × 4 : 3 × 5 or 8 : 15.

Duplicate Ratio

It is the compounded ratio of two equal ratios. Thus the duplicate ratio of

a : b is a^2 / b^2 or $a^2 : b^2$

Example: Find the duplicate ratio of 4 : 5.

Solution: The duplicate ratio of 4 : 5 is 16 : 25.

Triplicate Ratio

It is the compounded ratio of three equal ratios. Thus the triplicate ratio of a : b is a^3 / b^3 or $a^3 : b^3$

Example: Find the triplicate ratio of 4 : 5.

Solution: The triplicate ratio of 4 : 5 is 64 : 125.

Sub-Duplicate Ratio

For any ratio a : b its sub-duplicate ratio is defined as $\sqrt{a} : \sqrt{b}$.

Example: Find the sub-duplicate ratio of 16 : 25.

Solution: Sub-duplicate ratio of 16 : 25 is $\sqrt{16} : \sqrt{25}$, i.e. 4 : 5.

Sub-Triplicate Ratio

For any ratio a : b, its sub-triplicate ratio is defined as $\sqrt[3]{a} : \sqrt[3]{b}$.

Example:

Find the sub-triplicate ratio of 27 : 64.

Solution:

Sub-triplicate ratio of 27 : 64 is $\sqrt[3]{27} : \sqrt[3]{64}$ i.e. 3 : 4.

Comparing Ratios

To compare two ratios, we express them as fractions and then compare.

Example:

Which is greater 3 : 4 or 4 : 5?

Solution:

$3 : 4 = \frac{3}{4}$ and $4 : 5 = \frac{4}{5}$.

$\frac{3}{4} = \frac{3 \times 5}{4 \times 5} = \frac{15}{20}$

$\frac{4}{5} = \frac{4 \times 4}{5 \times 4} = \frac{16}{20}$

Since $\frac{16}{20} > \frac{15}{20}$ = or, $\frac{4}{5} > \frac{3}{4}$,

So, 4 : 5 > 3 : 4.

Note: A ratio is said to be in its simplest form if the HCF of the antecedent and the consequent is 1.

Example:

Divide 2400 in the ratio 3 : 5.

Solution:

The first part is 3 units and the second part is 5 units. The total of both the parts = 3 units + 5 units = 8 units.

Here, 8 units = 2400,

So, 1 unit = $\frac{2400}{8}$ = 300.

The first part = 3 units = 3 × 300 = 900.

The second part = 5 units = 5 × 300 = 1500.

Example:

A sum of money is divided between Vinod and Lokesh in the ratio of 3 : 7. Vinod gets Rs. 240. What does Lokesh get?

Solution:

Vinod gets 3 units = Rs. 240.

So, 1 unit = $\frac{240}{3}$ = 80.

Therefore, 7 units = 7 × 80 = 560.

Thus, Lokesh gets Rs. 560.

Problems Leading to the Application of Ratios

Example:

The ratio of the number of boys to the number of girls in a school of 1638 is 5 : 2. If the number of girls increased by 60, then what must be the decrease in the number of boys to make the new ratio of boys to girls as 4 : 3?

Solution:

Number of boys = $\frac{5}{7} \times 1638 = 1170$

Number of girls = $\frac{2}{7} \times 1638 = 468$

Number of girls after increase

= 468 + 60 = 528

Total number of boys as per new ratio

= 528 × $\frac{4}{3}$ = 704

∴ The number of boys to be decreased

= 1170 – 704 = 466.

Example:

If a : b = 3 : 4 and b : c = 6 : 13, then find a : b : c.

Solution:

The best way to solve such questions is to make b common in the two ratios.

Thus, we can write a : b = 9 : 12 and b : c = 12 : 26.

Now that b is equal in both the ratios, we can write the same as

a : b : c

9 : 12

12 : 26

Thus, we can write a : b : c = 9 : 12 : 26.

Using formula directly, we can get

a : b : c = (3 × 6) : (4 × 6) : (4 × 13) = 9 : 12 : 26

PROPORTION

When ratio of two terms is equal to the ratio of two other terms, then these four terms are said to be in **proportion,** i.e., if a : b = c : d, then a, b, c and d are in proportion.

a, b, c and d are called first, second, third and fourth proportionals respectively. The terms a and d are called **extremes,** while b and c are called **means.**

If a : b = c : d, then ad = bc i.e., product of extremes is equal to product of means.

Continued Proportion :

When $\frac{a}{b} = \frac{b}{c}$, then a, b and c are said to be in continued proportion and b is called *geometric mean* or *mean proportional* between a and c.

Also, $b^2 = a \times c$, $\Rightarrow$ $b = \sqrt{ac}$

If $\frac{a}{b} = \frac{b}{c} = \frac{c}{d} = \frac{d}{e} = ...$, then a, b, c, d, e ... are said to be in continued proportion.

Concept of continued proportion is useful in the following situation.

(*i*) If four quantities a, b, c and d and a : b, b : c and c : d, are know, then we can find a : d as $\frac{a}{d} = \frac{a}{b} \times \frac{b}{c} \times \frac{c}{d}$

(*ii*) If a, b, c and d are four quantities and a : b, b : c, c : d are known, then we can find a : b : c : d.

e.g., If a : b = 2 : 3; b : c = 4 : 5, c : d = 7 : 9, find a : b : c : d.

a : b = 8 : 12, b : c = 12 : 15 (LCM of 3 and 4 is 12)

∴ a : b : c = 8 : 12 : 15

a : b : c = 56 : 84 : 105, c : d = 105 : 135 (LCM of 7 and 15 is 105)

∴ a : b : c : d = 56 : 84 : 105 : 135

Example:

Find the fourth proportional to $12a^2$, $9a^2b$ and $6ab^2$.

Solution:

Let x be the fourth proportional, then

$$\frac{12a^2}{9a^2b} = \frac{6ab^2}{x} \Rightarrow x = \frac{6ab^2 \cdot 9a^2b}{12a^2} = \frac{9}{2}ab^3$$

Example:

Find the mean proportional between $9a^2b$ and $25b^3$.

Solution:

Let x be the mean proportional.

$$\therefore \frac{9a^2b}{x} = \frac{x}{25b^3} \Rightarrow x^2 = 9a^2b \cdot 25b^3$$

$$\Rightarrow x = \pm 15ab^2$$

VARIATION

Most of us would still remember statements like "The distance travelled varies directly as the speed of the vehicle." These and similar statements have precise mathematical meanings and they represent a specific type of function called **variation functions**.

The three general types of variation functions are direct variation, inverse variation and joint variation.

Direct Variation

If two quantities X & Y are related such that any increase or decrease in 'Y' produces a proportionate increase or decrease in 'X' or vice versa, then the two quantities are said to be in direct proportion.

X is directly proportional to Y is written as $X \propto Y$ or X = KY.

In other words $X : Y = \frac{X}{Y} = K$. Here K is a constant whose value for a particular variation is same.

Consider $X_1 = KY_1$ and $X_2 = KY_2$, dividing the two we get $\frac{X_1}{X_2} = \frac{Y_1}{Y_2}$.

Thus, the chances of your success in the test are directly proportional to the number of hours of sincere work devoted every day.

Example:

If $X \propto Y$ and x = 9 when y = 30, then find the relation between x and y. Find x when $y = 7\frac{1}{2}$ and y when x = 6.

Solution:

Let x = ky, then 9 = k(30), $k = \frac{3}{10}$, i.e. $x = \frac{3}{10}y$.

When $y = 7\frac{1}{2}$, $x = \frac{3}{10}\left(7\frac{1}{2}\right) = 2\frac{1}{4}$.

When x = 6, $x = \frac{3}{10}y$ ∴ y = 20.

Example:

Different sizes of the car have different models. The weight of a car model varies directly as the cube of its length. The weight of a car model of length 3 cm is 10 gm. What is the weight of a car model of length 12 cm?

Solution:

Let W gm be the weight of a car model and L cm be its length.

$\therefore W \propto L^3$ or $W = kL^3$ (where k is a constant)

$\Rightarrow 10 = k(3)^3. \Rightarrow k = \frac{10}{27}$ i.e. $W = \frac{10}{27}L^3$

When L = 12, $W = \frac{10}{27}(12)^3 = 640$ gm.

$\therefore$ The required weight is 640 gm.

Inverse Variation

Here two quantities X & Y are related such that, any increase in X would lead to a decrease in Y or any decrease in X would lead to an increase in Y. Thus the quantities X & Y are said to be inversely related and X is inversely proportional to Y is written as $X \propto \frac{1}{Y}$

or $X = \frac{k}{Y}$ or $XY = k$ (Constant)

Thus, $X_1Y_1 = X_2Y_2$

Or the product of two quantities remains constant. Thus, the chances that you will be able to cheat in a test are inversely proportional to the smartness of the invigilator.

Example:

If y varies inversely as x, and y = 3 when x = 2, then find x when y = 21.

Solution:

$y \propto \frac{1}{x}$ or $y = \frac{k}{x}$ (where k is a constant)

then $3 = \frac{k}{2}$, $k = 6$ i.e. $y = \frac{6}{x}$

When y = 21, $21 = \frac{6}{x}$, $x = \frac{2}{7}$

Joint Variation

(a) A varies jointly as B and C and is denoted by $A \propto BC$ Or $A = kBC$ (where k is a constant).

(b) A varies directly as B and inversely as C and is denoted by $A \propto \frac{B}{C}$ Or $A = \frac{kB}{C}$ (where k is a constant).

(c) If A varies as B when C is constant, and if A varies as C when B is constant then A varies as BC when B and C both vary.

$\therefore A \propto BC$ Or $A = kBC$ (where k is a constant).

Example: Given, a varies as b when c is constant, and as c^2 when b is constant. If a = 770, then b = 15 & c = 7, and when c = 3 & a = 132 find b.

Solution:

As $a \propto bc^2$ or $a = kbc^2$ (where k is a constant)

$\therefore 770 = k(15)(7)^2$

$\Rightarrow k = \frac{22}{21}$ i.e. $a = \frac{22}{21}bc^2$. When c = 3, a = 132.

$\therefore 132 = \frac{22}{21}(b)(3)^2$ $\therefore b = 14$

Note:

1. If $A \propto B$ and $B \propto C$, then $A \propto C$
2. If $A \propto C$ and $B \propto C$, then $(A \pm B) \propto C$ $\sqrt{(AB)} \propto C$
3. If $A \propto BC$, then $\frac{A}{C} \propto B$ and $\frac{A}{B} \propto C$.
4. If $A \propto B$ and $C \propto D$, then $AC \propto BD$.
5. If $A \propto B$, then $A^n \propto B^n$.
6. If $A \propto B$ and $A \propto C$, then $A \propto (B - C)$ and $A \propto (B + C)$
7. If $A \propto B$, then $AP \propto BP$ where P is any quantity, constant or variable.

MIXTURES AND ALLIGATIONS

Simple Mixture : When two different ingredients are mixed together, it is called *simple mixture.*

Compound Mixture : When two or more simple mixtures are mixed together to form another mixture, it is called compound mixture.

Alligation Rule

Alligation rule states that "When different quantities of the same or different ingredients, of different costs (one cheap and other dear) are mixed together to produce a mixture of a mean cost, then ratio of their quantities is inversely proportional to the difference in their cost from the mean cost."

$$\frac{\text{Quantity of Cheap}}{\text{Quantity of Dear}} = \frac{\text{Price of Dear} - \text{Mean Price}}{\text{Mean Price} - \text{Price of Cheap}}$$

Application of Alligation Rule :

(1) To find mean value of a mixture when prices of two or more ingredients, which are mixed together and the proportion in which they are mixed are given.

(2) To find proportion in which ingredients at given prices must be mixed to produce a mixture at a given price.

Some Important Results :

(1) When x_1 quantity of ingredient A of cost C_1 and x_2 quantity of ingredient B of cost C_2 are mixed, the,cost of the mixture,

$$C_m = \frac{C_1x_1 + C_2x_2}{x_1 + x_2}$$

(2) Similarly, when more than two ingredients are mixed, then $C_m = \dfrac{C_1x_1 + C_2x_2 + C_3x_3 + ...C_nx_n}{x_1 + x_2 + x_3 + ...x_n}$

(3) When two mixtures M_1 and M_2 each containing ingredient A and B in the ratio a : b and x : y, respectively, are mixed, then proportion of the ingredients A and B i.e., $q_A : q_B$, in the compound mixture is given by.

$$\frac{q_A}{q_B} = \frac{M_1 \times \left(\frac{a}{a+b}\right) + M_2 \times \left(\frac{x}{x+y}\right)}{M_1 \times \left(\frac{b}{a+b}\right) + M_2 \times \left(\frac{y}{x+y}\right)}$$

and, quantity in which M_1 and M_2 is to be mixed when quantity of A and B i.e., q_A and q_B in the compound mixture is known is given by

$$\frac{\text{Quantity of } M_1}{\text{Quantity of } M_2} = \frac{\left(\frac{x}{x+y}\right) - \left(\frac{q_A}{q_A + q_B}\right)}{\left(\frac{q_A}{q_A + q_B}\right) - \left(\frac{a}{a+b}\right)}$$

(4) When a mixture of three ingredients A, B and C is given, take any two ingredients such that cost of the mixture is between costs of the two chosen ones and find the ratio. Once again, take two more ingredients and find their ratio. Then find combined ratio. This will give infinite number of solutions.

(5) If a vessel contains 'a' litres of liquid A, and 'b' litres be withdrawn and replaced by liquid B, then if 'b' litres of mixture be withdrawn and replaced by liquid B, and operation repeated 'n' times in all, then

$$\frac{\text{Liquid A left after } n^{th} \text{ operation}}{\text{Initial quantity of liquid A in vessel}}$$

$$= \frac{\left(\frac{a-b}{a}\right)^n}{1 - \left(\frac{a-b}{a}\right)^n}$$

$$\frac{\text{Liquid A left after } n^{th} \text{ operation}}{\text{Liquid B left after } n^{th} \text{ operation}}$$

$$= \frac{\left(\frac{a-b}{a}\right)^n}{1 - \left(\frac{a-b}{a}\right)^n}$$

Example: In what ratio should tea @ 35 per kg be mixed with tea @27 per kg so that mixture may cost Rs .30 per kg?

Solution:

$$\frac{\text{Quantity of cheaper}}{\text{Quantity of dearer}} = \frac{35-30}{30-27} = \frac{5}{3}$$

Hence the two should be mixed in the ratio 3 : 5, the above ratio calculated is 3 : 5 and not 5 : 3, i.e. tea at 35 per kg to have 3 parts and tea at 27 per kg to have 5 parts. This may be checked mentally by the following simple rule: "If mean price is closer to cheaper C.P, then quantity of cheaper will be more and vice versa."

Example: In what ratio should two different types of mixtures containing milk and water in the ratio of 5 : 1 and 2 : 1 respectively be mixed to obtain a final mixture containing milk and water in the ratio 3 : 1?

Solution: Strength of milk in mixture 1 = 5/6 (because there are 5 parts of milk for every 1 part of water. Hence there are 5 parts of milk for a total of 6 parts of the mixture). Strength of milk in mixture 2 = 2/3 (because there are 2 parts of milk for every 1 part of water. Hence there are 2 parts of milk for total of 3 parts of the mixture).

Strength of milk in the final mixture required = 3/4 (because there are 3 parts of milk for every 1 part of water. Hence there are 3 parts of milk for a total of 4 parts of the mixture). As we can see, the mixture 1 is strong in milk as the dearer mixture, mixture 2 as the cheaper mixture and the final mixture as the mean mixture.

$$\frac{\text{Quantity of cheaper}}{\text{Quantity of dearer}} = \frac{\frac{5}{6} - \frac{3}{4}}{\frac{3}{4} - \frac{2}{3}} = \frac{1}{1}$$

Hence, both the types of mixtures will have to be mixed in the ratio of 1:1.

Two mixtures of same ingredients mixed (Compound mixture)

Mixture 1 has ingredients (A and B) in a : b.

Mixture 2 has same ingredients (A and B) in x: y.

$$\Rightarrow \frac{\text{quantity of ingredient A}}{\text{quantity of ingredient B}} = \frac{X}{Y}.$$

Now 'M' unit of Mixture 1 and 'N' units of Mixture 2 are mixed to form a resultant mixture with ingredients (A and B) in the ratio q_A:q_B

Case 1:

When q_A and q_B are to be found out. By alligation rule in the resultant mixture,

$$\frac{\text{Quantity of ingredient A}}{\text{Quantity of ingredient B}} = \frac{q_A}{q_B}$$

$$= \frac{M \times \left(\frac{a}{a+b}\right) + N \times \left(\frac{x}{x+y}\right)}{M \times \left(\frac{b}{a+b}\right) + N \times 8\left(\frac{y}{x+y}\right)}$$

Then, amount of ingredient A in the resultant mixture

$$= \frac{q_A}{(q_A + q_B)} \times (M + N)$$

Amount of ingredient B in the resultant mixture

$$= \frac{q_A}{(q_A + q_B)} \times (M \times N)$$

Case 2:

When M and N are to be found out (i.e. amount of mixtures)

Consider the quantity of any ingredients (Say, A) in all the three mixtures, i.e. quantity of ingredients a in

Mix 1 = $\frac{a}{a+b}$, i.e a out of (a + b)

Mix 2 = $\frac{x}{x+y}$, in the resultant mix = $\frac{q_a}{q_A + q_B}$

By alligation rule we get , in the resultant mixture,

$$\frac{\text{Quantity of 1st mixture}}{\text{Quantity of 2nd mixture}} = \frac{Q_1}{Q_2} = \frac{\left(\frac{x}{x+y}\right) - \left(\frac{q_A}{q_A + q_b}\right)}{\left(\frac{q_A}{q_A + q_B}\right) - \left(\frac{a}{a+b}\right)}$$

Then, amount of 1st mixture in the resultant mixture

$$= \frac{Q_1}{(Q_1 + Q_2)} \times (M + N)$$

Amount of 2nd mixture in the resultant mixture

$$= \frac{Q_2}{(Q_1 + Q_2)} \times (M + N)$$

Example: A tea merchant buys two kinds of tea, the price of the first kind being twice that of the second. He sells the mixture at Rs.14/kg there by making a profit of 40%. If the ratio of the first to second kind of tea in the mixture is 2 : 3, then find the cost price of each kind of tea.

Solution:

The cost of mixture = $\frac{14 \times 100}{140}$ = Rs.10kg.

Ratio in which the cheaper and dearer is mixed = 3:2.

Let the price of cheaper tea be Rs. x/kg and dearer tea be rs.2x/kg.

∴ Applying the alligation rule, we get

$$\frac{3}{2} = \frac{2x - 10}{10 - x}$$

$\therefore 30 - 3x = 4x - 20$

$\therefore 7c = 50$

$\therefore x = \frac{1}{7}$ and cost of dearer tea = Rs. $14\frac{2}{7}$

EXERCISE

MCQ TYPE QUESTIONS

1. A cask of wine when fully filled holds 10 litres. 2 litres of wine is removed and filled with water. Then 4 litres in the solution is replaced with water. Then, 6 and 8 litres respectively. At the end of the 4th operation, the ratio of wine to water is

   (a) $\frac{4!}{(5)^4}$ (b) $\frac{8!}{(10)^4}$

   (c) $\frac{4!}{5^4 - 4!}$ (d) $\frac{8!}{(10)^4 - 8!}$

2. A 20 litre vessel is filled with alcohol. Some of the alcohol is poured out into another vessel of an equal capacity, which is then completely filled by adding water. Then mixture obtained is poured into the first vessel to capacity. Then $6\frac{2}{3}$ litres is poured from the first vessel into second. Both vessels now contain an equal amount of alcohol. How much alcohol was originally poured from the first vessel into the second?

   (a) 9 litres
   (b) 10 litres
   (c) 12 litres
   (d) 12.5 litres

3. What is the ratio compounded of x – y : x + y and $y^2 + xy : x^2 - xy$?

   (a) $y^2 : x^2$ (b) xy : 1

   (c) y : x (d) $\sqrt{x} : \sqrt{y}$

4. If ratio of two natural numbers x and y is 'a' and that of y and x is 'b', then value of 'a + b' is

   (a) greater than 2 (b) lesser than 1

   (c) greater than 1 (d) data insufficient.

5. The charges of a goldsmith is partly fixed and partly variable with the amount in grams of gold. The charge is ₹ 300 for 20 grams of gold and ₹ 550 for 45 grams of gold. Find the charge of work for 100 grams of gold.

(*a*) ₹ 900 (*b*) ₹ 1000
(*c*) ₹ 1100 (*d*) ₹ 9000

Directions (Q.6–7) : *Refer to the data below and answer the questions that follow.*

The ratio of white collar to blue collar employees in a firm is 8 : 3. Also ratio of male employees to female employees is 7 : 4. It is observed that 60% of the white collar employees are males.

6. What is the ratio of female white collar employees to male blue collar employees?
(*a*) 16 : 11 (*b*) 24 : 11
(*c*) 3 : 1 (*d*) Data insufficient

7. If there are 48 female blue collar workers, what is the difference between male white collar employees and female blue collar employees?
(*a*) 288 (*b*) 240
(*c*) 220 (*d*) Data insufficient

8. In a chemical experiment, two NaOH solution bottles are used. Bottle A contains salt and water in the ratio 7 : 3 and bottle B contains salt and water in the ratio 4 : 3. In what proportion should the quantities be taken from A and B to give the 2 : 1 NaOH solution?
(*a*) 2 : 1 (*b*) 10 : 7
(*c*) 20 : 7 (*d*) 1 : 2

9. A triswitch mixer grinder operates in three stages of speed in revolutions per min i.e., rpm. The speed of the motor with some load in the second stage varies as a square of the speed in stage 1 and that in stage 3 varies as a cube of that in stage 2. If mixer motor makes 50 revolutions in 3 minutes in stage 1, its speed in stage 2 and 3 with the same load is 2500 and 12500 rpm respectively. What will be the speed of motor (in rpm) in stage 3, if at the same load motors speed in rpm is 25 revolutions in 6 minutes in stage 1?
(*a*) $\frac{3125}{16}$ rpm (*b*) $\frac{5^5}{2^{10}}$ rpm
(*c*) $\frac{5^{10}}{2^5}$ rpm (*d*) None of these

10. Two bags have certain number of mangoes. If half the mangoes from bag 1 are transferred to bag 2, the ratio of mangoes in bag 1 to bag 2 gets reversed. Now one-third and two-third of the mangoes from the first and the second bag are transferred to the second and first bag respectively. The ratio of mangoes in bag 1 to bag 2 now is
(*a*) 2 : 1 (*b*) 2 : 2
(*c*) 4 : 5 (*d*) 5 : 4

NUMERICAL TYPE QUESTIONS

1. Five litres are drawn from a cask full of wine and it is then filled with water. Five litres of the mixture are drawn and the cask is again filled with water. The quantity of wine now left in the cask to that of the water in it is in the ratio 361 : 39. The cask hold ______ litres.

2. In a mixture of wheat and barley, wheat is 60%. To 400 kg of the mixture a quantity of barley is added and then the wheat is $53\frac{1}{3}$%. The______kg of barley are added?

3. Two jars of capacity 3 and 5 litres are filled with mixtures of alcohol and water. In the smaller jar 25% of the mixture is alcohol and in the larger 25% of the mixture is water. The jars are emptied into a 9 litre cask and remaining volume is filled up with water. Then percentage of alcohol in the cask will be ________

4. A man buys milk at ₹ 5 a litre and after adding water, sells it at ₹ 6 a litre, thereby making a profit of $33\frac{1}{3}$%. The proportion of water to milk in the mixture is_______

5. The percentage of alcohol in two wine tanks is 20 and 40 respectively and rest is grape juice. If a wine glass is filled to its half by the tank 1 and 2 in the ratio 2 : 3 and the rest is filled again by pure alcohol, then the ratio of the grape juice to alcohol in the wine glass is _______

6. In ______ ratio must a grocer mix two varieties of tea worth ₹ 60 a kg and ₹ 65 a kg so that by selling the mixture at ₹ 68.20 a kg he may gain 10%?

7. A container contains 40 lit of milk. From this container 4. lit of milk was taken out and replaced by water. This process was repeated further two times. The ______litres of milk is now contained by the container?

8. A merchant has 1000kg of sugar,part of which he sells at 8% profit and the rest at 18% profit. He gains 14% on the whole. The quantity sold at 18% profit is_____ kg

9. Volumes in the proportion 3 : 4 : 7 of three different substances are mixed together. The densities of equal volume of the substances are in the ratio 5 : 2 : 6 respectively. The weight of the third substance contained in 130 lbs. of the mixture is _______ lbs

10. In a bag, there are 50 paise coins, 25 paise coins and rupee coins, proportional to the numbers 3, 4 and $2\frac{1}{2}$. If they amount to ₹ 210, then number of rupee coins is______

ANSWERS

MCQ Type Questions

1. (*c*) **2.** (*b*) **3.** (*c*) **4.** (*c*) **5.** (*c*) **6.** (*a*) **7.** (*b*) **8.** (*c*) **9.** (*b*) **10.** (*a*)

Numerical Type Questions

1. 100 **2.** 50 **3.** 50 **4.** 1 : 9 **5.** 17 : 33 **6.** 3 : 2 **7.** 29.16 **8.** 600 **9.** 84 **10.** 105

EXPLANATIONS

MCQ TYPE QUESTIONS

1.

Stage	$\frac{\text{Wine left}}{\text{Original quantity}}$
0	$\frac{10}{10}$
1	$\frac{8}{10}$
2	$\frac{8}{10}\times\frac{6}{10}$
3	$\frac{8}{10}\times\frac{6}{10}\times\frac{4}{10}$
4	$\frac{8}{10}\times\frac{6}{10}\times\frac{4}{10}\times\frac{2}{10} \Rightarrow \frac{4\times3\times2\times1}{5^4} = \frac{4!}{5^4}$

$$\therefore \quad \frac{\text{Wine}}{\text{Water}} = \frac{\frac{4!}{5^4}}{1-\frac{4!}{5^4}} = \frac{4!}{5^4-4!}$$

2.

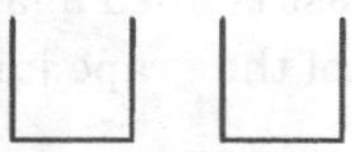

(1) (2)

	C_1	C_2
Operation1	20 litres alcohol	Empty
Operation 2	(20 – x) alcohol	x litres alcohol (20 – x) litres water
Operation 3	(20 – x) litres alcohol + xlitres mixture i.e., $\left(\frac{x^2}{20}\text{ litres alcohol}\right)$	(20 – x) litres mixture

Now $\frac{1}{3}$rd of C_1 poured into C_2

Quantity of alcohol in $C_1 = C_2$ = 10 litres

In C_1 : $\frac{2}{3}\left[20 - x + (x)\left(\frac{x}{20}\right)\right] = 10$

Solving we get, x = 10 litres

Hence 10 litres of alcohol was initially poured from C_1 to C_2.

3. Required ratio $= \frac{x-y}{x+y}\times\frac{y^2+xy}{x^2-xy} = \frac{y}{x} = y : x$

4. *Given* : $\frac{x}{y} = a$ and $\frac{y}{x} = b$

If $\frac{x}{y} > 1$, then $\frac{y}{x} < 1$,

$\Rightarrow \quad \frac{x}{y} + \frac{y}{x} > 1$.

Hence option (*c*) is correct.

5. Total charge = fixed + variable

$T_C = m + kx$

where m and k are constants.

$\therefore \quad 300 = m + k \times 20$

Also, $550 = m + k \times 45$

Solving for m and k; we get

m = 100; k = 10.

Hence, for 100 grams,

total charge = 100 + 10 × 100 = ₹ 1100.

Q. 6 and 7.

Using given data, we can draw following table

	Male	Female	
White Collar	4.8x	3.2x	8x
Blue Collar	2.2x	0.8x	3x
	7x	4x	11x

Now, all the questions can be answered.

6. $\dfrac{\text{Female white collar employess}}{\text{Male blue collar employees}} = \dfrac{3.2x}{2.2x} = \dfrac{16}{11}$

7. Since $0.8x = 48$

$\Rightarrow x = 60$

$\therefore$ Required difference $= 4.8x - 0.8x$

$= 4.8 \times 60 - 0.8x = 240$

8. Let 10x be taken from bottle A and 7y be taken from bottle B.

$$\text{Total salt taken} = \left(\frac{7}{10} \times 10x\right) + \left(\frac{4}{7} \times 7y\right)$$

$$= 7x + 4y$$

Total water taken = 3x + 3y.

$$\therefore \frac{7x + 4y}{3x + 3y} = \frac{2}{1}$$

$$\Rightarrow 7x + 4y = 6x + 6y$$

$$\Rightarrow x = 2y$$

$$\therefore \frac{x}{y} = \frac{2}{1}$$

$$\therefore \text{Ratio of quantities} = \frac{10x}{7y}$$

$$= \frac{10}{7} \times \frac{2}{1} = \frac{20}{7}$$

9. Let speed of the motor in stage 1 be x rpm. Then its speed in stage 2 be $(x^2) = ax^2$ (say) and in stage 3 be $(ax^2)^3 = b.a^3x^6$ (say)

Given that $x = \dfrac{50}{3}$ ...(i)

$ax^2 = 2500$ ...(ii)

$ba^3x^6 = 12500$ ...(iii)

Substituting value of x in (ii), we get

$$a = \frac{2500}{(50)^2} \times 3^2 = 9$$

Again, substituting values of a and x, we get

$$b = \frac{12500}{(2500)^3} = \frac{1}{125} \times 10^{-4}$$

Given : $x = \dfrac{25}{6}$ rpm

$$\therefore \text{Speed in stage 3} = \frac{10^{-4}}{125} \times (9)^3 \times \left(\frac{25}{6}\right)^6$$

$$= \frac{10^{-4}}{5 \times 25} \times (3)^6 \times \frac{(25 \times 5) \times (5 \times 25^4)}{(2)^6 \cdot (3)^6}$$

$$= \frac{5 \times 25^4}{2^6 \times 10^4} = \frac{5^5}{2^{10}} \text{ rpm}$$

10. Let bag 1 and 2 have

	x mangoes	y mangoes
Step 1	$\dfrac{x}{2}$	$y + \dfrac{x}{2}$

But $\dfrac{x}{2} : \ y + \dfrac{x}{2} = y : x$

Step 2	$\dfrac{2}{3}y + \dfrac{2}{3}x$	$\dfrac{1}{3}y + \dfrac{1}{3}x$

$$\therefore \text{Required ratio} = \frac{\frac{2}{3}y + \frac{2}{3}x}{\frac{1}{3}y + \frac{1}{3}x} = \frac{2\left(\frac{x+y}{3}\right)}{1\left(\frac{x+y}{3}\right)}$$

$$= 2 : 1$$

NUMERICAL TYPE QUESTIONS

1. Let quantity of wine be a, i.e., capacity of the cask be 'a'

$$\text{i.e., } \frac{\text{Quantity of wine}}{\text{Quantity of water}} = \frac{\left(1 - \frac{5}{a}\right)^2}{1 - \left(1 - \frac{5}{a}\right)^2}$$

$$\therefore \frac{361}{39} = \frac{\left(1 - \frac{5}{a}\right)^2}{1 - \left(1 - \frac{5}{a}\right)^2}$$

$$\Rightarrow \frac{361}{39} - \frac{361}{39}\left(1 - \frac{5}{a}\right)^2 = \left(1 - \frac{5}{a}\right)^2$$

$$\Rightarrow \left(1 - \frac{5}{a}\right) = \sqrt{\frac{361}{400}} = \frac{19}{20}$$

$$\Rightarrow \frac{5}{a} = \frac{1}{20}$$

$$\therefore a = 100 \text{ litres}$$

2. Quantity of wheat in 400 kg of mixture

$= 0.6 \times 400 = 240$ kg

Quantity of barley = 400 – 240 = 160 kg

Let x kg of barley be added to 400 kg of the mixture

$$\therefore \frac{240}{160 + x} = \frac{53\frac{1}{3}}{46\frac{2}{3}} = \frac{160}{140} = \frac{8}{7}$$

$$\Rightarrow 240 \times 7 = 8 \times 160 + 8x$$

$$\Rightarrow x = 50 \text{ kg}$$

3. Alcohol content in 1st jar = 0.25 × 3 = 0.75 litre

Alcohol content in 2nd jar = 0.75 × 5 = 3.75 litres

$\therefore$ Total alcohol in 9 litre cask = 0.75 + 3.75

= 4.5 litres

$\therefore$ Percentage of alcohol in the cask

$$= \frac{4.5}{9} \times 100 = 50\%$$

4. Cost price of milk = ₹ 5/litre

Cost price of mixture = $\frac{100}{133\frac{1}{3}} \times 6 =$ ₹ 4.5/litre

$$\therefore \quad \frac{\text{Quantity of water}}{\text{Quantity of milk}} = \frac{5-4.5}{4.5-0} = \frac{1}{9}$$

i.e.,

Milk ₹ 5.0   Water ₹ 0

₹. 4.5

₹ 4.5   ₹ 0.5

Milk : Water = 9 : 1

5.

	Alcohol	Grape juice	Alcohol : Grape
Tank 1	20%	80%	1 : 4
Tank 2	40%	60%	2 : 3

Ratio of grape juice to alcohol in half filled wine glass

$$= \frac{2\times\frac{4}{5}+3\times\frac{3}{5}}{2\times\frac{1}{5}+3\times\frac{2}{5}} = \frac{17}{8}$$

So, half capacity of glass = 17 + 8 = 25

So, full capacity = 25 × 2 = 50

Now rest 25 parts has to be filled with alcohol.

So, ratio of the grape juice to alcohol

= 17 : (8 + 25) = 17 : 33.

Alternatively,

As the wine glass is filled to half by taking two parts from tank 1 and three parts from tank 2 and second half is equivalent to five parts which is pure alcohol,

percentage concentration of grape juice

$$= \frac{2\times 80+3\times 60+5\times 10}{2+3+5} = 34\%$$

Hence ratio of grape juice to alcohol

= 34 : 66 = 17 : 33.

6. S.P of 1 kg of mixture = ₹ 68.20, Gain =10%

S.P of 1 kg of mixture = ₹ $\left(\frac{100}{110}\times 68.2\right)$ = ₹ 62.

Cost of 1 kg tea of 1st kind   Cost of 1 kg tea of 2nd kind

60   65

Mean Price ₹ 62

3   2

$\therefore$ Required ratio = 3 : 2.

7. Amount of milk left after 3 operations

$$= 40\left(1-\frac{4}{40}\right)^3$$

$$= 40 \times \frac{9}{10}\times\frac{9}{10}\times\frac{9}{10}$$

= 29.16 lit

8. Profit on 1st part 8%   Profit on 2nd part 18%

Mean Profit 14%

4   6

Ratio of 1st and 2nd parts = 4 : 6 = 2 : 3.

Quantity of 2nd part = $\frac{3}{5}\times 1000$ kg = 600 kg.

9. Ratio of volumes = $\frac{3}{14}:\frac{4}{14}:\frac{7}{14}$

Ratio of densities = $\frac{5}{13}:\frac{2}{13}:\frac{6}{13}$

$\therefore$ Ratio of weights = $\frac{3}{14}\times\frac{5}{13}:\frac{4}{14}\times\frac{2}{13}:\frac{7}{14}\times\frac{6}{13}$

= 15 : 8 : 42

... (Mass = Volume × Density)

Weight of the third substance = $\frac{42}{65}\times 130 = 84\ \ell$bs.

10. Let number of 50p coins be 3x, number of 25p coins be 4x and number of rupee coins be $2\frac{1}{2}$x.

$\therefore \quad 2\frac{1}{2}x+\frac{1}{2}\times 3x+\frac{1}{4}\times 4x = 210$

$\Rightarrow \quad \frac{5}{2}x+\frac{3}{2}x+x = 210$

$\Rightarrow \quad 5x = 210$

$\Rightarrow \quad x = 42$

$\therefore$ Number of rupee coins = $\frac{5}{2}\times 42$ = 105.

Number of goats = 416 × $\frac{13}{32}$ = 169

$\therefore$ Total number of hens = 65 + 65 = 130

Total number of ducks = 91 + 182 = 273

Total number of goats = 169 + 169 = 338

$\therefore$ The new ratio of hens: ducks: goats = 130: 273: 338 = 10: 21:26.

5 CHAPTER
Permutations-Combinations, Elementary Statistics and Probability

PERMUTATIONS-COMBINATION

Permutations

Each of the arrangements which can be made by taking some or all of a number of items is called a *Permutation*. Permutation implies *"arrangement"* or that *"order of the items"* is important.

Permutations of three items a, b and c taken two at a time are ab, ba, ac, ca, cb and bc. Since the order in which items are taken is important, ab and ba are counted as two different permutations. The words "permutation" and "arrangement" are synonymous and can be used interchangeably."

Number of permutations of n things taking r at a time is denoted by ^nP_r (and read as "nPr")

Combinations

Each of the groups or selections which can be made by taking some or all of a number of items is called *Combination*. In combinations, order in which the items are taken is not considered as long as the specific things are included.

Combinations of three items a, b and c taken two at a time are ab, bc and ca. Here, ab and ba are not considered separately because the order in which a and b are taken is not important but it is only required that a combination including a and b is what is to be counted. Words *"combination"* and *"selection"* are synonymous.

Number of combinations of n things taking r at a time is denoted by ^nC_r (and read as "nCr")

$$^nP_r = \frac{n!}{(n-r)!},\quad ^nC_r = \frac{n!}{r!(n-r)!}$$

Number of ways in which n things may be arranged taking them all at a time, when p of the things are exactly alike of one kind, q of them exactly alike of another kind, r of them exactly alike of a third kind, and the rest all distinct is $\frac{n!}{p!q!r!}$

Number of ways of selecting one or more items from n given items = $2^n - 1$

Number of ways of dividing (p + q) items into two groups of p and q items respectively is $\frac{(p+q)!}{p!q!}$.

Number of ways of dividing 2p items into two equal groups of p each is $\frac{(2p)!}{(p!)^2}$ where two groups have distinct identity.

Number of ways of dividing 2p items into two equal groups of p each is $\frac{(2p)!}{2!(p!)^2}$ where two groups do not have distinct identity.

Number of ways in which (p + q + r) things can be divided into three groups containing p, q and r things respectively is $\frac{(p+q+r)!}{p!q!r!}$

Number of circular arrangements of n distinct items is (n − 1)! If there is a DIFFERENCE between clockwise and anticlockwise arrangements and (n − 1)!/2 if there is NO DIFFERENCE between clockwise and anticlockwise arrangements.

For $x_1 + x_2 + x_3 + \text{-------} + x_n = s$ where $s \geq 0$, number of positive integral solutions (when $s \geq n$) is $^{s-1}C^{n-1}$ and number of non-negative integral solution is $^{n+s-1}C^{n-1}$.

NOTES:

The Number of Diagonals in an N–Sided Regular Polygon

An n-sided regular polygon has n vertices. Joining any two vertices we get a line of the polygon which are $^nC_2$ lines, n of them are sides. Hence diagonals are $^nC_2 = \frac{n(n-3)}{2}$

The following points are useful in solving the problems.

1) $^nC_0 = {^nC_n} = 1$
2) $^nC_1 = {^nC_{n-1}} = n$
3) $^nC_r = {^nC_{n-r}}$
4) if $^nC_r = {^nC_s}$ then r = s r n = r + s
5) $^nC_r + {^nC_{r-1}} = {^{n+1}C_r}$

Example:

Find the number of 4 –digit numbers that can be formed using the digits 0.2, 5, 7, 9, 6 where a digit can be repeated more than once.

Solution:

Among the given digits 0, 2, 5, 7, 9, 6 zero cannot be in thousands place 2, 5, 7, 9, 6 are to be used for the thousands place of a four digit number, which can be done in 5 ways.

Since the repetition of digits is allowed, for each of the hundreds, tens and the units place, all the given digits can be used which can be done in 6(6)(6) (6) or 216 ways. ∴ The total number of 4 digit numbers that can be formed is 5(216) or 1080

Example:

If $^nP_4 = 7920$, then find $^nC_4$

Solution:

$^nP_4 = n(n-4)(n-2)(n-3)$

$= 10(4)(198)$

$= 8(9)(10)(11)$

$$\therefore n = 11 \therefore {}^nP_4 = {}^{11}C_4 = {}^nP_4\left(\frac{1}{4!}\right)$$

$$= \frac{1}{4!}7920 = 330$$

Example:

In how many ways can the letters of the word RECRUDESCENCE be arranged?

Solution:

The given word contains 4E's ,2R's, 3C's ,1U, 1D, 1S,1N. Here the total numbers of letters =13

$$\therefore \text{Number of arrangements } \frac{13!}{1!2!3!}$$

Example:

Find the number of ways in which six boys and five girls can be seated in a row such that each girls has a boy sitting on either side.

Solution:

There are 6 boys and 5 girls

Possible seating arrangement : BGBGBGBGBGB

∴ Total number of sitting arrangements = 5!6!

Example:

Find the number of ways of drawing four cards, all of different suits from pack of 52 playing cards. (Diamonds, spades, hearts and clubs are the suits in a pack)

Solution :

A pack of 52 cards contains 4 different suits. Number of ways of drawing four cards each $= {}^{13}C_1 . {}^{13}C_1 . {}^{13}C_1 . {}^{13}C_1 = 13^4$

ELEMENTARY STATISTICS

If you find 'elementary' statistical formulae meaningless, or merely wish to refresh our understanding,

The term 'statistic' can refer to several rather different things.

A statistic summarizes or represents a set of information - most commonly as a single number. The term statistic is used both for the value and for the mathematical 'function' (usually an equation) used to obtain that value.

Many functions are available to summarize information. For example, a salesman could equally truthfully provide the most typical cost as 'on average' or give the maximum ('up to...') or the minimum ('from...') just $ 300. The 'average', 'maximum' and 'minimum' are all statistics.

Note, summary statistics of a sample are often used as estimates for the population at large - for instance when you are told 'the average man has 1.8 children' that result was found in a sample of men - it is usually impossible to check every man.

Statistics are also used to describe how sets of results varied, or to infer how they are liable to vary, or to infer how their summary statistic might be expected to vary. Inferential statistics are variously used to indicate how reliable an outcome is, or the probability it occurred by simple chance - given a simple (hopefully plausible) set of assumptions.

When used in the plural, 'statistics' also describes 'the study of the collection, organization, analysis, interpretation and presentation of data'.

Humans, of course, use non-numerical summaries all of the time. For example when you say 'cats are smaller than dogs' you are probably describing the average situation - however some people assume you mean every cat is smaller than every dog.

Humans also use non-numerical estimates of probability, using a simple scale, ranging from impossible to certain. Research shows most people divide that scale into surprisingly few levels - seldom more than 7 - and have problems in dealing with very small probabilities.

Statistics is a mathematical science including methods of collecting, organizing and analyzing data in such a way that meaningful conclusions can be drawn from them. In general, its investigations and analyses fall into two broad categories called descriptive and inferential statistics.

Descriptive statistics deals with the processing of data without attempting to draw any inferences from it. The data are presented in the form of tables and graphs. The characteristics of the data are described in simple terms. Events that are dealt with include everyday happenings such as accidents, prices of goods, business, incomes, epidemics, sports data, population data.

Inferential statistics is a scientific discipline that uses mathematical tools to make forecasts and projections by analyzing the given data.

Measures of Central Tendency in Elementary Statistics

Average of a distribution is the value of the variable which is representative of the entire distribution.

The following are the measures of central tendency.

(*i*) Arithmetic mean
(*ii*) Geometric mean
(*iii*) Harmonic Mean
(*iv*) Median
(*v*) Mode

(*i*) Arithmetic Mean: If $x_1, x_2, ..., x_n$ are n values of variable X, then the arithmetic mean of these values is given by

$$\bar{X} = \frac{1}{n}(x_1 + x_2 + ... x_n) = \frac{1}{n}\left[\sum_{i=1}^{h} x_i\right]$$

In case of a frequency distribution n_i/f_i, $i = 1, 2, ..., n$ where f_i is the frequency of the variable x:

$$\bar{X} = \frac{f_1x_1 + f_2x_2 + ... + f_nx_n}{f_1 + f_2 + ... + f_n} = \frac{1}{N}\left(\sum_{l=1}^{h} f_ix_i\right) ...(i)$$

$$\text{where } N = f_1 + f_2 + ... + f_n = \sum_{i=1}^{n} f_i$$

$$\boxed{\bar{X} = A + \frac{1}{N}\sum_{i=1}^{n} f_i\, d_j}$$

e.g., Find arithmetic mean of 10, 15, 17, 20.

$$x = \frac{10 + 15 + 17 + 20}{4} = 15.5$$

When large numbers are involved, an Arithmetic Mean is generally assumed to facilitate calculation. If a is assumed Arithmetic mean, then true Arithmetic mean $\bar{x}$ of group of n numbers $x_1, x_2, x_3 .... x_n$ is

$$\bar{x} = a + \frac{\sum_{i=1}^{n} d_i}{n},$$

where $d_i = x_i - a$ is deviation from the assumed mean.

e.g., Find the Arithmetic mean of 42, 46, 48, 50, 55, 60, 70

Let assumed mean be 50

x_i	d_i
42	42 – 50 = – 8
46	46 – 50 = – 4
48	48 – 50 = – 2
50	50 – 50 = 0
55	55 – 50 = 5
60	60 – 50 = 10
70	70 – 50 = 20

$$\therefore\ \bar{x} = 50 + \frac{\sum_{i=1}^{7} di}{7} = 50 + \frac{21}{7} = 53.$$

Weighted Arithmetic Mean: If values $x_1, x_2, ... x_n$ are assigned weights $w_1, w_2, ... w_n$ respectively, then

$$\text{weighted arithmetic mean},\ x_w = \frac{\sum_{i=1}^{n} w_ix_i}{\sum_{i=1}^{n} w_i}$$

The term '*weight*' stands for relative importance that is attached to the different values.

'w' can also be replaced by 'f' in a frequency distribution where 'f' denotes frequency or simply the number of times a particular observation occurs.

e.g.,

x_i	w_i	w_ix_i
20	3	60
30	1	30
40	5	200
50	1	50

$\Sigma w_i = 10$; $\Sigma x_i\ w_i = 340$

$$\therefore\quad \bar{x}_w = \frac{\sum_{i=1}^{4} w_ix_i}{\sum_{i=1}^{4} w_i} = \frac{340}{10} = 34$$

When two or more groups are combined, then combined average of the groups is

$$x_c = \frac{x_1n_1 + x_2n_2 + x_3n_3 + ... + x_nn_n}{n_1 + n_2 + n_3 + ... + n_n}$$

where $x_1, x_2, x_3....x_n$ individual averages of the groups and $n_1, n_2, ..., n_n$ are number of observations or values in each of the groups.

e.g., Mean marks scored by 50 boys are 70 and those scored by 100 girls are 55. Find average marks scored by this group.

$$x_c = \frac{n_1x_1 + n_2x_2}{n_1 + n_2} = \frac{70 \times 50 + 100 \times 55}{150} = \frac{9000}{150} = 60$$

Properties of Arithmetic Mean

(*i*) Algebraic sum of the deviations of a set of values from their arithmetic mean is zero.

$$\sum_{i=1}^{n} f_i(x_i - \bar{X}) = \sum_{i=1}^{n} f_i x_i - \bar{X}\sum_{i=1}^{n} f_i$$

$$= N\bar{X} - N\bar{X}\left[\because \bar{X} = \frac{1}{N}\sum_{i=1}^{n} f_i x_i\right] = 0$$

(*ii*) The sum of the squares of the deviations of a set of values is minimum when taken about mean.

(*iii*) Mean of the composite series: If $\bar{X}_1, \bar{X}_2, ...\bar{X}_K$ are the mean of K series of sizes $n_1, n_2, ...n_K$ respectively then the mean $\bar{X}$ of the composite series is given by

$$\bar{X} = \frac{n_1\bar{X}_1 + n_2\bar{X}_2 + ... + n_K\bar{X}_K}{n_1 + n_2 + ... + n_K}.$$

(*ii*) Geometric Mean: Geometric mean of a set of n observations is the nth root of their product.

$$\boxed{G = \text{anti}\log\left(\frac{1}{N}\sum_{i=1}^{n} f_i \log x_i\right)}$$

e.g., Geometric mean of 2, 3 and 4 is $\sqrt[3]{2\times3\times4}$ $= \sqrt[3]{24} = 2.\sqrt[3]{3}$

Geometric mean of a and b is $\sqrt{ab}$

e.g., Find Geometric mean of 9 and 4.

$$\text{G.M.} = \sqrt{9\times4} = 6$$

Geometric mean is generally used to calculate rate of growth.

If geometric mean of one group of 'a' numbers is 'x' and that of another group of 'b' numbers is 'y', then

Geometric mean of the combined groups

$$= (x_a + y_b)\frac{1}{a+b}$$

Propertie of Geometric Mean

If G_1 and G_2 are the geometric means of two series of sizes n_1 and n_2 respectively, then the geometric mean G of the combined series is given by

$$\boxed{\log G = \frac{n_1 \log G_1 + n_2 \log G_2}{n_1 + n_2}}$$

(*iii*) Harmonic Mean: Harmonic mean of a observations is the reciprocal of the AM of the reciprocals of the given values. Thus, harmonic mean H of n non-zero observations $x_1, x_2, ..., x_n$ is

$$H = \frac{1}{\frac{1}{n}\sum_{i=1}^{n}\left(\frac{1}{x_i}\right)} \quad ...(i)$$

In case of a frequency distribution $x_i | f_i\ i = 1, 2, ..., n$

$$H = \frac{1}{\frac{1}{N}\sum_{i=1}^{n}\left(\frac{f_i}{x_i}\right)} \quad ...(ii)$$

e.g., Find Harmonic mean of 1, 0.5, 10.

x_i	$\frac{1}{x_i}$
1	1
0.5	2
10	0.1

$$\sum\frac{1}{x_i} = 3.1$$

$$\text{H.M.} = \frac{3}{3.1} = \frac{30}{31}$$

This is an important result and is generally used to find average speed when equal distances are covered at different speeds. (when different distances are covered in the same time, arithmetic mean is used to determine the average speed).

e.g., A person travels from A to B at 6 kmph and returns at 4 kmph, find his average speed.

Average Speed= Harmonic Mean of 6 and 4

$$= \frac{2\times6\times4}{6+4} = \frac{48}{10} = 4.8 \text{ kmph}$$

Note: This concept will again be covered in time, speed and distance.

Weighted Harmonic Mean: Weighted Harmonic Mean of 3 numbers x_1, x_2 and x_3 with weights w_1, w_2 and w_3 respectively is $\frac{\sum_{i=1}^{3} w_i}{\sum_{i=1}^{3}\frac{w_i}{x_i}}$

This is used to calculate average speed when different distances are covered at different speeds.

e.g., A cyclist covers his first 5 km at an average speed of 10 kmph, another 3 km at 8 kmph and last 2 km at 5 kmph. Find his average speed during the journey.

Average Speed= Weighted HM

$$= \frac{5+3+2}{\frac{5}{10}+\frac{3}{8}+\frac{2}{5}} = 7.84 \text{ kmph}$$

(*iv*) Median and Other Partition Values

Median of a distribution is the value of the variable which divides it into equal parts.

$$\text{Median} = l + \left(\frac{\frac{N}{2} - F}{f}\right)h$$

where l = lower limit of the median class
f = frequency of the median class
h = size (width) of the median class
F = *c.f.* of the class preceding the median class.

1. $N = \sum_{i=1}^{n} f_i$
2. As median divides a distribution into two equal parts, similarly the quantities, deciles and percentiles divide the distribution respectively into 4, 5, 10 and 100 equal parts the ith quartile is given by.

$$Q_1 = l + \left[\frac{\frac{iN}{4} - F}{f}\right]h, i = 1, 2, 3$$

3. Q_1 is the lower quartile, Q_2 is the median and Q_3 is called the upper quartile.

The ith decile is given by

$$D_i = l + \left(\frac{\frac{iN}{10} - F}{f}\right)h, i = 1, 2, \ldots\ldots, 9$$

D_5 is the median.

e.g., Find median of the numbers 40, 41, 33, 48, 41, 39 and 38.

(v) **Mode :** The mode or modal value of a distribution is that value of the variable for which the frequency is maximum.

In case of a discrete frequency distribution the value of the mode is determined by the method of grouping.

In case of a grouped or continuous frequency distribution, mode is given by the formula.

$$\text{Mode} = l + \frac{f - f_1}{2f - f_1 - f_2} \times h$$

where l = lower limit of the modal class.
h = width of the modal class.
f_1 = frequency of the class proceeding the modal class
f_2 = frequency of the class following the modal class.
f = frequency of model class.

e.g., Find mode of the set of numbers 2, 4, 6, 6, 5, 4, 3, 6, 2, 6, 4, 1, 4, 5, 4.

Relation between A.M., G.M. and H.M.

1. Consider two number a and b.

Arithmetic Mean $= \frac{a+b}{2}$; Geometric mean $= \sqrt{ab}$ Harmonic mean $= \frac{2ab}{a+b}$

$\therefore$ $(GM)^2 = AM \times HM$

e.g., If Arithmetic mean of two numbers is 6.5 and their Geometric mean is 6, what is their Harmonic Mean?

$$HM = \frac{(G.M.)^2}{A.M.} = \frac{6 \times 6}{6.5} = 5.54$$

2. Arithmetic Mean > Geometric Mean > Harmonic Mean

e.g., Arithmetic Mean of 2 and 4 is 3.
Geometric Mean of 2 and 4 is 2.82.
Harmonic Mean of 2 and 4 is 2.66

$\therefore$ $3 > 2.82 > 2.66$

Symmetric Distribution

A distribution is a symmetric distribution if the value of mean, made and median coincide. In a symmetric distribution frequencies are symmetrically distributed on both side of the centre be point of the frequency curve.

(i)

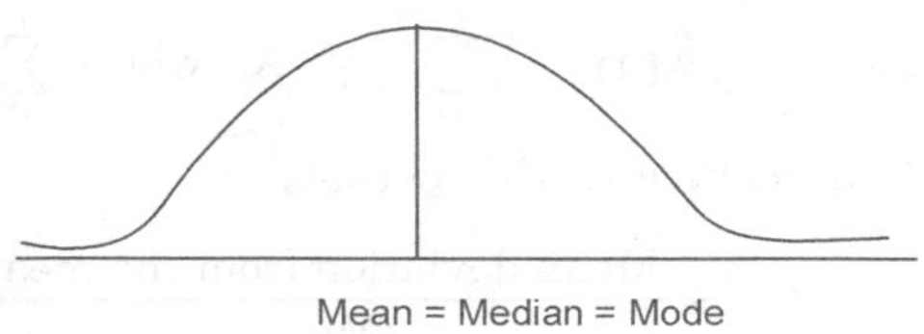

(ii)

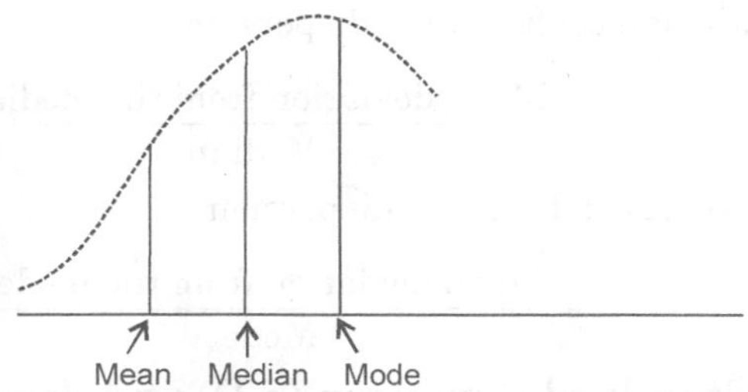

(iii)

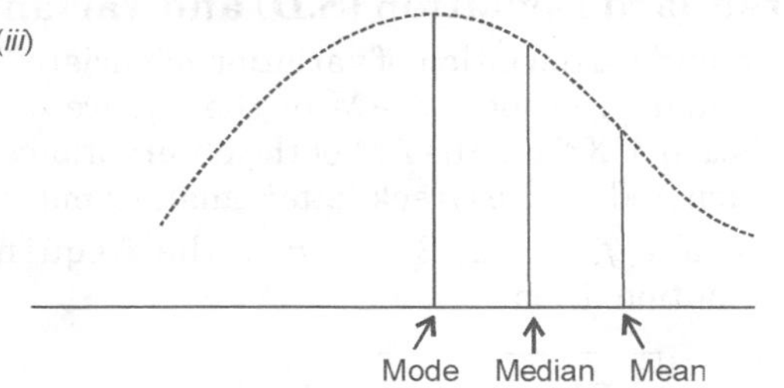

We have the following empirical relation between them:

$$\text{Mean} - \text{Mode} = 3(\text{Mean} - \text{Median})$$

$$\Rightarrow \quad \text{Mode} = 3\,\text{Median} - 2\,\text{Mean}$$

Measure of Dispersion

The measure of dispersion commonly used are:

(*i*) Range

(*ii*) Quartile deviation or the semi-interquartile range

(*iii*) Mean- deviation and

(*iv*) standard deviation.

(*i*) **Range:** If A and B are the greatest and the smallest observations respectively in a distribution, then its range = A – B.

The coefficient of range (or scatter) = $\frac{A-B}{A+B}$.

(*ii*) **Quartile Deviation or semi-interquartile range (Q.D.):** If Q_1 and Q_3 be the lower and upper quartiles.

Then quartile deviation $Q = \frac{1}{2}(Q_3 - Q_1)$

coefficient of quartile deviation

$$= \frac{(Q_3 - Q_1)/2}{(Q_3 + Q_1)/2} = \frac{Q_3 - Q_1}{Q_3 + Q_1}.$$

(*iii*)**Mean Deviation (M.D):** If x_1/f_i, i = 1, 2, ... n is the frequency distribution, then mean deviation from an average A (median, modern, mode, or Am)

is given by $M.D = \frac{1}{N}\sum_{i=1}^{n} f_i |x_i - A|$, where $\sum_{i=1}^{n} f_i = N$.

Mean coefficient of dispersion

$$= \frac{\text{Mean deviation from the mean}}{\text{Mean}}$$

Median coefficient of dispersion

$$= \frac{\text{Mean deviation from the median}}{\text{Median}}$$

Mode coefficient of dispersion

$$= \frac{\text{Mean deviation from the mode}}{\text{Mode}}.$$

(*iv*) Standard Deviation (S.D) and Variance

The standard deviation of variation of variate X is the square root of the AM of the square of all deviation of X from the AM of the observation and it is denoted by the Greek letter small sigma (σ).

Thus, if x_1/f_i, i = 1, 2, ... n is the frequency distribution, then,

$$\sigma = \sqrt{\frac{1}{N}\sum_{i=1}^{n} f_i \left(X_i - \bar{X}\right)^2} \qquad ...(i)$$

Where $\bar{X}$ is the A.M of the distribution and

$N = \sum_{i=1}^{n} f_i$.

Variance of the Combined Series

If n_1, n_2 are the sizes, $\bar{X}_1, \bar{X}_2$ the means and σ_1 σ_2 the standard deviation of two series, is given by

$$\sigma^2 = \frac{1}{n_1 + n_2}\left[n_1\left(\sigma_1^2 + d_1^2\right) + n_2\left(\sigma_2^2 + d\right)\right]$$

where $d_1 = \bar{X}_1 - \bar{X}$, $d_2 = \bar{X}_2 - \bar{X}$ and $\bar{X} = \frac{n_1\bar{X}_1 + n_2\bar{X}_2}{n_1 + n_2}$

Skewness: As we have seen that a distribution is skewness if (1) mean ≠ median ≠ mode.

(*i*) Quartiles are not equidistant from the median and

(*ii*) the frequency curve drawn is stretched more to one side than to the other.

Measures of Skewness

Measures of skewness tell us the direction and extent of asymmetry in a series. There are two types of measures of skewness (*i*) absolute (*ii*) relative.

Absolute Measures of Skewness

Various measures of skewness are

(*i*) $S_K = M - M_d$

(*ii*) $S_K = M - M_0$

(*iii*) $S_K = Q_3 + Q_1 - 2\,M_d$.

where M_d = median, M_0 = Mode, M = mean.

Relative Measures of Skewness

There are four relative measures of skewness.

(*i*) The karl person's coefficient of skewness.

(*ii*) The bowley's coefficient of skewness.

(*iii*) The kelly's coefficient of skewness and

(*iv*) Measures of skewness based upon moments.

PROBABILITY

If in an experiment there are 'n' occurrences out of which 'm' occurrences are favourable to a particular event E, then probability of the event E [written as P(E)] is defined as P(E) = m/n.

For the event E, non-occurrence of the event is denoted by $\bar{E}$. Out of a total n occurrences, if m occurrences are favourable to the event E, it means that (n – m) occurrences are not favourable to the event $\bar{E}$, i.e. (n - m) occurrences are favourable to the event $\bar{E}$.

Hence, probability of the event $\bar{E}$ will be $P(\bar{E})\ \frac{n-m}{n}$ (complementary event).

We can also see that $P(\bar{E}) + P(E) = 1$.

In some problems when we are asked to find out the probability of an event $\bar{E}$, it becomes easier to calculate probability of E and subtract that probability from 1 to get $P(\bar{E})$.

Note:

(*i*) Probability [P(E)] of the happening of an event E is called *probability* of success and probability $\left[P(\bar{E})\right]$ of non-happening of the event is called *probability of failure*.

(*ii*) If P(E) = 1, then event is called a *certain event* and if P(E) = 0, then event is called *impossible event*.

(*iii*) Instead of saying that the chance of happening of an event is m/n, we can also say that odds in favour of the event are m (to n – m). Similarly, if m/n is probability of the occurrences of an event, since (n – m)/n is probability of non-occurrence of the event, we can also say that the odds against the event are (n – m) to m.

Addition Theorem on Probability

If there are two sets A and B, we know that the number of elements in A ∪ B is given by

$$n(A \cup B) = n(A) + n(B) - n(A \cap B).$$

A similar relationship exists in probability theory (called Addition Theorem of Probability) as $P(A \cup B) = P(A) + P(B) - P(A \cap B)$, where A and B are two events.

If A and B are mutually exclusive events (i.e, A and B are disjoint sets), then $P(A \cap B) = 0$

If there are two mutually exclusive events A and B, then $P(A \cup B) = P(A) + P(B)$.

Independent Events

Two events, E_1 and E_2 are said to be independent, if occurrence of the event E_2 is not affected by the occurrence or non-occurrence of the event E_1.

If A and B are two independent events, then

$$P(A \cap B) = P(A) \times P(B)$$

Example:

If dice is rolled, then find the probability of getting a number divisible by 2.

Solution:

When a dice is rolled n(S) = 6

Favorable outcomes are {2, 4, 6} i.e, 3

∴ Required probability = 3/6 = 1/2

Example:

When a dice is rolled, what is the probability of getting a composite number?

Solution:

Number of favorable outcomes = 2({4, 6})

Total outcomes = 6

Required probability $= \frac{2}{6} = \frac{1}{3}$

Example:

When two coins are tossed together, find the probability of getting exactly two tails?

Solution:

Number of favorable outcomes = 1(TT)

Total outcomes = 4(HH, HT, TH, TT)

∴ Required probability $= \frac{1}{4}$

Example: If two dice are tossed, the probability that the sum is 8 is

Solution:

Two dice are rolled; the number of possibilities is 36.

The favorable cases are (2, 6), (6, 2), (3, 5), (5, 3), (4, 4)

∴ Required probability $= \frac{5}{36}$

Example:

Suppose, seven coins are tossed then the probability of getting atleast one head is

Solution:

Probablity of getting at least one head

= 1 – (probability of getting no heads).

$$= 1 - \frac{1}{2^7} = 1 - \frac{1}{128} = \frac{127}{128}$$

Example: A and B play a game where each is asked to selected a number from 1 to 8. If both of them select the same number, then they win a prize. Find the probability that they will not win the prize.

Solution:

The total number of ways of selecting numbers for both A and B is = 8(8) = 64

They win the prize if they select same numbers.

The total possibilities are 8, i.e. (1, 1), (2, 2), (3, 3), (4, 4), (5, 5), (6, 6), (7, 7), (8, 8)

∴ Probablity of winning prizes $= \frac{8}{64} = \frac{1}{8}$

The probability of not getting a prize $= 1 - \frac{1}{8} = \frac{7}{8}$.

EXERCISE

MCQ TYPE QUESTIONS

Directions (Q. 1 – 2) :

If A and B are two possible events of an experiment such that P (A ∩ B) = 0.6 and P(A) = 0.3 then find P (B) given that

1. A and B are mutually exclusive events

(a) $\frac{7}{10}$ (b) $\frac{4}{10}$

(c) $\frac{3}{10}$ (d) $\frac{1}{2}$

2. A and B are independent events

(a) $\frac{4}{7}$ (b) $\frac{9}{10}$ (c) $\frac{3}{10}$ (d) $\frac{3}{7}$

3. Raju throws a fair dice. He is promised an amount (in rupees), which is twice the number showing up if that number is odd and an amount thrice the number showing up, if it is even. What is the maximum amount Raju would be willing to pay each time to throw the dice, if in the long run he wants to make an average profit of ₹ 7 per throw?

(a) ₹ 6 (b) ₹ 7 (c) ₹ 3 (d) ₹ 2

Directions (Q. 4 – 5) :

7 letters are to be placed in seven addressed envelops. If the letters are placed at random into the envelopes, the probability that

4. All of them are placed in the corresponding envelops is

(a) 1

(b) $\frac{1}{6!}$

(c) $\frac{1}{7!}$

(d) None of these

5. Exactly six letters are placed in their corresponding envelopes is

(a) $1-\frac{1}{7!}$ (b) $\frac{1}{7!}$ (c) 1 (d) 0

Directions (Q. 6 – 8) :

If three cards are drawn at random, from a well shuffled pack of cards, then what is the probability that

6. All of them are from the same suit?

(a) $\frac{4\left({}^{13}C_3\right)}{{}^{52}C_3}$ (b) $\frac{\left({}^{13}C_3\right)^3}{{}^{52}C_3}$

(c) $\frac{4\left({}^{13}C_3\right)^2}{{}^{52}C_3}$ (d) $\frac{{}^{13}C_3}{{}^{52}C_3}$

7. All of them are from different suits?

(a) $\frac{13}{{}^{52}C_3}$ (b) $\frac{4(13)^3}{{}^{52}C_3}$

(c) $\frac{{}^{39}C_3}{{}^{52}C_3}$ (d) $\frac{(13)^4}{{}^{52}C_3}$

8. Two of them are number cards from the same suit and the remaining number card is from a different suit?

(a) $\frac{{}^{9}C_2\,{}^{9}C_1}{{}^{52}C_3}$ (b) $\frac{{}^{9}C_2\,{}^{27}C_1}{{}^{52}C_3}$

(c) $\frac{4\left({}^{9}C_2\,{}^{27}C_1\right)}{{}^{52}C_3}$ (d) $\frac{4\left({}^{9}C_2\,{}^{9}C_1\right)}{{}^{52}C_3}$

9. A bag contains 5 white balls and 7 red balls. If 5 balls are drawn at random, then probability that 3 balls are red and the rest are white is

(a) $\frac{175}{198}$ (b) $\frac{175}{396}$

(c) $\frac{38}{132}$ (d) $\frac{221}{396}$

10. A bag contains 6 five rupee coins, 5 two rupee coins and 4 one rupee coins. If 5 coins are selected at random from the bag, then odds in favour of the draw yielding minimum possible amount are

(a) 1:3002 (b) 5:2998

(c) 2:3003 (d) 5:3003

11. In how many ways can 6 girls and 6 boys sit around a circular table so that no two boys sit together?

(a) $(5!)^2$ (b) $(6!)^2$

(c) 5! 6! (d) 11!

12. How many ten-digit numbers can be formed using all the digits of 1324642119 such that the even digits appear only in odd places?

(a) $(5!)^2$ (b) $\frac{(5!)^2}{3!}$

(c) 5! 3! 2! (d) $\frac{(5!)^2}{(2!)^2\,3!}$

Directions (Q. 13 – 14) :

A man has 12 friends whom he wants to invite for lunch. In how many ways can he invite

13. At least one of them?

(*a*) 4096 (*b*) 4095
(*c*) 2047 (*d*) 2048

14. At least 10 of them?

(*a*) 66 (*b*) 140
(*c*) 153 (*d*) None of these

15. In how many ways can 6 prizes be distributed among 3 boys, if each boy is eligible to receive one or more prizes?

(*a*) 3^6 (*b*) 6^3
(*c*) ${}^6C_3$ (*d*) ${}^6P_3$

Directions (Q. 16 – 17) :

Find the number of ways of dividing 16 different books equally

16. Among 4 boys

(*a*) $\frac{16!}{(4!)^4}$ (*b*) $\frac{16!}{(4!)^3}$
(*c*) $\frac{16!}{(4!)^5}$ (*d*) $(4!)^4$

17. Into 4 parcels (Parcels are not distinct)

(*a*) $\frac{16!}{(4!)^4}$ (*b*) $\frac{16!}{(4!)^3}$
(*c*) $\frac{16!}{(4!)^5}$ (*d*) $(4!)^4$

18. How many linear arrangements can be made using all the letters of the word ILLUSTRATE which begin with S and end with a vowel?

(*a*) $\frac{8!}{(2!)^2}$ (*b*) 8!
(*c*) $\frac{8!}{2!}$ (*d*) $\frac{8!}{(2!)^3}$

19. From a group of 10 professors and 6 assistant professors, a management institute desires to send to delegation of 8 persons consisting of 5 professors and 3 assistant professors to the IIMs annual meet. If Prof. Balamurali, a science professor refuses to be in the delegation if assistant Prof. Sheshdari, an arts professor is included in the delegation, then in how many ways can the delegation be formed?

(*a*) ${}^9C_4\,{}^4C_3$
(*b*) ${}^9C_5\,{}^4C_2$
(*c*) ${}^{10}C_5\,{}^6C_3 - {}^9C_4\,{}^5C_2$
(*d*) ${}^9C_4\,{}^4C_3 + {}^9C_5\,{}^4C_2$

20. Two red pencils, three black pencils and two white pencils are to be arranged in a row such that

1. No two adjacent pencils are of the same colour and
2. Pencils at the two ends of the row are of same colour.

(*a*) 12 (*b*) 8 (*c*) 9 (*d*) 10

NUMERICAL TYPE QUESTIONS

Directions (Q. 1 – 2) :

The odds against an event are 4 to 5 and the odds in favour of another independent event are 3 to 7. The probability that

1. Exactly one of them will occur is _______

2. Neither of them will occur is _______

Directions (Q. 3 – 5) :

Three persons Shiva, Jagan and Rohit aim at a target. Their respective probabilities of hitting the target are 2/3, 5/7 and 3/8. The probability that

3. None of them hits the target is _______

4. At least two of them hit the target is _______

5. Exactly one of them hits the target si _______

6. If three consecutive letters are selected at random from the English alphabet, then the probability that at least one letter is a vowel is _______

7. Eight unbiased coins are tossed together. The probability that the number of heads is equal to the number of tails is _______

8. In a biased coin, Head occurs three times as frequently as tail occurs. If the coin is tossed 3 times, then the probability of getting two heads is _______

Directions (Q. 9 – 10) :

Two cards are drawn at random from a well shuffled pack of cards. Given that both are black, the probability that the cards have

9. Same honour on them is_______

10. Different honours and belong to different suits is _______

Directions (Q. 11 – 13) :

Three mountaineers Akil, Nikil, and Sunil are climbing up a mountain with their respective probability of reaching the summit being $\frac{2}{3}, \frac{5}{8}$ and $\frac{4}{7}$, respectively. The probability that

11. None of them reach the summit is _______

12. Exactly two of them reaches the summit is _______

13. Atleast two of them reach the summit is _______

Directions (Q. 14 – 15) :

There are 12 points in a plane of which 4 are on a straight line and no three of the other points lie on a straight line.

14. The _______number straight lines can be formed by joining these points

15. The _______number of triangles can be formed by joining these points

16. The _______ number of non-negative integral solutions does the equation $x_1 + x_2 + x_3 + x_4 = 10$ have?

17. A certain number of students of a school participated in the chess tournament of their annual sports meet. Each player played 1 game against each of the other player. It was found that in 66 games both the players were girls, and in 240 games one was a girl and the other was a boy. The number of games in which both the players were boys is _______

18. Nine points are marked on a straight line and 10 points are marked on another line which is parallel to the first line. The_______ number of triangles can be formed with these points as vertices?

19. The _______ number of positive integers less than 100,000 and divisible by 125 can be formed using the digits 0, 1, 2, 5 and 8, if repetition is allowed?

20. There is an unlimited supply of identical red, blue, and green coloured balls. In _______ number of ways can 12 balls be selected from the supply.

21. Number of positive integral solutions of the equation $a + b + c = 15$ is _______

22. If six unbiased coins are tossed together, then probability that the number of heads exceeds the number of tails is_______

ANSWERS

MCQ Type Questions

1. (*c*)	**2.** (*d*)	**3.** (*d*)	**4.** (*c*)	**5.** (*d*)	**6.** (*a*)	**7.** (*b*)	**8.** (*c*)	**9.** (*b*)	**10.** (*b*)
11. (*c*)	**12.** (*d*)	**13.** (*b*)	**14.** (*d*)	**15.** (*a*)	**16.** (*a*)	**17.** (*c*)	**18.** (*b*)	**19.** (*c*)	**20.** (*d*)

Numerical Type Questions

1. $\frac{47}{90}$	**2.** $\frac{14}{45}$	**3.** $\frac{5}{84}$	**4.** $\frac{107}{168}$	**5.** $\frac{17}{56}$	**6.** $\frac{13}{24}$	**7.** $\frac{35}{128}$	**8.** $\frac{27}{64}$	**9.** $\frac{4}{325}$	**10.** $\frac{12}{325}$
11. $\frac{3}{56}$	**12.** $\frac{37}{84}$	**13.** $\frac{19}{28}$	**14.** 61	**15.** 216	**16.** 286	**17.** 190	**18.** 765	**19.** 99	**20.** 91
21. 91	**22.** $\frac{57}{64}$								

EXPLANATIONS

MCQ TYPE QUESTIONS

1 and 2

Given : $P(A \cup B) = 0.6$

and $P(A) = 0.3$

1. If A and B are mutually exclusive events, then

$$P(A \cap B) = 0$$

$$\therefore \quad P(A \cup B) = P(A) + P(B)$$

$$0.6 = 0.3 + P(B)$$

$$\Rightarrow \quad 0.3 = P(B)$$

Hence $\quad P(B) = 0.3$

2. If A and B are independent events, then

$$P(A \cap B) = P(A).P(B)$$

$$\therefore \quad P(A \cup B) = P(A) + P(B) - P(A \cap B)$$

$$= P(A) + P(B) - P(A).\ P(B)$$

$$P(A \cap B) = P(A) + P(B)\ [1 - P(A)]$$

$$\Rightarrow \quad 0.6 = 0.3 + P(B)\ (1 - 0.3)$$

$$0.3 = P(B)(0.7)$$

$$\Rightarrow \quad \frac{0.3}{0.7} = P(B)$$

$$\therefore \quad P(B) = \frac{3}{7}$$

3. When a die is rolled, for each number the probability of getting it is $\frac{1}{6}$.

When number on the dice is 1, 3 or 5 Raju receives ₹ 2, ₹ 6 and ₹10 respectively.

When the number on the die is 2, 4 or 6, then Raju receives ₹ 6, ₹ 12 and ₹ 18 respectively.

$$\therefore \quad \text{Expected value} = \frac{1}{6}\ [2 + 6 + 10 + 6 + 12 + 18]$$

$$= \frac{54}{6} = ₹\, 9$$

Hence to make an average profit of ₹ 7 per throw, Raju must pay ₹ 2 (9 – 7) for each time to throw the die.

4. 7 letters can be arranged in 7 addressed envelopes in 7! ways.

$\therefore \quad n(S) = 7!$

We can arrange 7 letters into corresponding 7 addressed envelopes in only one way.

$\therefore$ Required probability $= \frac{1}{7!}$.

5. If six letters can be placed in their corresponding addressed envelopes, then seventh letter is also in the correct envelope.

Hence required probability is 0.

6, 7 and 8.

Three cards can be drawn from 52 cards in $^{52}C_3$ ways

$n(S) = 52c_3$

6. Since we have 4 suits i.e., Diamonds, Spades, Clubs and Hearts and each suit contains 13 cards.

Three cards can be selected from 13 cards in $^{13}C_3$ ways.

$\therefore$ Number of favorable outcomes $= 4.\, ^{13}C_3$

Required probability $= 4.\frac{^{13}C_3}{^{52}C_3}$

7. From four suits we can select 3 suits in $^{4}C_3$ ways.

One card can be selected from each suit in $^{13}C_1$. $^{13}C_1$. $^{13}C_1$

$\therefore$ Number of favorable outcomes

$= {}^{4}C_3 . {}^{13}C_1 . {}^{13}C_1 . {}^{13}C1$

$\therefore$ Required probability $= \frac{4(13)^3}{^{52}C_3}$

8. Each suit contains 9 number cards. 2 number cards can be drawn from 9 cards in $^{9}C_2$ ways..

Third card can be selected from the remaining 27 number cards in $^{27}C_1$ ways.

Hence total number of favorable outcomes is $4(^{9}C_2)(^{27}C_1)$

Required probability $= \frac{4\left(^{9}C_2\right)\left(^{27}C_1\right)}{^{52}C_3}$

9. Total number of balls in the bag = 5 + 7 = 12.

5 balls can be drawn from 12 balls in $^{12}C_5$ ways

Number of ways of drawing 3 red balls and 2 white balls from 5 white balls and 7 red balls is $^{5}C_2 . {}^{7}C_3$

$\therefore$ Required probability $= \frac{^{5}C_2 . {}^{7}C_3}{^{12}C_5}$

$$= \frac{10 \times 35}{792} = \frac{175}{396}$$

10. Total number of coins in bag = 6 + 5 + 4 = 15.

5 coins can be drawn from 15 coins in $^{15}C_5$ ways.

When 4 coins are one rupee coins and fifth coin is a two rupee coin the amount will be minimum.

$\therefore$ Number of favourable cases $= {}^{4}C_4 . {}^{45}C_1 = 5$

Probability that the amount will be minimum is

$$P(E) = \frac{5}{^{15}C_5} = \frac{5}{3003}$$

$$\therefore \quad P(\bar{E}) = 1 - P(E) = 1 - \frac{5}{3003} = \frac{2998}{3003}$$

$\therefore$ Odds in favour of the required event = 5: 2998

11.

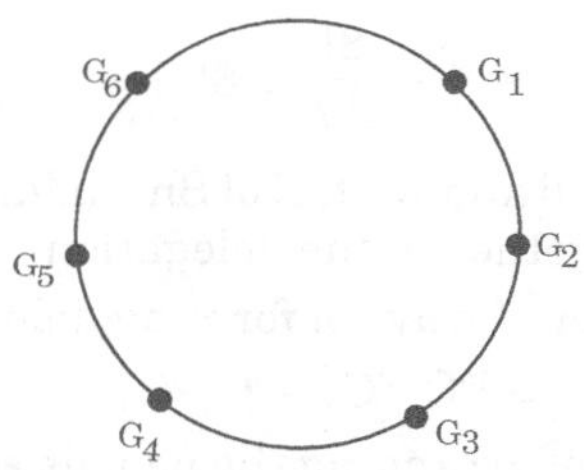

Six girls can be arranged around a circle in 5! ways.

Now, the six boys can be arranged in the six places in between the girls in 6! ways.

Hence, required number of ways = 5 ! 6!

12. Of the 10 places available, there are 5 even places.

Digits 2, 4, 6, 4 and 2 can be placed in these 5 even places in $\frac{5!}{2!2!}$ ways.

Remaining digits 1, 3, 1, 1 and 9 can be placed in the remaining 5 places in $\frac{5!}{3!}$ ways.

Hence, required number of ways

$$= \frac{5!}{2!2!} \frac{5!}{3!} = \frac{(5!)^2}{(2!)^2\, 3!}$$

13. Number of ways of inviting at least one friend is

$$2^{12} - 1 = 4095$$

14. Number of ways of inviting at least 10 friends

$= {}^{12}C_{10} + {}^{12}C_{11} + {}^{12}C_{12} = 66 + 12 + 1 = 79$

15. Each prize can be distributed in 3 ways, so, 6 prizes can be distributed in 3^6 ways.

16. Number of ways of dividing 16 books equally among 4 boys

$$=\frac{16!}{4!4!4!4!}=\frac{16!}{(4!)^4}$$

17. Number of ways of dividing 16 books equally into 4 parcels

$$=\frac{16!}{4!(4!)^4}=\frac{16!}{(4!)^5}$$

18. There are four vowels viz., U, A, I, E in the word "ILLUSTRATE".

First letter must be S and the last letter can be any one of the vowels U, A, I and E.

Remaining 8 places can be filled with the remaining 8 letters in $\frac{8!}{2!2!}$ ways.

Hence, required number of arrangements is

$$=4\times\frac{8!}{2!2!}=8!$$

19. Prof. Balamurali and Asst. B of Sheshadri cannot be included together in the delegation.

Hence, number of ways of forming delegation

$$={}^{10}C_5\,{}^6C_3-{}^9C_4\,{}^5C_2$$

20. Possible ways of arranging the pencils are: RBWBWBR, WBRBRBW, BWBRWRB, BRBWRWB, BWRBWRB, BWRRBRWB, BRWBRWB, BRWBWRB, BWRWBRB and BRWBBWB i.e. a total of 10 possible arrangements.

NUMERICAL TYPE QUESTIONS

1 and 2 :

Let A and B be the two events.

Given: $P(\bar{A}):P(A)=4:5$

$$\Rightarrow \quad P(\bar{A})=\frac{4}{9};\ P(A)=\frac{5}{9}$$

Also, $P(B):P(\bar{B})=3:7$

$$\Rightarrow \quad P(B)=\frac{3}{10} \text{ and } P(\bar{B})=\frac{7}{10}$$

1. $A\cap\bar{B}$ and $\bar{A}\cap B$ denote events that exactly one of them will occur.

$$P(A\cap\bar{B})=P(A)\,P(\bar{B})=\frac{5}{9}\cdot\frac{7}{10}=\frac{7}{18}$$

$$=P(\bar{A}\cap B)=P(\bar{A}).P(B)$$

$$=\frac{4}{9}\cdot\frac{3}{10}=\frac{2}{15}$$

Required probability is

$$=\frac{7}{18}+\frac{2}{15}=\frac{35+12}{90}=\frac{47}{90}$$

2. P (neither of them occurs)

$$=P(\bar{A}\cap\bar{B})=P(\bar{A}).P(\bar{B})=\frac{7}{10}\left(\frac{4}{9}\right)=\frac{14}{45}$$

3, 4 and 5.

Let A, B and C be the events that Shiva, Jagan and Rohit hit the target respectively.

$$\therefore \quad P(A)=\frac{2}{3};\quad P(\bar{A})=\frac{1}{3}$$

$$P(B)=\frac{5}{7};\quad P(\bar{B})=\frac{2}{7}$$

$$P(C)=\frac{3}{8};\quad P(\bar{C})=\frac{5}{8}$$

3. P (None of them hit the target)

$$=P(\bar{A}\cap\bar{B}\cap\bar{C})=P(\bar{A}).P(\bar{B}).P(\bar{C})$$

$$=\left(\frac{1}{3}\right)\cdot\left(\frac{2}{7}\right)\cdot\left(\frac{5}{8}\right)=\frac{5}{84}$$

4. Event $(A\cap B\cap\bar{C})\cup(A\cap\bar{B}\cap C)\cup(\bar{A}\cap B\cap C)$ denotes that exactly two persons hit the target, and $A\cap B\cap C$ is the event that all the three persons hit the target.

$$P(A\cap B\cap\bar{C})=P(A)\,P(B)\,P(\bar{C})$$

$$=\frac{2}{3}\cdot\left(\frac{5}{7}\right)\cdot\left(\frac{5}{8}\right)=\frac{25}{84}$$

$$P(A\cap\bar{B}\cap C)=P(A)\,P(\bar{B})\,P(C)$$

$$=\frac{2}{3}\cdot\left(\frac{2}{7}\right)\cdot\left(\frac{3}{8}\right)=\frac{1}{14}$$

$$P(\bar{A}\cap B\cap C)=P(\bar{A})\,P(B)\,P(C)$$

$$=\frac{1}{3}\cdot\left(\frac{5}{7}\right)\cdot\left(\frac{3}{8}\right)=\frac{5}{56}$$

$$P(A\cap B\cap C)=P(A)\,P(B)\,P(C)$$

$$=\frac{2}{3}\cdot\left(\frac{5}{7}\right)\cdot\left(\frac{3}{8}\right)=\frac{5}{28}$$

Hence required probability

$$= P(A \cap B \cap \bar{C}) + P(A \cap \bar{B} \cap C) + P(\bar{A} \cap B \cap C) + P(A \cap B \cap C)$$

$$= \frac{25}{84} + \frac{1}{14} + \frac{5}{56} + \frac{5}{28} = \frac{50+12+15+30}{168} = \frac{107}{168}$$

5. Exactly one of them hits the target

$$(A \cap \bar{B} \cap \bar{C}) \cup (\bar{A} \cap B \cap \bar{C}) + (\bar{A} \cap \bar{B} \cap C)$$

denotes that exactly one of them hits the target.

$$P(A \cap \bar{B} \cap \bar{C}) = P(A) \,.\, P(\bar{B}) \,.\, P(\bar{C})$$

$$= \frac{2}{3} \times \frac{2}{7} \times \left(\frac{5}{8}\right) = \frac{5}{42}$$

$$P(\bar{A} \cap B \cap \bar{C}) = P(\bar{A}).P(B).P(\bar{C}) = \frac{1}{3} \cdot \frac{2}{7} \cdot \frac{3}{8} = \frac{1}{28}$$

$$P(\bar{A} \cap \bar{B} \cap C) = P(\bar{A}).P(\bar{B})P(C) = \frac{1}{3} \cdot \frac{5}{7} \cdot \frac{5}{8} = \frac{25}{168}$$

Hence required probability is

$$= P(A \cap \bar{B} \cap \bar{C}) + P(\bar{A} \cap B \cap \bar{C}) + P(\bar{A} \cap \bar{B} \cap C)$$

$$= \frac{5}{42} + \frac{25}{168} + \frac{1}{28} = \frac{20 + 25 + 26}{168} = \frac{51}{168} = \frac{17}{56}$$

6. Three consecutive letters can be selected from English alphabet in 24 ways.

$\therefore \quad n(S) = 24$

Since, no two consecutive letters are vowels, the number of ways that among three letters one letter is vowel is 13.

$\therefore$ Required probability $= \dfrac{13}{24}$

7. Eight coins are tossed, so $n(S) = 2^8$

Let E be the favorable event. It is required to get 4 heads and 4 tails which is possible in the favorable out come HHHHTTTT.

This can be arranged in $\dfrac{8!}{4!4!}$ ways.

$$n(E) = \frac{8 \times 7 \times 6 \times 5 \times 4!}{4 \times 3 \times 2 \times 1 \times 4!} = 70$$

$\therefore$ Required probability $= \dfrac{n(E)}{n(S)} = \dfrac{70}{2^8} = \dfrac{35}{128}$

8. *Given* : $P(H) = 3P(T)$

We know

$$P(H) + P(T) = 1$$

$$3P(T) + P(T) = 1$$

$$P(T) = \frac{1}{4}$$

$$P(H) = \frac{3}{4}$$

Since the coin is tossed 3 times, two heads may occur in 1st and 2nd trial, 2nd and 3rd trial or 1st and 3rd trial.

$\therefore$ Required probability

$$= \frac{3}{4} \times \frac{3}{4} \times \frac{1}{4} + \frac{1}{4} \times \frac{3}{4} \times \frac{3}{4} + \frac{1}{4} \times \frac{3}{4} \times \frac{3}{4}$$

$$= \frac{9}{64} + \frac{9}{64} + \frac{9}{64} = \frac{27}{64}$$

9. Two black cards can be selected in $26c_2$ ways.

A pack contain 8 black honours.

Hence number ways of selecting two same honours is 4.

$\therefore$ Required probability $= \dfrac{4}{26c_2} = \dfrac{4}{325}$

10. Let one card is king, then second card can be any of the other 3 cards from the other suit and it can be selected in $4 \times 3 = 12$ ways. Since there are four honour cards, hence

$\therefore$ Required probability $= \dfrac{12}{325}$

11, 12 and 13

Probability that Akil reach the summit

$$= P(A) = 2/3$$

$\therefore \quad P(\bar{A}) = 1/3$

Probability that Nikil reach the summit is

$$P(N) = 5/8$$

$\therefore \quad P(\bar{N}) = 3/8$

Probability that Sunil reach the summit

$$P(S) = \frac{4}{7}$$

$\therefore \quad P(\bar{S}) = \dfrac{3}{7}$

11. Probability that none of them reaches the summit

$$= P(\bar{A} \cap \bar{N} \cap \bar{S}) = P(\bar{A})P(\bar{N})P(\bar{S})$$

$$= \frac{1}{3} \times \frac{3}{8} \times \frac{3}{7} = \frac{3}{56}$$

12. Probability that exactly two of them reaches the summit

$$= P(A \cap S \cap \bar{N}) + P(A \cap N \cap \bar{S}) + P(\bar{A} \cap N \cap S)$$

$$= P(A)P(S)P(\bar{N}) + P(A)P(N)P(\bar{S}) + P(N)P(S)P(\bar{A})$$

$$= \frac{2}{3}.\frac{4}{7}.\frac{3}{8} + \frac{2}{3}.\frac{5}{8}.\frac{3}{7} + \frac{1}{3}.\frac{5}{8}.\frac{4}{7} = \frac{37}{84}$$

13. Probability that atleast two of them reaches the summit

= Exactly two of them reach summit

+ three of them reach the summit

$$= \frac{37}{84} + \frac{5}{21} = \frac{57}{84} = \frac{19}{28}$$

14. 2 points determine a line.

Hence12 points can form $^{12}C_2$ straight lines.

But 4 points are collinear, i.e they give only one line.

Number of straight lines that can be formed

$$= {}^{12}C_2 - {}^{4}C_2 + 1 = 66 - 6 + 1 = 61$$

15. Three points determine a triangle.

∴ 12 non-collinear points can give $^{12}C_3$ triangles. But 4 points are collinear, so, these 4 points do not give any triangles.

Hence, number of triangles = $^{12}C_3 - {}^{4}C_3 = 216$

16. Number of non-negative integral solutions of the equation

$$x_1 + x_2 + ... + x_k = n \text{ is } {}^{(n+k-1)}C_{k-1}.$$

Hence number of non-negative integral solutions of

$$x_1 + x_2 + x_3 + x_4 = 10 \text{ is } {}^{13}C_3 = 286.$$

17. Let m, n be the number of girls and the number of boys respectively. Then

$$^{m}C_2 = 66$$

$$\Rightarrow \quad \frac{m(m-1)}{2!} = 66 \Rightarrow m = 12$$

Also, $^{m}C_1 \times {}^{n}C_1 = 240 \Rightarrow n = 20$

Hence, number of games in which both the players were boys = $^{20}C_2 = \dfrac{20(19)}{2} = 190$

18. Points lying on a straight line do not form triangles. Hence, number of triangles that can be formed by using the given points

$$= {}^{19}C_3 - ({}^{9}C_3 + {}^{10}C_3) = 765$$

19. A number is divisible by 125 if and only if number formed by its last three digits is divisible by 125.

So, possible combinations for last three digits are: 000, 125, 250 and 500.

Now, number of three-digit number divisible by 125 is 3.

Number of four-digit numbers divisible by 125

= 4 (4) = 16.

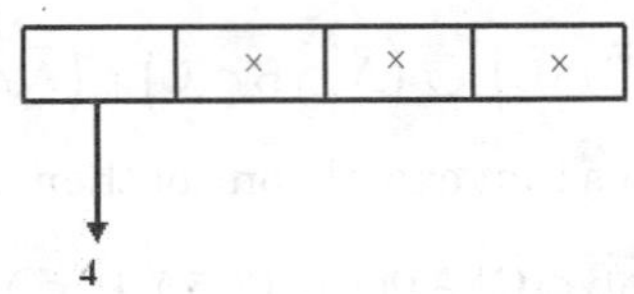

Number of five-digit numbers divisible by 125

= 4 (4) (5) = 80.

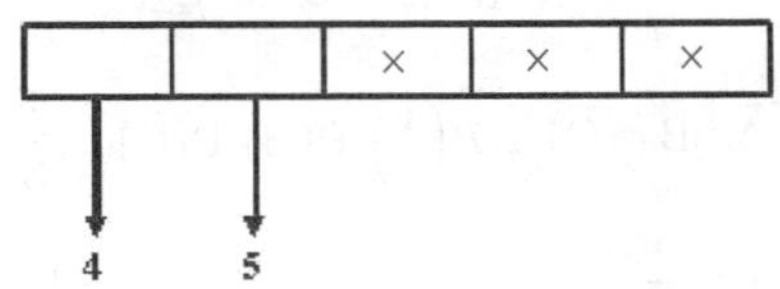

Hence, required number of numbers

= 3 + 16 + 80 = 99.

20. Let the number of red, blue and green balls selected from the supply be x_1, x_2 and x_3 respectively.

Then, $x_1 + x_2 + x_3 = 12$.

Number of non-negative integral solutions of the above equation

$$= {}^{(n+k-1)}C_{k-1} = {}^{14}C_2 = 91.$$

21. Number of positive integral solutions of equation

$$x_1 + x_2 + ... + x_k = n \text{ is } {}^{n-1}C_{k-1}.$$

Number of positive integral solutions of the equation

Hence $\quad a + b + c = 15$

$$^{15-1}C_{3-1} = {}^{14}C_2 = 91$$

22. When 'n' coins are tossed together, then

probability of getting exactly 'r' tails = $\dfrac{^{n}C_r}{2^n}$

Hence probability of getting atleast two tails

= 1 – (probability of getting no tail

+ probability of getting exactly one tail)

$$= 1 - \left\{\frac{^{6}C_0}{64} + \frac{^{6}C_1}{64}\right\} = 1 - \frac{7}{64} = \frac{57}{64}$$

CHAPTER 6

Miscellaneous

LINEAR EQUATIONS

An equation where the maximum power of any variable is unity (one) is a linear equation. A **linear equation** is of the form:

AX + BY + C = O

Where A and B are coefficient and C is a constant

The equation is called **LINEAR** because the graph of the equation on the X-Y Cartesian plane is a straight line. The sets of values of X & Y satisfying any equation(s) are called its solution(s).

Consider the equation $2x + y = 4$. Now, if we substitute $x = -2$ in the equation, we obtain $2.(-2) + y = 4$ or $-4 + y = 4$ or $y = 8$. Hence $(-2, 8)$ is a solution. If we substitute $x = 3$ in the equation, we obtain $2.3 + y = 4$ or $6 + y = 4$ or $y = -2$. Hence $(3, -2)$ is a solution. The following table lists six possible values for x and the corresponding values for y, i.e. six solutions of the equation.

x	–2	–1	0	1	2	3
y	8	6	4	2	0	–2

If we plot the solutions of the equation $2x + y = 4$ which appear in the table above then we see that they all lie on the same line. We call this line the graph of the equation since it corresponds precisely to the solution set of the equation.

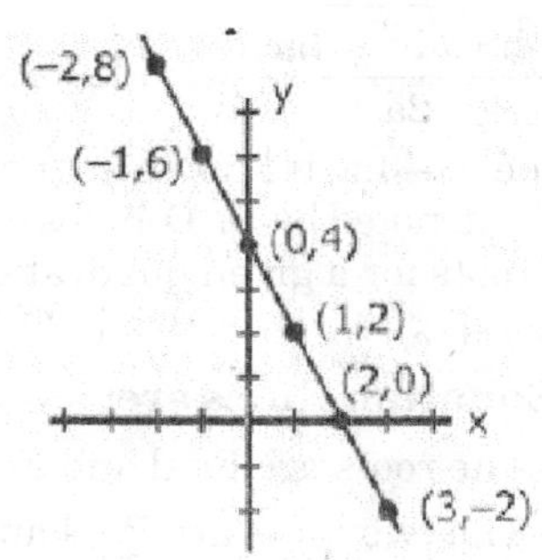

Two Linear Equations in two Unknowns Variables

We now consider a system of two linear equations, in two unknowns x and y:

$a_1x + b_1y = c_1$

$a_2x + b_2y = c_2$

A pair of numbers which satisfies both equations is called a simultaneous solution of the given equations or a solution of the system of equations. There are three cases which can be described geometrically. (Here we assume that the coefficients of x and y in each equation are not both zero.)

1. **The system has exactly one solution:** Here the lines corresponding to the linear equations intersect in one point as shown in fig. below.

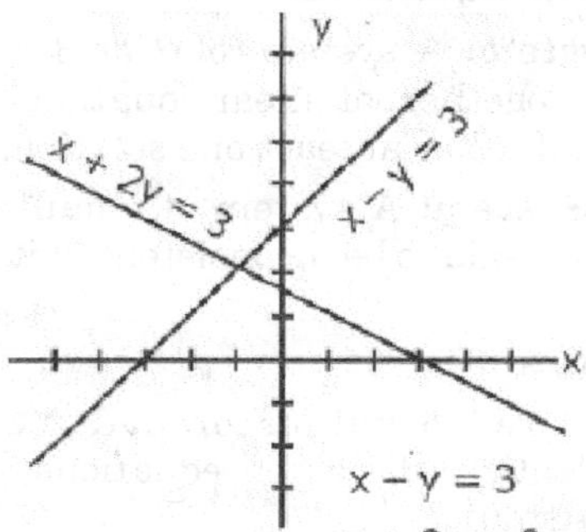

x – y = 3
x + 2y = 3

The system $a_1X + b_1Y = c_1$ and $a_2X + b_2Y = c_2$ has a unique solution, if $\frac{a_1}{a_2} \neq \frac{b_1}{b_2}$.

2. **The system has no solution:** Here the lines corresponding to the linear equations are parallel as shown in fig. below.

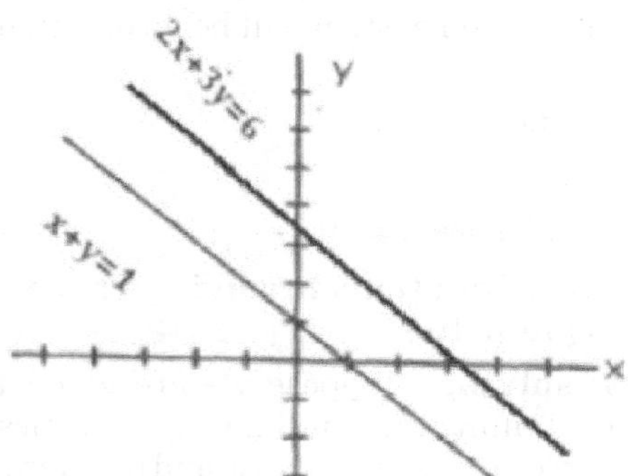

2x+3y=6
x+y=1

The system $a_1X + b_1Y = c_1$ & $a_2X + b_2Y = c_2$ has no solution if $\frac{a_1}{a_2} = \frac{b_1}{b_2} \neq \frac{c_1}{c_2}$.

3. **The system has an infinite number of solution:** Here the lines corresponding to the linear equations coincide as shown in fig. below.

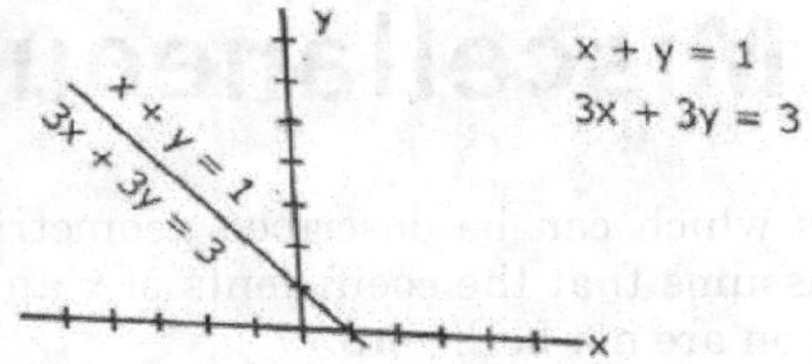

The system $a_1X + b_1Y = c_1$ & $a_2X + b_2Y = c_2$ has infinitely many solution if $\frac{a_1}{a_2} = \frac{b_1}{b_2} = \frac{c_1}{c_2}$

System of Linear Equations

Consistent System: A system (of 2 or 3 or more equations taken together) of linear equations is said to be consistent, if it has at least one solution.

Inconsistent System: A system of simultaneous linear equations is said to be inconsistent, if it has no solutions at all.

e.g. X + Y = 9; 3X + 3Y = 8

Clearly there are no values of X & Y which simultaneously satisfy the given equations. So the system is inconsistent.

Example: Find k for which the system 6x – 2y = 3, kx – y = 2 has a unique solution.

Solution: The given system will have a unique solution

if $\frac{a_1}{a_2} \neq \frac{b_1}{b_2}$; i.e. $\frac{6}{k} \neq \frac{-2}{-1}$ or $k \neq 3$.

Example: What is the value of k for which the system x + 2y = 3, 5x + ky = –7 is consistent?

Solution: The given system will be inconsistent if

$\frac{a_1}{a_2} = \frac{b_1}{b_2} \neq \frac{c_1}{c_2}$ i.e.

if $\frac{1}{5} = \frac{2}{k} \neq \frac{3}{-7}$. Hence k = 10.

Simultaneous Equations Involving Three Unknown Quantities

Method of solving: Suppose we are given three equations containing three unknown quantities X, Y and Z. Mark these equations (1), (2) and (3). Now, from any two equations say (1) and (2), try to eliminate (or get rid of) one unknown quantity, say Z, i.e., to say, multiply (1) and (2) by such suitable number which will make the coefficients of Z in (1) and (2) equal; then add or subtract to cancel Z. Mark this equation (4). After this, take any two equations, say (2) and (3), and from this, eliminate Z in a similar way. Mark this equation (5). Now, solving (4) and (5) by any of the methods given previously, we will find X and Y. Substituting the value of X and Y in any of the equations (1), (2) and (3), we will get the value of Z.

Example: Solve X + Y + Z = 6, 2X + 2Y + 3Z = 13, 3X + 4Y + 5Z = 22

Solution:

We have

X + Y + Z = 6 ... (1)
2X + 2Y + 3Z = 13 ... (2)
3X + 4Y + 5Z = 22 ... (3)

Multiply (1) by 3 and subtract (2) and we have

3(X + Y + Z) – (2X + 2Y + 3Z) = 18 – 13
or X + Y = 5 ... (4)

Again, multiply (1) by 5 and subtract (3), then

5(X + Y + Z) – (3X + 4Y + 5Z) = 30 – 22
or 2X + Y = 8 ... (5)

Subtracting (4) from (5), we have X = 3.

Hence, from (4), Y = 2.

Substituting these values of X and Y in (1), we have 3 + 2 + Z = 6 or Z = 1.

QUADRATIC EQUATION

A quadratic polynomial of the form $ax^2 + bx + c$, where $a \neq 0$ and a, b, c are real numbers, is called a quadratic equation

when $ax^2 + bx + c = 0$.

Here a and b are the coefficients of x^2 and x respectively and 'c' is a constant term.

Any value is a solution of a quadratic equation if and only if it satisfies the quadratic equation.

Quadratic formula: The roots, i.e., α and β of a quadratic equation $ax^2 + bx + c = 0$ are given by

$\frac{-b \pm \sqrt{D}}{2a}$ or $\frac{-b \pm \sqrt{b^2 - 4ac}}{2a}$ provided $b^2 - 4ac \geq 0$.

Here, the value $b^2 - 4ac$ is known as the discriminant and is generally denoted by D. 'D' helps us to determine the nature of roots for a given quadratic equation.

Hence, $D = b^2 - 4ac$.

Quadratic Equation Rules are:

1. If D = 0 ⇒ The roots are Real and Equal.
2. If D > 0 ⇒ The two roots are Real and Unequal.
3. If D < 0 ⇒ No Real roots exist.

If α and β are the roots of the quadratic equation, then Quadratic equation is $x^2 - (\alpha + \beta)x + \alpha\beta = 0$ Or x^2 – (sum of roots) x + product of roots = 0

where, Sum of roots (α + α) = –coefficientofx/ coefficiento fx^2 = –b/a

Product of roots (α × β) = coefficientterm /coefficientofx^2 = c/a

GEOMETRY AND MENSURATION

STRAIGHT LINE

Parallel lines: - Two straight lines are parallel if they lie on the same plane and do not intersect however far produced.

Transversal :- It is a straight line that intersects two parallel lines. When a transversal two parallel lines then

(1) Corresponding angles are equal, ? (that is : For the figure below)

(2) Alternate interior angles are equal, that is (Refer figure below.) 4 = 6; 5 = 3

(3) Alternate exciter or angles are equal, that is 2 = 8; 1 = 7

(4) Interior angles on the same side of transversal add up to 180°, that is 4 + 5 = 33 + 6 = 180°

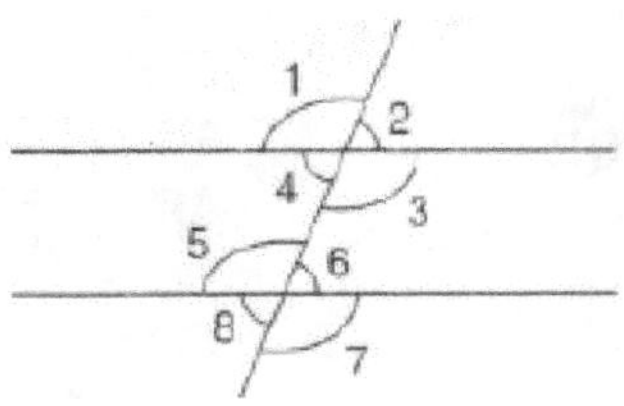

Polygons

Polygons are plane figures formed by a closed series of rectilinear (straight) segments.

Example:

Triangle, Rectangles.

Polygons can broadly be divided into two types:

(a) Regular polygons: Polygons with all the sides and angles equal.

(b) Irregular polygons: Polygons in which all the sides or angles are not of the same measure.

Triangles (Δ)

A triangle I a polygon having three sides. Sum of all the angles of a triangle =180°

Types:

(1) Acute angle triangle: Triangles with all the angles acute (less than 90°)

(2) Obtuse angle triangle: Triangles with one of the angles obtuse (more than 90°)

Note: We cannot have more than one obtuse angle triangle.

(3) Right angle triangle: Triangle with one of the angles equal to 90°

(4) Equilateral triangle: Triangle with all sides equal. All the angles in such a triangle measure 60°

(5) Isosceles triangle: Triangle with two of its sides equal and consequently the angles opposite the equal sides are also equal.

(6) Scalene triangle: Triangle with none of the sides equal to any other side.

Properties (General)

(1) Sum of the length of any two sides of a triangle has to be always greater than the third side.

(2) Difference between the lengths of any two sides of a triangle has to be always lesser than the third side.

(3) Side opposite to the greatest angle will be the greatest and the side opposite to the smallest angle the smallest.

(4) The exterior angle is equal to the sum of two interior angles not adjacent to it.

$\angle ACD = \angle BCE = \angle A + \angle B$

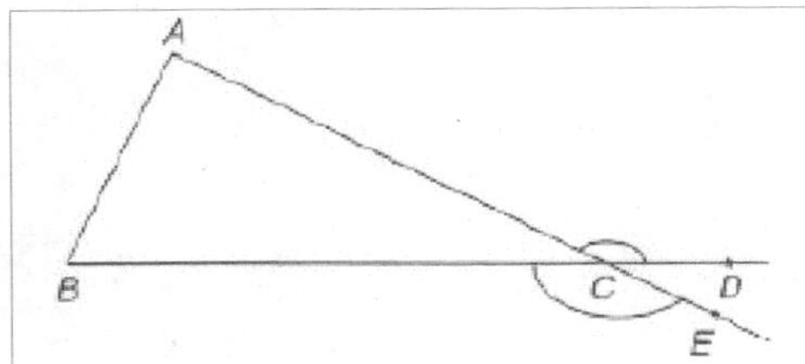

AREA:

(1) Area = 1/2 base × height or 1/2bh.

Height = Perpendicular distance between the base and vertex opposite to it

(2) Area $= \sqrt{s(s-a)(s-b)(s-c)}$ (Heros's formula)

Where $S = \frac{a+b+c}{2}$ (a, b and c being the length of the sides

(3) Area= r × S (where r is in radius)

(4) Area = 1/2 × product of two sides × sine of the included angle.

= 1/2 ac sin B

= 1/2 ab sin B

= 1/2 bc sin B

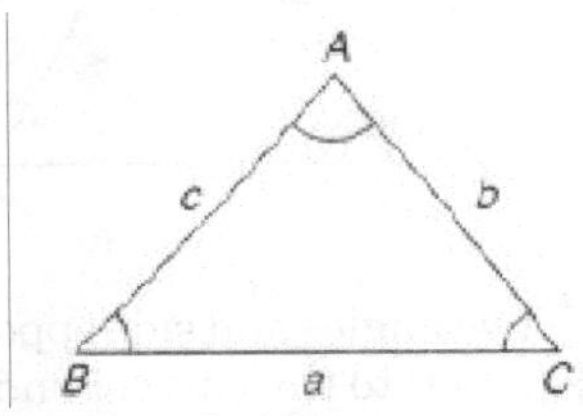

(5) Area = abc/4R

Where R = circum radius

CONGRUENCY OF TRIANGLES

Two triangles are congruent if all the sides of one are equal to the corresponding sides of another, It follows that all the angles of one are equal to the corresponding angles of another. The notation for congruency is ≅

1. **SAS congruency**: If two sides and an included angle of one triangle are equal to two sides and an included angle of another, the two triangles are congruent. (See figure below.)

   Here,

   AB = PQ

   BC = QR

   And ∠B = ∠Q

   So ΔABC ≅ ΔPQR

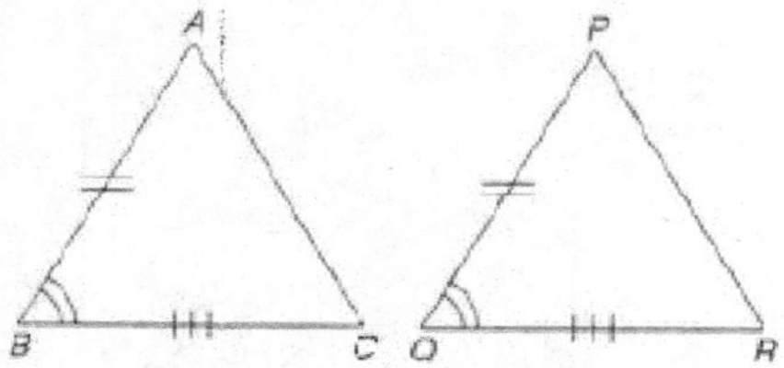

2. **ASA congruency**: If two angles and the included side of one triangle is equal to two angles and the included side of another, the triangles are congruent.

   (See figure below.)

   Here,

   ∠A = ∠P

   ∠B = ∠Q

   And AB = PQ

   So ΔABC ≅ ΔPQR

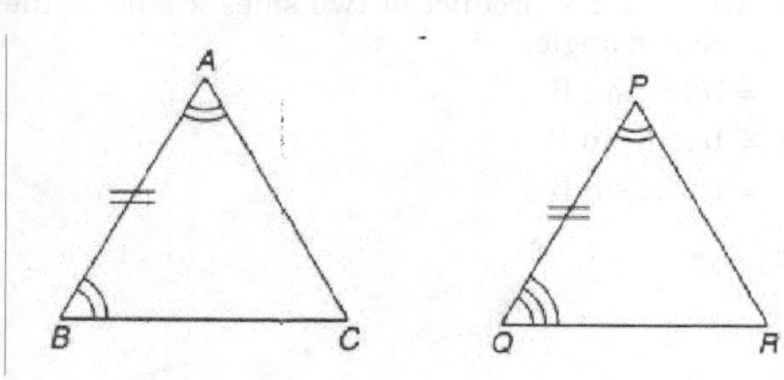

3. **AAS congruency**: If two angles and side opposite to one of the angle is equal to the corresponding angles and the side of another triangle, the triangles are congruent. In the figure below:

   ∠A = ∠P

   ∠B = ∠Q

   And AC = PR

   So ΔABC ≅ ΔPQR

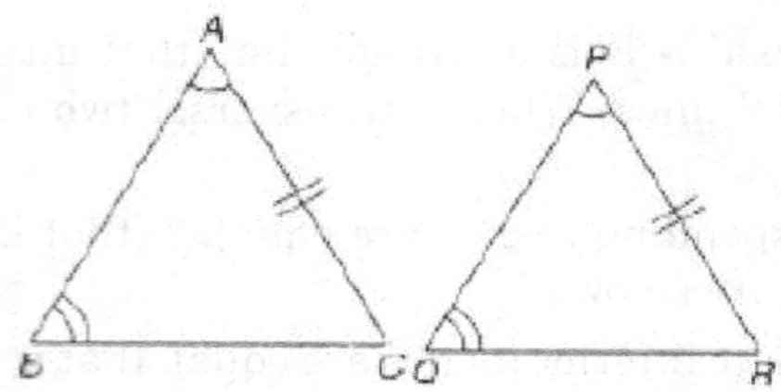

4. **SSS congruency**: If three sides of one triangle are equal to three sides of another triangle, the two triangles are congruent. In the figure below:

   AB = PQ

   BC = QR

   AC = PR

   ∴ ΔABC ≅ΔPQR

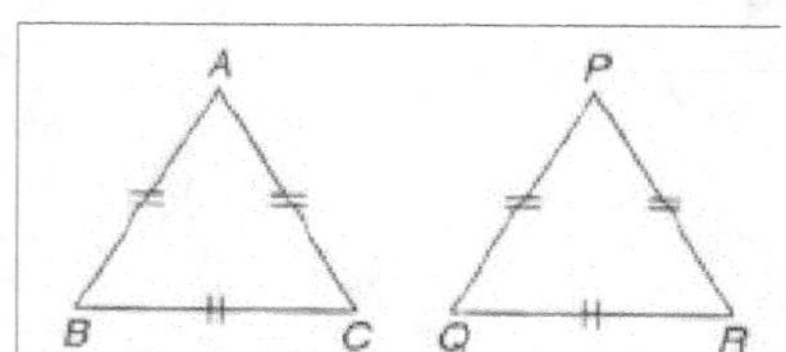

5. **SSA congruency** : If two sides and the angle opposite the greater side of one triangle are equal to the two sides and the angle opposite to the greater side of another triangle, then the triangles are congruent. The congruency doesn't hold if the equal angles lie opposite the shorter side. In the figure below, if the below, if

   AB = PQ

   AC = PR

   ∠B = ∠Q

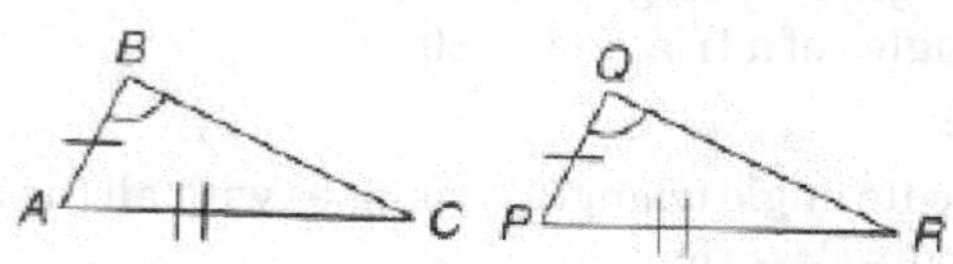

   Then the triangles are congruent.

   i.e., ΔABC ≅ΔPQR

SIMILARITY OF TRIANGLES

Similarity of triangles is a special case where if either of the conditions of similarity of polygons holds if either of the conditions of similarity of polygons holds, the other will hold automatically.

Type of Similarity

(1) AAA similarity: If in two triangles corresponding angles are equal, that is the two triangles are equiangular then triangles are similar. Corollary (AA similarity) If two angles of another triangle then the two triangles are similar. The reason being, the third angle becomes equal automatically.

(2) SSS similarity: If the corresponding sides of two triangles are proportional then they are similar. For ΔABC to be similar to ΔPQR, AB/PQ = BC/QR = AC/PR, must hold true.

(3) SAS similarity: If in two triangles, one pair of corresponding sides are proportional and the included angles are equal then the two triangles are similar.

$\Delta ABC \sim \Delta PQR$

If AB/BC = PQ/QR and $\angle B = \angle Q$

EQUILATERAL TRIANGLES (OF SIDE A)

(1) $h = \frac{a\sqrt{3}}{2} (\therefore \sin 60\sqrt{3}/2 = h/\text{side})$

(2) Area 1/2 (base) × (height)

$= \frac{1}{2} \times a \times \frac{a\sqrt{3}}{4} = a^2$

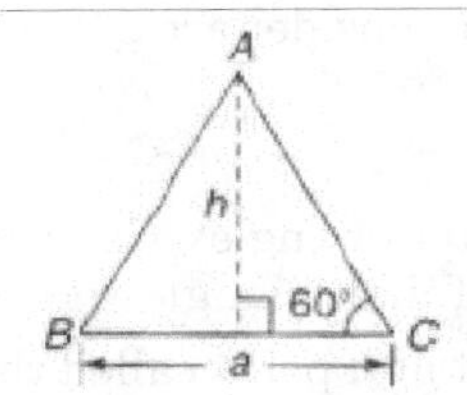

(3) $R\,(\text{circum radius}) = \frac{2h}{3} = \frac{a}{\sqrt{3}}$

(4) $R\,(\text{in radius}) = \frac{h}{3} = \frac{a}{2\sqrt{3}}$

Properties

(1) The in centre and circum centre lies at a point that divides the height in the ratio 2 : 1

(2) The circum radius is always twice the in radius . [R = 2r]

(3) Among all the triangles that can be formed with a given perimeter, the equilateral triangle will have the maximum area.

(4) An equilateral triangles in a circle will have the maximum area compared to other triangles inside the same circle.

ISOSCELES TRIANGLE

$\text{Area} = \frac{b}{4}\sqrt{4a^2 - b^2}$

In an isosceles triangle, the angles opposite to the equal sides are equal.

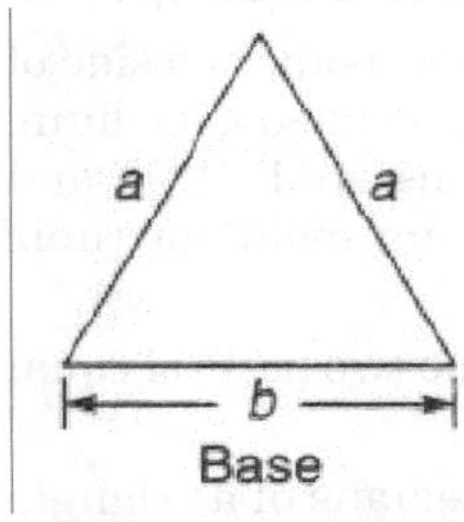

RIGHT- ANGLED TRIANGLE

PYTHAGORAS THEOREM

In the case of a right angled triangle, the square of the hypotenuse is equal to the sum of the squares of the other two sides. In the figure below, for triangle ABC, $a^2 = b^2 + c^2$

Area =1/2 (product of perpendicular sides)

$R\,(\text{circumradius}) = \frac{\text{hypotenuse}}{2}$

Area = rs

(Where r = in radius and s = (a + b + c)/2 where a, b, and c are sides of the triangle)

$\Rightarrow 1/2\ bc = r\,(a + b + c)/2$

$r = (bc)/(a + b + c)/2$

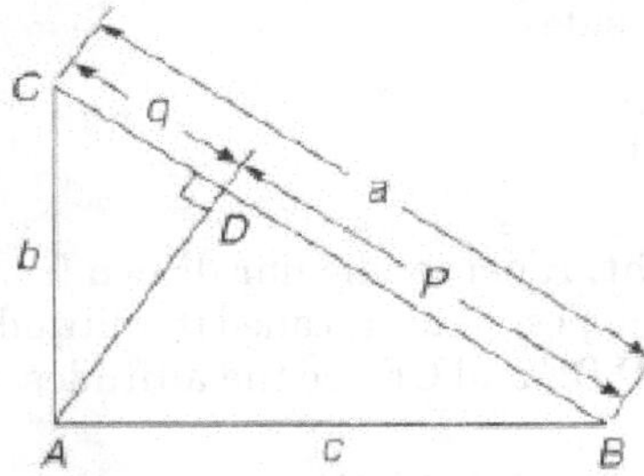

In the triangle ABC,

$\Delta ABC \sim \Delta DBA \sim \Delta DAC$

(1) $\Delta ABC \sim \Delta DBA$

$\therefore AB/BC = DB/BA$

$\Rightarrow AB^2 = DB \times BC$

$\Rightarrow c^2 = pa$

(2) $\Delta ABC \sim \Delta DAC$

$AC/BC = DC/AC$

$\Rightarrow AC^2 = DC \times BC$

$\Rightarrow b^2 = qa$

(3) $\Delta DBA \sim \Delta DAC$

$DA/DB = DC/DA$

$\Rightarrow DA^2 = DB \times DC$

$\Rightarrow AD^2 = pq$

IMPORTANT TERMS OF TRIANGLE

Important terms with respect to a triangle

1. **Median:** A line joining the mid- point of a side of triangle to the opposite vertex is called a medium. In the figure the three medians are PG, QF and RE where G, E and F are mid-point of their respective sides

(1) A median divides a triangle into two parts of equal area.

(2) The point where the three medians of a triangle meet is called the centroid of the triangle.

(3) The centroid of a triangle divides each median in the ratio 2:1

i.e., PC : CG = 2:1 = QC : CF = RC : CE

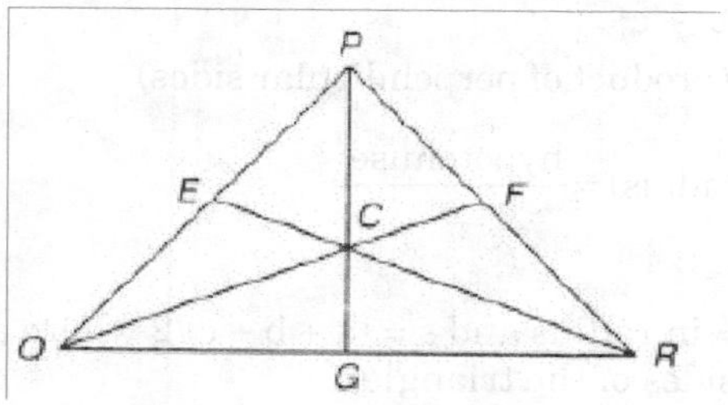

Important formula with respect to a median

→ 2 × (median)^2 + (1/2 the third side)^2 = Sum of the squares of other two sides

$$\Rightarrow 2(PG)^2 + 2\times\left(\frac{QR}{2}\right)^2$$

2. **Altitude/ Height:** A perpendicular drawn from any vertex to the opposite side is called the altitude (In the figure, AD BF and CE are the altitudes of the triangles)

(1) All the altitudes of a triangle met at a point called the orthocenter of the triangle.

(2) The angle made by any side at the orthocenter and the vertical angle make a supplementary pair (i.e., they both add up to 180°). In the figure below:

$\angle A + \angle BOC = 180° = \angle C + \angle AOB$

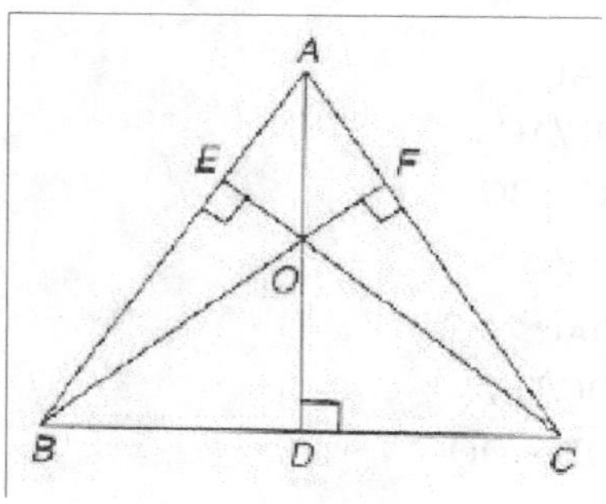

3. **Perpendicular Bisectors:** A line that is a perpendicular to a side and bisects it is the perpendicular bisector of the side.

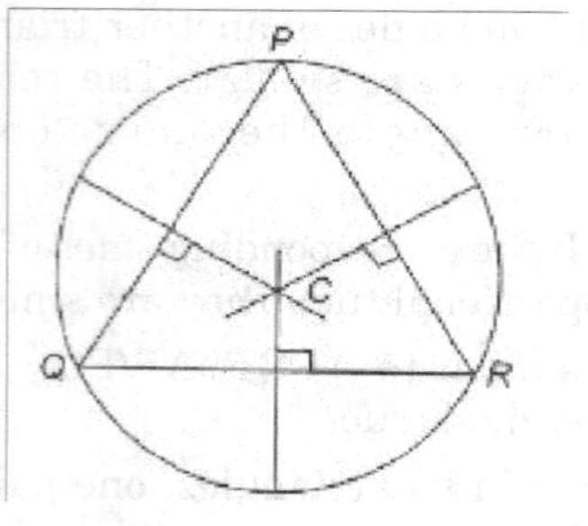

(1) The point at which the perpendicular bisectors of the sides meet is called the circumcentre of the triangle.

(2) The circumcentre is the centre of the circle that circumscribes the triangle. There can be only one such circle.

(3) Angle formed by any side at the circumcentre is two times the vertical angle opposite to the side. This is the property of the circle where by angles formed the property of the circle where by angles formed by an arc at the centre are twice that of the angle formed by the same arc in the opposite arc. Here we can view this as:

$\angle QCR = 2QPR$ (When we consider arc QR and its opposite arc QPR)

4. **In center:**

(1) The lines bisecting the interior angles of a triangle are the angle bisectors of that triangle.

(2) The angle bisectors meet at a point called the in centre of the triangle.

(3) The in centre is equidistant from all the sides of the triangle.

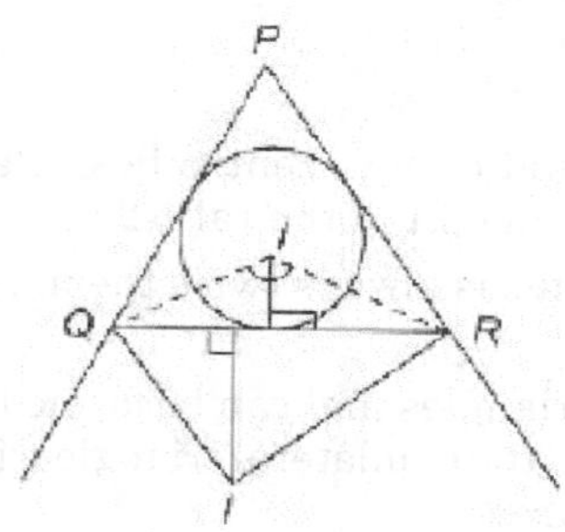

(4) From the in centre with a perpendicular drawn to any of the sides as the radius, a circle can be drawn touching all the three sides. This is called the in circle of the triangle. The radius of the circle is known as in radius.

(5) The angle formed by any side at the incentre is always a right angle more than half the angle opposite to the side. This can be illustrated as

$\angle QIR = 90 + 1/2\angle P$

(6) If QI and RI be the angle bisectors of exterior angles at Q and R then, QIR = 90+1/2 $\angle$P

QUADRILATERALS

A quadrilateral is a plane figure that has four side or ages, and also have four corners or vertices.

Types of Quadrilaterals

1. Parallelogram

A parallelogram is a quadrilateral with opposite sides parallel (as shown in the figure below)

(1) Area = Base (b) × Height (h)

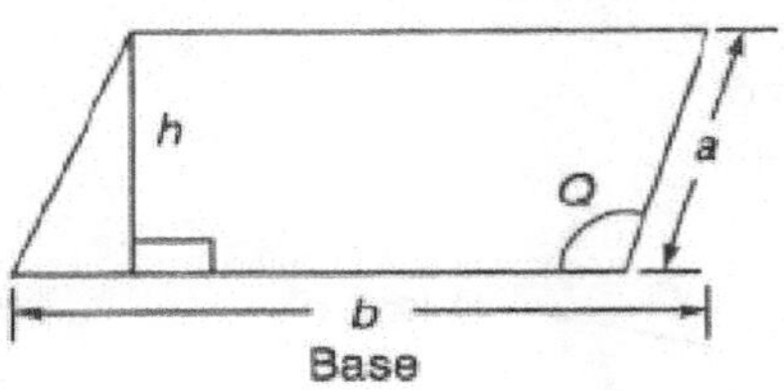

(2) Area = product of any two adjacent sides × sine of the included angle.

(3) perimeter = 2(a + b)

where a and b are any two adjacent sides.

Properties of Parallelogram

(1) Diagonals of a parallelogram bisect each other.

(2) Bisectors of the angles of a parallelogram form a rectangle.

(3) A parallelogram inscribed in a circle is a rectangle.

(4) A parallelogram circumscribed about a circle is a rhombus.

(5) The opposite angles I a parallelogram are equal.

(6) The sum of the squares of the diagonals is equal to the sum of the squares of the four sides in the figure:

$AC^2 + BD^2 = AB^2 + BC^2 + CD^2 + AD^2$

$= 2(AB^2 + BC^2)$

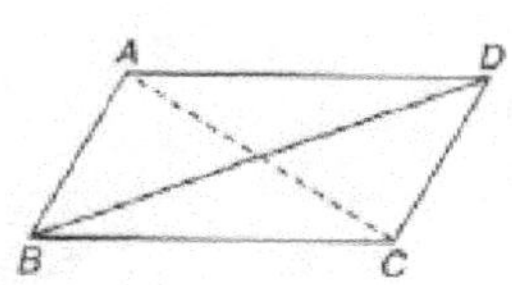

2. Rectangles:

A rectangle is a parallelogram with all angles 90°

(1) Area = Base × Height = b × h

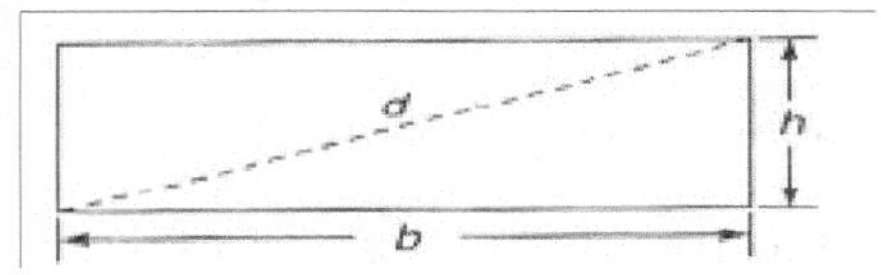

Note: Base and height are also referred to as the length and the breadth in a rectangle.

(2) Diagonal (d) = $\sqrt{b^2 + h^2}$ → by Pythagoras theorem

PROPERTIES OF A RECTANGLE

A parallelogram having all the sides equal is a rhombus.

(1) Area = 1/2 × product of diagonals × sine of the angles between them.

$1/2 \times d_1 \times d_2 \sin 90°$ (diagonals in a rhombus intersect at right angles) = $1/2 \times d_1 \times d_2$ (since sin 90° = 1)

(2) Area = Product of adjacent sides × sine of the angle between them.

(3) Diagonals bisect each other at right angles.

(4) All rhombuses are parallelograms by the reverse is not true.

(5) A rhombus may or may not be a square but all square are rhombus.

3. Properties of Square:

A square is a rectangle with adjacent sides equal or a rhombus with each angle 90°

(1) Area = base × height = a^2

(2) Area = 1/2 (diagonal)^2 = $\frac{1}{2}d^2$ (square s a rhombus too)

(3) Perimeter = 4a (a = side of the square)

(4) Diagonal = $\sqrt{2}$

(5) inradius = $\frac{a}{2}$

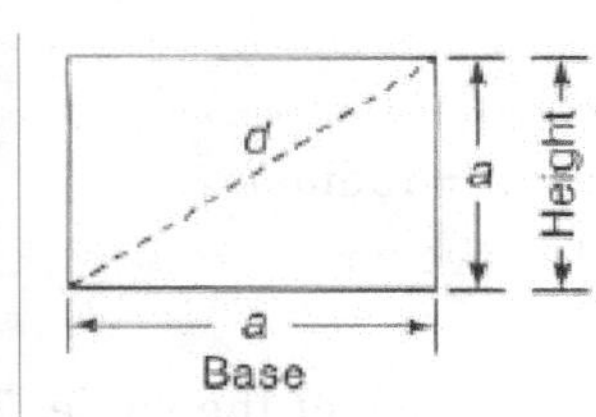

Trapezium of Square

(1) Diagonals are equal and bisect each other at right angles.

(2) Side is the diameter of the inscribed circle.

(3) Diagonal is the diameter of the circumscribing circle

$\Rightarrow$ Diameter $= a\sqrt{2}$

circumradius $= a / \sqrt{2}$

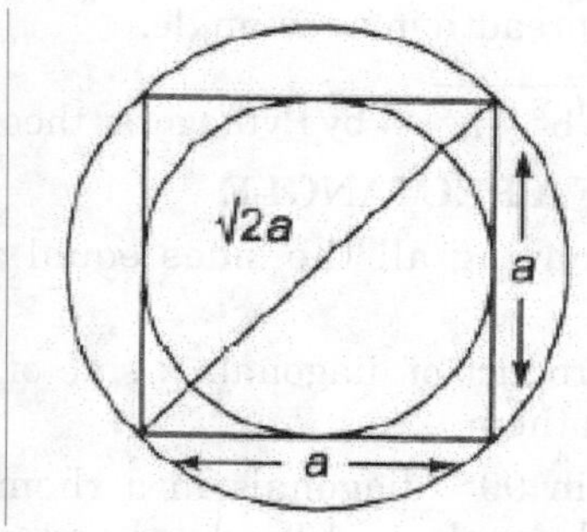

4. Trapezium:

A trapezium is a quadrilateral with only two sides parallel to each other.

(1) Area = 1/2 × sum of parallel sides × height = 1/2 (AB + DC) × h. For the figure below.

(2) Median = 1/2 × sum of the parallel sides*(median is the line equaidstant from the parallel sides)

For any line EF parallel to

$$EF = \frac{\{[P \times (AB)] + [Q \times (DC)]\}}{AD}$$

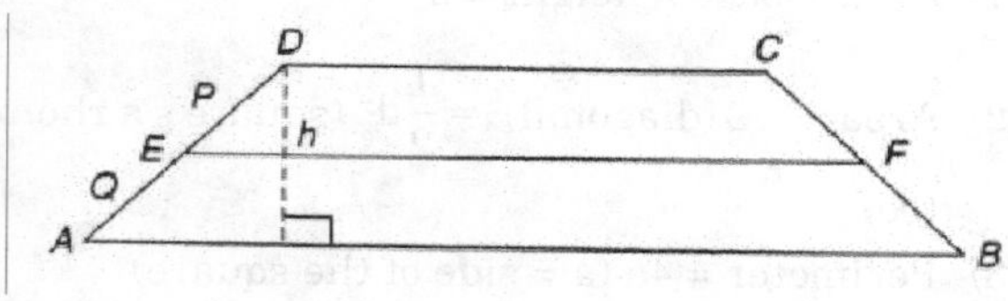

Properties of Trapezium

(1) If the non-parallel sides are equal then diagonals will be equal too.

CIRCLES

(1) Area = πr^2

(2) Circumference = $2\pi r$ = (r = radius)

(3) Area = 1/2 × circumference × r

Arc

It is a part of the circumference of the circle. The bigger one is called the major arc and the smaller one the minor arc.

(1) Length (ArcXY) $= \frac{\theta}{360} \times 2\pi r$

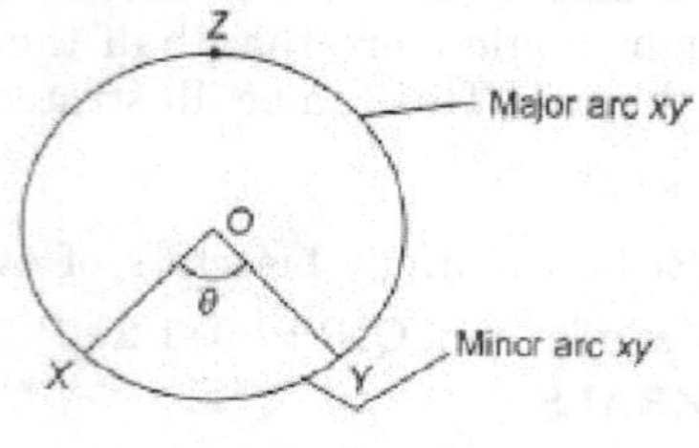

(2) Sector of a circle is a part of the area of a circle between two radii

(3) Area of a sector $= \frac{\theta}{360} \times \pi r^2$

(where θ is the angle between two radii)

= (1/2)r × length (arc xy)

($\therefore \pi r\theta / 180$ = length arc xy)

$$= \frac{1}{2} \times r \times \frac{\pi r \theta}{360}$$

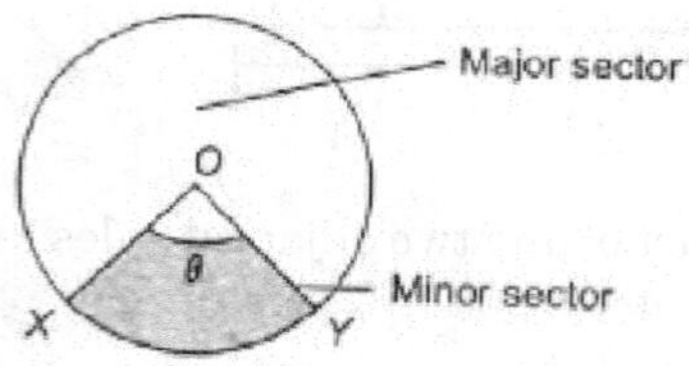

(4) **Segment:** A sector minus the triangle formed by the two radii is called the segment of the circle.

Area = Area of the sector–Area

$$\Delta OAB = \frac{\theta}{360} \times \pi r^2 - \frac{1}{2} \times r^2 \sin\theta$$

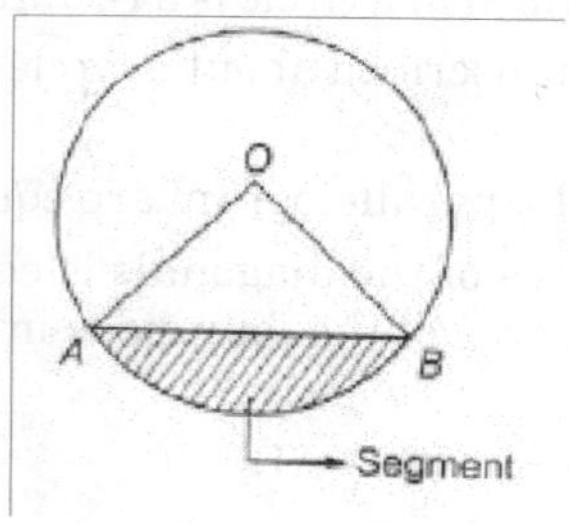

(5) Perimeter of segment = length of the arc + of segment AB

$$= \frac{\theta}{360} \times 2\pi r + 2r\sin\left(\frac{\theta}{2}\right)$$

$$= \frac{\pi r \theta}{180} + 2r\sin\left(\frac{\theta}{2}\right)$$

(6) **Congruency:**

Two circles can be congruent if and only if they have equal radii.

Properties of Circle

(1) The perpendicular from the centre of circle to a chord bisects the chord. The converse is also true.

(2) The perpendicular bisectors of two chords of a circle intersect at its centre.

(3) There can be one and only one circle passing through three of more non-collinear points.

(4) If two circles intersect in two points then the line through the centers is the perpendicular bisector of the common chord.

(5) If two chords of a circle are equal, then the centre of the circle lies on the angle bisector of the two chords.

(6) Equal chords of a circle or congruent circles are equidistant from the centre.

(7) Equidistant chords from the centre of a circle are equal to each other in terms of their length.

(8) The degree measure of an arc of a circle is twice the angle subtended by it at any point on the alternate segment of the circle. This can be clearly seen in the following figure:

With respect to the arc

AB, $\angle AOB = 2\angle ACB$

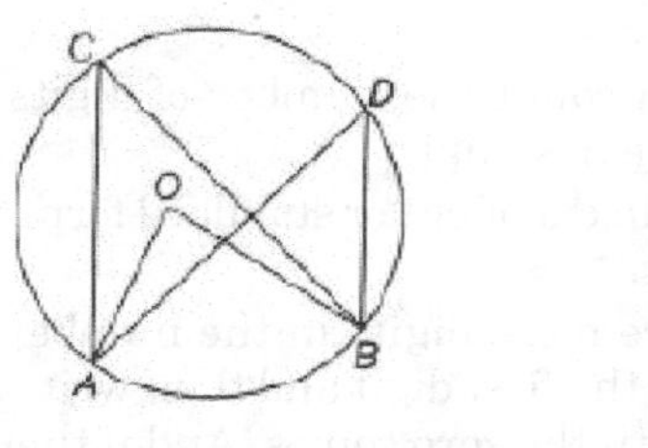

(9) Any two angles in the same segment are equal.

Thus $\angle ACB = \angle ADB$

(10) The angle subtended by a semi-circle is right angle. Conversely, the arc of a circle subtending a right angle at any point of the circle in its alternate segment is a semi-circle.

(11) Any angle subtended by a minor arc in the alternate segment is acute, and any angle subtended by a major arc in the alternate segment is obtuse. In the figure below.

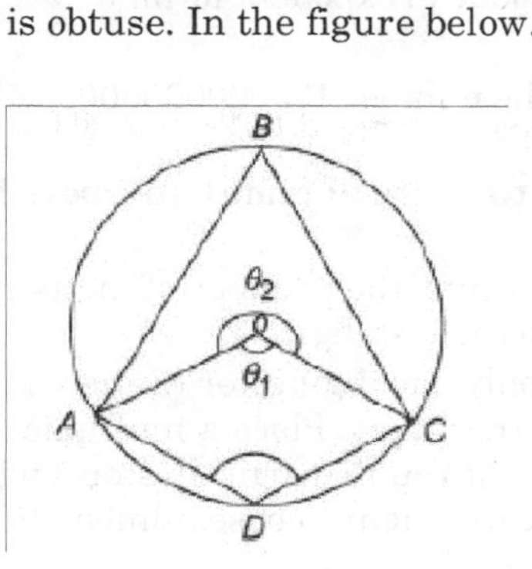

$\angle ABC$ is acute and

$\angle ADC$ = obtuse

Also $\theta_1 = 2\angle B$

And $\theta_2 = 2\angle D$

$\therefore \theta_1 + \theta_2 = 2(\angle B + \angle D)$

$= 360° = 2(\angle B + \angle D)$

or $\angle B + \angle D = 180°$

Or sum of opposite angles of a cyclic quadrilateral is 180°

MENSURATION

The following formulae hold true in the area of mensuration:

1. Cuboid

A cuboids is a three-dimensional box. Its volume is defined by the virtue of its length l breadth b and height h. It can be visualized as a room which has its length, breadth and height different from each other.

(1) Total surface area of a cuboid = 2(lb + bh + lh)

(2) Volume of cuboid = lbh

2. Cube of Side 'S'

A cube is cuboids which has all its edges equal i.e. length = breadth = height = s

(1) Total surface area of a cube = $6s^2$

(2) Volume of the cube = s^2

3. Prism

A prism is a solid which can have any polygon at both its ends. Its dimensions are defined by the dimensions of the polygon at its ends and its height.

(1) Lateral surface area of a tight prism = perimeter of base* height

(2) Volume of a right prism = area of base *height

(3) Whole surface of a right prism = Lateral surface of the prism + area of the two plane ends.

4. Cylinder

Is a solid which has both its ends in the form of a circle. Its dimensions are defined in the form of the radius of the base (r) and the height h. A gas cylinder is a close approximation of a cylinder.

(1) Curved surface of a right cylinder $= 2\pi rh$ where r is the radius of the base and h the height.

(2) Whole surface of a right circular cylinder

$= 2\pi rh + 2\pi r^2$

(3) Volume of a right circular cylinder $\pi r^2 h$

5. Pyramid

A pyramid is a solid which can have any polygon as its base and its edges converge to a single apex. Its dimensions are defined by the dimensions of the polygon at its base and the length of its lateral edges which lead to the apes. The Egyptian pyramids are example of pyramids.

(1) Slant surface of a pyramid = 1/2* Perimeter of the base* slant height

(2) Whole surface of a pyramid = Slant surface + area of the base

(3) Volume of a pyramid $\frac{\text{area of the base *}}{3}$ height

6. Cone

A cone is a solid which has a circle at its base and a slanting lateral surface that converges at the apex. Its dimensions are defined by the radius of the base? (r) the height (h) and the slant height (I) A structure similar to a cone is used in ice cream cones

(1) Curved surface of a cone = πrl where l is the height

(2) Whole surface of a cone $= \pi rl + \pi r^2$

(3) Volume of a cone $= \frac{\pi r^2 h}{3}$

7. Sphere

Is a solid in the form of a ball with radius r.

(1) Surface area of a sphere $= 4\pi^2$

(2) Volume of a sphere $= \frac{4}{3}\pi r^3$

POWER AND EXPONENTS:

The exponents tell us that how many times the number should be multiplied. It is called the Exponential form. Exponential formis written like this:

10^9 ⇒ Exponent

10 → Base

Power

Here 10 is the base and 9 is the exponent and this complete number is the power. We pronounce it as 10 raised to the power 9. The exponent could be positive or negative.

This tells us that the number 10 will be multiplied 9 times, like, 10 × 10 × 10 × 10 × 10 × 10 × 10 × 10 × 10

Powers with Negative Exponents

The exponents could be negative also and we can convert them in positive by the following method.

$$100^{-3} = \frac{1}{100^3}$$

This shows that for any non-zero negative integers a,

$$a^{-m} = \frac{1}{a^m}$$

where m is the positive integer and a^m is the multiplicative inverse of a^{-m}.

Laws of Exponents

If we have a and b as the base and m and n as the exponents, then

$a^m \times a^n = a^{m+n}$

$(a^m)^n$

$\frac{a^m}{a^n} = a^{m-n}, m > n$

$a^m b^m = (ab)^m$

$a^0 = 1$

$a^1 = a$

Use of Exponents to Express Small Numbers in Standard Form:

Sometimes we need to write the numbers in very small or large form and we can use the exponents to represent the numbers in small numbers.

1. Standard form to write the natural numbers like xyz000000......

Step 1: First of all count the number of digits from left leaving only the first digit.

Step 2: To write it in exponent or standard form, write down the first digit.

Step 3: If there are more digits in the number then put a decimal after the first digit and then write down the other digits until the zero comes. And if there are no digits after the first digit then skip this step.

Step 4: Now place a multiplication sign and then write down the counted digits in the first step as the exponent to the base number 10.

2×10^9

2.000000000 (1 2 3 4 5 6 7 8 9)

2,000,000,000

Example:

Express 1730000000000 in exponent form.

Solution:

In standard form, the number 1730000000000 will be written as 1.73×10^{12}.

2. Standard form to write decimal numbers like 0.00000.....xyz.

Step 1: First of all count the number of digits from the decimal point to the last digit.

Step 2: If there is only one digit after the zeros then simply write down that digit. Place a multiplication sign and write down the counted digits in step-1 with a negative sign as the exponent to base number 10.

Step 3: If there are two or more non-zero digits at the end of the number. Then, write down the digits followed by a decimal point after the first digit and the other non-zero digits.

Step 4: Now calculate the number of digits in the first step and minus the number of digits appearing after the decimal point.

Step 5: Place a multiplication sign and write down the counted digits in step-4 with a negative sign as an exponent to base number 10.

2×10^{-9}
0.000000002
9 8 7 6 5 4 3 2 1
0.000000002

Example:

Express 0.000000000000073 in exponent form.

Solution:

In standard form, the number will be written as 7.3×10^{-14}.

Comparing Very Large and Very Small Numbers:

To compare the very large or very small numbers we need to make their exponents same. When their exponents are the same then we can compare the numbers and check which number is large or small.

Example:

Compare the two numbers 4.56×10^8 and 392×10^7.

Solution:

To compare these numbers, we need to make their exponents same.

4.56×10^8

$392 \times 10^7 = 39.2 \times 10^8$

As the exponents are the same, we can easily see that which number is larger.

$392 \times 10^7 > 4.56 \times 10^8$

Note: To add and subtract also we need to make their exponents same and then they can be easily added or subtracted.

LOGARITHMS

Determinants

⇒ The logarithm of a given number base is then index of power to which the base must be raised in order to equal the given number.

⇒ If $a > 0$ and $\neq 1$, then logarithm of a positive number N is defined as the Index x of that power 'a' which equals N i.e., $\log_a N = x$ iff $a^x = N \Rightarrow a^{\log} a^N = N$, $a > 0, a \neq 1$ and $N > 0$

⇒ It is also known as fundamental logarithmic Identity.

⇒ Its domain is $(0, \infty)$ and range is R, a is called the base of the logarithmic function.

⇒ When base 'e' then the logarithmic function is called natural or Napierian logarithmic function and when is 10, then it is called common logarithmic function.

Characteristic and Mantissa

1. The integral part of a logarithm is called the characteristic and the fractional part is called mantissa.

$\log_{10} N$ = integer + fraction (+ve)

integer ↓ Characteristics; fraction ↓ Mantissa

2. If the characteristics of $\log_{10} N$ be n, then the number of digits in digits in N is $(n + 1)$.
3. The mantissa part of log of a number is always kept positive.
4. If the characteristics of $\log_{10} N$ be $(-n)$ then there exists $(n-1)$ number of zeros after decimal part of N.

Properties of Logarithms

1. $\log_a(mn) = \log_a m + \log a^n$
2. $\log_a\left(\frac{m}{n}\right) = \log_a m - \log a^n$
3. $\log_a\left(\frac{1}{n}\right) = -\log a^n$
4. $a \log_a m = m$
5. $\log a\left(\frac{1}{n}\right) = -\log a^n.$
6. $\log_a a = 1, \log_a 1 = 0$
7. $\log_a b . \log_b a = 1 \Rightarrow \log_a b = \frac{1}{\log b^a}$

Logarithmic Inequalities

1. If $a > 1, P > 1 \Rightarrow \log_a P > 0$
2. If $a > 1, 0 < P < 1 \Rightarrow \log_a P < 0$
3. If $0 < a < 1, P > 1 \Rightarrow \log_a P < 0$
4. If $0 < a < P < 1 \Rightarrow 0 < \log_a P < 1$
5. If $p > a > 1 \Rightarrow \log_a P > 1$
6. If $a > p > 1 \Rightarrow 0 < \log_a P < 1$
7. $\log_p a > \log_p b = a \geq b$ if base p is positived and > 1 or $a \leq b$ if base p is positive and < 1 i.e., $0 < p < 1$
8. $\log_m a < b \Rightarrow \begin{cases} a < m^b; \text{ if } m > 1 \\ a > m^b; \text{ if } 0 < m < 1 \end{cases}$
9. If $\log_m a > b \Rightarrow \begin{cases} a > m^b; \text{ if } m > 1 \\ a < m^b; \text{ if } 0 < m < 1 \end{cases}$
10. If $0 < p < a < 1 \Rightarrow \log_a p > 1$

EXERCISE

MCQ TYPE QUESTIONS

1. If one root of a quadratic equation is $3 + \sqrt{2}$, then quadratic equation is

(a) $x^2 - 6x - 7 = 0$ (b) $x^2 + 6x + 7 = 0$

(c) $x^2 + 6x - 7 = 0$ (d) $x^2 - 6x + 7 = 0$

2. Three consecutive positive integers such that square of their sum exceeds the sum of their squares by 214, are

(a) 5, 6, 7 (b) 10, 11, 12

(c) 9, 10, 11 (d) 7, 8, 9

3. A and B attempt to solve a quadratic equation of the form $ax^2 + bx + c = 0$. A starts with a wrong value of b and gets the roots as –3 and –5. B starts with a wrong value of C and gets the roots as 6 and 2. The correct roots are

(a) 3, 4 (b) 3, 5

(c) 4, 6 (d) 2, 3

4. What is the value of

$$\cfrac{1}{5+\cfrac{1}{5+\cfrac{1}{5+\dots}}}\ ?$$

(a) $\dfrac{5+\sqrt{29}}{2}$ (b) $\dfrac{5-\sqrt{29}}{2}$

(c) $\dfrac{5\pm\sqrt{29}}{2}$ (d) 7

5. What is the number of triplets (a, b, c) where a, b, c are positive integers, which satisfy the simultaneous equations ?

(i) $ab + bc = 44$

(ii) $ac + bc = 23$ is/are

(a) 1 (b) 2

(c) 4 (d) Indefinitely many

6. A man sitting beside me at a hotel ate idlis one after the other by ordering plate by plate. He said to me after drinking some water the last one I ate was my hundredth idli in last five days. Each day I ate 6 more than the previous day. How many he ate yesterday?

(a) 26 (b) 24 (c) 22 (d) 20

7. If log3 log4 $x > 0$, then

(a) $x > 1$ (b) $x > 4$

(c) $x > 64$ (d) none of these

8. log 1/4 $(a^2 - 1) < \log_{1/2} (a + 1)^2$ then:

(a) $a < 1$

(b) $a < -1$

(c) $a > 1$

(d) none of these

9. If $\log_e \left(\dfrac{a+b}{2}\right) = \dfrac{1}{2}(\log_e a + \log_e b)$, then:

(a) $a = b$ (b) $a = \dfrac{b}{2}$

(c) $2a = b$ (d) $a = \dfrac{b}{3}$

10. If 2 log $(x + 1)$– 10g $(x^2 - 1)$= log^2, then x equals

(a) 1 (b) 0 (c) 2 (d) 3

11. The number log 2^7 is:

(a) an integer (b) a rational number

(c) an irrational number (d) a prime number

12. If $y = 2^{1/\log x\,(8)}$, then x is equal to:

(a) y (b) y^2

(c) y^3 (d) none of these

13. If $\log_{05} \sin x = 1 - \log_{05} \cos x$, then number of solution of $x \in [-2\pi, 2\pi]$ is:

(a) 1 (b) 2 (c) 3 (d) 4

14. If a, b, c are in G.P. then $\log ax^x$, $\log_{bx} x$, $\log_{cx} x$ are in:

(a) GP (b) HP

(c) AP (d) none of these

15. Consider the following statements:

1. Solution of the inequality $\log_5(x^2 - 11x + 43) < 2$ is (0, 2)

2. If $[x-1]^{(\log_3 \cdot x^2 - 2\log x^9)} = (x-1)^7$ then x is 2 and 81.

Which of these is/are correct?

(a) only (1) (b) only 2

(c) both of these (d) none of these

16. log cos x sin $x \geq 2$ and $x \in [0, 3\pi]$, then sin x lines in the interval:

(a) $\left[0, \dfrac{1}{2}\right]$ (b) $\left[\dfrac{\sqrt{5}-1}{2}, 1\right]$

(c) $\left[0, \dfrac{\sqrt{5}-1}{2}\right]$ (d) none of these

17. If $\log_2 x + \log_2 y \geq 6$, then the least value of $(x + y)$ is

(a) 4 (b) 9

(c) 16 (d) 32

18. If $4^{\log 3\sqrt{3}} + 9^{\log 2^2} = 10^{\log_x 83}, (x \in R)$, then x is:

(a) 4 (b) 9

(c) 10 (d) none of these

19. Which of the following is not true?

(a) $\dfrac{1}{\log_3 \pi} + \dfrac{1}{\log_4 \pi} > 2$

(b) $\log_3 5$ is an irrational number

(c) $\log \sqrt{8x} = \dfrac{10}{3} \Rightarrow x = 16$

(d) If $\log_x (a^2 + 1) < 0$, $(a \neq 0)$then $0 < x < 1$

20. If $y = \dfrac{1}{a^{1-\log ax}}$ and $z = \dfrac{1}{a^{1-\log ay}}$ then x is equal to:

(a) $\dfrac{1}{a^{1+\log_a z}}$ (b) $\dfrac{1}{a^{z+\log_a z}}$

(c) $\dfrac{1}{a^{1-\log_a z}}$ (d) none of these

21. The least value of n in order that the sum of first n terms of an infinite series $1+\dfrac{3}{4}+\left(\dfrac{3}{4}\right)^2+\left(\dfrac{3}{4}\right)+\ldots$ should differ from the sum of the series by less then 10^{-6} is

(Given: $\log_{10} 2 = 0{\cdot}30103$, $\log_{10} 3 = 0.47712$)

(a) 14 (b) 27

(c) 53 (d) 57

22. Solution of the equation $x \log^{x^2} = \log 3 (x + y)$ and $x^2 + y^2 = 65$ is:

(a) $x = 8, y = 1$

(b) $x = 1, y = 8$

(c) $(x = 8, y = 1)$; $(x = 1, y = 8)$

(d) none of the above

23. The identity $\log_a n \log_b n + \log_b n \log_c n + \log_c n \log_a n$ is:

(a) $\dfrac{\log_a n \log_b n \log_c n}{\log_{abc} n}$

(b) $\dfrac{\log_{abc} n}{\log_a n}$

(c) $\dfrac{\log_b n}{\log_{abc} n}$

(d) none of these

24. The least value of the expression $2 \log_{10} x - \log x (0.01)$ for $x > 1$, is:

(a) 10 (b) 2

(c) -0.01 (d) none of the above

25. There are two numbers such that the sum of twice the first number and thrice the second number is 100 and sum of thrice the first number and twice the second number is 120. Which is the largest number?

(a) 32 (b) 12

(c) 14 (d) 35

26. Three mangoes, four guavas and five watermelons cost Rs. 750. Ten watermelon, six mangoes and nine guavas cost Rs.1580. What is the cost of six mangoes, ten watermelon and four guavas?

(a) 1280 (b) 1180

(c) 1080 (d) Cannot be determined

27. The sum of the digits of a two digit number is 9. The digit at the tens place is 4 less than 1/6th of the number. Find the number.

(a) 45 (b) 54

(c) 90 (d) None of these

28. In the figure, ΔABC is similar to ΔEDC

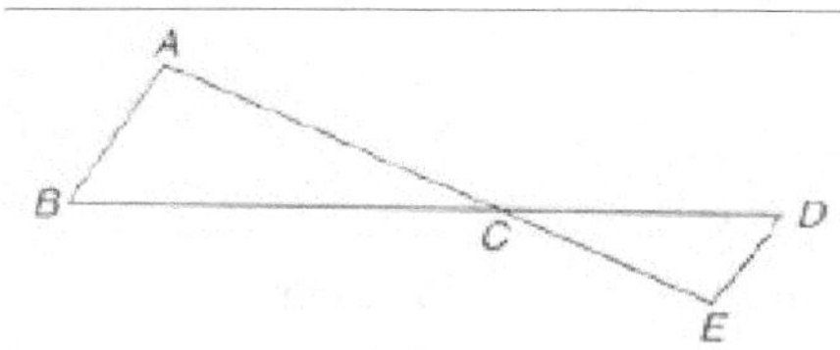

If we have AB = 4cm,
ED = 3cm, CE = 4.2 and
CD = 4.8 cm,
Find the value of CA and CB

(a) 6cm, 6.4 cm (b) 4.8 cm, 6.4 cm

(c) 5.4 cm, 6.4 cm (d) 5.6 cm, 6.4 cm

29. Two poles of height 6 m and 11 m stand vertically upright on a plane ground. If the distance between their foot is 12 m, find the distance between their tops.

(a) 12 cm (b) 14 cm

(c) 13 cm (d) 11 cm

30. If O is the centre of circle, find $\angle$x

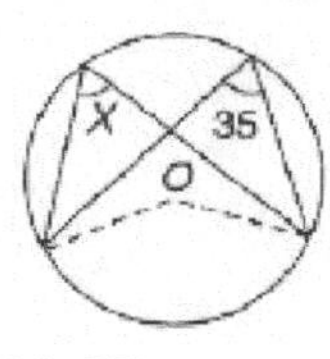

(a) 35° (b) 30°

(c) 39° (d) 40°

31. Find the value of x in the given figure.

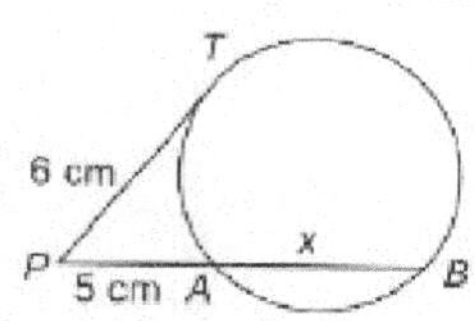

(a) 2.2 cm (b) 1.6 cm

(c) 3 cm (d) 2.6 cm

32. Find the value of x in the given figure.

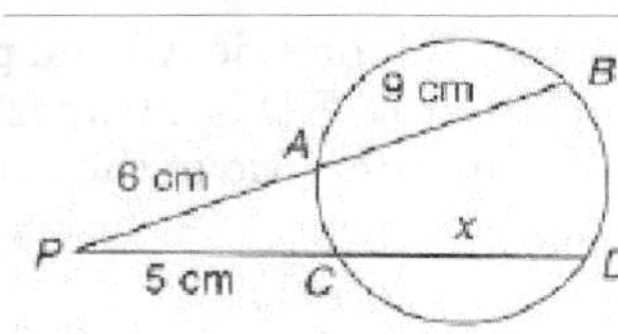

(a) 13 cm (b) 12 cm

(c) 16 cm (d) 15 cm

33. ABC is a right angled triangle with BC = 6cm & AB = 8cm. A circle with centre 0 and radius x has been inscribed in ΔABC. What is the value of x.

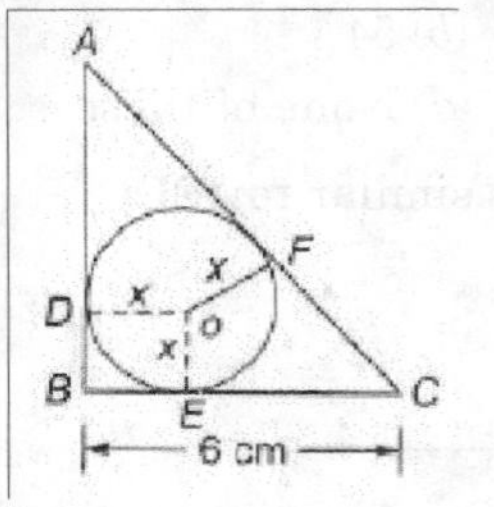

(*a*) 2.4 cm (*b*) 2 cm
(*c*) 3.6 cm (*d*) 4 cm

34. In the given figure find ∠ADB

(*a*) 132° (*b*) 144°
(*c*) 48° (*d*) 96°

35. In the given two straight line PQ and RS intersect each other at O. If ∠SOT = 75°, find the value of a, b, and c.

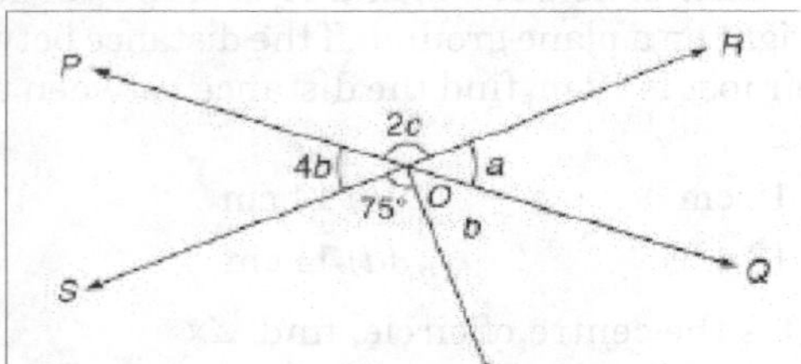

(*a*) a = 84°, b = 21°, c = 48°
(*b*) a = 48°, b = 20°, c = 50°
(*c*) a = 72°, b = 24°, c = 54°
(*d*) a = 64°, b = 28°, c = 45°

36. In the following figure, it is given that O is the centre of the circle and ∠AOC = 140°. Find ∠ABC

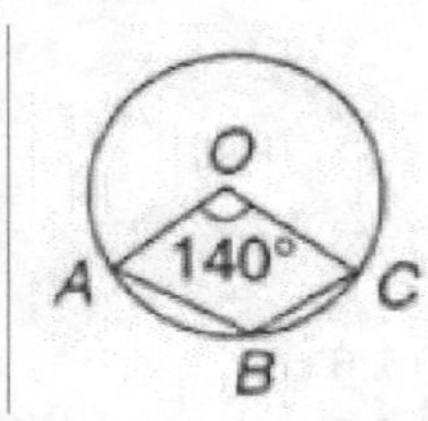

(*a*) 110° (*b*) 120°
(*c*) 115° (*d*) 130°

37. In a triangle ABC, point D is on side AB and point E is on side AC, such that BCED is a trapezium. DE : BC = 3:5. Calculate the ratio of the area of Δ ADE and the trapezium BCED

(*a*) 3:4 (*b*) 9:16
(*c*) 3:5 (*d*) 9:25

38. In the adjoining figure, ABCD is a trapezium in which AB | | DC and AB = 3DC. Determine the ratio of the areas of (ΔAOB and ΔCOD).

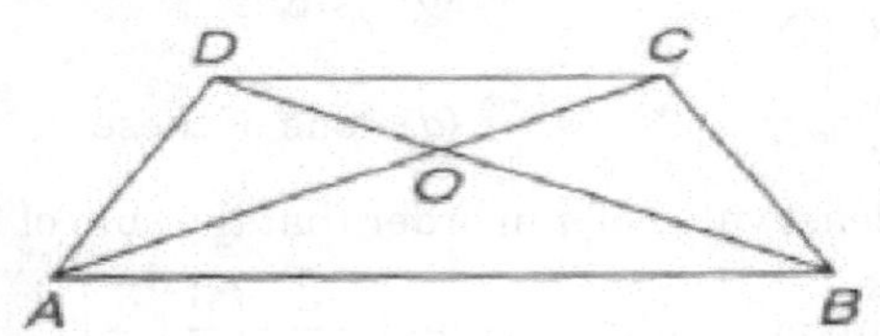

(*a*) 9:1 (*b*) 1:9
(*c*) 3:1 (*d*) 1:3

39. The diagram below represents three circular garbage cans, each of diameter 2 m. The three cans are touching as shown fig, in meters , the perimeter of the rope encompassing the three cans.

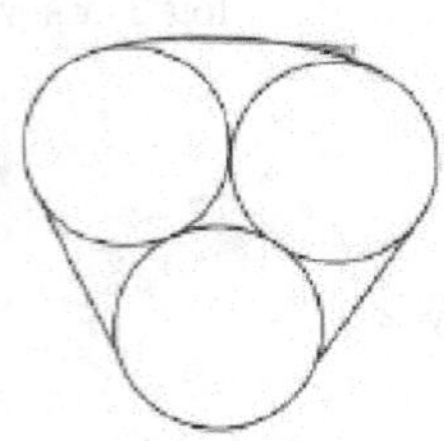

(*a*) $2\pi + 6$ (*b*) $3\pi + 6$
(*c*) $4\pi + 6$ (*d*) $6\pi + 6$

40. In the above figure, AD is the bisector of ∠BAC, AB = 6 cm, AC = 5 cm and BD = 3 cm. Find DC,

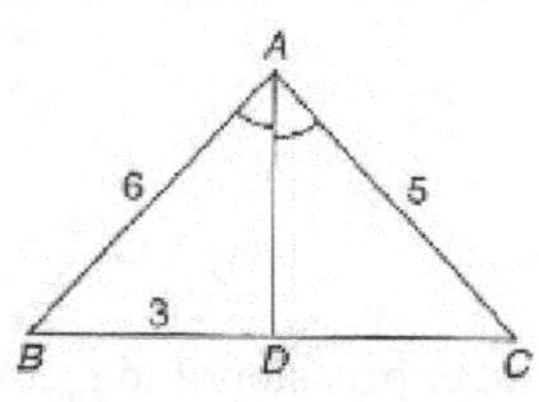

(*a*) 11.3 cm (*b*) 2.5 cm
(*c*) 3.5 cm (*d*) 4 cm

41. ABCD is a trapezium in which AB is parallel to DC, AD = BC, AB = 6 cm, AB = EF and DF = EC. If two lines AF and BE are drawn so that area of ABEF is half of ABCD. Find $\frac{DF}{CD}$.

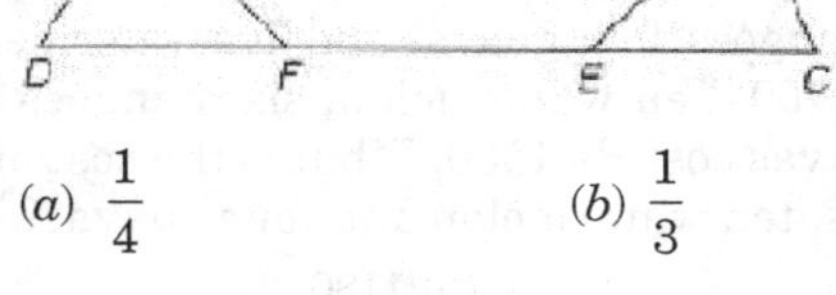

(*a*) $\frac{1}{4}$ (*b*) $\frac{1}{3}$
(*c*) $\frac{2}{5}$ (*d*) $\frac{1}{6}$

42. The diagram represents the area swept by the wiper of a car. With the dimensions given in the figure, calculate the shaded area swept by the wiper.

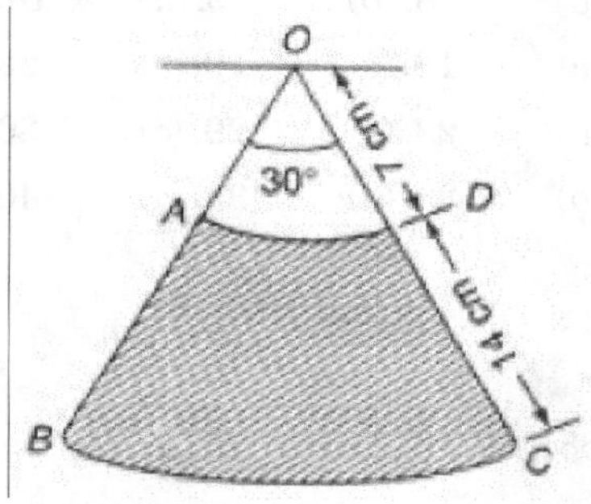

(*a*) 102.67 cm (*b*) 205.34 cm

(c) 51.33 cm (*d*) 208.16 cm

NUMERICAL TYPE QUESTIONS

1. When $x^5 + x^4 + 5x^2 - 3$ is divided by $(x + 2)$, the remainder will be______

2. Pressure of a gas is inversely proportional to the volume. If pressure of oxygen is 84 atm when the volume is 1000 cm^3, then volume when the pressure is 28 atm will be______ cm^3

3. The value of 'a' will make $x^2 - ax + 25$ a perfect square is ______

4. The value of $\alpha^3 + \beta^3$ if α and β are roots of the equation : $2x^2 - 7x - 4 = 0$ is ______

5. Sum of Anita's and Sunita's age is 43 years. 11 years hence, Anita's age will be $\frac{7}{6}$ times Sunita's age then. Sunita's present age is___ years

6. In a school trip, the students were accommodated in two rooms A and B. If 5 students are shifted from room A to room B then there will be equal number of students in both the rooms. But if 5 students from room B is shifted to room A, then the resulting number of students in room A is double the number of students left in room B. Total number of students are ______

7. A band of workers unloaded a certain number of boxes from their ship. The number of workers were between 50 and 60 (both inclusive). Each worker handled exactly 11 boxes, and each box was handled by exactly 7 workers. Exactly______number of boxes were there?

8. If ax = b,by = c ,cz = a, then value of xyz is______

9. The value of

$$\frac{1}{\left(1+\frac{xb}{xa}+\frac{xc}{xa}\right)}+\frac{1}{\left(1+\frac{xa}{xb}+\frac{xc}{xb}\right)}+\frac{1}{\left(1+\frac{xb}{xc}+\frac{xa}{xc}\right)}$$

is ______

10. A palm tree was 90 cm high, when it was planted. It grows by an equal number of cm each year, and at the end of the seventh year it was one ninth taller than at the end of the sixth year. The height of tree in cm at the end of the twelfth year is ______

11. A bottle and its cork together cost ₹ 1.10, and the bottle costs ₹ 1.00 more than its cork. The price of the bottle is ______

12. The value of is (100 – 1) (100 – 2) (100 – 3) .... (100 + 1) (100 + 2) (100 + 3) is ______

13. The value of $\dfrac{2+\dfrac{1}{3+\dfrac{4}{5}}}{2+\dfrac{1}{3+\dfrac{1}{1+\dfrac{1}{4}}}}$ is ______

14. Three integers, a, b and c, are such that $\frac{b}{c} = 3$. The fraction $\frac{b}{a}$ is in its lowest form. The______ is the possible value of a.

ANSWERS

MCQ Type Questions

1. (*d*)	**2.** (*a*)	**3.** (*b*)	**4.** (*a*)	**5.** (*b*)	**6.** (*a*)	**7.** (*b*)	**8.** (*b*)	**9.** (*a*)	**10.** (*d*)
11. (*c*)	**12.** (*c*)	**13.** (*b*)	**14.** (*b*)	**15.** (*b*)	**16.** (*c*)	**27.** (*c*)	**18.** (*c*)	**19.** (*c*)	**20.** (*c*)
21. (*c*)	**22.** (*a*)	**23.** (*a*)	**24.** (*d*)	**25.** (*a*)	**26.** (*b*)	**27.** (*b*)	**28.** (*d*)	**29.** (*c*)	**30.** (*a*)
31. (*a*)	**32.** (*a*)	**33.** (*b*)	**34.** (*a*)	**35.** (*a*)	**36.** (*a*)	**37.** (*b*)	**38.** (*a*)	**39.** (*a*)	**40.** (*b*)
41. (*b*)	**42.** (*a*)								

Numerical Type Questions

1. 1	**2.** 3000	**3.** 10	**4.** $\frac{511}{8}$	**5.** 19	**6.** 60	**7.** 88	**8.** 1	**9.** 1
10. 450	**11.** 1.05	**12.** 0	**13.** 1	**14.** 104				

EXPLANATIONS

MCQ TYPE QUESTIONS

1. If one root = $3 + \sqrt{2}$, then other root = $3 - \sqrt{2}$

$\therefore$ Sum of roots = $3 + \sqrt{2} + 3 - \sqrt{2} = 6$

$\therefore$ Product of roots

$= (3 + \sqrt{2})(3 - \sqrt{2}) = 9 - 2 = 7$

Quadratic equation is

x^2 – (Sum of roots) x + Product of roots = 0

i.e., $x^2 - 6x + 7 = 0$

2. Let consecutive positive integers be x – 1, x and x + 1

$\therefore [x - 1 + x + x + 1]^2$

$= (x - 1)^2 + x^2 + (x + 1)^2 + 214$

$(3x)^2 = x^2 + 1 - 2x + x^2 + x^2 + 2x + 1 + 214$

$9x^2 = 3x^2 + 216$

$6x^2 = 216$

$x^2 = 36$; $x = \pm 6$

But x is a positive integer, hence x = 6

So, numbers are 5, 6, and 7.

3. Equation formed by A with the roots –3 and –5 is

$x^2 - (-8x) + 15 = 0$

i.e., $x^2 + 8x + 15 = 0$

But value of B is wrong.

Hence coefficient of x is wrong.

Equation formed by B with roots 6 and 2 is

$x^2 - 8x + 12 = 0$

But the value of C is wrong.

hence constant term is wrong.

So, correct equation is

$x^2 - 8x + 15 = 0$.

Hence, real roots are 3 and 5.

4. Let

$$x = 5 + \cfrac{1}{\left[5 + \cfrac{1}{5 + \cfrac{1}{5 + \ldots}}\right]}$$

i.e., $x = 5 + \frac{1}{x} = \frac{5x + 1}{x}$

$\Rightarrow x^2 = 5x + 1$

$\Rightarrow x^2 - 5x - 1 = 0$

$$\Rightarrow x = \frac{5 \pm \sqrt{(-5)^2 - 4(-1)(1)}}{2}$$

$$= \frac{5 \pm \sqrt{25 + 4}}{2} = \frac{5 \pm \sqrt{29}}{2}$$

Since, all the terms are positive. The answer has to be $\frac{5 + \sqrt{29}}{2}$.

5. Equation (ii) is c(a + b) = 23, a prime number.

So, 2 factors must be 1, 23.

Since a, b are positive, therefore a + b > 1

$\Rightarrow$ c = 1; a + b = 23

Put c = 1; b = 23 – a in (i)

$\Rightarrow a^2 - 22a + 21 = 0$

$\Rightarrow$ a = 1 or 21

$\Rightarrow$ b = 22 or 2

$\therefore$ Solution sets are (1, 22, 1); (21, 2, 1).

6. First day the number of idlis he ate be x

Second day the count is (x+6)

Third day ---------------------- (x+12)

Fourth day -------------------(x+18)

Fifth day -----------------------(x+24)

Total is5x + 6(1+2+3+4) =100

5x + 60 =100

x = 8

Day	Idlis
1	8
2	14
3	20
4	26
5	32

So, on fourth day, number of idlis the man ate were 26.

7. We have log3 log4 $x > 0$

$\Rightarrow \log 4\, x > 1 \Rightarrow x > 4'$

$\Rightarrow x > 4$

8. $\log_{1/4}(a^2 - 1) < \log_{(1/4)}(a+1)^2$

$\Rightarrow a^2 - 1 > (a+1)^2$

$\Rightarrow a^2 - 1 > a^2 + 1 + 2a$

$\Rightarrow 2a + 2 < 0 \Rightarrow a < -1$

9. $\log\left(\frac{a+b}{2}\right) = \frac{1}{2}(\log_e a + \log_e b)$

$\Rightarrow \quad \frac{a+b}{2} = \sqrt{ab}$

$\Rightarrow \quad a + b - 2\sqrt{ab} = 0$

$\Rightarrow \quad \sqrt{a} = \sqrt{b} \Rightarrow a = b$

10. By definition $x \neq 1, = 1$

Given equation can be written as

$$\log\frac{(x+1)^2}{x^2-1} = \log 2 \Rightarrow \frac{x+1}{x-1} = 2 \Rightarrow x = 3$$

11. Let $\quad x = \log 2^7$

$\Rightarrow \quad 2^x = 7$

which is only possible for irrational number.

12. $y = 2^{1/\log x(8)}$

$\Rightarrow y = 2^{\log 8x} \Rightarrow y = 2^{\log 2\sqrt[3]{x}}$

$\Rightarrow x = y^3$

13. log 0.5 sin x + log 0.5 cos x = 1

$\Rightarrow$ log 0.5 sin x cos x = 1

$\Rightarrow \sin x \cos x = \frac{1}{2} \Rightarrow \sin 2x = 1$

$\Rightarrow 2x = n\pi + (-1)^n \frac{\pi}{2} \quad \Rightarrow x = \frac{n\pi}{2} + (-1)^n \frac{\pi}{4}$

Since $\log_{0.5} \sin x$ and $\log_{0.5} \cos x$ are real.

$\therefore$ sin x and cos x must lie in first quardrant

$\therefore \ x = \frac{\pi}{4}, -2\pi + \frac{\pi}{4}$

Hence, the number of solutions is 2.

14. $\log_{ax} x = \frac{1}{\log_x ax} \Rightarrow \log_{ax} x = \frac{1}{1 + \log_x a}$

If a, b, c are in G.P

$\Rightarrow \log_x a, \log_x b, \log_x c$ are in AP

$\Rightarrow 1 + \log_x a,\ 1 + \log_x b,\ 1 + \log_x C$ are in AP

$\Rightarrow \frac{1}{1+\log_x a}, \frac{1}{1+\log_x b}, \frac{1}{1+\log_x c}$ are in HP.

$\Rightarrow \log_{ax} x, \log_{bx} x \log_{cx} x$ are in HP.

15. (1) $a\log_5(x^2 - 11x + 43) < 2$

and $x^2 - 11x + 43 > 0$

$\Rightarrow x^2 - 11x + 43 < 5^2$

and $\left(x - \frac{11}{2}\right)^2 + \frac{51}{4} > 0$

$\Rightarrow x^2 - 11x + 18 < 0$

and $(x-2)(x-9) < 0$

solution is (2, 9)

(2) For domain $|x, -1| \neq 0, \neq -1$

Now, $|x - 1| = 1, \Rightarrow x - 1 = \pm 1$

$\Rightarrow \quad x = 0, 2$

$x = 0$ is not in the domain and $x = 2$ satisfies the given equation.

If $x - 1 > 0$ i.e., $x > 1$ then the given equation becomes

$$2\log_3 x - \frac{4}{\log_3 x} = 7 \Rightarrow x = 81, \frac{1}{\sqrt{3}}$$

But $\frac{1}{\sqrt{3}}$ being less than 1 is not valid.

Hence, $x = 2, 81$

16. log cos x sin $x \geq 2 \Rightarrow \sin x \leq \cos^2 x$

$\Rightarrow \sin x \leq 1 - \sin^2 x$

$\Rightarrow \sin^2 x + \sin x - 1 \leq 0$

$\Rightarrow \left(\sin x + \frac{1}{2}\right)^2 - \frac{5}{4} \leq 0$

Also by definition of logarithm

$\sin x > 0, \cos x > 0, \cos x \neq 1$

$\Rightarrow \sin x + \frac{1}{2} \leq \frac{\sqrt{5}}{2} \quad \Rightarrow 0 < \sin x \leq \frac{\sqrt{5}-1}{2}$

17. $\log_2 xy \geq 6 \Rightarrow xy \geq 2^6$

Now $\frac{x+y}{2} \geq \sqrt{xy}$

$\Rightarrow \frac{x+y}{2} \geq 2^3 \Rightarrow x + y \geq 2^4 \Rightarrow x + y \geq 16$

18. $4^{\log_3 31/2} + 9^{\log_2 2^2} = 10^{\log x^{83}}$

$\Rightarrow \quad 4^{1/2} + 9^2 = 10^{\log} x^{83}$

$\Rightarrow \quad 2 + 81 = 10^{\log} x^{83}$

$\Rightarrow \quad 83 = 10^{\log} x^{83}$

$\Rightarrow \quad x = 10$

19. (*a*) $\dfrac{1}{\log_3 \pi} + \dfrac{1}{\log_4 \pi} = \log_\pi 3 + \log_\pi 4 = \log_\pi 12 > 2$

$\Rightarrow \quad 12 > \pi^2$

(*b*) $\log_3 5$ is an irrational number

(*c*) $\log \sqrt{8} x = \dfrac{10}{3} \Rightarrow x = \left(\sqrt{8}\right)^{10/3} = 2^5$

$\Rightarrow \quad x = 32$

(*d*) $\log_x (a^2 + 1) < 0, a \neq 0 \Rightarrow a^2 + 1 > 1$

So, log is negative

Hence, base is (0, 1).

20. From the given relation, we have

$a = y^{1 - \log ax} = z^{1 - \log ay}$

$\therefore \quad \log_a a = (1 - \log_a x) \log_a y$

and $\quad \log_a a = (1 - \log_a y) \log_a z$

$\Rightarrow \log_a y(1 - \log_a x) = 1$

and $\log_a z\,(1 - \log_a y) = 1$

$\Rightarrow \quad \log_a y = \dfrac{1}{1 - \log_a x}$

and $\quad \log_a z = \dfrac{1}{1 - \log_a y}$

$\therefore \quad \log_a z = \dfrac{1}{1 - \log_a y} = \dfrac{1}{1 - \dfrac{1}{1 - \log_a x}}$

$= \dfrac{1 - \log_a x}{-\log_a x}$

Now $\dfrac{1}{1 - \log_a z} = \dfrac{1}{1 + \dfrac{1 - \log_a x}{\log_a x}} = \log_a x$

$\log_a x = \dfrac{1}{1 - \log_a z}$

$x = a^{\dfrac{1}{1 - \log_a z}}$

21. Series $1 + \dfrac{3}{4} + \left(\dfrac{3}{4}\right)^2 + \ldots + n \text{ terms} = \dfrac{1 - \left(\dfrac{3}{4}\right)^n}{1 - \dfrac{3}{4}}$

and $1 + \dfrac{3}{4} + \left(\dfrac{3}{4}\right)^2 + \ldots = \dfrac{1}{1 - \dfrac{3}{4}} = 4$

Now, $\dfrac{1 - \left(\dfrac{3}{4}\right)^n}{\dfrac{1}{4}} = 4 - 10^{-6}$

$\Rightarrow \quad 1 - \left(\dfrac{3}{4}\right)^n = 1 - \dfrac{1}{4}(10^{-6})$

$\Rightarrow \quad \left(\dfrac{3}{4}\right)^n = \dfrac{10^{-6}}{4}$

$\Rightarrow n \log_{10}\left(\dfrac{3}{4}\right) = \log_{10} 10^{-6} - \log_{10} 4$

$\Rightarrow n(0.47712 - 2 \times 0.30103) = -6 - 2 \times (0.30103)$

$\Rightarrow \quad n = \dfrac{6 \cdot 60206}{0 \cdot 12494} = 53$

22. Given that, $x^{\log x^2} = 2 = \log_3 (x + y)$

$\Rightarrow \quad x + y = 9$ and $x^2 + y^2 = 65$

$\Rightarrow \quad x = 8, y = 1$ or $x = 1, y = 8$

But $\quad x \neq 1$

$\therefore \quad x = 8, y = 1$

23. $\log_a n \log_b n + \log_b n \log_c n + \log_c n \log_a n$

$= \dfrac{1}{\log_n a \log_n b} + \dfrac{1}{\log_n b \log_n c} + \dfrac{1}{\log_n c \log_n a}$

$\left[\because \log_m n = \dfrac{1}{\log_n m}\right]$

$= \dfrac{\log_n c + \log_n a + \log_n b}{\log_n a \log_n b \log_n c}$

$= \dfrac{\log_n (abc)}{\log_n a \log_n b \log_n c} = \dfrac{\log_a n \log_b n \log_c n}{\log_{abc} n}$

24. Here, $2 \log_{10} x - \log_x (10)^{-2} = 2 \log_{10} x + 2 \log x 10$

$= 2 \log_{10} x + 2 \dfrac{1}{\log_{10} x}$

$= 2\left\{\log_{10} x + \dfrac{1}{\log_{10} x}\right\} \quad \ldots(i)$

Using AM $\geq$ GM we get

$\dfrac{\log_{10} x + \dfrac{1}{\log_{10} x}}{2} \geq \left(\log_{10} x \dfrac{1}{\log_{10} x}\right)^{1/2}$

$\Rightarrow \log_{10} x + \dfrac{1}{\log_{10} x} \geq 2$

$\therefore 2 \log_{10} x - \log_x (0.01) \geq 4$

$\therefore$ Least value is 4.

26. $3M + 4G + 5W = 750$ ...(i)

$6M + 9G + 10W = 1580$ ...(ii)

Now, if we look at the equation (i) and multiply it by 2, we get:

$6M + 8G + 10W = 1500$ ...(iii)

From equation (ii) and (iii) we get, $G = 80$

If we subtract the cost of 4 guavas from equation (iii) we would get:

$6M + 4G + 10W = 1500 - 320 = 1180$

27. Let, the two digits number be $10x + y$

$x + y = 9$ ...(1)

$x = \frac{1}{6}(10x + y) - 6$ ...(2)

Solving Eq (1) and (2) we get,

$x = 5$ and $y = 4$

Therefore, Number = 54

28. $\Delta ABC \sim \Delta CDE$

Then $\frac{AC}{CD} = \frac{BC}{CE} = \frac{AB}{DE}$

Then $\frac{AC}{4.8} = \frac{4}{3} = AC = 6.4$ cm and

$\frac{BC}{4.2} = \frac{4}{3} = BC = 5.6$ cm

29.

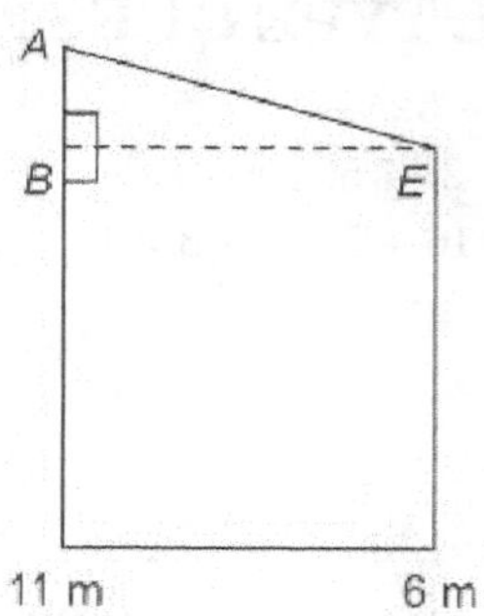

$BC = ED = 6$ m

$AB = AC - BC = 11 - 6 = 5$ m

$CD = BC = 12$ m

Then by Pythagoras theorem :

$AC^2 = AB^2 + BE^2 \Rightarrow AE = 13$ m

30. $\angle x = 35°$ because angles subtended by an arc anywhere on the circumference are equal.

31. By the rule of tangents, we know :

$6^2 = (5 + n)\ 5$

$\Rightarrow 36 = 25 + 5x$

$\Rightarrow 11 = 5x$

$\Rightarrow x = 2.2$ cm

32. By the rule of chords, cutting externally, we get

$\Rightarrow (9 + 6)6 = (5 + x)5$

$\Rightarrow 90 = 25 + 5x$

$\Rightarrow 5x = 65$

$\Rightarrow x = 13$ cm

33. In radius $= \frac{\text{area}}{\text{semi perimeter}}$

$= \frac{24}{12} = $ cm

34. ADBC is a cyclic quadrilateral as well its four vertices are on the circumference of the circle. Also, the opposite angles of the cyclic quadrilateral are supplementary.

Therefore, $\angle ADB = 180 - 48° = 132°$

35. $\angle POS = \angle QOR$

(Vertically opposite angles)

So $a = 4b$

$\angle SOT + \angle TOQ + \angle QOR = 180°$

(sum of angles on a line = 180°)

$4b + 2c = 180°$

$84 + 2c = 180°$

$\Rightarrow 2c = 96°$

$\Rightarrow c = 48°$

So $a = 84°$, $b = 21°$ $c = 48°$

36.

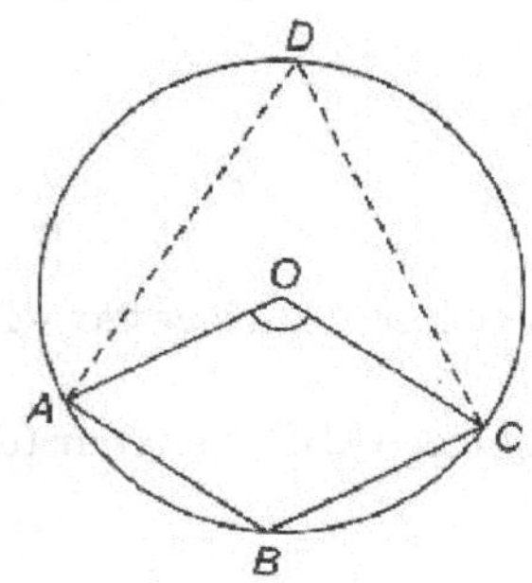

$\angle ADC = \frac{140}{2} 70°$ (because the angle subtended by an arc on the circumference is half of what it subtends at the centre.) ABCD one cyclic quadrilateral So $\angle ABC = 180° - 70° = 110°$ (because opposite angles of a cyclic quadrilateral are supplementary.)

37.

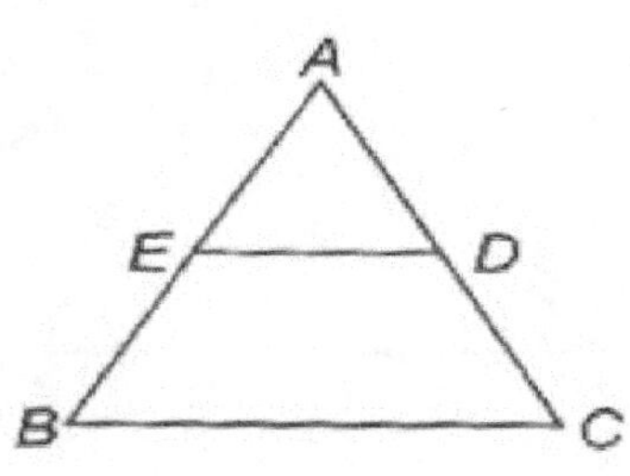

ΔADE is similar to ΔABC (AAA property)

ED : BC = 3 : 5

Area of ΔADE : Area of ΔABC = 9 : 25

Area of trapezium = area of ABC

Area of ΔADE = 25 – 9 = 16

Thus, Area of ΔADE : Area of trapezium EDBC = 9 : 16

38. $\angle DOC$ and ΔAOB are similar by AAA property)

AB : DC = 3 : 1

So area of AOB: Area of DOC = $(3 : 1)^2$

$\Rightarrow 9 : 1$

39. $\angle AOB = \angle COD = \angle FOE = 120°$

Distance between 2 centers = 2 m

$\therefore$ BC = DE = FA = 2m

Perimeter of the figure = BC + DE + FA + circumference of sector AB,CO'B and FO'' E.

But three equal sectors of 120° =1 full circle of same radius

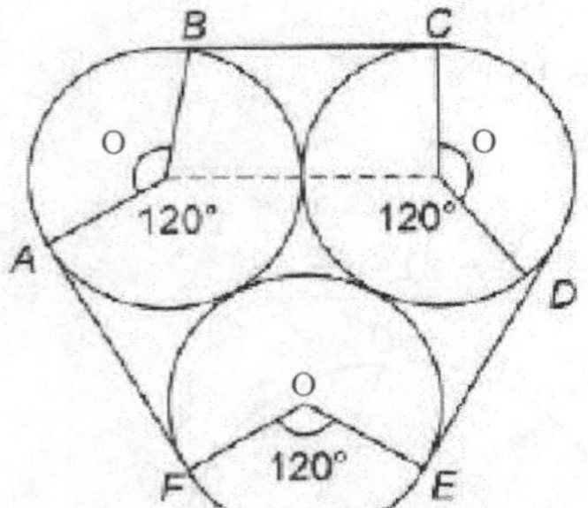

Therefore, perimeter of surface $2\pi r$ + BC + DE + FA = $(2\pi + 6)$m

40. In the given figure, ΔABD is similar to ΔADC

Then

$$\frac{AB}{BD} = \frac{AC}{DC}$$

$$\Rightarrow \frac{6}{3} = \frac{5}{DC}$$

$\Rightarrow$ DE = 2.5 cm

41.

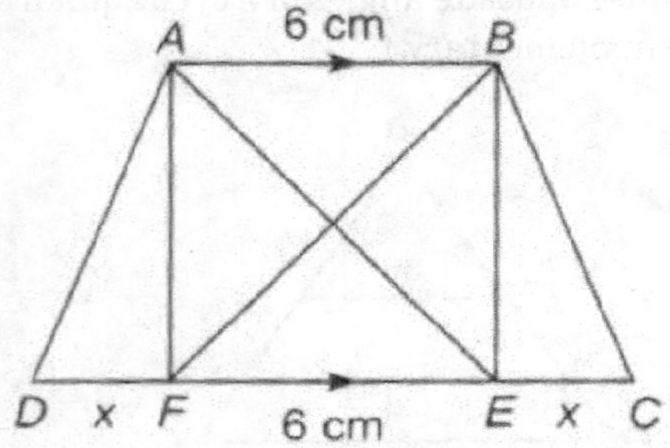

In the above question

FE = AB = 6 cm

$\Delta ADF \cong \Delta BEC$; so DF = EC

Let DF = EC = x

Solving through options; e.g. options

(b) $\frac{1}{3}$; x = 6

Then by Pythagoras triplet AF = 8

Area of ABEF = 8 × 6 = 48 cm^2

Area of $\Delta AFD + \Delta BEC$

$$= 2 \times \frac{1}{2} \times 6 \times 8$$

$\Rightarrow$ 48cm^2

$\therefore$ Area of ABCD = 48 + 48 = 96 cm^2

Hence the condition is proved.

42. Larger Radius (R) = 14 + 7 = 21 cm

Smaller Radius (π) = 7 cm

Area of shaded portion

$$\pi R^2 = \frac{8}{360} - \frac{\pi r^2 \theta}{360}$$

$$\Rightarrow \frac{\pi\theta}{360}\left(21^2 - 7^2\right)$$

$\Rightarrow$ 102.67 cm

NUMERICAL TYPE QUESTIONS

1. $p(x) = x^5 + x^4 + 5x^2 - 3$

$= (-2)^5 + (-2)^4 + 5(-2)^2 - 3$

$= (-32) + 16 + 5 \times 4 - 3 = 1$

2. $p \propto \frac{1}{v}$

$\therefore$ $p = \frac{k}{v}$

where, k = 84 × 1000

$\therefore$ $v = \frac{k}{p} = \frac{84 \times 1000}{28} = 3000$ cm^3

3. Middle term = $2\sqrt{\text{first term} \times \text{last term}}$

ax = 2 × x × 5 = 10x

$\therefore$ a = 10

4. $\alpha + \beta = \frac{7}{2}$

$\alpha\beta = -\frac{4}{2} = -2$

$(\alpha + \beta)^3 = \alpha^3 + \beta^3 + 3\alpha\beta(\alpha + \beta)$

$$\left(\frac{7}{2}\right)^3 = \alpha^3 + \beta^3 + 3(-2)\left(\frac{7}{2}\right)$$

$$\frac{343}{8} = \alpha^3 + \beta^3 - 21$$

$$\frac{343}{8} + 21 = \alpha^3 + \beta^3$$

$$\therefore \quad \frac{511}{8} = \alpha^3 + \beta^3$$

5. Let Anita's present age be 'x' years and that of Sunita by 'y' years.

$\therefore \quad x + y = 43 \quad ...(1)$

and $(x + 11) = \frac{7}{6}(y + 11)$

$\Rightarrow \quad 6x + 66 = 7y + 77$

$\Rightarrow \quad 6x - 7y = 11 \quad ...(2)$

Multiplying (1) by 6.

$6x + 6y = 258 \quad ...(3)$

Subtracting (3) from (2),

$-13y = -247$

$\therefore \quad y = 19$ years.

6. Let the number of students in room A be x and number of students in room B be y.

$\therefore \quad x - 5 = y + 5$

$x - y = 10 \quad ...(1)$

Also, $x + 5 = 2(y - 5)$

$\Rightarrow \quad x - 2y = -15 \quad ...(2)$

Solving (1) and (2) we get,

$x = 35$ and $y = 25$

$\therefore \quad x + y = 60$

7. Let P be the number of workers, and let C be the number of boxes. Then, by counting the number of boxes handling in 2 different ways yields $11P = 7C$.

Since P and C are integers, it follows that P must be divisible by 7.

Since $50 \le P \le 60$,

P must equal 56

$$\therefore \quad C = 11 \times \frac{56}{7} = 88$$

8. $ax = b$

$\Rightarrow \quad (cz)\ x = b \quad$ ....[since $cz = a$]

$\Rightarrow \quad (by)\ xz = b \quad$ ....[since $by = c$]

$\Rightarrow \quad xyz = 1$

9. $$\frac{1}{\left(1 + \frac{xb}{xa} + \frac{xc}{xa}\right)} + \frac{1}{\left(1 + \frac{xa}{xb} + \frac{xc}{xb}\right)} + \frac{1}{\left(1 + \frac{xb}{xc} + \frac{xa}{xc}\right)}$$

$$= \frac{xa}{(xa + xb + xc)} + \frac{xb}{(xa + xb + xc)} + \frac{xc}{(xa + xb + xc)}$$

$$= \frac{(xa + xb + xc)}{xa + xb + xc} = 1$$

10. Let tree grows x cm each year

Height of the tree at the end of the sixth year

$= (90 + 6x)$ cm

Growth in seventh year,

$$x = \frac{1}{9}(90 + 6x) \text{ cm}$$

$$x = 10 + \frac{2x}{3} \qquad \therefore \quad x = 30$$

Hence height of tree at the end of the twelfth year $= (90 + 12 \times 30) = 450$cm.

11. Let B = price of the bottle

C = price of the cork

Given : B + C = ₹ 1.10 ...(1)

and B – C = ₹ 1.00 ...(2)

From equations (1) and (2)

B = ₹ 1.05

C = ₹ 0.05.

12. $(100 - 1)(100 - 2)(100 - 3)$.........

$(100 - 100)(100+1)(100+2)(100+3)$

$\Rightarrow \quad (100 - 1)(100 - 2)(100 - 3)$.........

$(0)(100 + 1)(100 + 2)(100 + 3) = 0.$

13. $$\frac{2 + \frac{1}{\frac{19}{5}}}{2 + \frac{1}{3 + \frac{4}{5}}} = \frac{2 + \frac{5}{19}}{2 + \frac{5}{19}} = 1$$

14. $\frac{b}{c} = 3 \Rightarrow b = 3c$, i.e., b is divisible by 3.

Also, $\frac{b}{c}$ cannot be reduced,

i.e., a is not divisible by 3.

ENGINEERING MATHEMATICS

1

CHAPTER

Linear Algebra

VECTOR SPACES

Let V be a non-empty set of certain objects, which may be vectors, matrices, functions or some other objects. Each object is an element of V and is called a vector. The elements of V are denoted by a, b, c, u, v, etc. Assume that the two algebraic operations

(*i*) vector addition and

(*ii*) scalar multiplication

are defined on elements of V.

If the vector addition is defined as the usual addition of vectors, then

$$\mathrm{a} + \mathrm{b} = (a_1, a_2, ..., a_n) + (b_1, b_2, ..., b_n) = (a_1 + b_1, a_2 + b_2, ..., a_n + b_n).$$

If the scalar multiplication is defined as the usual scalar multiplication of a vector by the scalar α, then

$$\alpha\mathrm{a} = \alpha(a_1, a_2, ..., a_n) = (\alpha a_1, \alpha a_2, ..., \alpha a_n).$$

The set V defines a vector space if for any elements a, b, c in V and any scalars α, β the following properties (axioms) are satisfied.

Properties (axioms) with respect to vector addition

1. $\mathbf{a} + \mathbf{b}$ is in V.
2. $\mathbf{a} + \mathbf{b} = \mathbf{b} + \mathbf{a}$. (commutative law)
3. $(\mathbf{a} + \mathbf{b}) + \mathbf{c} = \mathbf{a} + (\mathbf{b} + \mathbf{c})$. (associative law)
4. $\mathbf{a} + \mathbf{0} = \mathbf{0} + \mathbf{a} = \mathbf{a}$. (existence of a unique zero element in V)
5. $\mathbf{a} + (-\mathbf{a}) = \mathbf{0}$. (existence of additive inverse or negative vector in V)

Properties (axioms) with respect to scalar multiplication

6. $\alpha\,\mathbf{a}$ is in V.
7. $(\alpha + \beta)\,\mathbf{a} = \alpha\mathbf{a} + \beta\mathbf{a}$. (left distributive law)
8. $(\alpha\beta)\,\mathbf{a} = \alpha(\beta\,\mathbf{a})$.
9. $\alpha(\mathbf{a} + \mathbf{b}) = \alpha\mathbf{a} + \alpha\mathbf{b}$. (right distributive law)
10. $1\mathbf{a} = \mathbf{a}$. (existence of multiplicative inverse)

The properties defined in 1 and 6 are called the *closure* properties. When these two properties are satisfied, we say that the vector space is closed under the vector addition and scalar multiplication. The vector addition and scalar multiplication defined above need not always be the usual addition and multiplication operators. Thus, *the vector space depends not only on the set V of vectors, but also on the definition of vector addition and scalar multiplication on V.*

If the elements of V are real, then it is called a *real vector space* when the scalars α, β are real numbers, whereas V is called a *complex vector space,* if the elements of V are complex and the scalars α, β may be real or complex numbers or if the elements of V are real and the scalars α, β are complex numbers.

Remark

(*a*) If even one of the above properties is not satisfied, then V is not a vector space. We usually check the closure properties first before checking the other properties.

(*b*) The concepts of length, dot product, vector product etc. are not part of the properties to be satisfied.

(*c*) The set of real numbers and complex numbers are called *fields* of scalars. We shall consider vector spaces only on the *fields* of scalars. In an advanced course on linear algebra, vector spaces over arbitrary fields are considered.

(*d*) The vector space V= {0} is called a trivial vector space.

The following are some examples of vector spaces under the usual operations of vector addition and scalar multiplication.

1. The set V of real or complex numbers.
2. The set of real valued continuous functions f on any closed interval $[a, b]$. The **0** vector defined in property **4** is the zero function.
3. The set of polynomials P_n of degree less than or equal to n.
4. The set V of n-tuples in R^n or C^n.
5. The set V of all $m \times n$ matrices. The element **0** defined in property **4** is the null matrix of order $m \times n$.

The following are some examples which are not vector spaces. Assume that usual operations of vector addition and scalar multiplication are being used.

1. The set V of all polynomials of degree n. Let P_n and Q_n be two polynomials of degree n in V. Then, $\alpha\mathrm{P}_n + \beta\mathrm{Q}_n$ need not be a polynomial of degree n and thus may not be in V. For example, if $\mathrm{P}_n = x^n + a_1x^{n-1} + ... + a_n$ and $\mathrm{Q}_n = -x^n + b_1x^{n-1} + ... + b_n$, then $\mathrm{P}_n + \mathrm{Q}_n$ is a polynomial of degree $(n - 1)$.

2. The set V of all real-valued functions defined and continuous on the closed interval $[a, b]$ such that the value of the function at b is some non-zero constant p. For example, let $f(x)$ and $g(x)$ be two elements in V. Now, $f(b) = g(b) = p$. Since $f(b) + g(b) = 2p$, $f(x) + g(x)$ is not in V. Note that if $p = 0$, then V forms a vector space.

Example. Let V be the set of all ordered pairs (x, y), where x, y are real numbers.

Let $a = (x_1, y_1)$ and $b = (x_2, y_2)$ be two elements in V. Define the addition as

$$a + b = (x_1, y_1) + (x_2, y_2) = (2x_1 - 3x_2, y_1 - y_2)$$

and the scalar multiplication as

$$\alpha(x_1, y_1) = (\alpha x_1/3, \alpha y_1/3).$$

Show that V is not a vector space. Which of the properties are not satisfied?

Solution :

We illustrate the properties that are not satisfied.

(i) $(x_2, y_2) + (x_1, y_1) = (2x_2 - 3x_1, y_2 - y_1) \neq (x_1, y_1) + (x_2, y_2)$

Therefore, property 2 (commutative law) does not hold.

(ii) $((x_1, y_1) + (x_2, y_2)) + (x_3, y_3) = (2x_1 - 3x_2, y_1 - y_2) + (x_3, y_3) = (4x_1 - 6x_2 - 3x_3, y_1 - y_2 - y_3)$

$(x_1, y_1) + ((x_2, y_2) + (x_3, y_3)) = (x_1, y_1) + (2x_2 - 3x_3, y_2 - y_3) = (2x_1 - 6x_2 + 9x_3, y_1 - y_2 + y_3).$

Therefore, property **3** (associative law) is not satisfied.

(iii) $1(x_1, y_1) = (x_1/3, y_1/3) \neq (x_1, y_1)$.

Therefore, property **10** (existence of multiplicative inverse) is not satisfied.

Hence, V is not a vector space.

Subspaces

Let V be an arbitrary vector space defined under a given vector addition and scalar multiplication. A non-empty subset W of V, such that W is also a vector space under the same two operations of vector addition and scalar multiplication, is called a *subspace* of V. Thus, W is also closed under the two given algebraic operations on V. As a convention, the vector space V is also taken as a subspace of V.

Remark

To show that W is a subspace of a vector space V, it is not necessary to verify all the 10 properties as given in section 3.3. If it is shown that W is closed under the given definition of vector addition and scalar multiplication, then the properties 2, 3, 7, 8, 9 and 10 are automatically satisfied because these properties are valid for all elements in V and hence are also valid for all elements in W. Thus, we need to verify the remaining properties, that is, the existence of the zero element and the additive inverse in W.

Example. Let F and G be subspaces of a vector space V. The direct sum of F and G is written as F + G and is defined by

$$F + G = \{\mathbf{f} + \mathbf{g} : \mathbf{f} \in F, \mathbf{g} \in G\}.$$

Show that F + G is a subspace of V assuming the usual definition of vector addition and scalar multiplication.

Solution :

Let W = F + G and $\mathbf{f} \in F$, $\mathbf{g} \in G$. Since $\mathbf{0} \in F$ and $\mathbf{0} \in G$, we have $\mathbf{0} + \mathbf{0} = \mathbf{0} \in W$. Let $\mathbf{f}_1 + \mathbf{g}_1$ and $\mathbf{f}_2 + \mathbf{g}_2$ belong to W where $\mathbf{f}_1, \mathbf{f}_2 \in F$ and $\mathbf{g}_1, \mathbf{g}_2 \in G$. Then

$$(\mathbf{f}_1 + \mathbf{g}_1) + (\mathbf{f}_2 + \mathbf{g}_2) = (\mathbf{f}_1 + \mathbf{f}_2) + (\mathbf{g}_1 + \mathbf{g}_2) \in F + G = W.$$

Also for any scalar α,

$$\alpha(\mathbf{f} + \mathbf{g}) = \alpha\mathbf{f} + \alpha\mathbf{g} \in F + G = W.$$

Therefore, W = F + G is a subspace of V.

We now, present an important result on subspaces.

Theorem

Let $v_1, v_2, \ldots, v_r$ be any r elements of a vector space V. Then, the set of all linear combinations of these elements, that is the set of all elements of the form

$$\alpha_1 v_1 + \alpha_2 v_2 + \ldots + \alpha_r v_r$$

is a subspace of V, where $\alpha_1, \alpha_2, \ldots, \alpha_r$ are scalars.

Proof

Let W be the set of all linear combinations of $v_1, v_2, \ldots, v_r$.

Let $w_1 = \sum_{i=1}^{r} a_i v_i$ and $w_2 = \sum_{i=1}^{r} b_i v_i$

be any two linear combinations (any two elements of W). Then,

$$w_1 + w_2 = (a_1 + b_1)v_1 + (a_2 + b_2)v_2 + \ldots + (a_r + b_r)v_r \in W$$

$$\alpha w_1 = (\alpha a_1)v_1 + (\alpha a_2)v_2 + \ldots + (\alpha a_r)v_r \in W$$

$$\alpha w_2 = (\alpha b_1)v_1 + (\alpha b_2)v_2 + \ldots + (\alpha b_r)v_r \in W$$

and $\alpha(w_1 + w_2) = \alpha w_1 + \alpha w_2$.

Taking $\alpha = 0$, we find that $0w_1 = 0 \in W$.

Taking $\alpha = -1$, we find that $(-1)w_1 = (-w_1) \in W$.

Therefore, W is a subspace of V.

The elements $v_1, v_2, \ldots, v_r$ are in the subspace W as $v_1 = 1v_1 + 0v_2 + \ldots + 0v_r$, $v_2 = 0v_1 + 1v_2 + \ldots + 0v_r, \ldots$

We say that the *subspace* W is *spanned* by the r elements $v_1, v_2, \ldots, v_r$. Also, any subspace that contains the elements $v_1, v_2, \ldots, v_r$ must contain every linear combination of these elements.

Spanning Set

Let S be a subset of a vector space V and suppose that every element in V can be obtained as a linear combination of the elements taken from S. Then S is said to be the *spanning set* for V. We also say that S spans V.

Example. Let V be the vector space of all 2 × 2 matrices. Show that the sets

(*i*) $S = \left\{\begin{pmatrix}1 & 0\\0 & 0\end{pmatrix}, \begin{pmatrix}0 & 1\\0 & 0\end{pmatrix}, \begin{pmatrix}0 & 0\\1 & 0\end{pmatrix}, \begin{pmatrix}0 & 0\\0 & 1\end{pmatrix}\right\}$

(*ii*) $S = \left\{\begin{pmatrix}1 & 0\\0 & 0\end{pmatrix}, \begin{pmatrix}1 & 1\\0 & 0\end{pmatrix}, \begin{pmatrix}1 & 1\\1 & 0\end{pmatrix}, \begin{pmatrix}1 & 1\\1 & 1\end{pmatrix}\right\}$

span V.

Solution :

Let $X = \begin{bmatrix}a & b\\c & d\end{bmatrix}$ be an arbitrary element of V.

(*i*) We write

$$\begin{bmatrix}a & b\\c & d\end{bmatrix} = a\begin{bmatrix}1 & 0\\0 & 0\end{bmatrix} + b\begin{bmatrix}0 & 1\\0 & 0\end{bmatrix} + c\begin{bmatrix}0 & 0\\1 & 0\end{bmatrix} + d\begin{bmatrix}0 & 0\\0 & 1\end{bmatrix}$$

Since every element of V can be written as a linear combination of the elements of S, the set S spans the vector space V.

(*ii*) We need to determine the scalars $\alpha_1, \alpha_2, \alpha_3, \alpha_4$ so that

$$\begin{bmatrix}a & b\\c & d\end{bmatrix} = \alpha_1\begin{bmatrix}1 & 0\\0 & 0\end{bmatrix} + \alpha_2\begin{bmatrix}1 & 1\\0 & 0\end{bmatrix} + \alpha_3\begin{bmatrix}1 & 1\\1 & 0\end{bmatrix} + \alpha_4\begin{bmatrix}1 & 1\\1 & 1\end{bmatrix}$$

Equating the corresponding elements, we obtain the system of equations

$$\alpha_1 + \alpha_2 + \alpha_3 + \alpha_4 = a,$$
$$\alpha_2 + \alpha_3 + \alpha_4 = b,$$
$$\alpha_3 + \alpha_4 = c,$$
$$\alpha_4 = d.$$

The solution of this system of equations is

$\alpha_4 = d, \alpha_3 = c - d, \alpha_2 = b - c, \alpha_1 = a - b.$

Therefore, we can write

$$\begin{bmatrix}a & b\\c & d\end{bmatrix} = (a-b)\begin{bmatrix}1 & 0\\0 & 0\end{bmatrix} + (b-c)\begin{bmatrix}1 & 1\\0 & 0\end{bmatrix} + (c-d)\begin{bmatrix}1 & 1\\1 & 0\end{bmatrix} + d\begin{bmatrix}1 & 1\\1 & 1\end{bmatrix}$$

Since every element of V can be written as a linear combination of the elements of S, the set S spans the vector space V.

Linear Independence of Vectors

Let V be a vector space. A finite set $\{v_1, v_2, ..., v_n\}$ of the elements of V is said to be *linearly dependent* if there exist scalars $\alpha_1, \alpha_2, \ldots, \alpha_n$, not all zero, such that

$$\alpha_1 v_1 + \alpha_2 v_2 + ... + \alpha_n v_n = 0. \qquad ...(i)$$

If Eq. (*i*) is satisfied only for $\alpha_1 = \alpha_2 = ... = \alpha_n = 0$, then the set of vectors is said to be *linearly independent*.

The above definition of linear dependence of $v_1, v_2, ..., v_n$ can be written alternately as follows.

Theorem

The set of vectors $\{v_1, v_2, ..., v_n\}$ is linearly dependent if and only if at least one element of the set is a linear combination of the remaining elements.

Proof

Let the elements $v_1, v_2, ..., v_n$ be linearly dependent. Then, there exist scalars $\alpha_1, \alpha_2,..., \alpha_n$, not all zero such that

$$\alpha_1 v_1 + \alpha_2 v_2 + ... + \alpha_{i-1} v_{i-1} + \alpha_i v_i + \alpha_{i+1} v_{i+1} + ... + \alpha_n v_n = 0.$$

Let $\alpha_i \neq 0$. Then, we can write

$$v_i = -\left(\frac{\alpha_1}{\alpha_i}\right)v_1 - \left(\frac{\alpha_2}{\alpha_i}\right)v_2 - ... - \left(\frac{\alpha_{i-1}}{\alpha_i}\right)v_{i-1} - .... - \left(\frac{\alpha_{i+1}}{\alpha_i}\right)v_{i+1} - ... - \left(\frac{\alpha_n}{\alpha_i}\right)v_n$$

$$= \overset{*}{\alpha}_1 v_1 + \overset{*}{\alpha}_2 v_2 + ... + \overset{*}{\alpha}_{i-1} v_{i-1} + \overset{*}{\alpha}_{i+1} v_{i+1} + ... + \overset{*}{\alpha}_n v_n$$

where $\overset{*}{\alpha}_1, \overset{*}{\alpha}_2, ..., \overset{*}{\alpha}_n$ are some scalars. Hence, the vector v_i is a linear combination of the vectors $v_1, v_2, ..., v_{i-1}, v_{i+1}, ..., v_n$.

Now let v_i be a linear combination of $v_1, v_2, ..., v_{i-1}, v_{i+1}, ..., v_n$. Therefore, we have

$$v_i = a_1 v_1 + a_2 v_2 + ...+ a_{i-1} v_{i-1} + a_{i+1} v_{i+1} + ...+ a_n v_n$$

where a_i's are scalars. Then

$$a_1 v_1 + a_2 v_2 + ...+ a_{i-1} v_{i-1} + (-1)v_i + a_{i+1} v_{i+1} + ...+ a_n v_n = 0$$

Since the coefficient of v_i is not zero, the elements are linearly dependent.

Remarks

Eq. (i) gives a homogeneous system of algebraic equations. Non-trivial solutions exist if *det* (coefficient matrix) = 0, that is the vectors are linearly dependent in this case. If the *det* (coefficient matrix) ≠ 0, then by Cramer's rule, $\alpha_1 = \alpha_2 = ... = \alpha_n = 0$ and the vectors are linearly independent.

Example. Let $v_1 = (1, -1, 0)$, $v_2 = (0, 1, -1)$ and $v_3 = (0, 0, 1)$ be elements of R_3. Show that the set of vectors $\{v_1, v_2, v_3\}$ is linearly independent.

Solution :

We consider the vector equation

$$\alpha_1 v_1 + \alpha_2 v_2 + \alpha_3 v_3 = 0.$$

Substituting for v_1, v_2, v_3, we obtain

$$\alpha_1(1, -1, 0) + \alpha_2(0, 1, -1) + \alpha_2(0, 0, 1) = 0$$

or $(\alpha_1, -\alpha_1 + \alpha_2, -\alpha_2 + \alpha_3) = 0.$

Comparing, we obtain $\alpha_1 = 0$, $-\alpha_1 + \alpha_2 = 0$ and $-\alpha_2 + \alpha_3 = 0$. The solution of these equations is $\alpha_1 = \alpha_2 = \alpha_3 = 0$. Therefore, the given set of vectors is linearly independent.

Alternately,

$$det(v_1, v_2, v_3) = \begin{vmatrix} 1 & 0 & 0 \\ -1 & 1 & 0 \\ 0 & -1 & 1 \end{vmatrix} = 1 \neq 0.$$

Therefore, the given vectors are linearly independent.

DETERMINANTS

A determinant of order n has n rows and n columns. It has $n \times n$ elements.

$$\Delta = \begin{vmatrix} a_{11} & a_{12} & a_{13} & \cdots & a_{1n} \\ a_{21} & a_{22} & a_{23} & \cdots & a_{2n} \\ \cdots & \cdots & \cdots & \cdots & \cdots \\ \cdots & \cdots & \cdots & \cdots & \cdots \\ a_{n1} & a_{n2} & a_{n3} & \cdots & a_{nn} \end{vmatrix}$$

MINOR

If a_{ij} is an element which is in the ith row and j^{th} column of a squre matrix A, then determinant of the matrix obtained by deleting i^{th} row and j^{th} column of A is called minor of a_{ij} and is denoted by M_{ij}

e.g. If $A = \begin{vmatrix} a_{11} & a_{12} & a_{13} \\ a_{21} & a_{22} & a_{23} \\ a_{31} & a_{32} & a_{33} \end{vmatrix}$, then

$$M_{11} = \text{minor of } a_{11} = \begin{vmatrix} a_{22} & a_{23} \\ a_{32} & a_{33} \end{vmatrix} = a_{22}\,a_{33} - a_{32}\,a_{23}$$

$$M_{12} = \text{minor of } a_{12} = \begin{vmatrix} a_{21} & a_{23} \\ a_{31} & a_{33} \end{vmatrix}$$

$$= a_{21}\,a_{33} - a_{31}\,a_{23} \text{ and so on.}$$

9 minors can be obtained corresponding to 9 elements of Δ.

e.g. Minor of the element

$$a_{11} = \begin{vmatrix} a_{22} & a_{23} \\ a_{32} & a_{33} \end{vmatrix} = M_{11}$$

$$a_{12} = \begin{vmatrix} a_{21} & a_{23} \\ a_{31} & a_{33} \end{vmatrix} = M_{12} \text{ and so on.}$$

Determinant of a Square Matrix

Only square matrix can have determinants because in a determinant, number of rows is equal to number of columns.

Let $A = [a_{ij}]_{max}$ be a square matrix of order n.

Then $$|A| = \begin{vmatrix} a_{11} & a_{12} & a_{13} & \cdots & a_{1n} \\ a_{21} & a_{22} & a_{23} & \cdots & a_{2n} \\ \cdots & \cdots & \cdots & \cdots & \cdots \\ a_{n1} & a_{n2} & a_{n3} & \cdots & a_{nn} \end{vmatrix}$$

is called *determinant of matrix A.*

Properties of Determinants

(*i*) If all the elements of one column or one row of a determinant are multiplied by the same number "C", then value of the new determinant is "C" times the value of the given determinant.

(*ii*) Value of a determinant does not change when rows and columns are interchanged.

(*iii*) If any two columns or two rows of a determinant are interchanged, then value of the determinant is multiplied by – 1.

(*iv*) In a determinant, sum of the products of the element of any column (or row) with the cofactors of the corresponding elements of any other column (or row) is zero.

(*v*) If two columns (or two rows) of a determinant are identical, then value of the determinant is zero.

(*vi*) If in a determinant each element in any column (or row) consists of sum of two terms, then determinant can be expressed as the sum of two determinants of the same order.

(*vii*) If the elements of a column (or row) of a determinant are added K times the corresponding elements of another column (or row), then value of the determinant obtained is equal to value of the original determinant.

CO-FACTOR

If a_{ij} is an element which is in the ith row and j^{th} column of a square matrix A, then product of $(-1)^{i+j}$ and minor of a_{ij} is called cofactor of a_{ij} and is denoted by A_{ij}.

e.g. If $A = \begin{bmatrix} a_{11} & a_{12} & a_{13} \\ a_{21} & a_{22} & a_{23} \\ a_{31} & a_{32} & a_{33} \end{bmatrix}$, then

$$A_{11} = \text{cofactor of } a_{11} = (-1)^{1+1} M_{11}$$

$$= +1 \begin{vmatrix} a_{22} & a_{23} \\ a_{32} & a_{33} \end{vmatrix} = a_{22}\,a_{33} - a_{32}\,a_{23}$$

$$A_{12} = \text{cofactor of } a_{12} = (-1)^{1+2} M_{11}$$
$$= -1\begin{vmatrix} a_{21} & a_{23} \\ a_{31} & a_{33} \end{vmatrix} = (a_{31}a_{23} - a_{21}a_{33})$$
and so on.

and if $A = \begin{bmatrix} a_1 & b_1 & c_1 \\ a_2 & b_2 & c_2 \\ a_3 & b_3 & c_3 \end{bmatrix}$, then

$$A_1 = \text{cofactor of } a_1 = (-1)^{1+1}\begin{vmatrix} b_2 & c_2 \\ b_3 & c_3 \end{vmatrix}$$
$$= b_2c_3 - b_3c_2$$
$$B_1 = \text{cofactor of } b_1 = (-1)^{1+2}\begin{vmatrix} a_2 & c_2 \\ a_3 & c_3 \end{vmatrix}$$
$$= -(a_2c_3 - a_3c_2) = a_3c_2 - a_2c_3$$
$$C_1 = \text{cofactor of } c_1$$
$$= (-1)^{1+3}\begin{vmatrix} a_2 & b_2 \\ a_3 & b_3 \end{vmatrix} = a_2b_3 - a_3b_2 \text{ and so on.}$$

Example. If $A = \begin{bmatrix} 1 & 2 & 3 \\ 4 & 5 & 6 \\ 7 & 8 & 9 \end{bmatrix}$, then find all cofactors of element of A.

Solution. $A_{11} = (-1)^{1+1}\begin{vmatrix} 5 & 6 \\ 8 & 9 \end{vmatrix} = +(45-48) = -3$

$$A_{12} = (-1)^{1+2}\begin{vmatrix} 4 & 6 \\ 7 & 9 \end{vmatrix} = -(36-42) = 6$$
$$A_{13} = (-1)^{1+3}\begin{vmatrix} 4 & 5 \\ 7 & 8 \end{vmatrix} = +(32-35) = -3$$
$$A_{21} = (-1)^{2+1}\begin{vmatrix} 2 & 3 \\ 8 & 9 \end{vmatrix} = -(18-24) = 6$$
$$A_{22} = (-1)^{2+2}\begin{vmatrix} 1 & 3 \\ 7 & 9 \end{vmatrix} = +(9-21) = -12$$
$$A_{23} = (-1)^{2+3}\begin{vmatrix} 1 & 2 \\ 7 & 8 \end{vmatrix} = -(8-14) = 6$$
$$A_{31} = (-1)^{3+1}\begin{vmatrix} 2 & 3 \\ 5 & 6 \end{vmatrix} = +(12-15) = -3$$
$$A_{32} = (-1)^{3+2}\begin{vmatrix} 1 & 3 \\ 4 & 6 \end{vmatrix} = -(6-12) = 6$$
$$A_{33} = (-1)^{3+3}\begin{vmatrix} 1 & 2 \\ 4 & 5 \end{vmatrix} = +(5-8) = -3$$

ALGEBRA OF MATRICES

Matrix

A set of m,n numbers (real or complex) arranged in the form of a rectangular may have m rows and n columns is called an $m \times n$ *matrix*.

An $m \times n$ matrix is written as

$$A = \begin{bmatrix} a_{11} & a_{12} & a_{13} & \cdots & a_{1n} \\ a_{21} & a_{22} & a_{23} & \cdots & a_{2n} \\ \cdots & \cdots & \cdots & \cdots & \cdots \\ a_{m1} & a_{m2} & a_{m3} & \cdots & a_{mn} \end{bmatrix}$$

Element a_{ij} belongs to i^{th} row and j^{th} column and is sometimes called as $(i, j)^{th}$ element of the matrix.

Note : In a matrix, number of rows and columns need not be equal.

Types of Matrix

1. Square Matrix

In this matrix, the number of rows is equal to the number of columns.

e.g. $\begin{bmatrix} 1 & 2 \\ 3 & 4 \end{bmatrix}_{2\times2}$ and $\begin{bmatrix} 1 & 1 & 2 \\ 0 & 1 & 2 \\ 1 & 0 & 3 \end{bmatrix}_{3\times3}$

In these square matrices, m = n

Elements a_{ij} of a square matrix $A = [a_{ij}]_{n \times n}$ for which $i = j$ are called *diagonal elements* and line along which they lie is called *principal diagonal* of the matrix.

2. Null Matrix or Zero Matrix

The $m \times n$ matrix whose all elements are 0 is called *null matrix* (or zero matrix) of the type $m \times n$. It is denoted by 0.

e.g. $\begin{bmatrix} 0 & 0 & 0 & 0 \\ 0 & 0 & 0 & 0 \\ 0 & 0 & 0 & 0 \end{bmatrix}_{3\times4}$ is zero matrix of type 3 × 4

Above matrix can be written as

$$A = [a_{11}\, a_{22}\, a_{33}\, a_{44}]$$

3. Diagonal Matrix

In this matrix, elements above and below the principal diagonal are all zero.

e.g. $A = \begin{bmatrix} a_{11} & 0 & 0 & 0 \\ 0 & a_{22} & 0 & 0 \\ 0 & 0 & a_{33} & 0 \\ 0 & 0 & 0 & a_{44} \end{bmatrix}$

4. Scalar Matrix

In this matrix, all diagonal elements are equal.

e.g. $A = \begin{bmatrix} C & 0 & 0 \\ 0 & C & 0 \\ 0 & 0 & C \end{bmatrix}$

- If A be a matrix and $A^2 = A$, then A is called *idempotent*.

  If AB = A and BA = B, then A and B are called *idempotent*.
- If A be a matrix and satisfies $A^2 = I$, then A is called *invalutory*.

5. Rectangular Matrix

A matrix of order m × n is called a *rectangular matrix* if m ≠ n

e.g. $\begin{bmatrix} 1 & 2 & 3 \\ 4 & 0 & 1 \end{bmatrix}_{2\times3}$, $\begin{bmatrix} 1 & 0 \\ 0 & 2 \\ 1 & 2 \end{bmatrix}_{3\times2}$, $\begin{bmatrix} 1 & 2 & 5 & 7 \\ 2 & 4 & 6 & 8 \end{bmatrix}_{2\times4}$

6. Unit Matrix or Identity Matrix

A square matrix each of whose diagonal elements is 1 and each of whose non-diagonal elements is equal to zero is called a *unit matrix or an identity matrix*. It is denoted by I_n when n is order of the matrix.

e.g. $I_2 = \begin{bmatrix} 1 & 0 \\ 0 & 1 \end{bmatrix}$, $I_3 = \begin{bmatrix} 1 & 0 & 0 \\ 0 & 1 & 0 \\ 0 & 0 & 1 \end{bmatrix}$, $I_4 = \begin{bmatrix} 1 & 0 & 0 & 0 \\ 0 & 1 & 0 & 0 \\ 0 & 0 & 1 & 0 \\ 0 & 0 & 0 & 1 \end{bmatrix}$

are unit matrices of order 2, 3, 4, respectively.

7. Column Matrix

Any $m \times 1$ matrix which has m rows and only one column is a column matrix or a column vector.

Row Matrix

Any $1 \times n$ matrix which has only one row and n columns is called a row matrix or as row vector.

e.g. $X_1 = [1\ 2\ 3\ 4\ 5\ 6]_{1 \times 6}$ is a row matrix of the type 1×6

$X_2 = \begin{bmatrix} 1 \\ 2 \\ 3 \\ 4 \end{bmatrix}_{4 \times 1}$ is a column matrix of type 4×1

8. Triangular Matrix

It is of two types

(*i*) **Lower triangular matrix.** In this matrix, all the elements above the principal diagonal are zero.

e.g. $\begin{bmatrix} a_{11} & 0 & 0 \\ a_{21} & a_{22} & 0 \\ a_{31} & a_{32} & a_{33} \end{bmatrix}_{3 \times 3}$

The diagonal and lower triangular elements may or may not be zero.

(*ii*) **Upper triangular matrix.** In this matrix, all the elements below the principal diagonal are zero.

e.g. $\begin{bmatrix} a_{11} & a_{12} & a_{13} & a_{14} \\ 0 & a_{22} & a_{23} & a_{24} \\ 0 & 0 & a_{33} & a_{34} \\ 0 & 0 & 0 & a_{44} \end{bmatrix}_{4 \times 4}$

The diagonal and upper triangular elements may or may not be zero.

Types of Square Matrices

(*i*) **Nilpotent Matrix**

A square matrix A is called a *nilpotent matrix* if there exists a positive integer n such that $A^n = 0$ and $A^{x-1} \neq 0$

If n is least positive integer such that $A^n = 0$, then n is called *index of the nilpotent* matrix A.

(*ii*) **Symmetrical Matrix**

It is a square matrix in which $a_{ij} = a_{ji}$ for all i, j. A symmetric matrix is necessarily a square one.

If A is symmetric, **then $A^T = A$**

e.g. $\begin{bmatrix} a & h & g \\ h & b & f \\ g & f & c \end{bmatrix}$

(*a*) A A^T is always symmetric matrix.

(*b*) $\dfrac{A + A^T}{2}$ is always symmetric matrix.

(c) If A and B are symmetric then A + B and A – B also symmetric AB, BA may or May not by symmetric.

(*iii*) **Skew-Symmetrical Matrix**

It is a square matrix in which $a_{ij} = -a_{ji}$ for all i and j.

In a skew symmetrical matrix, all elements along the diagonal are zero.

e.g. $\begin{bmatrix} 0 & h & g \\ h & 0 & f \\ g & f & 0 \end{bmatrix}$

For any matrix A, the matrix $\dfrac{A - A^T}{2}$ is always skew symmetric.

Theorem

Every square matrix can be uniquely expressed as a sum of a symmetric matrix and a skew symmetric matrix.

(*iv*) **Hermitian Matrix**

It is a square matrix A in which $(i, j)^{th}$ element is equal to conjugate complex of the $(j, i)^{th}$ element.

i.e. $a_{ij} = \bar{a}_{ji}$ for all i and j

e.g. $\begin{bmatrix} x & 3-4i & 1+2i \\ 3+4i & 0 & 5-4i \\ 1-2i & 5+4i & 1 \end{bmatrix}$

A necessary and sufficient condition for a matrix A to be hermitian is that

$$A = A^\theta$$

where A^θ is transposed conjugates of A.

(*v*) **Skew-Hermitian Matrix**

It is a square matrix $A = [a_{ij}]$ in which $a_{ij} = -\bar{a}_{ji}$ for all i and j.

The diagonal elements of a skew-Hermitian matrix must be pure imaginary numbers or zero.

A necessary and sufficient condition for a matrix A to be skew-Hermitian is that

$$A^\theta = -A$$

***(vi)* Orthogonal Matrix**

A square matrix A is called *orthogonal matrix if*

$$AA^T = A^TA = I$$

e.g. If $A = \begin{bmatrix} \cos\theta & -\sin\theta \\ \sin\theta & \cos\theta \end{bmatrix}$ and

$$A^T = \begin{bmatrix} \cos\theta & \sin\theta \\ -\sin\theta & \cos\theta \end{bmatrix}$$

then $AA^T = \begin{bmatrix} \cos\theta & -\sin\theta \\ \sin\theta & \cos\theta \end{bmatrix}\begin{bmatrix} \cos\theta & \sin\theta \\ -\sin\theta & \cos\theta \end{bmatrix}$

$$= \begin{bmatrix} \cos^2\theta + \sin^2\theta & 0 \\ 0 & \sin^2\theta + \cos^2\theta \end{bmatrix}$$

$$= \begin{bmatrix} 1 & 0 \\ 0 & 1 \end{bmatrix} = I$$

***(vii)* Adjoint Matrix**

Let A be the square matrix given by

$$A = \begin{bmatrix} a_{11} & a_{12} & a_{13} & \dots & a_{1n} \\ a_{21} & a_{22} & a_{23} & \dots & a_{2n} \\ \dots & \dots & \dots & \dots & \dots \\ a_{n1} & a_{n2} & a_{n3} & \dots & a_{nn} \end{bmatrix}$$

Let A_{ij} be the cofactor a_{ij} then transpose of the matrix $[A_{ij}]$ is called *adjoint of the matrix A.*

$$\text{Adjoint } A = \begin{bmatrix} A_{11} & A_{21} & \dots & \dots & A_{n1} \\ A_{12} & A_{22} & \dots & \dots & A_{n2} \\ \dots & \dots & \dots & \dots & \dots \\ A_{1n} & A_{2n} & \dots & \dots & A_{nn} \end{bmatrix}$$

If A be an n-rowed square matrix, then $(\text{Adj } A)A = A(\text{Adj } A) = |A|\, I_n$

where I_n is the $n \times n$ identity matrix.

Sub-Matrix

Any matrix obtained by omitting some rows and columns from a given $(m \times n)$ matrix B is called *submatrix of B.*

The matrix B itself is a sub-matrix of B as it can be obtained from B by omitting no rows or columns.

Principal Sub-Matrix

A square sub-matrix of a square matrix B whose diagonal elements are also diagonal elements of matrix B is called *principal sub-matrix.*

Addition of Matrices

Let A and B be two matrices of the same type $m \times n$.

If $A = [a_{ij}]_{m \times n}$, $B = [b_{ij}]_{m \times n}$, then

$$A + B = [a_{ij} + b_{ij}]_{m \times n}$$

e.g. If $A = \begin{bmatrix} a_{11} & a_{12} & a_{13} & \dots & a_{1n} \\ a_{21} & a_{22} & a_{23} & \dots & a_{2n} \\ \dots & \dots & \dots & \dots & \dots \\ a_{m1} & a_{m2} & a_{m3} & \dots & a_{mn} \end{bmatrix}$

and $B = \begin{bmatrix} b_{11} & b_{12} & b_{13} & \dots & b_{1n} \\ b_{21} & b_{22} & b_{23} & \dots & b_{2n} \\ \dots & \dots & \dots & \dots & \dots \\ b_{m1} & b_{m2} & b_{m3} & \dots & b_{mn} \end{bmatrix}$

then, $A + B = \begin{bmatrix} a_{11}+b_{11} & a_{12}+b_{12} & a_{13}+b_{13} & \dots\dots & a_{1n}+b_{1n} \\ a_{21}+b_{21} & a_{22}+b_{22} & a_{23}+b_{23} & \dots\dots & a_{2n}+b_{2n} \\ \dots\dots & & \dots\dots & & \dots\dots \\ a_{m1}+b_{m1} & a_{m2}+b_{m2} & a_{m3}+b_{m3} & \dots\dots & a_{mn}+b_{mn} \end{bmatrix}$

Properties of Matrix Addition

***(i)* Matrix addition is Commulative**

If A and B be two $m \times n$ matrices, then

$$A + B = B + A$$

***(ii)* Matrix addition is Associative**

If A, B, C be three matrices each of type $m \times n$, then

$$(A + B) + C = A + (B + C)$$

***(iii)* Existence of Additive Identity**

If 0 be the $m \times n$ matrix each of whose element is zero, then $A + 0 = A = 0 + A$

where A is a $m \times n$ matrix

***(iv)* Existence of Additive Inverse**

If $B = [b_{ij}]_{m \times n}$, then negative of matrix B is defined as he matrix $[-b_{ij}]_{m\times n}$ and is denoted by $-B$.

$$-B + B = 0 = B + (-B)$$

0 is null matrix of order $m\times n$

***(v)* Cancellation Law**

If A, B, C are three main matrices, then

$B + A = C + A$

$\Rightarrow$ $B = C$ (right cancellation law)

$A + B = A + C$

$\Rightarrow$ $B = C$ (left cancellation law)

These cancellation laws hold good in the case of addition of matrices.

(vi) The equation $B + X = 0$ has a unique solution in the set of all $m \times n$ matrices.

EQUALITY OF TWO MATRICES

Two matrices A = $[a_{ij}]$ and B = $[b_{ij}]$ are said to be equal if

(*i*) both are the same size, and

(*ii*) $a_{ij} = b_{ij}$ for each pair of subscripts i and j.

If two matrices A and B are equal, then we write

$$A = B$$

Multiplication of a Matrix

1. By a scalar

Let C any real number called scaler and B be any $m \times n$ matrix, then

$$CB = BC = [cb_{ij}]_{m \times n}$$

where, $[b_{ij}]_{m \times n} = B$

Properties

(*i*) If p and q are two scalars and B is any $m \times n$ matrix, then

$$p(qB) = (pq)B$$

(*ii*) $(p + q)B = pB + qB$

(*iii*) If A and B are two matrices each of the type $m \times n$, then

$$p(A + B) = p(A) + p(B)$$

where p is any scalar.

(*iv*) If B be any $m \times n$ matrix, then

$$(-p)B = (-pB) = p(-B)$$

(*a*) $1B = B$

(*b*) $(-1)B = -B$

2. By a matrix

Let A = $[a_{ij}]_{l \times m}$ and B = $[b_{jk}]_{m \times n}$ be two matrices such that number of columns in A is equal to number of rows in B. Then in $l \times n$ matrix

$$C = [C_{ik}]_{l \times n} \text{ such that } C_{ik} = \sum_{j=1}^{n} a_{ij} b_{jk}$$

is called *product of the matrices A and B*.

We can write C = AB

Properties

(*i*) **Associative laws**

If A, B, C are $l \times m$, $m \times n$ and $n \times p$ matrices respectively, then

$$A(BC) = (AB)C$$

(*ii*) **Distributive laws**

$$A(B + C) = AB + AC$$

where A, B, C are any three $l \times m$, $m \times n$ and $n \times p$ matrices respectively.

(*iii*) **Commutative**

Multiplication of matrices is not always commutative.

For commutative AB = BA

For anti commutative AB = – BA

(*iv*) Product of two matrices can be zero matrix while neither of them is a zero matrix.

(*v*) If A be an $l \times m$ matrix. and I_m denotes m rowed unit matrix, then

$$AI_m = A = I_l A$$

TRANSPOSE OF A MATRIX

If row and column are interchanged in a matrix A, then matrix obtained is called *transpose matrix A* and is denoted by A^T.

- If $A = \begin{bmatrix} a_{11} & a_{12} & a_{13} \\ a_{21} & a_{22} & a_{23} \\ a_{31} & a_{32} & a_{33} \end{bmatrix}$, then $A^T = \begin{bmatrix} a_{11} & a_{21} & a_{31} \\ a_{12} & a_{22} & a_{32} \\ a_{13} & a_{23} & a_{33} \end{bmatrix}$
- If $A = [a_{ij}]$, then $A^T = [a_{ji}]$
- If A is a square matrix, then $|A| = |A^T|$

Properties of transpose

Theorem 1

If A is any matrix, then $(A^T)^T = A$.

Theorem 2

If A and B are two matrices of same type, then

$$(A + B)^T = A^T + B^T.$$

Note : If A is any matrix and k is any complex number, then $(kA)^T = kA^T$.

Theorem 3

If A and B are two matrices for which conformability for multiplication is assured, then

$$(AB)^T = B^T A^T.$$

Corollary 1 : If $A_1, A_2, \ldots\ldots A_n$ are n matrices which are conformable for multiplication, then

$$(A_1, A_2, \ldots.. A_n)^T = A_n^T A_{n-1}^T \ldots\ldots\ldots\ldots A_1^T.$$

Note : If A is any matrix, then AA^T and A^TA are defined. They are square matrices but their order may be different.

e.g. If $A = \begin{bmatrix} 1 & 2 & 0 \\ -2 & 0 & 3 \end{bmatrix}$ then $A^T = \begin{bmatrix} 1 & -2 \\ 2 & 0 \\ 0 & 3 \end{bmatrix}$

and $AA^T = \begin{bmatrix} 5 & -2 \\ -2 & 13 \end{bmatrix}$

$$A^TA = \begin{bmatrix} 5 & 2 & -6 \\ 2 & 4 & 0 \\ -6 & 0 & 9 \end{bmatrix}$$

Inverse of Matrix

Let A and B are two square matrices of the same order, such that $AB = BA = I$

then B is called *inverse of A* and is denoted by A^{-1}.

$$A^{-1} = \frac{(adjA)}{|A|}$$

Properties

(*i*) A^{-1} exists only when A is non singular, i.e. $|A| \neq 0$

(*ii*) The inverse of a matrix is unique.

(*iii*) $(AB)^{-1} = B^{-1}A^{-1}$

(*iv*) $(A^T)^{-1} = (A^{-1})^T$

(*v*) $(A^{-1})^\theta = (A^\theta)^{-1}$

Inverse of a Square Matrix

A square matrix A is said to be an invertible matrix if there exists a square matrix B such that

AB = BA = I, then matrix B is called inverse of A.

- A rectangular matrix can not be invertible.
- Every square matrix need not be invertible.

Theorem 1

An invertible matrix has unique inverse.

If A is an invertible matrix, then its inverse is denoted by A^{-1}

Hence $AA^{-1} = A^{-1}A = I$.

Theorem 2

If A is an invertible matrix, and A^{-1} is also invertible then $(A^{-1})^{-1} = A$

Theorem 3

If A and B are two invertible matrices of same type, then AB is also invertible and $(AB)^{-1} = B^{-1}A^{-1}$.

Corollary 1: If $A_1, A_2, \ldots A_n$ are invertible matrices of same type, then $A_1A_2 \ldots A_n$ is also invertible and

$$(A_1 A_2 \ldots A_n)^{-1} = A_n^{-1} A_{n-1}^{-1} \ldots A_1^{-1}$$

If A is an invertible matrix and k is a non zero complex number, then kA is invertible and

$$(kA)^{-1} = k^{-1}A^{-1}$$

Theorem 4

If A is an invertible matrix, then A^T is also invertible and $(A^T)^{-1} = (A^{-1})^T$

Theorem 5

If A is a non singular matrix, then A is invertible and

$$A^{-1} = \frac{\text{Adj A}}{\det A}$$

If A is a square matrix, then

$$A\,(\text{Adj A}) = (\det A)\, I$$

Theorem 6

If A is an invertible matrix, then A is non singular.

If A is an invertible matrix, then

$$\det(A^{-1}) = \frac{1}{\det A}$$

Theorem 7

A square matrix A is non singular if A is invertible.

Corollary 2 : If $A = \begin{bmatrix} a & b \\ c & d \end{bmatrix}$ and $ad \neq bc$, then

$$A^{-1} = \frac{1}{ad - bc}\begin{bmatrix} d & b \\ c & a \end{bmatrix}$$

ORTHOGONAL MATRIX

An invertible square matrix is called orthogonal, if

$$A^T = A^{-1}$$

$\therefore \quad AA^T = I$

$\Rightarrow \quad |A| = |A^T|$

$\Rightarrow \quad |A| = \pm 1$

Example. If $A = \begin{bmatrix} 2 & 3 \\ 5 & 7 \end{bmatrix}$, then find A^{-1}.

Solution. $A^{-1} = \frac{1}{14 - 15}\begin{bmatrix} 7 & -3 \\ -5 & 2 \end{bmatrix}$

$$= -1\begin{bmatrix} 7 & -3 \\ -5 & 2 \end{bmatrix} = \begin{bmatrix} 7 & -3 \\ -5 & 2 \end{bmatrix}$$

Theorem 8

Let A, B, C be three square matrices of same type and A be a non-singular matrix. Then

$AB = AC \quad \Rightarrow B = C$

and $\quad BA = CA \quad \Rightarrow B = C$.

Theorem 9

If A and B are non singular matrices of same type, then

$$\text{Adj}(AB) = (\text{Adj B})(\text{Adj A})$$

Theorem 10

If A is a square matrix of type n, then

$\det(\text{Adj A}) = (\det A)^{n-1}$

Example. If A is a square matrix of type 3 and $\det A = 10$, then find (Adj A).

Solution. $\det(\text{Adj A}) = (\det A)^{3-1} = 10^2 = 100$

RANK OF A MATRIX

If A is not a null matrix; then rank of the matrix A is positive integer, if it possesses following two properties

(i) Each minor of order $n + 1$ of the matrix A vanishes (i.e. is zero).

(ii) There exists at least one minor of order n of the matrix A which is not zero.

If A is a null matrix, then rank of A is defined as *zero*.

A need not be square matrix.

Properties

- Rank of A is denoted by $p(A)$ and is UNIQUE.
- Every matrix will have a rank.
- If A is a matrix of order $m \times n$, then
  $p(A) \leq m$ and n (smaller of the two).
- If $P(A) = n$, then every minor of order $n + 1$, $n + 2$, etc. is zero.
- If A is a matrix of order $n \times n$, then
  A is non-singular $\Leftrightarrow p(A) = n$.
- Rank of $I_n = n$.
- A is a matrix of order $m \times n$. If every kth order minor $(k < m, k < n)$ is zero, then
  $$P(A) < k$$
- A is a matrix of order $m \times n$. If there is a minor of order $k(k < m, k < n)$ which is not zero, then
  $$p(A) \geq k$$
- Rank of a zero matrix is defined to be zero. The rank of a unit matrix is defined to be one.
- If x is a non-zero column matrix and y is a non-zero row matrix, then $p(x\,y) = 1$
- Rank of a matrix is greater than or equal to (not less than) rank of every sub-matrix thereof.
- If A is any n-rowed square matrix of rank, $n - 1$, then adj A $\neq 0$.
- Rank of the transpose of a matrix is equal to rank of the original matrix.
- Rank of a matrix does not change by pre-multiplication or post-multiplication with a non-singular matrix.
- If $A \sim B$, then $p(A) = p(B)$.
- Rank of a product of two matrices cannot exceed rank of either matrix.
- Rank of the sum of two matrices cannot exceed sum of their ranks.
- Elementary transformations do not change the rank of a matrix.

Echelon form of a Matrix

Let A be the given matrix of order m × n, then it can be reduced to Echelon form by row tranformation.

If we can make all the entries equal to zero, then

R (A) = Number of none zero rows

Consistency and Solution of Algebraic Equations

In argument matrix A*, we reduce echelon form.

(i) If Rank of A* = Rank of A,
then equation are consistent.

(ii) If Rank of A* ≠ Rank of A,
then system is inconsistent (i.e. no solution)

Homogenous System of Equations AX = 0

(i) If rank = number of unknowns,
then it has zero solution.

(ii) If rank < number of unknowns,
then it has infinitely may solution.

Elementary Transformation of a Matrix

Following operations on a matrix are called *elementary transformations*.

(i) Interchange of any two rows (or columns).

(ii) Multiplication of elements of any row (or column) by any non-zero number.

(iii) Addition to elements of any other row (or column) the corresponding elements of any other row (or column) multiplied by any number.

Solution of Simultaneous Linear Equations by Matrix

Simultaneous linear equations of two or three variables are solved by using matrices.

1. Homogeneous Linear Equations

Let system of m homogeneous equations in n unknown $x_1, x_2, \ldots, x_n$ be

$$\left.\begin{aligned} a_{11}x_1 + a_{12}x_2 + a_{13}x_3 + \ldots + a_{1n}x_n &= 0 \\ a_{21}x_1 + a_{22}x_2 + a_{23}x_3 + \ldots + a_{2n}x_n &= 0 \\ \ldots\ldots\ldots\ldots\ldots\ldots\ldots\ldots\ldots\ldots\ldots \\ a_{m1}x_1 + a_{m2}x_2 + a_{m3}x_3 + \ldots + a_{mn}x_n &= 0 \end{aligned}\right\} \quad \ldots(i)$$

$$\text{Let } A = \begin{bmatrix} a_{11} & a_{12} & \ldots & a_{1n} \\ a_{21} & a_{22} & \ldots & a_{2n} \\ \ldots & \ldots & \ldots & \ldots \\ \ldots & \ldots & \ldots & \ldots \\ a_{m1} & a_{m2} & \ldots & a_{mn} \end{bmatrix}_{m \times n}$$

$$X = \begin{bmatrix} x_1 \\ x_2 \\ x_3 \\ \ldots \\ \ldots \\ x_n \end{bmatrix}_{n \times 1}; \quad B = \begin{bmatrix} 0 \\ 0 \\ 0 \\ \ldots \\ \ldots \\ 0 \end{bmatrix}_{m \times 1}$$

where A, X and B are $m \times n$, $n \times 1$ and $m \times 1$ matrices respectively.

Equation *(i)* can be written in the form of a single matrix equation

$$AX = B \quad \ldots(ii)$$

The matrix A is called *coefficient matrix* of the system of equation *(i)*

Solutions of Equation AX = B

Working Rule

Reduce the coefficient matrix A to Echelon form by applying elementary row transformations only.

From this we know rank of the matrix A.

Let matrix A is of the type $m \times n$ and its rank is r.

- If $r < m$, then $(m - n)$ equations will be eliminated. Thus given system of m equations will be replaced by an equivalent system of r equations. Solving these r equations we can express values of some unknown in terms of remaining $n - r$ unknowns. These $n - r$ unknown can be given any arbitrarily chosen values.
- If $r = n$, then zero solution (trivial solution) will be the only solution.
- If $r < n$, there will be an infinity of solutions.

2. Linear Non-Homogeneous Equations

Let system of m simultaneous equations in n unknowns $x_1, x_2, \ldots, x_n$ be

$$a_{11}x_1 + a_{12}x_2 + a_{13}x_3 + \ldots + a_{1n}x_n = K_1$$
$$a_{21}x_1 + a_{22}x_2 + a_{23}x_3 + \ldots + a_{2n}x_n = K_2$$
$$a_{m1}x_1 + a_{m2}x_2 + a_{m3}x_3 + \ldots + a_{mn}x_n = K_m$$

Let $A = \begin{bmatrix} a_{11} & a_{12} & \ldots & a_{1n} \\ a_{21} & a_{22} & \ldots & a_{2n} \\ \ldots & \ldots & \ldots & \ldots \\ a_{m1} & a_{m2} & \ldots & a_{mn} \end{bmatrix}$, $X = \begin{bmatrix} x_1 \\ x_2 \\ \ldots \\ \ldots \\ x_n \end{bmatrix}_{n \times 1}$,

$K = \begin{bmatrix} K_1 \\ K_2 \\ \ldots \\ \ldots \\ K_m \end{bmatrix}_{m \times 1}$

where A, X, K are $m \times n$, $n \times 1$ and $m \times 1$ matrices respectively.

Above equations can be written in the form of a single matrix equation AX = K

Consistent and Inconsistent Equations

If above system has a solution, then equations are *consistent* otherwise equations are *called inconsistent equations*.

- The system of equation AX = K is consistent, i.e. possesses a solution, if and only if coefficient matrix A and augmented matrix [A K] are of the same rank.
- If A be an n-rowed non-singular matrix, X be an $n \times 1$ matrix, K be an $m \times 1$ matrix, then system of equation AX = K has a unique solution.

Solution of Equation AX = K

Working Rule

Let coefficient matrix A is of the type $m \times n$.

Write augmented matrix [A K] and reduce it to a Echelon form by applying only elementary row transformation on it. This Echelon form will give ranks of the augmented matrix [A K] and coefficient matrix A.

Different Cases

Case I : *Rank A < Rank [A K]*

In this case, equation AX = K are inconsistent, i.e. they have no solution.

Case II : *Rank A = Rank [A K] = r (suppose)*

In this case, equation AX = K are consistent.

- If $r < m$, then in the process of reducing the matrix [A K] to Echelon form, $(m - r)$ equations will be eliminated. Then given system of m equations will be replaced by an equivalent system of r equations. From these r equations we shall be able to express the values of some r unknowns in terms of the remaining $n - r$ unknowns which can be given any arbitrary chosen values.
- If $r = n$, there will be a unique solution.
- If $r < n$, then $n-r$ variables can be assigned arbitrary values. So there will be an infinite number of solutions. Only $n - r + 1$ solutions will be linearly independent and rest of the solutions will be linear combinations of them.
- If $m < n$, then $\quad r \le m < n$.

  Thus $\quad n - r > 0$.

Therefore when number of equations is less than the number of unknowns, equations will always have an infinite number of solutions, provided they are consistent.

Matrix Inversion Method

Theorem 1

If A is a non singular matrix, then $A^{-1}B$ is a solution AX = B.

Proof : A is non singular

$\Rightarrow$ A^{-1} exists and $AA^{-1} = A^{-1}A = I$

$\Rightarrow$ $A(A^{-1}B) = (AA^{-1})B = IB = B$

$\therefore$ $X = A^{-1}B$ is a solution of AX = B.

Theorem 2

If A is a non singular matrix, then AX = B has a unique solution.

Proof : A is non singular

$\Rightarrow$ A^{-1} exists and $AA^{-1} = A^{-1}A = I$

By theorem 1, $X = A^{-1}B$ is a solution of AX = B

Let $X = X_1$, $X = X_2$ be two solutions of AX = B.

$\therefore \quad AX_1 = B, AX_2 = B$

$\Rightarrow \quad AX_1 = AX_2$

$\Rightarrow \quad A^{-1}(AX_1) = A^{-1}(AX_2)$

$\Rightarrow \quad (A^{-1}A)X_1 = (A^{-1}A)X_2$

$\Rightarrow \quad IX_1 = IX_2$

$\Rightarrow \quad X_1 = X_2$

Hence AX = B has a *unique solution*.

Theorem 3

If A is a non singular matrix, then $X = A^{-1}B$ is the unique solution of AX = B.

Method of finding solution $X = A^{-1}B$ of AX = B is called **matrix inversion method**

Example. Solve following equations by matrix inversion method

$$x + y + z = 6$$
$$2x - y + z = 3$$
$$x + 2y - z = 2.$$

Solution :

Let $A = \begin{bmatrix} 1 & 1 & 1 \\ 2 & -1 & 1 \\ 1 & 2 & -1 \end{bmatrix}, X = \begin{bmatrix} x \\ y \\ z \end{bmatrix}, B = \begin{bmatrix} 6 \\ 3 \\ 2 \end{bmatrix}$

Given equations can be written as

$$AX = B$$

By matrix inversion method solution is,

$$X = A^{-1}B$$

$$\det A = 1(1-2) - 1(-2-1) + 1(4+1) = -1 + 3 + 5 = 7.$$

Cofactors of elements of A are

$A_{11} = +(1-2) = -1$

$A_{12} = -(-2-1) = 3$

$A_{13} = +(4+1) = 5$

$A_{21} = -(-1-2) = 3$

$A_{22} = +(-1-1) = -2$

$A_{23} = -(2-1) = -1$

$A_{31} = +(1+1) = 2$

$A_{32} = -(1-2) = 1$

$A_{33} = +(-1-2) = -3$

$$\therefore \quad \text{Adj } A = \begin{bmatrix} A_{11} & A_{21} & A_{31} \\ A_{12} & A_{22} & A_{32} \\ A_{13} & A_{23} & A_{33} \end{bmatrix} = \begin{bmatrix} -1 & 3 & 2 \\ 3 & -2 & 1 \\ 5 & -1 & -3 \end{bmatrix}$$

$$A^{-1} = \frac{\text{Adj } A}{\det A} = \frac{1}{7}\begin{bmatrix} -1 & 3 & 2 \\ 3 & -2 & 1 \\ 5 & -1 & -3 \end{bmatrix}$$

$$X = A^{-1}B = \frac{1}{7}\begin{bmatrix} -1 & 3 & 2 \\ 3 & -2 & 1 \\ 5 & -1 & -3 \end{bmatrix}\begin{bmatrix} 6 \\ 3 \\ 2 \end{bmatrix}$$

$$= \frac{1}{7}\begin{bmatrix} -6 & +9 & +4 \\ 18 & -6 & +2 \\ 30 & -3 & -6 \end{bmatrix} = \frac{1}{7}\begin{bmatrix} 7 \\ 14 \\ 21 \end{bmatrix} = \begin{bmatrix} 1 \\ 2 \\ 3 \end{bmatrix}$$

$\therefore \quad x = 1, y = 2, z = 3$ is the solution.

CRAMER'S RULE

Theorem 4

Let
$$a_1x + b_1y + c_1z = d_1$$
$$a_2x + b_2y + c_2z = d_2 \quad ...(i)$$
$$a_3x + b_3y + c_3z = d_3$$

be a system of linear equations.

If $\Delta = \begin{vmatrix} a_1 & b_1 & c_1 \\ a_2 & b_2 & c_2 \\ a_3 & b_3 & c_3 \end{vmatrix} \neq 0, \Delta_1 = \begin{vmatrix} d_1 & b_1 & c_1 \\ d_2 & b_2 & c_2 \\ d_3 & b_3 & c_3 \end{vmatrix},$

$\Delta_2 = \begin{vmatrix} a_1 & d_1 & c_1 \\ a_2 & d_2 & c_2 \\ a_3 & d_3 & c_3 \end{vmatrix}, \Delta_3 = \begin{vmatrix} a_1 & b_1 & d_1 \\ a_2 & b_2 & d_2 \\ a_3 & b_3 & d_3 \end{vmatrix}$

then solution of equation, (*i*) is

$$x = \frac{\Delta_1}{\Delta}, y = \frac{\Delta_2}{\Delta}, z = \frac{\Delta_3}{\Delta} \text{ is}$$

Proof : Let $A = \begin{bmatrix} a_1 & b_1 & c_1 \\ a_2 & b_2 & c_2 \\ a_3 & b_3 & c_3 \end{bmatrix}, X = \begin{bmatrix} x \\ y \\ z \end{bmatrix}, B = \begin{bmatrix} d_1 \\ d_2 \\ d_3 \end{bmatrix}.$

Then $\det A = \Delta \neq 0$

$\therefore \quad AX = B$ has unique solution

Hence equation (*i*) has unique solution.

Let, A_1, A_2, A_3 be the cofactors of a_1, a_2, a_3 respectively. Thus

$$(a_1x + b_1y + c_1z) A_1 + (a_2x + b_2y + c_2z) A_2 + (a_3x + b_3y + c_3z) A_3$$
$$= d_1A_1 + d_2A_2 + d_3A_3$$

$$\Rightarrow \quad x[a_1A_1 + a_2A_2 + a_3A_3] + y[b_1A_1 + b_2A_2 + b_3A_3] + z[c_1A_1 + c_2A_2 + c_3A_3] = \Delta_1$$

$$\Rightarrow \quad x\Delta + y0 + z0 = x = \Delta_1$$

$$\Rightarrow \quad x = \frac{\Delta_1}{\Delta}$$

Similarly, $y = \frac{\Delta_2}{\Delta}$, and $z = \frac{\Delta_3}{\Delta}$

Hence solution of (*i*) is $x = \frac{\Delta_1}{\Delta}, y = \frac{\Delta_2}{\Delta}, z = \frac{\Delta_3}{\Delta}$

- $\Delta_1, \Delta_2, \Delta_3$ are also denoted as $\Delta_x, \Delta_y, \Delta_z$ respectively.
- $\frac{x}{\Delta_1} = \frac{y}{\Delta_2} = \frac{z}{\Delta_3} = \frac{1}{\Delta}$

Example. Solve following equations by Cramer's rule.

$$x + y + z = 6$$
$$2x - y + z = 3$$
$$x + 2y - z = 2.$$

Solution : $\Delta = \begin{vmatrix} 1 & 1 & 1 \\ 2 & -1 & 1 \\ 1 & 2 & -1 \end{vmatrix}$

$= 1(1-2) - 1(-2-1) + 1(4+1)$

$= -1 + 3 + 5 = 7$

$\Delta_1 = \begin{vmatrix} 6 & 1 & 1 \\ 3 & -1 & 1 \\ 2 & 2 & -1 \end{vmatrix}$

$= 6(1-2) - 1(-3-2) + 1(6+2)$

$= -6 + 5 + 8 = 7$

$\Delta_2 = \begin{vmatrix} 1 & 6 & 1 \\ 2 & 3 & 1 \\ 1 & 2 & -1 \end{vmatrix}$

$= 1(-3-2) - 6(-2-1) + 1(4-3)$

$= -5 + 18 + 1 = 14$

$\Delta_3 = \begin{vmatrix} 1 & 1 & 6 \\ 2 & -1 & 3 \\ 1 & 2 & 2 \end{vmatrix}$

$= 1(-2-6) - 1(4-3) + 6(4+1)$

$= -8 - 1 + 30 = 21$

By Cramer's rule,

$$\frac{x}{\Delta_1} = \frac{y}{\Delta_2} = \frac{z}{\Delta_3} = \frac{1}{\Delta}$$

$$\therefore \quad \frac{x}{7} = \frac{y}{14} = \frac{z}{21} = \frac{1}{7}$$

Hence solution is, $x = 1, y = 2, z = 3$

Coefficient and Augmented matrix

Let $a_1x + b_1y + c_1z = d_1$

$a_2x + b_2y + c_2z = d_2$ ... (*i*)

$a_3x + b_3y + c_3z = d_3$

be a system of linear equations.

Then matrix, $A = \begin{bmatrix} a_1 & b_1 & c_1 \\ a_2 & b_2 & c_2 \\ a_3 & b_3 & c_3 \end{bmatrix}$ is called **Coefficient Matrix**

and matrix $K = \begin{bmatrix} a_1 & b_1 & c_1 & d_1 \\ a_2 & b_2 & c_2 & d_2 \\ a_3 & b_3 & c_3 & d_3 \end{bmatrix}$ is called **Augmented Matrix**

EIGENVALUES AND EIGENVECTORS

Let $A = [a_{ij}]$ be an $n \times n$ matrix.

- Matrix $A - \lambda I$ is called *characteristic matrix* of A, where I is identity matrix.
- Determinant $|A - \lambda I|$ is called *characteristic polynomial of A.*
- Equation $|A - \lambda I| = 0$ is called *characteristic equation* of A and its roots are called *eigen values.*
- Problem of finding the eigen values of a matrix is called *eigen-value problem.*
- If λ is a characteristic root or eigen values of an $n \times n$ matrix A, then a non-zero vector X such that $AX = \lambda X$ is called *characteristic vector* or *eigen vector of* A corresponding to the characteristic root λ.

  The set of eigen values of A is called the spectrum of A.

Relations between Eigenvalues and Eigenvectors

- λ is a eigen values of a matrix A if and only if there exists a non-zero vector X such that

$$AX = \lambda X$$

- If X is a characteristic vector of a matrix A corresponding to the characteristic value λ, then KX is also a characteristic vector of A corresponding to the same characteristic value λ. Here K is any non-zero scalar.
- If X is a eigen vector of a matrix A, then X cannot correspond to more than one eigen value of A.
- Eigen vectors corresponding to distinct characteristic roots of a matrix are linearly independent.

Properties of Eigenvalues

(*i*) Sum of all the eigen values of a matrix is equal to sum of the elements of the principal diagonal.

(*ii*) Product of the eigen values of a matrix is equal to its determinant.

(*iii*) If λ is an eigen value of a matrix A, then $\frac{1}{\lambda}$ is the eigen value of A^{-1}.

(*iv*) If λ is an eigen value of an orthogonal matrix, then $\frac{1}{\lambda}$ is also its eigen value.

(*v*) If $\lambda_1, \lambda_2 \ldots\ldots\ldots\ldots \lambda_m$ are the eigen values of a matrix A, then A^n has the eigen values $\lambda_1^n, \lambda_2^n \ldots\ldots\ldots\ldots \lambda_m^n$.

(*vi*) Eigen value of A = Eigen value of A^T

(*vii*) Maximum number of distinct eigen values = size of A

(*viii*) In a triangular and diagonal matrix, Eigen values are diagonal elements themselves.

(*ix*) If A and B are two matrices of same order then the matrix AB and BA will have same characteristic roots.

Cayley-Hamilton Theorem

Every square matrix satisfies its characteristic equation, i.e. if for a square matrix A of order n

$$|A - \lambda I| = (-1)^n [\lambda^n + a_1\lambda^{n-1} + a_2\lambda^{n-2} + \ldots\ldots + a_n]$$

then matrix equation

$$X^n + a_1 X^{n-1} + a_2 X^{n-2} + a_3 X^{n-3} + \ldots\ldots + a_n I = 0$$

is satisfied by X = A

i.e., $A^n + a_1 A^{n-1} + a_2 A^{n-2} + \ldots\ldots + a_n I = 0$

LU Decomposition Method

1. LU Decomposition Method

Let us suppose that 'A' is a matrix.

Now we can write A = LU

where 'L' is Lower Triangular matrix and 'U' is Upper Triangular matrix.

An LU decomposition of any given matrix A is product of a lower triangular matrix and an upper triangular matrix. The product of L and U is equal to A.

$$\text{Let } A = \begin{bmatrix} 25 & 5 & 1 \\ 64 & 8 & 1 \\ 144 & 12 & 1 \end{bmatrix} = LU$$

Our aim is to find L and U, So that product of it result into A.

$$\text{where} \quad L = \begin{bmatrix} 1 & 0 & 0 \\ L_{21} & 1 & 0 \\ L_{31} & L_{32} & 1 \end{bmatrix}$$

$$\text{and} \quad U = \begin{bmatrix} U_{11} & U_{12} & U_{13} \\ 0 & U_{22} & U_{23} \\ 0 & 0 & U_{33} \end{bmatrix}$$

Now $\quad A = LU$

$$\therefore \begin{bmatrix} 25 & 5 & 1 \\ 64 & 8 & 1 \\ 144 & 12 & 1 \end{bmatrix} = \begin{bmatrix} 1 & 0 & 0 \\ L_{21} & 1 & 0 \\ L_{31} & L_{32} & 1 \end{bmatrix} \begin{bmatrix} U_{11} & U_{12} & U_{13} \\ 0 & U_{22} & U_{23} \\ 0 & 0 & U_{33} \end{bmatrix}$$

$$\therefore \begin{bmatrix} 25 & 5 & 1 \\ 64 & 8 & 1 \\ 144 & 12 & 1 \end{bmatrix} =$$

$$\begin{bmatrix} U_{11} & U_{12} & U_{13} \\ L_{21}U_{11} & L_{21}U_{12} + U_{22} & L_{21}U_{13} + U_{23} \\ L_{31}U_{11} & L_{31}U_{12} + L_{32}U_{22} & L_{31}U_{13} + L_{32}U_{23} + U_{33} \end{bmatrix}$$

$\therefore \quad U_{11} = 25$

$U_{12} = 5$

$U_{13} = 1$

Now $\quad L_{21}U_{11} = 64$

$$\therefore \quad L_{21} = \frac{64}{U_{11}} = \frac{64}{25} = 2.56$$

Now $\quad L_{31}U_{11} = 144$

$$\therefore \quad L_{31} = \frac{144}{U_{11}} = \frac{144}{25} = 5.76$$

and $\quad L_{21}U_{12} + U_{22} = 8$

$\therefore \quad (2.56)(5) + U_{22} = 8$

$\therefore \quad U_{22} = -4.8$

and $\quad L_{21}U_{13} + U_{23} = 1$

$\therefore \quad (2.56)(1) + U_{23} = 1$

$\therefore \quad U_{23} = -1.56$

and $\quad L_{31}U_{12} + L_{32}U_{22} = 12$

$\therefore \quad (5.76)(5) + L_{32}(-4.8) = 12$

$\therefore \quad 28.8 + L_{32}(-4.8) = 12$

$$\therefore \quad L_{32} = \frac{12 - 28.8}{-4.8} = 3.5$$

and $\quad L_{31}U_{13} + L_{32}U_{23} + U_{33} = 1$

$\therefore \quad (5.76)(1) + (3.5)(-1.56) + U_{33} = 1$

$\therefore \quad U_{33} = 1 - 5.76 + 5.46$

$\therefore \quad U_{33} = 0.7$

Hence Lower triangular matrix

$$L = \begin{bmatrix} 1 & 0 & 0 \\ L_{21} & 1 & 0 \\ L_{31} & L_{32} & 1 \end{bmatrix}$$

$$= \begin{bmatrix} 1 & 0 & 0 \\ 2.56 & 1 & 0 \\ 5.76 & 3.5 & 1 \end{bmatrix}$$

$$\text{and} \quad U = \begin{bmatrix} U_{11} & U_{12} & U_{13} \\ 0 & U_{22} & U_{23} \\ 0 & 0 & U_{33} \end{bmatrix}$$

$$= \begin{bmatrix} 25 & 5 & 1 \\ 0 & -4.8 & -1.56 \\ 0 & 0 & 0.7 \end{bmatrix}$$

Cross check

$$[L][U] = \begin{bmatrix} 1 & 0 & 0 \\ 2.56 & 1 & 0 \\ 5.76 & 3.5 & 1 \end{bmatrix} \begin{bmatrix} 25 & 5 & 1 \\ 0 & -4.8 & -1.56 \\ 0 & 0 & 0.7 \end{bmatrix}$$

$$= \begin{bmatrix} 25 & 5 & 1 \\ 64 & 8 & 1 \\ 144 & 12 & 1 \end{bmatrix} = [A]$$

Hence Verified

Note : LU decomposition method is not possible for such matrix whose leading submatrices has zero value as determinant.

$$A = \begin{bmatrix} 1 & 2 & 3 \\ 2 & 4 & 5 \\ 1 & 3 & 4 \end{bmatrix}$$ does not have LU decomposition

as the leading submatrix has determinant equal to

$$\begin{vmatrix} 1 & 2 \\ 2 & 4 \end{vmatrix} = (4 \times 1) - (2 \times 2) = 0$$

Hence LU decomposition of the above matrix 'A' is not possible.

LU decomposition method to solve system of equation

Once a matrix A is decomposed into L and U we can obtain solution to

$$AX = B \quad \text{...}(i)$$

Steps to solve system of equation

1. For given matrix A, find L and U.

So that $A = LU$

$\therefore \quad LUX = B$ (from (i)) ...(ii)

2. Let $Y = UX$

$\therefore \quad LY = B$ (from (ii))

Solve the triangular system for Y.

3. Lastly solve the triangular system

$UX = Y$ for X

SOLVED PROBLEMS

1. Find inverse of $\begin{bmatrix} 1 & 1 & 3 \\ 1 & 3 & -3 \\ -2 & -4 & -4 \end{bmatrix}$

Solution :

Determinant of given matrix A is

$$\Delta = \begin{bmatrix} 1 & 1 & 3 \\ 1 & 3 & -3 \\ -2 & -4 & -4 \end{bmatrix} = \begin{bmatrix} a_1 & b_1 & c_1 \\ a_2 & b_2 & c_2 \\ a_3 & b_3 & c_3 \end{bmatrix}$$

If A_1, A_2 ... be the co-factors of a_1, a_2... in Δ, then

$$A_1 = -24, \quad A_2 = -8, \quad A_3 = -12,$$
$$B_1 = 10, \quad B_2 = 2, \quad B_3 = 6,$$
$$C_1 = 2, \quad C_2 = 2, \quad C_3 = 2$$

Thus, $D = a_1A_1 + a_2A_2 + a_3A_3 = -8$

$$\therefore \text{ Adj. } A = \begin{bmatrix} A_1 & A_2 & A_3 \\ B_1 & B_2 & B_3 \\ C_1 & C_2 & C_3 \end{bmatrix} = \begin{bmatrix} -24 & -8 & -12 \\ 10 & 2 & 6 \\ 2 & 2 & 2 \end{bmatrix}$$

Hence inverse of matrix A

$$= \frac{\text{adj.A}}{\Delta} = \frac{1}{-8} \begin{bmatrix} -24 & -8 & -12 \\ 10 & 2 & 6 \\ 2 & 2 & 2 \end{bmatrix}$$

$$A^{-1} = \begin{bmatrix} 3 & 1 & \frac{3}{2} \\ -\frac{5}{4} & -\frac{1}{4} & -\frac{3}{4} \\ -\frac{1}{4} & -\frac{1}{4} & -\frac{1}{4} \end{bmatrix}$$

2. Determine rank of the matrix $\begin{bmatrix} 1 & 2 & 3 \\ 1 & 4 & 2 \\ 2 & 6 & 5 \end{bmatrix}$

Solution :

Operate $R_2 - R_1$ and $R_3 - 2R_1$, therefore

$$\sim \begin{bmatrix} 1 & 2 & 3 \\ 0 & 2 & -1 \\ 0 & 2 & -1 \end{bmatrix} = A \text{ (say)}$$

Obviously, 3rd oder minor of A vanishes. All its second order minors formed by 2nd and 3rd rows are zero.

But another minor

$$\begin{bmatrix} 1 & 3 \\ 0 & -1 \end{bmatrix} = -1 \neq 0$$

Hence rank of A is 2

3. Investigate value of λ and μ, so that the equations

$$2x + 3y + 5z = 9$$
$$7x + 3y - 2z = 8$$
$$2x + 3y + \lambda z = \mu$$

have

(i) no solution,

(ii) a unique solution, and

(iii) an infinite number of solutions

Solution :

We have $\begin{bmatrix} 2 & 3 & 5 \\ 7 & 3 & -2 \\ 2 & 3 & \lambda \end{bmatrix} \begin{bmatrix} x \\ y \\ z \end{bmatrix} = \begin{bmatrix} 9 \\ 8 \\ \mu \end{bmatrix}$

The system admits of unique solution if, and only if, coefficent matrix is of rank 3. This requires in

$$\begin{vmatrix} 2 & 3 & 5 \\ 7 & 3 & -2 \\ 2 & 3 & \lambda \end{vmatrix} = 15(5-\lambda) \neq 0$$

Thus for a unique solution $\lambda \neq 5$ and m may have any value.

If $\lambda = 5$, the system wiil have no solution for those values of μ for which the matrices

$$A = \begin{bmatrix} 2 & 3 & 5 \\ 7 & 3 & -2 \\ 2 & 3 & \lambda \end{bmatrix}$$

and
$$K = \begin{bmatrix} 2 & 3 & 5 & 9 \\ 7 & 3 & -2 & 8 \\ 2 & 3 & 5 & \mu \end{bmatrix}$$

are not of the same rank. But A is of rank 2 and K is not of rank 2 unless $\mu = 9$.

Thus if $\lambda = 5$ and $\mu \neq 9$, the system will have no solution.

If $\lambda = 5$ and $\mu = 9$, the system will have an infinite number of solutions.

4. Find eigen values and eigen vectors of the matrix.

$$A = \begin{bmatrix} 8 & -6 & 2 \\ -6 & 7 & -4 \\ 2 & -4 & 3 \end{bmatrix}$$

Solution :

Characteristic equation is

$$|A-\lambda I| = \begin{bmatrix} 8-\lambda & -6 & 2 \\ -6 & 7-\lambda & 4 \\ 2 & -4 & 3-\lambda \end{bmatrix}$$

$$= \lambda^3 + 18\lambda^3 - 45\lambda = 0$$

$$\Rightarrow \quad \lambda(\lambda-3)(\lambda-15) = 0$$

$$\therefore \quad \lambda = 0, 3, 15$$

Thus eigen values of A are 0, 3, 15.

If x, y, z be the component of an eigen vectors corresponding to the eigen value λ, then

$$[A-\lambda I]X = \begin{bmatrix} 8-\lambda & -6 & 2 \\ -6 & 7-\lambda & -4 \\ 2 & -4 & 3-\lambda \end{bmatrix}\begin{bmatrix} x \\ y \\ z \end{bmatrix} = 0$$

Putting $\lambda = 0$, we have

$$8x - 6y + 2z = 0,$$
$$-6x + 7y + 4z = 0,$$
$$2x - 4y + 3z = 0$$

These equations determine a single linearly independent solution which may be takne as (1, 2, 2), so that every non-zero multiple of this vector is an eigen vector corresponding to $\lambda = 0$.

Similarly, eigen vectors corresponding to $\lambda = 3$ and $\lambda = 15$ are arbitrary non-zero multiples of the vectors (2, 1, –2) and (2, –2, 1) which are obtained from above equation.

Hence three vectors may be taken as (1, 2, 2), (2, 1, – 2), (2, –2, 1)

5. If $A\begin{bmatrix} -2 & 1 \\ 3 & 5 \end{bmatrix} = \begin{bmatrix} -1 & 7 \\ -1 & 20 \end{bmatrix}$, find matrix A

Solution :

$$A = \begin{bmatrix} -1 & 7 \\ -1 & 20 \end{bmatrix}\begin{bmatrix} -2 & 1 \\ 3 & 5 \end{bmatrix}^{-1}$$

$$= \begin{bmatrix} -1 & 7 \\ -1 & 20 \end{bmatrix}\frac{1}{-10-3}\begin{bmatrix} 5 & -1 \\ -3 & -2 \end{bmatrix}$$

$$= \frac{-1}{13}\begin{bmatrix} -1 & 7 \\ -1 & 20 \end{bmatrix}\begin{bmatrix} 5 & -1 \\ -3 & -2 \end{bmatrix}$$

$$= \frac{-1}{13}\begin{bmatrix} -26 & -13 \\ -65 & -39 \end{bmatrix}$$

$$= \begin{bmatrix} 2 & 1 \\ 5 & 3 \end{bmatrix}$$

6. If ω is cube root of –1, then find value of

$$\begin{bmatrix} 1 & -\omega & \omega^2 \\ -\omega & \omega^2 & 1 \\ \omega^2 & 1 & -\omega \end{bmatrix}$$

Solution :

Since ω is cube root of –1, therefore

$$\omega^3 = -1$$

i.e., $\omega^3 + 1 = 0 \Rightarrow (\omega+1)(\omega^2 - \omega + 1) = 0$

Hence either $\omega = -1$ or $\omega^2 - \omega + 1 = 0$

Let
$$A = \begin{bmatrix} 1 & -\omega & \omega^2 \\ -\omega & \omega^2 & 1 \\ \omega^2 & 1 & -\omega \end{bmatrix}$$

$$C_1 \to C_1 + C_2 + C_3$$

$$A = \begin{bmatrix} 1-\omega+\omega^2 & -\omega & \omega^2 \\ -\omega+\omega^2+1 & \omega^2 & 1 \\ \omega^2+1-\omega & 1 & -\omega \end{bmatrix}$$

$$= \begin{bmatrix} 0 & -\omega & \omega^2 \\ 0 & \omega^2 & 1 \\ 0 & 1 & -\omega \end{bmatrix}$$

$$= 0 \qquad (\text{since } 1 - \omega + \omega^2 = 0)$$

EXERCISE

MCQ TYPE QUESTIONS

1. Eigen values of the matrix $\begin{bmatrix} 3 & -1 & -1 \\ -1 & 3 & -1 \\ -1 & -1 & 3 \end{bmatrix}$ are

(a) 1, 1, 1 (b) 1, 1, 2 (c) 1, 4, 4 (d) 1, 2, 4

2. Rank of the following $(n+1)\times(n+1)$ matrix, where a is a real number, is

$$\begin{bmatrix} 1 & a & a^2 & . & . & . & a^n \\ a & a & a^2 & . & . & . & a^n \\ . & . & . & . & . & . & . \\ . & . & . & . & . & . & . \\ . & . & . & . & . & . & . \\ 1 & a & a^2 & . & . & . & a^n \end{bmatrix}$$

(a) 1
(b) 2
(c) n
(d) dependent upon the value of A

3. Matrices $\begin{bmatrix} \cos\theta & -\sin\theta \\ \sin\theta & \cos\theta \end{bmatrix}$ and $\begin{bmatrix} a & 0 \\ 0 & b \end{bmatrix}$ commute under multiplication
(a) if $a = b$ or $\theta = n\pi$, where n is an integer
(b) always
(c) never
(d) if $a \cos\theta \neq b \sin\theta$

4. If $A = \begin{bmatrix} 0 & 2 & 3 \\ -2 & 0 & 5 \\ -3 & -5 & 0 \end{bmatrix}$, then
(a) $A' = -A$
(b) $A' = A$
(c) $A' = -\dfrac{1}{A}$
(d) none of these

5. Eigen values of the system represented by

$$\dot{X} = \begin{bmatrix} 0 & 1 & 0 & 0 \\ 0 & 0 & 1 & 0 \\ 0 & 0 & 0 & 1 \\ 0 & 0 & 0 & 1 \end{bmatrix} X \text{ are}$$

(a) 0, 0, 0, 0 (b) 1, 1, 1, 1
(c) 0, 0, 0, –1 (d) 1, 0, 0, 0

6. Matrix, $A = \begin{bmatrix} \cos\theta & \sin\theta & 0 \\ \sin\theta & \cos\theta & 0 \\ 0 & 0 & 1 \end{bmatrix}$ is not

(a) orthogonal (b) non–singular
(c) singular (d) have A^{-1} exists

7. In the matrix equation $Px = q$, which of the following is a necessary condition for the existence of at least one solution for the unknown vector x?
(a) Augmented matrix $[Pq]$ must have the same rank as matrix P
(b) Vector q must have only non-zero elements
(c) Matrix P must be singular
(d) Matrix P must be square

8. Eigen vectors of a real symmetric matrix corresponding to different eigen values are
(a) orthogonal
(b) singular
(c) non-singular
(d) none of these

9. An arbitrary vector X is an eigen vector of the matrix $A = \begin{bmatrix} 1 & 0 & 0 \\ 0 & a & 0 \\ 0 & 0 & b \end{bmatrix}$, if (a, b) =

(a) (0, 0) (b) (1, 1) (c) (0, 1) (d) (1, 2)

10. The system of equations

$$a_1x + a_2y = 0$$
$$b_1x + b_2y = 0$$

where a_1, a_2, b_1, b_2, are real numbers, has a non-trivial solution if
(a) $a_1b_1 = a_2b_2$ (b) $a_1b_2 = b_1a_2$
(c) $a_1a_2 = b_1b_2$ (d) none of these

11. Solution of the given matrix equation is

$$\begin{bmatrix} 2 & 2 & 2 \\ 1 & 5 & -3 \\ -1 & -3 & 1 \end{bmatrix} \begin{bmatrix} x_1 \\ x_2 \\ x_3 \end{bmatrix} = \begin{bmatrix} 0 \\ 0 \\ 0 \end{bmatrix}$$

(a) $x_1 = 0, x_2 = 0, x_3 = 0$
(b) $x_1 = 1, x_2 = 1, x_3 = 0$
(c) $x_1 = 0, x_2 = 1, x_3 = 1$
(d) $x_1 = 2, x_2 = -5, x_4 = -1$

12. If product of matrix

$$A = \begin{bmatrix} \cos^2\theta & \cos\theta\sin\theta \\ \cos\theta\sin\theta & \sin^2\theta \end{bmatrix} \text{ and}$$

$$B = \begin{bmatrix} \cos^2\phi & \cos\phi\sin\phi \\ \cos\phi\sin\phi & \sin^2\phi \end{bmatrix}$$

is a null matrix, then θ and ϕ differ by an
(a) odd multiple of π (b) even multiple of π
(c) odd multiple of $\pi/2$ (d) even multiple of $\pi/2$

13. Inverse of the matrix $\begin{Bmatrix} a & 0 \\ 0 & b \end{Bmatrix}$ is

(a) $\begin{Bmatrix} a & 0 \\ 0 & a \end{Bmatrix}$ (b) $\begin{Bmatrix} b & 0 \\ a & b \end{Bmatrix}$

(c) $\begin{pmatrix} 1/a & 0 \\ 0 & 1/b \end{pmatrix}$ (d) $\begin{Bmatrix} a & 0 \\ 0 & 1/b \end{Bmatrix}$

14. The system of equations

$$2x + 4y = 10,$$
$$5x + 10y = 25$$

has

(a) no unique solution (b) only one solution
(c) only two solutions (d) infinite solutions

15. Sum of the eigen values of the matrix $\begin{pmatrix} 3 & 4 \\ x & 1 \end{pmatrix}$ for real and negative values of x is

(a) greater than zero
(b) less than zero
(c) zero
(d) dependent on the value of x

16. The system of equations

$$4x + 6y + 8z = 0$$
$$7x + 8y + 9z = 0$$
$$3x + 2y + z = 0$$

has

(a) no solution
(b) only one solution
(c) two solutions
(d) infinite number of solutions

17. Eigen values of $\begin{bmatrix} 1 & 1 & 1 \\ 1 & 1 & 1 \\ 1 & 1 & 1 \end{bmatrix}$ are

(a) 0, 0 0 (b) 0, 0 1
(c) 0, 0, 3 (d) 1, 1, 1

18. For following set of simulataneous equations

$$1.5x - 0.5y = 2$$
$$4x + 2y + 3z = 9$$
$$7x + y + 5z = 10$$

(a) solution is unique
(b) infinitely many solutions exist
(c) equations are incompatible
(d) finite number of multiple solutions exist

19. If $R = \begin{bmatrix} 1 & 0 & -1 \\ 2 & 1 & -1 \\ 2 & 3 & 2 \end{bmatrix}$, then top row of R^{-1} is

(a) [5 6 4] (b) [5 –3 1]
(c) [2 0 –1] (d) [2 –1 1/2]

20. The system of simultaneous equations

$$x + 2y + z = 6$$
$$2x + y + 2z = 6$$
$$x + y + z = 5$$

has

(a) unique solution
(b) infinite number of solutions
(c) no solution
(d) exactly two solutions

21. Eigen vectors of the matrix $\begin{bmatrix} 1 & -1 \\ -1 & 1 \end{bmatrix}$ is (are)

(a) (1, 0) (b) (0, 1)
(c) (1, 1) (d) (1, –1)

22. Inverse of the matrix $\begin{bmatrix} -3 & 5 \\ 2 & 1 \end{bmatrix}$ is

(a) $\begin{bmatrix} \frac{5}{13} & \frac{-1}{13} \\ \frac{2}{13} & \frac{3}{13} \end{bmatrix}$ (b) $\begin{bmatrix} \frac{2}{13} & \frac{5}{13} \\ \frac{-1}{13} & \frac{3}{13} \end{bmatrix}$

(c) $\begin{bmatrix} \frac{-1}{13} & \frac{5}{13} \\ \frac{2}{13} & \frac{3}{13} \end{bmatrix}$ (d) $\begin{bmatrix} \frac{1}{13} & \frac{-5}{13} \\ \frac{2}{13} & \frac{3}{13} \end{bmatrix}$

23. If A and B are square matrices of size $n \times n$, then which of the following statement is not true ?

(a) det (AB) = det (A) det (B)
(b) det (kA) = k^n det (A)
(c) det (A + B) = det (A) + det (B)
(d) det (A^T) = 1/det (A^{-1})

24. If matrix A is $m \times n$ and B is $n \times p$, then number of multiplication operations and addition operations needed to calculate the matrix AB, respectively are

(a) mn^2p, mpn (b) mpn, $(n - 1)$
(c) mpn, mpn (d) $mn^2 p$, $(m + p) n$

25. If A be an invertible matrix and inverse of 7A is $\begin{bmatrix} -1 & 2 \\ 4 & -7 \end{bmatrix}$, then matrix A is

(a) $\begin{bmatrix} 1 & 2/7 \\ 4/7 & 1/7 \end{bmatrix}$ (b) $\begin{bmatrix} 7 & 2 \\ 4 & 1 \end{bmatrix}$

(c) $\begin{bmatrix} 1 & -4/7 \\ -2/7 & 1/7 \end{bmatrix}$ (d) $\begin{bmatrix} 7 & 4 \\ 2 & 1 \end{bmatrix}$

26. Matrix B = A^T, where A is any matrix is

(a) skew symmetric
(b) symmetric about the secondary diagonal
(c) always symmetric
(d) another general matrix

27. If $A^T = A^{-1}$, where A is a real matrix, then A is

(a) normal (b) symmetric
(c) Hermitian (d) orthogonal

28. If A and B are non-zero square matrices, then AB = I implies

(a) A and B are orthogonal
(b) A and B are singular
(c) B is singular
(d) A is singular

29. Eigen values of a square symmetric matrix are always

(a) positive (b) real and imaginary
(c) negative (d) real

30. If $A = \begin{bmatrix} 1 & 0 \\ 0 & 1 \end{bmatrix}$, then state transition matrix e^{At} is

(a) $\begin{bmatrix} 0 & e^{-t} \\ e^{-t} & 0 \end{bmatrix}$ (b) $\begin{bmatrix} e^{t} & 0 \\ 0 & e^{t} \end{bmatrix}$

(c) $\begin{bmatrix} e^{-t} & 0 \\ 0 & e^{-t} \end{bmatrix}$ (d) $\begin{bmatrix} 0 & e^{t} \\ e^{t} & 0 \end{bmatrix}$

31. For matrix $p = \begin{bmatrix} 3 & -2 & 2 \\ 0 & -2 & 1 \\ 0 & 0 & 1 \end{bmatrix}$, if one of the eigen values is equal to –2, then which of the following is an eigen vector ?

(a) $\begin{bmatrix} 3 \\ -2 \\ 1 \end{bmatrix}$ (b) $\begin{bmatrix} -3 \\ 2 \\ -1 \end{bmatrix}$

(c) $\begin{bmatrix} 1 \\ -2 \\ 3 \end{bmatrix}$ (d) $\begin{bmatrix} 5 \\ 2 \\ 0 \end{bmatrix}$

32. If $A = \begin{bmatrix} 2 & -0.1 \\ 0 & 3 \end{bmatrix}$ and $A^{-1} = \begin{bmatrix} \frac{1}{2} & a \\ 0 & b \end{bmatrix}$, then (a + b) equals

(a) $\frac{7}{20}$ (b) $\frac{3}{20}$

(c) $\frac{19}{60}$ (d) $\frac{11}{20}$

33. Eigen values and corresponding eigen vectors of a 2 × 2 matrix are given by

Eigenvalue	Eigenvector
$\lambda_1 = 4$	$v_1 = \begin{bmatrix} 1 \\ 1 \end{bmatrix}$
$\lambda_2 = 8$	$v_2 = \begin{bmatrix} 1 \\ -1 \end{bmatrix}$

The matrix is

(a) $\begin{bmatrix} 6 & 2 \\ 2 & 6 \end{bmatrix}$ (b) $\begin{bmatrix} 4 & 6 \\ 6 & 4 \end{bmatrix}$

(c) $\begin{bmatrix} 2 & 4 \\ 4 & 2 \end{bmatrix}$ (d) $\begin{bmatrix} 4 & 8 \\ 8 & 4 \end{bmatrix}$

34. A second order system starts with an initial condition of $\begin{bmatrix} 2 \\ 3 \end{bmatrix}$ without any external input. The state transition matrix for the system is given by $\begin{bmatrix} e^{-2t} & 0 \\ 0 & e^{-t} \end{bmatrix}$. The state of the system at the end of 1 second is given by

(a) $\begin{bmatrix} 0.271 \\ 1.100 \end{bmatrix}$ (b) $\begin{bmatrix} 0.135 \\ 0.368 \end{bmatrix}$

(c) $\begin{bmatrix} 0.271 \\ 0.736 \end{bmatrix}$ (d) $\begin{bmatrix} 0.135 \\ 1.100 \end{bmatrix}$

NUMERICAL TYPE QUESTIONS

1. Rank of the matrix $A = \begin{bmatrix} 0 & 0 & 0 & 0 \\ 4 & 2 & 3 & 0 \\ 1 & 0 & 0 & 0 \\ 4 & 0 & 3 & 0 \end{bmatrix}$ is______

2. Rank of the diagonal matrix $\begin{pmatrix} 3 & & & 0 \\ & 3 & & \\ & & 4 & \\ 0 & & & 4 \end{pmatrix}$ is____

3. Rank of the matrix $A = \begin{bmatrix} 1 & 3 & 5 & 1 \\ 2 & 4 & 8 & 0 \\ 3 & 1 & 7 & 5 \end{bmatrix}$ is_______

4. Rank of the matrix $\begin{bmatrix} 1 & 1 \\ 0 & 0 \end{bmatrix}$ is________

5. For which value of k, following system is consistent?

$$2x - 5ky + 6z = 0$$
$$kx + 2y - 2z = 0$$
$$2x + 2y - kz = 0$$

6. The value of λ for which the equations

$$2x + y + 2z = 0$$
$$x + y + 3z = 0$$
$$4x + y + \lambda z = 0$$

have a non-zero solution, is______

7. The system of linear equations

$$(4d - 1)\,x + y + z = 0$$
$$-y + z = 0$$
$$(4d - 1)\,z = 0$$

has a non-trivial solution, if d equals _______

8. Rank of a 3 × 3 matrix C (= AB), found by multiplying a non-zero column matrix A of size 3 × 1 and a non-zero row matrix B of size 1 × 3, is_______

9. Sum of eigen values of the matrix $\begin{bmatrix} 1 & 1 & 3 \\ 1 & 5 & 1 \\ 3 & 1 & 1 \end{bmatrix}$ is_____

10. For which value of x, will the matrix $\begin{bmatrix} 8 & x & 0 \\ 4 & 0 & 2 \\ 12 & 6 & 0 \end{bmatrix}$ become singular?

11. Rank of the matrix, $A = \begin{bmatrix} 1 & -2 & -1 \\ -3 & 3 & 0 \\ 2 & 2 & 4 \end{bmatrix}$ is______

12. Solution(s) to the equations

$$2x + 3y = 1$$
$$x - y = 4$$
$$4x - y = a$$

will exist if a is equal to______

13. For matrix $\begin{bmatrix} 4 & 2 \\ 2 & 4 \end{bmatrix}$, eigen value corresponding to the eigen vector $\begin{bmatrix} 101 \\ 101 \end{bmatrix}$ is______

ANSWERS

MCQ Type Questions

1. (*c*)	**2.** (*a*)	**3.** (*a*)	**4.** (*a*)	**5.** (*d*)	**6.** (*b,d*)	**7.** (*a*)	**8.** (*a*)	**9.** (*b*)	**10.** (*b*)
11. (*a*)	**12.** (*c*)	**13.** (*c*)	**14.** (*d*)	**15.** (*a*)	**16.** (*b*)	**17.** (*c*)	**18.** (*c*)	**19.** (*b*)	**20.** (*c*)
21. (*c,d*)	**22.** (*c*)	**23.** (*c*)	**24.** (*b*)	**25.** (*a*)	**26.** (*c*)	**27.** (*d*)	**28.** (*a*)	**29.** (*d*)	**30.** (*b*)
31. (*d*)	**32.** (*a*)	**33.** (*a*)	**34.** (*a*)						

Numerical Type Questions

1. 3	**2.** 4	**3.** 3	**4.** 1	**5.** 2	**6.** 8	**7.** $\frac{1}{4}$	**8.** 1	**9.** 7	**10.** 4
11. 2	**12.** 12	**13.** 6							

EXPLANATIONS

MCQ TYPE QUESTIONS

1. Characteristic equation is

$$[A - \lambda I] = 0$$

$$\Rightarrow \begin{bmatrix} 3-\lambda & -1 & -1 \\ -1 & 3-\lambda & -1 \\ -1 & -1 & 3-\lambda \end{bmatrix} = 0$$

$$\Rightarrow (3-\lambda)\{(3-\lambda)^2 - 1\} - 1\{(3-\lambda)+1\} - 1\{(3-\lambda)+1\} = 0$$

$$\Rightarrow \{(3-\lambda)+1\}\{(3-\lambda)(3-\lambda-1)\} - 2\} = 0$$

$$\lambda = 1, 4, 4$$

$\therefore$ λ = **1, 4, 4** are eigen values.

4. $A' = \begin{bmatrix} 0 & -2 & -3 \\ 2 & 0 & -5 \\ 3 & 5 & 0 \end{bmatrix} = -\begin{bmatrix} 0 & 2 & 3 \\ -2 & 0 & 5 \\ -3 & -5 & 0 \end{bmatrix} = -A.$

This is a skew-symmetric matrix.

6. Determinant $A = 1(\cos^2\theta - \sin^2\theta)$

Hence A is non-singular and A^{-1} exists.

7. According to *Rouche's theorem,* the system is consistent if and only if coefficient matrix and augmented matrix K are of the same rank, otherwise the system is inconsistent.

8. Let A be a real symmetric matrix, therefore

$$A^T = A$$

Let α_1 and α_2 be different eigen values of the matrix A, and X_1 and X_2 be the corresponding vectors, then

$$AX_1 = \alpha_1 X_1 \quad \text{and} \quad AX_2 = \alpha_2 X_2$$

Taking transpose of the second equation

$$(AX_2)^T = (\alpha_2 X_2)^T$$

$$\Rightarrow \quad X_2^T A^T = \alpha_2 . X_2^T$$

But $\quad A^T = A,$

$$\therefore \quad X_2^T A = \alpha_2 . X_2^T$$

Post multiply by X_1, we get

$$X_2^T AX_1 = \alpha_2 X_2^T X_1$$

But $\quad AX_1 = \alpha_1 X_1$

$$\therefore \quad X_2^T \alpha_1 X_1 = \alpha_2 X_2^T X_1$$

$$\Rightarrow \quad (\alpha_1 - \alpha_2) X_2^T X_1 = 0$$

Since $\alpha_1 \neq \alpha_2, \alpha_1 - \alpha_2 \neq 0$

$$\therefore \quad X_2^T X_1 = 0$$

i.e. X_2 and X_1 are orthogonal.

9. Since matrix is triangular, hence eigen values are α, *a*, *b*.

If (X_1, X_2, X_3) is an arbitrary eigen vector, say corresponding to 1, then

$$\begin{bmatrix} 1 & 0 & 0 \\ 0 & a & 0 \\ 0 & 0 & b \end{bmatrix} \begin{bmatrix} X_1 \\ X_2 \\ X_3 \end{bmatrix} = 1 \begin{bmatrix} X_1 \\ X_2 \\ X_3 \end{bmatrix}$$

X_2, X_3 being not zero, we have

$$X_1 = X_1 \,;\, aX_2 = X_2$$

which gives $a = 1$

and $bX = X_3$

$\Rightarrow$ $b = 1$

$\therefore$ $(a, b) = (1, 1).$

10. $\dfrac{x}{y} = -\dfrac{a_2}{a_1}$ and $\dfrac{x}{y} = -\dfrac{b_2}{b_1}$

Equations are consistent only if

$$-\frac{a_2}{a_1} = -\frac{b_2}{b_1} \Rightarrow a_1b_2 = b_1a_2$$

11. Let matrics, $A = \begin{bmatrix} 2 & 2 & 2 \\ 1 & 5 & -3 \\ -1 & -3 & 1 \end{bmatrix}$

$$X = \begin{bmatrix} x_1 \\ x_2 \\ x_3 \end{bmatrix} \text{ and } B_1 = \begin{bmatrix} 0 \\ 0 \\ 0 \end{bmatrix}$$

$\therefore$ $AX = B$

Multiplying both sides by A^{-1},

$$X = A^{-1}B$$

But as $B = 0$; therefore $X = 0$

Hence $\begin{bmatrix} x_1 \\ x_2 \\ x_3 \end{bmatrix} = \begin{bmatrix} 0 \\ 0 \\ 0 \end{bmatrix}$

$\therefore$ $x_1 = 0, \quad x_2 = 0, \quad x_3 = 0.$

12. AB

$$= \begin{bmatrix} \cos\theta.\cos\phi\cos(\theta-\phi) & \cos\theta.\sin\phi.\cos(\theta-\phi) \\ \cos\phi\sin\theta.\cos(\theta-\phi) & \sin\theta.\sin\phi\cos(\theta-\phi) \end{bmatrix}$$

= A null matrix, when $\cos(\theta - \phi) = 0$

i.e. if $(\theta - \phi)$ is an odd multiple of $\left(\dfrac{\pi}{2}\right)$.

15. Eigen values are given by the solution of equation,

$$\begin{vmatrix} 3-\lambda & 4 \\ x & 1-\lambda \end{vmatrix} = 0$$

Since x is real and negative, put $x = -k$, where k is positive constant

$\therefore$ $(3-\lambda)(1-\lambda) + 4k = 0$

$\Rightarrow$ $\lambda^2 - 4\lambda + 3 + 4k = 0$

If λ_1 and λ_2 be the solutions of above equation, then λ_1 and λ_2 are eigen values.

Now, Sum of eigen values

= Sum of roots of above equation

i.e. $\lambda_1 + \lambda_2 = \dfrac{-(-4)}{1} = 4\ (> 0)$

16. For given system of equations,

$$\Delta = \begin{vmatrix} 4 & 6 & 8 \\ 7 & 8 & 9 \\ 3 & 2 & 1 \end{vmatrix}$$

$$= 4(8-18) - 6(7-27) + 8(14-24)$$

$$= -40 + 120 - 80 = 0$$

Since $\Delta = 0$, hence given system of equations has unique solution, *i.e.* only one solution.

17. Given matrix is $\begin{bmatrix} 1 & 1 & 1 \\ 1 & 1 & 1 \\ 1 & 1 & 1 \end{bmatrix}$

Now $\begin{bmatrix} (1-\lambda) & 1 & 1 \\ 1 & (1-\lambda) & 1 \\ 1 & 1 & (1-\lambda) \end{bmatrix} = 0$

$R_1 \to R_1 + R_2 + R_3$

$$\Rightarrow \begin{bmatrix} (3-\lambda) & (3-\lambda) & (3-\lambda) \\ 1 & (1-\lambda) & 1 \\ 1 & 1 & (1-\lambda) \end{bmatrix} = 0$$

$$\Rightarrow (3-\lambda)\begin{vmatrix} 1 & 1 & 1 \\ 1 & (1-\lambda) & 1 \\ 1 & 1 & (1-\lambda) \end{vmatrix} = 0$$

$C_2 \to C_2 - C_1$

$C_3 \to C_3 - C_1$

$$\Rightarrow (3-\lambda)\begin{vmatrix} 1 & 0 & 0 \\ 1 & -\lambda & 1 \\ 1 & 0 & -\lambda \end{vmatrix} = 0$$

Hence eigen values are 0, 0, 3

19. $\text{Adj } R = \begin{bmatrix} 5 & -6 & 4 \\ -3 & 4 & -3 \\ 1 & -1 & 1 \end{bmatrix}^T = \begin{bmatrix} 5 & -3 & 1 \\ -6 & 4 & -1 \\ 4 & -3 & 1 \end{bmatrix}$

Since, $|R| = 1$, therefore top row $\to [5\ -3\ 1]$

20. Given equations are :

$$x + 2y + z = 6$$
$$2x + y + 2z = 6$$
$$x + y + z = 5$$

Given system can be written as

$$\begin{bmatrix} 1 & 2 & 1 \\ 2 & 1 & 2 \\ 1 & 1 & 1 \end{bmatrix} . \begin{bmatrix} x \\ y \\ z \end{bmatrix} = \begin{bmatrix} 6 \\ 6 \\ 5 \end{bmatrix}$$

Applying row operation

$R_2 = R_2 - 2R_1$, $R_3 = R_3 - R_1$, we get

$$\begin{bmatrix} 1 & 2 & 1 \\ 0 & -3 & 0 \\ 0 & -1 & 0 \end{bmatrix} . \begin{bmatrix} x \\ y \\ z \end{bmatrix} = \begin{bmatrix} 6 \\ -6 \\ -1 \end{bmatrix}$$

Applying $R_3 = 3R_3 - R_2$, we get

$$\begin{bmatrix}1 & 2 & 1\\ 0 & -3 & 0\\ 0 & 0 & 0\end{bmatrix}.\begin{bmatrix}x\\ y\\ z\end{bmatrix} = \begin{bmatrix}6\\ -6\\ 3\end{bmatrix}$$

Since rank of co-efficient matrix is 2 and rank of argument matrix is 3, which is not equal. Hence system has no solution.

21. Eigen value are roots of the equation

$$\begin{bmatrix}1-\alpha & -1\\ -1 & 1-\alpha\end{bmatrix} = 0$$

i.e. $(1-\alpha)^2 - 1 = 0$

$\Rightarrow$ $\alpha^2 - 2\alpha = 0$

$\Rightarrow$ $\alpha = 0$ or $\alpha = 2$

When $\alpha = 0$, $\begin{bmatrix}1 & -1\\ -1 & 1\end{bmatrix}\begin{bmatrix}x\\ y\end{bmatrix} = 0$

$\Rightarrow$ $\begin{bmatrix}x\\ y\end{bmatrix} = \begin{bmatrix}0\\ 0\end{bmatrix}$

This gives, $x = y$

When $\alpha = 2$, $\begin{bmatrix}1 & -1\\ -1 & 1\end{bmatrix}\begin{bmatrix}x\\ y\end{bmatrix} = 2$

$\Rightarrow$ $\begin{bmatrix}x\\ y\end{bmatrix} = \begin{bmatrix}2x\\ 2y\end{bmatrix}$.

This gives, $x = -y$

Hence, eigen vectors corresponding to 0 and 2 are $\begin{bmatrix}x\\ x\end{bmatrix}$ and $\begin{bmatrix}x\\ -x\end{bmatrix}$, or $\begin{bmatrix}1\\ 1\end{bmatrix}$ and $\begin{bmatrix}1\\ -1\end{bmatrix}$

22. For matrix $A = \begin{bmatrix}-3 & 5\\ 2 & 1\end{bmatrix}$

cofactors are :

$a_{11} = 1, a_{21} = -5, a_{12} = -2, a_{22} = -3$

$\therefore$ Matrix by cofactors $= \begin{bmatrix}1 & -2\\ -5 & -3\end{bmatrix}$

$$\text{Adjoint A} = \begin{bmatrix}1 & -5\\ -2 & -3\end{bmatrix}$$

$$|A| = -3(1) - 5(2) = -3 - 10 = -13$$

$$\therefore A^{-1} = \frac{\text{Adjoint A}}{|A|} = \begin{bmatrix}-\frac{1}{13} & \frac{5}{13}\\ \frac{2}{13} & \frac{3}{13}\end{bmatrix}$$

25. Inverse of $7A = \begin{bmatrix}-1 & 2\\ 4 & -7\end{bmatrix}$

$$\det|\Delta| = \begin{vmatrix}-1 & 2\\ 4 & -7\end{vmatrix} = 7 - 8 = -1$$

$\therefore$ Matrices of cofactor $= \begin{bmatrix}-7 & -4\\ -2 & -1\end{bmatrix}$,

and it's transpose, $A^T = \begin{bmatrix}-7 & -2\\ -4 & -1\end{bmatrix}$

$\therefore$ Inverse $= \dfrac{A^T}{|\Delta|} = \begin{bmatrix}7 & 2\\ 4 & 1\end{bmatrix}$

$\therefore$ $A = \dfrac{1}{7}\left[\dfrac{A^T}{|\Delta|}\right] = \begin{bmatrix}1 & \frac{2}{7}\\ \frac{4}{7} & \frac{1}{7}\end{bmatrix}$

30. Here $[sI - A]^{-1} = \begin{bmatrix}s-1 & 0\\ 0 & s-1\end{bmatrix}^{-1}$

Then, $2^{-1}[sI - A]^{-1} = 2^{-1}\left\{\begin{bmatrix}\frac{1}{s-1} & 0\\ 0 & \frac{1}{s-1}\end{bmatrix}\right\} = \begin{bmatrix}e^t & 0\\ 0 & e^t\end{bmatrix}$

31. $|P - \lambda I| = \begin{vmatrix}3-\lambda & -2 & 2\\ 0 & -2-\lambda & 1\\ 0 & 0 & 1-\lambda\end{vmatrix} = 0$

$\therefore$ $\lambda = -2, 3, 1$

Putting $\lambda = -2$, we get

$5x - 2y + 2z = 0$

$z = 0$

So. $5x = 2y$ eigen vector $= \begin{bmatrix}5\\ 2\\ 0\end{bmatrix}$

32. We know, $AA^{-1} = I$

$\therefore$ $\begin{bmatrix}2 & -0.1\\ 0 & 3\end{bmatrix}\begin{bmatrix}\frac{1}{2} & a\\ 0 & b\end{bmatrix} = \begin{bmatrix}1 & 2a - 0.1b\\ 0 & 3b\end{bmatrix}$

$\Rightarrow$ $3b = 1$ $\Rightarrow$ $b = \dfrac{1}{3}$

or $2a - 0.1b = 0$

$\Rightarrow$ $2a = \dfrac{b}{10}$ $\Rightarrow$ $a = \dfrac{1}{60}$

$\therefore$ $a + b = \dfrac{1}{3} + \dfrac{1}{60} = \dfrac{7}{20}$

33. Given matrix : $A = \begin{bmatrix} 6 & 2 \\ 2 & 6 \end{bmatrix}$

We know $|\lambda I - A| = 0$

$$\Rightarrow \begin{vmatrix} \lambda-6 & -2 \\ -2 & \lambda-6 \end{vmatrix} = 0$$

$$\Rightarrow \lambda^2 - 12\lambda + 32 = 0$$

$$\Rightarrow \lambda = 4, 8 \text{ (eigen values)}$$

For $\lambda_1 = 4$,

$$(\lambda_1 I - A) = \begin{bmatrix} -2 & -2 \\ -2 & -2 \end{bmatrix}, v_1 = \begin{bmatrix} 1 \\ 1 \end{bmatrix}$$

For $\lambda_1 = 8$,

$$(\lambda_2 I - A) = \begin{bmatrix} 2 & -6 \\ -6 & 2 \end{bmatrix}, v_2 = \begin{bmatrix} 1 \\ -1 \end{bmatrix}$$

34. State of the system at time t is,

$$X(t) = [sI - A]^{-1} X(0) = \phi(t) X(0)$$

$$= \begin{bmatrix} e^{-2t} & 0 \\ 0 & e^{-t} \end{bmatrix} \begin{bmatrix} 2 \\ 3 \end{bmatrix} = \begin{bmatrix} 2e^{-2t} \\ 3e^{-t} \end{bmatrix}$$

At $t = 1$, $X(1) = \begin{bmatrix} 2e^{-2} \\ 3e^{-1} \end{bmatrix} = \begin{bmatrix} 0.271 \\ 1.100 \end{bmatrix}$

NUMERICAL TYPE QUESTIONS

1. Determinant A = 0

$$\Rightarrow \begin{vmatrix} 4 & 2 & 3 \\ 1 & 0 & 0 \\ 4 & 0 & 3 \end{vmatrix} = (-1) \times 6 = -6 \neq 0$$

Hence rank of A = 3

2. Since the number of non-zero elements on this diagonal matrix is four, hence the rank is four.

3. Given matrix A possesses a minor of order 3

viz. $\begin{vmatrix} 1 & 3 & 1 \\ 2 & 4 & 0 \\ 3 & 1 & 5 \end{vmatrix} = \begin{vmatrix} 0 & 0 & 1 \\ 2 & 4 & 0 \\ -2 & -14 & 5 \end{vmatrix}$

Replacing C_1 and C_2 by $C_1 - C_3$ and $C_2 - 3C_3$.

$$= \begin{vmatrix} 2 & 4 \\ -2 & -14 \end{vmatrix} \text{ expanding with respect to } R_1$$

$$= 2(-14) - (4)(-2) = -28 + 8 \neq 0$$

$\therefore \quad p(A) \geq 3 \qquad \ldots (i)$

Also A does not possess any minor of order 4, *i.e.* 3 + 1

$\therefore \quad p(A) \leq 3 \qquad \ldots (ii)$

From equations (*i*) and (*ii*), we get

$p(A) = 3$ *i.e.* rank of A is 3.

5. $A = \begin{bmatrix} 2 & -5k & 6 \\ k & 2 & -2 \\ 2 & 2 & -k \end{bmatrix}$

Equations are consistent, if rank of A and that of k are equal. But in this case it is always true.

Hence equations will have a trivial solution if $|A| \neq 0$. Therefore only non-trivial solution will exist if $|A| = 0$

i.e. $\begin{bmatrix} 2 & -5k & 6 \\ k & 2 & -2 \\ 2 & 2 & -k \end{bmatrix} = 0$

$$\Rightarrow 2(-2k + 4) + 5k(-k^2 + 4) + 6(2k - 4)$$

$$\Rightarrow -5k^3 + 20k - 4k + 8 + 12k - 24 = 0$$

$$\Rightarrow 5k^3 - 28k + 16 = 0$$

$$\Rightarrow 5k^3 - 10k^2 + 10k^2 - 20k - 8k + 16 = 0$$

$$\Rightarrow (5k^2 + 10k - 8)(k - 2) = 0$$

$$k = \frac{-10 \pm \sqrt{100 + 160}}{2} \text{ or } 2$$

$$= -1 \pm \sqrt{13/5} \text{ or } 2$$

6. Equivalent matrix equation is

$$\begin{bmatrix} 2 & 1 & 2 \\ 1 & 1 & 3 \\ 4 & 3 & \lambda \end{bmatrix} \begin{bmatrix} x \\ y \\ z \end{bmatrix} = 0$$

In order that given system of equations may have non-zero solution, rank of A should be less than 3. This requires that

$$\begin{bmatrix} 2 & 1 & 2 \\ 1 & 1 & 3 \\ 4 & 3 & \lambda \end{bmatrix} = 0$$

Interchanging R_1 and R_2,

$$\begin{bmatrix} 1 & 1 & 3 \\ 2 & 1 & 2 \\ 4 & 3 & \lambda \end{bmatrix} = 0$$

By $(R_2 - 2R_1)$ and $(R_3 - 4R_1)$,

$$\begin{bmatrix} 1 & 1 & 3 \\ 1 & -1 & -4 \\ 0 & -1 & \lambda - 12 \end{bmatrix} = 0$$

$$\Rightarrow \begin{bmatrix} -1 & -4 \\ -1 & \lambda - 12 \end{bmatrix}$$

$$\Rightarrow -1(\lambda - 12) - (-1)(4) = 0$$

$$\Rightarrow \lambda = 8$$

7. The system of homogeneous linear equations has a non-trivial solution if

$$\begin{bmatrix} 4d-1 & 1 & 1 \\ 0 & -1 & 1 \\ 0 & 0 & 4d-1 \end{bmatrix} = 0$$

$$\Rightarrow -(4d-1)^2 = 0 \qquad \Rightarrow d = \frac{1}{4}$$

8. Let $A = \begin{bmatrix} a_1 \\ b_1 \\ c_1 \end{bmatrix}$; $B = [a_2 \; b_2 \; c_2]$

$$\therefore \quad C = \begin{bmatrix} a_1a_2 & a_1b_2 & a_1c_2 \\ b_1a_2 & b_1b_2 & b_1c_2 \\ c_1a_2 & c_2b_2 & c_1c_2 \end{bmatrix}$$

$$= a_1\, a_2\, [b_1\, b_2\, c_1\, c_2 - b_1b_2\, c_1c_2] = 0$$

$$\text{Cofactor} = \begin{vmatrix} a_1a_2 & a_1b_2 \\ b_1a_2 & b_1a_b \end{vmatrix} = 0$$

So rank = 1

9. Sum of eigen values of given matrix
= Sum of element of diagonal of given matrix
= 1 + 5 + 1
= 7

10. For singular matrix,

$$\begin{vmatrix} 8 & x & 0 \\ 4 & 0 & 2 \\ 12 & 6 & 0 \end{vmatrix} = 0$$

$\Rightarrow \quad 8[0 - 12] - x\,[0 - 2 \times 12] = 0$

$\Rightarrow \quad -96 + 24x = 0$

$\Rightarrow \quad x = 4$

11. Since, *det* $A = \begin{vmatrix} 1 & -2 & -1 \\ -3 & 3 & 0 \\ 2 & 2 & 4 \end{vmatrix} = 12 - 24 + 12 = 0$

and $M_{11} = 12 \neq 0$. Hence rank is 2.

12. Equations are consistent when

$$\begin{vmatrix} 2 & 3 & -1 \\ 1 & -1 & -4 \\ 4 & -1 & -a \end{vmatrix} = 0$$

$\Rightarrow \quad 2(a - 4) - 3(-a + 16) - 1(-1 + 4) = 0$

$5a - 8 - 48 - 3 = 0$

$5a = 59$

$a = \dfrac{59}{5} \approx 12$

13. Given : Matrix $M = \begin{bmatrix} 4 & 2 \\ 2 & 4 \end{bmatrix}$

$\therefore \quad M - \lambda I = \begin{bmatrix} 4-\lambda & 2 \\ 2 & 4-\lambda \end{bmatrix}$

For the eigen vector $\begin{bmatrix} 101 \\ 101 \end{bmatrix}$

$(4 - \lambda)(101) + 2(101) = 0$

$\Rightarrow \quad 4 - \lambda + 2 = 0$

$\Rightarrow \quad \lambda = 6$ (eigen value)

32. $|AB| = |A| \cdot |B| = (5) \cdot (40) = 200$

33. General 2×2 real symmetric matrix is $\begin{bmatrix} y & x \\ x & z \end{bmatrix}$

$\Rightarrow \quad \det = yz - x^2$

and trace is $\quad y + z = 14$ (given)

$\Rightarrow \quad z = 14 - y$

...(*)

Let $\quad f = yz - x^2(\det)$

$= -x^2 - y^2 + 14y$ ($u \sin g$*)

Using maxima and minima of a function of two variables, we have f is maximum at $x = 0$, $y = 7$ and therefore, maximum value of the determinant is 49.

34. $A = \begin{bmatrix} 1 & 0 & 0 & 0 & 1 \\ 0 & 1 & 1 & 1 & 0 \\ 0 & 1 & 1 & 1 & 0 \\ 0 & 1 & 1 & 1 & 0 \\ 1 & 0 & 0 & 0 & 1 \end{bmatrix}$

Let $X = \begin{bmatrix} x_1 \\ x_2 \\ x_3 \\ x_4 \\ x_5 \end{bmatrix}$

$Am = Xm \Rightarrow x_1 + x_5 = m\,x_1 = m\,x_5$

$\Rightarrow x_2 + x_3 + x_4 = m\,x_2 = m\,x_4$

For $m \neq 0$

Say, $\quad x_1 = x_5 = a$

$x_2 = x_3 = x_4 = b$

$\Rightarrow \quad x_1 + x_5 = m\,x_1$

$\Rightarrow \quad 2a = m\,a$

$\Rightarrow \quad m = 2$

$\Rightarrow \quad x_2 + x_3 + x_4 = m\,x_2$

$\therefore \quad 3b = mb$

$\therefore \quad m = 3$

For $m = 0 \Rightarrow$ eigenvalue $m = 0$

$\therefore$ There are 3 distinct eigen value = 0, 2, 3.

Product of non zero eigen value = $2 \times 3 = 6$.

■■

2
CHAPTER

Calculus

FUNCTIONS OF SINGLE VARIABLE LIMITS

By limit we are able to find out behaviour of a function $f(x)$ near a point c, when distance between x and c is very small i.e. $|x-c|$ is small.

Specifically, we would like to know whether there is some real number k such that $f(x)$ approaches k as x approaches c.

We write as $\lim_{x \to c} f(x) = k$

Right Hand and Left Hand Limits

Left-hand limit : $\lim_{x \to c^-} f(x) = k$

Right-hand limit : $\lim_{x \to c^+} f(x) = k$

If $\lim_{x \to c^-} f(x) = \lim_{x \to c^+} f(x)$, then limit of $f(x)$ as $x \to c$ exists and is written as $\lim_{x \to c} f(x)$.

The limit of a function f(x) may not exist in any of the following situations :

(*i*) $\lim_{x \to c^-} f(x)$ does not exist

(*ii*) $\lim_{x \to c^+} f(x)$ does not exist

(*iii*) Both $\lim_{x \to c^-} f(x)$, and $\lim_{x \to c^+} f(x)$ exist but are unequal

Theorems on Limits

If $\lim_{x \to c} f(x)$ and $\lim_{x \to c} g(x)$ exist, then

(1) $\lim_{x \to c} (f(x) \pm g(x)) = \lim_{x \to c} f(x) \pm \lim_{x \to c} g(x)$

(2) $\lim_{x \to c} (f(x).g(x)) = \lim_{x \to c} f(x).\lim_{x \to c} g(x)$

(3) $\lim_{x \to c} \dfrac{f(x)}{g(x)} = \dfrac{\lim_{x \to c} f(x)}{\lim_{x \to c} g(x)}$ where $\left(\lim_{x \to c} g(x) \neq 0\right)$

Some useful Limits

(1) $\lim_{x \to 0} \dfrac{\sin x}{x} = 1 = \lim_{x \to 0} \cos x = \lim_{x \to 0} \dfrac{\tan x}{x}$

(2) $\lim_{x \to 0} (1+x)^{\frac{1}{x}} = e = \lim_{x \to \infty} \left(1+\dfrac{1}{x}\right)^x$

(3) $\lim_{x \to 0} \dfrac{\log_a (1+x)}{x} = \log_a e \quad (a > 0, a \neq 0)$

(4) $\lim_{x \to 0} \dfrac{\log(1+x)}{x} = 1$

(5) $\lim_{x \to 0} \dfrac{a^x - 1}{x} = \log a \quad (a > 0)$

(6) $\lim_{x \to 0} \dfrac{(1+x)^m - 1}{x} = m$

(7) $\lim_{x \to \infty} \dfrac{\log x}{x^m} = 0 \quad (m > 0)$

(8) $\lim_{x \to 0} \dfrac{x^n - a^n}{x - a} = na^{n-1}$

Theorem (L'Hospital's Rule) : Suppose f and g are differentiable and $g'(x) \neq 0$ near a (except possibly at a). Suppose that

$$\lim_{x \to a} f(x) = 0 \text{ and } \lim_{x \to a} g(x) = 0$$

or $\lim_{x \to a} f(x) = \pm\infty$ and $\lim_{x \to a} g(x) = \pm\infty$

(In other words, we have an indeterminate form of type $\frac{0}{0}$ or $\frac{\infty}{\infty}$.) Then

$$\lim_{x \to a} \frac{f(x)}{g(x)} = \lim_{x \to a} \frac{f'(x)}{g'(x)}$$

if the limit on the right side exists (or is ∞ or $-\infty$)

Indeterminate Forms of Type $\frac{0}{0}$ and $\frac{\infty}{\infty}$

Examples :

1. Find $\lim_{x \to \infty} \dfrac{5x-2}{7x+3}$

Solution :

We have, $\lim_{x \to \infty} \dfrac{5x-2}{7x+3} \quad \left[\dfrac{\infty}{\infty} \text{ form}\right]$

$$= \lim_{x \to \infty} \frac{\frac{5x-2}{x}}{\frac{7x+3}{x}} = \lim_{x \to \infty} \frac{\frac{5x}{x} - \frac{2}{x}}{\frac{7x}{x} + \frac{3}{x}}$$

$$= \lim_{x \to \infty} \frac{5 - \frac{2}{x}}{7 + \frac{3}{x}} = \frac{5-0}{7+0} = \frac{5}{7}$$

2. Find $\lim\limits_{x\to\infty}\dfrac{x^5+x^4+x^3+x^2+x+1}{2x^5+x^4+x^3+x^2+x+1}$

Solution : We have

$$\lim_{x\to\infty}\frac{x^5+x^4+x^3+x^2+x+1}{2x^5+x^4+x^3+x^2+x+1}\left[\frac{\infty}{\infty}\text{ form}\right]$$

$$=\lim_{x\to\infty}\frac{\dfrac{x^5+x^4+x^3+x^2+x+1}{x^5}}{\dfrac{2x^5+x^4+x^3+x^2+x+1}{x^5}}$$

$$=\lim_{x\to\infty}\frac{\dfrac{x^5}{x^5}+\dfrac{x^4}{x^5}+\dfrac{x^3}{x^5}+\dfrac{x^2}{x^5}+\dfrac{x}{x^5}+\dfrac{1}{x^5}}{\dfrac{2x^5}{x^5}+\dfrac{x^4}{x^5}+\dfrac{x^3}{x^5}+\dfrac{x^2}{x^5}+\dfrac{x}{x^5}+\dfrac{1}{x^5}}$$

$$=\lim_{x\to\infty}\frac{1+\dfrac{1}{x}+\dfrac{1}{x^2}+\dfrac{1}{x^3}+\dfrac{1}{x^4}+\dfrac{1}{x^5}}{2+\dfrac{1}{x}+\dfrac{1}{x^2}+\dfrac{1}{x^3}+\dfrac{1}{x^4}+\dfrac{1}{x^5}}$$

$$=\frac{1+0+0+0+0+0}{2+0+0+0+0+0}$$

$$=\frac{1}{2}$$

3. Find $\lim\limits_{x\to-2}\dfrac{x+2}{\ln(x+3)}$

Solution : We have

$$\lim_{x\to-2}\frac{x+2}{\ln(x+3)}\left[\frac{0}{0}\text{ form}\right]$$

$$=\lim_{x\to-2}\frac{(x+2)'}{(\ln(x+3))'}$$

$$=\left\{\begin{aligned}&\lim_{x\to-2}\frac{(x+2)'}{\dfrac{1}{x+3}\cdot(x+3)'}=\lim_{x\to-2}\frac{x'+2'}{\dfrac{1}{x+3}\cdot(x'+3')}\\&\qquad=\lim_{x\to-2}\frac{1+0}{\dfrac{1}{x+3}\cdot(1+0)}\end{aligned}\right\}$$

$$=\lim_{x\to-2}\frac{1}{\dfrac{1}{x+3}}$$

$$=\lim_{x\to-2}\frac{1\cdot(x+3)}{\dfrac{1}{x+3}\cdot(x+3)}$$

$$=\lim_{x\to-2}\frac{x+3}{1}=\lim_{x\to-2}(x+3)$$

$$=-2+3=1$$

4. Find $\lim\limits_{x\to\infty}\dfrac{3^x}{x^2+x-1}$

Solution : We have

$$\lim_{x\to\infty}\frac{3^x}{x^2+x-1}\left[\frac{\infty}{\infty}\text{ form}\right]$$

$$=\lim_{x\to\infty}\frac{(3^x)'}{(x^2+x-1)'}=\lim_{x\to\infty}\frac{3^x\ln 3}{2x+1}\left[\frac{\infty}{\infty}\text{ form}\right]$$

$$=\lim_{x\to\infty}\frac{(3^x\ln 3)'}{(2x+1)'}=\lim_{x\to\infty}\frac{\ln 3(3^x)'}{2}$$

$$=\lim_{x\to\infty}\frac{\ln 3\cdot 3^x\cdot\ln 3}{2}=\infty\text{ (D.N.E.)}$$

5. Find $\lim\limits_{x\to\infty}\dfrac{\ln x}{\sqrt{x}}$

Solution : We have

$$\lim_{x\to\infty}\frac{\ln x}{\sqrt{x}}\left[\frac{\infty}{\infty}\text{ form}\right]$$

$$=\lim_{x\to\infty}\frac{(\ln x)'}{(x^{1/2})'}=\lim_{x\to\infty}\frac{x^{-1}}{\dfrac{1}{2}x^{-1/2}}$$

$$=\lim_{x\to\infty}\frac{x^{-1}\cdot x}{\dfrac{1}{2}x^{-1/2}\cdot x}=\lim_{x\to\infty}\frac{1}{\dfrac{1}{2}x^{1/2}}=0$$

6. Find $\lim\limits_{x\to\infty}\dfrac{x+\sin x}{x+\cos x}$

Solution : We have

$$\lim_{x\to\infty}\frac{x+\sin x}{x+\cos x}\left[\frac{\infty}{\infty}\text{ form}\right]$$

$$=\lim_{x\to\infty}\frac{\dfrac{x+\sin x}{x}}{\dfrac{x+\cos x}{x}}$$

$$=\lim_{x\to\infty}\frac{\dfrac{x}{x}+\dfrac{\sin x}{x}}{\dfrac{x}{x}+\dfrac{\cos x}{x}}$$

$$=\lim_{x\to\infty}\frac{1+\dfrac{\sin x}{x}}{1+\dfrac{\cos x}{x}}$$

It is easy to show that

$$\lim_{x\to\infty}\frac{\sin x}{x}=0$$

and $$\lim_{x\to\infty}\frac{\cos x}{x}=0$$

$$\lim_{x\to\infty}\frac{x+\sin x}{x+\cos x}=\lim_{x\to\infty}\frac{1+\dfrac{\sin x}{x}}{1+\dfrac{\cos x}{x}}$$

$$=\frac{1+0}{1+0}=1$$

Indeterminate Forms of Type ∞ – ∞ and 0 ·∞

Examples :

1. Find $\lim_{x\to\infty}(x-\ln x)$.

Solution : We have

$$\lim_{x\to\infty}(x-\ln x)\ [(\infty-\infty)\text{ form}]$$

$$= \lim_{x\to\infty}\left(x\cdot 1 - x\cdot\frac{\ln x}{x}\right)$$

$$= \lim_{x\to\infty} x\left(1-\frac{\ln x}{x}\right)$$

Note that $\lim_{x\to\infty}\frac{\ln x}{x}\ \left[\frac{\infty}{\infty}\text{ form}\right]$

$$= \lim_{x\to\infty}\frac{(\ln x)'}{x'}$$

$$= \lim_{x\to\infty}\frac{x^{-1}}{1}$$

$$= \lim_{x\to\infty}\frac{1}{x} = 0$$

$$\therefore\quad \lim_{x\to\infty}(x-\ln x) = \lim_{x\to\infty} x\left(1-\frac{\ln x}{x}\right)$$

$$= \lim_{x\to\infty} x(1-0) = \infty$$

2. Find $\lim_{x\to 1}\left(\frac{1}{\ln x}-\frac{1}{x-1}\right)$.

Solution : We have

$$\lim_{x\to 1}\left(\frac{1}{\ln x}-\frac{1}{x-1}\right)[(\infty-\infty)\text{ form}]$$

$$= \lim_{x\to 1}\left(\frac{1\cdot(x-1)}{\ln x\cdot(x-1)}-\frac{\ln x\cdot 1}{\ln x\cdot(x-1)}\right)$$

$$= \lim_{x\to 1}\frac{x-1-\ln x}{\ln x(x-1)}\ \left[\frac{0}{0}\text{ form}\right]$$

$$= \lim_{x\to 1}\frac{(x-1-\ln x)'}{(\ln x(x-1))'}$$

$$= \lim_{x\to 1}\frac{x'-1'-(\ln x)'}{(\ln x)'\cdot(x-1)+\ln x\cdot(x-1)'}$$

$$= \lim_{x\to 1}\frac{1-\frac{1}{x}}{\frac{x-1}{x}+\ln x}$$

$$= \lim_{x\to 1}\frac{\left(1-\frac{1}{x}\right)x}{\left(\frac{x-1}{x}+\ln x\right)x}$$

$$= \lim_{x\to 1}\frac{1\cdot x-\frac{1}{x}\cdot x}{\frac{x-1}{x}\cdot x+\ln x\cdot x}$$

$$= \lim_{x\to 1}\frac{x-1}{x-1+x\ln x}\ \left[\frac{0}{0}\text{ form}\right]$$

$$= \lim_{x\to 1}\frac{(x-1)'}{(x-1+x\ln x)'}$$

$$= \lim_{x\to 1}\frac{x'-1'}{x'-1'+x'\ln x+x(\ln x)'}$$

$$= \lim_{x\to 1}\frac{1-0}{1-0+1\cdot\ln x+x\cdot\frac{1}{x}}$$

$$= \lim_{x\to 1}\frac{1}{2+\ln x}$$

$$= \frac{1}{2+0}=\frac{1}{2}$$

3. Find $\lim_{x\to\infty} x^{1/x}$

Solution :

Note that $\lim_{x\to\infty} x^{1/x}$ is ∞^0 type of an indeterminate form.

Put $\quad y = x^{1/x}$

then $\quad \ln y = \ln x^{1/x}$

$$= \frac{1}{x}\ln x = \frac{\ln x}{x}$$

We have $\lim_{x\to\infty}\frac{\ln x}{x}\ \left[\frac{\infty}{\infty}\text{ form}\right]$

$$= \lim_{x\to\infty}\frac{(\ln x)'}{x'}$$

$$= \lim_{x\to\infty}\frac{x^{-1}}{1}$$

$$= \lim_{x\to\infty}\frac{1}{x}=0$$

$$\therefore\quad \lim_{x\to\infty} x^{1/x} = e^0 = 1$$

4. Find $\lim_{x\to\pi/2}(\tan x)^{2x-\pi}$

Solution :

Note that $\lim_{x\to\pi/2}(\tan x)^{2x-\pi}$ is ∞^0 type of an indeterminate form. Put

$$y = (\tan x)^{2x-\pi}$$

Then $\ln y = \ln\left((\tan x)^{2x-\pi}\right)$

$$= (2x-\pi)\ln(\tan x)$$

$$= \left\{\frac{(2x-\pi)\ln(\tan x)}{1} = \frac{(2x-\pi)^{-1}\cdot(2x-\pi)\ln(\tan x)}{(2x-\pi)^{-1}\cdot 1}\right\}$$

$$= \frac{\ln(\tan x)}{(2x-\pi)^{-1}}$$

We have $\lim_{x\to\pi/2}\frac{\ln(\tan x)}{(2x-\pi)^{-1}}\left[\frac{\infty}{\infty}\text{form}\right]$

$$= \lim_{x\to\pi/2}\frac{[\ln(\tan x)]'}{[(2x-\pi)^{-1}]'}$$

$$= \lim_{x\to\pi/2}\frac{\frac{1}{\tan x}\cdot(\tan x)'}{(-1)(2x-\pi)^{-2}\cdot(2x-\pi)'}$$

$$= \lim_{x\to\pi/2}\frac{\frac{1}{\tan x}\cdot\sec^2 x}{(-1)(2x-\pi)^{-2}\cdot 2}$$

$$= -\frac{1}{2}\lim_{x\to\pi/2}\frac{\frac{1}{\frac{\sin x}{\cos x}}\cdot\frac{1}{\cos^2 x}}{(2x-\pi)^{-2}}$$

$$= -\frac{1}{2}\lim_{x\to\pi/2}\frac{\frac{1}{\frac{\sin x}{\cos x}\cdot\cos^2 x}}{(2x-\pi)^{-2}}$$

$$= -\frac{1}{2}\lim_{x\to\pi/2}\frac{\frac{1}{\sin x\cos x}}{(2x-\pi)^{-2}}$$

$$= -\frac{1}{2}\lim_{x\to\pi/2}\frac{\frac{1}{\sin x\cos x}\cdot\sin x\cos x(2x-\pi)^2}{(2x-\pi)^{-2}\cdot\sin x\cos x(2x-\pi)^2}$$

$$= -\frac{1}{2}\lim_{x\to\pi/2}\frac{(2x-\pi)^2}{\sin x\cos x}$$

$$= -\frac{1}{2}\lim_{x\to\pi/2}\frac{(2x-\pi)^2}{\sin\left(\frac{\pi}{2}\right)\cos x}$$

$$= -\frac{1}{2}\lim_{x\to\pi/2}\frac{(2x-\pi)^2}{\cos x}\left[\frac{0}{0}\text{form}\right]$$

$$= -\frac{1}{2}\lim_{x\to\pi/2}\frac{[(2x-\pi)^2]'}{(\cos x)'}$$

$$= -\frac{1}{2}\lim_{x\to\pi/2}\frac{2(2x-\pi)\cdot(2x-\pi)'}{-\sin x}$$

$$= -\frac{1}{2}\lim_{x\to\pi/2}\frac{2(2x-\pi)\cdot 2}{-\sin x}$$

$$= -\frac{1}{2}\cdot\frac{2\cdot\left(2\cdot\frac{\pi}{2}-\pi\right)\cdot 2}{-\sin\frac{\pi}{2}} = 0$$

Therefore $\lim_{x\to\pi/2}(\tan x)^{2x-\pi} = e^0 = 1$

5. Find $\lim_{x\to 0^+} x^x$

Solution :

Note that $\lim_{x\to 0^+} x^x$ is 0^0 type of indeterminate form.

Put $y = x^x$

then $\ln y = \ln x^x = x\ln x$

$$= \left\{\frac{x\ln x}{1} = \frac{x^{-1}\cdot x\ln x}{x^{-1}\cdot 1}\right\}$$

$$= \frac{\ln x}{x^{-1}}$$

We have $\lim_{x\to 0^+}\frac{\ln x}{x^{-1}}\left[\frac{\infty}{\infty}\text{form}\right]$

$$= \lim_{x\to 0^+}\frac{(\ln x)'}{(x^{-1})'}$$

$$= \lim_{x\to 0^+}\frac{x^{-1}}{-x^{-2}}$$

$$= \lim_{x\to 0^+}\frac{x^{-1}\cdot x^2}{-x^{-2}\cdot x^2}$$

$$= \lim_{x\to 0^+}\frac{x}{-1}$$

$$= \frac{0}{-1} = 0$$

Therefore $\lim_{x\to 0^+} x^x = e^0 = 1$

6. $\lim_{x\to 0}\frac{\sin x}{x} = \frac{0}{0}$

By L' Hospital Rule

$$\lim_{x\to 0}\frac{\sin x}{x} = \lim_{x\to 0}\frac{\cos x}{x} = \frac{1}{1} = 1$$

7. $\lim_{x\to 0}\frac{x^2+x-\sin x}{x^2} = \frac{0}{0}$

By L' Hospital Rule

$$\Rightarrow \lim_{x\to 0}\frac{x^2+x-\sin x}{x^2} = \lim_{x\to 0}\frac{2x+1-\cos x}{2x} = \frac{0}{0}$$

$$\Rightarrow \lim_{x\to 0}\frac{x^2+x-\sin x}{x^2} = \lim_{x\to 0}\frac{2+\sin x}{2} = 1$$

8. $\lim_{x \to 0} \frac{x^n}{x^m} = \frac{0}{0}$

By L' Hospital Rule

$$\lim_{x \to 0} \frac{x^n}{x^m} = \frac{nx^{n-1}}{mx^{m-1}} = \frac{0}{0}$$

$$\Rightarrow \quad \lim_{x \to 0} \frac{x^n}{x^m} = \frac{n(n-1)x^{n-2}}{m(m-1)x^{m-2}} = \frac{0}{0}$$

$$\lim_{x \to 0} \frac{x^n}{x^m} = \frac{n(n-1).......1}{m(m-1).....1} = \frac{n!}{m!}$$

so on

CONTINUITY AND DISCONTINUITY

Continuity

- The function f is said to be *continuous* at $x = c$ if $\lim_{x \to c} f(x) = f(c)$
- The function f is said to be continuous on an open interval (a, b) if it is continuous at each point of (a,b)
- The function is said to be continuous on a closed interval $[a, b]$ if
  (*i*) f is continuous at each point of (a, b)
  (*ii*) $\lim_{x \to a^+} f(x) = f(a)$
  (*iii*) $\lim_{x \to b^-} f(x) = f(b)$

Note : Graphically, $f(x)$ is continuous if its graph can be drawn on paper without raising the pencil.

Discontinuity

A function f is said to be discontinuous at c if it is not continuous at that point. A function f is said to be discontinuous on an interval if it is discontinuous at least one point of the interval.

A function is discontinuous at c under any of the following circumstances:

(*i*) f is not defined at c

(*ii*) $\lim_{x \to c} f(x)$ does not exist

(*iii*) f is defined at c and $\lim_{x \to c} f(x)$ exists but $\lim_{x \to c} f(x) \neq f(c)$

Note : If f and g are continuous at $x = c$, then $f + g$, $f.g$, αf, and $\frac{f}{g}$ is also continuous at $x = c$ $(g(c) \neq 0)$

DIFFERENTIABILITY

A function $f(x)$ is said to be differentiable at $x = a$ if

both $\lim_{h \to 0} \frac{f(a+h) - f(a)}{h}$, $h > 0$

and $\lim_{h \to 0} \frac{f(a-h) - f(a)}{-h}$, $h > 0$

exist and have common value (finite). This common value is called *derivative of* $f(x)$ at the point $x = a$ and is denoted by $f'(a)$.

MEAN VALUE THEOREMS

1. Rolle's theorem

If a function $f(x)$ is continuous in closed interval $[a, b]$, derivable in the open interval (a, b) and $f(a) = f(b)$, then there exists atleast one real c in (a, b) such that $f'(c) = 0$.

2. Lagrange's mean value theorem

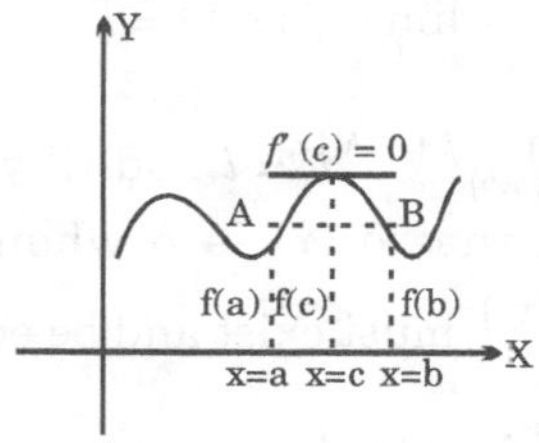

If $f(x)$ be a function such that $f(x)$ is continuous in $[a, b]$, and $f'(x)$ exists in (a, b), then there exist at least one c, such that

$$f'(c) = \frac{f(b) - f(a)}{b - a}.$$

3. Cauchy's Mean value theorem

If $f(x)$ and $g(x)$ be two functions such that

(*i*) $f(x)$ and $g(x)$ both are continuous in $[a \ b]$

(*ii*) $f'(x)$ and $g'(x)$ both exist in (a, b)

$g'(x) \neq 0 \ \forall \ x \in (a \ b)$

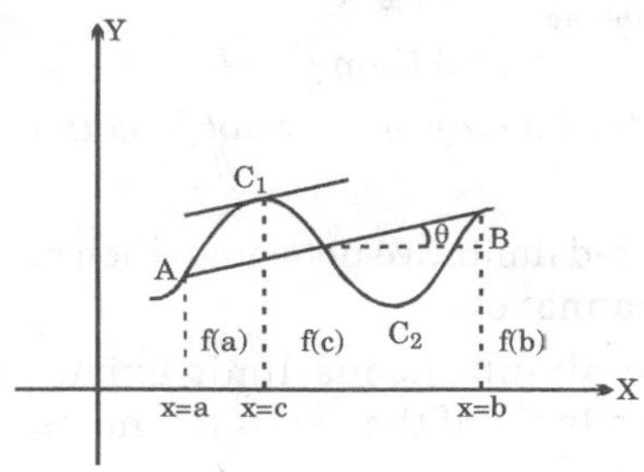

then there exist atleast one point c in the closed interval $(a \ b)$ such that

$$\frac{f'(c)}{g'(c)} = \frac{f(b) - f(a)}{g(b) - g(a)}$$

4. Taylor's theorem

If $f(x)$ be a continuous function such that

$f'(x), f''(x), f'''(x) ... f^{n-1}(x)$

all are continuous in $[a, a + h]$

$f^n(x)$ exist in $(a, \ a \ + h)$, where $h = b - a$

then $f(a + h) = f(a) + hf'(a) + \frac{h^2}{2!} f''(a)$

$$+ ..... \frac{h^{n-1}}{n-1!} f^{n-1}(a) + \frac{h^n}{n!} f^n(a)$$

If we put $a = 0$, $h = x$, then the series becomes

$$f(x) = f(0) + xf'(0) + \frac{x^2}{2!} f''(0) + ...$$

FUNCTIONS OF TWO VARIABLES

Let $f(x,y)$ be any function of two independent varibles x and y.

Limit

A function f tends to a limit l, when (x, y) tends to (a, b)

$$\lim_{(x,y)\to(a,b)} f(x,y) = l$$

Note : If $\lim_{(x,y)\to(a,b)} f(x,y) = l$ and if $y = \phi(x)$ is any function such that $\phi(x) \to b$ when $x \to a$, then $\lim_{(x,y)\to(a,b)} f(x,\phi(x))$ must exist and be equal to l.

Repeated Limits

If a function f is defined in some neigbourhood of (a, b), then limit $\lim_{y\to b} f(x,y)$ if exists, is a function of x, say $\phi(x)$.

If limit $\lim_{x\to a} \phi(x)$ exists and is equal to K, we write

$$\lim_{x\to a}\lim_{y\to b} f(x,y) = K$$

Here K is repeated limits of f as $y \to b, x \to a$

If we change order of taking the limit, we get other repeated limit.

$$\lim_{y\to b}\lim_{x\to a} f(x,y) = K'$$

when first $x \to a$ and then $y \to b$

These two limits may or may not be equal.

Note :

(i) If repeated limits are not equal, then simultaneous limits cannot exist.

(ii) In case simultaneous limit exists, these two repeated limits if they exist are necessarily equal but the converse is not true.

Continuity

A function f is said to be *continuous* at a point (a,b) of its domain to definition if

$$\lim_{(x,y)\to(a,b)} f(x,y) = f(a, b)$$

COMPUTING THE DERIVATIVE

Rules of Differentiation

$(f+g)' = f' + g'$ → Sum Rule

$(f-g)' = f' - g'$ → Difference Rule

$(f.g)' = fg' + gf'$ → Product Rule

$\left(\frac{f}{g}\right)' = \frac{gf' - fg'}{g^2}$ → Quotient Rule

$\frac{1}{dx}(f(g(x))) = \frac{df}{dg}\cdot\frac{dg}{dx}$ → Chain Rule

Some Standard Derivative

$f(x)$	$f'(x)$	$f(x)$	$f'(x)$
x^n	nx^{n-1}	$\sin h\, x$	$\cos h\, x$
$\ln x$	$1/x$	$\cos h\, x$	$\sin h\, x$
$\log_a x$	$\log_a e. \frac{1}{x}$	$\sin^{-1}x$	$\frac{1}{\sqrt{1-x^2}}$
e^x	e^x	$\cos^{-1}x$	$\frac{-1}{\sqrt{1-x^2}}$
a^x	$a^x \log_a e$	$\tan^{-1}x$	$\frac{1}{1+x^2}$
$\sin x$	$\cos x$	$\text{cosec}^{-1}x$	$\frac{-1}{x\sqrt{x^2-1}}$
$\cos x$	$-\sin x$		
$\tan x$	$\sec^2 x$	$\sec^{-1} x$	$\frac{1}{x\sqrt{x^2-1}}$
$\sec x$	$\sec x \tan x$	$\cot^{-1}x$	$\frac{1}{1+x^2}$
$\text{cosec}\, x$	$-\text{cosec}\, x \tan x$	$\lvert x \rvert$	$\frac{x}{\lvert x \rvert}$
$\cot x$	$-\text{cosec}^2 x$		$(x \neq 0)$

Differentiation by Substitution

If a function contain as expression of the form

(1) $a^2 - x^2$; put $x = a \sin t$ or $x = a$ cost

(2) $a^2 + x^2$; put $x = a \tan t$ or $x = a \cot t$

(3) $x^2 - a^2$; put $x = a \sec t$ or $x = a\, \text{cosec}\, t$

(4) $\sqrt{\frac{a-x}{a+x}}$ or $\sqrt{\frac{a+x}{a-x}}$; put $x = a \cos t$

(5) $a \cos x \pm b \sin x$; put $a = r \cos\theta$, and $b = r \sin\theta$, $r > 0$

PARTIAL DERIVATIVES

The partial differential coefficient of $z = f(x, y)$ with respect to x is defined as

$$\text{Lt}_{\delta x\to 0} \frac{f(x+\delta x, y) - f(x,y)}{\delta x}$$

provided this limit exists and is denoted by

$$\frac{\partial f}{\partial x} \text{ or } \frac{\partial z}{\partial x} \text{ or } f_x \text{ or } z_x$$

Similarly, partial derivatives of $f(x, y)$ with respect to y is defined as

$$\text{Lt}_{\delta y\to 0} \frac{f(x, y+\delta y) - f(x,y)}{\delta y}$$

provided this limit exists and is denoted by

$$\frac{\partial f}{\partial y} \text{ or } \frac{\partial z}{\partial y} \text{ or } f_y \text{ or } z_y$$

$\frac{\partial z}{\partial x}$ and $\frac{\partial z}{\partial y}$ are called *first order partial derivatives of z.*

- *The partial derivation of f(x, y) with respect to y is the ordinary derivative of f(x, y) when x is regarded as constant.*

Example. If $z = \log(x^2 + y^2)$, find $x\frac{\partial z}{\partial x} + y\frac{\partial z}{\partial y}$

Solution

Given : $z = \log(x^2 + y^2)$

$$\therefore \quad \frac{\partial z}{\partial x} = \frac{2x}{x^2 + y^2}$$

$$\Rightarrow \quad x\frac{\partial z}{\partial x} = \frac{2x^2}{x^2 + y^2}$$

$$\frac{\partial z}{\partial y} = \frac{2y}{x^2 + y^2}$$

$$\Rightarrow \quad y\frac{\partial z}{\partial y} = \frac{2y^2}{x^2 + y^2}$$

$$\therefore \quad x\frac{\partial z}{\partial x} + y\frac{\partial z}{\partial y} = \frac{2(x^2 + y^2)}{(x^2 + y^2)} = 2$$

Partial Derivatives of Higher Order

Since f_x and f_y are also functions of x and y and so we can further differentiate f_x and f_y partially with respect to x or y. The further partial derivatives of f_x or f_y are called *partial derivatives f higher order of u.*

- Partial derivatives of $\frac{\partial u}{\partial x}$ with respect to x and y are $\frac{\partial^2 u}{\partial^2 x}$ or $\frac{\partial^2 u}{\partial y\, \partial x}$ respectively.
- Partial derivatives $\frac{\partial^2 u}{\partial^2 x^2}, \frac{\partial^2 u}{\partial y\, \partial x}, \frac{\partial^2 u}{\partial xy}, \frac{\partial^2 u}{\partial y^2}$ are also denoted by $f_{xx}, f_{yx}, f_{xy}, f_{yy}$ respectively.

Example. If $u = e^{xyz}$, then find $\frac{\partial^3 u}{\partial x \partial y \partial z}$.

Solution

Given : $u = e^{xyz}$

$$\therefore \quad \frac{\partial u}{\partial z} = e^{xyz}\, xy = uxy, \quad \frac{\partial u}{\partial y} = uxz$$

and $$\frac{\partial u}{\partial x} = uyz$$

Now, $$\frac{\partial^2 u}{\partial y \partial z} = ux + xy$$

$$\frac{\partial u}{\partial y} = ux + (xy)(uxz) = ux + ux^2 yz$$

$$\therefore \quad \frac{\partial^3 u}{\partial x \partial y \partial z} = u + x\frac{\partial u}{\partial x} + 2u\, xyz + x^2 yz\, \frac{\partial u}{\partial x}$$

$$= u + x(uyz) + 2uxyz + x^2yz\,(uyz)$$

$$= e^{xyz}(1 + 3xyz + x^2 y^2 z^2)$$

MAXIMA AND MINIMA OF FUNCTIONS OF TWO VARIABLES

A function $f(x, y)$ is said to have a maximum or minimum at $x = a, y = b$, according as $f(a + h, b + k)$ is less or greater than $f(a, b)$ for all positive or negative small values of h and k.

In other words, if $\Delta = f(a + h, b + k) - f(a, b)$ is of the same sign for all small values of h, k and if this sign is negative, then $f(a, b)$ is a maximum. If this sign is positive, then $f(a, b)$ is a minimum.

A maximum or minimum value of a function is called its *extreme value.*

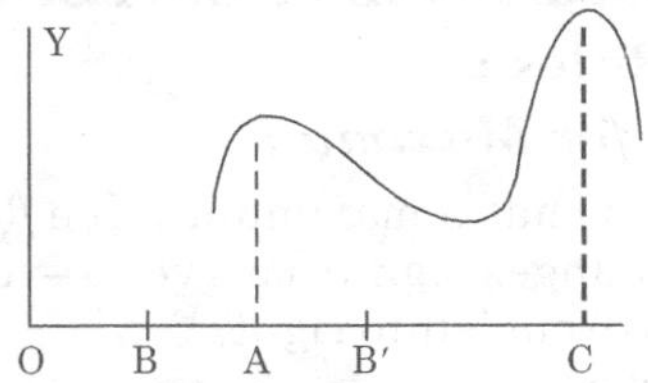

Above figure shows a graph of the function $f(x)$ and OA = a, *i.e.*, $f(x)$ has a maximum value for $x = a$ because $f(a)$ has a value more than the values of $f(x)$ for every value of x between B and B′. $f(x)$ is said to be a *maximum* at $x = a$, even though value of $f(x)$ at $x = a$ should be greater than all other values of $f(x)$ in some small neighbourhood. Thus a maximum value of $f(x)$ is not necessarily the greatest value of $f(x)$. In fact, a curve might have several maxima (and minima).

Condition for Maxima and Minima

Necessary conditions for existence of a maximum or a minimum of $f(x)$ at $x = a, y = b$ is

$$\left(\frac{\partial f}{\partial x}\right)_{\substack{x=a\\y=b}} = 0; \text{ and } \left(\frac{\partial f}{\partial y}\right)_{\substack{x=a\\y=b}} = 0$$

Sufficient Condition for Maxima and Minima

Let $$r = \left(\frac{\partial^2 f}{\partial x^2}\right)_{\substack{x=a\\y=b}}, \quad s = \left(\frac{\partial^2 f}{\partial x \partial y}\right)_{\substack{x=a\\y=b}}$$

and $$t = \left(\frac{\partial^2 f}{\partial y^2}\right)_{\substack{x=a\\y=b}}$$

Case I: $rt - s^2 > 0$

$f(x, y)$ will have maximum or a minimum of $x = a, y = b$.

Again, $f(x, y)$ is a maximum or a minimum according as r is negative or positive

Case II: $rt - s^2 < 0$

$f(x,y)$ will have neither a maximum nor a minimum at $x = a, y = b$.

Case III: $rt - s^2 = 0$

This case is doubtful and further investigation is needed to determine whether (x, y) is maximum or a minimum at $x = a, y = b$ or not.

Finding Absolute Maxima and Minima Values

Working Rules :

If f is a differentiable function in $[a\ b]$ except at finitely many points, then to find absolute maximum and absolute minimum values, adopt following procedure :

(*i*) Evaluate $f(x)$ at the points, where $f'(x) = 0$.

(*ii*) Evaluated $f(a)$ and $f(b)$.

Then maximum of these values is the **absolute maxima** and minimum of these values is called **absolute minima.**

Finding Local Maxima and Local Minima.

Working Rules :

Condition for Maxima :

- Function $f(x)$ has a maximum value $f(a)$ if $f'(a) = 0$ and $f'(x)$ changes sign from +ve to –ve as x passes through 'a' from left to right.
- In general, for any even number n,
  $f'(a) = f''(a) = f'''(a) \ \dots = f^{n-1}(a) = 0$ and $f^n(a) < 0$,
  then $f(a)$ is a maximum value of $f(x)$.
- In particular, if $n = 2$, then $f'(a) = 0$ and $f''(a) < (o)$, and the function is maximum at $x = a$.

Condition for minima :

- Function $f(x)$ has a minimum value $f(a)$ if $f'(a) = 0$ and $f'(x)$ changes sign from –ve to +ve as x passes through $x = a$ from left to right.
- In general, for any even number n, if
  $f'(a) = f''(a) = f'''(a) = \dots f^{n-1}(a) = 0$, and $f^n(a) > 0$,
  then $f(a)$ is minimum value of $f(x)$
- In particular, if $n = 2$, then $f'(a) = 0$ and $f''(a) > 0$, the function is minimum at $x = a$.

Note : If n is a even number such that

$$f'(a) = f''(a) = \dots f^n(a) = 0, f^{n+1}(a) \neq 0,$$

then $f(a)$ is neither maximum nor minimum of the function $f(x)$.

In particular if $n = 2$ and $f'(a) = f''(a) = 0$. $f'''(a) \neq 0$, then $f(a)$ is neither maximum nor minimum values of $f(x)$. Then the point $x = a$ is called **point of inflexion.**

Extreme Value

Either a maximum value or minimum value $f(a)$ of the function $f(x)$ is called *extreme value*.

Stationary Value

If $f'(a) = 0$, then $f(a)$ is called *stationary value.*

Every extreme value is stationary but every stationary value of the function need not to be extreme value

Example. Find the extreme value of

$$f(x) = x^5 - 5x^4 + 5x^3 - 1$$

Solution

$$f'(x) = 5x^4 - 20x^3 + 15x^2$$

and $\quad f'(0) = 0$

Hence, $f(0)$ is stationary value but $f(0)$ is not an extreme value because $f''(0) = 0$ but $f'''(0) \neq 0$.

Greatest Value

The greatest value of the function in an interval (a, b) is either a maximum value in an interval or end value (i.e. at $x = a$ or $x = b$) of $f(x)$ whichever is greater.

Least Values

The least value of $f(x)$ in an interval (a, b) is either a minimum value of $f(x)$ at a point inside the interval or an end value (i.e. at $x = a$ or $x = b$) of $f(x)$ whichever is less.

Example. Find greatest and least value of the function $2x^2 - 9x - 1$, in the internal $[2, 5]$

Solution

Let $\quad f(x) = 2x^2 - 9x - 1$

$\Rightarrow \quad f'(x) = 4x - 9$

At $x = \frac{9}{4} \quad f'(x) = 0$

$\therefore \quad f''(x) = 4 > 0$

$\therefore \quad f\left(\frac{9}{4}\right) = -\frac{89}{8}$ is a minimum value.

Also, $\quad f(2) = -11, f(5) = 4$

$\therefore$ Greatest value = 4, and least value = $-\frac{89}{8}$

TAYLOR SERIES

In the preceding section we were able to find power series representations for a certain restricted class of functions. Here we investigate more general problems : Which functions have power series representations? How can we find such representation?

We start by supposing that f is any function that can be represented by a power series :

$$f(x) = c_0 + c_1(x-a) + c_2(x-a)^2 + c_3(x-a)^3 + c_4(x-a)^4 + \dots \quad |x-a| < R \qquad (1)$$

Let's try to determine what the coefficients c_n must be in terms of f. To begin, notice that if we put $x = a$ in Equation 1, then all terms after the first one are 0 and we get

$$f(a) = c_0$$

By theorem 8.6.2, we can differentiate the series in Equation 1 term by term :

$$f'(x) = c_1 + 2c_2(x-a) + 3c_3(x-a)^2 + 4c_4(x-a)^3 + \dots \quad |x-a| < R \qquad (2)$$

and substitution of $x = a$ in Equation 2 gives

$$f'(a) = c_1$$

Now we differentiate both sides of Equation 2 and obtain

$$f''(x) = 2c_2 + 2 \cdot 3c_3(x-a) + 3 \cdot 4c_4 (x-a)^2 + \dots \quad |x-a| < R \quad (3)$$

Again we put $x = a$ in Equation 3. The result is

$$f''(a) = 2c_2$$

Let's apply the procedure one more time. Differentiation of the series in Equation 3 gives

$$f'''(x) = 2 \cdot 3c_3 + 2 \cdot 3 \cdot 4c_4(x-a) + 3 \cdot 4 \cdot 5c_5(x-a)^2 + \dots \quad |x-a| < R \quad (4)$$

and substitution of $x = a$ in Equation 4 gives

$$f'''(a) = 2 \cdot 3c_3 = 3!c_3$$

By now you can see the pattern. If we continue to differentiate and substitute $x = a$, we obtain

$$f^{(n)}(a) = 2 \cdot 3 \cdot 4 \cdot \dots\dots \cdot nc_n = n!c_n$$

Solving this equation for the nth coefficient c_n, we get

$$c_n = \frac{f^{(n)}(a)}{n!}$$

This formula remains valid even for $n = 0$ if we adopt the conventions that

$$0! = 1$$

and $\quad f^{(0)} = f.$

Thus we have proved the following theorem.

Theorem : If f has a power series representation (expansion) at a, that is, if

$$f(x) = \sum_{n=0}^{\infty} c_n (x-a)^n$$

and $\quad |x-a| < R,$

then $$c_n = \frac{f^{(n)}(a)}{n!} \quad (5)$$

Substituting this formula for c_n back into the series, we see that if f has a power series expansion at a, then it must be of the following form.

$$f(x) = \sum_{n=0}^{\infty} \frac{f^{(n)}(a)}{n!}(x-a)^n$$

$$= f(a) + \frac{f'(a)}{1!}(x-a) + \frac{f''(a)}{2!}(x-a)^2 + \frac{f'''(a)}{3!}(x-a)^3 + \dots \quad (6)$$

Theorem : If $f(x) = T_n(x) + R_n(x)$, where T_n is the nth-degree Taylor polynomial of f at a and

$$\lim_{n\to\infty} R_n(x) = 0 \quad (8)$$

for $|x-a| < R$, then f is equal to the sum of its Taylor series on the interval $|x-a| < R$.

Example : Find the Taylor series for

$$f(x) = e^x$$

at $\quad a = 2$

Solution : We have $f^{(n)}(2) = e^2$ and so, putting $a = 2$ in the definition of a Taylor series (6), we get

$$\sum_{n=0}^{\infty} \frac{f^{(n)}(2)}{n!}(x-2)^n = \sum_{n=0}^{\infty} \frac{e^2}{n!}(x-2)^n$$

As in Example 2 we can verify that $\lim_{n\to\infty} R_n(x) = 0$, so

$$e^x = \sum_{n=0}^{\infty} \frac{e^2}{n!}(x-2)^n \text{ for all } x \quad (9)$$

We have two power series expansions for e^x. The first is better if we are interested in values of x near 0 and the second is better if x is near 2.

TAYLOR AND MACLAURIN SERIES

In the preceding section we were able to find power series representations for a certain restricted class of functions. Here we investigate more general problems : Which functions have power series representations? How can we find such representation?

We start by supposing that f is any function that can be represented by a power series :

$$f(x) = c_0 + c_1(x-a) + c_2(x-a)^2 + c_3(x-a)^3 + c_4(x-a)^4 + \dots \quad |x-a| < R \quad (1)$$

Let's try to determine what the coefficients c_n must be in terms of f. To begin, notice that if we put $x = a$ in Equation 1, then all terms after the first one are 0 and we get

$$f(a) = c_0$$

By theorem 8.6.2, we can differentiate the series in Equation 1 term by term :

$$f'(x) = c_1 + 2c_2(x-a) + 3c_3(x-a)^2 + 4c_4(x-a)^3 + \dots \quad |x-a| < R \quad (2)$$

and substitution of $x = a$ in Equation 2 gives

$$f'(a) = c_1$$

Now we differentiate both sides of Equation 2 and obtain

$$f''(x) = 2c_2 + 2 \cdot 3c_3(x-a) + 3 \cdot 4c_4 (x-a)^2 + \dots \quad |x-a| < R \quad (3)$$

Again we put $x = a$ in Equation 3. The result is

$$f''(a) = 2c_2$$

Let's apply the procedure one more time. Differentiation of the series in Equation 3 gives

$$f'''(x) = 2 \cdot 3c_3 + 2 \cdot 3 \cdot 4c_4(x-a) + 3 \cdot 4 \cdot 5c_5(x-a)^2 + \dots \quad |x-a| < R \quad (4)$$

and substitution of $x = a$ in Equation 4 gives

$$f'''(a) = 2 \cdot 3c_3 = 3!c_3$$

By now you can see the pattern. If we continue to differentiate and substitute $x = a$, we obtain

$$f^{(n)}(a) = 2 \cdot 3 \cdot 4 \cdot \dots\dots \cdot nc_n = n!c_n$$

Solving this equation for the nth coefficient c_n, we get

$$c_n = \frac{f^{(n)}(a)}{n!}$$

This formula remains valid even for $n = 0$ if we adopt the conventions that

$$0! = 1$$

and $\qquad f^{(0)} = f.$

Thus we have proved the following theorem.

Theorem : If f has a power series representation (expansion) at a, that is, if

$$f(x) = \sum_{n=0}^{\infty} c_n (x-a)^n$$

and $\qquad |x-a| < R,$

then $$c_n = \frac{f^{(n)}(a)}{n!} \tag{5}$$

Substituting this formula for c_n back into the series, we see that if f has a power series expansion at a, then it must be of the following form.

$$f(x) = \sum_{n=0}^{\infty} \frac{f^{(n)}(a)}{n!}(x-a)^n$$

$$= f(a) + \frac{f'(a)}{1!}(x-a) + \frac{f''(a)}{2!}(x-a)^2 + \frac{f'''(a)}{3!}(x-a)^3 + \dots \tag{6}$$

The series in Equation 6 is called the **Taylor series of the function f at a (or about a or centered at a).** For the special case $a = 0$ the Taylor series becomes

$$f(x) = \sum_{n=0}^{\infty} \frac{f^{(n)}(0)}{n!} x^n$$

$$= f(0) + \frac{f'(0)}{1!}x + \frac{f''(0)}{2!}x^2 + \frac{f'''(0)}{3!}x^3 + \dots \tag{7}$$

This case arises frequently enough that is is given the special name **Maclaurin series**.

Note : We have shown that if f can be represented as a power series about a, then f is equal to the sum of its Taylor series. But there exist functions that are not equal to the sum of their Taylor series. For example, one can show that the function defined by

$$f(x) = \begin{cases} e^{-1/x^2} & \text{if } x \neq 0 \\ 0 & \text{if } x = 0 \end{cases}$$

is not equal to its Maclaurin series.

Example : Find the Maclaurin series of the function $f(x) = e^x$ and its radius of convergence.

Solution : If $\quad f(x) = e^x,$

then $\quad f^{(n)}(x) = e^x,$

so $\quad f^{(n)}(0) = e^0 = 1$

for all n.

Therefore, the Taylor series for f at 0 (that is, the Maclaurin series) is

$$\sum_{n=0}^{\infty} \frac{f^{(n)}(0)}{n!} x^n = \sum_{n=0}^{\infty} \frac{x^n}{n!}$$

$$= 1 + \frac{x}{1!} + \frac{x^2}{2!} + \frac{x^3}{3!} + \dots$$

To find the radius of convergence we let

$$a_n = \frac{x^n}{n!}.$$

Then $$\left|\frac{a_{n+1}}{a_n}\right| = \left|\frac{x^{n+1}}{(n+1)!} \cdot \frac{n!}{x^n}\right|$$

$$= \frac{|x|}{n+1} \to 0 < 1$$

So, by the Ratio Test, the series converges for all x and the radius of convergence is $R = \infty$.

The conclusion we can draw from (5) and is that if e^x has a power series expansion at 0, then

$$e^x = \sum_{n=0}^{\infty} \frac{x^n}{n!}$$

So how can we determine whether e^x does have a power series representation?

Let's investigate the more general question: Under what circumstances is a function equal to the sum of its Taylor series? In other words, if f has derivatives of all orders, when is it true that

$$f(x) = \sum_{n=0}^{\infty} \frac{f^{(n)}(a)}{n!}(x-a)^n$$

As with any convergent series, this means that f(x) is the limit of the sequence of partial sums.

In the case of the Taylor series, the partial sums are

$$T_n(x) = \sum_{i=0}^{n} \frac{f^{(i)}(a)}{i!}(x-a)^i$$

$$= f(a) + \frac{f'(a)}{1!}(x-a) + \frac{f''(a)}{2!}(x-a)^2 + \dots + \frac{f^{(n)}(a)}{n!}(x-a)^n$$

Theorem : If $f(x) = T_n(x) + R_n(x)$, where T_n is the nth-degree Taylor polynomial of f at a and

$$\lim_{n\to\infty} R_n(x) = 0 \tag{8}$$

for $|x-a| < R$, then f is equal to the sum of its Taylor series on the interval $|x-a| < R$.

Example : Find the Taylor series for

$$f(x) = e^x$$

at $$a = 2$$

Solution : We have $f^{(n)}(2) = e^2$ and so, putting $a = 2$ in the definition of a Taylor series (6), we get

$$\sum_{n=0}^{\infty} \frac{f^{(n)}(2)}{n!}(x-2)^n = \sum_{n=0}^{\infty} \frac{e^2}{n!}(x-2)^n$$

As in Example 2 we can verify that $\lim_{n\to\infty} R_n(x) = 0$, so

$$e^x = \sum_{n=0}^{\infty} \frac{e^2}{n!}(x-2)^n \text{ for all } x \qquad (9)$$

We have two power series expansions for e^x. The first is better if we are interested in values of x near 0 and the second is better if x is near 2.

Example : Find the Maclaurin series for sin x and prove that it represents sin x for all x.

Solution : We arrange our computation in two columns as follows :

$$\begin{aligned} f(x) &= \sin x \\ f(0) &= 0 \\ f'(x) &= \cos x \\ f'(0) &= 1 \\ f''(x) &= -\sin x \\ f''(0) &= 0 \\ f'''(x) &= -\cos x \\ f'''(0) &= -1 \\ f^{(4)}(x) &= \sin x \\ f^{(4)}(0) &= 0 \end{aligned}$$

Since the derivatives repeat in a cycle of four, we can write the Maclaurin series as follows :

$$f(0) + \frac{f'(0)}{1!} + \frac{f''(0)}{2!}x^2 + \frac{f'''(0)}{3!}x^3 + \frac{f^{(4)}(0)}{4!}x^4 + \ldots$$

$$= 0 + \frac{1}{1!}x + \frac{0}{2!}x^2 + \frac{-1}{3!}x^3 + \frac{0}{4!}x^4 + \ldots$$

$$= x - \frac{x^3}{3!} + \frac{x^5}{5!} - \frac{x^7}{7!} + \ldots$$

$$= \sum_{n=0}^{\infty} (-1)^n \frac{x^{2n+1}}{(2n+1)!}$$

Using the remainder term (10) with $a = 0$, we have

$$R_n(x) = \frac{f^{(n+1)}(z)}{(n+1)!}x^{n+1}$$

where $f(x) = \sin x$ and z lies between 0 and x.

But $f^{(n+1)}(z)$ is $\pm$ sin z or $\pm$ cos z.

In any case, $\left|f^{(n+1)}(z)\right| \le 1$ and so

$$0 \le \left|R_n(x)\right| = \frac{\left|f^{(n+1)}(z)\right|}{(n+1)!}\left|x^{n+1}\right| \le \frac{1}{(n+1)!}\left|x^{n+1}\right|$$

$$= \frac{\left|x^{n+1}\right|}{(n+1)!} \qquad (10)$$

Maclaurin Series Formula Table

$$\frac{1}{1-x} = \sum_{n=0}^{\infty} x^n = 1 + x + x^2 + x^3 + \ldots \qquad R = 1$$

$$e^x = \sum_{n=0}^{\infty} \frac{x^n}{n!} = 1 + \frac{x}{1!} + \frac{x^2}{2!} + \frac{x^3}{3!} + \ldots \qquad R = \infty$$

$$\sin x = \sum_{n=0}^{\infty} (-1)^n \frac{x^{2n+1}}{(2n+1)!}$$

$$= x - \frac{x^3}{3!} + \frac{x^5}{5!} - \frac{x^7}{7!} + \ldots \qquad R = \infty$$

$$\cos x = \sum_{n=0}^{\infty} (-1)^n \frac{x^{2n}}{(2n)!}$$

$$= 1 - \frac{x^2}{2!} + \frac{x^4}{4!} - \frac{x^6}{6!} + \ldots \qquad R = \infty$$

$$\tan^{-1} x = \sum_{n=0}^{\infty} (-1)^n \frac{x^{2n+1}}{2n+1}$$

$$= x - \frac{x^3}{3} + \frac{x^5}{5} - \frac{x^7}{7} + \ldots \qquad R = 1$$

$$\ln(1+x) = \sum_{n=1}^{\infty} (-1)^{n-1} \frac{x^n}{n}$$

$$= x - \frac{x^2}{2} + \frac{x^3}{3} - \frac{x^4}{4} + \ldots \qquad R = 1$$

$$(1+x)^k = \sum_{n=0}^{\infty} \binom{k}{n} x^n$$

$$= 1 + kx + \frac{k(k-1)}{2!}x^2 + \frac{k(k-1)(k-2)}{3!}x^3 + \ldots$$

$$R = 1$$

SOME STANDARD INTEGRATIONS

1. $\int x^n = \frac{x^{n+1}}{n+1}, \ (n \ne -1) + c$
2. $\int \frac{1}{x} dx = \log x + c$
3. $\int e^x dx = e^x + c$
4. $\int a^x dx = \frac{a^x}{\log a} + c$
5. $\int \sin x \, dx = -\cos x + c$
6. $\int \cos x \, dx = \sin x + c$
7. $\int \sec^2 x \, dx = \tan x + c$
8. $\int \text{cosec}^2 dx = -\cot x + c$
9. $\int \sec x \tan x \, dx = \sec x + c$
10. $\int \text{cosec } x \cot x \, dx = -\text{cosec } x + c$

11. $\int \tan x\, dx = \log \sec x = -\log \cos x + c$

12. $\int \cot x\, dx = \log \sin x + c$

13. $\int \sec x\, dx = \log (\sec x + \tan x) + c$

14. $\int \text{cosec}\, x\, dx = -\log(\text{cosec}\, x + \cot x) = \log \tan \frac{x}{2} + c$

15. $\int \frac{1}{\sqrt{1-x^2}} dx = \sin^{-1} x \text{ or } -\cos^{-1} x + c$

16. $\int \frac{1}{\sqrt{1+x^2}} dx = \sin h^{-1} x + c$

17. $\int \frac{1}{a^2+x^2} dx = \frac{1}{a} \tan^{-1}\left(\frac{x}{a}\right) + c$

18. $\int \frac{1}{x\sqrt{x^2-1}} dx = \sec^{-1} x = \cos^{-1} \frac{1}{x} + c$

19. $\int \sin hx\, dx = \cos hx + c$

20. $\int \cos hx\, dx = \sin hx + c$

21. $\int \frac{1}{\sqrt{x^2-a^2}} dx = \log\left(x + \sqrt{x^2-a^2}\right) = \cos h^{-1} \frac{x}{a} + c$

22. $\int \frac{1}{\sqrt{a^2+x^2}} dx = \log\left(x + \sqrt{a^2+x^2}\right) + c$

23. $\int \frac{1}{\sqrt{a^2+x^2}} dx = \sin h^{-1}\left(\frac{x}{a}\right) + c$

24. $\int \frac{dx}{x\sqrt{x^2-a^2}} = \frac{1}{a} \sec^{-1} \frac{x}{a} + c$

25. $\int \frac{dx}{x^2+a^2} = \frac{1}{a} \tan^{-1}\left(\frac{x}{a}\right) + c$

26. $\int \frac{dx}{x^2-a^2} = \frac{1}{2a} \log\left(\frac{x-a}{x+a}\right)$ if $x > a + c$

27. $\int \frac{dx}{a^2-x^2} = \frac{1}{2a} \log\left(\frac{a+x}{a-x}\right)$ if $x < a + c$

28. $\int u \cdot v\, dx = u \int v\, dx - \int \left[\left(\frac{du}{dx}\right)\left(\int v dx\right)\right] dx + c$

where u and v are the function of x. u is a first function and v as a second function.

29. $\int e^x [f(x) + f'(x)]\, dx = \boldsymbol{e^x f(x) + c}$

30. $\int \frac{f'(x)}{f(x)}$ **dx = *log f(x)* + *C***

Note: ILATE

I → Inverse

L → Logarithmic function

A → Algebraic

T → Trigonometric

E → Exponential

Example. Calculate $\int \frac{e^{\tan^{-1} x}}{1+x^2} dx$

Solution.

Let $\tan^{-1} x = y$

$\Rightarrow \frac{1}{1+x^2} dx = dy$

then $\int \frac{e^{\tan^{-1} x}}{1+x^2} dx = \int e^y dy = e^y + C = e^{\tan^{-1} x} + C$

Example. Calculate $\int 1. \sin^{-1} x\, dx$

Solution.

Take $u = \sin^{-1} x$ and $v = 1$

Then $\int \sin^{-1} x dx = \sin^{-1} x \int dx - \int \left[\frac{d(\sin^{-1} x)}{dx} \int dx\right] dx$

$= x \sin^{-1} x - \int \frac{x}{\sqrt{1-x^2}} x\, dx$

$= x \sin^{-1} x - \int \frac{x}{\sqrt{1-x^2}} dx$

$= x \sin^{-1} x + \sqrt{1-x^2} + C$

Example. Calculate $\int \frac{e^x + e^{-x}}{e^x - e^{-x}} dx$

Solution.

Let $y = e^x - e^{-x} \Rightarrow dy = (e^x + e^{-x})\, dx$

$\therefore \int \frac{(e^x + e^{-x})}{e^x - e^{-x}} = \int \frac{dy}{y} = \log y + C = \log (e^x + e^{-x}) + C$

DEFINITE INTEGRAL

If a function is defined in the interval $[a, b]$, then definite integral of $f(x)$ is

$$\int_a^b f(x)\, dx = [F(x)]_a^b = F(b) - F(a)$$

where F(x) is an integral of $f(x)$

$f(b)$ is the value of $f(x)$ at $x = b$, and

$f(a)$ is the value of $f(x)$ at $x = a$.

a and b are called *lower upper limit respectively.*

Geometrically, definite integral represents the area bounded by curve $y = f(x)$, x-axis and the lines $x = a$ and $x = b$.

Properties of Definite Integrals

Let $\int f(x)\,dx = F(x)$. Then

1. $\int_a^b f(x)\,dx = \int_a^b f(y)\,dy$

   i.e. in definite integral the variable of integration is dummy.

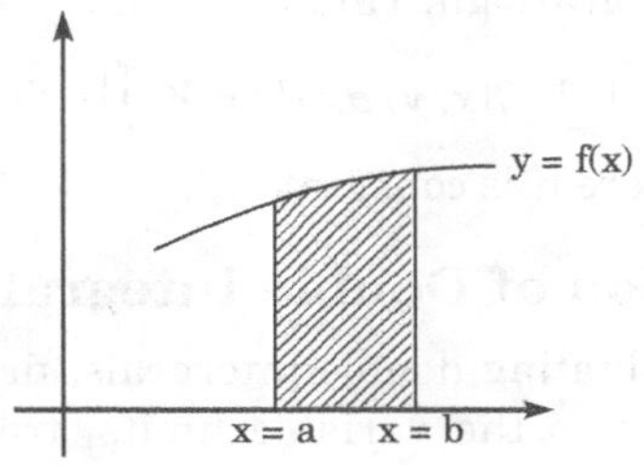

2. $\int_a^b f(x)\,dx = -\int_b^a f(x)\,dx$

3. $\int_a^b f(x)\,dx = \int_a^c f(x)\,dx + \int_c^b f(x)\,dx$

   where c is a point lies between the interval $[a, b]$. and $f(x)$ integrable over $[a, c]$ and $[c, b]$.

4. $\int_0^a f(x)\,dx = \int_0^a f(a-x)\,dx$

5. $\int_{-a}^a f(x)\,dx = \int_0^a f(x)\,dx + \int_0^a f(-x)\,dx$

   If function is even, i.e. $f(-x) = f(x)$, then

   $$\int_{-a}^a f(x)\,dx = 2\int_0^a f(x)\,dx$$

   If function is odd, i.e. $f(-x) = -f(x)$, then

   $$\int_{-a}^a f(x)\,dx = 0$$

6. $\int_0^{2a} f(x)\,dx = \int_0^a f(x)\,dx + \int_0^a f(2a-x)\,dx$

   If $f(x) = f(2a-x)$, then

   $$\int_0^{2a} f(x)\,dx = 2\int_0^a f(x)\,dx$$

   and if $f(x) = -f(2a-x)$, then

   $$\int_0^{2a} f(x)\,dx = 0$$

Definite Integral as Limit of Sum

If a function $f(x)$ is either increasing or decreasing on $[a, b]$, then

$$\int_a^b f(x)\,dx = \underset{h\to 0}{\text{Lt}}\; h[f(a) + f(a+h) + \ldots + f(a+(n-1)h]$$

$$= \underset{h\to 0}{\text{Lt}}\; h\,[f(a+h) + f(a+2h) + \ldots + f(a+nh)]$$

where, $h = \dfrac{b-a}{n}$

Example. From definition of a definite integral as the limit of a sum, evaluate $\int_1^2 (x + x^2)\,dx$.

Solution

$$\int_1^2 (x + x^2)\,dx$$

$$= \underset{h\to 0}{\text{Lt}} \sum_{i=1}^{n} h\,[(1+ih) + (1+ih)^2.....]\text{ where}\left(h = \frac{1}{n}\right)$$

$$= \underset{h\to 0}{\text{Lt}}\; h\sum_{i=1}^{n} [1 + ih + 1 + 2\,ih + i^2h^2]$$

$$= \underset{h\to 0}{\text{Lt}}\; h\sum_{i=1}^{n} (2 + 3hi) + i^2h^2$$

$$= \underset{h\to 0}{\text{Lt}}\left[2h\sum_{i=1}^{n} 1 + 3h^2\sum_{i=1}^{n} i + h^3\sum_{i=1}^{n} i^2\right]$$

$$= \underset{h\to 0}{\text{Lt}}\left[2nh + \frac{3}{2}h^2n\,(n+1) + \frac{1}{6}h^3n\,(n+1)(2n+1)\right]$$

$$= \underset{h\to 0}{\text{Lt}}\left[2 + \frac{3}{2}(1+h) + \frac{1}{6}(1+h)\,(2+h)\right] \ldots \text{ since } nh = 1$$

$$= 2 + \frac{3}{2} + \frac{1}{3} = \mathbf{\frac{23}{6}}$$

Improper Integral

Definite integral $\int_0^a f(x)\,dx$ *is called improper integral* if

(*i*) range of integration is infinite and integrand is bounded.

(*ii*) range of integration is definite and integrand is unbounded.

(*iii*) neither range of integration is finite nor integrand is bounded over it.

MULTIPLE INTEGRALS

Let a single-valued and bounded function $f(x,y)$ of two independent variables x,y be defined in a closed region R of the xy-plane. Divide the region R into sub-regions by drawing lines parallel to co-ordinate axes. Number the rectangles which lie entirely inside the region R from 1 to n.

Let (x_r, y_r) be any point inside the rth rectangle whose area is δA_r

$$f(x_1,y_1)\,\delta A_1 + f(x_2,y_2)\,\delta A_2 + ...f(x_n,y_n)\,\delta A_n$$
$$= \sum_{r=1}^{n} f(x_r,y_r)\delta A_r \quad .... \text{(i)}$$

Let number of these sub-regions increases indefinitely such that the largest linear dimension (i.e.diagonal) of δA_r approaches zero. The limit of the sum(*i*), if it exists, irrespective of the mode of sub-division, is called *double integral of f(x,y) over the region R* and is denoted by $\iint_R f(x,y)\,dA$.

In other words, $\underset{\substack{n\to\infty \\ \delta A_r \to 0}}{Lt} \sum_{r=1}^{n} f(x_r,y_r)\,\delta A_r = \iint_R f(x,y)\,dA$

which is also expressed as $\iint_R f(x,y)\,dxdy$

or $\iint_R f(x,y)\,dy\,dx$

Let the region R be bounded by the curves

$x = x_1$ *and* $y = y_2$

$x = x_2$ *and* $y = y_2$

Properties of a Double Integral

(i) When x_1, x_2 are functions of y and y_1, y_2 are constants

Let AB and CD be the curves $x_1 = \phi_1(y)$ and $x_2 = \phi_2(y)$. Then

$$\iint_R f(x,y)\,dxdy = \int_{y_1}^{y_2}\int_{x_1=\phi_1(y)}^{x_2=\phi_2(y)} f(x,y)\,dx\ \ dy$$

(ii) When y_1, y_2 are functions of x and x_1, x_2 are constants

Let AB and CD be the curves

$y_1 = \phi_1(x)$ and $y_2 = \phi_2(x)$. Then

$$\iint_R f(x,y)\,dx\,dy = \int_{x_1}^{x_2}\int_{y_1=\phi_1(x)}^{y_2=\phi_2(x)} f(x,y)\,dy\ \ dx$$

(iii) When x_1, x_2, y_1, y_2 are constants

Here the region of integration R is the rectangle. Therefore

$$\iint_R f(x,y)\,dxdy = \int_{y_1}^{y_2}\int_{x_1}^{x_2} f(x,y)\,dx\,dy$$
$$= \int_{x_1}^{x_2}\int_{y_1}^{y_2} f(x,y)\,dy\,dx$$

(*i*) If region R is partitioned into two parts, say R_1 and R_2, then

$$\iint_R f(x,y)\,dx\,dy = \iint_{R1} f(x,y)\,dx\,dy + \iint_{R2} f(x,y)\,dx\,dy$$

(*ii*) Double integral of the algebraic sum of a fixed number of functions is equal to algebraic sum of the double integrals taken far each term.

Thus

$$\iint_R [f_1(x,y) + f_2(x,y) + f_3(x,y) + ...]\,dx\,dy$$
$$= \iint_R f_1(x,y)\,dx\,dy + \iint_R f_2(x,y)\,dx\,dy + \iint_R f_3(x,y)\,dx\,dy + ...$$

(*iii*) A constant factor may be taken outside the integral sign. Thus

$$\iint_R K\,f(x,y)\,dx\,dy = K\iint_R f(x,y)\,dx\,dy$$

where K is constant

Evaluation of Double Integrals

During evaluating double intergrals, first integrate with respect to the variable limits, treating other varibles as constant, and then integrate with respect to the variable with constant limit.

CHANGE OF ORDER OF INTEGRATION

In a double integral, if limits of integration are constant, then order of integration is immaterial, provided limits of integration are changed accordingly. Thus

$$\int_c^d \int_a^b f(x,y)\,dx\,dy = \int_a^b \int_c^d f(x,y)\,dy\,dx$$

But if limits of integration are variable, a change in the order of integration changes in the limits of integration.

Triple Integrals

Consider a function $f(x,y,z)$ which is continuous at every point of the 3-dimensional finite region V.

Divide V into n elementary volumes $\delta V_1, \delta V_2, ... \delta V_n$.

Let (x_r, y_r, z_r) be any point within rth sub-division δV_r.

Consider the sum $\sum_{r=1}^{\infty} f(x_r,y_r,z_r)\,\delta V_r$

The limit of this sum, if it exists, as $n\to\infty$ and $\delta V_r \to 0$ is called *triple integral* of $f(x,y,z)$ over the region V and is denoted by $\iiint f(x,y,z)\,dV$

It can also be expressed as the repeated integral $\int_{x_1}^{x_2}\int_{y_1}^{y_2}\int_{z_1}^{z_2} f(x,y,z)\ dx\,dy\,dz$.

APPLICATIONS OF THE DEFINITE INTEGRAL

Area Between Curves

If $f(x)$ is continuous and nonnegative on $[a, b]$, then

$\int_a^b f(x)dx$ = the area below $y = f(x)$ and above x-axis ($y = 0$) from $x = a$ to $x = b$

Example :

1. Find the area between $y = x^2$ and the x-axis from $x = 0$ to $x = 2$
2. Find the area between $y = x$ and the x-axis from $x = -2$ to $x = 2$
3. Find the total area bounded by the curve $y = x^3 - 4x$ and the x-axis

Solutions

1. $\int_0^2 x^2 dx = \frac{8}{3}$ 2. $2\int_0^2 x\, dx = 4$
3. $2\int_{-2}^{0} \left(x^3 - 4x\right) dx = 8$

In general, if $f(x)$ and $g(x)$ are continuous on $[a, b]$ and $f(x) \le g(x)$ for all $x \in [a, b]$ then

$\int_a^b (f(x) - g(x))dx$ = the area between the curves $y = f(x)$ and $y = g(x)$ from $x = a$ to $x = b$

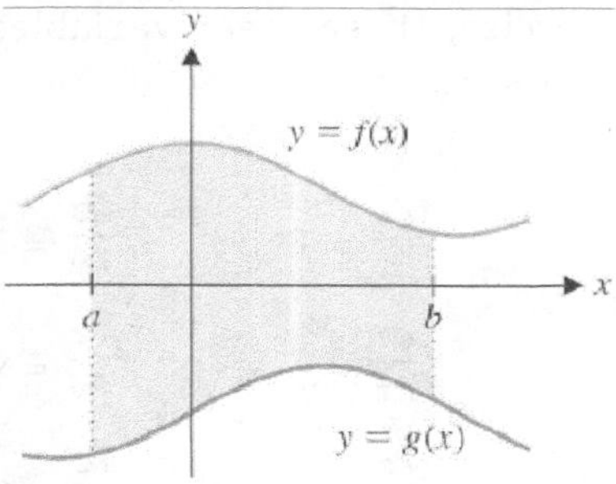

Example

1. Find the area bounded by the graphs of $y = 3 - x$ and $y = x^2 - 9$.
2. Find the area bounded by the graphs of $y = x^2$ and $y = 2 - x^2$ for $0 \ge x \ge 2$.
3. Find the area bounded by the graphs of $y = x^2$, $y = 2 - x$ and $y = 0$.
4. Find the area between the curves $y = 2 - x$ and $y = x^2$.
5. Find the area between the curves $y = 2 - x$, $y = x^2$ and x-axis.
6. Find the area bounded by the lines $y = x$ and $x = 2$ and the curve $y = \frac{1}{x^2}$.
7. Find the area between the curves $y = x^2$ and $y = x^3$ for $0 \ge x \ge 2$.

Solutions

1. $\int_{-4}^{3} \left((3 - x) - \left(x^2 - 9\right)\right) dx = \frac{343}{6}$
2. $\int_0^1 \left(\left(2 - x^2\right) - x^2\right) dx + \int_1^2 \left(x^2 - \left(2 - x^2\right)\right) dx = 4$
3. $\int_0^1 x^2 dx + \int_1^2 (2 - x) dx = \frac{5}{6}$
4. $\int_{-2}^{1} \left((2 - x) - x^2\right) dx = \frac{9}{2}$
5. $\int_0^1 x^2 dx + \int_1^2 (2 - x) dx = \frac{5}{6}$
6. $\int_1^2 \left(x - \frac{1}{x^2}\right) dx = 1$
7. $\int_0^1 \left(x^2 - x^3\right) dx + \int_1^2 \left(x^3 - x^2\right) dx = \frac{3}{2}$

VOLUME : SLICING AND DISKS

Volumes by Slicing : Suppose a solid object has boundaries extending from $x = a$, to $x = b$, and that its cross-section in a plane passing through $(x, 0, 0)$ and parallel to the yz-plane has area $A(x)$. To a first order approximation, the volume of the slice of the object on the right to the plane of thickness Δx is then $A(x)\ \Delta x$ so that the volume of the solid is the limit of the elemental sum $\Sigma A(x)\Delta x$.

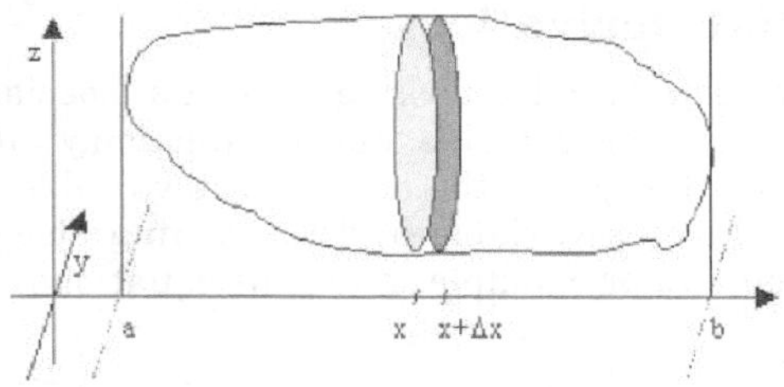

This being a Riemann sum, the volume is given by the formula

$$V = \int_a^b A(x)dx$$

The Method of Disks : Suppose that $f(x) \ge 0$ and f is continuous on the interval $[a, b]$. Take the region bounded by the curve $y = f(x)$ and the x-axis, for $a \ge x \ge b$, and revolve it about the x-axis, generating a solid. We can find the volume of this solid by slicing it perpendicular to the x-axis and recognizing that each cross section is a circular disk of radius $r = f(x)$,

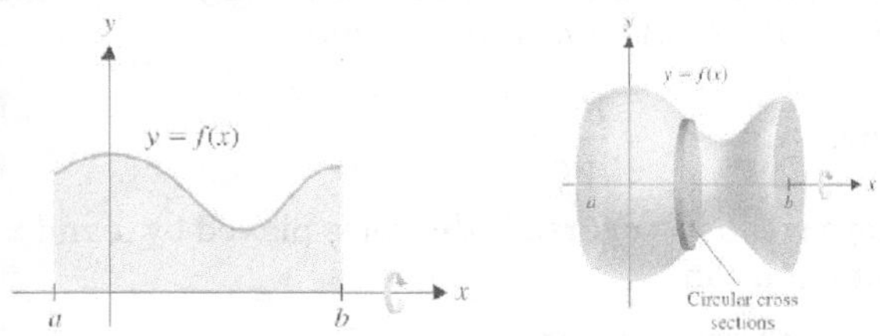

we then have that the volume of the solid is

$$V = \int_a^b \pi\left[f(x)\right]^2 dx$$

Since the cross sections of such a solid of revolution are all disks, we refer to this method of finding volume as the **method of disks**.

In a similar way, suppose that $g(y) \geq 0$ and g is continuous on the interval $[c, d]$. Then, revolving the region bounded by the curve $x = g(y)$ and the y-axis, for $c \geq y \geq d$, about the y-axis generates a solid. Notice that the cross sections of the resulting solid of revolution are circular disks of radius $r = g(y)$. All that has changed here is that we have interchanged the roles of the variables x and y.

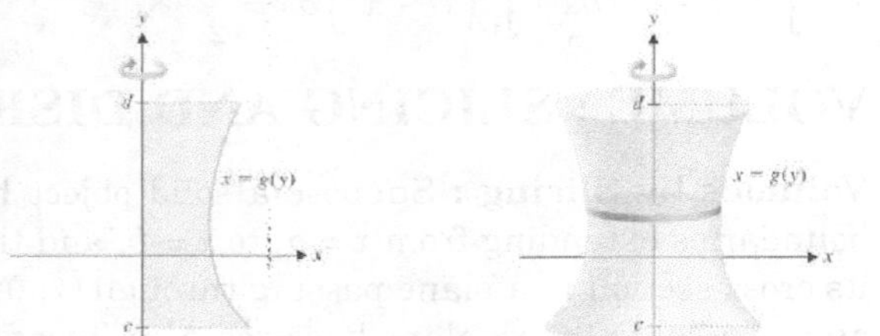

The volume of the solid is then given by

$$V = \int_c^d \pi\left[g(y)\right]^2 dy$$

Partial Derivatives

Single variable calculus is really just a "special case" of multivariable calculus. For the function $y = f(x)$, we assumed that y was the endogenous variable, x was the exogenous variable and everything else was a parameter. For example, given the equations

$$y = a + bx$$

or

$$y = ax^n$$

we automatically treated a, b, and n as constants and took the derivative of y with respect to $x(dy/dx)$. However, what if we decided to treat x as a constant and take the derivative with respect to one of the other variables? Nothing precludes us from doing this. Consider the equation

$$y = ax$$

where $\frac{dy}{dx} = x$

Now suppose we find the derivative of y with respect to a, but *TREAT x as the constant.*

Then $\frac{dy}{da} = x$

Here we just "reversed" the roles played by a and x in our equation.

Two Variable Case

let $z = f(x, y)$, which means **"z is a function of x and y"**. In this case z is the endogenous (dependent) variable and both x and y are the exogenous (independent) variables. To measure the the effect of a change in a single independent variable (x or y) on the dependent variable (z) we use what is known as the *PARTIAL DERIVATIVE*. The partial derivative of z with respect to x measures the instantaneous change in the function as x changes while *HOLDING* y constant. Similarly, we would hold x constant if we wanted to evaluate the effect of a change in y on z. Formally:

- $\frac{\partial z}{\partial x}$ is the **"partial derivative"** of z with respect to x, treating y as a constant. Sometimes written as f_x.
- $\frac{\partial z}{\partial y}$ is the **"partial derivative"** of z with respect to y, treating x as a constant. Sometimes written as f_y.

The "∂" symbol ("bent over" lower case D) is called the "partial" symbol. It is interpreted in exactly the same way as $\frac{dy}{dx}$ from single variable calculus. The ∂ symbol simply serves to remind us that there are other variables in the equation, but for the purposes of the current exercise, these other variables are held constant.

Examples

$z = x + y$	$\frac{\partial z}{\partial x} = 1$	$\frac{\partial z}{\partial y} = 1$
$z = xy$	$\frac{\partial z}{\partial x} = y$	$\frac{\partial z}{\partial y} = x$
$z = x^2y^2$	$\frac{\partial z}{\partial x} = 2(y^2)x$	$\frac{\partial z}{\partial y} = 2(x^2)y$
$z = x^2y^3 + 2x + 4y$	$\frac{\partial z}{\partial x} = 2xy^3 + 2$	$\frac{\partial z}{\partial y} = 3x^2y^2 + 4$

- **Remember:** When you are taking a partial derivative you treat the other variables in the equation as constants!

Rules of Partial Differentiation

Product Rule: given $z = g(x, y) \cdot h(x, y)$

$$\frac{\partial z}{\partial x} = g(x, y) \cdot \frac{\partial h}{\partial x} + h(x, y) \cdot \frac{\partial g}{\partial x}$$

$$\frac{\partial z}{\partial y} = g(x, y) \cdot \frac{\partial h}{\partial y} + h(x, y) \cdot \frac{\partial g}{\partial y}$$

Quotient Rule: given $z = \frac{g(x, y)}{h(x, y)}$ and $h(x, y) \neq 0$

$$\frac{\partial z}{\partial x} = \frac{h(x, y).\frac{\partial g}{\partial x} - g(x, y)\frac{\partial h}{\partial x}}{[h(x, y)]^2}$$

$$\frac{\partial z}{\partial y} = \frac{h(x, y).\frac{\partial g}{\partial y} - g(x, y)\frac{\partial h}{\partial y}}{[h(x, y)]^2}$$

Chain Rule: given $z = [g(x, y)]^n$

$$\frac{\partial z}{\partial x} = n[g(x, y)]^{n-1} \cdot \frac{\partial g}{\partial x}$$

$$\frac{\partial z}{\partial y} = n[g(x, y)]^{n-1} \cdot \frac{\partial g}{\partial y}$$

Further Examples

For the function U = U(x, y) find the partial derivates with respect to x and y for each of the following examples

Example

$$U = -5x^3 - 12xy - 6y^5$$

Answer

$$\frac{\partial U}{\partial x} = U_x - 15x^2 - 12y$$

$$\frac{\partial U}{\partial y} = U_y = -12x - 30y^4$$

Example

$$U = 7x^2y^3$$

Answer

$$\frac{\partial U}{\partial x} = U_x = 14xy^3$$

$$\frac{\partial U}{\partial y} = U_y = 21x^2y^2$$

Example

$$U = 3x^2(8x - 7y)$$

Answer

$$\frac{\partial U}{\partial x} = U_x = 3x^2(8) + (8x - 7y)(6x) = 72x^2 - 42xy$$

$$\frac{\partial U}{\partial y} = U_y = 3x^2(-7) + (8x - 7y)(0) = -21x^2$$

Example

$$U = (5x^2 + 7y)(2x - 4y^3)$$

Answer

$$\frac{\partial U}{\partial x} = U_x = (5x^2 + 7y)(2) + (2x - 4y^3)(10x)$$

$$\frac{\partial U}{\partial y} = U_y = (5x^2 + 7y)(-12y^2) + (2x - 4y^3)(7)$$

Example

$$U = \frac{9y^3}{x - y}$$

Answer

$$\frac{\partial U}{\partial x} = U_x = \frac{(x-y)(0) - 9y^3(1)}{(x-y)^2} = \frac{-9y^3}{(x-y)^2}$$

$$\frac{\partial U}{\partial y} = U_y = \frac{(x-y)(27y^2) - 9y^3(-1)}{(x-y)^2} = \frac{27xy^2 - 18y^3}{(x-y)^2}$$

Example

$$U = (x - 3y)^3$$

Answer

$$\frac{\partial U}{\partial x} = U_x = 3(x - 3y)^2(1) = 3(x - 3y)^2$$

$$\frac{\partial U}{\partial y} = U_y = 3(x - 3y)^2(-3) = -9(x - 3y)^2$$

A Special Function: Cobb-Douglas

The Cobb-douglas function is a mathematical function that is very popular in economic models. The general form is

$$z = x^a y^b$$

and its partial derivatives are

$$\frac{\partial z}{\partial x} = ax^{a-1}y^b$$

and $$\frac{\partial z}{\partial y} = bx^a y^{b-1}$$

Furthermore, the slope of the level curve of a Cobb-douglas is given by

$$\frac{\partial z / \partial x}{\partial z / \partial y} = \text{MRS} = \frac{a}{b}\frac{y}{x}$$

Differentials

Given the function

$$y = f(x)$$

the derivative is

$$\frac{dy}{dx} = f'(x)$$

However, we can treat dy/dx as a fraction and factor out the dx

$$dy = f'(x)dx$$

where dy and dx are called differentials. If dy/dx can be interpreted as "the slope of a function", then dy is the "rise" and dx is the "run". Another way of looking at it is as follows:

- dy = the change in y
- dx = the change in x
- $f'(x)$ = how the change in x causes a change in y

Example

If $y = x^2$

then $dy = 2xdx$

Lets suppose $x = 2$ and $dx = 0.01$. *What is the change in y(dy)?*

$$dy = 2(2)(0.01) = 0.04$$

Therefore, at $x = 2$, if x is increased by 0.01 then y will increase by 0.04.

The Two Variable Case

If $$z = f(x, y)$$

then the change in z is

$$dz = \frac{\partial z}{\partial x}dx + \frac{\partial z}{\partial y}dy$$

or $$dz = f_x dx + f_y dy$$

which is read as "the change in $z(dz)$ is due partially to a change in $x(dx)$ plus partially due to a change in $y(dy)$. For example, if

$$z = xy$$

then the total differential is

$$dz = ydx + xdy$$

and, if $$z = x^2y^3$$

then $$dz = 2xy^3dx + 3x^2y^2dy$$

Remember: When you are taking the total differential, you are just taking all the partial derivatives and adding them up.

Example

Find the total differential for the following utility functions

1. $U(x_1, x_2) = ax_1 + bx_2 \; (a, b > 0)$
2. $U(x_1, x_2) = x_1^2 + x_2^3 + x_1x_2$
3. $U(x_1, x_2) = x_1^a x_2^b$

Answers

1. $$\frac{\partial U}{\partial x_1} = U_1 = a$$

$$\frac{\partial U}{\partial x_2} = U_2 = b$$

$$dU = U_1dx_1 + U_2dx_2 = adx_1 + bdx_2$$

2. $$\frac{\partial U}{\partial x_1} = U_1 = 2x_1 + x_2$$

$$\frac{\partial U}{\partial x_2} = U_2 = 3x_2^2 + x_1$$

$$dU = U_1dx_1 + U_2dx_2 = (2x_1 + x_2)dx_1 + (3x_2^2 + x_1)dx_2$$

3. $$\frac{\partial U}{\partial x_1} = U_1 = ax_1^{a-1}x_2^b = \frac{ax_1^a x_2^b}{x_1}$$

$$\frac{\partial U}{\partial x_2} = U_2 = bx_1^a x_2^{b-1} = \frac{bx_1^a x_2^b}{x_2}$$

$$dU = \left(\frac{ax_1^a x_2^b}{x_1}\right)dx_1 + \left(\frac{bx_1^a x_2^b}{x_2}\right)dx_2$$

$$= \left[\frac{adx_1}{x_1} + \frac{bdx_2}{x_2}\right]x_1^a x_2^b$$

The Implicit Function Theorem

Suppose you have a function of the form

$$F(y, x_1, x_2) = 0$$

where the partial derivatives are $\partial F/\partial x_1 = Fx_1$, $\partial F/\partial x_2 = Fx_2$ and $\partial F/\partial y = F_y$. This class of functions are known as implicit functions where $F(y, x_1, x_2) = 0$ implicity define $y = y(x_1, x_2)$. What this means is that it is possible (theoretically) to rewrite to get y isolated and expressed as a function of x_1 and x_2. While it may not be possible to explicitly solve for y as a function of x, we can still find the effect on y from a change in x_1 or x_2 by applying the implicit function theorem:

Example

If a function

$$F(y, x_1, x_2) = 0$$

has well defined continuous partial derivatives

$$\frac{\partial F}{\partial y} = F_y$$

$$\frac{\partial F}{\partial x_1} = Fx_1$$

$$\frac{\partial F}{\partial x_2} = Fx_2$$

and if, at the values where F is being evaluated, the condition that

$$\frac{\partial F}{\partial y} = F_y \neq 0$$

holds, then y is implicitly defined as a function of x. The partial derivatives of y with respect to x_1 and x_2, are given by the ratio of the partial derivatives of F, or

$$\frac{\partial y}{\partial x_i} = -\frac{F_{x_i}}{F_y} \quad i = 1, 2$$

To apply the implicit function theorem to find the partial derivative of y with respect to x_1 (for example), first take the total differential of F

$$dF = F_y dy + F_{x_1}dx_1 + F_{x_2}dx_2 = 0$$

then set all the differentials except the ones in question equal to zero (i.e. set $dx_2 = 0$) which leaves

$$F_y dy + F_{x_1}dx_1 = 0$$

or

$$F_y dy = -F_{x_1}dx_1$$

dividing both sides by F_y and dx_1 yields

$$\frac{dy}{dx_1} = -\frac{F_{x_1}}{F_y}$$

which is equal to $\frac{\partial y}{\partial x_1}$ from the implicit function theorem.

Example

For each $f(x, y) = 0$, find dy/dx for each of the following:

1. $y - 6x + 7 = 0$
2. $3y + 12x + 17 = 0$
3. $x^2 + 6x - 13 - y = 0$
4. $f(x, y) = 3x^2 + 2xy + 4y^3$
5. $f(x, y) = 12x^5 - 2y$
6. $f(x, y) = 7x^2 + 2xy^2 + 9y^4$

Answer

1. $\frac{dy}{dx} = -\frac{f_x}{f_y} = -\frac{(-6)}{1} = 6$
2. $\frac{dy}{dx} = -\frac{f_x}{f_y} = -\frac{(-12)}{3} = 4$
3. $\frac{dy}{dx} = -\frac{f_x}{f_y} = \frac{-(2x+6)}{-1} = 2x+6$
4. $\frac{dy}{dx} = -\frac{f_x}{f_y} = -\frac{6x+2y}{12y^2+2x}$
5. $\frac{dy}{dx} = -\frac{f_x}{f_y} = -\frac{60x^4}{-2} = 30x^4$
6. $\frac{dy}{dx} = -\frac{f_x}{f_y} = -\frac{14x+2y^2}{36y^3+4xy}$

Example

For $f(x, y, z)$ use the implicit function theorem to find dy/dx and dy/dz :

1. $f(x, y, z) = x^2y^3 + z^2 + xyz$
2. $f(x, y, z) = x^3z^2 + y3 + 4xyz$
3. $f(x, y, z) = 3x^2y^3 + xz^2y^2 + y^3zx^4 + y^2z$

Answer

1. $\frac{dy}{dx} = -\frac{f_x}{f_y} = -\frac{2xy^3 + yz}{3x^2y^2 + xz}$

   $\frac{dy}{dz} = -\frac{f_z}{f_y} = -\frac{2z + xy}{3x^2y^2 + xz}$

2. $\frac{dy}{dx} = -\frac{f_x}{f_y} = -\frac{3x^2z^2 + 4yz}{3y^2 + 4xz}$

   $\frac{dy}{dz} = -\frac{f_z}{f_y} = -\frac{2x^3z + 4xy}{3y^2 + 4xz}$

3. $\frac{dy}{dx} = -\frac{f_x}{f_y} = -\frac{6xy^3 + z^2y^2 + 4y^3zx^3}{9x^2y^2 + 2xz^2y + 3y^2zx^4 + 2yz}$

   $\frac{dy}{dz} = -\frac{f_z}{f_y} = -\frac{2xzy^2 + y^3x^4 + y^2}{9x^2y^2 + 2xz^2y + 3y^2zx^4 + 2yz}$

Chain Rule and Total Differentials

1. Find the total differential of

$$w = x^3yz + xy + z + 3 \text{ at } (1, 2, 3).$$

Answer:

The total differential at the point (x_0, y_0, z_0) is

$$dw = w_x(x_0, y_0, z_0)dx + w_y(x_0, y_0, z_0)dy + w_z(x_0, y_0, z_0)dz.$$

In our case,

$$w_x = 3x^2yz + y,$$
$$w_y = x^3z + x,$$
$$w_z = x^3y + 1.$$

Substituting in the point (1, 2, 3) we get: $w_x(1, 2, 3) = 20$, $w_y(1, 2, 3) = 4$, $w_z(1, 2, 3) = 3$.

Thus, $dw = 20\,dx + 4\,dy + 3\,dz.$

2. Suppose $w = x^3yz + xy + z + 3$ and

$$x = 3\cos t, \quad y = 3\sin t, \quad z = 2t.$$

Compute $\frac{dw}{dt}$ and evaluate it at $t = \frac{\pi}{2}$.

Answer

We do not substitute for x, y, z before differentiating, so we can practice the chain rule.

$$\frac{dw}{dt} = \frac{\partial w}{\partial x}\frac{dx}{dt} + \frac{\partial w}{\partial y}\frac{dy}{dt} + \frac{\partial w}{\partial z}\frac{dz}{dt}$$
$$= (3x^2yz + y)(-3\sin t) + (x^3z + x)(3\cos t) + (x^3y + 1)(2).$$

At $t = \frac{\pi}{2}$

we have $x = 0$, $y = 3$, $z = \pi$, $\sin\frac{\pi}{2} = 1$, $\cos\frac{\pi}{2} = 0$.

Thus,

$$\left.\frac{\partial w}{\partial t}\right|_{\frac{\pi}{2}} = 3(-3) + 3(0) + (1)2 = -7.$$

3. Show how the tangent approximation formula leads to the chain rule that was used in the previous problem.

Answer

The approximation formula is

$$\Delta w \approx \left.\frac{\partial f}{\partial x}\right|_o \Delta x + \left.\frac{\partial f}{\partial y}\right|_o \Delta y + \left.\frac{\partial f}{\partial z}\right|_o \Delta z.$$

If x, y, z are functions of time then dividing the approximation formula by Δt gives

$$\frac{\Delta w}{\Delta t} \approx \left.\frac{\partial f}{\partial x}\right|_o \frac{\Delta x}{\Delta t} + \left.\frac{\partial f}{\partial y}\right|_o \frac{\Delta y}{\Delta t} + \left.\frac{\partial f}{\partial z}\right|_o \frac{\Delta z}{\Delta t}.$$

In the limit as $\Delta t \to 0$ we get the chain rule.

Note : We use the regular '*d*' for the derivative $\frac{dw}{dt}$ because in the chain of computations

$$t \to x, y, z \to w$$

the dependent variable w is ultimately a function of exactly one independent variable t. Thus, the derivative with respect to t is not a partial derivative.

LINE INTEGRAL

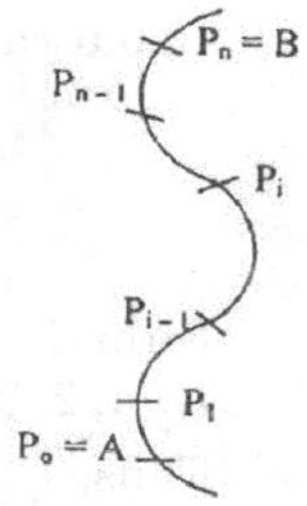

Consider a continuous vector function F(R) which is defined at each point of curve C in space. Divide C into n parts at the points $A = P_0, P_1, ..., -P_{i-1}, P_i ..., P_n = B$

Let their position vectors be

$$R_0, R_1, ... R_{i-1}, R_i, ... R_n.$$

Let U_i be the position vector of any point on the arc $P_{i-1} P_i$.

Now consider the sum $S = \sum_{i=0}^{n} F(U_i) . \delta R$

where $\delta R_i = R_i - R_{i-1}$

Limit of this sum as $n \to \infty$ in such a way that $\delta R_i | \to 0$, provided it exists, is called *tangential line integral of F(R)* along C and is symbolically written as

$$\int_C F(R) . dR \text{ or } \int_C F . \frac{dR}{dt} dt.$$

When path of integration is a closed curve, then this is denoted by using $\oint$ in place of $\int$.

If $F(R) = if(x, y, z) + j \phi (x, y, z) + k \psi (x, y, z)$

and $dR = idx + jdy + kdz$

then $\int_C F(R). dR = \int_C (fdx + \phi dy + \psi dz)$

Two other types of line integrals are $\int_C F \times dR$ and $\int_C fdR$ which are both vectors.

SURFACES

A surface S may be represented by $F(x, y, z) = 0$

The parametric representation of S is of the form

$$R (u, v) = x (u, v)i + y (u, v) j + z (u, v)k$$

and continous functions $u = \phi(t)$ and $v = \psi (t)$ of a real parameter represent a curve C on this surface S.

Parametric representation of the circular cylinder

$x^2 + y^2 = a^2, -1 \le z \le 1$, (radius a and height 2) is

$$R (u, v) = a \cos u\, i + a \sin u\, j + vk$$

where parameters u and v vary in the rectangle

$$0 \le u \le 2\pi \text{ and } -1 \le v \le 1.$$

Also u = t, v = bt represent a circular helix on this circular cylinder. The equation of the circular helix is

$$R = a \cos t\, i + a \sin t\, j + bt\, k.$$

Differentiating R = R (u, v), with respect to t, we get

$$\frac{dR}{dt} = \frac{\partial R}{\partial u} . \frac{du}{dt} + \frac{\partial R}{\partial v} . \frac{dv}{dt} .$$

The vectors $\left(\frac{\partial R}{\partial u}\right)$ and $\left(\frac{\partial R}{\partial v}\right)$ are tangential to S at P and determine tangent plane of S at P.N

$\left(\frac{\partial R}{\partial u}\right) \times \left(\frac{\partial R}{\partial v}\right) (\neq 0)$ gives a normal vector N of S at P.

Smooth Surface

If S has a unique normal at each of its points whose direction depends continuously on the points of S, then the surface S is called a *smooth surface*. If S is not smooth but can be divided into finitely many smooth portions, then it is called a *piecewise smooth surface*.

e.g. surface of sphere is smooth while the surface of a cube is piecewise smooth.

Orientable and Non-Orientable Surfaces

A surface S is said to be orientable or two sided if the positive normal direction at any point P of S can be continued in a unique and continuous way to the entire surface. If positive direction of the normal is reversed as we move around a curve on S passing through P, then the surface generate is *non-orientable*.

Surface Integral

Consider a continuous function F(R) and a surface S. Divide S into a finite number of sub-surfaces. Let the surface element surrounding any point P(R) be δS which can be regarded as a vector; its magnitude being the area and its direction that of the outward normal to the element.

Consider the sum $\Sigma F(R). \delta S$, where the summation extends over all the sub-surfaces. The limit of this sum as the number of sub-surfaces tends to infinity and the areas of each sub-surface tends to zero, is called *normal surface integral* of F(R) over S and is denoted by $\int_S F. dS$ or $\int_S F. Nds$, where N is a unit outward normal at P to S.

Other types of surface integrals are $\int_S F \times dS$ and $\int_S fds$ which are both vectors.

Notation: Only one integral sign when there is one differential (say dR or dS) and two (or three) signs when there are two (or three) differentials.

Flux across a surface

If F represents velocity of a fluid particle, then total outward flux of F across a closed surface S is the surface integral $\int_S F. dS$.

When flux of F across every surface S in a region E vanishes, F is said to be a solenoid vector point function in E.

F could equally well be taken as any other physical quantity, e.g. gravitational force, electric forces and magnetic force.

VOLUME INTEGRAL

Consider a continuous vector F(R) and surface S enclosing the region E.

Divide E into finite number of sub-regions $E_1, E_2, \ldots E_n$. Let δv_i be the volume of sub-region E_i enclosing any point whose position vector is R_i

Consider the sum $V = \sum_{i=1}^{n} F(R_i)\,\delta v_i$

The limit of this sum as $n \to \infty$ in such a way that $\delta v_i \to 0$, is called *volume integral* of F(R) over E and is symbolically written as $\int_E F dv$

If $\quad F(R) = f(x, y, z)\, i + \phi(x, y, z)\, j + \psi(x, y, z) K$

$\Rightarrow \quad \delta v = \delta x \delta y \delta z$

then $\int_E F dv = i \iiint_E f dxdydz + j \iiint_E \phi dxdydz + k \iiint_E \psi dxdydz$

FOURIER SERIES

If numbers $a_0, a_1, \ldots, a_n, \ldots, b_1, b_2, \ldots, b_n, \ldots$ are derived from a function *f* by means of Euler-Fourier formulas

$$a_n = \frac{1}{\pi}\int_{-\pi}^{\pi} f(x)\cos nx\, dx \qquad (n = 0,1,2,\ldots)$$

$$b_n = \frac{1}{\pi}\int_{-\pi}^{\pi} f(x)\sin nx\, dx \qquad (n = 1,2,\ldots)$$

then series $a_0 + \sum_{n=1}^{\infty} (a_n \cos nx + b_n \sin nx)$ is called *Fourier series*.

STOKE'S THEOREM

The close line integral of any vector $\overline{A}$ integral over any close curve C is always equal to the surface integral of the curl of vector A integrated over the surface area S which is enclosed by the closed curve C

$$\oint_C (\overline{A}) . \overline{dl} = \iint_S (\nabla \times \overline{A})\, ds$$

GREEN THEOREM

Let $\phi(x, y)$, $\psi(x, y)$, ϕy and ψx be continuous in a region E of the xy plane bound by a closed curve C, then

$$\int_C \phi dx + \psi dy = \iint_E \left(\frac{\partial \psi}{\partial x} - \frac{\partial \phi}{\partial y}\right) dx\, dy$$

This equation can be extended to regions which may be divided into finite number of sub-regions such that the boundary of each is cut atmost in two points by any line parallel to either axis apply in equation (i) to each of the sub-region and adding result , the surface integral combine into a integral over the whole region, the line integral over the common boundary cancel, whereas remaining line integrals combine into the line integral over the external curve C. This theorem convert line integrals around a closed curve into a double integrals and is a special case of Stroke's theorem.

GAUSS'S DIVERGENCE THEOREM

(Relation Between Surface and Volume Integrals)

"Volume integral of a vector field F taken over a any volume V is equal to the surface integral of F taken over the closed surface surrounding the volume V."

Mathematically, for any vector-field $\overline{F}$

$$\int_S F . N\, d_s = \int_E \text{div } F d_v$$

where N is the unit external vector

If $\quad F(R) = f(x, y, z)\, i + \phi(x, y, z)\, j + \psi(x, y, z)\, k$

then, $\iint (f\, dy\, dz + \phi\, dzdx + \psi\, dx\, dy)$

$$= \iiint_E \left(\frac{\partial f}{\partial x} + \frac{\partial \phi}{\partial y} + \frac{\partial \psi}{\partial z}\right) dx\, dy\, dx \qquad \ldots(i)$$

Consider a general region E assume that it can be split into finite number of sub region each of which is met by line parallel to any axis in only two points. Applying equation (i) to each of these sub-regions and adding results, the volume integral will be combine to give the volume integrals over the wholes region E. Also surface integrals over the common boundaries of two sub regions cancel because of each occur twice and having corresponding normal in opposite direction of whereas remaining surface integrals combine to give the surface integral over the entire – surface.

Consider a region E bounded by two closed surface S_1, S_2 (S, being within S_2). The outward normal at points S_1 is directed inwards (i.e. away from S_2) and introducing an adding surface cutting S_1, S_2 so that all parts of E are bounded by single closed surface. Thus the theorem also holds for regions enclosed by several surfaces.

EXERCISE

MCQ TYPE QUESTIONS

1. $\lim_{x\to\infty} x^n e^{-x}$ is equal to

   (a) ∞ (b) 1

   (c) $\lfloor n$ (d) 0

2. If $|x|$ denotes the greatest integer not greater than 2, then $\lim_{x\to 2}[x]$ is

   (a) 0 (b) 1

   (c) 2 (d) does not exist

3. If two functions $f(x) = \begin{cases} x\sin\frac{1}{x} & , x\neq 0 \\ 0 & , x\neq 0 \end{cases}$ $g(x) = |x|$ are given, then at $x = 0$

   (a) both f and g continuous

   (b) f is continuous and g is discontinuous

   (c) f is discontinuous and g is continuous

   (d) both f and g are discontinuous

4. If $f(x) = |x|$, then $f(x)$ is

   (a) discontinuous at $x = 0$

   (b) continuous only at $x = 0$

   (c) continuous at all values of x

   (d) discontinuous at $x = 1$

5. The function $f(x) = \begin{cases} ax^2 + b & if \quad x > 2 \\ 2 & if \quad x = 2 \\ 2ax - b & if \quad x < 2 \end{cases}$ is continuous at $x = 2$, if

   (a) $a = 0, b = \frac{1}{2}$ (b) $a = \frac{1}{2}, b = 0$

   (c) $a = -\frac{1}{2}, b = 0$ (d) $a = 0, b = -\frac{1}{2}$

6. If $f(x) = x - [x]$, then $f(x)$ is discontinuous

   (a) $\forall x \in Z$

   (b) $\forall x \in Q$

   (c) $\forall x \in R$

   (d) at very real number except integers

7. The interval in which Lagrange's theorem is applicable for the function $f(x) = \frac{1}{x}$ is

   (a) $[-3, 3]$ (b) $[-2, 2]$

   (c) $[2, 3]$ (d) $[-1, 1]$

8. If $f(x) = 3x^4 - 4x^2 + 5$, then the interval for which $f(x)$ satisfied all the condition of Rolle's theorem is

   (a) $[0, 2]$ (b) $[-1, 1]$

   (c) $[-1, 0]$ (d) $[1, 2]$

9. If $f(x) = |x|$, then In the interval $[-1, 1]$, $f(x)$ is

   (a) satisfied all the conditions of Rolle's theorem

   (b) satisfied all the conditions of Mean Value theorem

   (c) does not satisfied the conditions of Mean Value theorem

   (d) none of these

10. If f and F be both continuous in $[a, b]$ and are derivable in (a, b) and if $f'(x) = F'(x)$ for all x in (a, b), then $f(x)$ and $F(x)$ differ

    (a) by 1 in $[a, b]$ (b) by x in $[a, b]$

    (c) by constant in $[a, b]$ (d) none of these

11. $x^{1/x}$ is a decreasing function if

    (a) $x < e$ (b) $x > e$

    (c) $x = e$ (d) $x > \frac{1}{e}$

12. $1 - \frac{x^2}{\lfloor 2} + \frac{x^4}{\lfloor 4} + ... + (-1)^n \frac{x^{2n}}{\lfloor 2n} + (-1)^{n+1} \frac{x^{2n+1}}{\lfloor 2n+1} \sin\theta x$ is equal to

    (a) $\sin x$ (b) $\cos x$

    (c) $\tan x$ (d) $\log x$

13. Taylor's expansion of the function $f(x) = \frac{1}{1+x^2}$ is

    (a) $\sum_{n=0}^{\infty} (-1)^n x^{2n}$ where $-1 < x < 1$

    (b) $\sum_{n=0}^{\infty} (-1)^n x^{2n}$ for all real x

    (c) $\sum_{n=0}^{\infty} x^{2n}$ where $-1 < x < 1$

    (d) $\sum_{n=0}^{\infty} (-1)^n x^n$ where $-1 < x < 1$

14. If $f(x)$, $\phi(x)$ and $\psi(x)$ have derivatives when $a \le x \le b$, then $\exists$ a value of c of x lying between a and b such that

    (a) $\begin{vmatrix} f'(c) & \phi'(c) & \psi'(c) \\ f(b) & \phi(b) & \psi(b) \\ f(a) & \phi(a) & \psi(a) \end{vmatrix} = 0$

    (b) $\begin{vmatrix} f(c) & \phi(c) & \psi(c) \\ f'(a) & \phi'(a) & \psi'(a) \\ f'(b) & \phi'(b) & \psi'(b) \end{vmatrix} = 0$

    (c) $\begin{vmatrix} f(a) & \phi(a) & \psi(a) \\ f'(b) & \phi'(a) & \psi'(b) \\ f''(c) & \phi''(c) & \psi''(c) \end{vmatrix} = 0$

    (d) none of these

15. If $f(x) = ax + b, x \in [-1, 1]$, then the point $c \in (-1, 1)$ where $f'(c) = \dfrac{f(1) - f(-1)}{2}$

(a) does not exist (b) can be any $c \in (-1, 1)$

(c) can be only $\frac{1}{2}$ (d) can be only $\frac{-1}{2}$

16. Let $f(x) = x(x-1)(x-2)$, $a = 0$, $b = \frac{1}{2}$.

The value of 'c' in the Lagrange's mean value theorem is

(a) $\dfrac{6+\sqrt{11}}{6}$ (b) $\dfrac{6-\sqrt{11}}{6}$

(c) $\dfrac{6+\sqrt{-12}}{6}$ (d) $\dfrac{6-\sqrt{21}}{6}$

17. Maxima and Minima occur

(a) simultaneously (b) once

(c) alternately (d) rarely

18. If $x + y = k, x > 0, y > 0$, then xy is the maximum when

(a) $x = ky$ (b) $kx = y$

(c) $x = y$ (d) none of these

19. The triangle of maximum area inscribed in a circle of radius r is

(a) a right angled triangle with hypotenuse measuring $2r$

(b) an equilateral triangle

(c) an isosceles triangle of height r

(d) does not exist

20. Sum of the perimeters of a circle and a square is l. If the sum of area is least, then

(a) side of the square is double the radius of the circle

(b) side of the square is half of the radius of the circle

(c) side of the square is equal to radius of the circle

(d) none of these

21. Which of the following is false?

(a) $f(a)$ is an extreme value of $f(x)$ if $f'(a) = 0$

(b) If $f(a)$ is an extreme value of $f(x)$, then $f'(a) = 0$

(c) If $f'(a) = 0$, then f(a) is an extreme value of $f(x)$

(d) None of these

22. Let $f(x) = |x|$, then

(a) $f'(0) = 0$

(b) $f(x)$ is maximum at $x = 0$

(c) $f(x)$ is minimum at $x = 0$

(d) none of these

23. The greatest and the least value of $f(x) = x^4 - 8x^3 + 22x^2 - 24x + 1$ in $[0, 2]$ are

(a) 0, 8 (b) 0, – 8

(c) 1, 8 (d) 1, – 8

24. The function $f(x) = x^5 - 5x^4 + 5x^3 - 1$ has

(a) one minima and two maxima

(b) two minima and one maxima

(c) two minima and two maxima

(d) one minima and one maxima

25. The maximum value of $\dfrac{\log x}{x}$ in $(0, \infty)$ is

(a) e (b) $\frac{1}{e}$

(c) 1 (d) none of these

26. Maximum volume of a cylinder in a cone with semi vertical angle 'α' and height 'b' is

(a) $\frac{4}{27}\pi b^3 \tan^2\alpha$ (b) $\frac{4}{27}\pi b^2 \tan\alpha$

(c) $\frac{4}{27}\pi b \tan^3\alpha$ (d) none of these

27. The maximum point on the curve $x = e^x y$ is

(a) $(1, e)$ (b) $(1, e^{-1})$

(c) $(e, 1)$ (d) $(e^{-1}, 1)$

28. The value of function $f(x) = x + \frac{1}{x}$ at the points of minimum and maximum are respectively

(a) – 2 and 2 (b) 2 and – 2

(c) – 1 and 1 (d) 1 and – 1

29. The value of $\int_0^{5\pi} (2 - \sin x)dx$ is

(a) > 0 (b) 2

(c) 0 (d) undefined

30. The unit normal to the plane $2x + y + 2z = 6$ can be expressed in the vector form as

(a) $i3 + j2 + k2$ (b) $i\frac{2}{3} + j\frac{1}{3} + k\frac{2}{3}$

(c) $i\frac{1}{3} + j\frac{1}{2} + k\frac{1}{2}$ (d) $i\frac{2}{6} + j\frac{1}{6} + k\frac{2}{6}$

31. The function $f(x) = 3x(x-2)$ has a

(a) minimum at $x = 1$ (b) maximum at $x = 1$

(c) minimum at $x = 2$ (d) maximum at $x = 2$

32. The value of integral $\int_{-2}^{2} \frac{dx}{x^2}$ is

(a) 0 (b) 0.25

(c) 1 (d) ∞

33. The value of ξ in the mean value theorem of $f(b) - f(a) = (b - a)f'(\xi)$ for $f(x) = Ax^2 + Bx + C$ in (a, b) is

(a) $b + a$ (b) $b - a$

(c) $\frac{(b+a)}{2}$ (d) $\frac{(b-a)}{2}$

34. The function $f(x) = x^3 - 6x^2 + 9x + 25$ has

(a) a maxima at $x = 1$ and a minima at $x = 3$

(b) a maxima at $x = 3$ and a minima at $x = 1$

(c) no maxima, but a minima at $x = 1$

(d) a maxima at $x = 1$, but no minima

35. The temperature field in a body varies according to the equation $T(x, y) = x^3 + 4xy$. The direction of fastest variation in temperature at the point (1, 0) is given by

(a) $3\hat{i} + 8\hat{j}$ (b) $\hat{i}$

(c) $0.6\hat{i} + 0.8\hat{j}$ (d) $0.5\hat{i} + 0.866\hat{j}$

36. If $\Phi(x) = \int_0^{x^2} \sqrt{t}\, dt$, then $\frac{d\Phi}{dx}$ is

(a) $2x^2$ (b) $\sqrt{x}$

(c) 0 (d) 1

37. What is the derivative of $f(x) = |x|$ at $x = 0$?

(a) 1 (b) –1

(c) 0 (d) Does not exist

38. The minimum point of the function $f(x) = (x^3/3) - x$ is at

(a) $x = 1$ (b) $x = -1$

(c) $x = 0$ (d) $x = \frac{1}{\sqrt{3}}$

39. Volume of an object expressed in spherical coordinates is given by

$$V = \int_0^{2\pi} \int_0^{\pi/3} \int_0^1 r^2 \sin\phi \, dr \, d\phi \, d\theta .$$

Value of the integral is

(a) $\frac{\pi}{3}$ (b) $\frac{\pi}{6}$

(c) $\frac{2\pi}{3}$ (d) $\frac{\pi}{4}$

40. The function, $y = x^2 + \frac{250}{x}$ $x = 5$ attains

(a) maximum (b) minimum

(c) neither (d) 1

41. Volume generated by revolving the area bounded by the parabola $y^2 = 8x$ and the line $x = 2$ about y-axis is

(a) $\frac{128\pi}{5}$ (b) $\frac{5}{128\pi}$

(c) $\frac{127}{5\pi}$ (d) none of these

42. The integration of $\int \log x \, . \, dx$ has the value

(a) $(x \log x - 1)$ (b) $\log x - x$

(c) $x(\log x - 1)$ (d) None of these

43. Area bounded by the parabola $2y = x^2$ and the line $x = y - 4$ is equal to

(a) 6 (b) 18

(c) ∞ (d) none of these

44. If a function is continuous at a point, then its first derivative

(a) may or may not exist

(b) exists always

(c) will not exist

(d) has a unique value

45. The integral $\lim_{a \to \infty} \int_1^a x^{-4} dx$

(a) diverges (b) converges to $\frac{1}{3}$

(c) converges to $-\frac{1}{a^3}$ (d) converges to 0

NUMERICAL TYPE QUESTIONS

1. $\lim_{x \to 0} \frac{e^x - 1}{x}$ is equal to ________

2. If $\lim_{x \to 1} \frac{f(x) - 2}{f(x) + 2} = 0$, then $\lim_{x \to 1} f(x)$ is equal to ________

3. $\lim_{x \to 0} \frac{1 - \cos x}{2}$ is equal to ________

4. The value (N) of $\lim_{x \to 0} \frac{\sqrt{1+x} - \sqrt{1-x}}{x}$ is given by ________

5. $\lim_{n \to 0} e^{\left(\frac{1}{n \log n}\right)}$ is ________

6. $\lim_{x \to \infty} x \sin\frac{1}{x}$ is ________

7. $\lim_{n \to \infty} \left[\frac{1}{n^2} + \frac{2}{n^2} + \frac{3}{n^2} + \ldots + \frac{1}{n}\right]$ is

8. The 'c' of the mean value theorem for the function $f(x) = x(x - 2)$, when $a = 0$, $b = \frac{3}{2}$ is ________

9. The minimum value of $|x^2 - 5x + 2|$ is ________

10. The maximum value of $\frac{1}{\sqrt{2}}(\sin x - \cos x)$ is ________

11. Maximum slope of the curve $-x^3 + 6x^2 + 2x + 1$ is ________

12. A rectangular box with square base is open at the top. The maximum volume of the box made from 300 m^3 wood is ________

13. Given $f(x, y) = x^2 + y^2$. Then $\nabla^2 f$ is ________

14. The directional derivative of

$$f(x, y, z) = x^2 + y^2 + z^2$$

at the point (1, 1, 1) in the direction i − k is ________

15. A triangle ABC consists of vertex points

A (0, 0), B(1, 0) and C(0, 1).

Value of integral $\iint 2x\, dx\, dy$ over the triangle is ________

16. $\int_0^{\pi/2}\int_0^{\pi/2} \sin(x + y)\, dx\, dy$ is ________

ANSWERS

MCQ Type Questions

1. (d)	**2.** (d)	**3.** (a)	**4.** (c)	**5.** (b)	**6.** (d)	**7.** (c)	**8.** (b)	**9.** (c)	**10.** (c)
11. (b)	**12.** (b)	**13.** (b)	**14.** (a)	**15.** (b)	**16.** (d)	**17.** (c)	**18.** (d)	**19.** (b)	**20.** (a)
21. (d)	**22.** (c)	**23.** (d)	**24.** (d)	**25.** (b)	**26.** (a)	**27.** (b)	**28.** (d)	**29.** (a)	**30.** (b)
31. (a)	**32.** (d)	**33.** (c)	**34.** (a)	**35.** (c)	**36.** (a)	**37.** (b)	**38.** (a)	**39.** (a)	**40.** (b)
41. (a)	**42.** (c)	**43.** (b)	**44.** (d)	**45.** (a)					

Numerical Type Questions

1. (1)	**2.** (2)	**3.** (0)	**4.** (1)	**5.** (0)	**6.** (1)	**7.** (1/2)
8. (3/4)	**9.** (0)	**10.** (1)	**11.** (14)	**12.** (500)	**13.** (4)	**14.** (0)
15. (1/3)	**16.** (2)					

EXPLANATIONS

MCQ TYPE QUESTIONS

1. $$\lim_{x\to\infty} x^n e^{-x} = \lim_{x\to\infty}\frac{x^n}{e^x} \qquad \left[\text{form}\frac{\infty}{\infty}\right]$$

By L' Hospital Rule Differentiating denominator and numerator n times, we get

$$\lim_{x\to\infty}\frac{x^n}{e^x} = \lim_{x\to\infty}\frac{\lfloor n}{e^x} = 0$$

2. $$\text{L.}\lim_{x\to 2}[x] = \lim_{h\to 0}[2-h] = \lim_{h\to 0} 1 = 1$$

$$\text{R.}\lim_{x\to 2}[x] = \lim_{h\to 0}[2+h] = \lim_{h\to 0} 2 = 2$$

$$\therefore \quad \text{L.}\lim_{x\to 2}[x] \neq \text{R.}\lim_{h\to 2}[x]$$

3. $$\text{L.}\lim_{x\to 0} f(x) = \lim_{h\to 0} f(0-h) = \lim_{h\to 0} -h\sin\left(-\frac{1}{h}\right)$$

$$= \lim_{h\to 0} +h\sin\frac{1}{h} = 0$$

$$\text{R.}\lim_{x\to 0} f(x) = \lim_{h\to 0} f(0+h)$$

$$= \lim_{h\to 0} -h\sin\left(\frac{1}{h}\right) = 0$$

Also $f(0) = 0$

Since $\text{L.}\lim_{x\to 0} f(x) = \text{R.}\lim_{x\to 0} f(x) = f(0) = 0$

Hence $f(x)$ is continuous at $x = 0$

Now, $$\text{L.}\lim_{x\to 0} g(x) = \lim_{h\to 0} g(0-h)$$

$$= \lim_{h\to 0}|-h| = \lim_{h\to 0} h = 0$$

$$\text{R.}\lim_{x\to 0} g(x) = \lim_{h\to 0} g(0+h)$$

$$= \lim_{h\to 0}|h| = \lim_{h\to 0} h = 0$$

$$g(0) = |0| = 0$$

$$\therefore \text{L.}\lim_{x\to 0} g(x) = \text{R.}\lim_{x\to 0} g(x) = g(0) = 0$$

Hence $g(x)$ is also continuous at $x = 0$

4. Let $a \in$ R (real numbers)

$$\text{L.}\lim_{x\to a} f(x) = \lim_{h\to 0}|a-h| = \lim_{h\to 0}(a-h) = a$$

$$\text{R.}\lim_{x\to a} f(x) = \lim_{h\to 0}|a+h| = \lim_{h\to 0} a + h = a$$

and $f(a) = |a| = a$

$$\therefore \quad \lim_{x\to a}|x| = a = f(a)$$

Hence $\lim_{x\to a}|x|$ continuous for all values of x.

5. $$\text{L.}\lim_{x\to 2} f(x) = \lim_{h\to 0} 2a(2-h) - b = 4a - b$$

$$\text{R.}\lim_{x\to 2} f(x) = \lim_{h\to 0} a(2+h)^2 + b = 4a + b$$

The function will be continuous if

$$\text{L.}\lim_{x\to 2} f(x) = \text{R.}\lim_{x\to 2} f(x) = f(2)$$

$\therefore$ $4a - b = 2$

$4a + b = 2$

Solving we get, $a = \dfrac{1}{2}$, $b = 0$

6. Given function is, $f(x) = x - [x]$

Let a is a real number except integer and h be a very small real number, then

$[a - h] = b$ (an integer less than to a)

$[a + h] = b$ (an integer greater than to a)

Also $[a] = b$

$$\therefore \text{ L.}\lim_{x\to 0} f(x) = \lim_{h\to 0}(a-h) - [a-h]$$

$$= \lim_{h\to 0}(a-h) - b = a - b$$

Also $f(a) = (a) - [a] = a - b$

$$\therefore \text{ L.}\lim_{x\to a} f(x) = \text{R.}\lim_{x\to a} f(x) = f(a) = a - b$$

$\therefore$ $f(x) = x - [x]$

is continuous all real except integers.

7. Since $f(x) = \dfrac{1}{x}$ is not continuous in $[-3, 3]$ $[-2, 2]$ or, $[-1, 1]$, hence the point of discontinuity is '0'. Only in $[2, 3]$ the function is continuous, and differentiable. Hence mean value theorem is applicable in $[2, 3]$.

8. Given function is

$$f(x) = 3x^4 - 4x^2 + 5$$

(*i*) $f(x)$ is continuous in every real interval

(*ii*) $f'(x)$ exist in any real interval

(*iii*) $f(-1) = 3(-1)^4 - 4(-1)^2 + 5 = 4$

$f(1) = 3(1)^4 - 4(1)^2 + 5 = 4$

$f(-1) = f(1)$

Also $f'(c) = 12c^3 - 8c = 0$

$$\Rightarrow \quad c = 0, c = \pm\sqrt{\frac{2}{3}}$$

$\therefore$ $c \in [-1, 1]$

Hence all the condition of Rolle's Theorem are satisfied in the interval $[-1, 1]$

9. Since $f(x) = |x|$ is continuous in $[-1, 1]$ but it is not differentiable at $x = 0 \in (-1, 1)$

10. Since f and F are two continuous function in [a, b] such that

$$f'(x) = F'(x)$$

Let $\phi(x) = f(x) - F(x)$

$\therefore$ $\phi'(x) = f'(x) - F'(x)$

$= 0$ [from equation]

$\therefore$ $\phi(x) = \text{constant}$

11. Let $f(x) = x^{1/x}$

$$\therefore \quad f'(x) = x^{1/x}\left(\frac{1-\log_e x}{x^2}\right) < 0$$

$f(x)$ is decreasing if $f'(x) < 0$

$$i.e., \ x^{1/x}\left(\frac{1-\log x}{x^2}\right) < 0$$

$$\Rightarrow 1 - \log_e x < 0 \quad \Rightarrow \quad \log_e x > 1 \quad \Rightarrow \quad x > e$$

12. Let $f(x) = \cos x$

$$f''(x) = \cos\left(x + \frac{n\pi}{2}\right)$$

$\therefore$ $f'(0) = 0$

$f''(0) = -1$

$f'''(0) = 0$

$f''''(0) = 1$

....

....

and so on

By Maclaurin's theorem, we get

$$\cos(x) = f(0) + xf'(0) + \frac{x^2}{\lfloor 2}f''(0) + \frac{x^2}{\lfloor 3}f'''(0)$$

$$+ \ldots + \frac{x^{2n}}{\lfloor 2n}f^{2n}(0) + \frac{x^{2n+1}}{\lfloor 2n+1}f^{2n+1}(\theta x)$$

$$= 1 + 0 - \frac{x^2}{\lfloor 2} + 0 + \frac{1}{\lfloor 4}\ldots + (-1)^{n+1}\frac{x^{2n+1}}{\lfloor 2n+1}\sin\theta x$$

$$= 1 - \frac{x^2}{\lfloor 2} + \frac{1}{\lfloor 4}x^4 + \ldots + (-1)^{n+1}\frac{x^{2n+1}}{\lfloor 2n+1}\sin\theta x$$

13. Given: $f(x) = \dfrac{1}{1+x^2}$

$\therefore$ $(1 + x^2) f(x) = 1$; $f(0) = 1$

$(1 + x^2) f'(x) + 2x f(x) = 0$; $f'(0) = 0$

$(1 + x^2) f''(x) + 4x f'(x) + 2f(x) = 0$; $f''(0) = -2$

$(1 + x^2) f'''(x) + 6x f''(x) + 6f'(x) = 0$; $f'''(0) = 0$

$(1 + x^2) f^{iv}(x) + 8x f'''(x) + 12 f''(x) = 0$; $f^{iv}(0) = 24$

and so on

$$\therefore \ f(x) = f(0) + x f'(0) + \frac{x^2}{\lfloor 2}f'(0) + \ldots$$

$$\therefore \quad \frac{1}{1+x^2} = 1 + x.0 + \frac{x^2}{\lfloor 2}(-2) + \frac{x^2}{\lfloor 3}.0 + \frac{x^4}{\lfloor 4}.24$$

$$+ \ldots + (-1)^n.\frac{x^{2n}}{\lfloor 2n}.\lfloor 2n$$

$$= 1 - x^2 + x^4 + \ldots + (-1)^n.x^{2n}$$

$$= \sum_{n=0}^{\infty}(-1)^n.x^{2n} \ \text{ for all real } x.$$

19. Let ABC be a triangle inscribed in the circle with centre O and radius r.

If area of this triangle is maximum, then vertex C should be at a maximum distance from the base

AB i.e., CD must be perpendicular to AB. Hence ABC is an isosceles triangle.

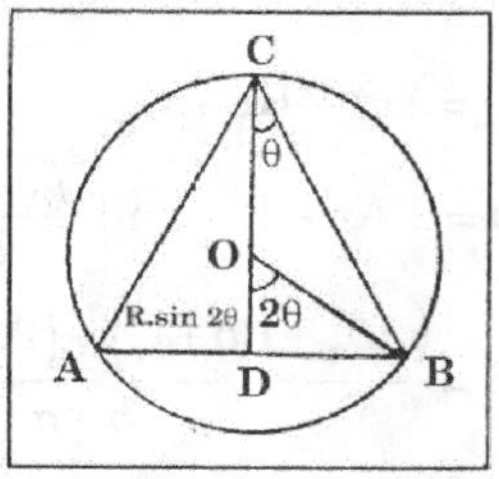

If $\angle BCD = \theta$,

where D is the mid-point of AB, then

$\angle BOD = 2\theta$

$\therefore$ $AB = 2BD = 2r \sin 2\theta$

$CD = CO + OD = r + r\cos 2\theta$

If S be the area of the triangle ABC, then

$$S = \frac{1}{2} AB \times CD = \frac{1}{2} \times 2r \sin 2\theta\,(r + r\cos 2\theta)$$

$$\therefore \quad \frac{ds}{d\theta} = r^2[\sin 2\theta\,(-2\sin 2\theta] + (1 + \cos 2\theta)(2\cos 2\theta)]$$

For maximum and minimum,

$$\frac{dS}{d\theta} = 0$$

$\therefore$ $\cos 4\theta + \cos 2\theta = 0$

$\Rightarrow$ $2\cos 3\theta \cos\theta = 0$

Hence either $\cos 3\theta = 0$, or $\cos\theta = 0$

If $\cos\theta = 0$, then $\theta = \frac{\pi}{2}$

If $\cos 3\theta = 0$, then $3\theta = \frac{\pi}{2}$

$\Rightarrow$ $\theta = \frac{\pi}{6}$

$\left(\frac{d^2S}{d\theta^2}\right)_{\theta=\pi/6}$ is negative.

Hence S is maximum for $\theta = \frac{1}{6}\pi$

$$\angle ACB = 2\theta = 2\left(\frac{\pi}{6}\right) = \frac{\pi}{3}$$

$$= \angle ABC = \angle BAC$$

Hence ΔABC is an equalateral triangle.

20. Let radius of circle is r and side of a sqaure is x.

$\therefore$ Perimeter $= 2\pi r + 4x = l$ ...(*i*)

Let area be A

$\therefore$ $A = \pi r^2 + x^2$

$$= \pi r^2 + \left(\frac{l - 2\pi r}{4}\right)^2 \quad \text{[from equation (i)]}$$

$$\frac{dA}{dr} = 2\pi r + \frac{2}{16}(1 - 2\pi r).\,(-2\pi)$$

and $\frac{d^2A}{dr^2} = 2\pi + \frac{\pi^2}{2}$

For maximum and minimum,

$$\frac{dA}{dr} = 0$$

$\Rightarrow$ $2\pi r - \frac{\pi}{4}(l - 2\pi r) = 0$

$\Rightarrow$ $l = 8r + 2\pi r$ ...(*ii*)

$\frac{d^2A}{dr^2} > 0$, hence area is least

From equations (*i*) and (*ii*),

$$2\pi r + 4x = 8r + 2\pi r$$

$\Rightarrow 4x = 8r \quad \Rightarrow \quad x = 2r$

Hence side of a square is double the radius of circle.

21. Since statements (*a*), (*b*), (*c*) are all correct, hence the statement (*d*) is false.

23. $f(x) = x^4 - 8x^3 + 22x^2 - 24x + 1$

$f(0) = 1$

$f(2) = 2^4 - 8.2^3 + 22.2^2 - 24.2 + 1 = -7$

Now $f'(x) = 4x^3 - 24x^2 + 44x - 24$

For maximum and minimum, $f'(x) = 0$

$\Rightarrow$ $4x^3 - 24x^2 + 44x - 24 = 0$

$\Rightarrow$ $4(x-1)(x-2)(x-3) = 0$

$\therefore$ $x = 1, 2, 3$

Since $x = 3$ does not lie in $[0, 2]$

Hence, consider only $x = 1$ and $x = 2$.

We have

$$f(1) = 1^4 - 8.1^3 + 22.1^2 - 24.1 + 1 = -8$$

Greatest of $f(x)$ = largest of $\{1, -7, -8\} = 1$

Least of $f(x)$ = smallest of $\{1, -7, -8\} = -8$

24. $f(x) = x^5 - 5x^4 + 5x^3 - 1$

$f'(x) = 5x^4 - 20x^3 + 15x^2$

$f''(x) = 20x^3 - 60x^2 + 30x$

$f'''(x) = 60x^2 - 120x + 30$

For maximum and minimum

$f'(x) = 0$

$\therefore$ $x^4 - 4x^3 + 3x^2 = 0$

$\Rightarrow$ $x^2(x-1)(x-3) = 0$

$\therefore$ $x = 0, 1, 3$

$f''(0) = 0$

$f'''(0) = 30 \neq 0$

$\therefore f(x)$ is neither maximum nor minimum at $x = 0$

$f''(1) = 20 - 60 + 30$

$= -10 < 0$

$\therefore$ $f(x)$ is maximum at $x = 1$

$f''(3) = 540 - 540 + 90 = 90 > 0$

$\therefore$ $f(x)$ is maximum at $x = 3$

Hence there is one maximum and one minimum for $f(x)$.

25. Given function is

$$f(x) = \frac{\log x}{x}$$

$$\therefore \quad f'(x) = \frac{x.\frac{1}{x} - \log x.1}{x^2} = \frac{1 - \log x}{x^2}$$

$$\text{and} \quad f''(x) = \frac{x^2.\left(-\frac{1}{x}\right) - (1 - \log x).2x}{x^4}$$

$$= \frac{2\log x - 3}{x^3}$$

For maximum and minimum of $f(x)$,

$f'(x) = 0 \Rightarrow 1 - \log x = 0 \quad \Rightarrow \quad x = e$

$$\therefore \quad f''(e) = \frac{2-3}{e^3} = -\frac{1}{e^3} < 0$$

Hence $f(x)$ is maximum at $x = e$

$$\therefore \text{ Maximum value, } f(e) = \frac{\log e}{e} = \frac{1}{e}$$

29. $\int_0^{5\pi} (2 - \sin x)dx = [2x + \cos x]_0^{5\pi}$

$= 10\pi - 1 - 1 = 10\pi - 2 > 0$

31. $f(x) = 3x(x - 2) = 3x^2 - 6x$

Differentiating, we get

$$\frac{df(x)}{dx} = f'(x) = 6x - 6$$

Again differentiating,

$$\frac{d^2 f(x)}{dx} = f''(x) = 6 > 0$$

For maxima or minima, $\quad f'(x) = 0$

$\therefore \quad 6x - 6 = 0,$

$\Rightarrow \quad x = 1$

As $f''(x) = 6 > 0$, $f(x)$ is minimum at $x = 1$

32.

$$\int_{-2}^{2} \frac{dx}{x^2} = \int_{-2}^{2} \frac{1}{x^2}\,dx$$

$$f(x) = \frac{1}{x^2} = \text{even function}$$

For even function,

$$\int_{-a}^{a} f(x)dx = 2\int_0^a f(x)\,dx = 2\int_0^2 \frac{1}{x^2}dx = 2\left[-1\frac{1}{x}\right]_0^2 = \infty$$

$$\therefore \quad \int_{-2}^{2} \frac{1}{x^2} = \infty$$

33. Given : $f(x) = Ax^2 + Bx + C$

$$\Rightarrow \quad f'(x) = 2\,Ax + B \quad \text{and} \quad \frac{f(b) - f(a)}{b - a} = f'(\xi)$$

$$\Rightarrow 2A\,\xi + B = \frac{(Ab^2 + Bb + C) - (Aa^2 + Ba + C)}{b - a}$$

$$= \frac{A(b^2 - a^2) + B(b - a)}{b - a} = A\,(b + a) + B$$

$$\therefore \quad \xi = \frac{b + a}{2}$$

34. $f'(x) = 3(x^2 - 4x + 3)$, $f''(x) = 6x - 12$

At $x = 1$; $f''(1) = -$ve; and, at $x = 3$, $f''(3) = +$ ve

35. Given $\quad T = x^3 + 4xy$

$$\Rightarrow \quad \frac{\partial T}{\partial x} = 3x^2 + 4y$$

$$\therefore \quad \left(\frac{\partial T}{\partial y}\right)_{(1,0)} = 3, \quad \Rightarrow \quad \left(\frac{\partial T}{\partial y}\right)_{(1,0)} = 4$$

Hence direction of fastest variation in temperature at (1, 0) is given by $3\hat{i} + 4\hat{j}$ or $0.6\hat{i} + 0.8\hat{j}$

36.

$$\phi(x) = \int_0^{x^2} \sqrt{t}\;dt = \left|\frac{2}{3}t^{3/2}\right|_0^{x^2} = \frac{2}{3}x^3$$

$$\therefore \quad \frac{d\phi}{dx} = \frac{2}{3} \times 3x^2 = 2x^2$$

37. $f(x) = |x|$, and $f(x) = x$ for $x > 0$

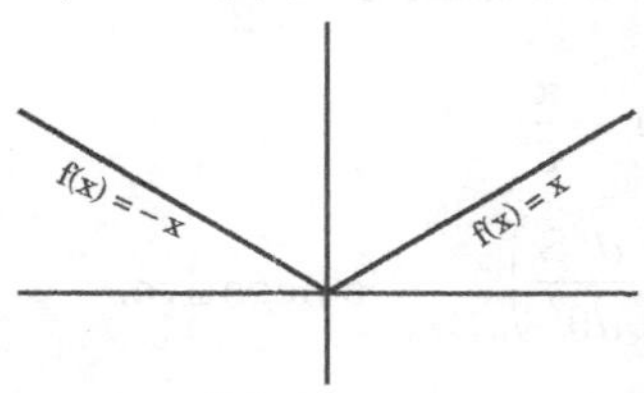

$\Rightarrow \quad f(x) = -x$ for $x < 0$

Hence from graph it is clear that derivative does not exist for $x = 0$.

38.

$$f(x) = \frac{x^3}{3} - x$$

$\Rightarrow f'(x) = x^2 - 1$

For maximum and minimum value, $f'(x) = 0$

$\therefore \quad x^2 - 1 = 0$

$\Rightarrow \quad x = \pm 1$

Again $\quad f''(x) = 2x$

At $x = 1$, $\quad f''(x) = 2 > 0$

$x = 1$, $\quad f''(x) = -2 < 0$

Hence minimum at $x = 1$

39. $V = \int_0^{2\pi}\int_0^{\pi/3}\int_0^1 r^2 \sin\phi \, dr\, d\phi\, d\theta$

$$= \int_0^{2\pi}\int_0^{\pi/3} \frac{1}{3}\sin\phi \, d\phi\, d\theta$$

$$= \frac{1}{3}\int_0^{2\pi} [-\cos\phi]_0^{\pi/3}\, d\theta$$

$$= \frac{1}{3}\int_0^{2\pi} \frac{1}{2}\, d\theta = \frac{1}{6}\times 2\pi = \frac{\pi}{3}.$$

40. Given: $y = x^2 + \frac{250}{x}$

$\therefore \quad \frac{dy}{dx} = 2x - \frac{250}{x^2}$,

and $\frac{dy^2}{dx^2} = 2 + \frac{250\times 2}{x^3}$

When $x = 5$, $\frac{dy}{dx} = 10 - 10 = 0$

and $\frac{d^2y}{dx^2}$ is positive

Therefore function is minimum at $x = 5$.

42. $\int \log x \, . \, dx = \log x \, . \, x - \int x \, . \, \frac{d}{dx}(\log x)\, dx$

$$= x\log x - \int 1 \, . \, dx = x\log x - x = x(\log x - 1)$$

43. Given : $2y = x^2$

$\Rightarrow \quad y = \frac{x^2}{2}$

and $\quad y = x + 4$

$\therefore \quad x^2 - 2x - 8 = 0$

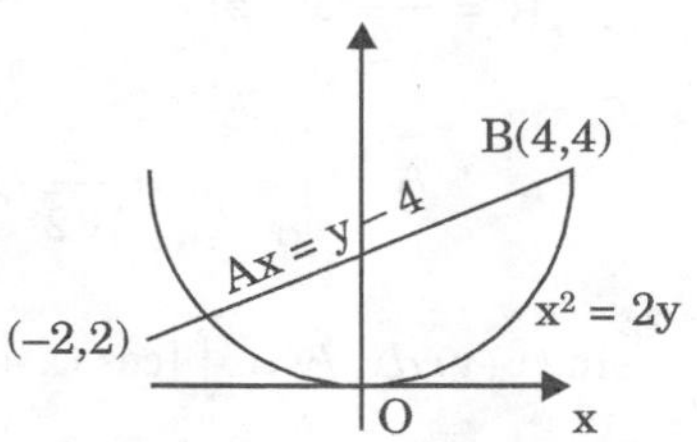

Given curve and the line intersect at A(–2, 2) and B (4, 8)

$$\therefore \text{ Area} = \int_{-2}^{4}(y_1 - y_2)\, dx = \int_{-2}^{4}\left\{(x+4) - \frac{x^2}{2}\right\} dx$$

$$= \left[\frac{x^2}{2} + 4x - \frac{x^3}{6}\right]_{-2}^{4}$$

$$= \left(8 + 16 - \frac{32}{3}\right) - \left(2 - 8 - \frac{4}{3}\right) = 18$$

45. $\lim_{a\to\infty}\int_1^a x^{-4}dx = \lim_{a\to\infty}\left[-\frac{1}{3}\left(x^{-3}\right)_1^a\right]$

$$= \lim_{a\to\infty} -\frac{1}{3}\left[a^{-3} - 1\right] = -\frac{1}{3}[0-1] = \frac{1}{3}$$

NUMERICAL TYPE QUESTIONS

1. $\lim_{x\to 0}\frac{e^x - 1}{x} = \lim_{x\to 0}\frac{\left(1 + x + \frac{x^2}{\lfloor 2} + \frac{x^3}{\lfloor 3} + \ldots\right) - 1}{x}$

$$= \lim_{x\to 0}\left(1 + \frac{x}{\lfloor 2} + \frac{x^2}{\lfloor 3} + \ldots\right) = 1$$

By L' Hospital's Rule

$$\lim_{x\to 0}\frac{e^x - 1}{x}\left[\text{form}\frac{0}{0}\right] = \lim_{x\to 0}\frac{e^x}{1} = 1$$

4. $\lim_{x\to 0}\frac{\sqrt{1+x} - \sqrt{1-x}}{x}$

$$= \lim_{x\to 0}\frac{\sqrt{1+x} - \sqrt{1-x}}{x}\times\left(\frac{\sqrt{1+x} + \sqrt{1-x}}{\sqrt{1+x} + \sqrt{1-x}}\right)$$

$$= \lim_{x\to 0}\frac{2x}{x(\sqrt{1+x} + \sqrt{1-x})}$$

$$= \frac{2}{\sqrt{1+0} + \sqrt{1-0}} = \frac{2}{1+1} = 1$$

5. Let $A = \lim_{n\to 0} e^{\frac{1}{n\log n}}$ $\quad\therefore \log_e A = \lim_{n\to 0}\frac{1}{n\log n}$.

$\Rightarrow \log_e A = \lim_{n\to 0}\frac{1/n}{\log n}$ $\quad\left[\text{form}\frac{\infty}{\infty}\right]$

$$= \lim_{n\to 0}\frac{-1/n^2}{1/n} = \lim_{n\to 0} -\frac{1}{n}$$

$\log_e A = -\infty \qquad A = e^{-\infty} = 0$

6. $\lim_{x\to\infty} x\sin\frac{1}{x}$ = Let $\frac{1}{x} = n$

So, $\quad x\to\infty,\ n\to 0$

$$\lim_{n\to 0}\frac{\sin n}{n} = 1$$

7. $\lim_{n\to\infty}\left[\frac{1}{n^2} + \frac{2}{n^2} + \frac{3}{n^2} + \ldots + \frac{1}{n}\right]$

$$= \lim_{n\to\infty}\left[\frac{1+2+3+\ldots+n}{n^2}\right] = \lim_{n\to\infty}\frac{n(n+1)/2}{n^2}$$

$$= \lim_{n\to\infty}\frac{1}{2}\left(1 + \frac{1}{n}\right) = \frac{1}{2}(1+0) = \frac{1}{2}$$

8. Given function is

$$f(x) = x(x-2) = x^2 - 2x$$

$f(x)$ is continuous function in $\left[0, \frac{3}{2}\right]$ and differentiable in $\left(0, \frac{3}{2}\right)$. Therefore, mean value thorem is applicable in $\left[0, \frac{3}{2}\right]$

$$f\left(\frac{3}{2}\right) = \left(\frac{3}{2}\right)^2 - 2\left(\frac{3}{2}\right) = \frac{-3}{4}$$

$$f(0) = 0$$

By Lagrange's mean value theorem, there exist $c \in \left(0, \frac{3}{2}\right)$ such that

$$\therefore \quad f'(c) = \frac{f\left(\frac{3}{2}\right) - f(0)}{\frac{3}{2} - 0}$$

$$\Rightarrow \quad 2c - 2 = \frac{-\frac{3}{4} - 0}{\frac{3}{2}} \text{ or } -\frac{1}{2}$$

$$\Rightarrow \quad c = +\frac{3}{4} \in \left(0, \frac{3}{2}\right)$$

9. Value of a mod function cannot be less than zero, therefore

$$f(x) = |x^2 - 5x + 2| \text{ is zero.}$$

10. Let $\quad f(x) = \frac{1}{\sqrt{2}}(\sin x - \cos x)$

$$= \sin x \cos \frac{\pi}{4} - \cos x \sin \frac{\pi}{4}$$

$$= \sin\left(x - \frac{\pi}{4}\right)$$

Since maximum value of sin θ is 1, hence maximum value of $f(x) = 1$

11. Let $\quad f(x) = -x^3 + 6x^2 + 2x + 1$

Slope of the function is,

$$f'(x) = -3x^2 + 12x + 2$$

$$= F(x) \text{ (say)}$$

Now to find minimum or maximum of F(x).

$$\therefore \quad F'(x) = -6x + 12$$

and $\quad F''(x) = -6$

For maxima and minima

$$F'(x) = 0$$

$$\Rightarrow \quad x = 2$$

$$\therefore \quad F''(2) = -6 < 0$$

Hence F(x) (slope) is maximum at $x = 2$

∴ Maximum slope = F(2)

$$= -3(2)^2 + 12(2) + 2 = 14$$

12. Volume is maximum at $x = 10$

At $x = 10$, $\quad y = \frac{300 - 100}{40}$

$$= 5$$

∴ Maximum volume $= x^2y = (10)^2 \times 5 = 500$

13. $f(x, y) = x^2 + y^2$

$$\nabla^2 f = \vec{\nabla}.\vec{\nabla} f$$

$$= \vec{z}\left(\vec{i}\frac{\partial}{\partial x} + \vec{j}\frac{\partial}{\partial y}\right)\left(\vec{i}\frac{\partial}{\partial x} + \vec{j}\frac{\partial}{\partial y}\right)\left[x^2 + y^2\right]$$

$$= \frac{\partial^2}{\partial x^2}(x^2) + \frac{\partial^2}{\partial y^2}(y^2)$$

$$= 4$$

14. $\quad f(x, y, z) = x^2 + y^2 + z^2$

$$\Delta f(x, y, z) = 2x\vec{i} + 2y\vec{j} + 2z\vec{k} = (\vec{A}) \text{ say}$$

Now, $\quad \vec{B} = \vec{i} - \vec{k}\,; \hat{B} = \frac{1}{\sqrt{2}}(\vec{i} - \vec{k})$

$$\therefore \quad \vec{A}.\hat{B} = \frac{2}{\sqrt{2}}(x - z)$$

$$\Rightarrow \quad \vec{A}.\hat{B}\Big|_{(1,1,1)} = \frac{2}{\sqrt{2}}(1-1) = 0$$

16. $\int_0^{\pi/2}\int_0^{\pi/2} \sin(x+y)\,dx.dy = \int_0^{\pi/2} [\cos(x+y)_0^{\pi/2}]\,dy$

$$= \left[\sin y - \sin\left(\frac{\pi}{2} + y\right)\right]_0^{\pi/2}$$

$$= (1 - 0) - (0 - 1) = 2$$

3 CHAPTER Ordinary and Partial Differential Equations

An equation which contains independent variable, dependent variables and their derivatives with respect to independent variables, is called *differential equation.*

I. ORDINARY DIFFERENTIAL EQUATION

A differential equation which involves only one independent variables is called *ordinary differential equation.*

e.g. $$\frac{dy}{dx} + xy = 0$$

and $$x\left(\frac{dy}{dx}\right)^2 - y\frac{dy}{dx} + 1 = 0$$

Order of Differential equation

It is the order of the highest derivative involving in the equation.

Degree of Differential equation

It is the degree of the highest order derivative involving in the equation, when the equation is free from radicals and fractional powers.

e.g. $$\frac{dy}{dx} + xy = 0$$

is equation of first order and first degree.

$$\frac{d^2y}{dx^2} + x\frac{dy}{dx} = 2$$

is equation of second order and first degree.

$$\frac{d^2y}{dx^2} = \left[1 + \left(\frac{dy}{dx}\right)\right]^{1/3}$$

is equation of second order and third degree.

Solution of a Differential equation

It is any relation between variables involved which satisfies the differential equation.

General or Complete Solution of a Differential equation

Solution of a differential equation in which number of arbitrary constant is equal to order of the differential equation is called *general solution.*

Variable Separables

$$f_1(x)\,dx = f_2(y)\,dy$$

It's solution is

$$\int f_1(x) = \int f_2(y)\,dy + c$$

Homogeneous equation

$$\frac{dy}{dx} = \frac{\phi_1(x, y)}{\phi_2(x, y)}$$

where, ϕ_1, ϕ_2 are of same degree

$$\frac{dy}{dx} = f\left(\frac{y}{x}\right)$$

Put $\frac{y}{x} = v$; $\frac{dy}{dx} = v + x\frac{dv}{dx}$

$$\therefore \quad \frac{dv}{f(v) - v} = \frac{dx}{x}, \quad \Rightarrow \int \frac{dv}{f(v) - v} = \log x + c$$

Equation Reducible to Homogeneous Equation

$$\frac{dy}{dx} = \frac{ax + by + c}{a'x + b'y + c'}$$

$$a : b \neq a' : b'$$

Putting $x = x + h,\ y = y + k$

$$\frac{dy}{dx} = \frac{ax + by + ah + bk + c}{a'x + b'y + a'h + b'k + c'}$$

Choosing h, k such that

$$ah + bk + c = a'h + b'k + c'$$

$$\frac{dy}{dx} = \frac{ax + by}{a'x + b'y} = \frac{y}{x} = v$$

Special Case

When $a : b = a' : b'$

then $ax + by = v$

Linear Differential Equation

$$\frac{dy}{dx} + py = Q$$

where P and Q are functions of x only.

Integrating factor I.F. $= e^{\int p dx}$

$$y.\, e^{\int p dx} = \int Q.\, e^{\int p dx} + C$$

Equations Reducible to Linear Form

$$\frac{dy}{dx} + Py = Qy^n$$

Extended form of linear equation are Bernoulli's equation

$$\frac{1}{y^n}\frac{dy}{dx} + \frac{P}{y^{n-1}} = Q$$

Put $y^{1-n} = t$

$$\Rightarrow \quad dt = (1-n)\, y^{-n} dy \quad \Rightarrow \quad \frac{dy}{y^n} = \frac{1}{(1-n)} dt$$

$$\therefore \quad \frac{dt}{dx} + P(1-n)t = Q(1-n)$$

$$\frac{dy}{dx} + y\frac{d\phi}{dx} = \phi(x)\frac{d\phi}{dx}$$

$$\Rightarrow \quad \frac{\dfrac{dy}{dx} + y\dfrac{d\phi}{dx} = \phi(x)\dfrac{d\phi}{dx}}{\left(\dfrac{d\phi}{dx}\right)}$$

$$\Rightarrow \quad \frac{dy}{d\phi} + y = \phi(x)$$

$$ye^x = \int \phi(x)e^x dx + C$$

Exact Differential Equation

$$Mdx + Ndy = 0$$

$$\frac{\partial M}{\partial Y} = \frac{\partial N}{\partial X}$$

(*a*) Integrate M with respect to x regarding Y as a constant.

(*b*) Integrate N with respect to Y and then retain only that term which have not been already obtained by integrating M, *i.e.* in (*a*) above.

(*c*) Add the two expression obtained in (*a*) and (*b*) above and equate the result to an arbitrary constant which gives the required solution.

Differential Equations of First Order and First Degree

Type I : Variable Separation Form

The equation of this type can be put in the form,

$$f(x)\, dx + g(y) dy = 0$$

Now, $\int f(x)\, dx + \int g(y)\, dy = C$

Example. Solve $(e^y + 1)\, dx + xe^y\, dy = 0$

Solution

Given equation can be written as

$$e^y\, dx + xe^y\, dy + dx = 0 \quad \Rightarrow \quad d\left(xe^y\right) + dx = 0$$

Integrating $\int d\left(xe^y\right) + \int dx = \int 0 \Rightarrow xe^y + x = C$

Type II : Linear Differential Equations

A linear equations is called linear if the dependent variable and its derivative occuring in it are of the first degree and are not multiplied together.

Linear differential equation is of the form

$$\frac{dy}{dx} + Py = Q,$$

where P and Q are functions of x only.

Solution of the linear differential equations is

$$y \, . \, e^{\int pdx} = \int \left(Q \, . \, e^{\int pdx} \right) dx + C$$

where, $e^{\int pdx}$ is called I.F

Example. Solve $(x + 2y^3)\, dy = ydx$

Solution

Given equation can be written as

$$y\frac{dx}{dy} = x + 2y^3 \quad \Rightarrow \quad \frac{dx}{dy} - \frac{1}{y}x = 2y^2$$

Here, $P = -\frac{1}{y}$ and $Q = 2y^2$

$$\therefore \quad \text{I.F.} = e^{\int -\frac{1}{y}dy} = e^{-\log y} = e^{\log (1/y)} = \frac{1}{y}$$

Solution of the equation is,

$$x(\text{IF}) = \int (Q \, . \, \text{IF} \, . \, dy)$$

$$\therefore \quad x\left(\frac{1}{y}\right) = \int 2y^2 \frac{1}{y} dy$$

$$\Rightarrow \quad \frac{x}{y} = y^2 + C \quad \Rightarrow \quad x = y\,(C + y^2)$$

Type III : Homogeneous Differential Equations

This equation is of the form, $\frac{dy}{dx} = \frac{f(x, y)}{\phi(x, y)}$

where $f(x, y)$ and $\phi(x, y)$ are homogeneous functions of same degree.

This can be solved by putting $y = vx$

$$\therefore \quad \frac{dy}{dx} = v + x\frac{dy}{dx}$$

Substitute the value of y and then solve by variable-separable method.

Example. Solve $x^2dy + y\,(x + y)\, dx = 0$

Solution

Here $$\frac{dy}{dx} = -\frac{y(x + y)}{x^2}$$

Let $y = vx,$

$\Rightarrow \quad \dfrac{dy}{dx} = v + x\dfrac{dv}{dx}$

Now $\quad v + x\dfrac{dv}{dx} = -\dfrac{x(vx) + v^2x^2}{x^2} = -(v^2 + v)$

$\Rightarrow \quad x\dfrac{dv}{dx} = -(v^2 + 2v)$

Integrating $\quad \displaystyle\int \frac{dv}{v^2 + 2v} = -\int \frac{dx}{x}$

$\Rightarrow \quad \displaystyle\int \frac{dv}{v(v+2)} = -\int \frac{dx}{x}$

$\Rightarrow \quad \displaystyle\int \frac{1}{2}\left(\frac{1}{v} - \frac{1}{v+2}\right) dv = -\int \frac{dx}{x}$

$\Rightarrow \quad \dfrac{1}{2}\ln v - \dfrac{1}{2}\ln(v + 2) = -\ln x + C$

$\Rightarrow \quad \dfrac{1}{2}\log \dfrac{v}{v+2} = -\log x + C$

$\Rightarrow \quad \dfrac{1}{2}\log \dfrac{(y/x)}{(y/x)+2} = -\log x + C$

$\Rightarrow \quad \log \dfrac{y}{2x + y} = \ln \dfrac{1}{x^2} + C$

$\Rightarrow \quad \dfrac{y}{2x + y} = \dfrac{C'}{x^2}$

$\Rightarrow \quad x^2y = K'(2x + y)$

Type IV : Exact Differential Equations

A differential equation is said to be exact if it can be derived from its primitive by direct differentiation without any further transformation such as elimination, etc.

e.g. $\quad (x^2 - ay)\,dx + (y^2 - ax)\,dy = 0$

is exact in as much as it can be derived from its primitive

$x^3 - 3ayx + y^3 = C$ by direct differentiation.

Theorem : The necessary and sufficient condition for the differential equation,

$$M + N\frac{dy}{dx} = 0$$

$$\Rightarrow \quad Mdx + Ndy = 0$$

to be exact is that $\quad \dfrac{\partial M}{\partial y} = \dfrac{\partial N}{\partial x}$

If equation $Mdx + Ndy = 0$ is exact, then it can be integrated as follows :

(*i*) Firstly integrate M with respect to x regarding y as a constant.

(*ii*) Then integrate with respect to y those of the terms in N which do not involve x.

(*iii*) Sum of the two expressions thus obtained equated to a constant is the required solution.

i.e. $\quad \displaystyle\int Mdx + \int Ndy = C$

$\therefore$ y = constant terms contain $x = 0$

Example. Show that

$$(x^4 - 2xy^2 + y^4)\,dx - (2x^2y - 4xy^3 + \sin y)\,dy = 0$$

is exact and also solve it.

Solution

Here, $M = x^4 - 2xy^2 + y^4$

and $\quad N = -(2x^2y - 4xy^3 + \sin y)$

$\therefore \quad \dfrac{\partial M}{\partial y} = -4xy + 4y^3$

and $\quad \dfrac{\partial N}{\partial x} = -4xy + 4y^3 \Rightarrow \dfrac{\partial M}{\partial y} = \dfrac{\partial N}{\partial x}.$

Thus the given equation is exact.

Now, $\quad \displaystyle\int \left(x^4 - 2xy^2 + y^4\right) dx - \int \left(2x^2y - 4xy^3 + \sin y\right) dy = C$

(since terms contains $x = 0$)

$\Rightarrow \quad \dfrac{1}{5}x^5 - x^2y^2 + xy^4 + \cos y = C$

Integrating Factors (I.F.)

Sometimes an equation which is not exact may become so on multiplication by some suitable function known as an *integrating factor.*

Consider an equation $xdy - ydx = 0$, which is not exact

It will become exact on multiplication by $\dfrac{1}{y^2}$. Then

$$\frac{x}{y^2}dy - \frac{1}{y}dx = 0$$

Hence solution is, $\quad -\displaystyle\int \frac{1}{y}dx + \int \frac{x}{y^2}dy = C$

$\Rightarrow \quad -\dfrac{x}{y} + 0 = C \quad \Rightarrow \quad \dfrac{x}{y} = C$

Rule 1 : If equation $Mdx + Ndy = 0$ is homogeneous and $Mx + Ny \neq 0$, then

$$\text{I.F.} = \frac{1}{Mx + Ny}$$

Rule 2 : If equation is of the form

$$f_1(x, y)\,ydx + f_2(x, y)\,xdy = 0$$

and $\quad Mx - Ny \neq 0$, then

$$\text{I.F.} = \frac{1}{Mx + Ny}$$

Rule 3 : If equation $Mdx + Ndy = 0$

and $\frac{1}{N}\left(\frac{\partial M}{\partial y} - \frac{\partial N}{\partial x}\right) = f(x)$, is a function of x, then

$$\text{I.F.} = e^{\int f(x)dx}$$

Rule 4 : If equation $Mdx + Ndy = 0$ and

$$\frac{1}{M}\left(\frac{\partial N}{\partial x} - \frac{\partial M}{\partial y}\right) = f(y), \text{ then}$$

$$\text{I.F.} = e^{\int f(y)dy}$$

Integrating Factors

- $d\left(\frac{y}{x}\right) = \frac{xdy - ydx}{x^2}$
- $d\left(\frac{x}{y}\right) = \frac{ydx - xdy}{y^2}$
- $d(xy) = xdy + ydx$
- $d\left(\frac{x^2}{y}\right) = \frac{2yxdx - x^2dy}{y^2}$
- $d\left(\frac{y^2}{x^2}\right) = \frac{2.y.x^2dy - 2xy^2dx}{x^4}$
- $d\left(\tan^{-1}\frac{x}{y}\right) = \frac{ydx - xdy}{x^2 + y^2}$
- $d\left(\tan^{-1}\left(\frac{y}{x}\right)\right) = \frac{xdy - ydx}{x^2 + y^2}$
- $d\left(\frac{1}{2}\log(x^2 + y^2)\right) = \frac{xdx + ydy}{x^2 + y^2}$
- $d\left(-\frac{1}{xy}\right) = \frac{xdy + ydx}{x^2y^2}$
- $d\log\left(\frac{x}{y}\right) = \frac{ydx - xdy}{xy}$
- $d\log\left(\frac{y}{x}\right) = \frac{xdy - ydx}{xy}$
- $d\left(\frac{e^x}{y}\right) = \frac{ye^x dx - e^x dy}{y^2}$

Example. Solve $(x^2 + y^2 + 2x)\, dx + 2ydy = 0$.

Solution

Here, $M = x^2 + y^2 + 2x$

$$\Rightarrow \quad \frac{\partial M}{\partial y} = 2y \text{ and, } N = 2y; \ \frac{\partial N}{\partial x} = 0$$

$$\therefore \quad \frac{1}{N}\left(\frac{\partial M}{\partial y} - \frac{\partial N}{\partial x}\right) = 1 \text{ which is function of } x$$

Now $\quad \text{I.F.} = e^{\int 1.dx} = e^x$

Multiplying by I.F, we get

$$e^x (x^2 + y^2 + 2x)\, dx + 2ye^x\, dy = 0$$

which is exact equation.

Hence the solution is,

$$\int_{y = \text{const}} e^x\left(x^2 + y^2 + 2x\right) dx + \int_{f(x)} 2ye^x dy = C$$

$$\Rightarrow \quad e^x (x^2 + y^2) = C$$

Example. Solve $\cos^2 x\left(\frac{dy}{dx}\right) + y = \tan x$ ...(*i*)

Solution

Given equation can be written as

$$\frac{dy}{dx} + \sec^2 x y = \sec^2 x \tan x$$

It is linear equation in y.

Here $P = \sec^2 x$

$$\therefore \quad \text{I.F.} = e^{\int pdx} = e^{\int \sec^2 xdx} = e^{\tan x}$$

Multiplying equation (*ii*) by I.F., we get

$$\left(\frac{dy}{dx} + y\sec^2 x\right)e^{\tan x} = \tan x \sec^2 x\, e^{\tan x}$$

Integrating, we get

$$y.\, e^{\tan x} = C + \int \tan x \sec^2 x.\, e^{\tan x}\, dx$$

where C is an arbitrary constant

Putting $t = \tan x$

$$dt = \sec^2 x dt$$

$$\therefore \quad y\, e^{\tan x} = C + \int t\, e^t\, dt$$

Integration by parts

$y\, e^{\tan x} = C + (te^t - e^t) \quad \Rightarrow y\, e^{\tan x} = C + e^{\tan x}(\tan x - 1)$

Clairaut's Equation

Equation of the form $y = Px + f(P)$ is called *Clairaut's equation.*

Here $\quad P = \frac{dy}{dx}$

Solution of the equation is,

$$y = cx + f(c)$$

Example. Solve $P = \tan (Px - y)$

Solution

$Px - y = \tan^{-1} P \quad \Rightarrow \quad y = Px - \tan^{-1} P$

Hence general solution of the equation is

$$y = cx - \tan^{-1} C$$

Singular Solution

Let $f(x, y, p) = 0$ be a given differential equation and its general solution is $\phi(x, y, c) = 0$.

Envelope of the family of curves $\phi(x, y, c) = 0$ is contained in the locus obtained by eliminating c between $\phi(x, y, c) = 0$ and $\frac{\partial}{\partial c}\phi(x, y, c) = 0$.

Example. Find singular solution of the differential equation,

$$P^3 + Px - y = 0$$

Solution

Given equation is, $y = Px + P^3$

Hence, general solution is,

$$y = cx + c^3$$

Let $\quad \phi(x, y, c) = cx + c^3 - y = 0 \qquad ...(i)$

$\therefore \quad \frac{\partial}{\partial c}\phi(x, y, c) = x + 3c^2 = 0 \qquad ...(ii)$

Eliminating c between equations (i) and (ii), we get

$$y^2 = c^2(x + c^2)^2$$

$$\Rightarrow \quad y^2 = -\frac{x}{3}\left(x - \frac{x}{3}\right)^2 \quad \Rightarrow \quad 27y^2 + 4x^3 = 0$$

Linear Differential Equation of nth Order (Higher Order)

Linear equations are the equations in which the dependent variable and its derivatives appear only in the first degree and are not multiplied together.

$$\frac{d^n y}{dx^n} + P_1\frac{d^{n-1}}{dx^{n-1}} + P_2\frac{d^{n-2}}{dx^{n-2}} + ... + P_n y = X$$

where $P_1, P_2, ... P_n$ are functions of x only.

The operator $\frac{d}{dx}$ is denoted by D.

$\therefore \quad D^n y + P_1 D^{n-1} y + ... + P_n y = X$

$\Rightarrow \quad f(D)y = X$

where, $\quad f(D) = D^n + P_1 D^{n-1} + ... + P_n = 0$

Case I : *When A.E.has Distinct and real roots, i.e.*

$$m_1 \neq m_2 \neq m_3 ....$$

$$y = c_1 e^{m_1 x} + c_2 e^{m_2 x} + c_3 e^{m_3 x} + ...$$

Case II : *When A.E. has real and some equal roots.*

If $m_1 = m_2 \neq m_3 \neq m_4 .....$

then $\quad y = (c_1 + c_2 x)\, e^{m1x} + c_3 e^{m3x} + ...,$

If $m_1 = m_2 = m_3 \neq m_4 \neq m_5 .....$

then $\quad y = (c_1 + c_2 x + c_3 x^2)\, e^{m1x} + c_4 e^{m4x} + ...$

Case III : *When A. E. has complex roots,*

i.e. $\quad m_1, m_2 = \alpha \pm i\beta, m_3, m_4 ......$

$$y = (A\cos\beta x + B\sin\beta x)\, e^{\alpha x} + C_3 e^{m3x} + C_4 e^{m4x} + ...$$

If complex roots are equal, then

$$y = [(A_1 + A_2 x)\cos\beta x + (B_1 + B_2 x)\sin\beta x]\, e^{\alpha x} + c_5 e^{m5x} + ...$$

Example. Solve $\frac{d^2y}{dx^2} - 5\frac{dy}{dx} + 6y = 0$

Solution

A.E. is, $\quad D^2 - 5D + 6 = 0$

$\therefore \quad D = 3, 2$

Hence solution is, $\quad y = c_1 e^{3x} + c_2 e^{2x}$

PARTICULAR INTEGRAL

When the equation is, $f(D)\, y = X$

then general solution is, $\quad y = (C. F.) + (P. I.)$

A particular integral is given by $\frac{1}{f(D)}X$

I. When $X = e^{ax}$, then

$$P. I. = \frac{1}{f(D)}e^{ax} = \frac{1}{f(a)}e^{ax}, \text{ if } f(a) \neq 0$$

II. When $f(a) = 0$, then

$$P.I. = \frac{1}{f(D)}e^{ax}$$

$$= \frac{x}{f'(a)}e^{ax}, \text{ if } f'(a) \neq 0$$

III. If $f'(a) = 0$, then

$$P.I. = \frac{1}{f(D)}e^{ax} = \frac{x^2}{f''(a)}e^{ax} \text{ if } f'(a) \neq 0$$

Example. Find P.I. when $(D^2 + 4)\, y = \sin 2x + \cos 3x$

Solution

$$P. I. = \frac{1}{D^2 + 4}\sin 2x + \frac{1}{D^2 + 4}\cos 3x$$

$$= \frac{x}{2D}\sin 2x + \frac{1}{-3^2 + 4}\cos 3x = \frac{x}{2D^2}D\sin 2x - \frac{1}{5}\cos 3x$$

$$= \frac{2x}{-8}\cos 2x - \frac{1}{5}\cos 3x = -\frac{x}{4}\cos 2x - \frac{1}{5}\cos 3x$$

Example. Find general solution of the equation

$$(D^2 + D - 2)\, y = e^x$$

Solution

A.E. is, $\quad D^2 + D - 2 = 0$

$\Rightarrow \quad D = -1, 2$

$$C.F. = C_1 e^{-x} + C_2 e^{2x}$$

Now $\quad P. I. = \frac{1}{D^2 + D - 2}e^x = x\frac{1}{2D + 1}e^x$

Hence general solution is,

$$y = c_1e^{-x} + c_2e^{2x} + \frac{x}{3}e^x$$

IV. $\text{P.I.} = \frac{1}{F(D^2)}.\sin ax$

$$= \frac{1}{F(-a^2)}\sin ax, \quad \text{if } F(-a^2) \neq 0$$

V. $\text{P.I.} = \frac{1}{F(D^2)}\cos ax$

$$= \frac{1}{F(-a^2)}\cos ax \quad \text{if} \quad F(-a^2) \neq 0$$

VI. $\text{P.I.} = \frac{1}{F(D)}x^m$

Expansion is to be carried upto the term D^m, since $D^{m+1}x^m = 0, D^{m+2}x^m = 0,$

VII. When $X = e^{ax}v(x)$, then

$$\text{P.I.} = \frac{1}{f(D)}e^{ax}v(x) = e^{ax}\frac{1}{f(D+a)}v(x)$$

Example. Solve $(D^2 - 4D + 1)y = e^{2x}\sin 2x$

Solution

A.E. is $m^2 - 4m + 1 = 0$

$\therefore$ $m = 2 \pm \sqrt{3}$

$\therefore$ $\text{C.F.} = c_1e^{(2+\sqrt{3})x} + c_2e^{(2-\sqrt{3})x}$

$$= e^{2x}\left[c_1e^{\sqrt{3}x} + c_2e^{-\sqrt{3}x}\right]$$

Now, $\text{P.I.} = \frac{1}{D^2 - 4D + 1}e^{2x}\sin 2x$

$$= e^{2x}\frac{1}{(D+2)^2 - 4(D+2) + 1}\sin 2x$$

$$= e^{2x}\frac{1}{D^2 + 4D + 4 - 4D - 8 + 1}\sin 2x$$

$$= e^{2x}\frac{1}{D^2 - 3}\sin 2x = e^{2x}\frac{1}{-4-3}\sin 2x$$

$$= \frac{-1}{7}e^{2x}\sin 2x$$

Hence general solution is

$$y = \left(c_1e^{\sqrt{3}x} + c_2e^{-\sqrt{3}x}\right)e^{2x} - \frac{1}{7}e^{2x}\sin 2x$$

VIII. When $X = x\,v(x)$, then

$$\text{P.I.} = \frac{1}{f(D)}xv(x) = \left[x - \frac{f'(D)}{f(D)}\right]\frac{1}{f(D)}v(x)$$

Example. Find PI of $(D^2 - 1)y = x\sin x$

Solution

$$\text{P.I.} = \frac{1}{D^2 - 1}x\sin x$$

$$= \left[x - \frac{2D}{D^2 - 1}\right]\frac{1}{D^2 - 1}\sin x$$

$$= \left(x - \frac{2D}{D^2 - 1}\right)\frac{\sin x}{-2}$$

$$= -\frac{1}{2}\left[x\sin x - \frac{2D\sin x}{-2}\right]$$

$$= -\frac{1}{2}[x\sin x + \cos x]$$

IX. $Q = e^{ax}$ and $F(a) = 0$

$$\text{P.I.} = \frac{1}{\phi(a)}e^{ax}.\frac{x^n}{n} = \frac{e^{ax}}{(D+a)}$$

X. To find P.I., when Q is of the form xV, where V is any function of x.

By Leibnitz's theorem,

$$\frac{1}{F(D)}(x.V) = x\frac{1}{F(D)}V - \frac{F'(D)}{\{F(D)\}^2}$$

XI. To find P.I. when Q is of the form x^m. V where V is any function of x.

(*i*) If $V = x^n$, then $x^m.\,x^n = x^{m+n}$

(*ii*) If $V = e^{ax}$, then $x^{mv} = x^me^{ax}$ [evaluate as (X)]

(*iii*) If $V = \cos ax$, then $x^{mv} = x^m\cos ax$

$$\text{P.I.} = \frac{1}{F(D)}x^m\cos ax = \frac{1}{F(D)}(\text{Real part of } x^me^{iax})$$

$$= \text{Real part of } \frac{1}{F(D)}x^me^{iax}$$

If $V = \sin ax$, then

$$\text{P.I.} = \text{I.P. of } \frac{1}{F(D)}x^me^{iax}$$

and calculate as above.

XII. $\frac{1}{D-a}Q = e^{ax}\int Qe^{-ax}dx$

XIII. $\frac{1}{(D^2 - a^2)} = \text{cosec } ax\frac{1}{2ia}\left[\left(\frac{1}{D-ia}\right) - \left(\frac{1}{D+ia}\right)\right]$

Method of Variation of Parameters

Finding Particular Integral

Consider linear differential equation

$$a\frac{d^2y}{dx^2}+b\frac{dy}{dx}+cy=X$$

Complementary function = $C_1y_1 + C_2y_2$

and Particular integral = $uy_1 + vy_2$

where $u=\int\frac{-y_2X}{y_1y_2'-y_1^1y_2'}dx$

and $v=\int\frac{y_1X}{y_1y_2'-y_1'y_2}dx$

Complete solution = Complementary function + Particular integral.

Method of Undetermined Coefficients to Find Particular Integral

Consider a linear differential equation

$$D^ny+p_1D^{n-1}y+\ldots+p_ny=X$$

For particular integral, assume a trial solution containing unknown constants which are determined by substitution in the given equation. The trial solution to be assumed in each case, depends on the form of X. Thus

(1) When $X = 2e^{3x}$

Trial solution = ae^{3x}

(2) When $X = 3\sin 2x$

Trial solution = $a_1\sin 2x + a_2\cos 2x$

(3) When $X = 2x^3$

Trial solution = $a_1x^3 + a_2x^2 + a_3x + a_4$

However when X = tan x or sec x, this method fails, since number of terms obtained by differentiating X = tan x or sec x is infinite.

This method holds as long as no term in the trial solution appears in the C.F.

If any term of the trial solution appears in the C.F., multiply this trial solution by the lowest positive integral power of x which is large enough so that none of the terms which are then present, appear in the C.F.

Example : Solve $\left(\frac{d^4y}{dx^4}+m^4\right)y=0$

Solution

Given equation can be written as

$(D^4 + m^4)y = 0$

here, $D=\frac{d}{dx}$

Its auxiliary equation is

$$M^4+m^4=0$$

$$\Rightarrow (M^2+m^2)^2-2M^2m^2=0$$

$$\Rightarrow (M^2+m^2)^2-(\sqrt{2}Mm)^2=0$$

$$\Rightarrow (M^2+m^2+\sqrt{2}Mm)(M^2+m^2)-\sqrt{2}Mm)=0$$

$$M=\frac{-\sqrt{2}m\pm\sqrt{2m^2-4m^2}}{2};\frac{\sqrt{2}m\pm\sqrt{2m^2-4m^2}}{2}$$

$$=\frac{-\sqrt{2}m\pm\sqrt{2}mi}{2};\frac{\sqrt{2}m\pm\sqrt{2}mi}{2}$$

$$=\frac{-m}{\sqrt{2}}\pm\frac{m}{\sqrt{2}}i;\frac{m}{\sqrt{2}}\pm\frac{m}{\sqrt{2}}i$$

Hence required solution,

$$e^{-mx/\sqrt{2}}\{C_1\cos(mx/\sqrt{2})+C_2\sin(mx/\sqrt{2})\}+e^{mx/\sqrt{2}}\{C_3\cos(mx/\sqrt{2})+C_4\sin(mx/\sqrt{2})\}$$

where C_1, C_2, C_3, C_4 are arbitrary constants.

Example : Solve $(D^2 - 2D + 5)y = e^{-x}$

Solution

Given equation is $(D^2 - 2D + 5)y = e^{-x}$

Auxiliary equation is $m^2 - 2m + 5 = 0$,

where roots are $m = -1 \pm 2i$

Complementary Function,

$$C.F. = e^{-x}(C_1\cos 2x + C_2\sin 2x)$$

where C_1 and C_2 are arbitrary constants.

and $$P.I.=\frac{1}{D^2-2D+5}e^{-x}=\frac{e^{-x_1}}{(-1)^2-2(-1)+5}$$

$$=\frac{e^{-x}}{1+2+5}=\frac{e^{-x}}{8}$$

Hence required solution is

$$y = C.F. + P.I$$

$$=e^{-x}(C_1\cos 2x+C_2\sin 2x)+\frac{e^{-x}}{8}$$

First Order Equations (Linear and Non-Linear)

Order of a differential equation means order of the highest differential coefficient which appears in it.

Linear Differential Equations

$$\frac{dy}{dx}+Py=Q$$

where P and Q are constant or functions of x alone.

This equation is called *linear differential* of the first order in y.

It's integrating factor is, I.F. $= e^{\int pdx}$

Multiplying both sides by I.F., we get

$$\left(\frac{dy}{dx} + Py\right)e^{\int pdx} = Q\, e^{\int pdx} \qquad ...(i)$$

Now, $\frac{d}{dx}(ye^{\int pdx}) = y.\frac{d}{dx}\left[e^{\int pdx}\right] + e^{\int pdx}.\frac{dy}{dx}$

$$= y.e^{\int pdx}.\, P + e^{\int pdx}.\frac{dy}{dx}$$

$$\frac{d}{dx}(ye^{\int pdx}) = \left(Py + \frac{dy}{dx}\right)e^{\int pdx}$$

From equation (i), $\frac{d}{dx}\left[ye^{\int pdx}\right] = Q\, e^{\int pdx}$

Integrating both sides with respect to x, we get

$$y\,.e^{\int pdx} = C + \int Q\,.e^{\int pdx}dx$$

where, C is a integration constant.

This equation is complete solution.

Cauchy's Homogeneous Linear Equation

$$x^n \frac{d^n y}{dx^n} + a_1 x^{n-1}\frac{d^{n-1}y}{dx^{n-1}} + a_2 x^{n-2}\frac{d^{n-2}y}{dx^{n-2}} + ... + a_n y = X(x)$$

where $a_1, a_2, ..., a_n$ are constants and X is a function of x. This can be reduced to linear differential equation with constant co-efficients by the substitution

$$x = e^t \quad \Rightarrow \quad \log x = t$$

which can be solved by above methods.

Legendre's Homogeneous Linear Equation

$$(ax+b)^n \frac{d^n y}{dx^n} + a_1(ax+b)^{n-1}\frac{d^{n-1}y}{dx^{n-1}}$$

$$+ a_2(ax+b)^{n-2}\frac{d^{n-2}y}{dx^{n-2}} + ... + a_n y = X$$

where $a_1, a_2, ... a_n$ are constants and X is a function of x. If we substitute $ax+b=t$, then above equation will reduce to the Cauchy's form and then a further substitution $t = e^z$ reduces it to the linear differential equation with constant co-efficients. A single substitution $ax+b=e^z$ will reduce the above equation to the linear differential equation with constant coefficients.

Example. Solve by the method of variation of parameters : $\frac{d^2y}{dx^2} + y = \text{cosec } x.$

Solution

Its A.E. is $D^2 + 1 = 0 \Rightarrow D = \pm i$

$\therefore$ C.F. $= C_1 \cos x + C_2 \sin x = C_1F_1 + C_2F_2$ (say)

where, $F_1 = \cos x$ and $F_2 = \sin x$

Now P.I. $= PF_1 + QF_2$

where, $P = -\int \frac{F_2X}{F_1F_2' - F_2F_1'}dx$

and, $Q = \int \frac{F_1X}{F_1F_2' - F_2F_1'}dx$

$F_1 = \cos x \Rightarrow F'_1 = -\sin x$

and $F_2 = \sin x \Rightarrow F'_2 = \cos x$

$\therefore$ $F_1F'_2 - F_2F'_1 = \cos^2 x - (\sin x)(-\sin x)$

$$= \cos^2 x + \sin^2 x = 1$$

$\therefore$ $P = -\int \frac{\sin x \text{ cosec } x}{1}dx = -x$

Similarly, $Q = \int \frac{F_1X}{F_1F_1' - F_2F_1'}dx$

$$= -\int \frac{\cos x}{1}\text{cosec } x\, dx = \int \frac{\cos x}{\sin x}dx$$

$$= \int \cot x\, dx = \log \sin x.$$

$\therefore$ P.I. $= PF_1 + QF_2$

$$= -x\cos x + \log(\sin x).\sin x$$

$$= -x\cos x + \sin x \log(\sin x)$$

Hence complete solution is,

$$y = C_1 \cos x + C_2 \sin x - x\cos x + \sin x.\log(\sin x).$$

Euler-Cauchy Equations

Homogeneous

If the ODE is of the form

$$ax^2y'' + bxy'' + cy = 0, \qquad ...(1)$$

where a, b and c are constants; then (1) is called homogeneous Euler-Cauchy equation. Two linearly independent solutions (i.e. basis) depend on the quadratic equation

$$am^2 + (b-a)m + c = 0. \qquad ...(2)$$

Equation (2) is called *characteristic equation* for (1). The ODE (1) is singular at $x = 0$. Hence, we solve (1) for $x \neq 0$. We consider the case when $x > 0$.

Theorem

(i) If the roots of (2) are real and distinct, say m_1 and m_2, then two linearly independent (LI) solutions of (1) are x^{m1} and x^{m2}. Thus, the general solution to (1) is

$$y = C_1x^{m1} + C_2x^{m2}.$$

(ii) If the roots of (2) are real and equal, say $m_1 = m_2 = m$, then two LI solutions of (1) are x^m and $x^m \ln x$. Thus, the general solution to (1) is

$$y = (C_1 + C_2 \ln x)x^m.$$

(iii) If the roots of (2) are complex conjugate, say $m_1 = \alpha + i\beta$ and $m_2 = \alpha - i\beta$, then two real LI solutions of (1) are $x^\alpha \cos(\beta \ln x)$ and $x^\alpha \sin(\beta \ln x)$. Thus, the general solution to (1) is

$$y = x^\alpha (C_1 \cos(\beta \ln x) + C_2 \sin(\beta \ln x)).$$

Example. *Solve $x^2y'' - xy' - 3y = 0$*

Solution

The characteristic equation is $m^2 - 2m - 3 = 0$

$\Rightarrow m = -1, 3$. The general solution is $y = C_1/x + C_2x^3$

(ii) If $p(m) = 0$ has real equal roots $m_1 = m_2 = m$, then x^m is a solution of (1). To find the other solution, note that if m is repeated root, then $p(m) = p'(m) = 0$. This suggests differentiating (3) w.r.t. m. Since L consists of differentiation w.r.t. x only,

$$\frac{\partial}{\partial m}\left(L(x^m)\right) = L\left(\frac{\partial}{\partial m}x^m\right)$$

$$= L(x^m \ln x).$$

Now $\quad L(x^m \ln x) = (p'(m) + p(m)\ln x)x^m$

where $'$ represents the derivative. Since, m is a repeated root, the RHS is zero. Thus, $x^m \ln x$ is also a solution to (1) and it is independent of x^m. Hence, the general solution to (1) is

$$y = (C_1 + C_2 \ln x)x^m.$$

Example. *Solve $x^2y'' - 3xy' + 4y = 0$*

Solution

The characteristic equation is $m^2 - 4m + 4 = 0$

$\Rightarrow m = 2, 2.$

The general solution is $y = (C_1 + C_2 \ln x)x^2$.

(iii) If $p(m) = 0$ has complex conjugate roots, say $m_1 = \alpha + i\beta$ and $m_2 = \alpha - i\beta$, then two LI solutions are

$$Y_1 = x^{(\alpha+i\beta)} = x^\alpha e^{i\beta \ln x},$$

and

$$Y_2 = x^\alpha e^{-i\beta \ln x}.$$

But these are complex valued. Note that if Y_1, Y_2 are LI, then so are $y_1 = (Y_1 + Y_2)/2$ and $y_2 = (Y_1 - Y_2)/2i$. Hence, two real LI solutions of (1) are $y_1 = x^\alpha \cos(\beta \ln x)$ and $y_2 = x^\alpha \sin(\beta \ln x)$. Thus, the general solution to (1) is

$$y = x^\alpha(C_1\cos(\beta \ln x) + C_2\sin(\beta \ln x)).$$

Nonhomogeneous

If the ODE is of the form

$$ax^2y'' + bxy'' + cy = \tilde{r}(x), \qquad \text{...(4)}$$

where a, b and c are constants; then (4) is called nonhomogeneous Euler-Cauchy equation. We can use the method of variation of parameters as follows. First divide (4) by ax^2 so that the coefficient of y'' becomes unity:

$$y'' + \frac{b}{ax}y'' + \frac{c}{ax^2}y = r(x) \qquad \text{...(5)}$$

where $r(x) = \tilde{r}(x)/ax^2$. Now we already know two LI solutions y_1, y_2 of the homogeneous part. Hence, the particular solution to (4) is given by

$$y_p(x) = -y_1(x)\int\frac{y_2(x)r(x)}{W(y_1,y_2)}dx + y_2(x)\int\frac{y_1(x)r(x)}{W(y_1,y_2)}dx.$$

Thus, the general solution to (4) is

$$y(x) = C_1y_1(x) + C_2y_2(x) + y_p(x).$$

Example. *Consider*

$$x^2y'' - xy' - 3y = \frac{\ln x}{x}, \; x > 0.$$

The characteristic equation is $m^2 - 2m - 3 = 0$

$\Rightarrow m = -1, 3.$

Hence $y_1 = 1/x$ and $y_2 = x^3$. Hence,

$$y_p(x) = y_1(x)u(x) + y_2(x)v(x)$$

where

$$u(x) = -\int\frac{y_2(x)r(x)}{W(y_1,y_2)}dx$$

$$v(x) = \int\frac{y_1(x)r(x)}{W(y_1,y_2)}dx.$$

Now $W(y_1, y_2) = 4x$ and $\boxed{r(x) = \ln x/x^3}$! Hence,

$$u(x) = -\int\frac{\ln x}{4x}dx$$

$$= -\frac{(\ln x)^2}{8}$$

$$v(x) = \int\frac{\ln x}{4x^5}dx$$

$$= -\frac{\ln x}{16x^4} - \frac{1}{16x^4}$$

Hence,

$$y_p(x) = -\frac{(\ln x)^2}{8x} - \frac{\ln x}{16x} - \frac{1}{64x}$$

Hence the general solution is $y = c_1y_1 + c_2y_2 + y_p$, i.e.

$$y(x) = \frac{A}{x} + Bx^3 - \frac{(\ln x)^2}{8x} - \frac{\ln x}{16x}.$$

Note that last term of y_p is absorbed with y_1.

Aliter: Let us make the transformation $t = \ln x$. Then the given transformed to

$$\ddot{y} - 2\dot{y} - 3y = te^{-t},$$

where $\dot{} = d/dt$. This is the same problem we have solved in lecture 9 using method of undetermined coefficients.

The solution is (see lecture 9)

$$y(t) = C_1 e^{-t} + C_2 e^{3t} - \frac{te^{-1}}{16}(2t+1).$$

which in terms of original x variable becomes

$$y(x) = \frac{C_1}{x} + C_2 x^3 - \frac{\ln x}{16x}(2\ln x + 1),$$

Solutions to Differential Equations Using Laplace Transformation

As promised, we can now solve differential equations using Laplace Transforms. What we do is transform the differential equation into an algebraic relation using Lapalce Transforms, solve for the Laplace Transform of the solution, then use inverse Laplace Transforms to bring the solution back into the time domain. The most important property of the Laplace transform in this case is the differentiation property:

$$\mathcal{L}\left\{\frac{d}{dt}f(t)\right\} = sF(s) - f(0^-),$$

$$\mathcal{L}\left\{\frac{d^2}{dt^2}f(t)\right\} = s\mathcal{L}\left\{\frac{d}{dt}f(t)\right\} - \frac{df}{dt}(0^-)$$

$$= s^2F(s) - sf(0^-) - \frac{dt}{dt}(0^-)$$

$$\mathcal{L}\left\{\frac{d^n}{dt^n}f(t)\right\} = s^nF(s) - s^{n-1}f(0^-) -$$

$$\ldots - s^n f^{(n-2)}(0^-) - f^{n-1}(0^-).$$

A good starting point to remember this is: if the initial conditions are zero, we can replace n^{th} order differentiation with s^n. Note that because we have defined the Laplace Transform with the lower limit of integration zero, we must always take the starting time as 0, which is no problem for linear constant coefficient differential equations.

Example. Here is a simple differential equation:

$$\frac{dx}{dt} = 1 \quad t \geq 0.$$

Since the functions are equal on the left and right, the Laplace Transforms will be equal:

$$sX(s) - x(0^-) = \frac{1}{s}.$$

Solve for $X(s)$:

$$X(s) = \frac{1}{s^2} + \frac{x(0^-)}{s}$$

We recognize the Laplace Transforms on the right hand side, resulting in the solution

$$x(t) = t + x(0^-) \quad t \geq 0.$$

A. Initial Value Problems

- These are the types of problems we have been solving with RK methods

$$\frac{dy_1}{dt} = f_1(t, y_1, y_2)$$

$$\frac{dy_2}{dt} = f_2(t, y_1, y_2)$$

Subject to :

$t = 0$

$y_1(t = 0) = Y_1$

$y_2(t = 0) = Y_2$

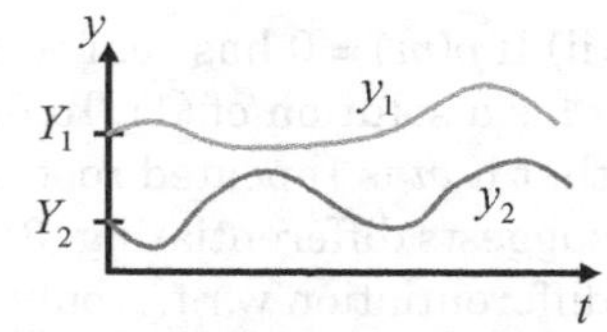

- These are the types of problems we have been solving with RK methods

$$\frac{dy_1}{dt} = f_1(t, y_1, y_2)$$

$$\frac{dy_2}{dt} = f_2(t, y_1, y_2)$$

All conditions are specified at the same value of the independent variable!

Subject to :

$t = 0$

$y_1(t = 0) = Y_1$

$y_2(t = 0) = Y_2$

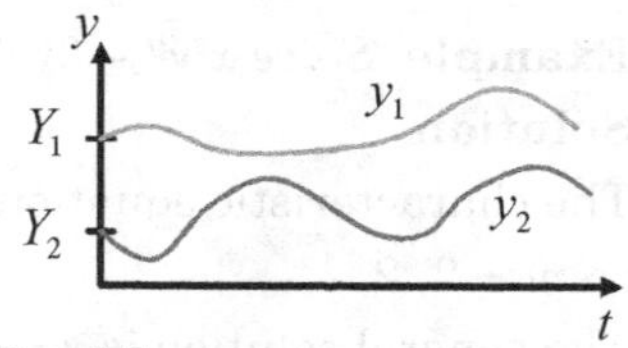

B. Boundary Value Problems

- Auxiliary conditions are specified at the boundaries (not just a one point like in initial value problems)

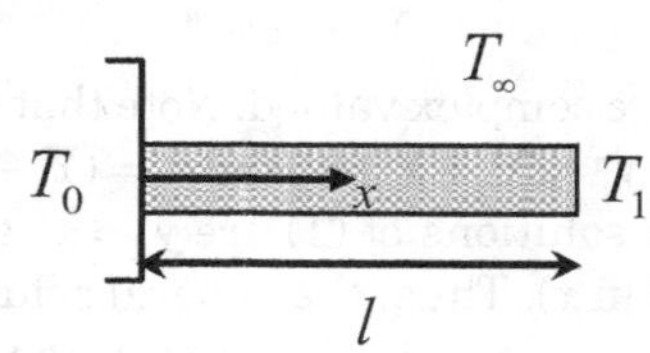

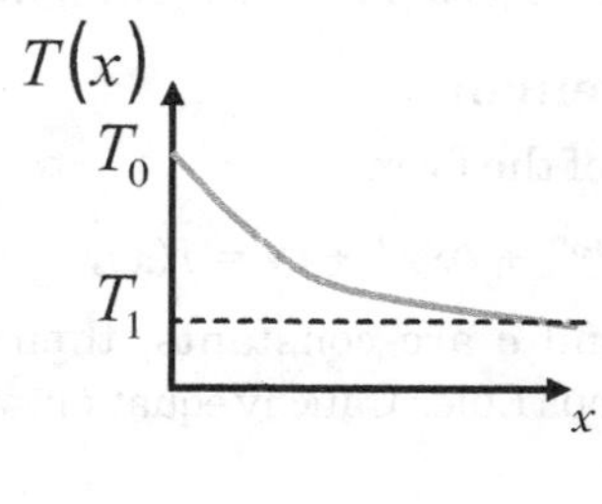

Two Methods: Shooting Method Finite Difference Method

- Auxiliary conditions are specified at the boundaries (not just a one point like in initial value problems)

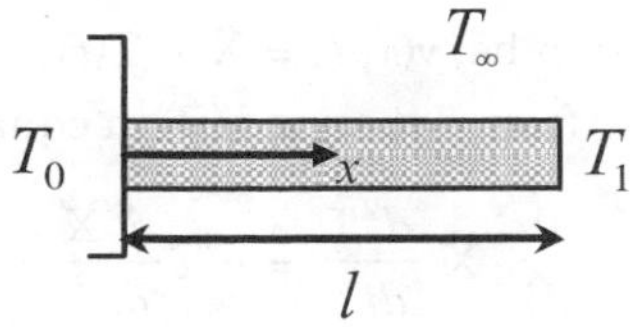

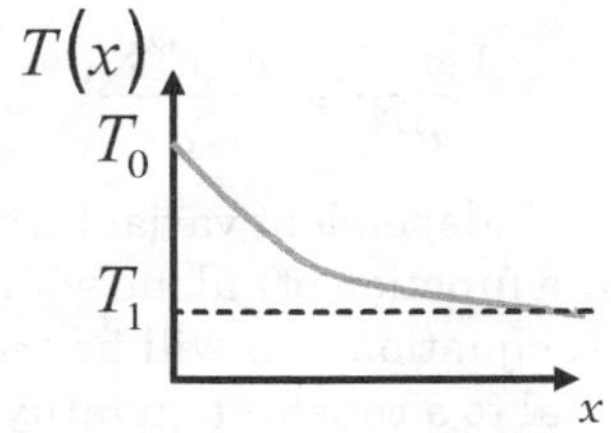

Conditions are specified at different values of the independent variable!

Two Methods: Shooting Method Finite Difference Method

C. Shooting Method

- Applicable to both linear & non-linear Boundary Value (BV) problems.
- Easy to implement
- No guarantee of convergence

D. Approach

- Convert a BV problem into an initial value problem
- Solve the resulting problem iteratively (trial & error)
- Linear ODEs allow a quick linear interpolation
- Non-linear ODEs will require an iterative approach similar to our root finding techniques.

II. PARTIAL DIFFERENTIAL EQUATIONS

If a dependent variable is a function of two or more independent variables, an equation involving partial differential coefficients is called *partial differential equation.* The order of a partial differential equation is the same as that of the highest order differential coefficient in it.

If $z = f(x, y)$ where x and y are independent variables, then partial differntial coefficients $\frac{\partial z}{\partial x}, \frac{\partial z}{\partial y}$ will be denoted by p and q respectively.

Formation of Partial Differential Equation

The partial differential equations can be formed either by elimination of arbitrary constants from a relation between x, y, z or by the elimination of arbitrary functions of these variables.

Solutions of Partial Differential Equations

- The solution $f(x, y, z, a, b) = 0$ of a first order partial differential equation, which contains two orbitrary constants is called *complete solution* or *complete integral.*
- If in this solution, we put $b = \phi(a)$ and find evnelope of the family of surfaces $f(x, y, z, a, \phi(a)) = 0$, we get a solution involving an arbitrary function ϕ. This is called *general solution* or *general integral.*
- A solution obtained from the complete integral by giving particular values to the arbitrary constant is called *particular solution* or *particular integral.*

Equations Solved by Direct Integration

Those equations which contain **only one** partial derivative can be solved by direct integration. However, in place of the constants of integration, we must use arbitrary functions of the variable kept constant.

Linear Partial Differential Equations of First Order

A differential equation involving first order partial derivatives p and q only is called *paritial diferential equation* of the first order. If p and q both occur in the first degree only and are not multiplied together, then it is called a *linear partial differential equation of the first order.*

Lagrange's Linear Equation

Partial differential equation of the form $Pp + Qq = R$, where P, Q and R are functions of x, y, z is the standard form of a linear partial differential equation of the first order and is called *Lagrange's linear equation.*

WORKING PROCEDURE TO SOLVE $P_p + Q_q = R$

(*i*) Form Auxiliary equations $\frac{dx}{P} = \frac{dy}{Q} = \frac{dz}{R}$

(*ii*) Find two independent solutions of the auxiliary equations.

Let $u = c_1$ and $v = c_2$ be two solutions of these equations. Then $f(u, v) = 0$ or $u = f(v)$ is solution of the given equation.

Example. The equation of any sphere of radius r having its centre in the XOY plane is

$$(x-a)^2+(y-b)^2+z^2=r^2$$

Form a partial differential equation by eliminating the constants a, b.

Solution

Given : $(x-a)^2+(y-b)^2+z^2=r^2$

Differentiating with respect to x and y, we get

$$2(x-a)+2z\frac{\partial z}{\partial x}=0 \quad\Rightarrow\quad x-a=-z.\frac{\partial z}{\partial x}$$

$$\text{and}\quad 2(y-b)+2z\frac{\partial z}{\partial y}=0 \quad\Rightarrow\quad y-b=-z.\frac{\partial z}{\partial y}$$

Substituting values, we get

$$z^2\left[\left(\frac{\partial z}{\partial x}\right)^2+\left(\frac{\partial z}{\partial y}\right)^2+1\right]=r^2$$

Applications of Partial Differential Equations

Method of Separation of Variables (Product Method)

This method describes a solution which breaks up into a product of functions each of which depends only on one variable.

i.e. if u is a dependent variable and depends on two independent variables x and t, *i.e.* $u(x, t)$, then we assume solution of the equation in the form

$$u = X(x)\,T(t)$$

where X is a function of x alone and T is a function of t alone.

Substitution for u from above equation in the given differential equation reduces to ordinary differential equations which can be conveniently studied.

Wave Equations

(*i*) One dimensional wave equation or equation of the vibration of a string.

$$\frac{\partial^2 u}{\partial t^2}=c^2\frac{\partial^2 u}{\partial x^2}$$

(*ii*) Two dimensional wave equation or equation of the vibration of a rectangular membrance.

$$\frac{\partial^2 u}{\partial t^2}=c^2\left(\frac{\partial^2 u}{\partial x^2}+\frac{\partial^2 u}{\partial y^2}\right)$$

(*iii*) Three dimensional wave equation is

$$\frac{\partial^2 u}{\partial t^2}=c^2\left(\frac{\partial^2 u}{\partial x^2}+\frac{\partial^2 u}{\partial y^2}+\frac{\partial^2 u}{\partial z^2}\right)$$

Solution of Wave Equation by Separation of Variables

Wave equation is, $$\frac{\partial^2 y}{\partial t^2}=c^2\frac{\partial^2 y}{\partial x^2} \quad ...(i)$$

Let its solution be, $y(x, t) = X(x)\,T(t)$ ...(*ii*)

Substituting from equations (*ii*) in equation (*i*), we get

$$X\frac{d^2T}{dt^2}=c^2T\frac{d^2X}{dx^2}$$

$$\Rightarrow \quad \frac{1}{T}\frac{d^2T}{dt^2}=\frac{c^2}{X}\frac{d^2X}{dx^2} \quad ...(iii)$$

Since x and t are independent variables and left side of equation (*iii*) is a function of t alone while right side is a function of x; equation (*iii*) will be true only when each side is equal to a constant quantity.

Following three possibilities arise:

(*i*) *Constant is negative ($-k^2$).*

General solution is,

$$y(x,t)=(A\cos kt+B\sin k)\left(C\cos\frac{k}{c}x+D\,sin\frac{k}{c}x\right)$$

(*ii*) *Constant is zero.*

General solution is,

$$y(x,t)=(At+B)(Cx+D)$$

(*iii*) *Constant is positive (k^2).*

General solution is

$$y(x,t)=(Ae^{kt}+Be^{-kt})(Ce^{(k/c)x}+De^{-(k/c)x})$$

Example. Solve by method of separation of variables the equation

$$\frac{\partial^2 u}{\partial t^2}-2\frac{\partial u}{\partial t}+\frac{\partial u}{\partial x}=0.$$

Solution

Let $u = XT$ ...(*i*)

$$\therefore \quad X\frac{d^2T}{dt^2}-2X\frac{dT}{dt}+T\frac{dX}{dx}=0$$

$$\Rightarrow \quad \frac{1}{T}\left(\frac{d^2T}{dt^2}-2\frac{dT}{dt}\right)=-\frac{1}{X}\frac{dX}{dx}=k\ \text{(say)}$$

[since x and t are independent of each other]

$$\frac{d^2T}{dt^2}-2\frac{dT}{dt}-kT=0 \quad ...(ii)$$

$$\text{and} \quad \frac{dX}{dx}+kX=0 \quad ...(iii)$$

Solutions of equations (*ii*) and (*iii*) are

$$T = A\,e^{\{1+\sqrt{(1+k)}\}t} + B\,e^{\{1-\sqrt{(1+k)}\}t} \quad \text{and} \quad X = C\,e^{-kx}$$

Substituting for X and T in equation (*i*), we get

$$u = \left[A'\,e^{\{1+\sqrt{(1+k)}\}t} + B'\,e^{\{1-\sqrt{(1+k)}\}t}\right] e^{-kx}$$

where A′ = AC and B′ = BC.

One Dimensional Diffusion Equation

Consider the problem where a Brownian particle gets absorbed once it reaches the position x_a. We want to solve for the probability distribution $P_a(x, t)$ in the presence of the absorbing boundary. Thus we need to solve the diffusion equation:

$$\frac{\partial P_a(x,t)}{\partial t} = D\frac{\partial^2 P_a(x,t)}{\partial x^2} \quad ...(1)$$

with the *boundary* condition

$$P_a(x = x_a, t) = 0 \quad ...(2)$$

and *initial* condition

$$P_a(x, t = 0) = \delta(x). \quad ...(3)$$

We first expand $P_a(x, t)$ using a basis set of functions $\{\phi_k(x)\}$ which satisfy the following equation and boundary condition (corresponding to Eq. 2) :

$$\frac{\partial^2\phi_k}{\partial^2 x} = -k^2\phi_k \text{ with } \phi_k(x = x_a) = 0.$$

One easily gets the following solution:

$$\phi_k(x) = \left(\frac{2}{\pi}\right)^{1/2} \sin k(x - x_a)$$

and these satisfy the orthonorraality condition $\int_{-\infty}^{x_a} dx\phi_k(x)\phi_{k'}(x) = \delta(k - k')$. Note that k is restricted to positive values only. Thus we get the following expansion and inverse transform:

$$P_a(x,t) = \int_0^{\infty} dk\tilde{P}_a(k,t)\phi_k(x) \quad ...(4)$$

$$\tilde{P}_a(k,t) = \int_{-\infty}^{x_a} dxP_a(x,t)\phi_k(x) \quad ...(5)$$

Substituting the expansion Eq. 4 in Eq. 1 we get the following equation for $\tilde{P}_a(k,t)$:

$$\frac{\partial\tilde{P}_a(k,t)}{\partial t} = -Dk^2\tilde{P}_a(k,t)$$

which is a simple differential equation of the form $dy/dt = -ay$ and has the solution $\tilde{P}_a(k,t) = \tilde{P}_a(k,0)e^{-Dk^2t}$. We can determine $\tilde{P}_a(k,0)$ from Eq. 5 and the initial condition $P_a(x, t = 0) = \delta(x)$.

This gives $\tilde{P}_a(k,0) = \phi_k(x_a) = -(2/\pi)^{1/2}\sin kx_a$. We thus finally get the following solution:

$$P_a(x, t) = -\int_0^{\infty} dk\frac{2}{\pi}\sin kx_a \sin k(x - x_a)e^{-Dk^2t}$$

$$= \frac{1}{2\pi}\int_{-\infty}^{\infty} dk[\cos kx - \cos k(2x_a - x)e^{-Dk^2t}$$

$$= P(x, t) - P(2x_a - x, t) \quad ...(6)$$

where $P(x) = e^{-x^2/(4Dt)}/(4\pi Dt)^{1/2}$ is the usual solution of the diffusion equation without any boundary conditions. Note that $P(2x_a - x,t)$ is also a solution of the diffusion equation but with the Brownian particle initially located at the miage point $2x_a$. You can thus directly verify that Eq. 6 is a solution of the diffusion equation and satisfies the boundary and initial conditions.

Note: When we think of the solution P(x,t) of the diffusion equation, we can either think of it as giving the probability of a single random walker to be between x and x + dx at time t, OR, alternatively we can imagine a gas of noninteracting brownian particles and then P(x)dx gives the fraction of particles in dx. Both viewpoints are valid and useful

Survival probaility of a particle. Since particles get continually absorbed at the boundary at x_a, the total probability is not conserved and the quantity

$$p_s(t) = \int_{-\infty}^{x_a} dxp_a(x,t) \quad ...(7)$$

infact gives the fraction of particles that have not been absorbed till time t. Also one can think of $P_s(t)$ as the probability that a diffusing particle, starting from the origin, has not hit the wall at x_a (and got killed!). Hence it is referred to as the survival probability of a random walker. After some simple manipulations we get:

$$P_s(t) = 2\int_0^{x_a} P(x)dx$$

$$= \frac{2}{(4\pi Dt)^{1/2}}\int_0^{x_a} e^{-\frac{x^2}{4Dt}}dx$$

$$= \frac{2}{\pi^{1/2}}\int_0^{\frac{x_a}{(4Dt)^{1/2}}} e^{-u^2}du \quad ...(8)$$

From the definition of the error function $erf(z) = \left(2/\sqrt{\pi}\right)\int_0^z e^{-u^2}du$ we then finally get:

$$P_s(t) = erf\left(\frac{x_a}{(4Dt)^{1/2}}\right) \sim \frac{1}{t^{1/2}} \text{ at large times} \quad ...(9)$$

Next let us ask the question "what is the probability $F(t)dt$ that a particle gets absorbed (at x_a) between times t and $t + dt$?"

By definition: $F(t)dt = P_s(t) - P_s(t + dt)$. Thus

$$F(t) = -\frac{dP_s(t)}{dt}$$

$$= \frac{x_a}{t}\frac{1}{(4\pi Dt)^{1/2}}e^{-\frac{x_a^2}{4Dt}} \sim \frac{1}{t^{3/2}} \text{ largo Ume,} \quad \text{...(10)}$$

A more direct way to obtain $F(t)$ is to note that it also gives the fraction of particles that get absorbed per unit time and so is simply the particle current evaluated at x_a. Thus

$$F(t) = -D\frac{\partial P_a}{\partial x}\bigg|_{x=x_a} \quad \text{...(11)}$$

Finally note that $F(t)$ also gives the first passage probability, ie., the probability that a particle, starting from the origin, reaches the point x_a for the first time between times t and $t + dt$. We can then ask "What is the probability that a random walker, starting from the origin, ever reaches the point x_a ?". This is simply given by $\int_0^\infty dtF(t)$ and evaluation of the integral gives the value one. Thus in one dimension, a random walker will eventually reach any given point. This is also true in two dimensions but not so in higher dimensions !!(This result is known as Polya's theorem).

First-Order One Dimension Wave Equation

First-order partial differential equations can be tackled with the **method of characteristics,** a powerful tool which also reaches beyond first-order. We'll be looking primarily at equations in two variables, but there is an extension to higher dimensions.

Wave Equation with Constant Speed

Consider the first-order wave equation with constant speed:

$$\frac{\partial u}{\partial t} + c\frac{\partial u}{\partial x} = 0.$$

It responds well to a change of variables:

$$\xi = x + ct \qquad \eta = x - ct$$

The chain rule gives us

$$\frac{\partial}{\partial x} = \frac{\partial \xi}{\partial x}\frac{\partial}{\partial \xi} + \frac{\partial \eta}{\partial x}\frac{\partial}{\partial \eta}$$

$$= \frac{\partial}{\partial \xi} + \frac{\partial}{\partial \eta}$$

$$\frac{\partial}{\partial t} = \frac{\partial \xi}{\partial t}\frac{\partial}{\partial \xi} + \frac{\partial \eta}{\partial t}\frac{\partial}{\partial \eta}$$

$$= c\left(\frac{\partial}{\partial \xi} - \frac{\partial}{\partial \eta}\right)$$

and so the wave equation is equivalent to

$$2c\frac{\partial u}{\partial \xi} = 0.$$

Integrating gives the general solution $u = F(\eta)$, $u = F(x - ct)$.

But where did we get the change of variables from? The line $x - ct$ = constant is a line in the $x - t$ plane along which u is constant. This means that if we parametrise this line

$$x = x(r) \qquad t = t(r)$$

then moving along the line by changing r will not change w, i.e.

$$\frac{du}{dr} = 0$$

This is the underlying principle of the characteristic.

Variable Speed

Let's look now at the variable speed case:

$$\frac{\partial u}{\partial t} + c(x,t)\frac{\partial u}{\partial x} = 0.$$

We would like again to find curves along which u is constant. Suppose such a curve is given by $x = x(r)$ and $t = t(r)$. Then, using the chain rule,

$$\frac{\partial u}{\partial r} = \frac{\partial u}{\partial t}\frac{dt}{dr} + \frac{\partial u}{\partial x}\frac{dx}{dr}.$$

We want this to be zero, which is easily achieved if we make this expression the same as the original linear operator:

$$\frac{dt}{dr}\frac{\partial u}{\partial t} + \frac{dx}{dr}\frac{\partial u}{\partial x} = \frac{\partial u}{\partial t} + c(x,t)\frac{\partial u}{\partial x} = 0.$$

This gives us the two parametric equations governing the shape of the characteristic curve:

$$\frac{dt}{dr} = 1, \qquad \frac{dx}{dr} = c(x, r)$$

These are both ODEs and straightforward to solve.

Example.

Look at the equation

$$2\sin\theta\cos 2\phi\frac{\partial u}{\partial \theta} - \frac{\cos\theta\sin 2\phi}{\sin\theta}\frac{\partial u}{\partial \phi} = 0.$$

Suppose that our characteristic is given by $\theta = \theta(r)$, $\phi = \phi(r)$. Then the requirement that u be constant along a characteristic becomes

$$\frac{\partial u}{\partial \theta}\frac{d\theta}{dr} + \frac{\partial u}{\partial \phi}\frac{d\phi}{dr} = 0.$$

A nai've attempt would be to look at the coupled ODEs

$$\frac{\partial\theta}{\partial r} = 2\sin\theta\cos 2\phi$$

$$\frac{\partial\phi}{\partial r} = -\frac{\cos\theta\sin 2\phi}{\sin\theta}$$

but we can uncouple them if, before we start, we multiply the original equation by $\sin\theta/\cos\theta\cos 2\phi$:

$$\frac{2\sin^2\theta}{\cos\theta}\frac{\partial u}{\partial\theta} - \frac{\sin 2\phi}{\cos 2\phi}\frac{\partial u}{\partial\phi} = 0$$

$$\frac{\partial\theta}{\partial r} = \frac{2\sin^2\theta}{\cos\theta}$$

$$\frac{\partial\phi}{\partial r} = -\frac{\sin 2\phi}{\cos 2\phi}$$

Now the equations are decoupled, and solving them in turn gives

$$\sin\theta = -\frac{1}{2r}$$

$$\sin 2\phi = \exp[C - 2r]$$

Note that we only use a constant of integration in one of these equations; since r is just a parameter, the point $r = 0$ is not defined *a priori*. Effectively, we are making a change of variables from x, t to r, C. We can invert the transformation:

$$C = \ln\sin 2\phi - \frac{1}{\sin\theta}$$

$$r = -\frac{1}{2\sin\theta}$$

and since u is constant on this curve, we can deduce the general solution

$$u = F(C) = F\left(\ln\sin 2\phi - \frac{1}{\sin\theta}\right)$$

More Than Two Dimensions

Now suppose we have the PDE

$$\frac{\partial u}{\partial x} + c_1(x,y,z)\frac{\partial u}{\partial y} + c_2(x,y,z)\frac{\partial u}{\partial z} = 0.$$

Again, we look for a curve on which u is constant; being a curve, it can still be described with a single variable r so we set $x = x(r)$, $y = y(r)$ and $z = z(r)$.

Then the chain rule gives

$$\frac{\partial u}{\partial r} = \frac{\partial u}{\partial x}\frac{dx}{dr} + \frac{\partial u}{\partial y}\frac{dy}{dr} + \frac{\partial u}{\partial z}\frac{dz}{dr}$$

and to make this equal to zero we choose

$$\frac{\partial x}{\partial r} = 1$$

$$\frac{\partial y}{\partial r} = c_1\left(x(r), y(r), z(r)\right)$$

$$\frac{\partial u}{\partial r} = c_2\left(x(r), y(r), z(r)\right)$$

The latter two are now coupled ODEs so we are not guaranteed to be able to find a solution; but sometimes you may be lucky.

Example.

Look at the equation

$$\frac{\partial u}{\partial x} + xy\frac{\partial u}{\partial y} + 2x^2 z\ln y\frac{\partial u}{\partial z} = 0$$

We set $x = x(r)$, $y = y(r)$ and $z = z(r)$ and the chain rule gives

$$\frac{\partial u}{\partial r} = \frac{\partial u}{\partial x}\frac{dx}{dr} + \frac{\partial u}{\partial y}\frac{dy}{dr} + \frac{\partial u}{\partial z}\frac{dz}{dr}$$

To match the three coefficients we set:

$$\frac{dx}{dr} = 1 \qquad x(r) = r$$

$$\frac{dy}{dr} = xy = ry \qquad y(r) = y_0\exp\left[r^2/2\right]$$

$$\frac{dz}{dr} = 2x^2 z\ln y = -r^4 z\ln y_0 \quad z(r) = z_0\exp\left[-r^5\ln y_0/5\right].$$

Now we have expressed all points in terms of the three parameters r, y_0 and z_0 and u is independent of r, so the solution is any function of y_0 and z_0. Reversing the change of variables gives

$$r = x \quad y_0 = y\exp[-x^2/2] \quad z_0 = z\exp\left[-x^7/10\right]y^{\left[x^5/5\right]}$$

and the full solution is

$$u = F(y\exp[-x^2/2];\ z\exp\left[-x^7/10\right]y^{\left[x^5/5\right]}$$

Inhomogeneous Case

The characteristic curve is just as fundamental if the equation is not homogeneous, although the function value is no longer constant along characteristics. The method is best seen by example:

$$\frac{\partial u}{\partial t} + 2xt\frac{\partial u}{\partial x} = u$$

over all x, with initial condition (at $t = 0$) $u = x$. We start by finding the characteristic. Here the characteristic is given by

$$\frac{dt}{dr} = 1 \qquad t = r$$

$$\frac{dx}{dr} = 2xr \qquad x = x_0 \exp[r^2].$$

This time our two new variables are r and XQ- Along a specific characteristic we have

$$\frac{du}{dr} = \frac{\partial u}{\partial t} + \frac{\partial u}{\partial x}\frac{dx}{dt} = u$$

$u = u_0 e^r$ or more exactly $u = F(u_0)e^r$.
We now have a one-parameter family of solutions (parameter x_0): on the curve
$x = x_0 \exp[t^2]$, $u = F(x_0)e^t$ so $u = F(x \exp(-t^2])e^t$.
We need to apply the initial conditions to determine the function F. At $t = 0$ we have $x = x_0$ and $u = F(x_0)$ so the initial condition gives $F(x_0) = x_0$:

$$u = x \exp(-t^2]e^t = x \exp[t - t^2].$$

Nonlinear Homogeneous Case

A general first-order homogeneous PDE in two variables can be written as

$$\frac{\partial u}{\partial t} + c(u, x, t)\frac{\partial u}{\partial x} = 0$$

and the method of characteristics still applies (but we expect an implicit solution in general). The characteristic curves are given by

$$\frac{dx}{dt} = c(u, x, t).$$

Again, we will write the curve parametrically as $t = r$, and x some function of r and a constant x_0 (that is, constant for a given characteristic). Along any characteristic we will have

$$\frac{du}{dr} = 0$$

and so u is constant along a characteristic. We can then make this into a solution everywhere by setting $u = F(x_0)$ on the characteristic specified by x_0.
Since u is constant on our characteristic, the equation of the curve is simply

$$\frac{dx}{dt} = c(F(x_0), x, t),$$

which is a straightforward ODE in x and t. Once we have solved it we have the characteristic curve

$$t = r$$
$$x = G(x_0, F(x_0), r)$$
$$= G(x_0, u, r).$$

The implicit form of the solution is now:

$$x = G(x_0, u, t)$$
$$u = f(x_0)$$

Which is a one-parameter family with parameter x_0. In many cases it is possible to rearrange the first equation to obtain x_0 in terms of u and t; then substituting this into the second equation gives the more standard form of the implicit solution.

Example.

Consider the advection equation

$$\frac{\partial u}{\partial t} + ux^2 t\frac{\partial u}{\partial x} = 0$$

Because it is homogeneous, we expect u to be constant along characteristics: so we parameterise with x_0 (constant on each characteristic) and r (which varies along the characteristic) and we can say $u = f(x_0)$.
Now our characteristic curve becomes

$$\frac{dx}{dt} = ux^2 t = f(x_0)x^2 t,$$

which we can solve:

$$\int \frac{dx}{x^2} = F(x_0)\int t\, dt$$

$$-\frac{1}{x} = \frac{1}{2}F(x_0)t^2 - \frac{1}{x_0}$$

$$x = \frac{2x_0}{2 - x_0 F(x_0)t^2}.$$

Thus the characteristic curve and implicit solution are:

$$t = r$$

$$x = \frac{2x_0}{2 - x_0 F(x_0)r^2}$$

$$u = F(x_0).$$

As above, we can rearrange to get x_0 in terms of x, t and u:

$$x = \frac{2x_0}{2 - x_0 u t^2}$$

$$x_0 = \frac{2x}{2 + uxt^2}$$

and the standard implicit form of the solution is

$$u = F\left(\frac{2x}{2+uxt^2}\right)$$

Nonlinear Inhomogeneous

We are now looking at the most complex of first-order PDEs: those of the form

$$\frac{\partial u}{\partial t} + c(u,x,t)\frac{\partial u}{\partial x} = f(u, x, t)$$

Characteristics still exist in these systems, and they may have important physical properties (for instance, discontinuities in the derivatives of the solution will propagate along them) but unfortunately, since u itself now varies along a characteristic, we can no longer solve even implicitly in general.

Example.

Here is a case where we can achieve a little:

$$\frac{\partial u}{\partial t} + u\frac{\partial u}{\partial x} = \cos t.$$

In tills case we can immediately spot one solution

$$u = A + \sin t$$

but can we show that this is not the most general solution?

The characteristics are defined by

$$\frac{du}{dt} = u;$$

let us suppose we know the family of curves $x = f(x_0, r)$, $t = r$. Then we have

$$\frac{du}{dr} = \frac{\partial u}{\partial t}\frac{dt}{dr} + \frac{\partial u}{\partial x}\frac{dx}{dr} = \frac{\partial u}{\partial t} = \frac{\partial}{\partial r}f(x_0,r)\frac{\partial u}{\partial x}$$

which gives us the two coupled equations

$$\frac{\partial}{\partial r}f(x_0,r) = u$$

$$\frac{du}{dr} = \cos r.$$

These are easy to solve in reverse order:

$$u = A(x_0) + \sin r$$

$$x = A(x_0)r - \cos r + x_0.$$

Unfortunately, we can't extract x_0 explicitly without choosing the function $A(x_0)$; but note that

$$A(x_0) = \alpha$$

$$\Rightarrow \quad x_0 = x + \cos t - \alpha t$$

$$u = a + \sin t$$

$$A(x_0) = \frac{x_0}{\beta}$$

$$\Rightarrow \quad x_0 = x + \frac{\beta\cos t}{t+\beta}$$

$$u = \frac{x}{\beta} + \frac{\cot}{t+\beta} + \sin t$$

so our "spotted" solution is only one of a family of possible solutions.

The Second-Order One-Dimensional Wave Equation

Homogeneous Wave Equation with Constant Speed

The simplest form of the second-order wave equation is given by:

$$\frac{\partial^2 u}{\partial t^2} - c^2\frac{\partial^2 u}{\partial x^2} = 0$$

Like the first-order wave equation, it responds well to a change of variables:

$$\xi = x + ct$$

$$\eta = x - ct$$

which reduces it to

$$\frac{\partial^2 u}{\partial\xi\partial\eta} = 0$$

which is solved by

$$u = p(\xi) + q(\eta)$$

$$= p(x + ct) + q(x - ct)$$

for any differentiate functions p and q. The lines ξ = constant and η = constant are the characteristics, exactly analogous to the characteristics for the first-order equation.

If we add initial conditions

$$u(x, 0) = f(x)$$

$$\frac{\partial u}{\partial t}(x, Q) = g(x)$$

then a little algebra gives us **d'Alembert's solution:**

$$u(x, t) = \frac{1}{2}[f(x+ct) + f(x-ct)] + \frac{1}{2c}\int_{x-ct}^{x+ct} g(x)dy.$$

Inhomogeneous Wave Equation

The inhoraogeneous wave equation:

$$\frac{\partial^2 u}{\partial t^2} - c^2\frac{\partial^2 u}{\partial x^2} = F(x, t) \quad ...(3)$$

can be solved in a very similar way. The change of variables results in:

$$-4c^2\frac{\partial^2 u}{\partial\xi\partial\eta} = F\left(\frac{\xi+\eta}{2}, \frac{\xi-\eta}{2c}\right)$$

which can be integrated directly for any specific function F; however (Wein-berger p. 25) it is also possible to carry out the integrals symbolically (paying particular attention to which variable is held constant when integrating with respect to another). The general result is

$$u = p(x + ct) + q(x - ct)$$

$$+\frac{1}{2c}\int_0^t \int_{x-c(t-t')}^{x+c(t-t')} F(x-t')dx'\,dt'. \quad ...(4)$$

As this is a linear PDE, the general solution to the inhomogeneous equation is the sum of the general solution to the homogeneous equation (the CF in the notation of ODEs) and one particular solution to the full equation (the PI). To verify the solution we simply check that the last term of (4) satisfies (3).

Example.

Let's consider the example inhoraogeneous wave equation

$$\frac{\partial^2 u}{\partial t^2} - c^2\frac{\partial^2 u}{\partial x^2} = 12xt.$$

We can find the solution using the formula (4) but it's not straightforward!

$$u = p(x + ct) + q(x - ct)$$

$$+\frac{1}{2c}\int_0^t \int_{x-c(t-t')}^{x+c(t-t')} 12x'\,t'\,dx'\,dt'.$$

Looking just at the integral and carrying out the x' integration first gives

$$\int_0^t \int_{x-c(t-t')}^{x+c(t-t')} 12x'\,t'\,dx'\,dt'$$

$$= \int_0^t \left[6(x')^2 t'\right]_{x'=x-c(t-t')}^{x+c(t-t')} dt'$$

$$= \int_0^t 6((x+c(t-t'))^2 t' - (x-c(t-t'))^2 t')dt'$$

$$= \int_0^t 24cxt'(t-t')dt' = \left[cx(12tt'^2 - 8tt'^3)\right]_{t'=0}^{t}$$

$$= 4cxt^3,$$

and the general solution is

$$u = p(x + ct) + q(x - ct) + 2xt^3.$$

For a specific case, though, it is usually more straightforward to work directly from the original equation (with change of variables). In this case when we put $\xi = x + ct$ and $\eta = x - ct$ we obtain

$$-4c^2\frac{\partial^2 u}{\partial\xi\partial\eta} = \frac{3}{c}(\xi+\eta)(\xi-\eta) = \frac{2}{c}(\xi^2-\eta^2)$$

Integrating gives

$$-4c^2\frac{\partial u}{\partial\xi} = p(\xi) + \frac{1}{c}(3\xi^2\eta - \eta^3);$$

$$-4c^2 u = p(\xi) + q(\eta) + \frac{\xi\eta}{c}(\xi+\eta)(\xi-\eta).$$

Converting the coordinates gives

$$u = f(x + ct) + g(x - ct) - \frac{(x+ct)(x-ct)}{4c^3}(2x)(2ct)$$

$$= f(x + ct) + g(x - ct) - \frac{xt(x^2 - c^2t^2)}{c^2}.$$

This doesn't immediately look the same; but note that the difference can be absorbed into $f(x + ct)$ and $g(x - ct)$:

$$2xt^3 + \frac{xt(x^2 - c^2t^2)}{c^2}$$

$$= \frac{(xt(x^2 - c^2t^2))}{c^2}$$

$$= \frac{(\xi+\eta)(\xi-\eta)((\xi+\eta)^2 + (\xi-\eta)^2)}{64c^3}$$

$$= \frac{1}{32c^3}((\xi^2-\eta^2)(\xi^2+\eta^2))$$

$$= \frac{1}{32c^3}(\xi^4 - \eta^4)$$

which is the sum of a function of ξ and a function of η.

EXERCISE

MCQ TYPE QUESTIONS

1. The solution for the differential equation

$$\frac{d^2y}{dx^2} + 5\frac{dy}{dx} + 6y = 0 \text{ is}$$

(a) $C_1 e^{-2t} + C_2 e^{3t}$ (b) $C_1 \sin 2t + C_2 \cos 2t$
(c) $C_1 e^{2t} + C_2 e^{-2t}$ (d) $C_1 e^{-2t} + C_2 e^{-3t}$

2. Integrating factor for the differential equation

$$\left(\cos^2 x\right)\frac{dy}{dx} + y = \tan x, \text{ is}$$

(a) $e^{\tan x}$ (b) $\cos 2x$
(c) $e^{-\tan x}$ (d) $\sin 2x$

3. Find range of values for a constant K to yield a stable system in the following set of time dependent differential equations

$$\frac{dy_1}{dt} = -5y_2 + (4-K)y_2; \quad \frac{dy_2}{dt} = y_1 - 2y_2$$

(a) $0 < K < 7$ (b) $6.25 < K < 10$
(c) $-6 < K \le 6.25$ (d) $0 \le K \le 7$

4. Differential equation,

$$\frac{d^2x}{dt^2} + 10\frac{dx}{dt} + 25x = 0$$

will have a solution of the form

(a) $(C_1 + C_2 t)e^{-5t}$ (b) $C_1 e^{-2t}$
(c) $C_1 e^{-5t} + C_2 e^{5t}$ (d) $C_1 e^{-5t} + C_2 e^{2t}$

where C_1 and C_2 are constants.

5. Differential equation for the variation of the amount of salt x in a tank with time t is given by

$$\frac{dx}{dt} + \frac{x}{20} = 10$$

where x is in kg and t is in minutes.

Assuming that there is no salt in the tank initially, the time (in min.) at which the amount of salt increases to 100 kg is

(a) 10 ln 2 (b) 20 ln 2
(c) 50 ln 2 (d) 100 ln 2

6. Solve the differential equation

$$xdx + ydy + 2(x^2 + y^2)\,dx = 0$$

(a) $x^2 - y^2 = ce^{-4x}$
(b) $x^2 + y^2 = ce^{+4x}$
(c) $x^2 + y^2 = ce^{-4x}$
(d) $x^2 - y^2 = ce^{4x}$

7. The differential equation

$$\frac{d^2y}{dx^2} + \sin x\frac{dy}{dx} + ye^x = \sinh x$$

is

(a) first order and linear
(b) first order and non-linear
(c) second order and linear
(d) second order and non-linear

8. For the differential equation $\frac{dy}{dt} + 5y = 0$ with $y(0) = 1$, the general solution is

(a) e^{5t} (b) e^{-5t}
(c) $5e^{-5t}$ (d) $e^{\sqrt{-5t}}$

9. The necessary and sufficient for the differential equation of the from

$$M(x, y)\,dx + N(x, y)\,dy = 0$$

to be exact is

(a) $M = N$ (b) $\frac{\partial M}{\partial x} = \frac{\partial N}{\partial y}$
(c) $\frac{\partial M}{\partial y} = \frac{\partial N}{\partial x}$ (d) $\frac{\partial^2 M}{\partial x^2} = \frac{\partial^2 N}{\partial y^2}$

10. The differential equation

$$y'' + (s^3 \sin x)^5 y' + y = \cos x^3 \text{ is}$$

(a) homogeneous
(b) nonlinear
(c) second order linear
(d) non-homogeneous with constant coefficients

11. For the differential equation,

$$f(x, y)\frac{dy}{dx} + g(x, y) = 0$$

to be exact

(a) $\frac{\partial f}{\partial y} = \frac{\partial g}{\partial x}$ (b) $\frac{\partial f}{\partial x} = \frac{\partial g}{\partial y}$
(c) $f = g$ (d) $\frac{\partial^2 f}{\partial x^2} = \frac{\partial^2 g}{\partial y^2}$

12. If $f(D)\,y = 0$ be a linear differential equation with constant coefficients, then its auxiliary equation is

(a) $f(m^3) = 0$
(b) $f(m) = 0$
(c) $f(D - m) = 0$
(d) $f(x) = 0$

13. The general solution of the linear differential equation $(D^4 - 81)y = 0$ is given by

(a) $(C_1 + C_2x)e^{3x} + (C_3 + C_4x)\sin 3x$

(b) $(C_1 + C_2x)e^{3x} + (C_3 + C_4x)e^{-3x} + (C_5 + C_6x)\cos 3x + (C_7 + C_8x)\sin 3x$

(c) $C_1e^{3x} + C_2e^{-3x} + C_3\cos 3x + C_4\sin 3x$

(d) $C_1e^{3x} + C_2e^{-3x} + e^{3x}(C_3\cos x + C_4\sin x)$

14. $\dfrac{1}{D-\alpha}Q(x)$ is equal to

(a) $e^{\alpha x}\int Q(x)dx$ (b) $e^{-\alpha x}\int e^{\alpha x}Q(x)dx$

(c) $e^{-\alpha x}\int Q(x)dx$ (d) $e^{\alpha x}\int e^{-\alpha x}Q(x)dx$

15. P.I. of $(D^2 - 2D)y = e^x \sin x$ is

(a) $-\dfrac{1}{2}e^x\sin x$ (b) $e^x\cos x$

(c) $-\dfrac{1}{2}e^x\cos x$ (d) none of these

16. General solution of the differential equation $(D^2 - m^2)y = 0$ is

(a) $y = (C_1 + C_2x)e^{mx}$

(b) $y = C_1\sin x + C_2\cos x$

(c) $y = C_1\cosh x + C_2\sinh x$

(d) none of these

17. $\dfrac{1}{\phi(D)(D-a)^n}e^{ax}$ is equal to

(a) $\dfrac{x^n e^{ax}}{\phi(a)}$ (b) $\dfrac{x^n e^{ax}}{n!\phi(a)}$

(c) $\dfrac{x^n e^{ax}}{\phi(a)}, \phi(a) \neq 0$ (d) $\dfrac{x^n e^{ax}}{n!\phi(a)}, \phi(a) \neq 0$

18. If V be any function of x, then $\dfrac{1}{f(D)}xV$ is equal to

(a) $\left[x + \dfrac{f(D)}{f'(D)}\right]f'(D).V$

(b) $\left[x - \dfrac{f'(D)}{f(D)}\right]f(D).V$

(c) $\left[x - \dfrac{f'(D)}{f(D)}\right]\dfrac{1}{f'(D)}.V$

(d) $\left[x - \dfrac{f'(D)}{f(D)}\right]\dfrac{1}{f(D)}.V$

19. If $y = 1 + \cos x$ and $y = 1 + \sin x$ solutions of the differential equation $\dfrac{d^2y}{dx^2} + y = 1$, then its solution will be also

(a) $y = 2(1 + \cos x)$

(b) $y = 2 + \cos x + \sin x$

(c) $y = \cos x - \sin x$

(d) $y = 1 + \cos x + \sin x$

20. Solution of the differential equation

$$\frac{d^2y}{dx^2} - \frac{dy}{dx} - 2y = 3e^{2x},$$

where $y(0) = 0$ and $y'(0) = -2$ is

(a) $y = e^{-x} - e^{2x} + xe^{2x}$ (b) $y = e^x - e^{-2x} - xe^{2x}$

(c) $y = e^{-x} + e^{2x} + \dfrac{x}{2}e^{2x}$ (d) $y = e^x - e^{-2x} - \dfrac{x}{2}e^{2x}$

21. The general solution of the differential equation $\dfrac{d^2y}{dx^2} + n^2y = 0$ is

(a) $c_1\sqrt{\cos nx} + c_2\sqrt{\sin nx}$

(b) $c_1\cos nx + c_2\sin nx$

(c) $c_1\cos^2 nx + c_2\sin^2 nx$

(d) $c_1\cos^3 nx + c_2\sin^3 nx$

22. The particular integral for the differential equation $\dfrac{d^2y}{dx^2} - 5\dfrac{dy}{dx} + 6y = e^{3x}$ is

(a) xe^{-x} (b) x^2e^{-x}

(c) $(x-1)e^{3x}$ (d) $(x-2)e^{4x}$

23. Solution of the differential equation $\dfrac{d^2y}{dx^2} = x$ is

(a) $y = \dfrac{x^3}{6} + c_1x^2 + c_2$

(b) $y = \dfrac{x^3}{6} + c_1x + c_2$

(c) $y = \dfrac{x^2}{6} + c_1x + c_2$

(d) $y = -\dfrac{x^2}{6} + c_1x + c_2$

24. Particular integral of $\dfrac{d^2y}{dx^2} + \dfrac{dy}{dx} = x^2 + 2x + 4$ is

(a) $\dfrac{x^2}{3} + 4x$ (b) $\dfrac{x^3}{3} + 4x$

(c) $\dfrac{x^3}{3} + 4$ (d) $\dfrac{x^3}{3} + 4x^2$

NUMERICAL TYPE QUESTIONS

1. The value of y as $t \to \infty$ for an initial value of $y(1) = 0$, for the differential equation

$$\left(4t^2 + 1\right)\frac{dy}{dt} + 8yt - t = 0 \text{ is } \_\_\_\_$$

2. The number of boundary conditions required to solve the following differential equation

$$\frac{\partial^2\phi}{\partial x^2} + \frac{\partial^2\phi}{\partial y^2} = 0 \text{ is/are } \_\_\_\_$$

ANSWERS

MCQ Type Questions

1. (*d*)	**2.** (*c*)	**3.** (*c*)	**4.** (*a*)	**5.** (*b*)	**6.** (*c*)	**7.** (*d*)	**8.** (*b*)	**9.** (*c*)	**10.** (*c*)
11. (*b*)	**12.** (*b*)	**13.** (*c*)	**14.** (*d*)	**15.** (*a*)	**16.** (*c*)	**17.** (*d*)	**18.** (*d*)	**19.** (*d*)	**20.** (*a*)
21. (*b*)	**22.** (*c*)	**23.** (*b*)	**24.** (*b*)						

Numerical Type Questions

1. (1/8) **2.** (2)

EXPLANATIONS

MCQ TYPE QUESTIONS

3. Given : $\frac{dy_1}{dt} = -5y_2 + (4-k)\ y_2$ ...(*i*)

and $\frac{dy_2}{dt} = y_1 - 2y_2$ ...(*ii*)

Putting $\frac{d}{dt}$ = D, we have from equation (*i*),

$$\mathrm{D}y_1 + 5y_2 - (4-\mathrm{K})\ y_2 = 0$$

$\Rightarrow$ $(\mathrm{D}+5)y_1 - (4-\mathrm{K})\ y_2 = 0$ ...(*iii*)

From equation (*ii*)

$$\mathrm{D}y_2 = y_1 - 2y_2$$

$\Rightarrow$ $(\mathrm{D}+2)y_2 - y_1 = 0$ ...(*iv*)

$\Rightarrow$ $y_1 = (\mathrm{D}+2)\ y_2$

Putting in equation (*iii*), we get

$$(\mathrm{D}+5)(\mathrm{D}+2)y_2 - (4-\mathrm{K})y_2 = 0$$

$\Rightarrow$ $[\mathrm{D}^2 + 7\mathrm{D} + (6+\mathrm{K})]y_2 = 0$

$\Rightarrow$ $\mathrm{D}^2 + 7\mathrm{D} + 6\mathrm{K} = 0$

$\Rightarrow$ $\mathrm{D} = \frac{-7 \pm \sqrt{49 - 4(6+\mathrm{K})}}{2\times 2}$

For real value of D,

$49 - 4(6+\mathrm{K}) \geq 0 \Rightarrow 49 - 24 - 4\mathrm{K} \geq 0 \Rightarrow 4\mathrm{K} \leq 25$

$\Rightarrow \mathrm{K} \leq \frac{25}{4} \Rightarrow \mathrm{K} \leq 6.25$

4. $\frac{d^2x}{dt^2} + 10\frac{dx}{dt} + 25x = 0$

$\Rightarrow (\mathrm{D}^2 + 10\mathrm{D} + 25)\ x = 0 \quad \Rightarrow (\mathrm{D}+5)^2 = 0$

$\Rightarrow \mathrm{D} = -5, -5$

Hence soloution is $(\mathrm{C}_1 + \mathrm{C}_2 t)\ e^{-5t}$

5. Given differential equation is,

$$\frac{dx}{dt} + \frac{x}{20} = 10$$

$$\Rightarrow \frac{dx}{dt} = 10 - \frac{x}{20} = \frac{200-x}{20}$$

Integrating,

$$20\int_0^{100} \frac{dx}{200-x} = \int_0^t dt$$

$$\therefore \quad t = 20\left[\ln\frac{1}{200-x}\right]_0^{100} = 20 \ln 2$$

6. $xdx + ydy + 2(x^2+y^2)dx = 0$

$$\frac{1}{2}d(x^2+y^2) + 2(x^2+y^2)dx = 0$$

$$\frac{d(x^2+y^2)}{(x^2+y^2)} = -4dx$$

$$x^2 + y^2 = ce^{-4x}$$

7. Given: $\frac{d^2y}{dx^2} + \sin x\frac{dy}{dx} + ye^x = \sinh x$

The above equation is second order and non linear.

8. Given : $\frac{dy}{dt} + 5y = 0$

$\Rightarrow$ $\frac{dy}{y} = -5\ dt$

Integrating, we get $\log_e y = c - 5t$

When $t = 0$, $y = 1$.

$\therefore$ $\log_e 1 = c - 5 \times 0$

$\Rightarrow$ $c = 0$

$\therefore$ $\log_e y = -5t$

$\Rightarrow$ $y = e^{-5t}$

10. y'' indicates that it is of second order. Degree of y'', y' and y is one. Hence given differential equation is second order linear equation.

11. For the differential equation Mdx + Ndy = 0, to be exact, $\frac{\partial \mathrm{M}}{\partial y} = \frac{\partial \mathrm{N}}{\partial x}$

Hence for the differential equation

$f(x, y)\frac{dy}{dx} + g(x, y) = 0$ to be exact $\frac{\partial f}{\partial x} = \frac{\partial g}{\partial y}$

13. $(\mathrm{D}^4 - 81)y = 0$

Auxiliary equation is

$m^4 - 81 = 0 \Rightarrow (m+3)(m-3)(m+3i)(m-3i) = 0$

Hence Solution is

$$y = C_1e^{3x} + C_2e^{-3x} + C_3 \cos 3x + C_4 \sin 3x$$

15. $\text{P.I.} = \frac{1}{D^2 - 2D} e^x \sin x = e^x \frac{1}{(D+1)^2 - 2(D+1)} \sin x$

$$= e^x \frac{1}{(-1^2) - 1} \sin x = -\frac{1}{2} e^x \sin x$$

16. Auxiliary equation of the given equation is,

$D^2 - m^2 = 0$

$\Rightarrow D = \pm m$

Hence C.F. is $y = C_1e^x + C_2e^{-x} = C_1 \cosh x + C_2 \sinh x$

17. $\frac{1}{\phi(D)(D-a)^n} e^{ax} = \frac{1}{\phi(a)} \cdot \frac{1}{(D-a)^n} e^{ax}$

$$= \frac{1}{\phi(a)} x . \frac{1}{\frac{d}{dD}(D-a)^n} e^{ax} \; (\because f(a) = 0)$$

$$= \frac{1}{\phi(a)} x^2 . \frac{1}{n\frac{d}{dD}(D-a)^{n-1}} e^{ax}$$

$(\because f'(a) = 0)$

$$= \frac{1}{\phi(a)} x^2 . \frac{1}{n(n-1)(D-a)^{n-2}} e^{ax}$$

$$= \left[\frac{1}{\phi(a)} . \frac{x^n e^{ax}}{n!}\right]$$

19. $y = 1 + \cos x + \sin x$

$\therefore \frac{dy}{dx} = -\sin x + \cos x$ and $\frac{d^2y}{dx^2} = -\cos x - \sin x$

$\therefore \quad y + \frac{d^2y}{dx^2} = 1$

Hence $y = 1 + \cos x + \sin x$ is solution of, $\frac{d^2y}{dx^2} + y = 1$

20. Auxiliary equation is,

$m^2 - m - 2 = 0$

$\therefore m = -1, 2$

$CF = c_1e^{-x} + c_2e^{2x}$

$$\text{P.I.} = \frac{1}{D^2 - D - 2} .3 . e^{2x}$$

$$= 3. \frac{1}{D^2 - D - 2} e^{2x}$$

$$= 3x \frac{1}{2D - 1} e^{2x} = 3x \frac{1}{3} . e^{2x} = xe^{2x}$$

$\therefore y = c_1e^{-x} + c_2e^{2x} + xe^{2x}$

$$\frac{dy}{dx} = -c_1e^{-x} + c_2e^{2x} + 2xe^{2x} + e^{2x}$$

$y(0) = c_1 + c_2 = 0$ (given)

$y'(0) = -c_2 + 2c_2 + 1 = -2$ (given)

$\therefore c_1 + c_2 = 0$ ...(i)

$c_1 - 2c_2 = 3$ ...(ii)

Solving equations (i) and (ii), we get

$$c_1 = 1, c_2 = -1$$

Hence required solution is

$$y = e^{-x} - e^{2x} + xe^{2x}$$

22. Given equation is,

$$(D^2 - 5D + 6)y = e^{3x}$$

$\therefore \quad \text{P.I.} = \frac{1}{D^2 - 5D + 6} e^{3x} = \frac{1}{(D-3)(D-2)} e^{3x}$

$$= x.e^{3x} - \frac{1}{3-2} e^{3x} = (x-1)\, e^{3x}$$

24. Given differential equation is, $D^2 + D = x^2 + 2x + 4$

$\therefore \quad \text{P.I.} = \frac{1}{D(D+1)} (x^2 + 2x + 4)$

$$= \frac{1}{D} (D+1)^{-1} (x^2 + 2x + 4)$$

$$= \frac{1}{D} (x^2 + 2x + 4 - 2x - 2 + 2) = \frac{x^3}{3} + 4x$$

NUMERICAL TYPE QUESTION

1. Given: $(4t^2 + 1) \frac{dy}{dt} + 8yt - t = 0$

$\Rightarrow \quad \frac{dy}{dt} + \frac{8t}{4t^2 + 1} y = \frac{t}{4t^2 + 1}$ ...(i)

$\therefore \quad \text{I.F.} = e^{\int .8t/(4t^2+1)} = (4t^2 + 1)$

Hence solution of equation (i) is

$$y\,(4t^2 + 1) = \int \frac{t}{(4t^2 + 1)} . (4t^2 + 1)\, dt + C = \frac{t^2}{2} + C$$

$\Rightarrow \quad y = \frac{t^2}{2\,(4t^2 + 1)} + \frac{C}{(4t^2 + 1)}$ ...(ii)

But $\quad y(1) = 0,$

$\therefore \quad 0 = \frac{1}{10} + \frac{C}{5}$

$\Rightarrow \quad C = -\frac{1}{2}$

From equation (ii), we have

$$y = \frac{t^2}{2\,(4t^2 + 1)} - \frac{1}{2(4t^2 + 1)}$$

$\therefore \quad y|_{t \to \infty} = \underset{t \to \infty}{\text{Lt}} \frac{1}{2\,(4 + 1/t^2)} - \frac{1}{2\,(4t^2 + 1)}$

$$= \frac{1}{2\,(4+0)} - 0 = \frac{1}{8}$$

4 CHAPTER Probability and Statistics

PROBABILITY

It is a concept of mathematics which measures the degree of certainty or uncertainty of the occurrence of events.

- If any event can happen in m ways and fails in n ways and each of the $(m + n)$ ways are equally likely to occur, then probability of the happening of the events is defined as the ratio, $\frac{m}{m+n}$ and that of its failing as $\frac{n}{m+n}$.

  If probability of the happening is denoted by p and not happening by q, then $p + q = 1$.
- If event is certain to happen, its probability is unity.
- If happening is impossible, then its probability is zero.

Sampling

A small section selected from the population is called a *sample* and the process of drawing a sample is called *sampling*.

Random Sampling

It is essential that the sample must be a random selection so that each member of the population has the same chance of being included in the sample. Thus the fundamental assumption underlying theory of sampling is *random sampling*

Simple Sampling

A special case of random sampling in which each event has the same probability p of success and the chance of success of different events are independent whether previous trials have been made or not, is called *simple sampling*.

Parameters

The statistical constants of the population such as mean (μ), standard deviation (σ) etc. are called *parameters*.

Statistics

Constants for the sample drawn from the given population, i.e. mean ($\bar{x}$), standard deviation (S) etc. are called *statistic*. The population parameters are in general, not known and their estimates given by the corresponding sample statistic are used. Greek letters are used to denoted population parameters and Roman letters for sample statistic.

Objectives of Sampling

Sampling aims at gathering the maximum information about the population with the minimum effort, cost and time. The object of sampling studies is to obtain the best possible value of the parameters under specific conditions. Sampling determines reliability of these estimates. The logic of the sampling theory is the logic of induction in which we pass from a particular (sample) to general (population). Such a generalization from sample to population is called *statistical inference*.

Sampling Distribution

Consider all possible sample of size n which can be drawn from a given population at random. For each sample, mean can be computed. The mean of the samples will not be identical. If these different means are grouped according of their frequencies, the frequency distribution so formed is called *sampling distribution of the mean*. Similarly sampling distribution of the standard deviation etc, can be obtained.

When drawing each sample, the previous sample is put back so that the parent population remains the same. This is called *sampling with replacement* and all subsequent formulae will pertain to sampling with replacements.

Standard Error

It is used to assess the difference between expected and observed values. Standard deviation of the sampling distribution is called *standard error* **(S.E.).** Thus standard error of the sampling distribution of means is called *standard error of means*. The reciprocal of the standard error is called *precision*.

If $n \geq 30$, a sample is called **large** otherwise **small**. The sampling distribution of large samples is assumed to be normal.

Population

Collection of all possible samples is population. It is impractical to collect data on each sample of a population. Statistics helps to determine best estimate of the population parameter from randomly selected samples. Because random erros are involved in determination of parameters, the estimate will represent a parameter with a given probability only.

Given an event B such that $P(B) > 0$ and any other event A, we define conditional probability of A given by

$$P(A|B) = \frac{P(A \cap B)}{P(B)}$$

Simple Multiplication Rule

$$P(A \cap B) = P(B)P(A|B) = P(A)P(B|A)$$

Example. Let a balanced die is tossed once. Use the definition to find the probability of getting 1, given that an odd number was obtained.

Solution

Probability of A given that the event B has occurred,

$$P(A|B) = \frac{P(A \cap B)}{P(B)} \quad \text{But} \quad P(B) = \frac{1}{2},$$

$$\therefore \quad P(A|B) = \frac{P(A \cap B)}{P(B)} = \frac{\frac{1}{6}}{\frac{1}{2}} = \frac{1}{3}$$

CONDITIONAL PROBABILITY

Consistency of the definition with the relative frequency concept of probability can be obtained from the following construction.

Let an experiment is repeated a large number, of times, N resulting in

both A and B, $A \cap B$, n_{11} times;

A, not B, $A \cap \bar{B}$, n_{21} times;

B, not A, $\bar{A} \cap B$, n_{12} times;

and neither A nor B, $\bar{A} \cap \bar{B}$, n_{22} times.

These results are:

	A	$\bar{A}$	
B	n_{11}	n_{12}	$n_{11} + n_{12}$
$\bar{B}$	n_{21}	n_{22}	$n_{21} + n_{22}$
	$n_{11} + n_{21}$	$n_{12} + n_{22}$	N

$$n_{11} + n_{12} + n_{21} + n_{22} = N.$$

Then it follows that with large N

$$P(A) \approx \frac{n_{11} + n_{21}}{N}, \quad P(B) \approx \frac{n_{11} + n_{12}}{N},$$

$$P(A \cup B) \approx \frac{n_{11} + n_{12} + n_{21}}{N}$$

and conditional probabilities

$$P(B|A) \approx \frac{n_{11}}{n_{11} + n_{21}} \quad \text{and} \quad P(A \cap B) \approx \frac{n_{11}}{N}$$

$$P(A/B) \approx \frac{n_{11}}{n_{11} + n_{12}}$$

Here definitions of P(B | A) and P(A | B) is verified.

Example. Let a box contains r red balls labeled 1, 2, 3, ..., r and b black balls labeled 1, 2, 3, ..., b.

Assume that the probability of drawing any particular ball is $\frac{1}{b+r}$

If a ball from the box is known to be red, what is the probability it is the red ball labeled 1?

Solution

First find probability that it is red (event A).

$$P(A) = \frac{r}{b+r}$$

Then compute probability that the ball is the red ball with the number 1 on it.

$$P(A \cap B) = \frac{1}{b+r}$$

where B is the event that the ball has a 1 on it.

Then probability that the ball is red and labeled 1 given that it is red is given by

$$P(B|A) = \frac{P(A \cap B)}{P(A)} = \frac{\frac{1}{b+r}}{\frac{r}{b+r}} = \frac{1}{r}$$

This differs from the unconditional probability of B, which is given by

$$P(B) = \frac{2}{b+r}$$

Summation Rule

Let $B_1, B_2, ..., B_n$ be (pairwise) disjoint events of positive probability whose union is $\cap$.

Then $$A = \bigcup_{j=1}^{n} (A \cap B_j)$$

Using multiplication rule, we have

$$P(A) = P\left(\bigcup_{j=1}^{n} (A \cap B_j)\right) = \sum_{j=1}^{n} P(A \cap B_j)$$

$$= \sum_{j=1}^{n} P(A|B_j)P(B_j)$$

e.g. Consider rolling a die. Let the sample space be given by $\Omega = \{1, 2, 3, 4, 5, 6\}$

Now consider a partition B_1, B_2, B_3, given by

$B_1 = \{1, 2\}$

$B_2 = \{3, 4\}$

$B_3 = \{5, 6\}$

Let A be the event that the number on the die is equal to one or greater than or equal to four. This gives

A = {1, 4, 5, 6}

$$P(A) = \frac{4}{6} = \frac{2}{3}.$$

This can also be computed using the formula as follows :

$$P(A) = P(A|B_1)P(B_1) + P(A|B_2)P(B_2) + P(A|B_3)P(B_3)$$

$$= \left(\frac{1}{2}\right)\left(\frac{1}{3}\right) + \left(\frac{1}{2}\right)\left(\frac{1}{3}\right) + (1)\left(\frac{1}{3}\right)$$

$$= \frac{1}{6} + \frac{1}{6} + \frac{1}{3} = \frac{2}{3}$$

Bayes Rule

$$P(B_i|A) = \frac{P(A|B_i)P(B_i)}{P(A)}$$

Now, if $B_1, B_2, ..., B_n$ are disjoint events of positive probability whose union is Ω, then

$$P(B_i|A) = \frac{P(A|B_i)P(B_i)}{\sum_{j=1}^{n} P(A|B_j)P(B_j)}$$

Consider same example as before.

Probability for B_1,

$$P(B_1|A) = \frac{P(A|B_1)P(B_1)}{P(A)} = \frac{\left(\frac{1}{2}\right)\left(\frac{1}{3}\right)}{\frac{2}{3}} = \frac{\frac{1}{6}}{\frac{2}{3}} = \frac{1}{4}$$

Probability for B_2,

$$P(B_2|A) = \frac{P(A|B_2)P(B_2)}{P(A)} = \frac{\left(\frac{1}{2}\right)\left(\frac{1}{3}\right)}{\frac{2}{3}} = \frac{\frac{1}{6}}{\frac{2}{3}} = \frac{1}{4}$$

Probability for B_3,

$$P(B_3|A) = \frac{P(A|B_3)P(B_3)}{P(A)} = \frac{(1)\left(\frac{1}{3}\right)}{\frac{2}{3}} = \frac{\frac{1}{3}}{\frac{2}{3}} = \frac{1}{2}$$

Conditional Probability for given Multiple events

Conditional probability of A given $B_1, B_2, ...B_n$, is written $P(A|B_1, B_2, ..., B_n)$ and is defined as

$$P(A|B_1, B_2, ..., B_n) = P(A|B_1 \cap B_2 \cap ... \cap B_n)$$

for any events A, $B_1, B_2, ..., B_n$ such that

$$P(B_1 \cap B_2 \cap ... \cap B_n) > 0$$

More Complex Multiplication Rule

$$P(B_1 \cap B_2 \cap B_3 ... \cap B_n)$$
$$= P(B_1)P(B_2 | B_1)P(B_3 | B_1B_2)...P(B_n | B_1, B_2, ...B_{n-1})$$

DISCRETE RANDOM VARIABLES

Definition, Mean and Variance

Let X be a (discrete) r.v. taking on the values x_i with corresponding probabilities

$$p_i = f(x_i),\ i = 1,...,n.$$

Function $f(x)$ is called **probability density function** (p.d.f.).

It is convenient to present the values of a (discrete) r.v. and the corresponding probabilities in a tabular form as follows.

x	x_1	....	x_n	Total
$f(x)$	p_1	....	p_n	1

The **cumulative distribution function** of a discrete random variable X, denoted as $F(x)$ is defined as

$$F(x) = P(X \le x) = \sum_{x_1 \le x} f(x_i)$$

For a discrete random variable X, $F(X)$ satisfies the following properties:

1. $0 \le F(x) \le 1$
2. If $x \le y$, then $F(x) \le F(y)$

The cumulative distribution function can be used to calculate the probabilities:

$$P(a < x \le b) = P(x \le b) - P(x \le a) = F(b) - F(a)$$

The **mathematical expectation** of r.v. X (or just expectation or mean value of X or just mean of X) is denoted by $E[X]$ and is defined by:

$$E[X] = \sum_{i=1}^{n} x_i f(x_i)$$

The alternative notations $\mu(X)$ or μ, are also often used.

Example.

Suppose an insurance company pays the amount of £1,000 for lost luggage on an airplane trip. From past experience, it is known that the company pays this amount in 1 out of 200 policies it sells. What premium should the company charge?

Solution

Define the r.v. X as follows: $X = 0$ if no loss occurs, which happens with probability $1 - (1/200) = 0.995$, and $X = -1,000$ with probability $\frac{1}{200} = 0.005$.

x	0	−1,000	Total
$f(x)$	0.995	0.005	1

Then the expected loss to the company is:

$$E[X] = 0 \times 0.995 + (-1,000 \times 0.005) = -5.$$

Thus, the company must charge £5 to break even.

Properties of mean:

1. $E[cX] = cE[X]$

Proof

$$E[cX] = \sum_{i=1}^{n} cx f(x_i) = c\sum_{i=1}^{n} x f(x_i) = cE[X]$$

2. $E[cX + d] = cE[X] + d$

Proof

$$E[cX + d] = \sum_{i=1}^{n}(cx_i + d)f(x_i) = \sum_{i=1}^{n} cx_i f(x_i) + \sum_{i=1}^{n} f(x_i)$$
$$= c\sum_{i=1}^{n} x_i f(x_i) + d\sum_{i=1}^{n} f(x_i) = cE[X] + d$$

The **variance** of a r.v. X is denoted by Var(X) and is defined by:

$$Var[X] = E[X - E[X]]^2.$$

The alternative notations $\sigma^2(X)$ and σ_X^2 are also often used for the $Var(X)$.

The variance account for the difference in size of the range of the distributions of r.v. More generally, for a r.v. X taking on finitely many values $x_i, \ldots, x_n$ with respective probabilities $f(x_1), \ldots, f(x_n)$, the variance is:

$$Var[X] = \sum_{i=1}^{n} (x_i - E[X])^2 f(x_i)$$

and represents the sum of the weighted squared distances of the points $x_i = 1, \ldots, n$ from the center of location of the distribution, $E[X]$. Thus, the further from $E[X]$ the x_is are located, the larger the variance, and vice versa. Because of this characteristic property of the variance, the variance is referred to as a **measure of dispersion** of the underlying distribution.

The positive square root of the $Var[X]$ is called the **standard deviation** (s.d.) of X. Unlike the variance, the s.d. is measured in the same units as X.

Properties of variance:

1. $Var[X] = E[X]^2 - (E[X])^2$
2. $Var[cX] = c^2 Var[X]$
3. $Var[cX + d] = c^2 Var[X]$
4. $Var[X] = E[X(X-1)] + E[X] - E[X]^2$

Example

The r.v. X has p.d.f. f given by:

$$f(x) = \begin{cases} 0, & x < 4 \\ 0.1, & x = 4 \\ 0.3, & x = 5 \\ 0.3, & x = 6 \\ 0.2, & x = 8 \\ 0.1, & x = 9 \\ 0, & x > 9. \end{cases}$$

1. Calculate the probabilities $P(X \le 6.5)$, $P(X > 8.1)$, $P(5 < X < 8)$.
2. Calculate and draw the graph of c.d.f. F.
3. Calculate mean $E[X]$.
4. Calculate variance $Var[X]$.
5. If we define new r.v. $Y = 5 — 2X$, calculate $E[Y]$ and Varf [Y"].

Solution

1. $P(X \le 6.5) = P(X = 4) + P(X = 5) + P(X = 6)$

$$= 0.1 + 0.3 + 0.3 = 0.7$$

$$P(X > 8.1) = P(X = 9) = 0.1$$

$$(P(5 < X < 8) = P(X = 6) = 0.3$$

2. $$F(x) = \begin{cases} 0, & x < 4 \\ 0.1, & 4 \le x < 5 \\ 0.4, & 5 \le x < 6 \\ 0.7, & 6 \le x < 8 \\ 0.9, & 8 \le x < 9 \\ 1.0, & x \ge 9 \end{cases}$$

3. $E[X] = 0.1*4 + 0.3*5 + 0.3*6 + 0.2*8 + 0.1*9 = 6.2$
4. $E[X^2] = 0.1*4^2 + 0.3*5^2 + 0.3*6^2 + 0.2*8^2 + 0.1*9^2 = 40.8$

$$Var[X] = E[X^2] - E[X]^2 = 40.8 - 6.2^2 = 2.36$$

5. $E[Y] = E[5 - 2X] = 5 - 2 * E[X] = 5 - 2 * 6.2 = -7.4$

$$Var[Y] = Var[5 - 2X] = (-2)^2 * Var[X] = 4 * 2.36 = 9.44$$

Example

The number of light switch turn-ons at which the first failure occurs is a r.v. X whose p.d.f. is given by :

$$f(x) = c\left(\frac{9}{10}\right)^{x-1}, \; x = 1, 2, \ldots \text{ (and 0 otherwise).}$$

(a) Determine the constant c.

(b) Calculate the probability that the first failure will not occur until the 10th turn on

(c) Determine the corresponding c.d.f. F

Solution

(a) The constant c is determined through the relationship :

$$\sum_{x=1}^{\infty} f(x) = 1 \text{ or } \sum_{x=1}^{\infty} c\left(\frac{9}{10}\right)^{x-1} = 1.$$

However, $\sum_{x=1}^{\infty} c\left(\frac{9}{10}\right)^{x-1}$

$$= c\sum_{x=1}^{\infty}\left(\frac{9}{10}\right)^{x-1} = c\frac{1}{1-\frac{9}{10}}$$

$$= 10\,c, \text{ so that } c = \frac{1}{10}$$

(b) $$P(X > 10) = P(X \ge 11) = c\sum_{x=11}^{\infty}\left(\frac{9}{10}\right)^{x-1}$$

$$= c\left[\left(\frac{9}{10}\right)^{10} + \left(\frac{9}{10}\right)^{11} + \ldots..\right]$$

$$= c\frac{\left(\frac{9}{10}\right)^{10}}{1-\frac{9}{10}} = c10\left(\frac{9}{10}\right)^{10} = \frac{1}{10}10\left(\frac{9}{10}\right)^{10}$$

$$= (0.9)^{10} \approx 0.349.$$

(c) First, for $x < 1$, $F(x) = 0$.

Next, for $x \ge 1$, $F(x) = \sum_{t=1}^{x} c\left(\frac{9}{10}\right)^{t-1}$

$$= 1 - \sum_{t=x+1}^{\infty} c\left(\frac{9}{10}\right)^{t-1} = 1 - c\sum_{t=x+1}^{\infty}\left(\frac{9}{10}\right)^{t-1}$$

$$= 1 - \frac{1}{10} \times \frac{\left(\frac{9}{10}\right)^{x}}{1-\frac{9}{10}} = 1 - \left(\frac{9}{10}\right)^{x}.$$

Thus, $F(x) = 0$ for $x < 1$, and $F(x) = 1 - \left(\frac{9}{10}\right)^{x}$ for $x \ge 1$.

ELEMENTS OF PROBABILITY

1. Experiment

It is a process in which a certain work is repeated under the same conditions, the outcomes (or results) of which need not be the same.

e.g. tossing a coin or rolling a die are experiments

2. Sample Space

The set S of all possible outcomes of a given experiment is called *sample space* for the experiment. An outcome an element of S, is called a *sample point*.

e.g. for experiment of tossing a fair (or unbiased) coin

sample space S = (H, T)

where H, T refer to head, tail respectively.

3. Event

A subset of the sample space S is called an *event*. The set (a) consisting of a single sample point a in S is called *elementary event*.

Since empty set ϕ and S are always subsets of S, they are also events ϕ is called *impossible event* or *null event*, and S is called *sure* or *certain event*.

Given two events A, B. form New events can be formed using the set operations of union, intersection and complementation.

(*i*) $A \cup B$ is the event that occurs if and only if A occurs or B occurs (or both).

(*ii*) $A \cap B$ is the event that occurs if and only if both A and B occurs.

(*iii*) A^c or $\overline{A}$, the complement of A, is the event that occurs if and only if A does not occur.

Mutually Exclusive Events

Two events A and B are said to be mutually exclusive if and only if they cannot occur simultaneously, i.e. if $A \cap B = \phi$.

Three or more events are called *mutually exclusive*, if every two of them are mutually exclusive.

Example. Find total number of possible events that can occur for an experiment.

Solution

Let sample space S consist of n sample points. Then total number of possible events

= total number of subsets of S

= $|p(S)|$

where $p(S)$ is power set of S = 2^n.

Example. In an experiment, a fair coin is tossed 4 times. Describe the sample

Solution

The sample space S consists of 16 (= 2^4) sample points.

S = [HHHH, HHHT, HHTH, HTHH, THHH, HHTT, HTTH, TTHH, THTH, HTHT, THHT, HTTT, THTT, TTHT, TTTH, TTTT]

Finite Sample Space

A sample space S is said to be a finite sample space, if S is a finite set.

Finite Probability Space

Let S = (a_1, a_2 ...., a_n) be a finite sample space. S is said to be a finite probability space or probability model, if each sample point a_i in S we can assign a real number p_i, called probability of a_i such that

(*i*) $p_i \geq 0$, for each i

(*ii*) $p_1 + p_2 + ... + p_n = 1$

If A is an event, then probability of A, denoted by P(A) is sum of the probabilities of the sample points in A. For the elementary event $\{a_i\}$ we write $P(a_i)$ instead of $P(a_i)$.

Equiprobable Space

A finite probability space S in which each sample point has the same probability is called an *equiprobable space*. If E is an event, then

$$P(E) = \frac{\text{number of elements in E}}{\text{number of elements in S}} = \frac{|E|}{|S|}$$

$$= \frac{\text{number of outcomes favourable to E}}{\text{total number of possible outcomes}}$$

- The word *"at random"* is used only when dealing with a equiprobable space. By the statement, *"a ball is drawn at random from a bag considering 10 balls"* we mean that each ball in the bag has the same probability of being chosen.
- If p = probability of happening of an event E = P(E),

q = probability of not happening of event E = $P(\overline{E})$

then $q = P(\overline{E}) = \frac{|\overline{E}|}{|S|} = \frac{|S|-|E|}{|S|} = 1 - \frac{|E|}{|S|} = 1 - p$

$\Rightarrow \quad p + q = 1$

DEPENDENT EVENTS

Events A and B in the space S are said to be independent if occurence of one of them does not influence the occurence of the other

i.e. B is dependent of A if $P(B/A) = P(B)$.

Now, $P(A \cap B) = P(A). P(B | A) = P(A) P(B)$

Thus events A and B are independent, if

$$P(A \cap B) = P(A).P(B)$$

Mutual Independence

Let (E_k : k is a positive integer) be a collection of events. These events are said to be mutually independent if for each finite non-empty subset $\{E_1, E_2, ...E_n\}$

$$P(E_1 \cap E_2 \cap .... \cap E_n) = P(E_1).\ P(E_2)\ ....\ P(E_n)$$

Pairwise Independence

Let $\{E_k$: k is a positive integer) be a collection of events. These events are said to be pairwise independent if

$$P(E_i \cap E_j) = P(E_i).\ P(E_j) \text{ for all } i \neq j.$$

Possible Event

An event E is said to be a possible event, if P (E) > 0.

- Possible independent events are not mutually exclusive.
- Possible mutually exclusive events are not independent.

Example. Let p be the probability that a man aged y years will meet with an accident in a year. What is the probability that a man among n men all aged y years will meet with an accident first ?

Solution

Probability that a man aged y years will not meet with an accident $= 1 - p$

P (none meets with an accident)

$= (1 - p)(1 - p) ......(1 - p)$ (n times) $= (1 - p)^n$

$\therefore$ P (atleast one man meets with an accident)

$= 1 - (1 - p)^n$

So, P (atleast one man meets with anaccident

[a person is chosen) $= \dfrac{1}{n} \times [1 - (1-p)^n]$

Example. A box contains 4 white, 3 blue and 5 green balls. Four balls are chosen. What is the probability that all three colours are represented ?

Solution

Total number of balls in the box is 12.

Hence total number of ways in which 4 balls can be chosen

$$= {}^{12}C_4 = \frac{12 \times 11 \times 10 \times 9}{4 \times 3 \times 2 \times 1} = 495$$

Each colour will be represented in the following mutually exclusive ways :

	White	Blue	Green
(i)	2	1	1
(ii)	1	2	1
(iii)	1	1	2

Hence number of ways of drawing four balls in the above fashion

$= {}^4C_2 \times {}^3C_1 \times {}^5C_1 + {}^4C_1 \times {}^3C_2 \times {}^5C_1 + {}^4C_1 \times {}^3C_1 \times {}^5C_2$

$= 90 + 60 + 120 = 270$

$\therefore$ Required probability $= \dfrac{270}{495}$.

LAW OF TOTAL PROBABILITY

Let B_1 B_2, ... B_n be mutually exclusive and let an event A occur only if anyone of B_i occurs. Then

$$P(A) = \sum_{i=1}^{n} P(A/B_i)P(B_i)$$

Baye's Theorm

Let E_1, E_2, ... E_n be mutually exclusive events such that $P(E_i) > 0$ for each i.

Then for any event $A \subseteq \bigcup_{1}^{n} E_i$ such that $P(A) > 0$, we have

$$P(E_i/A) = \frac{P(A/E_i)P(E_i)}{\sum_{i=1}^{n} P(A/E_i)P(E_i)}, \; i = 1, 2, .., n$$

Example. Three urns A, B and C have 1 white , 2 black, 3 red balls, 2 white 1 black, 1 red balls and 4 white, 5 black, 3 red balls respectively. One urn is chosen at random and two balls are drawn. They happen to be white and red balls. What is the probability that they came from urn B ?

Solution

Let E_1, E_2 and E_3 be the events of choosing urns A, B and C respectively.

Then $\quad P(E_1) = P(E_2) = P(E_3) = \dfrac{1}{3}$ ... (i)

Let X be the event of choosing two balls, white and red. To find $P(E_2/X)$.

By Baye's theorem, (E_2/X)

$$= \frac{P(X/E_2)P(E_2)}{P(X/E_1)P(E_1) + P(X/E_2)P(E_2) + P(X/E_3)P(E_3)}$$

$$= \frac{P(X/E_2)}{P(X/E_1) + P(X/E_2) + P(X/E_3)} \quad \{\text{from } (i)\}$$

Now, $\quad P(X/E_1) = \dfrac{{}^1C_1 \times {}^3C_1}{{}^6C_2} = \dfrac{1}{5}$;

$P(X/E_2) = \dfrac{{}^2C_1 \times {}^1C_1}{{}^4C_2} = \dfrac{1}{3}$ and $P(X/E_3)$

$= \dfrac{{}^4C_1 \times {}^3C_1}{{}^{12}C_2} = \dfrac{2}{11}$

From (ii), $\quad P(E_2/X) = \dfrac{\frac{1}{3}}{\frac{1}{5} + \frac{1}{3} + \frac{2}{11}} = \dfrac{55}{118}$

MEASURES OF CENTRAL TENDENCY

1. **Mean** (Arithmetic Mean A.M. or the average),
2. **Median** (positional average)
3. **Mode** (value of the variable which occurs most, frequently)

4. **Geometric Mean** (G.M. used extensively in finding the rate of population growth)
5. **Harmonic Mean** (H. M.)

1. Arithmetic Mean ($\bar{x}$)

Let $x_1, x_2, \ldots\ldots\ldots\ldots x_n$ be the observed values and let f_i be the frequency of x_i (t_i) value of x_i repeated f_i times.

i.e., if the data is 3, 1, 3, 3, 2, 1, 3, 2, 2, then

$x_1 = 1, x_2 = 2$ and $x_3 = 3$, and $f_1 = f_2 = 3$ and $f_3 = 4$.

$$\bar{x} = \frac{1}{n}(x_1 + x_2 + \ldots + x_n) = \frac{1}{n}\sum_{i=1}^{n} x_i$$

and for frequency distribution

$$\bar{x} = \frac{1}{N}(f_1x_1 + f_2x_2 + \ldots + f_nx_n)$$

$$\bar{x} = \frac{1}{N}\sum_{i=1}^{n} f_ix_i \quad \text{where} \quad N = \sum_{i=1}^{n} f_i$$

2. Median

Median is the middle score. If number of events are even then average of the two middles are taken. The median is better for describing the typical value.

Example. At a ski rental shop, data was collected on number of rentals on each of ten consecutive Saturdays:

44, 50, 38, 96, 42, 47, 40, 39, 46, 50.

To find sample mean, add them and divide by 10:

$$\frac{44+50+38+96+42+47+40+39+46+50}{10} = 49.2$$

Note: Mean value is not a value of the sample.

To find the median, first sort the data:

38, 39, 40, 42, 44, 46, 47, 50, 50, 96

There are two middle numbers 44 and 46. To find the median, we take average of the two.

$$\text{Median} = \frac{44+46}{2} = 45$$

Also mean is larger than all but three of the data points. The mean is influenced by outliers while the median is robust.

$$\text{Median} = L + \left(\frac{\frac{N}{2} - F_{i-1}}{f_m}\right)h$$

where, L = lower limited of Median class

N = Total Number of data items

h = width of Median class

f_m = frequency of Median class.

3. Mode

Mode is the value which occurs most frequently.

$$\text{Mode} = a + \frac{C(f_i - f_{i-1})}{2f_i - f_{i-1} - f_{i+1}}$$

where a = modal class

f_i = maximum frequency

C = constant difference for each class.

Example. Find median of the following data.

Cost	10–20	20–30	30–40	40–50	50–60
Items in a group	4	5	3	6	3

Solution

Cost	Number of items in the group	Cumulative frequency
10 – 20	4	4
20 – 30	5	9
30– 40	3	12
40 – 50	6	18
50 – 60	3	21

Here, $N = 21.$

$\therefore$ $\frac{N}{2} = 10.5$

The median class is 30 – 40.

$$\text{From formula, Median} = 30 + \frac{10}{12}(10.5 - 9) = 30 + 1.25 = \mathbf{31.25}$$

Example. Find mode for the following distribution:

Class interval	0 – 10	10 – 20	20 – 30	30 – 40	40 – 50	50 – 60	60 – 70	70 – 80
Frequency	5	9	8	12	28	20	12	11

Solution

Maximum frequency = 28

$\therefore$ Modal class = 40 – 50

$$\text{Mode} = a + \frac{C(f_i - f_{i-1})}{2f_i - f_{i-1} - f_{i+1}}$$

$$= 40 + \frac{10(28-12)}{(2\times 28)-12-20} = 40 + 6.666 = \mathbf{46.67}$$

Example. The expenditure of 100 families is given in the following frequency table.

Monthly Expenditure (in hundreds of rupees)	0 – 10	10 – 20	20 – 30	30 – 40	40 – 50
Number of families	14	?	27	?	15

If mode for the distribution is 24, calculate the missing frequencies.

Solution

Let missing frequencies for the classes 10 – 20 and 30 – 40 be f_1 and f_2 respectively.

Expenditure	**Number of families**	**Cumulative frequency**
0 – 10	14	14
10 – 20	f_1	$14 + f_1$
20 – 30	27	$41 + f_1$
30 – 40	f_2	$41 + f_1 + f_2$
40 – 50	15	$56 + f_1 + f_2$

$$N = 100 = 56 + f_1 + f_2$$

$$\Rightarrow \quad f_1 + f_2 = 100 - 56 = 44 \qquad \ldots(i)$$

Mode is given to be 24, which lies in the class 20 – 30. So 20 – 30 is the modal class.

$$\text{Mode} = a + \frac{C(f_i - f_{i-1})}{2f_i - f_{i-1} - f_{i+1}}$$

$$\therefore \quad 24 = 20 + \frac{10(27 - f_1)}{(2\times 27) - f_1 - f_2}$$

$$\Rightarrow \quad 24 - 20 = \frac{270 - 10f_1}{54 - f_1 - f_2} = \frac{270 - 10f_1}{54 - 44} \qquad (\because f_1 + f_2 = 44)$$

$$\therefore \quad f_1 = 23$$

Substituting in (*i*), we get $f_2 = 21$.

Relation between Mean, Mode and Median

For a symmetrical distribution

Mean = Median = Mode

For a moderately asymmetrical distribution

Mean – Mode = 3 (Mean – Median)

4. Geometric Mean (G)

$$G = \sqrt[n]{x_1 x_2 \ldots\ldots x_n} = (x_1 x_2 \ldots\ldots\ldots x_n)^{1/n}$$

For a frequency distribution,

$$G = \left(x_1^{f_1} x_2^{f_2} \ldots\ldots x_n^{f_n}\right)^{1/N}, \quad \text{where } N = \sum_{i=1}^{n} f_i$$

For both cases, taking log

$$\log G = \frac{1}{n}\sum_{i=1}^{n} \log(x_i)$$

For a frequency distribution

$$\log G = \frac{1}{N}\sum_{i=1}^{n} f_i \log(x_i)$$

- If n_1 and n_2 are sizes and G_1 and G_2 are geometric means of two series, then geometric mean of the combined series is given by

$$\log G = \frac{n_1 \log G_1 + n_2 \log G_2}{n_1 + n_2}$$

- Geometric mean is used to find the rate of population growth and the rate of interest and it is also used in the construction of index numbers.

5. Harmonic Mean

It is the reciprocal of mean of reciprocals, i.e.

$$H = \frac{1}{\left(\frac{1}{n}\sum_{i=1}^{n}(1/x_i)\right)}$$

For a frequency distribution,

$$H = \frac{1}{\left(\frac{1}{N}\sum_{i=1}^{n}(f_i / x_i)\right)}; \text{ where } N = \sum_{i=1}^{n} f_i$$

Skewness

It is defined as lack of symmetry.

$$\text{Coefficient of skewness} = \frac{\text{Mean} - \text{Mode}}{\text{Standard deviation}}$$

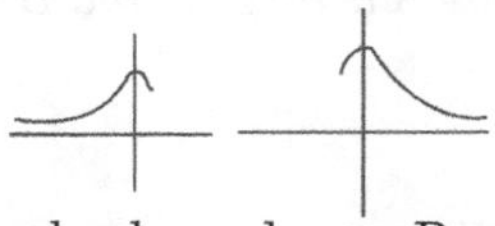

Negatively skewed Positively skewed

MEASURES OF DISPERSION

The set of constant which would in a concise way explain the "*variability*" or "*spread*" in a data is called *measures of dispersion or variability.*

The average for two groups of the same number of measurements may be equal, but one group may be more variable than the other.

e.g. set of five values 5, 6, 7, 8, 9 has the mean as 7; while another set of five values 1, 6, 4, 10, 14 also has the same mean 7. The second set has, obviously more variability than the first.

Usually four measures of dispersion or variability are defined:

1. Range

This is the difference between two extreme observations in the data given.

In a frequency distribution ,

R = (largest x value) – (smallest x value).

Obviously, this is not a very satisfactory measure except in cases where the number of observations are small and a quick calculation for the scatter is needed. It is used in statistical quality control studies rather widely.

2. Quartile Deviation (Q.D.)

Median bisects the distribution. If distribution divided into four parts, *quartiles* are obtained. *First quartile*, Q_1, would have 25 percent of the values below it and the rest above it;

Third quartile, Q_3 would have 75 percent of values below it and the rest above it. The method of calculation of quartiles is similar to that of the median with slight variations.

$$Q_1 = l + \frac{\frac{N}{4} - f_{Q_1}}{f} \times C; \qquad Q_3 = l + \frac{\frac{3N}{4} - f_{Q_3}}{f} \times C$$

where we locate Q_1 – class and Q_3 – class properly

l = lower limit of the quartile class

C = common factor

Quartile deviation is defined as

$$\text{Q.D.} = \frac{1}{2}(Q_3 - Q_1)$$

3. Average Deviation (A)

If average chosen is A (say), then average deviation about A is average deviation

$$\text{A.D. (A)} = \frac{1}{n}\sum |(x_i - A)| \text{ for discrete data}$$

$$= \frac{1}{n}\sum f_i |x_i - A| \text{ for a frequency distribution.}$$

Usually, either median or the mean are takes as the average.

4. Standard Deviation (Root Mean Square Deviation)

Standard deviation,

$$\sigma = \sqrt{\frac{1}{n}\Sigma(x_i - \bar{x})^2} \quad \text{for discrete data}$$

$$= \sqrt{\frac{1}{N}\Sigma f_i(x_i - \bar{x})^2} \quad \text{for frequency distribution.}$$

Square of the standard deviation , σ^2 is defined as the *variance* (V).

$$\therefore \quad V = \sigma^2 = \frac{1}{N}\Sigma f_i(x_i - \bar{x})^2$$

Out of these measures, σ is widely used as a companion to $\bar{x}$ on which is based, when dealing with dispersion or scatter.

Calculation of V or σ

$$V = \left[\frac{\Sigma f_i d_1^2}{N} - \frac{(\Sigma f_i d_1^2)^2}{N}\right] \times C^2$$

Some important results

- Average deviation is least when taken from the median.
- Standarsd deviation is not less than the average deviation from the mean.
- For raw data $x_1, x_2, \ldots\ldots x_n$,

$$V = \frac{\Sigma x^2}{N} - \left(\frac{\Sigma x}{N}\right)^2$$

$$\text{or} \quad V = \frac{\Sigma x^2}{n} - (\bar{x})^2$$

If n_1, n_2 are sizes of two groups, $\bar{x}_1$, $\bar{x}_2$ their means; σ_1, σ_2 their standard deviations, then standard deviation of the combined group is determined from

$$(n_1 + n_2)\,\sigma_2 = n_1\left(\sigma_1^2 + d_1^2\right) + n_2\left(\sigma_2^2 + d_2^2\right)$$

where, $d_1 = \bar{x} - \bar{x}_1$ and $d_2 = \bar{x} - \bar{x}_2$,

$\bar{x}$ being the combined mean.. This result can be extended to more than two groups.

Determination of Variance and Standard Deviation

Steps

1. Calculate the mean, $\bar{x}$.
2. Write a table that subtracts the mean from each observed value.
3. Square each of the differences.
4. Add this column.
5. Divide by n –1, where n is number of items in the sample This is the *variance*.
6. To get the *standard deviation,* take square root of the variance.

Example. The owner of a restaurant is interested in how much people spend at the restaurant. He examines 10 randomly selected receipts for parties of four and writes down the following data :

44, 50, 38, 96, 42, 47, 40, 39, 46, 50

Find mean, standard deviation and variance.

Solution

Mean is calculated by adding and dividing by 10

Mean, $\bar{x}$ = 49.2

Following table is used for getting standard deviation :

p	*x*–49.2	(*x*–49.2)²
44	–5.2	27.04
50	0.8	0.64
38	11.2	125.44
96	46.8	2190.24
42	–7.2	51.84
47	–2.2	4.84
40	–9.2	84.64
39	–10.2	104.04
46	–3.2	10.24
50	0.8	0.64
Total		**2600.4**

$$\text{Now } V = \left(\frac{2600.1}{10-1}\right) = 288.7 \approx 289$$

Hence variance is 289 and standard deviation is square root of 289 = 17.

Since standard deviation can be thought of measuring how far the data values lie from the mean, we take mean and move one standard deviation in either direction. The mean for this example was about 49.2 and the standard deviation was 17.

We have, 49.2 – 17 = 32.2

and 49.2 + 17 = 66.2

What is means is that most of the patrons probably spend between 32.20 and 66.20.

Coefficients to Dispersion

When two series of measurements have to be compared, then averages may be different and the units which the measurements are recorded may also be different. Hence, coefficients of dispersion are provided as

- $\dfrac{\text{Range}}{A+B}$, *where A is largest and B is smallest of the values.*
- $\dfrac{Q_3 - Q_1}{Q_3 + Q_1}$
- $\dfrac{\text{Average deviation about} A}{A}$
- *Each of these is free from units of measure*ment and is a pure number.

  Coefficient of variation (C.V.) $= \dfrac{\sigma}{x} \times 100$

  Thus coefficient of variation is a percentage.

RANDOM VARIABLE

Many variables of interest in a system are not expressible on explicit function of time, all such variable on explicit function of time, all such variable are classified as random variable. These are implying uncertainty about their exact nature.

Random Variable

Discrete | Continuous

Random variable deal with probability theory

Discrete Probability Theory | Continuous Probability Theory

Let X be a random variable.

- Function $F(x) = P(X \le x)$ is called *distribution function of X.*
- Mean or Expectation of $X = \mu = E(X)$

$$= X(a_1)P(a_1) + \ldots\ldots\ldots\ldots + X(a_n)P(a_n)$$

$$= \sum_{i=1}^{n} X(a_i)P(a_i),$$

where S is probability space $(a_1, a_2, \ldots.. a_n)$

- Variance of $X = Var(X) = \sigma^2 = E(X^2) - [E(X)]^2$

Standard deviation of $X = \sigma$

Example. Find expectation of the number of points when a fair die is rolled.

Solution

Let X be the random variable showing number of points.

Then $X = 1, 2, 3, 4, 5, 6$

a_i	$P(X = a_i) = P(a_i)$	Product
1	$\frac{1}{6}$	$\frac{1}{6}$
2	$\frac{1}{6}$	$\frac{2}{6}$
3	$\frac{1}{6}$	$\frac{3}{6}$
4	$\frac{1}{6}$	$\frac{4}{6}$
5	$\frac{1}{6}$	$\frac{5}{6}$
6	$\frac{1}{6}$	$\frac{6}{6}$
		$E(X) = \frac{21}{6} = \frac{7}{2}$

$\therefore$ Expectation $= \frac{7}{2}$.

Example. What is the expectation of the number of failure preceding the first success in an infinite series of independent trials with constant probability of success p in each trial ?

Solution

Let p = probability of succes

$\therefore$ q = probability of failure $= 1 - p$

Let X be the random variable representing the number of failure preceding the first success.

Since an infinite series of independent trials are conducted, X = 0 , 1, 2

$P(X = x)$ = probability that there are x failures preceding the first success.

$$= q \times q \times \ldots\ldots.. \times q \times p = q^x p$$

$$\therefore \quad E(X) = \sum_{x=0}^{\infty} xp(X = x) = \sum_{x=0}^{\infty} xq^x p$$

$$= 1\,qp + 2q^2p + 3q^3p + \ldots\ldots$$

$$= qp(1 + 2q + 3q^2 + \ldots\ldots)$$

$$= qp(1-q)^{-2}\ [\because (1-q)^{-2} \text{by Binomial theroem}]$$

$$= \frac{qp}{(1-q)^2} = \frac{qp}{p^2} = \frac{q}{p}$$

Example. The Bernouilli probability law with parameter p in which $0 \le P \le 1$ is defined by a random variable X taking the value 1 with probability P and the value 0 with probability q. Find the mean and variance of X.

Solution

a_i	$P(a_i)$	Product
1	p	p
0	q	0
		$E(X) = p$

$\therefore$ Mean $= E(X) = p$.

Now to find Var (X)

a_i	a_i^2	$P(a_i)$	Product $[a_i^2 \times P(a_i)]$
1	1	p	p
0	0	q	0
			$E(X^2) = p$

$$\therefore \quad Var(X) = E(X^2) - [E(X)]^2$$

$$= p - p^2$$

$$= (1 - p)\,p = pq$$

Absolute Frequency

Let a value x_i is repeated f_i times, then f_i is called *frequency or absolute frequency.*

Let N = sum of all frequencies f_i, then

$\frac{f_i}{N}$ is called *relative frequency* , $i = 1, 2 \ldots\ldots..n$.

Sum of all relative frequencies

$$= \frac{f_1}{N} + \frac{f_2}{N} + ...+ \frac{f_n}{N}$$

$$= \frac{f_1 + f_2 + ...+ f_n}{N} = \frac{N}{N} = 1.$$

PROBABILITY DISTRIBUTION

Let a random experiment is repeated n times. If an event A occurs n_A times, then its probability is defined as

$$p(A) = p(x_k) = \lim_{n\to\infty} \frac{n_A}{n}$$

= Probability of outcome A.

where, n_A = times occur A

n = total time experiment

- This limit may not exist.
- This definition presumes that all outcomes are equally likely to occur.

$$p(A + B) = p(A) + p(B) - p(AB)$$

Here, p(AB) = joint probability.

Type of Probability Distribution

It is of two types :

1. Discrete Distribution

It is of two types

(i) Binomial Distribution

It is concerned with trials of a repetitive nature in which only the occurrence or non-occurrence, success or failure, acceptance or erejection, yes or no of a particular event is of interest.

If a series of independent trials are perfomed such that for each trial, *p* is the probability of success and q that of a failure, then probability of r successes in a series of n trials is given by $^nC_r p^r q^{n-r}$, where r takes any integral value from 0 to n.

The probabilities of 0, 1, 2,......r,.....,n successes are therefore, given by

$$q^n, {}^nC_1 pq^{n-1}, {}^nC_2 p^2q^{n-2},....{}^nC_r p^r q^{n-r},......,p^n$$

Probability of the number of successes so obtained is called **binomial distribution** because probabilities are successive terms in the expansion of the binominal $(q + p)^n$.

Hence sum of the probabilities

$$= q^n + {}^nC_1 pq^{n-1} + {}^nC_2 p^2q^{n-2} +....+ p^n$$

$$= (q + p)^n = 1$$

Constants of Binomial distribution.

Moment generating function about the origin,

$$M_0(t) = E(e^{tx}) = \sum {}^nC_1 p^x q^{n-x} e^{tx}$$

$$= \sum {}^nC_x (pe^t)^x q^{n-x} = (q + pe^t)^n$$

Differentiating with respect to t and putting *t* = 0, we get mean given by

$$M_m(t) = e^{-npt} (q + pe^t)^n = (qe^{-pt} + pe^{qt})^n$$

$$= 1 + pq\frac{t^2}{2!} + pq(q^2 - p^2)\frac{t^3}{3!} + pq(q^3 + p^3)\frac{t^4}{4!} + ...$$

$$\Rightarrow \quad 1 + \mu_1 t + \mu_2\frac{t^2}{2!} + \mu_3\frac{t^3}{3!} + \mu_4\frac{t^4}{4!} + ...$$

$$= 1 + npq\frac{t^2}{2!} + npq(q - p)\frac{t^3}{3!} + npq\,[1 + 3(n - 2)\,pq]\frac{t^4}{4!} + ...$$

Equating coefficients of like power of t on either side, we have

$$\mu_2 = npq,\ \mu_3 = npq(q-p),\ \mu_4 = npq\,[1+ 3(n - 2)pq]$$

Also, $$\beta_1 = \frac{\mu_3^2}{\mu_2^3} = \frac{(q-p)^2}{npq} = \frac{(1-2p)^2}{npq}$$

$$\beta_1 = \frac{\mu_4}{\mu_2^2} = 3 + \frac{1 - 6pq}{npq}$$

∴ Mean = np

Standard deviation = √(npq)

Skewness = $\frac{1-2p}{\sqrt{npq}}$

Kurtosis = β_2

Obs. Skewness is

positive for $p < \left(\frac{1}{2}\right)$

negative for $p > \left(\frac{1}{2}\right)$

zero for $p = \left(\frac{1}{2}\right)$, i.e. when probability curve of binomial distribution will be symmetrical (bell-shaped)

As number of trials *n* increase indefinitely $\beta_1 \to 0$ and $\beta_2 \to 3$

Binomial Frequency Distribution

If n independent trials constitute one experiment and this experiment be repeated N times, then frequency of r successes = $N^n C_r p^r q^{n-r}$.

The possible number of successes together with these expected frequencies constitute binomial frequency distribution.

Applications of Binomial Distribution

This distribution is applied to problems concerning

(*a*) number of defectives in a sample from production line.

(*b*) estimation of reliability of system.

(*c*) number of rounds fired from a gun hitting a target.

(*d*) rader detection.

Bernouilli Trials

Independent repeated trials of an experiment with two outcomes only are called *Bernouilli trials,* call One of the outcomes is called *success* and the other outcome is called *failure*.

Let p = probability of success in a Bernouilli trial

q = probability of failure = $1 - p$.

A binomial experiment consisting of a fixed number n of trials is denoted by B (n, p).

Probability of r success in the experiment B (n, p) is given by

$$P(r) = {}^nC_r\, p^r\, q^{n-r}$$

The function $P(r)$ for $r = 0, 1, 2, \ldots, n$ for B (n, p) is called **binomial distribution.**

Example. If four coins are tossed , find the chance that there should be two heads and two tails.

Solution

$$p = \text{probability of a head} = \frac{1}{2}$$

$$q = \text{probability of a tail} = 1 - p = \frac{1}{2}$$

Hence probability of getting 2 heads (and 2 tails) when four coins are tossed

$$= P(2) = {}^4C_2\, p^2\, q^{4-2} = 6 \times \left(\frac{1}{2}\right)^2 \times \left(\frac{1}{2}\right)^2 = \frac{3}{8}$$

(*ii*) Poisson Distribution

It is a distribution related to the probabilities of events which are extremely rare, but which have a large number of independent opportunities for occurrence.

e.g. number of persons born blind per year in a large city is the phenomena, in which this law is followed.

This distribution can be derived as limiting case of the binomial distribution by making n very large and p very small keeping np fixed (= m, say)

The probability of r successes in a binomial distribution

$$P(r) = {}^nC_r p^r q^{n-r}$$

$$= \frac{n(n-1)(n-2)\ldots(n-r+1)\,p^r q^{n-r}}{r!}$$

$$= \frac{np(np-p)(np-2p)..(\overline{np-r}-1p)(1-p)^{n-r}}{r!}$$

As $n \to \infty$, $p \to 0$ $(np = m)$,

$$p(r) = \frac{m^r}{r!} \underset{n\to\infty}{\text{Lt}} \frac{(1-m/n)^2}{(1-m/n)^r} = \frac{m^r}{r!}\, e^{-m}$$

so that probabilities of 0, 1, 2, ..., r,...... success in poisson distributio is given by

$$e^{-m},\ me^{-m},\ \frac{m^2 e^{-m}}{2!}, \ldots \frac{m^r e^{-m}}{r!}, \ldots$$

Sum of these probabilities is unity as it should be.

2. Continuous Probability Distribution

When a variate X takes every value in an interval, it gives rise to continuous distribution of X. e.g. distributions defined by the variates like heights or weights are continuous distributions.

A major conceptual difference, however, exists between discrete and continuous probabilities. When thinking in discreate terms, the probability associated with an event is meaningful. With continuous event, however, where number of events is infinitely large, the probability that a specific event will occur is practically zero. For this reason, continuous probability statements must be worded somewhat, differently from discrete ones. Instead of finding the probability that z equals some value, we find the probability of x falling in a small interval.

Thus probability distribution of a continuous variate x is defined by a function f(x) such that probability of the variate x falling in the small interval

$$[x - \frac{1}{2}dx] \text{ to } [x + \frac{1}{2}dx] \text{ is } f(x)\,dx.$$

Let X be a continuous random variable. If $f(x)$ is probability density functions of X, then

- $P(X \le a) = \int_{-\infty}^{a} f(x)dx$
- $P(a \le X \le b) = \int_{a}^{b} f(x)dx$
- Expectation of X = E (X) = Mean = $\int_{-\infty}^{\infty} xf(x)dx$
- Variance of X = y (X) = $\int_{-\infty}^{b} [x - E(x)^2 f(x)dx]$
- $\int_{-\infty}^{\infty} f(x)dx = 1$

Types of Continuous Probability Distribution

(*i*) Uniform Distribution

A continuous random variable X is said to follow uniform distribution or rectangular distribution if its probability density function is given by

$$f(x) = \frac{1}{b-a},\ a \le x \le b = 0, \text{ otherwise}$$

The uniform distribution generally arises in the study of round off errors where measurements are recorded upto a certain level of accuracy.

Cumulative distribution function of X is given by

$f(x) = p\,(X \le x).$

$$= \int_{-\infty}^{x} f(x)dx = \begin{cases} 0 & if\ x \le a \\ \dfrac{x-a}{b-a}, & if\ a \le x \le b \\ 1 & if\ x > b \end{cases}$$

$$\text{Mean} = \frac{a+b}{2}, \text{ and Variance} = \frac{(b-a)^2}{12}$$

Example. The melting point X of a certain specimen be assumed to be a continuous random variable which is uniformly distributed over the interval [110, 120]. Find density function of X, mean of X, variance of X and P $(112 \le x \le 115.)$.

Solution

Here, $a = 100$, $b = 120$

$$\therefore \quad f(x) = \begin{cases} \dfrac{1}{120-110} & 110 \le x \le 120 \\ 0, & \text{otherwise} \end{cases}$$

$$= \begin{cases} \dfrac{1}{10} & 110 \le x \le 120 \\ 0, & \text{otherwise} \end{cases}$$

$$\text{Mean of X} = \frac{a+b}{2} = \frac{110+120}{2} = 115$$

$$\text{Variance of X} = \frac{(b-a)^2}{12} = \frac{(120-110)^2}{12} = \frac{25}{3}$$

$$P\,(112 \le x \le 115) = \int_{112}^{115} f(x)dx = \int_{112}^{115} \frac{1}{10}dx = \frac{3}{10}$$

(ii) Exponential Distribution

A continuous random variable X is said to have the exponential distribution if its probability density function is given by

$$f(x) = \begin{cases} ae^{-ax}, & x > 0 \\ 0, & \text{otherwise} \end{cases}$$

a being some positive number, called its *parameter*

This distribution arises in the study of life-length of electronic components.

Also the inter-arrival time and service time in queuing theory have exponential distributions.

$$\text{Mean} = \frac{1}{a}, \text{ Variance} = \frac{1}{a^2}$$

Example. The sales tax return of a salesman is exponentially distributed with parameter $\frac{1}{4}$. What is the probability that his sale will exceed ₹ 10,000 assuming that sales tax is charged at the rate of 5% on the sales ?

Solution

$$\text{Here,} \qquad f(x) = \begin{cases} \dfrac{1}{4}e^{-\frac{x}{4}}, & x > 0 \\ 0 \quad , & \text{otherwise} \end{cases}$$

Sales tax for the sale of ₹ 10,000

$$= 10000 \times \frac{5}{100} = ₹\ 500$$

Now, P $(x > 500)$ = 1 P $(x \le 500)$

$$= 1 - \int_0^{500} f(x)dx = 1 - \frac{1}{4}\int_0^{500} e^{-x/4}dx$$

$$= 1 + \left(e^{-x/4}\right)_0^{500} = e^{-125}$$

(iii) Normal Distribution

A continuous random variable X is said to follow normal distribution if its probability density function is

$$f(x) = \frac{1}{\sigma\sqrt{2\pi}} e^{-\frac{(x-\mu)^2}{2\sigma^2}}$$

$$-\infty \le x \le \infty,\ \sigma > 0,\ -\infty < \mu < \infty$$

In such a situation, write X – N (μ, σ^2)

The distribution involves two parameters μ and σ.

Properties

(a) Distribution is symmetrical.

(b) Mean = μ, Variance = σ^2.

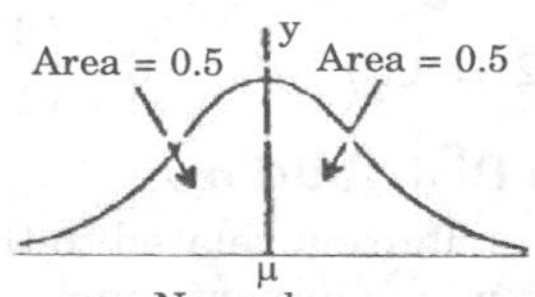

Normal curve

(c) Mean, median and mode coincide.

(d) $f(x) \ge 0$ for all x.

(e) $\int_{-\infty}^{\infty} f(x)dx = 1$, *i.e.* total area under the curve $y = f(x)$ bounded by the axis of x is 1.

Note

Curve $y = f(x)$, called *normal curve* is a *bell-shaped curve*. It is symmetrical about $x = \mu$. The two tails on the left and right sides of the mean extend to infinity.

Put $z = \dfrac{x-\mu}{\sigma}$. Then z is called a standard *normal variate* and its probability density function is given by

$$p(z) = \frac{1}{\sqrt{2\pi}} e^{-\frac{z^2}{2}}, \ -\infty \le z \le \infty.$$

Mean of the standard normal distribution is 0 and variance is 1. Write Z – N (0, 1).

Area under the standard normal curve

(a) Between $z = -1$ and $z = 1$ is 0.6827 (since total area under standard normal curve is 1)

i.e. $p(-1 < z < 1) = 0.6827$

(b) Between $z = -2$ and $z = 2$ is 0.9545

i.e., $p(-2 < z < 2) = 0.9545$

(c) Between $z = -3$ and $z = 3$ is 0.9973

i.e., $p(-3 < z < 3) = 0.9973$

In other words,

$p(\mu - \sigma < x < \mu + \sigma) = 0.6827$

$p(\mu - 2\sigma < x < \mu + 2\sigma) = 0.9545$

$p(\mu - 3\sigma < x < \mu + 3\sigma) = 0.9973$

$p(\mu - 1.96\,\sigma < x < \mu + 1.96\sigma) = 0.95$

$p(\mu - 2.58\sigma < x < \mu + 2.58\,\sigma) = 0.99$

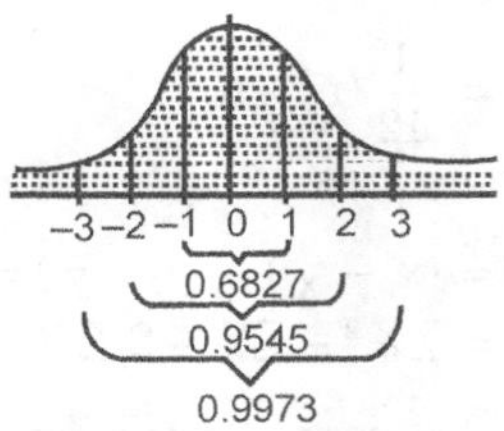

Since distribution is symmetric, we consider only positive values of z.

Then $$\phi(z) = \frac{1}{\sqrt{2\pi}} \int_0^z e^{-\frac{z^2}{2}}\, dz$$

gives area under the normal curve between $z = 0$ and $z = z$.

For different values of z, $\phi(z)$ can be calculated. These values of $\phi(z)$ have been tabulated.

There is another table called *table of ordinates*. We read

$$p(z) = \frac{1}{\sqrt{2\,\pi}} e^{-\frac{z^2}{2}}$$

corresponding to a value of z from this table.

Many distributions tend to a normal distribution in the limit.

When a variable is not normal, it can be made normal using some suitable transformation.

When the sample size is large, distributions of the sample mean, sample variance etc. approach normality. Thus distribution forms a basis for tests of significance.

Normal distribution is also called *distribution of errors*.

Normalized Gaussian Random Variable

It mean of the standard normal distribution is 0 and variance is 1, write $Z - N(0, 1)$.

The distribution function of X are plotted respectively. Mean and variance of X are

$$\mu_x = E[X] = \mu$$
$$\sigma_x^2 = E[(X - \mu_x)^2] = \sigma^2$$

Notation $N(\mu\,;\sigma^2)$ are used to denote that X is normal (or Gaussian) with mean m and variance σ^2.

In particular, $X = N(0\,;1)$, i.e. X with zero mean and unit variance is defined as a *normalized Gaussian random variable.*

Central Unit Theorem

Normal (or Gaussian) distribution has played a significant role in the study of random phenomena in nature. Many naturally occurring random phenomena are approximately normal. *Central–limit theore* states that *sum of a large number of independent random variables, under certain conditions can be approximated by a normal distribution.*

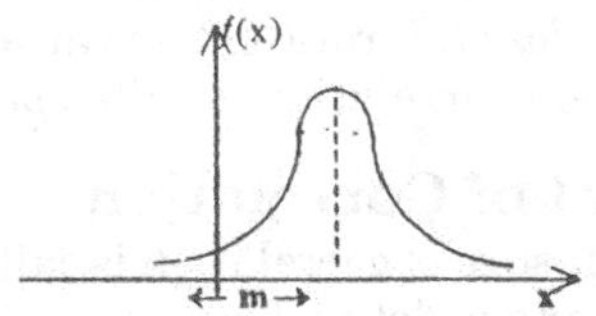

$$f(x) = \frac{1}{\sigma\sqrt{2\pi}} \exp - \frac{(x-\mu)^2}{2\sigma^2}$$

If mean = 0, i.e. $\mu = 0$,

then $$f(x) = \frac{1}{\sigma\sqrt{2\pi}} \exp - \frac{(x)^2}{2\sigma^2}$$

Gaussian Process

$$f(y) = \frac{1}{\sqrt{(2\,\pi\,\sigma_y^{\,2})}} \exp \frac{[-(y-\mu_y)^2}{2\sigma_y^{\,2}}$$

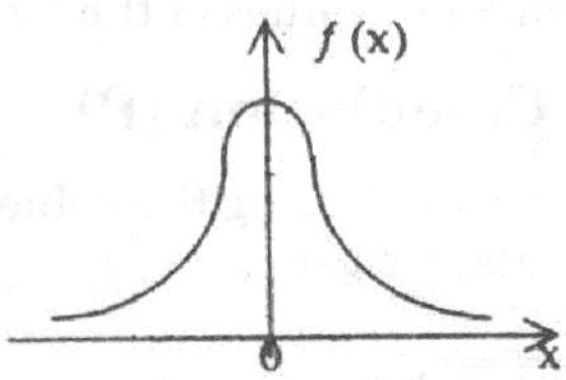

A Gaussian Pulse

where, μ_y = mean

$\sigma_y^{\,2}$ = variance

Properties of Gaussian Process

(*i*) If a Gaussian process x(t) is applied to a stable linear filter, then random process y(t) developed. The output of the filter is also Gaussian.

(*ii*) Consider the set of random variables or samples $x(t_1), x(t_2), x(t_3)$ ----- obtained by observing a random process x(t) at time t_1, t_2, -----.

Second Moment or Mean Square Value

Consider random variable X which assume the possible value x_1, x_2, x_3 ----- x_m. In a sequence of n experiment, let event x_1 occur n_1 time, x_2 occur n_2 times etc.

Arithmetic average or mean of x_k is

$$X_{avg} = \frac{n_1x_1 + n_2x_2 + - - - - - + x_m n_m}{n} = \frac{1}{n}\sum_{k=1}^{m} x_k n_k$$

If $n \to \infty$, then $\frac{n_1}{n} \to p(x_1)$

$$\therefore \quad X_{avg} = p_1x_1 + p_2x_2 + \text{----} \; p_m x_m = \sum_{k=1}^{m} p_k x_k$$

and $$X^2_{avg} = \frac{n_1{x_1}^2 + n_2\,{x_2}^2 + - - - -}{N}$$

CORRELATION

In a divariate distribution, if the classes in one variable are associated by clases in the other, then variables are called *correlated*. If ratio of two variable deviations is constant, then correlation is called *perfect*.

Co-efficient of Correlation

Numerical measure of correlation is called *co-efficient of correlation* and is defined as

$$r = \frac{\Sigma XY}{n\sigma_x\sigma_y}$$

$$= \frac{\Sigma XY}{\sqrt{\Sigma X^2 \Sigma Y^2}} \quad .... \left[\because \sigma_x^2 = \frac{\Sigma x^2}{n}; \sigma_y^2 = \frac{\Sigma y^2}{n}\right]$$

where, X = deviation from mean, $= x - \bar{x}$

Y = deviation from mean, $= y - \bar{y}$

σ_x = standard deviation of x series

σ_y = standard of y series

n = number of values of the two variables

Method of Calculation (P)

Direct method : Substituting the value of σ_x and σ_y in the above formula, we get

$$r = \frac{\Sigma XY}{\sqrt{\Sigma X^2 \Sigma Y^2}},$$

or $$\frac{n\sum XY}{\sqrt{\left[n\sum x^2 - \left(\sum x\right)^2 \times \left\{n\sum y^2 - \sum x^2\right\}\right]}}$$

Rank Correlation

A group of n individuals may be arranged in order of merit respect to some characteristic. The same group would give different order for different characteristics.

Consider order corresponding to two characteristic A and B. The correlation between these n pairs of ranks is called *rank correlation* in the *characteristics A and B* for that group of individuals.

Let x_i, y_i be the ranks of the ith individuals in A and B respectively. Assuming that no two individuals are bracketed equal in either case, each of the variables taking the values 1, 2, 3,....., n, we have

$$\bar{x} = \bar{y} = \frac{1+2+3+...+n}{n} = \frac{n(n+1)}{2n} = \frac{n+1}{2}$$

If X, Y be the deviation of x, y from their mean, then

$$\Sigma X_i^2 = \Sigma(x_i - \bar{x})^2$$

$$= \Sigma x_i^2 + n(\bar{x})^2 - 2\bar{x}\,\Sigma x_i$$

$$= \Sigma n^2 + \frac{n(n+1)^2}{4} - 2\frac{n+1}{2}.\Sigma n$$

$$= \frac{n(n+1)(2n+1)}{4} + \frac{n(n+1)^2}{4} - \frac{n(n+1)^2}{2}$$

$$= \frac{1}{12}(n^3 - n)$$

Similarly $\Sigma Y_i^2 = \frac{1}{12}(n^3 - n)$

Now let $d_i = x_i - y_i$

$\therefore \quad d_i = (x_i - \bar{x}) - (y_i - \bar{y}) = X_i - Y_i$

$$\Sigma d_i^2 = \Sigma X_i^2 + \Sigma Y^2 - 2\,\Sigma X_i Y_i$$

$$\Sigma X_i Y_i = \frac{1}{2}(\Sigma X_i^2 + \Sigma Y_i^2 - \Sigma d_i^2)$$

$$= \frac{1}{12}(n^3 - n) - \frac{1}{2}\Sigma d_i^2$$

Hence correlation coefficient between these variables

$$\rho = \frac{\sum X_i Y_i}{\sqrt{\sum X_i^2 \sum Y_i^2}} = \frac{\left(\frac{1}{12}\right)(n^3 - n) - \left(\frac{1}{2}\right)\Sigma d_i^2}{[1/12]\,(n^3 - n)}$$

$$= 1 - \frac{6\sum {d_i}^2}{n^3 - n}$$

This is called *rank correlation coefficient.*

Line of Regression

If dots of scatter diagram generally, tend to cluster along a well defined direction which suggests a linear relationship between variables x and y, such a line of best – fit for the given distribution of dots is called *line of regression*.

There are two such lines, one giving the best possible mean values of y for each specified value of x and the other giving the best possible mean values for x for given values of y. The former is called the *line for regression of y on x* and the later is called the *line of regression of x on y*.

First consider the line of regression of y on x.

Let straight line satisfying the general trend of n dots in a scatter diagram be

$$y = a + bx \quad ...(i)$$

We have to determine the constants a and b so that equation (*i*) gives for each value of x, the best estimate for the average value of y in accordance with the principle of least squares. Thus normal equations for a and b are

$$\sum y = na + b\sum x \qquad ...(ii)$$
$$\sum xy = a\sum x + b\sum x^2$$

Equation (*ii*) gives $\frac{1}{n}\sum y = a + b.\frac{1}{n}\sum x$

i.e. $\bar{y} = a + b\bar{x}$

This shows that $(\bar{x}, \bar{y})$, i.e. means of x and y lie on (*i*)

Shifting the origin to $(\bar{x}, \bar{y})$, equation (*iii*) takes the form

$$\sum(x-\bar{x})(y-\bar{y}) = a\sum(x-\bar{x} + b\sum(x-\bar{x})^2$$

But $\sum(x-\bar{x}) = 0$

$$\therefore \quad b = \frac{\sum(x-\bar{x})(y-\bar{y})}{\sum(x-\bar{x})^2} = \frac{\sum XY}{\sum X^2} = \frac{\sum XY}{n\sigma_x^2}$$

$$= r\frac{\sigma_y}{\sigma_x} \qquad ...\left(\because r = \frac{\sum XY}{n\sigma_x\sigma_y}\right)$$

Thus line of best fit becomes

$$(y-\bar{y}) = r\frac{\sigma_y}{\sigma_x}(x-\bar{x})$$

which is equation of the *line of regression of y on x*. Its slope is called *regression co-efficient of y on x*

Interchanging x and y, the line of regression of x on y is

$$(x-\bar{x}) = r\frac{\sigma_x}{\sigma_y}(y-\bar{y})$$

Thus regression co-efficient of y on $x = r\frac{\sigma_y}{\sigma_x}$

and regression co-efficient of x on $y = r\frac{\sigma_x}{\sigma_y}$

cor. Correlation co-efficient is the geometric mean between the two regression cofficients

$$r\frac{\sigma_y}{\sigma_x} \times r\frac{\sigma_x}{\sigma_y} = r^2$$

Regression Co-efficient

It is defined as the hope of lines of regression,

i.e. regression coefficient of line *x* on $y = r\frac{\sigma_x}{\sigma_y}$

A. Hypothesis Testing

Terms, Concepts

- In general, we do not know the true value of population parameters - they must be estimated. However, we do have hypotheses about what the true values are.
- The major purpose of hypothesis testing is to choose between two competing hypotheses about the value of a population parameter. For example, one hypothesis might claim that the wages of men and women are equal, while the alternative might claim that men make more than women.
- The hypothesis actually to be tested is usually given the symbol H_0, and is commonly referred to as the null hypothesis. As is explained more below, the null hypothesis is assumed to be true unless there is strong evidence to the contrary - similar to how a person is assumed to be innocent until proven guilty.
- The other hypothesis, which is assumed to be true when the null hypothesis is false, is referred to as the alternative hypothesis, and is often symbolized by H_A or H_1. *Both the null and alternative hypothesis should be stated before any statistical test of significance is conducted.* In other words, you technically are not supposed to do the data analysis first and then decide on the hypotheses afterwards.
- In general, it is most convenient to always have the null hypothesis contain an equals sign, e.g.

H_0: $\mu = 100$
H_A: $\mu > 100$

- The true value of the population parameter should be included in the set specified by H_0 or in the set specified by H_A- Hence, in the above example, we are presumably sure μ is at least 100.
- A statistical test in which the alternative hypothesis specifies that the population parameter lies entirely above or below the value specified in H_0 is a one-sided (or one-tailed) test, e.g.

H_0: $\mu = 100$
H_A: $\mu > 100$

- An alternative hypothesis that specified that the parameter can lie on either side of the value specified by H_0 is called a two-sided (or two-tailed) test, e.g.

H_0: $\mu = 100$
H_A: $\mu <> 100$

- Whether you use a 1-tailed or 2-tailed test depends on the nature of the problem.Usually we use a 2-tailed test. A 1-tailed test typically requires a little more theory.

For example, suppose the null hypothesis is that the wages of men and women are equal. A two-tailed alternative would simply state that the wages are not equal - implying that men could make more than women, or they could make less. A one-tailed alternative would be that men make more than women. The latter is a stronger statement and requires more theory, in that not only are you claiming that there is a difference, you are stating what direction the difference is in.

- *In practice, a 1-tailed test such as*

H_0: $\mu = 100$

H_A: $\mu > 100$

is tested the same way as

H_0: $\mu \le 100$

H_A: $\mu > 100$

For example, if we conclude that $\mu > 100$, we must also conclude that $\mu > 90$, $\mu > 80$, etc.

B. The decision problem

- How do we choose between H_0 and H_A? The standard procedure is to assume H_0 is true - just as we presume innocent until proven guilty. Using probability theory, we try to determine whether there is sufficient evidence to declare H_0 false.
- We reject H_0 only when the chance is small that H_0 is true. Since our decisions are based on probability rather than certainty, we can make errors.
- **Type I error** - We reject the null hypothesis when the null is true. The **probability of Type I error = α.** Put another way,

  α = Probability of Type I error = P(rejecting $H_0 \mid H_0$ is true)

  Typical values chosen for α are .05 or .01. So, for example, if $\alpha = .05$, there is a 5% chance that, when the null hypothesis is true, we will erroneously reject it.
- **Type II error** - we accept the null hypothesis when it is not true. **Probability of Type II error = β.** Put another way,

  β = Probability of Type II error = P(accepting $H_0 \mid H_0$ is false) E.

EXAMPLES of type I and type II error:

H_0: $\mu = 100$

H_A: $\mu <> 100$

Suppose u really does equal 100. But, suppose the researcher accepts H_A instead. A type I error has occurred. Or, suppose $\mu = 105$ -but the researcher accepts H_A. A type II error has occurred.

The following tables from Hamett help to illustrate the different types of error.

The true situation is

Action		H_0 is true	H_a is true
	Reject H_0 (Accept H_a)	Type I error	Correct decision
	Reject H_a (Accept H_0)	Correct decision	Type II error

The true situation is

Action		Not guilty (H_0)	Guilty (H_a)
	Jury finds guilty (Accept H_a)	Type I error	Correct decision
	Jury finds not guilty (Accept H_0)	Correct decision	Type II error

- α and β are not independent of each other - as one increases, the other decreases. However, increases in N cause both to decrease, since sampling error is reduced.
- In this class, we will primarily focus on Type I error. But, you should be aware that Type II error is also important. A small sample size, for example, might lead to frequent Type II errors, i.e. it could be that your (alternative) hypotheses are right, but because your sample is so small, you fail to reject the null even though you should.

SOLVED EXAMPLES

1. A bag contains 8 white and 6 red balls. Find the probability of drawing two balls of the same colour.

Solution

Two balls out of 14 can be drawn in $^{14}C_2$ ways which is total number of outcomes.

Two white balls out of 8 can be drawn in $^{8}C_2$ ways. Thus probability of drawing 2 white balls

$$= \frac{^{8}C_2}{^{14}C_2} = \frac{28}{91}$$

Similarly 2 red balls out of 6 can be drawn in $^{6}C_2$ ways.

Thus probability of drawing 2 red balls

$$= \frac{^{6}C_2}{^{14}C_2} = \frac{15}{91}$$

Hence probability of drawing 2 balls of the same colour (either both white or both red)

$$= \frac{28}{91} + \frac{15}{91} = \frac{43}{91}$$

2. Find probability of drawing an ace or a spade or both from a deck of cards?

Solution

Probability of drawing an ace from a deck of 52 cards $= \frac{4}{52}$

Similarly probability of drawing a card of spades

$$= \frac{13}{52}$$

and probability of drawing an ace of spades $= \frac{1}{52}$.

Since two events (i.e., a card being an ace and a card being of spades) are not mutually exclusive.

Hence probability of drawing an ace or a spade

$$= \frac{4}{52} + \frac{13}{52} - \frac{1}{52} = \frac{4}{13}.$$

3. A pair of dice is tossed twice. Find the probability of scoring 7 points

(a) once (b) at least once (c) twice.

Solution

In a single toss of two dice, the sum 7 can be obtained as (1, 6), (2, 5), (3, 4), (4, 3), (5, 2), (6, 1) i.e., in 6 ways

Hence probability of getting 7 = $\frac{6}{36} = \frac{1}{6}$.

Thus, probability of not getting 7 = $1 - \frac{1}{6} = \frac{5}{6}$.

(a) Probability of getting 7 in the first toss and getting 7 in the second toss

$$= \frac{1}{6} \times \frac{5}{6} = \frac{5}{36}.$$

Similarly, probability of not getting 7 in the first toss and getting 7 in the second toss

$$= \frac{5}{6} \times \frac{1}{6} = \frac{5}{36}.$$

Since these are mutually exclusive events, addition law of probability applies.

$\therefore$ Required probability = $\frac{5}{36} + \frac{5}{36} = \frac{5}{18}$

(b) Probability of not getting 7 in either toss

$$= \frac{5}{6} \times \frac{5}{6} = \frac{25}{36}$$

Hence probability of getting 7 at least once

$$= 1 - \frac{25}{36} = \frac{11}{36}.$$

(c) Probability of getting 7 twice

$$= \frac{1}{6} \times \frac{1}{6} = \frac{1}{36}.$$

4. There are two groups of subjects; one of which consists of 5 science and 3 engineering subjects, and the other consists of 3 science and 5 engineering subjects. An unbiased die is cast. If the number 3 or number 5 turns up, a subject is selected at randam from the first group, otherwise the subject is selected at randam from the second group. Find probability that an engineering subject is selected ultimately.

Solution

Probability of turning up 3 or 5 = $\frac{2}{6} = \frac{1}{3}$.

Probability of selecting an engineering subject from first group = $\frac{3}{8}$

Hence probability of selecting an engineering subject from first group on turning up 3 or 5

$$= \frac{1}{3} \times \frac{3}{8} = \frac{1}{8}$$

Now probability of not turning 3 or 5

$$= 1 - \frac{1}{3} = \frac{2}{3}$$

Probability of selecting an engineering subject from second group = $\frac{5}{8}$

Hence probability of selecting an engineering subject from second group on turning up 3 or 5

$$= \frac{2}{3} \times \frac{5}{8} = \frac{5}{12}$$

Thus probability of selecting an engineering subject

$$= \frac{1}{8} + \frac{5}{12} = \frac{13}{24}$$

5. A box A contains 2 white and 4 black balls. Another box B contains 4 white and 7 black balls. A ball is transferred from the box A to the box B. Then a ball is drawn from the box B. Find the probability that it is white.

Solution

Probability of drawing a white ball from box B will depend on whether the transferred ball is black or white.

If a black ball is transferred, its probability is $\frac{4}{6}$.

There are now 5 white and 8 black balls in the box B. Then probability of drawing white ball from box B is $\frac{5}{13}$.

Thus probability of drawing a white ball from urn B, if transferred ball is black $= \frac{4}{6} \times \frac{5}{13} = \frac{10}{39}$

Similarly the probability of drawing a white ball from urn B, if transferred ball is white

$$= \frac{2}{6} \times \frac{6}{13} = \frac{2}{13}$$

$\therefore$ Required probability = $\frac{10}{39} + \frac{2}{13} = \frac{16}{39}$.

6. Two persons A and B toss an unbiased coin alternately on the understanding that the first who gets the head wins. If A starts the game, find their respective chances of winning.

Solution

Probability of getting a head = $\frac{1}{2}$.

Then A can win in 1st, 3rd, 5th, ... throws.

Hence chances of A's winning

$$= \frac{1}{2} + \left(\frac{1}{2}\right)^2 \frac{1}{2} + \left(\frac{1}{2}\right)^4 \frac{1}{2} + \left(\frac{1}{2}\right)^6 \frac{1}{2} + ...$$

$$= \frac{\frac{1}{2}}{1 - \left(\frac{1}{2}\right)^2} = \frac{1}{2} \cdot \frac{4}{3} = \frac{2}{3} .....$$

Hence chance of B's winning = $1 - \frac{2}{3} = \frac{1}{3}$.

7. (*i*) Is the function defined as follows a density function?

$$f(x) = e^{-x}, x \ge 0 = 0, \quad x < 0$$

(*ii*) If so, determine probability that the variate having this density will fall in the interval (1, 2)?

(*iii*) Also find cumulative probability function F(2)?

Solution

(*i*) $f(x)$ is clearly ≥ 0 for every x in (1, 2) and

$$\int_{-\infty}^{\infty} f(x)dx = \int_{-\infty}^{0} 0.dx + \int_{0}^{\infty} e^{-x}dx = 1$$

Hence function $f(x)$ satisfies the requirements for a density function.

(*ii*) Required probability = P $(1 \le x \le 2)$

$$= \int_{1}^{2} e^{-x}dx$$

$$= e^{-1} - e^{-2}$$

$$= 0.368 - 0.135 = 0.233.$$

This probability is equal to the shaded area in Fig. (*a*).

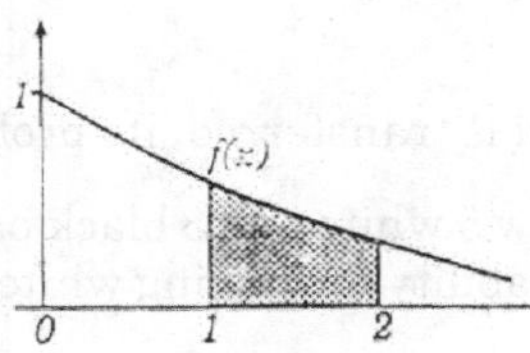

(a)

(*iii*) Cumulative probability function F(2)

$$\int_{-\infty}^{2} f(x)dx = \int_{-\infty}^{0} 0.dx + \int_{0}^{2} e^{-x}dx$$

$$= 1 - e^{-2}$$

$$= 1 - 0.135 = 0.865$$

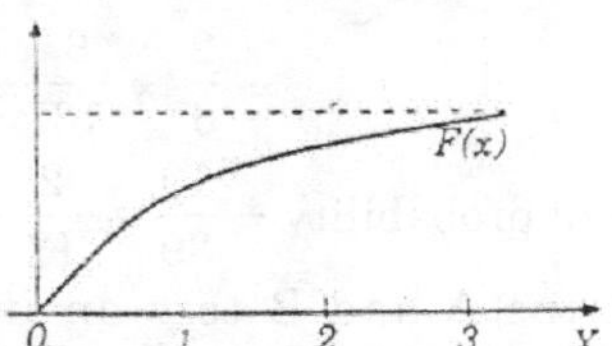

(b)

which is shown in Fig.(*b*).

8. X is a continuous random variable with probability density function given by

$$f(x) = kx \ (0 \le x < 2) = 2k \ (2 \le x < 4)$$
$$= -kx + 6k \ (4 \le x < 6)$$

Find k and mean value of X.

Solution

Since total probability is unity

$$\therefore \quad \int_{0}^{6} f(x)dx = 1$$

i.e., $\int_{0}^{2} kx\,dx + \int_{2}^{4} 2k\,dx + \int_{4}^{6} (-kx + 6k)dx = 1$

$$\Rightarrow k\left|\frac{x^2}{2}\right|_0^2 + 2k|x|_2^4 + \left(-k\frac{x^2}{2} + 6kx\right)_4^6 = 1$$

$$\Rightarrow 2k + 4k + (-10k + 12k) = 1$$

$$\Rightarrow k = \frac{1}{8}.$$

Mean of X

$$= \int_{0}^{6} xf(x)dx$$

$$= \int_{0}^{2} kx^2dx + \int_{2}^{4} 2kx\,dx + \int_{4}^{6} x(-kx + 6k)dx$$

$$= k\left|\frac{x^3}{3}\right|_0^2 + 2k\left|\frac{x^2}{2}\right|_2^4 + \left(-k\left|\frac{x^3}{3}\right|_4^6 + 6k\left|\frac{x^2}{2}\right|_4^6\right)$$

$$= k\left(\frac{8}{3}\right) + k(12) - k\left(\frac{152}{3}\right) + 3k(20) = \frac{1}{8}(24) = 3$$

9. The frequency distribution of a measurable characteristic varying between 0 and 2 is as under

$$f(x) = x^3, \ 0 \le x \le 1 = (2 - x)^3, \ 1 \le x \le 2.$$

Calculate standard deviation and also mean deviation about the mean.

Solution

Total frequency, $N = \int_{0}^{1} x^3\,dx + \int_{1}^{2} (2-x)^3\,dx$

$$= \frac{1}{4} + \frac{1}{4} = \frac{1}{2}$$

$\therefore$ μ_1' (about the origin)

$$= \frac{1}{N}\left[\int_{0}^{1} x.x^3\,dx + \int_{1}^{2} x(2-x)^3\,dx\right]$$

$$= 2\left\{\left|\frac{x^5}{5}\right|_0^1 + \left|-x.\frac{(2-x)^4}{4}\right|_1^2 - \left|\frac{(2-x)^5}{20}\right|_1^2\right\}$$

$$= 2\left(\frac{1}{5} + \frac{1}{4} + \frac{1}{20}\right) = 1$$

μ_2' (about the origin)

$$= \frac{1}{N}\left[\int_{0}^{1} x^2.x^3dx + \int_{1}^{2} x^2(2-x)^3\,dx\right]$$

$$= 2\left\{\left|\frac{x^6}{6}\right|_0^1 + \left|-x^2\frac{(2-x)^4}{4}\right|_1^2 + \frac{1}{2}\int_{1}^{2} x(2-x^4)dx\right\}$$

$$= 2\left\{\frac{1}{6} + \frac{1}{4} + \frac{1}{2}\left[\frac{1}{5} + \frac{1}{30}\right]\right\} = \frac{16}{15}$$

$$\therefore \quad \sigma^2 = \mu_2 = \mu_2' - (\mu_1')^2 = \frac{1}{15}$$

i.e., standard deviation $\sigma = \dfrac{1}{\sqrt{15}}$

Mean deviation about the mean

$$= \frac{1}{N}\left\{\int_{0}^{1}|x-1|x^3dx + \int_{1}^{2}|x-1|(2-x)^3\,dx\right\}$$

$$= 2\left\{\int_{0}^{1}(1-x)x^3dx + \int_{1}^{2}(x-1)(2-x)^3\,dx\right\}$$

$$= 2\left\{\left(\frac{1}{4} - \frac{1}{5}\right) + \left(0 + \frac{1}{20}\right)\right\} = \frac{1}{5}$$

10. The probability that a pen manufactured by a company will be defective is $\frac{1}{10}$. If 12 such pens are manufactured, find the probability that

(*a*) exactly two will be defective.

(*b*) at least two will be defective.

(*c*) none will be defective

Solution

Probability of a defective pen $= \frac{1}{10} = 0.1$

Hence probability of a non-defective pen

$$1 - 0.1 = 0.9$$

(*a*) Probability that exactly two will be defective

$= {}^{12}C_2\,(0.1)^2\,(0.9)^{10} = 0.2301.$

(*b*) Probability that at least two will be defective

= 1 – (probability that either none or one is non-defective)

$= 1 - [{}^{12}C_0(0.9)^{12} + {}^{12}C_1(0.1)(0.9)^{11}]$

$= 0.3412.$

(*c*) Probability that none will be defective

$= {}^{12}C_{12}\,(0.9)^{12} = 0.2833$

11. In 256 sets of 12 tosses of a coin, in how many cases one can expect 8 heads and 4 tails.

Solution

$$\text{P(head)} = \frac{1}{2} \text{ and P(tail)} = \frac{1}{2}$$

By Binomial distribution, probability of 8 heads and 4 tails in 12 trials,

$$P(X = 8) = {}^{12}C_8 \left(\frac{1}{2}\right)^8 \left(\frac{1}{2}\right)^4 = \frac{12!}{8!4!} \cdot \frac{1}{2^{12}} = \frac{495}{4096}$$

$\therefore$ Expected number of such cases in 256 sets

$$= 256 \times P(X = 8) = 256\ \frac{495}{4096} = 30.9 \approx 31$$

12. If probability of a bad reaction from a certain injection is 0.001, determine the chance that out of 2,000 individuals more than two will get a bad reaction.

Solution

It follows a Poisson distribution as probability of occurrence is very small.

Mean, $m = np = 2000\,(0.001) = 2$

Probability that more than 2 will get a bad reaction

= 1 – [probability that no one gets a bad reaction + probability that one gets a bad reaction + probability that two get bad reaction]

$$= 1 - \left[e^{-m} + \frac{m^1 e^{-m}}{1!} + \frac{m^2 e^{-m}}{2!}\right]$$

$$= 1 - \left[\frac{1}{e^2} + \frac{2}{e^2} + \frac{2}{e^2}\right] \quad \ldots[\because m = 2]$$

$$= 1 - \frac{5}{e^2} = 0.32 \quad [\because e = 2.718]$$

EXERCISE

MCQ TYPE QUESTIONS

1. If A, B, C are three events, then $P[A \cap (B \cup C)] =$

(*a*) $P(A) + P(B) + P(C) - P(A \cap B) - P(A \cap C)$

(*b*) $P(A) + P(B) + P(C) - P(B)\,P(C)$

(*c*) $P(A \cap B) + P(A \cap C) - P(A \cap B \cap C)$

(*d*) $P(B \cap C) + P(A \cap B) - P(A \cap B \cap C)$

2. If A and B are two arbitrary events, then $P(A \cap B)$ cannot be

(*a*) less than $P(A) + P(B) - 1$

(*b*) greater than $P(A) + P(B)$

(*c*) equal to $P(A) + P(B) - P(A \cup B)$

(*d*) equal to $P(A) + P(B) + P(A \cup B)$

3. If M and N are two events, then probability that exactly one of them occurs is not equal to

(*a*) $P(M) + P(N) - 2P(M \cap N)$

(*b*) $P(M) + P(N) - P(M \cap N)$

(*c*) $P(\overline{M}) + P(\overline{N}) - 2P(\overline{M} \cap \overline{N})$

(*d*) $P(M \cap \overline{N}) + P(\overline{M} \cap N)$

4. If A and B are any two events then which one of the following pairs of events are mutually exclusive ?

(*a*) $A \cup B$ and $A \cap \overline{B}$ (*b*) $A \cup B$ and $\overline{A} \cap B$

(*c*) $A \cup B$ and $\overline{A} \cap \overline{B}$ (*d*) $A \cup B$ and $A \cap B$

5. There are three events A, B, C one of which must, and only one can happen, the odds are 8 to 3 against A, 5 to 2 against B. The odds against C are

(*a*) 43 to 34 (*b*) 32 to 23

(*c*) 34 to 13 (*d*) none of these

6. A has one share in a lottery in which there is 1 prize and 2 blanks; B has three shares in a lottery in which there are 3 prizes and 6 blanks; compare the probability of A's success to that of B's success is

(*a*) 7 : 16

(*b*) 16 : 7

(*c*) 6 : 14

(*d*) 14 : 6

7. If A and B are independent, then A and $\overline{B}$ are also

(a) independent (b) dependent

(c) both (a) and (b) (d) none of these

8. If $E_1, E_2, ..., E_n$ are n mutually exclusive events, then the probability of the happening of any one of them is equal to the sum of the probabilities of the happening of separate events.

i.e., $P(E_1 \cup E_2 \cup E_3 ..... \cup E_n) = P(E_1) + P(E_2) + ... + P(E_n)$.

This is

(a) Law of total probability

(b) Law of probability

(c) Both (a) and (b)

(d) None of these

9. What is the probability of correctly choosing an unknown integer between 0 to 9 within three chances ?

(a) $\frac{963}{1000}$ (b) $\frac{973}{1000}$

(c) $\frac{983}{1000}$ (d) None of these

10. If $P(A/C) \geq P(B/C)$ and $P(A/\overline{C_1}) \geq P(B/\overline{C_1})$, then

(a) $P(A) \geq P(B)$ (b) $P(A) \leq P(B)$

(c) $P(A) < P(B)$ (d) $P(A) > P(B)$

11. Six dice are thrown simultaneously. The probability that all will show different faces is

(a) $\frac{5!}{6^5}$ (b) $\frac{5!}{6^4}$

(c) $\frac{5!}{6^3}$ (d) none of these

12. Given two events A and B and

$P(A) = \frac{1}{4}$, $P(B/A) = \frac{1}{2}$ and $P(A/B) = \frac{1}{4}$.

State that following is true?

(a) A is sub-event of B

(b) $P(A/B) = \frac{3}{4}$

(c) $P(A/B) + P(A/\overline{B}) = 1$

(d) None of these

13. Which of the following is correct?

(a) If $P(A) > 0$, $P(B) > 0$ and $P(A/B) = P(B/A)$, then

$$P(A) = P(B).$$

(b) If A and B mutually exclusive, then

$$P(A/\overline{B}) = P(A)/(1 - P(B)).$$

(c) If A and B are independent, then

$$P(A \cup B) = 1 - P(\overline{A}) . P(\overline{B})$$

(d) All of these

14. If $P(A) = \frac{3}{4}$ and $P(B) = \frac{5}{8}$, then

(a) $P(A \cup B) \geq \frac{3}{4}$ (b) $\frac{3}{8} < P(A \cap B) \leq \frac{5}{8}$

(c) $\frac{1}{8} \leq P(A \cap \overline{B}) \leq \frac{3}{8}$ (d) All of these

15. Three machines M_1, M_2 and M_3 produce identical items. Of their respective output 5%, 4% and 3% of items are faulty. On a certain day, M_1 has produced 25% of the total output, M_2 has produced 30% and M_3 the remainder. An item selected at random is found to be faulty. What are the chances that it was produced by the machine with the highest output?

(a) 0.155 (b) 0.255 (c) 0.355 (d) 0.455

16. For any two events A and B

(a) $P(\overline{A} \cap B) = P(B) - P(A \cap B)$

(b) $P(A \cup B) = P(A) + P(B) - P(A \cap B)$

(c) $P(A/B) \leq P(A)$.

(d) All of these

17. Which of the following is correct?

(a) $P(A \cup B/C) = P(A/C) + P(B/C) - P(A \cap B/C)$

(b) $P(A \cap B/C) + P(A \cap \overline{B}/C) = P(A/C)$

(c) If A and B are independent, then A and $\overline{B}$ are also independent.

(d) All of these

18. Which of the following is correct?

(a) If A, B and C are mutually independent, then $A \cup B$ and C are also independent.

(b) The events $E_1, E_2, ..., E_n$ are mutually exclusive. Let $E = \bigcup_{i=1}^{n} E_i$.

If $P(A/E_i) = P(B/E_i)$, i = 1, 2, ..., n, then $P(A/E) = P(B/E)$.

(c) Both (a) and (b)

(d) None of these

19. Let p be the probability that a man aged y will get into an accident in a year. What is the probability that a man among n men of all aged y will get into an accident first ?

(a) $\frac{1}{n}\left(1-(1-p)^n\right)$ (b) $\frac{1}{n}\left(1-(1+p)^n\right)$

(c) $n\left(1-(1+p)^n\right)$ (d) None of these

20. What would be the expectation of the number of failures preceding the first success in an infinite series of independent trials with the constant probability of success p ?

(a) $\frac{1}{p}$ (b) $\frac{1}{q}$

(c) $\frac{q}{p}$ (d) None of these

21. A restaurant serves two special dishes A and B to its customers consisting of 60% men and 40% women. 80% of men order dish A and the rest B and 70% women order dish A and the rest B. In what ratio of A to B should the restaurant prepare the dishes?

(a) 19 : 6 (b) 6 : 9 (c) 16 : 7 (d) 7 : 16

22. If A and B are mutually exclusive and $P(A \cup B) = P(A) + P(B)$, then $P(A/A \cup B) =$

(a) $\frac{P(A)}{P(A)+P(B)}$ (b) $\frac{P(A)}{P(A)-P(B)}$

(c) $\frac{P(B)}{P(A)+P(B)}$ (d) None of these

23. If $P(A) = P(\bar{B})$, then

(a) $A = \bar{B}$ (b) $B = 0$

(c) $A \neq B$ (d) None of these

24. If $P(A) = 0$, then

(a) $A = \phi$ (b) $A = 0$

(c) $A = \infty$ (d) none of these

25. The mean of a set of number is $\bar{x}$. If each number is increased by λ, then mean of the new set is

(a) $\bar{x}$ (b) $\bar{x} + \lambda$

(c) $\lambda\bar{x}$ (d) none of these

26. If each observation of raw data whose variance is σ^2, is increased by λ, then variance of the new set is

(a) σ^2 (b) $\lambda^2\sigma^2$

(c) $\lambda + \sigma^2$ (d) $\lambda^2 + \sigma^2$

27. If μ is mean of distribution, then $\sum f_i(\gamma_i - \mu)$ is equal to

(a) M.D (b) stamdard deviation

(c) 0 (d) none of these

28. $b_{xy} \times b_{yx}$ is equal to

(a) $\rho(X, Y)$ (b) cov (x, y)

(c) $\{\rho(x, y)\}^2$ (d) None of these

29. If cov (X, Y) = 0, then which of the following is negative?

(a) b_{yx} (b) b_{xy}

(c) $\rho(X, Y)$ (d) none of these

30. The mean of a set of observations is $\bar{x}$. If each observation is divided by α, $\alpha \neq 0$ and then is increased by 10, then the mean of new set is

(a) $\frac{\bar{x}}{\alpha}$ (b) $\frac{\bar{x}+10}{\alpha}$

(c) $\frac{\bar{x}+10\alpha}{\alpha}$ (d) $\alpha\bar{x} + 10$

31. S.D. of n observation $a_1, a_2, a_3, \ldots\ldots a_n$ is σ, then If S.D. of the observations $\lambda a_1, \lambda a_2, \lambda a_3, \ldots\ldots \lambda a_n$ is

(a) $\lambda\sigma$ (b) $-\lambda\sigma$

(c) $|\lambda|\sigma$ (d) σ

32. If two lines of regression are $Y = 3x - 5$ and $Y = 2x - 4$, then $\rho(X, Y)$ is equal to

(a) $\sqrt{\frac{2}{3}}$ (b) $\sqrt{\frac{1}{6}}$

(c) $\sqrt{\frac{3}{2}}$ (d) none of these

33. If two lines regression are $3x - y - 5 = 0$ and $2x - y - 4 = 0$, then $\bar{x}$ and $\bar{y}$ are respectively

(a) 1 and –2 (b) –1 and 2

(c) 2 and –1 (d) –2 and –1

34. If regression equation of Y on X is $Y = \lambda x + 4$ and that of X on Y be $4x = Y - 5$, then

(a) $0 < 4\lambda \leq 1$ (b) $0 \leq \lambda \leq 1$

(c) $0 \leq \lambda \leq 4$ (d) none of these

35. If cov (X,Y) = 0, then two lines of the regression are

(a) parallel (b) concident

(c) at right angles (d) none of these

36. If in a regression analysis problem, $b_{yx} = -0.9$ and $b_{xy} = 0.4$, then $\rho(x, y)$ is

(a) 0.6 (b) –0.6

(c) 0 (d) none of these

37. It b_{xy} and b_{yx} be the two regression coefficients, then coefficient of correlation ρ is given by

(a) $\rho = \sqrt{b_{xy}b_{yx}}$ (b) $\rho = -\sqrt{b_{yx}b_{yx}}$

(c) $\rho = (S_{gn}b_{yx})\sqrt{b_{xy}b_{yx}}$ (d) none of these

38. It standard deviation for two variables X and Y are 3 and 4 respectively and their covariance is 8, then correlation coefficient between them is

(a) $\frac{2}{3}$ (b) $\frac{8}{3\sqrt{2}}$ (c) $\frac{9}{8\sqrt{2}}$ (d) $\frac{2}{9}$

39. Mean deviation of the data $a, a + d, a + 2d, \ldots a + 2nd$ from the mean is equal to

(a) $\frac{n(n+1)|d|}{2n+1}$ (b) $\frac{n(n+1)d}{2n+1}$

(c) $\frac{(n+1)|d|}{2n}$ (d) none of these

40. If two variables X and Y have a perfect correlation (direct or indirect), then they may be connected by a relation of the type

(a) $xy = a^2$ (b) $\frac{a}{x} + \frac{b}{y} = 1$

(c) $\frac{x}{a} + \frac{y}{b} = 1$ (d) none of these

41. If $\rho > 0$ and $m = \frac{b_{yx} + b_{xy}}{2}$, then

(a) $m \geq \rho$ (b) $m = \rho$
(c) $m \leq \rho$ (d) none of these

42. If two lines of regression are $x + 2y = 7$ and $2x + y = 7$, then regression equation of Y and X is

(a) $2x + y = 7$ (b) $x + 2y = 7$
(c) $x + 2y = 0$ (d) $2x + y = 0$

43. Two variables X and Y are connected by the relation $ax + by + c = 0$ where $ab < 0$, then $\rho(x, y)$ is equal to

(a) $\frac{a}{b}$ (b) $\frac{b}{a}$ (c) -1 (d) 1

44. Mean of the numbers 0, 1, 2, 3, ...n with respective weight ${}^nC_0$, ${}^nC_1$, ${}^nC_2$,......, ${}^nC_n$ is

(a) $\frac{2^n}{n+1}$ (b) $\frac{2^{n+1}}{n+1}$
(c) $\frac{n}{2}$ (d) none of these

45. The minimum number of cards to be dealt from an arbitrarily shuffled deck of 52 cards to guarantee that three cards are from same suit is

(a) 3 (b) 8 (c) 9 (d) 12

46. A die is rolled three times. The probability that exactly one ODD number turns up among the three outcomes is

(a) $\frac{1}{6}$ (b) $\frac{1}{5}$ (c) $\frac{1}{8}$ (d) $\frac{1}{2}$

47. Two girls have picked 10 Roses, 15 Sunflowers and 14 Daffodils. What is the number of ways they can divide the flowers amongst themselves ?

(a) 1638 (b) 2100
(c) 2640 (d) None of these

48. The probability that it will rain today is 0.5. The probability that it will rain tomorrow is 0.6. The probability that it will rain either today or tomorrow is 0.7. What is the probability that it will rain today and tomorrow ?

(a) 0.3 (b) 0.25 (c) 0.35 (d) 0.4

49. The probability that top and bottom cards of a randomly shuffled deck are both aces is

(a) $\frac{4}{52} \times \frac{4}{52}$ (b) $\frac{4}{52} \times \frac{3}{52}$
(c) $\frac{4}{52} \times \frac{3}{51}$ (d) $\frac{4}{52} \times \frac{4}{51}$

50. The probability that a number selected at random between 100 and 999 (both inclusive) will not contain the digit 7 is

(a) $\frac{16}{25}$ (b) $\left(\frac{9}{10}\right)^3$ (c) $\frac{27}{75}$ (d) $\frac{18}{25}$

51. If 20 per cent managers are technocrats, the probability that a random committee of 5 managers consists of exactly 2 technocrats is

(a) 0.2048 (b) 0.4000 (c) 0.4096 (d) 0.9421

52. A regression moodel is used to express a variable Y as a function of another variable X. This implies that

(a) there is a causal relationship between Y and X
(b) a value of X may be used to estimate a value of Y
(c) values of X exactly determine values of Y
(d) there is no causal relationship between Y and X

53. Arrivals at a telophone booth are considered to be Poisson, with an average time of 10 minutes between successive arrivals. The length of a phone call is distributed exponentially with mean 3 minutes. The probability that an arrival does not have to wait before service is

(a) 0.3 (b) 0.5 (c) 0.7 (d) 0.9

54. From a pack of regular playing cards, two cards are drawn at random. What is the probability that both cards will be Kings, if the first card is NOT replaced?

(a) $\frac{1}{26}$ (b) $\frac{1}{52}$ (c) $\frac{1}{169}$ (d) $\frac{1}{221}$

55. A box contains 10 screws, 3 of which are defective. Two screws are drawn at random with replacement. The probability that none of the two screws is defective will be

(a) 100% (b) 50%
(c) 49% (d) None of these

56. A fair coin is tossed three times in succession. If the first toss poduces a head, then the probability of getting exactly two heads in three tosses is

(a) $\frac{1}{8}$ (b) $\frac{1}{2}$ (c) $\frac{3}{8}$ (d) $\frac{3}{4}$

57. In a frequencey distribution, mid value of a class is 15 and class interval is 4. The lower limit of the class is

(a) 14 (b) 13 (c) 12 (d) 10

58. Following marks were obtained by the students in a test :
81, 72, 90, 90, 86, 85, 92, 70, 71, 83, 89, 95, 85, 79, 62. Range of the marks is

(a) 9 (b) 17 (c) 27 (d) 33

59. Standard deviation for 7, 9, 11, 13, 15 is

(a) 2.4 (b) 2.5 (c) 2.7 (d) 2.8

60. If probabilities that A and B will die withina year are p and q respectively, then probability that only one of them will be alive at the end of the year is

(a) pq (b) $p(1-q)$
(c) $q(1-p)$ (d) $p + q - 2pq$

61. In a binomial distribution, mean is 4 and variance is 3. Then, its mode is

(*a*) 5 (*b*) 6
(*c*) 4 (*d*) none of these

62. If sum and product of the mean and variance of a binomial distribution are 24 and 18 respectively, then distriution is

(*a*) $\left(\frac{1}{7}+\frac{1}{8}\right)^{12}$ (*b*) $\left(\frac{1}{4}+\frac{3}{4}\right)^{16}$

(*c*) $\left(\frac{1}{6}+\frac{5}{6}\right)^{24}$ (*d*) $\left(\frac{1}{2}+\frac{1}{2}\right)^{32}$

63. Shen correlation coefficient $r=\pm 1$, then two regession lines

(*a*) are perpendicular to each other
(*b*) coincide
(*c*) are parallel to each other
(*d*) do not exist

NUMERICAL TYPE QUESTIONS

1. A card is taken out of a pack of 52 cards numbered 2 to 53. The probability that the numbers on the card is a prime number less than 20 is ______

2. The probability of getting a number between 1 and 100. Which is divisible by one and itself only is ______

3. If A and B are two elements such that $P(A) = \frac{2}{3}$, $P(B) = \frac{3}{4}$, $P(A \cap B) = \frac{1}{2}$, then $P(A \cup B)$ is ______

4. If A and B are two events such that $P(A \cup B) = 0.65$, $P(A \cap B) = 0.15$, then

$P(\overline{A}) + P(\overline{B}) =$ ______

5. There are two bags one of which contains 3 black and 4 white balls while other contains 4 black and 3 white balls. A die is cast, if the face 1 or 3 turns up, a ball is taken from first bag and if any other face turns up, a ball is chosen from second bag. The probability of choosing of black ball is______

6. An urn contains 5 red and 10 black balls. Eight of them are placed in another urn. The chance that the latter then contains 2 red and 6 black balls is______

7. The chance that a leap year selected at random will contain 53 Sundays is______

8. A five figure number is formed by the digits 0, 1, 2, 3, 4 without repetition. Then the probability that the number formed is divisible by 4 is______

9. A man alternately tosses a coin and throws a dice, beginning with the coin. Then the probability that he will get a head before he gets a 5 or 6 on dice is____

10. A box A contains 2 white and 4 black balls. Another box B contains 5 white and 7 black balls. A ball is transferred from the box A to the box B. Then a ball is drawn from the box B. The probability that it is white is______

11. If $P(A) = 0$, then $P(A \cap B)$ is ______

12. A problem in mechanics is given to three students A, B and C whose chances of slogging it are $\frac{1}{2}$, $\frac{1}{3}$ and $\frac{1}{4}$ respectively. Then probability that the problem will be solved is________

13. A and B throw alternately with a pair of dice. A wins if he throws 6 before B throws 7 and B wins if he throws 7 before A throws 6. If A begins, then his chance of winning is______

14. In an experiment a coin is tossed 4 times. The size of the sample space is________

15. If mode of a data is 18 and mean is 24, then median is______

16. Regression equation of Y on X is

$$8x - 10y + 66 = 0 \text{ and } \sigma_x = 3$$

Hence cov (X, Y) is equal is______

17. Coefficient of correlation between two variates X and Y is

X	1	2	3	4	5
Y	5	4	3	2	1

18. If mean = (3 median – mode) x, then value of x is______

19. If median = (mode + 2 mean) μ, then μ is equal to______

20. If two lines of regression are at right angles, then ρ (X, Y) is equal to ______

21. Given : $n = 10$, $\Sigma x= 4$, $\Sigma y = 3$, $\Sigma x^2 = 8$, $\Sigma y^2 = 9$ and $\Sigma xy = 3$, then coefficient of correlation is______

22. 25% of the items of data are less than 35 and 25% of the items are more than 75. Q.D. of the data is______

23. If X and Y are two independent variables such that $\bar{x} = 5$, $\bar{y} = 10$ $\sigma_x^2 = 4$ $\sigma_y^2 = 9$ and if $u = 3x + 4y$, $v = 3x - y$, then $\rho(u, v)$ is equal to____

24. The probability that two friends share the same birth-month is______

25. A box contains 5 black and 5 red balls. Two balls are randomly picked one after another from the box, without replacement. The probability for both balls being red is______

ANSWERS

MCQ Type Questions

1.(c)	2.(c)	3.(b)	4.(c)	5.(a)	6.(a)	7.(a)	8.(a)	9.(b)	10.(a)
11.(b)	12.(d)	13.(a)	14.(d)	15.(c)	16.(d)	17.(a)	18.(a)	19.(a)	20.(c)
21.(a)	22.(a)	23.(a)	24.(a)	25.(b)	26.(b)	27.(c)	28.(c)	29.(d)	30.(c)
31.(c)	32.(a)	33.(a)	34.(c)	35.(c)	36.(d)	37.(c)	38.(a)	39.(a)	40.(c)
41.(a)	42.(b)	43.(d)	44.(c)	45.(c)	46.(c)	47.(c)	48.(d)	49.(c)	50.(d)
51.(a)	52.(b)	53.(a)	54.(d)	55.(d)	56.(d)	57.(b)	58.(d)	59.(d)	60.(d)
61.(c)	62.(d)	63.(b)							

Numerical Type Questions

1. $\frac{2}{3}$	2. $\frac{25}{98}$	3. $\frac{11}{12}$	4. 1.2	5. $\frac{11}{21}$	6. $\frac{140}{429}$	7. $\frac{2}{7}$	8. $\frac{5}{16}$	9. $\frac{3}{4}$	10. $\frac{16}{39}$
11. ∞	12. $\frac{3}{4}$	13. $\frac{30}{61}$	14. 16	15. 22	16. 7.2	17. 1	18. $\frac{1}{2}$	19. $\frac{1}{3}$	20. 0
21. $\frac{1}{4}$	22. 50	23. $\frac{2}{3}$	24. $\frac{1}{6}$	25. $\frac{2}{9}$					

EXPLANATIONS

MCQ TYPE QUESTIONS

5. Since odds are 8 to 3 against A

$\therefore$ $P(A) = \frac{3}{11}$; $\left(P(B) = \frac{2}{7}\right)$

since odds are 5 to 2 aganist B

and $P(A) + P(B) + P(C) = 1$

$\therefore$ $P(C) = \frac{1}{1} - \frac{3}{11} - \frac{2}{7} = \frac{34}{77}$

Hence odds against C are 43 to 34.

6. A can draw a ticket in ${}^{3}C_1 = 3$ ways.

Number of cases in which A can get a prize is clearly 1.

$\therefore$ Probability of A's success $= \frac{1}{3}$.

Again B can draw a ticket in ${}^{9}C_3 = \frac{9\times8\times7}{3\times2\times1}$

$= 84$ way.

Number of ways in which B gets all blanks

$= {}^{6}C_3 = \frac{6\times5\times4}{3\times2\times1} = 20.$

$\therefore$ Number of ways of getting a prize $= 84 - 20 = 64$

Thus, probability of B's success $= \frac{64}{84} = \frac{16}{21}$

$\therefore$ A's probability of success : B's probability of success

$= \frac{1}{3} : \frac{16}{21} = 7 : 16$

7.
$$P(A \cap \overline{B}) = P(A) - P(A \cap B)$$
$$= P(A) - P(A)\,P(B)$$
$$= P(A)\,(1 - P(B)) = P(A)\,P(\overline{B})$$

Hence, **A** and $\overline{\mathbf{B}}$ are independent.

9. $S = \{0, 1, 2, ..., 9\}$

Suppose we want a particular integer "3" to be chosen.

Probability of not choosing "3" is $\frac{9}{10}$ (in one chance)

Probability of not choosing 3 in all the three chances

$= \frac{9}{10} \times \frac{9}{10} \times \frac{9}{10}$

Hence probability of choosing it in at least one of the three chances

$= 1 - \frac{27}{1000} = \frac{973}{1000}$

10. $$\frac{P(A\cap C)}{P(C)} > \frac{P(B\cap C)}{P(C)}$$

$\Rightarrow$ $P(A\cap C) > P(B\cap C)$

and $$\frac{P(A\cap \bar{C})}{P(\bar{C})} > \frac{P(B\cap \bar{C})}{P(\bar{C})}$$

$\Rightarrow$ $P(A\cap \bar{C}) > P(B\cap \bar{C})$

Adding

$P(A\cap C) + P(A\cap \bar{C}) > P(B\cap C) + P(B\cap \bar{C})$

$P(A) > P(B)$ since $(A\cap C)$ and $A\cap \bar{C}$ are disjoint.

11. Number of ways different numbers can occur on six dice = 6!

Total number of ways of occurrence = 6^5

$\therefore$ Required probability, $P = \frac{6!}{6^5} = \frac{5!}{6^4}$

12. $$P(A) = \frac{1}{4}$$

and $$P(B/A) = \frac{P(B\cap A)}{P(A)} = \frac{1}{2}$$

$\Rightarrow$ $P(A\cap B) = \frac{1}{8}$, $P(A/B) = \frac{1}{4}$

$\Rightarrow$ $\frac{P(A\cap B)}{P(B)} = \frac{1}{4}$ $\Rightarrow P(B) = \frac{1}{2}$

(*a*) If A is a sub-event of B, $A\subset B$, then $A\cap B = A$

$\therefore$ $P(A\cap B)$ should be equal to P(A). This is false.

(*b*) $P(A/B) = \frac{3}{4}$ is false.

(*c*) $P(A\cap \bar{B}) = P(A) - P(A\cap B) = \frac{1}{4} - \frac{1}{8} = \frac{1}{8}$

$$\therefore P(A/\bar{B}) = \frac{P(A\cap \bar{B})}{P(\bar{B})} = \frac{\frac{1}{8}}{1-\frac{1}{2}} = \frac{1}{4}$$

$$\therefore P(A/B) + P(A/\bar{B}) = \frac{1}{4} + \frac{1}{4} = \frac{1}{2}$$

Hence given statement is false.

14. $(A\cup B)\supset A$

(*a*) $P(A\cup B) \geq P(A) = \frac{3}{4}$

(*b*) $A\cap B\subset B$

$\therefore$ $P(A\cap B) \leq P(B) = \frac{5}{8}$

Also
$$\begin{aligned} P(A\cap B) &= 1 - P(A\cap B)^c \\ &= 1 - P(A^c\cup B^c) \\ &> 1 - \{P(A^c) + P(B^c)\} \\ &= 1 - \left\{\frac{1}{4}+\frac{3}{8}\right\} = \frac{3}{8} \end{aligned}$$

(*c*) $A\cap \bar{B} \subset \bar{B}$

$\therefore$ $P(A\cap \bar{B}) \leq P(\bar{B}) = \frac{3}{8}$

Also $A\cap \bar{B} = (\overline{\bar{A}\cup B})$

$$\therefore P(A\cap \bar{B}) = 1 - P(\bar{A}\cup B) \geq 1 - \{P(\bar{A}) + P(B)\}$$
$$= 1 - \left\{\frac{1}{4}+\frac{5}{8}\right\} = \frac{1}{8}$$

15. Let the event of drawing a faulty item from any of the machines be *A*, and the event that an item drawn at random was produced by M_i be B_i. To $P(B_i/A)$ proceed as follows :

	M_1	M_2	M_3	Remarks
$P(B_i)$	0.25	0.30	0.45	$\because$ sum = 1
$P(A/B_i)$	0.05	0.04	0.03	
$P(B_i)P(A/B_i)$	0.0125	0.012	0.0135	sum = 0.38
$P(B_i/A)$	$\frac{0.0125}{0.038}$	$\frac{0.0125}{0.038}$	$\frac{0.0135}{0.038}$	By Baye's theorem

The highest output being from M_3, required probability $= \frac{0\cdot 0135}{0\cdot 038} = 0.355$.

16. (*a*) $B = (A\cap B)\cup(\bar{A}\cap B)$

$\Rightarrow$ $P(B) = P(A\cap B) + P(\bar{A}\cap B)$ as $A\cap B$ and $\bar{A}\cap B$ are mutually exclusive.

$\Rightarrow$ $P(\bar{A}\cap B) = P(B) - P(A\cap B)$.

(*b*)
$$\begin{aligned} P(A\cup B) &= P[A\cup(\bar{A}\cap B)] \\ &= P(A) + P(\bar{A}\cap B) \\ &= P(A) + P(B) - P(A\cap B) \end{aligned}$$

(*c*)
$$\begin{aligned} P(A/B) &= P(A)\,P(B/A) \\ &\leq P(A) \quad \text{as } P(B/A) \leq 1. \end{aligned}$$

17. (*a*) $P(A\cup B) = P(A) + P(B) - P(A\cap B)$

$\therefore$ $P((A\cap C)\cup(B\cap C)) = P(A\cap C) + P(B\cap C) - P(A\cap C\cap B\cap C)$

and $P((A\cup B)\cap C) = P(A\cap C) + P(B\cap C) - P(A\cap B\cap C)$

$$\Rightarrow \frac{P((A\cup B)\cap C)}{P(C)} = \frac{P(A\cap C)}{P(C)} + \frac{P(B\cap C)}{P(C)} \frac{P(A\cap B\cap C)}{P(C)}$$

$\Rightarrow P(A\cup B/C) = P(A/C) + P(B/C) - P(A\cap B/C)$

(*b*) $P(A\cap \bar{B}/C) + P(A\cap B/C)$

$$= \frac{P(A\cap \bar{B}\cap C)}{P(C)} + \frac{P(A\cap B\cap C)}{P(C)}$$

$$= \frac{P(A\cap \bar{B}\cap C) + P(A\cap B\cap C)}{P(C)}$$

$$= \frac{P(A\cap\bar{B}\cap C)\cup(A\cap B\cap C)}{P(C)}$$

$$= \frac{P(A\cap C)}{P(C)} = \mathbf{P\ (A/C)}$$

(c) $P(A \cap \bar{B})$

$= P(A) - P(A\cap B) = P(A) - P(A)\,P(B)$

$= P(A)(1 - P(B)) = P(A)\,P(\bar{B})$

Hence A and $\bar{B}$ are independent.

18. (a) $P((A\cup B)\cap C) = P((A\cap C)\cup(B\cap C))$

$= P(A\cap C) + P(B\cap C) - P(A\cap B\cap C)$

$= P(A)\,P(C) + P(B)\,P(C) - P(A)\,P(B)\,P(C)$

$= P(C)\,(P(A) + P(B) - P(A\cap B))$

$= P(C)\,P(A\cup B)$

(b) $P(A/E) = P(B/E)$

$$\frac{P(A\cap E)}{P(E)} = \frac{P(B\cap E)}{P(E)}$$

if $P(A\cap E) = P(B\cap E)$

if $P\left(A\cap\bigcup_{i=1}^{n}E_i\right) = P\left(B\cap\bigcup_{i=1}^{n}E_i\right)$

if $P\bigcup_{i=1}^{n}(A\cap E_i) = P\left(\bigcup_{i=1}^{n}(B\cap E_i)\right)$

$$\sum_{i=1}^{n}P(A\cap E_i) = \sum_{i=1}^{n}P(B\cap E_i)$$

This is true since $P(A\cap E_i) = P(B\cap E_i)$ for each i.

19. Let E_i be the event of a person to get into an accident. Then

$P(E_i) = p\ \forall\ i$

P (at least one man meet with an accident)

$= P(E_1\cup E_2\cup ...\cup E_n)$

$= 1 - P(\bar{E}_1\cup\bar{E}_2\cup ...\cup\bar{E}_n)$

$= 1 - P(\bar{E}_1\cap\bar{E}_2\cap ...\cap\bar{E}_n)$

$= 1 - P(\bar{E}_1)\,P(\bar{E}_2)\,...\,P(\bar{E}_n)$

$= 1 - (1-p)(1-p)\,...\,(1-p)$

$= 1 - (1-p)^n$

Hence P (at least one man meets with an accident/a person is chosen) $= \frac{1}{n}(1-(1-p)^n)$.

20. Expectation of X = E(X) = average = mean

$$= \frac{1}{n}\sum_{i=1}^{n}x_i = \sum_{i=1}^{n}p_i\,x_i$$

where p_i denote probability of occurrence of x_i.

Let $q = 1 - p$.

Then probabilities of success in 1st, 2nd, 3rd trials are

$p, qp, q^2p, ...$

$\therefore\ E(X) = (0)\cdot p + (1)\cdot qp + (2)\cdot q^2p + ...$

$= qp\,(1 + 2q + 3q^2 + ... + nq^{n+1} + ...)$

$$= \frac{qp}{(1-q)^2} = \frac{qp}{p^2} = \frac{q}{p}$$

21. Ratio of A and B

$= 60\times 80 + 40\times 70 : 60\times 20 + 40\times 30$

$= 76 : 24 = \mathbf{19 : 6}$

22. $P(A/A\cup B) = \frac{P(A\cap(A\cup B))}{P(A\cup B)}$

$$= \frac{P(A)}{P(A\cup B)} = \frac{P(A)}{P(A)+P(B)}$$

23. $P(A) = P(\bar{B})$

$\Rightarrow P(A) = 1 - P(B) \Rightarrow P(A) + P(B) = 1$

$\Rightarrow P(A\cup B) - P(A\cap B) = 1 = P(S)$

Hence $A\cup B = S$ only if $A\cap B = \phi$

i.e. $A = \bar{B}$ only if $A\cap B = \phi$

24. $P(A) = 0$

$$\Rightarrow P\left(\frac{A}{S}\right) = 0 \Rightarrow \frac{n(A)}{n(S)} = 0 \Rightarrow n(A) = 0$$

$\Rightarrow A = \phi$

25. When each item is increased by λ, the mean is also increased by λ.

26. When each item is increased by λ, the S.D. is multiplied by $|\lambda|$ and variance is multiplied by λ^2.

27. $\Sigma f_i\,(y_i - \mu) = f_i\,y_i - M\,\Sigma f_i = \Sigma f_i\,y_i - \Sigma f_i\,Y_i = 0$

28. $$b_{yx}\,b_{xy} = \frac{\text{cov}(X,Y)}{\sigma_x^2}\,\frac{\text{cov}(X,Y)}{\sigma_y^2}$$

$$= \left\{\frac{\text{cov}(X,Y)}{\sigma_x\sigma_y}\right\}^2 = \{\rho\,X, Y\}^2$$

29. Cov (X, Y), b_{xy}, b_{xy} and ρ(X, Y) all are either positive, or 0 or negative simultaneously.

30. When each item is divided by α, *i.e.* it becomes $\frac{\bar{x}}{\alpha}$

When each item is increased by 10, mean also increased by 10.

$$\therefore\ \text{New mean} = \frac{\bar{x}}{\alpha} + u = \frac{x+10\alpha}{\alpha}$$

31. Standard Deviation of new data

$$= \sqrt{\frac{\Sigma(\lambda a_1)^2}{n} - \left(\frac{\Sigma\lambda a_i}{n}\right)^2} = \sqrt{\lambda^2}\,\sqrt{\frac{\Sigma a_1^2}{n} - \left(\frac{\Sigma a_i}{n}\right)^2} = |\lambda|\sigma$$

32. If we take $y = 3x - 5$ as regression equation of Y on X and $Y = 2x - 4$ as that of X on Y, then

$$b_{yx} = 3\ b_{xy} = \frac{1}{2}$$

$$\Rightarrow b_{yx}\,b_{xy} = \frac{3}{2}, \text{ which is not possible.}$$

Hence, equation $y = 3x - 5$, *i.e.* $x = \frac{1}{3}y + \frac{5}{3}$ is the regression equation of X on Y

and equation, $Y = 2x - 4$ is that Y on X.

Then $b_{yx} = 2$ and $b_{xy} = \frac{1}{3}$

$$\Rightarrow b_{xy}\, b_{xy} = \frac{3}{2}$$

$$\Rightarrow \{\rho(X, Y)\}^2 = \frac{3}{2}$$

$$\Rightarrow \rho(XY) = \sqrt{\frac{2}{3}}$$

33. Solving equations

$3x - y - 5 = 0$ and $2x - y - 4 = 0$,

we get $x = 1$ and $y = -2$.

34. Here $b_{yx} = \lambda$ and $b_{xy} = \frac{1}{4}$.

Since $0 \le b_{yx}\, b_{xy} \le 1$, therefore we must have

$$0 \le \lambda\left(\frac{1}{4}\right) \le 1 \quad \Rightarrow \quad 0 \le \lambda \le 4$$

As $b_{xy} \ne 0$, $\therefore\ b_{yx} \ne 0 \quad \Rightarrow \lambda \ne 0$.

Hence, we must have $0 < \lambda \le 4$.

35. $\text{Cov}(X, Y) = 0 \quad \Rightarrow \quad b_{yx} = b_{xy} = 0$

So, two lines of regression are $y - \bar{y} = 0$ and $x - \bar{x} = 0$, which are at right angles.

36. Given : $b_{yx} = -0.9$ is negative and $b_{xy} = 0.4$ is positive which is not possible as b_{yx} and b_{xy} are always of same sign. Hence given data is inconsistent.

37. We know $b_{yx}\, b_{xy} = \rho^2 \quad \Rightarrow \quad \rho = \pm\sqrt{b_{yx}\, b_{xy}}$

But b_{yx} and b_{xy} ρ and cov (X, Y) are all of the same sign

$$\therefore \qquad \rho = (\text{sgn}\, b_{yx})\sqrt{b_{yx}\, b_{xy}}$$

38. $$\rho(X, Y) = \frac{\text{cov}(X, Y)}{\sigma_x \sigma_y} = \frac{8}{3 \times 4} = \frac{2}{3}$$

39. Mean, $$\bar{x} = \frac{a + (a + d) + (a + 2d) + \ldots + (a + 2nd)}{2n + 1}$$

$$= \frac{2n + 1}{2}\{a + (a + (a + 2nd))\} = a + nd$$

Sum of numerical deviations from the mean

$$= \sum_{i=1}^{2n+1} |\{a + (i - 1)d\} - \{a + nd\}|$$

$$= |a - a - nd| + |a + d - a - nd| + \ldots + |a + nd - a - nd| + |a + (n + 1)\, d - a - nd| + \ldots + |a + 2nd - a - nd|$$

$$= |d|\{n + (n - 1) + (n - 2) + \ldots + 1 + 0 + 1 + 2 + \ldots + n\}$$

$$= |d| \times 2\{1 + 2 + 3 + \ldots + n\} = n(n + 1)|d|$$

$$\therefore \quad \text{M.D. (about mean)} = \frac{n(n + 1)|d|}{2n + 1}$$

40. If two variables X and Y are in perfect correlation (direct or indirect), then there is a linear relationship between the two variables.

41. Since $\rho > 0$, therefore b_{yx} and b_{xy} are also positive

Hence $$\frac{b_{yx} + b_{xy}}{2} \ge \sqrt{b_{yx}\, b_{xy}}$$

$$\Rightarrow \frac{b_{yx} + b_{xy}}{2} \ge \sqrt{\rho^2}$$

$$\Rightarrow \frac{b_{yx} + b_{xy}}{2} \ge \rho \ \Rightarrow\ m \ge \rho$$

42. If first time $x + 2y = 7$ is taken as regression equation of Y on X, then we write it as

$$y = -\frac{1}{2}x + \frac{7}{2} \quad \Rightarrow \quad b_{yx} = -\frac{1}{2}$$

Also the second equation (which will be regression equation X on Y) can be written as

$$x = -\frac{1}{2}y + \frac{7}{2} \quad \Rightarrow \quad b_{xy} = -\frac{1}{2}$$

$$\therefore \quad b_{yx}\, b_{xy} = \left(\frac{-1}{2}\right)\left(\frac{-1}{2}\right) = \frac{1}{4} \le 1$$

Hence regression equation of Y on X is

$$x + 2y = 7$$

43. Since X and Y connected by linear relationship, therefore they are in perfect correlation.

Also $ab < 0$

$$\frac{a}{b} < 0.$$

$\therefore\ y = \left(-\frac{a}{b}\right)x - \frac{c}{b}$ shows that when x increases, y also increases; so X and Y are in perfect direct correlation

$$\Rightarrow \qquad \rho(X, Y) = 1.$$

44. Required mean

$$= \frac{0^n C_0 + 1^n C_1 + 2^n C_2 + \ldots + n^n C_n}{{}^nC_0 + {}^nC_1 + {}^nC_2 + \ldots + {}^nC_n} = \frac{n2^{n-1}}{2^n}$$

$$= \frac{n}{2} \qquad (\because C_1 + 2C_2 + 3C_3 + \ldots nC_n = n2^{n-1})$$

51. Probability of technocrat manager $= \frac{20}{100} = \frac{1}{5} = p$

Probability of non-technocrat manager

$$= \frac{80}{100} = \frac{4}{5} = q$$

Probability of a random committee of 5 with exactly 2 technocrats

$$= {}^5C_2p^2q^3 = \frac{5\times 4}{1\times 2}\left(\frac{1}{5}\right)^2\left(\frac{4}{5}\right)^3 = 0.2048$$

54. Probability of both cards being kings

$$= \frac{4}{52}\times\frac{3}{51} = \frac{1}{221}$$

56. After first heads in first toss probability of tails in 2nd and 3rd toss

$$= \frac{1}{2}.\frac{1}{2} = \frac{1}{4}$$

$\therefore$ Probability of exactly two heads $= 1 - \frac{1}{4} = \frac{3}{4}$

57. Let lower limit be x.

Then, upper limit $= x+4$

$$\therefore \quad \frac{x+(x+4)}{2} = 15$$

$$\Rightarrow \quad x = 13$$

58. Range = Difference between largest values $= 95 - 62 = 33$

59. $m = \frac{7+9+11+13+15}{5} = \frac{55}{5} = 11$

$$\sum \delta^2 = |7-11|^2 + |9-11|^2 + |11-11|^2 + |13-11|^2 + |15-11|^2$$
$$= 40$$

$$\therefore \quad \sigma = \sqrt{\frac{\sum\delta^2}{n}} = \sqrt{\frac{40}{5}} = \sqrt{8}$$

$$= 2\sqrt{2} = 2\times 1.41 = 2.8$$

60. Required probability

= P [(A dies and B is alive) or (A is alive and B dies)]

$= p(1-q) + (1-p)q$

$= p + q - 2pq$

61. We have $np = 4$ and $npq = 3$

$$\Rightarrow \quad q = \frac{3}{4} \text{ and } p = (1-q) = \frac{1}{4}$$

Mode is an integer such that

$$np + p > x > np - q$$

$$\Rightarrow \quad 4 + \frac{1}{4} > x > 4 - \frac{3}{4}$$

$$\Rightarrow \quad \frac{13}{4} < x < \frac{17}{4}$$

$$\Rightarrow \quad 3.25 < x < 4.25$$

$$\Rightarrow \quad x = 4$$

62. Here, $m + \sigma^2 = 24$

and $m\sigma^2 = 128$

Solving we get

$m = 16$ or 8.

If $m = 16$ then

$\sigma^2 = 8$

If $m = 8$, then

$\sigma^2 = 56$.

Case I : $np = 16$ and $npq = 8$

$$\Rightarrow \quad p = \frac{1}{2},\ q = \frac{1}{2} \text{ and } n = 32.$$

Case II : $np = 8$ and $npq = 56$

$\Rightarrow \quad q = 7$, which is not possible.

Hence distribution is $(q+p)^n = \left(\frac{1}{2}+\frac{1}{2}\right)^{32}$.

NUMERICAL TYPE QUESTIONS

5. Let E_1 be the event that a ball is drawn from first bag, E_2 the event that a ball is drawn from second bag and E the event a black ball is chosen, therefore

$$P(E) = P(E_1)\,P\left(\frac{E}{E_1}\right) + P(E_2)\,P\left(\frac{E}{E_2}\right)$$

$$= \frac{2}{6}\times\frac{3}{7} + \frac{4}{6}\times\frac{4}{7} = \frac{11}{21}$$

6. Number of ways in which 8 balls can be drawn out of 15 is ${}^{15}C_8$.

Number of ways of drawing 2 red balls is ${}^5C_2$ and corresponding to each of these ${}^5C_2$ ways of drawing a red ball, there are ${}^{10}C_6$ ways of drawing 6 black balls.

Hence total number of ways in which 2 red and 6 black balls can be drawn $= {}^5C_2 \times {}^{10}C_6$.

$$\therefore \quad \text{Required probability} = \frac{{}^5C_2 \times {}^{10}C_6}{{}^{15}C_8}$$

$$= \frac{140}{429}$$

7. A leap year consists of 366 days, so that there are 52 full weeks (and hence 52 Sundays) and two extra days. These two days can be

(i) Monday, Tuesday
(ii) Tuesday, Wednesday,
(iii) Wednesday, Thursday
(iv) Thursday, Friday
(v) Friday, Saturday
(vi) Saturday, Sunday
(vii) Sunday, Monday

Of these 7 cases, the last two are favourable

$\therefore$ Required probability $= \frac{2}{7}$.

8. The five digits can be arranged in 5! ways, out of which 4! will begin with zero.

$\therefore$ Total number of 5-figure numbers formed
$= 5! - 4! = 96.$

Those numbers formed will be divisible by 4 which will have two extreme right digits divisible by 4,

i.e., numbers ending in 04, 12, 20, 24, 32, 40.

Now, numbers ending in 04 = 3! = 6,
numbers ending in 12 = 3! – 2! = 4,
numbers ending in 20 = 3! = 6,
numbers ending in 24 = 3! – 2! = 4,
numbers ending in 32 = 3! – 2! = 4,
andnumbers ending in 40 = 3! = 6.

[Numbers having 12, 24, 32 in the extreme right are (3! – 2!), since the numbers having zero on the extreme left are to excluded.]

$\therefore$ Total number of favourable ways
$= 6 + 4 + 6 + 4 + 4 + 6 = 30$

Hence, required probability $= \frac{30}{96} = \frac{5}{16}$.

9. $P(H) = \frac{1}{2}, P(T) = \frac{1}{2}$

Let A be the event of getting 5 or 6.

$\therefore$ $P(A) = \frac{1}{3}$ and B be the event of getting 1, 2, 3, 4;

$\therefore$ $P(B) = \frac{2}{3}$

In order to get a 'Head' before "5 or 6", the following events have to occur:

H, TBH, TBTBH, TBTBTBH, ...

$$\therefore \quad P = \frac{1}{2} + \frac{1}{2} \times \frac{2}{3} \times \frac{1}{2} + \left(\frac{1}{3}\right)^2 \cdot \frac{1}{2} + \left(\frac{1}{3}\right)^3 \cdot \frac{1}{2}$$

$$= \frac{\frac{1}{2}}{1 - \frac{1}{3}} = \mathbf{\frac{3}{4}}$$

10. Probability of drawing a white ball from box B will depend on whether transferred ball is black or white.

If a black ball is transferred, then its probability is $\frac{4}{6}$.

There are now 5 white and 8 black balls in the box B. Then probability of drawing white ball from box $B = \frac{5}{13}$

Hence probability of drawing a white ball from urn B, if the transferred ball is black

$$= \frac{4}{6} \times \frac{5}{13} = \frac{10}{39}$$

Similarly, probability of drawing a white ball from urn B, if transferred ball is white

$$= \frac{2}{6} \times \frac{6}{13} = \frac{2}{13}.$$

$$\therefore \quad \text{Required probability} = \frac{10}{39} + \frac{2}{13} = \mathbf{\frac{16}{39}}.$$

11.
$$P(A) = 0$$
$$\Rightarrow \quad A = \phi$$
$$\therefore \quad A \cap B = \phi$$
$$\therefore \quad P(A \cap B) = P(\phi) = 0$$

13. The sum 6 can be obtained as follows :
(1, 5), (2, 4), (3, 3), (4, 2), (5, 1), *i.e.* in 5 ways.

Probability of A's throwing 6 with 2 dice $= \frac{5}{36}$

$\therefore$ Probability of A's not throwing 6 $= \frac{31}{36}$

Similarly, probability of B's throwing 7 $= \frac{6}{36}$, i.e. $\frac{1}{6}$

$\therefore$ Probability of B's not throwing 7 $= \frac{5}{6}$

Now A can win, if he throws 6 in the first, third, fifth, seventh etc. throws.

$\therefore$ Chance of A's winning

$$= \frac{5}{36} + \frac{31}{36} \times \frac{5}{6} \times \frac{5}{36} + \frac{31}{36} \times \frac{5}{6} \times \frac{31}{36} \times \frac{5}{6} \times \frac{5}{36} + \ldots..$$

$$= \frac{5}{36} \times \left[1 + \left(\frac{31}{36} \times \frac{5}{6}\right) + \left(\frac{31}{36} \times \frac{5}{6}\right)^2 + \left(\frac{31}{36} \times \frac{5}{6}\right)^3 + \ldots..\right]$$

$$= \frac{5}{36} \times \frac{1}{1 - (31/36) \times (5/6)} = \frac{5}{36} \times \frac{36 \times 6}{61} = \frac{30}{61}$$

14. Sample space, S = {HHHH, HHHT, HHTH, HHTT, ... , TTTT}

Clearly, $|S| = 2^4 = 16.$

15. 18 = 3 Median –2 × 24

16. Given regression limit is

$$8x - 10y + 66 = 0$$

$$\Rightarrow \quad y = \frac{8}{10}x + \frac{66}{10} = \frac{4}{5}x + \frac{33}{5}$$

$$\text{Hence} \quad b_{yx} = \frac{4}{5}$$

$$\Rightarrow \quad \frac{\text{cov}(X, Y)}{\sigma_x^2} = \frac{4}{5}$$

$$\Rightarrow \quad \text{cov}(X, Y) = \frac{9 \times 4}{5}$$

17. Since, $x + y = 6$ for all the pairs of observations,

$$\therefore \quad b_{xy} = b_{yx} = -1$$
$$\Rightarrow \quad \rho(X, Y) = -1$$

18. Mode = 3 Median – 2 Mean

$$\Rightarrow \quad \text{Mean} = \frac{3}{2}\text{Median} - \frac{1}{2}\text{Mode}$$

$$= \frac{1}{2}(3\,\text{Median} - \text{Mode})$$

19. Mode= 3 Median – 2 Mean

$\Rightarrow$ 3 Median= Mode + 2 Mean

$$\Rightarrow \quad \text{Median} = \frac{1}{3}(\text{Mode} + 2\,\text{Mean})$$

$$\Rightarrow \quad M = \frac{1}{3}$$

20. The two lines of regression are at right angles only when

$$b_{yx}\left(\frac{1}{b_{xy}}\right) = -1 \Rightarrow b_{yx} = -b_{xy}$$

But both b_{yx} and b_{xy} are of same sign, therefore we must have

$$b_{yx} = b_{xy} = 0 \Rightarrow \rho(X, Y) = 0$$

21.
$$\rho(X, Y) = \frac{\Sigma xy - \frac{1}{n}\Sigma x \Sigma y}{\sqrt{\Sigma x^2 - \frac{1}{n}(\Sigma x)^2}\sqrt{\Sigma y^2 - \frac{1}{n}(\Sigma y)^2}}$$

$$= \frac{3 - \frac{1}{10}\times 4\times 3}{\sqrt{8 - \frac{1}{10}(4)^2 \times \left\{9 - \frac{1}{10}(3)^2\right\}}} = \frac{1}{4}$$

22. Given : Lower quartile, $Q_1 = 35$
Upper quartile $Q_3 = 75$.

$$\therefore \text{ Quartile deviation} = \frac{Q_3 - Q_1}{2} = \frac{75 - 35}{2} = 20$$

23. Here, $\bar{u} = 3\bar{x} + 4\bar{y}$ and $\bar{v} = 3\bar{x} - \bar{y}$

$$u - \bar{u} = 3(x - \bar{x}) + 4(y - \bar{y})$$

$$v - \bar{v} = 3(x - \bar{x}) - (y - \bar{y})$$

$$\Rightarrow \quad \text{cov}(u, v) = \frac{1}{n}\Sigma(u - \bar{u})(v - \bar{v})$$

$$= \frac{1}{n}\Sigma\{9(x - \bar{x})^2 - 4(y - \bar{y})^2 + 9(x - \bar{x})(y - \bar{y})\}$$

$$= 9\left(\frac{\Sigma(x - \bar{x})^2}{n}\right) - 4\left(\frac{\Sigma(y - \bar{y})^2}{n}\right) + 9\left(\frac{\Sigma(x - \bar{x})(y - \bar{y})}{n}\right)$$

$$= 9\,\sigma_x^2 - 4\sigma_y^2 + 9\text{ cov. }(x, y)$$

$$= 9 \times 4 - 4 \times 9 + 9(0) = 0$$

{since X, Y are independent, therefore cov (X, Y) = 0}

$$\Rightarrow \quad \rho(u, v) = \frac{\text{cov}(u, v)}{\sigma_u \sigma_v} = 0$$

24. Probability that first friend is born in any month = 100% = 1

Probability that second friend is born in the same month as that of first friend $= 1 \times \frac{1}{12} = \frac{1}{12}$

25. Probability of drawing a red ball $= \frac{5}{10}$

If the ball is not replaced, then box will have a ball, so probability of drawing the red ball in next chance $= \frac{4}{9}$

Hence probability of drawing 2 balls

$$= \frac{5}{10} \times \frac{4}{9} = \frac{2}{9}$$

5 CHAPTER

Numerical Methods

ACCURACY AND PRECISION

Accuracy is the agreement between an experimental value, or the average of several determinations of the value, with an accepted or theoretical ("true") value for a quantity.

Accuracy is usually expressed as a percent difference:

$$\% \text{ difference} = \frac{(\text{experimental} - \text{true}) \times 100\%}{\text{true}}$$

The accepted value for the density of gold metal is 19.31 g/cm^3. If a student measured the mass and volume of a sample of gold, and obtained a value of 19.03 g/cm^3, the percent difference would be calculated as

$$\% \text{ difference} = \frac{(19.03 - 19.31)\cancel{g} \times 100\%}{19.31\cancel{g}}$$

$$= -1.5\%$$

A percent difference can be positive or negative; the sign shows whether the experimental value is higher or lower than the actual or theoretical value. This distinction is not used in precision, since all values are experimental.

If one has high precision, or reproducibility, and poor accuracy, there is usually a systematic error of some sort. This can involve improper calibration or mishandling of a measuring device (very consistent mishandling, but mishandling just the same!).

Good accuracy and poor precision can result from a combination of sloppy experimental procedure and dumb luck. However, some quantities are difficult to determine with high precision, such as those involving lab animals. In such cases, the experimenter must make many determinations of the value and average them.

The goal of any experimenter should be to achieve both high precision and high accuracy in his or her work. Your lab work will often be graded on the basis of your precision, accuracy, or both. Pay attention to technique in the lab and your work will be of high quality.

ERROR ANALYSIS

The Uncertainty of Measurements

Some numerical statements are exact: Mary has 3 brothers, and 2 + 2 = 4. However, all *measurements* have some degree of uncertainty that may come from a variety of sources. The process of evaluating this uncertainty associated with a measurement result is often called *uncertainty analysis* or *error analysis*.

The complete statement of a measured value should include an estimate of the level of confidence associated with the value. Properly reporting an experimental result along with its uncertainty allows other people to make judgments about the quality of the experiment, and it facilitates meaningful comparisons with other similar values or a theoretical prediction. Without an uncertainty estimate, it is impossible to answer the basic scientific question: "Does my result agree with a theoretical prediction or results from other experiments?" This question is fundamental for deciding if a scientific hypothesis is confirmed or refuted.

When we make a measurement, we generally assume that some exact or true value exists based on how we define what is being measured. While we may never know this true value exactly, we attempt to find this ideal quantity to the best of our ability with the time and resources available. As we make measurements by different methods, or even when making multiple measurements using the *same* method, we may obtain slightly different results. So how do we report our findings for our best estimate of this elusive *true value*? The most common way to show the range of values that we believe includes the true value is:

measurement = best estimate ± uncertainty (units)

Let's take an example. Suppose you want to find the mass of a gold ring that you would like to sell to a friend. You do not want to jeopardize your friendship, so you want to get an accurate mass of the ring in order to charge a fair market price. By simply examining the ring in your hand, you estimate the mass to be between 10 and 20 grams, but this is not a very precise estimate. After some searching, you find an electronic balance that gives a mass reading of 17.43 grams. While this measurement is much more **precise** than the original estimate, how do you know that it is **accurate**, and how confident are you that this measurement represents the true value of the ring's mass? Since the digital display of the balance is limited to 2 decimal places, you could report the mass as m = 17.43 ± 0.01 g. Suppose you use the same electronic balance and obtain several more readings: 17.46 g, 17.42 g, 17.44 g, so that the average mass appears to be in the range of 17.44 ± 0.02 g. By now you may feel confident that you know the mass of this ring to the nearest hundredth of a gram, but how do you know that the true value definitely lies between 17.43 g and 17.45 g? Since you want to be honest, you decide to use another balance that gives a reading of 17.22 g. This value is clearly below the range of values found on the first balance, and under normal circumstances, you might not care, but you want to be fair to your friend. So what do you do now?

To help answer these questions, we should first define the terms *accuracy* and *precision*:

> **Accuracy is the closeness of agreement between a measured value and a true or accepted value. Measurement error is the amount of inaccuracy.**
>
> **Precision is a measure of how well a result can be determined (without reference to a theoretical or true value). It is the degree of consistency and agreement among independent measurements of the same quantity; also the reliability or reproducibility of the result.**
>
> **The statement of uncertainty associated with a measurement should include factors that affect both the accuracy and precision of the measurement.**

Caution: Unfortunately the terms *error* and *uncertainty* are often used interchangeably to describe both imprecision and inaccuracy. This usage is so common that it is impossible to avoid entirely. Whenever you encounter these terms, make sure you understand whether they refer to accuracy or precision, or both.

Notice that in order to determine the *accuracy* of a particular measurement, we have to know the ideal, true value, which we really never do. Sometimes we have a "textbook" measured value, which is known *precisely*, and we assume that this is our "ideal" value, and use it to estimate the *accuracy* of our result. Other times we know a theoretical value, which is calculated from basic principles, and this also may be taken as an "ideal" value. But physics is an empirical science, which means that the theory must be validated by experiment, and not the other way around. We can escape these difficulties and retain a useful definition of *accuracy* by assuming that, even when we do not know the true value, we can rely on the best available *accepted* value with which to compare our experimental value.

For our example with the gold ring, there is no accepted value with which to compare, and both measured values have the same precision, so we have no reason to believe one more than the other. The only way to assess the accuracy of the measurement is to compare with a known **standard.** For this situation, it may be possible to calibrate the balances with a standard mass that is accurate within a narrow tolerance and is traceable to a **primary mass standard** at the National Institute of Standards and Technology (NIST). Calibrating the balances should eliminate the discrepancy between the readings and provide a more *accurate* mass measurement.

Precision is often reported quantitatively by using **relative** or **fractional uncertainty**:

$$\text{Relative Uncertainty} = \left|\frac{\text{uncertainty}}{\text{measured quantity}}\right| \quad (1)$$

For example, $m = 75.5 \pm 0.5$ g has a fractional uncertainty of:

$$\frac{0.5g}{75.5g} = 0.0066 = 0.66\%$$

Accuracy is often reported quantitatively by using **relative error**:

$$\text{Relative Error} = \frac{\text{measured value} - \text{expected value}}{\text{expected value}} \quad (2)$$

If the expected value for m is 80.0 g, then the relative error is:

$$\frac{75.5 - 80.0}{80.0} = -0.056 = -5.6\%$$

Note: The minus sign indicates that the measured value is *less* than the expected value.

When analyzing experimental data, it is important that you understand the difference between precision and accuracy. Precision indicates the quality of the measurement, without any guarantee that the measurement is "correct." Accuracy, on the other hand, assumes that there is an ideal value, and tells how far your answer is from that ideal, "right" answer. These concepts are directly related to *random* and *systematic* measurement errors.

ALGEBRAIC EQUATIONS

An equation of type $y = f(x)$ is said to be algebraic if it can be expressed in the form

$$f_n y_n + f_{n-1} y_{n-1} + ... + f_1 y_1 + f_0 = 0$$

where f_1 is an ith order polynomial in x.

This equation can be thought of as having a general form

$$f(x, y) = 0$$

This implies that above equation portrays a dependence between the variables x and y.

e.g. $3x + 5y - 21 = 0$ (linear)

$2x + 3xy - 25 = 0$ (non-linear)

$x^3 - xy - 3y^3 = 0$ (non-linear)

These equations have an infinite number of pairs of values of x and y which satisfy them.

POLYNOMIAL EQUATIONS

Polynomial equations are a simple class of algebraic equations that are represented as follows:

$$a_n x^n + a_{n-1} x^{n-1} + \ldots + a_1 x + a_0 = 0$$

This is called ***n*th degree polynomial** and has n roots. The roots may be

(*i*) real and different

(*ii*) real and repeated

(*iii*) complex numbers

Since complex roots appear in pairs, if n is odd, then polynomial has at least one real root.

e.g. a cubic equation of the type $a_3x^3 + a_2x^2 + a_1x + a_0 = 0$ will have at least one real root and the remaining two may be real or complex roots.

Specific examples of polynomial equations:

(*i*) $5x^5 - x^3 + 3x^2 = 0$

(*ii*) $x^3 - 4x^2 + x + 6 = 0$

(*iii*) $x^2 - 4x + 4 = 0$

TRANSCENDENTAL EQUATIONS

A non-algebraic equation is called *transcendental equation.* These include trigonometric, exponential and logarithmic functions.

e.g. $2\sin x - x = 0,\ e^x \sin x - \dfrac{1}{2x} = 0, \log x^2 - 1 = 0,\ x - e^{1/x} = 0$

A transcendental equation may have finite or infinite number of real roots or may not have real root at all.

METHODS OF SOLVING NON-LINEAR EQUATIONS

(1) Direct Analytical Methods

In certain cases, roots can be found by using direct analytical methods.

e.g. consider a quadratic equation $ax^2 + bx + c = 0$

Solution of this equation is

$$x = \frac{-b \pm \sqrt{b^2 - 4ac}}{2a} \quad (x\text{: gives roots of equation})$$

This equation gives the two roots of quadratic equation. However, there are equations that cannot be solved by analytical methods.

e.g. simple transcendental equation

$$2 \sin x - x = 0$$

cannot be solved analytically. Direct methods for solving non-linear equations do not exist except for certain simple cases.

(2) Graphical Methods

These are useful when we are satisfied with approximate solution for a problem. This method involves plotting the given function and determining the points where it crosses the x-axis. These points represent approximate values of the roots of the function.

(3) Trial and Error Methods

This method involves a series of guesses for x, each time evaluating the function to see whether it is close to zero. The value of x that causes the function value closer to zero is one of the approximate roots of the equation.

Although graphical and trial and error methods provide satisfactory approximations for many problem situations, they become cumbersome and time consuming. Moreover, accuracy of the results are inadequate for the requirements of many engineering and scientific problems.

(4) Iterative Methods

It usually begins with an approximate value of the root called *initial guess*, which is then successively corrected iteration by iteration. The process of iteration stops when desired level of accuracy is obtained. Since iterative methods involve a large number of iterations and arithmetic operations to reach a solution, the use of computers has become inevitable to make the task simple and efficient.

Before initiating an iterative process, we have to determine either an approximate value of root or a *search interval* that contains a root.

One simple method of guessing starting points is to plot the curve of $f(x)$ and to identify a search interval near the root of interest. Graphical representation of a function cannot only provide rough estimates of the roots, but also help in understanding the properties of the function, thereby identifying possible problems in numerical computing.

A plot of $f(x) = x^3 - x - 1$ is shown in the figure.

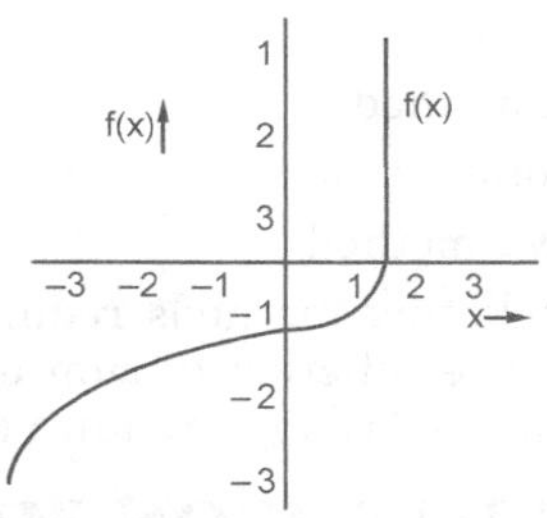

Fig. Plot of $f(x) = x^3 - x - 1$

Although $f(x)$ is a cubic function, it intersects x-axis at only one point. This suggests that remaining two roots are imaginary ones.

In the case of polynomials, many theoretical relation-ships between roots and coefficients are available.

Relation Used for Making Initial Guesses.

For a polynomial represented by

$$f(x) = a_n x^n + a_{n-1} x^{n-1} + \dots a_1 x \; a_0$$

the *largest possible* root is given by $f(x_1) = \dfrac{a_{n-1}}{a_n}$

This value is taken as the initial approximation when no other value is suggested by the knowledge of the problem at hand.

Types of Iterative Methods

There are a number of iterative methods that have been tried and used successfully in various problem situations. All these methods typically generate a sequence of estimates of the solution which is expected to converge to the true solution.

All iterative methods begin their process of solution with one or more guesses at the solution being sought.

Types of iterative methods based on number of guesses they use are

(*i*) Bracketing methods

These are also called *interpolation methods.* It start with two initial guesses that 'bracket' the root and then systematically reduce the width of bracket until the solution is reached.

Types of Braketing methods:

(*a*) Bisection method

(*b*) False position method

These methods are based on the assumption that the function changes sign in the vicinity of a root.

(*ii*) Open end methods

These are also called *extrapolation methods.* It uses a single starting value or two values that do not necessarily bracket the root.

Types of open end methods:

(*a*) Newton-Raphson method

(*b*) Secant method

(*c*) Muller's method

(*d*) Fixed-point method

(*e*) Bairstow's method

Note : Bracketing methods require to find sign changes in the function during every iteration. Open end methods do not require this.

NEWTON-RAPHSON METHOD

If point x_0 is close to the root α, then tangent line to the graph of $f(x)$ at x_0 is a good approximation to $f(x)$ near α. So, root of the tangent line, where line cuts the x-axis ; x_1 is a better approximation to α than x_0 is.

$$\text{Slope of the tangent} = \frac{f(x_0)}{x_0 - x_1} = f'(x_0)$$

$$\therefore \quad x_0 - x_1 = \frac{f(x_0)}{f'(x_0)}$$

$$\Rightarrow \quad x_1 = x_0 - \frac{f(x_0)}{f'(x_0)}$$

Repeating process, we obtain a better approximation

$$x_2 = x_1 - \frac{f(x_1)}{f'(x_1)}$$

and, in general, for $n = 0, 1, 2, \dots,$

$$x_{n+1} = x_n - \frac{f(x_n)}{f'(x_n)}$$

If x_0 is sufficiently close to α, $x_n \to \alpha$ as $n \to \infty$.

Order of Convergence

Let $x_0, x_1, x_2, \dots, x_n, \dots$ be a sequence of approximations to a root α produced by a numerical method, where $\lim\limits_{n\to\infty} x_n = \alpha$.

Let $\varepsilon_n = \alpha - x_n$

If $\lim\limits_{n\to\infty} \dfrac{|\varepsilon_{n+1}|}{|\varepsilon_n|^p} = C$, for some p and some *non-zero* constant C, then method has *order of convergence p*, and C is asymptotic error constant (bigger p is, faster the convergence).

Now consider Newton Raphson method, where $f(x)$ is assumed to be twice differentiable, and α is a simple root.

$$\text{Then } f(\alpha) = f(x_n) + (\alpha - x_n) f'(x_n) + \frac{(\alpha - x_n)^2}{2} f''(\xi) = 0$$

where ξ is an unknown point between x_n and α;

$$i.e. \quad \alpha - x_n + \frac{f(x_n)}{f'(x_n)} = \frac{-(\alpha - x_n)^2 f''(\xi)}{2f'(x_n)}$$

$$\Rightarrow \quad \alpha - x_{n+1} = \frac{-(\alpha - x_n)^2 f''(\xi)}{2f'(x_n)}$$

$$\Rightarrow \quad \frac{|\varepsilon_{n+1}|}{|\varepsilon_n|^2} = \frac{|f''(\xi)|}{2|f'(x_n)|}$$

$$\Rightarrow \quad \lim_{n\to\infty} \frac{|\varepsilon_{n+1}|}{|\varepsilon_n|^2} = \frac{|f''(\alpha)|}{2|f'(\alpha)|} \text{ if } x_n \to \alpha \text{ as } n \to \infty$$

Thus, in case of a simple root α, Newton's method has second order quadratic—convergence, with asymptotic error constant $\dfrac{|f''(\alpha)|}{2|f'(\alpha)|}$

Note. If α is a root of multiplicity m, then

$$f(x) = (x - \alpha)^m g(x)$$

$$\Rightarrow \quad f_0'(x) = (x - \alpha)^m g'(x) + m(x - \alpha)^{m-1} g(x)$$

$$= (x - \alpha)^{m-1}[(x - \alpha)g'(x) + mg(x)]$$

So, $f'(\alpha) = 0$ unless $m = 1$,
in which case $f'(\alpha) = g(\alpha) \neq 0$.
Hence requirement that α be simple for the method to be quadratically convergent.
Otherwise, if α is not simple,

$$x_{n+1} - \alpha = x_n - \alpha - \frac{f(x_n)}{f'(x_n)}$$

$$= x_n - \alpha - \frac{(x_n - \alpha) g(x_n)}{(x_n - \alpha) g'(x_n) + m g(x_n)}$$

So, dividing across by $x_n - \alpha$ and taking absolute values of both sides, we have

$$\frac{|x_{n+1} - \alpha|}{|x_n - \alpha|} = \left|1 - \frac{g(x_n)}{(x_n - \alpha) g'(x_n) + m g(x_n)}\right|$$

$$\therefore \lim_{n\to\infty} \frac{|\varepsilon_{n+1}|}{|\varepsilon_n|} = \left|1 - \frac{g(\alpha)}{m g(\alpha)}\right|$$

$$= 1 - \frac{1}{m} = \frac{m-1}{m}; m \geq 2$$

assuming $\lim_{n\to\infty} x_n = \alpha$.

So, order of convergence for multiple roots is only 1—linear and as m, multiplicity of the root gets bigger, convergence slows down since asymptotic error constant $\frac{m-1}{m} \to 1$.

However, function $q(x) = \frac{f(x)}{f'(x)}$ always has simple roots :

$$q(x) = \frac{(x - \alpha) g(x)}{(x - \alpha) g'(x) + m g(x)}$$

So $q(\alpha) = 0$

$$\therefore q'(x) = \frac{[(x-\alpha)g'(x) + mg(x)][(x-\alpha)g'(x) + g(x)] - (x-\alpha)g(x)[(x-\alpha)g''(x) + (m+1)g'(x)]}{[(x-\alpha)g'(x) + mg(x)]^2}$$

Hence, $q'(\alpha) = \frac{mg(\alpha)^2}{m^2 g(\alpha)^2} = \frac{1}{m} \neq 0$

Hence, Newton's method applied to $q(x)$ will be quadratically convergent:

$$x_{n+1} = x_n - \frac{q(x_n)}{q'(x_n)}$$

$$q'(x) = \frac{f'(x)^2 - f(x) f''(x)}{f'(x)^2}$$

$$\Rightarrow \frac{q(x)}{q'(x)} = \frac{f(x) f'(x)}{f'(x)^2 - f(x) f''(x)}$$

Thus, we have *modified Newton method:*

$$x_{n+1} = x_n - \frac{f(x_n) f'(x_n)}{f'(x_n)^2 - f(x_n) f''(x_n)}$$

which will be quadratically convergent to a root of any multiplicity; it is more unwieldy than Newton's method and requires calculation of $f''(x_n)$.

Alternatively

If $f(x) = k(x - \alpha)^m$ for some constant k, then

$$f'(x) = km(x - \alpha)^{m-1}$$

$$\therefore \frac{f(x)}{f'(x)} = \frac{(x - \alpha)}{m}$$

So, $\alpha = x - m\frac{f(x)}{f'(x)}$ for any x.

In general case of a multiple root, this becomes an approximation rather than an equality, and we have another modified Newton method:

$$x_{n+1} = x_n - m\frac{f(x_n)}{f'(x_n)}$$

This method does not require calculation of $f''(x_n)$; but it does require calculation of the multiplicity m from

$$\frac{|\varepsilon_{n+1}|}{|\varepsilon_n|} \approx \frac{m-1}{m}$$

where ε_n and ε_{n+1} are errors in Newton's method.

Example. $f(x) = x - \cos x$ ($\alpha = 0.739085$)

$$f'(x) = 1 + \sin x$$

$$x_{n+1} = x_n - \frac{x_n - \cos x_n}{1 + \sin x_n}$$

$$= \frac{x_n \sin x_n + \cos x_n}{1 + \sin x_n}$$

n	x_n	$\varepsilon_n = \alpha - x_n$	$x_{n+1} - x_n$
0	0	0.739085	1
1	1	–0.260915	–0.249636
2	0.750364	–0.011279	–0.011251
3	0.739113	–0.000028	–0.000028
4	0.739085	0	0
5	0.739085	0	

In this case, $\frac{|\varepsilon_1|}{|\varepsilon_0|^2} = 0.47765$; $\frac{|\varepsilon_2|}{|\varepsilon_1|^2} = 0.165681$;

$$\frac{|\varepsilon_3|}{|\varepsilon_2|^2} = 0.220098$$

Note : $x_{n+1} - x_n$ is a good approximation for $\alpha - x_n$ as x_n approaches α.

Example. $f(x) = x^2 - 4x + 4 = 0$;

$\alpha = 2\ m = 2$

$f'(x) = 2x - 4$

Solution

From Newton's method :

$$x_{n+1} = x_n - \frac{x_n^2 - 4x_n + 4}{2x_n - 4} = \frac{x_n^2\ 4}{2x_n - 4} = \frac{x_n + 2}{2}$$

n	x_n	$\varepsilon_n = \alpha - x_n$	$x_{n+1} - x_n$
0	1	1	0.5
1	1.5	0.5	0.25
2	1.75	0.25	0.125
3	1.875	0.125	0.0625
4	1.9375	0.0625	

Here,

$$\frac{|\varepsilon_{n+1}|}{|\varepsilon_n|} = \frac{|\alpha - x_{n+1}|}{|\alpha - x_n|} = \frac{1}{2} = \frac{|x_{n+1} - x_n|}{|x_n - x_{n+1}|} \text{ for all } n.$$

Thus convergence is only linear :

$$\frac{m-1}{m} = \frac{1}{2}$$

$$\Rightarrow \quad 2m - 2 = m$$

$$\Rightarrow \quad m = 2$$

So an alternative method should be used.

Using first modified method :

$$f(x) = x^2 - 4x + 4$$

$$\Rightarrow \quad f'(x) = 2(x - 2)$$

$$\Rightarrow \quad f''(x) = 2$$

$$\therefore \quad x_{n+1} = x_n - \frac{2(x_n - 2)^3}{4(x_n - 2)^2 - 2(x_n - 2)^2}$$

$$= x_n - (x_n - 2) = 2$$

Using second method :

$$x_{n+1} = x_n - 2\,\frac{f(x_n)}{f'(x_n)}$$

$$\Rightarrow \quad x_{n+1} = x_n - \frac{2(x_n - 2)^3}{2(x_n - 2)} = 2$$

Example. $f(x) = x^4 - 2x^3 + 2x - 1 = 0$; & $\alpha = 1\ f'(x) = 4x^3 - 6x^2 + 2$

Using Newton's method,

$$x_{n+1} = x_n - \frac{x_n^4 - 2x_n^3 + 2x_n - 1}{4x_n^3 - 6x_n^2 + 2} = \frac{3x_n^4 - 4x_n^3 + 1}{4x_n^3 - 6x_n^2 + 2}$$

n	x_n	$\varepsilon_n = \alpha - x_n$	$x_{n+1} - x_n$	$\frac{\lvert\alpha - x_{n+1}\rvert}{\lvert\alpha - x_n\rvert}$	$\frac{\lvert x_{n+2} - x_{n+1}\rvert}{\lvert x_{n+1} - x_n\rvert}$
0	0	1	0.5	0.5	0.375
1	0.5	0.5	0.1875	0.625	0.59227
2	0.6875	0.3125	0.11102	0.64474	0.62836
3	0.79852	0.20148	0.06976	0.65376	0.64464
4	0.86828	0.13172	0.04497	0.65859	0.65288
5	0.91325	0.08675	0.02936	0.66156	
6	0.94261	0.05739			

Here $\frac{m-1}{m} \approx \frac{2}{3} \Rightarrow 3m - 3 = 2m \Rightarrow \quad m = 3$

Since root is multiplicity 3, we use modified method.

$$x_{n+1} = x_n - 3\,\frac{f(x_n)}{f'(x_n)} = x_n - \frac{3(x_n^4 - 2x_n^3 + 2x_n - 1)}{4x_n^3 - 6x_n^2 + 2} = \frac{x_n^4 - 4x_n^3 + 3}{4x_n^3 - 6x_n^2 + 2}$$

Using this method, we have

n	x_n
0	0
1	1.5
2	1.03125
3	1.00016
4	0.99993
5	0.99961
6	1.00006
7	1.00000

So, there is a root of multiplicity 3 at $x = 1$.

In practice, Newton's method would be terminated if $\frac{|x_{n+2} - x_{n+1}|}{|x_{n+1} - x_n|}$ approached a non-zero limit after a preset number of steps and whatever the most recent value of x_n was would be used as an initial guess for the modified method.

Limitations of Newton-Raphson Method

This method fails in the following situations :

(*i*) Division by zero may occur if $f'(x_i)$ is zero or very close to zero.

(*ii*) If initial guess is too far away from the required root, the process may converge to some other root.

(*iii*) A particular value in the iteration sequence may repeat, resulting in an infinite loop. This occurs when to the curve $f(x)$ at $x = x_{i+1}$ cuts the x-axis again at $x = x_i$.

Example. Find by Newton Raphson's method, real root of equation $xe^x - 2 = 0$ correct to three decimal places.

Solution

Let $\quad f(x) = xe^x-2 \quad ...(i)$

Then $\quad f(0) = -2$ and $f(1) = e - 2 = 0.7183$

So a root of (*i*) lies between 0 and 1 and nearer to 1.

Let us take $\quad x_0 = 1.$

Also, $\quad f'(x) = xe^x + e^x$

and $\quad f'(1) = e + e = 5.4366$

By Newton's rule,

first appoximation, $\quad x_1 = x_0 - \dfrac{f(x_0)}{f'(x_0)}$

$$= 1 - \frac{0.7183}{5.4366} = 0.8679$$

$\therefore \quad f(x_1) = 0.0672,$

and $\quad f'(x_1) = 4.4491$

second approximation,

$$x_2 = x_1 - \frac{f(x_1)}{f'(x_1)} = 0.8679 - \frac{0.0672}{4.4491} = 0.8528$$

Thus, required root is 0.0853 correct to 3 decimal places.

SECANT METHOD

This method may be regarded as an approximation to Newton's method where, instead of $f'(x_n)$, the quotient $\dfrac{f(x_n)-f(x_{n-1})}{x_n - x_{n-1}}$ is used; *i.e.* instead of tangent to the graph of $f(x)$ at x_n, we use *secant joining* the points $(x_{n-1}, f(x_{n-1}))$ and $(x_n, f(x_n))$.

Equating two expressions for the slope of the secant, we have

$$\frac{f(x_n)}{x_n - x_{n+1}} = \frac{f(x_n)-f(x_{n-1})}{x_n - x_{n-1}}$$

$$\Rightarrow \quad x_{n+1} = x_n - \frac{(x_n - x_{n-1})f(x_n)}{f(x_n)-f(x_{n-1})}$$

Convergence rate of the secant method is 1.618, *i.e. almost quadratic,* if α is a simple root.

Advantage

$f'(x)$ need not be evaluated; so the number of function evaluations is half that of Newton's method. However, its convergence rate is slightly less than Newton and it requires two initial guesses x_0 and x_1.

In common with Newton's method, its convergence rate deteriorates for multiple roots, and if m can be estimated, the modification

$$x_{n+1} = x_n - m\frac{(x_n - x_{n-1})f(x_n)}{f(x_n)-f(x_{n-1})} \text{ should be used.}$$

Also in common with Newton's method, convergence is not guaranteed at all unless x_0 and in case of secant method x_1 also is a sufficiently good initial guess. For this reason a starting method such as Bisection method or *Regula Falsi* is used where applicable for a few iterations to obtain a good initial guess for one of the more accurate methods.

SOLUTION OF NON-LINEAR SIMULTANEOUS EQUATIONS

1. Newton-Raphson Method

Consider equations, $f(x, y) = 0$ and $g(x, y) = 0$

If an initial approximation (x_0, y_0) to a solution has been found by graphical method or otherwise, then a better approximation (x_1, y_1) can be obtained as follows :

Let $\quad x_1 = x_0 + h \quad$ and $\quad y_1 = y_0 + k$

$\therefore \quad f(x_0 + h, y_0 + k) = 0; \quad g(x_0 + h, y_0 + k) = 0$

Expanding each of the functions by Taylor's series to first degree terms, we get approximately

$$f_0 + h\frac{\partial f}{\partial x_0} + k\frac{\partial f}{\partial y_0} = 0$$

$$g_0 + h\frac{\partial g}{\partial x_0} + k\frac{\partial g}{\partial y_0} = 0$$

where, $\quad f_0 = f(x_0\, y_0), \quad \dfrac{\partial f}{\partial x_0} = \left(\dfrac{\partial f}{\partial x}\right)_{x_0, y_0} \quad$ etc.

Solving above equations for h and k, we get a new approximation to the root as

$$x_1 = x_0 + h,\ y_1 = y_0 + k$$

This process is repeated till we get the values to the desire accuracy.

Example. Solve the system of non-linear equations

$$x^2 + y = 11$$
$$y^2 + x = 7$$

Solution

An initial approximation to the solution is obtained from a rough graph of above equations, as

$$x_0 = 3.5 \text{ and } y_0 = -1.8$$

We have, $\quad f = x^2 + y - 11$

and $$g = y^2 + x - 7$$

$\therefore$ $$\frac{\partial f}{\partial x} = 2x, \frac{\partial f}{\partial y} = 1$$

and $$\frac{\partial g}{\partial x} = 1, \frac{\partial g}{\partial y} = 2y$$

Newton-Raphson's equations above will be

$$7h + k = 0.55$$

and $$h - 3.6k = 0.26$$

Solving these, we get

$$h = 0.0855$$

Repeating the above process, replacing (x_0, y_0) by (x_1, y_1), we get

$$x_2 = 3.5844,$$
$$y_2 = -1.8482$$

2. Bisection Method

It is one of the simplest and most reliable of iterative methods for the solution of nonlinear equations. This method, also called *binary chopping or half-interval method*, relies on the fact that if $f(x)$ is real and continuous in the interval $a < x < b$, and $f(a)$ and $f(b)$ are of opposite signs, i.e.

$$f(a)f(b) < 0$$

then there is at least one real root in the interval between a and b (there may be more than one root in the interval).

Let $x_1 = a$ and $x_2 = b$.

Let us also define another point x_0 to be the midpoint between a and b, i.e.

$$x_0 = \frac{x_1 + x_2}{2}$$

Now, there exists following three conditions:

(i) If $f(x_0) = 0$, we have a root at x_0.

(ii) If $f(x_0)f(x_1) < 0$, there is a root between x_0 and x_1.

(iii) If $f(x_0)\, f(x_2) < 0$, there is a root between x_0 and x_2.

It follows that by testing sign of the function at midpoint, we can deduce which part of the interval contains the root.

Figure shows that, since $f(x_0)$ and $f(x_2)$ are of opposite signs, a root lies between x_0 and x_2.

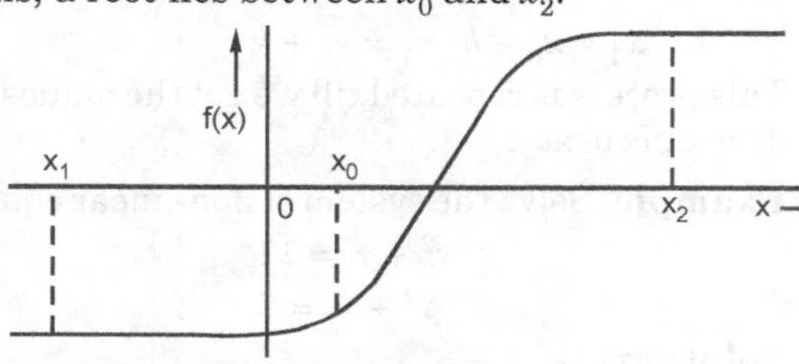

This subinterval can be further divided into two halves to locate a new sub-interval containing the root. This process can be repeated until interval containing the root is as small as needed.

Example. Find a root of the equation,
$x^2 - 4x - 10 = 0$ using bisection method.

Solution

First step is to guess two initial values that would bracket a root.

Using equation

$$|x^*_{\max}| = \sqrt{\left(\frac{a_n - 1}{a_n}\right)^2 - 2\left(\frac{a_n - 2}{a_n}\right)}$$

maximum absolute of the solution can be decided.

Thus $$x_{\max} = \sqrt{\left(\frac{-4}{1}\right)^2 - 2\left(\frac{-10}{1}\right)} = 6$$

Therefore, both the roots in the interval (– 6, 6). The table below gives values of $f(x)$ between –6 and 6 and shows that there is a root in the interval (– 2, – 1) and another in (5, 6).

x	–6	–5	–4	–3	–2	–1	0	1	2	3	4	5	6
f(x)	50	35	22	11	2	–4	–10	–13	–14	–13	–10	–5	2

Now, take $x_1 = -2$ and $x_2 = -1$

Then $$x_0 = \frac{-2-1}{2} = -1.5$$

$$f(-2) = 2 \text{ and } f(-1.5) = -1.75$$

Since $$f(-2)\, f(-1.5) < 0$$

the root must be in the interval (– 2, – 1.5).

The next step begins.

$$x_1 = -1,\ x_2 = -1.5 \text{ and } x_0 = -1.75$$
$$f(-1.75) = 0.0625$$

Since $f(-1.75)$ and $f(-1.5)$ are of opposite sign, the root lies in the interval (– 1.75, – 1.5), another iteration begins.

$$x_1 = -1.75, x_2 = -1.5 \text{ and } x_0 = -1.625$$
$$f(-1.625) = -0.859$$

Now, root lies in the interval (– 1.75, – 1.625)

$\therefore$ $$x_0 = -1.6875$$
$$f(-1.6875) = -0.40$$

Again $$x_0 = -\frac{1.75 + 1.6875}{2} = -1.72$$
$$f(-1.72) = -0.1616$$

Again, $$x_0 = -\frac{1.75 + 1.72}{2} = -1.735$$
$$f(-1.735) = -0.05$$

Again, $$x_0 = 1.7425$$
$$f(-1.7425) = +0.0063$$

The root lies between – 1.735 and – 1.7425.

Approximate root is – 1.7416.

3. Regula Falsi Method

This is a variation of the bisection method, where m_{k+1}, instead of being midpoint of $I_k = (a_k, b_k)$ is the point of intersection with the x-axis of the secant joining $(a_k, f(a_k))$ and $(b_k, f(b_k))$.

$$m_{k+1} = \frac{a_k f(b_k) - b_k f(a_k)}{f(b_k) - f(a_k)}$$

Like bisection method, it only work for odd roots, and has linear convergence rate.

Advantages

(*i*) It always converges to odd roots.

(*ii*) Error bounds are easily obtained.

(*iii*) Convergence is unaffected by the multiplicity of the root.

Disadvantages

(*i*) Slow to converge.

(*ii*) Skips even roots.

LEAST SQUARES APPROXIMATION

Another approach to approximating a function $f(x)$ on an interval $a \le x \le b$ is to seek an approximation $p(x)$ with a small 'average error' over the interval of approximation. A convenient definition of the average error of the approximation is given by

$$E(p; f) \equiv \left[\frac{1}{b-a}\int_a^b [f(x) - p(x)]^2\, dx\right]^{\frac{1}{2}} \quad (1)$$

This is also called the root-mean-square-error (denoted subsequently by RMSE) in the approximation of $f(x)$ by $p(x)$. Note first that choosing $p(x)$ to minimize $E(p; f)$ is equivalent to minimizing

$$\int_a^b [f(x) - p(x)]^2\, dx$$

thus dispensing with the square root and multiplying fraction (although the minimums are generally different). The minimizing of (1) is called the least squares approximation problem.

Example. Let $f(x) = e^x$, let $p(x) = \alpha_0 + \alpha_1 x$, α_0, α_1 unknown. Approximate $f(x)$ over $[-1, 1]$. Choose α_0, α_1 to minimize

$$g(\alpha_0, \alpha_1) \equiv \int_{-1}^{1} [e^x - \alpha_0 - \alpha_1 x]^2\, dx \quad (2)$$

$$g(\alpha_0, \alpha_1) = \int_{-1}^{1} \{e^{2x} + \alpha_0^2 + \alpha_1^2 x^2 - 2\alpha_0 e^x - 2\alpha_1 x e^x + 2\alpha_0\alpha_1 x\} dx$$

Integrating,

$$g(\alpha_0, \alpha_1) = c_1\alpha_0^2 + c_2\alpha_1^2 + c_3\alpha_0\alpha_1 + c_4\alpha_0 + c_5\alpha_1 + c_6$$

with constants $\{c_1, ..., c_6\}$, e.g.

$$c_1 = 2,\ c_6 = \frac{(e^1 - e^{-1})}{2}.$$

g is a quadratic polynomial in the two variables α_0, α_1. To find its minimum, solve the system

$$\frac{\partial g}{\partial \alpha_0} = 0,\quad \frac{\partial g}{\partial \alpha_1} = 0$$

It is simpler to return to (2) to differentiate, obtaining

$$2\int_{-1}^{1} [e^x - \alpha_0 - \alpha_1 x]\,(-1)\, dx = 0$$

$$2\int_{-1}^{1} [e^x - \alpha_0 - \alpha_1 x]\,(-x)\, dx = 0$$

This simplifies to

$$2\alpha_0 = \int_{-1}^{1} e^x dx = e - e^{-1}$$

$$\frac{2}{3}\alpha_1 = \int_{-1}^{1} x e^x dx = 2e^{-1}$$

$$\alpha_0 = \frac{e - e^{-1}}{2} = 1.1752$$

$$\alpha_1 = 3e^{-1} = 1.1036$$

Using these values for α_0 and α_1, we denote the resulting linear approximation by

$$\ell_1(x) = \alpha_0 + \alpha_1 x$$

It is called the best linear approximation to e^x in the sense of least squares. For the error,

$$\max_{-1 \le x \le 1} |e^x - \ell_1(x)| = 0.439$$

Errors in linear approximations of e^x:

Approximation	Max Error	RMSE
Taylor $t_1(x)$	0.718	0.246
Least squares $\ell_1(x)$	0.439	0.162
Chebyshev $c_1(x)$	0.372	0.184
Minimax $m_1(x)$	0.279	0.190

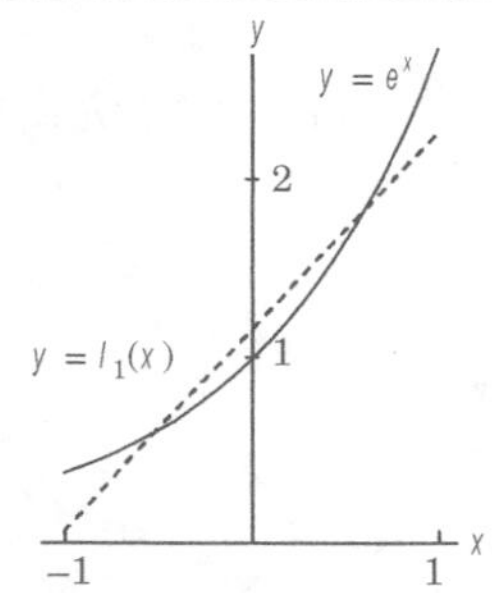

The linear least squares approximation to e^x

The General Case

Approximate $f(x)$ on $[a, b]$, and let $n \ge 0$. Seek $p(x)$ to minimize the RMSE. Write

$$p(x) = \alpha_0 + \alpha_1 x + ... + \alpha_n x^n$$

$$g(\alpha_0, \alpha_1, ..., \alpha_n) \equiv \int_{-1}^{1} \left[f(x) - \alpha_0 - \alpha_1 x - ... - \alpha_n x^n\right]^2 dx$$

Find coefficients $\alpha_0, \alpha_1, ..., \alpha_n$ to minimize this integral. The integral $g(\alpha_0, \alpha_1, ..., \alpha_n)$ is a quadratic polynomial in the $n + 1$ variables $\alpha_0, \alpha_1, ..., \alpha_n$.

To minimize $g(\alpha_0, \alpha_1, ..., \alpha_n)$, invoke the conditions

$$\frac{\partial g}{\partial \alpha_i} = 0, \quad i = 0, 1, ..., n$$

This yields a set of $n + 1$ equations that must be satisfied by a minimizing set $\alpha_0, \alpha_1, ..., \alpha_n$ for g. Manipulating this set of conditions leads to a simultaneous linear system.

To better understand the form of the linear system, consider the special case of $[a, b] = [0, 1]$. Differentiating g with respect to each α_i, we obtain

$$2\int_{-1}^{1} [e^x - \alpha_0 - \ldots - \alpha_n x^n]\,(-1)\,dx = 0$$

$$2\int_{-1}^{1} [e^x - \alpha_0 - \ldots - \alpha_n x^n]\,(-x)\,dx = 0$$

$$\vdots$$

$$2\int_{-1}^{1} [e^x - \alpha_0 - \ldots - \alpha_n x^n]\,(-x^n)\,dx = 0$$

Then the linear system is

$$\sum_{j=0}^{n} \frac{\alpha_j}{i+j+1} = \int_0^1 x^i f(x)dx\,,\; i = 0, 1, \ldots, n$$

We will study the solution of simultaneous linear systems in Chapter 6. There we will see that this linear system is 'ill-conditioned' and is difficult to solve accurately, even for moderately sized values of n such as $n = 5$. As a consequence, this is not a good approach to solving for a minimizer of $g(\alpha_0, \alpha_1, \ldots, \alpha_n)$.

Legendre Polynomials

Define the Legendre polynomials as follows.

$$P_0(x) = 1$$

$$P_n(x) = \frac{1}{n!2^n}\cdot\frac{d^n}{dx^n}\left[(x^2-1)^n\right],\; n = 1, 2, \ldots$$

For example,

$$P_1(x) = x$$

$$P_2(x) = \frac{1}{2}(3x^2 - 1)$$

$$P_3(x) = \frac{1}{2}(5x^2 - 3x)$$

$$P_4(x) = \frac{1}{2}(35x^4 - 30x^2 + 3)$$

The Legendre polynomials have many special properties, and they are widely used in numerical analysis and applied mathematics.

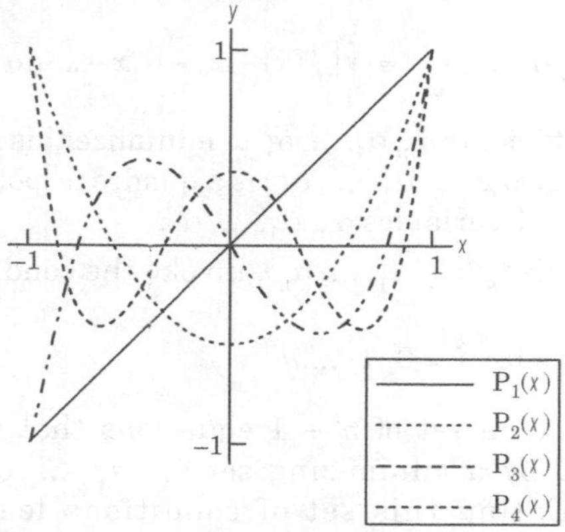

Legendre polynomials of degrees 1, 2, 3, 4

Properties

Introduce the special notation

$$(f, g) = \int_a^b f(x)g(x)dx$$

for general functions f (x) and g(x).

- Degree and normalization:

  $$\deg P_n = n,\; P_n(1) = 1,\; n \geq 0$$

- Triple recursion relation: For $n \geq 1$,

  $$P_{n+1}(x) = \frac{2n+1}{n+1} x P_n(x) - \frac{n}{n+1} P_{n-1}(x)$$

- Orthogonality and size:

  $$(P_i, P_j) = \begin{cases} 0, & i \neq j \\ \dfrac{2}{2j+1}, & i = j \end{cases}$$

- Zeroes:

  All zeroes of $Pn(x)$ are located in $[-1, 1]$; all zeroes are simple roots of $P_n(x)$

- Basis: Every polynomial $p(x)$ of degree $\leq n$ can be written in the form

  $$p(x) = \sum_{j=0}^{n} \beta_j P_j(x)$$

  with the choice of $\beta_0, \beta_1, \ldots, \beta_n$ uniquely determined from $p(x)$:

  $$\beta_j = \frac{(p, P_j)}{(P_j, P_j)},\; j = 0, 1, \ldots, n$$

Finding the Least Squares Approximation

We solve the least squares approximation problem on only the interval $[-1, 1]$. Approximation problems on other intervals $[a, b]$ can be accomplished using a linear change of variable.

We seek to find a polynomial $p(x)$ of degree n that minimizes

$$\int_a^b [f(x) - p(x)]^2\,dx$$

This is equivalent to minimizing

$$(f - p, f - p) \qquad (3)$$

We begin by writing $p(x)$ in the form

$$p(x) = \sum_{j=0}^{n} \beta_j P_j(x)$$

$$p(x) = \sum_{j=0}^{n} \beta_j P_j(x)$$

Substitute into (3), obtaining

$$g(\beta_0, \beta_1, \ldots \beta_n) \equiv (f - p, f - p) = \left(f - \sum_{j=0}^{n} \beta_j P_j,\, f - \sum_{i=0}^{n} \beta_i P_i\right)$$

Expand this into the following:

$$\tilde{g} = (f,f) - \sum_{j=0}^{n}\frac{(f,P_j)^2}{(P_j,P_j)} + \sum_{j=0}^{n}(P_j,P_j)\left[\beta_j - \frac{(f,P_j)}{(P_j,P_j)}\right]^2$$

Looking at this carefully, we see that it is smallest when

$$\beta_j = \frac{(f,P_j)}{(P_j,P_j)}, j = 0, 1, ..., n$$

Looking at this carefully, we see that it is smallest when

$$\beta_j = \frac{(f,P_j)}{(P_j,P_j)}, j = 0, 1, ..., n$$

The minimum for this choice of coefficients is

$$\tilde{g} = (f,f) - \sum_{j=0}^{n}\frac{(f,P_j)^2}{(P_j,P_j)}$$

We call

$$\ell_n(x) = \sum_{j=0}^{n}\frac{(f,P_j)^2}{(P_j,P_j)}P_j(x) \qquad (4)$$

the least squares approximation of degree n to $f(x)$ on $[-1, 1]$.

If $\beta_n = 0$, then its actual degree is less than n.

Example. Approximate $f(x) = e^x$ on $[-1, 1]$. We use (4) with $n = 3$:

$$\ell_3(x) = \sum_{j=0}^{3}\beta_j P_j(x),\ \beta_j = \frac{(f,P_j)^2}{(P_j,P_j)} \qquad (5)$$

The coefficients $\{\beta_0, \beta_1, \beta_2, \beta_3\}$ are as follows.

j	0	1	2	3
β_j	2.35040	0.73576	0.14313	0.02013

Using (5) and simplifying,

$$\ell_3(x) = .996294 + .997955x + .536722x^2 + .176139x^3$$

The error in various cubic approximations

Approximation	Max Error	RMSE
Taylor $t_3(x)$	.0516	.0145
Least squares $\ell_3(x)$	.0112	.00334
Chebyshev $c_3(x)$	.00666	.00384
Minimax $m_3(x)$	.00553	.00388

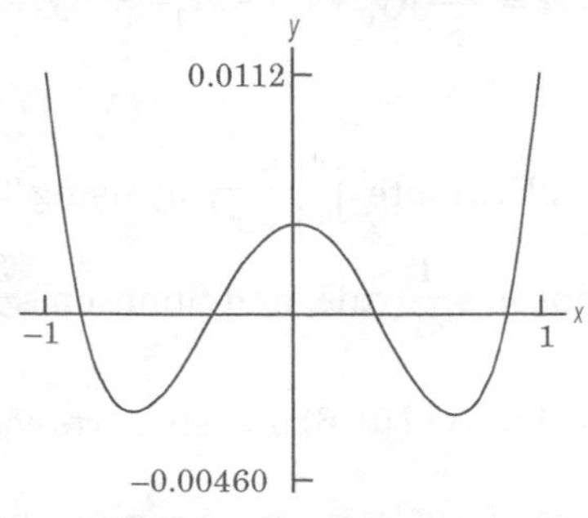

Error in the cubic least squares approximation to e^x.

NUMERICAL INTERGRATION

The process of evaluating a definite integral from a set of tabulated values of the integrand $f(x)$ is called *numerical integration.* This process when applied to a function of a single variable is called *quadrature.*

1. Newton-Cote's quadrature formula

Let $$I = \int_a^b f(x)\,dx$$

where $f(x)$ takes the values $y_0, y_1, y_2, ..., y_n$,

For $x = x_0, x_1, x_2...x_n$ divide the interval (a, b) into n sub-intervals of width h, so that

$$x_0 = a, x_1 = x_0 + h, x_2 = x_0 + 2h, ..., x_n = x_0 + nh = b.$$

Then $$I = \int_{x_0}^{x_0+nh} f(x)\,dx$$

Putting $$x = x_0 + rh$$

$\Rightarrow$ $$dx = h\,dr$$

$\therefore$ $$I = h\int_0^n f(x_0 + rh)\,dr$$

By Newton's forward interpolation formula

$$I = h\int_0^n\left[y_0 + r\Delta y_0 + \frac{r(r-1)}{2!}\Delta^2 y_0 + \frac{r(r-1)(r-2)}{3!}\Delta^3 y_0 + ...\right]dr$$

Integrating term by term, we get

$$\int_{x_0}^{x_0+nh} f(x)\,dx = nh\left[y_0 + \frac{n}{2}\Delta y_0 + \frac{n(2n-3)}{12}\Delta^2 y + \frac{n(n-2)^2}{24}\Delta^3 y + ...\right] \quad ...(i)$$

2. Trapezoidal rule

For definite integrals such as

$$\int_0^1 \sqrt{1-x^3}\,dx \quad \text{or} \quad \int_0^1 e^{-x^2}\,dx$$

fundamental theorem of calculus cannot be used to evaluate them since there are no elementary functions that are antiderivatives of $\sqrt{1-x^3}$ or e^{-x^2}. *Approximation methods are used for such integrals. Trapezoidal rule* is a numerical method that approximates value of a definite integral.

Consider the definite integral $\int_a^b f(x)\,dx$.

Assume that $f(x)$ is continuous on $[a, b]$ and divide $[a, b]$ into n subintervals of equal length

$$\Delta x = \frac{b-a}{n}$$

using $n + 1$ points $x_0 = a, x_1 = a + \Delta x, x_2 = a + 2\Delta x, \ldots,$ $x_n = a + n\Delta x = b.$

Compute value of $f(x)$ at these points.

$$y_0 = f(x_0), y_1 = f(x_1), y_2 = f(x_2), \ldots , y_n = f(x_n)$$

Approximate integral by using n trapezoids formed by using straight line segments between points (x_{i-1}, y_{i-1}) and (x_i, y_i) for $1 \le i \le n$ as shown in the figure below.

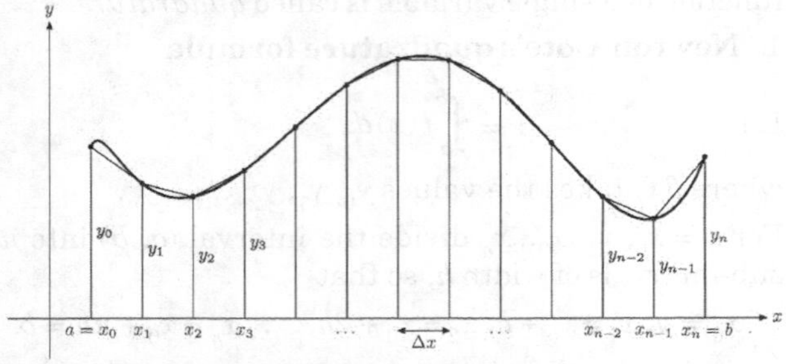

Area of a trapezoid is obtained by adding areas of a rectangle and triangle.

$$A = y_0\,\Delta x + \frac{1}{2}(y_1 - y_0)\Delta x = \frac{(y_0 + y_1)\Delta x}{2}$$

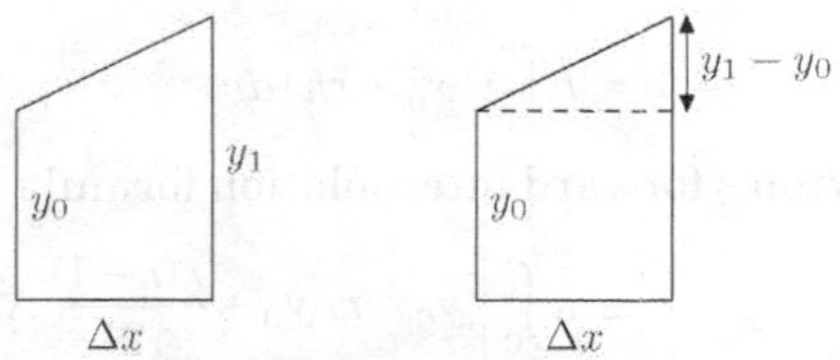

Adding area of the n trapezoids, the approximation obtained is

$$\int_a^b f(x)\,dx \approx \frac{(y_0 + y_1)\Delta x}{2} + \frac{(y_1 + y_2)\Delta x}{2} + \frac{(y_2 + y_3)\Delta x}{2} + \cdots + \frac{(y_{n-1} + y_n)\Delta x}{2}$$

which simplifies to the trapezoidal rule formula.

$$\int_a^b f(x)dx \approx \frac{\Delta x}{2}(y_0 + 2y_1 + 2y_2 + \ldots + 2y_{n-1} + y_n)$$

Example 1. Use trapezoidal rule with $n = 8$ to estimate

$$\int_1^5 \sqrt{1+x^2}\,dx.$$

Solution For $n = 8, \Delta x = \dfrac{5-1}{8} = 0.5.$

Compute the values of $y_0, y_1, y_2, \ldots , y_8$.

x	1	1.5	2	2.5	3	3.5	4	4.5	5
$y = \sqrt{1+x^2}$	$\sqrt{2}$	$\sqrt{3.25}$	$\sqrt{5}$	$\sqrt{7.25}$	$\sqrt{10}$	$\sqrt{13.25}$	$\sqrt{17}$	$\sqrt{21.25}$	$\sqrt{26}$

Therefore,

$$\int_1^5 \sqrt{1+x^2}\,dx \approx \frac{0.5}{2}\left(\sqrt{2} + 2\sqrt{3.25} + 2\sqrt{5} + 2\sqrt{7.25} + 2\sqrt{10} + 2\sqrt{13.25} + 2\sqrt{17} + 2\sqrt{21.25} + \sqrt{26}\right)$$

$$\approx 12.76$$

Example 2. Following points were found empirically.

x	2.1	2.4	2.7	3.0	3.3	3.6
y	3.2	2.7	2.9	3.5	4.1	5.2

Use the trapezoidal rule to estimate $\displaystyle\int_{2.1}^{3.6} y\,dx.$

Solution

By inspection, $\Delta x = 0.3$

$$\therefore \int_{2.1}^{3.6} y\,dx \approx \frac{0.3}{2}(3.2 + 2(2.7) + 2(2.9) + 2(3.5) + 2(4.1) + 5.2)$$

$$\approx 5.22$$

3. Simpson's one-third rule

Putting $n = 2$ in equation (i) above and taking the curve through (z_0, z_0) (z_1, y_1) and (z_2, y_2) as a parabola, *i.e.* a polynomial of second order so that differences of order higher than second vanish, we get

$$\int_{x_0}^{x_0+2h} f(x)dx = 2h\left(y_0 + \Delta y_0 + \frac{1}{6}\Delta^2 y_0\right)$$

$$= \frac{h}{3}(y_0 + 4y_1 + y_2)$$

Similarly, $\displaystyle\int_{x_0+2h}^{x_0+4h} f(x)dx = \frac{h}{3}(y_2 + 4y_3 + y_4)$

$$\int_{x_{0(n-2)h}}^{x_0+nh} f(x)dx = \frac{h}{3}(y_{n-2} + 4y_{n-1} + y_n)$$

n being even.

Adding all these integrals, when n is even,

$$\int_{x_0}^{x_0+nh} f(x)dx = \frac{h}{3}[(y_0 + y_n) + 4(y_1 + y_3 + \ldots + y_{n-1}) + 2(y_2 + y_4 + \ldots + y_{n-2})]$$

4. Simpson's three-eighth rule.

Putting $n = 3$ in equation (i) above and taking the curve through (x_i, y_i) $i = 0, 1, 2, 3$ as a polynomial of third order so that differences above the third order vanish, we get

$$\int_{x_0}^{x_0+3h} f(x)dx = 3h\left(y_0 + \frac{3}{2}\Delta y_0 + \frac{3}{4}\Delta^2 y_0 + \frac{1}{8}\Delta^3 y_0\right)$$

$$= \frac{3h}{8}(y_0 + 3y_1 + 3y_2 + y_3)$$

Similarly, $\displaystyle\int_{x_0+3h}^{x_0+6h} f(x)dx = \frac{3h}{8}(y_3 + 3y_4 + 3y_5 + y_6)$

Adding all such expressions from x_0 to $x_0 + nh$, where n is a multiple of 3, we get

$$\int_{x_0}^{x_0+nh} f(x)dx = \frac{3h}{8}[(y_0 + y_n) + 3(y_1 + y_2 + y_4 + y_5 + \ldots + y_{n-1}) + 2(y_3 + y_6 + y_{n-3})]$$

Example. Evaluate $\displaystyle\int_0^6 \frac{dx}{1+x^2}$ by using Trapezoidal rule, Simpson's $\frac{1}{3}$ rule, and Simpson's $\frac{3}{8}$ rule,

Solution:

Divide the interval (0, 6) into six parts each of width $h = 1$.

The values of $f(x) = \dfrac{1}{1+x^2}$ are given below :

x	0	1	2	3	4	5	6
$f(x)$	1	0.5	0.2	0.1	0.0588	0.0385	0.027
	y_0	y_1	y_2	y_3	y_4	y_5	y_6

(*i*) By Trapezoidal rule,

$$\int_0^6 \frac{dx}{1+x^2} = \frac{h}{2}[(y_0+y_6)+2(y_1+y_2+y_3+y_4+y_5)]$$

$$= \frac{1}{2}[(1+.027)+2(.5+.2+.1+0588+.0385)]$$

$$= \mathbf{1.4108}$$

(*ii*) By Simpson's $\frac{1}{3}$ rule

$$\int_0^6 \frac{dx}{1+x^2} = \frac{h}{3}[(y_0+y_6)+4(y_1+y_3+y_5)+2(y_2+y_4)]$$

$$= \frac{1}{3}[(1+.027)+4(0.5+0.1+0.0385)+2(0.2+0.0588)]$$

$$= \mathbf{1.3662}$$

(*iii*) By Simpson's $\frac{3}{8}$ rule,

$$\int_0^6 \frac{dx}{1+x^2} = \frac{3h}{8}[(y_0+y_6)+3(y_1+y_2+y_4+y_5)+2y_3]$$

$$= \frac{1}{8}[(1+0.027)+3(0.5+0.2+0.0588+0.0385+2(0.1)]$$

$$= \mathbf{1.3571}$$

SINGLE AND MULTI-STEP METHOD FOR FIRST ORDER DIFFERENTIAL EQUATION

1. Runge-Kutta Method

This methods do not require calculations of higher order derivatives. These methods agree with Taylor's series solution upto the terms in h^r, where r differs from method to method and is called *order of that method*. Euler's method, Modified Euler's method and Runge's method are Runge-Kutta methods of the first, second and third order respectively.

The fourth-order Runge-Kutta method is most commonly used and is often referred to as 'Runge-Kutta method only'.

Working rule for finding increment k of y corresponding to an increment h of x by Runge-Kutta method from, $\frac{dy}{dx} = f(x, y)$, $y(x_0) = y_0$, is as follows :

Calculate successively,

$$k_1 = hf(x_0, y_0)$$

$$k_2 = hf\left(x_0 + \frac{1}{2}h, y_0 + \frac{1}{2}k_1\right)$$

$$k_3 = hf\left(x_0 + h, y_0 + \frac{1}{2}k_2\right)$$

$$k_4 = hf(x_0 + h, y_0 + k_3)$$

and finally compute,

$$k = \frac{1}{6}(k_1 + 2k_2 + 2k_3 + k_4)$$

which gives required approximate value

$$y_1 = y_0 + k.$$

(*Note : k is weighted mean of k_1, k_2, k_3 and k_4*)

Example. Apply Runge-Kutta fourth order method, to find an approximate value of y when $h = 0.2$. Given: $\frac{dy}{dx} = x + y$ and $y = 1$, when $x = 0$.

Solution

$$x_0 = 0, \quad y_0 = 1, \quad h = 0.2, \quad f(x_0, y_0) = 1$$

$$\therefore \quad k_1 = hf(x_0, y_0) = 0.2 \times 1 = 0.2$$

$$k_2 = hf\left(x_0 + \frac{1}{2}h, y_0 + \frac{1}{2}k_1\right)$$

$$= 0.2 \times f(0.1, 1.1) = 0.2400$$

$$k_3 = hf\left(x_0 + \frac{1}{2}h, y_0 + \frac{1}{2}k_2\right)$$

$$= 0.2 \times f(0.1, 1.12) = 0.2440$$

$$\text{and} \quad k_4 = hf(x_0 + h, y_0 + k_3)$$

$$= 0.2 \times f(0.2, 1.244) = 0.2888$$

$$\therefore \quad k = \frac{1}{6}(k_1 + 2k_2 + 2k_3 + k_4)$$

$$= \frac{1}{6}(0.2000 + 0.4880 + 0.2888) = \mathbf{0.2428}$$

Hence, required approximate value of

$$Y_1 = Y_0 + K$$

$$= 1 + 0.2428 = 1.2428$$

2. Taylor's Series Method

Given initial value problem

$$y' = \frac{dy}{dx} = f(x, y), \text{ with } y(x_0) = y_0 \quad ...(i)$$

$$\text{Then} \quad y'' = \frac{d^2y}{dx^2} = f_x + f_y y' = f_x + f_y f$$

$$y''' = \frac{d^3y}{dx^3} = f_{xx} + 2f_{xy}f + f_{yy}f^2 + f_x f_y + f_y^2 f \quad ...(ii)$$

By Taylor's theorem, series about a point $x = x_0$

$$y = y_0 + (x - x_0)(y')_0 + \frac{(x-x_0)^2}{2!}(y'')_0 + \frac{(x-x_0)^3}{3!}(y''')_0 + \ldots \quad ...(iii)$$

From equation (*iii*) we can find the value y_1 of y for $x = x_1$ and y', y'', y''',... can be found at $x = x_1$ with equations (*i*) and (*ii*) and so on.

3. Euler's Method

Given initial value problem

$$y' = \frac{dy}{dx} = f(x, y), \text{ with } y(x_0) = y_0$$

By Taylor series, for $h \to 0$,

$$y(x + h) = y(x) + hy'(x) = y(x) + hf(x, y)$$

This gives, $y_{n+1} = y_n + hf(x_n, y_n)$

where $h = \frac{x_n - x_0}{n}$ (*i.e.*, $x_n = x_0 + nh$)

EXERCISE

MCQ TYPE QUESTIONS

1. The order of error is the Simpson's rule for numerical integration with a step size h is
(a) h (b) h^2
(c) h^3 (d) h^4

2. Four arbitrary points (x_1, y_1), (x_2, y_2), (x_3, y_3), (x_4, y_4) are given in the x, y-plane. Using the method of least squares, if regressing y upon x gives the fitted line $y = ax + b$; and regressing x upon y gives the fitted line $x = cy$ + d, then
(a) two fitted lines must coincide
(b) two fitted lines need not coincide
(c) it is possible that $ac = 0$
(d) a must be $1/c$

3. The accuracy of Simpson's rule quadrature for a step size h is
(a) $O(h^2)$ (b) $O(h^3)$
(c) $O(h^4)$ (d) $O(h^5)$

4. The Newton's-Raphson iterative formula for finding $f(x) = x^2 - 1$, is
(a) $x_{i+1} = \dfrac{x_i^2 - 1}{2x_i}$ (b) $x_{i+1} = \dfrac{x_i^2 + 1}{2x_i}$
(c) $x_{i+1} = \dfrac{2x_i^2 + 1}{2x_i}$ (d) $x_{i+1} = \dfrac{2x_i}{2x_i^2 + 1}$

5. The convergence of which of the following method is sensitive to starting value?
(a) False position
(b) Gauss seidal method
(c) Newton-Raphson method
(d) All of these

6. Newton-Raphson method of solution of numerical equation is not preferred when
(a) graph of A(B) is vertical
(b) graph of $x(y)$ is not parallel
(c) graph of $f(x)$ is nearly horizontal where it crosses the x-axis.
(d) none of these

7. In which of the following methods, proper choice of initial value is very important ?
(a) Bisection method (b) False position
(c) Newton-Raphson (d) Bairsto method

8. Newton-Raphson method is applicable to the solution of
(a) both algebraic and trans-cendental equations
(b) both algebraic and trans-cendental and also used when the roots are complex
(c) algebraic equations only
(d) transcendental equations only

NUMERICAL TYPE QUESTIONS

1. Solving $x^2 - 2 = 0$ by Newton Raphson technique, when initial guess $x_0 = 1.0$, then subsequent estimate of x (*i.e.* x_1) will be _____

2. Following are the values of a function
$y(x) : y(-1) = 5, y(0), y(1) = 8$
$\dfrac{dy}{dx}$ at x = 0 as per Newton's central difference scheme is _____

3. The root of $x^3 - 2x - 5 = 0$ correct to three decimal places by using Newton-Raphson method is _____

4. Using Newton-Raphson method, a root correct to three decimal places of the equation $e^x = 1 + 2x$ is _____

5. Using Newton-Raphson method, a root correct to three decimal places of the equation $x^3 - 3x - 5 = 0$ is _____

6. Using Newton-Raphson method, a root correct to three decimal places of the equation sin $x = 1 - x$ is _____

7. A root of the equation $x^3 - x - 11 = 0$ correct to four decimals using bisection method, is _____

8. Using Bisection method, the negative of $x^3 - x + 11 = 0$ is _____

9. A real root of equation $x^3 + x^2 - 1 = 0$ by iteration (method of successive approximation) method is _____

10. A real root of equation $\cos x = 3x - 1$ correct to seven decimal places by the method of successive approximation is _____

11. A real root of the equation $x^3 - 2x - 5 = 0$ by the method of false position correct to three decimal places is _____

12. Using Bisection method, the negative root of $x^3 - 4x + 9 = 0$ correct to three decimal places is _____

13. A real root of the equation $x - \cos x = 0$ by the method of false position correct to four decimal places is _____

14. From certain experiment following data has been obtained:

x	1	3	4
y	4	12	19

Using Langrange's inverse formula, the value of x for which y = 7, will be _____

15. The double (repeated) root of $4x^3 - 8x^2 - 3x + 9 = 0$ by Newton-raphson method will be _____

16. Using trapezoidal rule, and the table given below, $\int_4^{5.2} \ln x \, dx$ will be _____

x :	4	4.2	4.4	4.6	4.8	5	5.2
In x:	1.39	1.44	1.48	1.53	1.57	1.61	1.65

17. Using trapezoidal rule, taking 10 equal interval $\int_0^{\pi} \sin x \, dx$ will be _____

18. The Lagrange interpolating polynomial of degree 2 approximating the function y = ln x defined by the following table of values. The value of ln 2.7 will be ___

x	2	2.5	3.0
y = ln x	0.69315	0.91629	1.09861

19. If $y_1 = 4$, $y_3 = 12$, $y_4 = 19$ and $y_x = 7$, then x will be _____

20. The function y = sin x is tabulated below

x	0	π/4	π/2
y = sin x	0	0.70711	1.0

Using Lagrange's interpolation formula, the value of sin (π/6) will be _____

ANSWERS

EXERCISE

MCQ Type Questions

1. (*b*) **2.** (*d*) **3.** (*d*) **4.** (*b*) **5.** (*c*) **6.** (*c*) **7.** (*c*) **8.** (*a*)

Numerical Type Questions

1. (1.5) **2.** (1.5) **3.** (2.0946) **4.**(1.256) **5.** (2.279) **6.** (0.511) **7.** (2.3736)
8. (–2.3736) **9.** (0.7548776) **10.** (0.6071016) **11.** (2.0944) **12.** (–2.706) **13.** (0.7391) **14.**(1.857)
15. (1.5) **16.** (1.8277) **17.** (1.984) **18.** (0.9941164) **19.** (1.86) **20.** (0.51743)

EXPLANATIONS

EXERCISE

MCQ TYPE QUESTION

2.
$$y = ax + b \quad \text{....}(i)$$
$$x = cy + d$$
$$\Rightarrow \quad x - d = cy$$
$$\Rightarrow \quad y = \frac{1}{c}x - d$$

From equatons (*i*),
$$a = \frac{1}{c} \text{ and } b = -d$$

NUMERICAL TYPE QUESTIONS

1. $x^2 - 2 = 0$
$$x_1 = x_0 - \frac{f(x_0)}{f'(x_0)} = 1 + \frac{1}{2} = 1.5$$
$$f(x_0) = 1^2 - 2 = -1$$
$$f(x_0) = 2x$$
$$f(x_0) = 2x_0 = 2(1) = 2$$

2. $$\left(\frac{dy}{dx}\right)_{at\,x=0} = \frac{y_2 - y_1}{x_2 - x_1} = \frac{y(1) - y(-1)}{1-(-1)} = \frac{8-5}{2} = 1.5$$

18. $$L_0(x) = \frac{(x-2.5)(x-3.0)}{(-0.5)(-1.0)} = 2x^2 - 11x + 15$$

Similarly,
$$L_1(x) = -(4x^2 - 20x + 24)$$
$$L_2(x) = 2x^2 - 9x + 10$$
$$\therefore \quad L_2(x) = (2x^2 - 11x + 15)(0.69315) - (4x^2 - 20x + 24)(0.091629) + (2x^2 - 9x + 10)(1.09861)$$
$$= -0.08164x^2 + 0.81366x - 0.60761$$

which is the required quadratic polynomial.
Putting $x =$ 2.7 in the above polynomial, we get
$$\ln 2.7 \approx L_2(2.7)$$
$$= -0.08164(2.7)^3 + 0.81366(2.7) - 0.60761$$
$$= 0.9941164$$

19. $$x = \frac{(-5)(-12)}{(-8)(-15)}(1) + \frac{(3)(-12)}{(8)(-7)}(3) + \frac{(3)(-5)}{(15)(7)}(4)$$
$$= \frac{1}{2} + \frac{27}{14} - \frac{4}{7} = \mathbf{1.86}$$

20. $$\sin\frac{\pi}{6} \approx \frac{\left(\frac{\pi}{6}-0\right)\left(\frac{\pi}{6}-\frac{\pi}{2}\right)}{\left(\frac{\pi}{4}-0\right)\left(\frac{\pi}{4}-\frac{\pi}{2}\right)}(0.70711) + \frac{\left(\frac{\pi}{6}-0\right)\left(\frac{\pi}{6}-\frac{\pi}{4}\right)}{\left(\frac{\pi}{2}-0\right)\left(\frac{\pi}{2}-\frac{\pi}{4}\right)}(1)$$
$$= \frac{8}{9}(0.70711) - \frac{1}{9} = \mathbf{0.51743}$$

6

CHAPTER

Complex Analysis

LIMIT OF A COMPLEX FUNCTION

Let f be a complex valued function of the complex variable z defined on a set E_2 (where E_2 is a two dimensional euclidean space) and be single valued for all values of z in the neighbourhood of $z = z_0$ with the possible exception of $z = z_0$ itself.

Then number l is limit of $f(z)$ as z approaches z_0 and write $\lim_{Z \to Z_0} f(z) = l$ if for any positive number $\in$, however small, some positive number δ (usually depending on $\in$) can be found such that

$$|f(z) - l| < \in \text{ whenever } 0 < |z - z_0| < \delta.$$

Note that limit is independent of the manner in which $z \to z_0$.

CONTINUITY OF A COMPLEX FUNCTION

Let f be defined and single valued on a set in E_2 for all values of a complex variable z in a neighbourhood of $z = z_0$ as well as at $z = z_0$.

The function $f(z)$ is said to be continuous at the point z_0 provided that

$$\operatorname{Lim}_{z \to z_0} f(z) = f(z_0)$$

Note :

- If a function $f(z)$ is continuous in a region, then real and imaginary parts of the function $f(z)$ are also continuous in that region.
- If function $f(z)$ is continuous in a closed region is bounded in that region, i.e. there exists a constant M such that $|f(z)| < M$ for all points in z of the region.

DERIVATIVES OF A COMPLEX FUNCTION.

- If $f(z)$ is a complex single-valued function in some region R of the z-plane, then derivative of $f(z)$ is defined as

$$f'(z) = \lim_{\delta z \to 0} \frac{f(z+\delta z) - f(z)}{\delta z}$$

provided that this limit exists and independent of the manner in which $\delta z \to 0$.

- If f is a complex valued function defined on an open set $S \in E_2$ and it is assumed that $z_0 \in S$, then f is said to have a derivative $f'(z_0)$ at z_0 if the limit

$$\lim_{\delta z \to z_0} \frac{f(z) - f(z_0)}{\delta z} = f'(z_0) \text{ exists.}$$

Note : If a function $f(z)$ is differentiable at any point z_0, then it is not enough that

$$\lim_{\delta z \to z_0} \frac{f(z) - f(z_0)}{z - z_0}$$

exists, but it is also required to prove that this limit must be unique, i.e. it must have the same value along whatever path z approaches z_0.

ANALYTIC FUNCTIONS

A function f(z) which is single-valued and possesses a unique derivative with respect to z at all points of regions R, is called *analytic* or *regular function* of z in that region.

A point at which an analytic function cease to possess a derivative is called *singular point of the function*. Thus if u and v are real single – valued functions of x and y such that $\left[\frac{\partial u}{\partial x}\right], \left[\frac{\partial u}{\partial y}\right], \left[\frac{\partial v}{\partial x}\right]\left[\frac{\partial v}{\partial x}\right]$ are continuous thoughtout a region R, then Cauchy-Riemann equation

$$\frac{\partial u}{\partial x} = \frac{\partial v}{\partial y}; \quad \frac{\partial u}{\partial y} = \frac{-\partial v}{\partial x}$$

are both necessary and sufficient conditions for the function $f(z) = u + iv$ to be analytic in R.

The real and imaginary parts of an analytic function are called *conjugate functions*. The relation between two conjugate functions is given by C-R equation.

- Analytic function is called *regular function* or holomorphic function.
- A point at which an analytic function ceases to possesses a derivative is called *singular point of the function*.

Cauchy's Theorem

If f(z) is an analytic function and f(z) is continuous at each point within or on a closed curve C, then

$$\int_C f(z)\, dz = 0$$

Writing $f(z) = u(x, y) + iv(x, y)$ and noting that $dz = dx + idy$

$$\int_C f(z) = dz \int_C = (udx - vdy) + i\int_C (vdx + udy) \quad ...(i)$$

Since $f'(z)$ is continuous, therefore

$$\frac{\partial u}{\partial x}, \frac{\partial u}{\partial y}, \frac{\partial v}{\partial x}, \frac{\partial v}{\partial y}$$

are also continuous in the region D enclosed by C. Hence Green's theorem can be applied to integral (*i*), giving

$$\int_C f(z)\, dz = -\iint_D \left(\frac{\partial v}{\partial x} + \frac{\partial u}{\partial y}\right) dx\, dy + i\iint_D \left(\frac{\partial u}{\partial x} - \frac{\partial v}{\partial y}\right) dx\, dy \quad ...(ii)$$

Now f(z) being analytic, u and v necessarily satisfy the Cauchy-Riemann equation and thus the integrands of the two double integrals in integral (*ii*) vanish identically

Hence $\int_C f(z)\, dz = 0$

Obs.1. Cauchy-Riemann equations are precisely the conditions for the two real integrals in integral (*i*) to be independent of the path. Hence line integral of a function f(z) which is analytic in the region D, is independent of the path joining any two points of D.

Obs. 2. Extension of Cauchy's Theorem. If f(z) is analytic in the region D between two simple closed curves C and C_1, then

$$\int_C f(z)\, dz = \int_{C_1} f(z)\, dz + \int_{BA} f(z) dz = 0$$

To prove this, we need to introduce the cross-cut AB. Then $\int f(z)\, dz = 0$

where path is as indicated by arrows in Fig. (a), i.e. along AB-along C_1 in clock-wise sense and along BA–along C_1 in anti-clockwise sense

i.e. $$\int_{AB} f(z)\, dz + \int_{C_1} f(z)\, dz + \int_C f(z)\, dz \int_{BA} f(z) dz = 0$$

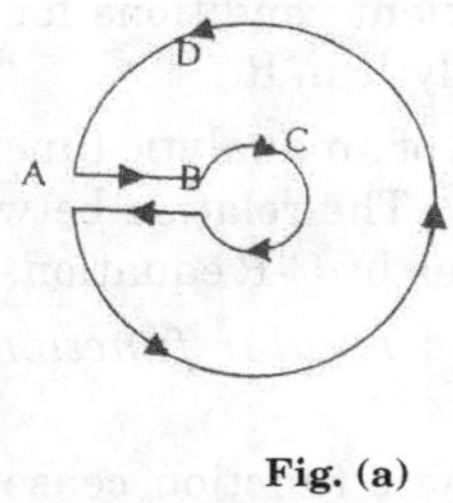

Fig. (a)

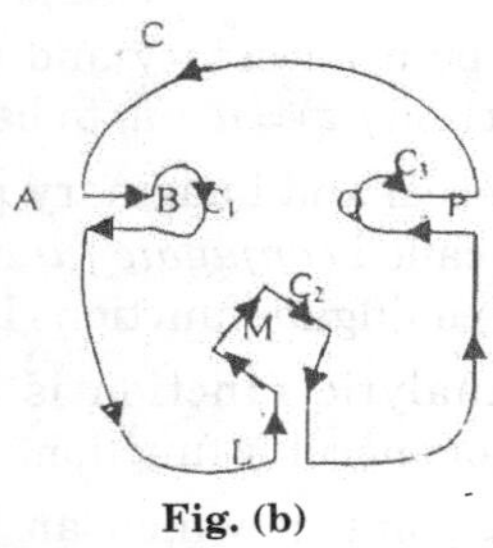

Fig. (b)

But, since integrals along AB and BA cancel,

$$\int_C f(z)\, dz + \int_{C_1} f(z)\, dz = 0$$

Reversing the direction or integral around C_1 and transposing, we get

$$\int_C f(z)\, dz = \int_{C_1} f(z)\, dz$$

Each integration being taken in the anti-clockwise sense.

If C_1, C_2, C_3 .... be any number of closed curves with C in Fig. (*b*), then

$$\int_C f(z)\, dz = \int_{C_1} f(z)\, dz + \int_{C_2} f(z)\, dz + \int_{C_3} f(z)\, dz$$

Cauchy's Integral Formula

If f(z) is analytic within and on a closed curve and if a is any point with C, then

$$f(a) = \frac{1}{2\pi i}\int_C \frac{f(z)dz}{z-a}$$

Consider the function $\frac{f(z)}{(z-a)}$ which is analytic at all points within C except at z = a. With the point 'a' as center and radius r, draw a small circle C_1 lying entirely within C.

Now $\frac{f(z)}{(z-a)}$ being analytic in the region enclosed by C and C_1, we have by Cauchy's therorem

$$\int_C \frac{f(z)}{z-a} dz = \int_{C_1} \frac{f(z)}{z-a} dz$$

For any point on C_1,

$$z - a = re^{i\theta} \Rightarrow dz = ire^{i\theta}\, d\theta$$

$$\therefore \quad = \int_{C1} \frac{f(a + re^{i\theta})}{re^{i\theta}} . ire^{i\theta}\, d\theta = i\int_{C1} f(a + re^{i\theta})\, d\theta \quad ...(i)$$

In the limiting form, as the circle C_1 shrinks to the point 'a', i.e. as $r \to 0$, the integral (*i*) will approach to

$$i\int_{C1} f(a)\, d\theta = if(a)\int_0^{2\pi} d\theta = 2\pi i f(a)$$

$$\therefore \quad \int_C \frac{f(z)}{z-a} dz = 2\pi i f(a)$$

$$\Rightarrow \quad f(a) = \frac{1}{2\pi i}\int_C \frac{f(z)}{z-a} dz \quad ...(ii)$$

which is desired Cauchy's integral formula.

Cor. Differentiation both sides of equation (*ii*) with respect to a, we get

$$f'(a) = \frac{1}{2\pi i}\int_C \frac{\partial}{\partial a}\left(\frac{f(z)}{z-a}\right) dz = \frac{1}{2\pi i}\int_C \frac{f(z)}{(z-a)^2} dz \quad ...(iii)$$

Similarly, $$f''(a) = \frac{2!}{2\pi i}\int_C \frac{f(z)}{(z-a)^3} dz \quad ...(iv)$$

and in general, $$f^n(a) = \frac{n!}{2\pi i}\int_c \frac{f(z)}{(z-a)^{n+1}} dz \quad ...(v)$$

Thus from results (*ii*) to (*v*) if a function f(z) is known to be analytic on the simple closed curve C, then values of the functions and all its derivative can be found at any point of C. Thus *an analytic function possesses derivative of all orders and these are themselves all analytic.*

Residue Theorem

If f(z) is analytic in a closed curve C except at a finite number of singular point within C, then

$$\int_C f(z)\,dz = 2\pi i \times \text{(sum of residues at the singular points within C)}$$

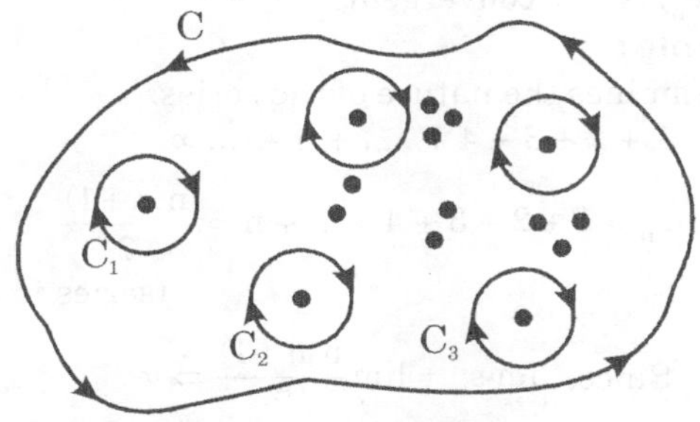

Let us surround each of the singular points $a_1, a_2, \ldots a_n$ by a small circle such that it encloses no other singular point. Then these circles $C_1, C_2, \ldots, C_n$ together with C form a multiply connected region in which f(z) is analytic.

Applying Cauchy's theorem, we have

$$\int_C f(z)\,dz = \int_{C_1} f(z)dz + \int_{C_2} f(z)\,dz + \text{-----} + \int_{C_n} f(z)\,dz$$

$$= 2\pi i\,[\text{Res } f(a_1) + \text{Res } f(a_2) + \ldots\ldots + \text{Res } f(a_n)]$$

Taylor's Series

If f(z) is analytic inside a circle C with center at 'a', then for z inside C

$$f(z) = f(a) + f'(a)(z-a) + \frac{f''(a)}{2!}(z-a)^2 + \ldots\ldots \frac{f^n(a)}{n!}(z-a)^n + \ldots\ldots \quad \ldots(i)$$

Obs. Complex analytic functions can always represented by power series of the form (*i*).

Laurent's Series

If f(z) is analytic in the ring-shaped region R bounded by two concentric circles C and C_1 of radii r and r_1 $(r > r_1)$ and with centre at 'a', then for all z in R

$$f(z) = a_0 + a_1(z-a) + a_2(z-a)^2 + \ldots\ldots + a_{-1}(z-1)^{-1} + a_{-2}(z-a)^{-2} + \ldots\ldots$$

where, $$a_n = \frac{1}{2\pi i}\int_r \frac{f(t)}{(t-a)^{n+1}}\,dt$$

r being any curve is R, encircling C_1

Obs. Laurent series of a given analytic function f(z) in its annulus of convergence is unique. There may be different Laurent series of f(z) in two annuli with the same center.

SEQUENCE

A sequence is a succession of numbers or terms formed according to some definite rule. The nth term in a sequence in denoted by u_n.

For example, If $u_n = 2_n + 1$.

By giving different values of n in u_n, we get different terms of sequence.

Thus, $u_1 = 3, u_2 = 5, u_3 = 7$

A sequence having unlimited number of term's know as an infinite sequence.

LIMIT

It is a sequence tends to a limit *l*, then we write

$$\lim_{n\to\infty}(u_n) = l$$

Convergent sequence

If the limit of a sequence is finite, the sequence is convergent. If the limit of a sequence does not lend to a finite number, the sequence is said to be divergent.

e.g. $1, \frac{1}{4}, \frac{1}{9}, \frac{1}{16} \ldots \frac{1}{n^2}$ (is a convergent sequence)

3, 5, 7.... (n + 1) (is divergent sequence)

Bounded Sequence

u_1, u_2, u_3, u_n is a bounded sequence if $\boxed{u_n < k}$ (for every n).

Monotonic Sequence

The sequence is either immersing or decreasing, such sequences are called monotonic.

e.g. 1, 4, 7, 10 is a monotonic sequence.

1, –1, 1, –1, 1..... is not a monotonic sequence.

A sequence which is monotonic and bounded is a convergent sequence.

Ex. Determine the general term of each of the following sequence. Prove that the following sequence are convergent.

1. $\frac{1}{2}, \frac{1}{4}, \frac{1}{8}, \frac{1}{16} \ldots$

Ans. $\frac{1}{2^n}$

Limits :

(i) $\lim_{n\to\infty} x^n = 0$ if $x < 1$ and $\lim_{n\to\infty} x^n = \infty$ if $x > 1$

(ii) $\lim_{n\to\infty} \frac{x^n}{n!} = 0$ for all the values of x

(iii) $\lim_{n\to\infty} \frac{\log n}{n} = 0$

(iv) $\lim_{n\to\infty} \left[\frac{n!}{n}\right]^{\frac{1}{n}} = \frac{1}{e}$

(v) $\lim_{n\to\infty} (n)^{\frac{1}{n}} = 1$

(vi) $\lim_{n\to\infty} [n!]^{\frac{1}{n}} = \infty$

(vii) $\lim_{n\to\infty} \left[1+\frac{1}{n}\right]^n = e$

(viii) $\lim_{n\to\infty} nx^n = 0$ if $x < 1$

(ix) $\lim_{n\to\infty} n^n = \infty$

(x) $\lim_{n\to\infty} \frac{1}{n^n} = 0$

(xi) $\lim_{n\to\infty} \left[\frac{a^n - 1}{n}\right] = \log a$ or $\lim_{n\to\infty} \left(\frac{a^{\frac{1}{n}} - 1}{\frac{1}{n}}\right) = \log a$

(xii) $\lim_{x\to\infty} \frac{\sin x}{x} = 1$

(xiii) $\lim_{x\to\infty} \frac{\tan x}{x} = 1$

SERIES

- A series is the sum of sequence.
  Let $u_1, u_2, u_3, \ldots u_n$ be a given sequence. Then, the expression $u_1 + u_2 + u_3 + \ldots + u_n + \ldots$ is called the series associated with the given sequence.
- For example, 1 + 3 + 5 + 7 .... is a series.
  If the number of terms of a series is limited, the series is called finite. When the number of terms of a series are unlimited, it is called an infinite series $u_1 + u_2 + u_3 + u_4 + \ldots u_n + \ldots \infty$ is called an infinite series and it is denoted by $\sum_{n=1}^{\infty} u_n$ or $\sum u_n$.
  The sum of the first n terms of a series is denoted by s_n.

Convergent, Divergent and Oscillatory Series :

Consider the infinite series,

$$\Sigma u_n = u_1 + u_2 + u_3 + \ldots u_n + \ldots \infty$$
$$s_n = u_1 + u_2 + u_3 + \ldots u_n$$

The cases arise :

(i) If s_n tends to a finite number as $n \to \infty$, the series Σu_n is said to be convergent.

(ii) If s_n tends to infinity as $n \to \infty$, the series Σu_n is said to be diverged.

(iii) If s_n does not tend to a unique limit, finite or infinite the series Σu_n is called oscillatory.

Properties of intinite seres :

1. The nature of an infinite series does not change.
   * By multiplication of all terms by a constant k.
   * By addition or deletion of a finite number of terms.
2. If tow series Σu_n and Σv_n are convergent, then $\Sigma(u_n + v_n)$ is also convergent.

Example :

1. Examines the nature of the series.
   $1 + 2 + 3 + 4 + \ldots + n + \ldots \infty$.

Sol. $s_n = 1 + 2 + 3 + 4 + \ldots + n = \frac{n(n+1)}{2}$ (series in AP)

Since, $\lim_{n\to\infty} s_n = \lim_{n\to\infty} \frac{n(n+1)}{2} \Rightarrow \infty$

Hence, the series is divergent.

2. Test the convergence of the series
   $1 + \frac{1}{2} + \frac{1}{4} + \frac{1}{8} - \infty$

Sol. $s_n = 1 + \frac{1}{2} + \frac{1}{4} + \frac{1}{8} + \ldots \infty$

$= \frac{1}{1-\frac{1}{2}} = 2$ (series in GP)

$\lim_{n\to\infty} s_n = 2$ $\left(\because s_n = \frac{a}{1-r}\right)$

Hence, the series is convergent.

3. Discuss the name of the series
   $2 - 2 + 2 - 2 + 2 - \infty$

$s_n = 2 - 2 + 2 - 2 + 2 + \ldots \infty$
$= 0$ if n is even
$= 2$ if n is odd.

Hence, s_n does not tend to a unique limit, and therefore, the given series is oscillatory.

Properties of Geometric Series :

The serie $1 + r + r^2 + r^3 + \ldots \infty$ is

(1) Convergent if $|r| < 1$
(2) Divergent if $r \geq 1$
(3) Oscillatory if $r \leq -1$.

Proof:

$$s_n = 1 + r + r^2 + \ldots r^{n-1} = \frac{1-r^n}{1-r}$$

(i) When $|r| < \lim_{n\to\infty} r^n = 0$

$$\lim_{n\to\infty} s_n = \lim_{n\to\infty} \frac{1-r^n}{1-r} = \frac{1-0}{1-r} = \frac{1}{1-r}$$

Hence, the series is convergent.

(ii) (a) When $r > 1$ $\lim_{n\to\infty} r^n = \infty$

$$\therefore \quad \lim_{n\to\infty} s_n = \lim_{n\to\infty} \frac{r^n - 1}{r - 1} \Rightarrow \infty$$

Hence the series is convergent.

(b) When $r = 1$, the series becomes

$1 + 1 + 1 + 1 + ... \infty$

$s_n = 1 + 1 + 1 + ... = n$

$\lim_{n\to\infty} s_n = \lim_{n\to\infty} n = \infty$

Hence, the series is divergent

(iii) (a) When $r = -1$, the series becomes

$1 - 1 + 1 - 1 + 1 - .... \infty$

$s_n = 0$ if n is even

$= 1$ if n is odd

Hence, the series is oscillatory.

(b) When $r < -1$, let $r = -k$, where $k > 1$

$r^n = (-k)^n = (-1)^n k^n$

$$\lim_{n\to\infty} s_n = \lim_{n\to\infty} \frac{1 - r^n}{1 - r} = \lim_{n\to\infty} \frac{1 - (-1)^n k^n}{1 - (-k)}$$

$= +\infty$ if n is odd

$= -\infty$ if n is even

Positive Term Series :

If all the terms after new negative terms in an infinite series are positive, such a series is a positive terms series.

e.g., $-10 - 6 - 1 + 5 + 12 + 20 + ...$ is a positive term series.

By omitting the negative terms, the nature of a positive term series remained unchanged.

Necessary Condition for Convergent Series :

For every convergent series $\sum u_n$,

$$\lim_{n\to\infty} u_n = 0$$

Solution : Let $s_n = u_1 + u_2 + u_3 + ... + u_n$

$\Rightarrow \lim_{n\to\infty} s_n = k$ (a finite quantity)

$\lim_{n\to\infty} s_{n-1} = k$ (a finite quantity)

$s_n = s_{n-1} + u_n$

$u_n = s_n - s_{n-1}$

$\lim_{n\to\infty} u_n = \lim_{n\to\infty}[s_n - s_{n-1}] = 0$

$\lim_{n\to\infty} u_n = 0$

Corollary : converse of the above theorem is not true.

e.g, $1 + \frac{1}{\sqrt{2}} + \frac{1}{\sqrt{3}} + \frac{1}{\sqrt{4}} + ... + \frac{1}{\sqrt{n}} + ...\infty$
(is divergent)

$$S_n = 1 + \frac{1}{\sqrt{2}} + \frac{1}{\sqrt{3}} + \frac{1}{\sqrt{4}} + ... + \frac{1}{\sqrt{n}} + \frac{1}{\sqrt{n}} + \frac{1}{\sqrt{n}} + \frac{1}{\sqrt{n}} + ... + \frac{1}{\sqrt{n}} > \frac{n}{\sqrt{n}} > \sqrt{n}$$

$\lim_{n\to\infty} s_n = \lim_{n\to\infty} \sqrt{n} = \infty$

Thus, the series is divergent, although,

$$\lim_{n\to\infty} u_n = \lim_{n\to\infty} \frac{1}{\sqrt{n}} = 0$$

so $\lim_{n\to\infty} u_n = 0$ is a necessary condition but not a sufficient condition for convergence.

Note : Test for divergence

If $\lim_{n\to\infty} u_n \neq 0$ the series is divergent.

Example : Test for convergence of the series

$$1 + \frac{2}{3} + \frac{3}{4} + \frac{4}{5} + ... + \frac{n}{n+1} + ... \infty$$

Solution : $\lim_{n\to\infty} u_n = \lim_{n\to\infty} \frac{n}{n+1} = \lim_{n\to\infty} \frac{1}{1 + \frac{1}{n}} = 1 \neq 0$

Hence, by Cauchy's Fundamental test for divergence the series is divergent.

Example : Test for convergence the series :

$$1 + \frac{3}{5} + \frac{8}{10} + \frac{15}{17} + ... + \frac{2^n - 1}{2^n + 1} + ...\infty$$

Solution : $\lim_{n\to\infty} u_n = \lim_{n\to\infty} \frac{2^n - 1}{2^n + 1} = \lim_{n\to\infty} \frac{1 - \frac{1}{2^n}}{1 + \frac{1}{2^n}} = 1 \neq 0$

Hence, by Cauchy's fundamental test for divergence the series is divergent.

Example : Examine the series for convergence.

CONVERGENT TESTS

List by the tests for convergene :

(1) Cauchy's Fundamental Test

$\lim_{n\to\infty} u_n \neq 0$, $\sum u_n$ is divergent.

(2) Comparison Test :

$\lim_{n\to\infty} \frac{u_n}{v_n} = k$, $\sum u_n$ and $\sum v_n$ are of the same nature.

$\sum \frac{1}{nP}$, $P > 1$, convergent; $P \leq 1$, divergent.

(3) D' Alembert's Test :

$\lim_{n\to\infty} \frac{u_n + 1}{u_n} = k$, if $k = 1$, $\sum u_n$ is convergent,

if $k > 1$, Σ is divergent.

(4) Raaber's Test :

$$\lim_{n\to\infty} n\left(\frac{u_n}{u_{n+1}}-1\right) = k, \text{ if } k > 1, \Sigma u_n \text{ is convergent,}$$

if k < 1, Σu_n is divergent.

(5) Cauchy's Root Test :

$$\lim_{n\to\infty} (u_n)^{\frac{1}{n}} = k, \text{ if } k < 1, \Sigma u_n \text{ is convergent;}$$

if k > 1, is divergent.

(6) Logarithmic Test :

$$\lim_{n\to\infty} n \log \frac{u_n}{u_n+1} = k, \text{ if } k > 1, \Sigma u_n \text{ is convergent; if}$$

k < 1, Σu_n is divergent.

(7) De Morgan's and Bertrand's Test :

$$\lim_{n\to\infty} \left[\left\{n\left(\frac{u_n}{u_{n+1}}-1\right)-1\right\}\log n\right] = k,$$

If k > 1, convergent; k < divergent.

(8) Gauss's Test :

$$\frac{u_n}{u_{n+1}} = \propto + \frac{\beta}{n} + \frac{\gamma}{n^2}$$

(i) (a) If $\propto > 1$, Σu_n is convergent.
(b) If $\propto < 1$, Σu_n is divergent.

(ii) If $\propto = 1$ and $\begin{bmatrix} \beta > 1, \Sigma u_n \text{ is convergent} \\ \beta < 1, \Sigma u_n \text{ is divergent} \end{bmatrix}$

(9) Cauchy's condenation Test :

$\Sigma\phi(n)$ and $\Sigma a^n\phi(a^n)$ are either both convergent or both divergent.

(10) Leibnitz's test:

Alternately positive and negative is convergent if

(i) $|u_{n+1}| > |u_n|$

(ii) $\lim_{n\to\infty} u_n = 0$

SOLVED EXAMPLES

Example. Evaluate $\int_C \frac{z^2-z+1}{z-1} dx$,

where C is the circle

(*i*) $|z| = 1$,

(*ii*) $|z| = \frac{1}{2}$

Solution :

(*i*) Here $f(z) = z^2 - z + 1$ and $a = 1$.

Since $f(z)$ is analytic within and on circle C : $|z| = 1$ and $a = 1$ lies on C.

By Cauchy's integral formula

$$\frac{1}{2\pi i}\int_C \frac{f(z)}{z-a} = f(a) = 1 \text{ i.e. } \int_C \frac{z^2-z+1}{z-1} dz = 2\pi i.$$

(*ii*) Here, $a = 1$ lies outside the circle

$$C : |z| = \frac{1}{2}.$$

So $(z^2 - z + 1)(z - 1)$ is analytic everywhere within C.

By Cauchy's theorem,

$$\int_C \frac{z^2-z+1}{z-1} dx = 0$$

Example. Evaluate, using Cauchy's integral formula:

(*i*) $\int_C \frac{e^{2z} dz}{(z-1)(z-2)}$, where C is the circle $|z| = 3$.

(*ii*) $\int_C \frac{\cos \pi z}{z^2-1} dz$ around a rectangle with vertices $2 \pm i, -2 \pm i$.

Solution :

(*i*) $f(z) = e^{2z}$ is analytic within the circle C : $|z| = 3$ and two singular points $a = 1$ and $a = 2$ lie inside C.

$$\therefore \int_C \frac{e^{2z}}{(z-1)(z-2)} dz = \int_C e^{2z}\left(\frac{1}{z-2}-\frac{1}{z-1}\right) dz$$

$$= \int_C \frac{e^{2z}}{z-2} dz - \int_C \frac{e^{2z}}{z-1} dz = 2\pi i e^4 - 2\pi i e^2$$

$= 2\pi i\,(e^4 - e^2)$ [By Cauchy's integral formula]

(*ii*) $f(z) = \cos \pi z$ is analytic in the region bounded by given rectangle and two singular points $a = 1$ and $a = -1$ lie inside this rectangle.

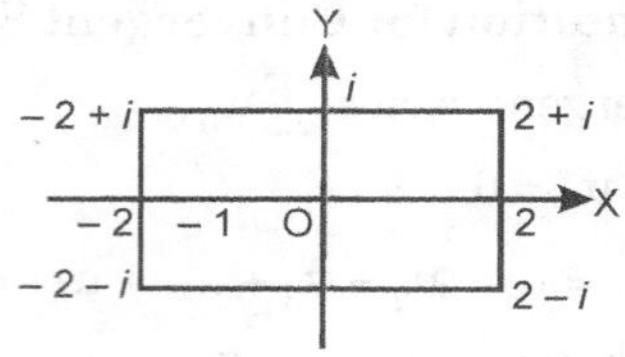

$$\therefore \int_C \frac{\cos \pi z}{z^2-1} dz = \frac{1}{2}\int_C \left(\frac{1}{z-1}-\frac{1}{z+1}\right)\cos \pi z\, dz$$

$$= \frac{1}{2}\int_C \frac{\cos \pi z}{z-1} dz - \int_C \frac{\cos \pi z}{z+1} dz$$

$$= \frac{1}{2}\{2\pi i \cos \pi(1)\} - \frac{1}{2}\{2\pi i \cos \pi(-1)\}$$

$= 0$

[By Cauchy's integral formula]

Example. Evaluate

(*i*) $\int_C \frac{\sin^2 z}{(z-\pi/6)^3}dz$, where C is the circle $|z| = 1$.

(*ii*) $\int_C \frac{e^{2z}}{(z+1)^4}dz$, where C is the circle $|z| = 2$.

(*iii*) $\int_C \frac{e^z}{(z^2+\pi^2)^2}dz$ where C is $|z| = 4$.

Solution :

(*i*) $f(z) = \sin^2 z$ is analytic inside the circle C : $|z| = 1$ and point $a = \frac{\pi}{6}$ (≈ 0.5) lies within C.

By Cauchy's integral formula

$$f''(a) = \frac{2!}{2\pi i}\int_C \frac{f(z)}{(z-a)^3}dz,$$

$$\therefore \int_C \frac{\sin^2 z}{(z-\pi/6)^3}dz = \pi i\left[\frac{d^2}{dz^2}(\sin^2 z)\right]_{z=\pi/6}$$

$$= \pi i(2\cos 2z)_{z=\pi/6}$$

$$= 2\pi i \cos \pi/3 = \pi i$$

(*ii*) $f(z) = e^{2z}$ is analytic within the circle C : $|z| = 2$.

Also $(a) = -1$ lies inside C.

By Cauchy's integral formula,

$$f'''(a) = \frac{3!}{2\pi i}\int_C \frac{f(z)dz}{(z-a)^4}$$

$$\therefore \int_C \frac{e^{2z}}{(z+1)^4}dz = \frac{2\pi i}{6}\left|\frac{d^3(e^{2z})}{dz^3}\right|_{z=-1}$$

$$= \frac{\pi i}{3}\left[8e^{2z}\right]_{z=-1}$$

$$= \frac{8\pi i}{3}e^{-2}.$$

(*iii*) $\frac{e^z}{(z^2+\pi^2)^2} = \frac{e^z}{(z+\pi i)^2(z-\pi i)^2}$ is not analytic at $z = \pm\pi i$.

However both $z = \pm\pi i$ lie within the circle $|z| = 4$.

Now $\frac{1}{(z+\pi i)^2(z-\pi i)^2}$

$$= \frac{A}{z+\pi i} + \frac{B}{(z+\pi i)^2} + \frac{C}{z-\pi i} + \frac{D}{(z-\pi i)^2}$$

where, $A = \frac{7}{2\pi^3 i}$,

$$C = -\frac{7}{2\pi^3 i},$$

$$B = D = -\frac{1}{4\pi^2}$$

$$\therefore \int_C \frac{e^z}{(z^2+\pi^2)^2}dz$$

$$= \frac{7}{2\pi^3 i}\left\{\int_C \frac{e^z}{z+\pi i}dz - \int_C \frac{e^z}{z-\pi i}dz\right\}$$

$$-\frac{1}{4\pi^2}\left\{\int_C \frac{e^z}{(z+\pi i)^2}dz + \int_C \frac{e^z}{(z-\pi i)^2}dz\right\}$$

$$= \frac{7}{2\pi^3 i}[2\pi i f(-\pi i) - 2\pi i f(\pi i)]$$

$$-\frac{1}{4\pi^2}[2\pi i f'(-\pi i) + 2\pi i f'(\pi i)]; \text{ where } f(z) = e^z$$

$$= \frac{7}{\pi^2}(e^{-\pi i} - e^{\pi i}) - \frac{i}{2\pi}(e^{-\pi i} + e^{\pi i})$$

$$= -\frac{14i}{\pi^2}\left(\frac{e^{\pi i} - e^{-\pi i}}{2i}\right) - \frac{i}{\pi}\left(\frac{e^{\pi i} + e^{-\pi i}}{2}\right)$$

$$= -\frac{14i}{\pi^2}\sin\pi - \frac{i}{\pi}\cos\pi = \frac{i}{\pi}.$$

Example. Expand $f(z) = \frac{1}{[(z-1)(z-2)]}$ in the region:

(*a*) $|z| < 1$,

(*b*) $1 < |z| < 2$,

(*c*) $|z| > 2$,

(*d*) $0 < |z-1| < 1$.

Solution :

(*a*) By partial fractions

$$\frac{1}{(z-1)(z-2)} = \frac{1}{z-2} - \frac{1}{z-1} \quad ...(i)$$

$$= -\frac{1}{2}\left(1-\frac{z}{2}\right)^{-1} + (1-z)^{-1} \quad ...(ii)$$

For $|z| < 1$, both $\left|\frac{z}{2}\right|$ and $|z|$ are less than 1.

Hence gives on expansion

$$f(z) = -\frac{1}{2}\left(1+\frac{z}{2}+\frac{z^2}{4}+\frac{z^3}{8}+...\right)$$

$$+ (1 + z^2 + z^2 + z^3 + ...)$$

$$= \frac{1}{2} + \frac{3}{4}z + \frac{7}{8}z^2 + \frac{15}{16}z^3 + ...$$

which is a *Taylor's series*.

(*b*) For $1 < |z| < 2$, write (*i*) as

$$f(z) = -\frac{1}{2}\frac{1}{(1-z/2)} - \frac{1}{z(1-z^{-1})} \quad ...(iii)$$

and both $\left|\frac{z}{2}\right|$ and $|z^{-1}|$ are less than 1, hence (*iii*) on expansion gives

$$f(z) = -\frac{1}{2}\left(1+\frac{z}{2}+\frac{z^2}{4}+\frac{z^3}{8}+...\right)$$

$$-\frac{1}{z}(1 + z^{-1} + z^{-2} + z^{-3} + ...)$$

$$= ... - z^{-4} - z^{-3} - z^{-2} - z^{-1} - \frac{1}{2} - \frac{1}{4}z$$

$$-\frac{1}{8}z^2 - \frac{1}{16}z^3 - ...$$

which is a *Laurent's series.*

(*c*) For $|z| > 2$, write (*i*) as

$$f(z) = \frac{1}{z(1-2z^{-1})} - \frac{1}{z(1-z^{-1})}$$

$$= z^{-1}(1 - 2z^{-1} + 4z^{-2} + 8z^{-3} + ...)$$

$$- z^{-1}(1 + z^{-1} + z^{-2} + z^{-3} + ...)$$

$$= ... + 7z^{-4} + 3z^{-3} + z^{-2} + ...$$

(*d*) For $0 < |z - 1| < 1$, write (*i*) as

$$f(z) = \frac{1}{(z-1)-1}, -\frac{1}{z-1}$$

$$= -(z-1)^{-1} - [1-(z-1)]^{-1}$$

$$= -(z-1)^{-1} - [1 + (z-1) + (z-1)^2 + (z-1)^3 + ...]$$

Example. Find sum of the residues of $f(z) = \frac{\sin z}{z\cos z}$ at its poles inside the circle $|z| = 2$.

Solution :

$f(z)$ has simple poles at $z = 0, \pm\frac{\pi}{2}, \pm 3\frac{\pi}{2}, - ...$

Only poles $z = 0$ and $z = \pm\frac{\pi}{2}$ lies inside $|z| = 2$.

$$\therefore \quad \text{Res } f(0) = \underset{z\to 0}{\text{Lt}}[z.f(z)]$$

$$= \underset{z\to 0}{\text{Lt}}\left(\frac{\sin z}{\cos z}\right) = 0$$

$$\text{Res } f(\frac{\pi}{2}) = \underset{z\to\pi/2}{\text{Lt}}\left[\left(z-\frac{\pi}{2}\right)f(z)\right]$$

$$= \underset{z\to\pi/2}{\text{Lt}}\left\{\frac{(z-\pi/2)\sin z}{z\cos z}\right\}$$

$$= \underset{z\to\pi/2}{\text{Lt}}\frac{(z-\pi/2)\cos z + \sin z}{\cos z - z\sin z} \quad \left[\text{being } \frac{0}{0} \text{ form}\right]$$

$$= \frac{1}{-\pi/2}$$

$$= -\frac{2}{\pi}$$

and $$\text{Res } f(-\pi/2) = \underset{z\to-\pi/2}{\text{Lt}}\left\{\frac{(z+\pi/2)\sin z}{z\cos z}\right\}$$

$$= \underset{z\to-\pi/2}{\text{Lt}}\frac{(z+\pi/2)\cos z + \sin z}{\cos z - z\sin z}$$

$$= \frac{-1}{-\pi/2}$$

$$= \frac{2}{\pi}$$

Hence sum of residues $= 0 - \frac{2}{\pi} + \frac{2}{\pi} = 0.$

Example. Evaluate $\int_C \frac{z-3}{z^2+2z+5}dz$, where C is the circle

(*i*) $|z| = 1$

(*ii*) $|z + 1 - i| = 2$

(*iii*) $|z + 1 + i| = 2$

Solution :

Poles of $f(z) = \frac{z-3}{z^2+2z+5}$ are given by

$$z^2 + 2z + 5 = 0$$

i.e. $$z = \frac{-2 \pm \sqrt{(4-20)}}{2} = -1 \pm 2i.$$

(*i*) Both poles $z = -1 + 2i$ and $z = -1 - 2i$ lie outside the circle $|z| = 1$. Therefore, $f(z)$ is analytic everywhere within C.

Hence by Cauchy's theorem,

$$\int_C \frac{z-3}{z^2+2z+5}dz = 0.$$

(*ii*) Here only one pole $z = -1 + 2i$ lies inside the circle C : $|z + 1 - i| = 2$.

Therefore, $f(z)$ is analytic within C except at this pole.

$$\therefore \text{ Res } f(-1+2i) = \underset{z\to-1+2i}{\text{Lt}}[\{z-(-1+2i\}f(z)]$$

$$= \underset{z\to-1+2i}{\text{Lt}}\frac{(z+1-2i)(z-3)}{z^2+2z+5}$$

$$= \underset{z\to-1+2i}{\text{Lt}}\frac{z-3}{z+1+2i}$$

$$= \frac{-4+2i}{4i}$$

$$= i + \frac{1}{2}$$

Hence by residue theorem

$$\int_C f(z)dz = 2\pi i \operatorname{Res} f(-1+2i)$$

$$= 2\pi i(i + 1/2)$$

$$= \pi(i - 2).$$

(*iii*) Here only pole $z = -1 - 2i$ lies inside the circle C : $|z + 1 + i| = 2$.

Therefore, $f(z)$ is analytic within C except at this pole.

$$\therefore \quad \operatorname{Res} f(-1-2i) = \operatorname*{Lt}_{z\to -1-2i} \frac{(z+1+2i)(z-3)}{z^2+2z+5}$$

$$= \operatorname*{Lt}_{z\to -1-2i} \frac{z-3}{z+1-2i}$$

$$= \frac{-4-2i}{-4i}$$

$$= \frac{1}{2} - i$$

Hence by residue theorem,

$$\int_C f(z)dz = 2\pi i \operatorname{Res} f(-1-2i)$$

$$= 2\pi i\left(\frac{1}{2} - i\right) = \pi(2+i)$$

Example. Evaluate $\oint_C \frac{e^z}{\cos \pi z} dz$, where C is the unit circle $|z| = 1$.

Solution :

$f(z) = \frac{e^x}{\cos \pi z}$ has simple poles at

$$z = \pm \frac{1}{2}, \pm \frac{3}{2}, \pm \frac{5}{2}, \ldots$$

Out of these only poles at $z = 1/2$ and $z = -\frac{1}{2}$ lie inside the given circle $|z| = 1$.

$$\therefore \quad \operatorname{Res} f(1/2) = \operatorname*{Lt}_{z\to 1/2}\left[\left(z - \frac{1}{2}\right)f(z)\right]$$

$$= \operatorname*{Lt}_{z\to 1/2}\left\{\frac{\left(z-\frac{1}{2}\right)e^z}{\cos \pi z}\right\} \quad \left[\frac{0}{0}\text{ form}\right]$$

$$= \operatorname*{Lt}_{z\to 1/2} \frac{e^z + \left(z - \frac{1}{2}\right)e^z}{-\pi \sin \pi z}$$

$$= \frac{e^{1/2}}{-\pi}$$

and $$\operatorname{Res} f(-1/2) = \operatorname*{Lt}_{z\to -1/2}\left\{\frac{\left(z-\frac{1}{2}\right)e^z}{\cos \pi z}\right\} \quad \left[\frac{0}{0}\text{ form}\right]$$

$$= \operatorname*{Lt}_{z\to -1/2} \frac{e^z + \left(z - \frac{1}{2}\right)e^z}{-\pi \sin \pi z}$$

$$= \frac{e^{-1/2}}{\pi}$$

Hence $$\oint_C \frac{e^z}{\cos \pi z} dz = 2\pi i\left\{\operatorname{Res} f\left(\frac{1}{2}\right) + \operatorname{Res} f\left(-\frac{1}{2}\right)\right\}$$

$$= 2\pi i\left(-\frac{e^{1/2}}{\pi} + \frac{e^{-1/2}}{\pi}\right)$$

$$= -4i\left(\frac{e^{1/2} - e^{-1/2}}{2}\right)$$

$$= -4i \sinh \frac{1}{2}.$$

Example. Evaluate $\int_C \tan z \, dz$ where C is the circle $|z| = 2$

Solution :

Poles of $f(z) = \sin z/\cos z$ are given by $\cos z = 0$

i.e. $z = \pm \frac{\pi}{2}, \pm \frac{3\pi}{2}, \pm \frac{5\pi}{2}$

Of these poles, $z = \frac{\pi}{2}$, and $-\frac{\pi}{2}$ only are within the given circle.

$$\therefore \quad \operatorname{Res} f(\pi/2) = \operatorname*{Lt}_{z\to \pi/2} \frac{\sin z}{\frac{d}{dz}(\cos z)}$$

$$= \operatorname*{Lt}_{z\to \pi/2}\left(\frac{\sin z}{-\sin z}\right) = -1$$

Similarly $$\operatorname{Res} f(-\pi/2) = \operatorname*{Lt}_{z\to -\pi/2} \frac{\sin z}{\frac{d}{dz}(\cos z)} = -1.$$

Hence by residue theorem,

$$\int_C f(z)dz = 2\pi i \{\operatorname{Res} f(\pi/2) + \operatorname{Res} f(-\pi/2)\}$$

$$= 2\pi i(-1-1) = 4\pi i.$$

EXERCISE

MCQ TYPE QUESTIONS

1. Taylor series expansion of the function,

$$F(x) = \frac{x}{1+x} \text{ around } x = 0 \text{ is}$$

(a) $x + x^2 + x^3 + x^4$............

(b) $1 = x + x^2 + x^3 + x^4$..............

(c) $2x + 4x^2 + 8x^3 + 16x^4$..............

(d) $x - x^2 + x^3 - x^4$..............

2. If $u = x^3 - 3xy^2 + 3x^2 - 3y^2 + 1$, then analytic function $f(z) = u(x, y) + iv(x, y)$ is

(a) $3xy - y^3 + 6xy^2 + C$

(b) $3x^2y - y^3 + 6xy + C$

(c) $3x^2y^2 - y^3 + 6xy + C$

(d) $3x^2y - y^3 + 6xy^2 + C$

3. The function $f(z) = \frac{xy^2(x+iy)}{x^2+y^4}, z \neq 0$

and $f(0) = 0$

(a) is not analytic at the origin

(b) is analytic at the origin

(c) can't be said

(d) none of these

4. $\oint_C \frac{\sin \pi z^2 + \cos \pi z^2}{(z-1)(z-2)}$, where C is the circle and $|z| = 3$, is equal to

(a) $2\pi i$ (b) $3\pi i$

(c) $4\pi i$ (d) $5\pi i$

5. The residue of $f(z) = \frac{z^2 - 2z}{(z+1)^2(z^2+4)}$ is

(a) $\frac{1 \pm i}{25}$ (b) $\frac{3 \pm i}{25}$

(c) $\frac{5 \pm i}{25}$ (d) $\frac{7 \pm i}{25}$

6. For $|z| = 1, \int_C \frac{z-3}{z^2+2z+5} dz$, where C is the circle, is

(a) 0 (b) 1

(c) 2 (d) none of these

7. For $|z + 1 - i| = 2, \int_C \frac{z-3}{z^2+2z+5} dz$, where C is the circle, is

(a) $\pi(i - 2)$ (b) $\pi(i - 3)$

(c) $\pi(i - 4)$ (d) $\pi(i - 5)$

8. Expansion of sin z about $z = \frac{\pi}{2}$ is

(a) $1 + 0 - \frac{\left(z - \frac{\pi}{2}\right)^2}{(2!)} + \frac{\left(z - \frac{\pi}{2}\right)^4}{(4!)} + ...$

(b) $1 + 0 + \frac{\left(z - \frac{\pi}{2}\right)^2}{(2!)} + \frac{\left(z - \frac{\pi}{2}\right)^4}{(4!)} + ...$

(c) $1 - 0 + \frac{\left(z - \frac{\pi}{2}\right)^3}{(2!)} + \frac{\left(z - \frac{\pi}{2}\right)^3}{(3!)} + ...$

(d) $1 - \frac{(z-\pi)^2}{(2!)} + \frac{(z-\pi)^3}{(3!)} + ...$

9. The most general complex analytical function $f(z) = u(x, y) + iv(x, y)$ for $u = x^2 - y^2$ is

(a) z (b) $2z$

(c) z^3 (d) $1/z^2$

10. If $f(z) = u + iv$, then solution for $f'(z)$ is

(a) $\frac{\partial u}{\partial x} + i\frac{\partial v}{\partial x}$ (b) $\frac{\partial v}{\partial y} - i\frac{\partial u}{\partial y}$

(c) $\frac{\partial u}{\partial x} - i\frac{\partial v}{\partial x}$ (d) both (a) and (b)

11. If $v = 2xy$, then analytic function $f(z) = u + iv$ is

(a) $z^2 + c$

(b) $z^{-2} + c$

(c) $z^3 + c$

(d) $z^{-3} + c$

12. If $v = \frac{x-y}{x^2+y^2}$, then analytic function $f(z) = u + iv$ is

(a) $z + c$ (b) $z^{-1} + c$

(c) $(1-i)\frac{1}{z} + c$ (d) $(1+i)\frac{1}{z} + c$

13. If $u = \frac{\sin 2x}{\cosh 2y - \cos 2x}$, then analytic function $f(z) = u + iv$ is

(a) $\cot z + ic$

(b) $\operatorname{cosec} z + ic$

(c) $\sinh z + ic$

(d) $\cosh z + ic$

14. If C is the circle $|z| = 3$, then $\int_C \frac{\cos \pi z}{z-1} dz$ equals

(a) $i2\pi$ (b) $-i2\pi$

(c) $i6\pi^2$ (d) $-i6\pi^2$

15. Laurent's series of $f(z) = \dfrac{z}{(z+1)(z^2+4)}$ is, where $|z| < 1$

(a) $\frac{1}{4}z - \frac{5}{16}z^3 + \frac{21}{64}z^5 \ldots\ldots\ldots$

(b) $\frac{1}{2} + \frac{1}{4}z^2 + \frac{5}{16}z^4 + \frac{21}{64}z^6 \ldots\ldots\ldots$

(c) $\frac{1}{2}z - \frac{3}{4}z^3 + \frac{15}{8}z^5 \ldots\ldots\ldots$

(d) $\frac{1}{2} + \frac{1}{2}z^2 + \frac{3}{4}z^4 + \frac{15}{8}z^6 \ldots\ldots\ldots$

16. If C is $|z-1| = 1$, then $\int_C z^2 e^{\frac{1}{z}} dz$ is

(a) $i3\pi$ (b) $-i3\pi$

(c) $\frac{i\pi}{3}$ (d) none of these

17. Test the following series :

$$1 - \frac{1}{2^2} - \frac{1}{3^2} - \frac{1}{4^2} - \ldots\ldots\ldots\infty$$

(a) divergent
(b) convergence
(c) both convergence and divergence
(d) None of the above

18. Test the following series :

$$1 - \frac{1}{2} + \frac{1}{3} - \frac{1}{4} + \frac{1}{5} - \frac{1}{6} + \ldots..$$

(a) absolute convergence
(b) Divergence
(c) Convergent but not absolutely convergent
(d) None of the above

19. Test the convergence of the series :

$$\frac{1}{6} - \frac{2}{11} + \frac{3}{16} - \frac{4}{21} + \frac{5}{26} \ldots\ldots\infty$$

(a) Convergent series
(b) Divergent series
(c) oscillatory
(d) both convergent and Divergent

20. Check the series :

$$x + \frac{2^2.x^2}{\sqrt{2}} + \frac{3^3.x^3}{\sqrt{3}} + \frac{4^4.x^4}{\sqrt{4}} + \ldots..\infty$$

(a) Convergent series
(b) Divergence series
(c) Oscillatory series
(d) None of the above

21. State which one of the alternatives in the following is correct :

The series 1 – 1 + 1 – 1 + 1 ..... is

(a) Convergent with its sum equal to 0
(b) Convergent with its sum equal to 1
(c) Divergent
(d) oscillatory,

22. In a series $1 + r + r^2 + r^3 + \ldots..\infty$

If r = 1 then the series is :

(a) Divergent with its sum equal to n
(b) Convergent with its sum equal to 0
(c) Oscillatory with its some is equal to 0
(d) Both oscillatory and convergent

23. The nature of the series :

2 – 2 + 2 – 2 + 2 .... ∞ is :

(a) $S_n = 0$ if n is even, $S_n = 1$ if n is odd
(b) $S_n = \infty$ if n is even, $S_n = -10$ if n is odd
(c) $S_n = -n$ if is even, $S_n = -1$ if n is odd
(d) $S_n = -\infty$ if n is even, $S_n = \infty$ if n is odd

24. The sum of the series $1 + \frac{1}{2} + \frac{1}{4} + \frac{1}{6} \ldots..\infty$ is:

(a) $\lim_{n\to\infty} S_n = 10$ (b) $\lim_{n\to\infty} S_n = 2$

(c) $\lim_{n\to\infty} S_n = 15$ (d) $\lim_{n\to\infty} S_n = 20.4$

25. In the infinite series :

$\sum u_n = u_1 + u_2 + u_3 \ldots u_n$, $S_n = u_1 + u_2 + u_3 + \ldots\ldots.. u_n$, which one of the statement are correct:

(a) if S_n tends to a finite number as $n \to \infty$, the series $\sum u_n$ is equal to be convergent.

(b) if S_n tends to infinity as $n \to \infty$, the series $\sum u_n$ is equal to be divergent

(c) if S_n does not tends to a unique limit, finite or infinity, the series the series $\sum u_n$ called oscillatory

(d) All are correct.

NUMERICAL TYPE QUESTIONS

1. Residue of the function $\dfrac{1-e^{2z}}{z^4}$ at its pole is______.

2. Residue of $z\cos\frac{1}{z}$ at $z = 0$ is______.

ANSWERS

MCQ Type Questions

1. (*d*)	**2.** (*b*)	**3.** (*a*)	**4.** (*c*)	**5.** (*d*)	**6.** (*a*)	**7.** (*a*)	**8.** (*a*)	**9.** (*b*)	**10.** (*d*)
11. (*a*)	**12.** (*d*)	**13.** (*a*)	**14.** (*b*)	**15.** (*a*)	**16.** (*c*)	**17.** (*b*)	**18.** (*c*)	**19.** (*c*)	**20.** (*c*)
21. (*d*)	**22.** (*a*)	**23.** (*a*)	**24.** (*b*)	**25.** (*d*)					

Numerical Type Questions

1. $\frac{-4}{3}$ **2.** $\frac{-1}{2}$

EXPLANATIONS

MCQ TYPE QUESTIONS

1. $f(x) = \frac{x}{1+x}$

$f(x) = f(0) + f'(0)(x-0) + \frac{f''(0)}{\lfloor 2}(x-a)^2 + \frac{f'''(0)}{\lfloor 3}x^3 + ...$

$= 0 + x + (-x^2) + x^3 + ...$

$= x - x^2 + x^3 + ...$

2. Given: $u = x^3 - 3xy^2 + 3x^2 - 3y^2 + 1$

$\therefore \quad \frac{\partial u}{\partial x} = 3x^2 - 3y^2 + 6x$

and $\quad \frac{\partial u}{\partial y} = -6xy - 6y$

By C.R. equations,

$\frac{\partial u}{\partial x} = 3x^2 - 3y^2 + 6x$

$\therefore \quad V = 3x^2y - y^3 + 6xy + \phi(x) + C \quad ...(i)$

$\frac{\partial u}{\partial x} = -\frac{\partial v}{\partial x} = -6xy - 6y$

$\Rightarrow \quad \frac{\partial v}{\partial x} = 6xy + 6y$

$\therefore \quad V = 3x^2y + 6xy + \phi(y) + C \quad ...(ii)$

From equations (*i*) and (*ii*), we get

$3x^2y - y^3 + 6xy + \phi(x) + C = 3x^2y + 6xy + \phi(y) + C.$

$\Rightarrow \quad \phi(x) - \phi(y) = y^3$

$\therefore \quad \phi(x) = 0, \quad$ and $\quad \phi(y) = -y^3.$

$\therefore \quad V = 3x^2y - y^3 + 6xy + C.$

3. We know that given function is not analytic at (0, 0) if derivative $f'(z)$ does not exists at (0, 0)

Now, $\quad f'(z) = \lim_{z\to 0}\frac{f(z)-f(0)}{z-0}$

$$= \lim_{z\to 0}\frac{f(z)}{z} = \lim_{\substack{x\to 0\\ y\to 0}}\left[\frac{xy^2}{x^2+y^4}\right]$$

provided this limit exists independent of the manner in which $z \to 0$.

Consider two ways in which $z \to 0$.

Case I : $x = 0, y \to 0$

Then $\lim_{x\to 0}\frac{0}{0+y^4} = 0$

Case II : $x = y^2, x \to 0$

Then $\lim_{x\to 0}\frac{x.x}{x^2+x^2}\lim_{x\to 0}\frac{x^2}{2x^2} = \frac{1}{2}$

This shows that above limit is not unique, i.e. it depends on the manner in which $z \to 0$. So derivative $f'(z)$ does not exists at (0, 0) and hence given function $f(z)$ is not analytic at (0, 0).

4. Since $\frac{1}{(z-1)(z-2)} = \frac{1}{z-2} - \frac{1}{z-1}, ...$

$$\therefore \oint \frac{\sin \pi z^2 + \cos \pi z^2}{(z-1)(z-2)}$$

$$= +\int_c \frac{\sin \pi z^2 + \cos \pi z^2}{z-2}dz$$

$$- \int_c \frac{\sin \pi z^2 + \cos \pi z^2}{z-1}dz$$

By Cauchy integral formula.

$$\int_c \frac{\sin \pi z^2 + \cos \pi z^2}{z-2}\, dz = 2\pi i[\sin \pi(2)^2 + \cos \pi(2)^2]$$

$$= 2\pi i\,[0+1]$$

$$= 2\pi i$$

and $\int_c \frac{\sin \pi z^2 + \cos \pi z^2}{z-1}\, dz$

$$= 2\pi i\,[\sin \pi + \cos \pi]$$

$$= -2\pi i$$

$\therefore$ Required integral $= 2\pi i - (-2\pi i) = 4\pi i$

5. $f(z)$ has a double pole at $z = -1$ and simple poles at $z = \pm 2i$

$$\text{Res}(-1) = \lim_{z\to(-1)} \frac{1}{1!}\frac{d}{dz}\left\{(z+1)^2 \frac{z^2-2z}{(z+1)^2(z^2+4)}\right\}$$

$$= \lim_{z\to -1} \frac{(z^2+4)(2z-2)-(z^2-2z)(2z)}{(z^2+4)^2}$$

$$= \frac{14}{25}$$

$$\text{Res}(2i) = \lim_{z\to -2} \frac{(z+2i)(z^2-2z)}{(z^2+4)(z+1)^2}$$

$$= \frac{z^2-2z}{(z+2i)(z+1^2)}$$

$$= \frac{(2i)^2-4i}{4i(2i+1)^2}$$

$$= \frac{7+i}{25}$$

$$\text{Res}(-2i) = \lim_{z\to -2} \frac{(z+2i)(z^2-2z)}{(z^2+4)(z+1)^2}$$

$$= \frac{(-2i)^2+2(2i)}{(-2i-2i)(-2i+1)^2}$$

$$= \frac{7-i}{25}$$

6. Poles of $f(z) = \frac{z-3}{z^2+2z+5}$ are given by

$z^2 + 2z + 5 = 0$

$$\therefore \quad z = \frac{-2\pm 4i}{2} \quad 1 \pm 2i$$

Since, both the poles lie outside the circle $|z| = 1$, therefore $f(z)$ is analytic inside the circle

$$\therefore \quad \oint \frac{z-3}{z^2+2z+5}\, dz = 0$$

7. When $z = -1 - 2i$,

$$|-1-2i+1-1| = |-3i| = 3 > 2.$$

When $z = -1 + 2i$,

$$|-1+2i+1-i| = |i| = 1 < 2.$$

Therefore only $-1 + 2i$ lie inside the closed curve.

Res at $(-1+2i)$

$$= \lim_{x\to -1+2i} [z-(-1+2i)]\frac{z-3}{z^2+2z+5}$$

$$= \lim_{x\to 1+2i} \frac{z-3}{(z+1+2i)}$$

$$= \frac{-1+2i-3}{-1+2i+1+2i}$$

$$= \frac{-4+2i}{4i}$$

$$= \left(i+\frac{1}{2}\right)$$

Hence by residue theorem,

$$\oint_c \frac{z-3}{z^2+2z+5}\, dz = 2\pi i\left(i+\frac{1}{2}\right) = \pi(i-2)$$

8. Here $a = \pi/2$

$$f(z) = \sin z,\ f\left(\frac{\pi}{2}\right) = 1$$

$$f'(z) = \cos z,\ f'\left(\frac{\pi}{2}\right) = 0$$

$$f''(z) = -\sin z,\ f''\left(\frac{\pi}{2}\right) = -1$$

$$f'''(z) = -\cos z,\ f'''\left(\frac{\pi}{2}\right) = 0$$

$$\therefore \quad f(z) = f(a) + f'(a)(z-a) + \frac{f''(a)}{2i}(z-a)^2 + \ldots.$$

$$= 1 + 0 - \frac{\left(z-\frac{\pi}{2}\right)^2}{2!} + \frac{\left(z-\frac{\pi}{2}\right)^4}{4!} + - \ldots.$$

9. Here, $f(z) = u(x, y) + iv(x, y)$ and $u = x^2 - y^2$

From Cauchy-Riemann equation

$$\frac{\partial u}{\partial x} = \frac{\partial v}{\partial y} = 2y$$

$$\Rightarrow \quad \frac{\partial u}{\partial y} = -\frac{\partial v}{\partial x} = -2y$$

Now, $\frac{df(z)}{dz} = \frac{\partial u}{\partial x} + i\frac{\partial v}{\partial y} = 2x + i2y = 2(x+iy) = 2z$

10. $$f'(z) = \frac{\partial u}{\partial x} + i\frac{\partial v}{\partial x}$$

$$= \frac{\partial v}{\partial y} - i\frac{\partial u}{\partial y}$$

11. $$\frac{\partial v}{\partial x} = 2y = h(x,y),$$

$$\frac{\partial v}{\partial y} = 2x = g(x,y)$$

By Milne's method

$$f'(z) = g(z,0) + ih(z,0)$$

$$= 2z + i0 = 2z$$

Integrating, we get

$$f'(z) = z^2 + c$$

12. $$\frac{\partial v}{\partial y} = \frac{-(x^2+y^2)-(x-y)2y}{(x^2+y^2)^2}$$

$$= \frac{y^2-x^2-2xy}{(x^2+y^2)^2} = g(x,y)$$

and $$\frac{\partial v}{\partial x} = \frac{(x^2+y^2)-(x-y)2x}{(x^2+y^2)^2}$$

$$= \frac{y^2-x^2+2xy}{(x^2+y^2)^2} = h(x,y)$$

By Milne's method

$$f'(z) = g(z,0) + ih(z,0)$$

$$= -\frac{1}{z^2} + i\left(-\frac{1}{z^2}\right) = -(1+i)\frac{1}{z^2}$$

Integrating, we get

$$f'(z) = -(1+i)\int \frac{1}{z^2}dz + c = (1+i)\frac{1}{z} + c$$

13. $$\frac{\partial u}{\partial x} = \frac{2\cos 2x(\cosh 2y - \cos 2x) - 2\sin^2 2x}{(\cosh 2y - \cos 2x)^2}$$

$$= \frac{2\cos 2x \cosh 2y - 2}{(\cosh 2y - \cos 2x)^2} = \phi(x,y)$$

$$\frac{\partial u}{\partial x} = \frac{2\sin 2x \sinh 2y}{(\cosh 2y - \cos 2x)^2} = \psi(x,y)$$

By Milne's method

$$f'(z) = \phi(z,0) - i\psi(z,0)$$

$$= \frac{2\cos 2z - 2}{(1-\cos 2z)^2} - i(0)$$

$$= \frac{-2}{1-\cos 2z} = -\text{cosec}^2 z$$

Integrating, we get

$$f'(z) = -\int \text{cosec}^2 z\, dz + ic = \cot z + ic$$

14. Let $f(z) = \cos \pi z$

Then $f(z)$ is analytic within and on $|z| = 3$

By Cauchy's integral formula

$$f(z_0) = \frac{1}{2\pi i}\int_c \frac{f(z)}{z-z_0}dz \Rightarrow \int_c \frac{f(z)dz}{z-z_0} = 2\pi i f(z_0)$$

Taking $f(z) = \cos \pi z$ and $z_0 = 1$, we have

$$\int_{|z|=3} \frac{\cos \pi z}{z-1}dz = 2\pi i f(1) = 2\pi i \cos \pi = -2\pi i$$

15. $$f(z) = \frac{z}{(z^2+1)(z^2+4)}$$

$$= \frac{z}{3(z^2+1)} - \frac{z}{3(z^2+4)}, |z| < 1 \Rightarrow |z^2| < 1$$

$$f(z) = \frac{z}{3}(1+z^2)^{-1} - \frac{z}{12}\left(1+\frac{z^2}{4}\right)^{-1}$$

$$= \frac{z}{3}(1-z^2+z^4-\ldots)^{-1} - \frac{z}{12}\left(1-\frac{z^2}{4}+\frac{z^4}{16}-\ldots\right)$$

$$= \frac{1}{4}z - \frac{5}{16}z^3 + \frac{21}{64}z^5 \ldots..$$

16. $f(z) = z^2 e^{\frac{1}{z}}$

$$= z^2\left(1+\frac{1}{z}+\frac{1}{2!z^2}+\frac{1}{3!z^3}+\ldots..\right)$$

$$= z^2 + z + \frac{1}{2} + \frac{1}{6z} + \ldots..$$

The only pole of $f(z)$ is at $z = 0$, which lies within the circle $|z| = 1$

$$\int_c f(z)dz = 2\pi i \text{ (residue at z = 0)}$$

Residue of $f(z)$ at $z = 0$ is coefficient of $\frac{1}{z}$ i.e., $\frac{1}{6}$

$$\therefore \quad \int_c f(z)dz = 2\pi i \times \frac{1}{6} = \frac{1}{3}\pi i$$

17. $1 - \frac{1}{2^2} - \frac{1}{3^2} - \frac{1}{4^2} - \ldots$

$$|u_n| = \frac{1}{n^2}, |u_{n+1}| = \frac{1}{(n+1)^2}$$

(i) $|u_{n+1}| < |u_n|$

(ii) $\lim\limits_{n\to\infty} u_n = \lim\limits_{n\to\infty} \frac{1}{n^2} = 0$

Thus, the given series is absolutely convergent by Leibnitz rule and $1 + \frac{1}{2^2} + \frac{1}{3^2} + \frac{1}{4^2} + \ldots..$ is also convergent

18. $1 - \frac{1}{2} + \frac{1}{3} - \frac{1}{4}$

The terms of the series are alternatively positive and negative

(i) $|u_{n+1}| < |u_n|$ as $\frac{1}{n+1} < \frac{1}{n}$

(ii) $\lim\limits_{n\to\infty} u_n = 0$

Thus the given series is convergent

But $1 + \frac{1}{2} + \frac{1}{3} + \frac{1}{4} + \ldots\infty$ is divergent (p series, p=1)

Hence, the series is conditionally convergent

19. The terms of the given series are alternately positive and negative

$$u_n = (-1)^n \left(\frac{n}{5n+1}\right)$$

$$|u_n| - |u_{n-1}| = \frac{n}{5n+1} - \frac{n+1}{5(n+1)+1}$$

$$= \frac{5n^2 + 6n - 5n^2 - 5n - n - 1}{(5n+1)(5n+6)}$$

$$= \frac{-1}{(5n+1)(5n+6)}$$

$$|u_n| > |u_{n+1}|$$

$$\lim_{n\to\infty} = \lim_{n\to\infty} \frac{b}{5n+1} = \lim_{n\to\infty} \frac{1}{5+\frac{1}{n}}$$

$\frac{1}{5} \neq 0$ (Hence, the series is not convergent. It is oscillatory)

20. $x + \frac{2^2.x^2}{\lfloor 2} + \frac{3^3.x^3}{\lfloor 3} + \frac{4^4.x^4}{\lfloor 4} + \ldots. \frac{n^n.x^n}{\lfloor n} \ldots.\infty$

$$u_n = \frac{n^n x^n}{\lfloor n}, u_{n+1} = \frac{(n+1)^{n+1}.x^{n+1}}{\lfloor n+1}$$

$$\frac{u_n}{u_n+1} = \frac{n^n x^n}{\lfloor n} \frac{\lfloor n+1}{(n+1)^{n+1} x^{n+1}}$$

$$= \frac{n^n 1}{(n+1)^n x} = \frac{1}{\left(1+\frac{1}{n}\right)^n} \cdot \frac{1}{x}$$

$$\lim_{n\to\infty} \frac{u_n}{u_n+1} = \lim_{n\to\infty} \frac{1}{\left(1+\frac{1}{n}\right)^n} \cdot \frac{1}{x} = \frac{1}{e} \cdot \frac{1}{x}$$

If, $\frac{1}{ex} > 1$ or $x < \frac{1}{e}$, the series is convergent

If, $\frac{1}{ex} < 1$ or $\frac{1}{e} < x$, the series is divergent.

If $\frac{1}{ex} = 1$ or $x = \frac{1}{e}$, the test fails

$$\log \frac{u_n}{u_n+1} = \log \frac{1}{\left(1+\frac{1}{n}\right)} e = \log e - \log\left(1+\frac{1}{n}\right)^n$$

$$= 1 - n \log\left(1+\frac{1}{n}\right) = 1 - n\left[\frac{1}{n} - \frac{1}{2n^2} + \frac{1}{3n^3} + \ldots.\right]$$

$$= \left[1 - 1 + \frac{1}{2n} - \frac{1}{3n} + \ldots \frac{1}{2n} - \frac{1}{3n^2}\right]$$

$$\lim_{n\to\infty} n \log \frac{u_n}{u_n+1} = \lim_{n\to\infty}\left[\frac{1}{2} - \frac{1}{3n}\right] = \frac{1}{2} < 1$$

Thus, the series is divergent

21. The series $1 + r + r^2 + r^3 \ldots. \infty$ is oscillatory if $r \leq -1$, Divergent if $r \geq 1$ and convergent if $|r| < 1$

(a) When $r = -1$, the series becomes: $1 - 1 + 1 - 1 + 1 \ldots. \infty$

$S_n = 0$ if n is even

$= 1$ if n is odd

Hence, the series is oscillatory

22. When r = 1, the series becomes 1 + 1 + 1 + 1 ---- ∞.

$S_n = 1 + 1 + 1 + 1 + 1$ ------ n

$$\lim_{n\to\infty} S_n = \lim_{n\to\infty} \infty$$

Hence, the series is divergent.

23. 2 – 2 + 2 – 2 + 2 ------ ∞

$S_n = 2 - 2 + 2 - 2 +$ ------ ∞

$S_n = 0$, if n is odd

$S_n = 1$, if n is even.

Hence, S_n does not tent to a unique limit, and therefore, the given series is oscillatory.

24. $S_n = 1 + \frac{1}{2} + \frac{1}{4} + \frac{1}{8} + - - - - - \infty$

Series is in G.P, so

$$S_n = \left(\frac{a}{1-r}\right)$$

$$S_n = \left(\frac{1}{1-\frac{1}{2}}\right) \left(a = 1, r = \frac{1}{2}\right)$$

$S_n = 2$

$\lim_{n\to\infty} S_{n=2}$ (Hence, the series is convergent)

25. $\sum u_n = u_1 + u_2 + u_3 + - - - - u_n + - - - - \infty$

$S_n = u_1 + u_2 + u_3 + - - - - + u_n$

(i) S_n $\lim \to \infty$ = Finite number

$\left(\sum u_n \text{ in convergent}\right)$

(ii) S_n $_{n\to\infty}$ $= \infty \left(\sum u_n \text{ is divergent}\right)$

(iii) $S_n = 0, S_n = 1$ or $0\left(\sum u_n \text{ is oscillatry}\right)$

Hence, S_n, does not tend to a unique limit and, therefore, the given series is oscillatory.

NUMERICAL TYPE QUESTIONS

1. Let $f(z) = \frac{1-e^{2z}}{z^4}$

then $f(z)$ has a pole at $z = 0$ of order 4.

Residue of $f(z)$ at $z = 0$

$$= \frac{1}{3!}\lim_{z\to z_0}\frac{d^3}{dz^3}(1-e^{2z})$$

$$= \frac{1}{(m-1)!}\lim_{z\to z_0}\frac{d^{m-1}}{dz^{m-1}}(z-z_0)^m f(z)$$

$$= \frac{1}{(4-1)!}\lim_{z\to z_0}\frac{d^3}{dz^3}\left[z^4.\left(\frac{1-e^{2z}}{z^4}\right)\right]$$

$$= \frac{-1}{3!}\lim_{z\to z_0} 8e^{2z}$$

$$= \frac{-8}{6}(e^0)$$

$$= \frac{-8}{6} = \frac{-4}{3}$$

2. Putting $z = 0 + t$ we have

$$f(z) = z\cos\frac{1}{z}$$

$$= t\cos\frac{1}{t}$$

$$= t\left(1 - \frac{1}{2!}\frac{1}{t^2} + \frac{1}{4!}\frac{1}{t^4}.....\right)$$

$$= t - \frac{1}{2t} + \frac{1}{24t^3} - .....$$

Residue of $f(z)$ at $z = 0$ is coefficient of $\frac{1}{t}$ i.e. $-\frac{1}{2}$

7

CHAPTER

Discrete Mathematics

MATHEMATICAL LOGIC

STATEMENTS

Atomic statements.

The simple statements without connectives are called *atomic or primary statements*.

Molecular statements.

The new statements formed by joining atomic statements through connectives are called *molecular or compound statements.*

Truth table.

A table showing the truth value of a statement formula for each possible combinations of truth values of component statements is called its *truth table*.

A statement formula having n distinct components will have 2^n rows in its Truth table.

Truth values.

A statement is a declarative sentence which has one and only one of two possible values, called *truth values*. These two truth values are *true* and *false* denoted by the symbol T and F respectively; sometimes it is also denoted by the symbols 1 and 0.

The object language contains such declarative sentences. Since we allow only two possible truth values in our study, the logic is called *two-valued logic*.

The sentence, " *This statement is false*" is not a statement because we cannot assign a definite truth value to it. If we assign the value T, then sentence says that the '*statement is false*'; if we assign F, then the 'statement is true'. This is an example of *semantic paradox.* The sentence like

$$101 + 1 = 110$$

is a statement. It has a truth value, but the truth value depends on the context. If we take decimial system, the statement is false, but in binary system the statement is true. Even then we consider such sentences as statements.

CONNECTIVES

If we form new sentences from given sentences using 'and', 'but', 'if' etc., we can get new propositions from given propositions using *connectives*.

(1) Propositional connectives.

The new sentence obtained will be a proposition only when the new sentence has a truth value T or F (but not both).

(2) Logical connectives.

Truth value of the new sentence depends on the (logical) connectives and the truth value of the given propositions.

SOME CONNECTIVES.

(1) Negation (NOT) .

If P is a proposition, then negation P or NOT P (denoted by P) is a proposition whose truth value is T if P has truth value F, and F if P has truth value T.

Table. Truth Table for Negation

P	P
T	F
F	T

(2) Conjunction (AND).

If P and Q are two propositions, then conjunction of P and Q (read as "P and Q") is a proposition whose truth values are given in the Truth Table and denoted by $P \wedge Q$.

Table. Truth Table for Conjunction

P	Q	$P \wedge Q$
T	T	T
T	F	F
F	T	F
F	F	F

(3) Disjunction (OR).

If P and Q are two propositions, then disjunction of P and Q (read as P or Q) is a proposition whose truth values are given in the truth table and denotd by $P \vee Q$.

$P \vee Q$ is true if P is true or Q is true or both are true.

Table. Truth Table for Disjunction

P	Q	$P \vee Q$
T	T	T
T	F	T
F	T	T
F	F	F

Example. If P represents *"This book is nice"* and Q represents *"This book is cheap"*, write the following sentences in symbolic form

(*a*) This book is good and cheap.

(*b*) This book is not good but cheap.

(*c*) This book is costly but good.
(*d*) This book is neither good nor cheap.
(*e*) This book is good or cheap.

Solution:

(*a*) $P \wedge Q$ (*b*) $(\neg P) \wedge Q$ (*c*) $(\neg Q) \wedge P$ (*d*) $(\neg P \wedge (\neg Q))$ (*e*) $P \vee Q$

Table. Representation and meaning of Logical connectives

Connective	Resulting proposition	Read as
Negation $\neg$	$\neg P$	NOT P
Conjunction $\wedge$	$P \wedge Q$	P AND Q
Disjunction $\vee$	$P \vee Q$	P OR Q (or both)
Implication $\rightarrow$	$P \rightarrow Q$	IF P THEN Q (P Implies Q)
IF AND ONLY IF $\rightleftarrows$	$P \rightleftarrows Q$	P IF AND ONLY IF Q

(4) Implication (IF ... THEN .....).

If P and Q are two propositions, then "IF P THEN Q" is a proposition whose truth values are given in the truth table and is denoted by $P \rightarrow Q$. We also read $P \rightarrow Q$ as "*P implies Q*".

Table. Truth Table for Implication

P	Q	$P \rightleftarrows Q$
T	T	T
T	F	F
F	T	T
F	F	T

$P \rightarrow Q$ assumes the truth value F only if P has truth value T and Q has truth value F. In all the other cases, $P \rightarrow Q$ assumes the truth value T. In the case of natural languages, we are concerned about the truth values of the sentence "IF P THEN Q" only when P is true. When P is false, we are not concerned about the truth value of "IF P THEN Q". But in the case of mathematical logic, we have to definitely specify the truth value of $P \rightarrow Q$ in all cases. So truth value of $P \rightarrow Q$ is defined as T when P has truth value F (irrespective of the truth value of Q).

(5) IF and Only IF.

If P and Q are two statements, then P if and only if Q (denoted by $P \rightleftarrows Q$) is a statement whose truth value is T when the truth values of P and Q are the same and whose truth value is F when the statements differ. It is also called B conditional and means that P implies Q and Q implies P

i.e. $(P \rightarrow Q) \wedge (Q \rightarrow P)$

Table. Truth Table for IF and Only IF

P	Q	$P \rightleftarrows Q$
T	T	T
T	F	F
F	T	F
F	F	T

Example. Translate the following sentences into propositional forms :

(*a*) If it is not raining and I have time then I will go to a movie.
(*b*) It is raining and I will not go to a movie.
(*c*) It is not raining.
(*d*) I will not go to a movie.
(*e*) I will go to movie only if it is not raining.

Solution:

Let P be proposition *"It is raining"*,
Q be the proposition *"I have time"*,
R be proposition *"I will go to a movie"*.

Then

(*a*) $\neg P \wedge Q \rightarrow R$ (*b*) $P \wedge \neg R$ (*c*) $\neg P$
(*d*) $\neg R$ (*e*) $R \rightleftarrows \neg P$

WELL-FORMED FORMULAS

Definition 1.

A propositional variable is a symbol representing any proposition. Usually a real variable is represented by the symbol x. This means that x is not a real number but can take a real value. Similarly, a propositional variable is not a proposition but can be replaced by a proposition.

Definition 2.

A well-formed formula (wff) is defined recursively as follows :

I. *If P is a propositional variable, then it is a wff.*

II. *If α is a wff, then $\neg\alpha$ is a wff.*

III. *If α and β are well-formed formulas, then*

$(\alpha \vee \beta), (\alpha \wedge \beta), (\alpha \rightarrow \beta), (\alpha \rightleftarrows \beta)$ *are well-formed formulas.*

IV. *A string of symbols is a wff if and only if it is obtained by finitely many applications of (I)–(III).*

A well formed formula is not a proposition, but if we substitute the proposition in place of propositional variable, we get a proposition.

e.g.(*i*) $[\neg(P \vee Q) \wedge (\neg Q \wedge R) \rightarrow Q)$ is a wff.

(*ii*) $(\neg P \wedge Q) \rightleftarrows Q$ is a wff.

We can drop parentheses when there is no ambiguity.
e.g. in propositions we can remove the outermost parentheses. We can also specify hierarchy of connectives and avoid parentheses.

For the convenience, we can refer to a wff as a formula.

Duality law.

Two formulas A & A′ are said to be duals of each other if either can be obtained form the other by replacing $\wedge$ by $\vee$ and vice-versa and T by F and vice-versa.

e.g. Dual of $(P \wedge Q) \vee T$ is $(P \vee Q) \wedge F$

Let A and A′ are duals consisting of $P_1, P_2 \ldots\ldots P_n$ propositional variables. By repeated application of De Morgans Law, it can be shown that

$$\neg A(P_1, P_2 \ldots P_n) = A'(\neg P_1, \neg P_2 \ldots\ldots \neg P_n)$$

Theorem.

If two formulas are equivalent, then their duals are also equivalent.

Truth Table For Well-Formed Formula.

If we replace the propositional variables in a formula α by propositions, we get a proposition involving connectives. The table giving truth value of such proposition obtained by replacing the propositional variables by arbitrary propositions is called *truth table of α.*

If α involves n propositional constants, we have 2^n possible combinations of truth values of propositions replacing the variables.

Example. Obtain the truth table for $\alpha = (P \vee Q) \wedge (P \rightarrow Q) \wedge (Q \rightarrow P)$.

Solution:

P	Q	$P \vee Q$	$P \rightarrow Q$	$(P \vee Q) \wedge (P \rightarrow Q)$	$(Q \rightarrow P)$	α
T	T	T	T	T	T	T
T	F	T	F	F	T	F
F	T	T	T	T	F	F
F	F	F	T	F	T	F

Example. Construct the truth table for $\alpha = (P \vee Q) \rightarrow ((P \vee R) \rightarrow (R \vee Q))$.

Solution:

P	Q	R	$P \vee R$	$R \vee Q$	$(P \vee R) \rightarrow (R \vee Q)$	$(P \vee Q)$	α
T	T	T	T	T	T	T	T
T	T	F	T	T	T	T	T
T	F	T	T	T	T	T	T
T	F	F	T	F	F	T	F
F	T	T	T	T	T	T	T
F	T	F	F	T	T	T	T
F	F	T	T	T	T	F	T
F	F	F	F	F	T	F	T

Some formulas have truth value T for all possible assignments of truth values to the propositional variables. e.g., $P \vee \neg P$ has the truth value T irrespective of the truth value of P. Such formulas are called *tautologies.*

Definition 3.

A tautology or a universally true formula is a well-formed formula whose truth value is T for all possible assignments of truth values to the propositional variables.

e.g. $P \vee \neg P$, $(P \wedge Q) \rightarrow P$ and $((P \rightarrow Q) \wedge (Q \rightarrow R)) \rightarrow (P \rightarrow R)$ *are tautologies.*

Example. Show that $\alpha = (P \rightarrow (Q \rightarrow R)) \rightarrow ((P \rightarrow Q) \rightarrow (P \rightarrow R))$ is a tautology.

Solution:

Truth Table for α

P	Q	R	$Q \rightarrow R$	$P \rightarrow (Q \rightarrow R)$	$P \rightarrow Q$	$P \rightarrow R$	$(P \rightarrow Q) \rightarrow (P \rightarrow R)$	α
T	T	T	T	T	T	T	T	T
T	T	F	F	F	T	F	F	T
T	F	T	T	T	F	T	T	T
T	F	F	T	T	F	F	T	T
F	T	T	T	T	T	T	T	T
F	T	F	F	T	T	T	T	T
F	F	T	T	T	T	T	T	T
F	F	F	T	T	T	T	T	T

Definition 4.

A contradiction (or absurdity) is a wff whose truth value is F for all possible assignments of truth values to the propositional variables.

e.g. $P \wedge \rceil P$ and $(P \wedge Q) \wedge \rceil Q$ are contradictions.

Note : α is a contradiction if and only $\rceil\alpha$ is a tautology.

Equivalence of Well Formed Formulas.

Definition 5.

Two wffs α and β in propositional variables $P_1, P_2, \ldots P_n$ are equivalent (or logically equivalent) if the formula $\alpha \rightleftarrows \alpha$ is a tautology. When α and β are equivalent, we write $\alpha \equiv \beta$.

α and β are equivalent if the truth tables for α and β are the same.

e.g. $P \wedge Q \equiv Q \wedge P$ and $P \wedge P \equiv P$

Note : The difference betwen $\alpha \rightleftarrows \beta$ and $\alpha \equiv \beta$, $\alpha \rightleftarrows \beta$ is a formula, where as $\alpha \equiv \beta$ is not a formula, but it denotes the relation between α and β.

Example. Show that $(P \to (Q \vee R)) \equiv ((P \to Q) \vee (P \to R))$.

Solution:

Let $\alpha = (P \to Q \vee R)$ and $\beta = ((P \to Q) \vee (P \to R))$.

We construct the truth values of α and β for all assignments of truth values to the variables P, Q and R, given in the table below.

P	Q	R	$Q \vee R$	$P \to Q \vee R$ (α)	$P \to Q$	$P \to R$	$(P \to Q) \vee (P \to R)(\beta)$
T	T	T	T	T	T	T	T
T	T	F	T	T	T	F	T
T	F	T	T	T	F	T	T
T	F	F	F	F	F	F	F
F	T	T	T	T	T	T	T
F	T	F	T	T	T	T	T
F	F	T	T	T	T	T	T
F	F	F	F	T	T	T	T

As the columns corresponding to α and β coincide, $\alpha \equiv \beta$.

As truth value of a tautoloy is T, irrespective of the truth values of propositional variables, we denote any tautology by T. Similarly, we denote any contradiction by F.

LOGICAL IDENTITIES

Some equivalences are useful for deducing other equivalences, and are called *identities*

The identities $I_1 - I_{12}$ can be used to simplify formulas. If a formula β is part of another formula α and β is equivalent to β', then we can replace β and β' in α and resulting wff is equivalent to α.

Table : Logical identities

I_1	**Idempotent laws**	$P \vee P \equiv P$, $P \wedge P \equiv P$
I_2	**Commutative laws**	$P \vee Q \equiv Q \vee P$, $P \wedge Q \equiv Q \wedge P$
I_3	**Associative laws**	$P \vee (Q \vee R) \equiv (P \vee Q) \vee R$, $P \wedge (Q \wedge R) \equiv (P \wedge Q) \wedge R$
I_4	**Distributive laws**	$P \vee (Q \wedge R) \equiv (P \vee Q) \wedge (P \vee R)$, $P \wedge (Q \vee R) \equiv (P \wedge Q) \vee (P \wedge R)$
I_5	**Absorption laws**	$P \vee (P \wedge Q) \equiv P$, $P \wedge (P \vee Q) \equiv P$
I_6	**De Morgan's laws**	$\rceil(P \vee Q) \equiv \rceil P \wedge \rceil Q$, $\rceil(P \wedge Q) \equiv \rceil P \vee \rceil Q$
I_7	**Involution law**	$P \equiv \rceil(\rceil P)$
I_8	$P \vee \rceil P \equiv T$, $P \wedge \rceil P \equiv F$	
I_9	$P \vee T \equiv T$, $P \wedge T \equiv P$, $P \vee F \equiv P$	$P \wedge F \equiv F$
I_{10}	$(P \to Q) \wedge (\rceil Q) \equiv \rceil P$	
I_{11}	**Contrapositive**	$P \to Q \equiv \rceil Q \to \rceil P$
I_{12}	$P \to Q \equiv (\rceil P \vee Q)$	

Example. Show that $(P \wedge Q) \vee (P \wedge \neg Q) \equiv P$

Solution:

$$\begin{aligned} \text{L.H.S.} &= (P \wedge Q) \vee (P \wedge \neg Q) \\ &\equiv P \wedge (Q \vee \neg Q) && \text{using Distributive law (i.e. } I_4) \\ &\equiv P \wedge T && \text{using } I_8 \\ &\equiv P && \text{using } I_8 \end{aligned}$$

Example. Show that $(P \to Q) \wedge (R \to Q) \equiv (P \vee R) \to Q$

Solution:

$$\begin{aligned} \text{L.H.S.} &= (P \to Q) \wedge (R \to Q) \\ &\equiv (\neg P \vee Q) \wedge (\neg R \vee Q) && \text{using } I_{12} \\ &\equiv (Q \vee \neg P) \wedge (Q \vee \neg R) && \text{using Commutative law} \\ &\equiv Q \vee (\neg P \wedge \neg R) && \text{using Distributive law} \\ &\equiv Q \vee (\neg(P \vee R)) && \text{using De Morgan's law} \\ &\equiv (\neg(P \vee R)) \vee Q && \text{using Commutative law} \\ &\equiv (P \vee R) \to Q && \text{using } I_{12} \\ &= \text{R.H.S.} \end{aligned}$$

Normal Forms Of Well-Formed Formulas.

Various well-formed formulas are seen in terms of two propositional variables, say P and Q. We also know that two such formulas are equivalent if and only if they have the same truth table. The number of distinct truth tables for formulas in P and Q is 2^4 (as possible combinations of truth values of P and Q are TT, TF, FT, FF, truth table of any formula in P and Q has 4 rows. So number of distinct truth tables is 2^4). Thus there are only 16 distinct (nonequivalent) formulas, and any formula in P and Q is equivalent to one of these 16 formulas.

Here we give a method of reducing a given formula to an equivalent form called a *normal form*. We also use '*sum*' for disjunction. '*product*' for conjunction, and '*literal*' either for P or for $\neg P$, where P is any propositional variable.

Definition 6 .

An elementary product is a product of literals. An elementary sum is a sum of literals.

e.g. $P \wedge \neg Q$, $\neg P \wedge \neg Q$, $P \wedge Q$, $\neg P \wedge Q$ *are elementary products.*

$P \vee \neg Q$, $P \vee \neg R$ *are elementary sums.*

Definition 7.

A formula is in disjunctive normal form if it is a sum of elementary products.

e.g. $P \vee (Q \wedge R)$ and $P \vee (\neg Q \wedge R)$ *are in disjunctive normal form.*

$P \wedge (Q \vee R)$ *is not in disjunctive normal form.*

Construction to obtain Disjunctive Normal form of a given formula.

Step 1: Eliminate → and → using logical identities. (We can use I_{12}, *i.e.*, $P \to Q \equiv (\neg P \vee Q)$.

Step 2: Use De Morgan's law (I_6) to eliminate ¬ before sums or products. The resulting formula has ¬ only before propositional variables, *i.e.* it involves sum, product and literals.

Step 3: Apply distributive laws (I_4) repeatedly to eliminate product of sums. The resulting formula will be a sum of products of literals, *i.e.* sum of elementary products.

Example. Obtain a disjunctive normal form of

$$P \vee (\neg P \to (Q \vee (Q \to \neg R)))$$

Solution:

$$\begin{aligned} & P \vee (\neg P \to (Q \vee (Q \to \neg R))) \\ &\equiv P \vee (\neg P \to (Q \vee (\neg Q \vee \neg R))) && \text{(step 1 using } I_{12}) \\ &\equiv P \vee (P \vee (Q \vee (\neg Q \vee \neg R))) && \text{(step 1 using } I_{12} \text{ and } I_7) \\ &\equiv P \vee P \vee Q \vee \neg Q \vee \neg R && \text{using } I_3 \\ &\equiv P \vee Q \vee \neg Q \vee \neg R && \text{using } I_1 \end{aligned}$$

Thus, $P \vee Q \vee \neg Q \vee \neg R$ is a disjunctive normal form of the given formula.

Advantages of constructing principal disjunctive normal forms

(*i*) For a given formula, its principal disjunctive normal form is unique.

(*ii*) Two formulas are equivalent if and only if their principal disjunctive normal forms coincide.

Definition 8.

A min term in n propositional variables $P_1, \ldots, P_n$ is $Q_1 \wedge Q_2 \ldots \wedge Q_n$, where each Q_i is either P_i or $\neg P_i$ ....

e.g. min terms in P_1 and P_2 are $P_1 \wedge P_2$, $\neg P_1 \wedge P_2 \wedge \neg P_2$, $P_1 \wedge \neg P_2$, $\neg P_1 \wedge \neg P_2$. The number of min terms in *n* variables is 2^n.

Definition 9.

A formula α is in principal disjunctive normal form if α is a sum of min terms.

Construction to obtain principal Disjunctive Normal form of a given formula.

Step 1 : Obtain a disjunctive normal form.

Step 2 : Drop elementary products which are contradictions (such s $P \wedge \rceil P$).

Step 3 : If P_i and $\rceil P_i$ are missing in an elementary product α replace α by

$$(\alpha \wedge P_i) \vee (\alpha \wedge \rceil P_i).$$

Step 4 : Repeat step 3 until all elementary products are reduced to sum of min terms. Use idempotent laws to avoid repetition of min terms.

Example. Obtain the canonical sum-of-products form (i.e. principal disjunctive normal form) of

$$\alpha = P \vee (\rceil P \wedge \rceil Q \wedge R).$$

Solution:

α is already in disjunctive normal form. There are no contradictions. So we have to introduce missing variables (step 3). $\rceil P \wedge \rceil Q \wedge R$ in α is already a min term.

$$\text{Now, } P \equiv (P \wedge Q) \vee (P \wedge \rceil Q)$$
$$\equiv ((P \wedge Q \wedge R) \vee (P \wedge Q \wedge \rceil R)) \vee (P \wedge \rceil Q \wedge R) \vee (P \wedge \rceil Q \wedge \rceil R)$$
$$\equiv ((P \wedge Q \wedge R) \vee (P \wedge Q \wedge \rceil R)) \vee (P \wedge \rceil Q \wedge R) \vee (P \wedge \rceil Q \wedge \rceil R)$$

Therefore, canonical sum-of-products form of α is

$$(P \wedge Q \wedge R) \vee (P \wedge Q \wedge \rceil R) \vee (P \wedge \rceil Q \wedge R) \vee (P \wedge \rceil Q \wedge \rceil R) \vee (\rceil P \wedge \rceil Q \wedge R)$$

Example. Obtain the principal disjunctive normal form of

$$\alpha = (\rceil P \vee \rceil Q) \to (\rceil P \wedge R)$$

Solution.

$$\alpha = (\rceil P \vee \rceil Q) \to (\rceil P \wedge R)$$
$$\equiv (\rceil(\rceil P \vee \rceil Q)) \vee (\rceil P \wedge R) \quad \text{using } I_{12}$$
$$\equiv (P \wedge Q) \vee (\rceil P \wedge R) \quad \text{using De Morgan's law}$$
$$\equiv ((P \wedge Q \wedge R) \vee (P \wedge Q \wedge \rceil R)) \vee ((\rceil P \wedge R \wedge Q) \vee (\rceil P \wedge R \wedge \rceil Q))$$
$$\equiv (P \wedge Q \wedge R) \vee (P \wedge Q \wedge \rceil R) \vee (\rceil P \wedge Q \wedge R) \vee (\rceil P \wedge \rceil Q \wedge R)$$

So principal disjunctive normal form of α is

$$(P \wedge Q \wedge R) \vee (P \wedge Q \wedge \rceil R) \vee (\rceil P \wedge Q \wedge R) \vee (\rceil P \wedge \rceil Q \wedge R)$$

A min term of the form $Q_1 \wedge Q_2 \dots \wedge Q_n$ can be represented by $a_1 a_2 \dots a_n$,

where, $a_i = 0$ if $Q_i = \rceil P_i$ and $a_i = 1$ if $Q_i = P_i$.

So, principal disjunctive normal form can be represented by a 'sum' of binary strings.

e.g. $(P \wedge Q \wedge R) \vee (P \wedge Q \wedge \rceil R) \vee (\rceil P \wedge \rceil Q \wedge R)$ is represented by $111 \vee 110 \vee 001$.

The main terms in two variables P and Q are 00, 01, 10, and 11. Each wff is equivalent to its principal disjunctive normal form. Every principal disjunctive normal form corresponds to the min terms in it, and hence to a subset of {00, 01, 10, 11}. As the number of subsets is 2^4, the number of distinct formulas is 16.

Functionally complete set of connectives.

Any set of connectives in which every formula can be expressed in terms of as equivalent formula containing connectives from this set is called *functionally complete set of connectives*.

e.g. $\{\wedge, \rceil\}$ and $\{\vee, \rceil\}$ are functionally complete

$$\therefore \quad P \vee Q = \rceil(\rceil P \wedge \rceil Q)$$

$$\text{and} \quad P \wedge Q = \rceil(\rceil P \vee \rceil Q)$$

$\{\rceil\}, \{\wedge\}, \{\vee\}, \{\wedge, \vee\}$ and not functionally complete.

Other connectives.

(1) Exclusive OR ($\overline{\vee}$)

$P \,\overline{\vee}\, Q$ is true when either of P and Q is true, but not both

$\overline{\vee}$ is associative and commulative

$$P \,\overline{\vee}\, Q \equiv \rceil(P \rightleftarrows Q)$$

(2) NAND ($\uparrow$)

$$P \uparrow Q \equiv \rceil(P \wedge Q)$$
$$\rceil P \equiv P \uparrow P$$
$$P \vee Q \equiv (P \uparrow P) \uparrow (Q \uparrow Q)$$
$$P \wedge Q \equiv (P \uparrow Q) \uparrow (P \uparrow Q)$$

(3) NOR ($\downarrow$)

$$P \downarrow Q = \rceil(P \vee Q)$$
$$\rceil P = P \downarrow P$$
$$P \vee Q = (P \downarrow Q) \downarrow (P \downarrow Q)$$
$$P \wedge Q = (P \downarrow P) \downarrow (Q \downarrow Q)$$

$\therefore \quad \{\uparrow\}, \{\downarrow\}$ are functionally complete.

Example. For a given formula α, truth table is given in the Table below. Find the principal disjunctive normal form.

P	Q	R	α
T	T	T	T
T	T	F	F
T	F	T	F
T	F	F	T
F	T	T	T
F	T	F	F
F	F	T	F
F	F	F	T

Solution:

We have T in the α-column corresponding to the rows 1, 4, 5 and 8.

The min term corresponding to the first row is

$$P \wedge Q \wedge R.$$

Similarly, main terms corresponding to rows 4, 5 and 8 are

$$P \wedge \rceil Q \wedge \rceil R,\ \rceil P \wedge Q \wedge R \text{ and } \rceil P \wedge \rceil Q \wedge \rceil R.$$

Therefore, principal disjunctive normal form of α is

$$(P \wedge Q \wedge R) \vee (P \wedge \rceil Q \wedge \rceil R) \vee (\rceil P \wedge Q \wedge R) \vee (\rceil P \wedge \rceil Q \wedge \rceil R)$$

We can form 'dual' of disjunctive normal form which is termed as **conjunctive normal form.**

Definition 10.

A formula is in conjunctive normal form if it is a product of elementary sums.

If α is in disjunctive normal form, then $\rceil\alpha$ is in conjunctive normal form. (This can be seen by applying De Morgan's laws). So to obtain conjunctive normal form of α, we construct the disjunctive normal of $\rceil\alpha$ and use negation.

Definition 11.

A max term in n propositional variables $P_1, P_2, \ldots, P_n$ is

$Q_1 \vee Q_2 \ldots \vee Q_n$, *where each Q_i is either P_i or $\rceil P_i$.*

Definition 12.

A formula α is in principal conjunctive normal form if α is a product of max terms. For obtaining principal conjunctive normal form of α, we can construct principal disjunctive normal form of $\rceil\alpha$ and apply negation ($\rceil$).

Example. Find the principal conjunctive normal form of

$$\alpha = P \vee (Q \to R).$$

Solution:

$$\begin{aligned}
\rceil\alpha &= \rceil(P \vee (Q \to R)) \\
&\equiv \rceil(P \vee (\rceil Q \vee R)) && \text{using } I_{12} \\
&\equiv \rceil P \wedge (\rceil(\rceil Q \vee R)) && \text{using De Morgan's law} \\
&\equiv \rceil P \wedge (Q \wedge \rceil R) && \text{using De Morgan's law and } I_7
\end{aligned}$$

$\rceil P \wedge Q \wedge \rceil R$ is the principal disjunctive normal form of $\rceil\alpha$.

Hence, principal conjunctive normal form of α is

$$\rceil(\rceil P \wedge Q \wedge \rceil R) = P \vee \rceil Q \vee R$$

The logical identities given in Table above and normal forms of well-formed formulas bear a close resemblance of identities in Boolean algebras and normal forms of Boolean functions. Actually, the propositions under $\vee, \wedge$ and $\rceil$ form a Boolean algebra if equivalent propositions are identified. T and F act as bounds (*i.e.* 0 and 1 of a Boolean algebra). Also, statement formulas form a Boolean algebra under $\vee, \wedge$ and $\rceil$ if equivalent formulas are identified.

The normal forms of well-formed formulas correspond to normal form of Boolean functions and we can 'minimise' a formula in a similar manner.

PROPOSITIONAL CALCULUS

(Statement calculus)

In a logical reasoning, certain propositions are assumed to be true, and based on that assumption some other propositions are derived (deduced or inferred).

Hypothesis or Premises.

The propositions that are assumed to be true are called *hypotheses* or *premises*.

Conclusion.

The proposition derived by using the rules of inference is called a *conclusion.*

Valid argument.

The process of deriving conclusions based on the assumption of premises is called *a valid argument.*

So in a valid argument, we are concerned with the process of arriving at the conclusion rather than obtaining the conclusion.

Rules of inference.

These are simply tautologies in the form of implication (i.e. $P \to Q$).

e.g. $P \to (P \vee Q)$ is such a tautology, and it is a rule of inference.

We write this in the form $\dfrac{P}{P \vee Q}$. Here P denotes a premise. The proposition below the line, *i.e.* $P \vee Q$ is the conclusion.

For valid arguments, we can use the rules of inference given in the table below.

Rules of Inference		*Implication Form*
RI_1 :	Addition $\dfrac{P}{\therefore P \vee Q}$	$P \rightarrow (P \vee Q)$
RI_2 :	Conjunction $\dfrac{Q}{\therefore P \wedge Q}$	$Q \rightarrow P \wedge Q$
RI_3 :	Simplification $\dfrac{P \wedge Q}{\therefore Q}$	$(P \wedge Q) \rightarrow Q$
RI_4 :	Modus ponens P $\dfrac{P \rightarrow Q}{\therefore Q}$	$(P \wedge (P \rightarrow Q)) \rightarrow Q$
RI_5 :	Modus tollens $\rceil Q$ $\dfrac{P \rightarrow Q}{\therefore \rceil P}$	$(\rceil Q \wedge (P \rightarrow Q)) \rightarrow \rceil P$
RI_6 :	Disjunctive syllogism $\rceil P$ $\dfrac{P \vee Q}{\therefore Q}$	$(\rceil P \wedge (P \vee Q)) \rightarrow Q$
RI_7 :	Hypothetical syllogism $P \rightarrow Q$ $\dfrac{Q \rightarrow R}{\therefore P \Rightarrow R}$	$((P \rightarrow Q) \wedge (Q \rightarrow R)) \rightarrow (P \rightarrow R)$
RI_8 :	Constructive dilemma $(P \rightarrow Q) \wedge (R \rightarrow S)$ $\dfrac{P \vee R}{\therefore Q \vee S}$	$(P \rightarrow Q) \wedge (R \rightarrow S) \wedge (P \vee R) \rightarrow (Q \vee S)$
RI_9 :	Constructive dilemma $(P \rightarrow Q) \wedge (R \rightarrow S)$ $\dfrac{\rceil Q \vee \rceil S}{\therefore \rceil P \vee \rceil R}$	$(P \rightarrow Q) \wedge (R \rightarrow S) \wedge (\rceil Q \vee \rceil S) \rightarrow (\rceil P \vee \rceil R)$

Example. Can we conclude S from the following premises ?

I. $P \rightarrow Q$ II. $P \rightarrow R$

III. $\rceil(Q \wedge R)$ IV. $S \vee P$

Solution:

The valid argument for deducing S from the given for premises is given as a sequence. On the left, well-formed formula are given and on the right indicate whether the proposition is a premises (hypothesis) or a conclusion. If it is a conclusion, indicate premises and the rules of inference or logical identities used for deriving the conclusion.

1. $P \rightarrow Q$	Premise I
2. $P \rightarrow R$	Premise II
3. $(P \rightarrow Q) \wedge (P \rightarrow R)$	Lines 1, 2 and RI_2
4. $\rceil(Q \wedge R)$	Premise III
5. $\rceil Q \vee \rceil R$	Line 4 and De Morgan's law (I_6)
6. $\rceil P \vee \rceil P$	Line 3, 5 and destructive dilemma (RI_9)
7. $\rceil P$	Idempotent law I_1
8. $S \vee P$	Premise IV
9. S	Lines 7, 8 and disjunctive syllogism RI_6

Thus S can be concluded from the given premises.

Exmaple. Check validity of the following argument:

If Ram has completed B.E. Computer Science or M.B.A., then he is assured of a good job. If Ram is assured of a good job, he is happy. Ram is not happy. So Ram has not completed M.B.A.

Solution:

The propositions can be named in the following ways:

P denotes "Ram has completed B.E. Computer Science".

Q denotes "Ram has completed M.B.A.".

R denotes "Ram is assured of a good job".

S denotes "Ram is happy".

The given premises are :

I. $(P \vee Q) \rightarrow R$

II. $R \rightarrow S$

III. $\urcorner S$

The conclusion is $\urcorner Q$

1. $(P \vee Q) \rightarrow R$ Premise (I)

2. $R \rightarrow S$ Premise (II)

3. $(P \vee Q) \rightarrow S$ Lines 1, 2 and hypothetical syllogism R I_7

4. $\urcorner S$ Premise (III)

5. $\urcorner(P \vee Q)$ Lines 3, 4 and modus tollens

6. $\urcorner P \wedge \urcorner Q$ De Morgan's law I_6

7. $\urcorner Q$ Line 6 and simplification RI_3

Thus the argument is valid.

Example. Test validity of the following argument :

If milk is black, then every crow is white. If every crow is white, then it has four legs. If every crow has four legs, then every buffalo is white and brisk. The milk is black'. Therefore, the buffalo is white.

Solution:

We name the proposition in the following way :

P denotes "The milk is black"

Q denotes "Every crow is white"

R denotes "Every crow has four legs"

S denotes ""Every buffalo is white"

T denotes "Every buffalo is brisk"

The given premises are:

I. $P \rightarrow Q$

II. $Q \rightarrow R$

III. $R \rightarrow S \wedge T$

IV. P

The conclusion is S

1. P Premise (IV)

2. $P \rightarrow Q$ Premise (I)

3. Q Modus ponens RI_4

4. $Q \rightarrow R$ Premise (II)

5. R Modus ponens RI_4

6. $R \rightarrow S \wedge T$ Premise (III)

7. $S \wedge T$ Modus ponens RI_4

8. S Simplification RI_3

Thus the argument is valid.

PREDICATE CALCULUS

Consider two propositions "*Rita is a student*", and "*Sita is a student*".

As propositions, there is no relation between them, but we know they have something in common. Both Rita and Sita share the property of being a student. We can replace the two propositions by a single statement "x is a student". The common feature expressed by "is a student" is called a *predicate*. In predicate calculus we deal with sentences involving predicates.

Statements involving predicates occur in Mathematics and programming languages. *e.g.* "$2x + 3y = 4z$", "IF (D.GE. 0.0) GO TO 20" are statements in Mathematics and FORTRAN, respectively involving predicates. Some logical deductions are possible only by separating predicates'.

Predicates.

Sentences involving predicates describing property of the objects are denoted by $P(x)$, where P denotes the predicates and x is a variable denoting any object.

e.g. $P(x)$ can denote "x is a student".

In this sentence, x is a variable and P denotes the predicate "is a student".

The sentence "x is the father of y" also involves a predicate "is the father of". Here the predicate describes the relation between two persons. We can write this sentence as $F(x, y)$.

Similarly, $2x + 3y = 4z$ can be described by $S(x, y, z)$.

Note : Although, $P(x)$, involving a predicate looks like a proposition, it is not a proposition. As $P(x)$ involves a variable x, we cannot assign a truth value to $P(x)$. However, if we replace x by an individual object, we get a proposition.

e.g. if we replace x by Rita in $P(x)$, we get the proposition "Rita is a student". (we can denote this proposition by P(Rita).

If we replace x by 'A can', then also we get a proposition (whose truth value is F).

S (2, 0, 1) is the proposition 2.2 + 3.0 = 4.1 (whose truth value is T).S (1, 1, 1) is the proposition 2.1 + 3.1 = 4.1 (whose truth value is F).

The following definition is regarding possible 'values" which can be assigned to variables.

Definition 13.

For a declarative sentence involving a predicate, the universe discourse, or simply the universe, is the set of all possible values which can be assigned to variables.

e.g. universe of discourse for $P(x)$: "x is a student", can be taken as the set of all human names; the universe of discourse for $E(n)$; "n is an even integer", can be taken as the set of all integers (or the set of all real numbers).

Quantifier.

Universal Quantifier:

The phrase 'for all' (denoted by ∀) is called *universal quantifier.*

Using this symbol, we can write

"For all x, $x^2 = (-x)^2$" as $\forall x\, Q(x)$, where $Q(x)$ is" $x^2 = (-x)^2$"

Existential Quantifier:

The phrase 'there exists' (denoted by ∃) is called *existential quantifier*.

The sentence. "There exists x such that $x^2 = 5$" can be written as $\exists x R(x)$, where $R(x)$ is '$x^2 = 5$'.

Scope of the Quantifier:

$P(x)$ in $\forall x P(x)$ or in $\exists x P(x)$ is called the *scope of the quantifier* ∀ or ∃.

Note : The symbol ∀ can be read as 'for every', 'for any', 'for each', 'for arbitrary'. The symbol ∃ can be read as 'for some' for 'at least one'.

When we use quantifiers, we should specify the universe of discourse. If we change universe of discourse, the truth value may change.

e.g. consider $\exists x R(x)$ where $R(x)$ is $x^2 = 5$.

If universe of discourse is the set of all integers, then $\exists x R(x)$ is false.

If universe of discourse is the set of all real numbers, then $\exists x R(x)$ is true (when $x = \pm\sqrt{5}, x^2 = 5$).

The logical connectives involving predicates can be used for declarative sentences involving predicates.

Example. *Express the following sentences involving predicates in symbolic form :*

(*a*) All students are clever.

(*b*) Some students are not successful.

(*c*) Every clever student is successful.

(*d*) There are some successful students who are not clevers.

(*e*) Some students are clever and successful.

Solution:

As quantifiers are involved, we have to specify the universe of discourse. we can take the universe of discourse as the set of all students.

Let C (x) denote "x is clever"

Let $S(x)$ denote "x is successful".

Then sentence 1 can be written as $\forall x C(x)$. Sentences (*b*) – (*e*) can be written as

$$\exists x(\urcorner S(x))$$
$$\forall x\,(C(x) \to S(x))$$
$$\exists x(S(x) \wedge \urcorner C(x))$$
$$\exists x\,(C(x) \wedge S(x))$$

Well-formed formulas of predicate calculus.

A well-formed formula (wff) of predicate calculus is a string of variables such as x_1, x_2, ...x_n, connectives, parentheses and quantifiers defined recursively by following rules :

I. P $(x_1, ..., x_n)$ is a wff, when P is a predicate involving n variables $x_1\, x_2, ..., x_n$.

II. If α is a wff, then $\urcorner\alpha$ is a wff.

III. If α and β are wffs, then $\alpha \vee \beta$, $\alpha \wedge \beta$, $\alpha \to \beta$, $\alpha \to \beta$ are also wffs.

IV. If α is a wff and x is any variable, then $\forall x(\alpha)$, $\exists x(\alpha)$ are wffs.

V. A string is a wff if and only if it is obtained by finitely many applications of rules (I) – (IV).

Note : A proposition can be viewed as a sentence involving a predicate with 0 variables. So propositions are wffs of predicate calculus by rule (I).

Definition 14.

Let α and β be two predicate formulas in variables x_1,.... x_n, and let U be a universe of discourse of α and β. Then α and β are equivalent to each other over U if for every possible assignment of values to each variable in α and β, the resutling statements have the same truth values. We can write $\alpha \equiv \beta$ over U.

We say that α and β are equivalent to each other ($\alpha \equiv \beta$) if $\alpha \equiv \beta$ over U for every universe of discourse U.

Remark : In predicate formulas, the predicate variables may or may not be quantified. We can classify the predicate variables in a predicate formula, depending on whether they are quantified or not. This gives to the following definitions.

Definition 15.

If a formula of the form $\exists x(Px)$ or $\forall x P(x)$ occurs as part of a predicate formula α, then such part is called an x-bound part of α, and occurrence of x is called bound occurrence of x. An occurrence of x is free if it is not a bound occurrence. A predicate variable in α is free if its occurrence is free in any part of α.

e.g. in $\alpha = (\exists x_1 P(x_1, x_2)) \wedge (\forall x_2 Q(x_2, x_3))$, the occurrence of x_1 in $\exists x_1 P(x_1, x_2)$ is a bound occurrence and that of x_2 is free.

In $\forall x_2 Q(x_2, x_3)$, the occurrence of x_2 is a bound occurrence the occurrence of x_3 in α is free.

Note : Quantified parts of a predicate formula such as $\forall x P(x)$ or $\exists x P(x)$ are propositions. We can assign values from the universe of discourse only to free variables in a predicate formula α.

Definition 16.

A predicate formula is valid if for all possible assignments of values from any universe of discourse to free variables, the resulting propositions have truth value T.

Definition 17.

A predicate formula is satisfiable if for some assignment of values to predicate variables the resulting proposition has ruth value T.

Definition 18.

A predicate formula is unsatisfiable if for all possible assignments of values from any universe of discourse to predicate variables the resulting propositions have truth value F.

Note: valid predicate formulas correspond to tautologies among proposition formulas and unsatisfiable predicate formulas correspond to contradictions.

Rules of inference for predicate calculus.

(1) Proposition formulas are also predicate formulas:

(2) Predicate formulas where all the variables are quantified are proposition formulas. Therefore, all the rules of inference for proposition formulas are also applicable for predicate calculus wherever necessary.

For predicate formulas not involving connectives such as A(x), P(x, y), we can get equivalences and rules of inference.

e.g. corresponding to I_6 in Table of logical identities, we get

$$\rceil(P(x) \vee Q(x)) \equiv \rceil(P(x)) \wedge \rceil(Q(x)).$$

Corresponding to RI_3 in Table of rules of reference $P \wedge Q \to R$, we get

$$P(x) \wedge Q(x) \to P(x).$$

Thus we can replace propositional variables by predicate variables in previous tables.

Table. Equivalences involving two quantifiers and valid implications

I_{13}	**Distributivity of ∃ over ∨**
	$\exists x(P(x) \vee Q(x)) = \exists xP(x) \exists xQ(x)$
	$\exists x(P \vee Q(x)) \equiv P \vee (\exists xQ(x))$
I_{14}	**Distributivity of ∀ over ∧**
	$\forall x(P(x) \wedge Q(x)) \equiv \forall xP(x) \wedge \forall xQ(x)$
	$\forall x(P \wedge Q(x)) \equiv P \wedge (\forall xQ(x))$
I_{15}	$\rceil(\exists xP(x) \equiv \forall x \rceil P(x))$
I_{16}	$\rceil(\forall xP(x)) \equiv \exists x \rceil(P(x))$
I_{17}	$\exists x(P \wedge Q(x)) \equiv P \wedge (\exists xQ(x))$
I_{18}	$\forall x(P \vee Q(x)) \equiv P \vee (\forall xQ(x))$
RI_{10}	$\forall xP(x) \to \exists xP(x)$
RI_{11}	$\forall xP(x) \vee \forall xQ(x) \to \forall x(P(x) \vee Q(x))$
RI_{12}	$\exists x(P(x) \wedge Q(x) \to \exists xP(x) \wedge \exists xQ(x)$

Sometimes it is required to derive some conclusions from a given set of premises involving quantifiers, we may have to eliminate the quantifiers before applying the rules of inference for proposition formulas. Also, when the conclusion involves quantifiers, we may have to introduce quantifiers. The necessary rules of inference for addition and deletion of quantifiers are given in Table below.

Table. Rules of inference for addition and deletion of quantifiers

RI_{13} : Universal instantiation

$$\frac{\forall xP(x)}{\therefore P(c)}$$

c is some element of the universe.

RI_{14} : Existential instantiation

$$\frac{\exists xP(x)}{\therefore P(c)}$$

c is some element for which P(c) is true

RI_{15} : Universal generalisation

$$\frac{P(x)}{\forall xP(x)}$$

x should not be free in any of the given premises.

RI_{16} : Existential generalisation

$$\frac{P(c)}{\therefore \exists xP(x)}$$

c is some element of the universe.

Example. Discuss the validity of the following arguments

I. All graduates are educated.

II. Rita is a graduate. Therefore,

III. Rita is educated.

Solution:

Let G(x) denote "x is a graduate".

Let E (x) denote "x is educated".

Let R denote "Rita".

So the premises are $\forall x(G(x) \Rightarrow E(x))$ and G(R). The conclusion is E(R).

1. $\forall x(G(x) \Rightarrow E(x))$	Premise (I)
2. $G(R) \Rightarrow E(R)$	Universal instantiation RI_{13}
3. G(R)	Premise (II)
4. $\therefore$E(R)	Modus ponens RI_{14}

Thus the conclusion, viz. (III), is valid.

Example. Discuss the validity of the following argument :

All educated persons are well behaved. Rita is educated. No well-behaved person is quarrelsome. Therefore, Rita is not quarrelsome.

Solution:

Let the universe of discourse be the set of all educated persons.

Let P(x) denote x is well-behaved".

Let y denote "Rita".

Let Q(x) denote "x is equarrelsome".

So the premises are :

I. $\forall xP(x)$

II. y is a particular element of the universe of discourse.

III. $\forall x(P(x) \Rightarrow \rceil Q(x))$.

To obtain the conclusion, we have the following arguments :

1. $\forall xP(x)$	Premise (I)
2. P(y)	Universal instantiation RI_{13}
3. $\forall x(P(x) \Rightarrow \rceil Q(x))$	Premise (III)
4. $P(y) \Rightarrow \rceil Q(y)$	Universal instantiation RI_{13}
5. P(y)	Line 2
6. $\rceil Q(y)$	Modus ponens RI_{14}

SOLVED EXAMPLES

1. Find infix, prefix and suffix notation of

$$P \wedge \neg R \rightleftarrows P \vee Q$$

Solution:

Let $f = P \wedge \neg R \rightleftarrows P \vee Q$.

Connectives must be considered in the following hierarchy

$$\neg, \wedge, \vee, \longrightarrow \text{ and } \rightleftarrows$$

Infix : $f = P \wedge \neg R \rightleftarrows P \vee Q$

$\equiv (P \wedge \neg R \rightleftarrows P \vee Q$

$\equiv (P \wedge \neg R) \rightleftarrows (P \vee Q)$.

Prefix : Consider $f =$

$(P \wedge \neg R) \rightleftarrows (P \vee Q)$.

It is given by $(P \wedge \neg R) \rightleftarrows (P \vee Q)$

$\equiv \rightleftarrows \wedge P \neg R \vee PQ$ is prefix notation of f

Suffix : It is given by

$(P \wedge \neg R) \rightleftarrows (P \vee Q)$

$\equiv PR\neg \wedge PQ \vee \rightleftarrows$ is suffix notation of f.

2. Conclude $\neg P$ from $\neg P \vee Q$, $\neg(Q \wedge \neg R)$ and $\neg R$.

Solution:

The premises are $\neg P \vee Q$, $\neg(Q \wedge \neg R)$ and $\neg R$.

(1)	$\neg R$	P
(2)	$\neg(Q \wedge \neg R)$	P
(3)	$\neg Q \vee R$	(2) De Morgans Law
(4)	$\neg Q$	(1), (3), Disjunctive syllogisn
(5)	$\neg P \vee Q$	P
(6)	$\neg P$	(4), (5), Disjunctive syllogisn

3. Find the equivalent formula for $(P \longrightarrow Q) \vee R$ contains $\uparrow$ only.

Solution:

We know, $P \uparrow P = \neg(P \wedge P) = \neg P$.

We can write $\neg P \equiv P \uparrow P$ ... (*i*)

$P \wedge Q \equiv (P \uparrow Q) \uparrow (P \uparrow Q)$ ... (*ii*)

and $P \vee Q \equiv (P \uparrow P) \uparrow (Q \uparrow Q)$ ... (*iii*)

Hence $(P \to Q) \vee R \equiv \neg P \vee Q \vee R$

$\equiv (\neg P \vee Q) \vee R$

$\equiv ((\neg P \vee Q) \uparrow (\neg P \vee Q)) \uparrow (R \uparrow R)$ [by equation (*iii*)]

$\equiv ((\neg P \uparrow \neg P) \uparrow (Q \uparrow Q)) \uparrow ((\neg P \uparrow \neg P) \uparrow (Q \uparrow Q)) \uparrow (R \uparrow R)$ [by equation (*iii*)]

$= ((P \uparrow (Q \uparrow Q)) \uparrow (P \uparrow (Q \uparrow Q)) \uparrow (R \uparrow R)$.

4. $\{\neg, \longrightarrow\}$ is functionally complete.

Solution:

A set of connectives is said to be functionally complete if any formula can be written as an equivalent formula containing only these connectives. Obviously $\{\neg, \vee, \wedge\}$ is functionally complete.

Since $P \wedge Q = \neg(\neg P \vee \neg Q)$, hence $\{\neg, \vee\}$ is also functionally complete.

Since $P \vee Q \equiv \neg P \to Q$, wherever $\vee$ comes we can substitute $\neg$ and $\to$ for $\vee$, hence $\{\neg, \vee\} = \{\neg, \to\}$ in turn implies, $\{\neg, \to\}$ is functionally complete.

5. Show that R can be concluded from $P \vee Q$, $P \longrightarrow R$ and $Q \longrightarrow R$.

Solution:

Take $\neg R$ as a premise. Therefore

{1}	(1) $\neg R$	P
{2}	(2) $Q \to R$	P
{1,2}	(3) $\neg Q$	T since $Q \longrightarrow R$, $\neg R \to \neg Q$
{3}	(4) $P \to R$	P
{1,3}	(5) $\neg P$	T since $P \longrightarrow R$, $\neg R \to \neg P$
{1, 2, 3}	(6) $\neg P \wedge \neg Q$	T since P, $Q \to P \wedge Q$
{1, 2, 3}	(7) $\neg(P \vee Q)$	T
{4}	(8) $(P \vee Q)$	P
{1, 2, 3, 4}		(9) F

T since P, $\neg P \to F$.

Hence $\neg R$ can't be a premise, *i.e.* $\neg R \equiv F$ leads a contradiction. Hence $R \equiv T$ and hence R can be concluded.

6. Show that $(A \vee C)$ is not a valid conclusion of

$A \rightleftarrows (B \longrightarrow C)$, $(B \rightleftarrows (\neg A \vee \neg C))$, $C \rightleftarrows (A \vee \neg B)$ and B

Solution:

Consider $A \equiv C \equiv F$ and $B \equiv T$. Therefore

$A \rightleftarrows (B \longrightarrow C) \equiv F \rightleftarrows (T \longrightarrow F) \equiv F \rightleftarrows F \equiv T$,

$B \rightleftarrows (\neg A \vee \neg C) \equiv T \rightleftarrows (T \vee T) \equiv T \rightleftarrows T \equiv T$,

$C \rightleftarrows (A \vee \neg B) \equiv F \rightleftarrows (F \vee F) \equiv F \rightleftarrows F \equiv T$,

$B \equiv T$.

But $A \vee C \equiv F \vee F \equiv F$

Hence above premises can't imply $A \vee C$.

(If $P \equiv T$ and $Q \equiv F$, then Q can't be a valid conclusion of P or P can't imply Q).

7. If $H_1, H_2, \ldots, H_n$ and P imply Q, then $H_1, H_2, \ldots, H_n$ imply $P \longrightarrow Q$.

Solution:

Given $(H_1 \wedge H_2 \wedge \ldots \wedge H_n \wedge P) \to Q$

To prove that $(H_1 \wedge H_2 \wedge \ldots \wedge H_n) \to P \to Q$.

i.e. to prove that if $H_1 \wedge H_2 \wedge \ldots \wedge H_n \equiv T$

then $P \to Q \equiv T$

Let $(H_1 \wedge H_2 \wedge \ldots \wedge H_n \equiv T)$ and let $(P \to Q \equiv F)$

$\Rightarrow (H_1 \equiv H_2 \equiv \ldots \equiv H_n \equiv T)$ and $(P \equiv T$ and $Q \equiv F)$

$\Rightarrow (H_1 \equiv H_2 \equiv \ldots \equiv H_n \equiv P \equiv T)$ and $(Q \equiv F)$

$\Rightarrow\ (H_1 \wedge H_2 \wedge \ldots \wedge H_n \wedge P) \rightarrow Q$, a contradiction to (1)

Hence the statement (2) is false.

i.e. $(H_1 \wedge H_2 \wedge \ldots \wedge H_n) \rightarrow (P \rightarrow Q)$.

8. Show that $P \rightarrow (Q \rightarrow R) \Rightarrow (P \rightarrow Q) \rightarrow (P \rightarrow R)$.

Solution:

As $P \rightarrow Q$ is false only when P = T and Q = F. Therefore if we want to show $P \rightarrow Q$ it is enough to show that $Q \equiv F$ implies $P \equiv F$.

Consider RHS

$F \Rightarrow (P \rightarrow Q) \rightarrow (P \rightarrow R) \equiv F$

$\Rightarrow (P \rightarrow Q) \equiv T$ and $(P \rightarrow R) \equiv F$

$\Rightarrow (P \rightarrow Q) \equiv T$ and $(P \equiv T$ and $R \equiv F)$

$\Rightarrow\ T \rightarrow Q \equiv T,\ P \equiv T$ and $R \equiv F$

$\Rightarrow\ Q \equiv T,\ P \equiv T$ and $R \equiv F$

Consider LHSP $\rightarrow (Q \rightarrow R) \equiv T \rightarrow (T \rightarrow F)$ by (1)

$\equiv T \rightarrow F \equiv F$.

Thus when RHS is F, then LHS is also F.

Hence **LHS ≡ RHS.**

9. Prove substitution instance of tautology is a tautology.

Solution:

The basic structure of a tautology is not going to change by substitution of a formula.

e.g. $Q \vee \neg Q$ is going to become

$(\neg P \longrightarrow Q) \vee \neg(\neg P \longrightarrow Q)$

10. Show that truth value of the following formula is independent of its components.

$$(P \longrightarrow Q) \rightleftarrows (\neg P \vee Q)$$

Solution:

Method 1: Let us construct the truth table.

P	Q	(P ⟶ Q)	¬P	(¬P ∨ Q)	(P ⟶ Q) ⇄ (¬P ∨ Q)
T	T	T	F	T	T
T	F	F	F	F	T
F	T	T	T	T	T
F	F	T	T	T	T

Hence $(P \longrightarrow Q) \rightleftarrows (\neg P \vee Q)$ is a tautology and independent of components.

Method 2 :

Consider $(P \longrightarrow Q) \rightleftarrows (\neg P \vee Q)$

$\equiv (P \longrightarrow Q) \rightleftarrows (P \longrightarrow Q)$

$\equiv T$ as $P \rightleftarrows P$ is always true.

Hence truth value of above formula is independent of its components.

11. Find equivalent formula consisting of $\wedge, \vee, \neg$ for $\bar{\vee}$.

Solution:

The connective exclusive OR, $\bar{\vee}$ is given by the truth table.

P	Q	P $\bar{\vee}$ Q	P ∧ ¬Q	¬P ∧ Q	(P ∧ ¬Q) ∨ (¬P ∧ Q)
T	T	F	F	F	F
T	F	T	T	F	T
F	T	T	F	T	T
F	F	F	F	F	F

$\Rightarrow\ P \,\bar{\vee}\, Q = (P \wedge \neg Q) \vee (\neg P \wedge Q)$.

$= x\bar{y} + \bar{x}y$.

Note. $P \rightleftarrows Q$ if and only if $P \rightleftarrows Q \equiv T$. (That is P and Q have the same truth value.)

12. $\{\wedge, \vee\}$ is not functionally complete.

Solution:

Consider $(P \vee \neg P = T)$. This formula can't be written as combination of $\vee$ and $\wedge$ only.

13. $\{\uparrow\}$ is functionally complete.

Solution:

We know $P \uparrow P \equiv \neg(P \wedge P) \equiv \neg P$

$(P \uparrow Q) \uparrow (P \uparrow Q) \equiv \neg(\neg(P \wedge Q) \wedge \neg(P \wedge Q))$

$\equiv \neg(\neg(P \wedge Q)) = P \wedge Q$

and $(P \uparrow P) \uparrow (Q \uparrow Q) \equiv \neg(\neg(P \wedge \neg Q) = P \vee Q$.

Hence any formula consisting $\wedge$, $\vee$, and $\neg$ can be written as a formula consisting of $\uparrow$ only.

14. Express $P \longrightarrow (\rceil P \longrightarrow Q)$ in terms of $\uparrow$ only.

Solution:

Consider

$$\begin{aligned} P \longrightarrow (\rceil P \longrightarrow Q) &\equiv \rceil P \vee (\rceil(\rceil P) \vee Q)) \\ &\equiv \rceil P \vee (P \vee Q) \\ &\equiv \rceil P \vee P \vee Q \\ &\equiv T \vee Q \\ &\equiv T \end{aligned}$$

15. Express $P \uparrow Q$ in terms of $\downarrow$ only.

Solution: $$\begin{aligned} P \uparrow Q &= \rceil(P \wedge Q) \\ &= (P \wedge Q) \downarrow (P \wedge Q) \\ &= ((P \downarrow P) \downarrow (Q \downarrow Q)) \downarrow ((P \downarrow P) \downarrow (Q \downarrow Q)) \end{aligned}$$

16. Find simplest form of $((P \to Q) \rightleftarrows (\rceil Q \to \rceil P)) \wedge R$

Solution:

Expression is $((P \to Q) \rightleftarrows (P \to Q)) \wedge R \equiv T \wedge R \equiv F.$

17. Obtain principal conjunctive normal form and principal disjunctive normal form of

$$P \vee (\rceil P \to (Q \vee (\rceil Q \to R)))$$

Solution:

Consider $A \equiv P \vee (\rceil P \longrightarrow (Q \vee (\rceil Q \to R))$

$\equiv P \vee (P \vee (Q \vee (Q \vee R))$ as $P \to Q \equiv \rceil P \vee Q$

$\equiv P \vee P \vee Q \vee Q \vee R$

$A \equiv (P \vee Q \vee R)$ = PCNF

$\Rightarrow$ $\bar{A}$ $\equiv$ product of remaining 7 max terms

$= (P \vee Q \vee \bar{R}) \wedge (P \vee \bar{Q} \vee R) \wedge (P \vee \bar{Q} \vee \bar{R})$

$\wedge (\bar{P} \vee Q \vee R) \wedge (\bar{P} \vee Q \vee \bar{R})$

$\wedge (\bar{P} \vee \bar{Q} \vee R) \wedge (\bar{P} \vee \bar{Q} \vee \bar{R})$

$\Rightarrow A = \bar{\bar{A}} \equiv (\bar{P} \wedge \bar{Q} \wedge R) \vee (\bar{P} \wedge Q \wedge \bar{R})$

$\vee (\bar{P} \wedge Q \wedge R) \vee (P \wedge \bar{Q} \wedge \bar{R})$

$\vee (\bar{P} \vee \bar{Q} \wedge R) \vee (P \wedge Q \wedge \bar{R})$

$\vee (P \wedge Q \wedge R)$

$\equiv$ PDNF

18. Convert the following into infix notation.

$$P \rceil P \longrightarrow P \longrightarrow P \longrightarrow.$$

Solution:

This is in postfix notation. Hence let us go from left to right.

$$\begin{aligned} A &\equiv P \rceil P \longrightarrow P \longrightarrow P \longrightarrow \\ &\equiv (\rceil P)\, P \longrightarrow P \longrightarrow P \longrightarrow \\ &\equiv (\rceil P \longrightarrow P)\, P \longrightarrow P \longrightarrow \\ &\equiv ((\rceil P \longrightarrow P) \longrightarrow P)\, P \longrightarrow \\ &\equiv ((\rceil P \longrightarrow P) \longrightarrow P) \longrightarrow P) \end{aligned}$$

19. Show that C follows from H_1 and H_2.

$H_1 : P \longrightarrow (Q \longrightarrow R)$ $H_2 : P \wedge Q$ $C : R$

Solution:

We have to show $(H_1 \wedge H_2) \Rightarrow C$,

i.e. if $(H_1 \wedge H_2) \equiv T$, then $C \equiv T$.

Let $H_1 \equiv T$ and $H_2 \equiv T$.

$\Rightarrow P \longrightarrow (Q \longrightarrow R) \equiv T$ and $P \wedge Q \equiv T$

$\Rightarrow P \longrightarrow (Q \longrightarrow R) \equiv T$ and $P \equiv T$ and $Q \equiv T$

$\Rightarrow T \longrightarrow (T \longrightarrow R) \equiv T$

$\Rightarrow$ $T \longrightarrow R \equiv T \Rightarrow R \equiv T.$

20. Without constructing a truth table, show that $A \wedge E$ is not a valid consequence of

$H_1 : A \rightleftarrows B$, $H_2 : B \rightleftarrows (C \wedge D)$,

$H_3 : C \rightleftarrows (A \vee E)$, $H_4 : A \vee E$

Solution:

If we want to prove Q is not valid consequence of P, then it is sufficient to prove $P \equiv T$ and $Q \equiv F$ (hence $P \longrightarrow Q \equiv F$).

For the combinations $A \equiv F$, $B \equiv F$, $C \equiv T$, $D \equiv F$, $E \equiv T$, we get

$$H_1 : A \rightleftarrows B \equiv F \rightleftarrows F \equiv T,$$

$$H_2 : B \rightleftarrows (C \wedge D) \equiv F \rightleftarrows (T \wedge F) \equiv T.$$

$$H_3 : C \rightleftarrows (A \vee E) \equiv T \rightleftarrows (F \vee T) \equiv T$$

and $H_4 : A \vee E \equiv F \vee T \equiv T.$

$A \wedge E \equiv F \wedge T \equiv F.$

Hence $(H_1 \wedge H_2 \wedge H_3 \wedge H_4) \equiv T$ and $(A \wedge E) \equiv F$.

$\Rightarrow$ $(A \wedge E)$ is not valid conclusion of H_1, H_2, H_3 and H_4.

21. Show that following are inconsistent

$$P \longrightarrow Q,\ P \longrightarrow R,\ Q \longrightarrow \rceil R \text{ and } P$$

Solution:

Let the consistent,

$P \longrightarrow Q \equiv T$	$P \longrightarrow R \equiv T$	$Q \longrightarrow \rceil R \equiv T$	$P \equiv T$
(1) $Q \equiv T$	—	—	—
—	—	(2) $\rceil R \equiv T$	—
—	—	(3) $R \equiv F$	—
—	(4) $P \equiv F$	—	—

Hence we get $P \equiv F \equiv T$, a contradiction.

22. Show that the formula $[(\sim p \vee q) \Rightarrow (q \Rightarrow p)]$ is not a tautology.

Solution:

Boolean expression can be simplified as below :

$[(\sim p \vee q) \Rightarrow (q \Rightarrow p)]$

i.e. $[(\sim p \vee q) \Rightarrow (\sim q \vee p)]$

i.e. $[\sim(\sim p \vee q) \vee (\sim q \vee p)]$

i.e. $[(p \wedge \sim q) \vee \sim q \vee p]$

i.e. $[\sim p \vee q]$

Above expression is false for $p = 0$, $q = 1$.

So, it is not a tautology.

23. Let A be a tautology and B be any other formula. Prove that $(A \vee B)$ is a tautology.

Solution:

Since A is a tautology, it is true for all possible inputs. So, $A \vee B$ will also be true for all possible inputs, regardless of whether B is true or not.

24. Show that proposition C is a logical consequence of the formula using truth tables.

$A \wedge (A \to (B \vee C)) \wedge (B \to \sim A)$

Solution:

If we want to show a formula Y is logical consequence of another formula X, we have to prove $X \to Y$. i.e we have to prove $X \to Y$ is a tautology. Hence $X \to Y \equiv T$ implies Y is logical consequence of X.

Let $X = A \wedge (A \to (B \vee C)) \wedge (B \to \sim A)$

Let us show $X \to C \equiv T$.

Consider following truth table.

A	B	C	$B \vee C$	C_1	$\sim A$	C_2	X	$X \to C$
				$A \to (B \vee C)$		$B \to \sim A$	$A \wedge C_1 \wedge C_2$	
T	T	T	T	T	F	F	F	T
T	T	F	T	T	F	F	F	T
T	F	T	T	T	F	T	T	T
T	F	F	F	F	F	T	F	T
F	T	T	T	T	T	T	F	T
F	T	F	T	T	T	T	F	T
F	F	T	T	T	T	T	F	T
F	F	F	F	T	T	T	F	T

In the above truth table, the last column shows that $X \to C \equiv T$. i.e. $X \Rightarrow C$.

Hence C is a logical consequence of $X \equiv A \wedge (A \to (B \vee C) \wedge (B \to \sim A))$.

SET THEORY

SET

S*et is any collection of distinct and distinguishable objects around us.*

By the form *distinct*, we mean that no object is repeated and by the term *distinguishable*, we mean that whether that object is in our collection or not. The objects belonging to a set are called *elements* or *members* of that set.

e.g. A is a set of stationary used by an student, i.e.

A = {Pen, Pencil, Eraser, Paper}

B is a set of Capitals of State, i.e. B = {Lucknow, Patna, Bhopal}

A set is represented by using all its elements between braces { } and by separating them from each other by commas (if there are more than one element).

Sets are denoted by capital letters of English alphabet while the elements are divided in general, but small letters.

If x is an element of a set A, we write $x \in A$ (read as *x belongs to A*).

If x is not an element of A, we write $x \notin A$ (read as x *does not belong to A*).

e.g. Let A = {4, 2, 8, 2, 6}. Elements of this collection are disinguishable but not distinct.

Hence A is not a set.

Let B = {*a, e, i, o, u*} is set of vowels in English.

Here elements of B are distinguishable as well as distinct. Hence B is a set.

FORMS OF REPRESENTATION OF A SET

(1) Set-builders form.

In this representation of set, we write between the braces { } a variable x which stands for each of the elements of the set, then we state the properties possessed by x.

We denote this property by $p(x)$ and separate x and $p(x)$ by a symbol : or (read as *such that*).

$A = \{x : p(x)\}$

$A = \{x : x$ is capital of States$\}$

or $A = \{x : x$ is a natural number and $2 < x < 11\}$

(2) Tabular form or Roaster form.

In this representation of set, elements of a set are listed one by one within braces { } and one separaed by each other by commas.

B = {Lucknow, Patna, Bhopal}

or B = (3, 4, 5, 6, 7, 8, 9, 10}

TYPES OF SETS

(1) Null set (or Empty set or Void set).

A set having no elements is called *empty set or void set.* It is denoted by ϕ or{ }

e.g. $A = \{x : x$ is an even number not divisible by $2\}$

B = {the set of immortal men}

(2) Singleton set.

A set having single element is called *singleton set.*

e.g. $A + \{x : x$ is present Prime Minister of India$\}$

$B = \{2\}$

(3) Pair set.

A set having two elements is called *pair set.*

e.g. {1, 2}, {0, 3}, {4, 9} etc.

(4) Finite set.

A set having finite number of elements, i.e. a set where counting of elements is possible is called *finite set.*

e.g. A = {1, 2, 4, 6} is a finite set because it has four elements.

B = a null set ϕ is also a finite set because it has zero number of elements.

(5) Infinite set.

A set having infinite number of elements, i.e. a set where counting of elements is impossile, is called *infinite set.*

e.g. A = $\{x : x$ is a set of all natural numbers$\}$

B = set of all points in a given plane.

SUBSETS

Set A is called subset of a set B if each element of set A is also an element of set B.

If A is a subset of set B, we represent it as $A \subseteq B$.

So, if $A \subseteq B \Leftrightarrow [x \in A \Rightarrow x \in B]$

e.g. Let A = {4, 5, 6} and B = {4, 5, 7, 8, 6}, then we write $A \subseteq B$.

Proper subset.

Set A is said to be a proper subset of a set B if

(*a*) every element of set A is an element of set B, and

(*b*) set B has at least one element which is not an element of set A.

This is expressed by writing $A \subset B$ or $B \subset A$ and read as A is a proper subset of B or B is a proper subset of A. If A is not a proper subset of B then we write it as $A \not\subset B$.

e.g. Let A = {4, 5, 6} and B = { 4, 5, 6, 7, 8}

Let A = {1, 2, 3} and B = {3, 2, 9}

$\therefore \quad A \not\subset B$ and $B \not\supset A$

SUPERSET

Let $\quad A = \{1, 2, 3\}, B = \{2, 3, 1\}$

$\Rightarrow \quad A \subseteq B$ Also $B \subseteq A$

Here $A \subseteq B$ can also be expressed equivalently by writing $B \supseteq A$, read as B is *superset of another set B*, if and only if set B contains all the elements of set A.

COMPARABILITY OF SETS

Two sets A and B are called *comparable* if either of these happens

(*i*) $A \subset B$

(*ii*) $B \subset A$

(*iii*) $A = B$

Similarly if neither of these above three exists, i.e. $A \not\subset B$, $B \not\supset A$ and $A \neq B$, then let A and B are called *incomparable.*

e.g. set A = {1, 2, 3} and B = {1, 2} are comparable.

But A = {1, 2, 3} and B = {2, 3, 6, 7} are incomparable.

UNIVERSAL SET

Any set which is superset of all the sets under consideration is called *universal set* and is either denoted by Ω or S or U.

e.g. Let A = {1, 2, 3}, B = {3, 4, 6, 9} and C = {0, 1}

We can take, S = {0, 1, 2, 3, 4, 5, 6, 7, 8, 9} as universal set.

POWER SET

The set or family of all the subsets of a given set A is said to be power set of A and is expressed by P(A).

Mathematically $\quad P(A) = \{x : x \subseteq A\}$

So, $\quad x \in P(A)$

$\Rightarrow \quad x \subseteq A$

Note :

(*i*) $\phi \in P(A)$ and $A \in P(A)$ for all sets A.

(*ii*) Elements of P(A) are subsets of A.

e.g. If A = {1}, then $\quad P(A) = \{\phi, \{1\}\}$

If A {1, 2}, then $\quad P(A) = \{\phi, \{1\}, \{2\}, \{1, 2\}\}$

Similarly if A = {1, 2, 3}, then

$P(A) = \{\phi, \{1\}, \{2\}, \{3\}, \{1, 2\}, \{1,3\}, \{2,3\}, \{1, 2, 3\}\}$

So trends show that if A has n elements, then P(A) has 2^n elements. It is not only true for $n = 1, 2, 3$, but it is true of all n.

OPERATIONS ON SETS

(1) Union of sets.

The union of two sets A and B is the set of all those elements which are either in A or in B or in both. This set is denoted by $A \cup B$ and read as 'A union B'.

Symbolically, $\quad A \cup B = \{x : x \in A$ or $x \in B\}$

e.g. If A = {4, 5, 6}, and B = {2, 1, 3, 8},

then $A \cup B = \{1, 2, 3, 4, 5, 6, 8\}$

(2) Intersection of sets.

Intersection of two sets A and B is the set of all the elements, which are common in A and B. This set is denoted by $A \cap B$ and read as '*A intersection of B*'.

Symbolically, $A \cap B = \{x : x \in A$ and $x \in B\}$

or $\quad A \cap B = \{x : x \in A \cap x \in B\}$

e.g. If A = {1, 2, 3}, and B = {2, 1, 5, 6}, then $A \cap B = \{1, 2\}$

(3) Difference of sets.

The difference of two sets A and B is the set of all those elements of A which are not elements of B. Sometimes, we call difference of sets as the relative components of B in A. It is denoted by $A \cap B$.

Symbolically, $\quad A - B = \{x : x \in A$ and $x \notin B\}$

Similarly $\quad B - A = \{x : x \in B$ and $x \notin A\}$

e.g. If A = {4, 5, 6, 7, 8, 9) and B = {3, 5, 2, 7}, then $A - B = \{4, 6, 8, 9\}$

Similarly, $B - A = \{3, 2\}$.

Here, since $A - B \neq B - A$, so difference of two sets is not commutative.

(4) Disjoint sets.

Two sets A and B are called disjoint sets, if they have no element in common

Symbolically, $A \cap B = \phi$

(5) Symmetric difference of sets.

It is defined as set of elements that are member of either A or B but not both

Symbolically, $A \oplus B = (A \cup B) - (A \cap B)$

$= (A - B) \cup (B - A)$

(6) Complements of set.

Complement of a set A, also known as *absolute complement' of A* is the sets of all those elements of the universal sets which are not element of A. It is denoted by A° or A^{-1} or A' or A^c.

Symbolically, A^{-1} or $A^c = S - A$.

or $A^{-1} = \{x : x \in S \text{ and } x \in A\}$.

LAWS AND THEOREMS

(1) $A \cup A = A, \quad A \cap A = A$

(2) $A \cup S = S, \quad A \cap S = A$

(3) $A \cup \phi = A, \quad A \cap \phi = \phi$

(4) $A - A = \phi, \quad \phi - A = \phi, \quad A - \phi = A.$

Theorem 1.

For any two sets A and B

(*a*) $A \cup B = B \cup A$ (*Commutative law of union*)

(*b*) $A \cap B = B \cap A$ (*Commutative law of intersection*)

Also

(*i*) $A \cup (B \cup C) = (A \cup B) \cup C$

(Associative law for union)

(*ii*) $A \cap (B \cap C) = (A \cap B) \cap C.$

(Associative law for intersection)

Theorem 2.

For any three sets A, B and C, following *disributive laws hold :*

(*i*) $A \cup (B \cap C) = (A \cup B) \cap (A \cup C)$

(*ii*) $A \cap (B \cup C) = (A \cap B) (A \cap C).$

DeMorgan's law.

For any two sets A and B

(*i*) $(A \cup B)' = A' \cap B'$

(*ii*) $(A \cap B)' = A' \cup B'$

For any three sets A, B and C, Demorgan's law can also be expressed as

(*i*) $A - (B \cup C) = (A - B) \cap (A - C)$

(*ii*) $A - (B \cap C) = (A - B) \cup (A - C)$

VENN DIAGRAM

It is a pictorial representation of sets.

A set is represented by circles or a closed geometrical figure inside the universal set. The universal set S is represented by a rectangular region.

Firstly we represent the set or a statement regarding sets with the help of diagram or venn diagram. The shaded area represents the set written.

1. Subset.

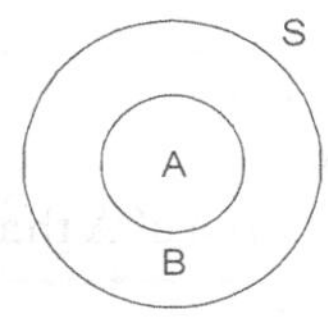

$A \subset B.$

2. Union of sets.

Let $A \cup B = B$. Here, whole area represented by B represents $A \cup B$.

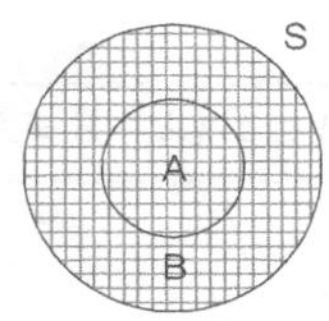

$A \cup B$ when $A \subset B$

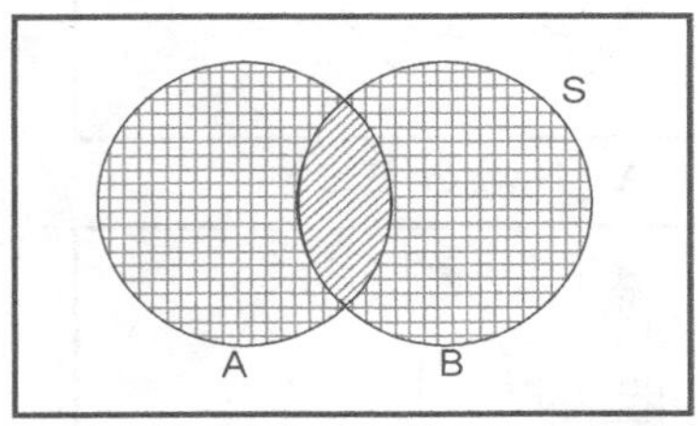

$A \cup B$ when neither $A \subset B$ nor $B \subset A$

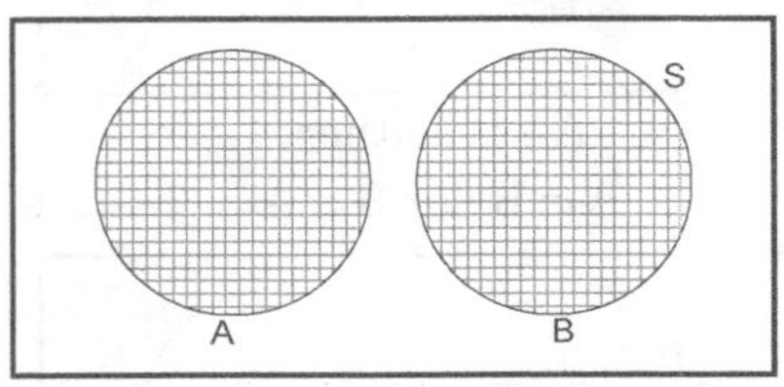

$A \cup B$ when A and B are disjoint events

3. Intersection of sets.

$A \cap B$ represents the common area of A and B.

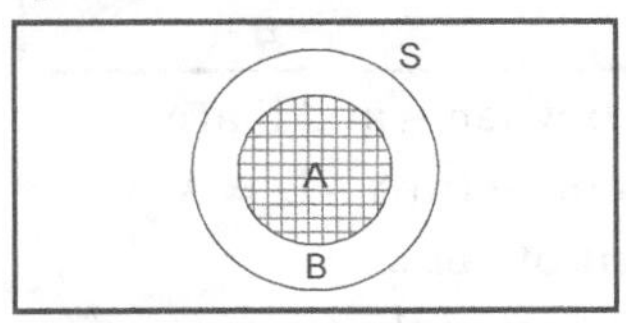

$A \cap B$ when $A \subset B$ $(A \cap B = A)$

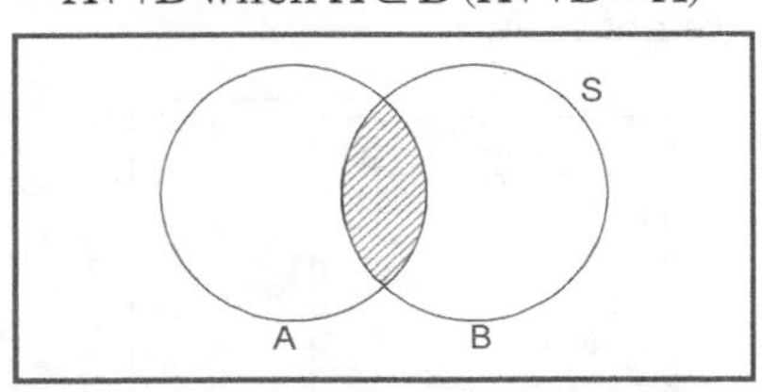

$A \cap B$ when neither $A \subset B$ or $B \subset A$

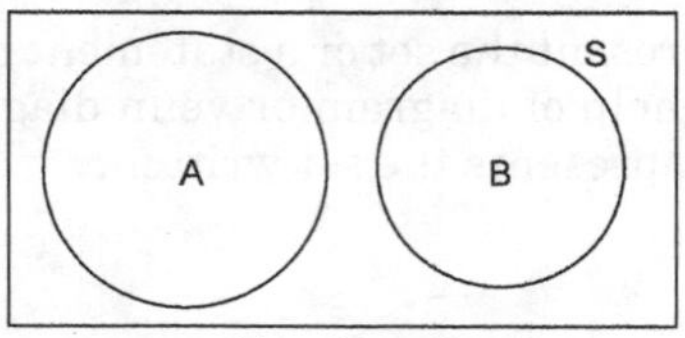

$(A \cap B = \phi)$

4. Difference of sets.

A – B represents the area of A that is not in B.

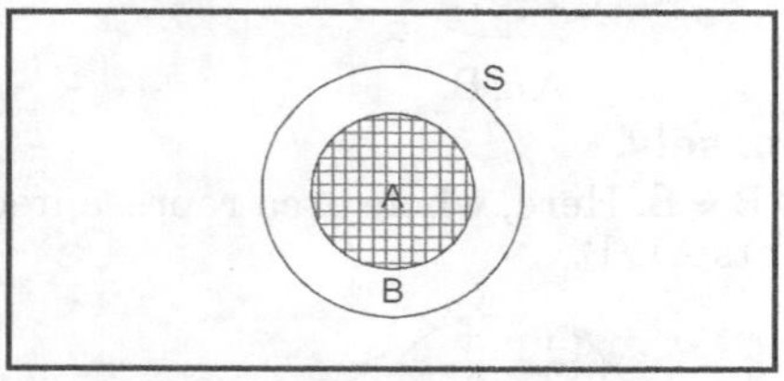

A – B when $A \subset B$ $(A - B = A)$

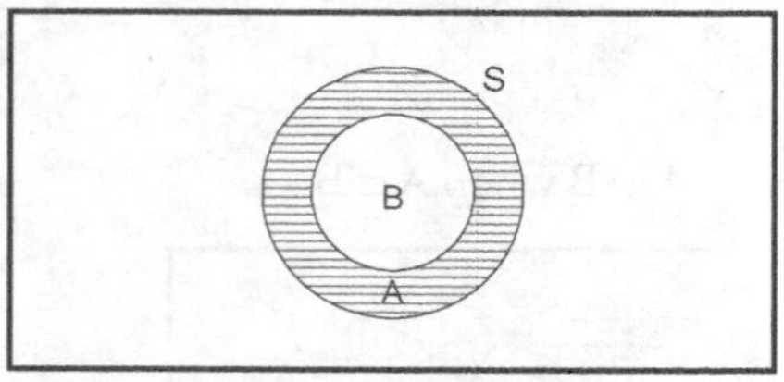

A – B when $B \subset A$

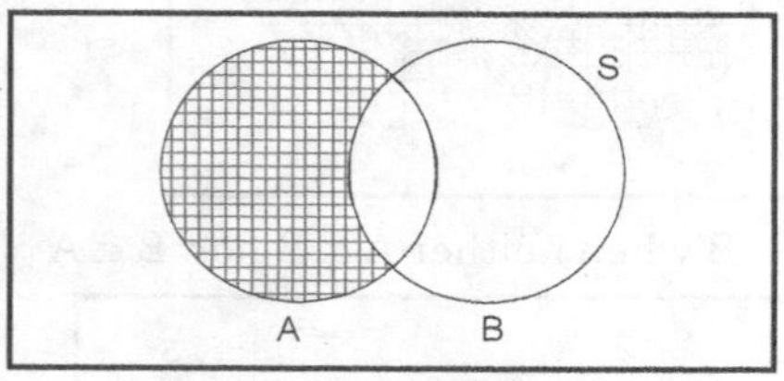

A – B neither

$A \subset B$ nor $B \subset A$

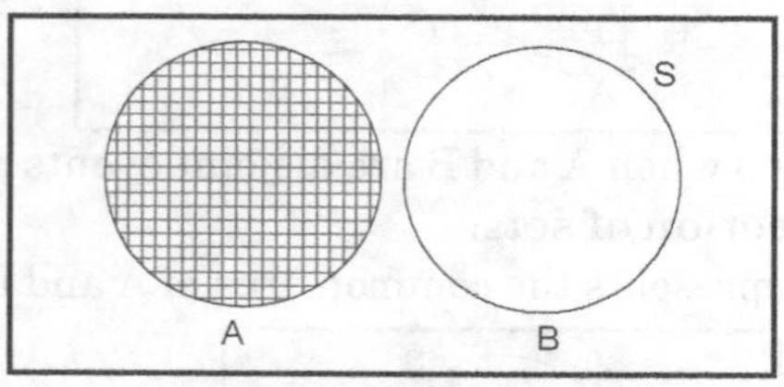

A – B when A and B are disjoint sets $(A - B = A)$

5. Complement of sets.

A' or A^c is the set of those points of universal set S which are not in A.

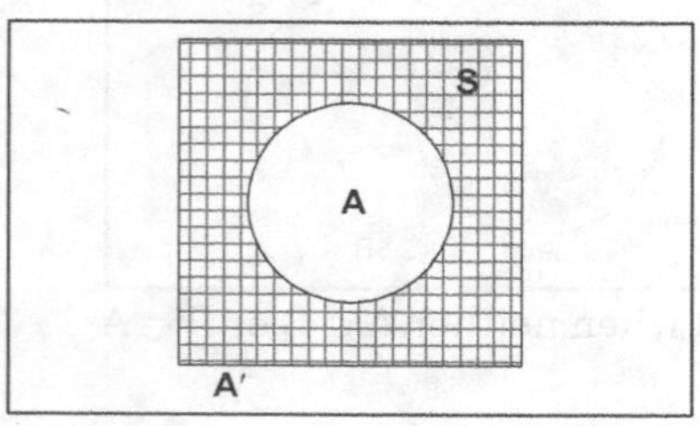

APPLICATION OF SET THEORY

The number of distinct elements of a finite set A is denoted by n (A).

Let $n(A)$, $n(B)$ and $N(A \cap B)$, where A and B are non-empty sets.

Then $n(A \cup B) = n(A) + n(B) - n(A \cap B)$

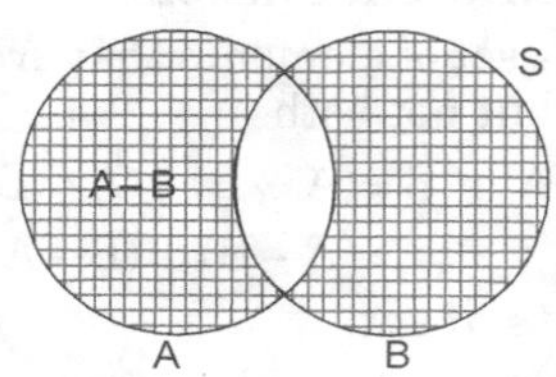

In case A and B are disjonit sets, so we have

$$A \cap B = \phi$$

$\therefore \quad n(A \cup B) = n(A) + n(B)$

From above Venn-diagram, following results are obtained :

(*i*) $n(A) = n(A - B) + n(A \cap B)$

(*ii*) $n(B) = n(B - A) + n(A \cap B)$

(*iii*) $n(A \cup B) = n(A - B) + n(B - A) + n(A \cap B)$

The result, $n(A \cap B) = n(A) + n(B) - n(A \cap B)$ can be generalised as,

$n(A \cup B \cup C) = n(A) + n(B) + n(C) - n(A \cap B) - n(B \cap C) - n(A \cap C) + n(A \cap B \cap C)$

Example. If 63% of persons like oranges while 76% like apples, what can be said about the percentage of persons who like both oranges and apples ?

Solution:

Let $\quad n(S)$ = total number of persons = 100

A = {x : x likes oranges}

B = {x : x likes apples}

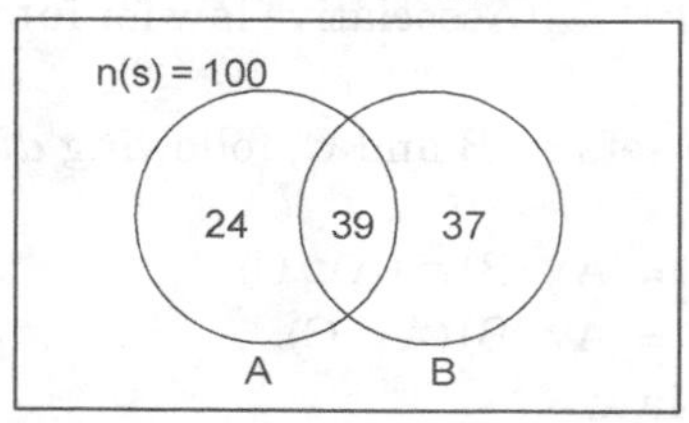

$n(A) = 63$

$n(B) = 76$

$A \cap B$ = {x : x likes oranges and apples both}

Now, $n(A \cup B) = n(A) + n(B) - n(A \cap B)$

$n(A \cap B) = n(A) + n(B) - n(A \cup B)$

$= 63 + 76 - 100 = 39$

$\therefore \quad n(A \cap B) = 39$

Hence, 39% people like both oranges and apples.

Example. In a group of 1000 people, there are 750, who can speak Hindi and 400, who can speak Bengali. How many can speak Hindi only ? How many can speak Bengali only ? How many can speak both ?

Solution:

Let $A = \{x : x \text{ speaks Hindi}\}$

and $B = \{x : x \text{ speaks Bengali}\}$

$\therefore$ $A - B = \{x : x \text{ speaks Hindi and can not speak Bengali}\}$

$B - A = \{x : x \text{ speaks Bengali and can not speak Hindi}\}$

$A \cap B = \{x : x \text{ speaks Hindi and Bengali both}\}$

Given : $n(A) = 750$, $n(B) = 400$

$\therefore$ $n(A \cup B) = 1000$

Now, $n(A \cup B) = n(A) + n(B) - n(A \cap B)$

$\therefore$ $n(A \cap B) = n(A) + n(B) - n(A \cup B)$

$= 750 + 400 - 1000 = 150$

So, 150 people are speaking Hindi and Bengali both.

Again, $n(A) = n(A - B) + n(A \cap B)$

$\therefore$ $n(A - B) = n(A) - n(A \cap B)$

$= 750 - 150 = 600$

Hence, 600 people are speaking Hindi only

Finally, $n(S - A) = n(B) - n(A \cap B)$

$= 400 - 150 = \mathbf{250}$

So, 250 people are speaking Bengali only.

CARTESIAN PRODUCT OF SETS.

Ordered pair.

It is usually denoted by (x, y) is a pair of elements x and y of some sets, which is ordered in the sense that $(x, y) \neq (y, x)$ whenever $x \neq y$

Here x is called *first co-ordinate* and y is called *second co-ordinate* of the ordered pair (x, y).

e.g. ordered pairs (1, 2) and (2, 1) though consist of same elements 1 and 2, are different because they represent different points in the co-ordinate plane.

Cartesian product.

Cartesion product of two sets A and B is the set of all those pairs whose first co-ordinate is an element of A and second co-odinate is an element B.

The set is denoted by $A \times B$ and is read as '*A cross B*' or '*product set of A and B*'

i.e. $A \times B = \{(x, y) : x \in A \wedge y \in B\}$

e.g. Let $A = \{1, 2, 3\}$, and $B = \{3, 5\}$

$\therefore$ $A \times B = \{1, 2, 3\} \times \{3, 5\}$

$= \{(1, 3), (1, 5), (2, 3), (2, 5), (3, 3), (3, 5)\}$

and, $B \times A = \{(3, 1), (3, 2), (3, 3), (5, 1), (5, 2), (5, 3)\}$

$\therefore$ $A \times B \neq B \times A$

Similarly, we can define Cartesian product for n sets $A_1, A_2, \ldots A_n$ as

$$A_1 \times A_2 \times A_3 \times \ldots \times A_n = \{(x_1, x_2, x_3, \ldots x_n) \; x_1 \in A \wedge x_2 \in A_2 \wedge x_3 \in A_3 \ldots \wedge x_n \in A_n\}$$

The element $(x_1, x_2, \ldots x_n)$ is called as n-type of $x_1, x_2 \ldots x_n$.

RELATIONS

Consider a set, $A = \{1, 2, 3\}$

Cartesian product set is,

$$A \times A = \{(1, 1), (1, 2), (1, 3), (2, 1), (2, 2), (2, 3), (3, 1), (3, 2), (3, 3)\}$$

Consider a subset from this set in such a way that first coordinate is '*less than*' the second coordinate.

If this relation is denoted by R, then R is described by

$$R = \{(x, y) \in A \times A : x < y\}$$

in which (x, y) is a member of the relation R.

Relation in a set.

A relation between two sets A and B is a subset of $A \times B$ and is denoted by R.

Thus $R \subseteq A \times B$

We write xRy, if and only if $(x, y) \in R$. xRy is read as '*x is related to y*'

If A = B, then we say that *R is a relation in the set A*.

e.g. Consider the relation

$R = \{(x, y) \in N \times N : x > y$, N is a set of natural numbers.$\}$

Obviously, $R \subseteq N \times N$ and so R is relation N.

Binary relation in a set.

A binary relation R is called *defined in set* A if for any ordered pair $(x, y) \in A \times A$, it is meaningful to say that xRy is true or false.

In other words, $R = \{(x, y) \in A \times A : xRy \text{ is true}\}$.

i.e., a relation R in a set A is a subset of $A \times A$. Binary is two. So binary relation is a relation between two sets, which may be different or identical. For the sake of convenience a binary relation will be written as only as relation.

Domain and Range of relation.

Domain D of the relation R is defined as the set of all first elements of the ordered pairs which belong to R

i.e. $D = \{x : (x, y) \in R, \text{ for } x \in A\}$

Range E of the relation R is defined as the set of all second elements of the ordered pairs which belong to R

i.e. $E = \{y : (x, y) \in R, \text{ for } y \in B\}$

Obviously, $D \subseteq A$ and $E \subseteq B$

e.g. Let $A = \{1, 2, 3, 4\}$ and $B = \{a, b, c\}$.

Every subset of $A \times B$ is a relation from A to B. So, if $R = \{(2, a), (4, a), (4, c)\}$, then domain of R is the set $\{2, 4\}$ and range of R is the set $\{a, c\}$.

Total number of distinct relations from set A to set B.

Let number of elements of A and B be m and n respectively.

Then number of elements of $A \times B$ is mn.

Therefore number of elements of the power set of $A \times B$ is 2^{mn}.

Thus, $A \times B$ has 2^{mn} different subsets.

Now every subset of $A \times B$ is a relation from A to B. Hence number of different relations A to B is 2^{mn}.

TYPES OF RELATIONS IN A SET

(1) Inverse relation.

Let R be a relation from set A to set B.

Then inverse relation R^{-1} from set B to set A is defined by

$$\{(b, a) : (a, b) \in R\}$$

In other words, inverse relation R^{-1} consists of

those ordered pairs which when reversed belong to R. Thus every relation R from set A to set B has an inverse relation R^{-1} from B to A.

e.g. Let A = {1, 2, 3}, B = {a, b} and R = {(1, a), (1, b), (3, a), (2, b)} be a relation from A to B.

The inverse relation R is

R^{-1} = {(a, 1), (b, 1), (a, 3), (b, 2)}

Let A = {2, 3, 4}, B = {2, 3, 4} and R = {(x, y) : $|x - y| = 1$} be a relation from A to B, i.e.

R = {(3, 2), (2, 3), (4, 3), (3, 4)}.

Inverse relation of R is,

R^{-1} = {(3, 2), (2, 3), (4, 3), (3, 4)}

It may be noted that $R = R^{-1}$.

Here every relation has an inverse relation.

If R be a relation from A to B, then $(R^{-1})^{-1} = R$.

Theorem.

If R be a relation from A toB, then domain of R is the range of R^{-1} and range of R is the domain of R^{-1}.

(2) Identity relation.

A relation R in a set A is said to be identity relation, generally denoted by I_A, if

I_A = {(x, x) : $x \in$ A}.

e.g. Let A = {2, 4, 6}, then

I_A = {(2, 2), (4, 4), (6, 6)} is an identity relation in A.

(3) Universal relation.

A relation R in a set A is said to be universal relation if R is equal to A × A i.e. if R = A × A.

e.g. Let A = {1, 2, 3}, then

R = A × A = {(1, 1), (1, 2), (1, 3), (2, 1), (2, 2), (2, 3), (3, 1), (3, 2), (3, 3)} is a universal relation in A.

(4) Void relation.

A relation R in a set A is said to be a void relation if R is a null set, i.e. if R = ϕ.

e.g. Let A = {2, 3, 7} and let R be defined as 'aRb if and only if a divides b', then

R = $\phi \subset$ A × A is a void relation.

PROPERTIES OF RELATIONS IN A SET

(1) Reflexive relations.

Let R be a relation in a set A. Then R is called a *reflexive relation,* if (a, a) $\in$ R, for all $a \in$ A.

In other words, R is reflexive if every element in A is related to itself. Thus, R is reflexive if aRa holds for all $a \in$A.

As relation R in a set A is not reflexive if there is atleast one element such that (a, a) $\notin$ R.

e.g. Let A = {1,2, 3, 4}. Then the relation

R_1 = {(1,1), (2, 4), (3, 3), (4, 1), (4, 4)} in A is not reflexive since 2 $\in$ A but (2, 2) $\notin R_1$.

(2) Anti-reflexive relations.

Let R be a relation in a set A. Then R is called an *anti-reflexive relation,* if (a, a) $\notin$ R for every $a \in$A, i.e. a/Ra for all $a \in$ A.

e.g. Relation in the set of natural numbers N defined by '$x < y$' is anti-reflexive, since a is not less than a for any natural number a.

Let A = {1, 2, 3, 4}. Then relation R = {(1, 1), (2, 3), (3, 4)} is not anti-reflexive since, (1, 1) $\in$ R.

(3) Symmetric relations.

Let R be a relation in a set A. Then R is said to be symmetric relation if (a, b) $\in$ R $\Rightarrow$ (b, a) $\in$ R.

Thus R is symmetric if bRa holds whenever aRb holds.

A relation R in set A is not symmetric if there exist two distinct elements a, b $\in$A, such that aRb, but b/Ra.

e.g. Let L be the set of all straight lines in a plane. The relation R in L defined by '*x is parallel to y*' is symmetric, since if a straight line a is parallel to a straight line b, then b is also parallel to a.

(a, b) $\in$ R $\Rightarrow$ (b, a) $\in$ R.

Let R be relation in the natural numbers N which is defined by '$x - y > 0$'. Then R is not symmetric since, 4 – 2 > 0 but 2 – 4 > 0.

Thus, (4, 2) $\in$ R but (2, 4) $\notin$R.

Remarks : Since (a, b) $\in$ R $\Rightarrow$ (b, a) $\in R^{-1}$, thus R is a symmetric relation if and only if $R = R^{-1}$.

(4) Anti-symmetric relations.

Let R be a relation in a set A, i.e.let R be a subset of A × A. Then R is said to be an anti-symmetric relation if (a, b) $\in$ R and (b, a) $\in$R $\Rightarrow a = b$.

Thus R is anti-symmetric if $a \neq b$, then possibly (a, b) $\in$ R or possibly (b, a) $\in$R, but never both.

A relation R in a set A is not anti-symmetric if there exist elements a, $b \in$ A, $a \neq b$ such that (a, b) $\in$R and (b, a) $\notin$R.

e.g. Let A be a family of sets and let R be the relation in A defined by 'x is a subset of y'. Then R is anti-symmetric since A $\subseteq$ B and B $\subseteq$ A $\Rightarrow$ A = B.

(5) Transitive relations.

Let R be a relation in a set, i.e. let R be a subset of A × A. Then R is called *transitive relation,* if

(a, b) $\in$ R and (b, c) $\in$ R $\Rightarrow$ (a, c) $\in$ R.

A relation R in a set A is *not transitive* if there exist elements a, b, $c \in$ A, not necessarily distinct, such that (a, b) $\in$ R, (b, c) $\in$R but (a, c) $\notin$ R.

e.g. Let L be the set of all straight lines in a plane and R be the relation in L defined by 'x is parallel to y'.

If a is parallel to b and b is parallel to c, then obviously a is parallel to c.

Thus (a, b) $\in$ R and (b, c) $\in$ R $\Rightarrow$ (b, c) $\in$ R. Hence R is transitive.

Let A be set of all Indians. Let R be the relation in A defined by '*x loves y*' if a loves b and b loves c, it does not necessarily follow that a loves c, so R is not a transitive relation.

(6) Partial order.

A relation R on a Set A is said to be partial order if

(*i*) R is reflexive, *i.e.* $(a, a) \in R\ \forall\ a \in A$

(*ii*) R is anti-symmetric, *i.e.* $\forall\ a, b \in A\ (a, b) \in R \wedge (b, a) \in R \Rightarrow a = b$

(*iii*) R is transitive, *i.e.* $\forall\ a, b, c \in A\ (a, b) \in R$ and $(b, c) \in R \Rightarrow (a, c) \in R$

(7) Equivalence relations.

Let R be a relation in a set A. Then R is an equivalence relation in A if and only if

(*i*) R is reflexive, i.e. for all $a \in R$, $(a, a) \in R$.

(*ii*) R is symmetric, i.e. $(a, b) \in R \Rightarrow (b, a) \in R$, for all $a, b \in A$.

(*iii*) R is transitive, i.e. $(a, b) \in R$ and $(b, c) \in R \Rightarrow (a, c) \in R$ for all $a, b, c \in A$.

e.g. Most trivial example of an equivalence relation is that of *equality* for any elements in any set

I. $a = a$, i.e. reflexive.

II. $a = b, \;\Rightarrow\; b = a$, i.e. symmetric

III. $a = b$ and $b = c \;\Rightarrow\; a = c$, i.e. transitive.

Let R be the relation in the real numbers defined by '$x \le y$'. Then

I. $x \le x$ i.e. $(x, x) \in R$, i.e., R is reflexive.

II. Let $x \le y$ but $y \not\le x$ i.e. $(x, y) \in R$ but not necessarily, $(y, x) \in R$, i.e. R is not symmetric. Thus R is not an equivalence relation.

Theorem.

If R be an equivalence relation in a set A, then R^{-1} is also an equivalence relation in A.

FUNCTION

Let A and B be two non-empty sets and *f* be a relation from A to B. If for each element $a \in A$, there exists a unique $b \in B$ (denoted by $f(a)$) such that $(a, b) \in f$, then *f* is called a *function (a mapping) from A to B.*

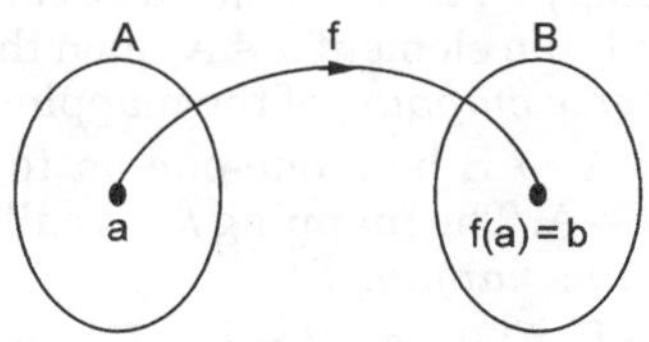

It is denoted by $f : A \to B$.

Take any element $a \in A$.

Since $f : A \to B$ is a function, there exists a unique element $b \in B$, such that $f(a) = b$.

It may be written as follows.

$$a \in A.\ f : A \to B \Rightarrow f(a) = b \in B, \text{ or}$$

$$f : A \to B,\ a \in A \Rightarrow \exists \text{ unique } (a, b) \in f.$$

Note:

For all elements of A, images should exist in B.

All elements in B may or may not have pre-images in A

There may be one or more pre-images in A for elements in B.

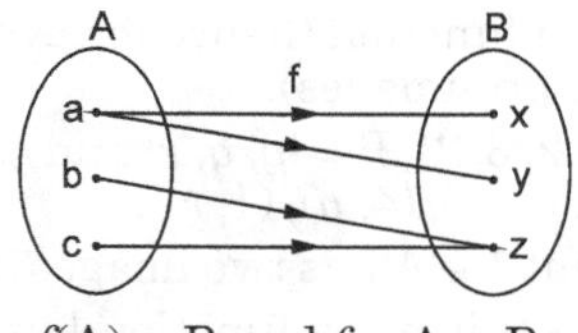

$f(A) \subseteq B$ and $f \subseteq A \times B$.

Every function from A to B is a relation from A to B but every relation from A to B need not be a function from A to B.

If an element in A has more than one image in B, *f* is not a function.

If $n(A) = m$, $n(B) = k$, then number of functions that can be defined from A to B is k^m.

IMAGE AND PRE-IMAGE

If $f : A \to B$ is a function and $(a, b) \in f$, then b is called *f-image of a.*

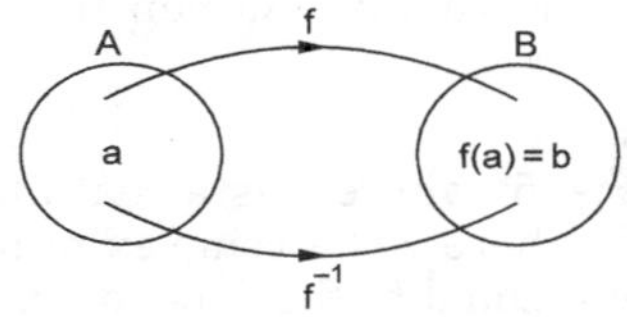

It is written as $f(a) = b$.

'a' is called a *pre-image of* 'b'.

We write this as

$$a \in f^{-1}(b) : f^{-1}(b) = \{x \in A \mid f(x) = b\}.$$

Every element of A has one and only one image in B. Every element of B may or may not have a pre-image in A. But one or more elements of A, may have the same image in B.

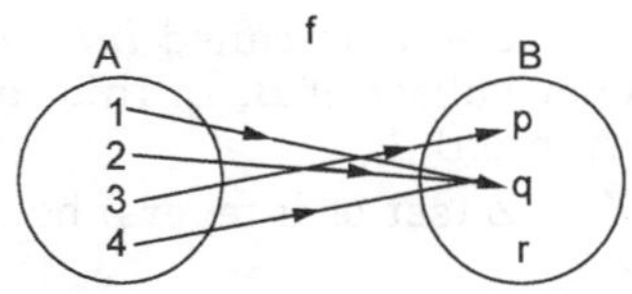

In above figure, there are more than one pre-images for $q \in B$.

In fact, $\quad f^{-1}(q) = \{a \in A \mid f(a) = q\} = \{1, 2, 4\}$

DOMAIN, CO-DOMAIN AND RANGE

If $f\ A \to B$ is a function, then A is called *domain*, B is called *co-domain of f.*

Range.

If $f : A \to B$ is a function, then set of all images of the elments of A is called *range of f.*

It is denoted by $f(A) = \{f(a)/a \in A\}$.

Range is a subset of the co-domain, i.e. $f(A) \subseteq B$.

e.g. If $A = \{1, 2, 3, 4\}$, $B = \{p, q, r\}$ and $f = \{(1, q), (2, q), (3, q), (4, q)\}$, then *f* is a function.

Domain = A, Co-domain = B.

$$\text{Range of } f = f(A) = \{f(1), f(2), f(3), f(4)\} = \{q, q, p, q\} = \{p, q\}$$

$$\therefore \quad f(A) \subseteq B$$

There is no pre-image for the element $r \subseteq B$. Even

then $f : A \rightarrow B$ is a function (hence for every element there may not be pre-images).

e.g. (*i*) Let A = {1, 2, 3, 4}, B = {*p, q, r*} and *g* = {(1, *p*), (2, *q*), (3, *p*), (4, *p*), (1, *r*)}.

The element $1 \in A$ has two images *p* and *r* in B. So, $g : A \rightarrow B$ is not a function but it is only a relation.

- If N is the set of natural numbers and $f = \{(a, b) \mid a, b \in N, b = 2a\}$, then $f : N \rightarrow N$ is a function.
- If $f(x) = |x|$, $\forall x \in R$, then $f : R \rightarrow R$ is a function, i.e.

$$f = \{(x, |x|) x \in R\}$$

$$f(x) = x \text{ if } x \geq 0, f(x) = -x \text{ if } x < 0.$$

EQUALITY OF TWO FUNCTIONS

If *f* and *g* are functions defined on the same domain A and if $f(a) = g(a)$ for every $a \in A$, then functions *f* and *g* are equal,

i.e. $f = g$

e.g. Let $f(x) = x^2 + 5$, where *x* is a real number and $g(x) = x^2 + 5$, where *x* is a complex number. Then function is not equal to the function *g*, since they have different domains although $f(x) = x^2 + 5 = g(x)$.

Let a function *g* be defined by the formula $g(x) = x^2$, where domain of *g* is the set {1, 2}.

Then $f = g$, since they both have same domain and since *f* and *g* assign same image of each element in the domain.

TYPES OF FUNCTIONS

(1) Into mappings.

A mapping $f : A \rightarrow B$ is called *into mapping* if $f(A)$ is a proper subset of B. In this case we say that *f* maps A into B.

e.g. Let $f : Z \rightarrow Z$ (set of integers) be defined by $f(x) = 2x$, $x \in Z$.

Then *f* is an into mapping, because $f(Z)$ (set of all even integers) is a proper subset of the codomain set Z.

(2) Onto mappings.

A mapping $f : A \rightarrow B$ is called *onto mapping* if $f(A) = B$.

In this case we say that *f* maps onto B.

e.g. Let $f : Z \rightarrow Z$ be defined by $f(x) = x + 1, x \in Z$.

Then every element *y* in the codomain set Z has a pre-image *y*–1 in the domain set Z.

Therefore, $f(z) = Z$ and *f* is an onto mapping.

(3) Injective mapping.

A mapping $f : A \rightarrow B$ is called *injective* (or one-to-one or injection) if for each pair of distinct elements of A, their *f*-images are distinct.

Thus $f : A \rightarrow B$ is injective if $x_1 \neq x_2$ in A implies $f(x_1) \neq (x_1)$ in B.

e.g. Let $f : R \rightarrow R$ be defined by $f(x) = 2x + 1, x \in R$. Then for $x_1, x_2 \in R$ $(x_1 \neq x_2)$, we have $f(x_1) \neq f(x_2)$.

(4) Surjective mapping.

A mapping $f : A \rightarrow B$ is called *surjective* (or onto or surjection) if $f(A) = B$.

(5) Bijective mapping.

A mapping $f A \rightarrow B$ is called *bijective* (or bijection) if *f* is both injective and surjective.

e.g. Let $f : Z \rightarrow Z$ be defined by $f(x) = x + 2$, $x \in 1 \in Z$.

The mapping *f* is both injective and surjective. So it is bijective.

Note : If *f* is injective, each element of B has at most one *pre-image*. If *f* is surjective, each element of B has at least one *pre-image*.

(6) Many-one mapping.

A mapping $f : A \rightarrow A$ is called *many-one* if two or more distinct elements in A have the same image, i.e. if $f(x) = f(x)'$ implies $x \neq x'$.

(7) Constant mapping.

A function *f* of A into B is called *constant mapping* if same element $b \in B$ is assigned to every element in A.

In other words, $f : A \rightarrow B$ is a constant mapping if range of *f* consists of only one element.

e.g. Let $f : R \rightarrow R$ be defined by the formula $f(x) = 50$. Then *f* is a constant function since 50 is assigned to every element.

(8) Identity mapping.

Let A be a set and the function $f : A \rightarrow A$ be defined by the formula $f(x)$.

Then *f* is called *identity function* or *identity mapping* or *identity transformation* on A.

(9) Inverse mapping.

Let $f A \rightarrow B$ be a one-one onto mapping.

Then mapping $f^{-1} : B \rightarrow A$ which associates to each element $b \in B$ the element $a \in A$, such that $f(a) = b$, is called *inverse mapping* of the mapping $f : A \rightarrow B$.

Hence if $f : A \rightarrow B$ be a one-one on to mapping, then $f^{-1} : B \rightarrow A$. The mapping f^{-1} is called *inverse mapping of the mapping f*.

e.g. Let A = R – {3} and B = R – {1} and function $f : A \rightarrow B$ be defined by

$$f(x) = \frac{x-2}{x-3}.$$

Then inverse function $f^{-1} : B \rightarrow A$ is given by

$$y = \frac{x-2}{x-3} \quad \text{or} \quad x = \frac{2-3y}{1-y}.$$

Hence $f^{-1}(x) = \dfrac{2-3x}{1-x}$.

SOME IMPORTANT THEOREMS

Theorem 1.

Only bijective mapping possess inverse mappings.

Theorem 2.

If $f : A \rightarrow B$ is bijective, then $f^{-1} : B \rightarrow A$ is also bijective.

Theorem 3.

If $f : A \to B$ be bijective, then inverse mapping of f is unique.

PRODUCT OR COMPOSITIONS OF MAPPINGS

Let $f : A \to B$ and $g : B \to C$.

Then composite of functions f and g denoted by (fog) or fg is a mapping of $A \to C$ given by $(fog) : A \to C$ such that

$$(gof)(x) = gf(x) = g[f(x)] = \forall x \in A$$

It may be noted that domain of g is equal to codomain of f.

e.g. Let $f(x) = 2x + 1$ and $g(x) = 3x$ be two functions.

$$\therefore \quad (gof)(x) = g[f(x)] = g(2x + 1)$$
$$= g(y) \text{ (where } y = 2x + 1)$$
$$= 3y \quad [\text{since } g(x) = 3x]$$
$$= 3(2x + 1)$$
$$= 6x + 3.$$

Note : If $f A \to B$ and $g : A \to A$, then we can find both the composite mapping $go f$ and fog.

But in general $(gof) \neq (fog)$.

e.g. Let $f(x) = 2x - 1$, $g(x) = 5x$

$$\therefore \quad (fog)(x) = f[g(x)] = f(5x)$$
$$= 2(5x) - 1$$
$$= 10x - 1$$

and

$$(gof)(x) = g[f(x)]$$
$$= g(2x - 1)$$
$$= 5(2x - 1)$$
$$= 10x - 5.$$

i.e. $fog \neq gof$.

Properties of Composites of mappings.

(1) If $f : A \to B$ be a bijective mapping, then

$$of fof^{-1} = I_B \text{ and } f^{-1} of = I_A$$

where I_A and I_B are identity mappings of the sets A.

(2) If $f : A \to B$, then

$$I_B \text{ of } = f \text{ and of } f_0 I_A = f$$

(3) If $f : A \to B$, $g : B \to C$, $h : C \to D$ be three mappings, then

$$ho(gof) = (hog) of.$$

(4) If $f : A \to B$ and $g : B \to C$ be both injective mappings, then composite mapping $gof : A \to C$ is injective.

(5) If $f : A \to B$ and $g : B \to C$ be two mappings such that $gof : A \to C$ is injective, then f is injective.

(6) If $f : A \to B$ and $g : B \to C$ be both surjective, then composite mapping $gof : A \to C$ is surjective.

(7) If $f : A \to B$ and $g : B \to C$ be two mappings such that $gof : A \to C$ is surjective, then g is surjective.

(8) If $f : A \to B$ and $g : B \to C$ be both bijective, then composite mapping $go f : A \to C$ is bijective.

REAL VALUED FUNCTIONS

R is the set of real numbers. If $A \subseteq R$, then a function, $f : A \to R$ is called *real valued function.*

Example. Find domain of the real valued function $f(x) = \sqrt{49 - x^2}$.

Solution:

$$\sqrt{49 - x^2} \in R$$
$$\Rightarrow \quad 49 - x^2 \geq 0$$
$$\Rightarrow \quad x^2 \leq 49$$
$$\Rightarrow \quad |x| \leq 7$$
$$\Rightarrow \quad -7 \leq x \leq 7$$

Thus domain of f is $[-7, 7]$

ALGEBRA

INTRODUCTION

Group is a mathematical system having an algebraic structure.

Groups are sets with an algebraic operation. Sets under an algebraic operation satisfy certain rules which we classify and put them under the name of axioms.

Notations used for some standard Sets.

N = set of all natural numbers (positive integers)

Z = set of all integers

Q = set of all rational numbers

R = set of all real numbers

C = set of all complex numbers

A^* = set of all nonzero numbers of A where $A \subseteq R$

A^+ = set of all positive numbers of A where $A \subseteq R$

A^- = set of all negative numbers of A where $A \subseteq R$.

BINARY OPERATIONS

Definition 1.

Let A be a nonempty set. Any mapping from $A \times A$ into A is called binary operation or binary composition on A.

- If f is a binary operation on a nonempty set A and $(x, y) \in A \times A$, then image $f(x, y)$ of (x, y) under f is denoted by xfy.
- The symbols +, ..., 0, * etc., is used oftenly to denoted binary operations on a nonempty set.
- If * is a binary operation on a nonempty set A, then for all $x, y \in A$, $x * y$ is a unique element of A.

Definition 2.

*Let * be a binary operation on a nonempty set A and $B \subseteq A$. Then B is said to be closed with respect to the binary operation * if $x * y \in B$ for all $x, y \in B$.*

- If * is a binary operation on a nonempty set A, then A is closed with respect to *.

  If $x, y \in A$ then $x * y$ is called *product of x and y with respect to* *.
- If A is closed with respect to *, then we also say that * *is closed on A.*

Symbols used for some usual operations

+ = usual addition
− = usual subtraction
. = usual multiplication
÷ = usual division

e.g. + and . are binary operations on N, Z, Q, R and C.

BINARY OPERATION TABLES.

Definition 3.

*If * is a binary operation on a finite set A, then products of elements of A can be represented in a table. These tables are called Cayley binary operation tables or Cayley multiplication tables or Cayley composition tables.*

If $x, y \in A$, then $x * y$ is shown as follows :

*		y	
x		x * y	

Example. Write down binary operation table for usual multiplication '.' on the set A = {1, −1}.

Solution:

·	1	−1
1	1	−1
−1	−1	1

Example. Write down composition table for the set A = {1, ω, ω²}, where ω is complex cube roots of unity, under usual multiplication.

Solution:

·	1	ω	ω^2
1	1	ω	ω^2
ω	ω	ω^2	1
ω^2	ω^2	1	ω

TYPES OF BINARY OPERATIONS.

Definition 4.

*A binary operation * on a nonempty set A is said to be associative if $(x * y) * z = x * (y * z)$, $\forall x, y, z \in A$.*

e.g. + and ·are associative binary operations on N, Z, Q, R and C.

Defintion 5.

*A binary operation * on a nonempty set A is said to be commutative if*

$$x * y = y * x,\ \forall x, y \in A.$$

e.g. + and ·are commutative binary operations on N,Z, Q, R and C.

Example. Which of the following are binary operations on N. Also verify which binary operations are associative and commutative ?

(*i*) $a * b = a + b$
(*ii*) $a * b = a - b$
(*iii*) $a * b = a + b + 1$
(*iv*) $a * b = ab$
(*v*) $a * b = a \div b$
(*vi*) $a * b = ab + 1$
(*vii*) $a * b = a + b + ab$.

Solution:

(***i***) If a, b are natural numbers, then $a + b$ is a natural number.
Therefore * is a binary operation on N in (I).
Addition of natural numbers is both associative and commutative.

(***ii***) * is not a binary operation on N, since $3 \in N$, $5 \in N$; but $3 * 5 = 3 - 5 \notin N$.

(***iii***) If a, b are natural numbers, then $a + b + 1$ is also a natural number.
Therefore * is a binary operation on N in (III).

$$a * b = a + b + 1 = b + a + 1 = b * a$$

Therefore * is commutative.

$$\begin{aligned}(a * b) * c &= (a + b + 1) * c\\ &= a + b + 1 + c + 1\\ &= a + (b + c + 1) + 1\\ &= a * (b + c + 1) = a * (b * c)\end{aligned}$$

Therefore * is associative.

(***iv***) If a, b are natural numbers, then ab is a natural number.
Therefore * is a binary operation on N in (IV).
Multiplication of natural numbers is both associative and commutative.

(***v***) * is not a binary operation in (V),
since, $3 \in N$, $5 \in N$,
but $3 * 5 = 3 \div 5 \notin N$.

(***vi***) If a, b are natural numbers, then $ab + 1$ is a natural number
Therefore * is a binary operation on N in (VI).

$$a * b = ab + 1 = ba + 1 = b * a$$

Therefore * is commutative.

$$\begin{aligned}(a * b) * c &= (ab + 1) * c = (ab + 1)\,c + 1\\ &= abc + c + 1\\ a * (b * c) &= a * (bc + 1) = a(bc + 1) + 1\\ &= abc + a + 1.\end{aligned}$$

Therefore * is not associative.

(***vii***) If a, b are natural numbers, then $a + b + ab$ is also a natural number.
Therefore * is a binary operation on N.

$$a * b = a + b + ab = b + a + ba = b * a$$

Therefore * is commutative.

$$\begin{aligned}(a * b) * c &= (a + b + ab) * c\\ &= a + b + ab + c + (a + b + ab)\,c\\ &= a + b + c + ab + bc + ac + abc\\ a * (b * c) &= a * (b + c + bc)\\ &= a + b + c + bc + a(b + c + bc)\\ &= a + b + c + ab + bc + ac + abc\end{aligned}$$

Therefore * is associative.

ALGEBRAIC SYSTEMS OR ALGEBRAIC STRUCTURES

Definition 6.

If A is a nonempty set and B is a set of binary operations on A, then (A, B) is called an algebraic system or an algebraic structure.

*If * is a binary operation on A, then (A, *) is an algebraic system.*

e.g. (N, +), (N, .), (Z, +), (Z, .), (Q, +), (Q, .), (R, +), (R, .) are algebraic systems.

SEMIGROUPS

Definition 7.

*If S is a nonempty set and * is a binary operation on S, then algebraic system (S, *) is called a semigroup if*

$$a * (b * c) = (a * b) * c, \ \forall\, a, b, c \in S.$$

(*i*) If S is a nonempty set and * is an associative binary operation on S, then (S, *) is a semigroup.

(*ii*) If (S, *) is a semigroup, then element $a \cdot b$ is denoted by ab.

(*iii*) If (S, *) is a semigroup, then S is a semigroup under * or a semigroup with respect to *.

e.g. (N, +), (N, ·), (Z, +), (Z, ·), (Q, +), (Q, ·), (R, +), (R, ·), (C, +), (C, ·), are all semigroups.

Definition 8.

Let (S, *) be a semigroup.

An element $e \in S$ is called

I. *left identity of S if* $e * a = a, \ \forall a \in S$

II. *right identity of S if* $a * e = a, \ \forall a \in S$

III. *two sided identity or simple identity of S if* $a * e = e * a = a, \ \forall a \in S.$

e.g. semigroups (Z, +), (Q, +), (R, +), (C, +) contain '0' as identity.

semigroups (N, .), (Z, .), (Q, .), (R, .), (C, .) contain '1' as identity.

Theorem 1.

Let (S, *) be a semigroup. If e is a left identity and f is a right identity of S, then $e = f$.

Theorem 2.

If (S, *) is a semigroup with identity, then identity in S is unique.

Definition 9.

A semigroup with identity is called a monoid.

e.g. (Z, +) is a monoid and (N, +) is a semigroup but not a monoid.

Definition 10.

*Let (S, *) be a semigroup with identity e.*

I. *An element $a \in S$ is called left invertible if there exists an element $b \in S$ such that $b * a = e$. The element b is called left inverse of a.*

II. *An element $a \in S$ is called right invertible if there exists an element $c \in S$ such that $a * c = e$. The element c is called right inverse of a.*

III. *An element $a \in S$ is called invertible if a is both left invertible and right invertible.*

e.g. 1 is invertible in the semigroup (Z, .).

Theorem 3.

Let (S, *) be a semigroup with identity e and $a \in S$. If b is a left inverse and c is a right inverse of a, then $b = c$.

Definition 11.

*Let (S, *) be a semigroup with identity e and $a \in S$. If a is invertible, then element which is both left inverse and right inverse of a is called inverse of a.*

Theorem 4.

Let (S, *) be a semigroup with identity e. If a is an invertible element, then inverse of a is unique.

Let (S, *) be a semigroup with identity e. If a is an invertible in S element, then inverse of a is denoted by a^{-1}.

Now, $$a * a^{-1} = a^{-1} * a = e.$$

Theorem 5.

Let (S, *) be a semigroup with identity e. If a is an invertible element, then a^{-1} is also invertible and

$$\left(a^{-1}\right)^{-1} = a.$$

Theorem 6.

Let (S, *) be a semigroup with identity e. If a, b are two invertible elements in S, then $a * b$ is also invertible and

$$(a * b)^{-1} = b^{-1} * a^{-1}$$

Example. Show that $S\left\{\begin{bmatrix} a & b \\ 0 & 0 \end{bmatrix} : a, b \in R\right\}$ is a semigroup under matrix multiplication. Verify that $\begin{bmatrix} 1 & 0 \\ 0 & 0 \end{bmatrix}$ is a left identify in S.

Solution:

Let $\begin{bmatrix} a & b \\ 0 & 0 \end{bmatrix}, \begin{bmatrix} c & d \\ 0 & 0 \end{bmatrix} \in S$

$$\begin{bmatrix} a & b \\ 0 & 0 \end{bmatrix}, \begin{bmatrix} c & d \\ 0 & 0 \end{bmatrix} = \begin{bmatrix} ac & ad \\ 0 & 0 \end{bmatrix} \in S.$$

Hence matrix multiplication is a binary operation on S. Since matrix multiplication is associative (S, .) is a semigroup.

Let $$\begin{bmatrix} a & b \\ 0 & 0 \end{bmatrix} \in S$$

$$\begin{bmatrix} 1 & 0 \\ 0 & 0 \end{bmatrix}\begin{bmatrix} a & b \\ 0 & 0 \end{bmatrix} = \begin{bmatrix} a & b \\ 0 & 0 \end{bmatrix}$$

$\therefore \quad \begin{bmatrix} 1 & 0 \\ 0 & 0 \end{bmatrix}$ is a left identity of S.

Example. If S is a nonempty set and P(S) is power set of S, then show that (P(S), $\cup$), (P(S), $\cap$) are monoids.

Solution:

Let A, B $\in$ P(S).

Now $A \subseteq S, B \subseteq S$ and therefore $A \cup B \subseteq S, A \cap B \subseteq S$.

Hence, $$A \cup B \in P(S), A \cap B \in P(S)$$

We know, $(A \cup B) \cup C = A \cup (B \cup C)$,
$(A \cap B) \cap C = A \cap (B \cap C)$,
$\forall$ A, B, C $\in p(S)$.

Again $A \cup \phi = \phi \cup A = A$

and $A \cap S = S \cap A = A$.

Hence (P(S), $\cup$) and (P(S), $\cap$) are monoids.

Example. If S = set of all odd integers, then show that (S, .) is a semigroup with identity (monoid)

Solution:

Now $S = \{2n + 1 : n \in z\}$.

Let $2m + 1, 2n + 1 \in S$

$(2m + 1)(2n + 1) = 2(2mn + m + n) + 1 \in S$.

Since multiplication is associative in Z, multiplication is associative in S.

Hence (S, .) is a semigroup and clearly 1 is the identity of S.

Definition 12.

*A semigroup (S, *) is called*

I. *left cancellative semigroup* if $a, b, c \in S$
$a * b = a * c$
$\Rightarrow b = c$

II. *right cancellative semigroup* if $a, b, c \in S$
$b * a = c * a$
$\Rightarrow b = c$

III. *cancellative semigroup* if it is both left cancellative and right cancellative semigroup.

- $a * b = a * c$
  $\Rightarrow b = c$ is called *left cancellation law.*
- $b * a = c * a$
  $\Rightarrow b = c$ is called *right cancellation law.*

e.g. (Z, +), (Z, *), (Q, +*), (Q, *.), (R, +), (R, .) are cancellative semigroups.

GROUP

A semigroup with identity and in which every element is invertible is called a *group.*

Definition 1 .

*Let G be a nonempty set and * be a binary operation on G. Then algebraic system (G, *) is called a group if*

I. $a * (b * c) = (a * b) * c, \forall a, b, c \in G$

II. $\exists e \in G \Rightarrow a * e = e * a = a, \forall a \in G$

III. $a \in G$
$\Rightarrow \exists b \in G \Rightarrow a * b = b * a = e$

- The element *e* in condition II of the definition of a group is called *identity element of the group.*
- The element *b* corresponding to a in condition III of the definition of a group is called *inverse of a* in the group.

Definition 2 .

*A group (G, *) is called an abelian group or commutative group if*

$$a * b = b * a, \forall a, b \in G.$$

e.g. (Z, +), (Q, +), (R, +), (C, +) are all abelian groups.

Example. If G = Q – {1}, if * is defined on G as $a * b = a + b - ab$, show that (G, *) is an abelian group.

Solution:

Clearly G is non-empty.

Let, $a, b \in G$

$a \in G \Rightarrow a \in Q, a \neq 1$

$b \in G \Rightarrow b \in Q, b \neq 1$

$a \in Q, b \in Q \Rightarrow a + b \in Q$,

$ab \in Q \Rightarrow a + b - ab \in Q$

$\Rightarrow a * b \in Q$

If $a * b = 1$

then $a + b - ab = 1$

$\Rightarrow a - 1 + b - ab = 0$

$\Rightarrow 1(a - 1) - b(a - 1) = 0$

$\Rightarrow (a - 1)(1 - b) = 0$

$\Rightarrow a = 1$ or $b = 1$

Contradiction : $a * b \neq 1$

$a * b \in Q, a * b \neq 1$,

$\Rightarrow a * b \in G$

Therefore* is a binary operation on G.

Let, $a, b, c \in G$.

$$\begin{aligned} a * (b * c) &= a * (b + c - bc) \\ &= a * (b + c) - bc) - a(b + c - bc) \\ &= a + b + c - bc - ab - ac + abc \\ (a * b) * c &= (a + b - ab) * c \\ &= a + b - ab + c - (a + b - ab) c \\ &= a + b + c - ab - ac - bc + abc \end{aligned}$$

$\therefore a * (b * c) = (a * b) * c$

*Therefore * is an associative binary operation on G.*

Let *e* be the identity element of G, then

$a * e = a, \forall a \in G$

$a * e = a$

$\Rightarrow a + e - ae = a$

$\Rightarrow e(1 - a) = 0$

$\Rightarrow e = 0$

$\therefore e = 0 \in G$ is identity element in G.

Let $a \in G$. If *b* is inverse of *a*, then

$a * b = 0$

$\Rightarrow a + b - ab = 0$

$\Rightarrow b(a - 1) = a$

$\Rightarrow b = \dfrac{a}{a-1} \in G$.

Therefore $b = \dfrac{a}{a-1} \in G$ is inverse of *a* in G.

Let $a, b \in G$, then $a * b = a + b - ab$
$= b + a - ba = \mathbf{b * a}$

*Therefore * is a commutative in G.*

Therefore (G, *) is an abelian group.

Definition 3.

A group (G, ∗) is said to be a nonabelian group, if (G, ∗) is not abelian.

PROPERTIES OF A GROUP.

Theorem 1.

In a group, identity element is unique.

Theorem 2.

In a group, inverse of every element is unique.

Theorem 3.

If (G, .) be a group and $a \in G$, then $\left(a^{-1}\right)^{-1} = a$.

Theorem 4.

If (G, .) be a group and $a, b \in G$, then $(ab)^{-1} = b^{-1}a^{-1}$

Theorem 5.

If (G, .) be a group and $a_1, a_2, \ldots.. a_n \in G$, then $(a_1a_2 \ldots.. a_n)^{-1} = a_n^{-1}\, a_{n-1}^{-1} \ldots.. a_1^{-1}$

Theorem 6.

Cancellation laws hold in a group, i.e. If (G, .) is a group, then

I. $a, b, c \in G$, $\quad ab = ac$

$\Rightarrow \quad b = c$

II. $a, b, c \in G$, $\quad ba = ca$

$\Rightarrow \quad b = c.$

Definition 4.

If (G, .) be a group and $a \in G$, $n \in Z$, then a^n is defined as follows :

I. $a^0 = e$

II. If $n > 0$, then $\quad a^1 = a;\ a^{n+1} = a^n \cdot a$

III. If $n < 0$, then $\quad a^n = \left(a^{-n}\right)^{-1}$

Theorem 7.

Let (G, .) be a group and $a \in G$.

If $m, n \in Z$, then

I. $a^m \cdot a^n = a^{m+n} = a^n a^m$

II. $\left(a^m\right)^n = a^{mn}$

Definition 5.

Let (G, .) be a group. An element $a \in G$ is called idempotent if $a^2 = a$.

Definition 6.

A group (G, .) is called finite group if G is a finite set. The number of different elements in G is called order of the finite group (G, .). It is denoted by O(G).

Definition 7.

A group (G, .) is called infinite group if G is an infinite set. The order of an infinite group is defined to be ∞.

Example. Let G = {1, – 1, i, – i} where $i^2 = -1$. Show that G is a finite abelian group under multiplications of complex numbers.

Solution:

The composition table for multiplication is as follows:

·	1	– 1	i	– i
1	1	– 1	i	– i
– 1	– 1	1	– i	i
i	i	– i	– 1	1
– i	– i	i	1	– 1

The multiplication of complex numbers is associative and commutative.

By observing the table it follows that 1 is the identity element in G and the inverses of 1, – 1, i, – i are 1, – 1, – i, i respectively.

Thus (G, .) is a *finite abelian group*.

Definition 8.

If S is a finite set containing n elements, then group of all bijections on S is called a permutation group or symmetric group. It is denoted by P_n or S_n.

Note : $o(S_n) = n!$

RESIDUE CLASSES

Definition 1.

Let n be a positive integer and $a, b \in Z$. Then a is called congruent to b modulo n if n divides a – b. It is denoted by $a \equiv b$ (mod n).

e.g. 5 divides 32 – 2 $\Rightarrow$ $32 \equiv 2$ (mod 5)

9 divides – 47 – 7 $\Rightarrow$ $47 \equiv 7$ (mod 9)

- The relation in the above definition is referred to as congruent modulo n.

Theorem 1.

The relation congruent modulo n is an equivalence relation on Z.

Definition 2.

The equivalence classes under the relation congruent modulo a positive integer n or Z are called residue classes modulo n.

- The residue class containing an integer a is denoted by $[a]$ or $\overline{a}$.
- The set of all residue classes modulo a positive integer n is denoted by Z_n.

Theorem 2 .

If n is a positive integer, then

$$Z_n = \left\{\overline{0}, \overline{1}, \overline{2}, \ldots, \overline{n-1}\right\}, \text{ and } o(Z_n) = n.$$

Theorem 3.

If $a \equiv b \pmod n$, $c \equiv d \pmod n$, then $a + c \equiv b + d$ (mod n), and $ac \equiv bd \pmod n$

Residue class Addition.

Definition 3.

The operation ⊕ defined on Z_n as, $\overline{a} \oplus \overline{b} = \overline{a+b}$, $\forall\ \overline{a}, \overline{b} \in Z_n$

is called *residue class addition* or addition modul n.

It is also denoted by $\oplus_n$ or $+_n$.

- If $\overline{a}, \overline{b} \in Z_n$, then $\overline{a} \oplus \overline{b} = \overline{r}$, where r is remainder of $a + b$ when divided by n.
- The operation, addition modulo n is also defined in $\{0, 1, 2, ..., n - 1\}$ as $a \oplus b = r$, where r is remainder of $a + b$ when divided by n.

e.g. $\overline{5} \oplus_6 \overline{4} = \overline{9} = \overline{3}$; $4 \oplus_7 6 = 3$

Theorem 4.

$(Z_n, \oplus)$ is an abelian group.

Residue class Multiplication.

Definition 4.

The operation. defined on Z_n, as $\overline{a}\,\overline{b} = \overline{ab}$ $\forall\ \overline{a}, \overline{b} \in Z_n$ is called *residue class multiplication* or multiplication modulo n.

It is also denoted by $\cdot_n$ or $\cdot_n$ or $\otimes$ or $\otimes_n$ or $\times_n$.

- If $\overline{a}, \overline{b} \in Z_n$, then $\overline{a} \otimes \overline{b} = \overline{r}$

  where r is remainder of ab when divided by n.
- The operation, multiplication modulo n is also defined in $\{0, 1, 2, ..., n - 1\}$ as $a \otimes b = r$

  where r is remainder of ab when divided by n.

  e.g. $\overline{5} \cdot_6 \overline{4} = \overline{20} = \overline{2}$; $4 \cdot_7 6 = 3$

Theorem 5.

$(Z_n, \otimes)$ is a commutative semigroup with identity.

Theorem 6.

$(Z^*_n, .)$ is an abelian group iff n is prime.

PARTIAL ORDERING

A relation ~ on a set P is called partial order if it is

(*a*) reflexive $x \sim x,\ \forall\ x \in P$:

(*b*) antisymmetric : $(x \sim y) \wedge (y \sim x) \Rightarrow x = y$

(*c*) transitive : $(x \sim y) \wedge (y \sim z) \Rightarrow x \sim z$

A partial order is always denoted as $\leq$.

The set P with a partial order $\leq$ is called *partially ordered set*, denoted by $(P, \leq)$ and it is also called *poset*.

Totally ordered Poset.

A poset $(P, \leq)$ is called *totally ordered set*, if for all $x, y \in P$, $(x \leq y) \vee (y \leq x)$.

Cover of x.

Let $x \in (P, \leq)$. An element $\in (P, \leq)$ is called *cover of* x if $x \leq y$ and there exists no element $z \in P$ such that $x \leq z$ and $z \leq y$.

Hasse diagram.

It is a graph whose points are points of P, where points x and y are connected by an edge if y covers x. Any finite poset can be represented by a Hasse diagram.

e.g. Let P = {1, 2, 3, 4, 5, 6, 25, 100}.

Hasse graph is given in the figure.

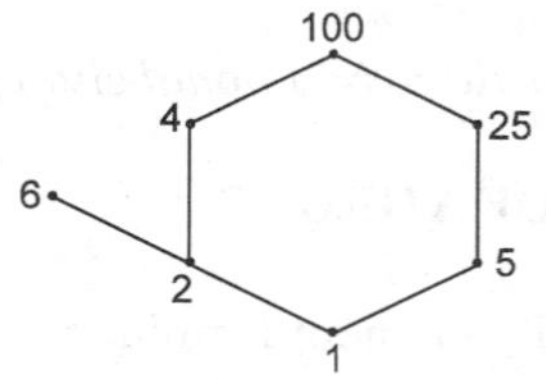

Fig. Poset

Least element.

y is least element of P if $y \leq x$ for all $x \in P$.

Greatest element.

y is greatest element of P if $x \leq y$ for all $x \in P$.

Upper bound.

Let $A \subseteq P$. An element $x \in P$ is called *upper bound of A* if $a \leq x$ for all $a \in A$.

Lower bound.

Let $A \subseteq P$. An element $y \in P$ is called *upper bound of A* if $y \leq a$ for all $a \in A$.

Least upper bound (LUB).

LUB of A is x is an upper bound and if y is any other upper bound, then $x \leq y$.

Greatest lower bound (GLB).

GLB of A is y if y is a lower bound for A and if b is any other lower bound, then $b \leq y$.

Sometimes GLB is called *infimum* and LUB is called *supremum*.

LATTICE

It is a partially ordered set $(P, \leq)$ in which any two elements $\in$ P has single GLB and single LUB.

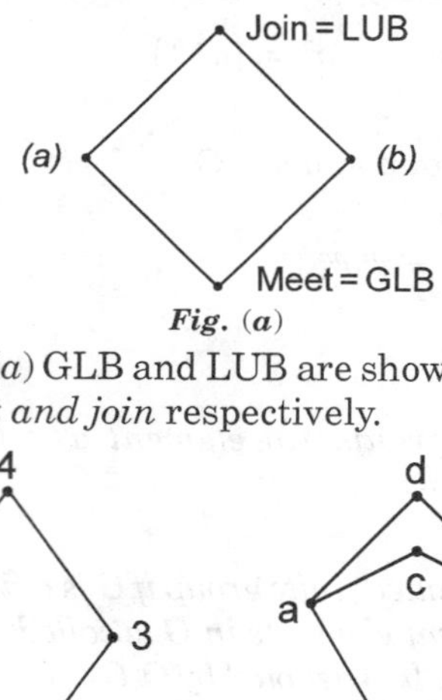

Fig. (*a*)

In the Fig.(*a*) GLB and LUB are shown, which are also called *meet and join* respectively.

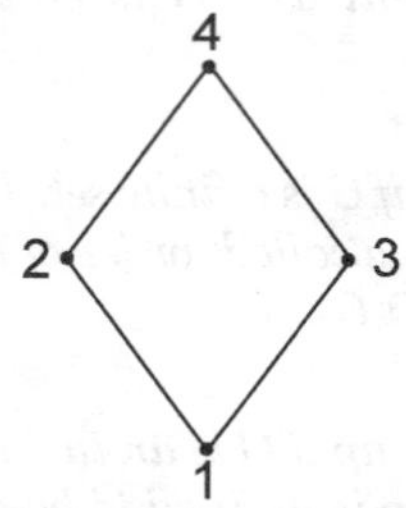

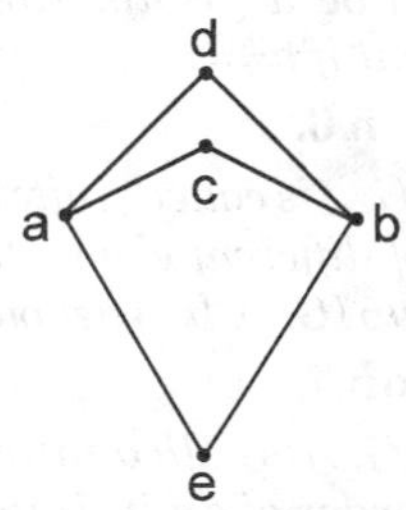

Fig. (*b*) *Lattice Fig.* (*c*) *Non-Lattice*

The poset shown in Fig (*c*) is not a lattice since a and b have two LUBs namely c and d.

From first lattice, following relations are obtained :

$$1 \leq 2,\ 1 \leq 3,\ 2 \leq 4 \text{ and } 3 \leq 4.$$

Here 2 and 3 are not compared.

If (P, ≤) is a lattice, then (P, ≥) is also a lattice, we define ≥ as follows

$$(x \leq y) \Rightarrow y \geq x.$$

In other words, GLB and LUB are interchanged if we interchange relations ≤ and ≥.

In terms of lattices, operations of meet and join on (L, ≤) become the operations of join and meet on (L, ≤).

DUAL LATTICE

For the lattice (P, ≤) the dual is (P, ≥).

The duals are shown in the figure below. The diagram of (P, ≥) can be obtained from that of (P, ≤) by simply turning it upside down.

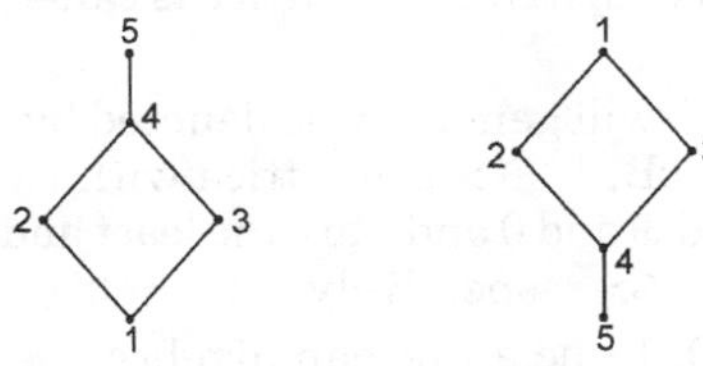

$1 \leq 2 \leq 4 \leq 5$ $\quad$ $5 \geq 4 \geq 2 \geq 1$

$1 \leq 3 \leq 4 \leq 5$ $\quad$ $5 \geq 4 \geq 3 \geq 1$

Let $a * b$ = meet of a and b = GLB, and $a \oplus b$ = join of a and b = LUB

Example. For $a, b \in$ a lattice (P, ≤), prove $a \leq b \Leftrightarrow a * b = a$

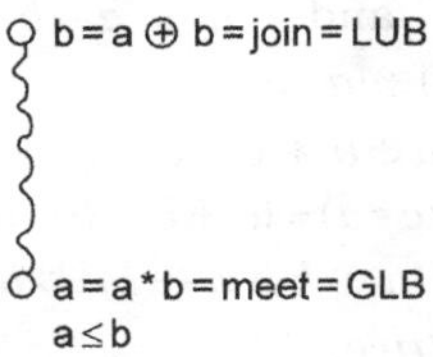

Solution:

Let $a \leq b$.

Since $a \leq a$, we have $a \leq a * b$.

Since $a * b$ is lower bound for both a and b, we have

$$a * b \leq a.$$

Hence $a = a * b$.

Example. Show that a lattice with less than 4 elements is a chain.

Solution:

Let (L, ≤) be a lattice.

Obviously if | L | = 1 or 2 it must be a chain.

Figure given below shows all possible combinations on three elements.

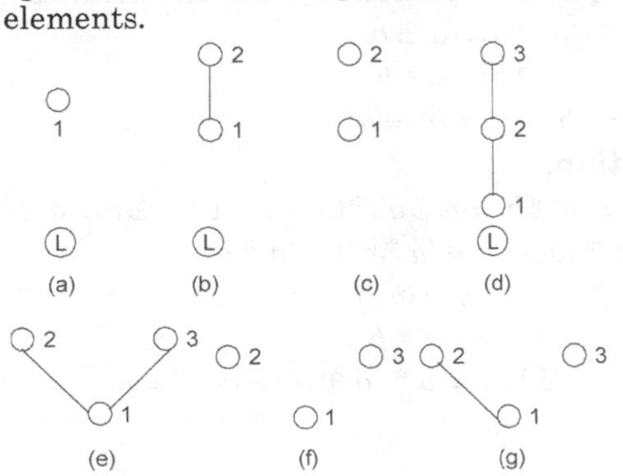

The posets shown on Figures (*c*), (*e*), (*f*) and (*g*) are not lattices as no two elements have GLB or LUB. Other posets are lattices and they are chains.

SUBLATTICE.

Let (L, *, ⊕) be a lattice and let S ⊆ L. The set (S, *, ⊕) is called *sublattice* iff it is closed under * and ⊕.

Sublattice is itself a lattice.

Example. Show that poset in Fig. (a) is a lattice and it is not a sublattice of the lattice in the Fig. (b).

Solution:

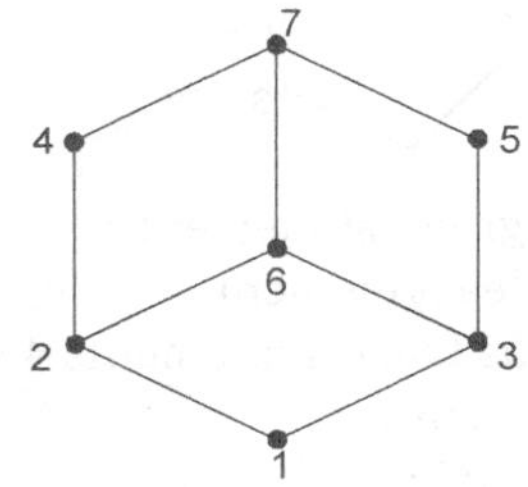

Fig. (*a*) $(P_1, \leq)$

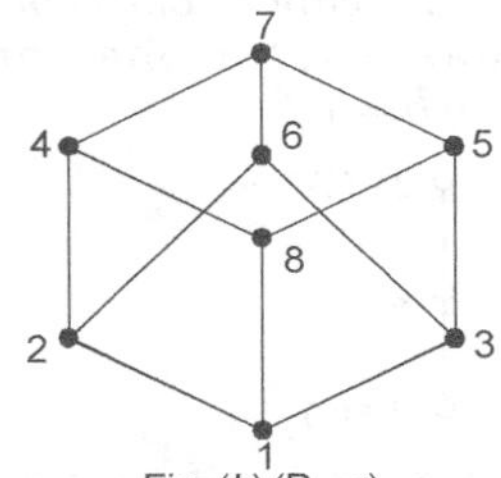

Fig. (*b*) $(P_2, \leq)$

In poset $(P_1, \leq)$, any two elements have unique GLB and LUB.

Hence $(P_1, \leq)$ is a lattice.

Since GLB (4, 5) in $P_1 = 1$ and LUB (4, 5) in $P_2 = 8 \notin P_1$.

Hence P_1 is not sublattice of P_2.

Closed interval.

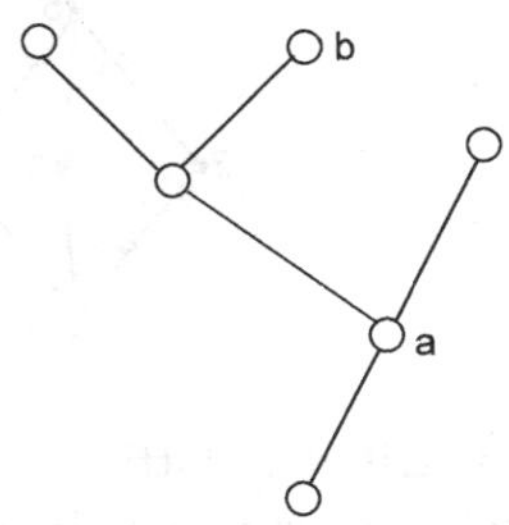

Fig. Interval [a,b]

Let $a \leq b$.

Then closed interval of a and b is defined as

$$[a, b] = [x / a \leq x \text{ and } x \leq b\}.$$

Clearly any closed interval is a chain.

Complete lattice.

A lattice is complete if each of its non-empty subsets has a least upper bound and a greatest lower bound.

Every finite lattice is *complete*.

BOUNDS

The least and greatest elements of lattice are called *bounds*, and denoted by 0 and 1.

Complement.

In a bounded lattice (L, *, ⊕, 0, 1), if $a * b = 0$ and $a \oplus b = 1$, then b is called *complement of a*.

If b is complement of a, then a is complement of b.

Consider lattice shown in the figure.

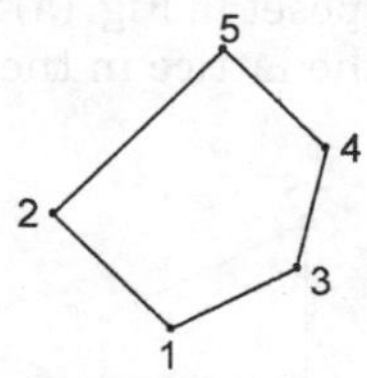

5 = (1) = greatest element

1 = (0) = least element

In that lattice, 2 has two complements 4 and 3.

0 and 1 are complement.

Complemented lattice.

A lattice (L, *, ⊕, 0, 1), is called *complemented* if every element a of L has atleast one complement b (*i.e.* $a * b = 0$ and $a \oplus b = 1$).

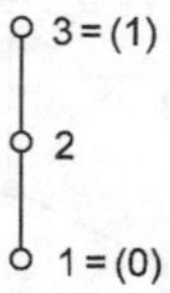

Fig. Lattice but not complemented

e.g. consider a chian of 3 elements show in the figure.

Here $2 * 3 = 2 \neq (0)$ = least element

Distributive lattice.

A lattice is distributive if

$$a * (b \oplus c) = (a * b) * (a * c)$$

and
$$a \oplus (b * c) = (a \oplus b) * (a \oplus c)$$

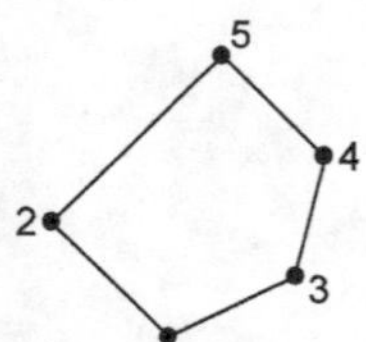

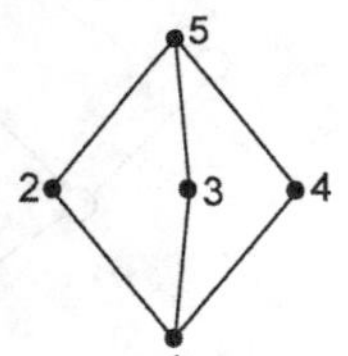

e.g.

- Every chain is a distributive lattice.
- Cancellation law can be stated as in a distributive lattice.

$$(a * b = a * c) \wedge (a \oplus b = a \oplus c) \Rightarrow b = c$$

Modular lattice.

A lattice is modular if $a \leq c$, then $a \oplus (b * c) = (a \oplus b) * c$

- We can show every distributive lattice is modular but the converse is not true.
- In a bounded distributive lattice, elements which have complements form a sublattice.

Sublattice.

Let <L, *, ⊕ > be a lattics and S ⊆ L.

Then <S, *, ⊕ >is a sublattics of <L, *, ⊕ > iff S is closed under operations * and ⊕.

Homomorphisn.

Let < L, *, ⊕ > and < S, ∧, ∨ > be two lattices.

A mapping $g : L \rightarrow S$ is called Homomorphisn if for any $a, b \in L$

$$g(a * b) = g(a) \wedge g(b)$$
$$g(a \oplus b) = g(a) \vee g(b)$$

BOOLEAN ALGEBRA

A complemented, distributive lattice is called *Boolean Algebra.*

Boolean algebra will generally be denoted by < B, *, ⊕, 0, 1> in which <B, *, ⊕ > is a lattice with two binary operations * and ⊕ and 0 and 1 are the least and greatest element of < B, ≤ > respectively.

Let < B, *, ⊕, 0, 1> be a Boolean algebra.

Following statements in which a, b and c denote any elements of the set B are true.

(1) $a * a = a$, and $a \oplus a = a$

(2) $a * b = b * a, a \oplus b = b \oplus a$

(3) $a * (b * c) = (a * b) * c$ and $a \oplus (b \oplus c) = (a \oplus b) \oplus c$

(4) $a * (a \oplus b) = a$ and $a \oplus (a * b) = a$

(5) $a * (b \oplus c) = (a * b) \oplus (a * c)$
and $a \oplus (b * c) = (a \oplus b) * (a \oplus c)$

(6) $(a * b) \oplus (b * c) \oplus (c * a) = (a \oplus b) * (b \oplus c) * (c \oplus a)$

(7) $(a * b = a * c) \wedge (a \oplus b = a \oplus c) \Rightarrow b = c$

If B is a bounded lattice, then

(8) $0 \leq a \leq 1$

(9) $a * 0 = 0, a \oplus 1 = 1$

(10) $a * 1 = a, a \oplus 0 = a$

If B is uniquely complemented, then

(11) $a * a' = 0, a \oplus a' = 1$

(12) $0' = 1, 1' = 0$

(13) $(a * b)' = a' \oplus b', (a \oplus b)' = a' * b'$

(14) $a ** b$ = GLB $\{a, b\}$, $a \oplus b$ = LUB $\{a, b\}$

(15) $a \leq b \Leftrightarrow a * b = a \Leftrightarrow a \oplus b = b$

(16) $a \leq b \Leftrightarrow a * b' = 0 = b' \leq a' \Leftrightarrow a' \oplus b = 1$

Example. Prove following Boolean identities

(*i*) $a \oplus (a' * b) = a \oplus b$

(*ii*) $a * (a' \oplus b) = a * b$

(*iii*) $(a * b) \oplus (a * b') = a$

Solution:

(*i*) $a \oplus (a' * b) = (a \oplus a') * (a \oplus b) = 1 * (a \oplus b) = a \oplus b$

(*ii*) $a * (a' + b) = (a * a') \oplus (a * b)$
$= 0 \oplus (a * b)$
$= a * b$

(*iii*) LHS $= a * (b \oplus b') = a * 1 = a$

SOLVED EXAMPLES

1. Let A = {– 2, – 1, 0, 1, 2}, B = {0, 1, 4}, and f : A → B is defined as f(x) = x^2 is a function. Find whether it is one-one or bijection ?

Solution: $f(x) = x^2$

$f(-2) = 4, \; f(-1) = 1$

$f(0) = 0, \; f(1) = 1, \; f(2) = 4.$

$f : A \to B$ is a function,

and $f : A \to B$ is onto (since $f(A) = B$).

$f(-1) = f(1) = 1$

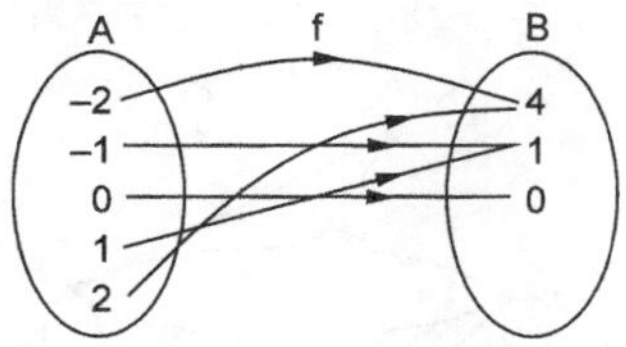

hence $f : A \to B$ is not one-one, since $f : A \to B$ is not a bijection.

2. If A = {p, q, r}, B = {2, 1, 0} and f : A → B, find whether it is one-one, onto or bijection ?

(i) f = {(p, 2), (q, 0), (r, 2)}

(ii) f = {(p, 1), (q, 0), (r, 2)}

Solution:

(i) $(p, 2), (r, 2) \in f$

$\Rightarrow$ p and r have the same image

$\therefore$ $f : A \to B$ is not one-one.

$f(A) = \{0, 2\} \neq B$. Therefore not onto.

$\therefore$ $f : A \to B$ is not a bijection.

(ii) $f = \{(p, 1), (q, 0), (r, 2)\}$

Different elements in A have different f – images in B and hence f is one-one.

$f(A) = \{1, 0, 2\} = B,$

$\therefore$ $f : A \to B$ is onto

$\therefore$ $f : A \to B$ is a bijection

3. Find whether function f : [0, ∞) → R defined as f(x) = x^2 is one-one, onto or bijection.

Solution: $f(x) = x^2$

$f : [0, \infty) \to R$ is a function.

Let $a_1, a_2 \in [0, \infty)$ and $f(a_1) = f(a_2)$

Now $f(a_1) = f(a_2)$

$\Rightarrow$ $a_1^2 = a_2^2$

$\Rightarrow$ $a_1 = a_2$

$\therefore$ $[0, \infty) \to R$ is one-one.

There is no pre-image in $[0, \infty]$ to (-1)

$\therefore$ f is not onto.

$\therefore$ $f : [0, \infty) \to R$ is not a bijection.

4. Find whether each of the following functions is one-one or onto or one-one and onto.

(i) $f : R \to R$ is defined as $f(x) = \begin{cases} x, & x > 2 \\ 5x - 2 & x \leq 2 \end{cases}$

(ii) $f(x) = 2x - 3$

(iii) $f(x) = x^2$

Solution:

(i) $3 > 2$, then $f(3) = 3$

$1 < 2$, then $f(10) = 5(1) - 2 = 3$

Thus 1 and 3 have same image. Hence it is not one-one.

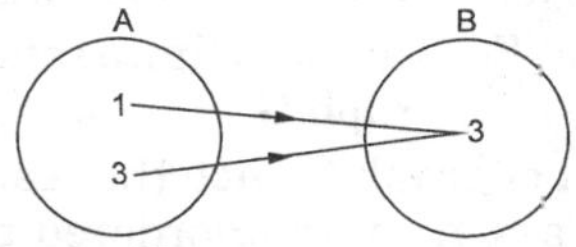

Let $y \in R$

If $y > 2$, take $x = y$

then $f(x) = f(y) = y.$

If $y \leq 2$, take $x = \dfrac{y+2}{5}$

Then $x \leq \dfrac{4}{5}$

$\therefore$ $f(x) = 5x - 2 = 5\left(\dfrac{y+2}{5}\right) - 2 = y.$

Thus f is onto since f is not one-one. f is not a bijection.

(ii) If $f : R \to R$, then $f(x) = 2x - 3 \in R$

$a_1, a_2 \in R$ $\quad f(a_1) = f(a_2)$

$\therefore$ $2a_1 - 3 = 3a_2 - 3$

$\Rightarrow$ $a_1 = a_2$

Hence f is one-one.

If $y \in R$, then $x = \dfrac{y+3}{2}$

$f(x) = 2x - 3 = 2\left(\dfrac{y+3}{2}\right) - 3 = y$

Thus f is onto. Hence f is a bijection.

(iii) $f(x) = x^2, \; \forall x, y \in R$

$f(-1) = (-1)^2 = 1$

and $f(1) = 1^2 = 1$

Hence 1 and – 1 have same image, hence f is not one-one.

– 1 has no pre-image, hence f is not onto.

Thus f is not a bijection.

5. Find inverse functions of the following of $f(x) = 4x + 7$

Solution: $y = f(x) = 4x + 7$

$\Rightarrow$ $x = \dfrac{y-7}{4}$

$\therefore$ Inverse function $f^{-1}(y) = \dfrac{y-7}{4}$, and inverse function $f^{-1}(x) = \dfrac{x-7}{4}$

6. Find inverse function of $f(x) = \dfrac{2x+1}{3}$

Solution:

Let $y = f(x) = \dfrac{2x+1}{3}$

$\therefore$ $f^{-1}(y) = x = \dfrac{3y-1}{2}$, and $f^{-1}(x) = \dfrac{3x-1}{2}$

GRAPH THEORY

GRAPH

Graph is a drawing or a diagram consisting of a collection of vertices together with edges joining certain pairs of these vertices.

Mathematically, A graph G = [V(G), E(G)]

where V(G) and E(G) are two finite sets defined as

V(G) = vertex set of graph G

E(G) = edge set of graph G such that each element e of E(G) is assigned an unordered pair of vertices (u, v) called *end vertices of e*.

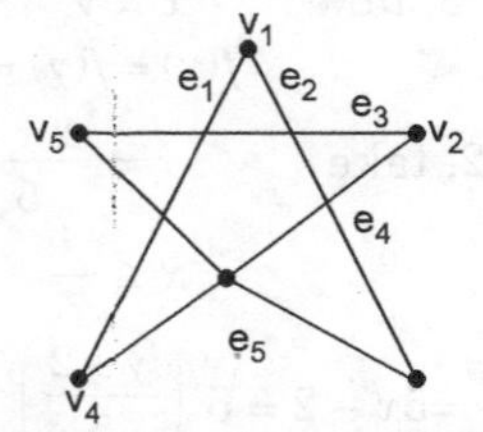

Vertices are also some times called *points or nodes*.

If e an edge with end vertices u and v, then e is said to join u and v. It is also possible to have a vertex u joined to itself by an edge, such an edge is called a *loop.*

Parallel edges.

If two (or more) edges of a graph G have the same end vertices, then these edges are called *parallel.* e.g.

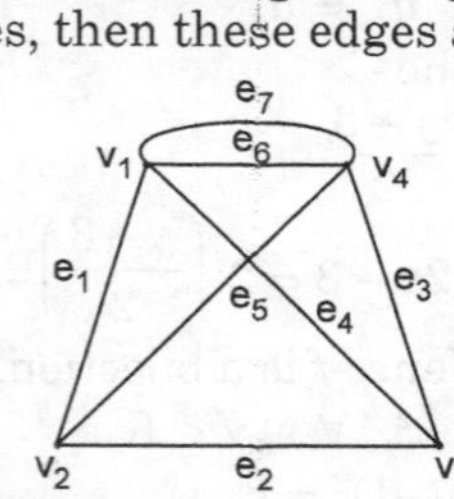

Isolated vertex.

A vertex of graph G, which is not the end of any edge is called *isolated.*

Neighbours or adjacent vertex.

Two vertices which are joined by an edge are called *'adjacent'* or *'neighbours'*. The set of all such neighbours of vertex V is called *neighbourhood set of V* and is denoted by N(v).

Incidence.

An edge e of a graph G is called *incident* with the vertex v if v is an end vertex of e (or v is incident with e). Two edges e and f which are incident with a common vertex v are called *adjacent*. An edge could be incident with either 1 or 2 vertices (1 in case of loop), whereas a vertex could be incident on any finite number of edges.

MULTIGRAPH

A multigraph G = G(V, E) also consists of a set V of vertices and a set E of edges except that E may contain *multiple* edges, i.e., edges connecting the same endpoints, and E may contain one or more *loops* i.e., an edge whose endpoints are the same vertex.

Diagram of a Multigraph.

Graphs (multigraphs) G = G(V, E) are pictured by diagrams in the plane as follows.

Each vertex v in V is represented by a dot (or small circle) and each edge $e = \{u, v\}$ is represented by a curve which connects its *endpoints* u and v. (we usually denote a graph, when possible, by drawing its diagram rather than explicitly listing its vertices and edges).

Example. Describe formally the multigraph shown below.

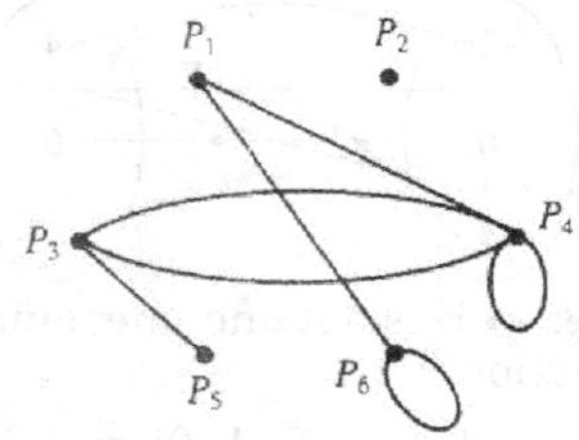

Solution.

There are six vertices, so

$$V = \{P_1, P_2, P_3, P_4, P_5, P_6\}$$

There are seven edges (of which two are multiple edges and two are loops) and thus seven pairs of vertices; hence

$E = [\{P_1, P_4\}, \{P_1, P_6\}, \{P_3, P_4\}, \{P_3, P_4\}, \{P_4, P_4\}, \{P_3, P_5\}, \{P_6, P_6\}]$

Finite multigraph.

A multigraph G = G(V, E) is *finite* if both V is finite and E is finite.

Note that a graph G with a finite number of vertices V must automatically have a finite number of edges and so must be finite.

DEGREE OF A VERTEX

Relation of adjacency and incidence in a graph G.

Let $e = \{u, v\}$ is an edge in G, i.e. u and v are endpoints of e. Then vertex u is said to be adjacent to the vertex v, and edge e is said to be incident on u and on v.

Degree and parity (even or odd) of a vertex.

Degree of a vertex v in a graph G, written deg (v), is equal to number of edges which are incident on v or, in other words, the number of edges which contain v as an endpoint. The vertex v is said to be even or odd according as deg (v) is even or odd.

Theorem : Sum of the degrees of the vertices of a graph is equal to twice the number of edges.

It follows directly from the fact that, each edge is counted twice in counting degrees of the vertices of a graph G.

First theorem of graph theory.

For any graph G with e edges and n vertices $v_1 .... v_n$,

$$\sum_{i=1}^{n} d(v_i) = 2e$$

Odd or Even vertex.

Vertex of a graph is called *odd or even* depending on whether its degree is even or odd.

PATHS, CONNECTIVITY

Path and its length in a graph (multigraph) G.

A *path* α in G with origin v_0 and end v_n is an alternating sequence of vertices and edges of the form

$$v_0, e_1, v_1, e_2, v_2, \dots, e_{n-1}, v_{n-1}, e_n, v_n$$

where each edge e_i is incident on vertices v_{i-1} and v_i.

The number of edges, n is called *length* of α. When there is no ambiguity, we denote α by its sequence of edges, $\alpha = (e_1, e_2, \dots, e_n)$, or by its sequence of vertices $\alpha = (v_0, v_1, \dots, v_n)$.

Simple path and trail in a graph (multigraph) G.

A path $\alpha = (v_0, v_1, \dots, v_n)$ is simple if all the vertices are distinct. The path is a *trail* if all the edges are distinct.

Closed path and a cycle in a graph (multigraph) G.

A path $\alpha = (v_0, v_1, \dots, v_n)$ is *closed* if $v_0 = v_n$, i.e. if origin (α) = end (α). The path α is a *cycle* if it is closed and if all vertices are distinct except $v_0 = v_n$. A cycle of length k is called k-cycle. A cycle in a graph must therefore have length three or more.

SPECIAL GRAPHS

Types of graphs.

(Here the term graph does not include multigraphs.)

(1) Complete graph.

A graph G is *complete* if each vertex is connected to every other vertex. The complete graph with n vertices is denoted by K_n.

Complete graphs K_1, K_2, K_3, and K_4.

First draw the appropriate number n of vertices. Then draw an edge from each vertex to every other vertex. The required diagrams are given below.

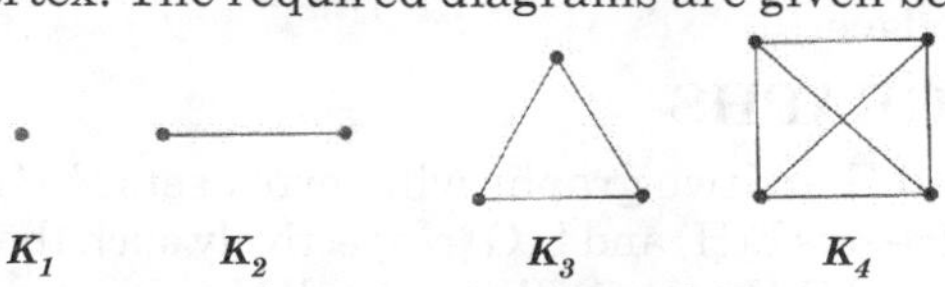

K_1 K_2 K_3 K_4

e.g. complete graphs K_5 and K_6.

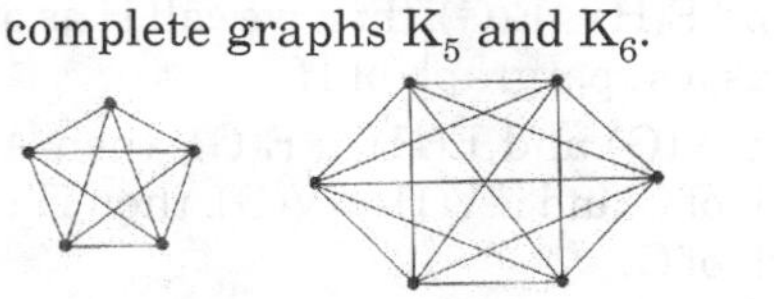

K_5 K_6

Number of edges, m in complete graph K_n.

Each pair of vertices determines an edge. Thus

$m = C(n, 2) = \dfrac{n(n-1)}{2}$ since there are $C(n, 2)$ ways of selecting two vertices out of n vertices.

Example. Find the number of edges, m in the graphs (*i*) K_8, (*ii*) K_{12}, and (*iii*) K_{15}.

Solution.

(*i*) $m = \dfrac{8 \cdot 7}{2} = 28$, (*ii*) $m = \dfrac{12 \cdot 11}{2} = 66$

(*iii*) $m = \dfrac{15 \cdot 14}{2} = 105$.

(2) Regular graph.

A graph G is *regular of degree* or *k-regular* if every vertex has degree k. In other words, a *graph is regular if every vertex has the same degree.*

Connected regular graphs of degrees 0, 1 and 2:

The connected 0-regular graph is the *trivial graph* with one vertex and no edges.

The connected 1-regular graph is the graph with two vertices and one edge connecting them.

The connected 2-regular graph with n vertices is the graph which consists of a single n-cycle.

Figure below shows the connected 0-regular and 1-regular graphs and some of the connected 2-regular graphs.

(a) 0-regular *(b) 1-regular*

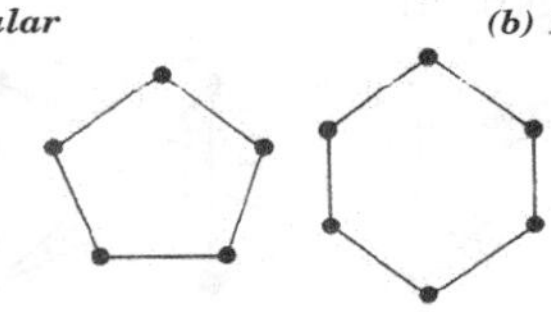

(c) 2-regular

Draw two-3 regular graphs with six vertices.

e.g.

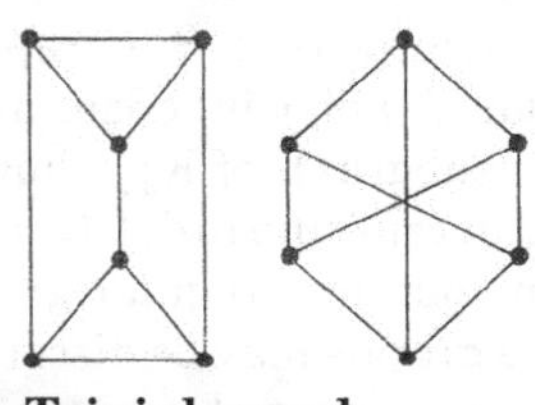

(3) Empty or Trivial graph.

A graph with no edges.

(4) Bipartite graphs.

If vertex set V of a graph G can be partitioned into two non-empty disjoint subsets X and Y in such a way that each edge of G has one end in X and one end in Y, then G is called *bipartite*.

The partition $V = X \cup Y$ is called a *bipartition of G.*

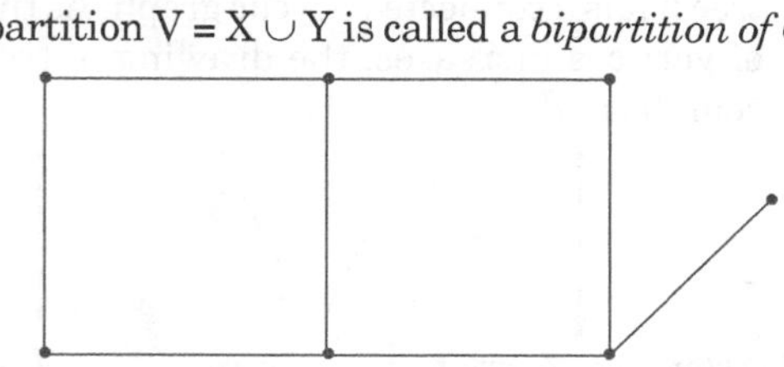

Complete bipartite graph :

If every vertex in X is joined to every vertex of Y, then it is called *complete bipartite graph* denoted by $k_{m,\,n}$, where m and n are vertices in X and Y respectively.

Any complete bipartite graph with a bipartition into two sets of m and n vertices is isomorphic to $k_{m,n}$.

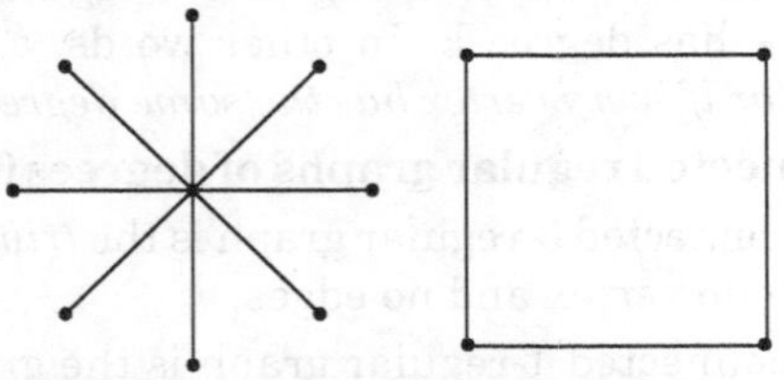

Diameter of any complete bipartite graph :

The diameter of $K_{1,\,1}$ will be one since there are only two vertices and the shortest path between them is length one. All other bipartite graphs will have diameter two since any two points in either M or N will be exactly distance 2 apart. (One edge to reach the other subgroup of vertices and one to return.)

Example. Which connected graphs can be both regular and bipartite?

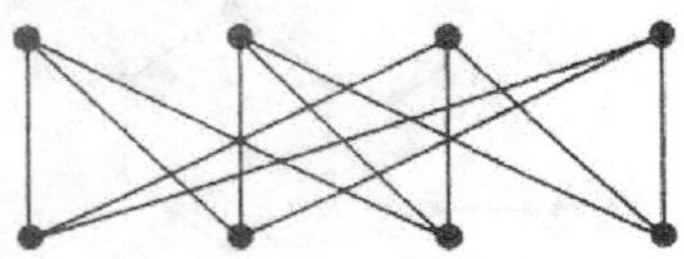

Solution.

The bipartite graph $K_{m,m}$ is regular of degree m since each vertex is connected to m other vertices and hence its degree is m. Subgraphs of $K_{m,m}$ can also be regular if m disjoint edges are deleted. For example, the subgraph of $K_{4,4}$ shown above is 3-regular. We can continue to delete m disjoint edges and each time obtain a regular graph of one less degree. These graphs may be disconnected, but in any case their connected components have the desired properties.

(5) Tree graph.

A graph is called a **cycle** if it contains no cycles. Since loops are cycles of length one while a pair of parallel edges produces a cycle of length two, any acyclic graph must be simple. A graph G is called a tree if it is a connected acyclic graph. As the number of vertices increases, the drawing of tree become complicated.

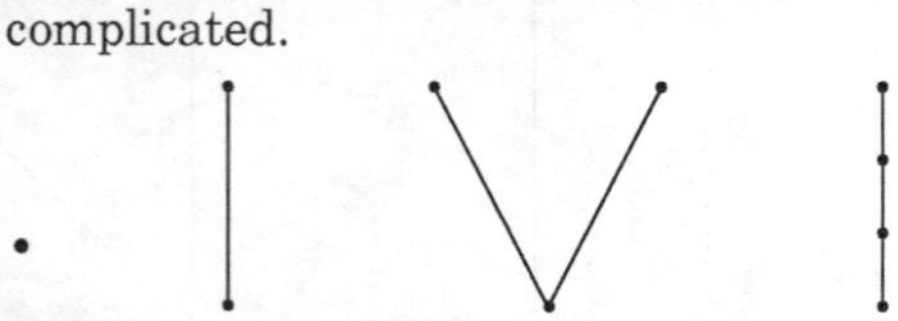

1 vertex *2 vertex* *3 vertex* *4 vertex*

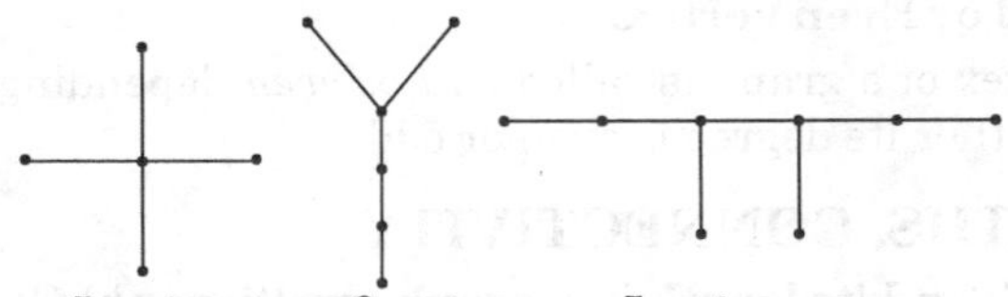

5 vertex *6 vertex* *7 vertex*

Fig. Trees with different vertices

Forest.

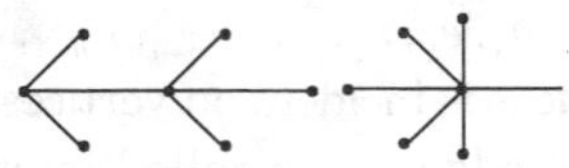

Fig. : Forest

Let G be an acyclic graph. Then any sub graph of G must also contains no cycles. The connected components of G are also acyclic and so they are trees, and therefore an acyclic graph is also called a *Forest.*

Theorem 1.

- Let u and v be distinct vertices of a Tree T, then there is precisely one path from u to v.
- Let G be a graph without any loops. If for every pair of distinct vertices u and v of G there is precisely one path from u to v, then G is a tree.

Theorem 2.

Let T be a tree with at least two vertices and let P $= u_0, u_1 \ldots.. u_n$ be a longest path in T. Then both u_0 and u_n have degree 1.

Corollary : Any tree T with atleast two vertices has more than one vertex of degree 1.

Theorem 3.

If T is a tree with n vertices, then it has precisely $(n - 1)$ edges.

Theorem 4.

If G be an acyclic graph with n vertices and k connected components, i.e. $w(G) = k$, then G has $n-k$ edges.

SUBGRAPHS

If G and H are two graphs will vertex sets V(H), V(G) and edge sets E(H) and E(G) respectively such that $V(H) \leq V(G)$ and $E(H) \leq E(G)$, then we call H as a subgraph of G (or G as a supergraph of H).

If $V(H) \subset V(G)$ and $E(H) \subset E(G)$, then A is a proper subgraph of G and if $V(H) = V(G)$, then H is a spanning subgraph of G.

Vertex deleted subgraphs.

If from a graph G with vertex set V, a subset U of V is deleted and all the edges which have a vertex in U as an end point, then (G-U) is called *vertex deleted subgraph.*

Edge deleted subraphs.

If a subset F of E is deleted from G, then (G-F) denotes the subgraph of G with vertex set V and edge set E-F, then (G-F) is called *edge deleted subgraph.*

Underlying simple graph.

By deleting from a graph G all loops and in each collection of parallel edges all edges but one a simple spanning subgraph of G is obtained, called *underlying simple graph of G.*

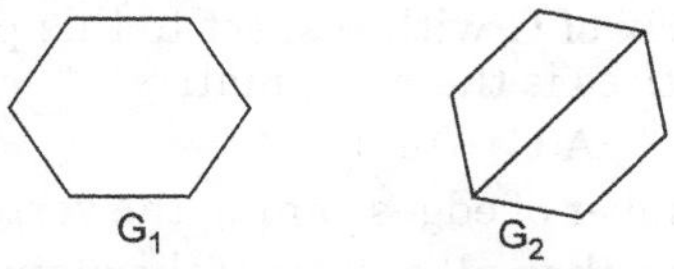

Fig. G_1 is a proper spanning subgraph of G_2.

If G is a graph with vertex set V, and U is a subset of V, then subgraph G(U) of G induced by U is defined to be the graph having vertex set U and edge set consisting of those edges of G that have both ends in U. Similarly if F is a non-empty subset of the edge set E of G, then subgraph G(F) of G induced by F is the graph whose vertex set of ends of edges in F and whose edge set is F.

PATHS AND WALKS

A walk in a graph G is a finite sequence

$$W = v_0e_1, v_1e_2, ... v_{k-1}e_kv_k$$

whose terms are alternately vertices and edges such that, for $1 \le i \le k$, the edge has ends v_{i-1} and v_i.

The vertex V_0 is called *origin* and V_k as the terminus of walk W. They need not be distinct.

Vertics v_1 ..... v_{k-1} are called *internal vertices.*

The number of edges in W are called *length of the walk.* Vertices and edges can be repeated in a walk. A trivial walk is one containing no edges and has length zero. A walk is closed or open depending on whether origin and terminus are same or different.

If edges $e_1, e_2, ... e_k$ of walk W are distinct, then such walk is called *trail.* So a trail is a walk if no edge is repeated.

If vertices $v_0, v_1, ... v_k$ of the walk $W = v_0e_1, v_1e_2, ... e_kv_k$ are distinct, then W is called *a path.* A path is a walk if no vertex is repeated. So every path is a trail and not every trail is a path.

CONNECTED GRAPHS AND CYCLES

Concatenation.

A vertex u is said to be connected to a vertex v in a graph G if there is a path in G from u to v. The converse is also true as v is also connected with u by the reverse path. If u is connected with v and v is connected with w, then u is connected with w.

This process of joining two walks which have a common end point to form a longer walk is called *concatenation (stringing together).*

Connected graphs.

A graph G is called **connected** if every two of its vertices are connected.

Given any vertex u of a graph G, let C(u) denote the set of all vertices in G that are connected to u. Then subgraph of G induced by u is called *connected component containing u.*

Diameter of a connected graph G.

The diameter of G, written diam (G), is the maximum distance between any two of its vertices.

Example. Determine whether or not each of the graphs given below is connected.

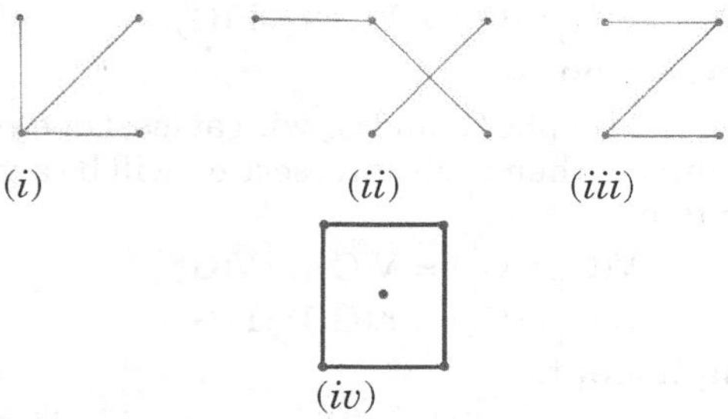

Solution.

(*i*) Yes. There is a path between any two vertices of the graph.

(*ii*) No. Here A, B, and Y are connected, and C and X are connected, but there is no path from A, B, to either C or X.

(*iii*) Yes. There is a path between any two vertices of the graph.

(*iv*) No. There is no path from C to any other vertex of the graph.

Cycle.

A non trivial closed trail in a graph a is called a **cycle**, if its origin and internal vertices are distinct.

The closed trail C = $v_1, v_2, ... v_n, v_1$ is cycle, if C has atleast 1 edge and $v_1, v_2, ... v_n$ are n distinct vertices.

A cycle with K edges is called **K-cycle.** A K-cycle is called **odd or even** depending on whether k is odd or even. Any two cycles with same length are isomorphic. An cycle is denoted by C_n.

Some cycles are defined as

1 cycle → Loop

2 cycle → Pair of parallel edges

3 cycle → triangle

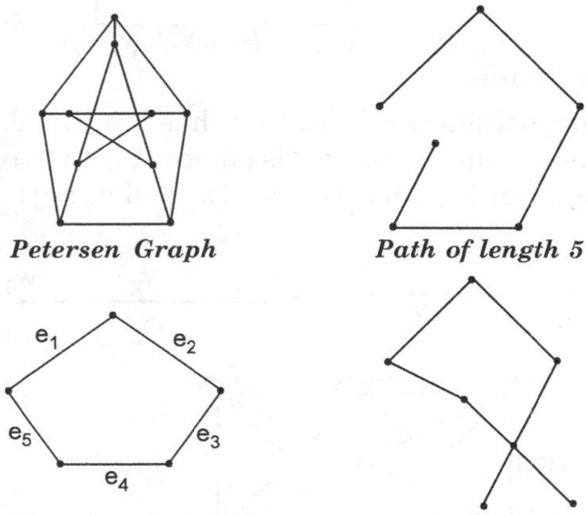

Fig. Path, Cycle and Trails

OPERATIONS ON GRAPHS

(1) Union.

Given two graphs G_1 and G_2. Their union will be a graph such that

$$V(G_1 \cup G_2) = V(G_1) \cup V(G_2)$$

and $$E(G_1 \cup G_2) = E(G_1) \cup E(G_2)$$

(2) Intersection.

When two graphs G_1 and G_2 with at least one vertex in common, then their intersection will be a graph such that

$$V(G_1 \cap G_2) = V(G_1) \cap V(G_2)$$
$$E(G_1 \cap G_2) = E(G_1) \cap E(G_2)$$

(3) Complement.

Complement $\overline{G}$ of G is defined as a simple graph with the same vertex set as G and where two vertices u and v are adjacent only when they are not adjacent in G.

e.g.

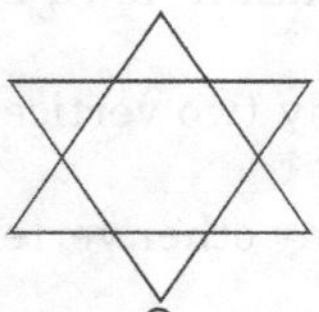

G

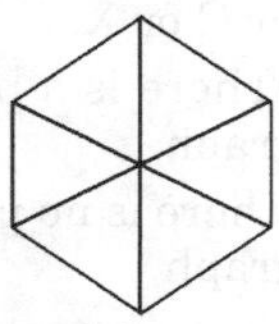

Complement of G

(4) Product of graphs.

To define product $G_1 \times G_2$ of two graphs, consider any two points $u = (u_1, u_2)$ and $v = (v_1, v_2)$ in $V = V_1 \times V_2$.

Then u and v are adjacent in $G_1 \times G_2$ whenever $[u_1 = v_1$ and u_2 adj $v_2]$ or $[u_2 = v_2$ and u_1 adj $v_1]$.

e.g.

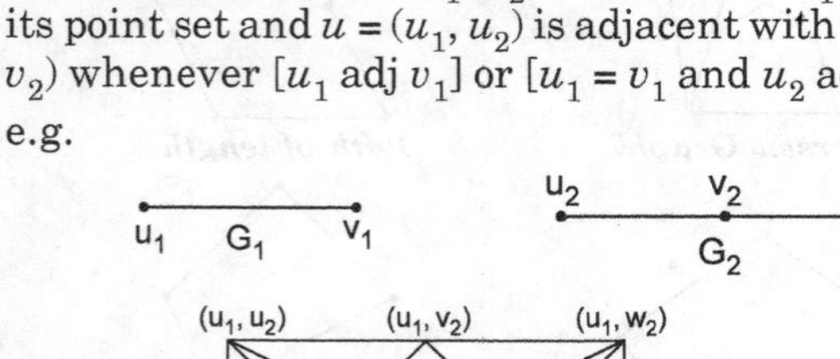

(5) Composition.

The composition $G = G_1[G_2]$ also has $V = V_1 \neq V_2$ as its point set and $u = (u_1, u_2)$ is adjacent with $v = (v_1, v_2)$ whenever $[u_1$ adj $v_1]$ or $[u_1 = v_1$ and u_2 adj $v_2]$.

e.g.

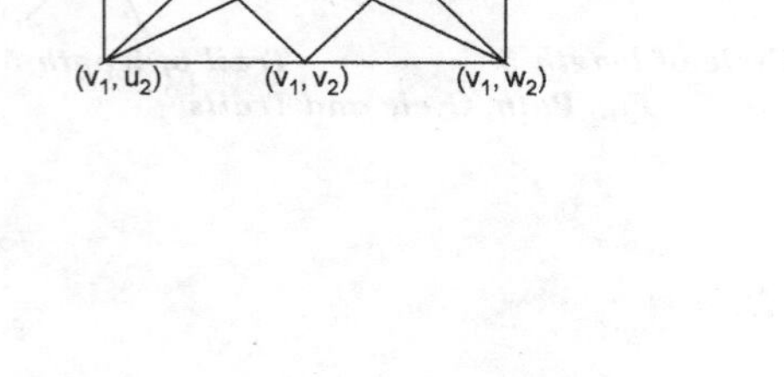

MATRIX REPRESENTATION OF GRAPHS

A graph can be represented inside a computer by using adjacency matrix or incidence matrix of a graph.

Let G be a graph with n vertices v_1 .... v_n. The adjacency matrix of G with respect to this particular listing of n vertices is the $n \times n$ matrix

$$A(G) = (a_{ij}),$$

where, a_{ij} = number of edges joining the vertex v_i to v_j.

If G has no loops, then all entries of the main diagonal will be 0 and if G has no parallel edges, then entries of A(G) are either 0 or 1.

Given a $n \times n$ symmetric matrix $A = (a_{ij})$, in which all the entries are non-negative integers, one can associate with it a graph G whose adjacency matrix is A by letting G have n vertices and joining vertex i to vertex j by a_{ij} edges.

Theorem 1.

Let G be a graph with n vertices v_i .... v_n and let A denote adjacency matrix of G with respect to this listing of the vertices. Let k be any positive integer and let A^k denote matrix multiplication of A, k times. Then $(ij)^{th}$ element of A^k is the number of different $v_i - v_j$ walks in G of length K.

This result can be used to check whether a graph is connected or not.

Theorem 2.

Let G be a graph with n vertices v_1, ... v_n and let A denote adjacency matrix of G with respect to this listing of the vertices.

Let $B = (b_{ij})$ be the matrix.

$\therefore$ $B = A + A^2 + \ldots\ldots + A^{n-1}$

Then G is a connected graph if for every pair of distinct indices i and j, we have $b_{ij} \neq 0$ i.e., iff B has no zero entries of the main diagonal.

This result can also be used to check the connect-edness of a graph by using its adjacency matrix.

Example. A graph G has the following adjacency matrix. Check whether it is connected.

$$A(G) = \begin{bmatrix} 0 & 0 & 1 & 0 & 0 \\ 0 & 0 & 0 & 1 & 0 \\ 0 & 0 & 0 & 0 & 1 \\ 0 & 1 & 0 & 0 & 1 \\ 0 & 0 & 1 & 1 & 0 \end{bmatrix}$$

Solution.

Since here, $n = 5$, therefore $B = A + A^2 + A^3 + A^4$

We find, A^2, A^3 and A^4 by matrix multiplication

$$A^2 = \begin{bmatrix} 1 & 0 & 0 & 0 & 1 \\ 0 & 1 & 0 & 0 & 1 \\ 0 & 0 & 2 & 1 & 0 \\ 0 & 0 & 1 & 2 & 1 \\ 1 & 1 & 0 & 0 & 1 \end{bmatrix},$$

$$A^3 = \begin{bmatrix} 0 & 0 & 2 & 1 & 0 \\ 0 & 0 & 1 & 2 & 0 \\ 2 & 1 & 0 & 0 & 3 \\ 1 & 2 & 0 & 0 & 3 \\ 0 & 0 & 3 & 3 & 0 \end{bmatrix}, A^4 = \begin{bmatrix} 2 & 1 & 0 & 0 & 3 \\ 1 & 2 & 0 & 0 & 3 \\ 0 & 0 & 5 & 4 & 0 \\ 0 & 0 & 4 & 5 & 0 \\ 3 & 3 & 0 & 0 & 6 \end{bmatrix}$$

Now $B = A + A^2 + A^3 + A^4 = \begin{bmatrix} 3 & 1 & 3 & 1 & 4 \\ 1 & 3 & 1 & 3 & 4 \\ 3 & 1 & 7 & 5 & 4 \\ 1 & 3 & 5 & 7 & 4 \\ 4 & 4 & 4 & 4 & 8 \end{bmatrix}$

Since B has no non zero entry off the main diagonal. The graph is connected.

Incidence matrix.

Another matrix associated with graph G is the incidence matrix.

Let G has n vertices listed as v_1 ..... v_n and t edges listed as e_1 ..... e_i.

The incidence matrix of a is the $n \times t$ matrix

$$M(G) = (m_{ij}),$$

where m_{ij} is number of times the vertex V_i is incident with the edge e_j,

i.e.

$$m_{ij} \begin{cases} 0 \text{ if } v_i \text{ is not an end of } e_j \\ 1 \text{ if } v_i \text{ is an end of non loop } e_j \\ 2 \text{ if } v_i \text{ is an end of the loop } e_j \end{cases}$$

Sum of elements in the ith row of MG gives degree of the vertex v_i.

BRIDGE OR CUT EDGE

On removing an edge from a graph, the number of connected components of the graph either remains unchanged or it increases by exactly 1. So, an edge e of a graph G is called *bridge or cut edge*, if the subgraph G-e has more connected components than G has.

Theorem 1.

Let e be an edge of the graph G and, as usual let G–e be the subgraph obtained by deleting e.

Then $\omega(G) \leq \omega(G-e) \leq \omega(G) + 1$

where $\omega(a)$ is number of connected components of G.

Theorem 2.

An edge e of a graph G is a bridge if e is not a part of any cycle in G.

Theorem 3.

Let G be a graph with n vertices, and q edges and, let $\omega(G)$ denote number of connected components of G. Then G has at least n-$\omega(G)$ edges, i.e. $q \geq n-\omega(G)$.

Corollary : A graph with n vertices less than $(n-1)$ edges can not be connected.

Theorem 4.

Let G be a graph with n vertices, then following three statements are equivalent.

I. G is tree

II. G is an acyclic graph witn $(n-1)$ edges.

III. G is a connected graph witn $(n-1)$ edges.

SPANNING TREES AND CONNECTOR PROBLEMS

Let G be a graph. A subgraph H of G is called a **spanning subgraph** of G if vertex set of H is same as the vertex set of G. Similarly, a spanning tree of a graph G is a spanning subgraph of G, that is a tree.

Theorem.

A graph G is connected if it has a spanning tree.

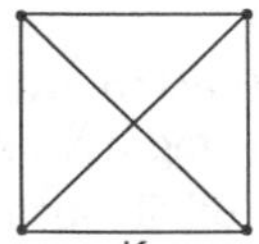

K_4

Cayley's theorem.

The complete graph K_n has n^{n-2} different spanning trees.

Spanning trees of K_4.

Since, here, $n = 4$, so, there will be $4^{4-2} = 16$ spanning trees.

A weighted graph is a graph G in which each edge e has been assigned a real number $\omega(e)$, called *weight of e*.

If H is a subgraph of a weighted graph, weight $\omega(H)$ of H is sum of the weights $\omega(e_1)$ + ... + $\omega(e_k)$, where $[e_1, \ldots\ldots, e_k]$ is set of edges of H.

Since many optimisation problems could be solved by using weighted graphs, hence try to find a subgraph of the weighted graph, with minimum (or maximum) weight.

Now look at a connected spanning subgraph of G which should not have a cycle, i.e. look at a spanning tree of G and finally look for a spanning tree with minimum weights, such a tree is called *minimal spanning tree or an optimal tree for G.*

Thus there are two algorithms for finding such a tree. A necessary condition for these algorithms is that no weights should be negative.

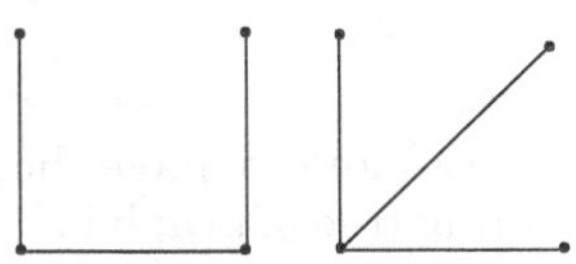

Fig. spanning trees for k_4

Kruskal's Algorithm.

In this algorithm, choose an edge of G which has smallest weight among the edges of G which are not loops. This algorithm gives an acyclic subgraph T of G and theorem given below proves that T is a minimal spanning tree of G.

Following steps are required.

Step 1 : Choose e_1, an edge of G, such that weight of e_1, $\omega(e_1)$ is as small as possible and e_1 is not a loop.

Step 2 : If edges $e_1, e_2, ... e_i$ have been chosen, then choose an edge e_{i+1} not already chosen, such that

(*i*) induced subgraph $G[(e_1, ... e_{i+1})]$ is acyclic, and

(*ii*) $\omega(e_{i+1})$ is as small as possible.

Step 3 : If G has n vertices, stop after n-1 edges have been chosen. Otherwise repeat step 2.

Theorem.

Let G be a weighted connected graph in which weights of the edges are all non-negative numbers.

Let T be a subgraph of G obtained by Kruskal's algorithm. Then T is a minimal spanning tree of G.

Example. Given below is a weighted graph G with 5 vertices. Apply Kruskal's Algorithm to determine a minimal spanning tree.

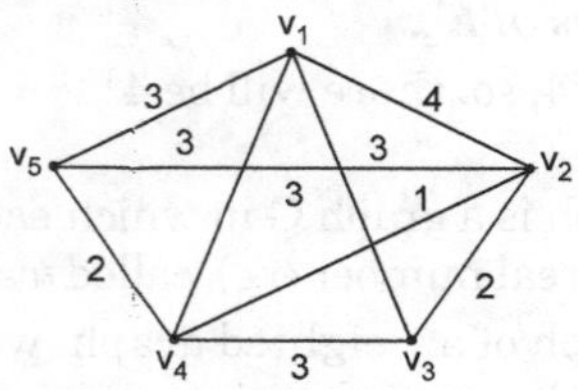

Solution.

According to step 1, choose, $e = v_2v_4$ as it has minimum weight.

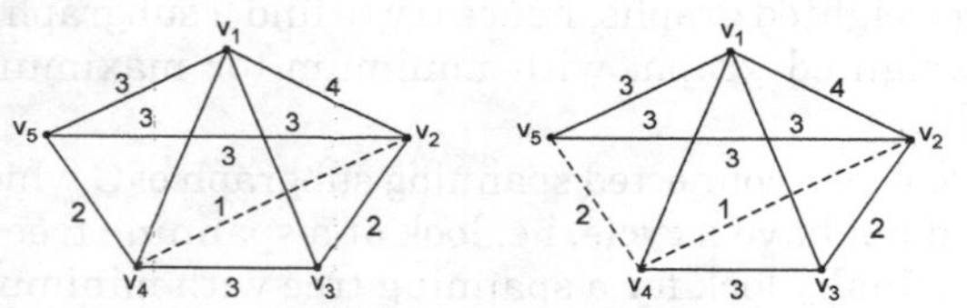

Choose $e = v_2\ v_3$

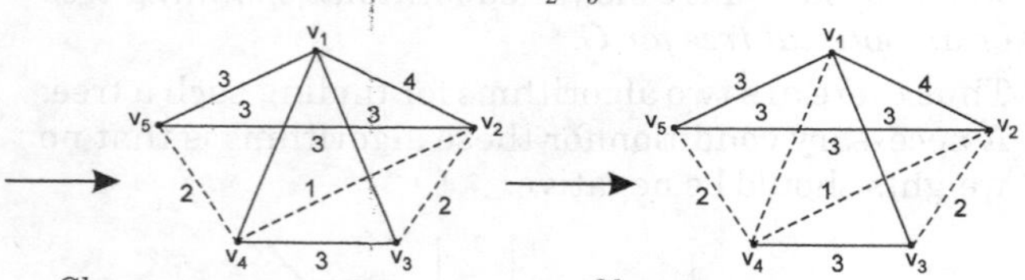

Choose $e = v_4\ v_5$ *Choose* $e = v_1\ v_4$

Since vertices are 5 and we have choosen 4 edges stop, minimal spanning tree of graph is

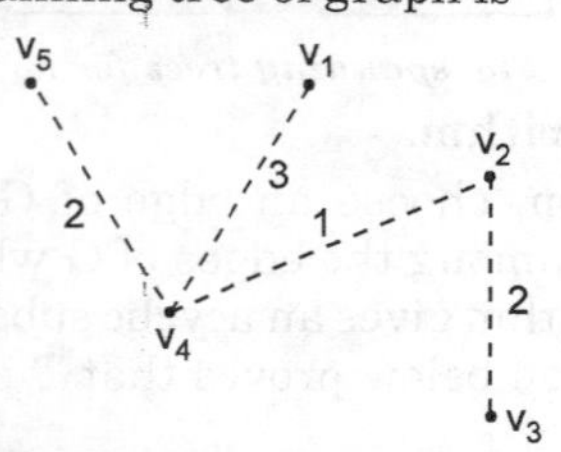

Greedy Algorithms.

Kruskal's algorithm is an example of a type of algorithm known as **Greedy**. Greedy algorithms are essentially algorithms that proceed by selecting the choice that looks best at the moment. All processes cannot be handled so simply.

Prim's Algorithm.

Another algorithm used for finding a minimal spanning tree is *Prim's algorithm.* It chooses a vertex first and chooses and edge with smallest weight incident on that vertex.

The algorithm involves following steps.

Step 1 : Choose any vertex v_1 of G.

Step 2 : Choose an edge $e_1 = v_1v_2$ of G such that $v_2 \neq v_1$ and e_1 has smallest weight among the edges of G incident with v_1.

Step 3 : If edges $e_1, e_2, ..., e_i$ have been chosen involving end points $v_1, v_2, ... v_{i+1}$.
Choose an edge $e_{i+1} = v_jv_k$ with $v_j \in \{v_i ...... v_{i+1}\}$ and $v_k \notin [v_1, ... v_{i+1}]$ such that e_{i+1} has smallest weight among the edges of G with precisely one end in $[v_1, ... v_{i+1}]$.

Step 4 : Stop after $n - 1$ edges have been chosen. Otherwise go to step 3.

Theorem.

Let G be a weighted connected graph in which weights of the edges are all non-negative numbers.

Let T be a subgraph of G obtained by Prim's algorithm. Then T is a minimal spanning tree of G.

Example. Find minimal spanning tree of the weighted graph in previous example using Prim's algorithm.

Solution.

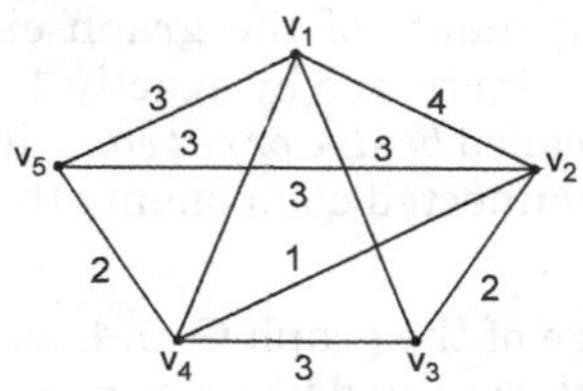

According to step 1, we choose vertex v_1.

Now edge with smallest weight incident on v_1 is $e = v_1, v_3$ so we choose $e = v_1, v_3$ (or $e = v_1, v_5$).

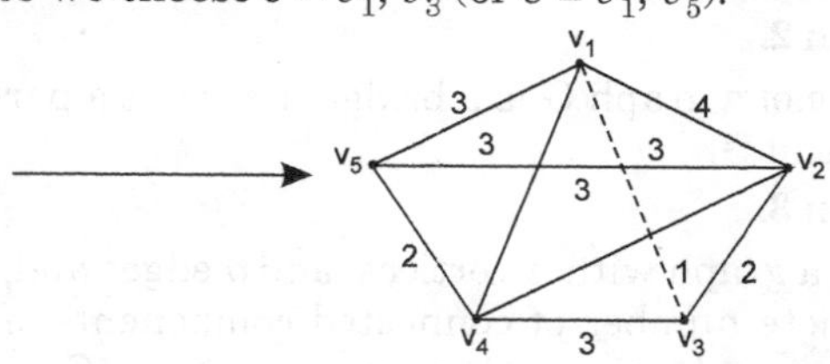

Now, we look on to the weights,

$w(v_1, v_2) = 4$, $w(v_1, v_4) = 3$, $w(v_1, v_5) = 3$, $w(v_3, v_2) = 2$, $w(v_3, v_4) = 3$

Since minimum is $w(v_3, v_2)$, we choose, $e = v_3, v_2$

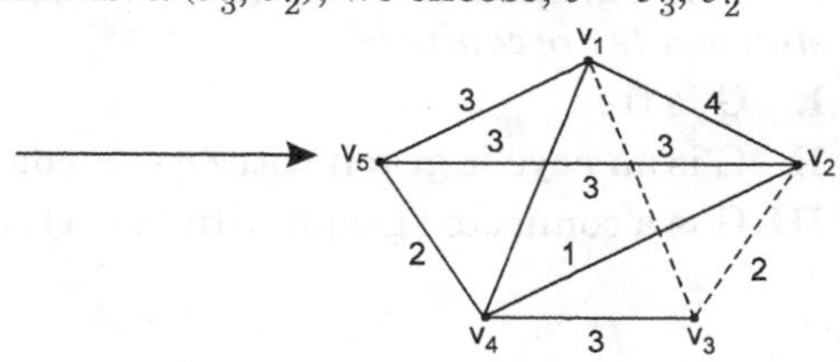

Again, $w(v_2, v_5) = 3$, $w(v_2, v_4) = 1$ and $w(v_3, v_4) = 3$.
We choose $e = v_2, v_4$

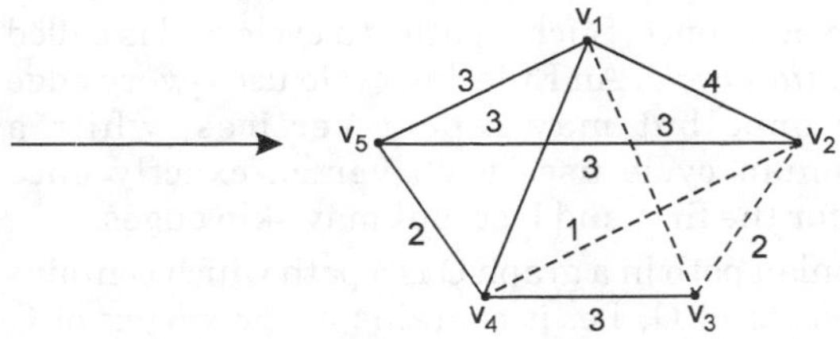

Now, $w(v_4, v_5) = 2$. We choose $e = v_4, v_5$

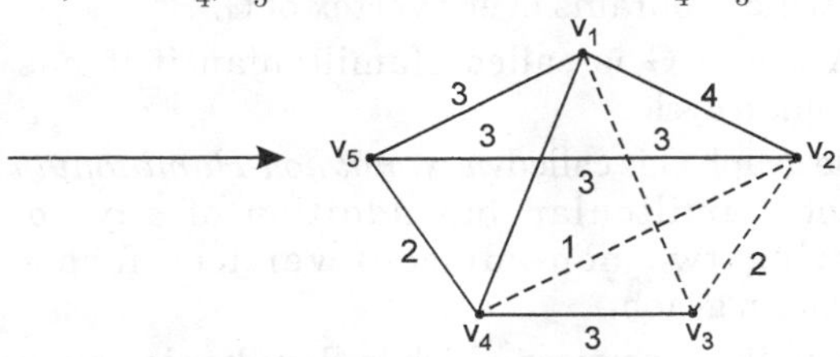

The minimal spanning tree is

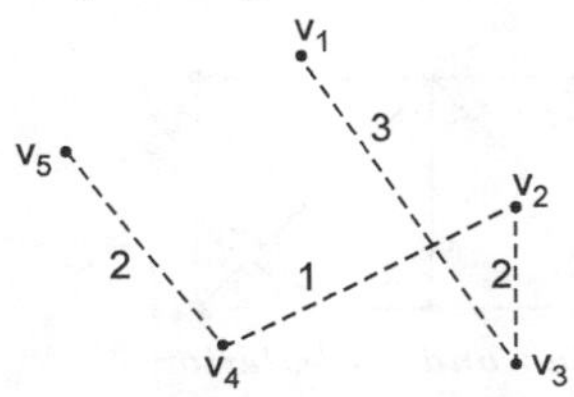

CUT VERTICES AND CONNECTIVITY

Cut vertex is analogue of a bridge. A vertex v of a graph is called a *cut vertex of G* if

$$\omega(G - v) > \omega(G),$$

where $\omega(G)$ are number of components in graph G, i.e. a cut vertex breaks a graph into a subgraph having more connected components, then G has

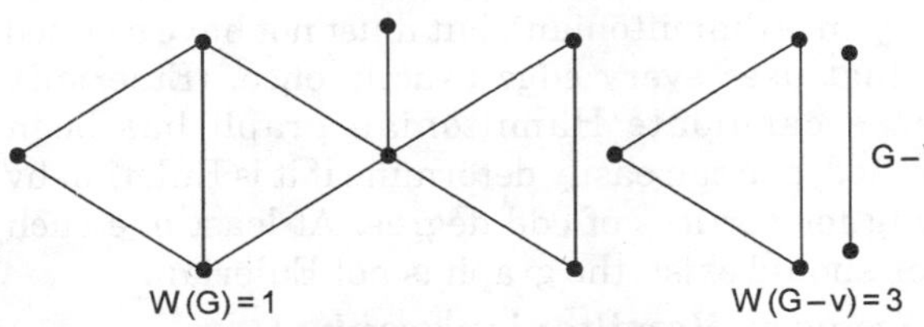

Fig. Graph after deleting cut vertex v

Theorem.

Let G be a graph with n vertices, when $n \geq 2$. Then G has at least two vertices which are not cut vertices.

If a connected graph G has a cut vertex v, then connectedness of G is vulnerable at v. Let G be a simple graph. The connectivity of G, denoted by $k(G)$, is the smallest number of vertices in G whose deletion from G leaves either a disconnected graph or k_1.

$$k(k_n) = n - 1.$$

Connected graph has

$k(G) = 1$, iff either $G = k_2$ or G has a cut vertex.

$k(G) = 0$, iff either $G = k_1$ or G is disconnected.

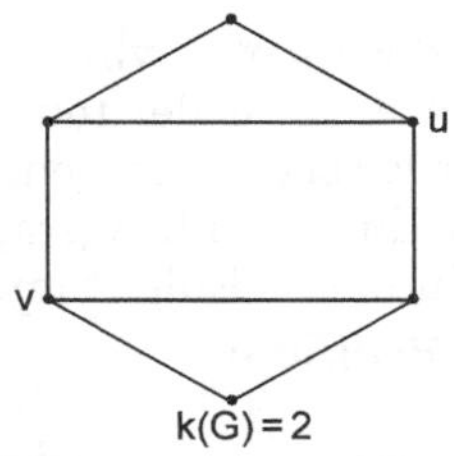

Fig. Connectivity of graph (by deleting u and v)

A simple graph G is called *n-connected* $(n \geq 1)$ if

$$k(G) \geq n.$$

Let u and v be two vertices of a graph G. A collection $\{p_{(1)} \ldots\ldots p_{(n)}\}$ of u-v paths is said to be *internally disjoint* if, given any distinct pair $P_{(i)}$ and $P_{(j)}$ in the collection, u and v are the only vertices $P_{(i)}$ and $P_{(j)}$ have in common.

BINARY TREES

Trees are principally used in the data representation. Here, all trees will be rooted, *i.e.* they will have a distinguished vertex r (called **root**) with the property that for every vertex $v \in V$, there is a directed r-v graph. The trees will also be levelled, i.e. root r will constitute level 0, neighbours of r will constitute level 1 and neighbours of vertices in level 1 will have level 2 and so on. The neighbours on next level are called *children (descendent)* and previous level are called *Father (predecessor).*

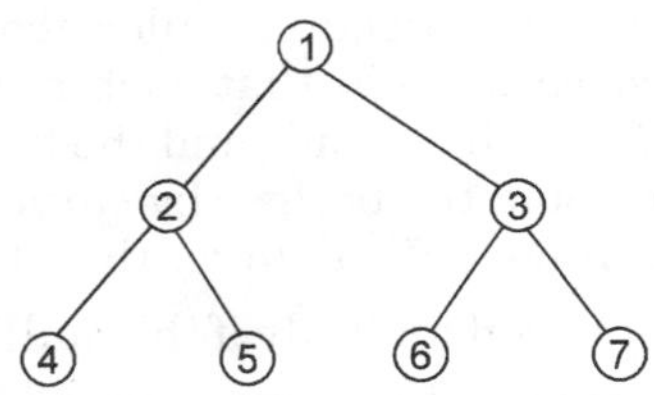

Fig. Full binary tree

A binary tree is a rooted, levelled tree in which any vertex has at most two children. If every vertex of binary tree has either two children or no children, then we say the tree is a *full binary tree*.

Use of binary tree in data representation

We have to express $a + 3b = a + 3 \times b$, then we will draw following tree :

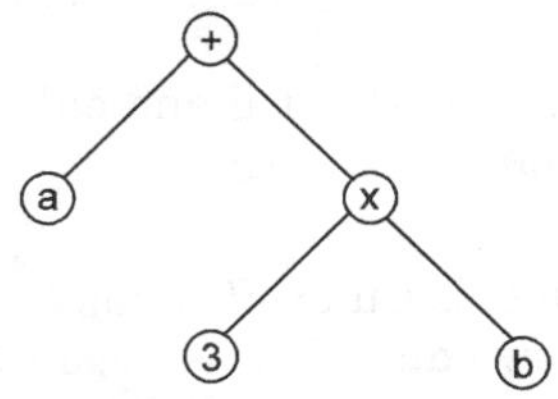

EULER TOURS.

A trail is defined as Euler trail if it includes every edge of G. So, a trail is Euler if each edge G is in the trail exactly once.

A tour of G is a closed walk of G which includes every edge of G at least once. An Euler tour of G is a tour which includes each edge of G exactly once. So, an Euler tour is just a closed Euler trail. A graph G is called *Eulerian or Euler* if it has as Euler tour.

Konisberg Bridge Problem

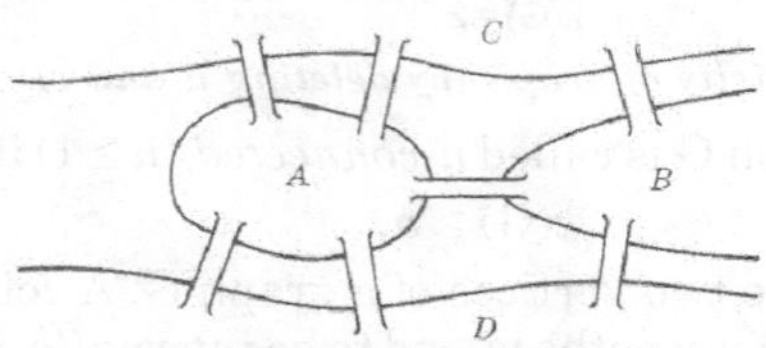

(a) Konigsberg in 1736

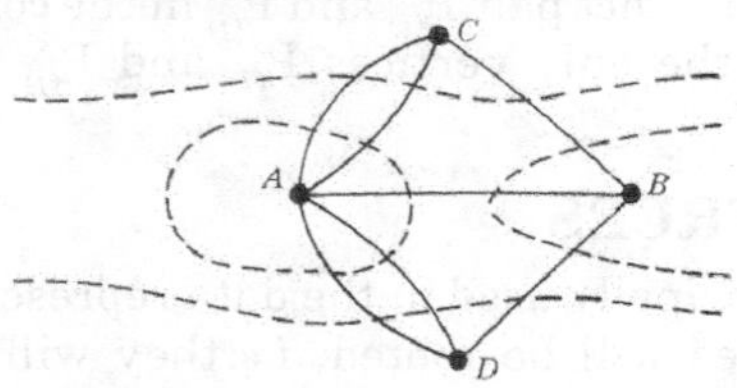

(b) Euler's graphical representation

The eighteenth-century East Prussian town of Königsberg included two islands and seven bridges as shown in above in Fig (*a*). Beginning anywhere and ending anywhere, can a person walk through town crossing all seven bridges but not crossing any bridge twice. The people of Königsberg wrote to the celebrated Swiss mathematician L. Euler about this question. Euler proved in 1736 that such a walk is impossible. He replaced the islands and the two sides of the river by points and the bridges by curves, thus obtaining Fig. (*b*). It is not difficult to see that the walk in Königsberg is possible if and only if the multigraph in Fig. (*b*) is traversable. But this multigraph has four odd vertices, and hence it is not traversable. Thus one cannot walk through Königsberg so that each bridge is crossed exactly once.

Theorems on Euler graph.

Theorem 1.

Let G be a graph in which degree of every vertex is at least two. Then G contains a cycle.

Theorem 2.

A connected graph G is Euler if and only if degree of every vertex is even.

Theorem 3.

A connected graph G is Euler iff G has cycles $C_{(1)}$,..... $C_{(n)}$ such that every edge of G belongs to exactly on cycle $C_{(i)}$, i.e. G is union of edge disjoint cycles.

Theorem 4.

A connected graph G has an Euler trail iff it has at most two odd vertices, i.e. it has either no vertices of odd degree or exactly two vertices of odd degree.

HAMILTONIAN GRAPHS

It is a graph with a closed path that includes every vertex exactly once. Such a path is a cycle and is called *Hamiltonian cycle*. An Eulerian cycle uses every edge exactly once but may repeat vertices, while a Hamiltonian cycle uses each vertex exactly once (except for the first and last) but may skip edges.

Hamiltonian path in a graph G is a path which contains every vertex of G, i.e. it contains every vertex of G once and only once. Hamiltonian cycle in a graph G is a cycle which contains every vertex of G.

Now, A graph G is called Hamiltonian if it has a Hamiltonian cycle.

A simple graph G is called *maximal non-Hamiltonian* if it is not Hamiltonian but addition of any edge connecting two non-adjacent vertices forms a Hamiltonian graph.

A graph with six vertices which is Hamiltonian but not Eulerian.

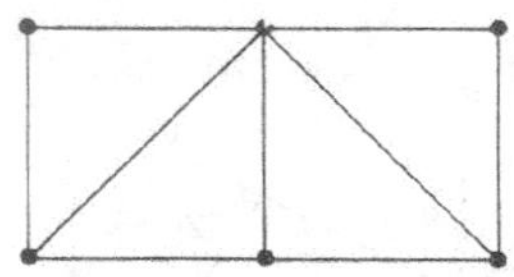

(a) Hamiltonian and noneulerian

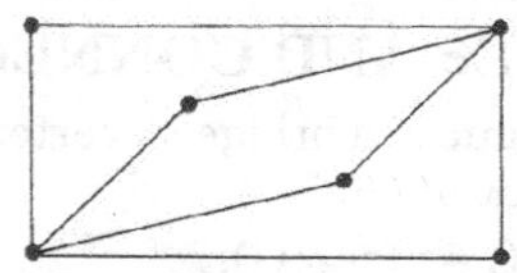

(b) Eulerian and nonhamiltonian

There are many possible solutions to this problem and one of these is shown above in Fig. (*a*). However every solution must have a cycle that includes every vertex exactly once (hamiltonian), but must not have a closed trail that uses every edge exactly once. (Eulerian). When a candidate Hamiltonian graph has been identified, one can easily determine if it is Eulerian by looking for vertices of odd degree. At least one such vertex should exist, the graph is not Eulerian.

Theorems on Hamiltonian graph.

Theorem 1 (Dirac).

If G is a simple graph with n vertices where $n \geq 3$, and the degree $d(v) \geq \frac{n}{2}$ for every vertex v of G, then G is Hamiltonian.

Theorem 2.

Let G be a simple graph with n vertices and u and v be non-adjacent vertices in G such that

$$d(u) + d(v) \geq n$$

If G + uv denote the subgraph of G obtained by joining u and v by an edge, then G is Hamiltonian iff G + uv is Hamiltonian.

CLOSURE OF A GRAPH

Let G be a simple graph.

If there are two non-adjacent vertices u_1 and v_1 in G such that $d(u_1) + d(v_1) \geq n$ in G, join u_1 and v_1 by an edge to form the supergraph G_1. If two vertices u_2 and v_2 are there such that $d(u_2) + d(v_2) \geq n$ in G_1, join u_2 and v_2 by an edge to form the supergraph G_2.

Continue in this was to join pairs of non adjacent vertices whose degree sum is at least n. The final supergraph thus obtained is called *closure of G* and is denoted by C(G).

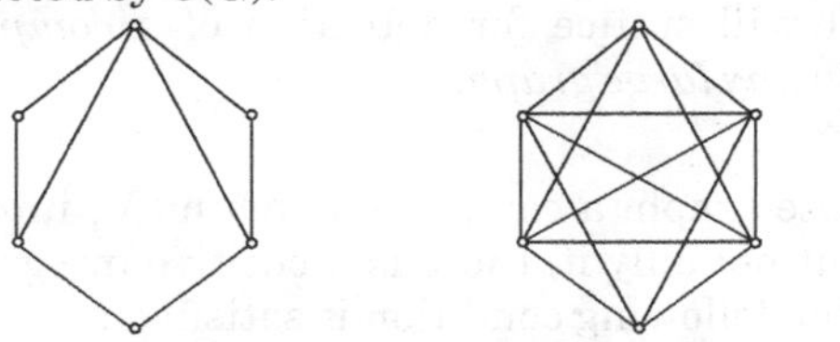

Fig. Graph and its closure C(G)

Theorem due to Bondy and Chvatal.

A simple graph is Hamiltonian iff its closure C(G) is Hamiltonian.

ISOMORPHISM OF GRAPHS

Let G(V, E) and G*(V*, E*) are graphs and $f: V \to V^*$ is a one-to-one correspondence beta sets of vertices such that $\{u, v\}$ is an edge of G if and only if $\{f(u), f(v)\}$ is an edge of G*. Then f is *isomorphism* between G and G*, and G and G* are called *isomorphic* graphs. Normally, we do not distinguish between isomorphic graphs (even though their diagrams may "look different").

Example. Figure shows ten graphs pictured as letters. Which of the ten graphs are isomorphic to M?

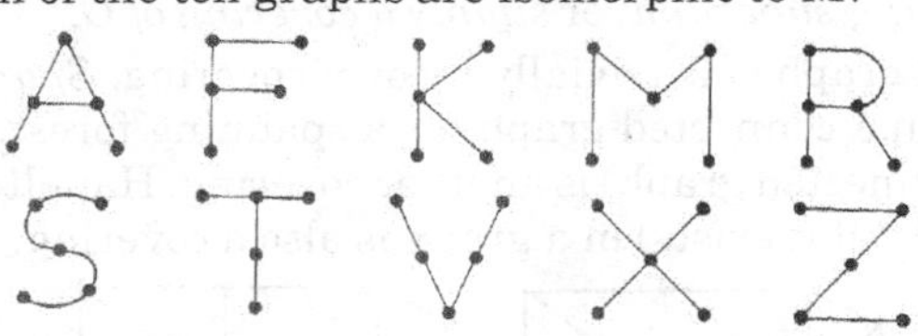

Solution.

M consists of five vertices in a single line. Thus S, V, and Z (and M itself) are isomorphic to M.

HOMEOMORPHIC GRAPHS

Given any graph G, we can get a new graph by dividing an edge of G with additional vertices. Two graphs G and G* are called *homeomorphic* if they can be obtained from isomorphic graphs by this method.

Graphs which are called homeomorphic, but not isomorphic.

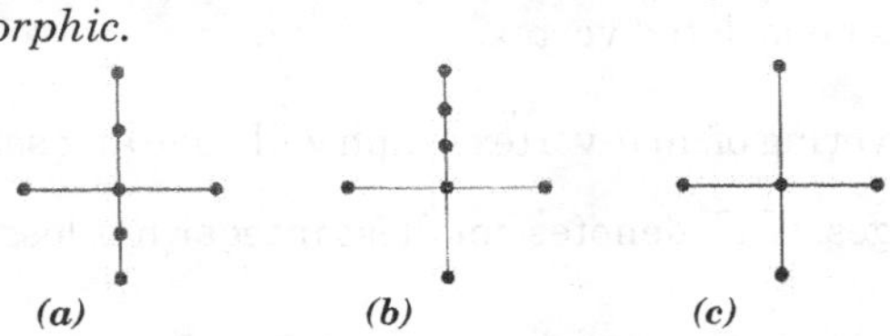

(a) *(b)* *(c)*

Graphs (a) and (b) in figure are not isomorphic; but they are homeomorphic since each can be obtained from (c) by adding appropriate vertices.

PLANAR GRAPHS

It is a graph drawn in the plane in such a way that any pair of edges meet only at their end vertices.

Planar graph is a graph which is isomorphic to a plane graph, *i.e.* it can be redrawn as a plane graph.

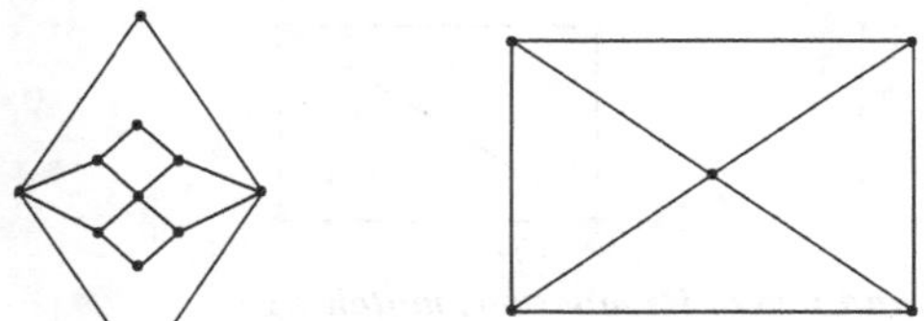

Jordan Curve in the plane is a continuous non-interesecting curve whose origin and terminus coincide.

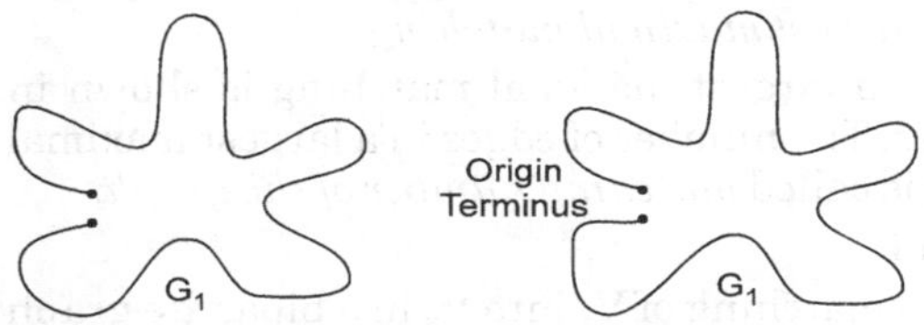

Fig. G_1 is Jordan Curve but G_2 is not

MATCHINGS

Let four applicants a_1, a_2, a_3, and a_4 are available to fill six vacant positions p_1, p_2, p_3, p_4, p_5 and p_6.

Applicant a_1 is qualified to fill position p_2 or p_5.

Applicant a_1 is qualified to fill position p_2 or p_5. Applicant a_2 can fill p_2 or p_5.

Applicant a_3 is qualified for p_1, p_2, p_3, p_4, p_5 or p_6.

Applicant a_4 can fill jobs p_2 or p_5.

This situation is represented by the graph in figure below.

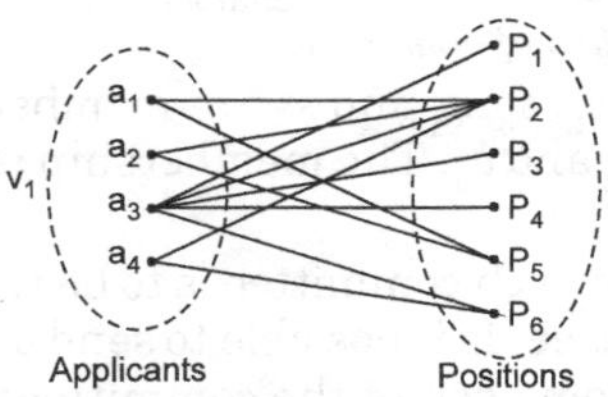

Fig. (a) Bipartite graph

The vacant positions and applicants are represented by vertices. The edges represent the qualifications of each applicant for filling different positions. The graph clearly is bipartite, the vertices falling into two sets $V_1 = \{a_1, a_2, a_3, a_4\}$ and $V_2 = \{p_1, p_2, p_3, p_4, p_5, p_6\}$.

The most likely questions one is to ask in this situation are :

Is it possible to hire all the applicants and assign each a position for which he is suitable ?

If the answer is no, what is the maximum number of positions that can be filled from the given set of applicants ?

This is a problem of matching (or assignment) of one set of vertices into another. More formally, a matching in a graph is a subset of edges in which no two edges are adjacent. A single edge in a graph is obviously a matching.

Maximal maching.

It is a matching to which no edge in the graph can be added.

e.g. in a complete graph of three vertices (*i.e.* a triangle) any single edge is a maximal matching. The edges shown by heavy lines.

(a) *(b)*

Fig. Graph and two of its maximal matchings.

Clearly, a graph may have many different maximal matchings and of different sizes. Among these, the maximal matchings with the largest number of edges are called *largest maximal matchings.*

In Fig. *(b)*, a largest maximal matching is shown in heavy lines. The number of edges in a largest maximal matching is called *matching number of the graph.*

Theorem 1.

A complete matching of V_1 into V_2 in a bipartite graph exists if and only if every subset of r vertices in V_1 is collectively adjacent to or more vertices in V_2 for all values of r.

Problem of distinct representatives.

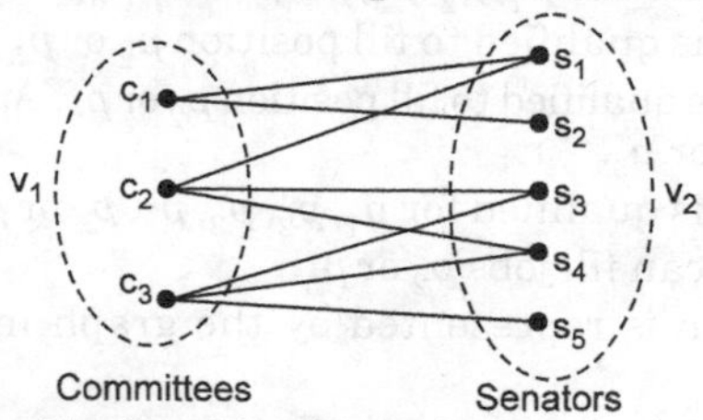

Fig. Membership of committees

Five senators s_1, s_2, s_3, s_4 and s_5 are members of three committees, c_1, c_2 and c_3. The membership is shown in the figure.

One member from each committee is to be represented in a super committee. Is it possible to send one distinct representative from each of the committees ?

This problem is one of finding a complete matching of a set V_1 into set V_2 in a bipartite graph.

Now, use Theorem 1 and check if r vertices from V_1 are collectively adjacent to at least r vertices from V_2, for all values of r.

Result is shown in the table below (ignore last column for the time being).

	V_1	V_2	$r - q$
$r = 1$	$\{c_1\}$	$\{s_1, s_2\}$	-1
	$\{c_2\}$	$\{s_1, s_3, s_4\}$	-2
	$\{c_3\}$	$\{s_3, s_4, s_5\}$	-2
$r = 2$	$\{c_1, c_2\}$	$\{s_1, s_2, s_3, s_4\}$	-2
	$\{c_2, c_3\}$	$\{s_1, s_3, s_4, s_5\}$	-2
	$\{c_3, c_1\}$	$\{s_1, s_2, s_3, s_4, s_5\}$	-3
$r = 3$	$\{c_1, c_2, c_3\}$	$\{s_1, s_2, s_3, s_4, s_5\}$	-2

Thus for this example, condition for the existence of a complete matching is satisfied as stated in Theorem 1. Hence, it is possible to from the super committee with one distinct representative from each committee.

The problem of distinct representatives just solved was a small one.

A larger problem would have become unwieldy, if there are M vertices in V_1. Theorem 1 requires that we take all $2^M - 1$ nonempty subsets of V_1 and find the number of vertices of V_2 adjacent collectively to each of these. In most cases, however, following simplified version of Theorem 1 will suffice for detection of a *complete matching in any large graph.*

Theorem 2.

In a bipartite graph, a complete matching V_1 into V_2 exists if (but not only if) there is a positive integer m for which the following condition is satisfied :

Degree of every vertex in $V_1 \geq m \geq$ degree of every vertex in V_2.

In the bipartite graph of above figure,

Degree of every vertex in $V_1 \geq 2 \geq$ degree of every vertex in V_2.

Therefore, there exists a complete matching.

In the bipartite graph, no such number is found, because

degree of $p_2 = 4 >$ degree of a_1.

COVERINGS

In a graph G, a set g of edges is said to cover G if every vertex in G is incident on at least one edge in g. A set of edges that covers a graph G is called *edge covering, covering subgraph*, or *simply a covering of G.*

e.g. a graph G is trivially its own covering. *Spanning tree* in a connected graph (or a spanning forest in an unconnected graph) is another covering. Hamiltonian circuit (if it exists) in a graph is also a covering.

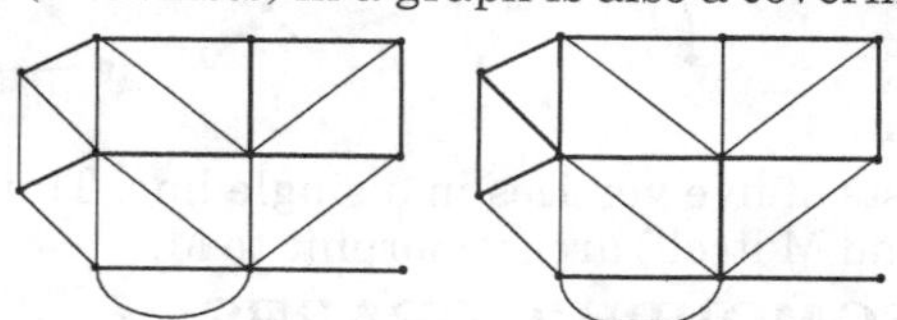

(a) (b)

Fig. Graph and two of its minimal coverings

Here we shall investigate minimal covering-a covering from which no edge can be removed without destroying its ability to cover the graph.

- Covering exists for a graph if and only if the graph has no isolated vertex.
- Covering of an n-vertex graph will have at least $\left\lceil \frac{n}{2} \right\rceil$ edges. ($\lceil x \rceil$ denotes smallest integer not less than x.)
- Every pendant edge in a graph is included in every covering of the graph.

- Every covering contains a minimal covering.
- If we denote remaining edges of a graph by $(G - g)$, the set of edges g is a covering if and only if, for every vertex v,

  degree of vertex in $(G - g) \leq$ (degree of vertex v in G) – 1
- No minimal covering can contain a circuit, for which we can always remove an edge from a circuit without leaving any of the vertices in the circuit uncovered. Therefore a minimal covering of an n-vertex graph can contain no more than n-edges.
- A graph in general has many minimal coverings, and they may be of different sizes (i.e. consisting of different numbers of edges). The number of edges in a minimal covering of the smallest size is called *covering number of the graph.*

Theorem 3.

A covering g of a graph is minimal if and only if g contains no paths of length three or more.

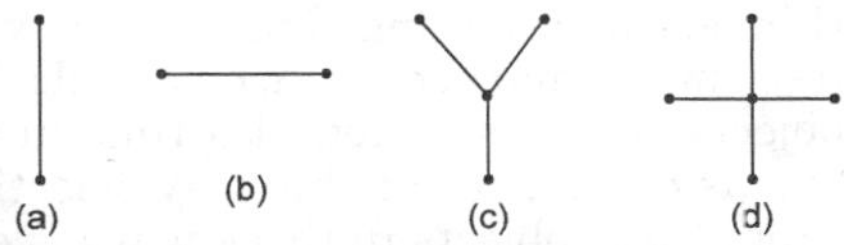

Fig. Star graphs of one, two, three, and four edges.

Let graph in the above figure represents street map of a part of a city, each of the vertices is a potential trouble spot and must be kept under the surveillance of a patrol car. How will you assign a minimum number of patrol cars to keep every vertex covered?

The answer is a smallest minimal covering. The covering shown in Fig. (a) is an answer, and it requires six patrol cars. Clearly, since there are 11 vertices and no edge can cover more than two, less than six edges cannot cover the graph.

SOLVED EXAMPLES

1. Prove by induction that expression for the number of diagonals in a polygon of n sides is

$$\left[\frac{n(n-3)}{2}\right]$$

Solution.

The result holds for $n = 4$, i.e. for quadrilateral.

Assume that it holds for $n = m$, if there are $\left[\frac{m(m-3)}{2}\right]$ diagonals of m sides polygon.

Now, increase the sides to $m + 1$, which results in addition of one corner to the polygon. To these edges diagonals from all other edges can be added except the neighbouring edges.

$\therefore$ Number of diagonals added $= (m + 1) - 2$

$= m - 1$

$\therefore$ Total number of diagonals

$$= \left[\frac{m(m-3)}{2}\right] + (m-1) = \frac{[((m+1)(m+1)-3)]}{3}$$

Hence the result is proved by induction.

2. Consider the multigraph G = G (V, E) shown in the figure.

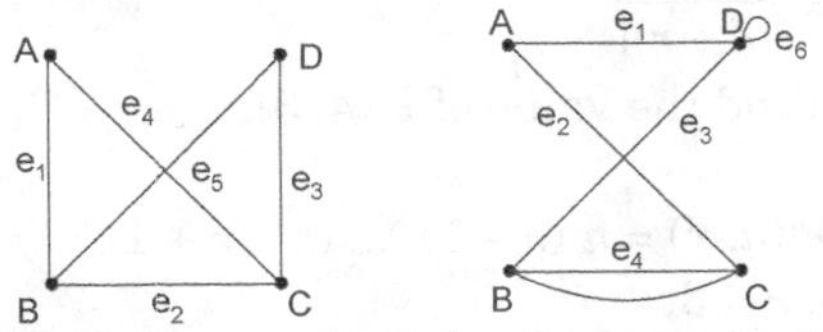

(i) Find the number of vertices and edges.

(ii) Are there any multiple edges or loops ? If So, what are they ?

Solution.

(i) G contains four vertices A, B, C, D; and six edges, $e_1, e_2, \ldots\ldots\ldots, e_6$ (although the edges e_2 and e_3 across at a point, the diagram does not indicate that the intersection point is a vertex of G.)

(ii) The edges e_4 and e_5 are multiple edges since they both have the same endpoints B and C. The edge e_6 is a loop.

3. Draw a diagram for each of the following graphs G = G (V, E) :

V = {A, B, C, D}, E = [{A, B}, {D, A}, {C, A}, {C, D}]

Solution.

First draw vertices of the graph, and then connect the appropriate vertices to indicate the edges of the graph, as shown in the figure.

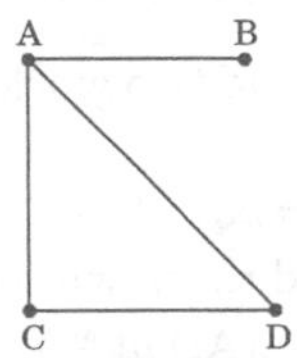

4. Draw a diagram for the graphs G = G (V, E)

V = {a, b, c, d, e, f}, E = [{a, d}, {a, f}, {b, c}, {b, f}, {c, e}]

Solution.

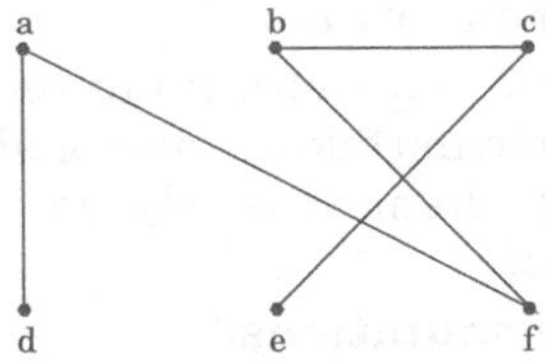

5. Let G = G (V, E) has five vertices. Find the maximum number of m of edges in E if

(i) G is a graph, and

(ii) G is a multigraph.

Solution.

(i) There are C (5, 2) = 10 ways of choosing two vertices from V ; hence $m = 10$.

(ii) Since multiple edges are peritted, G can have any number of edges (and loops), finite or infinite; hence no such maximum number m exists.

COMBINATORIES

INTRODUCTION

Objects (or things) can be arranged in many ways. Let there are three objects marked a, b, c on a table. From these, two objects can be selected at a time in three different ways as ab, ac, bc. In this way selection of two objects from three objects in three ways is called *combination.*

The above selections ab, ac, bc can also be arranged as ba, ca, cb. So, we can understand that two objects can be selected from three objects and arranged in six ways. These arrangements are called *permutations.*

FUNDAMENTAL PRINCIPLE

If A is a finite set, then number of different elements in A is denoted by $n(A)$.

Here $n(A)$ is non-negative integer.

e.g. If $A = \{2, 5, 7\}$, then $n(A) = 3$

If $A = \left\{a, \frac{2}{3}, 5, 0, \frac{1}{3}\right\}$, then $n(A) = 5$

If $C = \phi$ (empty set), then $n(C) = 0$.

Fundamental principle of counting.

If a man can do a work in p different ways and another work in q different ways, then he can do the two works one after the other in pq ways.

Now treating one way of doing a work as one element, we form the two sets A and B.

We denote the set of ways of doing the first work by A and set of ways of doing the second work by B. Then

$n(A) = p$ and $n(B) = q$.

Number of ways one can do the two works in an order is

$$n(A \times B) = n(A) \times n(B) = pq.$$

If same principle is extended to 'm' works, then

$$n(A_1 \times A_2 \times \ldots\ldots \times A_m) = n(A_1)\, n(A_2) \ldots\ldots n(A_m).$$

PERMUTATIONS

(1) Linear permutations.

From the elements of a finite set A (in general distinct elements) taking some or all elements and arranging them linearly is called *linear permutation.*

(2) Similar permutations.

If in a given set of things, some elements are alike, permutations formed from them are called *similar permutations.* In this case, number of permutations will change because we can not identify the change among the like things.

(3) Circular permutations.

If 'n' distinct things are arranged in a circular manner giving preference only to relative positions, then it is called *circular permutation.*

In a circular permutation, if places are changed circularly, the permutations will not change.

e.g. If $A = \{4, 5, 6\}$, then permutations obtained by arranging all elements in A in an order are:

456, 465, 546, 564, 645, 654.

- From a set A of n elements, number of permutations taken $r\ (\le n)$ at a time is denoted by n_{P_r}, or $P(n, r)$.

Notation.

The product of first 'n' natural numbers is called *factorial n* and is denoted by $n!$.

We define $0! = 1$ and we observe that $1! = 1$.

e.g. (*i*) $6! = 6 \times 5 \times 4 \times 3 \times 2 \times 1 = 720$

(*ii*) Taking all the letters at a time from A, A, B, the possible number of permutations is 3. They are AAB, ABA, BAA.

Here we can not find difference between A's but if you find it, then number of permutations is

$$3P_3 = 3! = 6.$$

Theorem.

The number of permutations of 'n' dissimilar things taken $r\ (r \le n)$ at a time is

$$n_{P_r} = n\,(n-1)\,(n-2) \ldots\ldots (n-r+1)$$

A finite bijective mapping is a permutation.

If $S = \{a, b, c, d, e\}$ and defining a permutation f as

$$f(a) = b;\quad f(b) = d;\quad f(c) = a;\quad f(d) = e;\quad f(e) = c$$

then we denote the permutation f as $f = \begin{bmatrix} a\,b\,c\,d\,e \\ b\,d\,a\,e\,c \end{bmatrix}$

$$n_{P_n} = n(n-1)\,(n-2) \ldots\ldots (n-n+1)$$
$$= n\,(n-1)\,(n-2) \ldots\ldots 2 \cdot 1$$
$$= 1.2.3. \ldots\ldots\ldots (n-2)\,(n-1)n = n!$$

i.e. number of permutations containing all objects from n distinct things is $n_{P_n} = n!$.

Note : $n_{P_r} = n(n-1)\,(n-2) \ldots\ldots (n-r+1)$

$$= \frac{n(n-1)\,(n-2) \ldots\ldots (n+r+1)\{(n-r)!\}}{(n-r)!}$$

$$= \frac{1 \cdot 2 \cdot 3 \ldots\ldots\ldots (n-1) \cdot n}{(n-r)!} = \frac{n!}{(n-r)!}$$

$$\therefore \quad n_{P_r} = \frac{n!}{(n-r)!}$$

Example. Find the value of P (4, 3)

Solution:

We know, $P(n, r) = n\,(n-1) \ldots\ldots (n-r+1)$

Here $n = 4$, $r = 3$,

$\therefore \quad n - r + 1 = 4 - 3 + 1 = 2.$

Hence $P(n, r) = P(4, 3) = 4.3.2 = 24$

Alternately : $P(n, r) = \dfrac{n!}{(n-r)!}$

$\therefore \quad P(4, 3) = \dfrac{4!}{(4-3)!} = \dfrac{4!}{1!} = 4 \times 3 \times 2 = 24$

Theorem. $^nP_r = {}^{(n-1)}P_r + r \cdot {}^{(n-1)}P_{(r-1)}$

CIRCULAR PERMUTATIONS

Types of circular permutations

Circular permutations are of two types.

(*i*) Clockwise (*ii*) Anti-clockwise.

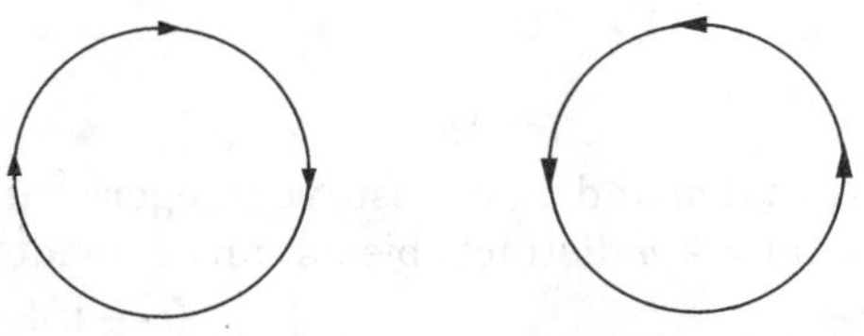

Clockwise ***Anti-clockwise***

Considering direction, the above two permutations are different, otherwise they are the same.

Number of Circular permutations.

Theorem : The number of circular permutations of n different things taken all at a time is $(n-1)!$

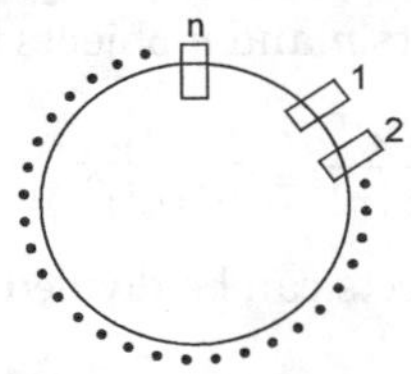

From the required n objects, object a can be put in any one of n places. After filling a place by a the remaining $(n-1)$ vacancies can be filled by $(n-1)$ objects in $(n-1)!$ ways. So the number of circular permutations formed with 'n' objects is $(n-1)!$.

In circular permutations, direction of arrangements of the objects is also important. The above mentioned circular permutations should be considered as total number of clock-wise and anti-clockwise permutations. Since number of permutations are equal in both the directions and if the direction is not considered, then number of circular permutations = $\dfrac{(n-1)!}{2}$

Example. Find the number of ways that a garland is made with 18 flowers such that two specified flowers should be side by side in the garland.

Solution:

First we should treat that two specified flowers as one, hence total 17 (18 – 1) flowers can be arranged circularly in 17! ways.

Two specified flowers can be arranged in 2! = 2 ways

Thus, based on the given condition, 18 flowers can be arranged as a garland in $2 \times 17!$ ways.

But when preparing a garland, we should not consider the direction, so flowers in a garland can be arranged in $\dfrac{2 \times 17!}{2} = 17!$ ways.

Theorem.

The number of permutations of 'n' dissimilar things, taken r at a time, when the repetition of things is allowed any number of times, is n^r.

Example. Using 9 digits 1, 2, 3, 4, 5, 6, 7, 8, 9, how many 3 digits numbers can be written

(*i*) without repetition of digit?

(*ii*) using a digit any number of times?

Solution:

(***i***) Above 3 blank places are to be filled with 3 different digits without repetition of a digit. This number is equivalent to the number of permutations formed from 9 digits which are taken 3 at a time,

$$i.e.\ {}^9P_3 = 9 \times 8 \times 7 = 504 \text{ ways.}$$

(***ii***) If a digit can be used in any number of times, then number of permutations obtained is

$$ {}^9P_3 = 9 \times 9 \times 9 = 729.$$

RESTRICTED PERMUTATION.

The number of permutations of n different objects take r at time in which m particular objects are

(*i*) never included will be $P(n-m, r)$

(*ii*) always included will be $C(n-mr-m)\,P(r, r)$, where $(n \geq m, r \geq m)$

Example. There are 9 objects and 9 boxes. Out of 9, 5 objects cannot fit to be put in 3 small boxes. How many arrangements can be made, such that each object can be put in one box only?

Solution:

Since 5 objects cannot fit to be put 3 small boxes, these 5 objects can be put in the remaining 6 boxes in 6_{P_5} ways.

The remaining 4 objects can be put in 4_{P_4} ways.

$\therefore$ Total number of arrangements

$= {}^6P_5 \times {}^4P_4 = (6 \times 5 \times 4 \times 3 \times 2) \times (4 \times 3 \times 2 \times 1)$

$= 720 \times 24 =$ **17280.**

Theorem.

There are n objects. Out of which m $(< n)$ objects are identical and the remaining are dissimilar. Using all objects, the number of permutations from n objects is $\dfrac{n!}{m!}$.

In n objects, if m objects are of one type, p objects are of another type and q objects are of different type etc., then number of permutations formed from $n(= m + p + \ldots.. + q)$ objects $= \dfrac{n!}{m!\,p!\ldots\ldots.q!}$

- In n objects, if m objects are of one type and $(n-m)$ objects are of another type, then number of permutations using all objects at a time $= \dfrac{n!}{m!(n-m)!}$
- In n objects, if m objects are of one type, q objects are of another type and $(n-p-q)$ objects are of different type, then using all the objects, number of permutations formed $= \dfrac{n!}{p!q!(n-p-q)!}$

Example. Find the number of permutations formed by using all the letters of the word 'ENGINEERING' when

(*i*) *all E's come together.*

(*ii*) *all the N's and all the E's come together.*

(*iii*) all N's come together and E's do not come together.

Solution:

In the given word 'ENGINEERING', there are 11 letters out, of which there are 3 E's, 3 N's, 2 G's, 2 I's and 1 R.

(*i*) Number of permutations in which all the E's come together $= \dfrac{9!}{3!2!2!1!1!}$

(*ii*) Number of permutations that can be formed with all the N's and all the E's which come together

$$= \frac{7!}{2!2!1!1!1!}$$

(*iii*) Number of permutations that can be formed in which all the N's that come together and E's do not come together

$$= \frac{9!}{3!2!2!1!1!} - \frac{7!}{2!2!1!1!1!}$$

COMBINATIONS

n, r are integers. If $n \geq r$, for the set of n elements a subset of r elements is called a *combination*. Therefore a combination is an unordered selection of r elements from a set of n elements. The number of combinations of r elements selected from n elements is denoted by ^nC_r or $\binom{n}{r}$ or C(n, r).

e.g. Out of 40 students in a class any 4 students can sit on a bench without considering the order in $^{40}C_4$ ways.

Theorem.

Number of combinations of n dissimilar things, taken r at a time is

$$^nC_r = \frac{n!}{r!\,(n-r)!}$$

There is no importance for the order of arrangements in combinations.

Theorem.

- $^nC_r = {}^nC_{n-r}$
- $^nC_n = \dfrac{n!}{n!\,(n-n)!} = \dfrac{n!}{n!\,0!} = 1$ (since $0! = 1$)
- $^nC_0 = \dfrac{n!}{0!(n-0)!} = \dfrac{n!}{n!} = 1$

Theorem.

If $^nC_r = {}^nC_s$, then either $r = s$, or $r + s = n$.

Theorem.

$$^{(n-1)}C_{(r-1)} + {}^{(n-1)}C_r = {}^nC_r$$

Theorem.

If n, r are positive integers with $1 \leq r \leq n$, then

- $^nC_r + {}^nC_{(r-1)} = {}^{(n+1)}C_r$
- $^{(n+1)}C_r - {}^nC_r = {}^nC_{(r-1)}$

or $^{(n+1)}C_r - {}^nC_{(r-1)} = {}^nC_r$

Corollary : If m and n are positive integers with $m \neq n$, then total $m + n$ distinct objects can be divided into two groups of m objects and n objects in $\dfrac{(m+n)!}{m!n!}$ ways.

Note : Let $2n$ objects are to be divided into 2 equal groups. Applying above corollary, there will not be any change in the division, if bags A and B are filled with the objects n and n instead of m and n objects. But interchange can be done in 2! ways. Hence number of dividing $2n$ objects n and n objects is

$$\frac{(2n)!}{2n!n!} = \frac{(2n!)}{2!\,(n!)^2}$$

- $(m + n + p)$ objects can be divided into 3 groups of m objects, n objects and p objects in $\dfrac{(m+n+p)!}{m!n!\,p!}$ ways.
- $3n$ objects can be divided into 3 equal groups in $\dfrac{(3n)!}{3!(n!)^3}$ ways.

Corollary : Let there are p objects of one type and q objects are of another type, then number of combinations selected from the total $(p + q)$ objects = $[(p + 1)(q + 1) - 1]$.

If there are p objects of one type q objects of 2nd type and so on.... and t objects of another type, then total number of combinations formed from $(p + q + ... + t)$ objects = $\{[(p + 1)(q + 1) ..... (t + 1)] - 1\}$

Here if $p = q = ... = t = 1$, and they are 'n' in number, then

total number of combinations formed

$= (1 + 1)(1 + 1) .... (1 + 1) - 1$ (n times)

$= (2, 2, ... 2) - 1 = 2^n - 1$ (n times) $= 2^n - 1$

Remember that number of subsets in a power set formed from the 'n' elements is $2^n - 1$ (except the empty set).

Example. In how many ways, a team of 3 boys and 3 girls can be selected from 6 boys and 4 girls?

Solution:

3 boys can be selected from 6 boys in

$$^6C_3 = \frac{6!}{3!3!} = \frac{6 \times 5 \times 4}{3 \times 2 \times 1} = 20 \text{ ways}$$

3 girls can be selected from 4 girls in $^4C_3 = {}^4C_1 = 4$ ways

Hence from fundamental principle, required number of ways of selection = 20 × 4 = 80

- *Number of straight lines drawn through 'n' points on a circle*

$$= {}^nC_2 = \frac{n(n-1)}{2 \times 1} = \frac{n(n-1)}{2}$$

Example. How many straight lines can be drawn through 10 points on a circle?

Solution:

Since two points are necessary to draw a straight line, then

number of straight lines drawn through 10 points

$${}^{10}C_2 = \frac{10 \times 9}{2 \times 1} = \mathbf{45.}$$

Example. In an election, 4 batches of people containing 15, 20, 18, 22 have to convass in 4 districts. If 75 people themselves come forward to convass, in how many ways can they distribute among different districts ?

Solution:

Since 4 groups are distinct, number of ways of sending

75 people to 4 districts = $\frac{75!}{15!\,20!\,18!\,22!}$

RESTRICTED COMBINATIONS

The number of combinations of n objects taken r at a time in which m particular objects are

(*i*) never included in C $(n - m, r)$ ways,

(*ii*) always included in C $(n - m, r - m)$ ways where $(n \geq m$ and $n \geq m)$.

Example. Find the diagonals of a polygon on n sides.

Solution:

Number of straight lines formed by joining the 2 vertices from 'n' vertices of the polygon.

Number of ways of selecting 2 vertices from 'n' vertices

$$= {}^nC_2 = \frac{n(n-1)}{2 \times 1}$$

Out of these straight lines, 'n' lines are the sides of polygon, therefore

$$\text{number of diagonals} = \frac{n(n-1)}{2 \times 1} - n = \frac{n^2 - n - 2n}{2}$$

$$= \frac{n^2 - 3n}{2} = \frac{n(n-3)}{2}$$

Example. Find the number of triangles formed by joining the vertices of the polygon.

Solution:

Since a triangle is formed with any three diagonals of the polygon, therefore number of triangles formed with 3 vertices

$$= {}^nC_3 = \frac{n(n-1)(n-2)}{3 \times 2 \times 1} = \frac{n(n-1)(n-2)}{6}$$

Example. In how many ways, can a committee of 5 members be formed from 8 men and 5 women such that at least 2 women should be there in each committee ?

Solution:

If there are atleast 2 women in each 5 member committee, then the committees are

(*a*) 3 men, 2 women

Number of selections $= {}^8C_3 \times {}^5C_2 = 56 \times 10 = \mathbf{560}$

(*b*) 2 men, 3 women

Number of selections $= {}^8C_2 \times {}^5C_3 = 28 \times 10 = \mathbf{280}$

(*c*) 1 man, 4 women

Number of selections $= {}^8C_1 \times {}^5C_4 = 8 \times 5 = \mathbf{40}$

(*d*) no man, 5 women

Number of selections $= 8c_0 \times 5c_5 = 1 \times 1 = \mathbf{1}$

∴ Number of ways of selecting a 5 member committee = 560 + 280 + 40 + 1 = **881.**

Example. There are 3 questions in 1st section, 3 questions in the 2nd section and 2 questions in the 3rd section in a question paper of an exam. In how many ways can a student who appeared for the exam select to answer any 5 questions of the paper choosing at least 1 question from each section.

Solution:

Out of 8 questions, student can select 5 questions in following ways :

Number of ways of selection in 1st way

$$= {}^3C_3 \times {}^3C_1 \times {}^2C_1 = 6$$

Number of ways of selection in 2nd way

$$= {}^3C_2 \times {}^3C_2 \times {}^2C_1 = 18$$

Number of ways of selection in 3rd way

$$= {}^3C_2 \times {}^3C_1 \times {}^2C_2 = 9$$

Number of ways of selection in 4th way

$$= {}^3C_1 \times {}^3C_2 \times {}^2C_2 = 9$$

COUNTING

Basic counting rules.

There are two basic counting rules which can be used to solve many different counting problems.

(*i*) **Sum rule :** If first task can be done in n_1 ways and a second task in n_2 ways, and if these tasks are such that they cannot be done at the same time, then there are $n_1 + n_2$ ways to do either task.

(*ii*) **Product rule :** Whenever a precedure can be broken down into two tasks and then there are n_1 ways to do the first task and n_2 ways to do the next task after the first task has been done, then there are $n_1 n_2$ ways to do the procedure.

Example. How many one-one functions are there from a set with m elements to one with n elements ?

Solution:

Whenever $m > n$, there are no one-one functions from a set with m elements to a set with n elements.

Let us say $m \le n$. If elements in the domain are assumed to be $a_1, a_2, \ldots a_n$, then there are n ways to choose value of the function at a_1.

Since the function is one-one, there are only $(n-1)$ ways to pick up for a_2, as value of the function. In function general value can be chosen in $n-k+1$ ways for a_k. Applying product rule, there are $n(n-1) \ldots (n-k+1)$ one-one functions from a set with m elements to one with n elements.

Example. How many bit strings of length nine either start with 1 bit or end with the two bits 00 ?

Solution:

Precedure can be divided into two tasks.

First task is constructing a bit string of length nine beginning with a 1 bit. This can be done in

2^8 ways = 256 ways.

Second task of constructing a bit string of length nine ending with two bits 00. This can be done in

$2^7 = 128$ ways.

From this we have to remove the number of ways of choosing 1 in the first bit and corresponding 00 in the last two bits. This can be done in

2^6 ways = 64 ways.

Therefore, total number of bit strings of length nine that can be constructed, that begin with 1 and end with a 00, in

$256 + 128 - 64 = 320$ ways.

EXERCISE

MCQ TYPE QUESTIONS

1. $(P \vee Q) \wedge (P \to R) \wedge (Q \to S)$ is equivalent to
(*a*) $S \wedge R$ (*b*) $S \to R$
(*c*) $S \vee R$ (*d*) none of these

2. The functionally complete set is
(*a*) $\{\rceil, \wedge, \vee\}$ (*b*) $\{\downarrow, \wedge\}$
(*c*) $\{\uparrow\}$ (*d*) $\{\to, \wedge\}$

3. $(P \vee Q) \wedge (P \to R) \wedge (Q \to R)$ is equivalent to
(*a*) P (*b*) Q
(*c*) R (*d*) True $\equiv$ T

4. $\rceil(P \to Q)$ is equivalent to
(*a*) $P \wedge \rceil Q$ (*b*) $\rceil P \wedge Q$
(*c*) $\rceil P \vee Q$ (*d*) $P \vee Q$

5. Logical expression $(A \wedge B) \to (C' \wedge A) \to (A \equiv 1)$ is
(*a*) contradiction
(*b*) tautology
(*c*) well-formed formula
(*d*) none of these

6. Which one of the following propositional forms is a tautology?
(*a*) $p \wedge \rceil q$
(*b*) $p \Rightarrow (p \vee q)$
(*c*) $p \Rightarrow (p \wedge q)$
(*d*) $(p \vee q) \Rightarrow (p \wedge q)$

7. The translation of the sentence (given in English): '*Some cats are black but all buffaloes are black*' using the following notations :
B(x) : x is a black
C(x) : x is cat
Bf(x) : x is a buffalo
($\forall x$) : for all x
($\exists x$: for some x.
in logical symbolism is given by
(*a*) $(\exists x)(C(x) \wedge B(x)) \wedge (\forall x)(Bf(x) \to B(x))$
(*b*) $(\exists x)(C(x) \wedge B(x)) \wedge (\forall x)(Bf(x) \wedge B(x))$
(*c*) $(\exists x)(C(x) \to B(x)) \wedge (\forall x)(Bf(x) \wedge B(x))$
(*d*) $(\exists x)(C(x) \to B(x)) \wedge (\forall x)(Bf(x) \to B(x))$

8. In propositional logic, which of the following is equivalent to $p \to q$?
(*a*) $\sim p \to q$ (*b*) $\sim p \vee q$
(*c*) $\sim p \vee \sim q$ (*d*) $p \to \sim q$

9. In predicate logic, $\sim \forall P(x)$ is equivalent to
(*a*) $\exists x P(x)$ (*b*) $\exists x \sim P(x)$
(*c*) $\forall x \sim P(x)$ (*d*) $\sim \exists x P(x)$

10. Which of the following are tautologies ?
(*a*) $((P \vee Q) \wedge Q) \leftrightarrow Q$
(*b*) $(P \vee (P \to Q)) \to P$
(*c*) $((P \vee Q) \wedge P) \to Q$
(*d*) $((P \vee Q) \wedge \rceil P) \to Q$

11. Identify the valid conclusion from the premises $P \vee Q$, $Q \to R$, $P \to M$, $\rceil M$
(*a*) $P \wedge (Q \vee R)$ (*b*) $P \wedge (Q \wedge R)$
(*c*) $R \wedge (P \vee Q)$ (*d*) $Q \wedge (P \vee R)$

12. Which one of the following statements is true when exactly two of P, Q, or R are TRUE ?
(*a*) $(P \wedge \overline{Q} \wedge R) \vee (P \wedge \overline{Q} \wedge \overline{R}) \wedge (\overline{P} \wedge Q \wedge \overline{R})$
(*b*) $(P \vee Q \vee R) \wedge (P \vee \overline{Q} \vee R) \wedge (P \vee Q \vee \overline{R})$
(*c*) $(\overline{P} \wedge Q \wedge R) \vee (P \wedge \overline{Q} \wedge R) \vee (P \wedge Q \wedge \overline{R})$
(*d*) $(P \wedge \overline{Q} \wedge R) \vee (P \wedge \overline{Q} \wedge \overline{R}) \vee (\overline{P} \wedge Q \wedge \overline{R})$

13. $P \to (Q \to R)$ is equivalent to
(*a*) $(P \wedge Q) \to R$
(*b*) $(P \vee Q) \to R$
(*c*) $(P \vee Q) \to \rceil R$
(*d*) $(P \wedge Q) \to \rceil R$

14. What can we correctly say about proposition P1 :

$$P1 : (p \vee \neg q) \wedge (q \rightarrow r) \vee (r \vee p)$$

(*a*) P1 is tautology

(*b*) P1 is a contradiction

(*c*) If p is true and q is false and r is false, then P1 is true

(*d*) If p is true and q is true and r is false, then P1 is true

15. Which of the following is a declarative statement?

(*a*) It is beautiful

(*b*) He says "*It is correct*"

(*c*) Two may not be an even integer

(*d*) I love you

16. The correct prefix formula is

(*a*) $\rightarrow \neg$,P$\vee$Q $\Longleftrightarrow$ $\neg$R $\neg$S

(*b*) $\rightarrow$ P$\neg$Q$\vee$ $\rightleftarrows$ $\neg$RS

(*c*) $\rightarrow \rightarrow$ PQ $\rightarrow \rightarrow$ QR$\rightarrow$ PR

(*d*) $\rightarrow \neg$PV $\rightleftarrows$ QSP

17. The negation of the statement $\exists x \forall y, P(x, y)$ is

(*a*) $\forall x \exists y, P(x, y)$

(*b*) $\forall x \exists y, \neg P(x, y)$

(*c*) $\exists x \exists y, \neg P(x, y)$

(*d*) $\forall y \forall x, \exists z \neg P(x, y, z)$

18. Which of the following sets are null sets ?

(*a*) $\{0\}$ (*b*) $\{\phi\}$

(*c*) $\{\ \}$ (*d*) ϕ

19. Which of the following set (*s*) are empty ?

(*a*) $\{x : x = x\}$ (*b*) $\{x : x \neq x\}$

(*c*) $\{x : x = x^2\}$ (*d*) $\{x : x \neq x^2\}$

20. If A = $\{x, y\}$, then power set of A is

(*a*) $\{\{x\}, \{y\}\}$ (*b*) $\{\{\phi\}, \{x, y\}\}$

(*c*) $\{\phi, \{x\}, \{y\}\}$ (*d*) none of these

21. If A = $\{x, y\}$, then proper subsets of A are

(*a*) $\{x\}, \{y\}$ (*b*) $\phi, \{x, y\}$

(*c*) $\phi, \{x\}, \{y\}$ (*d*) none of these

22. If A and B are sets and $A \cup B = A \cap B$, then

(*a*) $A = \phi$ (*b*) $B = \phi$

(*c*) $A = B$ (*d*) none of these

23. If X, Y, Z are sets such that $X \subset Y$ and $Y \subset Z$, then

(*a*) $X \cup Y = Z$ (*b*) $X \cap Y = Z$

(*c*) $X \cap Y \cap Z = Z$ (*d*) none of these

24. If X and Y are two sets, then $X \cap (Y \cup X)^c$ equals

(*a*) X (*b*) Y

(*c*) ϕ (*d*) none of these

25. If P, Q, R are subsets of a set A, then

$R \times (P^c \cup Q^c)^c = (R \times P) \cap (R \times Q)$

(*a*) True

(*b*) False

26. For any two subsets X and Y of a set A, define X o Y = $(X^c \cap Y) \cup (X \cap Y^c)$. Then for any three subsets X, Y, Z of the set A is

$$X \text{ o } (Y \text{ o } Z) = (X \text{ o } Y) \text{ o } Z$$

(*a*) True

(*b*) False

27. Relation R defined on the set A = {1, 2, 3, 4}, by R = {(1, 1), (2, 2) {3, 3)} is

(*a*) reflexive (*b*) symmetric

(*c*) transitive (*d*) none of these

28. If relation R is defined on N by

R = {(a, b): a divides b; $a, b \in$ N}.

Then R is

(*a*) reflexive (*b*) symmetric

(*c*) transitive (*d*) none of these

29. Relation R defined on a set N by

R = $\{(a, b) : |a - b|$ is divisible by 5}, is

(*a*) reflexive (*b*) symmetric

(*c*) transitive (*d*) all of these

30. Relation R is defined on the set N as $\{(a, b) : a, b$ are both odd}, is

(*a*) reflexive (*b*) symmetric

(*c*) transitive (*d*) none of these

31. If R is an equivalence relation on a set A, then R^{-1} is

(*a*) reflexive (*b*) symmetric

(*c*) transitive (*d*) all of these

32. "n/m" means that n is a factor of m, then the relation 'I' is

(*a*) reflexive and symmetric.

(*b*) transitive and symmetric

(*c*) reflexive, transitive and symmetric

(*d*) reflexive, transitive and not symmetric

33. The function $f : N \rightarrow N$ defined by $f(n) = 2n + 3$ is

(*a*) surjective (*b*) not surjective

(*c*) injective (*d*) none of these

34. Given : A = $\{x, y, z\}$; B = $\{u, v, w\}$

The function $f : A \rightarrow B$ defined by $f(x) = u$, $f(y) = v$, $f(z) = w$, is

(*a*) surjective (*b*) bijective

(*c*) injective (*d*) none of these

35. If $f : X \rightarrow Y$ and $a, b \subseteq X$, then $f(a \cap b)$ is equal to

(*a*) $f(a) - f(b)$

(*b*) $f(a) \cap f(b)$

(*c*) a proper subset of $f(a) \cap f(b)$

(*d*) $f(b) - f(a)$

36. If A and B are sets, then which of the following is FALSE ?

(*a*) $A - B' = A \cap B$

(*b*) $A \subset B \Rightarrow B' \subset A'$

(*c*) $A - (A - B) = A \cap B$

(*d*) None of these

37. A = {string of 0's and 1's}.
Relation d on A is defined as xdy if x is a sub string of y (Ex. 01 d 101), Then d is
(*a*) symmetric
(*b*) anti symmetric
(*c*) equivalence
(*d*) reflexive and symmetric

38. If relation R over {a, b, c} is given by R = {(a, a), (a,b), (b,a), (b,b), (c,c)},
then which of the following properties does R have?
(*a*) Symmetry (*b*) Reflexivity
(*c*) Transitivity (*d*) All of these

39. If R = {(a, b)/a, b are positive integers such that $a - b$ is on ODD positive integer}, then relation R is
(*a*) reflexive (*b*) symmetric
(*c*) transitive (*d*) equivalence

40. A – (B ∪ C) is
(*a*) (A – B) ∪ (A – C) (*b*) A – B – C
(*c*) (A – B) ∩ (A – C) (*d*) $A - (\overline{B \cap C})$

41. A ∪ B is NOT equivalent to
(*a*) $\left(\overline{\overline{A} \cap \overline{B}}\right)$ (*b*) $\left(\overline{\overline{A} - B}\right)$
(*c*) $\left(\overline{\overline{B} - A}\right)$ (*d*) $\left(\overline{A - B}\right)$

42. A ∪ B = A ∩ B if and only if
(*a*) A is empty set
(*b*) B is empty set
(*c*) A and B are non-empty sets
(*d*) A and B are empty sets

43. If A = {x | –1 < x < 1} = B, then function $f(x) = x/2$ from A to B is
(*a*) injective
(*b*) surjective
(*c*) both injective and surjective
(*d*) neither injective nor surjective

44. The domain and range are same for
(*a*) constant function
(*b*) identity function
(*c*) absolute value function
(*d*) greatest integer function

45. The function $f : Z \to Z$ given by $f(x) = x^2$ is
(*a*) one-one
(*b*) onto
(*c*) one-one and onto
(*d*) none of these

46. The number of functions
f : (1, 2, 3, ..., n), → {1, 2, ..., m},
which are one-to-one is
(*a*) m^n
(*b*) $m(m-1)(m-2)\ldots\ldots\ldots(m-n+1)$
(*c*) n^m
(*d*) $n(n-1)(n-2)\ldots\ldots\ldots(n-m+1)$

47. Which of the following relations may be characterised as a function defined on the set I = {1, 2, 3, 4, 5} ?
(*a*) {(x, y) | x, y ∈ I, x < y/
(*b*) {(x, y)| x, y ∈ I, x > y}
(*c*) {(x, y) | x, y ∈ I, x = y/2}
(*d*) None of these

48. Which of the following is a partition of the set S = {1, 2, 3, 4, 5, 6} ?
(*a*) [{(1, 3, 5), {2, 4}, {3, 6}]
(*b*) [{1, 5}, {2}, {3, 6}]
(*c*) [{(1, 5}, {2}, {4}, {1, 5}, {3, 6}]
(*d*) [{1, 2, 3, 4, 5, 6}]

49. When A = {1, 2, 3, 4, 5} and r defined on A by xry if $x + 1 = y$, then r^3 is
(*a*) {(1, 3) (2, 4)} (*b*) {(1, 3) (2, 5)}
(*c*) {(1, 4) (2, 5)} (*d*) {(1, 4) (4, 5)}

50. If R = {(1, 1), (3, 1), (2, 3), (4, 2)}, then which of the following represents R^2, where R^2 is R composite R ?
(*a*) {(1, 1), (3, 1), (2, 3), (4, 2)}
(*b*) {(1, 1), (9, 1), (4, 9), (16, 4)}
(*c*) {(1, 3), (3, 3), (3, 4), (3, 2)}
(*d*) {(1, 1), (2, 1), (4, 3), (3, 1)}

51. If I is set of positive integers and relation R is defined over the set I by xRy if $x^y = y^x$, then R is
(*a*) reflexive but neither symmetric nor transitive
(*b*) reflexive and transitive but not symmetric
(*c*) reflexive and symmetric but not transitive
(*d*) reflexive, symmetric and transitive

52. 'Subset' relation on a set of sets is
(*a*) a partial ordering
(*b*) an equivalence relation
(*c*) transitive and symmetric only
(*d*) transitive and anti-symmetric only

53. The correspondence $f : N \to N$ is such that
$f(x) = y$ and for $x = p_1^{e1}, p_2^{e2}, \ldots p_k^{ek}$
where p_i are distinct primes and
integers $e_i \geq 1$, if $y = e_1 + e_2 + \ldots. + e_k$, then
(*a*) f is not a function
(*b*) f is an onto function
(*c*) f is an 1 – 1 function
(*d*) f is neither 1 – 1 nor onto function

54. The power set of the set
$S = \{3, \{1, 4\}, 5\}$ is
(*a*) $\{S, 3, 1, 4, \{1, 3, 5\}, \{1, 4, 5\} \{3, 4\}, \phi\}$
(*b*) $\{S, 3, \{1, 4,\}, 5\}$
(*c*) $\{S, \{3\}, \{3, \{1, 4\}\}, \{3, 5\}, \phi\}$
(*d*) none of these

55. If $f(x)$ and $g(x)$ are defined on domains A, B respectively, then domain of $f(x) + g(x)$ is
(*a*) $A \cup B$ (*b*) $A \cap B$
(*c*) $A \Delta B$ (*d*) $A - B$

56. The function $f : R - \{2\} \to R$ defined by
$f(x) = (x^2 + 2x)/(x - 2)$ is
(*a*) one-one and onto
(*b*) one-one but not onto
(*c*) neither one-one nor onto
(*d*) not one-one but onto

57. If set $S = \{x \mid x$ is a student in your class$\}$ and relation $x \text{ R } y \leftrightarrow x$ sits in the same row as y, then which of the following statements about R is TRUE ?
(*a*) Reflexive only
(*b*) Transitive only
(*c*) Refleixve, transitvie and asymmetric
(*d*) Reflexive, transitive and symmetric

58. If set A has n elements, then number of functions that can be defined from A into A is
(*a*) n^2 (*b*) n !
(*c*) n^n (*d*) n

59. If $n \geq 2$, then number of surjections that can be defined from $\{1, 2, 3, ...n\}$ onto $\{1, 2\}$ is
(*a*) $2n$ (*b*) ${}^nP_2$
(*c*) 2^n (*d*) 2^n-2

60. Number of bijective functions from set A to itself when A contains 106 elements, is
(*a*) 106 (*b*) 106^2
(*c*) 106 ! (*d*) 2^{106}

61. If $f : Z \to Z$ be defined as $f(x) = x^2$, $x \in Z$, then function f is
(*a*) bijection (*b*) injection
(*c*) surjection (*d*) none of these

62. A relation over the set $S = \{x, y, z\}$ is defined by :
$\{(x, x), (x, y), (y, x), (x, z), (y, z), (y, y), (z, z)\}$.
What properties hold for this relation?
(*a*) Symmetric
(*b*) Reflexive
(*c*) Antisymmetric
(*d*) Irreflexive

63. A relation on the integers 0 through 4 is defined by :
$R = \{(x, y): x + y \leq 2x\}$.
Which of the properties listed below applies to this relation?
I. Tronsitivity
II. Symmetry
III. Reflexivity
(*a*) I only (*b*) III only
(*c*) both I and III (*d*) both II and III

64. The symmetric difference of sets A and B is
(*a*) $(A - B) \cup (B - A)$
(*b*) $(A \cup B) - (A \cap B)$
(*c*) $(A - B) \cap (B - A)$
(*d*) $(A \cap B) - (A \cup B)$

65. Order of the power set of a set of order n is
(*a*) n (*b*) $2n$
(*c*) n^2 (*d*) 2^n

66. If x, y are real numbers such that ordered pairs $(x + y, x - y)$ and $(2x + 3y, 3x - 2y)$ are equal, then (x, y) is equal to
(*a*) (1, 1) (*b*) (2, 3)
(*c*) (3, – 2) (*d*) (0, 0)

67. $f : R \to R$ is a function defined by $f(x) = 10x - 7$. If $g = f^{-1}$, then $g(x) =$
(*a*) $\dfrac{1}{10x-7}$ (*b*) $\dfrac{1}{10x+7}$
(*c*) $\dfrac{x+7}{10}$ (*d*) $\dfrac{x-7}{10}$

68. If $|A| = m$ and $|B| = n$, then number of possible relations $R : A \to B$ is
(*a*) mn (*b*) $(mn)^2$
(*c*) 2^{m+n} (*d*) 2^{mn}

69. Assume R and S are (non empty) relations in a set A. Which of the following given relations is FALSE ?
(*a*) If R and S are transitive, then $R \cup S$ is transitive
(*b*) If R and S are transitive, then $R \cap S$ is transitive
(*c*) If R and S are symmetric, then $R \cup S$ is symmetric
(*d*) If R and S are reflexive, then $R \cup S$ is reflexive

70. If R be a relation from
$A = \{1, 2, 3, 4\}$ to $B = \{1, 3, 5\}$.
i.e. $(a, b) \in R$ iff $a < b$, then RoR^{-1} is
(*a*) $\{(1, 3); (1, 5), (2, 3), (2, 5), (3, 5), (4, 5)\}$
(*b*) $\{(3, 1), (5, 1) (3,2), (5,2), (5,3), (5,4)\}$
(*c*) $\{(3, 3), (3, 5), (5, 3), (5, 5)\}$
(*d*) $\{(3, 3), (3, 4), (4, 5)\}$

71. If $f : R \to R$ defined by $f(x) = x^2 + 1$, then values of $f^{-1}(17)$ and $f^{-1}(-3)$ are respectively

(a) $\{\phi\}, (4, -4\}$ (b) $\{3, -3\}, \{\phi\}$

(c) $\{\phi\}, \{3, -3\}$ (d) $\{4, -4\}, \phi$

72. If function $f : R \to R$ is given by $f(x) = x^2 + 2x - 3$ and function $g : R \to R$ is gien by $g(x) = 3x - 4$,

then $(gof)(x)$ is given by

(a) $9x^2 + 18x + 5$ (b) $3x^2 + 6x - 13$

(c) $x^2 + x - 7$ (d) $x^2 - 5x - 1$

73. With respect to the relation '*x divides y*', which of the following sets are totally ordered ?

I. $\{36, 3, 9\}$

II. $\{7, 77, 11\}$

III. $(3, 6, 24, 12\}$

IV. $\{1, 2, 3, ...\}$

(a) I, II and III only (b) II and III only

(c) I and III only (d) All of these

74. If Q be the set of non-zero rational numbers and the relation R be defined over the set Q by

$x R y$ if $x = 1/y$, $x, y \in Q$, then

(a) R is an equivalence relation

(b) R is reflexive

(c) R is symmetric

(d) R is transitive

75. Let $n(A)$ denotes the number of elements in set A. If $n(A) = p$ and $n(B) = q$, then how many ordered pairs (a, b) are there with $a \in A$ and $b \in B$?

(a) p^2 (b) $p \times q$

(c) $p + q$ (d) $2pq$

76. Which of the following relations are functions?

I. $N = \{(x, y)/y = x^2, x \in \{-1, 0, 1, 2, 3\}\}$

II. $P = \{(x, y)/y^2 = x, x \in \{4, 9, 16\}\}$

III. $Q = \{(x, y)/y = 4x^2 - 14, x \in \{-1, 1, 2, 3\}\}$

(a) I only (b) I and II only

(c) I and III only (d) III only

77. In a group of 72 students, 47 have background is Electronics, 59 have background in Mathematics and 42 have background in both the subjects. How many students do not have background in any of the subjects ?

(a) 8 (b) 13

(c) 25 (d) 34

78. If $[x]$ denotes integer part of the real number, then the function $f(x) = x - [x]$ is/an

(a) even function (b) odd function

(c) periodic function (d) constant

79. To have inverse for the function f, f is

(a) one one (b) onto

(c) one one onto (d) identity function

80. Which of the following set is closed under addition, subtraction and multiplication?

(a) $A = \{x \mid x = 2^n, n \in N\}$

(b) $B = \{x \mid x = 3n, n \in N\}$

(c) $C = \{x \mid x = 3n, n \in Z\}$

(d) All of these

81. The set of all Equivalence classes of a set A of cardinality C

(a) has the same cardinality as A

(b) forms a partition of A

(c) is of cardinality 2C

(d) is of cardinality C^2

82. Which of the following sets is a null set ?

I. $X = \{x \mid x = 9, 2x = 4\}$

II. $Y = \{x \mid x = 2x. x \neq 0\}$

III. $Z = \{x \mid x - 8 = 4\}$

(a) I and II only (b) I, II and III

(c) I and III only (d) II and III only

83. Let R be a relation "$(x - y)$ is divisible by m", where x, y, m are integers and $m > 1$, then R is

(a) symmetric but not transitive

(b) partial order

(c) equivalence relation

(d) anti symmetric and not transitive

84. Let $s(w)$ denote the set of all the letters in w where w is an English word. Let us denote set equality, subset and union relations by $=$, $\subset$ and $\cup$ respectively.

Which of the following is NOT true?

(a) $s(\text{ten}) \subset s(\text{twenty})$

(b) $s(\text{stored}) = s(\text{sorted})$

(c) $s(\text{sixty}) \subset (s(\text{six}) \cup s(\text{twenty})$

(d) None of these

85. A relation R is defined on the set of positive integers as xRy if $2x + y \leq 5$.

The relation R is

(a) reflexive (a) symmetric

(a) transitive (d) none of these

86. Let R be an equivalence relation on the set $\{1, 2, 3, 4, 5, 6\}$ given by

$\{(1, 1), (1, 5), (2, 2), (2, 3), (2, 6) (3, 2), (3, 3), (3, 6), (4, 4), (5, 1), (5, 5), (6, 2), (6, 3), (6, 6)\}$.

The partition induced by R is

(a) $\{1, 2, 3, 4, 5, 6\}$ (b) $\{\{1, 3, 5, 6\}, \{2, 4\}\}$

(c) $\{\{1, 5\}, \{2, 3, 6\}, \{4\}\}$ (d) $\{\{1, 2, 3, 4\}, \{5, 6\}\}$

87. Let Z denote the set of all integers.
Define $f: Z \to Z$ by

$$f(x) = \begin{cases} x/2 & (x \text{ is even}) \\ 0 & (x \text{ is odd}) \end{cases},$$

then f is
(*a*) onto but not one-one
(*b*) one-one but not onto
(*c*) one-one and onto
(*d*) neither one-one nor-onto

88. If $f: A \to B$ is a bijective function, then f^{-1} of f =
(*a*) fof^{-1}
(*b*) f
(*c*) f^{-1}
(*d*) I_A (Identity map of the set A)

89. The matrices $\begin{bmatrix} \cos\theta & -\sin\theta \\ \sin\theta & \cos\theta \end{bmatrix}$ and $\begin{bmatrix} a & 0 \\ 0 & b \end{bmatrix}$ commute under the multiplication
(*a*) if $a = b$ (or) $\theta = n\pi$, n is an integer
(*b*) always
(*c*) never
(*d*) if $a \cos\theta \neq b \sin\theta$

90. The set of all real numbers under the usual multiplication operation is not a group since
(*a*) multiplication is not a binary operation
(*b*) multiplication is not associative
(*c*) identity element does not exist
(*d*) zero has no inverse

91. The set of integers Z with the binary operation '*' defined as $a * b = a + b + 1$ for $a, b \in Z$, is a group. The identity element of this group is
(*a*) 0 (*b*) 1
(*c*) -1 (*d*) 12

92. The identity element in the group $G = \left\{\begin{bmatrix} x & x \\ x & x \end{bmatrix} : x \in R, x \neq 0\right\}$ with respect to matrix multiplication is
(*a*) $\begin{pmatrix} 1 & 0 \\ 0 & 1 \end{pmatrix}$ (*b*) $\begin{pmatrix} -1 & 0 \\ 0 & -1 \end{pmatrix}$
(*c*) $\begin{pmatrix} \frac{1}{2} & \frac{1}{2} \\ \frac{1}{2} & \frac{1}{2} \end{pmatrix}$ (*d*) $\begin{pmatrix} 1 & 1 \\ 1 & 1 \end{pmatrix}$

93. (Z, *) is a group with $a * b = a + b + 1$, $\forall a, b \in Z$. The inverse of a is
(*a*) 0 (*b*) -2
(*c*) $a - 2$ (*d*) $-a - 2$

94. In the group (G, .), the value of $(a^{-1} b)^{-1}$ is
(*a*) ab^{-1} (*b*) $b^{-1}a$
(*c*) $a^{-1}b$ (*d*) ba^{-1}

95. If (G, .) is a group such that $a^2 = e$, $\forall a \in G$, then G is
(*a*) semi group (*b*) abelian group
(*c*) non-abelian group (*d*) none of these

96. If (G, .) is a group, such that $(ab)^2 = a^2b^2$ $\forall a, b \in G$, then G is a/an
(*a*) commutative semi group
(*b*) abelian group
(*c*) non-abelian group
(*d*) none of these

97. If (G, .) is a group such that $(ab)^{-1} = b^{-1} a^{-1}$, $\forall a, b \in G$, then G is a/an
(*a*) commutative semi group
(*b*) abelian group
(*c*) non-abelian group
(*d*) none of these

98. Let G denoted the set of all $n \times n$ non-singular matrices with rational numbers as entries. Then under multiplication G is a/an
(*a*) subgroup
(*b*) finite abelian group
(*c*) infinite, non abelian group
(*d*) infinite, abelian

99. The set of all n^{th} roots of unity under multiplication of complex numbers form a/an
(*a*) semi group with identity
(*b*) commutative semigroups with identity
(*c*) group
(*d*) abelian group

100. If the binary operation * is defined on a set of ordered pairs of real numbers as
$(a, b) * (c, d) = (ad + bc, bd)$
and is associative, then
(1, 2) * (3, 5) * (3, 4) equals
(*a*) 74, 40) (*b*) (32, 40)
(*c*) (23, 11) (*d*) (7, 11)

101. If every element of a group G is its own inverse, then G is
(*a*) finite (*b*) infinite
(*c*) cyclic (*d*) abelian

102. A subset H of a group (G, *) is a group if
(*a*) $a, b \in H \Rightarrow a * b \in H$
(*b*) $a \in H \Rightarrow a^{-1} \in H$
(*c*) $a, b \in H \Rightarrow a * b^{-1} \in H$
(*d*) H contains the identity element.

103. G{e, a, b, c} is an abelian group with 'e' as identity element. The order of the other elements are
(*a*) 2, 2, 3 (*b*) 3, 3, 3
(*c*) 2, 2, 4 (*d*) 2, 3, 4

104. Which of the following statements is true?
(*a*) Every equivalence relation is a partial-ordering relation.
(*b*) Number of relations form A = {x, y, z} to B = {1, 2}, is 64.
(*c*) Empty relation ϕ is reflexive.
(*d*) Properties of a relation being symmetric and being ant-symmetric are negative of each other.

105. Which of the following statements is false ?
(*a*) If R is reflexive, then $R \cap R^{-1} \neq \phi$.
(*b*) $R \cap R^{-1} \neq \phi \Rightarrow$ R is anti-symmetric.
(*c*) If R, R′ are equivalence relations in a set A, then $R \cap R'$ is also an equivalence relation in A.
(*d*) If R, R′ are reflexive relations in A, then R – R′ is reflexive.

106. Let A = {1, 2, 3.........}.
Define ~ by $x \sim y \Leftrightarrow x$ divides y. Then ~ is
(*a*) reflexive, but not a partial-ordering
(*b*) symmetric
(*c*) an equivalence relation
(*d*) a partial-ordering relation

107. If A = {1, 2, 3}, then relation S = {(1, 1), (2, 2) is
(*a*) symmetric only
(*b*) anti-symmetric only
(*c*) both symmetric and anti-symmetric
(*d*) an equivalence relation

108. If A = {1, 2, 3, 4}. Let ~ = {(1, 2), (1, 3), (4, 2). Then ~ is
(*a*) not anti-symmetric
(*b*) transitive
(*c*) reflexive
(*d*) symmetric

109. If R = {(1, 2),(2, 3),(3, 3)} be a relation defined on A = {1, 2, 3}, then R . R (= R^2) is
(*a*) R itself
(*b*) {(1, 2),(1, 3),(3, 3)}
(*c*) {(1, 3),(2, 3),(3, 3)}
(*d*) {(2, 1),(1, 3),(2, 3)}

110. The universal relation A × A on A is
(*a*) an equivalence relation
(*b*) anti-symmetric
(*c*) a partial ordering relation
(*d*) not symmetric and not anti-symmetric

111. Total number of different partitions of a set having four elements is
(*a*) 16 (*b*) 8
(*c*) 15 (*d*) 4

112. A partition of {1, 2, 3, 4, 5} is the family
(*a*) {(1, 2),(3, 4),(3, 5)} (*b*) {ϕ(1, 2),(3, 4),(5)}
(*c*) {(1, 2, 3),{5}} (*d*) {{1, 2,} {3, 4, 5}}

113. The less than relation, <, on reals is
(*a*) a partial ordering since it is asymmetric and reflexive.
(*b*) a partial ordering since it is anti-symmetric and reflexive.
(*c*) not a partial ordering because it is not asymmetric and not reflexive.
(*d*) not a partial ordering because it is not anti-symmetric and not reflexive.

114. Hasse diagrams are drawn for
(*a*) partially ordered sets
(*b*) lattices
(*c*) boolean Algebra
(*d*) none of these

115. A self-complemented, distributive lattice is called
(*a*) Boolean algebra
(*b*) Modular lattice
(*c*) Complete lattice
(*d*) Self dual lattice

116. Which of the following lattices are Boolean ?
(*a*) 
(*b*) 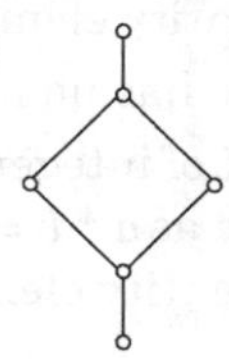
(*c*) 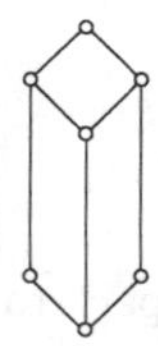
(*d*) None of these

117. Which of the following is not a "lattice" ?
(*a*) 
(*b*) 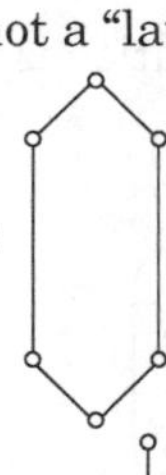
(*c*) 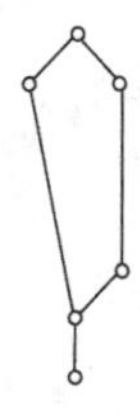
(*d*) 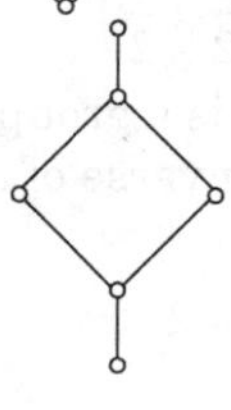

118. Let D_{30} = {1, 2, 3, 4, 5, 6, 10, 15, 30} and relation I be partial ordering on D_{30}. The all lower bounds of 10 and 15 respectively are

(*a*) 1, 3 (*b*) 1, 5

(*c*) 1, 3, 5 (*d*) none of these

119. Complement of each element of D_6 in $[D_6 ; \vee, \wedge]$, is

(*a*) $\overline{1} = 2, \overline{2} = 3, \overline{3} = 2, \overline{6} = 1$

(*b*) $\overline{1} = 6, \overline{2} = 3, \overline{3} = 6, \overline{6} = 1$

(*c*) $\overline{1} = 6, \overline{6} = 1, \overline{3} = 2, \overline{2} = 3$

(*d*) none of these

120. Maximal and minimal elements of the Poset are

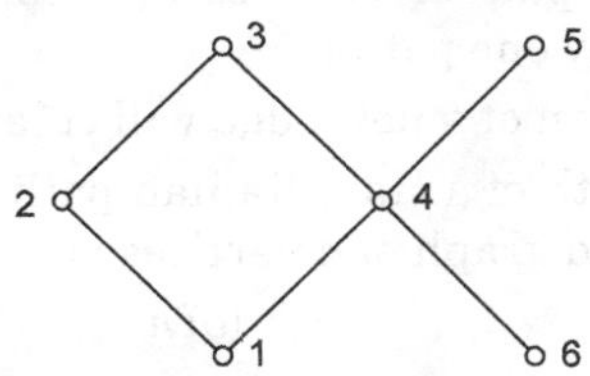

(*a*) Maximal 5, 6; Minimal 2

(*b*) Maximal 5, 6; Minimal 1

(*c*) Maximal 3, 5; Minimal 1, 6

(*d*) None of these

121. The greatest and least elements of the Poset are

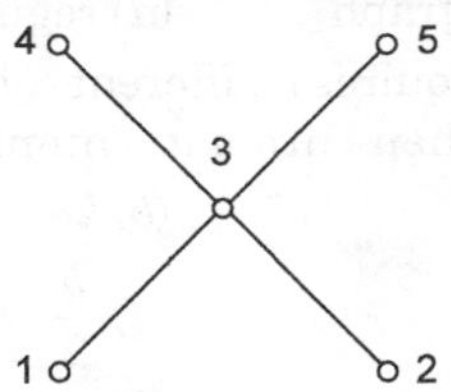

(*a*) greatest 4, 5; least 1, 2

(*b*) greatest 5; least 1

(*c*) greatest None; least none

(*d*) none of these

122. Different partially ordered sets may be represented by the same Hasse diagram if they are

(*a*) same

(*b*) lattices with same order

(*c*) isomorphic

(*d*) order-isomorphic

123. The absorption law is defined as

(*a*) $a * (a * b) = b$

(*b*) $a * (a \oplus b) = b$

(*c*) $a * (a * b) = a \oplus b$

(*d*) $a * (a \oplus b) = a$

124. If lattice (C, ≤) is a complemented chain, then

(*a*) $|C| \leq 1$ (*b*) $|C| \leq 2$

(*c*) $|C| > 1$ (*d*) C doesn't exist

125. If set A = {1, 2, 3, 4}, then ordered pairs in the relation determined by the Hasse diagram in the figure given below, are described as

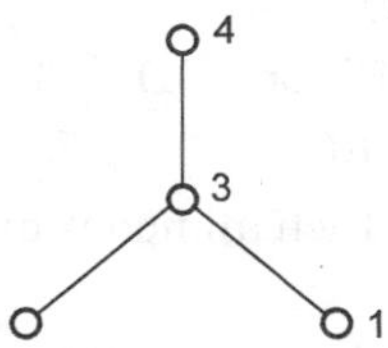

(*a*) {(1, 1), (2, 2), (3, 3), (4, 4), (1, 3), (1, 4), (2, 3), (2, 4), (3, 4)}

(*b*) (1, 1), (2, 2), (3, 3), (4, 4), (1, 2), (1, 3), (1, 4), (2, 3), (2, 4), (3, 4)}

(*c*) both (*a*) and (*b*)

(*d*) none of these

126. Matrix of the partial order whose Hasse diagram is given below is

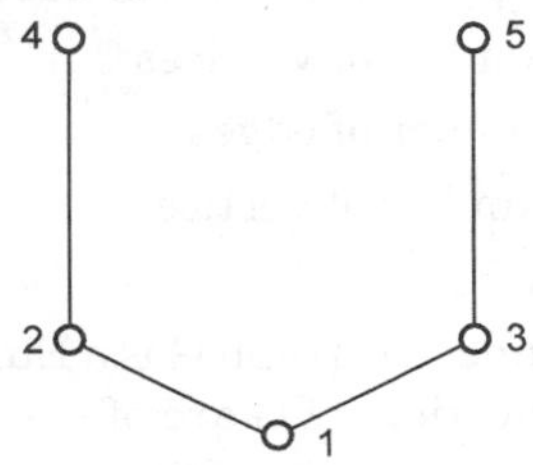

(*a*) $\begin{bmatrix} 0 & 1 & 1 & 1 & 1 \\ 0 & 1 & 0 & 0 & 0 \\ 0 & 0 & 0 & 0 & 0 \\ 0 & 0 & 0 & 0 & 1 \\ 0 & 0 & 0 & 0 & 1 \end{bmatrix}$ (*b*) $\begin{bmatrix} 1 & 1 & 1 & 1 & 1 \\ 0 & 1 & 0 & 0 & 0 \\ 0 & 0 & 1 & 0 & 0 \\ 0 & 0 & 0 & 1 & 0 \\ 0 & 0 & 0 & 0 & 1 \end{bmatrix}$

(*c*) $\begin{bmatrix} 1 & 1 & 1 & 1 & 1 \\ 0 & 1 & 0 & 1 & 0 \\ 0 & 0 & 1 & 0 & 1 \\ 0 & 0 & 0 & 1 & 0 \\ 0 & 0 & 0 & 0 & 1 \end{bmatrix}$ (*d*) None of these

127. Principle of duality is defined as

(*a*) ≤ is replaced by ≥

(*b*) LUB becomes GLB

(*c*) all properties are unaltered when ≤ is replaced by ≥

(*d*) all properties are unaltered when ≤ is replaced by ≥ other than 0 and 1 element.

128. Every finite subset of a lattice has

(*a*) a LUB and a GLB

(*b*) many LUBs and a GLB

(*c*) many LUBs and many GLBs

(*d*) either some LUBs or some GLBs

129. In any undirected graph, the sum of degrees of all the nodes

(*a*) must be even

(*b*) is twice the number of edges

(*c*) must be ODD

(*d*) need not be EVEN

130. Number of vertices of ODD degree in a graph is

(*a*) always EVEN
(*b*) always ODD
(*c*) either EVEN or ODD
(*d*) always ZERO

131. A graph in which all nodes are of equal degree is called

(*a*) multi graph (*b*) non regular graph
(*c*) regular graph (*d*) complete graph

132. Maximum degree of any node in a simple graph with *n* vertices is

(*a*) $n - 1$ (*b*) n
(*c*) $\frac{n}{2}$ (*d*) $n - 2$

133. Two isomorphic graphs must have

(*a*) same number of vertices
(*b*) same number of edges
(*c*) equal number of vertices
(*d*) all of these

134. A given connected graph G is a Eular graph if and only if all vertices of G are of

(*a*) same degree (*b*) even degree
(*c*) odd degree (*d*) different degree

135. Total number of edges in a complete graph of n vertices is

(*a*) n (*b*) $\frac{n}{2}$
(*c*) $\frac{n(n-1)}{2}$ (*d*) $\frac{n(n+1)}{2}$

136. A tree with '*n*' nodes has

(*a*) $\frac{n}{2}$ edges (*b*) n edges
(*c*) $n - 1$ edges (*d*) $n + 1$ edges

137. The number of circuits in a tree with '*n*' nodes is

(*a*) Zero (*b*) One
(*c*) $n - 1$ (*d*) $\frac{n}{2}$

138. A graph is a tree if and only if it

(*a*) is completely connected
(*b*) is minimally connected
(*c*) contains a circuit
(*d*) is planar

139. A graph with '*n*' vertices and $n - 1$ edges that is not a tree, is

(*a*) connected (*b*) disconnected
(*c*) eular (*d*) a circuit

140. A complete graph with "*n*" vertices is

(*a*) 2-chromatic (*b*) $\frac{n}{2}$-chromatic
(*c*) $(n-1)$-chromatic (*d*) n-chromatic

141. Number of different rooted labeled trees with *n* vertices is

(*a*) 2^{n-1} (*b*) 2^n
(*c*) n^{n-1} (*d*) n^n

142. T is a graph with *n* vertices. T is connected and has exactly $n - 1$ edges, then

(*a*) T is a tree
(*b*) T contains no cycles
(*c*) every pair of vertices in T is connected by exactly one path
(*d*) addition of a new edge will create a cycle.

143. The length of a Hamiltanian path (if exists) in a connected graph of *n* vertices is

(*a*) $n - 1$ (*b*) n
(*c*) $n + 1$ (*d*) $\frac{n}{2}$

144. A simple graph in which there exists an edge between every pair of vertices is called

(*a*) complete graph (*b*) eular graph
(*c*) planner graph (*d*) regular graph

145. If a graph requires k different colours for its proper colouring, then chromatic number of the graph is

(*a*) 1 (*b*) k
(*c*) $k - 1$ (*d*) $\frac{k}{2}$

146. Degree of each vertex in K_n is

(*a*) n (*b*) $n - 1$
(*c*) $n - 2$ (*d*) $2n - 1$

147. The parity of vertex D will be

(*a*) odd (*b*) even
(*c*) uncertain (*d*) none of these

148. The number of ways in which a team of eleven players can be selected from 22 players including 2 of them and excluding 4 of them is

(*a*) ${}^{16}C_{11}$ (*b*) ${}^{16}C_5$
(*c*) ${}^{16}C_9$ (*d*) ${}^{20}C_9$

149. In how many ways can 5 red and 4 white balls be drawn from a bag containing 10 red and 8 white balls ?

(*a*) ${}^{8}C_5 \times {}^{10}C_4$ (*b*) ${}^{10}C_5 \times {}^{8}C_4$
(*c*) ${}^{18}C_9$ (*d*) None of these

150. How many 10 digits numbers can be written by using the digits 1 and 2 ?

(*a*) ${}^{10}C_1 + {}^{9}C_2$ (*b*) 2^{10}
(*c*) ${}^{10}C_2$ (*d*) 10 !

151. The number of words that can be formed out of the letters of the word 'COMMITTEE' is

(a) $\frac{9!}{(2!)^3}$ (b) $\frac{9!}{(2!)^2}$

(c) $\frac{9!}{2!}$ (d) 9 !

152. How many different committees of 5 can be formed from 6 men and 4 women on which exact 3 men and 2 women serve ?

(a) 6 (b) 20

(c) 60 (d) 120

153. m men and n women are to be seated in a row so that no two women sit together. If $m > n$, then number of ways in which they can be seated is

(a) $\frac{m!n!}{(m+n)!}$ (b) $\frac{m!(m+1)!}{(m-n+1)!}$

(c) $\frac{m!n!}{(m-n+1)!}$ (d) none of these

154. There are $(n + 1)$ white and $(n + 1)$ black balls each set numbered 1 to $n + 1$. The number of ways in which the balls can be arranged in a row so that adjacent balls are of different colours, is

(a) $(2n + 2)!$ (b) $(2n + 2)! \times 2$

(c) $(n + 1)! \times 2$ (d) $2\{(n + 1)!\}^2$

155. The number of ways in which 10 persons can sit around a circular table so that none of them has the same neighbours in any two arrangements, is

(a) 9 ! (b) $\frac{1}{2}(9!)$

(c) 10 ! (d) $\frac{1}{2}(10!)$

156. In the above problem, if the first digit cannot be 0, then

(a) 42100 (b) 421200

(c) 42400 (d) 421600

157. Six teachers and six students have to sit round a circular table such that there is a teacher between any two students. The number of ways in which they can sit is

(a) $6! \times 6!$ (b) $5! \times 6!$

(c) $5! \times 5!$ (d) none of these

158. In how many ways can 10 people be seated in a row so that a given pair is not next to each other?

(a) $10! - 9!$

(b) $10! - 2(9!)$

(c) $9!$

(d) none of these

159. The number of ways in which three girls and nine boys can be seated in two vans each having numbered seats with four in the back and three in the front ?

(a) 7! (b) 2 (7!)

(c) 7 (13!) (d) 12!

160. Suppose a licence plate contains two letters followed by three digits with the first digit not zero. How many different licence plates can be printed ?

(a) 608000 (b) 608200

(c) 608400 (d) 608600

NUMERICAL TYPE QUESTIONS

1. Number of subsets of a set of order three is ________
2. Number of proper subsets of a set of order three is ________
3. The number of distinct relations on a set of 3 elements is ________
4. The number of elements in the power set of the set $\{\{a, b\}, \{c\}\}$ is ________
5. The number of distinct reflexive and symmetric relations that can be defined on a 3 element set is ________
6. If P(A) be the collection of all subsets of A = $\{a, b, c\}$ and R be a relation defined as "*x is disjoint from y*" over P(A), then number of elements in R is ________
7. If set A has 3 elements and set B has 4 elements, then number of injections that can be defined from A into B is ________
8. If sets A and B have 3 and 6 elements each, then minimum number of elements in $A \cup B$ is ________
9. In a beauty contest, half the number of experts voted for Mr. A and two thirds voted for Mr. B. 10 voted for both and 6 did not vote for either. How many experts were there in all ?
10. In a classroom containing 28 people, there are 18 people who speak English, 15 people who speak Hindi and 22 people who speak Telugu, 9 people speak both English and Hindi, while 1 people speak both Hindi and Telugu. Further 13 persons speak both Telugu and English. How many people speak all three languages ?
11. Let $f : R \to R$ be defined by

$$f(x) = \begin{cases} x+2 & (x \le -1) \\ x^2 & (-1 \le x \le 1) \\ 2-x & (x \ge 1) \end{cases}.$$

Then value of $f(-1.75) + f(0.5) + f(1.5)$ is ________

12. If * is defined on R* as $a * b = \frac{ab}{2}$, then identity element in the group (R*, *) is _______

13. The inverse of $-i$ in the multiplicative group, $\{1, -1, i, -i\}$ is _______

14. In the group G = {2, 4, 6, 8} under multiplication modulo 10, the identity element is _______

15. If a, b are positive integers, define $a * b = \alpha$ where $ab \equiv \alpha$ (modulo 7), with this * operation, then inverse of 3 in group G {1, 2, 3, 4, 5, 6} is _______

16. Let D_{30} = {1, 2, 3, 5, 6, 10, 15, 30} and relation I be a partial ordering on D_{30}. The all upper bounds of 10 and 15 respectively is _______

17. Let D_{30} = {1, 2, 3, 5, 6, 10, 15, 30} and relation I be a partial ordering on D_{30}. The lub of 10 and 15 respectively is _______

18. The minimum number of spanning trees in a connected graph with 'n' nodes is _______

19. The number of colours required to properly colour the vertices of every planar graph is _______

20. If G and G* are isomorphic graphs, then number of connected components of G* if G has connected components, are _______

21. If v is an isolated vertex in a graph (multigraph) G, then the degree is _______

22. The degree of vertex D in G will be _______

23. The number of different permutations of the word BANANA is _______

24. The number of diagonals that can be drawn by joining the vertices of an octagon is _______

25. In an examination there are three multiple choice questions and each question has 4 choices. Number of ways in which a student can fail to get all answers correct is _______

26. Ramesh has 6 friends. In how many ways can he invite one or more of them at a dinner? _______

27. The number n of ways that an organization consisting of twenty-six members can elect a president, treasurer, and secretary (assuming no person is elected to more than one position) is _______

28. Total number of words formed by 2 vowels and 3 consonants taken from 4 vowels and 5 consonants is equal to _______

29. The ______ number of ways can 5 prizes be distributed among 4 boys when every boy can take one or more prizes

30. The number of different signals which can be given from 6 flags of different colours taking one or more at a time, is _______

31. The number of ways to arrange the letters of the word CHEESE are_______

32. The number of words from the letters of the word 'BHARAT' in which B and H will never come together, is_______

33. Ten different letters of an alphabet are given. Words with five letters are formed from these given letters. Then the number of words which have at least one letter repeated is _______

34. If a five digit number divisible by 3 is to be formed using the numerals 0, 1, 2, 3, 4 and 5, without repetition, then total number of ways this can be done is _______

35. If sides AB, BC, CA of a triangle ABC have 3, 4 and 5 interior points respectively on them, then total number of triangles that can be constructed by using these points as vertices, is _______

36. A student can take one of four Mathematics sections and one of five English sections. The number n of ways he can register for the two courses, is _______

37. In a football championship, 153 matches were played. Every team played one match with each other. The number of teams participating in the champion-ship is _______

38. There are 10 lamps in a hall. Each one of them can be switched on independently. The number of ways in which the hall can be illuminated is _______

39. Three persons enter a railway compartment. If there are 5 seats vacant, _____ number of ways can they take these seats

40. Number n of license plates that can be made where each plate contains two distinct letters followed by three different digits is _______

41. The number of ways to cut a six sided convex polygon whose vertices are labeled into four triangles using diagonal lines that do not cross is _______

42. There are three identical red balls and four identical blue balls in a bag. Three balls are drawn. The number of different colour combinations is _______

43. The ______ number of five different letter words can be formed out of the word "LOGARITHMS"

44. The ______ number of ways can 5 men and 3 women be seated in a row so that two women cannot sit side by side

45. If 6 identical coins are arranged in a row, then number of arrange-ments in which 4 are heads and 2 are tails are _______

ANSWERS

MCQ Type Questions

1. (*c*)	**2.** (*c*)	**3.** (*c*)	**4.** (*a*)	**5.** (*d*)	**6.** (*b*)	**7.** (*d*)	**8.** (*b*)	**9.** (*b*)	**10.** (*a,d*)
11. (*c,d*)	**12.** (*c*)	**13.** (*a*)	**14.** (*c*)	**15.** (*b*)	**16.** (*c*)	**17.** (*b*)	**18.** (*c, d*)	**19.** (*b*)	**20.** (*d*)
21. (*c*)	**22.** (*c*)	**23.** (*d*)	**24.** (*c*)	**25.** (*a*)	**26.** (*a*)	**27.** (*b,c*)	**28.** (*a,c*)	**29.** (*d*)	**30.** (*b,d*)
31. (*d*)	**32.** (*d*)	**33.** (*b,c*)	**34.** (*b*)	**35.** (*c*)	**36.** (*d*)	**37.** (*b*)	**38.** (*d*)	**39.** (*b*)	**40.** (*c*)
41. (*d*)	**42.** (*d*)	**43.** (*a*)	**44.** (*b*)	**45.** (*b*)	**46.** (*b*)	**47.** (*d*)	**48.** (*d*)	**49.** (*c*)	**50.** (*d*)
51. (*d*)	**52.** (*a*)	**53.** (*b*)	**54.** (*d*)	**55.** (*b*)	**56.** (*c*)	**57.** (*d*)	**58.** (*c*)	**59.** (*d*)	**60.** (*c*)
61. (*d*)	**62.** (*b*)	**63.** (*c*)	**64.** (*a,b*)	**65.** (*d*)	**66.** (*d*)	**67.** (*c*)	**68.** (*d*)	**69.** (*a*)	**70.** (*c*)
71. (*d*)	**72.** (*b*)	**73.** (*c*)	**74.** (*c*)	**75.** (*b*)	**76.** (*c*)	**77.** (*a*)	**78.** (*b*)	**79.** (*b*)	**80.** (*c*)
81. (*b*)	**82.** (*a*)	**83.** (*c*)	**84.** (*d*)	**85.** (*c*)	**86.** (*c*)	**87.** (*a*)	**88.** (*d*)	**89.** (*a*)	**90.** (*d*)
91. (*c*)	**92.** (*c*)	**93.** (*d*)	**94.** (*b*)	**95.** (*b*)	**96.** (*b*)	**97.** (*b*)	**98.** (*c*)	**99.** (*d*)	**100.** (*a*)
101. (*d*)	**102.** (*c*)	**103.** (*a*)	**104.** (*b*)	**105.** (*d*)	**106.** (*d*)	**107.** (*c*)	**108.** (*b*)	**109.** (*c*)	**110.** (*a*)
111. (*c*)	**112.** (*d*)	**113.** (*d*)	**114.** (*a, d*)	**115.** (*a*)	**116.** (*d*)	**117.** (*b*)	**118.** (*b*)	**119.** (*c*)	**120.** (*c*)
121. (*c*)	**122.** (*d*)	**123.** (*d*)	**124.** (*b*)	**125.** (*a*)	**126.** (*c*)	**127.** (*d*)	**128.** (*a*)	**129.** (*b*)	**130.** (*a*)
131. (*c*)	**132.** (*a*)	**133.** (*d*)	**134.** (*b*)	**135.** (*c*)	**136.** (*c*)	**137.** (*a*)	**138.** (*b*)	**139.** (*b*)	**140.** (*d*)
141. (*c*)	**142.** (*a*)	**143.** (*a*)	**144.** (*a*)	**145.** (*d*)	**146.** (*b*)	**147.** (*b*)	**148.** (*c*)	**149.** (*b*)	**150.** (*b*)
151. (*a*)	**152.** (*d*)	**153.** (*b*)	**154.** (*d*)	**155.** (*b*)	**156.** (*b*)	**157.** (*b*)	**158.** (*b*)	**159.** (*c*)	**160.** (*c*)

Numerical Type Questions

1. 8	**2.** 6	**3.** 512	**4.** 4	**5.** 8	**6.** 11	**7.** 24
8. 6	**9.** 24	**10.** 6	**11.** 1	**12.** 2	**13.** *i*	**14.** 6
15. 5	**16.** 30	**17.** 30	**18.** 2	**19.** 2	**20.** 8	**21.** 0
22. 4	**23.** 60	**24.** 20	**25.** 63	**26.** 63	**27.** 15600	**28.** 7200
29. 1024	**30.** 1956	**31.** 120	**32.** 240	**33.** 69760	**34.** 216	**35.** 205
36. 20	**37.** 18	**38.** 1023	**39.** 60	**40.** 468000	**41.** 12	**42.** 30
43. 30240	**44.** 14400	**45.** 15				

EXPLANATIONS

NUMERICAL TYPE QUESTIONS

20. The graphs G* must also have eight connected components.

21. The vertex v is isolated if it does not belong to any edge. Thus v is isolated if and only if deg $(v) = 0$.

Consider the multigraph G where

V(G) = {A, B, C, D} and

E(G) = [{A, C}, {A, D}, {B, B}, {B, C}, {C, A}, {C, B}, {D, B}, {D, D}]

27. The president can be elected in twenty-six different ways; following this, the treasurer can be elected in twenty-five different ways (since the person chosen president is not eligible to be treasurer); and following this, the secretary can be elected in twenty-four different ways.

Thus, by above principle of counting, there are $n = 26 \times 25 \times 24 = 15\,600$ different ways in which the organization can elect the officers.

36. There are four choices for Mathematics and five choices for English.

$$\therefore \quad n = 4 \times 5 = 20$$

40.

$$n = 26 \times 25 \times 10 \times 9 \times 8 = 468000.$$

i.e. there are twenty-six choices for the first letter, but only twenty-five choices for the second letter which must be different from the first letter. Similarly, choices for the digits are 10, 9 and 8 since the digits must be distinct.

43. Total letters in the word "LOGARITHMS" is 10,

i.e. $n = 10$ and we have to form five-letter words $r = 5$

$\therefore$ Required number of words,

$$P(10, 5) = \frac{10!}{(10-5)!} = \frac{10!}{5!} = \frac{10 \times 9 \times 8 \times 7 \times 6 \times 5!}{5!} = 10 \times 9 \times 8 \times 7 \times 6 = \mathbf{30240}$$

44. Since there is no restriction on men, they can sit in a row in 5! = 120 ways.

Since women cannot sit side by side they can sit in between two men or in the first place or in the last place as shown below :

X | M | X | M | X | M | X | M | X | M | X

If sign 'X' shows the women place, there are 6 vacancies for women to sit. In these 6 vacancies 3 women can sit in

$${}^6P_3 = 120 \text{ ways.}$$

Hence number of ways where 5 men and 3 women can sit with the given condition

$$= 120 \times 120 = \mathbf{14400.}$$

45. Here we find the number of permutations from 6 objects in which 4 are of one type (heads) and 2 are of another type (tails).

$\therefore$ Required number of ways

$$= \frac{6!}{4!2!} = \frac{6 \times 5 \times 4 \times 3 \times 2 \times 1}{(4 \times 3 \times 2 \times 1)(2 \times 1)} = \mathbf{15}$$

8

CHAPTER

Transform Theory

FOURIER TRANSFORM

If a non-periodic signal has finite total energy, its frequency domain representation will be a continuous spectrum obtained from the fourier transform. Therefore an aperiodic signal $f(t)$ can be represented by a Fourier Integral (rather than Fourier Series).

Let, $f(t)$ denote a non periodic deterministic signal, expressed as some function of time t. Then fourier transform of signal $f(t)$ is given by

$$F(\omega) = \lim_{T\to\infty} F_n = \int_{-\infty}^{\infty} f(t)e^{-j\omega t}dt$$

The original signal $f(t)$ is recovered using the formula for inverse fourier transform.

i.e., $$f(t) = \frac{1}{2\pi}\int_{-\infty}^{\infty} F(\omega)e^{j\omega t}d\omega = \int_{-\infty}^{\infty} F(f)e^{j\omega t}df$$

In general, the fourier transform $F(\omega)$ is a complex function of frequency ω, so that we may express it in the form

$$F(\omega) = |F(\omega)|e^{j\theta(\omega)}$$

where $|F(\omega)|$ is called the continuous amplitude spectrum of $f(t)$ and $\theta(\omega)$ is called the continuous phase spectrum of $f(t)$.

For special case of real valued function $f(t)$, we have

$$F(-\omega) = F^*(\omega)$$

Therefore it follows that if $f(t)$ is a real valued function of time t, then

$$|F(-\omega)| = |F(\omega)|$$

and $$\theta(-\omega) = -\theta(\omega)$$

We may make the following statements on the spectrum of a real valued signals.

(*i*) The amplitude spectrum of the signal is an even function of the frequency, i.e., the amplitude spectrum is symmetric about the vertical axis.

(*ii*) The phase spectrum of the signal is an odd function of the frequency i.e., the phase spectrum is antisymmetric about the vertical axis.

Hence we conclude that the Fourier transform of a real valued signal exhibits Conjugate symmetry.

Dirichlet Conditions

For the fourier transform of a signal $f(t)$ to exist, it is sufficient that $f(t)$ will satisfy the three conditions known as Dirichlet's conditions. These are :

(*i*) The function $f(t)$ is a single valued with a finite number of maxima and minima.

(*ii*) The function $f(t)$ should have a finite number of discontinuities in any finite interval of time.

(*iii*) The function $f(t)$ is absolutely integrable

i.e., $$\int_{-\infty}^{\infty} |f(t)|dt < \infty$$

In other words, physical realizability is a sufficient condition for the fourier transform to exist.

Fourier Transform Operations

Operation	$f(t)$	$F(\omega)$
Addition	$f_1(t) + f_2(t)$	$F_1(\omega) + F_2(\omega)$
Scalar Multiplication	$Kf(t)$	$KF(\omega)$
Symmetry	$F(t)$	$2\pi f(-\omega)$
Scaling	$f(bt)$	$\frac{1}{\lvert b\rvert}F\left(\frac{\omega}{b}\right)$
Time Shift	$f(t-b)$	$F(\omega)e^{-j\omega b}$
Frequency shift	$f(t)e^{j\omega_c t}$	$F(\omega - \omega_c)$
Time Convolution	$f_1(t) * f_2(t)$	$F_1(\omega)\,F_2(\omega)$
Frequency Convolution	$f_1(t)f_2(t)$	$\frac{1}{2\pi}F_1(\omega) * F_2(\omega)$
Time Differentiation	$\frac{d^n f(t)}{dt^n}$	$(j\omega)^n F(\omega)$
Time Integration	$\int_{-\infty}^{t} f(t)d(t)$	$\frac{1}{j\omega}F(\omega) + \pi F(0)\delta(\omega)$

Short Table of Fourier Transform

$f(t)$	Property	$F(\omega)$	Property
1. $e^{-at}u(t)$	Asymmetric	$\frac{1}{a+j\omega}$	Complex
2. $e^{at}u(-t)$	Asymmetric	$\frac{1}{a-j\omega}$	Complex

$f(t)$	Property	$F(\omega)$	Property
3. $e^{-a\|t\|}$	Real and even symmetric	$\frac{2a}{a^2+\omega^2}$	Real and symmetric
4. $te^{-at}u(t)$		$\frac{1}{(a+j\omega)^2}$	$a > 0$
5. $t^n e^{-at}u(t)$		$\frac{n!}{(a+j\omega)^{n+1}}$	$a > 0$
6. $\delta(t)$	Real and Even symmetric	1	Real and Even symmetric
7. A	Real and Even symmetric	$2\pi A\delta(\omega)$ or $A\delta(f)$	Real and Even symmetric
8. $e^{jw_0 t}$		$2\pi\delta(\omega-\omega_0)$	
9. $A\cos\omega_0 t$		$A\pi[\delta(\omega-\omega_0)+\delta(\omega+\omega_0)]$ or, $\frac{A}{2}[\delta(f-f_0)+\delta(f+f_0)]$	
10. $A\sin\omega_0 t$		$Aj\pi[\delta(\omega+\omega_0)-\delta(\omega-\omega_0)]$ or, $\frac{jA}{2}[\delta(f+f_0)-\delta(f-f_0)]$	
11. $u(t)$		$\pi\delta(\omega)+\frac{1}{j\omega}$	
12. $\sin t$	Real and Anti-symmetric	$\frac{2}{j\omega}$	Imaginary and odd symmetric
13. $\cos\omega_0 t\, u(t)$		$\frac{\pi}{2}[\delta(\omega-\omega_0)+\delta(\omega+\omega_0)]+\frac{j\omega}{\omega_0^2-\omega^2}$	
14. $\sin\omega_0 t\, u(t)$		$\frac{\pi}{2j}[\delta(\omega+\omega_0)-\delta(\omega-\omega_0)]+\frac{\omega_0}{\omega_0^2-\omega^2}$	
15. $e^{-at}\sin\omega_0 t\, u(t)$		$\frac{\omega_0}{(a+j\omega)^2+\omega_0^2}$	
16. $e^{-at}\cos\omega_0 t\, u(t)$		$\frac{a+j\omega}{(a+j\omega)^2+\omega_0^2}$	
17. $\Delta\left(\frac{t}{\tau}\right)$	Even symmetric	$\frac{\tau}{2}Sa^2\left(\frac{\omega\tau}{4}\right)$	Even symmetric

Fourier Integral Theorem

Fourier integral of $f(x)$ is defined as

$$f(x) = \frac{1}{\pi}\int_0^\infty\int_{-\infty}^\infty f(t)\cos\lambda(t-x)\,dt\,d\lambda$$

$$= \frac{1}{\pi}\int_0^\infty\left[\int_{-\infty}^\infty f(t)\cos\lambda(t-x)\,dt\right]d\lambda.$$

The equation is true at a point of continuity. At a point of discontinuity, value of the integral on the right is

$$\frac{1}{2}\left[\lim_{\substack{x\to 0^+\\ x>0}} f(x) + \lim_{\substack{x\to 0^-\\ x<0}} f(x)\right].$$

Fourier Sine Integral

$$f(x) = \frac{1}{\pi}\int_0^\infty\left[\int_{-\infty}^\infty f(t)\cos(\lambda t-\lambda x)\,dt\right]d\lambda$$

$$= \frac{1}{\pi}\int_0^\infty\left[\int_{-\infty}^\infty f(t)\left[\cos\lambda t.\cos\lambda x+\sin\lambda t.\sin\lambda x\right]dt d\lambda\right] \quad ...(i)$$

$$= 0 + \frac{1}{\pi}\int_0^\infty \sin\lambda x\left(\int_{-\infty}^\infty f(t)\sin\lambda t\,dt\right)d\lambda$$

When $f(t)$ is an odd function, $f(t)\cos\lambda t$ is an odd while $f(t)\sin\lambda t$ is even. Thus the first integral in equation *(i)* vanishes and we get

$$f(x) = \frac{2}{\pi}\int_0^{\infty}\sin\lambda x\left[\int_0^{\infty}f(t)\sin\lambda t\,dt\right]d\lambda$$

Fourier Cosine Integral

When $f(t)$ is an even function, $f(t)\cos\lambda t$ is an even while $f(t)\sin\lambda t$ is odd. Thus second integral in equation *(i)* vanishes and we get

$$f(x) = \frac{2}{\pi}\int_0^{\infty}\cos\lambda x\left[\int_0^{\infty}f(t)\cos\lambda t\,dt\right]d\lambda\,.$$

Complex Fourier Integral

Since $\cos\lambda(t-x)$ is an even function of λ and $\sin(t-x)$ is an odd function of λ

$$f(x) = \frac{1}{2\pi}\int_{-\infty}^{\infty}\int_{-\infty}^{\infty}f(t)\cos\lambda(t-x)\,dt\,d\lambda$$

$$\therefore\quad 0 = \frac{1}{2\pi}\int_{-\infty}^{\infty}\int_{-\infty}^{\infty}f(t)\sin\lambda(t-x)\,dt\,d\lambda$$

Now $\quad f(x) = f(x) + i\,0$

$$= \frac{1}{2\pi}\int_{-\infty}^{\infty}\int_{-\infty}^{\infty}f(t)\,[\cos\lambda(t-x) + i\sin\lambda(t-x)]\,dt\,d\lambda$$

$$= \frac{1}{2\pi}\int_{-\infty}^{\infty}\int_{-\infty}^{\infty}f(t)\,e^{i\lambda(t-x)}\,d\lambda$$

Inversion Formula of Fourier Transform

Complex form of Fourier integral is

$$f(x) = \frac{1}{2\pi}\int_{-\infty}^{\infty}e^{-i\lambda x}\int_{-\infty}^{\infty}f(t)\,e^{+i\lambda t}\,dt\,d\lambda$$

Replacing λ by s, we get

$$f(x) = \frac{1}{\sqrt{2\pi}}\int_{-\infty}^{\infty}e^{-isx}\left[\frac{1}{\sqrt{2\pi}}\int_{-\infty}^{\infty}f(t)\,e^{ist}\,dt\right]ds\,.$$

Denoting the value of inner integral by F $\{f(x)\}$, we get

$$\mathrm{F}\,\{f(x)\} = \frac{1}{\sqrt{2\pi}}\int_{-\infty}^{\infty}f(x)\,e^{isx}\,dx = \mathrm{F}(s)$$

This is called Fourier transform of $f(x)$, and is denoted by F $\{f(x)\}$ or F(s).

$$\therefore\quad f(x) = \frac{1}{\sqrt{2\pi}}\int_{-\infty}^{\infty}\mathrm{F}(s)\,e^{-isx}\,ds$$

This is called **inversion formula** of F(s)

Fourier Sine Transform and its Inversion Formula

We have Fourier sine integral as

$$f(x) = \frac{2}{\pi}\int_0^{\infty}\sin\lambda x\int_0^{\infty}f(t)\sin\lambda t\,dt\,d\lambda$$

Replacing λ by s, we get

$$f(x) = \sqrt{\frac{2}{\pi}}\int_0^{\infty}\sin sx\left[\sqrt{\frac{2}{\pi}}\int_0^{\infty}f(t)\sin st\,dt\right]ds$$

Denoting value of the inner integral by $\mathrm{F}_s\,\{fx)\}$ or $\mathrm{F}_s\,(s)$, we get

$$f(x) = \sqrt{\frac{2}{\pi}}\int_0^{\infty}\mathrm{F}_s\,(s)\sin sx\,ds \qquad ...(i)$$

where $$\mathrm{F}_s\,(s) = \sqrt{\frac{2}{\pi}}\int_0^{\infty}f(x)\sin sx\,dx = \mathrm{F}_s\,\{f(x)\} \qquad ...(ii)$$

Equation *(i)* is called *Fourier sine transform* of $f(x)$ and equation *(ii)* is called *inversion formula of* $\mathrm{F}_s\,(s)$.

Fourier Cosine Transform and its Inversion Formula

Fourier cosine integral is

$$f(x) = \frac{2}{\pi}\int_0^{\infty}\cos\lambda x\int_0^{\infty}f(t)\cos\lambda t\,dt\,d\lambda$$

$$f(x) = \sqrt{\frac{2}{\pi}}\int_0^{\infty}\cos sx\left[\sqrt{\frac{2}{\pi}}\int_0^{\infty}f(t)\cos st\,dt\right]ds$$

Denoting inner integral by $\mathrm{F}_c\,(s)$ or $\mathrm{F}_c\,\{f(x)\}$, we get

$$f(x) = \sqrt{\frac{2}{\pi}}\int_0^{\infty}\mathrm{F}_c\,(s)\cos sx\,ds \qquad ...(i)$$

where $$\mathrm{F}_c\,(s) = \sqrt{\frac{2}{\pi}}\int_0^{\infty}f(t)\cos st\,dt = \mathrm{F}_c\,\{f(x)\} \qquad ...(ii)$$

Equation *(ii)* is called Fourier cosine transform of $f(x)$ and equation *(i)* is called inversion formula of $\mathrm{F}_c\,(s)$.

Finite Fourier Sine and Cosine Transforms

The finite Fourier sine transform of $f(x)$, in $0 < x < c$, is defined as

$$\mathrm{F}_s\,(n) = \int_0^{c}f(x)\sin\frac{n\pi x}{c}\,dx$$

where n is an integer.

The function $f(x)$ is called *inverse finite Fourier sine transform* of $\mathrm{F}_s\,(n)$ which is given by

$$f(x) = \frac{2}{c}\sum_{n=1}^{\infty}\mathrm{F}_s\,(n)\sin\frac{n\pi x}{c}$$

Finite Fourier Cosine Transforms

The finite Fourier cosine transform of $f(x)$ in $0 < x < c$, is defined as

$$\mathrm{F}_c\,(n) = \int_0^{c}f(x)\cos\frac{n\pi x}{c}\,dx$$

where n is an integer.

The function $f(x)$ is called *inverse finite Fourier cosine transform* of $\mathrm{F}_c\,(n)$ which is given by

$$f(x) = \frac{1}{c}f_c\,(0) + \frac{2}{c}\sum_{n=1}^{\infty}\mathrm{F}_c\,(n)\cos\frac{n\pi x}{c}$$

Properties of Fourier Transforms

1. **Linearity property**

If $F_1(s)$ and $F_2(s)$ are Fourier transforms of $f_1(x)$ and $f_2(x)$ respectively, then

$$F\{af_1(x) + bf_2(x)\} = a F_1(s) + b F_2(s)$$

where a and b are constants.

2. **Change of scale property**

If $F(s)$ is Fourier transform of $f(x)$, then

$$F\{f(a)\} = \frac{1}{a} F\left(\frac{s}{a}\right)$$

3. **If $F_s(s)$ and $F_c(s)$** are Fourier sine and cosine transform of $f(x)$, then

$$F_s\{f(ax)\} = \frac{1}{|a|} F_s\left(\frac{s}{a}\right)$$

and $$F_c\{f(ax)\} = \frac{1}{|a|} F_c\left(\frac{s}{a}\right).$$

4. **Shifting property**

If $F(s)$ is Fourier transform of $f(x)$, then

$$F\{f(x-a)\} = e^{isa} F(s)$$

5. **Modulation theorem**

If $F(s)$ is Fourier transform of $f(x)$, then

$$F\{f(x)\cos ax\} = \frac{1}{2}\{F(s+a) + F(s-a)\}.$$

6. If $F_s(s)$ and $F_c(s)$ are Fourier sine and cosine transform of $f(x)$ respectively, then

(i) $F_s\{f(x)\cos ax\} = \frac{1}{2}[F_s(s+a) + F_s(s-a)]$

(ii) $F_s\{f(x)\sin ax\} = \frac{1}{2}[F_s(s+a) - F_s(s-a)]$

(iii) $F_c\{f(x)\sin ax\} = \frac{1}{2}[F_c(s-a) - F_c(s+a)]$

CONVOLUTION

The convolution of two functions $f(x)$ and $g(x)$ over the interval $(-\infty, \infty)$ is defined as

$$f * g = \int_{-\infty}^{\infty} f(u)\, g(x-u)\, du.$$

Convolution theorem for Fourier transforms

The Fourier transform of the convolution of $f(x)$ is the product of their Fourier transforms, *i.e.*

$$F\{f(x) * g(x)\} = F(s) \cdot G(s).$$

Parseval's identity for fourier transforms

If $F(s)$ is Fourier transform of $f(x)$, then

$$\int_{-\infty}^{\infty} |f(x)|^2\, dx = \int_{-\infty}^{\infty} |F(s)|^2\, ds.$$

SOLVED EXAMPLES

1. Find the Fourier transforms of $f(t) = e^{-a|t|}$, where $a > 0$.

Solution.

Given $f(t) = e^{-a|t|}$ where $a > 0$

Since $|t| = t,\ t > 0 \quad = -t,\ t < 0$

$\therefore$ $f(t) = e^{-at} \quad t > 0 \quad = e^{at} \quad t < 0$

Fourier transform of $f(t)$ is given by

$$F(s) = \frac{1}{\sqrt{2\pi}} \int_{-\infty}^{\infty} f(t)\, e^{ist}\, dt$$

$$= \frac{1}{\sqrt{2\pi}} \left[\int_{-\infty}^{0} e^{at} e^{ist}\, dt + \int_{0}^{\infty} e^{-at} e^{ist}\, dt\right]$$

$$= \frac{1}{\sqrt{2\pi}} \left[\left(\frac{e^{(a+is)t}}{a+is}\right)_{-\infty}^{0} + \left(\frac{e^{-(a-is)t}}{-(a-is)}\right)_{0}^{\infty}\right]$$

$$= \frac{1}{\sqrt{2\pi}} \left[\frac{1}{a+is} - 0 - 0 + \frac{1}{a-is}\right]$$

$$= \frac{1}{\sqrt{2\pi}} \frac{a - is + a + is}{(a+is)(a-is)}$$

$$= \frac{1}{\sqrt{2\pi}} \cdot \frac{2a}{a^2 + s^2}$$

2. If Fourier sine transform of $f(x)$ is

$$\frac{1-\cos n\pi}{n^2\pi^2},\ 0 < x < \pi;\ \text{find } f(x).$$

Solution.

Given $$F_s(n) = \frac{1-\cos n\pi}{n^2\pi^2},\ 0 < x < \pi$$

Inverse finite Fourier sine transform of $F_s(n)$ is given by

$$f(x) = \frac{2}{\pi} \sum_{n=1}^{\infty} F_s(n) \sin \frac{n\pi x}{\pi}$$

$$= \frac{2}{\pi} \sum_{n=1}^{\infty} \frac{1-\cos n\pi}{n^2\pi^2} \sin nx$$

$$= \frac{2}{\pi^3} \sum_{n=1}^{\infty} \frac{1-\cos n\pi}{n^2} \sin nx.$$

3. Solve the integral equation

$$\int_0^{\infty} f(\theta) \cos \alpha\theta\, d\theta = \begin{cases} 1-\alpha, & 0 \le \alpha \le 1 \\ 0, & \alpha > 1 \end{cases}$$

Solution.

Since $$F_c = \int_0^{\infty} \sqrt{\frac{2}{\pi}} f(x) \cos sx\, dx$$

$\therefore$ $$F_c(\alpha) = \sqrt{\frac{2}{\pi}} \int_0^{\infty} f(\theta) \cos \alpha\theta\, d\theta.$$

$$= \begin{cases} \sqrt{\frac{2}{\pi}}, (1-\alpha) & 0 \le a \le 1 \\ 0, & \alpha > 1 \end{cases}$$

By inversion formula of $F_c(\alpha)$, we get

$$f(\theta) = \sqrt{\frac{2}{\pi}} \int_0^{\infty} F_c(\theta) \cos\theta \, d\alpha$$

$$= \sqrt{\frac{2}{\pi}} \left[\int_0^1 \sqrt{\frac{2}{\pi}} (1 - 0 \cos\alpha\theta \, d\alpha + \int_1^{\infty} 0 \, d\alpha \right]$$

$$= \frac{2}{\pi} \left[(1-\alpha)\left(\frac{\sin\alpha\theta}{\theta}\right) - (-1)\left(-\frac{\cos\alpha\theta}{\theta^2}\right) \right]_0^1$$

$$= \frac{2}{\pi} \left[0 - \frac{\cos\theta}{\theta^2} - 0 + \frac{1}{\theta^2} \right]$$

$$f(\theta) = \frac{2(1-\cos\theta)}{\pi\theta^2}$$

4. Solve the integral equation

$$\int_0^{\infty} f(x) \cos sx \, dx = e^{-s}$$

Solution.

Since $\quad F_c(s) = \int_0^{\infty} \sqrt{\frac{2}{\pi}} f(x) \cos sx \, dx$

$$F_c(s) = \sqrt{\frac{2}{\pi}} e^{-s}$$

By using inversion formula, we get

$$f(x) = \sqrt{\frac{2}{\pi}} \int_0^{\infty} F_c(s) \cos sx \, ds = \frac{2}{\pi} \int_0^{\infty} e^{-s} \cos sx \, ds$$

$$= \frac{2}{\pi} \left[\frac{e^{-s}}{1+x^2} (-\cos sx - x \sin sx) \right]_0^{\infty}$$

$$= \frac{2}{\pi} \left[0 - \frac{1}{1+x^2} (-1-0) \right]$$

$$\therefore \quad f(x) = \frac{2}{\pi} \cdot \frac{1}{1+x^2}$$

5. Find the Fourier cosine transform of e^{-mx} $(m > 0)$.

Solution.

We have $\quad f(x) = e^{-mx}$ $(m > 0)$.

Fourier cosine transform of $f(x)$ is given by

$$F_c(s) = \sqrt{\frac{2}{\pi}} \int_0^{\infty} e^{-mx} \cos sx \, dx$$

$$= \sqrt{\frac{2}{\pi}} \int_0^{\infty} \left[\frac{-e^{-mx}}{m^2+s^2} (-m \cos sx + s \sin sx) \right]$$

$$= \sqrt{\frac{2}{\pi}} \left[0 - \frac{1}{m^2+s^2} (-m+0) \right]$$

$$= \sqrt{\frac{2}{\pi}} \cdot \frac{m}{s^2+m^2}$$

6. Find Fourier sine transform of $e^{-|x|}$

Hence show that

$$\int_0^{\infty} \frac{x \sin mx}{1+x^2} dx = \frac{\pi e^{-m}}{2}, \; m > 0$$

Solution.

x being positive in the interval $(0, \infty)$,

$$e^{-|x|} = e^{-x}$$

Hence Fourier sine transform of $f(x) = e^{-|x|}$ is given by

$$F_s\{f(x)\} = \int_0^{\infty} f(x) \sin sx \, dx$$

$$= \int_0^{\infty} e^{-x} \sin sx \, dx$$

$$= \left| \frac{e^{-x}}{1+s^2} (-\sin sx - s \cos sx) \right|_0^{\infty}$$

$$= \frac{s}{1+s^2}$$

Using Inversion formula for Fourier sine transforms, we get

$$f(x) = \frac{2}{\pi} \int_0^{\infty} F_s\{f(s)\} \sin sx \, dx$$

$$\Rightarrow \quad e^{-x} = \frac{2}{\pi} \int_0^{\infty} \frac{s}{1+s^2} \sin sx \, dx$$

Changing x to m, we get

$$e^{-m} = \frac{2}{\pi} \int_0^{\infty} \frac{s \sin ms}{1+s^2} ds$$

$$= \frac{2}{\pi} \int_0^{\infty} \frac{x \sin mx}{1+m^2} dx$$

$$\therefore \int_0^{\infty} \frac{x \sin mx}{1+m^2} dx = \frac{\pi e^{-m}}{2}.$$

7. Find Fourier cosine transform of

$$f(x) = \begin{cases} x, & \text{for } 0 < x < 1 \\ 2-x, & \text{for } 1 < x < 2 \\ 0, & \text{for } x > 2 \end{cases}$$

Solution.

Fourier cosine transform of $f(x)$ *i.e.*,

$$F_c[f(x)] = \int_0^{\infty} f_c(x) \cos sx \, dx$$

$$= \int_0^1 x \cos sx \, dx + \int_1^2 (2-x) \cos sx \, dx + \int_0^{\infty} 0 . dx$$

$$= \left| x \frac{\sin sx}{s} - \left(\frac{-\cos sx}{s^2} \right) \right|_0^1 + \left| (2-x) \frac{\sin sx}{s} - (-1) \frac{-\cos sx}{s^2} \right|_1^2$$

$$= \left(\frac{\sin s}{s} + \frac{\cos s}{s^2} - \frac{1}{s^2} \right) + \left(-\frac{\cos 2s}{s^2} - \frac{\sin s}{s} + \frac{\cos s}{s^2} \right)$$

$$= \frac{2 \cos s}{s^2} - \frac{\cos 2s}{s^2} - \frac{1}{s^2}$$

8. Find Fourier sine transform of $\frac{e^{-ax}}{x}$.

Solution.

Let $f(x) = \frac{e^{-ax}}{x}$

then its Fourier sine transform *i.e.*,

$$F_s\{f(x)\} = \int_0^\infty f(x)\sin sx\,dx$$

$$= \int_0^\infty \frac{e^{-ax}}{x}\sin sx\,dx = F(s), \text{ say}$$

Differentiating w.r.t. s, we get

$$\frac{d}{ds}\{F(s)\} = \int_0^\infty \frac{xe^{-ax}\cos sx}{x}\,dx$$

$$= \int_0^\infty e^{-ax}\cos sx\,dx = \frac{a}{s^2+a^2}$$

Integrating w.r.t. s, we get

$$F(s) = \int \frac{a}{s^2+a^2}ds = \tan^{-1}\frac{s}{a} + c$$

But When $s = 0$, $F(s) = 0$

$\Rightarrow c = 0$

$$\therefore \quad F(s) = \tan^{-1}\left(\frac{s}{a}\right).$$

9. Find Fourier cosine transform of

$$f(x) = \frac{1}{(1+x^2)}$$

Hence derive Fourier sine transform of

$$\phi(x) = \frac{x}{(1+x^2)}.$$

Solution.

$$F_c[f(x)] = \int_0^\infty \frac{\cos sx}{1+x^2}dx = I \text{ (say)} \quad ...(i)$$

$$\therefore \quad \frac{dI}{ds} = \int_0^\infty \frac{-x\sin sx}{1+x^2}dx = -\int_0^\infty \frac{x^2\sin sx}{x(1+x^2)}dx \quad ...(ii)$$

$$= -\int_0^\infty \frac{\left[(1+x^2)-1\right]\sin sx}{x(1+x^2)}dx$$

$$= -\int_0^\infty \frac{\sin sx}{x}dx + \int_0^\infty \frac{\sin sx}{x(1+x^2)}dx$$

$$\Rightarrow \quad \frac{dI}{ds} = -\frac{\pi}{2} + \int_0^\infty \frac{\sin sx}{x(1+x^2)}dx \quad ...(iii)$$

$$\therefore \quad \frac{d^2I}{ds^2} = \int_0^\infty \frac{x\cos sx}{x(1+x^2)}dx = I$$

$$\Rightarrow \quad \frac{d^2I}{ds^2} - I = 0 \quad \Rightarrow \quad (D^2-1)\,I = 0$$

where, $D = \frac{dI}{ds}$

Its solution is

$$I = c_1e^s + c_2e^{-s} \quad ...(iv)$$

$$\therefore \quad \frac{dI}{ds} = c_1e^s - c_2e^{-s} \quad ...(v)$$

When $s = 0$, equations (*i*) and (*iv*) give

$$c_1 + c_2 = \int_0^\infty \frac{dx}{1+x^2} = \frac{\pi}{2} \quad ...(vi)$$

When $s = 0$, equations (*iii*) and (*v*) give

$$c_1 - c_2 = -\frac{\pi}{2} \quad ...(vii)$$

Solving equations (*vi*) and (*vii*), we get

$$c_1 = 0,\ c_2 = \frac{\pi}{2}$$

Thus from equations (*i*) and (*iv*), we have

$$F_c[f(x)] = I = \left(\frac{\pi}{2}\right)e^{-s}$$

Now $F_s[\phi(x)] = \int_0^\infty \frac{x\sin sx}{1+x^2}dx$

From equation (*ii*)

$$F_s[\phi(x)] = -\frac{dI}{ds}$$

From equation (*v*)

$$F_s[\phi(x)] = \left(\frac{\pi}{2}\right)e^{-s}$$

with $c_1 = 0$, $c_2 = \frac{\pi}{2}$

10. Find Fourier sine and cosine transform of x^{n-1}, $n > 0$.

Solution.

We know $F_s(x^{n-1}) = \int_0^\infty x^{n-1}\sin sx\,dx \quad ...(i)$

and $F_c(x^{n-1}) = \int_0^\infty x^{n-1}\cos sx\,dx \quad ...(ii)$

$$F_c(x^{n-1}) + iF_s(x^{n-1})$$

$$= \int_0^\infty (\cos sx + i\sin sx)x^{n-1}dx$$

$$= \int_0^\infty e^{isx}x^{n-1}dx$$

$$= \int_0^\infty e^{-t}\left(-\frac{t}{is}\right)^{n-1}\left(-\frac{dt}{is}\right) \quad \text{where } is = -t$$

$$= \left(\frac{-1}{(i)}\right)^n \int_0^\infty e^{-t}t^{n-1}dt$$

$$= \frac{(i)^{2n}}{(i)^n s^n}\Gamma(n) = \frac{(i)^n}{s^n}\Gamma(n)$$

$$= \left(\cos\frac{\pi}{2} + i\sin\frac{\pi}{2}\right)^n \Gamma\frac{(n)}{s^n}$$

$$= \left(\cos\frac{n\pi}{2} + i\sin\frac{n\pi}{2}\right)\Gamma\frac{(n)}{s^n}$$

Equating real and imaginary parts, we get

$$F_c(x^{n-1}) = \frac{\Gamma(n)}{s^n}\cos\frac{n\pi}{2}$$

and $$F_s(x^{n-1}) = \frac{\Gamma(n)}{s^n}\sin\frac{n\pi}{2}$$

LAPLACE TRANSFORM

Let f(t) be a function of t defined for all positive values of t. Then **Laplace transforms** *of f (t), is defined by*

$$L\{f(t)\} = \int_0^{\infty} e^{-st} f(t)dt$$

provided that the integral exists. s is a parameter which may be a real or complex number.

$L\{f(t)\}$ is a function of s is written as $\bar{f}(s)$

and $$L\{f(t)\} = \bar{f}(s),$$

When $$f(t) = L^{-1}\{\bar{f}(s)\}$$

then $f(t)$ is called **inverse Laplace transform** of $\bar{f}(s)$.

The symbol L, which transform $f(t)$ into $\bar{f}(s)$, is called *Laplace transformation operator.*

Transforms of Elementary Function

Direct application of the definition gives the following formulae :

- $L(1) = \frac{1}{s}$ $(s > 0)$
- $L(t^n) = \frac{n!}{s^{n+1}}$, when $n = 0, 1, 2, 3, ...$
- $L(e^{at}) = \frac{1}{s-a}$ $(s > a)$
- $L(\sin at) = \frac{a}{s^2 + a^2}$ $(s > 0)$
- $L(\cos at) = \frac{s}{s^2 + a^2}$ $(s > 0)$
- $L(\sinh at) = \frac{a}{s^2 - a^2}$ $(s > |a|)$
- $L(\cosh at) = \frac{s}{s^2 - a^2}$ $(s > |a|)$

Properties of Laplace Transform

(1) Linearity Property

If a, b, c, be any constants and f, g, h any functions of t, then

$$L[a\,f(t) + b\,g(t) - c\,h(t)] = a\,L\{f(t)\} + b\,L\{g(t)\} - c\,L\{h(t)\}$$

(2) First Shifting Property

If $$L\{f(t)\} = \bar{f}(s),$$

then $$L\{e^{at}f(t)\} = \bar{f}(s-a)$$

Application of this property gives following useful results

- $L(e^{at}t^n) = \frac{n!}{(s-a)^{n+1}}$ ... $\left[\because \; L(t^n) = \frac{n!}{s^{n+1}}\right]$
- $L(e^{at}\sin bt) = \frac{b}{(s-a)^2 + b^2}$
- $L(e^{at}\cos bt) = \frac{s-a}{(s-a)^2 + b^2}$
- $L(e^{at}\sinh bt) = \frac{b}{(s-a)^2 - b^2}$
- $L(e^{at}\cosh bt) = \frac{s-a}{(s-a)^2 - b^2}$

where in each case s 79

(3) Change of Scale Property

If $L\{f(t)\} = \bar{f}(s)$, then

$$L\{f(at)\} = \frac{1}{a}\bar{f}\left(\frac{s}{a}\right)$$

Existence conditions

$\int_0^{\infty} e^{-st} f(t)\,dt$ exists if $\int_0^{\lambda} e^{-st} f(t)\,dt$ can actually be evaluated and its limit as $\lambda \to \infty$ exists.

If f (t) is continuous and $\underset{t\to\infty}{Lt}\{e^{-at} f(t)\}$ *is finite; then Laplace transform of f(t), i.e.* $\int_0^{\infty} e^{-st} f(t)\,dt$ exists for $s > a$.

Inverse Transforms Standard Forms

- $L^{-1}\left[\frac{1}{s}\right] = 1$
- $L^{-1}\left[\frac{1}{s-a}\right] = e^{at}$
- $L^{-1}\left[\frac{1}{s^n}\right] = \frac{t^{n-1}}{(n-1)!}$, $n = 1, 2, 3, ...$
- $L^{-1}\left[\frac{1}{(s-a)^n}\right] = \frac{e^{at}t^{n-1}}{(n-1)!}$
- $L^{-1}\left(\frac{1}{s^2 + a^2}\right) = \frac{1}{a}\sin at$
- $L^{-1}\left(\frac{s}{s^2 + a^2}\right) = \cos at$
- $L^{-1}\left(\frac{1}{s^2 - a^2}\right) = \frac{1}{a}\sinh at$

- $L^{-1}\left(\frac{s}{s^2-a^2}\right) = \cosh at$

- $L^{-1}\left[\frac{1}{(s-a)^2+b^2}\right] = \frac{1}{b}e^{at}\sin bt$

- $L^{-1}\left[\frac{s-a}{(s-a)^2+b^2}\right] = e^{at}\cos bt$

- $L^{-1}\left[\frac{s}{(s^2-a^2)^2}\right] = \frac{1}{2a}t\sin at$

- $L^{-1}\left[\frac{1}{(s^2-a^2)^2}\right] = \frac{1}{2a^3}(\sin at - at\cos at)$

Transforms of Derivatives

(*i*) *If $f'(t)$ be continuous and $L\{f(t)\} = \bar{f}(s)$, then*

$$L\{f'(t)\} = s\bar{f}(s) - f(0)$$

(*ii*) *If $f'(t)$ and its first $(n-1)$ derivatives be continuous, then*

$$L\{f^n(t)\} = s^n\bar{f}(s) - s^{n-1}f(0) - s^{n-2}f'(0) - \ldots\ldots - f^{n-1}(0)$$

Transforms of Integrals

If $L\{f(t)\} = \bar{f}(s)$, *then*

$$L\left\{\int_0^t f(u)du\right\} = \frac{1}{s}\bar{f}(s)$$

$$\Rightarrow \quad L^{-1}\left\{\frac{1}{s}\bar{f}(s)\right\} = \int_0^t f(u)du$$

and $$L^{-1}\left\{\frac{\bar{f}(s)}{s^2}\right\} = \int_0^t\left(\int_0^t f(t)dt\right)dt$$

$$L^{-1}\left\{\frac{\bar{f}(s)}{s^3}\right\} = \int_0^t\left\{\int_0^t\left(\int_0^t f(t)dt\right)dt\right\}dt$$

and so on.

Multiplication by t^n

If $L\{f(t)\} = \bar{f}(s)$, *then*

$$L\{t^n f(t)\} = (-1)^n\frac{d^n}{ds^n}[\bar{f}(s)]$$

where $n = 1, 2, 3....$

Division by *t*.

If $L\{f(t)\} = \bar{f}(s)$, *then*

$$L\left\{\frac{1}{t}f(t)\right\} = \int_s^\infty \bar{f}(s)\,ds$$

provided the integral exists.

Convolution Theorem

If $L^{-1}\{\bar{f}(s)\} = f(t)$ and $L^{-1}\{\bar{g}(s)\} = g(t)$, then

$$L^{-1}\{\bar{f}(s)\times\bar{g}(s)\} = \int_0^t f(u)\,g(t-u)\,du = F * G$$

[*$F * G$ is called* **convolution** *of F and G*]

Applications to Differential equations

The Laplace transform method of solving differential equations gives particular solutions without the necessity of first finding the general solution and then evaluating the arbitrary constants. This method is, shorter and is useful for solving linear differential equations with constant coefficients.

Working procedure *to solve a linear differential equation with constant coefficient by transform method* :

(*i*) Take Laplace transform of both sides of the differential equation and using the given initial conditions.

(*ii*) Transpose the terms with minus signs to the right.

(*iii*) Divide by the coefficient of $\bar{y}$, getting $\bar{y}$ as a known function of *s*.

(*iv*) Resolve this function of s into partial fractions and take the inverse transform of both sides. This gives *y* as a function of *t* which is the desired solution satisfying the given conditions.

Simultaneous Linear equations with constant coefficients

The Laplace transfrom method can also be applied with advantage to the solution of simultaneous linear differential equations.

Periodic functions.

If $f(t)$ is a periodic function with period T,

i.e $\quad f(t+T) = f(t)$

then $$L\{f(t)\} = \frac{\int_0^T e^{-st}f(t)dt}{1-e^{-sT}}$$

SOLVED EXAMPLES

1. Find Laplace transforms of

$e^{2t} + 4t^3 - 2\sin 3t + 3\cos 3t$.

Solution.

$L[e^{2t} + 4t^3 - 2\sin 3t + 3\cos 3t]$

$$= L[e^{2t}] + 4L[t^3] - 2L[\sin 3t] + 3L[\cos 3t]$$

$$= \frac{1}{s-2} + 4.\frac{3!}{s^4} - 2.\frac{3}{s^2+3^2} + 3.\frac{s}{s^2+3^2}$$

$$= \frac{1}{s-2} + \frac{24}{s^4} - \frac{6}{s^2+9} + \frac{3s}{s^2+9}$$

2. Find Laplace transform of $1 + 2\sqrt{t} + \frac{3}{\sqrt{t}}$.

Solution.

$$L\left[1 + 2\sqrt{t} + \frac{3}{\sqrt{t}}\right] = L[1] + 2L[t^{1/2}] + 3L[t^{-1/2}]$$

$$= \frac{1}{s} + 2.\frac{\Gamma 3/2}{s^{3/2}} + 3.\frac{\Gamma 1/2}{s^{1/2}}$$

$$= \frac{1}{s} + 2.\frac{\sqrt{\pi}}{2\,s^{3/2}} + 3.\frac{\sqrt{\pi}}{s^{1/2}}$$

$$= \frac{1}{s} + .\frac{\sqrt{\pi}}{s^{3/2}} + 3.\sqrt{\frac{\pi}{s}}$$

3. Find Laplace transform of cosh at – cos at

Solution.

$L[\cosh at - \cos at] = L[\cosh at] - L[\cos at]$

$$= \frac{s}{s^2-a^2} - \frac{s}{s^2+a^2} = s.\frac{s^2+a^2-s^2+a^2}{s^4-a^4}$$

$$= \frac{2a^2 s}{s^4-a^4}$$

4. Find Laplace transform of sin *at* sin *bt*.

Solution.

$L[\sin at \sin bt]$

$$= \frac{1}{2} L[2\sin at \sin bt]$$

$$= \frac{1}{2} L[\cos(a-b)t - \cos(a+b)t]$$

$$= \frac{1}{2}.\left[\frac{s}{s^2+(a-b)^2} - \frac{s}{s^2+(a+b)^2}\right]$$

$$= \frac{1}{2}.s.\left[\frac{s^2+(a+b)^2-s^2-(a-b)^2}{\{s^2+(a-b)^2\}.\{s^2+(a+b)^2\}}\right]$$

$$= \frac{s}{2}\left[\frac{a^2+b^2+2ab-a^2-b^2+2ab}{[s^2+(a-b)^2][s^2+(a+b)^2]}\right]$$

$$= \left[\frac{2abs}{[s^2+(a-b)^2][s^2+(a+b)^2]}\right]$$

5. Find Laplace transform of $\sin^2 3t$.

Solution.

$L[\sin^2 3t]$

$$= \frac{1}{2} L[2\sin^2 3t]$$

$$= \frac{1}{2} L[1-\cos 6t] = \frac{1}{2}L[1] - \frac{1}{2}L[\cos 6t]$$

$$= \frac{1}{2}.\frac{1}{s} - \frac{1}{2}.\frac{s}{s^2+36}$$

$$= \frac{1}{2}\left[\frac{s^2+36-s^2}{s(s^2+36)}\right] = \frac{18}{s(s^2+36)}$$

6. Find Laplace transform of t – sinh 2t.

Solution.

$L[t - \sinh 2t]$

$$= L[t] - L[\sinh 2t]$$

$$= \frac{1}{s^2} - \frac{2}{s^2-4} = \frac{s^2-4-2s^2}{s^2(s^2-4)}$$

$$= \frac{-s^2-4}{s^2(s^2-4)} = \frac{s^2+4}{s^2(4-s^2)}$$

7. Find Laplace transform of $(t+2)^2 e^t$.

Solution.

$L[(t+2)^2 e^t] = L[(t^2+4t+4)e^t]$

$$= L[t^2 e^t] + 4[te^t] + 4L[e^t]$$

$$= \frac{2!}{(s-1)^3} + 4.\frac{1}{(s-1)^2} + 4.\frac{1}{s-1}$$

$$= 2\left[\frac{1+2(s-1)+2(s-1)^2}{(s-1)^3}\right]$$

$$= 2.\frac{1+2s-2+2s^2+2-4s}{(s-1)^3}$$

$$= \frac{2(2s^2-2s+1)}{(s-1)^3}$$

8. Find Laplace transform of $e^{-t}\sin^2 t$.

Solution.

$$= \frac{1}{2} L[e^{-t}(1-\cos 2t)]$$

$$= \frac{1}{2} L[e^{-t}] - \frac{1}{2} L[e^{-t}\cos 2t]$$

$$= \frac{1}{2}.\frac{1}{s+1} - \frac{1}{2}.\frac{s+1}{(s+1)^2+4}$$

$$= \frac{1}{2} \cdot \frac{4}{(s+1)(s^2+2s+5)}$$

$$= \frac{\mathbf{2}}{\mathbf{(s+1)(s^2+2s+5)}}$$

9. Find Laplace transform of $f(t) = \begin{cases} e^t, & 0 < t < 1 \\ 0, & t > 1 \end{cases}$.

Solution.

$$L[f(t)] = \int_0^\infty e^{-st} f(t)\, dt$$

$$= \int_0^1 e^{-st} f(t)\, dt + \int_1^\infty e^{-st} f(t)\, dt$$

$$= \int_0^1 e^{-st} e^t\, dt + \int_1^\infty e^{-st}\, dt$$

$$= \int_0^1 e^{-(s-1)t}\, dt + 0 = \left| \frac{e^{-(s-1)t}}{-(s-1)} \right|_0^1$$

$$= \left[\frac{e^{1-s}}{1-s} - \frac{1}{1-s} \right] = \frac{\mathbf{e^{1-s} - 1}}{\mathbf{1-s}}$$

10. Find Laplace transform of $f(t) = te^{-t} \sin 2t$.

Solution.

$$L[\sin 2t] = \frac{2}{s^2+4}$$

$$L[e^{-t} \sin 2t] = \frac{2}{(s+1)^2+4} = F(s) \text{ (say)}$$

$$L(t\, e^{-t} \sin 2t) = -F'(s)$$

$$= -\frac{d}{ds}\left[\frac{2}{(s+1)^2+4} \right]$$

$$= \frac{4(s+1)}{[(s+1)^2+4]^2}$$

11. Find Laplace transform of

$$f(t) = \begin{cases} \cos\left(t - \dfrac{2\pi}{3}\right), & t > \dfrac{2\pi}{3} \\ 0, & t < \dfrac{2\pi}{3} \end{cases}$$

Solution.

$$L[f(t)] = \int_0^{2\pi/3} e^{-st} . 0 . dt + \int_{2\pi/3}^\infty e^{-st} . \cos\left(t - \frac{2\pi}{3}\right) dt$$

$$= \int_{2\pi/3}^\infty e^{-st} \cos\left(t - \frac{2\pi}{3}\right) dt$$

$$= \frac{s^2}{s^2+1} \left[\frac{1}{s^2} e^{-st} \sin\left(t - \frac{2\pi}{3}\right) - \frac{e^{-st}}{s} \cos\left(t - \frac{2\pi}{3}\right) \right]_{\frac{2\pi}{3}}^\infty$$

$$= \frac{s^2}{s^2+1} \left[0 + \frac{e^{-\frac{2\pi s}{3}}}{s} \right] = \mathbf{e^{-\frac{2\pi s}{3}} . \frac{s}{s^2+1}}$$

12. Find L(cos at) and deduce from it L(sin at).

Solution.

$$L(\cos at) = \int_0^\infty e^{-st} \cos at\, dt$$

$$= \int_0^\infty e^{-st} \frac{e^{iat} + e^{-iat}}{2} dt$$

$$= \frac{1}{2} \int_0^\infty [e^{-(s-ia)t} + e^{-(s+ia)t}]\, dt$$

$$= \frac{1}{2} \left[\frac{e^{-(s-ia)t}}{-(s-ia)} + \frac{e^{-(s+ia)t}}{-(s+ia)} \right]_0^\infty$$

$$= \frac{1}{2} \left[\frac{1}{s-ia} + \frac{1}{s+ia} \right]$$

$$= \frac{1}{2} \frac{s+ia+s-ia}{s^2+a^2} = \frac{s}{s^2+a^2}$$

and $$L(\sin at) = -\frac{1}{a} L\left(\frac{d}{dt}(\cos at)\right)$$

By transforms of derivatives

$$L(\sin at) = -\frac{1}{a}\left[s . \frac{s}{s^2+a^2} - 1 \right]$$

$$= -\frac{1}{a}\left[\frac{s^2 - s^2 - a^2}{s^2+a^2} \right]$$

$$= \frac{\mathbf{a}}{\mathbf{s^2+a^2}}$$

13. If $L\left\{2\sqrt{\left(\dfrac{t}{\pi}\right)}\right\} = \dfrac{1}{s^{3/2}}$, evaluate $L\left\{\dfrac{1}{\sqrt{\pi t}}\right\}$.

Solution.

$$L\left\{\frac{1}{\sqrt{\pi t}}\right\} = L\left[\frac{d}{dt}\left(2\sqrt{\frac{t}{\pi}}\right)\right]$$

$$= s . \frac{1}{s^{3/2}} - 2\sqrt{\frac{0}{\pi}}$$

$$= \frac{\mathbf{1}}{\mathbf{\sqrt{s}}}$$

14. Evaluate L[t sin3t cos2t].

Solution.

$$L[\sin 3t \cos 2t] = \frac{1}{2} L[2 \sin 3t \cos 2t]$$

$$= \frac{1}{2} L[\sin 5t + \sin t] = \frac{1}{2}\left[\frac{5}{s^2+25} + \frac{1}{s^2+1}\right]$$

$\therefore$ $L[t \sin 3t \cos 2t]$

$$= -\frac{d}{ds}\left\{\frac{1}{2}\left(\frac{5}{s^2+25} + \frac{1}{s^2+1}\right)\right\}$$

$$= -\frac{5}{2}\left[-\frac{2s}{(s^2+25)^2}\right] - \frac{1}{2}\left[-\frac{2s}{(s^2+1)^2}\right]$$

$$= \frac{5s}{(s^2+25)^2} + \frac{s}{(s^2+1)^2}$$

15. Evaulate L[t sin hat].

Solution.

Since $L[\sin hat] = \frac{a}{s^2 - a^2}$

$$\therefore \quad L[t \sin hat] = -\frac{d}{ds}\left[\frac{a}{s^2-a^2}\right]$$

$$= -\left[\frac{-2s \, . \, a}{(s^2-a^2)^2}\right]$$

$$= \mathbf{\frac{2as}{(s^2-a^2)^2}}$$

16. Evaluate $L\left[\frac{e^{-at} - e^{-bt}}{t}\right]$.

Solution.

Since $L[e^{-at} - e^{-bt}] = \frac{1}{s+a} - \frac{1}{s+b}$

$$\therefore \quad L\left[\frac{e^{-at} - e^{-bt}}{t}\right] = \int_s^\infty \left[\frac{1}{s+a} - \frac{1}{s+b}\right] ds$$

$$= [\log(s+a) - \log(s+b)]_s^\infty$$

$$= \left[\log \frac{s+a}{s+b}\right]_s^\infty$$

$$= \underset{s \to \infty}{Lt} \log \frac{s\left(1+\frac{a}{s}\right)}{s\left(1+\frac{b}{s}\right)} - \log \frac{s+a}{s+b}$$

$$= 0 - \log \frac{s+a}{s+b}$$

$$= \mathbf{\log \frac{s+b}{s+a}}$$

17. Evaluate $L\left[\frac{e^{-t} \sin t}{t}\right]$.

Solution.

Since $L[e^{-t} \sin t] = \frac{1}{(s+1)^2+1}$

$$\therefore \quad L\left[\frac{e^{-t} \sin t}{t}\right] = \int_s^\infty \frac{1}{(s+1)^2+1} ds$$

$$= \left[\tan^{-1} \frac{s+1}{1}\right]_s^\infty$$

$$= \frac{\pi}{2} - \tan^{-1}(s+1) = \cot^{-1}(s+1).$$

18. Evaluate $L\left(\int_0^t \frac{\sin t}{t} dt\right)$.

Solution.

$$L\left[\frac{\sin t}{t}\right] = \int_s^\infty \frac{1}{s^2+1} ds = \frac{\pi}{2} - \tan^{-1} s$$

$$= \cot^{-1} s$$

By transform of integrals,

$$L\left[\int_0^t \frac{\sin t}{t} dt\right] = \frac{1}{s} \cot^{-1} s$$

19. Evaluate $L\left(\int_0^t e^{-t} \cos dt\right)$.

Solution.

Since $L[e^{-t} \cos t] = \frac{s+1}{(s+1)^2+1} = \frac{s+1}{s^2+2s+2}$

$$\therefore \quad L\left(\int_0^t e^{-t} \cos t \, dt\right) = \mathbf{\frac{1}{s} \cdot \frac{s+1}{s^2+2s+2}}$$

20. Find the inverse Laplace transforms of $\frac{3(s^2-2)^2}{2s^5}$.

Solution.

Since $\frac{3(s^2-2)^2}{2s^5} = \frac{3(s^4 - 4s^2 + 4)}{2s^5}$

$$= \frac{3}{2}\left[\frac{1}{s} - 4 . \frac{1}{s^3} + 4 \frac{1}{s^5}\right]$$

$$\therefore \quad L^{-1}\left[\frac{3(s^2-2)^2}{2s^5}\right]$$

$$= \frac{3}{2} L^{-1}\left[\frac{1}{s}\right] - 6L^{-1}\left[\frac{1}{s^3}\right] + 6L^{-1}\left[\frac{1}{s^5}\right]$$

$$= \frac{3}{2} . 1 - 6 . \frac{t^2}{2!} + 6 . \frac{t^4}{4!}$$

$$= \mathbf{\frac{3}{2} - 3t^2 + \frac{1}{4} t^4}$$

21. Find the inverse Laplace transform of

$$F(s) = \log \frac{s+a}{s+b}.$$

Solution.

$$L^{-1} \log \left(\frac{s+a}{s+b}\right)$$

$$= -\frac{1}{t} L^{-1}\left[\frac{d}{ds} \log \frac{s+a}{s+b}\right]$$

$$= -\frac{1}{t} L^{-1}\left[\frac{d}{ds} \log (s+a) - \frac{d}{ds} \log (s+b)\right]$$

$$= -\frac{1}{t} L^{-1}\left[\frac{1}{s+a} - \frac{1}{s+b}\right]$$

$$= -\frac{1}{t}[e^{-at} - e^{-bt}] = \frac{1}{t}(e^{-bt} - e^{-at})$$

22. Obtain the inverse Laplace transform of $\cot^{-1}\left(\frac{s+3}{2}\right)$.

Solution.

$$L^{-1}[F(s)] = -\frac{1}{t} L^{-1}\left[\frac{d}{ds} F(s)\right]$$

$$\therefore\ L^{-1}\left[\cot^{-1}\left(\frac{s+3}{2}\right)\right]$$

$$= -\frac{1}{t} L^{-1}\left[\frac{d}{ds}\cot^{-1}\left(\frac{s+3}{2}\right)\right]$$

$$= -\frac{1}{t} L^{-1}\left\{\frac{-\frac{1}{2}}{1+\left(\frac{s+3}{2}\right)^2}\right\}$$

$$= \frac{1}{2t} L^{-1}\left\{\frac{4}{4+(s+3)^2}\right\}$$

$$= \frac{1}{t} L^{-1}\left\{-\frac{2}{2^2+(s+3)^2}\right\}$$

$$= \frac{1}{t} e^{-3t} L^{-1}\left(\frac{2}{2^2+s^2}\right)$$

$$= \frac{e^{-3t}}{t} \sin 2t$$

23. Evaluate $L^{-1} \frac{1}{s^2(s^2+a^2)}$.

Solution.

$$L^{-1}\left[\frac{1}{s^2(s^2+a^2)}\right] = \int_o^t\left\{\int_o^t \frac{\sin at}{a} dt\right\} dt$$

$$= \frac{1}{a}\int_o^t\left[-\frac{\cos at}{a}\right]_o^t dt$$

$$= \frac{1}{a^2}\int_o^t (-\cos at + 1)\, dt$$

$$= \frac{1}{a^2}\left[t - \frac{\sin at}{a}\right]_o^t = \frac{1}{a^2}\left[t - \frac{\sin at}{a} + 0\right]$$

$$= \frac{1}{a^2}\left[t - \frac{1}{a}\sin at\right]$$

24. Evaluate $L^{-1} \log \left(\frac{1+s}{s}\right)$.

Solution.

Let $f(t) = L^{-1}\left[\log\left(\frac{1+s}{s}\right)\right]$

then $t f(t) = L^{-1}\left\{-\frac{d}{ds}\left[\log\frac{(1+s)}{s}\right]\right\}$

$$= L^{-1}\left\{\frac{d}{ds}[\log s] - \frac{d}{ds}[\log(1+s)]\right\}$$

$$= L^{-1}\left\{\frac{1}{s} - \frac{1}{1+s}\right\} = 1 - e^{-t}$$

$$\therefore\ f(t) = \frac{1-e^{-t}}{t}$$

25. Evaluate $L^{-1} \log \frac{s(s+1)}{s^2+4}$.

Solution.

Let $f(t) = L^{-1}\left[\log\frac{s(s+1)}{s^2+4}\right]$

$$= L^{-1}[\log s + \log(s+1) - \log(s^2+4)]$$

Then

$$t f(t) = L^{-1}\left[\frac{d}{ds}\{\log(s^2+4) - \log s - \log(s+1)\}\right]$$

$$= L^{-1}\left[\frac{2s}{s^2+2^2}\right] - L^{-1}\left[\frac{1}{s}\right] - L^{-1}\left[\frac{1}{s+1}\right]$$

$$= 2 . \cos 2t - 1 - e^{-t}$$

$$\therefore\ f(t) = \frac{2\cos 2t - e^{-t} - 1}{t}$$

26. Evalulate $L^{-1}\cot^{-1}(s+1)$.

Solution.

Let $f(t) = L^{-1}[\cot^{-1}(s+1)]$

Then $$t\,f(t) = L^{-1}\left[-\frac{d}{ds}\cot^{-1}(s+1)\right]$$

$$= L^{-1}\left[-\frac{-1}{1+(s+1)^2}\right]$$

$$= L^{-1}\left[\frac{1}{(s+1)^2+1}\right] = e^{-t}\sin t$$

$\therefore$ $$f(t) = \frac{1}{t}e^{-t}\sin t$$

Z-TRANSFORM

The z-transform of a discrete time signal $x(n)$ is defined as the power series.

$$X(z) = \sum_{n=-\infty}^{\infty} x(n)z^{-1} \quad ...(i)$$

where z is a complex variable.

z-transform of a signal $x(n)$ is denoted by

$$X(z) \equiv Z[x(n)]$$

or, $$x(n) \xleftrightarrow{z} X(z)$$

Region of Convergence

From equation (i), z-transform is an infinite power series and it exists only for those values of z for which this series converges.

The region of z-plane for which this series converges is called *Region of Convergence* (ROC). Hence ROC of $X(z)$ is the set of all values of z for which $X(z)$ attains a finite value. Thus any time we find z-transform we should also indicate its ROC.

Example. Determine z-transform of the following finite duration signals :

(i) $x(n) = \{2, 5, 7, 6, 3, 1, 4\}$

(ii) $x(n) = \{0, 0, 1, 6, 2, -3, 4\}$

(iii) $x(n) = [2, 8, \underset{\uparrow}{1}, 2, -3, 4\}$

(iv) $x(n) = \delta(n)$

(v) $x(n) = \delta(n-k), k > 0$

(vi) $x(n) = \delta(n+k), k > 0$

Solution.

(i) $$X(z) = \sum_{n=-\infty}^{\infty} x(n)z^{-n} = \sum_{n=0}^{6} x(n)z^{-n}$$

$$= x(0)z^0 + x(1)z^{-1} + x(2)z^{-2} + x(3)z^{-3} + x(4)z^{-4} + x(5)z^{-5} + x(6)z^{-6}$$

$$= 2 + 5z^{-1} + 7z^{-2} + 6z^{-3} + 3z^{-4} + z^{-5} + 4z^{-6}$$

ROC : Entire z-plane except $z = 0$

(ii) $$X(z) = \sum_{n=-\infty}^{\infty} x(n)z^{-n} = \sum_{n=0}^{6} x(n)z^{-n}$$

$$= x(0)z^0 + x(1)z^{-1} + x(2)z^{-2} + x(3)z^{-3} + x(4)z^{-4} + x(5)z^{-5} + x(6)z^{-6}$$

$$= z^{-2} + 6z^{-3} + 2z^{-4} - 3z^{-5} + 4z^{-6}$$

ROC : Entire z-plane except $z = 0$

(iii) $$X(z) = \sum_{n=-\infty}^{\infty} x(n)z^{-n} = \sum_{n=-3}^{2} x(n)z^{-n}$$

$$= x(-2)z^2 + x(-1)z^1 + x(0)z^0 + x(1)z^{-1} + x(2)z^{-2} + x(3)z^{-3}$$

$$= 2z^2 + 8z + 1 + 2z - 3z^{-2} + 4z^{-3}$$

ROC : Entire z-plane except $z = 0$ and $z = \infty$.

(iv) $$x(n) = \begin{cases} 1, & n = 0 \\ 0, & \text{elsewhere} \end{cases}$$

z-transform is given by

$$X(z) = \sum_{n=-\infty}^{\infty} x(n)z^{-n}$$

$$= .... + x(-3)z^3 + x(-2)z^2 + x(-1)z^1 + x(0)z^0 + x(1)z^{-1} + x(2)z^{-2} + x(3)z^{-3}$$

$$= .... + 0\times z^3 + 0\times z^2 + 0\times z^1 + 1\times z^0 + 0\times z^{-1} + 0\times z^{-2} + 0\times z^{-3} = 1$$

ROC : Entire z-plane.

(v) $$X(z) = \sum_{n=-\infty}^{\infty} x(n)z^{-n} = \sum_{n=-\infty}^{\infty} \delta(n-k)z^{-n}$$

Put $m = n - k \Rightarrow n = m + k$

$$X(z) = \sum_{n=-\infty}^{\infty} \delta(m)z^{-(m+k)}$$

$$= \sum_{n=-\infty}^{\infty} \delta(m)z^{-m}z^{-k}$$

$$= Z^{-k}\sum_{n=-\infty}^{\infty} \delta(m)z^{-m} = z^{-k}\times 1 = z^{-k}$$

ROC : Entire z-plane except z = 0

(vi) $$X(z) = \sum_{n=-\infty}^{\infty} x(n)z^{-n} = \sum_{n=-\infty}^{\infty} \delta(n+k)z^{-n}$$

Put, $m = n + k \Rightarrow n = m - k$

$$X(z) = \sum_{n=-\infty}^{\infty} \delta(m)z^{-(m-k)}$$

$$= \sum_{n=-\infty}^{\infty} \delta(m)z^{-m}z^{k}$$

$$= z^{k}\sum_{n=-\infty}^{\infty} \delta(m)z^{-m} = z^{k}\times 1 = z^{k}$$

ROC : Entire z-plane except $z = \infty$.

Example. Determine z-transform of the discrete time signal given by $x(n) = \alpha^n u(n)$

Solution.

z-transform is given by

$$X(z) = \sum_{n=-\infty}^{\infty} x(n) z^{-n} = \sum_{n=-\infty}^{\infty} \alpha^n u(n) z^{-n}$$

$$= \sum_{n=0}^{\infty} \alpha^n z^{-n} = \sum_{n=0}^{\infty} \left(\alpha z^{-1}\right)^n$$

Expanding above series, we get

$$X(z) = \left(\alpha z^{-1}\right)^0 + \left(\alpha z^{-1}\right)^1 + \left(\alpha z^{-1}\right)^2 + \left(\alpha z^{-1}\right)^3 + ...$$

This is a geometric series for infinite points and whose sum is given by

$$S = \frac{\text{First term}}{1 - \text{common ratio}} \quad i.e. \quad X(z) = \frac{1}{1 - \alpha z^{-1}}$$

ROC is given as

$$\left|\alpha z^{-1}\right| < 1 \quad \Rightarrow \quad |z| > |\alpha|$$

Hence this is a causal sequence.

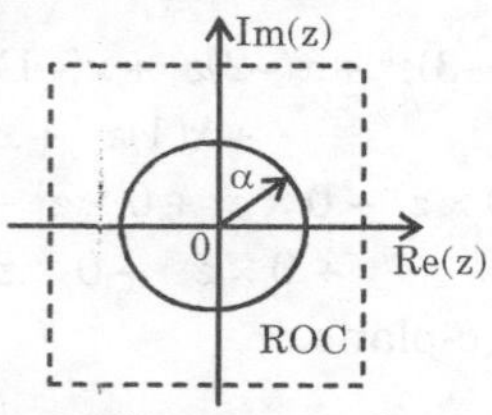

Example. Determine z-transform of the discrete time signal given by

$$x(n) = -\alpha^n u(-n-1)$$

Solution.

z-transform is given by

$$X(z) = \sum_{n=-\infty}^{\infty} x(n) z^{-n}$$

$$= \sum_{n=-\infty}^{\infty} -\alpha^n u(-n-1) z^{-n}$$

$$= \sum_{n=-\infty}^{-1} -\alpha^n z^{-n}$$

$$= -\sum_{n=-\infty}^{-1} \left(\alpha^{-1} z^{1}\right)^{-n}$$

Put $m = -n$

$$\therefore \quad X(z) = -\sum_{m=\infty}^{-1} \left(\alpha^{-1} z\right)^m$$

$$= [(\alpha^{-1}z)^1 + (\alpha^{-1}z)^2 + (\alpha^{-1}z)^3 + ...]$$

This is a geometric series for infinite points and whose sum is given by

$$X(z) = \frac{-\alpha^{-1} z}{1 - \alpha^{-1} z} = \frac{z}{1 - \alpha z^{-1}}$$

ROC is given as

$$|\alpha^{-1} z| < 1 \quad \Rightarrow \quad |z| < |\alpha|$$

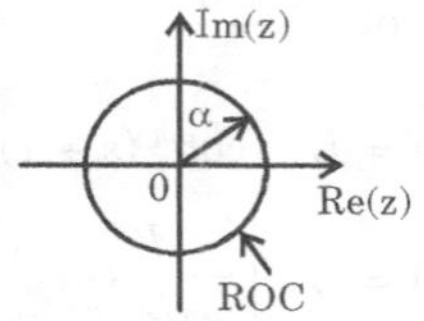

$x(n)$ is a non-causal sequence.

Example. Determine z transform of the signal

$$x(n) = \alpha^n u(n) + \beta^n u(-n-1)$$

Solution.

As calculated above, their z transform is given by

$$X(z) = \frac{1}{1 - \alpha z^{-1}} - \frac{\beta^{-1} z}{1 - \beta^{-1} z}$$

First power series converges, if

$$|z| > |\alpha|$$

Second power series converges, if

$$|z| < |\beta|$$

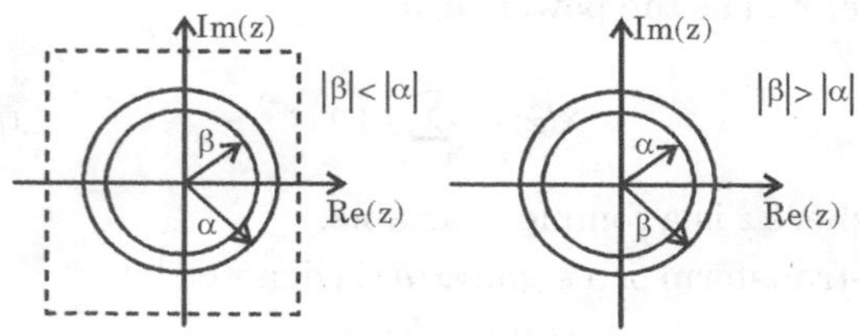

Fig (a) *Fig.(b)*

Hence two different cases are possible.

Case 1 : $|\beta| < |\alpha|$

In this case, two ROC do not overlap as shown in Fig.(a). Consequently values of z for which both power series converge simultaneously cannot be obtained. In this case, X(z) does not exist.

Case 2 : $|\beta| > |\alpha|$

In this case, there is a ring in the z-plane where both power series converge simultaneously as shown in Fig. (b).

In this case, X(z) is given by

$$X(z) = \frac{1}{1 - \alpha z^{-1}} - \frac{1}{1 - \beta z^{-1}} = \frac{\beta - \alpha}{\alpha + \beta - z - \alpha\beta z^{-1}}$$

Its ROC is given by

$$|\alpha| < |z| < |\beta|$$

PROPERTIES OF Z-TRANSFORM

1. Linearity

A discrete time signal is said to be linear, if system is homogenous and satisfies the principle of superposition.

If z-transform of two discrete sequences is given by

$$x_1(n) \xleftrightarrow{z} X_1(z)$$

and, $$x_2(n) \xleftrightarrow{z} X_2(z)$$

Then

$$A_1x_1(n) + A_2x_2(n) \xleftrightarrow{z} A_1X_1(z) + A_2X_2(z)$$

where, A_1 and A_2 are constants.

The linearity property helps to find z-transform of a signal expressing the signal as a sum of elementary signals.

Example. Determine z-transform and ROC of the signal

$$x(n)= 4(3)^n\, u(n) - 3(2)^n - u(n)$$

Solution.

If we define the signals,

$$x_1(n) = (3)^n\, u(n)$$

Its z transform is

$$X_1(z) = \frac{1}{1-3z^{-1}}$$

ROC : $|z| > 3$

$$x_2(n) = (2)^n u(n)$$

and its z transform is

$$X_2(z) = \frac{1}{1-2z^{-1}}$$

ROC : $|z| > 2$

Then, $x(n) = 4x_1(n) - 3x_2(n)$

and its z transform is given by

$$X(z) = 4X_1(z) - 3X_2(z)$$

Intersection of ROCs of both signals $X_1(z)$ and $X_2(z)$ is $|z| > 3$.

Thus overall z-transform is given by

$$X(z) = \frac{4}{1-3z^{-1}} - \frac{3}{1-2z^{-1}}$$

ROC : $|z| > 3$

2. Time Shifting

To see a particular signal in previous time or in future time, time shifting is used. The signal is delayed in time to see it in previous time and the signal is advances to see it in future time.

Consider a discrete time sequence $x(n)$ whose z transform is $X(z)$.

i.e. $\qquad x(n) \xleftrightarrow{z} X(z)$

If signal is delayd by k units of time, then

$$x(n-k) \xleftrightarrow{z} z^{-k}\, X(z)$$

and if signal is advances by k units of time, then

$$x(n+k) \xleftrightarrow{z} z^k.\, X(z)$$

ROC of $z^{-k}.X(z)$ is the same as that of $X(z)$ except for $z = 0$ and ROC of z^k. $X(z)$ is the same as that of $X(z)$ except at $z = \infty$.

3. Scaling in Z-Domain

Scaling in z-domain is used to increase or decrease the value of z. So, we multiply or divide z-domain parameter by any constant A. Scaling in z-domain by a scalar quantity A is equivalent to scaling in time domain by a factor of $\frac{1}{A}$.

If $\qquad x(n) \xleftrightarrow{z} X(z)$

ROC : $R_1 < |z| < R_2$

then, $\qquad A^n x(n) \xleftrightarrow{z} X\left(\frac{z}{A}\right)$

ROC : $|A|R_1 < |z| < |A|R_2$

$\Rightarrow \qquad A^{-n}x(n) \xleftrightarrow{z} X(Az)$

ROC : $\left|\frac{1}{A}\right| R_1 < |z| < \left|\frac{1}{A}\right| R_2$

4. Time Reversal

It means taking the mirror image of the signal.

If $\qquad x(n) \xleftrightarrow{z} X(z)$

ROC : $R_1 < |z| < R_2$

then, $\qquad x(-n) \xleftrightarrow{z} X\left(\frac{1}{z}\right)$

ROC : $\frac{1}{R_2} < |z| < \frac{1}{R_1}$

Thus folding a signal is equivalent to replacing z by z^{-1} in the z transform formula.

Example. Determine z transform of the signal

$$x(n) = u(-n)$$

Solution.

z transform of $u(n)$ is given by

$$u(n) \xleftrightarrow{z} \frac{1}{1-z^{-1}}$$

ROC : $\qquad |z| > 1$

Using time reversal property.

$$u(-n) \xleftrightarrow{z} \frac{1}{1-z} \quad \text{ROC} : |z| < 1$$

5. Differentiation in the Z-Domain

If we want to differentiate z transform of discrete time sequence with respect to z, then following relations are used :

If $\qquad x(n) \xleftrightarrow{z} X(z)$

then $\qquad nx(\text{n}) \xleftrightarrow{z} -z\frac{dX(z)}{dz}$

There is change in ROC and both the transforms have the same ROC.

Example. Determine z transform of the signal

$$x(n) = na^n u(n)$$

Solution.

First calculate $x_1(n)$

$$x_1(n) = a^n u(n)$$

and we know, $X_1(z) = \frac{1}{1-az^{-1}}$

ROC : $|z| > |a|$

Using differentiating property

$$na^n u(n) \xleftrightarrow{z} -z\frac{dX_1(z)}{dz} = \frac{az^{-1}}{\left(1-az^{-1}\right)^2}$$

ROC : $|z| > |a|$

6. Convolution of Two Sequences

Convolution of two discrete time signals in time domain is equivalent to multiplication of their z-transforms.

Consider two descrete time sequences $x_1(n)$ and $x_2(n)$.

If $x_1(n) \xleftrightarrow{z} X_1(z)$ and $x_2(n) \xleftrightarrow{z} X_2(z)$

Then, $x(n) = x_1(n) * x_2(n)$

$$= \sum_{k=-\infty}^{\infty} x_1(k)x_2(n-k) \xleftrightarrow{z} X(z) = X_1(z).X_2(z)$$

Computation of convolution of two signals using z-transform requires following steps :

(*i*) Compute z-transform of the two signals to be convolved.

(*ii*) Multiply two z-transforms

(*iii*) Find inverse z-transform of the multiplied signal.

Example. Find convolution of sequences given below :

$$x_1(n) = \{\underset{\uparrow}{1}, 2, -4\} \text{ and } x_2(n) = \{1, \underset{\uparrow}{2}, -1\}$$

Solution.

Step 1 : z-transform of individual sequences is

$$X_1(z) = 1 + 2z^{-1} - 4z^{-2}$$

and $X_2(z) = z + 2 - z^{-1}$

Step 2 : Multiply two z-transforms.

$$\begin{aligned} X(z) &= X_1(z)\,X_2(z) \\ &= (1 + 2z^{-1} - 4z^{-2})\,(z + 2 - z^{-1}) \\ &= z + 4 - z^{-1} - 10z^{-2} + 4z^{-3} \end{aligned}$$

Step 3 : Take inverse z-transform of the multiplied signal

$$\begin{aligned} x(n) &= Z^{-1}[X(z)] \\ &= Z^{-1}[z + 4 - z^{-1} - 10z^{-2} + 4z^{-3}] \\ x(n) &= \{1, \underset{\uparrow}{4}, -1, -10, 4\} \end{aligned}$$

7. Correlation of Two Sequences

Correlation is used to check the similarity between two discrete time signals.

Now consider two discrete time sequences $x_1(n)$ and $x_2(n)$ as given below :

If $x_1(n) \xleftrightarrow{z} X_1(z)$

and, $x_2(n) \xleftrightarrow{z} X_2(z)$

Then,

$$R_{x_1x_2}(\ell) = \sum_{k=-\infty}^{\infty} x_1(n)x_2(n-\ell) \xleftrightarrow{z} R_{x_1x_2}(z)$$

$$= X_1(z).X_2(z^{-1})$$

Computation of correlation of two signals using z-transform requires following steps :

(*i*) Compute z-transform of the two signals to be convolved.

If $x_1(n) \xleftrightarrow{z} X_1(z)$ and, $x_2(n) \xleftrightarrow{z} X_2(z)$

(*ii*) Calculate $X_2(z^{-1})$

(*iii*) Multiply two z-transforms

(*iv*) Find inverse z-transform of the multiplied signal.

INVERSE Z-TRANSFORM

The method of transforming $X(z)$ back to a discrete time sequence $x(n)$ is called *inverse z-transform*.

The inverse z-transform is formally given by

$$x(n) = \frac{1}{2\pi j}\oint_c X(z)z^{n-1}dz$$

where integral is a contour integration over a closed path C that encloses origin and lies within the ROC of $X(z)$.

Inverse *Z*-Transform by Power Series Expansion Method

Power series expansion of rational z-transforms are obtained by long division method.

z-transform $X(z)$ is expanded into power series of the form

$$X(z) = \sum_{n=-\infty}^{\infty} C_n z^{-1}$$

which converges in the given ROC.

Example. Determine inverse z-transform of

$$X(z) = \frac{3}{3 + 3z^{-1} - 2z^{-2}}$$

when,

(*i*)ROC : $|z| > 1$; and

(*ii*)ROC : $|z| < \frac{1}{2}$

Solution.

(*i*) Since, ROC is exterior of the circle and we expect $x(n)$ to be causal. For a causal sequence, the power series expansion results in negative powers of z.

Now dividing numerator with the denominator of $X(z)$, we get

$$
\begin{array}{r}
1 - z^{-1} + \frac{5}{3} z^{-2} + \ldots\ldots \\
3 + 3z^{-1} - 2z^{-2} \overline{)\, 3 \qquad\qquad\qquad} \\
3 + 3z^{-1} - 2z^{-2} \\
- \quad + \\
\hline
-3z^{-1} + 2z^{-2} \\
-3z^{-1} - 3z^{-2} + 2z^{-3} \\
+ \quad + \quad - \\
\hline
5z^{-2} - 2z^{-3} \\
5z^{-2} + 5z^{-3} - \frac{10}{3} z^{-4} \\
- \quad - \quad + \\
\hline
-7z^{-3} + \frac{10}{3} z^{-4}
\end{array}
$$

$$X(z) = \frac{3}{3 + 3z^{-1} - 2z^{-2}} = 1 - z^{-1} + \frac{5}{3} z^{-2} + \ldots$$

Writing above equation in sequence representation, we get

$$x(n) = \left\{1, -1, \frac{5}{3}\right\}$$

(*ii*) In this case ROC is interior of circle. The signal $x(n)$ is anti-causal.

For anti-causal sequence, power series expansion results in positive powers of z.

To obtain positive powers of z, we perform long division in the following ways :

$$
\begin{array}{r}
-\frac{3}{2} z^{2} - \frac{9}{4} z^{3} - \frac{65}{8} z^{4} + \ldots\ldots \\
-2z^{-2} + 3z^{-1} + 3 \overline{)\, 3 \qquad\qquad\qquad\qquad\qquad} \\
3 - \frac{9}{2} z - \frac{9}{2} z^{2} \\
- \quad + \quad + \\
\hline
\frac{9}{2} z + \frac{9}{2} z^{2} \\
\frac{9}{2} z - \frac{27}{4} z^{2} - \frac{27}{4} z^{3} \\
- \quad + \quad + \\
\hline
\frac{65}{4} z^{2} + \frac{27}{4} z^{3} \\
\frac{65}{4} z^{2} - \frac{195}{8} z^{3} - \frac{195}{8} z^{4} \\
- \quad + \quad + \\
\hline
+\frac{247}{8} z^{3} + \frac{195}{8} z^{4}
\end{array}
$$

$$X(z) = -\frac{3}{2} z^{2} - \frac{9}{4} z^{3} - \frac{65}{8} z^{4} + \ldots$$

Expressing above equation in sequence representation, we get

$$x(n) = \left\{\ldots, -\frac{65}{8}, -\frac{9}{4}, -\frac{3}{2}, 0, \underset{\uparrow}{0}\right\}$$

Inverse Z-Transform using Partial Fraction Expansion Method

In partial fraction expansion method, we attempt to express function $X(z)$ as a linear combination.

$$X(z) = AX_1(z) + BX_2(z) + CX_3(z) + \ldots$$

where $X_1(z)$, $X_2(z)$ .... are expression with inverse z-transforms $x_1(n)$, $x_2(n)$ ..... .

Then inverse z-transform of $X(z)$,

$$
\begin{aligned}
x(n) &= Z^{-1}[X(z)] \\
&= Z^{-1}[AX_1(z) + BX_2(z) + CX_3(z) + \ldots] \\
&= Z^{-1}[AX_1(z)] + Z^{-1}[BX_2(z)] + Z^{-1}[CX_3(z)] + \ldots\ldots \\
&= Ax_1(n) + Bx_2(n) + Cx_3(n) + \ldots
\end{aligned}
$$

Example. Determine inverse z-transform by using partial fraction expansion method for the z-transform shown below :

$$X(z) = \frac{4z^2 - 2z}{z^3 - 5z^2 + 8z - 4}$$

Solution.

$$
\begin{aligned}
X(z) &= \frac{4z^2 - 2z}{z^3 - 5z^2 + 8z - 4} \\
&= \frac{4z^2 - 2z}{(z-1)(z-2)^2}
\end{aligned}
$$

The expression of $X(z)$ is expanded using partial fraction as

$$
\begin{aligned}
X(z) &= \frac{4z^2 - 2z}{(z-1)(z-2)^2} \\
&= \frac{A}{z-1} + \frac{B}{z-2} + \frac{C}{(z-2)^2}
\end{aligned}
$$

Calculating co-efficients A, B and C and putting in above equation, we get

$$
\begin{aligned}
X(z) &= \frac{2}{z-1} + \frac{2}{z-2} + \frac{12}{(z-2)^2} \\
&= \frac{2z^{-1}}{1 - z^{-1}} + \frac{2z^{-1}}{1 - 2z^{-1}} + \frac{12z^{-2}}{(1 - 2z^{-1})^2}
\end{aligned}
$$

Taking inverse z-transform on both sides of above equation, we get

$$x(n) = 2(1)^{n-1} + 2(2)^{n-1} + 6(n-1)(2)^{n-1}$$

System Function

It is defined as the ratio of z-transform of output sequence to the z-transform of the input sequence keeping all the initial conditions zero.

i.e., $$H(z) = \frac{z - \text{transform of output}}{z - \text{transform of input}}$$

z-transform of unit sample response gives the system function. It is given as

$$H(z) = \sum_{n=-\infty}^{\infty} h(n) z^{-n}$$

Output of a relaxed LTI system is obtained by taking convolution of input sequence $x(n)$ and $h(n)$ of the system

i.e., $$y(z) = \sum_{n=-\infty}^{\infty} x(k) h(n-k) = x(n) * h(n)$$

In z-domain, above relation becomes

$$Y(z) = X(z)\, H(z)$$

Example. Calculate system function, $H(z)$ and unit sample response $h(n)$ of a system whose differential equation is given by

$$y(n) = 2y(n-1) + 5x(n)$$

where, $y(n)$ and $x(n)$ are output and input of the system respectively.

Solution.

Taking z-transform of the above differential equation, we get

$$Y(z) = 2z^{-1} Y(z) + 5X(z)$$

$$\Rightarrow \quad H(z) = \frac{Y(z)}{X(z)} = \frac{5}{1-2z^{-1}}$$

Taking, inverse z-transform of $H(z)$

$$h(n) = 5(2)^n u(n)$$

Solution of differential equations using z-transform

The one sided z-transform is a very efficient tool for the solution of differential equations with initial conditions. The difference equations relating two time domain signals to an equivalent algebraic equation relating their one sided z-transforms. This algebraic equation can be easily solved to obtain z-transform of the desired signal.

Example. Determine step response of a system given by

$$y(n) = Ay(n-1) + x(n); \; -1 < A < 1$$

Given initial condition is $y(-1) = 1$

Solution.

Taking one sided z-transform on both sides of the above difference equation, we get

$$Y(z) = A[z^{-1} Y(z) + y(-1)] + X(z)$$
$$= A[z^{-1} Y(z) + 1] + X(z)$$

$$\Rightarrow \quad Y(z)[1 - Az^{-1}] = A + X(z)$$

We know input is a step response,

i.e., $$x(n) = u(n)$$

$$\Rightarrow \quad X(z) = \frac{1}{1-z^{-1}}$$

Putting value of $X(z)$, above differential equation becomes

$$Y(z)[1 - Az^{-1}] = A + \frac{1}{1-z^{-1}}$$

$$\Rightarrow \quad Y(z) = \frac{A}{1-Az^{-1}} + \frac{1}{(1-z^{-1})(1-Az^{-1})}$$

Expanding above equation, using partial fraction expansion, it becomes

$$Y(z) = \frac{A}{1-Az^{-1}} + \frac{\left(\frac{1}{1-A}\right)}{1-z^{-1}} + \frac{\left(\frac{-A}{1-A}\right)}{1-Az^{-1}}$$

Taking inverse z-transform of above equation, A^{n+1}we get

$$y(n) = A^{n+1} u(n) + \frac{1}{1-A} u(n) + \frac{-1}{1-A} A^{n+1} u(n)$$

$$= A^{n+1} u(n) + \left(\frac{1-A^{n+1}}{1-A}\right) u(n)$$

$$= \frac{1}{1-A}(1 - A^{n+2})\, u(n)$$

Some Common Z-Transform Pairs

S.N.	Signal, $x(n)$	z-transform, $X(z)$	ROC
1.	$\delta(n)$	1	All values of z
2.	$u(n)$	$\frac{1}{1-z^{-1}}$	$\|z\| > 1$
3.	$a^n u(n)$	$\frac{1}{1-az^{-1}}$	$\|z\| > \|a\|$
4.	$nu(n)$	$\frac{z^{-1}}{(1-z^{-1})^2}$	$\|z\| > 1$

S.N.	Signal, $x(n)$	z-transform, X(z)	ROC
5.	$n\,a^n u(n)$	$\frac{az^{-1}}{(1-az^{-1})^2}$	$\lvert z\rvert > \lvert a\rvert$
6.	$-a^n u(-n-1)$	$\frac{1}{1-az^{-1}}$	$\lvert z\rvert < 1$
7.	$-n\,a^n u(-n-1)$	$\frac{az^{-1}}{(1-az^{-1})^2}$	$\lvert z\rvert < \lvert a\rvert$
8.	$(\cos\omega_0 n)\,u(n)$	$\frac{1-z^{-1}\cos\omega_0}{1-2z^{-1}\cos\omega_0+z^{-2}}$	$\lvert z\rvert > 1$
9.	$(\sin\omega_0 n)\,u(n)$	$\frac{z^{-1}\sin\omega_0}{1-2z^{-1}\cos\omega_0+z^{-2}}$	$\lvert z\rvert > 1$
10.	$(a^n\cos\omega_0 n)\,u(n)$	$\frac{1-az^{-1}\cos\omega_0}{1-2az^{-1}\cos\omega_0+a^2z^{-2}}$	$\lvert z\rvert > \lvert a\rvert$
11.	$(a^n\sin\omega_0 n)\,u(n)$	$\frac{az^{-1}\sin\omega_0}{1-2az^{-1}\cos\omega_0+a^2z^{-2}}$	$\lvert z\rvert > \lvert a\rvert$
12.	$u(n-1)$	$z^{-1}\left(\frac{1}{1-z^{-1}}\right)$	$\lvert z\rvert > 1$
13.	$\delta(n-k)$	z^{-k}	All values of z except $z = 0$
14.	$\delta(n+k)$	z^{k}	All values of z except $z = \infty$

Properties of Z-Transform

Property	Time domain	z-domain	ROC
Linearity	$Ax_1(n) + Bx_2(n)$	$AX_1(z) + BX_2(z)$	At least the intersection of both ROCs
Time shifting	$x(n-k)$	$z^{-k}\,X(z)$	Same as X(z), except $z = 0$ for $k > 0$ and $z = \infty$ for $k < 0$
Scaling in z-domain	$A^n x(n)$	$x\left(\frac{z}{A}\right)$	$\lvert A\rvert R_2 < \lvert z\rvert < \lvert A\rvert R_1$
Time Reversal	$x(-n)$	$X\left(\frac{1}{z}\right)$	$\frac{1}{R_1} < \lvert z\rvert < \frac{1}{R_2}$
Time expansion	$x(mn)$	$X(z^m)$	$R^{1/m}$
Differentiation in z-domain	$nx(n)$	$-z\frac{dX(z)}{dz}$	$R_2 < \lvert z\rvert < R_1$
Convolution	$x_1(n) * x_2(n)$	$X_1(z)\,X_2(z)$	At least the intersection of ROCs of $X_1(z)$ and $X_2(z)$
Correlation	$R_{x_1x_2}(\ell) = x_1(\ell) * x_2(-\ell)$	$X_1(z)\,X_2(z^{-1})$	At least the intersection of ROCs of $X_1(z)$ and $X_2(z^{-1})$
Initial value	If $x(n)$ is causal	$x(0) = \lim_{z\to\infty} X(z)$	
Theorem Conjugation	$x^{\circ}(n)$	$X^{\circ}(z^{\circ})$	ROC

EXERCISE

MCQ TYPE QUESTIONS

1. The fourier transform of the function sgn(t) defined in the figure is

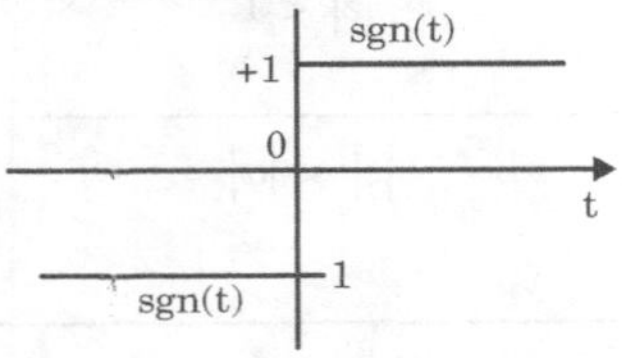

(a) $\frac{1}{j\omega}$ (b) $-\frac{1}{j\omega}$

(c) $\frac{2}{j\omega}$ (d) $-\frac{2}{j\omega}$

2. Which one of the following is the fourier transform of the signal given in figure (b), if the fourier transform of the signal in figure (a) is given by

$$\frac{2\sin\omega T_1}{\omega}$$

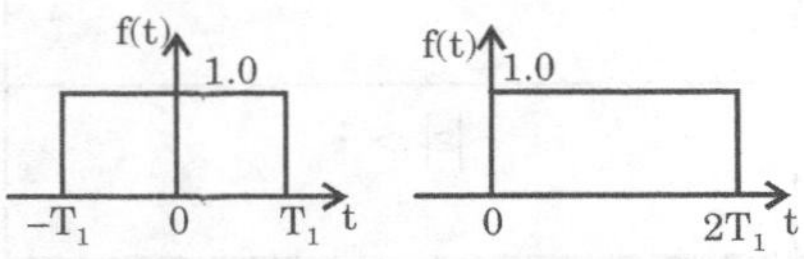

Fig. (a) *Fig. (b)*

(a) $\frac{2\sin\omega T_1}{\omega}e^{j\omega T_1}$ (b) $\frac{2\sin\omega T_1}{\omega}e^{-j\omega T_1}$

(c) $\frac{\sin\omega T_1}{\omega}e^{-j\omega T_1}$ (d) $\frac{\sin\omega T_1}{\omega}e^{j\omega T_1}$

3. Fourier transform of an arbitrary signal has the property.

(a) $F(j\omega) = F(-j\omega)$ (b) $F(j\omega) = -F(-j\omega)$

(c) $F(j\omega) = F^*(-j\omega)$ (d) $F(j\omega) = -F^*(-j\omega)$

4. The two inputs to an analog multiplier are $x(t)$ and $y(t)$ with fourier transforms X(f) and Y(t) respectively. The output $z(t)$ will have a transform Z(f) given by

(a) X(f). Y(f)

(b) X(F) + Y(f)

(c) $\frac{1}{2\pi}\int_{-\lambda}^{\lambda} x(\lambda)y(\lambda - f)d\lambda$

(d) $\int_{-\infty}^{\infty} x(\lambda)y(f-\lambda)d\lambda$

5. The fourier transform of a Gaussian time pulse is

(a) uniform (b) a pair of impulse

(c) gaussian (d) rayleigh

6. The response of a system is $h(t)$ when the input is $x(t)$. The output $y(t)$ will be [H(f) is the fourier transform of $h(t)$]

(a) $h(t)x(t)$ (b) $[h(t)]\,x(t)$

(c) $h(t) * x(t)$ (d) $h(t)$

7. Which one of the following is the correct fourier transform of the unit step signal

$$u(t) = 1 \text{ for } t > 0$$
$$= 0 \text{ for } t \le 0$$

(a) $\pi\delta(\omega)$ (b) $\frac{1}{j\omega}$

(c) $\frac{1}{j\omega} + \pi\delta(\omega)$ (d) $-2\pi\delta(\omega)$

8. The fourier transform of $v(t) = \cos\omega_0 t$ is given by

(a) $v(t) = \frac{1}{2}\delta(f - f_0)$

(b) $v(t) = \frac{1}{2}\delta(f + f_0)$

(c) $v(t) = \frac{1}{2}[\delta(f - f_0) + \delta(f + f_0)]$

(d) $v(t) = \frac{1}{2}[\delta(f - f_0) - \delta(f + f_0)]$

9. If $g(t) \leftrightarrow G(f)$ represents a fourier transform pair then according to the duality property of fourier transform.

(a) $G(t) \leftrightarrow g(f)$ (b) $G(t) \leftrightarrow g^*(f)$

(c) $G(t) \leftrightarrow g(-f)$ (d) $G(t) \leftrightarrow g^*(-f)$

10. If F(f) be the fourier transform of a function $f(t)$, then F(0) is

(a) $\int_{-\infty}^{\infty} f(t)dt$ (b) $\int_{-\infty}^{\infty} |f(t)|^2 dt$

(c) $\int_{-\infty}^{\infty} |tf(t)|^2 dt$ (d) $\int_{-\infty}^{\infty} tf(t)^2 dt$

11. The fourier transform of $x(f)$ is the periodic delta function

$$x(t) = \sum_{k=-\infty}^{\infty} \partial(t - kT) \text{ is}$$

(a) $T\sum_{k=-\infty}^{\infty} \partial(f - kT)$ (b) $T\sum_{k=-\infty}^{\infty} \partial\left(f - \frac{k}{T}\right)$

(c) $\frac{1}{T}\sum_{k=-\infty}^{\infty} \partial\left(f - \frac{k}{T}\right)$ (d) $\frac{1}{T}\sum_{k=-\infty}^{\infty} \partial(f - kT)$

12. z-transform of $x(n) = u(n)$ and ROC is

(a) $X(z) = \frac{1}{1 - z^{-1}}$ $|z| < 1$

(b) $X(z) = \frac{z^{-1}}{1 - z^{-1}}$ $|z| > 1$

(c) $X(z) = \frac{1}{1 - z^{-1}}$ $|z| > 1$

(d) $X(z) = \frac{z}{1 - z}$ $|z| < 1$

13. If $x(n) = -a^n u(-n-1)$, then z-transform and its ROC is

(a) $X(z) = \dfrac{z}{z-a} \quad |z| < a$

(b) $X(z) = \dfrac{1}{1+az^{-1}} \quad |z| < a$

(c) $X(z) = \dfrac{1}{1-a^{-1}z} \quad |z| < a$

(d) $X(z) = \dfrac{z}{z-a} \quad |z| > a$

14. If $X(z) = \dfrac{1}{1-az^{-1}}$ then two discrete time signals are

(a) $x(n) = a^n u(n)$ and $x(n) = a^n u(-n)$

(b) $x(n) = a^n u(n)$ and $x(n) = a^n u(-n-1)$

(c) $x(n) = a^n u(n)$ and $x(n) = -a^n u(-n-1)$

(d) none of these

15. If $x(n) = \left[7\left(\frac{1}{3}\right)^n u(n) - 6\left(\frac{1}{2}\right)^n u(n)\right]$, then ROC for the discrete time signal is

(a) $|z| < \frac{1}{3}$ (b) $|z| < \frac{1}{2}$

(c) $|z| > \frac{1}{2}$ (d) $|z| > \frac{1}{3}$

16. If $x(n) = b^{|n|}$, $b>0$, then z-transform and its ROC is

(a) $X(z) = \left(\dfrac{b^2-1}{b}\right)\dfrac{z}{(z-b)(z-b^{-1})}; \quad b < |z| < \dfrac{1}{b}$

(b) $X(z) = \left(\dfrac{b^2-1}{b}\right)\dfrac{z}{(z-b)(z+b^{-1})}; \quad b < |z| < \dfrac{1}{b}$

(c) $X(z) = \left(\dfrac{b}{b^2-1}\right)\dfrac{z}{(z+b)(z+b^{-1})}; \quad b < |z| < \dfrac{1}{b}$

(d) none of these

17. For the z-transform

$X(z) = 4z^2 + 2 + 3z^{-1}; \quad 0 < |z| < \infty$ $x(n)$ is

(a) $x(n) = 4\delta[n-2] + 2\delta[n] + 3\delta[n+1]$

(b) $x(n) = 4\delta[n+2] + 2 + 3\delta[n-1]$

(c) $x(n) = 4\delta[n+2] + 2\delta[n] + 3\delta[n-1]$

(d) none of these

18. If $X(z) = \log(1 + az^{-1})$, find $X(n)$ if ROC is $|z| > a$.

(a) $x(n) = \dfrac{-(a)^n}{n}u(n-1)$

(b) $x(n) = \dfrac{-(-a)^n}{n}u(n+1)$

(c) $x(n) = \dfrac{-(-a)^n}{n}u(n-1)$

(d) $x(n) = \dfrac{-(a)^n}{n}u(n+1)$

19. An LTI system is stable if and only if ROC of its system function H(z)

(a) inside the unit circle $|z| = 1$

(b) includes the unit circle $|z| = 1$

(c) outside the unit circle $|z| = 1$

(d) none of these

20. If $x(n) = \{1, 2, 5, 7, 0, 1\}$, then ROC of $X(z)$ is entire z-plane except

(a) $z = 0$

(b) $z = \infty$

(c) $z = 0$ and $z = \infty$

(d) none of these

21. z-transform and ROC of $x(n) = [3(2)^n - 4(3)^n]u(n)$ is

(a) $X(z) = \dfrac{3}{1-2z^{-1}} - \dfrac{4}{1-3z^{-1}}; \quad |z| > 2$

(b) $X(z) = \dfrac{3}{1-2z^{-1}} - \dfrac{4}{1-3z^{-1}}; \quad |z| > 3$

(c) $X(z) = \dfrac{3}{1-2z^{-1}} - \dfrac{4}{1-3z^{-1}}; \quad |z| < 2$

(d) $X(z) = \dfrac{3}{1-2z^{-1}} - \dfrac{4}{1-3z^{-1}}; \quad |z| < 3$

22. If $x(n) = \cos \omega_0 n u(n)$, then $X(z)$ and its ROC is

(a) $X(z) = \dfrac{1}{2}\left[\dfrac{1}{1-e^{+j\omega_0}z^{-1}} + \dfrac{1}{1-e^{-j\omega_0}z^{-1}}\right]; \quad |z| < 1$

(b) $X(z) = \dfrac{1}{2}\left[\dfrac{1}{1-e^{+j\omega_0}z^{-1}} + \dfrac{1}{1-e^{-j\omega_0}z^{-1}}\right]; \quad |z| > -1$

(c) $X(z) = \dfrac{1}{2}\left[\dfrac{1}{1-e^{+j\omega_0}z^{-1}} + \dfrac{1}{1-e^{-j\omega_0}z^{-1}}\right]; \quad |z| > 1$

(d) None of these

23. If $x(n) = u(-n)$, then $X(z)$ and ROC is

(a) $\dfrac{1}{1-z^{-1}} \quad |z| < 1$ (b) $\dfrac{1}{1-z} \quad |z| > 1$

(c) $\dfrac{z}{1-z} \quad |z| < 1$ (d) $\dfrac{1}{1-z} \quad |z| < 1$

24. If $x(n) = n\left(\frac{1}{3}\right)^n u(n)$, then $X(z)$ and its ROC is

(a) $X(z) = \dfrac{\frac{1}{3}z^{-1}}{(1-z^{-1})^2} \quad |z| > \dfrac{1}{3}$

(b) $X(z) = \dfrac{\frac{1}{3}z^{-1}}{\left(1-\frac{1}{3}z^{-1}\right)^2} \quad |z| > \dfrac{1}{3}$

(c) $X(z) = \dfrac{z^{-1}}{\left(1-\frac{1}{3}z^{-1}\right)^2} \quad |z| > \dfrac{1}{3}$

(d) None of these

25. What should be the ROC of H(z) and $h(n)$ the system is anti-casual if

(a) $h(n) = \left(\frac{1}{2}\right)^n u(n) - 2(3)^n(-n-1)$ $\frac{1}{2} < |z| < 3$

(b) $h(n) = -\left[\left(\frac{1}{2}\right)^n u(n) + 2(3)^n u(-n-1)\right]$ $|z| < \frac{1}{2}$

(c) $h(n) = \left(\frac{1}{2}\right)^n u(n) + 2(3)^n u(n)$ $|z| > 3$

(d) None of these

26. Inverse z-transform of 1 is

(a) $\delta(n+1)$ (b) $\delta(n)$

(c) $\delta(n-1)$ (d) $u(n)$

27. z-transform of $y(n) = n[u(n)]$ is

(a) $\dfrac{z}{(1-z^{-1})^2}$ (b) $\dfrac{z^{-1}}{(1-z)^2}$

(c) $\dfrac{z}{(1-z)^2}$ (d) $\dfrac{z^{-1}}{(1-z^{-1})^2}$

28. The region of convergence of X(z) consists of a ring in the z-plane centered about

(a) origin (b) pole

(c) Γ (d) curve

29. If $x(n)$ is anticausal (*i.e.* if it is left sided and equal to 0 for for $n > 0$), then ROC also includes as

(a) $z = 0$ (b) $z = \infty$

(c) $z = -\infty$ (d) $z = 1$ (pole)

30. For z-transform

$$X(z) = \frac{3 - \frac{5}{6}z^{-1}}{\left(1-\frac{1}{4}z^{-1}\right)\left(1-\frac{1}{3}z^{-1}\right)}, \quad |z| > \frac{1}{3}$$

if one pole is $z = \frac{1}{3}$, then other pole is

(a) $z = -\frac{1}{3}$ (b) $z = \frac{1}{4}$

(c) $z = -\frac{1}{4}$ (d) $z = 0$

31. What is the value of $z\{\cos(\omega_0 n)\, u(n)\}$ when ROC $|z| > 0$?

(a) $\dfrac{1-z^{-1}\cos\omega_0}{1-2z^{-1}\cos\omega_0 + z^{-2}}$ (b) $\dfrac{1-z^{-1}\cos\omega_0}{1+z^{-1}+z^{-2}}$

(c) $\dfrac{1-z^{-1}}{1-2z^{-1}\cos\omega_0}$ (d) None of these

32. z-transform of $\delta(n-k)$ is

(a) z^{-k} (b) z^k

(b) $z^{\frac{1}{k}}$ (d) z^{-k+1}

33. z-transform of $x(n) = [4(2^n) - 5(3^n)]\, u(n)$ is

(a) $\dfrac{4}{1-2z^{-1}} - \dfrac{5}{1-3z^{-1}}$ (b) $\dfrac{4}{1-3z^{-1}} + \dfrac{5}{1-3z^{-1}}$

(c) $\dfrac{5}{1-2z^{-1}} - \dfrac{4}{1-3z^{-1}}$ (d) $\dfrac{6}{1-2z} \times \dfrac{4}{1-3z^{-1}}$

34. Inverse of z-transform of $\dfrac{Az}{(z-A)^2}$ is

(a) $nA^n u(n-1)$

(b) $nA^n u(n+1)$

(c) $(n-1)A^{(n-1)} u(n)$

(d) $nA^n u(n)$

35. A LTI system is stable if and only if ROC of its system function H(z) includes the unit circle, *i.e.*

(a) $|z| > 0$ (b) $|z| \geq \pm 1$

(c) $|z| = 1$ (d) $|z| = \dfrac{1}{z^{-1}k}$

36. The inverse z-transform of

$$X(z) = \frac{z}{\left(z-\frac{1}{2}\right)\left(z-\frac{1}{4}\right)} \text{ is}$$

(a) $4\left[\left(\frac{1}{2}\right)^n - \left(\frac{1}{4}\right)^n\right]u(n)$

(b) $4\left[\left(\frac{1}{2}\right)^n - \left(\frac{1}{4}\right)^n\right]u(n-1)$

(c) $\left[\left(\frac{1}{2}\right)^n - \left(\frac{1}{4}\right)^n\right]u(n-1)$

(d) $4\left[\left(\frac{1}{2}\right)^n - \left(\frac{1}{4}\right)^n\right]$

37. If $|\alpha| < 1$, then value of $\sum_{n=0}^{n} \alpha^n$ is

(a) $\dfrac{1}{1+\alpha}$ (b) $\dfrac{1}{1+\alpha^2}$

(c) $\dfrac{1}{1-\alpha}$ (d) $\dfrac{1}{\alpha}$

38. The value of

$$\lim_{x\to\infty} X(z) = \lim_{z\to\infty} \frac{1-\frac{3}{2}z^{-1}}{\left(1-\frac{1}{3}z^{-1}\right)\left(1-\frac{1}{2}z^{-1}\right)} \text{ is equal to}$$

(a) 0 (b) 2

(c) -1 (d) 1

39. If z-transform, X(z) of $x(n)$ is rational and if $x(n)$ is right hand sided, then ROC is region of the z-plane outside outermost pole, *i.e* outside the circle of radius equal to the largest magnitude of the poles of

(a) X(n) (b) $x(n)$
(c) $z = 0$ (d) X(z)

40. z-transform of a discrete-time signal is defined as

(a) $X(z) = \sum_{n=-\infty}^{\infty} x(n)z^{-n}$ (b) $X(z) = \sum_{n=-\infty}^{0} x(n)z^{-n}$

(c) $X(z) = \sum_{n=0}^{\infty} x(n)z^{-n}$ (d) $X(z) = \sum_{n=0}^{1} x(n)z^{-n}$

41. If $x(n) \overset{z}{\longleftrightarrow} X(z)$ $x(n-k) \overset{z}{\longleftrightarrow} z^{-k}X(z)$, then ROC of z^{-k} X(z) is same as that of X(z) except for $z = 0$ if $K > 0$ and $z = \infty$ if

(a) $k = 0$ (b) $k \geq 0$
(c) $k < 0$ (d) $k \leq 0$

42. Unit impulse signal $d(n)$ z-transform of

$\delta(n) \overset{z}{\longleftrightarrow} \sum_{n=-\infty}^{\infty} \delta(n)z^{-n}$ is equal to

(a) 1 (b) 0
(c) ∞ (d) ± 1

43. For an LTI system with input $x(n)$ and unit impulse response $h(x)$ specified as $x(n) = 2^n u(-n)$ and

$h(n) = u(n)$, the value of $\sum_{h=-\infty}^{0} 2^k$ will be

(a) 1
(b) 2
(c) 3
(d) 4

44. A causal LTI system with rational system function H(z) is stable if and only if all the poles of H(z) lie inside the unit circle *i.e.* they must all have magnitude

(a) greater than 1
(b) smaller than 1
(c) greater than equal to 1
(d) equal to zero

45. A stable and casual system is described by the difference equation

$$y(n) + \frac{1}{4}y(n-1) - \frac{1}{8}y(n-2) = -2x(n) + \frac{5}{4}x(x-1)$$

The system impulse response is

(a) $h(n) = \left(\frac{1}{4}\right)^n u(n) - 3\left(-\frac{1}{2}\right)^n u(n)$

(b) $h(n) = \left(\frac{1}{4}\right)^n u(n) - 3u(n)$

(c) $h(n) = u(n)$

(d) None of these

46. Consider discrete time singal $x(n)$ defined by

$$x(n) = \begin{cases} 1, & -2 \leq n \leq 2 \\ 0, & |n| > 2 \end{cases}$$

then $y(n) = x(3n - 2)$ is equal to

(a) $x(n) = \begin{cases} 1, & n = 0,1 \\ -1, & \text{otherwise} \end{cases}$

(b) $x(n) = \begin{cases} 1, & n = 1 \\ -1, & n = -1 \end{cases}$

(c) $x(n) = \begin{cases} 1, & n = 0,1 \\ 0, & \text{otherwise} \end{cases}$

(d) None of these

NUMERICAL TYPE QUESTIONS

1. Consider a continuous-time signal x(t) defined by x(t) = 0 for $|t| > 1$, and x(t) = 1 – $|t|$ for $|t| \leq 1$. Let the Fourier transform of x(t) be defined as $X(\omega) = \int_{-\infty}^{\infty} x(t)e^{-j\omega t}dt$. The maximum magnitude of X(ω) is _____.

2. The Fourier transform of a continuous time signal x(t) is given by

$$X(\omega) = \frac{1}{(10 + j\omega)^2}, -\infty < \omega < \infty, \text{ where } j = \sqrt{-1}$$

and ω denotes frequency. Then the value of | Inx(t) | at t = 1 is _______ (upto 1 decimal place). (In denotes the logarithm to base e).

ANSWERS

MCQ Type Questions

1. (*c*)	**2.** (*b*)	**3.** (*c*)	**4.** (*d*)	**5.** (*c*)	**6.** (*c*)	**7.** (*c*)	**8.** (*c*)	**9.** (*c*)	**10.** (*a*)
11. (*d*)	**12.** (*c*)	**13.** (*a*)	**14.** (*c*)	**15.** (*c*)	**16.** (*a*)	**17.** (*b*)	**18.** (*c*)	**19.** (*c*)	**20.** (*a*)
21. (*b*)	**22.** (*c*)	**23.** (*d*)	**24.** (*b*)	**25.** (*c*)	**26.** (*b*)	**27.** (*d*)	**28.** (*a*)	**29.** (*a*)	**30.** (*b*)
31. (*a*)	**32.** (*a*)	**33.** (*a*)	**34.** (*d*)	**35.** (*c*)	**36.** (*a*)	**37.** (*c*)	**38.** (*d*)	**39.** (*d*)	**40.** (*a*)
41. (*c*)	**42.** (*a*)	**43.** (*b*)	**44.** (*b*)	**45.** (*b*)	**46.** (*c*)				

Numerical Type questions

1. (1 to 1) **2.** 10

EXPLANATIONS

MCQ TYPE QUESTIONS

1. We know that, $\text{sgn}(t) = 2u(t) - 1$ ...(*i*)

Let, $\text{sgn}(t) = s(t)$

$$\frac{ds(t)}{dt} \leftrightarrow j\omega S(\omega)$$

Now differentiate equation (*i*)

$$\frac{ds(t)}{dt} = 2\delta(t) - 0$$

Fourier transform of above equation is

$$F\left[\frac{ds(t)}{dt}\right] = 2$$

Therefore, $2 = j\omega S(\omega)$

$$S(\omega) = \frac{2}{j\omega}$$

2. If $f(t) \leftrightarrow F(\omega)$

then by $f(t-b) \leftrightarrow e^{-j\omega b}F(\omega)$ time shifting property

$$\frac{2\sin\omega T_1}{\omega}e^{-jwT_1}$$

4. Frequency Convolution Theorem : Multiplication of two signals in time domain is equivalent of their convolution in frequency domain.

i.e. $x(t).\,y(t) \leftrightarrow X(f) * 2\pi\, Y(f)$

5. Let $f(t)$ denote a function of time and $F(f)$ its fourier transform. Then differentiating the fourier transform $F(f)$ with respect to f, we get,

$$-j2\pi t f(f) \leftrightarrow \frac{d}{df}F(f)$$

Similarly differentiating $f(t)$ w.r.t. we get,

$$\frac{d}{dt}f(t) = -j2\pi f(f)$$

then, $$\frac{d}{df}F(f) = -2\pi f F(f)$$

This shows that the gaussian pulse and its fourier transform are the same function.

6. Using convolution theorem,

$$y(t) = h(t) * x(t)$$

7. $$u(t) = \frac{1}{2}[1 + \text{sgn}(t)]$$

Taking the fourier transform both sides

$$F[u(t)] = \frac{1}{2}\left[2\pi\delta(\omega) + \frac{2}{j\omega}\right]$$

$$F[u(t)] = \pi\delta(\omega) + \frac{1}{j\omega}$$

8. $$v(t) = \cos\omega_0 t$$

$$v(t) = \frac{1}{2}\left[e^{j\omega_0 t} + e^{-j\omega_0 t}\right]$$

Taking the fourier transform,

$$V(\omega) = \frac{1}{2}[2\pi\delta(\omega - \omega_0) + 2\pi\delta(\omega + \omega_0)]$$

$$V(\omega) = \pi[\delta(\omega - \omega_0) + \delta(\omega + \omega_0)]$$

or $$V(f) = \frac{1}{2}[\delta(f - f_0) + \delta(f + f_0)]$$

9. If, $g(t) \leftrightarrow G(f)$

then from the Duality property

$$G(t) \leftrightarrow g(-f)$$

10. Area under the curve is

$$\int_{-\infty}^{\infty} f(t)d(t) = F(0)$$

11. Fourier transform of periodic delta function $x(t)$ is also a periodic delta function with same period.

14. $x(n) = -a^n u(-n-1)$

$$x(z) = \sum_{-\infty}^{\infty} -a^n u(-n-1)z^{-n}$$

$$= -\sum_{-\infty}^{-1} a^n z^{-n} = -\sum_{1}^{\infty} a^{-n} z^n = -\sum_{-\infty}^{-1}\left(a^{-1}z\right)^n$$

$$= -[a^{-1}z + (a^{-1}z)^2 + (a^{-1}z)^3 + ...]$$

$$= -a^{-1}z[1 + (a^{-1}z) + (a^{-1}z)^2 + ...]$$

$$= -a^{-1}z \times \frac{1}{1-a^{-1}z} = \frac{1}{1-az^{-1}} \quad |a^{-1}z| < a$$

or $|z| < a$

15. $z > \frac{1}{3}, z > \frac{1}{2}$

24. $x(n) = a^n u(n) \leftrightarrow \frac{1}{1-az^{-1}}$

$$nx(n) \leftrightarrow -z\frac{dX(z)}{dz}$$

$$n\left(\frac{1}{3}\right)^n u(n) \leftrightarrow -z\frac{(-az^{-2})}{(1-az^{-1})^2} = \frac{az^{-1}}{(1-az^{-1})^2}$$

$$n\left(\frac{1}{3}\right)^n u(n) \leftrightarrow \frac{\frac{1}{3}z^{-1}}{\left(1-\frac{1}{3}z^{-1}\right)^2}$$

ROC: $|z| > \frac{1}{3}$

NUMERICAL TYPE QUESTIONS

1. As given that,

Signal : CTFT

$$x(t) = \begin{Bmatrix} 1-|t|; |t| \le 1 \\ 0; \quad |t| > 1 \end{Bmatrix}$$

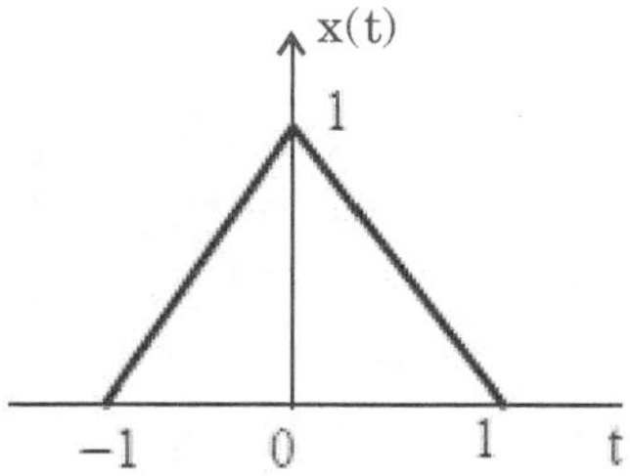

$$X(\omega) = (\text{Area of triangle}) \text{ X } Sa^2\left(\omega\frac{T}{2}\right)$$

$$= \left(\frac{1}{2}\times 2\times 1\right)Sa^2\left(\frac{\omega}{2}\right) = Sa^2\left(\frac{\omega}{2}\right) \quad [\because T = 1]$$

$\Rightarrow X(0) = Sa^2(0)$

$\Rightarrow X(0) = 1$

Hence, the maximum magnitude of x(W) is '1'.

2. Lapalce transform,

$$x(t) \Longleftrightarrow X(j\omega) = \frac{1}{(10+j\omega)^2}$$

Now by taking inverse Fourier transform, we get,

$$x(t) = te^{-10t}u(t)$$

Now, $x(t)|_{t=1} = 1 \times e^{-10} \times 1 = e^{-10}$

Thus, $|\ln\{x(t)\}| = |\ln(e^{-10})|$

$$= |-10| = 10$$

9 CHAPTER Vector Analysis

VECTORS

In chemical engineering one is interested in the concentrations of a chemical species at different levels of an absorption column. To record the concentrations at a certain time, we can write

$$C = (c_1, c_2 \ldots, c_n)$$

where c_k is the concentration at the kth level. We can call C a **concentration vector**; it is just a $1 \times n$ matrix.

Definition : An n-vector is a $1 \times n$ matrix.

Before developing the algebra of n-vectors we study two-vectors and three-vectors because their structure can be visualized geometrically.

Definition : A plane vector (or two-vector) is a 1×2 matrix with real entries. A **space vector** (or three-vector) is a 1×3 matrix with real entries. The set of all plane vectors is denoted $\mathbb{R}^2$; the set of all space vectors, $\mathbb{R}^3$.

The elements of the matrix are called the **components** of the vector. Two vectors are equal if they are **equal** as matrices.

Example.

The following are space vectors: (1, 1, –1), (2, 0, 1), (–1, 1, 1). We associate these vectors with points in the standard xyz coordinate system.

A plane vector (a, b) is also associated with any arrow in the standard xy plane which begins at a point (x_0, y_0) and ends at $(x_0 + a, y_0 + b)$. In fig. we have drawn several arrows associated with the vector (1, 2).

In general, when we write a vector (a, b), we assume that the initial point of the associated vector is (0, 0) unless specified otherwise .

Example.

Find the vector associated with the arrow with initial point (– 2, 4) and terminal point (– 7, 3).

Solution :

Now (–7, 3) must be equal to $(x_0 + a, y_0 + b)$.

Since $(x_0, y_0) = (-2, 4)$,

we have $(-7, 3) = (-2 + a, 4 + b)$.

Thus $a = -5$, $b = -1$, so $(a, b) = (-5, -1)$.

The easy way to solve this problem is to "subtract points": terminal point - initial point. In this way (just subtracting matrices)

$$(a, b) = (-7, 3) - (-2, 4) = (-5, -1)$$

For space vectors we have similar methods and interpretations in the xyz coorindate system.

Definition :

The **length** (or **magnitude**) of the vector $A = (a, b)$ is denoted $|A|$ or $|(a, b)|$ and is defined by

$$|A| = \sqrt{a^2 + b^2}$$

For three-vectors

$$|(a, b, c)| = \sqrt{a^2 + b^2 + c^2}$$

Example.

Calculate $|(0, 0)|$, $|(1, -1)|$, $|(3, 4)|$, $|(0, 1, 0)|$, and $|(\sqrt{11}, 3, 4)|$.

Solution :

$$|(0,0)| = \sqrt{0^2 + 0^2} = \sqrt{0} = 0$$
$$|(1,-1)| = \sqrt{1^2 + (-1)^2} = \sqrt{2}$$
$$|(3,4)| = \sqrt{3^2 + 4^2} = \sqrt{25} = 5$$
$$|(0,1,0)| = \sqrt{0^2 + 1^2 + 0^2} = \sqrt{1} = 1$$
$$|(\sqrt{11},3,4)| = \sqrt{(\sqrt{11})^2 + 3^2 + 4^2} = \sqrt{36} = 6$$

Example.

Let A = (1, 0, –1), B = (3, –4, 2), C = (π, 0, 4), D = (0, 0, 0), and E = (–1, 0, 1). Calculate A + B, B + D, A + B + C, and A + E.

Solution :

$$A + B = (1 + 3, 0 + -4, -1 + 2) = (4, -4, 1)$$
$$B + D = (3 + 0, -4 + 0, 2 + 0)$$
$$= (3, -4, 2) = B$$
$$A + B + C = (1 + 3 + \pi, 0 + (-4) + 0, -1 + 2 + 4)$$
$$= (4 + \pi, -4, 5)$$
$$A + E = (1 + -1, 0 + 0, -1 + 1) = (0, 0, 0)$$

The definition of addition corresponds to the "addition" of forces represented by the vectors. The sum of the two vectors is called the resultant force. The length of each vector corresponds to the magnitude of the force, and the direction to the direction of application of the force. The resultant vector is just the diagonal (having initial point in common with the two forces) of the parallelogram generated by the two forces.

Because vectors represent forces, some other concepts regarding forces translate to operations with vectors. For instance, if we will "twice as hard" on an object, we think of doubling the force we are applying. If the original force is (a, b), the doubled force is $(2a, 2b)$. This leads to the definition of scalar multiple.

Definition

Let r be a real number and (a, b) be a vector. The scalar multiple $r(a, b)$ is defined as

$$r(a, b) = (ra, rb)$$

The reader should write a definition for a scalar multiple of (a, b, c). Note that this multiplication is the same as multiplication of a matrix by a scalar.

The idea of reversing a force or changing it to the opposite direction leads to a definition of the negative of a vector.

Definition

The negative of a vector V is written –V and is defined as (–1)V. The difference A – B of two vectors A and B is defined by

$$\text{A} - \text{B} = \text{A} + (-1)\text{B}$$

Example :

Let A = (1, –1) and B = (2, 4).

Calculate 4A +B, A – B, and A – A.

Solution :

$$\begin{aligned} 4\text{A} + \text{B} &= 4(1, -1) + (2, 4) \\ &= (4, -4) + (2, 4) = (6, 0) \\ \text{A} - \text{B} &= \text{A} + (-1)\text{B} \\ &= (1, -1) + (-2, -4) = (-1, -5) \end{aligned}$$

or more simply

$$\begin{aligned} &= (1, -1) - (2, 4) \\ &= (1 - 2, -1 - 4) = (-1, -5) \end{aligned}$$

Finally,

$$\begin{aligned} \text{A} - \text{A} &= (1, -1) - (1, -1) \\ &= (1 - 1, -1 - (-1)) = (0, 0) \end{aligned}$$

Example :

Let A be 2×2 matrix and v a two vector. Show that $\text{A}v_s = (v\text{A}^{\text{T}})_s$.

Solution :

For any vector w with $w = (a, b)$, we have

$$w_s = \begin{pmatrix} a \\ b \end{pmatrix}$$

so that $w_s = w^{\text{T}}$.

Therefore $(v\text{A}^{\text{T}})_s = (v\text{A}^{\text{T}})^{\text{T}} = \text{A}^{\text{TT}} v^{\text{T}} = \text{A}v^{\text{T}} = \text{A}v_s$.

Introduction of standard coordiante matrices allows us to "stand (a, b, c) up" as

$$\begin{pmatrix} a \\ b \\ c \end{pmatrix}$$

The purpose is to allow multiplication of a vector on the left by a matrix. If A is 3×3 and $v = (a, b, c)$, then Av makes no sense, but Av_s does. In the next section we will use this multiplication. Note that $v_s \neq v$ but v_s represents v in a natural way.

THE ANGLE BETWEEN VECTORS: PROJECTIONS

One of the most important problems in the analysis of vectors is the angle problem:

Given two vectors A and B, find the angle θ, $0 \leq \theta \leq \pi$, between A and B.

If we can solve this problem, then we know whether A is parallel to B (θ is 0 or π) or A is perpendicular to B($\theta = \pi/2$).

The solution to this problem for plane vectors can be found by using the law of cosines: For a triangle with sides of length a, b, and c as shown

we have

$$c^2 = a^2 + a^2 + b^2 - 2ab \cos \theta \qquad \text{...(1)}$$

From Eq. (1) we find

$$\cos \theta = \frac{a^2 + b^2 - c^2}{2ab} \qquad \text{...(2)}$$

Therefore in any triangle with lengths of sides given, we can find the cosine of any of the angles. And this is just as good as finding θ—especially for the purposes of determining whether θ equals 0, $\pi/2$, or π.

To solve the angle problem, we consider the triangle formed by A, B, and A – B and use the law of cosines. We have from (1)

$$\cos \theta = \frac{|\text{A}|^2 + |\text{B}|^2 - |\text{A} - \text{B}|^2}{2|\text{A}|\,|\text{B}|} \qquad \text{...(3)}$$

Suppose now that $\text{A} = (a_1, a_2)$ and $\text{B} = (b_1, b_2)$. Then $\text{A} - \text{B} = (a_1 - b_1, a_2 - b_2)$ and

$$\begin{aligned} |\text{A}|^2 + |\text{B}|^2 &= |\text{A} - \text{B}|^2 \\ &= a_1^2 + a_2^2 + b_1^2 + b_2^2 - [(a_1 - b_1)^2 + (a_2 - b_2)^2] \\ &= a_1^2 + a_2^2 + b_1^2 + b_2^2 \\ &\qquad - (a_1^2 - 2a_1b_1 + b_1^2 + a_2^2 - 2a_2b_2 + b_2^2) \\ &= 2(a_1b_1 + a_2b_2) \end{aligned}$$

This means that

$$\cos \theta = \frac{a_1b_1 + a_2b_2}{|\text{A}|\,|\text{B}|} \qquad \text{...(4)}$$

and the angle problem is solved for two-vectors.

Example :

Find the cosine of the angle between A = (3, 4) and B = (5, 12).

Solution :

Since $a_1 = 3$, $a_2 = 4$, $b_1 = 5$, and $b_2 = 12$, we have

$$\cos \theta = \frac{3 \cdot 5 + 4 \cdot 12}{\sqrt{3^2 + 4^2}\sqrt{5^2 + 12^2}} = \frac{63}{5 \cdot 13} = \frac{63}{65}$$

Example :

Find the cosine of the angle between A = (1, –1, 4) and B = (4, 0, –1).

Solution :

$$\cos\theta = \frac{AB^T}{|A||B|} = \frac{4+0-4}{\sqrt{18}\sqrt{17}} = 0$$

Hence, these vectors are perpendicular.

Since the matrix product AB^T is used to solve the angle problem for vectors, we give it a special name, the **dot product**.

Definition :

Given two plane or space vectors A and B, the dot product A . B is defined by A . B = AB^T. Because A . B is a real number, $(A . B)^T$ = A . B. Therefore B . A = $BA^T = (AB^T)^T = (A . B)^T$ = A . B. This means that the dot product is commutative.

Theorem

Let A, B, and C be vectors of the same order. Let *r* and *s* be real numbers.

(*a*) A . B = |A| |B| cos θ, where θ is the angle between A and B.

(*b*) A . (B + C) = A . B + A . C

(*c*) (A + B) . C = A . C + B . C

(*d*) *r*(A . B) = (*r*A) . B = A . (*r*B)

(*e*) (*r*A) – (*s*B) = (*rs*)(A . B)

(*f*) $\sqrt{A.A} = |A|$

(*g*) A . A = 0 if and only if A = 0 (the zero vector).

(*h*) Let |A| ≠ 0 and |B| ≠ 0. Then A . B = 0 if and only if A ⊥ B.

(*i*) |A . B| ≤ |A| |B| (Cauchy-Schwarz inequality)

(*j*) |*r*A| = |*r*| |A|

(*k*) |A . B| ≤ |A| + |B|

Projections are an important geometric idea which we can now discuss by using the dot product. Geometrically, the projection of a vector B on a vector A. Roughly speaking, the projection of B on A is the shadow which B casts on A due to light rays which hit A, the light rays being perpendicular to A. In mechanics, the projection is the component of the force B in the direction of A, and B . A is the work done by the force B in the direction of A.

We can determine the vector projection of B on A (written $B_{proj\,A}$) by finding its length; its direction is the same as A. Once we find $|B_{proj\,A}|$, we can multiply this by (1/|A|)A (the unit vector in the direction of A) to get. We find $|B_{proj\,A}|$ by using trigonometry. $|B_{proj\,A}|$ = |B| cosθ, where ____ is the angle between A and B. Therefore, since

$$\frac{A.B}{|A||B|} = \cos\theta$$

we know that

$$|B_{proj\,A}| = |\,|B|\cos\theta| = |B|\frac{A.B}{|A||B|} = \frac{A.B}{|A|}$$

and so $B_{proj\,A} = |B_{proj\,A}|\,\frac{1}{|A|}A = \frac{A.B}{|A|^2}A$

Sometimes $|B_{proj\,A}|$ is called the scalar projection of B on A.

Example :

Let A = (1, 0, 1) and B = (1, –1, 2). Find $B_{proj\,A}$.

Solution

$$B_{proj\,A} = \frac{A.B}{|A|^2}A = \frac{3}{2}(1,0,1) = \left(\frac{3}{2},0,\frac{3}{2}\right)$$

A vector product which is important in three-dimensional mechanics is the cross **product**. This product is specific to three-vectors and does not generalize in a natural way to higher dimensions. Because of this, our treatment of the cross product is brief.

Definition

Let A = (a_1, a_2, a_3) and B = (b_1, b_2, b_3). Then the cross product of A and B, written A × B, is a vector defined by

$$A \times B = (a_2b_3 - b_2a_3,\ a_1b_3 + b_1a_3,\ a_1b_2 - a_2b_1)$$

Example :

Calculate A × B for A = (1, –1, 2) and B = (2, 2, 7).

Solution :

One way to do this is to form a symbolic determinant

$$\begin{matrix} & 1st & 2d & 3d & \text{componet of } A\times B \\ & i & j & k & \\ \det & 1 & -1 & 2 & \leftarrow A \text{ is row } 2 \\ & 2 & 2 & 7 & \leftarrow B \text{ is row } 3 \end{matrix}$$

and to expand by the first row:

$$A \times B = i\det\begin{pmatrix}-1 & 2\\ 2 & 7\end{pmatrix} - j\det\begin{pmatrix}1 & 2\\ 2 & 7\end{pmatrix} + k\det\begin{pmatrix}1 & -1\\ 2 & 2\end{pmatrix} = (-11,-3,4)$$

Notice that A × B is perpendicular to both A and B since

(A × B) . A = (–11, –3,4) . (1, –1, 2) = –11 + 3 + 8 = 0

(A × B) . B = (–11, –3,4) . (2, 2, 7) = –22 – 6 + 28 = 0

In fact, this is always true.

Theorem

Let A and B be nonzero vectors.

Then (A × B) . A = 0 = (A × B) . B.

Proof. For (A × B) . A we have

$$\begin{aligned}(A\times B).A &= (a_2b_3 - b_2a_3 - a_1b_3 + b_1a_3,\ a_1b_2 - a_2b_1).(a_1, a_2, a_3)\\ &= a_1a_2b_3 - a_1b_2a_3 - a_2a_1b_3 + a_2b_1a_3 + a_3a_1b_2 + a_2b_1a_3\\ &= 0\end{aligned}$$

Geometrically A × B is the vector of length |A| |B|sin θ, perpendicular to both A and B and pointing in the direction dictated by the right-hand rule: Using the right hand, curve your fingers from A to B; A × B points in the direction of the thumb. This can be shown rigorously, but we omit the proof.

From the geometric meaning of A × B it is not hard to see that B × A = −(A × B) and A × B = 0 if sin θ = 0, that is, A × B = 0 if A is parallel to B (the directions can be opposite).

VECTORS IN THE PLANE

The advantage gained in using vectors is that they are moveable, and not tied to any particular coordinate system. As we have seen in the examples of the previous section, geometric facts can be easily derived using vectors while working in coordinates may be cumbersome. However, it is often the case, that in working with vectors we must do calculations in a particular coordinate system. It is important to realize that it is the worker who gets to choose the coordinates; it is not necessarily inherent in the problem.

We now explain how to move back and forth between vectors and coordinates. Suppose, then, that a coordinate system has been chosen: a point O, the origin, and two perpendicular lines through the origin, the *x*- and *y*-axes. A vector V is determined by its length, |V| and its direction, which we can describe by the angle θ that V makes with the horizontal (see figure). In figure, we have realized V as the vector $\vec{OP}$ from the origin to P. Let (*a*, *b*) be the cartesian coordinates of P. Note that V can be realized as the sum of a vector of length *a* along the x-axis, and *a* vector of length *b* along the *y*-axis. We express this as follows.

Definition

We let I represent the vector from the origin to the point (1, 0), and J the vector from the origin to the point (0,1). These are the **basic** unit vectors (a unit vector is a vector of length 1). The unit vector in the direction θ is cos θI + sin θ J.

If V is a vector of length *r* and angle θ, then V = *r*(cosθI + cosθJ). If V is the vector from the origin to the point (*a*, *b*); *r* is the length of V, and cosθI + cosθJ is its direction. If P(*a*, *b*) is the endpoint of V, then V = $\vec{OP}$ = *a*I + *b*J. *a* and *b* are called the components of V.

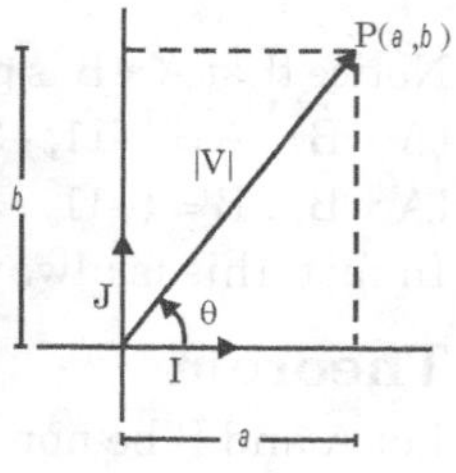

Of course, *r* and θ are the usual polar coordinates, and we have these relations:

$$|V| = \sqrt{a^2 + b^2}\ , \theta = \arctan \frac{b}{a},$$

$$a = |V| \cos \theta, b = |V| \sin \theta.$$

We add vectors by adding their components, and multiply a vector by a scalar by multiplying the components by the scalar.

Example :

An object on the plane is subject to the three forces F = 2I + J, G = − 8J. H. Assuming the object doesn't move, find H. At what angle to the horizontal is H directed?

By Newton's law, the sum of the forces must be zero. Thus

$$H = - F - G = - 2I - J + 8J = = - 2I + 7J.$$

If α is the angle from the positive *x*-axis to H, tan α = −7/2, so α = 105.95°, since H points upward and to the left.

Since vectors represent magnitude and length, we need a computationally straightforward way of determining lengths and angles, given the components of a vector.

The **dot product** of two vectors V_1 and V_2 is defined by the equation

$$V_1 \cdot V_2 = |V_1|\, |V_2| \cos \beta,$$

where β is the angle between the two vectors.

Note that since the cosine is an even function, it does not matter if we take β from V_1 to V_2, or in the opposite sense. In particular, we see that $V_1 \cdot V_2 = V_2 \cdot V_1$. Now, we see how to write the dot product in terms of the components of the two vectors.

Proposition :

Let $V_1 = a_1I + b_1J$ and $V_2 = a_2I + b_2J$.

Then $V_1 \cdot V_2 = a_1a_2 + b_1b_2$

with equality holding only when the vectors are parallel.

To see this, we use the polar representation of the vectors:

$$V_1 = r_1(\cos\theta_1 I + \sin\theta_1 J),$$

$$V_2 = r_2(\cos\theta_2 I + \sin\theta_2 J).$$

Then

$$V_1 \cdot V_2 = r_1r_2 \cos(\theta_1 - \theta_2)$$

$$= r_1r_2 \cos\theta_1 \cos\theta_2 + r_1r_2 \sin\theta_1 \sin\theta_2$$

by the addition formula for the cosine. This is the same as

$$V_1 \cdot V_2 = (r_1 \cos\theta_1)(r_2 \cos\theta_2) + (r_1 \sin\theta_1)(r_2 \sin\theta_2)$$

which is equation in Cartesian coordinates. As for the last statement, we have strict inequality unless cosβ = 1, that is β = 0 or π, in which case the vectors are parallel.

Proposition :

(*a*) Two vectors V and W are orthogonal if and only if V . W = 0.

(*b*) If L and M are two unit vectors with L . M = 0, then for any vector V, we can write

$$V = aL + bM, \text{ with } a = V \cdot L,$$

$$b = V\,M, \text{ and } |V| = \sqrt{a^2 + b^2}.$$

We shall say that a pair of unit vectors L, M with L . M = 0 form a **base** for the plane. This statement just reiterates that we can put cartesian coordinates on the plane with any point as origin and coordinate axes two orthogonal lines through the origin; that is the lines in the directions of L and M. To show part b) we start with figure given below.

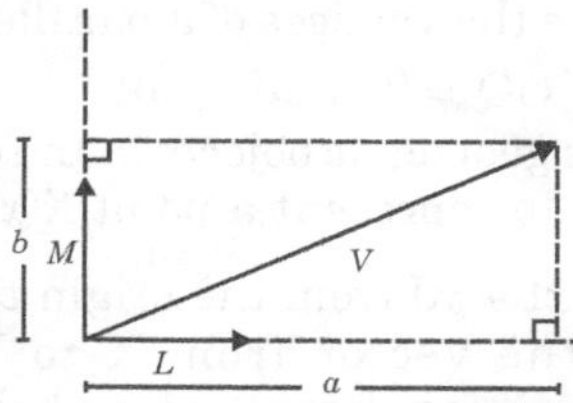

From that figure, we see that we can write any vector as a sum V = aL + bM with (by the Pythagorean theorem) |V| = $\sqrt{a^2+b^2}$. We now show that a, b are as described;

V . L = (aL + bM).L = aL . L + bM.L = a.

Similarly V . M = b.

Example :

Find the angle β between the vectors V = 2I – 3J and W = I + 2J.

We have $|V| = \sqrt{2^2+3^2} = \sqrt{13}$,

$|W| = \sqrt{1^2+2^2} = \sqrt{5}$

and V . W = 2(1) + (–3)(2) = –4. Thus

$$\cos\beta = \frac{V \cdot W}{|V||W|} = \frac{-4}{\sqrt{65}} = -0.496$$

so β = –119.7°.

Example :

Suppose we have put cartesian coordinates on the plane, with I, J the standard base.

Let $L = \dfrac{I+J}{\sqrt{2}}$, $M = \dfrac{-I+J}{\sqrt{2}}$

be a new base. Given the point P(5, 2), write $\vec{OP}$ in terms of L and M.

By the preceding proposition,

$$\vec{OP}.L = (5I+2J).\left(\frac{I+J}{\sqrt{2}}\right) = \frac{7}{\sqrt{2}},$$

$$\vec{OP}.M = (5I+2J).\left(\frac{-I+J}{\sqrt{2}}\right) = \frac{3}{\sqrt{2}},$$

so $\vec{OP} = (7I-3J)/\sqrt{2}$.

Example :

Show, using vectors, that the interior angles of an isosceles triangle are equal.

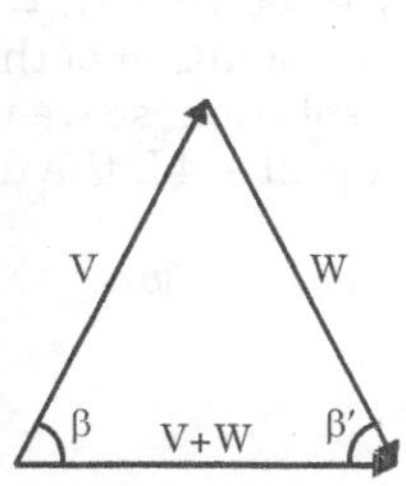

In the figure given above we have labelled the sides of equal length as V and W.

Thus, the base of the triangle is V+W. First of all, since | V | = | W |, we have (V + W).V = V.V+W.V = W.W + V.W = (V + W).W. Thus, by (2),

$$\cos\beta = \frac{(V+W).V}{|V+W||V|} = \frac{(V+W).W}{|V+W||W|} = \cos\beta'.$$

Since both angles are acute, β = β′.

Example :

Find a vector orthogonal to V = 31 + 4J and of the same length.

The vectors V = aI + bJ, W = cI + dJ, are orthogonal precisely when $ac + bd = 0$. Thus, if we are given a,b, we take $c = -b$, $d = a$ to get an orthogonal vector. So for this example, we can take W = –4I + 3J. Clearly, since the coefficients are the same but for sign, |W| = |V|. We could also take the vector in the opposite direction: –W = 4I – 3J

In general, if V = cI + dJ then both – dI + cJ and dI – cJ are orthogonal to V and of the same length. The first is counterclockwise to V, and the second, clockwise.

Definition

Given the vector V, we shall denote by $V^{\perp}$ that vector which is orthogonal to, of the same length as, and counterclockwise to V. In components, we have:

If V = aI + bJ, then $V^{\perp}$ = – bI + aJ

See figure given below to see that $V^{\perp}$ is counterclockwise to V (at least in the case where both a and b are positive).

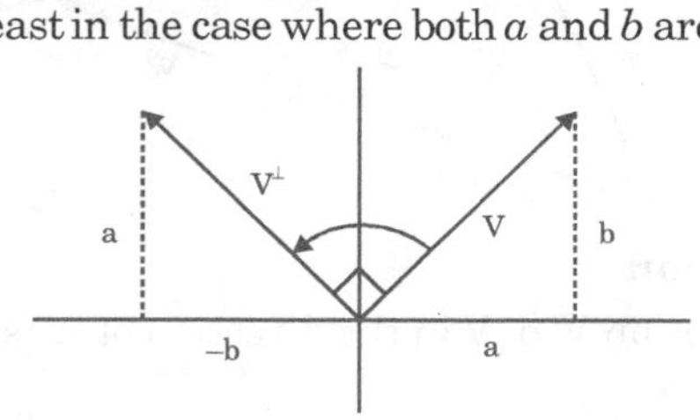

Definition

Given two vectors V and W, we define the determinant det(V, W) of the two vectors as the signed area of the parallelogram spanned by the two vectors. The sign is positive if W is counterclockwise from V; otherwise negative.

In figure given below, α is the angle from V to W.

Thus det(V, W) = |V| |W| sin α.

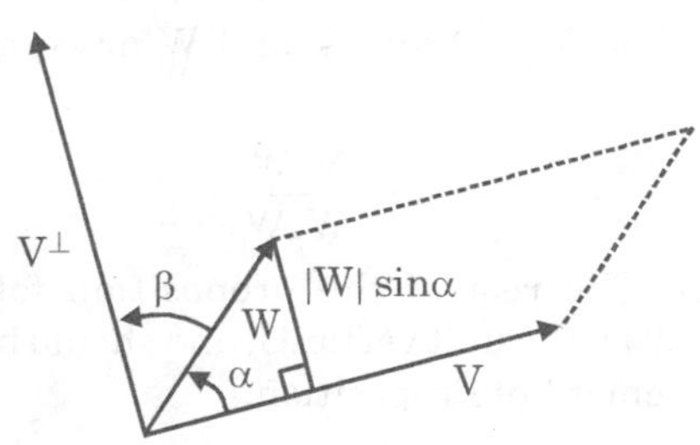

Now, let β be the angle from W to $V^{\perp}$ so that (in figure), $\alpha + \beta = \pi/2$, and we have $\sin \alpha = \cos\beta$. Since $|V| = |V^{\perp}|$, we can rewrite as

$$\det(V, W) = |V^{\perp}|\,|W| \cos \beta = V^{\perp} . W$$

This gives us the following.

Proposition

The determinant of the two vectors $V = aI + bJ$ and $W = cI + dJ$ is the determinant of the matrix whose rows are the vectors V and W:

$$\det(V, W) = ad - bc$$

For, $V^{\perp} = -bI + aJ$, and from,

$$\det(V, W) = V.W^{\perp} = -bc + ad = ad - bc.$$

The vectors V and W are parallel (or collinear) if and only if det (V, W) = 0, for in this case there is no parallelogram. We also have the inequality

$$|\det(V, W)| \le |V|\,|W|,$$

with equality holding if and only if V and W are orthogonal.

Definition

Given two vectors V and W, the projection of V′ in the direction of W is that vector V′ parallel to W such that V - V′ is orthogonal to V′ (See Figure).

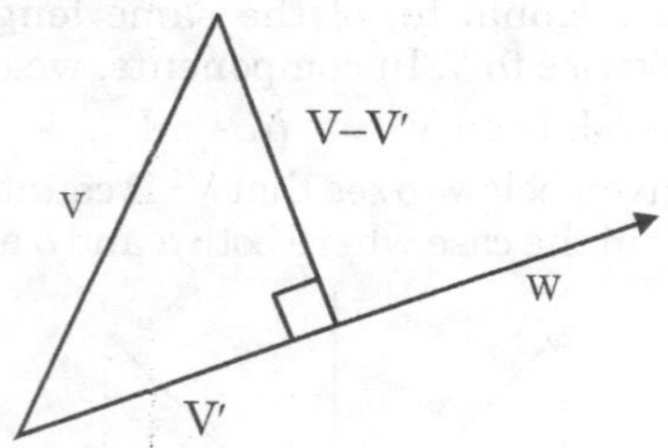

Proposition

The projection V′ of V in the direction of W is given by the formula

$$V' = pr_w(V) = \frac{V.W}{W.W}W.$$

If U is a unit vector in the direction of W, then

$$V' = (V . U)U, \text{ and } V = (V . U)U + (V.U^{\perp})U^{\perp}.$$

To show this we start with the equation $(V - V').V' = 0$. Since $V' = aW$ for some a, this gives us

$$(V - aW).\, aW = 0, \text{ or } a^2 W.W = aV.W$$

If $a = 0$, then $V' = 0$ and V and W are orthogonal. Otherwise

$$a = \frac{V.W}{W.W},$$

giving us. The rest of the proposition follows by replacing W by the unit vector U, and should be viewed as a restatement of Proposition.

Example :

Find the area of the parallelogram whose vertices are at O(0, 0), P(4, –2), Q(5, 8), R(9, 6).

This is the parallelogram determined by the vectors from the origin O to the points P and Q: $\vec{O}P = 4I - 2J$, $\vec{O}Q = 5I - 8J$, so has signed area $4(-8) - (-2)(5) = -22$. We verify these are the vertices of a parallelogram by calculating $\vec{O}P + \vec{O}Q = 9I + 6J = \vec{O}R$.

In order to discuss geometric objects in the coordinate plane, it is useful to represent a point X(x, y) by the vector $X = \vec{O}X - xI + yJ$ from the origin to X. For Y another point, the vector from X to Y is thus represented by Y – X (see figures given below).

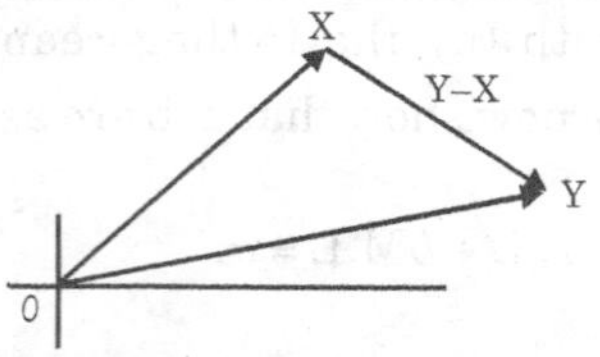

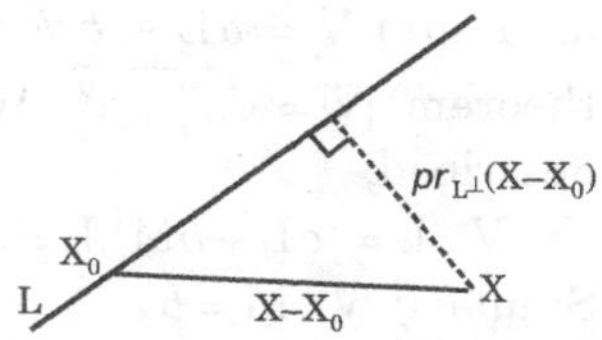

A line L is determined by its direction and a point on the line. let X_0 be a point on L, and L a vector parallel to the line L. Then, for any point X, it is on the line if and only if $X - X_0$ is parallel to L, or, what is the same, orthogonal to $L^{\perp}$. This leads to these two equations, called the equation of the line:

$$(X - X_0).L^{\perp} = 0 \text{ or } \det(X - X_0, L) = 0.$$

Also, since $X - X_0$ is parallel to L if and only if $X - X_0$ is a scalar multiple of L, we have the parametric form of the equation of the line:

$$L : X = X_0 + tL.$$

A line is also determined by two points X_0, X_1 on the line. Given that information, we find the equations of the line by taking $L = X_1 - X_0$.

Now, suppose L is a line and X is a point not on the line. We seek a formula for the distance from the point X to the line. We see from figure that this is the length of the projection in the direction perpendicular to L, of a vector from X to any point X_0 on L. This leads to the formula for the distance from X to L

$$d(X, L) = |pr_{L^{\perp}}(X - X_0)|$$

Example :

Let L be the line given by the equation $3x - y = 7$. Find the distance from (2, 4) to L.

We see that $L^{\perp} = 3I - J$, we need a point on the line; any solution of the equation $3x - y = 7$ will do. (3, 2) is a solution, so we take $X_0 = 3I + 2J$. Thus, for our point, $X = 2I + 4J$, the distance is

$$|pr_{L^{\perp}}(X - X_0)| = \frac{|(X - X_0).L^{\perp}|}{|L^{\perp}|}$$

$$= \frac{|(-I + 2J).(3I - J)|}{|3I - J|} = \frac{5}{\sqrt{10}}$$

Example :

Find the distance from X(3, 1) to the line through $X_0(2, -3)$ and parallel to V = –I + 4J. The vector $L^{\perp} = -4I - J$ is orthogonal to the line. Thus the distance is

$$|pr_{L^{\perp}}(X - X_0)| = \frac{|(I+4J).(-4I-J)|}{|-4I-J|} = \frac{8}{\sqrt{17}}$$

Example :

Find the point on the line L : $2x - 3y = 17$ which is closest to the origin.

Let X be the vector from the origin to the desired point. Then X is orthogonal to the line, so is parallel to the vector $L^{\perp} = 2I - 3J$. Writing $X = t(2I - 3J)$, since X ends on the line we have $2(2t) - 3(-3t) = 17$, so $t = 17/13$, and X = (34/13) I – (51/13)J.

VECTORS IN SPACE

In a Cartesian coordinate system for space, the vectors I, J, K are the vectors from the origin to the points (1, 0, 0), (0, 1, 0), (0, 0, 1) respectively. These are unit vectors, mutually orthogonal, and form the **standard base** for space. We always take a coordinatization so that {I, J, K} is a right-handed system. More precisely, if we situate I and J on the horizontal plane, then I is a unit vector, J is a unit vector perpendicular to I and counterclockwise from I, and K is a unit vector orthogonal to the horizontal plane, pointing upwards, see figure given below.

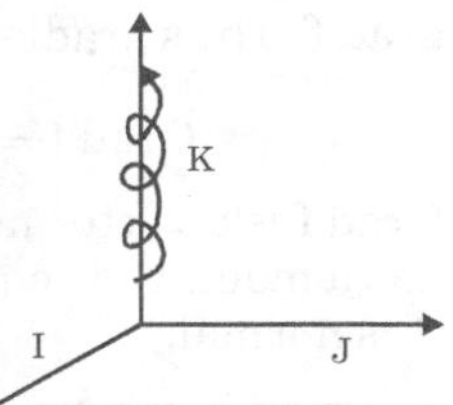

Any vector V can be written uniquely as

$$V = aI + bJ + cK,$$

where a, b, c are called the **components** of V. To add two vectors, add the components; to multiply a vector by a scalar, multiply the components by the scalar. If V is given as in equation above, its length is

$$|V| = \sqrt{a^2 + b^2 + c^2}.$$

The direction of V is determined by the cosines of the angles between V and the coordinate axes. Thus, for any vector V we can write

$$V = |V|(\cos\alpha I + \cos\beta J + \cos\gamma K)$$

where α, β, γ are those angles. The components of the unit vector in equation above are called the **direction cosines** of the vector V. Note that $\cos^2\alpha + \cos^2\beta + \cos^2\gamma = 1$.

Definition

The dot product of two vectors V, W is defined as

$$V . W = |V|\,|W| \cos\theta,$$

where θ is the angle between V and W.

As for plane vectors, this has an easy formulation in terms of the components of the vectors.

Proposition

Let $V = a_1I + b_1J + c_1K$, $W = a_2I + b_2J + c_2K$, in components. Then

$$V . W = a_1a_2 + b_1b_2 + c_1c_2$$

To see this, we start with the Law of Cosines for the triangle whose sides are the vectors V, W, W –V (see figure given below):

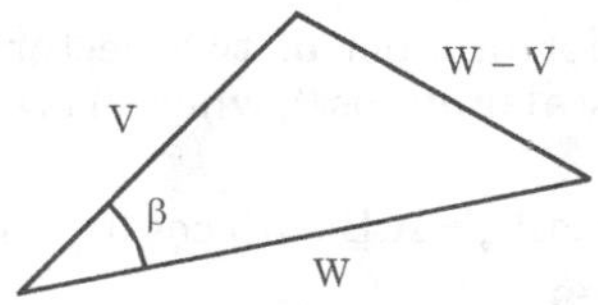

$$|W - V|^2 = |V|^2 + |W|^2 - 2|V||W|\cos\beta$$

$$|W - V|^2 = |W|^2 + |V|^2 - 2|V|\,|W| \cos\beta$$
$$= |W|^2 + |V|^2 - 2(W . V), \text{ so that}$$

$$W . V = \frac{1}{2}(|W|^2 + |V|^2 - |W - V|^2).$$

Now, writing the right hand side in terms of components, after some cancellation. In particular, just as in two dimensions, two vectors V, W are **orthogonal** if V . W=0.

Example :

Find the angle between the vectors

$$V = 2I - 3J + K, \quad W = 6I + J - 2K.$$

We have V . W = 12 – 3 – 2 = 7

and $|V| = \sqrt{2^2 + 3^2 + 1^2} = 3.74$,

$|W| = \sqrt{6^2 + 1^2 + 2^2} = 6.40$.

Thus

$$\cos\alpha = \frac{7}{(3.75)(6.40)} = 0.2923$$

VECTOR

Vector is completely specified by its magnitude and direction.

Vector is represented by a directed line segment. Thus $\overline{PQ}$ represents a vector where magnitude is the length PQ and direction is form P (starting point) to Q (end point)

Unit Vector

A vector of unit magnitude is called *unit vector.* Unit vector corresponding to vector A is written as $\overline{A}$.

Null Vector

A vector of zero magnitude, which can have no direction associated with it is called *zero (or null) vector* and is denoted by O-a thick zero.

PRODUCT OF TWO VECTORS

1. Scalar or Dot Product

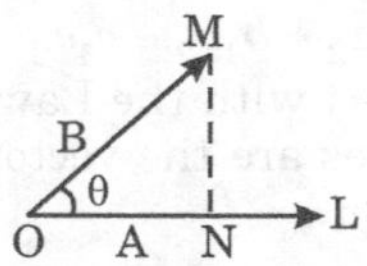

Scalar or dot product of two vectors A and B is defined as scalar ab cos θ, where θ is angle between A and B.

Mathematically, $A.B = ab \cos\theta$

Properties

(*i*) For two vectors to be perpendicular, their scalar product should be zero, i.e.

$$A.\,B = ab \cos 90° \qquad ...[\theta = 90°]$$

$$\Rightarrow \quad A.\,B = 0$$

(*ii*) Mutually perpendicular unit vectors i, j, k, are have relation

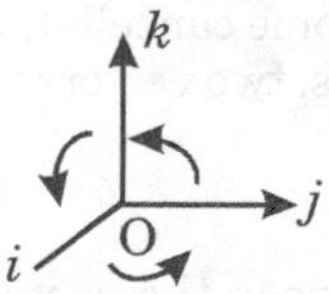

$$i.j = j.k = k.i = 0 \qquad ...(i = 0)$$

and $\quad i^2 = j^2 = k^2 = 1$

2. Vector or Cross Product

Vector or cross product of two vectors A and B is defined as a vector ab sin θ, where θ is angle between A and B.

Mathematically, $A \times B = ab \sin\theta\, N$

N is unit vector normal to the plane A and B.

Properties

(*i*) For two non zero vectors to be parallel, their vector product should be zero.

$$\theta = 0° \text{ or } 180°$$

$$\therefore \quad A \times B = \sin 0° \text{ or } \sin 180°$$

$$\Rightarrow \quad A \times B = 0$$

(*ii*) For the orthonormal vectors i, j and k, we have the relation

$$i \times i = j \times j = k \times k = 0,$$
$$i \times j = k, \; j \times i = -k$$
$$j \times k = i, \; k \times j = -i$$
$$k \times i = j, \; i \times k = -j$$

Differentiation of Vector

If a vector R varies continuously as a scalar variable t changes, then R is said to be function of t and is written as R = f(t)

In scalar calculus, we define derivative of vector function R = f(t) as

$$\underset{\delta t \to 0}{\text{lt}} = F\frac{(t + dt) - F(t)}{dt}$$

and it is written as $\frac{dR}{dt}$ or $\frac{df}{dt}$ or $F'(t)$

- If F(t) has constant magnitude, then $F.\frac{df}{dt} = 0$

  For $\quad F(t).\,F(t) = [F(t)]^2 = \text{constant}$

  $$F.\frac{dF}{dt} = 0, \text{ i.e. } \frac{dF}{dt} \text{ is perpendicular to F}$$

- If F(t) has constant or fixed direction, then

  $$F \times \frac{dF}{dt} = 0$$

VECTOR CALCULUS

Consider a vector operator ∇ (del) defined by

$$\nabla = i\frac{\partial}{\partial x} + j\frac{\partial}{\partial y} + k\frac{\partial}{\partial z}$$

Gradient

Del (∇) applied to vector function ∇ f is defined as the gradient of the scalar point function f and written as grad f. Thus gradient f defined as

$$\text{Grad } f = \nabla f = i\frac{\partial f}{\partial x} + j\frac{\partial f}{\partial y} + k\frac{\partial f}{\partial z}$$

Grad f is a vector normal to surface f = constant, and has a magnitude equal to the rate of change of f along this normal.

DIRECTIONAL DERIVATIVE

If δr denotes the length PP′ and N′ is a unit vector in the direction PP′, then limiting value of $\frac{\delta f}{\delta r}$ as $\delta r \to 0$ is called *directional derivative of f* at P along the direction PP′.

Since $\quad \delta r = \frac{\delta N}{\cos\alpha} = \frac{\delta N}{N}.\,N'$

$$\therefore \quad \frac{\partial f}{\partial r} = \underset{\delta r \to 0}{\text{Lt}} \left[N.\,N\frac{\partial f}{\partial n}\right] = \vec{N}\frac{\partial f}{\partial n}\vec{N} = N'\,\nabla f$$

Thus directional derivative of *f* in the direction of N′ is the resolved part of ∇*f* in the direction of N′

Since $\nabla f\, N' = |\nabla f| \cos\alpha \le |\nabla f|$.

therefore ∇*f* gives the maximum rate of change of *f*.

Vector Identities

1. grad (U + V) = grad U + grad V,

   i.e. $\nabla(U + V) = \nabla U + \nabla V$

2. $\text{div}\left(\vec{A} + \vec{B}\right) = \text{div } A + \text{div } B,$

   i.e. $\nabla.A + \nabla.B$

3. Curl $\left(\vec{A}+\vec{B}\right)$ = Curl A + Curl B,

i.e. $\nabla \times (A + B) = \nabla \times A + \nabla \times B$

4. $\nabla . \left(U\vec{A}\right) = (\nabla U) . A + U (\nabla . A)$

5. $\nabla \times \left(U\vec{A}\right) = (\nabla U) \times A + U (\nabla \times A)$

6. $\nabla . \left(\vec{A} \times \vec{B}\right) = B (\nabla \times A) - A . (\nabla \times B)$

7. $\nabla \times \left(\vec{A} \times \vec{B}\right) = (B.\nabla) A - B (\nabla.A) - (A.\nabla) B + A (\nabla.B)$

8. $\nabla (A.B) = (B.\nabla) A + (A.\nabla) B + B \times (\nabla\times A) + A \times (\nabla \times B)$

9. $\nabla . (\nabla U) = \nabla^2 U$

10. $\nabla \times (\nabla U) = 0$, i.e. curl of grad of U = 0

11. $\nabla (\nabla \times A) = 0$, i.e. div of the curl A = 0

12. $\nabla \times(\nabla \times A) = \nabla (\nabla . A) - \nabla^2 A$

Note : (*i*) $\nabla\left(\frac{f}{g}\right) = \frac{g\nabla f - f\nabla g}{g^2}$; $\nabla (f.g) = f\nabla g - g\nabla f$

(*ii*) div $\left(\vec{r} \times \vec{a}\right) = 0$, if $\vec{a}$ is a constant vector

(*iii*) curl $\left(\vec{r} \times \vec{a}\right) = -2\vec{a}$, if $\vec{a}$ is a constant vector

(*iv*) If div $f = 0$, then f is called **solenoidal.**

Example. Find the unit vector normal to the surface $xy^3z^2 = 4$ at the point (–1, –1, 2).

Solution

A vector normal to the surface is given by

$$\nabla(xy^3z^2) = i\frac{\partial}{\partial x}(xy^3z^2) + j\frac{\partial}{\partial y}(xy^3z^2) + \hat{k}\frac{\partial}{\partial z}(xy^3z^2)$$

$$= i(y^3z^2) + j(3xy^2z^2) + \hat{k}(2xy^3z)$$

$$= -4i - 12j + 4\hat{k} \text{ at } (-1, -1, 2)$$

Hence desired unit normal to the surface

$$= \frac{-4i - 12j + 4k}{\sqrt{(-4)^2 + (-12)^3 + 4^2}} = -\frac{1}{\sqrt{11}}(i + 3j - k)$$

Example. Find the directional derivative of $f(x, y, z) = xy^2 + yz^3$ at the point (2, –1, 1) in the direction of vector $i + 2j + 2k$.

Solution. $\nabla f = y^2 i + (2xy + z^3) j + (3yz^2)\hat{k}$

$$= i - 3j - 3k \quad \text{at} \quad (2, -1, 1)$$

Hence, directional derivative of f in the direction $i + 2j + 2k$

$$= (i - 3j - 3k) . \frac{(i + 2j + 2k)}{\sqrt{1^2 + 2^2 + 2^2}} = \frac{(1 - 6 - 6)}{3} = -\frac{11}{3}$$

Example. Prove that curl $r = 0$

Solution

$$\text{Curl } \vec{r} = \nabla \times \vec{r}$$

$$= \left(i\frac{\partial}{\partial x} + j\frac{\partial}{\partial y} + k\frac{\partial}{\partial z}\right) \times \left(xi + yj + z\hat{k}\right)$$

$$= i \times i + j \times j + k \times k = 0 + 0 + 0 = 0$$

DIVERGENCE

The divergence of a continuously differentiable vector point function F is denoted by divergence F and is define by the equation

$$\text{div F} = \nabla F = i.\frac{\partial F}{\partial x} + j\frac{\partial F}{\partial y} + k.\frac{\partial F}{\partial z}$$

If $F = fi + \phi j + \psi k$, then

$$\text{div F} = \nabla.F = \left(i.\frac{\partial}{\partial x} + j\frac{\partial}{\partial y} + k.\frac{\partial}{\partial z}\right)(fi + \phi j + \psi k)$$

$$\text{div F} = \left(i.\frac{\partial f}{\partial x} + j\frac{\partial \phi}{\partial y} + k.\frac{\partial \psi}{\partial z}\right)$$

CURL

The curl of a continuously differentiable vector point function F is defined by the equation

$$\text{Curl F} = \nabla \times F = i \times \frac{\partial F}{\partial x} + j \times \frac{\partial F}{\partial y} + k \times \frac{\partial F}{\partial x}$$

If $F = f i + \phi j + \psi k$, then $F = \nabla \times F$

$$= \left(i.\frac{\partial}{\partial x} + j\frac{\partial}{\partial y} + k.\frac{\partial}{\partial z}\right) \times (fi + \phi j + \psi k)$$

$$= \begin{vmatrix} i & j & k \\ \frac{\partial}{\partial x} & \frac{\partial}{\partial y} & \frac{\partial}{\partial z} \\ f & \phi & \psi \end{vmatrix}$$

$$= i\left(\frac{\partial \psi}{\partial y} - \frac{\partial \phi}{\partial z}\right) + j\left(\frac{\partial \psi}{\partial y} - \frac{\partial f}{\partial z}\right) + k\left(\frac{\partial \phi}{\partial x} - \frac{\partial f}{\partial y}\right)$$

Del (∇) Applied Twice to Point Function

- Div grad f = $\nabla. \nabla f = \nabla^2 f = \frac{\partial^2 f}{\partial x^2} + \frac{\partial^2 f}{\partial y^2} + \frac{\partial^2 f}{\partial z^2}$
- Curl grad f = $\nabla \times \nabla f = 0$
- Div curl F = $\nabla. \nabla \times F = 0$
- Curl Curl F = grad div F – ∇^2 f
  i.e. $\nabla \times (\nabla \times F) = \nabla(\nabla. f) - \nabla^2 f$
- Grad div F = curl Curl F + ∇^2 F
  i.e. $\nabla(\nabla.F) = \nabla \times (\nabla \times F) + \nabla^2 F$

Del(∇) Applied to Product of Point Functions

- $\nabla (fg) = f \nabla g + g \nabla f$
- $\nabla.(fG) = \nabla f.G + f \nabla.G$
- $\nabla \times (fG) = \nabla f \times G + f\nabla \times G$
- $\nabla(F.G) = (F.\nabla)G + (G.\nabla)F + F \times (\nabla \times G) + G \times (\nabla \times F)$
- $\nabla.(F \times G) = G.(\nabla \times F) - F.(\nabla \times F)$
- $\nabla \times (F \times G) = F(\nabla.G) - G(\nabla.F) + G. \nabla F - F.\nabla G$

STOKE'S THEOREM

The close line integral of any vector $\overline{A}$ integral over any close curve C is always equal to the surface integral of the curl of vector A integrated over the surface area S which is enclosed by the closed curve C

$$\oint_C (\overline{A}).\, \overline{dl} = \iint_S (\nabla \times \overline{A})\, ds$$

GREEN THEOREM

Let ϕ (x, y), ψ (x, y), ϕy and ψx be continuous in a region E of the xy plane bound by a closed curve C, then

$$\int_C \phi dx + \psi dy = \iint_E \left(\frac{\partial \psi}{\partial x} - \frac{\partial \phi}{\partial y} \right) dx\, dy$$

This equation can be extended to regions which may be divided into finite number of sub-regions such that the boundary of each is cut atmost in two points by any line parallel to either axis apply in equation (i) to each of the sub-region and adding result, the surface integral combine into a integral over the whole region, the line integral over the common boundary cancel, whereas remaining line integrals combine into the line integral over the external curve C. This theorem convert line integrals around a closed curve into a double integrals and is a special case of Stroke's theorem.

GAUSS'S DIVERGENCE THEOREM

(Relation Between Surface and Volume Integrals)

"Volume integral of a vector field F taken over a any volume V is equal to the surface integral of F taken over the closed surface surrounding the volume V."

Mathematically, for any vector-field $\overline{F}$

$$\int_S F.N\, d_s = \int_E \text{div } F d_v$$

where N is the unit external vector

If $\quad F(R) = f(x, y, z)\, i + \phi(x, y, z)\, j + \psi(x, y, z)\, k$

then, $\quad \iint (f\, dy\, dz + \phi\, dzdx + \psi\, dx\, dy)$

$$= \iiint_E \left(\frac{\partial f}{\partial x} + \frac{\partial \phi}{\partial y} + \frac{\partial \psi}{\partial z} \right) dx\, dy\, dx \quad ...(i)$$

Consider a general region E assume that it can be split into finite number of sub region each of which is met by line parallel to any axis in only two points. Applying equation (i) to each of these sub-regions and adding results, the volume integral will be combine to give the volume integrals over the wholes region E. Also surface integrals over the common boundaries of two sub regions cancel because of each occur twice and having corresponding normal in opposite direction of whereas remaining surface integrals combine to give the surface integral over the entire – surface.

Consider a region E bounded by two closed surface S_1, S_2 (S, being within S_2). The outward normal at points S_1 is directed inwards (i.e. away from S_2) and introducing an adding surface cutting S_1, S_2 so that all parts of E are bounded by single closed surface. Thus the theorem also holds for regions enclosed by several surfaces.

EXERCISE

MCQ TYPE QUESTIONS

1. The unit normal to the plane $2x + y + 2z = 6$ can be expressed in the vector form as

(a) $i3 + j2 + k2$ (b) $i\frac{2}{3} + j\frac{1}{3} + k\frac{2}{3}$

(c) $i\frac{1}{3} + j\frac{1}{2} + k\frac{1}{2}$ (d) $i\frac{2}{6} + j\frac{1}{6} + k\frac{2}{6}$

2. If $\vec{V}$ is a differentiable vector function and f is a sufficient differentiable scalar function, then curl $\left(f\vec{V}\right)$ is equal to

(a) $(\text{grad } f) \times \left(\vec{V}\right) + (f \text{ curl } \vec{V})$

(b) $\vec{O}$

(c) $f \text{ curl } \left(\vec{V}\right)$

(d) $(\text{grad } f) \times \left(\vec{V}\right)$

3. Among the following, the pair of vectors orthogonal to each-other is

(a) [3, 4, 7], [3, 4, 7] (b) [1, 0, 0], [1, 2, 0]

(c) [1, 0, 2], [0, 5, 0,] (d) [1, 1, 1], [–1, –1, –1]

4. The temperature field in a body varies according to the equation $T(x, y) = x^3 + 4xy$. The direction of fastest variation in temperature at the point (1, 0) is given by

(a) $3\hat{i} + 8\hat{j}$ (b) $\hat{i}$

(c) $0.6\hat{i} + 0.8\hat{j}$ (d) $0.5\hat{i} + 0.866\hat{j}$

5. If velocity vector in a two dimensional flow field is given by $\vec{v} = 2xy\,\vec{i} + \left(2y^2 - x^2\right)\vec{j}$, then vorticity vector, curl $\vec{v}$ will be

(a) $2y^2\,\vec{j}$ (b) $6y\,\vec{k}$

(c) zero (d) $-4x\,\vec{k}$

6. Divergence of the vector $\vec{r} = x\,\vec{i} + y\,\vec{i} + z\,\vec{i}$ is

(a) $\vec{i} + \vec{j} + \vec{k}$ (b) 3

(c) 0 (d) 1

7. The derivative of f (x, y) at point (1, 2) in the direction of vector i + j is $2\sqrt{2}$ and in the direction of the vector –2j is –3. Then derivative of f (x, y) in the direction –i –2j is

(a) $2\sqrt{2} + 3/2$ (b) $-7/\sqrt{5}$

(c) $-2\sqrt{2} - 3/2$ (d) $1/\sqrt{5}$

NUMERICAL TYPE QUESTIONS

Directions (Q. 1-6) : A pair of points is given. Find the vector beginning at the first point and ending at the second.

1. (1, –1), (3, 5)
2. (3, 2), (–3, –7)
3. (–1, –6), (2, –8)
4. (1, 1), (7, –8)
5. (1, –1, 2), (0, –2, –6)
6. (–3, 0, 1), (1, 1, 2)

Directions (Q. 7-11) : Let A = (1, –1, 2), B = (3, 0, 4), C = (4, 2), D = (–1, 1), i = (1, 0), and j = (0, 1). Calculate

7. A + B
8. A – B
9. 2C – 3D
10. $4i + 2j$
11. $|i|$, $|j|$, $|i + j|$
12. What value should k have to make $|(k, 3k, 4k)| = 1$?

Directions (Q. 13-15) : Find the angle between A and B.

13. A = (5, 0), B = (2, 2)
14. A = (2, –2), B = (–1, 1)
15. A = (3, 0, 4), B = (–6, 0, –8)
16. Given $f(x, y) = x^2 + y^2$. Then $\nabla^2 f$ is ________
17. The directional derivative of
$$f(x, y, z) = x^2 + y^2 + z^2$$
at the point (1, 1, 1) in the direction i – k is _____
18. The magnitude of the gradient of the function $f = xyz^3$ at (1, 0, 2) is ________
19. For the function, $\phi = ax^2y - y^3$ to represent the velocity potential of an ideal fluid, $\nabla^2 \phi$ should be equal to zero. In that case, the value of 'a' has to be ________
20. Maximum value of the directional derivative of the function $\phi = 2x^2 + 3y^2 + 5z^2$ at a point (1, 1, –1) is ________

ANSWERS

MCQ Type Questions

1. (b) 2. (a) 3. (c) 4. (c) 5. (d) 6. (b) 7. (b)

Numerical Type Questions

1. $2\hat{i} + 6\hat{j}$ 2. $-6\hat{i} - 9\hat{j}$ 3. $3\hat{i} - 2\hat{j}$ 4. $6\hat{i} - 9\hat{j}$ 5. $-\hat{i} - \hat{j} - 8\hat{k}$ 6. $4\hat{i} + 1\hat{j} + 1\hat{k}$ 7. (4, –1, 6)

8. (–2, –1, –2) 9. (11, 1) 10. (4, 2) 11. $(1, 1, \sqrt{2})$ 12. $(1/\sqrt{26})$ 13. (45°) 14. (0°)

15. (0°) 16. (4) 17. (0) 18. (8) 19. (3) 20. $\left(\sqrt{152}\right)$

EXPLANATIONS

MCQ TYPE QUESTIONS

2. $\text{curl}(f\vec{V}) = i \times \frac{\partial}{\partial x}\left(f\vec{V}\right) + j \times \frac{\partial}{\partial y}\left(f\vec{V}\right) + k \times \frac{\partial}{\partial z}\left(f\vec{V}\right)$

$$= i \times \left[f\frac{\partial \vec{V}}{\partial x} + \frac{\partial f}{\partial x}\vec{V} \right] + \text{two similar terms}$$

$$= i \times \left[f\frac{\partial \vec{V}}{\partial x} + j \times \partial \vec{V} + k \times \frac{\partial \vec{V}}{\partial z} \right] + \left[i \times \frac{\partial f}{\partial x} + j\frac{\partial f}{\partial y} + k\frac{\partial f}{\partial x} \right] \times \vec{V}$$

$$= f \,\text{curl}\, \vec{V} + (\text{grad}\, f) \times \vec{V}$$

3. (a_1, b_1, c_1) and (a_2, b_2, c_2) are orthogonal, only if

$$a_1a_2 + b_1 b_2 + c_1c_2 = 0$$

Only C fits, i.e. $[1, 0, 2] \times [0, 5, 0] = 0 + 0 + 0 = 0$

4. Given $T = x^3 + 4xy \Rightarrow \frac{\partial T}{\partial x} = 3x^2 + 4y$

$\therefore \left(\frac{\partial T}{\partial y}\right)_{(1,0)} = 3, \Rightarrow \left(\frac{\partial T}{\partial y}\right)_{(1,0)} = 4$

Hence direction of fastest variation in temperature at (1, 0) is given by $3\hat{i} + 4\hat{j}$ or $0.6\hat{i} + 0.8\hat{j}$

5. Given: $\vec{v} = 2xy\,\vec{i} + (2y^2 - x^2)\vec{j}$

$$\therefore \text{Curl}\ \vec{v} = \begin{vmatrix} i & j & k \\ \frac{\partial}{\partial x} & \frac{\partial}{\partial k} & \frac{\partial}{\partial z} \\ 2xy & 2y^2 - x^2 & 0 \end{vmatrix}$$

$$= i(0) - j(0) + k \left[\frac{\partial}{\partial x}(2y^2 - x^2) - \frac{\partial}{\partial y}(2xy) \right]$$

$$= (-2x - 2x)\,k = -4xk.$$

6. $\text{div.}\ \vec{r} = \vec{\Delta}.\vec{r}$

$$= \left(\frac{\partial}{\partial x}\hat{i} + \frac{\partial}{\partial y} j + \frac{\partial}{\partial z}\hat{k} \right)\left(x\hat{i} + y\hat{j} + z\hat{k}\right)$$

$$= 1 + 1 + 1 = 3$$

7. $\frac{1}{\sqrt{2}}\frac{\partial f}{\partial x} + \frac{1}{\sqrt{2}}\frac{\partial f}{\partial y} = 2\sqrt{2},$

$\Rightarrow \frac{\partial f}{\partial x} + \frac{\partial f}{\partial y} = 4$ and $\frac{\partial f}{\partial y} = -3 \Rightarrow \frac{\partial f}{\partial y} = 3.$

Hence $\frac{\partial f}{\partial x} = 1$ and $\frac{\partial f}{\partial y} = 3$

$\therefore$ Derivative in the direction $-i - 2j$

$$= \left(\frac{-1}{\sqrt{5}}\right)(1) + \left(\frac{-2}{\sqrt{5}}\right)3 = \frac{-7}{\sqrt{5}}$$

NUMERICAL TYPE QUESTIONS

16. $f(x, y) = x^2 + y^2$

$$\nabla^2 f = \vec{\nabla}.\vec{\nabla} f$$

$$= \vec{z}\left(\vec{i}\frac{\partial}{\partial x} + \vec{j}\frac{\partial}{\partial y} \right)\left(\vec{i}\frac{\partial}{\partial x} + \vec{j}\frac{\partial}{\partial y} \right)\left[x^2 + y^2\right]$$

$$= \frac{\partial^2}{\partial x^2}(x^2) + \frac{\partial^2}{\partial y^2}(y^2) = 4$$

17. $f(x, y, z) = x^2 + y^2 + z^2$

$$\Delta f(x, y, z) = 2x\,\vec{i} + 2y\,\vec{j} + 2z\,\vec{k} = (\vec{A}) \text{ say}$$

Now, $\vec{B} = \vec{i} - \vec{k}\,;\ \hat{B} = \frac{1}{\sqrt{2}}(\vec{i} - \vec{k})$

$\therefore \vec{A}.\hat{B} = \frac{2}{\sqrt{2}}(x - z)$

$\Rightarrow \vec{A}.\hat{B}\Big|_{(1,1,1)} = \frac{2}{\sqrt{2}}(1 - 1) = 0$

18. $\nabla . f = i\frac{\partial}{\partial x} + j\frac{\partial}{\partial y} + k\frac{\partial}{\partial z}$

$$= iyz^3 + jxz^3 + k \times 3xyz^2$$

$$= i \times 0 + j \times 1 \times 2^3 + + k \times 0 = j \times 8$$

$\therefore |\nabla . f| = 8$

19. Given: $\phi = ax^2y - y^3$

$\therefore \frac{\partial \phi}{\partial y} = ax^2 - 3y^2;$ and $\frac{\partial^2 \phi}{\partial y^2} = -6y$

$\frac{\partial \phi}{\partial x} = 2\,axy;$ and $\frac{\partial^2 \phi}{\partial x^2} = 2\,ay$

Now $\frac{\partial^2 \phi}{\partial x^2} + \frac{\partial^2 \phi}{\partial y^2} = 0$

$\therefore 2ay - 6y = 0$

$\Rightarrow a = 3$

20. $\phi = 2x^2 + 3y^2 + 5z^2$

At point (1, 1, –1), $\nabla\phi = i.\,4x + j.\,6y + k.\,10\,z$

Given : $x = 1, y = 1, z = -1$

$\therefore |\nabla\phi| = \sqrt{(4)^2 + (6)^2 + (-10)^2} = \sqrt{152}$

$\cos^2 \omega t = \cos 90°$

or $\theta = \omega t = 90°$

So, the value of θ is 90°

SOLVED PAPER – 2016

GENERAL APTITUDE

1. Based on the given statements, select the appropriate option with respect to grammar and usage.

   Statements :

   I. The height of Mr. X is 6 feet.

   II. The height of Mr. Y is 5 feet.

   (*a*) Mr. X is longer than Mr. Y.

   (*b*) Mr. X is more elongated than Mr. Y

   (*c*) Mr. X is taller than Mr. Y

   (*d*) Mr. X is lengthier than Mr. Y

2. The students _____ the teacher on teachers' day for twenty years of dedicated teaching.

   (*a*) facilitated (*b*) felicitated

   (*c*) fantasized (*d*) facillitated

3. After India's cricket world cup victory in 1985, Shrotria who was playing both tennis and cricket till then, decided to concentrate only on cricket. And <u>the rest is history</u>.

   What does the underlined phrase mean in this context?

   (*a*) history will rest in peace

   (*b*) rest is recorded in history books

   (*c*) rest is well known

   (*d*) rest is archaic

4. Given (9 inches)$^{1/2}$ = (0.25 yards)$^{1/2}$, which one of the following statements is TRUE?

   (*a*) 3 inches = 0.5 yards

   (*b*) 9 inches = 1.5 yards

   (*c*) 9 inches = 0.25 yards

   (*d*) 81 inches = 0.0625 yards

5. S, M, E and F are working in shifts in a team to finish a project. M works with twice the efficiency of others but for half as many days as E worked. S and M have 6 hour shifts in a day whereas E and F have 12 hours shifts. What is the ratio of contribution of M to contribution of E in the project?

   (*a*) 1 : 1

   (*b*) 1 : 2

   (*c*) 1 : 4

   (*d*) 2 : 1

6. The Venn diagram shows the preference of the student population for leisure activities.

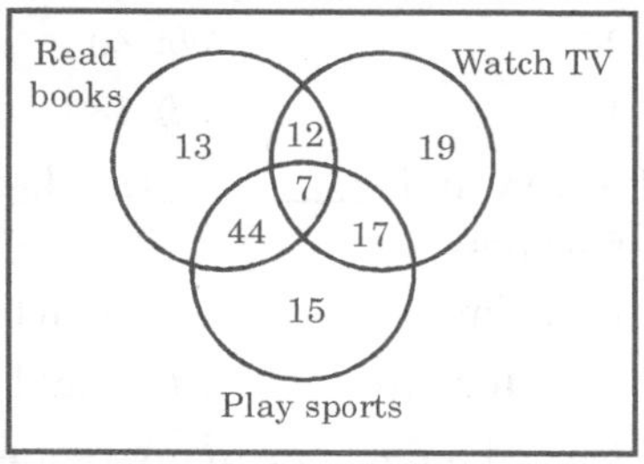

   From the data given, the number of students who like to read books or play sports is ____.

   (*a*) 44 (*b*) 51

   (*c*) 79 (*d*) 108

7. Social science disciplines were in existence in an amorphous form until the colonial period when they were institutionalized. In varying degrees, they were intended to further the colonial interest. In the time of globalization and the economic rise of postcolonial countries like India, conventional ways of knowledge production have become obsolete.

   Which of the following can be logically inferred from the above statements?

   I. Social science disciplines have become obsolete.

   II. Social science disciplines had a pre-colonial origin.

   III. Social science disciplines always promote colonialism.

   IV. Social science must maintain disciplinary boundaries.

   (*a*) II only (*b*) I and III only

   (*c*) II and IV only (*d*) III and IV only

8. Two and a quarter hours back, when seen in a mirror, the reflection of a wall clock without number markings seemed to show 1 : 30. What is the actual current time shown by the clock?

   (*a*) 8:15 (*b*) 11:15

   (*c*) 12:15 (*d*) 12:45

9. M and N start from the same location. M travels 10 km East and then 10 km North-East. N travels 5 km South and then 4 km South-East. What is the shortest distance (in km) between M and N at the end of their travel?

   (*a*) 18.60 (*b*) 22.50

   (*c*) 20.61 (*d*) 25.00

10. A wire of length 340 mm is to be cut into two parts. One of the parts is to be made into a square and the other into a rectangle where sides are in the ratio of 1:2. What is the length of the side of the square (in mm) such that the combined area of the square and the rectangle is a MINIMUM?

(*a*) 30 (*b*) 40

(*c*) 120 (*d*) 180

11. If I were you, I ________ that laptop. It's much too expensive.

(*a*) won't buy (*b*) shan't buy

(*c*) wouldn't buy (*d*) would buy

12. He <u>turned a deaf ear</u> to my request. What does the underlined phrasal verb mean?

(*a*) ignored (*b*) appreciated

(*c*) twisted (*d*) returned

13. Choose the most appropriate set of words from the options given below to complete the following sentence.

_______________ is a will, ________ is a way.

(*a*) Wear, there, their

(*b*) Were, their, there

(*c*) Where, there, there

(*d*) Where, their, their

14. (x % of y) + (y % of x) is equivalent to ______.

(*a*) 2 % of xy (*b*) 2 % of (xy/100)

(*c*) xy % of 100 (*d*) 100 % of xy

15. The sum of the digits of a two digit number is 12. If the new number formed by reversing the digits is greater than the original number by 54, find the original number.

(*a*) 39 (*b*) 57

(*c*) 66 (*d*) 93

16. Two finance companies, P and Q, declared fixed annual rates of interest on the amounts invested with them. The rates of interest offered by these companies may differ from year to year. Year-wise annual rates of interest offered by these companies are shown by the line graph provided below.

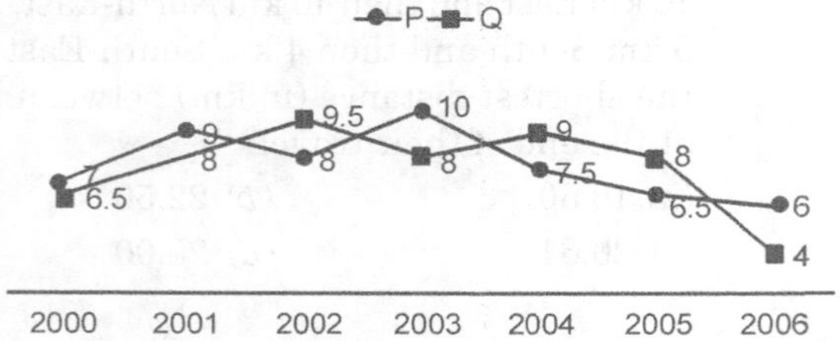

If the amounts invested in the companies, P and Q, in 2006 are in the ratio 8 : 9, then the amounts received after one year as interests from companies P and Q would be in the ratio:

(*a*) 2 : 3 (*b*) 3 : 4

(*c*) 6 : 7 (*d*) 4 : 3

17. Today, we consider Ashoka as a great ruler because of the copious evidence he left behind in the form of stone carved edicts. Historians tend to correlate greatness of a king at his time with the availability of evidence today. Which of the following can be logically inferred from the above sentences?

(*a*) Emperors who do not leave significant sculpted evidence are completely forgotten.

(*b*) Ashoka produced stone carved edicts to ensure that later historians will respect him.

(*c*) Statues of kings are a reminder of their greatness.

(*d*) A king's greatness, as we know him today, is interpreted by historians.

18. **Fact 1 :** Humans are mammals.

Fact 2 : Some humans are engineers.

Fact 3 : Engineers build houses.

If the above statements are facts, which of the following can be logically inferred ?

I. All mammals build houses.

II. Engineers are mammals.

III. Some humans are not engineers.

(*a*) II only (*b*) III only

(*c*) I, II and III (*d*) I only

19. A square pyramid has a base perimeter x, and the slant height is half of the perimeter. What is the lateral surface area of the pyramid?

(*a*) x^2 (*b*) 0.75 x^2

(*c*) 0.50 x^2 (*d*) 0.25 x^2

20. Ananth takes 6 hours and Bharath takes 4 hours to read a book. Both started reading copies of the book at the same time. After how many hours is the number of pages to be read by Ananth, twice that to be read by Bharath? Assume Ananth and Bharath read all the pages with constant pace.

(*a*) 1

(*b*) 2

(*c*) 3

(*d*) 4

21. Which of the following is **CORRECT** with respect to grammar and usage?

Mount Everest is ____________.

(*a*) the highest peak in the world

(*b*) highest peak in the world

(*c*) one of highest peak in the world

(*d*) one of the highest peak in the world

22. The policeman asked the victim of a theft, "What did you______ ?"

(*a*) loose (*b*) lose

(*c*) loss (*d*) louse

23. Despite the new medicine's _____________ in treating diabetes, it is not_____________widely.

(*a*) effectiveness --- prescribed

(*b*) availability --- used

(*c*) prescription --- available

(*d*) acceptance --- proscribed

24. In a huge pile of apples and oranges, both ripe and unripe mixed together, 15% are unripe fruits. Of the unripe fruits, 45% are apples. Of the ripe ones, 66% are oranges. If the pile contains a total of 5692000 fruits, how many of them are apples?

(*a*) 2029198 (*b*) 2467482

(*c*) 2789080 (*d*) 3577422

25. Michael lives 10 km away from where I live. Ahmed lives 5 km away and Susan lives 7 km away from where I live. Arun is farther away than Ahmed but closer than Susan from where I live. From the information provided here, what is one possible distance (in km) at which I live from Arun's place?

(*a*) 3.00 (*b*) 4.99

(*c*) 6.02 (*d*) 7.01

26. A person moving through a tuberculosis prone zone has a 50% probability of becoming infected. However, only 30% of infected people develop the disease. What percentage of people moving through a tuberculosis prone zone remains infected but does not show symptoms of disease?

(*a*) 15 (*b*) 33

(*c*) 35 (*d*) 37

27. In a world filled with uncertainty, he was glad to have many good friends. He had always assisted them in times of need and was confident that they would reciprocate. However, the events of the last week proved him wrong.

Which of the following inference(s) is/are logically valid and can be inferred from the above passage?

(*i*) His friends were always asking him to help them.

(*ii*) He felt that when in need of help, his friends would let him down.

(*iii*) He was sure that his friends would help him when in need.

(*iv*) His friends did not help him last week.

(*a*) (*i*) and (*ii*) (*b*) (*iii*) and (*iv*)

(*c*) (*iii*) only (*d*) (*iv*) only

28. Leela is older than her cousin Pavithra. Pavithra's brother Shiva is older than Leela. When Pavithra and Shiva are visiting Leela, all three like to play chess. Pavithra wins more often than Leela does.

Which one of the following statements must be **TRUE** based on the above?

(*a*) When Shiva plays chess with Leela and Pavithra, he often loses.

(*b*) Leela is the oldest of the three.

(*c*) Shiva is a better chess player than Pavithra.

(*d*) Pavithra is the youngest of the three.

29. If $q^{-a} = \frac{1}{r}$ and $r^{-b} = \frac{1}{s}$ and $s^{-c} = \frac{1}{q}$, the value of abc is ____ .

(*a*) $(rqs)^{-1}$ (*b*) 0

(*c*) 1 (*d*) $r + q + s$

30. P, Q, R and S are working on a project. Q can finish the task in 25 days, working alone for 12 hours a day. R can finish the task in 50 days, working alone for 12 hours per day. Q worked 12 hours a day but took sick leave in the beginning for two days. R worked 18 hours a day on all days. What is the ratio of work done by Q and R after 7 days from the start of the project?

(*a*) 10:11 (*b*) 11 :10

(*c*) 20:21 (*d*) 21:20

31. The chairman requested the aggrieved shareholders to _________ him.

(*a*) bare with (*b*) bore with

(*c*) bear with (*d*) bare

32. Identify the correct spelling out of the given options :

(*a*) Managable (*b*) Manageable

(*c*) Mangaeble (*d*) Managible

33. Pick the odd one out in the following :
13, 23, 33, 43, 53

(*a*) 23 (*b*) 33
(*c*) 43 (*d*) 53

34. R2D2 is a robot. R2D2 can repair aeroplanes. No other robot can repair aeroplanes. Which of the following can be logically inferred from the above statements?

(*a*) R2D2 is a robot which can only repair aeroplanes.
(*b*) R2D2 is the only robot which can repair aeroplanes.
(*c*) R2D2 is a robot which can repair only aeroplanes.
(*d*) Only R2D2 is a robot.

35. If $|9y-6| = 3$, then $y^2 - \frac{4y}{3}$ is ________.

(*a*) 0 (*b*) $+\frac{1}{3}$
(*c*) $-\frac{1}{3}$ (*d*) Undefined

36. The following graph represents the installed capacity for cement production (in tonnes) and the actual production (in tonnes) of nine cement plants of a cement company. Capacity utilization of a plant is defined as ratio of actual production of cement to installed capacity. A plant with installed capacity of at least 200 tonnes is called a large plant and a plant with lesser capacity is called a small plant. The difference between total production of large plants and small plants, in tonnes is ________.

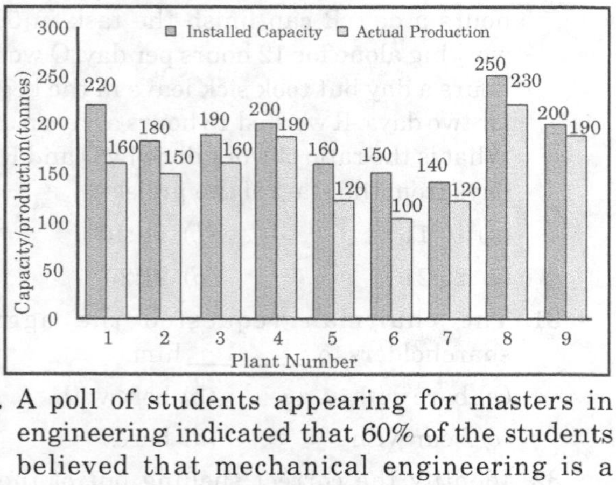

37. A poll of students appearing for masters in engineering indicated that 60% of the students believed that mechanical engineering is a profession unsuitable for women. A research study on women with masters or higher degrees in mechanical engineering found that 99% of such women were successful in their professions.

Which of the following can be logically inferred from the above paragraph?

(*a*) Many students have misconceptions regarding various engineering disciplines.
(*b*) Men with advanced degrees in mechanical engineering believe women are well suited to be mechanical engineers.
(*c*) Mechanical engineering is a profession well suited for women with masters or higher degrees in mechanical engineering.
(*d*) The number of women pursuing higher degrees in mechanical engineering is small.

38. Sourya committee had proposed the establishment of Sourya Institutes of Technology (SITs) in line with Indian Institutes of Technology (IITs) to cater to the technological and industrial needs of a developing country.

Which of the following can be logically inferred from the above sentence?

Based on the proposal,

(*i*) In the initial years, SIT students will get degrees from IIT.
(*ii*) SITs will have a distinct national objective.
(*iii*) SIT like institutions can only be established in consulation with IIT.
(*iv*) SITs will serve technological needs of a developing country.

(*a*) (*iii*) and (*iv*) only (*b*) (*i*) and (*iv*) only
(*c*) (*ii*) and (*iv*) only (*d*) (*ii*) and (*iii*) only

39. Shaquille O' Neal is a 60% career free throw shooter, meaning that he successfully makes 60 free throws out of 100 attempts on average. What is the probability that he will successfully make exactly 6 free throws in 10 attempts?

(*a*) 0.2508 (*b*) 0.2816
(*c*) 0.2934 (*d*) 0.6000

40. The numeral in the units position of $211^{870} + 146^{127} \times 3^{424}$ is ________.

41. Out of the following four sentences, select the most suitable sentence with respect to grammar and usage.

(*a*) I will not leave the place until the minister does not meet me.
(*b*) I will not leave the place until the minister doesn't meet me.
(*c*) I will not leave the place until the minister meet me.
(*d*) I will not leave the place until the minister meets me.

42. A rewording of something written or spoken is a__________.

(*a*) paraphrase (*b*) paradox

(*c*) paradigm (*d*) paraffin

43. Archimedes said, "Give me a lever long enough and a fulcrum on which to place it, and I will move the world."

The sentence above is an example of a ________ statement.

(*a*) figurative (*b*) collateral

(*c*) literal (*d*) figurine

44. If 'relftaga' means carefree, 'otaga' means careful and 'fertaga' means careless, which of the following could mean 'aftercare'?

(*a*) zentaga (*b*) tagafer

(*c*) tagazen (*d*) relffer

45. A cube is built using 64 cubic blocks of side one unit. After it is built, one cubic block is removed from every corner of the cube. The resulting surface area of the body (in square units) after the removal is _______.

(*a*) 56 (*b*) 64

(*c*) 72 (*d*) 96

46. A shaving set company sells 4 different types of razors, Elegance, Smooth, Soft and Executive. Elegance sells at Rs. 48, Smooth at Rs. 63, Soft at Rs. 78 and Executive at Rs. 173 per piece. The table below shows the numbers of each razor sold in each quarter of a year.

Quarter \product	Elegance	Smooth	Soft	Executive
Q1	27300	20009	17602	9999
Q2	25222	19392	18445	8942
Q3	28976	22429	19544	10234
Q4	21012	18229	16595	10109

Which product contributes the greatest fraction to the revenue of the company in that year?

(*a*) Elegance (*b*) Executive

(*c*) Smooth (*d*) Soft

47. Indian currency notes show the denomination indicated in at least seventeen languages. If this is not an indication of the nation's diversity, nothing else is.

Which of the following can be logically inferred from the above sentences?

(*a*) India is a country of exactly seventeen languages.

(*b*) Linguistic pluralism is the only indicator of a nation's diversity.

(*c*) Indian currency notes have sufficient space for all the Indian languages.

(*d*) Linguistic pluralism is strong evidence of India's diversity.

48. Consider the following statements relating to the level of poker play of four players P, Q, R and S.

I. P always beats Q

II. R always beats S

III. S loses to P only sometimes

IV. R always loses to Q

Which of the following can be logically inferred from the above statements?

(*i*) P is likely to beat all the three other players

(*ii*) S is the absolute worst player in the set

(*a*) (*i*) only (*b*) (*ii*) only

(*c*) (*i*) and (*ii*) (*d*) neither (*i*) nor (*ii*)

49. If $f(x) = 2x^7 + 3x - 5$, which of the following is a factor of $f(x)$?

(*a*) $(x^3 + 8)$ (*b*) $(x - 1)$

(*c*) $(2x - 5)$ (*d*) $(x + 1)$

50. In a process, the number of cycles to failure decreases exponentially with an increase in load. At a load of 80 units, it takes 100 cycles for failure. When the load is halved, it takes 10000 cycles for failure. The load for which the failure will happen in 5000 cycles is _____.

(*a*) 40.00 (*b*) 46.02

(*c*) 60.01 (*d*) 92.02

ENGINEERING MATHEMATICS

1. The value of x for which the matrix

$$A = \begin{bmatrix} 3 & 2 & 4 \\ 9 & 7 & 13 \\ -6 & -4 & -9+x \end{bmatrix}$$ has zero as an eigen value

is ______.

2. Consider the complex valued function $f(z) = 2z^3 + b|z|^3$ where z is a complex variable. The value of b for which the function $f(z)$ is analytic is _____

3. As x varies from -1 to $+3$, which one of the following describes the behaviour of the function $f(x) = x^3 - 3x^2 + 1$?

(*a*) $f(x)$ increases monotonically.

(*b*) $f(x)$ increases, then decreases and increases again.

(*c*) $f(x)$ decreases, then increases and decreases again.

(*d*) $f(x)$ increases and then decreases

Then the given function $f(x)$ increases, then decreases and increase again.

4. How many distinct values of x satisfy the equation $\sin(x) = x/2$, where x is in radians?

(*a*) 1 (*b*) 2

(*c*) 3 (*d*) 4 or more

5. Consider the time-varying vector

$I = \hat{x}15\cos(\omega t) + \hat{y}5\sin(\omega t)$ in Cartesian coordinates, where $\omega > 0$ is a constant.

When the vector magnitude $|I|$ is at its minimum value, the angle θ that I makes with the x axis (in degrees, such that $0 \le \theta \le 180°$) is _____.

6. The ordinary differential equation

$$\frac{dy}{dx} = -3y + 2, \text{ with } y(0) = 1$$

is to be solved using the forward Euler method. The largest time step that can be used to solve the equation without making the numerical solution unstable is _______.

7. Suppose C is the closed curve defined as the circle $x^2 + y^2 = 1$ with C oriented anti clockwise. The value of $\oint (xy^2 dx + x^2 y dy)$ over the curve C equals _______.

8. Two random variables X and Y are distributed according to

$$f_{X,Y}(x, y) = \begin{cases} (x + y) & 0 \le x \le 1,\ 0 \le y \le 1 \\ 0 & \text{otherwise} \end{cases}$$

The probability $P(X + Y \le 1)$ is _______

9. The matrix $A = \begin{bmatrix} a & 0 & 3 & 7 \\ 2 & 5 & 1 & 3 \\ 0 & 0 & 2 & 4 \\ 0 & 0 & 0 & b \end{bmatrix}$ has det(A)=100 and

trace (A) =14. The value of $|a - b|$ is _______.

10. The spot speeds (expressed in km/hr) observed at a road section are 66, 62, 45, 79, 32, 51, 56, 60, 53, and 49. The median speed (expressed in km/hr) is _______

(*Note : answer with one decimal accuracy*)

11. The optimum value of the function

$$f(x) = x^2 - 4x + 2 \text{ is}$$

(*a*) 2 (maximum) (*b*) 2 (minimum)

(*c*) – 2 (maximum) (*d*) – 2 (minimum)

12. The Fourier series of the function,

$$f(x) = 0, \quad -\pi < x \le 0$$
$$= \pi - x, \quad 0 < x < \pi$$

in the interval $[-\pi, \pi]$ is

$$f(x) = \frac{\pi}{4} + \frac{2}{\pi}\left[\frac{\cos x}{1^2} + \frac{\cos 3x}{3^2} + \ldots\right] + \left[\frac{\sin x}{1} + \frac{\sin 2x}{2} + \frac{\sin 3x}{3} + \ldots\right].$$

The convergence of the above Fourier series at $x = 0$ gives

(*a*) $\sum_{n=1}^{\infty} \frac{1}{n^2} = \frac{\pi^2}{6}$ (*b*) $\sum_{n=1}^{\infty} \frac{(-1)^{n+1}}{n^2} = \frac{\pi^2}{12}$

(*c*) $\sum_{n=1}^{\infty} \frac{1}{(2n-1)^2} = \frac{\pi^2}{8}$ (*d*) $\sum_{n=1}^{\infty} \frac{(-1)^{n+1}}{2n-1} = \frac{\pi}{4}$

13. X and Y are two random independent events. It is known that $P(X) = 0.40$ and $P(X \cup Y^C) = 0.7$. Which one of the following is the value of $P(X \cup Y)$?

(*a*) 0.7 (*b*) 0.5

(*c*) 0.4 (*d*) 0.3

14. What is the value of $\lim_{\substack{x \to 0 \\ y \to 0}} \frac{xy}{x^2 + y^2}$?

(*a*) 1 (*b*) – 1

(*c*) 0 (*d*) Limit does not exist

15. Consider the following linear system.

$$x + 2y - 3z = a$$
$$2x + 3y + 3z = b$$
$$5x + 9y - 6z = c$$

This system is consistent if *a*, *b* and *c* satisfy the equation

(*a*) $7a - b - c = 0$ (*b*) $3a + b - c = 0$

(*c*) $3a - b + c = 0$ (*d*) $7a - b + c = 0$

16. If $f(x)$ and $g(x)$ are two probability density functions,

$$f(x) = \begin{cases} \frac{x}{a} + 1 & : -a \le x < 0 \\ -\frac{x}{a} + 1 & : 0 \le x \le a \\ 0 & : \text{otherwise} \end{cases}$$

$$g(x) = \begin{cases} -\dfrac{x}{a} & : \ -a \le x < 0 \\ \dfrac{x}{a} & : \ 0 \le x \le a \\ 0 & : \ \text{otherwise} \end{cases}$$

Which one of the following statements is true?

(a) Mean of $f(x)$ and $g(x)$ are same; Variance of $f(x)$ and $g(x)$ are same

(b) Mean of $f(x)$ and $g(x)$ are same; Variance of $f(x)$ and $g(x)$ are different

(c) Mean of $f(x)$ and $g(x)$ are different; Variance of $f(x)$ and $g(x)$ are same

(d) Mean of $f(x)$ and $g(x)$ are different; Variance of $f(x)$ and $g(x)$ are different

17. The angle of intersection of the curves $x^2 = 4y$ and $y^2 = 4x$ at point (0, 0) is

(a) 0° (b) 30°

(c) 45° (d) 90°

18. The area between the parabola $x^2 = 8y$ and the straight line $y = 8$ is ________.

19. The quadratic approximation of $f(x) = x^3 - 3x^2 - 5$ at the point $x = 0$ is

(a) $3x^2 - 6x - 5$ (b) $-3x^2 - 5$

(c) $-3x^2 + 6x - 5$ (d) $3x^2 - 5$

20. The solution to the system of equations

$$\begin{bmatrix} 2 & 5 \\ -4 & 3 \end{bmatrix} \begin{Bmatrix} x \\ y \end{Bmatrix} = \begin{Bmatrix} 2 \\ -30 \end{Bmatrix}$$

is

(a) 6, 2 (b) –6, 2

(c) –6, –2 (d) 6, –2

21. If $f(t)$ is a function defined for all $t \ge 0$, its Laplace transform F(s) is defined as

(a) $\int_0^\infty e^{st} f(t)\,dt$ (b) $\int_0^\infty e^{-st} f(t)\,dt$

(c) $\int_0^\infty e^{ist} f(t)\,dt$ (d) $\int_0^\infty e^{-ist} f(t)\,dt$

22. $f(z) = u(x, y) + i\,v(x, y)$ is an analytic function of complex variable $z = x + i\,y$ where $i = \sqrt{-1}$. If $u(x, y) = 2xy$, then $v(x, y)$ may be expressed as

(a) $-x^2 + y^2$ + constant

(b) $x^2 - y^2$ + constant

(c) $x^2 + y^2$ + constant

(d) $-(x^2 + y^2)$ + constant

23. Consider a Poisson distribution for the tossing of a biased coin. The mean for this distribution is μ. The standard deviation for this distribution is given by

(a) $\sqrt{\mu}$ (b) μ^2

(c) μ (d) $1/\mu$

24. Solve the equation $x = 10\cos(x)$ using the Newton-Raphson method. The initial guess is $x = \pi/4$. The value of the predicted root after the first iteration, up to second decimal, is ________

25. Consider the function $f(x) = 2x^3 .- 3x^2$ in the domain [–1, 2]. The global minimum of $f(x)$ is ________

26. If $y = f(x)$ satisfies the boundary value problem $y'' + 9y = 0$, $y(0) = 0$, $y(\pi/2) = \sqrt{2}$, then $y(\pi/4)$ is ________.

27. The value of the integral

$$\int_{-\infty}^{\infty} \frac{\sin x}{x^2 + 2x + 2}\,dx$$

evaluated using contour integration and the residue theorem is

(a) $-\pi \sin(1)/e$ (b) $-\pi \cos(1)/e$

(c) $\sin(1)/e$ (d) $\cos(1)/e$

28. Gauss-Seidel method is used to solve the following equations (as per the given order):

$$x_1 + 2x_2 + 3x_3 = 5$$
$$2x_1 + 3x_2 + x_3 = 1$$
$$3x_1 + 2x_2 + x_3 = 3$$

Assuming initial guess as $x_1 = x_2 = x_3 = 0$, the value of x_3 after the first iteration is ________.

29. Consider the function $f(z) = z + z^*$ where z is a complex variable and z^* denotes its complex conjugate. Which one of the following is true?

(a) $f(z)$ is both continuous and analytic

(b) $f(z)$ is continuous but not analytic

(c) $f(z)$ is is not continuous but is analytic

(d) $f(z)$ is neither continuous nor analytic

30. A 3 × 3 matrix P is such that, $P^3 = P$. Then the eigenvalues of P are

(a) 1, 1, – 1

(b) 1, $0.5 + j0.866$, $0.5 - j0.866$

(c) 1, $-0.5 + j0.866$, $-0.5 - j0.866$

(d) 0, 1, – 1

31. The solution of the differential equation, for $t > 0$, $y''(t) + 2y'(t) + y(t) = 0$ with initial conditions $y(0) = 0$ and $y'(0) = 1$, is ($u(t)$ denotes the unit step function),

(*a*) $te^{-t}u(t)$

(*b*) $(e^{-t} - te^{-t})u(t)$

(*c*) $(-e^{-t} - te^{-t})u(t)$

(*d*) $e^{-t}u(t)$

32. The value of the line integral

$$\int_C \left(2xy^2\,dx + 2x^2 y\,dy + dz\right)$$

along a path joining the origin (0, 0, 0) and the point (1, 1, 1) is

(*a*) 0 (*b*) 2

(*c*) 4 (*d*) 6

33. The value of the integral $2\int_{-\infty}^{\infty}\left(\frac{\sin 2\pi t}{\pi t}\right) dt$ is equal to

(*a*) 0 (*b*) 0.5

(*c*) 1 (*d*) 2

34. Let $y(x)$ be the solution of the differential equation $\frac{d^2y}{dx^2} - 4\frac{dy}{dx} + 4y = 0$ with initial conditions $y(0) = 0$ and $\left.\frac{dy}{dx}\right|_{x=0} = 1$. Then the value of $y(1)$ is _____.

35. The line integral of the vector field

$$F = 5xz\hat{i} + \left(3x^2 + 2y\right)j + x^2 zk$$

along a path from (0, 0, 0) to (1, 1, 1) parameterized by (t, t^2, t) is ________.

36. Let $P = \begin{bmatrix} 3 & 1 \\ 1 & 3 \end{bmatrix}$. Consider the set S of all vectors $\begin{pmatrix} x \\ y \end{pmatrix}$ such that $a^2 + b^2 = 1$ where $\begin{pmatrix} a \\ b \end{pmatrix} = P\begin{pmatrix} x \\ y \end{pmatrix}$.

Then S is

(*a*) a circle of radius $\sqrt{10}$

(*b*) a circle of radius $\frac{1}{\sqrt{10}}$

(*c*) an ellipse with major axis along $\begin{pmatrix} 1 \\ 1 \end{pmatrix}$

(*d*) an ellipse with minor axis along $\begin{pmatrix} 1 \\ 1 \end{pmatrix}$

37. Let the probability density function of a random variable, X, be given as :

$$f_x(x) = \frac{3}{2}e^{-3x}u(x) + ae^{4x}u(-x)$$

where $u(x)$ is the unit step function.

Then the value of 'a' and Prob $\{X \le 0\}$, respectively, are

(*a*) $2, \frac{1}{2}$

(*b*) $4, \frac{1}{2}$

(*c*) $2, \frac{1}{4}$

(*d*) $4, \frac{1}{4}$

38. $\lim_{x \to 4} \frac{\sin(x-4)}{x-4} =$ ______.

39. A probability density function on the interval [a, 1] is given by $1/x^2$ and outside this interval the value of the function is zero. The value of a is ______ .

40. Two eigenvalues of a 3 × 3 real matrix P are $\left(2 + \sqrt{-1}\right)$ and 3. The determinant of P is______.

41. The coefficient of x^{12} in $(x^3 + x^4 + x^5 + x^6 + ...)^3$ is ______.

42. Consider the recurrence relation $a_1 = 8$, $a_n = 6n^2 + 2n + a_{n-1}$. Let $a_{99} = K \times 10^4$. The value of K is ______.

43. A function $f: \mathbb{N}^+ \to \mathbb{N}^+$, defined on the set of positive integers $\mathbb{N}^+$, satisfies the following properties:

$f(n) = f(n/2)$ if n is even

$f(n) = f(n + 5)$ if n is odd

Let $R = \{i \mid \exists j : f(j) = i\}$ be the set of distinct values that f takes. The maximum possible size of R is ______.

ANSWERS

General Aptitude

1. (c)	2. (b)	3. (c)	4. (c)	5. (b)	6. (d)	7. (a)	8. (d)	9. (c)	10. (b)
11. (c)	12. (a)	13. (c)	14. (a)	15. (a)	16. (d)	17. (d)	18. (b)	19. (d)	20. (c)
21. (a)	22. (b)	23. (a)	24. (a)	25. (c)	26. (c)	27. (b)	28. (d)	29. (c)	30. (c)
31. (c)	32. (b)	33. (b)	34. (b)	35. (c)	36. (120)	37. (a)	38. (c)	39. (a)	40. (7)
41. (d)	42. (a)	43. (a)	44. (c)	45. (d)	46. (b)	47. (d)	48. (a)	49. (b)	50. (b)

Engineering Mathematics

1. (1)	2. (0)	3. (b)	4. (c)	5. (90°)	6. (0.66)	7. (0)	8. (0.33)	9. (3)	10. (54.5)
11. (d)	12. (c)	13. (a)	14. (d)	15. (b)	16. (b)	17. (d)	18. (85.33)	19. (b)	20. (d)
21. (b)	22. (a)	23. (a)	24. 1.5641	25. – 5	26. –1	27. (a)	28. –6	29. (b)	30. (d)
31. (a)	32. (b)	33. (d)	34. (7.38)	35. (4.41)	36. (c)	37. (a)	38. (1)	39. (0.5)	40. (15)
41. (10)	42. (198)	43. (3)							

EXPLANATIONS

General Aptitude

1.

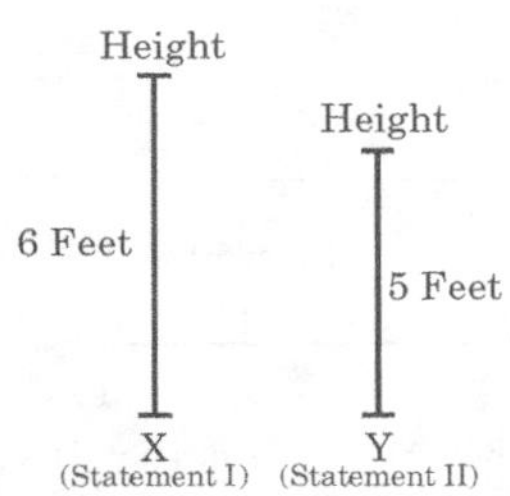

Hence from the given figure Mr. X is taller than Mr. Y by 1 foot.

4. $(9 \text{ inches})^{1/2} = (0.25 \text{ yards})^{1/2}$,

Solving we get 9 inch = 0.25 yards

(since 1 inch = 0.028 yard)

5. M is twice as efficient as E but worked for half as many days. So in this case they will do equal work if their shifts had same timings. But M's shift is for 6 hours, while E's shift for 12 hours. Hence, E will do twice the work as M.

Ratio of contribution of M : E in work is 1 : 2.

6. Given Venn diagram is

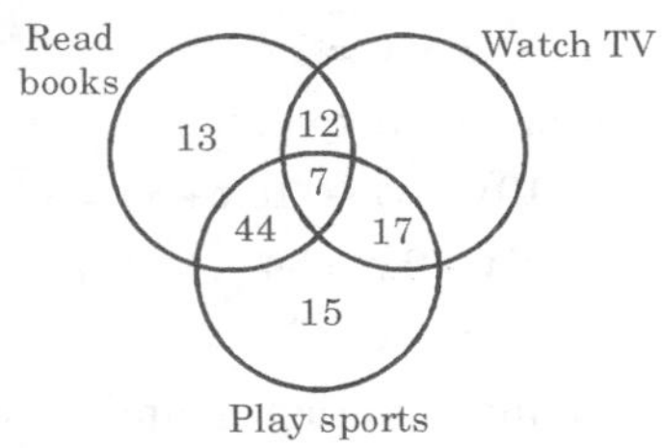

The number of students who like to read books or play sports

$= 13 + 12 + 44 + 7 + 15 + 17$

$= 108$

7. Social science disciplines had a pre-colonial origin.

8.

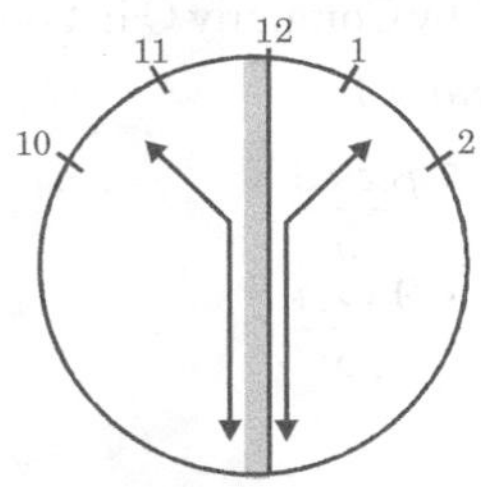

Mirror image of 1 : 20 is 10 : 30

10 : 30 was the time two and quarter hour back so time now will be 12 : 45

9.

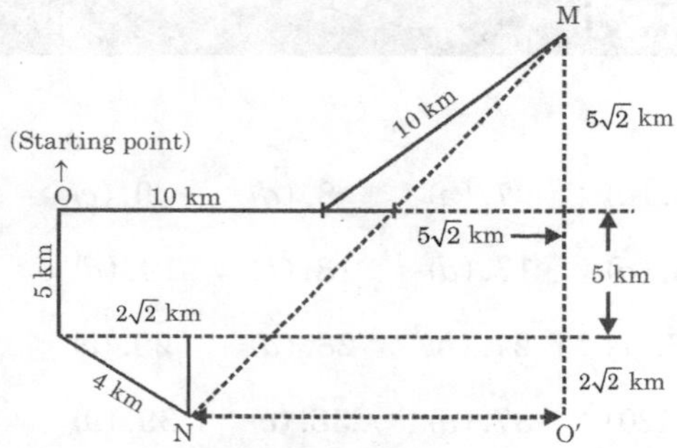

From the given figure

$MN = \sqrt{(O'M)^2 + (O'N)^2}$

$O'M = 5\sqrt{2} + 5 + 2\sqrt{2} = 5 + 7\sqrt{2}$

$O'N = 10 + 5\sqrt{2} - 2\sqrt{2} = 10 + 3\sqrt{2}$

$MN = \sqrt{(5+7\sqrt{2})^2 + (10+3\sqrt{2})^2}$

$= \sqrt{25 + 98 + 70\sqrt{2} + 100 + 18 + 60\sqrt{2})}$

≈ 20.61 km

14. (x% of y) + (y% of x)

$\frac{x}{100} \times y + \frac{y}{100} \times x = \frac{2xy}{100} = 2\%$ of xy

$\therefore$ Option (a) is correct.

15. Let the two digit number be xy

Now sum of the digit of a two digit number is 12

$\therefore \quad x + y = 12 \quad ...(i)$

Now according to question,

$(10y + x) - (10x + y) = 54$

$\therefore \quad 9y - 9x = 54$

$\therefore \quad y - x = 6 \quad ...(ii)$

Now solving equation (i) and (ii) we get,

$x = 3$ and $y = 9$

Hence the original number is 39.

$\therefore$ Option (a) is correct.

16. Amount invested by Company P in 2006 = ₹ 8x and amount invested by Company Q in 2006 = ₹ 9x

Interest from Company P

$$= \frac{\frac{₹\, 8x \times 6}{100}}{\frac{₹\, 9x \times 4}{100}}$$

$$= \frac{4}{3}$$

$\therefore$ Option (d) is correct.

21. Before superlative article 'the' has to be used. "One of" the expression should take plural noun and so option 'C' and 'D' can't be the answer.

22. 'lose' is verb; to complete that sentence a verb is required so 'B' is correct.

23. 'Effectiveness' is noun and 'prescribed' is verb. These words are apt and befitting with the word 'medicine'.

24. Total no. of fruits = 5692000

Unripe type of apples = 45% of 15% of 5692000

$= \frac{45}{100} \times \frac{15}{100} \times 5692000$

$= 384210$

Ripe type of apples $= \frac{34}{100} \times \frac{85}{100} \times 5692000$

$= 1644988$

$\therefore$ Total number of apples

$= 384210 + 1644988$

$= 2029198$

25.

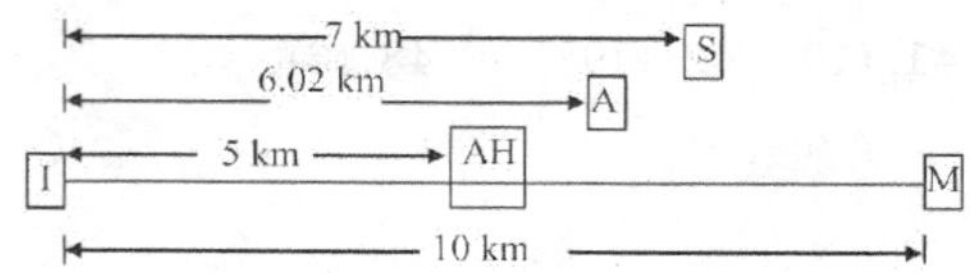

I = I live

AH = Ahmed lives

M = Michel lives

S = Susan lives

A = Arun lives

26.

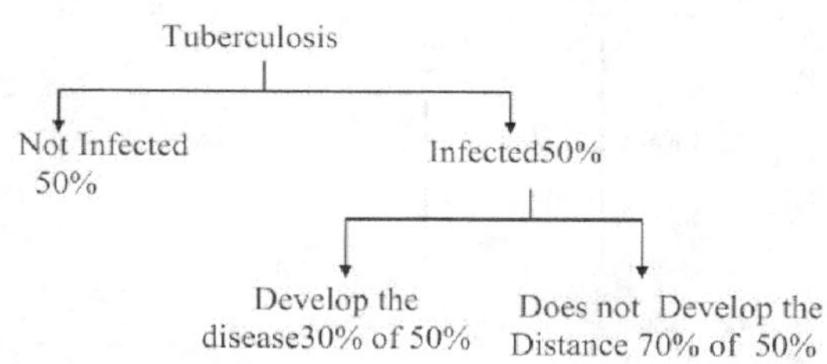

$\frac{70}{100} \times \frac{50}{100} \Rightarrow \frac{35}{100} = 35\%$

$\therefore$ Option (c) is correct

27. The words 'was confident that they would reciprocate' and 'last week proved him wrong' lead to statements iii and iv as logically valid inferences.

28. From given data, the following arrangement is possible

Shiva

Leela

Pavithra

Among four alternatives, option D is TRUE.

29. $q^{-a} = \frac{1}{r} \Rightarrow \frac{1}{q^a} = \frac{1}{r} \Rightarrow q^a = r$

$r^{-b} = \frac{1}{s} \Rightarrow \frac{1}{r^b} = \frac{1}{s} \Rightarrow s = r^b$

$s^{-c} = \frac{1}{q} \Rightarrow \frac{1}{s^c} = \frac{1}{q} \Rightarrow s^c = q$

$q^a = r \Rightarrow \left(s^c\right)^a = r \Rightarrow s^{ac} = r$

$\left(s^{ac}\right)^b = s \Rightarrow s^{abc} = s^1$

$\therefore$ $abc = 1$

$\therefore$ Option (*c*) is correct

30. Q can finish the task = 25 days, 12 hrs/day = 300 hrs

$\Rightarrow$ 1 hr = $\frac{1}{300}$th

R can finish the task = 50 days, 12 hrs/day
= 50 × 12 = 600 hrs

$\Rightarrow$ 1 hr = $\frac{1}{600}$th

Q working hours $\Rightarrow (7-2) \times 12 = 60$ hrs

R working hours $\Rightarrow 7 \times 18 = 126$ hrs

After 7 days, the ratio of work done by Q and R

Q	:	R
$\frac{60}{300}$		$\frac{126}{600}$
20	:	21

33. The given number 33 is odd one out, because the remaining numbers are prime number.

35. Given, $|9y - 6| = 3$

It has two values either

$(9y - 6) = 3$ ...(*i*)

Or $-(9y - 6) = 3$

$9y - 6 = -3$ ...(*ii*)

From equation (*i*),

$9y = 9$

$\therefore$ $y = 1$

Or from equation (*ii*),

$9y = 3$

$\therefore$ $y = \frac{1}{3}$

Now $y^2 = \frac{4y}{3}$

put $y = 1$,

we get $= 1 - \frac{4}{3}$

$= -\frac{1}{3}$

and also put $y = \frac{1}{3}$

we get, $\frac{1}{9} - \frac{4}{9} = \frac{-3}{9} = \frac{-1}{3}$

So, the required value is $\frac{-1}{3}$.

36. From the given graph large plants which are having installed capacity of at least 200 tonnes are 1, 4, 8 and 9.

So, the total production of large plants

= 160 + 190 + 230 + 190

= 770

Now the remaining plants with installed capacity is less then 200 tonnes are 2, 3, 5, 6 and 7.

So, total production of small plants

= 150 + 160 + 120 + 100 + 120

= 650

Then difference = 770 – 650

= 120

39. The question clearly explains that shaquille makes 60 free throws out of 100

Hence, Probability of free throw = $\frac{60}{100} = 0.6$

And probability of NOT free throw

= 1 – 0.6 = 0.4

So required probability of exactly 6 throws in 10 attempts will be given by

$^{10}C_6\, 0.6^6 \times 0.4^4 = 0.2508$

40. Unit digit of $211^{870} + 146^{127} \times 3^{424}$ is $1 + 6 \times 1 = 7$

41. I will not leave the place until the minister meets me.

42. Rewording of something written or spoken is PARAPHRASE.

43. Figurative statement

44.

relftaga ⟶ carefree

otaga ⟶ careful

fertaga ⟶ careless

taga ⟶ care

aftercare ⟶ taga____.

Option (*c*) matches i.e. tagazen.

45.

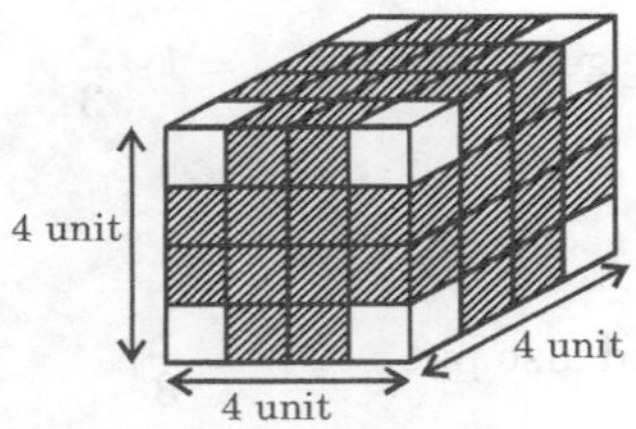

Required Surface Area = $(6 \times 4) + (8 \times 6) + (3 \times 8)$

$= 24 + 48 + 24$

$= 96$ units.

46. Elegance $\Rightarrow (27300 + 25222 + 28976 + 21012) \times 48$

Executive $\Rightarrow (9999 + 8942 + 10234 + 10109) \times 173$

Smooth $\Rightarrow (20009 + 19392 + 22429 + 18229) \times 63$

Soft $\Rightarrow (17602 + 18445 + 19544 + 16595) \times 78$

Executive has the highest revenue.

48. (*i*) is followed.

(*ii*) is not followed.

49. $f(x) = 2x^7 + 3x - 5$

Putting $x = 1$

$f(1) = 2 + 3 - 5 = 0$

So, $(x - 1)$ is the factor of $f(x)$.

Engineering Mathematics

1. From the given question,

A has an eigen value is zero

$$|A| = 0$$

$$\begin{vmatrix} 3 & 2 & 4 \\ 9 & 7 & 13 \\ -6 & -4 & -9+x \end{vmatrix} = 0$$

$3(-63 + 7x + 52) - 2(-81 + 9x + 78) + 4(-36 + 42) = 0$

$3(7x - 11) - 2(9x - 3) + 4(6) = 0$

$21x - 33 - 18x + 6 + 24 = 0$

$3x - 3 = 0$

$x = 1$

2. From the question, the complex variable function is:

$$f(z) = 2z^3 + b_1 |z|^3$$

Given $f(z)$ is analytic.

which is possible only when $b = 0$

since $|z^3|$ is differentiable at the origin but not analytic.

$2z^3$ is analytic everywhere

$\therefore \; f(z) = 2z^3 + b\,|z^3|$ is analytic

only when $b = 0$

3. Given $f(x) = x^3 - 3x^2 + 1$

$f'(x) = 3x^2 - 6x$

$f'(x) = 0$

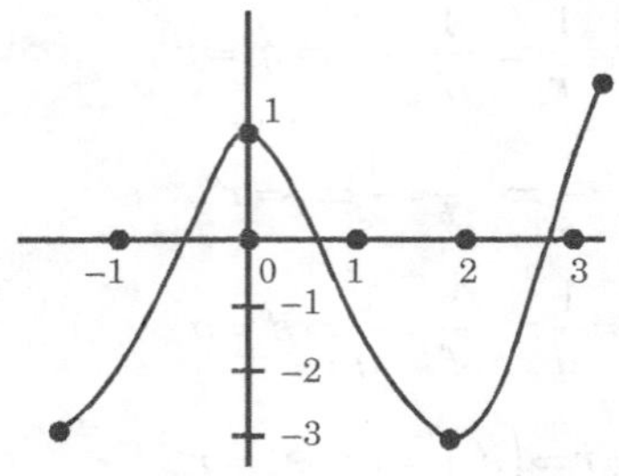

$3x^2 - 6x = 0$

$3x(x - 2) = 0$

$x = 0, 2$

$f''(x) = 6x - 6$

At $x = 0$

$f''(0) = -6$ maxima

$x = 2$

$f''(2) = 6$ minima

Then the given function $f(x)$ increases, then decreases and increase again.

4. Here the given equation is

$$\sin(x) = \frac{x}{2} \text{ (where } x \text{ is in radians)}$$

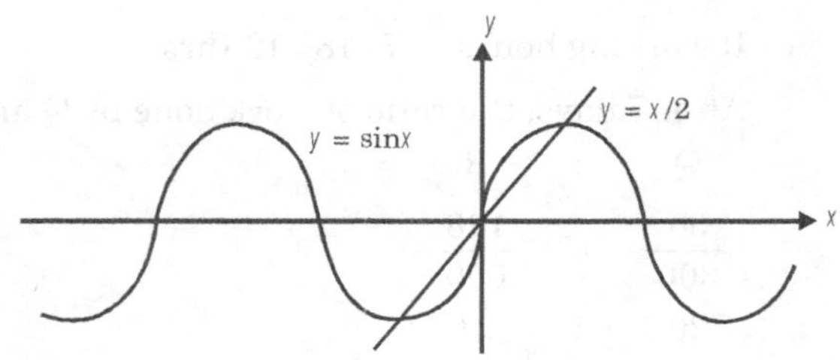

Hence there are 3 distinct values.

5. Here the given time-varying vector

$$I = \hat{x}15\cos\omega t + \hat{y}5\sin\omega t$$

$$|I| = \sqrt{(15\cos\omega t)^2 + (5\sin\omega t)^2}$$

$$= \sqrt{225\cos^2\omega t + 25\sin^2\omega t}$$

$$= \sqrt{225\cos^2\omega t + 25(1 - \cos^2\omega t)}$$

$$= \sqrt{25 + 200\cos^2\omega t}$$

$|I|$ is minimum when $\cos^2\omega t = 0$

$\cos^2\omega t = \cos 90°$

or $\theta = \omega t = 90°$

So, the value of θ is $90°$

6. Given differential equations

$$\frac{dy}{dx} = -3y + 2, \; y(0) = 1$$

If $|1 - 3h| < 1$, then solution of differential equation is stable.

$$-1 < 1 - 3h < 1$$
$$-2 < -3h < 0$$
$$0 < h < 2/3$$

$$h_{max} = \frac{2}{3} = 0.66$$

Hence, the largest time step that can be used to solve the equation is 0.66.

7. By Green's theorem

$$C = \oint xy^2\,dx + x^2 y\,dy \text{ (where C = Closed curve)}$$

$$= \iint_R \left(\frac{d}{dx}(x^2 y) - \frac{d}{dy}(xy^2) \right) dxdy$$

$$= \iint_R (2xy - 2xy) = 0$$

So, the value of closed curve C is 0.

8. The probability $P(X + Y \leq 1)$

$$= \int_{x=0}^{1} \int_{y=0}^{(1-x)} f_{xy}(x, y)\,dx\,dy$$

$$= \int_{x=0}^{1} \int_{y=0}^{(1-x)} (x + y)\,dx\,dy = \int_{x=0}^{1} \left(xy + \frac{y^2}{2} \right)_0^{1-x}$$

$$= \int_{x=0}^{1} \left(x(1-x) + \frac{(1-x)^2}{2} \right) dx$$

$$= \int_{x=0}^{1} (x - x^2)\,dx + \int_{x=0}^{1} \frac{(1-x)^2}{2}\,dx$$

$$= \int_{x=0}^{1} \left(\frac{1}{2} - \frac{x^2}{2} \right) dx = \left(\frac{x}{2} - \frac{x^3}{6} \right)_0^1$$

$$= \frac{1}{2} - \frac{1}{6} = \frac{1}{3} = 0.33$$

9. Trace of A = 14

$a + 5 + 2 + b = 14$

(Taking the diagonal element and then adding)

$$a + b = 7 \quad \text{...}(i)$$

$$\det(A) = 100$$

$$5\begin{vmatrix} a & 3 & 7 \\ 0 & 2 & 4 \\ 0 & 0 & b \end{vmatrix} = 100$$

$$5 \times 2 \times a \times b = 100$$
$$10\,ab = 100$$
$$ab = 10 \quad \text{...}(ii)$$

From equation (*i*) and (*ii*)

either $a = 5, b = 2$

or $a = 2\ b = 5$

$$|a - b| = |5 - 2| = 3$$

10. Median speed calculation = ~~32~~, ~~45~~, ~~49~~, ~~51~~, 53, 56 ~~60~~, ~~62~~, ~~66~~, ~~79~~

$$\text{Median speed} = \frac{53 + 56}{2} = 54.5 \text{ km/hr}$$

19. R tan 30→ = 0.577 R

20. Equations from this matrix are:

$2x + 5y = 2$ and $-4x + 3y = -30$

By solving these two equations, we get

$x = 6;\ y = -2$

21. According to the Laplace transformation of

$f(t) \forall \in t \geq 0$ is as $F(s) = L\{f(t)\} = \int_0^{\infty} e^{-st} f(t)\,dt$

22. For a function of complex variable to be an analytic function Cauchy-Riemann equation has to satisfies which states as

$$\frac{\partial u}{\partial x} = \frac{\partial v}{\partial y} \text{ and } \frac{\partial u}{\partial y} = -\frac{\partial v}{\partial x}$$

Only option (A) $-x^2 + y^2 + \text{constant}$ fulfills this.

23. In the case of Poisson distribution mean = variance

Given, mean = μ

$\therefore$ variance = μ

$\Rightarrow$ standard deviation = $\sqrt{\mu}$

24. Let $f(x) = x - 10\cos(x)$

and $x_0 = \frac{\pi}{4}$

$$f'(x) = 1 + 10\sin(x)$$

$$x_1 = x_0 - \frac{f(x_0)}{f'(x_0)} = \frac{\pi}{4} - \frac{\left(\frac{\pi}{4} - \frac{10}{\sqrt{2}}\right)}{\left(1 + \frac{10}{\sqrt{2}}\right)}$$

$$\Rightarrow \frac{\pi}{4} + \frac{(6.2857)}{(8.0711)}$$

$$\therefore x_1 = 1.5641$$

25.
$$f(x) = 2x^3 - 3x^2$$
$$f'(x) = 0$$
$$\Rightarrow 6x^2 - 6x = 0$$
$$6x(x - 1) = 0$$
$$x = 0 \text{ and } 1$$
$$f(-1) = -5,\ f(1) = -1,\ f(2) = -4$$

Global minimum = –5

26. $y'' + 9y = 0$

A.E. is $m^2 + 9 = 0$

$$m = \pm 3i$$

$$y = y_c + y_p$$

$$y = C_1 \cos 3x + C_2 \sin 3x \qquad (\because y_p = 0)$$

If $x = 0$, $y = 0$

$$0 = C_1(1) + C_2(0) \Rightarrow C_1 = 0$$

If $x = \pi/2$, $y = \sqrt{2}$

$$\sqrt{2} = C_1(0) + C_2 \sin\left(\tfrac{3\pi}{2}\right) \Rightarrow C_2 = -\sqrt{2}$$

$\therefore \; y = -\sqrt{2}\sin 3x$

If $x = \pi/4$

$$y\left(\tfrac{\pi}{4}\right) = -\sqrt{2}\sin\left(\tfrac{3\pi}{4}\right) = -\sqrt{2}\left(\frac{1}{\sqrt{2}}\right) = -1$$

27. $$I = \int_{-\infty}^{\infty} \frac{\sin(x)}{x^2 + 2x + 2}$$

let $f(x) = \dfrac{I_m\left(e^{iz}\right)}{z^2 + 2z + 2}$

Then poles of $f(z)$ are given by $z^2 + 2z + 2 = 0$

$\therefore \; z = -1 \pm i$

$$R_1 = \operatorname{Res}\left(f(z) : z = -1 + i\right)$$

$$= \underset{z \to -1 \pm i}{\text{Lt}} \left[z - (-1+i)\right] \frac{e^{iz}}{\left[z - (-1+i)\right]\left[z - (-1-i)\right]}$$

$$= \frac{e^{i(-1+i)}}{-1 + i + 1 + i} = \frac{e^{-i-1}}{2i}$$

$$\int_c f(z)\,dz = \int_c \frac{I_m\left(e^{iz}\right)}{z^2 + 2z + 2}\,dz = I_m\left[2\pi i(R_1)\right]$$

$$= I_m\left[2\pi i\left(\frac{e^{-i-1}}{2i}\right)\right]$$

$$= I_m\left[\pi e^{-1}\left(\cos(1) - i\sin(1)\right)\right]$$

$$= -\frac{\pi \sin(1)}{e}$$

28. Let x + 2y + 3z = 5

2x + 3y + z = 1

3x + 2y + z = 3 and $x_0 = 0$, $y_0 = 0$, $z_0 = 0$

Then first iteration will be

$$x_1 = x_1 = 5 - 2y_0 - 3z_0 = 5 - 0 - 0 = 5$$

$$x_0 = x_0 = \frac{1}{3}(1 - 2x_1 - z_0)$$

$$= \frac{1}{3}(1 - 10 - 0) = -3$$

$$x_3 = z_1 = 3 - 3x_1 - 2y_1$$
$$= 3 - 15 + 6 = -6$$

$\therefore \; x_3 = -6$

29. Here the given function

$$f(z) = z + z^*$$

where $\quad z$ = complex variable

$f(z) = 2x$ is continuous (polynomial)

$$u = 2x$$
$$v = 0$$
$$u_x = 2$$
$$u_y = 0$$
$$v_x = 0$$
$$v_y = 0$$

C.R. equation is not satisfied,

Hence there is no analytic is $f(z)$ is continuous but not analytic.

30. According to Calyey Hamilton theorem

$$\lambda^3 = \lambda$$
$$\lambda = 0, 1, -1$$

So, the eigen values of λ are 0, 1, – 1.

31. The differential equation is

$$y''(t) + 2y'(t) + y(t) = 0$$

$$\left(s^2 Y(s) - sy(0) - y'(0)\right) + 2\left[sY(s) - y(0)\right] + Y(s) = 0$$

$$Y(s) = \frac{sy(0) + y'(0) + 2y(0)}{\left(s^2 + 2s + 1\right)}$$

Given that $\quad y'(0) = 1,$

$\quad y(0) = 0$

Hence $\quad Y(s) = \dfrac{1}{(s+1)^2}$

$\therefore \quad y(t) = te^{-t}u(t)$

32. Let $\quad I = \int_C \bar{F} \cdot \bar{dr}$

where $\quad \bar{F} = xy^2\bar{i} + 2x^2y\bar{j} + \bar{k}$

Now $\quad \nabla \times F = 0$

(F is irrotational $\Rightarrow$ F is conservative)

$$F = \nabla\phi$$

(ϕ is scalar potational function)

$$\phi_x = 2xy^2$$
$$\phi_y = 2x^2y$$
$$\phi_z = 1$$
$$\Rightarrow \quad \phi = x^2y^2 + z + C$$

where, F is conservative

$$\int_C \bar{F}.\bar{d}r = \int_{(0,0,0)}^{(1,1,1)} d\phi = \left[x^2y^2 + z\right]_{(0,0,0)}^{(1,1,1)}$$
$$= 2$$

33. Let $\quad I = 2\int_{-\infty}^{\infty}\left(\frac{\sin 2\pi t}{\pi t}\right)dt \quad ...(i)$

As per the fourier transform of

$$\frac{2\sin(t\pi/2)}{t} \longrightarrow 2\pi\, \text{rect}\left(\frac{\omega}{t}\right)$$
$$\frac{\sin(2\pi t)}{\pi t} \longrightarrow \text{rect}\left(\frac{\omega}{4\pi}\right)$$

So, $\int_{-\infty}^{\infty}\frac{\sin(2\pi t)}{\pi t}e^{-j\omega t}dt = \text{rect}\left(\frac{\omega}{4\pi}\right) \quad ...(ii)$

Putting $\omega = 0$ in above equation, we get

$$\int_{-\infty}^{\infty}\frac{\sin(2\pi t)}{\pi t}dt = 1$$

Now from equation (*i*)

$$\therefore \quad I = 2\int_{-\infty}^{\infty}\frac{\sin(2\pi)}{\pi t}dt$$
$$= 2 \times 1 = 2$$

34. A.E. $\quad m^2 - 4m + 4 = 0$

$$m^2 - 2m - 2m + 4 = 0$$
$$m(m-2) - 2(m-2) = 0$$
$$(m-2)(m-2) = 0$$
$$\therefore \quad m = 2, 2$$
$$y = (C_1 + C_2x)e^{2x}$$

At $\quad y(0) = 0$

$$\Rightarrow \quad C_1 = 0$$
$$y = C_2xe^{2x}$$
$$y' = C_2e^{2x} + 2C_2xe^{2x}$$
$$y'(0) = 1$$
$$\Rightarrow \quad C_2 = 1$$
$$y = xe^{2x}$$
$$y(1) = e^2$$
$$= 7.38$$

So, the value of $y(1)$ is 7.38.

35. $\quad E = \int_C \bar{F}\cdot\bar{d}r$

Here $\quad \bar{F} = 5xz\hat{i} + (3x^2 + 2y) - \hat{j} + x^2 2\hat{k}$

$$= \int_C 5xzdx + (3x^2 + 2y)dy + x^2zdz$$

Put $\quad x = t,$

$$y = t^2,$$
$$z = t,$$
$$t = 0 \text{ to } 1$$
$$dx = dt$$
$$dy = 2t\,dy,$$
$$dz = dt$$
$$= \int_0^1 5t^2dt + (3t^2 + 2t^2)2tdt + t^3dt$$
$$= \int_0^1 (5t^2 + 11t^3)dt$$
$$= \left[\frac{5t^3}{3} + \frac{11t^4}{4}\right]_0^1 = \frac{5}{3} + \frac{11}{4} = \frac{53}{12} = 4.41$$

36. Given $\quad P = \begin{bmatrix}3 & 1\\1 & 3\end{bmatrix}$

According to question

$$P\begin{bmatrix}x\\y\end{bmatrix} = \begin{bmatrix}a\\b\end{bmatrix}$$
$$\begin{bmatrix}3 & 1\\1 & 3\end{bmatrix}\begin{bmatrix}x\\y\end{bmatrix} = \begin{bmatrix}a\\b\end{bmatrix}$$
$$3x + y = a$$
$$x + 3y = b$$
$$a^2 + b^2 = 1$$
$$\Rightarrow \quad 10x^2 + 10y^2 + 12xy = 1$$

Ellipse with major axis along $\begin{bmatrix}1\\1\end{bmatrix}$.

37. Given, $\quad f_x(x) = \begin{cases} ae^{4x} & x < 0 \\ \frac{3}{2}e^{-3x} & x \geq 0 \end{cases}$

$$\therefore \quad \int_{-\infty}^{\infty} f_x(x) = 1$$
$$\int_{-\infty}^{0} ae^{4x}dx + \int_0^{\infty}\frac{3}{2}e^{-3x}dx = 1$$
$$a\int_{-\infty}^{0} e^{4x}dx + \frac{3}{2}\int_0^{\infty} e^{-3x}dx = 1$$

$$\left[\frac{ae^{4x}}{4}\right]_{-\infty}^{0} + \left[\frac{\frac{3}{2}e^{-3x}}{-3}\right]_{0}^{\infty} = 1$$

$$\frac{a}{4}+\frac{3}{6} = 1$$

$$a = 2$$

$$P(x \le 0) = \int_{-\infty}^{0} 2e^{4x}dx$$

$$= \left[\frac{e^{4x}}{2}\right]_{-\infty}^{0} = \frac{1}{2}$$

So, the value of 'a' and 'P' are 2, $\frac{1}{2}$

38. $\lim_{x\to 4}\frac{\sin(x-4)}{(x-4)}$ $\left(\frac{0}{0}\text{form}\right)$

Apply L' Hospital Rule :

$$\lim_{x\to 4}\left[\frac{\cos(x-4)}{1}\right]$$

$$\Rightarrow \frac{\cos 0}{1} \Rightarrow 1.$$

39. Total probability = 1

So $\int_{a}^{1}\frac{1}{x^2} = 1$

$$\left[-\frac{1}{x}\right]_a^1 = 1$$

$$\frac{1}{a} = 2$$

$$a = \frac{1}{2} = 0.5$$

40. $(2 + i)$ is an eigenvalue

So $(2 - i)$ will be other.

So Determinant will be product of eigenvalues.

So $|A| = (2 + i)(2 - i) \times 3$

$$(2^2 - i^2) \times 3 = 15$$

41. $(x^3 + x^4 + x^5 + x^6 \ldots\ldots)^3$

$$\Rightarrow \left(\frac{x}{1-x} - x - x^2\right)^3$$

$$\Rightarrow \left(\frac{x - x(1-x) - x^2(1-x)}{(1-x)}\right)^3$$

$$\Rightarrow \left(\frac{x - x - x^2 + x^2 + x^3}{(1-x)}\right)^3$$

$$\Rightarrow \left(\frac{x^3}{1-x}\right)^3$$

$\Rightarrow$ Coefficient of x^{12} in $\left(\frac{x^3}{1-x}\right)^3$

$\Rightarrow$ Coefficient of x^3 in $\left(\frac{1}{1-x}\right)^3$

$\Rightarrow$ Coefficient of x^3 in $(1 - x)^{-3}$

$$f(0) + xf'(0) + \frac{x^2}{2!}f''(0) + \frac{x^3}{3!}f'''(0)$$

$f(x) = (1 - x)^{-3}$

$f'(x) = +3(1 - x)^{-4}$

$f''(x) = +12(1 - x)^{-5}$

$f'''(x) = +60(1 - x)^{-6}$

Coefficient of x^3 in $(1 - x)^{-3} = \frac{60}{3!} = 10$

42. $a_n = 6n^2 + 2n + a_{n-1}$, $a_1 = 8$.

$a_2 = 6 \times 2^2 + 2 \times 2 + a_1$

$a_3 = 6 \times 3^2 + 2 \times 3 + a_2$

$= 6 \times 3^2 + 2 \times 3 + 6 \times 2^2 + 2 \times 2 + a_1$

$\vdots$

$a_{99} = 6 \times 99^2 + 2 \times 99 + 6 \times 98^2 + 2 \times 98 + \ldots. + 6 \times 2^2 + 2 \times 2 + 8$

$= 6[1^2 + 2^2 + 3^2 + \ldots. + 99^2] + 2[1 + 2 + \ldots 99]$

$$= 6.\frac{(99(99+1)(2\times 99+1))}{6} + 2\frac{(99(99+1))}{2}$$

$= 99 \times 100 \times 199 + 99 \times 100$

$= 100 \times 99\ (199 + 1)$

$= 2 \times 99 \times 10^4 = \boxed{198} \times 10^4$

⇓ value of K

43. Maximum number of distinct values f takes is only 3.

1st $\Rightarrow f(1) = f(2) = f(3) = f(4) = f(8)$

2nd $\Rightarrow f(5) = f(10)$

3rd $\Rightarrow f(6) = f(7) = f(9)$

Remaining all other n values will provide only one of these three values.

SOLVED PAPER – 2017

GENERAL APTITUDE

1. What is the value of x when

$$81 \times \left(\frac{16}{25}\right)^{x+2} \div \left(\frac{3}{5}\right)^{2x+4} = 144?$$

(*a*) 1

(*b*) – 1

(*c*) – 2

(*d*) Cannot be determined

2. There was no doubt that their work was <u>thorough</u>

Which of the words is closest in meaning to the underlined word above?

(*a*) pretty (*b*) complete

(*c*) sloppy (*d*) haphazard

3. Four cards lie on a table. Each card has a number printed on the one side and a color on the other. The faces visible on the cards are 2, 3 red and blue.

Proposition. If a card has an even value on one side then its opposite face is red.

The cards which MUST be turned over to verify the above porposition are

(*a*) 2, red (*b*) 2, 3, red

(*c*) 2, blue (*d*) 2, red, blue

4. The event would have been successful if you _____ able to come.

(*a*) are (*b*) had been

(*c*) have been (*d*) would have been

5. Two dice are thrown simultaneously. The probability that the product of the numbers appearing on the top faces of the dice is a perfect square is

(*a*) 1/9

(*b*) 2/9

(*c*) 1/3

(*d*) 4/9

6. Bhaiclung was observing the pattern of people entering and leaving a car service centre. There was a single window where customers were being served. He saw that people inevitably came out of the centre in the order that they went in. However, the time they spent inside seemed to vary a lot some people came out in a matter of minutes while for others it took much longer.

From this what can one conclude?

(*a*) The centre operates on a first-come-first-served basis, but with variable service times depending on specific customer needs.

(*b*) Customers were served in an arbitrary order since they took varying amounts of time for service completion in the centre.

(*c*) Since some people came out within a few minutes of entering the centre, the system is likely to operate on a last-come-first-served basis.

(*d*) Entering the centre early ensured that one would have shorter service times and most-people attempted to do this.

7. The points in the graph below represent the halts of a lift for duration of 1 minute, over a period of 1 hour.

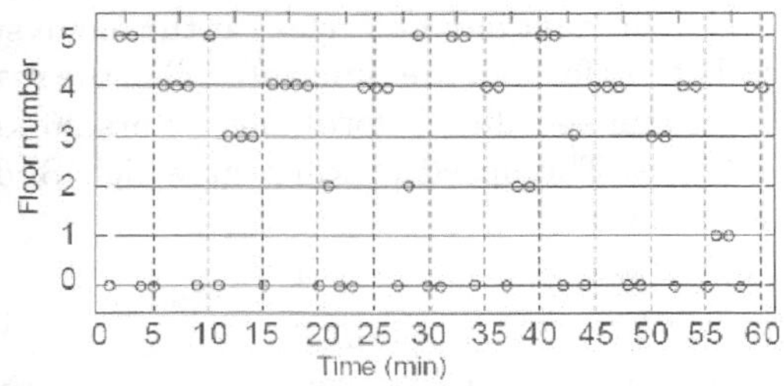

Which of the following statements are correct?

(*i*) The elevator never moves directly from any non-ground floor to another non-ground floor over the one hour period.

(*ii*) The elevator stays on the fourth floor for the longest duration over the one hour period.

(*a*) only (*i*) (*b*) only (*ii*)

(*c*) Both (*i*) and (*ii*) (*d*) Neither (*i*) nor (*ii*)

8. A map shows the elevations of Darjeeling, Gangtok, Kalimpong, Pelling and Siliguri. Kalimpong is at a lower elevation than Gangtok. Pelling is at a lower elevation than Gangtok Pelling is at a higher elevation than Siliguri. Darjeeling is at a higher elevation than Gangtok.

Which of the following statements can be inferred from the paragraph above?

(*i*) Pelling is at a higher elevation than Kalimpong

(*ii*) Kalimpong is at a lower elevation than Darjeeling

(*iii*) Kalimpong is at a higher elevation than Siliguri

(*iv*) Siliguri is at a lower elevation than Gangtok.

(*a*) Only (*ii*)

(*b*) Only (*ii*) and (*iii*)

(*c*) Only (*ii*) and (*iv*)

(*d*) Only (*iii*) and (*iv*)

9. P, Q, R, S, T and U are sealed around a circular table. R is seated two places to the right of Q. P is seated three places to the left of R. S is seated opposite U. If P and U now switch seats which of the following must necessarily be true.

(*a*) P is immediately to the right of R

(*b*) T is immediately to the left of P

(*c*) T is immediately to the left of P or P is immediately to other right of Q.

(*d*) U is immediately to the right of R or P is immediately to the left of T.

10. Budhan covers a distance of 19 km in 2 hours by cycling one fourth of the time and walking the rest. The next day he cycles (at the same speed as before) for half the time and walks the rest (at the same speed as before) and covers 26 km in 2 hours. The speed in km/h at which Budhan walks is

(*a*) 1 (*b*) 4

(*c*) 5 (*d*) 6

11. The probability that a k-digit number does NOT contain the digits 0, 5 or 9 is

(*a*) 0.3^k (*b*) 0.6^k

(*c*) 0.7^k (*d*) 0.9^k

12. Find the smallest number y such that that y x 162 is a perfect cube.

(*a*) 24 (*b*) 27

(*c*) 32 (*d*) 36

13. After Rajendra Chola returned from his voyage to Indonesia, he ______ to visit the temple in Thanjavur.

(*a*) was wishing (*b*) is wishing

(*c*) wished (*d*) had wished

14. Research in the workplace reveal that people work for many reasons________.

(*a*) money beside

(*b*) beside money

(*c*) money besides

(*d*) besides money

15. Rahul, Murali, Srinivas and Arul are seated around a square table. Rahul is sitting to the left of Murali. Srinivas is sitting to the right of Arul. Which of the following pairs are seated opposite each other?

(*a*) Rahul and Murali

(*b*) Srinivas and Arul

(*c*) Srinivas and Murali

(*d*) Srinivas and Rahul

16. "The hold of the nationalist imagination on our colonial past is such that anything inadequately or improperly nationalist is just not history."

Which of the following statements best reflects the author's opinion?

(*a*) Nationalists are highly imaginative

(*b*) History is viewed through the filter of nationalism

(*c*) Our colonial past never happened

(*d*) Nationalism has to be both adequately and properly imagined.

17. The expression $\frac{(x+y)-|x-y|}{2}$ is equal to

(*a*) the maximum of x and y

(*b*) the minimum of x and y

(*c*) 1

(*d*) none of the above

18. Arun, Gulab, Neel and Shweta must choose one shirt each from a pile of four shirts coloured red, pink, blue and white respectively. Arun dislikes the colour red and Shweta dislikes the colour white. Gulab and Neel like all the colours. In how many different ways can they choose the shirts so that no one has a shirt with a colour he or she dislikes?

(*a*) 21 (*b*) 18

(*c*) 16 (*d*) 14

19. Six people are seated around a circular table. There are at least two men and two women. There are at least three right-handed persons. Every woman has a left-handed person to her immediate right. None of the women are right-handed. The number of women at the table is

(*a*) 2

(*b*) 3

(*c*) 4

(*d*) Cannot be determined

20. A test has twenty questions worth 100 marks in total. There are two types of questions. Multiple choice questions are worth 3 marks each and essay questions are worth 11 marks each. How many multiple choice questions does the exam have ?

(*a*) 12 (*b*) 15

(*c*) 18 (*d*) 19

21. Choose the option with words that are not synonyms.

(*a*) aversion, dislike (*b*) luminous, radiant

(*c*) plunder, loot (*d*) yielding, resistant

22. There are five buildings called V, W, X, Y and Z in a row (not necessarily in that order). V is to the West of W. Z is to the East of X and the West of V. W is to the West of Y. Which is the building in the middle ?

(*a*) V

(*b*) W

(*c*) X

(*d*) Y

23. There are 3 red socks, 4 green socks and 3 blue socks. You choose 2 socks. The probability that they are of the same colour is

(*a*) 1/5

(*b*) 7/30

(*c*) 1/4

(*d*) 4/15

24. Saturn is _____ to be seen on a clear night with the naked eye.

(*a*) enough bright

(*b*) bright enough

(*c*) as enough bright

(*d*) bright as enough

25. "We lived in a culture that denied any merit to literary works, considering them important only when they were handmaidens to somethings seemingly more urgent – namely ideology. This was a country where all gestures, even the most private, were interprated in political terms."

The author's belief that ideology is not as important as literature is revealed by the word :

(*a*) 'culture'

(*b*) 'seemingly'

(*c*) 'urgent'

(*d*) 'political'

26. An air pressure contour line joins locations in a region having the same atmospheric pressure. The following is an air pressure contour plot is a geographical region. Contour lines are shown at 0.05 bar intervals in this plot.

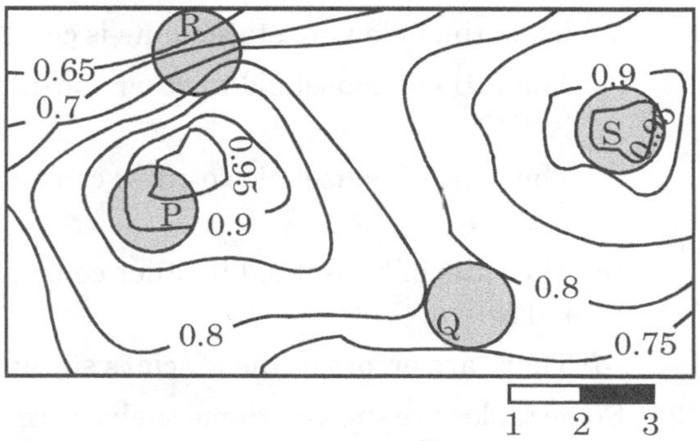

If the possibility of a thunderstorm is given by how fast air pressure rises or drops over a region, which of the following regions is most likely to have a thunderstorm ?

(*a*) P (*b*) Q

(*c*) R (*d*) S

27. The number of roots of $e^x + 0.5x^2 - 2 = 0$ in the range [–5, 5] is

(*a*) 0

(*b*) 1

(*c*) 2

(*d*) 3

28. There are three boxes. One contains apples another contains oranges and the last one contains both apples and oranges. All three are known to be incorrectly labelled. If you are permitted to open just one box and then pull out and inspect only one fruit, which box would you open to determine the contents of all three boxes?

(*a*) The box labelled 'Apples'

(*b*) The box labelled 'Apples and Oranges'

(*c*) The box labelled 'Oranges'

(*d*) Cannot be determined

29. X is a 30 digit number starting with the digit 4 followed by the digit 7. Then the number X^3 will have

(*a*) 90 digits

(*b*) 91 digits

(*c*) 92 digits

(*d*) 93 digits

30. In the summer, water consumption is known to decrease overall by 25%. A Water Board official states that in the summer household consumption decreases by 20%, while other consumption increases by 70%.

Which of the following statements is correct?

(*a*) The ratio of household to other consumption is 8/17.

(*b*) The ratio of household to other consumption is 1/17.

(*c*) The ratio of household to other consumption is 17/8.

(*d*) There are errors in the official's statement.

31. Some tables are shelves. Some shelves are chairs. All chairs are benches. Which of the following conclusions can be deduced from the preceding sentences?

(*i*) At least one bench is a table

(*ii*) At least one shelf is a bench

(*iii*)At least one chair is a table

(*iv*)All benches are chairs

(*a*) only (*i*) (*b*) only (*ii*)

(*c*) only (*ii*) and (*iii*) (*d*) only (*iv*)

32. I _______ made arrangements had I _______ informed earlier.

(*a*) could have, been (*b*) would have, being

(*c*) had, have (*d*) had been, been

33. 40% of deaths on city roads may be attributed to drunken driving. The number of degrees needed to represent this as a slice of a pie chart is

(*a*) 120 (*b*) 144

(*c*) 160 (*d*) 212

34. She has a sharp tongue and it can occasionally turn ________.

(*a*) hurtful (*b*) left

(*c*) methodical (*d*) vital

35. Trucks (10 m long) and cars (5 m long) go on a single lane bridge. There must be a gap of at least 20 m after each truck and a gap of at least 15 m after each car. Trucks and cars travel at a speed of 36 km/h. If cars and trucks go alternately. What is the maximum number of vehicles that can use the bridge in one hour?

(*a*) 1440 (*b*) 1200

(*c*) 720 (*d*) 600

36. "If you are looking for a history of India, or for an account of the rise and fall of the British Raj, or for the reason of the cleaving of the subcontinent into two mutually antagonistic parts and the effects this mutilation will have in the respective sections, and ultimately on Asia, you will not find it in these pages; for though I have spent a lifetime in the country, I lived too near the seat of events, and was too intimately associated with the actors, to get the perspective needed for the impartial recording of these matters".

Here, the word 'antagonistic' is closest in meaning to

(*a*) impartial (*b*) argumentative

(*c*) separated (*d*) hostile

37. A contour line joins locations having the same height above the mean sea level. The following is a contour plot of a geographical region. Contour lines are shown at 25 m intervais in this plot.

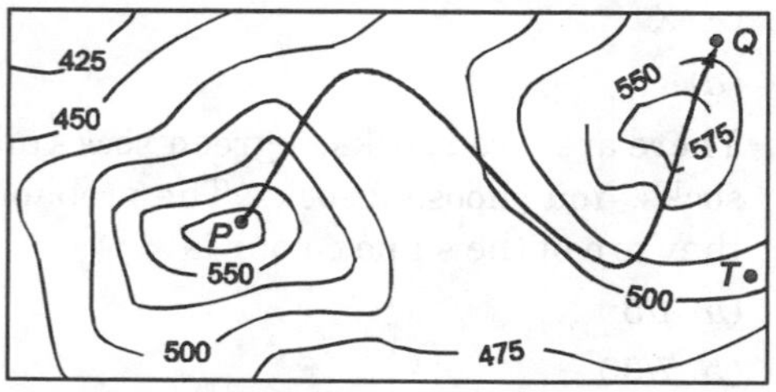

The path from P to Q is best described by

(*a*) Up-Down-Up-Down

(*b*) Down-Up-Down-Up

(*c*) Down-Up-Down

(*d*) Up-Down-Up

38. There are 3 Indians and 3 Chinese in a group of 6 people. How many subgroups of this group can we choose so that every subgroup has at least one Indian?

(*a*) 56 (*b*) 52

(*c*) 48 (*d*) 44

39. S, T, U, V, W, X, Y and Z are seated around a circular table. T's neighbours are Y and V. Z is seated third to the left of T and second to the right of S. U's neighbours are S and Y; and T and W are not seated opposite each other. Who is third to the left of V?

(*a*) X

(*b*) W

(*c*) U

(*d*) T

40. He was one of my best__________ and I felt his loss________

(a) Friend, keenly (b) Friends, keen

(c) Friend, keener (d) Friends, keenly

41. A right-angled cone (with base radius 5 cm and height 12 cm), as shown in the figure below , is rolled on the ground keeping the point P fixed until the point Q (at the base of the cone, as shown) touches the ground again.

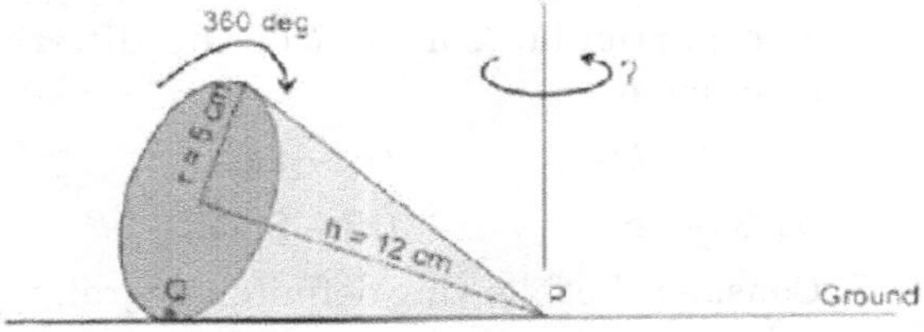

(a) $\frac{5\pi}{12}$ (b) $\frac{5\pi}{24}$

(c) $\frac{24\pi}{5}$ (d) $\frac{10\pi}{13}$

42. In a company with 100 employees, 45 earn ₹ 20,000 per month, 25 earn ₹ 30,000, 20 earn ₹ 40,000, 8 earn ₹ 60,000 and 2 earn ₹ 150,000. The median of the salaries is______

(a) ₹ 20,000 (b) ₹ 30,000

(c) ₹ 32,300 (d) ₹ 40,000

43. P, Q and R talk about S's car collection. P states that S has at least 3 cars. Q belives that S has less than 3 cars R indicates that to his knowledge, S has at least one car. Only one of P,Q and R is right. The number of cars owned by S is

(a) 0

(b) 1

(c) 3

(d) Cannot be determined

44. As the two speakers became increasingly agitated, the debate became__________.

(a) Lukewarm (b) Poetic

(c) Forgiving (d) Heated

45. The growth of bacteria (lactobacillus) in milk leads to curd formation. A minimum bacterial population density of 0.8 (in suitable units) is needed to form curd. In the graph below, the population density of lactobacillus in 1 litre of milk is plotted as a function of time, at two different temperature, 25°C and 37°C.

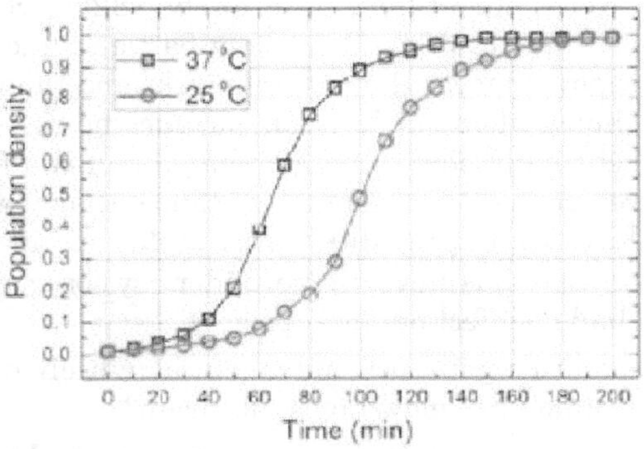

Consider the following statements based on the data shown above:

i. The growth in bacterial population stops earlier at 37°C as compared to 25°C

ii. The time taken for curd formation at 25°C is twice the time taken at 37°C Which one of the following options is correct?

(a) Only i (b) Only ii

(c) Both i and ii (d) Neither i nor ii

46. What is the sum of the missing digits in the subtraction problem below?

$$\begin{array}{r} 5\,\_\,\_\,\_\,\_ \\ -\,4\,8\,\_\,8\,9 \\ \hline 1\,1\,1\,1 \end{array}$$

(a) 8

(b) 10

(c) 11

(d) Cannot be determined

47. Let S_1 be the plane figure consisting of the points (x, y) given by the inequalities $|x - 1| \le 2$ and $|y + 2| \le 3$. Let S_2 be the plane figure given by the inequalities $x - y \ge -2$, $y \ge 1$, and $x \le 3$ Let S be the union of S_1 and S_2. The area of S is

(a) 26 (b) 28

(c) 32 (d) 34

48. Two very famous sportsmen Mark and Steve happened to be brothers, and played for country K. Mark teased James, an opponent from country E, "There is no way you are good enough to play for your country." James replied, "Maybe not, but at least I am the best player in my own family."

Which one of the following can be inferred from this conversation?

(a) Mark was known to play better than James

(b) Steve was known to play better than Mark

(c) James and Steve were good friends

(d) James played better than Steve

49. "Here, throughout the early 1820s, Stuart continued to fight his losing battle to allow his sepoys to wear their caste- marks and their own choice of facial hair on parade, being again reprimanded by the commander-in-chief. His retort that "A stronger instance than this of European prejudice with relation to this country has never come under my observations "had no effect on his superiors"

According to this paragraph, which of the statements below I most accurate?

(*a*) Start's commander-in-chief was moved by this demonstration of his prejudice

(*b*) The Europeans were accommodating of the sepoys' desire to wear their caste-marks

(*c*) Stuart's losing battle' refers to his inability to succeed in enabling sepoys to wear cast-marks

(*d*) The commander-in-chief was exempt from the European prejudice that dictated how the sepoys were to dress.

ENGINEERING MATHEMATICS

1. A two-faced fair coin has its faces designated as head (H) and tail (T). This coin is tossed three times in succession to record the following outcomes: H.H.H. If the coin is tossed one more time, the probability (up to one decimal place) of obtaining H again given the previous realizations of H, H and H. would be______

2. The divergence of the vector field $V = x^2\,i + 2y^3\,j + z^4\,k$ at x = 1, y = 2, z = 3 is __________

3. Consider the following simultaneous equation (with c_1 and c_2 beings constants):

$$3x_1 + 2x_2 = c_1$$
$$4x_1 + x_2 = c_2$$

The characteristic equation for these simultaneous equations is

(*a*) $\lambda^2 - 4\lambda - 5 = 0$ (*b*) $\lambda^2 - 4\lambda + 5 = 0$

(*c*) $\lambda^2 + 4\lambda - 5 = 0$ (*d*) $\lambda^2 + 4\lambda + 5 = 0$

4. Let w = (f(x, y), where x and y are functions of t. Then according to the chain rule $\frac{dw}{dt}$ is equal to

(*a*) $\frac{dw}{dx}\frac{dx}{dt} + \frac{dw}{dy}\frac{dt}{dt}$

(*b*) $\frac{\partial w}{\partial x}\frac{\partial x}{\partial t} + \frac{\partial w}{\partial y}\frac{\partial y}{\partial t}$

(*c*) $\frac{\partial w}{\partial x}\frac{dx}{dt} + \frac{\partial w}{\partial y}\frac{dy}{dt}$

(*d*) $\frac{dw}{dx}\frac{\partial x}{\partial t} + \frac{dw}{dy}\frac{\partial y}{\partial t}$

5. The tangent to the curve represented by $y = x \ln x$ is required to have 45° inclination with the x-axis. The coordinates of the tangent point would be

(*a*) (1, 0) (*b*) (0, 1)

(*c*) (1, 1) (*d*) $(\sqrt{2}, \sqrt{2})$

6. Consider the following second-order differential equation:

$$y'' - 4y' + 3y = 2t - 3t^2$$

The particular solution of the differential equation is

(*a*) $-2 - 2t - t^2$ (*b*) $-2t - t^2$

(*c*) $2t - 3t^2$ (*d*) $-2 - 2t - 3t^2$

7. Consider the following definite integral:

$$I = \int_0^1 \frac{(\sin^{-1} x)^2}{\sqrt{1-x^2}}\,dx$$

The value of the integral is

(*a*) $\frac{\pi^3}{24}$ (*b*) $\frac{\pi^3}{12}$

(*c*) $\frac{\pi^3}{48}$ (*d*) $\frac{\pi^3}{64}$

8. If $A = \begin{bmatrix} 1 & 5 \\ 6 & 2 \end{bmatrix}$ and $B = \begin{bmatrix} 3 & 7 \\ 8 & 4 \end{bmatrix}$. AB^T is equal to

(*a*) $\begin{bmatrix} 38 & 28 \\ 32 & 56 \end{bmatrix}$ (*b*) $\begin{bmatrix} 3 & 40 \\ 42 & 8 \end{bmatrix}$

(*c*) $\begin{bmatrix} 43 & 27 \\ 34 & 50 \end{bmatrix}$ (*d*) $\begin{bmatrix} 38 & 32 \\ 28 & 56 \end{bmatrix}$

9. The matrix $A = \begin{bmatrix} \frac{3}{2} & 0 & \frac{1}{2} \\ 0 & -1 & 0 \\ \frac{1}{2} & 0 & \frac{3}{2} \end{bmatrix}$ has three distinct eigenvalues and one of its eigenvectors is $\begin{bmatrix} 1 \\ 0 \\ 1 \end{bmatrix}$.

Which one of the following can be another eigenvector of A?

(*a*) $\begin{bmatrix} 0 \\ 0 \\ -1 \end{bmatrix}$ (*b*) $\begin{bmatrix} -1 \\ 0 \\ 0 \end{bmatrix}$

(*c*) $\begin{bmatrix} 1 \\ 0 \\ -1 \end{bmatrix}$ (*d*) $\begin{bmatrix} 1 \\ -1 \\ 1 \end{bmatrix}$

10. For a complex number z, $\lim_{z \to i} \frac{z^2+1}{z^3+2z-i(z^2+2)}$ is

(a) $-2i$ (b) $-i$

(c) i (d) $2i$

11. Only one of the real roots of $f(x) = x^6 - x - 1$ lies in the interval $1 \le x \le 2$ and bisection method is used to find its value. For achieving an accuracy of 0.001, the required minimum number of iterations is______. (Give the answer up to two decimal places.)

12. Consider the differential equation $(t^2-81)\frac{dy}{dt} + 5ty = \sin(t)$ with $y(1) = 2\pi$. There exists of a unique solution for this differential equation when t belongs to the interval

(a) (–2, 2) (b) (–10,10)

(c) (–10, 2) (d) (0,10)

13. A function $f(x)$ is defined as

$$f(x) = \begin{cases} e^x & x < 1 \\ \ln x + ax^2 + bx, & x \ge 1 \end{cases},$$

where $x \in R$ which one of the following statements is TRUE?

(a) $f(x)$ is NOT differentiable at $x = 1$ for any values of a and b.

(b) $f(x)$ is differentiable at $x = 1$ for the unique values of a and b.

(c) $f(x)$ is differentiable at $x = 1$ for all the values of a and b such that $a + b = e$.

(d) $f(x)$ is differentiable at $x = 1$ for all values of a and b.

14. Consider the line integral $I = \int_C (x^2 + iy^2)\,dz$, where $z = x + iy$. The line c is shown in the figure below.

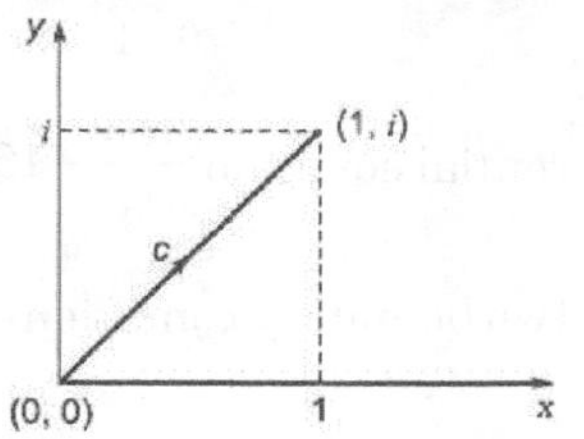

The value of I is

(a) $\frac{1}{2}i$ (b) $\frac{2}{3}i$

(c) $\frac{3}{4}i$ (d) $\frac{4}{5}i$

15. If $f(x) = R\sin\left(\frac{\pi x}{2}\right) + S$, $f'\left(\frac{1}{2}\right) = \sqrt{2}$ and $\int_0^1 f(x)\,dx = \frac{2R}{\pi}$, then the constants R and S are. respectively.

(a) $\frac{2}{\pi}$ and $\frac{16}{\pi}$

(b) $\frac{2}{\pi}$ and 0

(c) $\frac{4}{\pi}$ and 0

(d) $\frac{4}{\pi}$ and $\frac{16}{\pi}$

16. Consider a quadratic equation $x^2 - 13x + 36 = 0$ with coefficients in a base b. The solution of this equation in the same base b are $x = 5$ and $x = 6$. Then b = _____ .

17. G is an undirected graph with n vertices and 25 edges such that each vertex of G has degree at least 3. Then the maximum possible value of n is _____.

18. Consider the set X = {a, b, c, d, e} under the partial ordering

R = {(a, a), (a, b), (a, c), (a, d), (a, e), (b, b), (b, c), (b, e), (c, c), (c, e), (d, d), (d, e), (e, e)}.

The Hase diagram of the partial order (X, R) is shown below :

List II

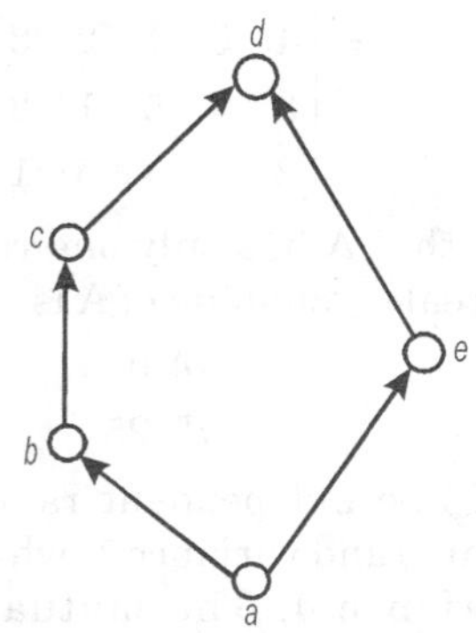

The minimum number of ordered pairs that need to be added to R to make (X, R) a lattice is______.

19. Let $P = \begin{bmatrix} 1 & 1 & -1 \\ 2 & -3 & 4 \\ 3 & -2 & 3 \end{bmatrix}$ and

$Q = \begin{bmatrix} -1 & -2 & -1 \\ 6 & 12 & 6 \\ 5 & 10 & 5 \end{bmatrix}$ be two matrices.

Then the rank of P + Q is ______ .

20. P and Q are considering to apply for a job. The probability that P applies for the job is $\frac{1}{4}$, the probability that P applies for the job given that Q applies for the job is $\frac{1}{2}$, and the probability that Q applies for the job given that P applies for the job is $\frac{1}{3}$. Then the probability that P does not apply for the job given that Q does not apply for the job is

(*a*) $\frac{4}{5}$ (*b*) $\frac{5}{6}$

(*c*) $\frac{7}{8}$ (*d*) $\frac{11}{12}$

21. If the ordinary generating function of a sequence $\{a_n\}_{n=0}^{\infty}$ is $\frac{1+z}{(1-z)^3}$. then $a_3 - a_0$ is equal to ______.

22. If a random variable X has a Poisson distribution with mean 5, then the expectation $E[(X + 2)^2]$ equals ______.

23. If the characteristic polynomial of a 3 × 3 matrix M over $\mathbb{R}$ (the set of real numbers) is $\lambda^3 - 4\lambda^2 + a\lambda + 30$, $a \in \mathbb{R}$, and one eigenvalue of M is 2, then the largest among the absolute values of the eigenvalues of M is ______.

24. Consider the 5 × 5 matrix

$$A = \begin{bmatrix} 1 & 2 & 3 & 4 & 5 \\ 5 & 1 & 2 & 3 & 4 \\ 4 & 5 & 1 & 2 & 3 \\ 3 & 4 & 5 & 1 & 2 \\ 2 & 3 & 4 & 5 & 1 \end{bmatrix}$$

It is given that A has only one real eigenvalue. Then the real eigenvalue of A is

(*a*) – 2.5 (*b*) 0

(*c*) 15 (*d*) 25

25. Let (X_1, X_2) be independent random variables. X_1 has mean 0 and variance 1, while X_2 has mean 1 and variance 4. The mutual information $I(X_1 : X_2)$ between X_1 and X_2 in bits is ______

26. The rank of the matrix $M = \begin{bmatrix} 5 & 10 & 10 \\ 1 & 0 & 2 \\ 3 & 6 & 6 \end{bmatrix}$ is

(*a*) 0 (*b*) 1

(*c*) 2 (*d*) 3

27. Three fair cubical dice are thrown simultaneously. The probability that all three dice have the same number of dots on the faces showing up is (up to third decimal place) ______.

28. Starting with $x = 1$, the solution of the equation $x^3 + x = 1$, after two iterations of Newton-Raphson's method (up to two decimal places) is ______.

29. Let $I = \int_C (2zdx + 2ydy + 2xdz)$ where x, y, z are real, and let C be the straight line segment from point A : (0, 2, 1) to point B : (4, 1, – 1). The value of I is ______.

30. Let $f(x) = e^{x + x^2}$ for real x. From among the following, choose the Taylor series approximation of $f(x)$ around $x = 0$, which includes all powers of x less than or equal to 3,

(*a*) $1 + x + x^2 + x^3$ (*b*) $1 + x + \frac{3}{2}x^2 + x^3$

(*c*) $1 + x + \frac{3}{2}x^2 + \frac{7}{6}x^3$ (*d*) $1 + x + 3x^2 + 7x^3$

31. A three dimensional region R of finite volume is described by

$$x^2 + y^2 \le z^3 \; ; 0 \le z \le 1,$$

where x, y, z are real. The volume of R (up to two decimal places) is ______.

32. Which one of the following is the general solution of the first order differential equation

$$\frac{dy}{dx} = (x + y - 1)^2,$$

where x, y are real?

(*a*) $y = 1 + x + \tan^{-1}(x + c)$, where c is a constant.

(*b*) $y = 1 + x + \tan(x + c)$, where c is a constant.

(*c*) $y = 1 - x + \tan^{-1}(x + c)$, where c is a constant.

(*d*) $y = 1 - x + \tan(x + c)$, where c is a constant.

33. The value of $\lim_{x \to 0} \frac{x^3 - \sin(x)}{x}$ is

(*a*) 0 (*b*) 3

(*c*) 1 (*d*) –1

34. The differential equation $\frac{d^2y}{dx^2} + 16y = 0$ for $y(x)$ with the two boundary conditions $\left.\frac{dy}{dx}\right|_{x=0} = 1$ and $\left.\frac{dy}{dx}\right|_{x=\frac{\pi}{2}} = -1$ has

(*a*) No solution

(*b*) Exactly two solutions

(*c*) Exactly one solution

(*d*) Infinitely many solutions

35. Consider the following partial differential equation for u(x,y) with the constant c>1:

$$\frac{\partial u}{\partial y} + c\frac{\partial u}{\partial x} = 0$$

Solution of this equation is

(a) $u(x, y) = f(x + cy)$ (b) $u(x, y) = f(x - cy)$

(c) $u(x, y) = f(cx + y)$ (d) $u(x, y) = f(cx - y)$

36. Consider the two-dimensional velocity field given by $\vec{V} = (5 + a_1x + b_1y)\hat{i} + (4 + a_2x + b_2y)\hat{j}$, where a_1, b_1, a_2 and b_2 are constants. Which one of the following conditions needs to be satisfied for the flow to be incompressible?

(a) $a_1 + b_1 = 0$ (b) $a_1 + b_2 = 0$

(c) $a_2 + b_2 = 0$ (d) $a_2 + b_1 = 0$

37. A six-face fair dice is rolled a large number of times. The mean value of the outcomes is ____

38. The product of eigenvalues of the matrix P is

$$P = \begin{bmatrix} 2 & 0 & 1 \\ 4 & -3 & 3 \\ 0 & 2 & -1 \end{bmatrix}$$

(a) – 6 (b) 2

(c) 6 (d) – 2

39. P(0, 3), Q(0.5, 4) and R(1, 5) are three points on the curve defined by $f(x)$. Numerical integration is carried out using both Trapezoidal rule and Simpson's rule within limits $x = 0$ and $x = 1$ for the

(a) 0

(b) 0.25

(c) 0.5

(d) 1

40. For the vector $\vec{V} = 2yz\hat{i} + 3xz\hat{j} + 4xy\hat{k}$, the value of $\nabla.(\nabla \times \vec{V})$ is ________

41. For a steady flow, the velocity field is $\vec{V} = (-x^2 + 3y)\hat{i} + (2xy)\hat{j}$. The magnitude of the acceleration of a particle at (1, –1) is

(a) 2

(b) 1

(c) $2\sqrt{5}$

(d) 0

ANSWERS

General Aptitude

1. (b)	2. (b)	3. (a)	4. (b)	5. (b)	6. (a)	7. (d)	8. (c)	9. (c)	10. (d)
11. (c)	12. (d)	13. (c)	14. (d)	15. (d)	16. (b)	17. (b)	18. (d)	19. (a)	20. (b)
21. (d)	22. (a)	23. (d)	24. (b)	25. (c)	26. (c)	27. (c)	28.(a)	29. (a)	30. (d)
31. (b)	32. (a)	33. (b)	34. (a)	35. (a)	36. (d)	37. (c)	38. (a)	39. (a)	40. (d)
41. (d)	42. (b)	43. (a)	44. (d)	45. (a)	46. (8)	47. (c)	48. (b)	49. (c)	

Engineering Mathematics

1. (0.5)	2. (134)	3. (a)	4. (c)	5. (a)	6. (a)	7. (a)	8. (a)
9. (c)	10. (d)	11. (10)	12. (a)	13. (a)	14. (b)	15. (c)	16. (8)
17. (16)	18. (0)	19. (2)	20. (a)	21. (15)	22. (54)	23. (5)	24. (c)
25. 0	26. (c)	27. 0.028	28. 0.68	29. –11	30. (c)	31. 0.7853	32. (d)
33. (d)	34. (a)	35. (b)	36. (b)	37. 3.5	38. (b)	39. (a)	40. 0
41. (c)							

EXPLANATIONS

1. $81 \times \left(\frac{16}{25}\right)^{x+2} \div \left(\frac{3}{5}\right)^{2x+4} = 144$

$$\Rightarrow \left(\frac{4}{5}\right)^{2x+4} \times \left(\frac{5}{3}\right)^{2x+4} = \frac{144}{81}$$

$$\Rightarrow \left(\frac{4}{3}\right)^{2x+4} = \frac{144}{81}$$

$$\Rightarrow \left(\frac{4}{3}\right)^{x+2} = \sqrt{\frac{144}{81}}$$

$$\Rightarrow \left(\frac{4}{3}\right)^{x} \times \frac{16}{9} = \frac{12}{9}$$

$$\Rightarrow \left(\frac{4}{3}\right)^{x} = \frac{3}{4}$$

$$\Rightarrow \left(\frac{4}{3}\right)^{x} = \left(\frac{4}{3}\right)^{-1}$$

$$x = -1$$

3. Card with visible face 2 should have red on the other side.

Card with visible face 3 may have any color on other side.

Card with visible face red should have even number on other side.

Card with visible face blue may have any number on other side.

Hence, to verify the above proposition, cards with face 2 and Red must be turned.

5. Favourable outcomes = {(1, 1), (1, 4), (2, 2), (3, 3), (4, 1), (4, 4), (5, 5), (6, 6)}

No. of favourable outcomes = 8

Total outcomes = 36

$\therefore$ Probability $= \frac{8}{36} = \frac{2}{9}$

7. The elevators has moved from 2nd to 5th floor between time 25 min to 30 min.

Elevation stayed at 4th floor for

= 3 + 4 + 3 + 2 + 3 + 2 + 2 = 19 min

Elevation stayed at ground floor for

= 1 + 2 + 1 + 1 + 1 + 1 + 2 + 1 + 2 + 1 + 1 + 1 + 1 + 2 + 1 + 1 + 1

= 21 min.

Hence, (*i*) and (*ii*) both are false.

8. K < G, P < G, P > S, G < D

if K < G and G < D

Hence, K < D

If S < P and P < G

Hence, S < G

Hence, (*ii*) and (*iv*) are correct.

9.

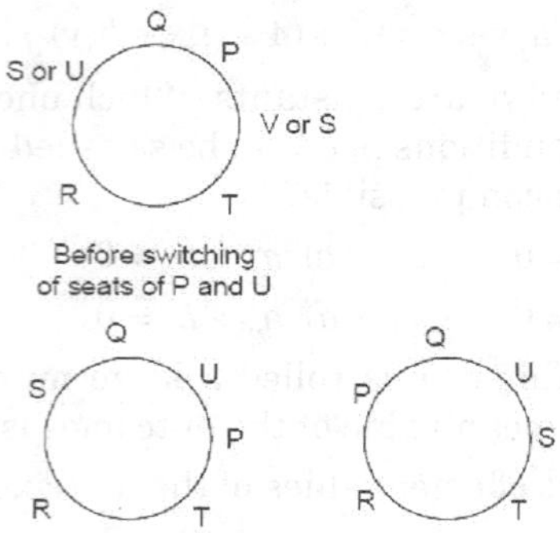

Before switching of seats of P and U

Ist possibility   2nd possibility

10. Let C = Speed by cycling (km/h)

W = Speed of walking (km/h)

Now, by according to questions,

$$C \times \frac{2}{4} + W \times 2 \times \frac{3}{4} = 19$$

$$\Rightarrow C + 3W = 38 \quad ...(i)$$

and, $$C \times \frac{2}{2} + W \times \frac{2}{2} = 26$$

$$\Rightarrow C + W = 26 \quad ...(ii)$$

By solving equations (*i*) and (*ii*) we get,

W = 6 km/h

11.

Each digit can be filled in 7 ways as 0, 5 and 9 is not allowed, so each of these places can be filled by 1, 2, 3, 4, 6, 7, 8 so required probability is $\left(\frac{7}{10}\right)^{K}$ or 0.7^{K}

$$\frac{fC}{TC} = \frac{7^{K}}{10^{K}} = 0.7^{K}$$

12. $y \times 162$ = Perfect cube

Option, $y = 24 \Rightarrow 2^3 \times 3\,(2 \times 81) \neq$ Not perfect cube

$y = 27 \Rightarrow 3^3 \times (2 \times 3^4) \neq$ Not perfect cube

$y = 32 \Rightarrow 2^5 \times 2 \times 3^4 \neq$ Not perfect cube

$y = 36 \Rightarrow 2^2 \times 3^2 \times 2 \times 3^4$

$= 2^3 \times 3^6 = (2 \times 3^2)^3$

is a perfect cube Hence the answer is, $y = 36$

13. Correct option is; Wished

After Rajender Chola **retured** from his voyage to Indonesia, he **wished** to visit the temple in Thanjavur.

Both are events of past. Use of past perfect form is unwarranted as it reflects part of past.

14. Besides money

Research in the work place reveals the people works for many reasons **besides money.** Besides conveys the meaning of 'in addition'

Beside means 'next to'

15. Following seating arrangement can be drawn

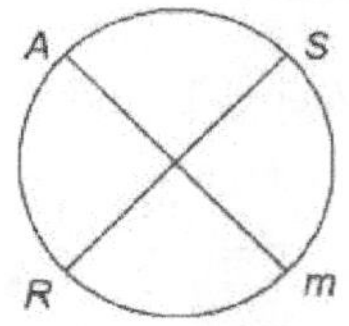

Therefore correct option is (*d*).

16. History is viewed through the filter of rationalism.

17. $$\frac{(x+y)-|x-y|}{2} \quad ...(i)$$

If $x > y$ $\quad |x-y| = x-y;$

If $x < y$ $\quad |x-y| = y-x$

Now if $x > y$, above expression (*i*) becomes

$$\frac{(x+y)-(x-y)}{2} = \frac{2x}{2}$$

$$= y = \text{minimum of } (x, y) \text{ as } x > y$$

Now if $x < y$,

$$\frac{x+y-(y-x)}{2} = \frac{2y}{2}$$

$$= x = \text{minimum of } (x, y) \text{ as } x < y$$

Therefore correct answer is option (*b*).

Alternate solution:

Use easy values,

$$x = 1$$

and $\quad y = -2$

Now, $$\frac{(1-2)-|1-(-2)|}{2} = -2$$

or $\quad x = 2$

and $\quad y = -1$

$$\frac{(2-1)-|2-(-1)|}{2} = \frac{1-3}{2} = -1$$

which is minimum of (x, y).

Therefore, correct answer is option (*b*).

18. As there are 4 people A, G, N, S and four colours so without any restriction total ways have to be $4 \times 4 = 16$.

Now, Arun → dislikes Red and Shweta dislikes white so $16 - 2 = 14$ ways.

Therefore correct answer should be option (*d*).

Alternate solution:

Only one option is less than 16.

Therefore correct answer should be option (d).

19. Out of six people. 3 place definitely occupied by right handed people as at least 2 women are there so these two will sit adjacently. Now as only on seat is left it will be occupied by a left handed man because on right side of this seat is sitting an right handed man.

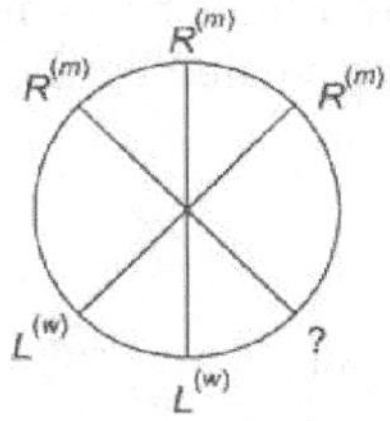

? have to be a left handed man.

Here, $R^{(m)}$indicates right handed man and $L^{(W)}$ indicates left handed women. Therefore answer should be only 2 women.

20. Let number of multiple choice question be 'x'

∴ Total number of essay type question = $20 - x$

Now according to question

$$x \times 3 + (20-x)11 = 100$$

$$\therefore \quad 3x + 220 - 11x = 100$$

$$\therefore \quad 220 - 100 = 8x$$

$$\therefore \quad x = \frac{120}{8}$$

$$\therefore \quad x = 15$$

∴ Opation (*b*) is correct.

21. Aversion means a strong dislike or dinclination. Hence aversion is synonyms of dislike.

Aversion means giving of light which is synonym of radiant plunder means loot while yielding and resistant are not synonyms of each other.

∴ Option (*d*) is correct.

22. From the given condition of sequence we get following order of building in

X, Z, V, W, Y

Hence 'V' building is in the middle

∴ Option (*a*) is correct

23. $$n(\text{R}) = 3$$

$$n(\text{G}) = 4$$

$$n(\text{B}) = 3$$

Probability that 2 socks choosen are of both Red or Both Green or Both blue

$$\text{Same colour} = \frac{3}{10}\times\frac{2}{9}+\frac{4}{10}\times\frac{3}{9}+\frac{3}{10}\times\frac{2}{9}$$

$$= \frac{6}{90}+\frac{12}{90}+\frac{6}{10}$$

$$= \frac{24}{90}$$

$$= \frac{4}{15}$$

$\therefore$ Option (*d*) is correct.

24. Saturn is bright enough to be seen on a clear night with the naked eye.

$\therefore$ Option (*b*) is correct

25. The author's belief that ideology is not as important as literature is revealed by the word 'urgent' in above paragraph.

$\therefore$ Option (*c*) is correct.

27. $e^x + 0.5x^2 - 2 = 0$

Differention w.r.t to 'x' we get

$$e^x + x = 0$$

$\therefore$ Hence '*x*' has total number of root which lies in the range [– 5,5] is '2'.

30. Let H is household consumption and P be other consumption

Now according to question,

$$H \times 0.8 + P \times 1.7 = (H + P) \times 0.75$$

From this it is concluded that ratio is negative.

31. From given condition

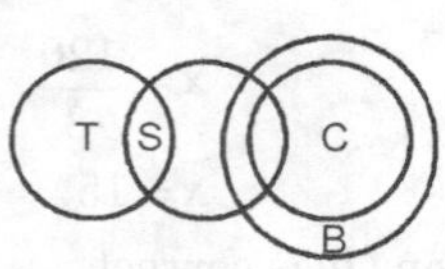

Only conclusion (*ii*) follows

32. Use of conditional sentence based on past participle form.

I could have made arrangements had I been informed earlier.

33. $40\% \text{ of } 360° = \frac{40}{100}\times 360°$

$$= 144°$$

So, the angle subtended on pie chart will be 144°

34. Hurtful means causing pain or suffering or something that is damaging or harmful. The expression 'sharp tongue' defines a bitter or critical manner of speaking.

35. According to question,

Length of Truck + gap required = 10 + 20 = 30 m and

Length of Car + gap required = 5 + 15 = 20 m

Alternative pairs of Truck and Car needs

$$30 + 20 = 50 \text{ m}$$

Let '*n*' be the number of repetition of (Truck + Car) in 1 hour (3600 secs)

Given speed = 36 km/hr

= 10 m/sec

$$\therefore \quad \frac{50m \times n}{3600 \text{ secs}} = 36 \text{ km/hr}$$

$$\therefore \quad \frac{50n}{3600} \text{ m/sec} = 10 \text{ m/sec}$$

$$\therefore \quad n = \frac{36000}{50}$$

= 720 (Trucks + Cars)

So, 720 (Trucks + Cars) passes

= (720 × 2)

= 1440 vehicles

36. Antagonist is a adversary or one who opposes/contends against another. Hence the closest meaning to word "antagonistic" is hostile.

37. Contour lines can be observed to cross region with height

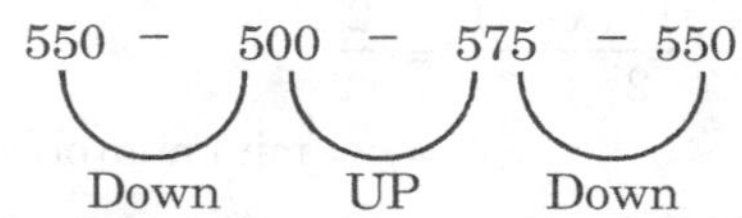

Down-Up-Down satisfies.

38. Subgroups containing only Indians

$$= {}^3C_1 + {}^3C_2 + {}^3C_3$$

$$= 3 + 3 + 1 = 7$$

and Subgroups containing one Indian and rest Chinese

$$= {}^3C_1[{}^3C_1 + {}^3C_2 + {}^3C_3]$$

$$= 3[3 + 3 + 1]$$

$$= 21$$

and Subgroups containing two Indian and remaining Chinese

$$= {}^3C_2[{}^3C_1 + {}^3C_2 + {}^3C_3]$$

$$= 21$$

and Subgroups containing three Indian and remaining Chinese

$$= {}^3C_3[{}^3C_1 + {}^3C_2 + {}^3C_3]$$

$$= 7$$

Hence total number of subgroups

$= 7 + 21 + 21 + 7$

$= 56.$

39. Following circular seating arrangement can be drawn from the given data

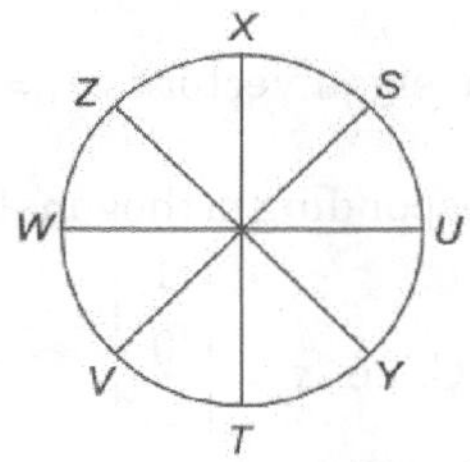

From the given arrangement 'X' is the third to the left of 'V'.

42.

Salary	No. of employees	Cumulative
2000	45	45
30000	25	70
40000	20	90
60000	8	98
150000	2	100

Median $= \dfrac{100+1}{2} = 50.5$

From cumulative column,

Median salary = 30000

46.

$$\begin{array}{r} 5\,\underline{0}\,\underline{0}\,\underline{0}\,\underline{0} \\ 4\,8\,\underline{8}\,8\,9 \\ \hline 1\,1\,1\,1 \end{array}$$

$sum = 0 + 0 + 0 + 0 + 8 = 8$

47.

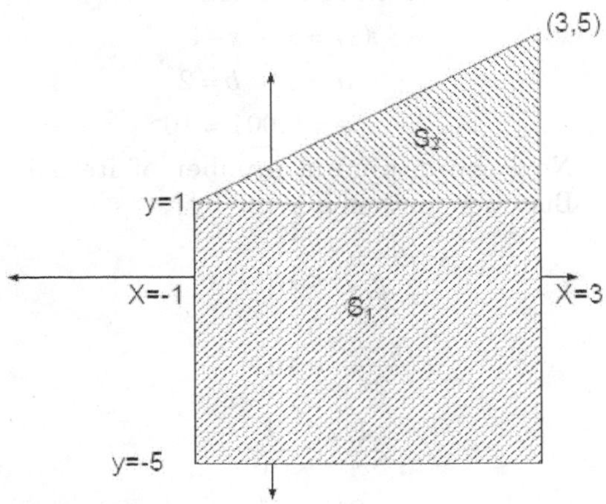

$S = S_1 + S_2$

$= 6 \times 4 + \dfrac{1}{2} \times 4 \times 4$

$= 24 + 8 = 32$

Engineering Mathematics

1. Probability of getting (H.H.H)

$$= \frac{1}{2} \times \frac{1}{2} \times \frac{1}{2}$$

$$= \frac{1}{8}$$

Probability of getting (H.H.H.H)

$$= \frac{1}{2} \times \frac{1}{2} \times \frac{1}{2} \times \frac{1}{2}$$

$$= \frac{1}{16}$$

$\because$ Now, Conditional probability of getting next H after (H.H.H)

$$= \frac{\frac{1}{16}}{\frac{1}{8}}$$

$$= \frac{1}{16} \times \frac{8}{1} = \frac{1}{2} = 0.5$$

2. $V = x^2\hat{i} + 2y^3\hat{j} + z^4\hat{k}$

Divergence $(\forall)$

$$= \frac{\partial}{\partial x}(x^2)\hat{i} + \frac{\partial}{\partial y}(2y^3)\hat{j} + \frac{\partial}{\partial x}(z^4)\hat{k}$$

$2x + 6y^2 + 4z^3$

$[\text{Divergence (V)}]_{(x=1,\, y=2,\, z=3)}$

$= 2 \times 1 + 6 \times 2^2 + 4 \times 3^3$

$= 2 + 24 + 108 = 134$

3.

$3x_1 + 2x_2 = c_1$

$4x_1 + x_2 = c_2$

$$\begin{bmatrix} 3 & 2 \\ 4 & 1 \end{bmatrix} = [A]$$

$$|[A] - \lambda[1]| = 0$$

$$\begin{bmatrix} 3-\lambda & 4 \\ 2 & 1-\lambda \end{bmatrix} = 0$$

$3 + \lambda^2 - 4\lambda - 8 = 0$

$\lambda^2 - 4\lambda - 5 = 0$

4. $w = f(x, y)$

$$\frac{dw}{dt} = \frac{\partial w}{\partial x}\frac{dx}{dt} + \frac{\partial w}{\partial y}\frac{dy}{dt}$$

5. Target is having inclination of 45° with x-axis

$$\Rightarrow \quad \frac{dy}{dx} = \tan 45°$$

$$\Rightarrow \quad \frac{d(x \ln x)}{dx} = 1$$

$\Rightarrow \quad ln\, x + \frac{x}{x} = 1$

$\therefore$ At $x = 1, y = 1 \times ln'\, 1 = 0$

(1, 0)

6. Given second order differential equation is

$$y'' - 4y' + 3y = 2t - 3t^2$$
$$f(D) = D^2 - 4D + 3$$

$$\text{P.I.} = \frac{I}{f(D)}(2t - 3t^2)$$

$$= \frac{1}{D^2 - 4D + 3}(2t - 3t^2)$$

$$= \left[\frac{1}{1-D} - \frac{1}{3\left(1 - \frac{D}{3}\right)}\right]\left(t - \frac{3t^2}{2}\right)$$

$$(1 + D + D^2 + ...)\left(t - \frac{3t^2}{2}\right) - \frac{1}{3}\left(t - \frac{D}{3} + \frac{D^2}{9} + ...\right) \times \left(t - \frac{32t^2}{2}\right)$$

$$= \left(t - \frac{3t^2}{2} + 1 - 3t - 3\right) - \frac{1}{3}\left(1 - \frac{3t^2}{2} + \frac{1}{3} - 8 - \frac{1}{3}\right)$$

$$= -2 - 2t - t^2$$

7. Given, $\quad I = \int_0^1 \frac{(\sin^{-1} x)^2}{\sqrt{1 - x^2}}\, dx$

Put $\sin^{-1} x = t$

Now, $\quad \frac{dx}{\sqrt{1 - x^2}} = dt = \int_0^{\pi/2} t^2 dt$

$$= \left[\frac{t^3}{3}\right]_0^{\pi/2} = \frac{\pi^3}{24}$$

8. $\quad A = \begin{bmatrix} 1 & 5 \\ 6 & 2 \end{bmatrix} [B]$

$$= \begin{bmatrix} 3 & 7 \\ 8 & 4 \end{bmatrix}$$

$$AB^T = \begin{bmatrix} 1 & 5 \\ 6 & 2 \end{bmatrix}\begin{bmatrix} 3 & 8 \\ 7 & 4 \end{bmatrix}$$

$$= \begin{bmatrix} 3 + 35 & 8 + 20 \\ 18 + 14 & 48 + 8 \end{bmatrix}$$

$$= \begin{bmatrix} 38 & 28 \\ 32 & 56 \end{bmatrix}$$

9. From the given question

The given matrix is symmetric and all its eigen values are distinct.

Therefore all its eigen vectors are orthogonal

one of the eigen vector is $x_1 = \begin{bmatrix} 1 \\ 0 \\ 1 \end{bmatrix}$

The corresponding orthogonal vector in the given option is C. i.e. $x_2 = \begin{bmatrix} 1 \\ 0 \\ -1 \end{bmatrix}$

$$x_1^T . x_2 = [101]\begin{bmatrix} 1 \\ 0 \\ -1 \end{bmatrix}$$

$$= 1 \times 1 + 0 \times 0 + 1 \times (-1)$$
$$= 1 + 0 - 1 = 0$$

10. For a complex number z

$$\underset{z \to i}{lt} \frac{z^2 + 1}{z^3 + 2z - i(z^2 + 2)} \quad \left(\frac{0}{0}\text{form}\right)$$

$$\underset{z \to i}{lt} \frac{2z}{3z^3 + 2 - i(2z)} = \frac{2i}{3i^2 + 2 - i(2i)}$$

$$= \frac{2i}{3i^2 + 2 - 2i^2}$$

$$= \frac{2i}{-3 + 2 + 2}$$

$$= \frac{2i}{-3 + 4} = 2i$$

So the value of complex number z is $2i$.

11. Given one of the real roots is

$$f(x) = x^e - x - 1$$
$$a = 1 \quad b = 2$$
$$\in = 0.001 = 10^{-3}$$

Now the minimum number of iterations by Bisection method is given by

$$\frac{|b - a|}{2^n} < \in$$

$$\frac{2 - 1}{2^n} < 10^{-3}$$

$$\frac{1}{2^n} < \frac{1}{10^3}$$

$$2^n > 10^3$$

$$2^n > 1000$$

$$\frac{n}{n_2} > ln\, 1000$$

$$n > \frac{ln1000}{ln2}$$

$$n > \frac{ln10^3}{ln2}$$

$$n > \frac{3\log 10}{0.3010}$$

$$n > \frac{3}{0.3010}$$

$$n > 9.96$$

$$n = 10$$

So the minimum no. of iterations is 10.

12. The differential equation is given by

$$(t^2 - 81)\frac{dy}{dt} + 5 + y = \sin t$$

$$\frac{dy}{dt} + \frac{5t}{t^2-81}y = \frac{\sin t}{t^2-81}$$

$$P = \frac{5t}{t^2-81}$$

$$Q = \frac{\sin t}{t^2-81}$$

$$\text{Integrating factor (I.F)} = e^{\int \rho dt}$$

$$= e^{\int \frac{5t}{t^2-81}}dt$$

$$= e^{\frac{5}{2}\int \frac{2t}{t^2-8t}}dt$$

$$= e^{\frac{5}{2}\ln(t^2-81)}$$

$$= e^{\ln(t^2-81)^{5/2}}$$

$$= (t^2-81)^{\frac{5}{2}}$$

Solution is given by

$$y(t^2-81)^{\frac{5}{2}} = \int \frac{\sin t}{t^2-81}.(t^2-81)^{\frac{5}{2}}dt$$

$$y = \frac{\int(\sin t)(t^2-81)^{3/2}dt}{(t^2-81)^{5/2}} + \frac{C}{(t^2-81)^{5/2}}$$

The solution exists for $t \neq +9$

Therefore option (*a*) is correct.

13. A function $f(x) = \begin{cases} e^x & x < 1 \\ \ln x + ax^2 + bx & x \geq 1 \end{cases}$

$$\text{LHD} = \underset{x\to 1}{\text{Lt}}\frac{f(x)-f(1)}{x-1}$$

$$= \underset{x\to 1}{\text{Lt}}\frac{e^x-(a+b)}{x-1}$$

$$= \underset{x\to 1}{\text{Lt}}\frac{e^x}{1} = e$$

$$\text{R.H.D.} = \underset{x\to 1}{\text{Lt}}\frac{f(x)-f(1)}{x-1}$$

$$= \underset{x\to 1}{\text{Lt}}\frac{\ln x + ax^2 + bx - a + b}{x-1}$$

$$= \underset{x\to 1}{\text{Lt}}\frac{\frac{1}{x}+2ax+b}{1}$$

$$= 1 + 2s + b$$

$$\therefore \quad \text{L.H.D} \neq \text{R.H.D.}$$

Hence $f(x)$ is not derivable at $x = 1$

14. From the diagram line C is given by $y = x$

$$\text{Line integral (I)} = \int_C (x^2 + iy^2)dz$$

$$= \int_C (x^2+iy^2)(dx+idy)$$

$$= \int_C (x^2+iy^2)(dx+idx)$$

$$= \int x^2dx + ix^2dx + ix^2dx - x^2dx$$

$$= 2i\int_0^1 x^2dx$$

$$= 2i\left(\frac{x^3}{3}\right)\Big|_0^1$$

$$= 2i\left(\frac{1}{3}-0\right)$$

$$= \frac{2i}{3}$$

15. Given,

$$f(x) = R\sin\left(\frac{\pi x}{2}\right) + S, \quad ...(1)$$

$$\text{Given} \quad f'\left(\frac{1}{2}\right) = \sqrt{2} \quad ...(2)$$

$$\int_0^1 f(x)dx = \frac{2R}{\pi} \quad ...(3)$$

Now we need to find R and S.

$$f'(x) = R\cos\left(\frac{\pi x}{2}\right)\frac{\pi}{2}$$

$$f'\left(\frac{1}{2}\right) = R\cos\left(\frac{\pi}{4}\right)\times\frac{\pi}{2}$$

$$= \sqrt{2}$$

$$\Rightarrow \quad \frac{R}{\sqrt{2}} \times \frac{\pi}{2} = \sqrt{2}$$

$$\Rightarrow \quad R = \frac{4}{\pi}$$

Now, $\int f(x)dx = \int R \sin\frac{\pi x}{2} + S$

Putting $R = \frac{4}{\pi}$ we get

$$\int f(x)dx = \int \frac{4}{\pi}\sin\left(\frac{\pi x}{2}\right)dx + \int S\, dx$$

$$= \frac{4}{\pi} \times -\frac{\cos\left(\frac{\pi x}{2}\right)}{\frac{\pi}{2}} + Sx$$

$$= \frac{-8}{\pi^2}\cos\left(\frac{\pi x}{2}\right) + Sx$$

Putting limit 0 and 1

$$\int_0^1 f(x)dx = \frac{-8}{\pi^2}\left(\cos\frac{\pi}{2} - \cos(0)\right) + S(1-0)$$

$$= \frac{2R}{\pi}$$

$$\Rightarrow \quad \frac{-8}{\pi^2}(0-1) + S = \frac{2R}{\pi}$$

Put $R = \frac{4}{\pi}$ and solve for S

$$\Rightarrow \quad S = 0$$

So, $R = \frac{4}{\pi}$

and $S = 0$ is answer.

16. $x^2 - 13x + 36 = 0$

In base b $13 = 1 \times b^1 + 3 \times b^0 = b + 3$

In base b $36 = 3 \times b^1 + 6 \times b^0$

$= 3b + 6$

So the equation becomes

$x^2 - (b+3)x + (3b+6) = 0$

Now since it is given that x = 5 is a solution , so

$5^2 - (b+3)5 + (3b+6) = 0$

$\Rightarrow \quad -2b + 16 = 0$

$\Rightarrow \quad b = 8$

Same can be obtained by putting $x = 6$ also.

17. In the undirected graph G, we have n vertices and 25 edges

Hence

$n \leq ?$

$e = 25$

As each vertex has at least 3 degree

and $2e = \Sigma\text{degree}$

i.e., $2e \geq 3n$

$n \leq 2e/3$

$$\Rightarrow \quad n \leq \frac{2\times 25}{3} \leq 16.66$$

$\therefore$ n is at the maximum possible value of 16.

18. There is no need to add any ordered pair as the given hasse diagram is already a lattice.

19. By adding both P and Q

$$P + Q = \begin{bmatrix} 0 & -1 & -2 \\ 8 & 9 & 10 \\ 8 & 8 & 8 \end{bmatrix}$$

Magnitude of P & Q

$|P+Q| = -16 + 16 = 0$

So rank $\neq 3$

rank < 3

Take the 2×2 minor $\begin{bmatrix} 0 & -1 \\ 8 & 9 \end{bmatrix} = 8 \neq 0$

So, rank of P + Q is 2

20. Given that

$$p(P) = \frac{1}{4} \quad \text{...}(i)$$

$$p(P \mid Q) = \frac{1}{2} \quad \text{...}(ii)$$

$$p(Q \mid P) = \frac{1}{3} \quad \text{...}(iii)$$

$$p(\overline{P} \mid \overline{Q}) = ?$$

First solve for $p(Q)$ and $p(P \cap Q)$ from equation (*ii*) and (*iii*) as follows :

From equation (*ii*)

$$p(P \mid Q) = \frac{p(P \cap Q)}{p(Q)} = \frac{1}{2} \quad \text{...}(iv)$$

From equation (*iii*) $p(Q \mid P) = \frac{p(P \cap Q)}{p(P)} = \frac{1}{3}$

$$\Rightarrow \quad p(P \cap Q) = \frac{1}{3} \times p(P)$$

$$= \frac{1}{3} \times \frac{1}{4} = \frac{1}{12}$$

Substitute in equation (*iv*) get

$$p(Q) = \frac{p(P \cap Q)}{1/2}$$

$$= \frac{1/12}{1/2}$$

$$= \frac{2}{12} = \frac{1}{6}$$

$$p(P) = \frac{1}{4}$$

$$p(Q) = \frac{1}{6}$$

and $p(P \cap Q) = \frac{1}{12}$

we need to find

$$p(\bar{P} \mid \bar{Q}) = \frac{p(\bar{P} \cap \bar{Q})}{p(\bar{Q})}$$

$$= \frac{1-(P \cup Q)}{1-p(Q)}$$

$$= 1 - \frac{[p(P) + p(Q) - p(P \cap Q)]}{1-p(Q)}$$

$$= \frac{1-\left[\frac{1}{4}+\frac{1}{6}-\frac{1}{12}\right]}{1-\frac{1}{6}}$$

$$= \frac{\frac{2}{3}}{\frac{5}{6}} = \frac{4}{5}$$

$$p(\bar{P} \mid \bar{Q}) = \frac{4}{5}$$

21. Then generating function $\{a_n\}_{n=0}^{\infty}$ is $\frac{1+z}{(1-z)^3}$

$$A(z) = \sum a_r z^r$$

$$= \frac{1+z}{(1-z)^3}$$

We can replace z by x, we get

$$A(x) = \sum a_r x^r$$

$$= \frac{1+x}{(1-x)^3}$$

$$= \frac{1}{(1-x)^3} + \frac{x}{(1-x)^3}$$

$$= \sum_{r=0}^{\infty} {}^{3-1+r}C_r x^r + x\sum_{r=0}^{\infty} {}^{3-1+r}C_r x^r$$

$$= \sum_{r=0}^{\infty} {}^{r+2}C_2 x^r + \sum_{r=0}^{\infty} {}^{r+2}C_2 x^{r+1}$$

$${}^2C_2 x^0 + {}^3C_2 x^1 + {}^4C_2 r^2 + {}^5C_2 x^3 \ldots\ldots\ldots$$
$$+ {}^2C_2 x^1 + {}^3C_2 x^2 + {}^4C_2 x^3 + \ldots\ldots\ldots$$

Now we read to find a_0 and a_3 which are coefficient of x^0 and x^3 respectively.

$$a_0 = \text{Coefficient } x^0 = {}^2C_2 = 1$$

$$a_3 = \text{Coefficient } x^3 = {}^5C_2 + {}^4C_2 = 16$$

So, $a_3 - a_0 = 16 - 1 = 15$

22. We know that in Poisson distribution representation

$$E(X) = V(X) = \lambda$$

Given $\lambda = 5$

$$E(X) = V(X) = 5$$

We need $E[(X + 2)^2]$

$$= E(X^2 + 4x + 4)$$

$$= E(X^2) + 4E(X + 4)$$

To find $E(X^2)$ we write,

$$V(X) = E(X^2) - (E(X))^2$$

$$5 = E(X^2) - 5^2$$

So, $E(X^2) = 5^2 + 5 = 30$

$$\text{required value} = 30 + 4 \times 5 + 4 = 54$$

23. $f(\lambda) = \lambda^3 - 4\lambda^2 + a\lambda + 30 = 0$

Given $m = 2$ will be one root of polynomial

So, $2^3 - 4 \times 2^2 + a \times 2 + 30 = 0$

$\Rightarrow$ $8 - 16 + 2a + 30 = 0$

$\Rightarrow$ $a = -11$ (satisfy the equation)

So, the equation is $\lambda^3 - 4\lambda^2 - 11\lambda + 30 = 0$

Divided polynomials by $(\lambda - 2)$

$$\frac{\lambda^3 - 4\lambda^2 - 11\lambda + 30}{\lambda - 2} = \lambda^2 - 2\lambda - 15$$

roots of $\lambda^2 - 2\lambda - 15 = 0$ are

$$\lambda = \frac{2 \pm \sqrt{4+60}}{2}$$

$$= \frac{2+8}{2} = 5 \text{ and } -3$$

$\lambda_1 = 2$ $\lambda_2 = 5$ $\lambda_3 = 3$ the maximum absolute eigen value is 5.

24. $|A - \lambda I| = 0$

$$\therefore \begin{bmatrix} 1 & 2 & 3 & 4 & 5 \\ 5 & 1 & 2 & 3 & 4 \\ 4 & 5 & 1 & 2 & 3 \\ 3 & 4 & 5 & 1 & 2 \\ 2 & 3 & 4 & 5 & 1 \end{bmatrix} - \lambda \begin{bmatrix} 1 & 0 & 0 & 0 & 0 \\ 0 & 1 & 0 & 0 & 0 \\ 0 & 0 & 1 & 0 & 0 \\ 0 & 0 & 0 & 1 & 0 \\ 0 & 0 & 0 & 0 & 1 \end{bmatrix} = 0$$

$$\begin{bmatrix} 1-\lambda & 2 & 3 & 4 & 5 \\ 5 & 1-\lambda & 2 & 3 & 4 \\ 4 & 5 & 1-\lambda & 2 & 3 \\ 3 & 4 & 5 & 1-\lambda & 2 \\ 2 & 3 & 4 & 5 & 1-\lambda \end{bmatrix} = 0$$

Sum of all elements in any one row must be zero,

i.e., $15 - \lambda = 0$

$\therefore \lambda = 15$

25. Mutual information of two random variables is a measure of the mutual dependence of the two variables.

Given that, X and Y are independent.

Hence, I(X : Y) = 0

26. $$M = \begin{bmatrix} 5 & 10 & 10 \\ 1 & 0 & 2 \\ 3 & 6 & 6 \end{bmatrix}$$

Now applying, $R_1 \leftrightarrow R_2$ in above matrix we get

$$\begin{bmatrix} 1 & 0 & 2 \\ 5 & 10 & 10 \\ 3 & 6 & 6 \end{bmatrix}$$

Applying $R_2 \leftarrow R_2 - 5R_1$ and $R_3 \leftarrow R_3 - 3R_1$, we get

$$\begin{bmatrix} 1 & 0 & 2 \\ 0 & 10 & 0 \\ 0 & 6 & 0 \end{bmatrix}$$

Now applying, $R_3 \leftarrow R_3 - \frac{6}{10} R_2$, we get

$$\begin{bmatrix} 1 & 0 & 2 \\ 0 & 10 & 0 \\ 0 & 0 & 0 \end{bmatrix}$$

The above matrix is in Echelon form

Rank of matrix M is given by

$\rho(M) = 2$

27. When three dice are thrown

Total number of possible cases

$= (6)^3 = 216$

Outcomes for all three dice having same number are,

$$\begin{Bmatrix} (1,1,1) & (2,2,2) & (3,3,3) \\ (4,4,4) & (5,5,5) & (6,6,6) \end{Bmatrix}$$

Hence number of favourable cases = 6

$\therefore$ Required probability

$$= \frac{\text{Number of favourable cases}}{\text{Total number of outcomes}}$$

$$= \frac{6}{216}$$

$$= \frac{1}{36}$$

$= 0.028$

28. $f(x) = x^3 + x -$

Now $f(1) = (1)^3 + 1 - 1 = 1$

and $f'(x) = \frac{d}{dx}(x^3 + x - 1)$

$= 3x^2 + 1$

$\therefore f'(1) = 4$

Now by Newton-Raphson method,

$$x_{i+1} = x_i - \frac{f(x_i)}{f'(x_i)}$$

For $x_0 = 1$, $x_1 = x_0 - \frac{f(x_0)}{f'(x_0)}$

$$= 1 - \frac{1}{4} = 0.75$$

and for $x_1 = 0.75$

$$x_2 = x_1 - \frac{f(x_1)}{f'(x_1)}$$

$$= 0.75 - \frac{f(0.75)}{f'(0.75)}$$

$$= 0.75 - \frac{0.171875}{2.6875}$$

$= 0.686 \approx 0.68$

29. A(0, 2, 1) and 6(4, 1, –1)

The equation of the line AB is

$$\frac{x-0}{4-0} = \frac{y-2}{1-2}$$

$$= \frac{z-1}{-1-1} = t(\text{say})$$

$\therefore x = 4t;\quad y = -t + 2;\quad z = -2t + 1$

$\therefore dx = 4dt;\quad dy = -dt;\quad dz = -2dt$

Also t varies from 0 to 1

$$\therefore I = \int_0^1 2(-2t+1)4\,dt + 2(-t+2)(-dt) + 2(4t)(-2\,dt)$$

$$= \int_0^1 (-16t + 8 + 2t - 4 - 16t)\,dt$$

$$= \int_0^1 (-30t + 4)\,dt$$

$$= \left(-30\frac{t^2}{2} + 4t\right)\Bigg|_0^1$$

$= (-15 + 4) = -11$

30. At $x = 0$,

$f(x) = e^x e^{x^2}$

$$= \left(1 + x + \frac{x^2}{2!} + \frac{x^3}{3!} + \ldots\right)\left(1 + x^2 + \frac{x^4}{2!} + \frac{x^6}{3!} + \ldots\right)$$

$$= 1 + x^2 + \frac{x^4}{2} + \frac{x^6}{6} + \ldots + x + x^3 + \frac{x^5}{2} + \frac{x^7}{6} + \ldots$$

$$+ \frac{x^2}{2} + \frac{x^4}{2} + \frac{x^6}{4} + \frac{x^8}{12} \ldots + \frac{x^3}{6} + \frac{x^5}{6} + \frac{x^7}{12} + \frac{x^9}{36} + \ldots$$

$$= 1 + x + \frac{3}{2}x^2 + \frac{7}{6}x^3$$

31. Let $x^2 + y^2 = t^2$

and $t^2 = z^3$

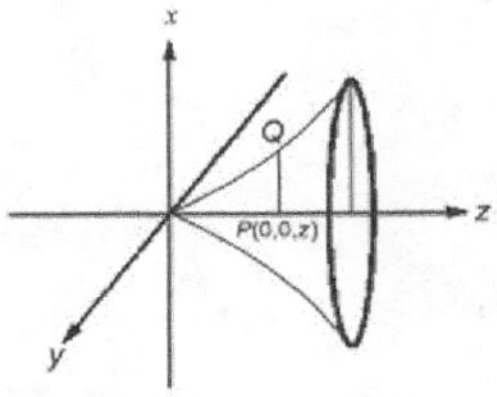

Here revolution is about z axis

Now volume of region R = $\int_0^1 \pi(\text{PQ})^2\, dz$

Where PQ is radius of circle at some z.

Now $\text{PQ} = \sqrt{x^2 + y^2}$

$\therefore$ $(\text{PQ})^2 = x^2 + y^2 = z^3$

$\therefore$ Volume of region R = $\int_0^1 \pi t^2\, dz$

$$= \int_0^1 \pi z^3\, dz$$

$$= \frac{\pi}{4}$$

$$= 0.7853$$

32. $\dfrac{dy}{dx} = (x + y - 1)^2$ ...(*i*)

Let $x + y - 1 = t$ ...(*ii*)

Now differiating equation (*ii*) w.r.t to '*x*' we get

$$1 + \frac{dy}{dx} = \frac{dt}{dx}$$

$$\therefore \quad \frac{dy}{dx} = \frac{dt}{dx} - 1$$

Substituting equations (*ii*) and (*iii*) in equation (*i*) we get

$$\frac{dt}{dx} - 1 = t^2$$

$$\frac{dt}{dx} = t^2 + 1$$

$$\frac{dt}{t^2 + 1} = dx$$

Now integrating on both side we get,

$$\int \frac{1}{t^2 + 1}\, dt = \int dx$$

$\therefore$ $\tan^{-1} t = x + c$

$\therefore$ $x + y - 1 = \tan(x + c)$ $(\because t = x + y - 1)$

$\therefore$ $y = 1 - x + \tan(x + c)$

33. From L – hospital rule

$$\lim_{x \to 0} \frac{x^3 - \sin(x)}{x} = \frac{3x^2 - \cos x}{1}$$

$$= \frac{3 \times 0^2 - \cos 0}{1}$$

$$= \frac{0 - 1}{1} = -1$$

34. $(D^2 + 16)y = 0$

A.E. $\to D^2 + 16 = 0$

$\Rightarrow$ $D = \pm 4i$

$y = C_1 \cos 4x + C_2 \sin 4x$

$$\frac{dy}{dx} = -4C_1 \sin 4x + 4C_2 \cos 4x$$

$$\left.\frac{dy}{dx}\right|_{x=0} = -4C_1 \times 0 + 4C_2 \times 1 = 4C_2$$

$$4C_2 = 1$$

$$C_2 = \frac{1}{4}$$

Now, $\left.\dfrac{dy}{dx}\right|_{x=\pi/2} = -4C_1 \times \sin 4/2 + 4C_2 \times \cos 4/2$

$$\left.\frac{dy}{dx}\right|_{x=\pi/2} = 0 + 4C_2 = 4\left(\frac{1}{4}\right) = 1$$

But $\left.\dfrac{dy}{dx}\right|_{x=\pi/2} = -1$ (given)

Hence no solution.

36. For continuous and incompressible flow

$$u_x + u_y = 0$$

$$a_1 + b_2 = 0$$

37.

x.	1	2	3	4	5	6
p(x)	1/6	1/6	1/6	1/6	1/6	1/6

We know that

$$\mu = \sum x.p(x)$$

$$= (1 + 2 + 3 + 4 + 5 + 6) \times \frac{1}{6}$$

$$= 21 \times \frac{1}{6} = 3.5$$

38. Product of eigen value = |p|

$$= \begin{vmatrix} 2 & 0 & 1 \\ 4 & -3 & 3 \\ 0 & 2 & -1 \end{vmatrix} = 2$$

39. $y_0 = 3, y_1 = 4, y_2 = 5$

$h = 0.5$

From Trapezoidal rule, we know that

$$I_T = \frac{h}{2}[(y_0 + y_2) + 2y_1]$$

$$= \frac{0.5}{2}[(3+5) + 2 \times 4] = 4$$

Also by Simpson's $\frac{1}{3}$ *rd* rule

$$I_S = \frac{h}{3}[(y_0 + y_2) + 4y_1]$$

$$= \frac{0.5}{3}[(3+5) + 4 \times 4] = 4$$

$$I_T - I_S = 4 - 4 = 0$$

40. Given vector $\vec{V} = 2yz\hat{i} + 3xz\hat{j} + 4xy\hat{k}$

Now $\nabla \times \vec{V} = \begin{vmatrix} i & j & k \\ \frac{\partial}{\partial x} & \frac{\partial}{\partial y} & \frac{\partial}{\partial z} \\ 2yz & 3xz & 4xy \end{vmatrix} = xi - 2yj + zk$

$\nabla.(\nabla \times \vec{V}) = 1 - 2 + 1 = 0$

41. $\vec{V} = (-x^2 + 3y)i + (2xy)j$

$u = -x^2 + 3y, v = 2xy$

$$\frac{\partial u}{\partial x} = -2x, \quad \frac{\partial v}{\partial x} = 2y$$

$$\frac{\partial u}{\partial y} = 3,$$

$$\frac{\partial v}{\partial y} = 2x$$

Now magnitude of particle at (1, –1)

$u = -4,$

$v = -2,$

$$\frac{\partial u}{\partial x} = -2, \quad \frac{\partial u}{\partial y} = 3, \quad \frac{\partial v}{\partial x} = -2, \quad \frac{\partial v}{\partial y} = 2$$

$$a_x = u\frac{\partial u}{\partial x} + v\frac{\partial u}{\partial y}$$

$$= (-4) \times (-2) + (-2) \times 3$$

$$a_x = 8 - 6 = 2$$

$$a_y = u\frac{\partial v}{\partial x} + v\frac{\partial v}{\partial y}$$

$$= [(-4) \times (-2) + (-2) \times 2]$$

$$a_y = 8 - 4 = 4$$

$$a = \sqrt{a_x^2 + a_y^2}$$

$$= \sqrt{2^2 + 4^2}$$

$$= 2\sqrt{5}$$

■■

SOLVED PAPER - 2018

GENERAL APTITUDE

SET-1

1. "When she fell down the _______. She received many _________ but little help." The words that best fill the blanks in the above sentence are

(*a*) stairs, stares (*b*) stairs, stairs

(*c*) stares, stairs (*d*) stares, stares

2. "In spite of being warned repeatedly, he failed to correct his __________ behaviour." The word that best fills that blank in the above sentence is

(*a*) rational (*b*) reasonable

(*c*) errant (*d*) good

3. For $0 \leq x \leq 2\pi$, sinx and cosx are both decreasing functions in the interval _________.

(*a*) $\left\{0, \frac{\pi}{2}\right\}$ (*b*) $\left\{\frac{\pi}{2}, \pi\right\}$

(*c*) $\left\{\pi, \frac{\pi}{2}\right\}$ (*d*) $\left\{\frac{3\pi}{2}, 2\pi\right\}$

4. The area of an equilateral triangle is $\sqrt{3}$. What is the perimeter of the triangle?

(*a*) 2 (*b*) 4

(*c*) 6 (*d*) 8

5. Arrange the following three-dimensional objects in the descending order of their volumes:

(i) A cuboid with dimensions 10 cm, 8 cm and 6 cm

(ii) A cube of side 8 cm

(iii) A cylinder with base radius 7 cm and height 7 cm

(iv) A sphere of radius 7 cm

(*a*) (i), (ii), (iii), (iv) (*b*) (ii), (i), (iv), (iii)

(*c*) (iii), (ii), (i), (iv) (*d*) (iv), (iii), (ii), (i)

6. An automobile travels from city A to city B and returns to city A by the same route. The speed of the vehicle during the onward and return journeys were constant at 60 km/h and 90 km/h, respectively. What is the average speed in km/h for the entire journey?

(*a*) 72 (*b*) 73

(*c*) 74 (*d*) 75

7. A set of 4 parallel lines intersect with another set of 5 parallel lines. How many parallelograms are formed?

(*a*) 20 (*b*) 48

(*c*) 60 (*d*) 72

8. To pass a test, a candidate needs or answer at least 2 out of 3 questions correctly. A total of 6,30,000 candidates appeared for the test. Question A was correctly answered by 3,30,000 candidates. Question B was answered correctly by 2,50,000 candidates. Question C was answered correctly by 2,60,000 candidates. Both questions A and B were answered correctly by 1,00,000 candidates. Both questions B and C were answered correctly by 90,000 candidates. Both questions A and C were answered correctly by 80,000 candidates. If the number of students answering all questions correctly is the same as the number answering none, how many candidates failed to clear the test?

(*a*) 30,000 (*b*) 2,70,000

(*c*) 3,90,000 (*d*) 4,20,000

9. If $x^2 + x - 1 = 0$ what is the value of $x^4 + \frac{1}{x^4}$?

(*a*) 1 (*b*) 5

(*c*) 7 (*d*) 9

10. In a detailed study of annual crow births in India, it was found that there was relatively no growth during the period 2002 to 2004 and a sudden spike from 2004 to 2005. In another unrelated study, it was found that the revenue from cracker sales in India which remained fairly flat from 2002 to 2004, saw a sudden spike in 2005 before declining again in 2006. The solid line in the graph below refers to annual sale of crackers and the dashed line refers to the annual crow births in India. Choose the most appropriate inference from the above data.

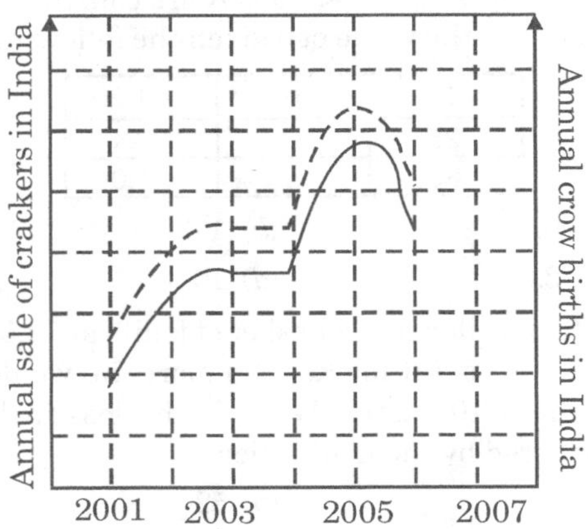

(*a*) There is a strong correlation between crow birth and cracker sales.

(*b*) Cracker usage increases crow birth rate.

(*c*) If cracker sale declines, crow birth will decline.

(*d*) Increased birth rate of crows will cause an increase in the sale of crackers.

SET-2

1. "The driver applied the ________ as soon as she approached the hotel where she wanted to take a ________."

The words that best fill the blanks in the above sentence are

(*a*) brake, break (*b*) break, break

(*c*) brake, brake (*d*) break, brake

2. "It is no surprise that every society has had codes of behavior; however, the nature of these codes is often ________."

The word that best fills the blank in the above sentence is

(*a*) unpredictable (*b*) simple

(*c*) expected (*d*) strict

3. Hema's age is 5 years more than twice Hari's age. Suresh's age is 13 years less than 10 times Hari's age. If Suresh is 3 times as old as Hema. How old is Hema?

(*a*) 14 (*b*) 17

(*c*) 18 (*d*) 19

4. Tower A is 90 m tall and tower B is 140 m tall. They are 100 m apart. A horizontal skywalk connects the floors at 70 m in both the towers. If a taut rope connects the top of tower A to the bottom of tower B, at what distance (in meters) from tower A will the rope intersect the skywalk?

5. The temperature T in a room varies as a function of the outside temperature T_0 and the number of persons in the room p, according to the relation $T = K(\theta p + T_0)$, where θ is K are constants. What would be the value of θ given the following data?

T_0	p	T
25	2	32.4
30	5	42.0

(*a*) 0.8 (*b*) 1.0

(*c*) 2.0 (*d*) 10.0

6. A fruit seller sold a basket of fruits at 12.5% loss. Had he sold it for Rs. 108 more, he would have made a 10% gain. What is the loss in Rupees incurred by the fruit seller?

(*a*) 48 (*b*) 52

(*c*) 60 (*d*) 108

7. The price of a wire made of a superalloy material is proportional to the square of its length. The price of 10 m length of the wire is Rs. 1600. What would be the total price (in Rs.) of two wires of lengths 4 m and 6 m?

(*a*) 768 (*b*) 832

(*c*) 1440 (*d*) 1600

8. What of the following function(s) in an accurate description of the graph for the range(s) indicated?

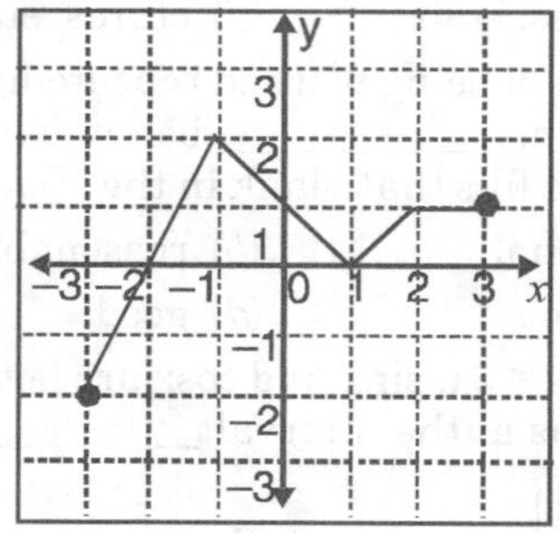

(*i*) $y = 2x + 4$ for $-3 \le x \le -1$

(*ii*) $y = |x - 1|$ for $-1 \le x \le 2$

(*iii*) $y = ||x| - 1|$ for $-1 \le x \le 2$

(*iv*) $y = 1$ for $2 \le x \le 3$

(*a*) (*i*), (*ii*) and (*iii*) only

(*b*) (*i*), (*ii*) and (*iv*) only

(*c*) (*i*) and (*iv*) only

(*d*) (*ii*) and (*iv*) only

9. Consider a sequence of number $a_1, a_2, a_3, \ldots, a_n$ where $a_n = \frac{1}{n} - \frac{1}{n+2}$, for each integer $n > 0$. What is the sum of the first 50 terms?

(*a*) $\left(1+\frac{1}{2}\right)-\frac{1}{50}$ (*b*) $\left(1+\frac{1}{2}\right)+\frac{1}{50}$

(*c*) $\left(1+\frac{1}{2}\right)-\left(\frac{1}{51}+\frac{1}{52}\right)$ (*d*) $1-\left(\frac{1}{51}+\frac{1}{52}\right)$

10. Each of the letters arranged as below represents a unique from 1 to 9. The letters are positioned in the figure such that (A × B × C), (B × G × E) and (D × E × F) are equal. Which integer among the following choices cannot be represented by the letters A, B, C, D, E, F or G?

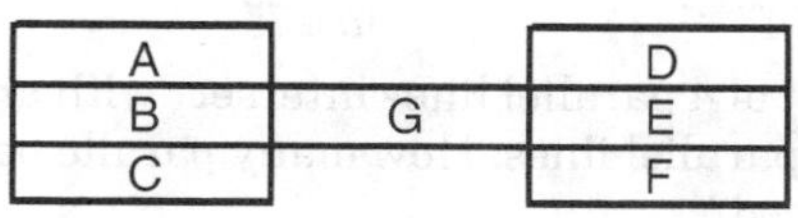

(*a*) 4 (*b*) 5

(*c*) 6 (*d*) 9

SET-3

1. "From where are they bringing their books? ______ bringing ______ books from ______."
The words that best fill the blanks in the above sentence are
(a) Their, they're, there
(b) They're, their, there
(c) There, their, they're
(d) They're, there, there

2. "A ______ investigation can sometimes yield new facts, but typically organized ones are more successful."
The word that best fills the blank in the above sentence is
(a) meandering (b) timely
(c) consistent (d) systematic

3. The area of a square is d. What is the area of the circle which has the diagonal of the square as its diameter?
(a) πd (b) πd^2
(c) $\frac{1}{4}\pi d^2$ (d) $\frac{1}{2}\pi d$

4. What would be the smallest natural number which when divided either by 20 or by 42 or by 76 leaves a remainder of 7 in each case?
(a) 3047 (b) 6047
(c) 7987 (d) 63847

5. What is the missing number in the following sequence?
2, 12, 60, 240, 720, 1440, ______, 0
(a) 2880 (b) 1440
(c) 720 (d) 0

6. In appreciation of the social improvements completed in a town, a wealthy philanthropist decided to gift Rs. 750 to each male senior citizen in the town and Rs. 1000 to each female senior citizen. Altogether, there were 300 senior citizens eligible for this gift. However, only 8/9th of the eligible men and 2/3rd of the eligible women claimed the gift. How much money (in Rupees) did the philanthropist give away in total?
(a) 1,50,000 (b) 2,00,000
(c) 1,75,000 (d) 1,51,000

7. In $pqr \neq 0$ and $p^{-x} = \frac{1}{q}, q^{-y} = \frac{1}{r}, r^{-z} = \frac{1}{p}$, what is the value of the product xyz?
(a) –1 (b) $\frac{1}{pqr}$
(c) 1 (d) pqr

8. In a party, 60% of the invited guests are male and 40% are female. If 80% of the invited guests attended the party and if all the invited female guests attended, what would be the ratio of males of females among the attendees in the party?
(a) 2 : 3 (b) 1 : 1
(c) 3 : 2 (d) 2 : 1

9. In the figure below, $\angle DEC + \angle BFC$ is equal to ___.

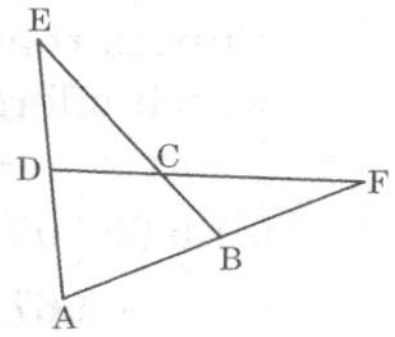

(a) $\angle BCD - \angle BAD$
(b) $\angle BAD + \angle BCF$
(c) $\angle BAD + \angle BCD$
(d) $\angle CBA + \angle ADC$

10. A six sided unbiased die with four green faces and two red faces is rolled seven times. Which of the following combinations is the most likely outcome of the experiment?
(a) Three green faces and four red faces.
(b) Four green faces and three red faces.
(c) Five green faces and two red faces.
(d) Six green faces and one red face.

SET-4

1. "By giving him the last ______ of the cake, you will ensure lasting ______ in our house today."
The words that best fill the blanks in the above sentence are
(a) peas, piece (b) piece, peace
(c) peace, piece (d) peace, peas

2. "Even though there is a vast scope for its ______, tourism has remained a/an ______ area."
The words that best fill the blanks in the above sentence are
(a) improvement, neglected
(b) rejection, approved
(c) fame, glum
(d) interest, disinterested

3. If the number 715 ? 423 is divisible 3 (? denotes the missing digit in the thousandths place), then the smallest whole number in the place of ? is ______
(a) 0 (b) 2
(c) 5 (d) 6

4. What is the value of $1 + \frac{1}{4} + \frac{1}{16} + \frac{1}{64} + \frac{1}{256} + ...?$
(a) 2 (b) $\frac{7}{4}$
(c) $\frac{3}{2}$ (d) $\frac{4}{3}$

5. As 1.5 m in tall person is standing at a distance of 3 m from a lamp post. The light from the lamp at the top of the post casts her shadow. The length of the shadow is twice her height. What is the height of the lamp post in meters?

(*a*) 1.5 (*b*) 3
(*c*) 4.5 (*d*) 6

6. Leila aspires to buy a car worth Rs. 10,00,000 after 5 years. What is the minimum amount in Rupees that she should deposit now in a bank which offers 10% annual rate of interest, if the interest was compounded annually?

(*a*) 5,00,000 (*b*) 6,21,000
(*c*) 6,66,667 (*d*) 7,50,000

7. Two alloys A and B contain gold and copper in the ratios of 2 : 3 and 3 : 7 by mass, respectively. Equal masses of alloys A and B are melted to make an alloy C. The ratio of gold to copper in alloy C is ______.

(*a*) 5 : 10 (*b*) 7 : 13
(*c*) 6 : 11 (*d*) 9 : 13

8. The Cricket Board has long recognized John's potential as a leader of the team. However, his on-field Temper has always been a matter of concern for them since his junior days. While this aggression has filled stadia with die-hard fans, it has taken a toll on his own batting. Until recently, it appeared that he found it difficult to convert his agression into big scores. Over the past three seasons though, that picture of John has been replaced by a cerebral, calculative and successful batsman-captain. After many years, it appears that the team has finally found a complete captain.

Which of the following statements can be logically inferred from the above paragraph?

(i) Even as a junior cricketer, John was considered a good captain.
(ii) Finding a complete captain is a challenge.
(iii) Fans and the Cricket Board have differing views on what they want in a captain.
(iv) Over the past three seasons John has accumulated big scores.

(*a*) (i), (ii) and (iii) only (*b*) (iii) and (iv) only
(*c*) (ii) and (iv) only (*d*) (i), (ii), (iii) and (v)

9. A cab was involved in a hit and run accident at right you are given the following data about the cabs in the city and the accident.

(i) 85% of cabs in the city are green and the remaining cabs are blue.
(ii) A witness identified the cab involved in the accident as blue.
(iii) It is known that a witness can correctly identify the cab colour only 80% of the time.

Which of the following options is closest to the probability that the accident was caused by a blue cab?

(*a*) 12% (*b*) 15%
(*c*) 41% (*d*) 80%

10. A coastal region with unparalleled beauty is home to many species of animals. It is dotted with coral reefs and unspoilt white sandy beaches. It has remained inaccessible to tourists due to poor connectivity and lack of accommodation. A company has spotted the opportunity and is planning to develop a luxury resort with helicopter service to the nearest major city airport. Environmentalists are upset that this would lead to the region becoming crowded and polluted like any other major beach resorts.

Which one of the following statements can be logically inferred from the information given in the above paragraph?

(*a*) The culture and tradition of the local people will be influenced by the tourists.
(*b*) The region will become crowded and polluted due to tourism.
(*c*) The coral reefs are on the decline and could soon vanish.
(*d*) Helicopter connectivity would lead to an increase in tourists coming to the region.

SET-5

1. Going by the ______ that many hands make light work, the school ______ involved all the students in the task."

(*a*) principle, principal
(*b*) principal, principle
(*c*) principle, principle
(*d*) principal, principal

2. "Her ____ should not be confused with miserliness; she is ever willing to assist those in need."

The word that best fills the blank in the above sentence is

(*a*) cleanliness (*b*) punctuality
(*c*) frugality (*d*) greatness

3. Seven machines take 7 minutes to make 7 identical toys. At the same rate, how many minutes would it take for 100 machines to make 100 toys?

(*a*) 1 (*b*) 7
(*c*) 100 (*d*) 700

4. A rectangle becomes a square when its length and breadth are reduced by 10m and 5 m, respectively, during this process, the rectangle loses 650 m^2 of area. What is the area of the original rectangle in square meters?

(*a*) 1125 (*b*) 2250

(*c*) 2924 (*d*) 4500

5. A number consists of two digits. The sum of the digits is 9. If 45 is subtracted from the number, its digits are interchanged. What is the number?

(*a*) 63 (*b*) 72

(*c*) 81 (*d*) 90

6. For integers a, b and c, what would be the minimum and maximum values respectively of a + b + c if $\log |a| + \log |b| + \log |c| = 0$

(*a*) –3 and 3 (*b*) –1 and 1

(*c*) –1 and 3 (*d*) 1 and 3

7. Given that a and b are integers and $a + a^2b^3$ is odd, which one of the following statements is correct?

(*a*) a and b are both odd

(*b*) a and b are both even

(*c*) a is even and b is odd

(*d*) a is odd and b is even

8. From the time the front of train enters a platform, it takes 25 seconds for the back of the train to leave the platform, while travelling at a constant speed of 54 km/h. At the same speed, it takes 14 seconds to pass a man running at 9 km/h in the same direction as the train. What is the length of the train and that of the platform in meters, respectively?

(*a*) 210 and 140 (*b*) 162.5 and 187.5

(*c*) 245 and 130 (*d*) 175 and 200

9. Which of the following functions describe the graph shown in the below figure.

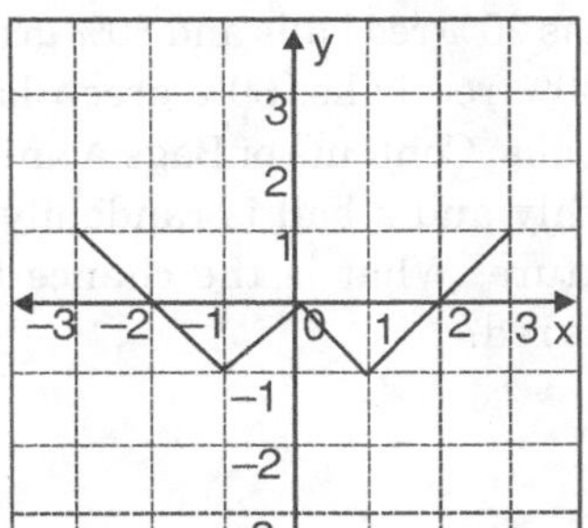

(*a*) $y = ||x| + 1| - 2$ (*b*) $y = ||x| - 1| - 1$

(*c*) $y = ||x| + 1| - 1$ (*d*) $y = ||x - 1| - 1|$

10. Consider the following three statements:

(i) Some roses are red

(ii) All red flowers fade quickly

(iii) Some roses fade quickly

Which of the following statements can be logically inferred from the above statements?

(*a*) If (i) is true and (ii) is false, then (iii) is false.

(*b*) If (i) is true and (ii) is false, then (iii) is true.

(*c*) If (i) and (ii) are true, then (iii) is true.

(*d*) If (i) and (ii) are false, then (iii) is false.

SET-6

1. "Since you have gone off the ______, the ______ sand is likely to damage the car". The words that best fill the blanks in the above sentence are

(*a*) course, coarse (*b*) course, course

(*c*) coarse, course (*d*) coarse, coarse

2. "A common misconception among writers is that sentence structure mirrors thought; the more ______ the structure, the more complicated the ideas".

(*a*) detailed (*b*) simple

(*c*) clear (*d*) convoluted

3. The three roots of the equation $f(x) = 0$ are x = {–2, 0, 3}. What are the three values of x for which $f(x - 3) = 0$?

(*a*) –5, –3, 0 (*b*) –2, 0, 3

(*c*) 0, 6, 8 (*d*) 1, 3, 6

4. For what values of k given below is $\dfrac{(k+2)^2}{k-3}$ and integer?

(*a*) 4, 8, 18 (*b*) 4, 10, 16

(*c*) 4, 8, 28 (*d*) 8, 26, 28

5. Functions, F(a, b) and G(a, b) are defined as follows: $F(a, b) = (a - b)^2$ and $G(a, b) = |a - b|$, where $|x|$ represents the absolute value of x. What would be the value of G(F(1, 3), G(1, 3))?

(*a*) 2 (*b*) 4

(*c*) 6 (*d*) 36

6. An e-mail password must contain three characters. The password has to contain one numeral from 0 to 9, one upper case and one lower case character from the English alphabet. How many distinct passwords are possible?

(*a*) 6,760 (*b*) 13,520

(*c*) 40,560 (*d*) 1,05,456

7. In a certain code AMCF is written as EQGJ and NKUF is written as ROYJ. How will DHLP be written in the code?

(*a*) RSTN (*b*) TLPH

(*c*) HLPT (*d*) XSVR

8. A class of twelve children has two more boys than girls. A group of three children are randomly picked from this class to accompany the teacher on a field trip. What is the probability that the group accompanying the teacher contains more girls than boys?

(a) 0 (b) $\frac{325}{864}$

(c) $\frac{525}{864}$ (d) $\frac{5}{12}$

9. A designer uses marbles of four different colours for his designs. The cost of each marble is the same, irrespective of the colour. The table below shows the percentage of marbles of each colour used in the current design. The cost of each marble increased by 25%. Therefore, the designer decided to reduce equal numbers of marbles of each colour to keep the total cost unchanged. What is the percentage of blue marbles in the new design?

Blue	Black	Red	Yellow
40%	25%	20%	15%

(a) 35.75 (b) 40.25

(c) 43.75 (d) 46.25

10. P, Q, R and S crossed a lake in a boat that can hold a maximum of two persons, with only one set of oars. The following additional facts are available

(i) The boat held two persons on each of the three forward trips across lake and one person on each of the two return trips.

(ii) P is unable is row when someone else is in the boat.

(iii) Q is unable to row with anyone else except R.

(iv) Each person rowed for at least one trip.

(v) Only one person can row during a trip.

Who rowed twice?

(a) P (b) Q

(c) R (d) S

ENGINEERING MATHEMATICS

SET-1

1. Let N be a 3 by 3 matrix with real number entries. The matrix N is such that $N^2 = 0$. The eigen values of N are

(a) 0, 0, 0 (b) 0, 0, 1

(c) 0, 1, 1 (d) 1, 1, 1

2. Let $f_1(z) = z^2$ and $f_2(z) = \bar{Z}$ be two complex variable function. Here $\bar{Z}$ is the complex conjugate of z. Choose the correct answer.

(a) Both $f_1(z)$ and $f_2(z)$ are analytic

(b) Only $f_1(z)$ is analytic

(c) Only $f_2(z)$ is analytic

(d) Both $f_1(z)$ and $f_2(z)$ are not analytic

3. X and Y are two independent random variables with variances 1 and 2, respectively. Let Z = X – Y. The variance of Z is

(a) 0 (b) 1

(c) 2 (d) 3

4. Consider two functions $f(x) = (x - 2)^2$ and $g(x) = 2x - 1$, where x is real. The smallest value of x for which f(x) = g (x) is ________.

5. Considers a sequence of tossing of a fair coin where the outcomes of tosses are independent. The probability of getting the head for the third time in the fifth toss is

(a) $\frac{5}{16}$ (b) $\frac{3}{16}$

(c) $\frac{3}{5}$ (d) $\frac{9}{16}$

6. Given $\vec{F} = (x^2 - 2y)\hat{i} - 4yz\hat{j} + 4xz^2\hat{k}$, the value of the line integral $\int_c \vec{F} \cdot dl$ along the straight line c from (0, 0, 0) to (1, 1, 1) is

(a) $\frac{3}{16}$ (b) 0

(c) $\frac{-5}{12}$ (d) –1

7. Two bags A and B have equal number of balls. Bag A has 20% red balls and 80% green balls. Bag B has 30% red balls. 60% green balls and 10% yellow balls. Contents of Bags A and B are mixed thoroughly and a ball is randomly picked from the mixture. What is the chance that the ball picked is red?

(a) 20%

(b) 25%

(c) 30%

(d) 40%

8. Consider the following system of linear equations:

$$3x + 2ky = -2$$
$$kx + 6y = 2$$

Here, x and y are the unknown and k is a real constant. The value of k for which there are infinite number of solutions is

(*a*) 3 (*b*) 1

(*c*) –3 (*d*) –6

9. Consider the following equations

$$\frac{\partial V(x,y)}{\partial x} = px^2 + y^2 + 2xy$$

$$\frac{\partial V(x,y)}{\partial y} = x^2 + qy^2 + 2xy$$

where p and q are constants. V(x, y) that satisfies the above equations is

(*a*) $p\frac{x^3}{3} + q\frac{y^3}{3} + 2xy + 6$

(*b*) $p\frac{x^3}{3} + q\frac{y^3}{3} + 5$

(*c*) $p\frac{x^3}{3} + q\frac{y^3}{3} + x^2y + xy^2 + xy$

(*d*) $p\frac{x^3}{3} + q\frac{y^3}{3} + x^2y + xy^2$

SET-2

1. Which one of the following matrices is singular?

(*a*) $\begin{bmatrix} 2 & 5 \\ 1 & 3 \end{bmatrix}$ (*b*) $\begin{bmatrix} 3 & 2 \\ 2 & 3 \end{bmatrix}$

(*c*) $\begin{bmatrix} 2 & 4 \\ 3 & 6 \end{bmatrix}$ (*d*) $\begin{bmatrix} 4 & 3 \\ 6 & 2 \end{bmatrix}$

2. For the given orthogonal matrix Q,

$$Q = \begin{bmatrix} \frac{3}{7} & \frac{2}{7} & \frac{6}{7} \\ -\frac{6}{7} & \frac{3}{7} & \frac{2}{7} \\ \frac{2}{7} & \frac{6}{7} & -\frac{3}{7} \end{bmatrix}$$

The inverse is

(*a*) $\begin{bmatrix} \frac{3}{7} & \frac{2}{7} & \frac{6}{7} \\ -\frac{6}{7} & \frac{3}{7} & \frac{2}{7} \\ \frac{2}{7} & \frac{6}{7} & -\frac{3}{7} \end{bmatrix}$ (*b*) $\begin{bmatrix} -\frac{3}{7} & -\frac{2}{7} & -\frac{6}{7} \\ \frac{6}{7} & -\frac{3}{7} & -\frac{2}{7} \\ -\frac{2}{7} & -\frac{6}{7} & \frac{3}{7} \end{bmatrix}$

(*c*) $\begin{bmatrix} \frac{3}{7} & -\frac{6}{7} & \frac{2}{7} \\ \frac{2}{7} & \frac{3}{7} & \frac{6}{7} \\ \frac{6}{7} & \frac{2}{7} & -\frac{3}{7} \end{bmatrix}$ (*d*) $\begin{bmatrix} -\frac{3}{7} & \frac{6}{7} & -\frac{2}{7} \\ -\frac{2}{7} & -\frac{3}{7} & -\frac{6}{7} \\ -\frac{6}{7} & -\frac{2}{7} & \frac{3}{7} \end{bmatrix}$

3. At the point $x = 0$, the function $f(x) = x^3$ has

(*a*) local maximum

(*b*) local minimum

(*c*) both local maximum and minimum

(*d*) neither local maximum nor local minimum

4. The value of the integral $\int_0^x x\cos^2 x dx$ is.

(*a*) $\frac{\pi^2}{8}$ (*b*) $\frac{\pi^2}{4}$

(*c*) $\frac{\pi^2}{2}$ (*d*) π^2

5. The solution at $x = 1$, $t = 1$ of the partial differential equation $\frac{\partial^2 u}{\partial x^2} = 25\frac{\partial^2 u}{\partial t^2}$ subject to initial conditions of $u(0) = 3x$ and $\frac{\partial u}{\partial x}(0) = 3$ is ______.

(*a*) 1 (*b*) 2

(*c*) 4 (*d*) 6

6. The solution (up to three decimal places) at $x = 1$ of the differential equation $\frac{d^2y}{dx^2} + 2\frac{dy}{dx} + y = 0$ subject to boundary conditions $y(0) = 1$ and $\frac{dy}{dx} = (0) = -1$ is _________.

SET-3

1. Which one of the following is a closed form expression for the generating function of the sequence $\{a_n\}$, where $a_n = 2n + 3$ for all $n = 0, 1, 2, ...$?

(*a*) $\frac{3}{(1-x)^2}$ (*b*) $\frac{3x}{(1-x)^2}$

(*c*) $\frac{2-x}{(1-x)^2}$ (*d*) $\frac{3-x}{(1-x)^2}$

2. Two people. P and Q, decide to independently roll two identical dice, each with 6 faces, Numbered 1 to 6. The person with the lower number wins. In case of a tie, they roll the dice repeatedly until there is no tie. Define a trial a throw of the dice by P and Q, Assume that all 6 numbers on each dice are equi-probable and that all trials are independent. The probability (rounded to 3 decimal places) that one of them wins on the third trial is _____.

3. The value of $\int_0^{\pi/4} x\cos(x^2)dx$ correct to three decimal places (assuming that $\pi = 3.14$) is ______.

4. Consider a matrix $A = uv^T$ where $u = \begin{pmatrix}1\\2\end{pmatrix}, v = \begin{pmatrix}1\\1\end{pmatrix}$.

Note that v^T denotes the transpose of v. The largest eigenvalue of A is ______.

5. Consider a matrix P whose only eigenvectors are the multiples of $\begin{bmatrix}1\\4\end{bmatrix}$.

Consider the following statements:

I. P does not have an inverse.

II. P has a repeated eigenvalue.

III. P cannot be diagonalized.

Which one of the following options is correct?

(*a*) Only I and III are necessarily true

(*b*) Only II is necessarily true

(*c*) Only I and II are necessarily true

(*d*) Only II and III are necessarily true

6. Let N be the set of natural numbers. Consider the following sets:

P. Set of rational numbers (positive and negative).

Q. Set of functions from {0, 1} to N.

R. Set of functions from N to {0, 1}.

S. Set of finite subsets of N.

Which of the sets above are countable?

(*a*) Q and S only (*b*) P and S only

(*c*) P and R only (*d*) P, Q and S only

SET-4

1. Let $f(x,y) = \frac{ax^2 + by^2}{xy}$, where a and b are constant. If $\frac{\partial f}{\partial x} = \frac{\partial f}{\partial y}$ at x = 1 and y = 2, then the relation between a and b is

(*a*) $a = \frac{b}{4}$ (*b*) $a = \frac{b}{2}$

(*c*) a = 2b (*d*) a = 4b

2. Consider matrix $A = \begin{bmatrix}k & 2k\\ k^2 - k & k^2\end{bmatrix}$ and vector $X = \begin{bmatrix}x_1\\x_2\end{bmatrix}$. The number of distinct real values of k for which the equation AX = 0 has infintely many solution is ______.

3. Let X_1, X_2, X_3 and X_4 be independent normal random variables with zero mean and unit variance. The probability that X_4 is the smallest among the four is ________.

4. Taylor series expansion of $f(x) = \int_0^x e^{-\left(\frac{t^2}{2}\right)} dt$ around x = 0 has the form

$$f(x) = a_0 + a_1x + a_2x^2 + ...$$

The coefficient a_2 (correct to two decimal places) is equal to ________.

5. A curve passes through the point (x = 1, y = 0) and satisfies the differential equation $\frac{dy}{dx} = \frac{x^2 + y^2}{2y} = \frac{y}{x}$. The equation that describes the curve is

(*a*) $In\left(1 + \frac{y^2}{x^2}\right) = x - 1$ (*b*) $\frac{1}{2}In\left(1 + \frac{y^2}{x^2}\right) = x - 1$

(c) $In\left(1 + \frac{y}{x}\right) = x - 1$ (*d*) $\frac{1}{2}In\left(1 + \frac{y}{x}\right) = x - 1$

6. A random variable X takes values –0.5 and 0.5 with probabilities $\frac{1}{4}$ and $\frac{3}{4}$, respectively. The noisy observation of X is Y = X + Z, where Z has uniform probability density over the interval (–1, 1). X and Z are independent. If the MAP rule based detector outputs $\hat{X}$ as

$$\hat{X} = \begin{cases} -0.5, & Y < \alpha \\ 0.5, & Y \geq \alpha, \end{cases}$$

then the value of α (accurate to two decimal places) is ________.

7. The contour C given below is on the complex plane $z = x + jy$, where $j = \sqrt{-1}$.

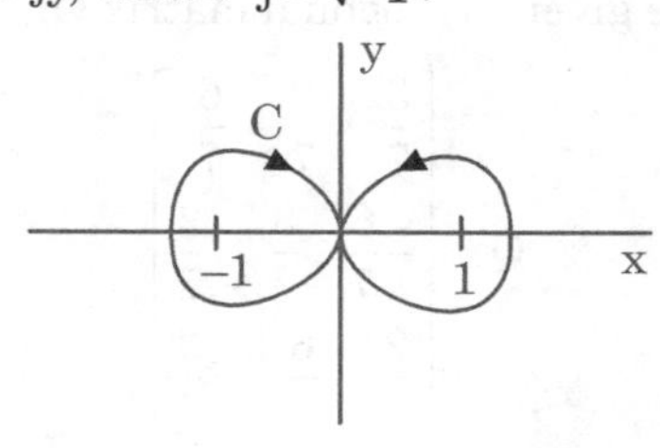

The value of the integral $\frac{1}{\pi j}\oint_C \frac{dz}{z^2 - 1}$ is ________.

8. Let $r = x^2 + y - z$ and $z^3 - zy + yz + y^3 = 1$. Assume that x and y are independent variables. At (x, y, z) = (2, –1, 1), the value (correct to two decimal places) of $\frac{\partial r}{\partial x}$ is ________.

SET-5

1. Four red balls, four green balls and four blue balls are put in a box. Three balls are pulled out of the box of random one after another without replacement. The probability that all the three balls are red is

(a) 1/72 (b) 1/55
(c) 1/36 (d) 1/27

2. The rank of the matrix $\begin{bmatrix} -4 & 1 & -1 \\ -1 & -1 & -1 \\ 7 & -3 & 1 \end{bmatrix}$ is

(a) 1 (b) 2
(c) 3 (d) 4

3. According to the Mean Value Theorem, for a continuous function $f(x)$ in the interval $[a, b]$, there exists a value ξ in this interval such that

$\int_n^b f(x)\,dx =$

(a) $f(\xi)(b-a)$ (b) $f(b)(\xi-a)$
(c) $f(a)(b-\xi)$ (d) 0

4. $F(z)$ is a function of the complex variable $z = x + iy$ given by

For what value of k will $F(z)$ satisfy the Cauchy-riemann equations?

$F(z) = iz + k\,Re(z) + i\,Im(z)$

(a) 0 (b) 1
(c) –1 (d) y

5. A six-faced fair dice is rolled five times. The probability (in%) of obtaining "ONE" at least four times is

(a) 33.3 (b) 3.33
(c) 0.33 (d) 0.0033

6. Let X_1, X_2 be two independent normal random variables with means μ_1, μ_2 and standard deviations σ_1, σ_2 respectively. Consider $Y = X_1 - X_2$; $\mu_1 = \mu_2 = 1$, $\sigma_1 = 1$, $\sigma_2 = 2$, Then,

(a) Y is normal distributed with mean 0 and variance 1

(b) Y is normally distributed with mean 0 and variance 5

(c) Y has mean 0 and variance 5, but is NOT normally distributed

(d) Y has mean 0 and variance 1, but is NOT normally distributed

7. The value of integral $\oiint_s \vec{r}.\vec{n}\,ds$ over the closed surface S bounding a volume, where $\vec{r} = x\hat{i} + y\hat{i} + z\hat{k}$ is the position vector and $\vec{n}$ is the normal to the surface S, is

(a) V (b) $2V$
(c) $3V$ (d) $4V$

8. $F(s)$ is the Laplace transform of the function $f(t) = 2t^2 e^{-t}$

F(1) is ______(correct to two decimal places).

9. The minimum value of 3x + 5y

such that : $3x + 5y \le 15$

$4x + 9y \le 8$

$13x + 2y \le 2$

$x \ge 0, y \ge 0$

SET-6

1. Let f be a real-valued function of a real variable defined as $f(x) = x^2$ for $x \ge 0$, and $f(x) = -x^2$ for $x < 0$. Which one of the following statements is true?

(a) f(x) is discontinuous at x = 0.

(b) f(x) is continuous but not differentiable at x = 0.

(c) f(x) is differentiable but its first derivative is not continuous at x = 0.

(d) f(x) is differentiable but its first derivative is not differentiable at x = 0.

2. The value of the directional derivative of the function $\phi(x, y, z) = xy^2 + yz^2 + zx^2$ at the point (2, –1, 1) in the direction of the vector p = i + 2j + 2k is

(a) 1 (b) 0.95
(c) 0.93 (d) 0.9

3. The value of the integral $\oint \frac{z+1}{z^2-4}dz$ in counter clockwise direction around a circle C of radius 1 with center at the point z = –2 is

(a) $\frac{\pi i}{2}$

(b) $2\pi i$

(c) $-\frac{\pi i}{2}$

(d) $-2\pi i$

4. A continuous-time out signal x(t) is an eigen function of an LTI system, if the output is

(*a*) k x(t), where k is an eigenvalue.

(*b*) $ke^{j\omega t}$ x(t), where k is an eigenvalue and $e^{j\omega t}$ is a complex exponential signal.

(*c*) x(t) $e^{j\omega t}$. where $e^{j\omega t}$ is a complex exponential signal.

(*d*) k H(ω), where k is an eigenvalue and H(ω) is frequency response of the system.

5. Consider a non-singular 2 × 2 square matrix A. If trace (A) = 4 and trace (A^2) = 5, the determinant of the matrix A is _______ (up to 1 decimal place).

6. Let f be a real valued function of a real variable defined as f(x) = x –[x], where [x] denotes the largest integer less than or equal to x. The value of $\int_{0.25}^{1.25} f(x)\,dx$ is ________ (up to 2 decimal places).

7. If C is a circle |z| = 4 and $f(z) = \frac{z^2}{(z^2 - 3z + 2)^2}$, then $\oint_C f(z)\,dz$ is

(*a*) 1 (*b*) 0

(*c*) –1 (*d*) –2

8. Let f(x) = $3x^3 - 7x^2 + 5x + 6$. The maximum value of f(x) over the interval [0, 2] is ________ (upto 1 decimal place).

9. Let $A = \begin{bmatrix} 1 & 0 & -1 \\ -1 & 2 & 0 \\ 0 & 0 & -2 \end{bmatrix}$ and $B = A^3 - A^2 - 4A + 5I$, where I is the 3 × 3 identify matrix. The determinant of B is ________ (upto 1 decimal place).

ANSWERS

GENERAL APTITUDE

SET-1

1. (*a*)	**2.** (*c*)	**3.** (*b*)	**4.** (*c*)	**5.** (*d*)
6. (*a*)	**7.** (*c*)	**8.** (*d*)	**9.** (*c*)	**10.** (*a*)

EXPLANATIONS

1. Stares - To look at someone for a long time

Stairs - A construction designed to bridge a large vertical distance by dividing it into smaller vertical distances, called steps

2. Errant means misbehaving, exhibiting inappropriate behaviour / offending conduct.

3.

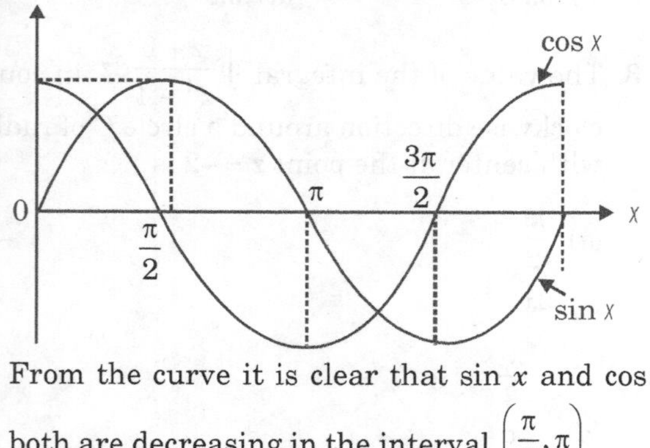

From the curve it is clear that sin x and cos x both are decreasing in the interval $\left(\frac{\pi}{2}, \pi\right)$

4. Area of equilateral triangle = $\frac{\sqrt{3}}{4}a^2$

$\therefore \quad \sqrt{3} = \frac{\sqrt{3}}{4}a^2$

$\therefore \quad a^2 = 4$

$\therefore \quad a = 2$

Now perimeter of equilateral triangle

= 3a = 3 × 2 = 6

5. (i) Volume of cuboid = (8 × 10 × 6) cm^3 = 480 cm^3

(ii) Volume of cube = (8 × 8 × 8) cm^3 = 512 cm^3

(iii) Volume of cylinder = $\pi r^2 h$

$\left(\frac{22}{7} \times 7 \times 7 \times 7\right) cm^3 = 1078\ cm^3$

(iv) Volume of Sphere $= \frac{4}{3}\pi r^3 = \left(\frac{4}{3}\pi \times (7)^2\right) cm^3$

$= 1436.75\ cm^3$

Hence, the descending order of their volumes is (iv), (iii), (ii), (i).

6. Average speed $= \frac{2(S_1 \times S_2)}{S_1 + S_2}$

$= \frac{2(60 \times 90)}{(60+90)}$ km/h

$= \frac{2 \times 60 \times 90}{150} = 72$ km / h

7. Number of parallelogram $= {}^4C_2 \times {}^5C_2$

$= \frac{4!}{2! \times 2!} \times \frac{5!}{3! \times 2!}$

$= 6 \times \frac{5 \times 4}{2}$

$= 60$

8.

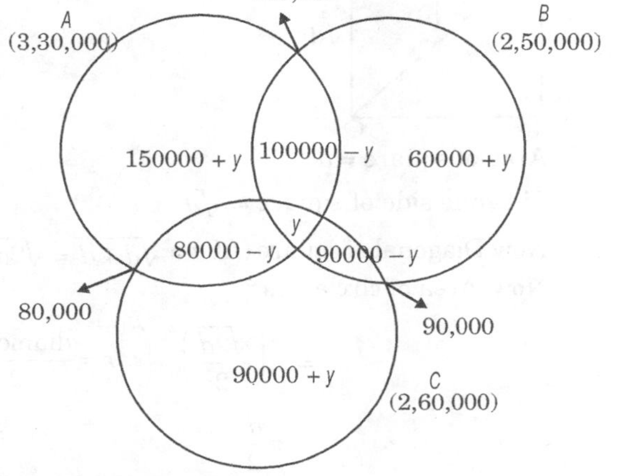

From above diagram,

$\therefore$ 6,30,000 = 2y + 1,50,000 + 100000 + 80000 + 60000 + 90000 + 90000

$\therefore$ 63000 = 2y + 57000

$\therefore$ 6,30,000 – 5,70,000 = 2y

$\therefore$ 60000 = 2y

$\therefore$ y = 30000

Students who failed to clear the test

= 150000 + 60000 + 90000 + 4y

= 300000 + 4 × 30000 ($\because$ y = 30000)

= 420000

Students who failed to clear the test = 420000

9. $x^2 - 1 = -x$

$x - \frac{1}{x} = -1$

$x^2 + \frac{1}{x^2} = 0.2$

Now, $x^4 + \frac{1}{x^4} = \left(x^2 + \frac{1}{x^2}\right)^2 - 2$

$= (3)^2 - 2$

$= 9 - 2$

$= 7$

ANSWERS

SET-2

1. (*a*)	**2.** (*a*)	**3.** (*d*)	**4.** 22.22	**5.** (*b*)
6. (*c*)	**7.** (*b*)	**8.** (*b*)	**9.** (*c*)	**10.** (*b*)

EXPLANATIONS

3. Let the age of Hari be '*x*'

$\therefore$ Hema age $= 5 + 2x$

and Suresh age $= 10x - 13$

Now Suresh is 3 time old as Hema

$\therefore$ $(10x - 13) = 3(5 + 2x)$

$\therefore$ $10x - 13 = 15 + 6x$

$\therefore$ $4x = 28$

$\therefore$ $x = 7$

$\therefore$ Hema's age $= 5 + 2x$

$= 5 + 14$

$= 19$

4.

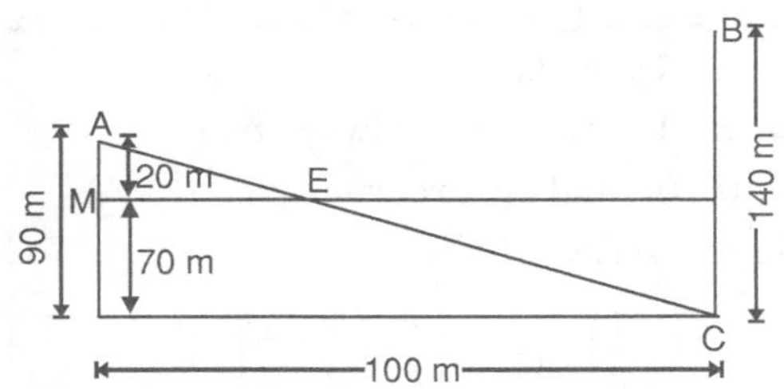

$\Delta AME \approx \Delta APC$

$\frac{AM}{AP} = \frac{ME}{PC}$

$\therefore$ $\frac{20}{90} = \frac{ME}{100}$

$\therefore$ $ME = \frac{20 \times 100}{90} = 22.22$

5. For $T = 32.4$,

$p = 2$

and $T_0 = 25$

$T = K(\theta p + T_0)$

$\therefore$ $32.4 = K(2\theta + 25)$ ...(*i*)

Now for $T = 42$, $p = 5$ and $T_0 = 30$

$T = K(\theta p + T_0)$

$\therefore$ $42 = K(5\theta + 30)$ ...(*ii*)

Solving equation (*i*) and (*ii*) we get

$$\theta = 1$$

and $K = \dfrac{78}{65}$

6. $12.5\%x + 10\%x = 108$

$$x = \frac{108}{22.5}$$

So loss $108 \times \dfrac{12.5}{22.5} = 60$

7. Total price $(p) = p_1 + p_2$

$$= 16(4)^2 + 16(6)^2$$
$$= 16[16 + 36]$$
$$= 16[52]$$
$$= 832$$

Price $\propto L^2$

Price $= KL^2$ (where K is constant)

Now according to question

$$1600 = K(10)^2$$
$$\therefore \quad K = 16$$

8. Put value and verify

(*i*) $y = 2x + 4$ is true in $-3 \le x \le -1$

On putting $x = -3, y = -2$ and $x = -2, y = 0$ and $x = 1, y = 2$

(*ii*) $y = |x - 1|$ is also true $(x = -1, y = 2)$, $(x = 0, y = 1)$ and $(x = 1, y = 0)$

(*iv*) $y = 1$ in $(2 \le x \le 3)$ always true

(*i*), (*ii*) and (*iv*) are true.

9. Sum of series will be

$$= \left(1 - \frac{1}{3}\right) + \left(\frac{1}{2} - \frac{1}{4}\right) + \left(\frac{1}{3} - \frac{1}{5}\right) \ldots \left(\frac{1}{48} - \frac{1}{50}\right) + \left(\frac{1}{49} - \frac{1}{51}\right) + \left(\frac{1}{50} - \frac{1}{52}\right)$$

All like terms will cancel out and we will be left with

$$= \left(1 + \frac{1}{2}\right) - \left(\frac{1}{51} + \frac{1}{52}\right)$$

10.

$$A \times B \times C = B \times G \times E$$
$$= D \times E \times F$$
$$= 72$$
$$8 \times 9 \times 1 = 9 \times 2 \times 4$$
$$= 3 \times 4 \times 6$$
$$= 72$$

Any of A, B, C, D, E, F, G cannot be 5.

ANSWERS

SET-3

1. (*b*)	**2.** (*a*)	**3.** (*d*)	**4.** (*c*)	**5.** (*b*)
6. (*b*)	**7.** (*c*)	**8.** (*b*)	**9.** (*a*)	**10.** (*c*)

EXPLANATIONS

1. They're used for pointing group.

Their is pointing people.

There is used for place.

2. Meandering : wandering aimlessly/indirect.

3.

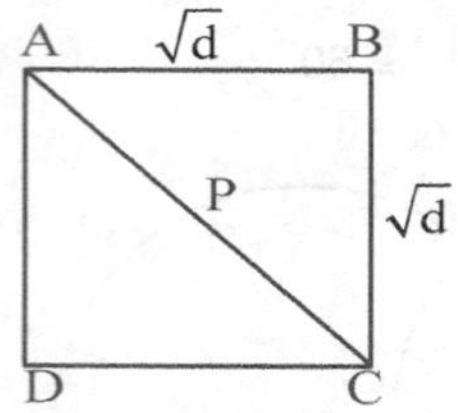

Area of square $= d$

Side one side of square $= \sqrt{d}$

Now Diagonal of square (AC) $= \sqrt{d + d} = \sqrt{2d}$

Now, Area of circle $= \pi r^2$

$$= \pi\left(\frac{\sqrt{2d}}{2}\right)^2 \quad \left(\because r = \frac{\text{diameter}}{2}\right)$$
$$= \pi\frac{d}{2}$$
$$= \frac{1}{2}\pi d$$

4. Number is divided by either by 20 or 42 or by 76

K × LCM (20, 42, 76) + constant difference

= 7890 K + 7 (K is natural number)

Least number will be 7890 + 7 = 7897.

5. 2, 12, 60, 240, 720, 1440, 1440, 0

(2×6), (12×5), (60×4), (240×3), (720×2), (1440×1), (1440×0)

−1, −1, −1, −1, −1, −1

So, 1440 must be answer.

6. Male (M) + Female (F) = 300

$$\text{Total money} = \frac{8}{9}M \times 750 + \frac{2}{3}F \times 1000 \quad \ldots(i)$$
$$= \frac{6000}{9}M + \frac{6000}{9}F$$
$$= \frac{6000}{9}(M + F)$$

From equation (i) we get,

$$\text{Total money} = \frac{6000 \times 300}{9} = 2{,}00{,}000$$

7. $p^x = q \Rightarrow x \log p = \log q$

and $q^y = r \Rightarrow y \log q = \log r$

and $z^2 = p \Rightarrow z \log r = \log p$

$\therefore \quad xyz = \dfrac{\log q}{\log p} \times \dfrac{\log r}{\log q} \times \dfrac{\log p}{\log r} = 1$

8. Let total number of people are 100.

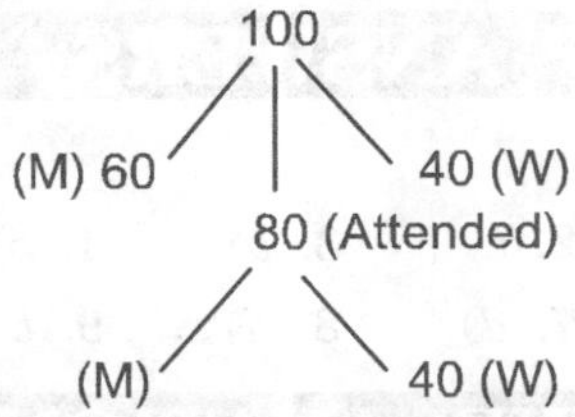

So, male attended the party 80 – 40 = 40

Ratio of male to female who attended the party

M : W = 40 : 40

= 1 : 1

9.

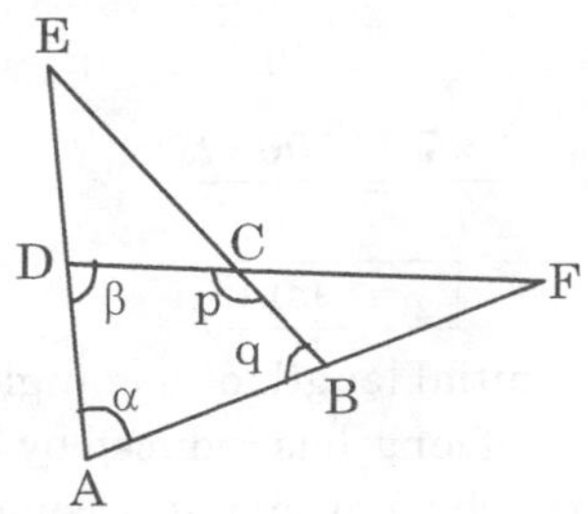

$\angle E + \angle F = ?$

$\alpha + q + E = 180$ ....(i)

and $\alpha + \beta + F = 180$ ....(ii)

and $\alpha + \beta + p + q = 360$ ....(iii)

Adding equations (i) and (ii) and equating it to (iii) we get

$\alpha + q + E + \alpha + \beta + F = \alpha + \beta + p + q$

$\therefore \quad E + F = p - \alpha$

10. Four green, two red face

$P(G) = \dfrac{4}{6} = \dfrac{2}{3}$

$q(R) = 1 - P(G)$

$= \dfrac{1}{3}$

$n = 7$

Now from options

From option (a), $P(G = 3) = {}^7C_3 \left(\dfrac{2}{3}\right)^3 \left(\dfrac{1}{4}\right)^4$

$\dfrac{35 \times 2^3}{(3)^7} = \dfrac{35 \times 2^3}{(3)^7}$

From option (b), $P(G = 4) = {}^7C_4 \left(\dfrac{2}{3}\right)^4 \left(\dfrac{1}{3}\right)^3$

$\dfrac{35 \times 2^4}{(3)^7} = \dfrac{35 \times 2^4}{(3)^7}$

From option (c), $P(G = 5) = {}^7C_5 \left(\dfrac{2}{3}\right)^5 \left(\dfrac{1}{3}\right)^2$

$\dfrac{21 \times 2^5}{(3)^7} = \dfrac{42 \times 2^4}{(3)^7}$

From option (d), $P(G = 6) = {}^7C_6 \left(\dfrac{2}{3}\right)^6 \times \left(\dfrac{1}{3}\right)$

$\dfrac{7 \times 2^6}{(3)^7} = \dfrac{28 \times 2^4}{(3)^7}$

Option c is maximum value.

ANSWERS

SET-4

1. (*b*)	**2.** (*a*)	**3.** (*b*)	**4.** (*d*)	**5.** (*b*)
6. (*b*)	**7.** (*b*)	**8.** (*c*)	**9.** (*a*)	**10.** (*b*)

EXPLANATIONS

3. Rule of divisibility by 3 : Sum of all digits should be divisible by 3

715 ? 423

7 + 1 + 5 + 4 + 2 + 3 = 22

So, the next number after sum 22 which are divisible by 3 are 24, 27, 30 etc.

So, 22 + 2 = 24

22 + 5 = 27

But according to given condition 2 is right answer.

4. The given series is in G.P,

where a = 1 and r = $\dfrac{1}{4}$

Now, in G.P, $s_n = \dfrac{a}{1 - r}$

$\dfrac{1}{1 - \frac{1}{4}} = \dfrac{1}{3/4} = \dfrac{4}{3}$

5.

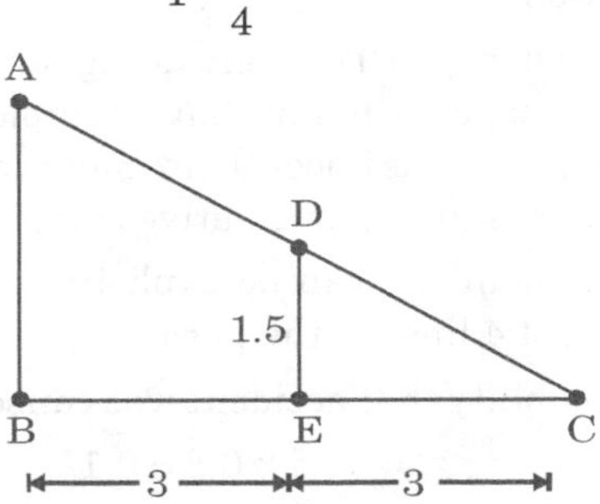

Let 'AB' be the height of lamp post

Here ΔABC and ΔDEC are similar triangles,

$$\therefore \quad \frac{AB}{DE} = \frac{BC}{EC}$$

$$\therefore \quad \frac{AB}{1.5} = \frac{6}{3}$$

$$\therefore \quad \frac{AB}{1.5} = 2$$

$$AB = 3 \text{ meter}$$

Hence height of lamp post is 3 m.

6. P : A

$(10)^5 : (11)^5$

$$\therefore \quad P = \frac{(10)^6 \times (10)^5}{(11)^5}$$

$$= \frac{(10)^{11}}{(11)^5}$$

$$= 620921.323 \approx 621000$$

7.

A	B
G : C	G : C
2 : 3	3 : 7

Now making the mass equal of alloy A and B.

A	B
G : C	G : C
4 : 6	3 : 7

$\therefore$ Alloy of C contain

$$G : C = 4 + 3 : 6 + 7 = 7 : 13$$

8. Statement (i) is not true as nowhere it is mentioned that John was a captain in junior team. The introductory line emphasizes on the board recognizing John's potential (Latent quality/ possibility) as leader of the team.

Statement (iii) also manipulates the facts mentioned in the argument.

The 3rd statement of the argument while this aggression has filled stadia with die-hard fans does not indicate fans expectations from John as a caption.

Statement (ii) The concluding statement of the para suggests that finding a completer captain is a tough task as it look John many years to become a successful and calculative batsman - captain.

Statement (iv) can be explicity concluded from the last 4 lines of the para.

9. Probability that accident was caused by blue cab

$$= 0.15 \times 0.8 = 0.12 = 12\%$$

10. (A) is beyond the scope of given information option (c) can also be discarded on the same grounds.

The argument deals with the coastal region becoming crowded and polluted because of the upcoming luxury resort. Option (b) precisely underlines the theme of the para.

ANSWERS

SET-5

1. (*a*)	**2.** (*c*)	**3.** (*b*)	**4.** (*b*)	**5.** (*b*)
6. (*a*)	**7.** (*d*)	**8.** (*d*)	**9.** (*b*)	**10.** (*c*)

EXPLANATIONS

3. $$\frac{\text{Machine} \times \text{time}}{\text{Number of toys}} = \frac{M_1 \times t_1}{T_1}$$

$$= \frac{M_2 \times t_2}{T_2}$$

$$\therefore \quad \frac{7 \times 7}{7} = \frac{100 \times t_2}{100}$$

$$\therefore \quad \boxed{t_2 = 7 \text{ min}}$$

4. Let the initial length of rectangle is l and breath is b. Now if length is reduced by 10 m and breath is reduced by 5 m then it becomes square.

$$\therefore \quad l - 10 = b - 5$$

[$\because$ In square length = breadth]

$$\therefore \quad l - b = 5$$

Given, initial, Area (A) = $l \times b$

Now according to question,

$$(l - 10)(b - 5) = A - 650$$

$$\therefore \quad lb - 10b - 5l + 50 = A - 650$$

$$\therefore \quad A - 10b - 5l = A - 700$$

$$\therefore \quad 10b + 5l = 700$$

$$\therefore \quad 10b + 5(b + 5) = 700$$

$$\therefore \quad 15b + 25 = 700$$

$$\therefore \quad 15b = 700 - 25$$

$$\therefore \quad 15b = 675$$

$$\Rightarrow \quad b = 45 \text{ m}$$

$$\therefore \quad l = 45 + 5 = 50 \text{ m}$$

$\therefore$ Area of original rectangle (A) = $(45 \times 50)\text{m}^2$

$$= 2250 \text{ m}^2.$$

5. Let two digit number be xy.

Now sum of digit is 9

$$\therefore \quad x + y = 9 \quad \text{....}(i)$$

and when '45' is subtracted from number then digit gets interchanged

$\therefore \quad 10x + y - 45 = 10y + x$

$\therefore \quad x - 5 = y$

$x - y = 5 \quad \quad ....(ii)$

Now solving equation (*i*) and (*ii*) we get

$x = 7$ and $y = 2$

Hence the number is 72.

6. $\log|a| + \log|b| + \log|c| = 0$

It is possible only,

when $|a|$, $|b|$ and $|c|$ all are equal to 1.

$\therefore$ a, b, c may be $\pm 1, \pm 1, \pm 1$ respectively.

Now for minimum value of all three will be negative.

$\therefore$ minimum value = – 3

and maximum value of all three will be positive.

$\therefore$ maximum value = + 3

7. From option (*a*),

Let $\quad a = 1, b = 3$

$\therefore \quad a + a^2b^3 = 1 + 27$

$= 28$ (Even)

$\therefore$ Option (*a*) is wrong

From option (*b*)

Let $\quad a = 2, b = 4$

$\therefore \quad a + a^2b^3 = 2 + 4 \times 64$

$= 258$ (Even)

From option (*c*),

Let $\quad a = 2, b = 1$

From option (*d*),

for $x = 1$

$$y = ||x-1|-1|$$

$$= |-1| = 1$$

$\therefore$ Option (*c*) is wrong.

Now from option (*b*),

for $\quad x = 1$

$$y = ||x|-1|-1$$

$$= |1-1|-1 = -1$$

$\therefore$ Option (*b*) is correct.

8. Train speed (S_T) = 54 km/h

Time = 25 sec for travelling length of train and length of platform

Man speed (S_M) = 9 km/h

Speed of train to man = 45 km/h

Time = 14 sec

So, length of train = time × speed

$$= 14 \times 45 \times \frac{5}{18}$$

Length of train (L_T) = (35 × 5 m) = 175 m

Length of platform (L) + length of train (L_T) = speed × time

$$= 54 \times \frac{5}{18} \times 25 = 15 \times 25 = 375\text{m}$$

$\therefore$ Length of platform (L) = 375 – 175 = 200 m

9. Here from graph for x = 1, y = – 1

Checking the same from options.

From option (*a*)

for $x = 1$

$$y = ||1|+1|-2$$

$$= 2 - 2 = 0$$

$\therefore$ Option (*a*) is wrong

From option (*c*)

for $\quad x = 1$

$$y = ||x|+1|-1$$

$$= |1+1|-1 = 1$$

$\therefore$ Option (*c*) is wrong.

10.

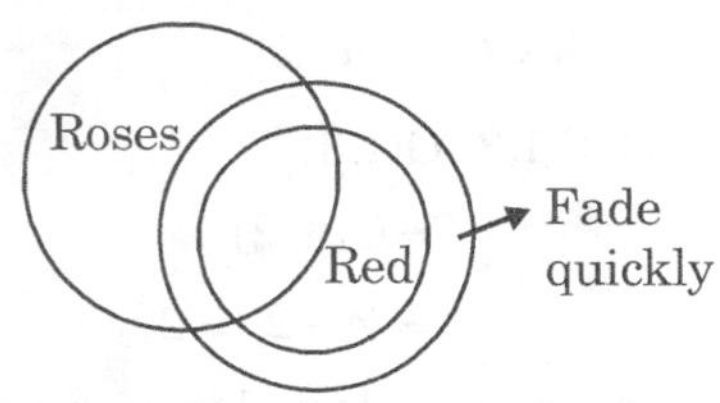

Case-I

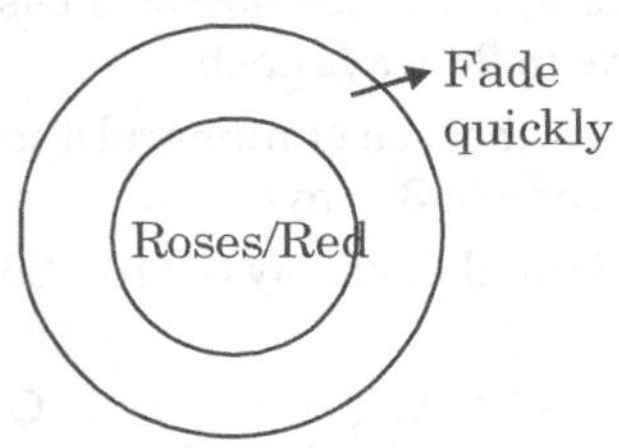

Case-II

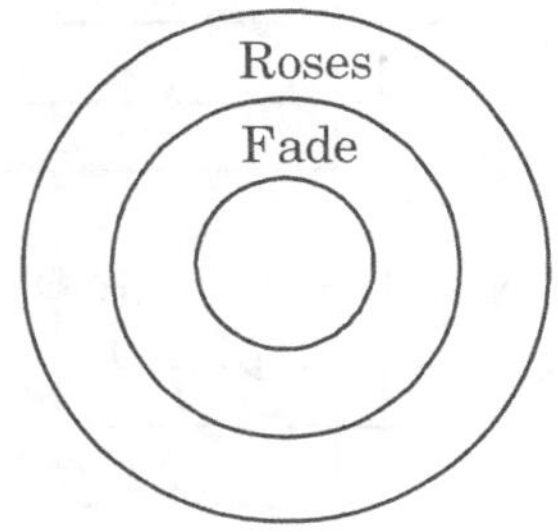

Case-III

ANSWERS

SET-6

1. (*a*)	**2.** (*d*)	**3.** (*d*)	**4.** (*c*)	**5.** (*a*)
6. (*c*)	**7.** (*c*)	**8.** (*b*)	**9.** (*c*)	**10.** (*c*)

EXPLANATIONS

3. $f(x) = 0$

for, $x = \{-2, 0, 3\}$

$\therefore$ $f(-2) = 0, f(0) = 0, f(3) = 0$

$f(x - 3) = 0$

Now from option (*d*),

$f(1 - 3) = f(-2) = 0$

$f(3 - 3) = f(0) = 0$

$f(6 - 3) = f(3) = 0$

4. For $K = 4 \Rightarrow \frac{(4+2)^2}{4-3} = 36$

$K = 8 \Rightarrow \frac{(8+2)^2}{8-3} = \frac{100}{5} = 20$

For $K = 28 \Rightarrow \frac{(28+2)^2}{28-3} = \frac{900}{25} = 36$

} Integers

$\therefore$ Option (*c*) is correct.

5. Here $F(a, b) = (a - b)^2$

and $G(a, b) = |a - b|$

Now $G[F(1,3), G(1,3)]$

$= G[4, 2]$

$= |4 - 2| = 2$

6. Numeral can be selected in 10 ways (0 - 9).

Each of upper case and lower case alphabets can be done in 26 ways each.

All three choosen (1 numeral and 2 alphabet) can be arrange in 3! ways.

$\therefore$ total number of ways – $(10 \times 26 \times 26)\, 3! = 40560$ ways

7.

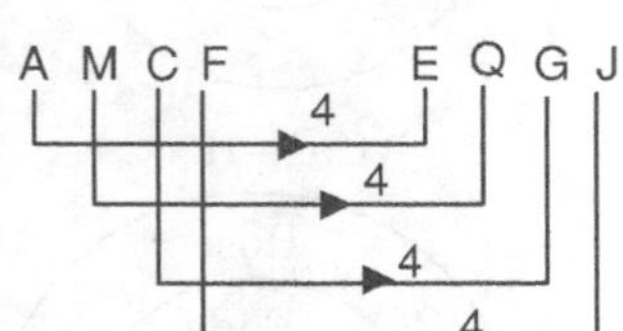

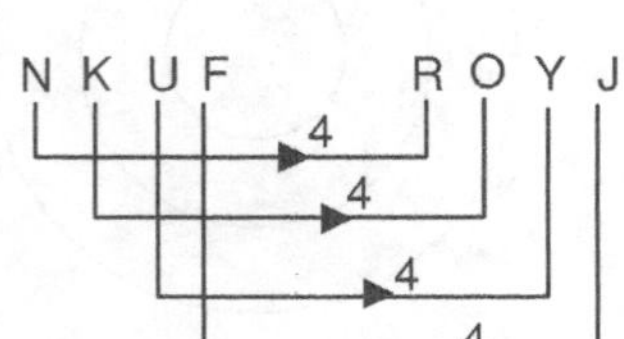

$\therefore$

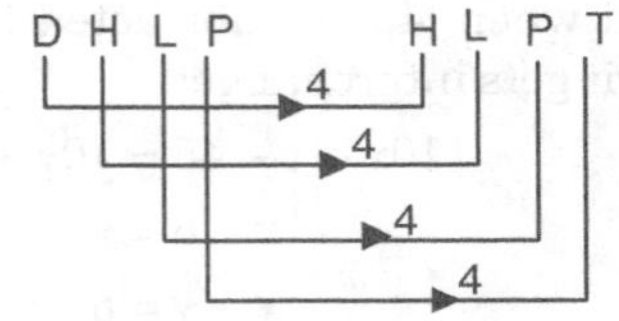

8. There are 7B and 5G in class

Now according to question 3 students are taken at random.

Hence possible ways are

G	B
2	1
3	0

$\therefore$ Required probability $= \frac{{}^5C_2 + {}^7C_1 + {}^5C_3 + {}^7C_0}{{}^{12}C_3}$

$= \frac{10 \times 7 + 10 \times 1}{\frac{12 \times 10 \times 11}{6}}$

$= \frac{80}{220}$

$= \frac{4}{11} \approx 0.3636$

9. Let the number of marbles be 100

Now Cost of marbles increased by Rs. 25

$\therefore$ New cost = Rs. 125

Number of marbles in Rs. 100 $= \frac{100}{125} \times 100 = 80$ marbles

Now according to question

$(40 - x) + (25 - x) + (20 - x) + (15 - x) = 80$

$\therefore$ $100 - 4x = 80$

$\therefore$ $x = 5$

% blue marbles in new design $= \left[\frac{(40-5)}{80} \times 100\right]\%$

$= \left(35 \times \frac{5}{4}\right)\% = 43.75\%$

10. (i) Q and R moves first.

In forward trip Q rowed.

In return trip R rowec

(ii) P and R moves in second trip.

R rowed in forward trip

P rowed in return trip

(iii) P and S moves in last trip

S rowed in forward trip.

R rowed twice.

ENGINEERING MATHEMATICS

ANSWERS

SET-1

1. (*a*)	**2.** (*b*)	**3.** (*d*)	**4.** 1	**5.** (*b*)
6. (*d*)	**7.** (*b*)	**8.** (*c*)	**9.** (*d*)	

EXPLANATIONS

1. $N^2 = 0$

Let the eigen values of N are $\lambda^1, \lambda^2, \lambda^3$:

Now eigen values of N^2 will be $\lambda^2{}_1, \lambda^2{}_2, \lambda^2{}_3$

But $N^2 = 0$

$\Rightarrow \lambda_1^2 = 0, \lambda_2^2 = 0, \lambda_3^2 = 0$

$\therefore$ Eigen values of N are 0, 0, 0

2. Only $f_1(z) = z^2$ is analytic

$f_2(z) = \bar{z}$ is not analytic

3. $Var(aY) = a^2 Var(Y)$

So, $Var(-Y) = (-1)^2 Var(Y)$

$= var(Y) \quad (\because a = -1)$

Since X and Y are independent random variables,

$\therefore Var(Z) = Var(X) + Var(-Y)$

$= Var(X) + Var(Y)$

$= 1 + 2 = 3$

4. $(x - 2)^2 = 2x - 1$

$x^2 - 6x + 5 = 0$

$x = 1, 5$

$\therefore$ Smallest value of 'x' is 1

5. Probability of getting head in a toss $= p = \frac{1}{2}$

Probability of getting tail in a toss $= q = 1 - p = \frac{1}{2}$

Required probability = (Probability of getting 2 heads in first 4 tosses) × (Probability of getting head in 5th toss)

$$= \left({}^4C_2 p^2 q^2\right) \times \left(\frac{1}{2}\right)$$

$$= \frac{4!}{(4-2)!\,2!} \times \frac{p^2 + q^2}{2}$$

$$= \frac{4!}{2! \times 2!} \times \frac{p^2 q^2}{2}$$

$$= 6 \times \frac{1}{4} \times \frac{1}{4} \times \frac{1}{2}$$

$$= \frac{6}{32} = \frac{3}{16}$$

6. $x = y = z \Rightarrow dx = dy = dz$

$$\int_C \vec{F} \cdot \vec{dl} = \int_0^1 \left(x^2 - 2x - 4x^2 + 4x^3\right) dx$$

$$= -3\left[\frac{x^3}{3}\right]_0^1 - 2\left[\frac{x^2}{2}\right]_0^1 + 4\left[\frac{x^2}{4}\right]_0^1$$

$$= -1 - 1 + 1 = -1$$

7.

A	B
	3 R
2 R	6 G
8 G	1 Y

Required probability $= \left[\frac{5}{20} \times 100\right]\%$ or 25%

8. Now from option (c),

Putting k = –3 in options,

$3x + 2ky = -2$

$\Rightarrow 3x - 6y = -2$

$Kx + 6y = 2$

$\Rightarrow 3x - 6y = -2$

Both the equations are reducing in the same form, thus the system will have infinite number of solutions. $\left(\because \frac{d_1}{d_2} = \frac{b_1}{b_2} = \frac{c_1}{c_2}\right)$

ANSWERS

SET-2

1. (*c*)	**2.** (*c*)	**3.** (*d*)	**4.** (*b*)	**5.** (*d*)
6. 0.36				

EXPLANATIONS

1. If $|A| = 0$, then the given matrix is singular.

Now from options.

From Option (a) : $|A| = 6 - 5 = 1$

From Option (b) : $|A| = 9 - 4 = 5$

From Option (c) : $|A| = 12 - 12 = 0$

From Option (d) : $|A| = 8 - 18 = -10$

Hence matrix (c) is singular.

2.

$$|Q| = \frac{3}{7}\left(-\frac{9}{49}-\frac{12}{49}\right)-\frac{2}{7}\left(\frac{18}{49}-\frac{4}{49}\right)+\frac{6}{7}\left(\frac{-36}{49}-\frac{6}{49}\right) = -1$$

$$(\text{Adj. Q}) = \begin{bmatrix} -\frac{21}{49} & \frac{42}{49} & -\frac{14}{49} \\ -\frac{14}{49} & -\frac{21}{49} & -\frac{42}{49} \\ -\frac{42}{49} & -\frac{14}{49} & \frac{21}{42} \end{bmatrix}$$

$$= \begin{bmatrix} -\frac{3}{7} & \frac{6}{7} & -\frac{2}{7} \\ -\frac{2}{7} & -\frac{3}{7} & -\frac{6}{7} \\ -\frac{6}{7} & -\frac{2}{7} & \frac{3}{7} \end{bmatrix}$$

Now, $$Q^{-1} = \frac{\text{AdjQ}}{|Q|} = \begin{bmatrix} \frac{3}{7} & -\frac{6}{7} & \frac{2}{7} \\ \frac{2}{7} & \frac{3}{7} & \frac{6}{7} \\ \frac{6}{7} & \frac{2}{7} & -\frac{3}{7} \end{bmatrix}$$

3. $f(x) = x^3$ at $x = 0$

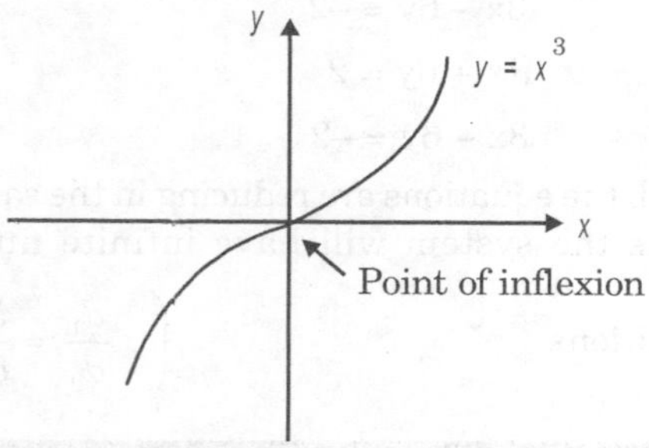

$F(x) = x^3$

Differentiating $f(x)$ w.r.t. to 'x' we and equating it to zero we get

$$f'(x) = 3x^2 = 0$$

and $$f''(x) = 6x$$

Now $$f''(x)\big|_{x=0} = 0$$

4. The value of $\int_0^{\pi} x\cos^2 x dx$

$$= \int_0^{\pi} x\left(\frac{1+\cos 2x}{2}\right)dx \quad [\because \cos 2\theta = 2\cos^2\theta - 1]$$

$$= \int_0^{\pi}\left(\frac{x}{2}+\frac{x\cos 2x}{2}\right)dx$$

$$= \left[\frac{x^2}{4}\right]_0^{\pi} + \frac{1}{2}\left[\frac{x\sin 2x}{2}+\frac{\cos 2x}{4}\right]_0^{\pi}$$

$$= \frac{\pi^2}{4}+\frac{1}{2}\left[\frac{\pi}{2}\times\sin(2\pi)+\frac{\cos 2\pi}{4}-\frac{1}{4}\right]$$

$$= \frac{\pi^2}{4}+\frac{1}{2}\left[\frac{1}{4}-\frac{1}{4}\right] = \frac{\pi^2}{4}$$

6. $(D^2 + 2D + 1)y = 0$

$$CF = (C_1 + C_2x)e^{-x}$$

$\therefore$ $$y = C_1e^{-x} + C_2xe^{-x} \quad ...(i)$$

Now $y(0) = 1$ $\quad 1 = C_1 \quad ...(ii)$

Now differenting equation (*i*) w.r.t to 'x' we get

$$y' = C_1^{e-x} + C_2(e^{-x} - xe^{-x})$$

Now $y'(0) = -1,$ $\quad -1 = -C_1 + C_2 \quad ...(iii)$

From equation (*ii*) and (*iii*) we get

$$C_1 = 1, C_2 = 0$$

$\therefore$ $$y = e^{-x}$$

Now $x = 1, y = e^{-1} = \frac{1}{e} = 0.368$

ANSWERS

SET-3

1. (*d*) **2.** 0.023 **3.** 0.289 **4.** 3 **5.** (*d*) **6.** (*d*)

EXPLANATIONS

1. Given, $a_n = 2n + 3$

Since generating function for 1 is $\frac{1}{1-x}$ and n is $\frac{x}{(1-x)^2}$, the generating function for a_n is

$$A(x) = \frac{2x}{(1-x)^2}+\frac{3}{1-x}$$

$$\frac{2x+3(1-x)}{(1-x)^2} = \frac{3-x}{(1-x)^2}$$

which is option (d).

2. P(one of them wins in 3rd trial)

= P(1st trial is Tie) × P (IInd trial is Tie) × P(one of them wins 3rd trial)

P(Tie in any trial)

= P(P = 1 and Q = 1) + P(P = 2 and Q = 2) + ... + P(P = 6 and Q = 6)

$$= \frac{6}{36} = \frac{1}{6}$$

$\therefore$ P(one of them wins) = 1 – P(Tie)

$$= 1 - \frac{1}{6} = \frac{5}{6}$$

So, required probability

$$= \frac{1}{6} \times \frac{1}{6} \times \frac{5}{6} = \frac{5}{216} = 0.023$$

(rounded to 3 decimal places)

3. $\int_0^{\pi/4} x \cos(x^2)\, dx$

Let, $t = x^2$

$\therefore \quad \frac{dt}{dx} = 2x$

$dt = 2x\, dx$

$\Rightarrow \quad xdx = xdx = \frac{dt}{2}$

when $x = 0 \Rightarrow t = 0$ and when $x = \frac{\pi}{4} \Rightarrow t = \left(\frac{\pi}{4}\right)^2$

So required integral reduce to

$$\int_0^{(\pi/4)^2} \cos t\, dt = [\sin t]_0^{(\pi/4)^2}$$

$$= \sin\left(\frac{\pi}{4}\right)^2 - \sin(0)$$

$$= \sin\left(\frac{\pi}{4}\right)^2 = 0.28898$$

$$\simeq 0.289$$

(rounded to 3 decimal places)

4. $u = \begin{pmatrix} 1 \\ 2 \end{pmatrix}, v = \begin{pmatrix} 1 \\ 1 \end{pmatrix}$

$$A = uv^T$$

$$= \begin{pmatrix} 1 \\ 2 \end{pmatrix} [1 \quad 1]$$

$$= \begin{bmatrix} 1 & 1 \\ 2 & 2 \end{bmatrix}$$

Now $|A - \lambda I| = 0$

$$\left\| \begin{bmatrix} 1 & 1 \\ 2 & 2 \end{bmatrix} - \begin{bmatrix} \lambda & 0 \\ 0 & \lambda \end{bmatrix} \right\| = 0$$

$$\begin{vmatrix} 1-\lambda & 1 \\ 2 & 2-\lambda \end{vmatrix} = 0$$

$\therefore \quad (1-\lambda)(2-\lambda) - 2 = 0$

$$\lambda^2 - 3\lambda = 0$$

$$\lambda(\lambda - 3) = 0$$

$$\lambda = 0 \text{ or } \lambda = 3$$

$\therefore$ The largest eigen value is 3.

5. Only Eigen vector is $\begin{bmatrix} 1 \\ 4 \end{bmatrix}$ multiples means that eigen value is repeated since if eigen values were distinct we will get one more independent eigen vector. So, II P has repeated eigen values is true. I need not be true since $\begin{bmatrix} 2 & 0 \\ 0 & 2 \end{bmatrix}$ has repeated eigen values and yet it is invertible. III is true since if matrix has repeated eigen values then it cannot be diagonalizable.

6. P : Set of rational number $\rightarrow$ countable

Q : Set of functions from {0, 1} to N $\rightarrow$ N

0 1

$N \times N$

0 can be assigned in N ways

1 can be assigned in N ways

There are N × N functions, cross product of countable set in countable.

R : Set of functions from N to {0, 1}

1 2 3 4 5 6

Each of thus boxes can be assigned to 0 or 1 so each such function is a binary number with infinite number of bits.

Example: 0000 ..... is the binary number corresponding to 0 is assigned to all boxes and so on.

Since each such binary number represents a subset of N (the set of natural numbers) by characteristic function method, therefore, the set of such function is same as power set of N which is uncountable due to Cantor's theorem, which says that power set of a countably infinite set is always uncountably infinite.

S : Set of finite subsets of N $\rightarrow$ countably infinite since we are counting only finite subsets.

So, P, Q and S are countable.

ANSWERS

SET-4

1. (*d*)	**2.** 2	**3.** 0.25	**4.** 0	**5.** (*a*)
6. –0.50	**7.** 2	**8.** 4.50		

EXPLANATIONS

1. Here, $f(x,y) = \frac{ax^2 + by^2}{xy} = a\left(\frac{x}{y}\right) + b\left(\frac{y}{x}\right)$

Now $\frac{\partial f}{\partial x} = \frac{a}{y} - \frac{by}{x^2}$

Now $\frac{\partial f}{\partial x}$ at x = 1 and y = 2

$$\left.\frac{\partial f}{\partial x}\right|_{(1,2)} = \left[\frac{a}{y} - \frac{by}{x^2}\right]_{(1,2)} \left[\frac{a}{2} - \frac{b(2)}{1}\right] = \frac{a}{2} - 2b$$

Now $\frac{\partial f}{\partial y} = \left(-\frac{ax}{y^2} + \frac{b}{x}\right)$

Now $\frac{\partial f}{\partial y}$ at x = 1 and y = 2

$$\therefore \left.\frac{\partial f}{\partial y}\right|_{(1,2)} = \left[-\frac{ax}{y^2} + \frac{b}{x}\right]_{(1,2)} = -\frac{a}{4} + b$$

Now according to question,

$$\frac{\partial f}{\partial x} = \frac{\partial f}{\partial y}$$

$$\therefore \frac{a}{2} - 2b = -\frac{a}{4} + b$$

$$\therefore \frac{3a}{4} = 3b$$

$$\therefore a = 4b$$

2. AX = 0 has infinitely many solutions (given)

$\Rightarrow |A| = 0$

$$\therefore \begin{vmatrix} k & 2k \\ k^2 - k & k^2 \end{vmatrix} = 0$$

$\therefore k(k^2) - 2k(k^2 - k) = 0$

$\therefore k^3 - 2k^3 + 2k^2 = 0$

$\therefore k^2(2 - k) = 0$

$\therefore k = 0, 2 \Rightarrow$ "two" distinct values of k

3. $P(X_4 \text{ is smallest}) = \frac{3!}{4!} = \frac{3 \times 2}{4 \times 3 \times 2} = \frac{1}{4} = 0.25$

4. $$f(x) = \int_0^x e^{-t^2/2} dt$$

Differentiating $f(x)$ w.r.t. 'x' we get

$$f'(x) = e^{-x^2/2} \text{ and } f''(x) = e^{-x^2/2}(-x)$$

Now $f''(0) = 0$

$$a_2 = \frac{f''(0)}{2!} = 0$$

5. $$\frac{dy}{dx} = \frac{x^2}{2y} + \frac{y}{2} + \frac{y}{x}$$

Let $y = xt$

$$\therefore \frac{dy}{dx} = t + x\frac{dt}{dx}$$

$$\therefore t + x\frac{dt}{dx} = \frac{x}{2t} + \frac{tx}{2} + t$$

$$\therefore x\frac{dt}{dx} = x\left(\frac{1}{2t} + \frac{t}{2}\right)$$

$$\therefore x\frac{dt}{dx} = x\left(\frac{1+t^2}{2t}\right)$$

Now $\int \frac{2t}{1+t^2} dt = \int dx + C$

$In(1 + t^2) = x + C$

$$\therefore t = \frac{y}{x}$$

So, $\text{In}\left(1 + \frac{y^2}{x^2}\right) = x + C$

At x = 1, y = 0

$$\text{In}\left(1 + \frac{0}{1}\right) = \text{In}(1) = 0 = 1 + C$$

C = –1

So, $\text{In}\left(1 + \frac{y^2}{x^2}\right) = x - 1$

6.

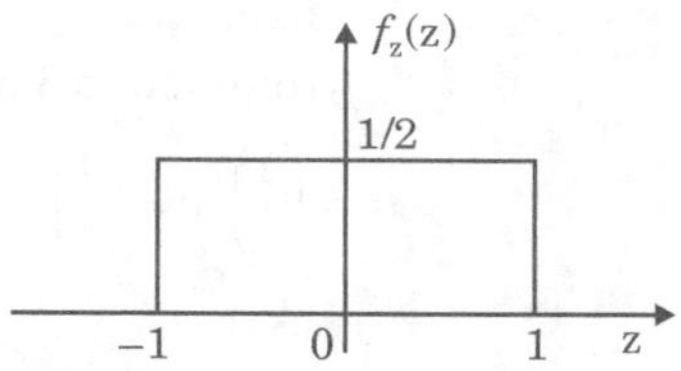

$$P(x_0) = \frac{1}{4}$$

$$P(x_1) = \frac{3}{4}$$

MAP criteria,

$$f_Y(y \mid x_0)P(x_0) \underset{x_1}{\overset{x_0}{\gtrless}} f_Y(y \mid x_1)P(x_1)$$

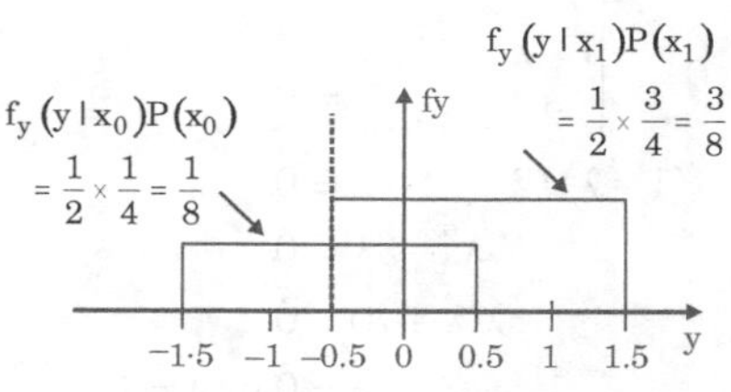

Optimum threshold exists here

So, $\alpha = -0.50$

7.

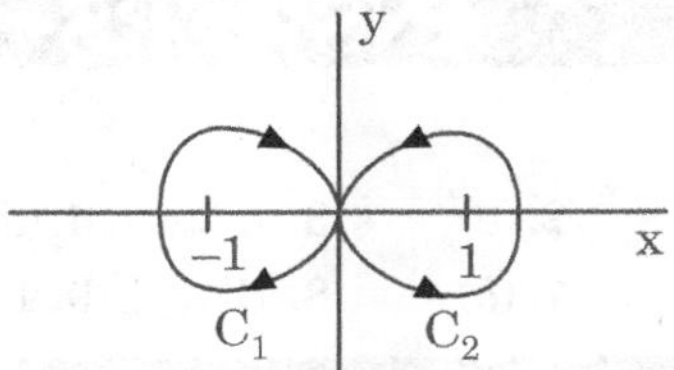

$$\frac{1}{\pi i}\oint_C \frac{dz}{z^2-1}$$

$$= 2\left[\frac{1}{2\pi i}\oint_{C_1}\frac{dz}{(z+1)(z-1)} + \frac{1}{2\pi i}\oint_{C_2}\frac{dz}{(z+1)(z-1)}\right]$$

$$= 2\left[-\left(\frac{1}{z-1}\right)\Big|_{z=-1} + \left(\frac{1}{z+1}\right)\Big|_{z=1}\right]$$

$$= 2\left[-\left(-\frac{1}{2}\right) + \left(\frac{1}{2}\right)\right] = 2$$

8. $$r = x^2 + y - z \quad ...(i)$$

$$\therefore \quad \frac{\partial r}{\partial x} = 2x - \frac{\partial z}{\partial x} \quad ...(ii)$$

Now, $z^3 - xy + yz + y^3 = 1$ ...(iii)

$$\therefore \quad 3z^2\frac{\partial z}{\partial x} - y + y\frac{\partial z}{\partial x} = 0$$

$$\therefore \quad \frac{\partial z}{\partial x} = \frac{y}{3z^2 + y}$$

By substituting the value of $\frac{\partial z}{\partial x}$ in equation (ii), we get,

$$\frac{\partial r}{\partial x} = 2x - \frac{y}{3z^2 + y}$$

Now $$\frac{\partial r}{\partial x}\Big|_{(2,-1,1)} = 2(2) - \frac{(-1)}{3(1)^2 + (-1)}$$

$$= 4 + \frac{1}{3-1} = 4.50$$

ANSWERS

SET-5

1. (*b*)	**2.** (*b*)	**3.** (*a*)	**4.** (*b*)	**5.** (*c*)
6. (*b*)	**7.** (*c*)	**8.** 0.5	**9.** 0	

EXPLANATIONS

1. (4R, 4G, 4B)

P(RRR) (without replacement)

$$= \frac{^4C_1}{^{12}C_1} \times \frac{^3C_1}{^{11}C_1} \times \frac{^2C_1}{^{10}C_1} = \frac{1}{55}$$

2. Let $A = \begin{bmatrix} -4 & 1 & -1 \\ -1 & -1 & -1 \\ 7 & -3 & 1 \end{bmatrix}$

Applying $R_1 \longleftrightarrow R_2$ $\begin{bmatrix} -1 & -1 & -1 \\ -4 & 1 & -1 \\ 7 & -3 & 1 \end{bmatrix}$

Applying $R_2 - 4R_1, R_3 + 7R_1$ $\begin{bmatrix} -1 & -1 & -1 \\ 0 & 5 & 3 \\ 0 & -10 & -6 \end{bmatrix}$

Now applying $R_3 + 2R_2$ $\begin{bmatrix} -1 & -1 & -1 \\ 0 & 5 & 3 \\ 0 & 0 & 0 \end{bmatrix}$

$\therefore$ No. of non zero rows = 2

rank = 2

3. $\int_a^b f(x)\,dx = f(x)(b-a)$

4. $$F(z) = iz + k\,Re(z)\;i\,Im(z)$$

$$u + iv = i(x+iy) + kx + iy$$

$$u + iv = (kx - y) + i(x + y)$$

Now comparing real and imaginary part we get

$u = kx - y$

and $v = x + y$

Now $u = kx - y$

Now $\frac{du}{dx} = k$

and $\frac{dv}{dx} = -1$

Now $V = x + y$

$\therefore$ $\frac{dv}{dx} = 1$

and $\frac{dv}{dy} = 1$

$\Rightarrow$ $u_x = v_y$

$\therefore$ $K = 1$

5. A dice is rolled 5 times

$n = 5$

P = (Probability of getting 1) = $\frac{1}{6}$

0% $q = 1 - p = 1 - \frac{1}{6} = \frac{5}{6}$

Probability of getting 1 at least 4 times is

$$P(x \geq 4) = P(x = 4) + P(x = 5)$$

$$= {}^nC_4 p^4 q^{n-4} + {}^nC_5 p^5 q^{n-5} \quad \left[\because {}^nC_r p^x q^{n-x}\right]$$

$$= {}^5C_4p^4q^1 + {}^5C_5p^5q^0$$

$$= 5\times\left(\frac{1}{6}\right)^4\left(\frac{5}{6}\right)^1 + 1\times\left(\frac{1}{6}\right)^5\times\left(\frac{5}{6}\right)^0$$

$$= \frac{25}{(6)^5} + \frac{1}{(6)^5} = \frac{26}{(6)^5} = 0.0033$$

Required probability = 0.33%

6. $Y = X_1 - X_2$ → $E_{(y)} = E(x_1 - x_2) = E(x_1) - E(x_2) = 1 - 1 = 0$; $V_{(y)} = V(x_1 - x_2) = V(x_1) + V(x_2) = 1 + 4 = 5$

Y is normally distributed with mean 0 and variance 5.

7. By Gauss Divergence Theorem

$$\iint_s \bar{r}\cdot\hat{n}ds = \iiint_v \bar{V}\cdot\hat{r}dV = \iiint_v 3\, dV = 3V$$

8. Applying Laplace transform

$$L(t^2) = \frac{L^2}{S^3}$$

$$L(2t^2) = \frac{4}{S^3}$$

$$F(s) = L\left[e^{-t}\cdot 2t^2\right] = \frac{4}{(S+1)}$$

Putting s = 1 we get

$$F(1) = \frac{4}{(1+1)^3} = \frac{4}{8} = 0.5$$

9. $3x + 5y \le 15 \rightarrow \frac{x}{b} + \frac{y}{3} \le 1$

$4x + 9y \le 8 \rightarrow \frac{x}{2} + \frac{y}{(8/9)} \le 1$

$13x + 2y \le 2 \rightarrow \frac{x}{(2/13)} + \frac{y}{1} \le 1$

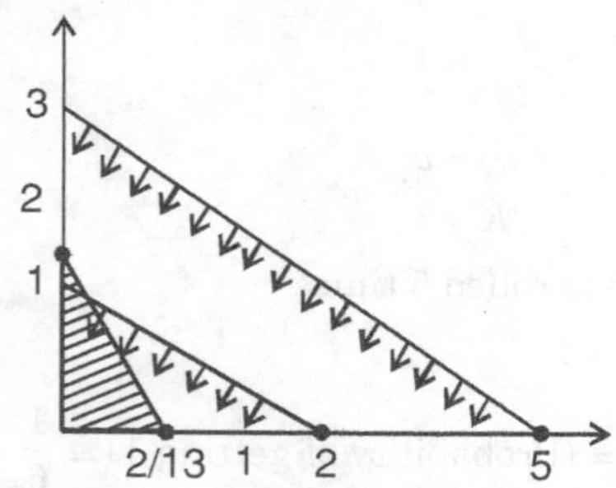

The minimum value of $3x + 5y$.

for minimum value the value of x & y in common area is - so respectively it means. The minimum value of 3x + 5y = 3(0) + 5(0) = 0

ANSWERS

SET-6

1. (*d*)	**2.** (*a*)	**3.** (*a*)	**4.** (*a*)	**5.** 5.5
6. 0.5	**7.** (*b*)	**8.** 12	**9.** 1	

EXPLANATIONS

1. Here, $f(x) = \begin{cases} x^2, & x \ge 0 \\ -x^2, & x < 0 \end{cases}$

Now differentiating f(x) w.r.t to 'x' we get,

$$f'(x) = \begin{cases} 2x, & x \ge 0 \\ -2x, & x < 0 \end{cases}$$

Again differentiating w.r.t. to 'x' we get,

$$f''(x) = \begin{cases} 2, & x \ge 0 \\ -2, & x < 0 \end{cases}$$

The first derivation of f (i.e) f'(x) is not derivable at x = 0.

2. $\phi = xy^2 + yz^2 + zx^2$

Now $\nabla = \frac{\partial}{\partial x}\hat{i} + \frac{\partial}{\partial y}\hat{j} + \frac{\partial}{\partial z}\hat{k}$

Also $\frac{\partial\phi}{\partial x} = y^2 + 2xz$

$$\frac{\partial\phi}{\partial y} = 2xy + z^2$$

and $\frac{\partial\phi}{\partial z} = 2yz + x^2$

$$\therefore \quad \nabla\phi = \bar{i}\frac{\partial\phi}{\partial x} + \bar{j}\frac{\partial\phi}{\partial y} + \bar{k}\frac{\partial\phi}{\partial z}$$

$$= \bar{i}(y^2 + 2xz) + \bar{j}(2xy + z^2) + \bar{k}(2yz + x^2)$$

$$\nabla\phi_{(2,-1,1)} = \bar{i}(1+4) + \bar{j}(-4+1) + \bar{k}(-2+4)$$

$$= 5\bar{i} - 3\bar{j} + 2\bar{k}$$

Now $\bar{P} = \bar{i} + 2\bar{j} + 2\bar{k}$

$$\therefore \quad |\bar{P}| = \sqrt{1+4+4} = 3$$

The directional derivative of $\phi(x, y, z)$ at (2, –1, 1) in the direction of $\bar{P}$ is $\nabla\phi_{\text{at P}}\cdot\frac{\bar{P}}{|\bar{P}|}$

$$= (5\bar{i} - 3\bar{j} + 2\bar{k})\cdot\left(\frac{\bar{i} + 2\bar{j} + 2\bar{k}}{3}\right)$$

$$= \frac{5 - 6 + 4}{3} = 1$$

3. Here $f(z) = \dfrac{z+1}{z-2}$

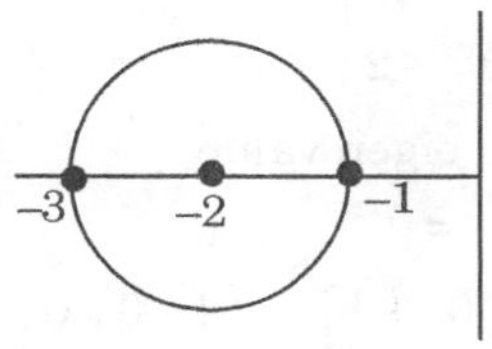

$$\int \frac{z+1}{z^2-4}\,dz$$

$$= \int \frac{z+1}{(z-2)(z+2)}\,dz$$

$$= \int \frac{\left(\dfrac{z+1}{z-2}\right)}{(z+2)}\,dz = 2\pi i\, f(-2)$$

$$= 2\pi i\left(\frac{-2+1}{-2-2}\right) = 2\pi i\left(\frac{-1}{-4}\right) = \frac{\pi i}{2}$$

4. Eigen-function is a type of input for which output is constant times of input.

i.e.

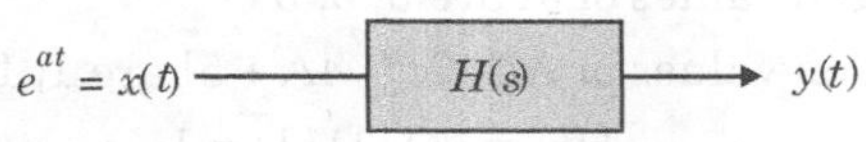

where

x(t) = system input = eigen-function

H(s) = transfer function of system

y(t) = sytem output

Here,

$y(t) = H(s)|_{s=a}$ (eigen value) e^{at} (eigen-function)

$= K \,.\, x(t)$

where,

K = eigen-value

x(t) = eigen-function input

5. Here matrix A is 2 × 2 matrix

$$Tr\,(A) = 4$$

$$\therefore \quad \lambda_1 + \lambda_2 = 4 \qquad \text{...(i)}$$

and $\quad Tr\,(A^2) = 5$

$$\therefore \quad {\lambda_1}^2 + {\lambda_2}^2 = 5 \qquad \text{...(ii)}$$

From equation (i) and (ii) we get

$$\therefore \quad \left(\lambda_1 + \lambda^2\right)^2 = {\lambda_1}^2 + {\lambda_2}^2 + 2\lambda_1\lambda_2$$

$$\therefore \quad 16 = 5 + 2\lambda_1\,\lambda_2$$

$$2\lambda_1\lambda_2 = 16 - 5 = 11$$

$$\therefore \quad \lambda_1\lambda_2 = \frac{11}{2}$$

$$\therefore \quad |A| = \frac{11}{2} = 5.5$$

6.

$$\int_{0.25}^{1.25} f(x)\,dx = \int_{0.25}^{1} (x-[x])\,dx + \int_{1}^{1.25} (x-[x])\,dx$$

$$= \int_{0.25}^{1.25} (x\,dx) - \left[\int_{0.25}^{1} [x]\,dx + \int_{1}^{1.25} [x]\,dx\right]$$

$$\left(\frac{x^2}{2}\right)\Bigg|_{0.25}^{1.25} - \left[\int_{0.25}^{1} 0\,dx + \int_{1}^{1.25} 1\,dx\right]$$

$$= \frac{(1.25)^2}{2} - \left(\frac{(0.25)^2}{2}\right) - (0 + 0.25)$$

$$= \frac{1}{2}\left[(1.5625 - 0.0625)\right] - 0.25 = 0.5$$

7.

$$\int \frac{z^2}{\left(z^2 - 3z + 2\right)}\,dz$$

$$\int \frac{z^2}{(z-1)^2(z-2)^2}\,dz$$

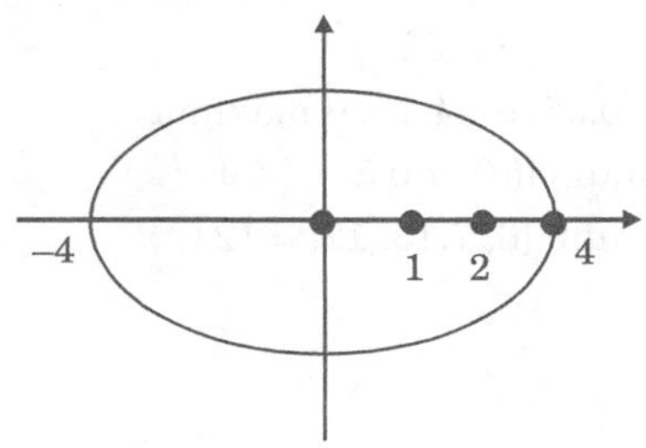

Residue of f(z) at z = 1,

$$\underset{z=1}{\text{Res.}}\, f(z) = \underset{Z\to 1}{\text{Lt}} \frac{1}{1!}\frac{d}{dz}\left((z-1)^2 \cdot \frac{z^2}{(z-1)^2(z-2)^2}\right)$$

$$= \underset{Z\to 1}{\text{Lt}} \left(\frac{2z(z-2)^2 - 2z^2(z-2)}{(z-2)^4}\right)$$

$$= \underset{Z\to 1}{\text{Lt}} \left(\frac{2z(z-2) - 2z^2}{(z-2)^3}\right)$$

$$= \frac{2(1)(1-2) - 2(1)^2}{(1-2)^3}$$

$$= \frac{-4}{-1} = 4$$

and residue of f(z) at z = 2

$$\underset{z=2}{\text{Res.}}\, f(z) = \underset{Z\to 2}{\text{Lt}} \frac{1}{1!}\frac{d}{dz}\left((z-2)^2 \frac{z^2}{(z-1)^2(z-2)^2}\right)$$

$$= \underset{Z\to 2}{\text{Lt}} \left(\frac{(z-1)^2 . 2z - z^2 2(z-1)}{(z-1)^4}\right)$$

$$= \underset{Z \to 2}{Lt}\left(\frac{2z(z-1)-2z^2}{(z-1)^3}\right)$$

$$= \frac{4-8}{1} = -4$$

By residue theorem, $I = 2\pi i\,(4-4) = 0$

8. $f(x) = 3x^3 - 7x^2 + 5x + 6$ in[0, 2]

Now, differentiating f(x) w.r.t to 'x'

$$f'(x) = 9x^2 - 14x + 5$$

$$f''(x) = 18x - 14$$

Again differentiating w.r.t. to 'x' we get

$$f'(x) = 0$$

Now equating f'(x) to zero, we get

$$\therefore \quad x^2 - 14x + 5 = 0$$

$$x = 1, 0.55$$

For $x = 1$

$$f''(1) = 18 - 14$$

$$= 4 > 0 \text{ minima}$$

For $x = 0.55$

$$f''(0.55) = -4.1 < 0 \text{ maxima}$$

Maximum {f(0), f(0.55), f(2)}

Maximum {6, 7.13, 12} = 12

9. $A = \begin{bmatrix} 1 & 0 & -1 \\ -1 & 2 & 0 \\ 0 & 0 & -2 \end{bmatrix}$

Now to find eigen value

$$|A - \lambda I| = 0$$

$$= \begin{bmatrix} 1 & 0 & -1 \\ -1 & 2 & 0 \\ 0 & 0 & -2 \end{bmatrix} - \lambda \begin{bmatrix} 1 & 0 & 0 \\ 0 & 1 & 0 \\ 0 & 0 & 1 \end{bmatrix} = 0$$

$$\therefore \begin{vmatrix} 1-\lambda & 0 & -1 \\ -1 & 2-\lambda & 0 \\ 0 & 0 & -2-\lambda \end{vmatrix} = 0$$

$$\therefore \quad (1-\lambda)((2-\lambda)(-2-\lambda)) - 1(0-0) = 0$$

$$\therefore \quad \lambda = 1, 2 \text{ and } -2$$

Eigen values of A are 1, 2, –2

Eigen values of A^3 are 1, 8, –8

Eigen values of A^2 are 1, 4, 4

Eigen values of 4A are 4, 8, –8

Eigen values of 5I are 5, 5, 5

Eigen values of $A^3 - A^2 - 4A + 5I$ are 1, 1, 1

$$\therefore \quad |B| = (1)(1)(1) = 1$$

SOLVED PAPER – 2019

GENERAL APTITUDE

SET-1

(Q.1 – 5) : Carry One Mark Each

1. The lecture was attended by quite _______ students, so the hall was not very _______.

   (*a*) few, quite (*b*) few, quiet
   (*c*) a few, quiet (*d*) a few, quite

2. They have come a long way in _________ trust among the users.

   (*a*) creating (*b*) create
   (*c*) created (*d*) creations

3. If E = 10; J = 20; O = 30; and T = 40, what will be P + E + S + T?

   (*a*) 51 (*b*) 120
   (*c*) 82 (*d*) 164

4. The CEO's decision to quit was as shocking to the Board at it was to ________.

   (*a*) myself (*b*) I
   (*c*) my (*d*) me

5. On a horizontal ground, the base of a straight ladder is 6m away from the base of a vertical pole. The ladder makes an angle of 45° to the horizontal. If the ladder is resting at a point located at one-fifth of the height of the pole from the bottom, the height of the pole is _______ meters.

   (*a*) 35
   (*b*) 30
   (*c*) 15
   (*d*) 25

(Q.6 – 10) : Carry Two Marks Each

6. "The increasing interest in tribal characters might be a mere coincidence, but the timing is of interest. None of this, though, is to say that the tribal here has arrived in Hindi cinema, or that the new crop of characters represents the acceptance of the tribal character in the industry. The films and characters are too few to be described as a pattern."

   What does the word „arrived. mean in the paragraph above?

   (*a*) went to a place
   (*b*) came to a conclusion
   (*c*) attained a status
   (*d*) reached a terminus

7. In a sports academy of 300 people, 105 play only cricket, 70 play only hockey, 50 play only football, 25 play both cricket and hockey, 15 play both hockey and football and 30 play both cricket and football. The rest of them play all three sports. What is the percentage of people who play at least two sports?

   (*a*) 28.00 (*b*) 23.30
   (*c*) 50.00 (*d*) 25.00

8. P, Q, R, S and T are related and belong to the same family. P is the brother of S. Q is the wife of P. R and T are the children of the siblings P and S respectively. Which one of the following statements is necessarily FALSE?

   (*a*) S is the aunt of T
   (*b*) S is the aunt of R
   (*c*) S is the sister-in-law of Q
   (*d*) S is the brother of P

9. The new cotton technology, Bollgard-II, with herbicide tolerant traits has developed into a thriving business in India. However, the commercial use of this technology is not legal in India. Now with standing that, reports indicate that the herbicide tolerant Bt cotton had been purchased by farmers at an average of Rs 200 more than the control price of ordinary cotton, and planted in 15% of the cotton growing area in the 2017 Kharif season.

   Which of the following statements can be inferred from the given passage?

   (*a*) Farmers want to access the new technology by paying high price
   (*b*) Farmers want to access the new technology if India benefits from it
   (*c*) Farmers want to access the new technology for experimental purposes
   (*d*) Farmers want to access the new technology even if it is not legal

10. A square has sides 5 cm smaller than the sides of a second square. The area of the larger square is four times the area of the smaller square. The sided of the larger square is _________ cm.

    (*a*) 18.50 (*b*) 10.00
    (*c*) 15.10 (*d*) 8.50

SET-2

(Q.1 – 5) : Carry One Mark Each

1. Ten friends planned to share equally the cost of buying a gift for their teacher. When two of them decided not to contribute, each of the other friends had to pay Rs. 150 more. The cost of the gift was Rs. _______.

   (*a*) 666 (*b*) 6000
   (*c*) 3000 (*d*) 12000

2. A court is to a judge as _______ is to a teacher.

   (*a*) a student (*b*) a syllabus
   (*c*) a school (*d*) a punishment

3. The search engine's business model _______ around the fulcrum of trust.

   (*a*) bursts (*b*) revolves
   (*c*) plays (*d*) sinks

4. Two cars start at the same time from the same location and go in the same direction. The speed of the first car is 50 km/h and the speed of the second car is 60 km/h. The number of hours it takes for the distance between the two cars to be 20 km is ______.

   (*a*) 2 (*b*) 1
   (*c*) 6 (*d*) 3

5. The expenditure on the project _______ as follows: equipment Rs. 20 lakhs, salaries Rs. 12 lakhs, and contingency Rs. 3 lakhs.

   (*a*) break (*b*) break down
   (*c*) breaks down (*d*) breaks

(Q.6 – 10) : Carry Two Marks Each

6. In the given diagram, teachers are represented in the triangle, researchers in the circle and administrators in the rectangle. Out of the total number of the people, the percentage of administrators shall be in the range of _______.

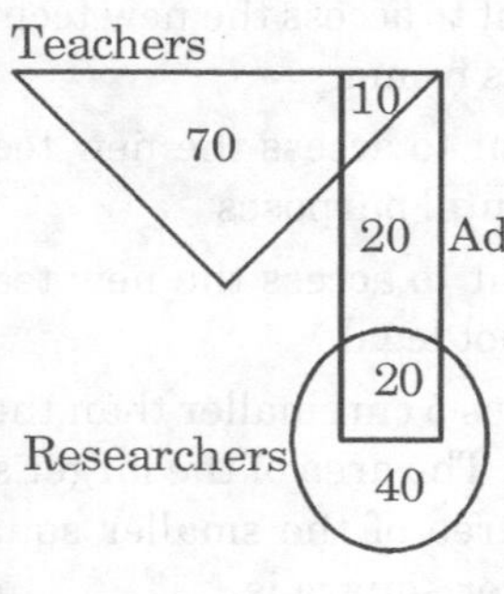

   (*a*) 0 to 15 (*b*) 31 to 45
   (*c*) 46 to 60 (*d*) 16 to 30

7. In a college, there are three student clubs. Sixty students are only in the Drama club, 80 students are only in the Dance club, 30 students are only in the Maths club, 40 students are in both Drama and Dance clubs, 12 students are in both Dance and Maths clubs, 7 students are in both Drama and Maths clubs, and 2 students are in all the clubs. If 75% of the students in the college are not in any of these clubs, then the total number of students in the college is _______.

   (*a*) 975 (*b*) 1000
   (*c*) 225 (*d*) 900

8. Three of the five students allocated to a hostel put in special requests to the warden. Given the floor plan of the vacant rooms, select the allocation plan that will accommodate all their requests.

   Request by X : Due to pollen allergy, I want to avoid a wing next to the garden.

   Request by Y : I want to live as far from the washrooms as possible, since I am very sensitive to smell.

   Request by Z : I believe in Vaastu and so want to stay in the South-west wing.

   The shaded room are already occupied. WR is washroom.

(*a*)

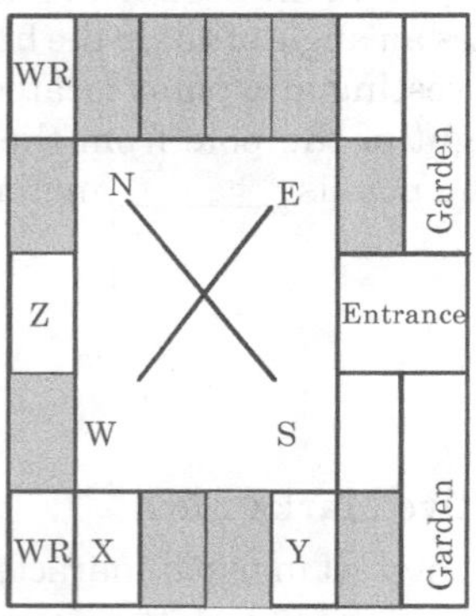

(*b*)

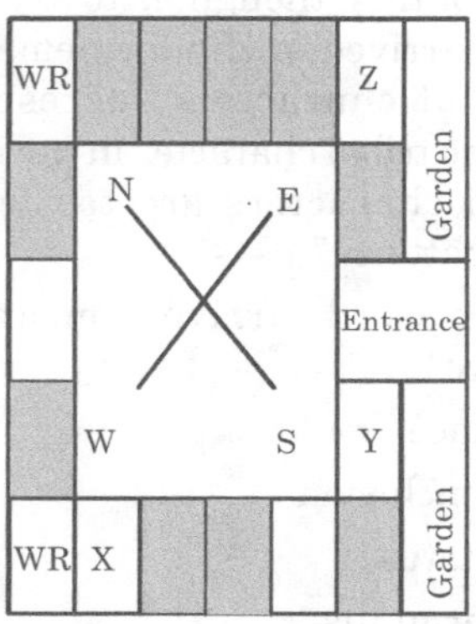

(*c*)

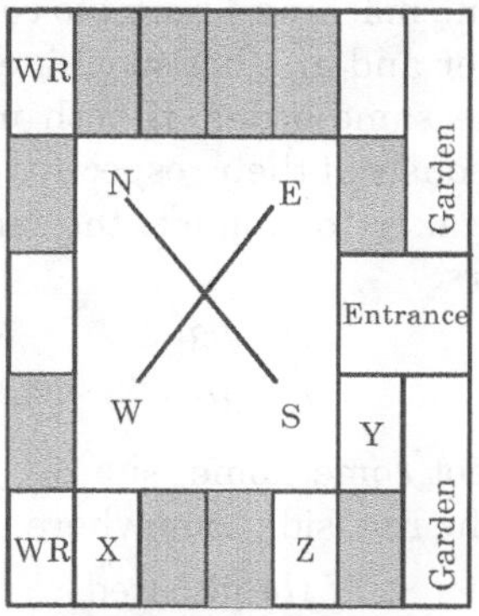

(*d*)

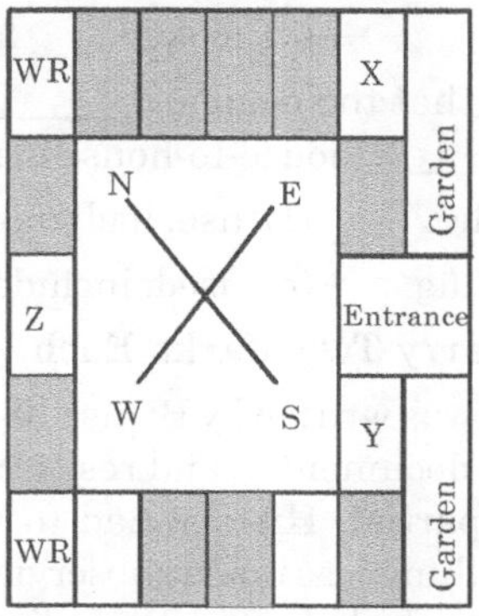

9. The police arrested four criminals – P, Q, R and S. The criminals knew each other. They made the following statements :

P says "Q committed the crime."

Q says "S committed the crime."

R says "I did not do it."

S says "What Q said about me is false."

Assume only one of the arrested four committed the crime and only one of the statements made above is true. Who committed the crime?

(*a*) R (*b*) P

(*c*) Q (*d*) S

10. "A recent High Court judgement has sought to dispel the idea of begging as a disease — which leads to its stigmatization and criminalization — and to regard it as a symptom. The underlying disease is the failure of the state to protect citizens who fall through the social security net."

Which one of the following statements can be inferred from the given passage?

(*a*) Begging is an offence that has to be dealt with firmly

(*b*) Beggars are lazy people who beg because they are unwilling to work

(*c*) Beggars are created because of the lack of social welfare schemes

(*d*) Beggings has to be banned because it adversely affects the welfare of the state

SET-3

Q. 1 to Q. 5 : Carry One Mark Each.

1. The passengers were angry _______ the airline staff about the delay.

(*a*) towards (*b*) on

(*c*) with (*d*) about

2. The missing number in the given sequence 343, 1331, ________, 4913 is

(*a*) 4096 (*b*) 3375

(*c*) 2744 (*d*) 2197

3. Newspapers are a constant source of delight and recreation for me. The _______ trouble is that I read ______ many of them.

(*a*) only, too (*b*) only, quite

(*c*) even, too (*d*) even, quite

4. It takes two hours for a person X to mow the lawn. Y can mow the same lawn in four hours. How long (in minutes) will it take X and Y, if they work together to mow the lawn?

(*a*) 60 (*b*) 80

(*c*) 120 (*d*) 90

5. I am not sure if the bus that has been booked will be able to _____ all the students.

(*a*) deteriorate (*b*) sit

(*c*) accommodate (*d*) fill

Q. 6 to Q. 10 : Carry Two Marks Each.

6. Given two sets X = {1, 2, 3} and Y = {2, 3, 4}, we construct a set Z of all possible fractions where the numerators belong to set X and the denominators belong to set Y. The product of elements having minimum and maximum values in the set Z is _________.

(*a*) $\frac{1}{12}$ (*b*) $\frac{3}{8}$

(*c*) $\frac{1}{8}$ (*d*) $\frac{1}{6}$

7. Consider five people-Mita, Ganga, Rekha, Lakshmi and Sana. Ganga is taller than both Rekha and Lakshmi. Lakshmi is taller than Sana. Mita is taller than Ganga.

Which of the following conclusions are TRUE?

1. Lakshmi is taller than Rekha
2. Rekha is shorter than Mita
3. Rekha is taller than Sana
4. Sana is shorter than Ganga

(*a*) 3 only (*b*) 1 only

(*c*) 2 and 4 (*d*) 1 and 3

8. How many integers are there between 100 and 1000 all of whose digits are even?

(a) 60 (b) 100
(c) 90 (d) 80

9. An award-winning study by a group researchers suggests that men are as prone to buying on impulse as women but women feel more guilty about shopping.

Which one of the following statements can be inferred from the given text?

(a) Many men and women indulge in buying on impulse
(b) All men and women indulge in buying on impulse
(c) Few men and women indulge in buying on impulse
(d) Some men and women indulge in buying on impulse

10. The ratio of the number of boys and girls who participated in an examination is 4 : 3. The total percentage of candidates who passed the examination is 80 and the percentage of girls who passed is 90. The percentage of boys who passed is ________.

(a) 90.00
(b) 80.50
(c) 55.50
(d) 72.50

SET-4

Q. 1 to Q. 5 : Carry One Mark Each.

1. Five different books (P, Q, R, S, T) are to be arranged on a shelf. The books R and S are to be arranged first and second, respectively from the right side of the shelf. The number of different order in which P, Q and T may be arranged is ________.

(a) 2
(b) 120
(c) 6
(d) 12

2. The boat arrived ________ dawn.

(a) on
(b) at
(c) under
(d) in

3. It would take one machine 4 hours to complete a production order and another machine 2 hours to complete the same order. If both machines work simultaneously at their respective constant rates, the time taken to complete the same order is ______ hours.

(a) 2/3 (b) 7/3
(c) 4/3 (d) 3/4

4. When he did not come home, she ______ him lying dead on the roadside somewhere

(a) concluded (b) pictured
(c) notice (d) looked

5. The strategies that the company ______ to sell its products ________ house-to-house marketing.

(a) uses, include (b) use, includes
(c) uses, including (d) used, includes

Q. 6 to Q. 10 : Carry Two Marks Each.

6. "Indian history was written by British historians– extremely well documented and researched, but not always impartial. History had to serve its purpose: Everything was made subservient to the glory of the Union Jack. Latter-day Indian scholar presented a contrary picture."

From the text above, we can infer that :

Indian history written by British historians ____.

(a) was well documented and not researched but was always biased
(b) was not well documented and researched and was sometimes biased
(c) was well documented and researched but was sometimes biased
(d) was not well documented and researched and was always biased

7. Two design consultants, P and Q, started working from 8 AM for a client. The client budgeted a total of USD 3000 for the consultants. P stopped working when the hour hand moved by 210 degrees on the clock. Q stopped working when the hour hand moved by 240 degrees. P took two tea breaks of 15 minutes each during her shift, but took no lunch break. Q took only one lunch break for 20 minutes, but no tea breaks. The market rate for consultants is USD 200 per hour and breaks are not paid. After paying the consultants, the client shall have USE _____ remaining in the budget.

(a) 000.00 (b) 433.33
(c) 166.67 (d) 300.00

8. Five people P, Q, R, S and T work in a bank. P and Q don't like each other but have to share an office till T gets a promotion and moves to the big office next to the garden. R, who is currently sharing an office with T wants to move to the adjacent office with S, the handsome new intern.

Given the floor plan, what is the current location of Q, R and T?

(O = Office, WR = Washroom)

(*a*)

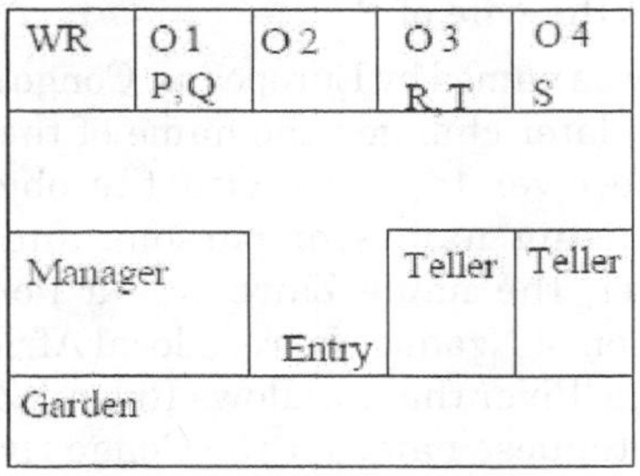

(*b*)

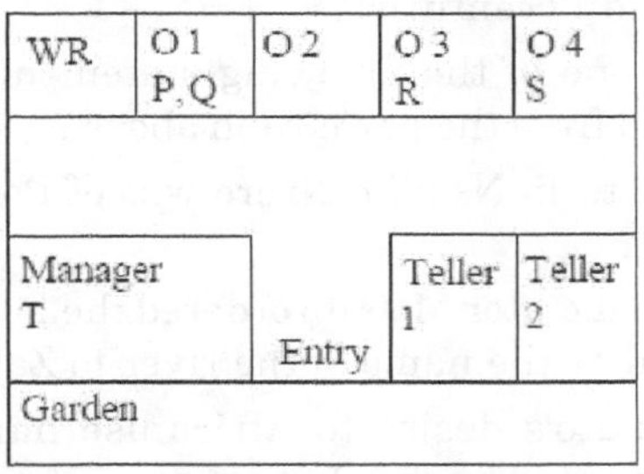

(*c*)

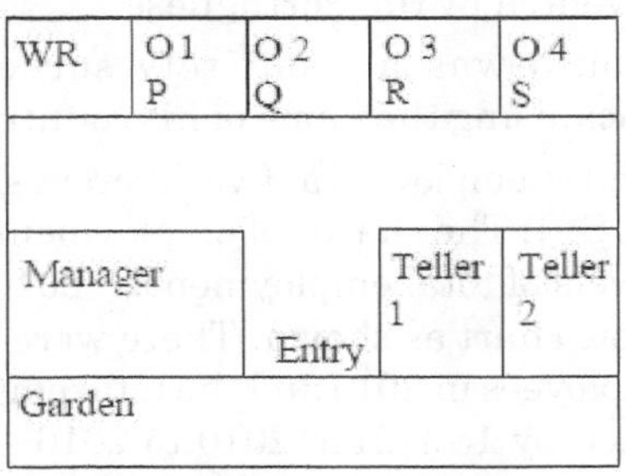

(*d*)

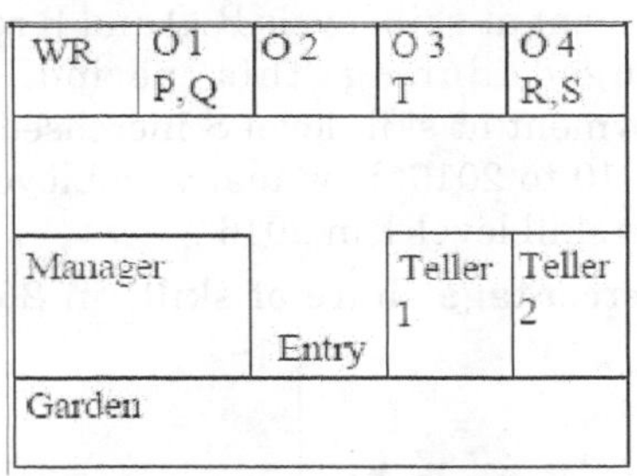

9. Four people are standing in a line facing you. They are Rahul, Mathew, Seema and Lohit. One is an engineering, one is a doctor, one a teacher and another a dancer. You are told that:

1. Mathew is not standing next to Seema
2. There are two people standing beween Lohit and the engineer
3. Rahul is not a doctor
4. The teacher and the dancer are standing next to each other.
5. Seema is turning to her right to speak to the doctor standing next to her.

Who amongst them is an engineer?

(*a*) Rahul

(*b*) Mathew

(*c*) Seema

(*d*) Lohit

10. The bar graph in Panel (a) shows the proportion of male and female illiterates in 2001 and 2011. The proportions of males and females in 2001 and 2011 are given in Panel (b) and (c), respectively. The total population did not change during this period.

The percentage increase in the total number of literate from 2001 to 2011 is ________.

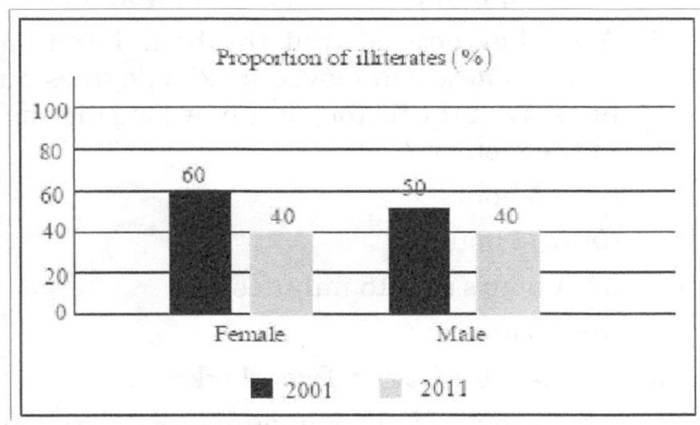

Panel (a)

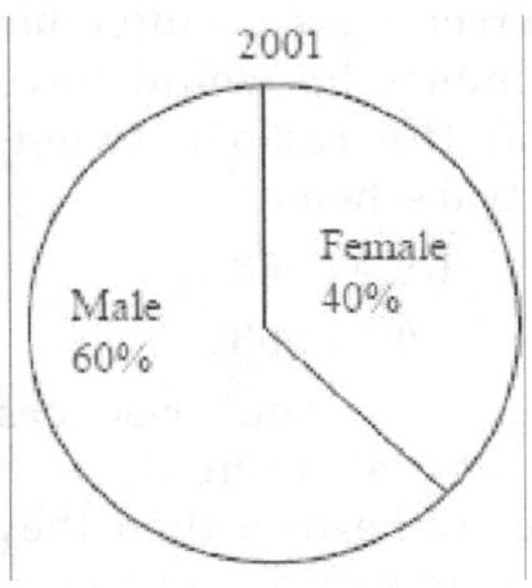

Panel (b)

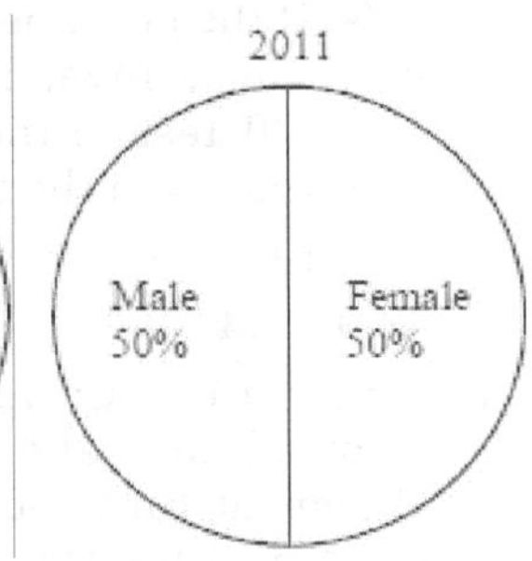

Panel (c)

(*a*) 33.43

(*b*) 35.43

(*c*) 34.43

(*d*) 30.43

SET-5

Q. 1 to Q. 5 : Carry One Mark Each.

1. John Thomas, an ___ writer, passed away in 2018.

(*a*) imminent (*b*) dominant

(*c*) eminent (*d*) prominent

2. The minister avoided any mention of the issue of women's reservation in the private sector. He was accused of _____ the issue.

(*a*) collaring (*b*) belting

(*c*) tying (*d*) skirting

3. The sum and product of two integers are 26 and 165 respectively. The difference between these two integers is ______.

(*a*) 4 (*b*) 2

(*c*) 6 (*d*) 3

4. _____ I permitted him to leave, I wouldn't have had any problem with him being absent _____ I?

(*a*) Have, would (*b*) Have, wouldn't

(*c*) Had, would (*d*) Had, wouldn't

5. A worker noticed that the hour hand on the factory clock had moved by 225 degrees during her stay at the factory. For how long did she stay in the factory ?

(*a*) 8.5 hours

(*b*) 3.75 hours

(*c*) 4 hours and 15 minutes

(*d*) 7.5 hours

Q. 6 to Q. 10 : Carry Two Marks Each.

6. A person divided an amount of Rs. 100,000 into two parts and invested in two different schemes. In one he got 10% profit and in the other he got 12%. If the profit percentages are interchanged with these investments he would have got Rs. 120 less. Find the ratio between his investments in the two schemes.

(*a*) 9 : 16 (*b*) 47 : 53

(*c*) 11 : 14 (*d*) 37 : 63

7. Under a certain legal system, prisoners are allowed to make one statement. If their statement turns out to be true then they are hanged. If the statement turns out to be false then they are shot. One prisoner made a statement and the judge had no option but to set him free. Which one of the following could be that statement ?

(*a*) I committed the crime

(*b*) You committed the crime

(*c*) I will be shot

(*d*) I did not commit the crime

8. M and N had four children P, Q, R and S. Of them, only P and R were married. They had children X and Y respectively. If Y is a legitimate child of W, which one of the following statements is necessarily FALSE ?

(*a*) R is the father of Y

(*b*) W is the wife of R

(*c*) M is the grandmother of Y

(*d*) W is the wife of P

9. Congo was named by Europeans. Congo's dictator Mobuto later changed the name of the country and the river to Zaire with the objective of Africanising names of persons and spaces. However, the name Zaire was a Portuguese alteration of Nzadi o Nzere, a local African term meaning 'River that swallows Rivers'. Zaire was the Portuguese name for the Congo river in the 16th and 17 centuries.

Which one of the following statements can be inferred from the paragraph above ?

(*a*) The term Nzadi o Nzere was of Portuguese origin

(*b*) As a dictator Mobuto ordered the Portuguese to alter the name of the river to Zaire

(*c*) Mobuto's desire to Africanise names was prevented by the Portuguese

(*d*) Mobuto was not entirely successful in Africanising the name of his country

10. A firm hires employees at five different skill levels P, Q, R, S, T. The shares of employment at these skill levels of total employment in 2010 is given in the pie chart as shown. There were a total of 600 employees in 2010 and the total employment increased by 15% from 2010 to 2016. The total employment at skill levels P, Q and R remained unchanged during this period. If the exmployment at skill level S increased by 40% from 2010 to 2016, how many employees were there at skill level T in 2016 ?

Percentage share of skills in 2010

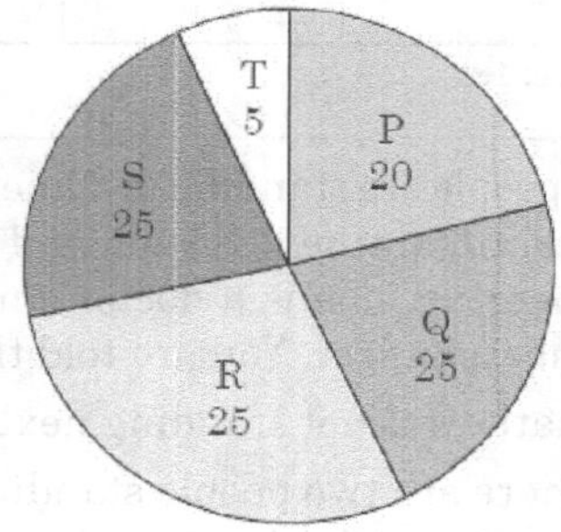

(*a*) 72 (*b*) 60

(*c*) 30 (*d*) 35

ENGINEERING MATHEMATICS

SET-1

1. For a small value of h, the Taylor series expansion for f(x+h) is

(a) $f(x) - hf'(x) + \frac{h^2}{2}f''(x) - \frac{h^3}{3}f'''(x) + ...\infty$

(b) $f(x) - hf'(x) + \frac{h^2}{2!}f''(x) - \frac{h^3}{3!}f'''(x) + ...\infty$

(c) $f(x) + hf'(x) + \frac{h^2}{2}f''(x) + \frac{h^3}{3}f'''(x) + ...\infty$

(d) $f(x) + hf'(x) + \frac{h^2}{2!}f''(x) + \frac{h^3}{3!}f'''(x) + ...\infty$

2. Which of one of the following is correct?

(a) $\lim_{x\to 0}\left(\frac{\sin 4x}{\sin 2x}\right) = 1$ and $\lim_{x\to 0}\left(\frac{\tan x}{x}\right) = 1$

(b) $\lim_{x\to 0}\left(\frac{\sin 4x}{\sin 2x}\right) = \infty$ and $\lim_{x\to 0}\left(\frac{\tan x}{x}\right) = 1$

(c) $\lim_{x\to 0}\left(\frac{\sin 4x}{\sin 2x}\right) = 2$ and $\lim_{x\to 0}\left(\frac{\tan x}{x}\right) = \infty$

(d) $\lim_{x\to 0}\left(\frac{\sin 4x}{\sin 2x}\right) = 2$ and $\lim_{x\to 0}\left(\frac{\tan x}{x}\right) = 1$

3. Which one of the following is NOT a correct statement?

(a) The function $|x|$ has the global minima at $x = 0$

(b) The function $\sqrt[x]{x}$, $(x > 0)$, has the global minima at $x = e$

(c) The function $\sqrt[x]{x}$, $(x > 0)$, has the global maximum at $x = e$

(d) The function x^3 has neither global minima nor global maxima

4. Consider the ordinary differential equations $x^2\frac{d^2y}{dx^2} - 2x\frac{dy}{dx} + 2y = 0$ Given the values of $y(1) = 0$ and $y(2) = 2$, the value of $y(3)$ (round off to 1 decimal place), is _______.

5. A one-dimensional domain is discretized into N sub-domains of width Δx with node numbers $i = 0, 1, 2, 3, ..., N$. If the time scale is discretized in steps of Δt the forward-time and centered-space finite difference approximation at i^{th} node and n^{th} time step, for the partial differential equation $\frac{\partial v}{\partial t} = \beta\frac{\partial^2 v}{\partial x^2}$ is

(a) $\frac{v_{i+1}^{(n+1)} - v_i^{(n)}}{\Delta t} = \beta\left[\frac{v_{i+1}^{(n)} - 2v_i^{(n)} + v_{i-1}^{(n)}}{2\Delta x}\right]$

(b) $\frac{v_i^{(n)} - v_i^{(n-1)}}{2\Delta t} = \beta\left[\frac{v_{i+1}^{(n)} - 2v_i^{(n)} + v_{i-1}^{(n)}}{2\Delta x}\right]$

(c) $\frac{v_i^{(n+1)} - v_i^{(n)}}{\Delta t} = \beta\left[\frac{v_{i+1}^{(n)} - 2v_i^{(n)} + v_{i-1}^{(n)}}{(\Delta x)^2}\right]$

(d) $\frac{v_i^{(n)} - v_i^{(n-1)}}{\Delta t} = \beta\left[\frac{v_{i+1}^{(n)} - 2v_i^{(n)} + v_{i-1}^{(n)}}{(\Delta x)^2}\right]$

6. Consider two functions: $x = \psi\ell n\phi$ and $y = \phi\ell n\psi$. Which one of the following is the correct expression for $\frac{\partial\psi}{\partial x}$?

(a) $\frac{\ell n\psi}{\ell n\phi\ell n\psi - 1}$

(b) $\frac{\ell n\phi}{\ell n\phi\psi - 1}$

(c) $\frac{x\ell n\psi}{\ell n\phi\psi - 1}$

(d) $\frac{x\ell n\phi}{\ell n\phi\ell n\psi - 1}$

SET-2

1. Compute $\lim_{x\to 3}\frac{x^4 - 81}{2x^2 - 5x - 3}$

(1) 53/12

(2) 1

(3) Limit does not exist

(4) 108/7

2. Let X be a square matrix. Consider the following two statements on X.

I. X is invertible

II. Determinant of X is non-zero

Which one of the following is TRUE?

(1) I and II are equivalent statements.

(2) I implies II; II does not imply I.

(3) I does not imply II; II does not imply I.

(4) II implies I; I does not imply II.

3. Let U = {1, 2, ..., n}. Let A = {(x, X | x ∈ X, X ⊆ U}. Consider the following two statements on |A|.

I. $|A| = n2^{n-1}$

II. $|A| = \sum_{k=1}^{n} k\binom{n}{k}$

Which of the above statements is/are TRUE?

(1) Neither I nor II

(2) Only I

(3) Only II

(4) Both I and II

4. Let G be an undirected complete graph on n vertices, where n > 2. Then, the number of different Hamiltonian cycles in G is equal to

(1) n! (2) $\frac{(n-1)!}{2}$

(3) 1 (4) (n – 1)!

5. Let G be an arbitrary group. Consider the following relations on G:

R_1: ∀ a, b ∈ G, a R_1b if and only if ∃ g ∈ G such that $a = g^{-1}bg$

R_2: ∀ a, b ∈ G, aR_2b if and only if $a = b^{-1}$

Which of the above is/are equivalence relation/ relations?

(1) Neither R_1 nor R_2

(2) R_2 only

(3) R_1 only

(4) R_1 and R_2

6. Suppose Y is distributed uniformly in the open interval (1, 6). The probability that the polynomial $3x^2 + 6xY + 3Y + 6$ has only real roots is (rounded off to 1 decimal place)______.

7. Consider the following matrix:

$$R = \begin{bmatrix} 1 & 2 & 4 & 8 \\ 1 & 3 & 9 & 27 \\ 1 & 4 & 16 & 64 \\ 1 & 5 & 25 & 125 \end{bmatrix}$$

The absolute value of the product of Eigen values of R is ______.

SET-3

1. If $f = 2x^3 + 3y^2 + 4z$, the value of line integral $\int_C \text{grad}\, f.dr$ evaluated over contour C formed by the segments (–3, –3, 2) → (2, –3, 2) → (2, 6, 2) → (2, 6, –1) is ________.

2. The inverse Laplace transform of

$$H(s) = \frac{s+3}{s^2+2s+1} \text{ for } t \geq 0 \text{ is}$$

(A) $3te^{-t} + e^{-t}$ (B) $3e^{-t}$

(C) $4te^{-t} + e^{-t}$ (D) $2te^{-t} + e^{-t}$

3. The partial differential equation

$$\frac{\partial^2 u}{\partial t^2} - C^2\left(\frac{\partial^2 u}{\partial x^2} + \frac{\partial^2 u}{\partial y^2}\right) = 0;$$

where C ≠ 0 is known as

(A) Wave equation

(B) Poisson's equation

(C) Laplace equation

(D) Heat equation

4. M is a 2 × 2 matrix with eigen values 4 and 9. The eigen values of M^2 are

(A) 16 and 81

(B) 2 and 3

(C) – 2 and – 3

(D) 4 and 9

5. The rank of the matrix, $M = \begin{bmatrix} 0 & 1 & 1 \\ 1 & 0 & 1 \\ 1 & 1 & 0 \end{bmatrix}$, is ____.

6. If A = 2xi + 3yj + 4zk and $u = x^2 + y^2 + z^2$, then div(uA) at (1, 1, 1) is _______.

7. The closed loop line integral $\oint_{|z|=5} \frac{z^3 + z^2 + 8}{z+2} dz$ evaluated counter-clockwise, is

(a) +4 jπ

(b) – 4 jπ

(c) +8 jπ

(d) – 8 jπ

8. A periodic function f(t), with a period of 2π, is represented as its Fourier series,

If $f(t) = a_0 \sum_{n=1}^{\infty} a_n \cos nt + \sum_{n=1}^{\infty} b_n \sin nt$

$$f(t)\begin{cases} A \sin t, & 0 \leq t \leq \pi \\ 0, & \pi \leq t < 2\pi' \end{cases}$$

The fourier series coefficients a_1 and b_1 of f(t) are

(A) $a_1 = 0; b_1 = A/\pi$

(B) $a_1 = \frac{A}{\pi}; b_1 = 0$

(C) $a_1 = \frac{A}{2}; b_1 = 0$

(D) $a_1 = 0; b_1 = \frac{A}{2}$

9. Consider a 2 × 2 matrix $M = [v_1\ v_2]$, where v_1 and v_2 are the column vectors. Suppose $M^{-1} = \begin{bmatrix} u_1^T \\ u_2^T \end{bmatrix}$ where u_1^T and u_2^T are the row vectors. Consider the following statements:

Statement 1: $u_1^T v_1 = 1$ and $u_2^T v_2 = 1$

Statement 2: $u_1^T v_2 = 0$ and $u_2^T v_1 = 0$

Which of the following options is CORRECT ?

(A) Statement 2 is true and statement 1 is false
(B) Statement 1 is true and statement 2 is false
(C) Both the statements are false
(D) Both the statements are true

SET-4

1. Which one of the following functions is analytic over the entire complex plane?

(a) $\ell n(z)$

(b) $\cos(z)$

(c) $e^{1/z}$

(d) $\frac{1}{1-z}$

2. The value of integral $\int_0^{\pi}\int_y^{\pi} \frac{\sin x}{x} dx\, dy$, is equal to ________.

3. The number of distinct eigen values of the matrix

$$A = \begin{bmatrix} 2 & 2 & 3 & 3 \\ 0 & 1 & 1 & 1 \\ 0 & 0 & 3 & 3 \\ 0 & 0 & 0 & 2 \end{bmatrix}$$ is equal to ________.

4. The families of curves represented by the solution of the equation $\frac{dy}{dx} = -\left(\frac{x}{y}\right)^n$

For n = – 1 and n = + 1, respectively, are

(a) Hyperbolas and Circles
(b) Circles and Hyperbolas
(c) Hyperbolas and Parabolas
(d) Parabolas and Circles

5. If X and Y are random variables such that E[2X + Y] = 0 and E[X + 2Y] = 33, the E[X] + E[Y] = ______.

6. Let Z be an exponential random variable with mean 1. That is, the cumulative distribution function of Z is given by

$$F_z(x) = \begin{cases} 1 - e^{-x} & \text{if } x \ge 0 \\ 0 & \text{if } x < 0 \end{cases}$$

The $\Pr(Z > 2 \mid Z > 1)$, rounded off two decimal places, is equal to ______.

7. The value of the contour integral

$$\frac{1}{2\pi j}\oint\left(z + \frac{1}{z}\right)^2 dz$$

Evaluated over the unit circle |Z| = 1 is ______.

8. Consider the line integral $\int_C (x\,dy - y\,dx)$. The integral being taken in a counterclockwise direction over the closed curve C that forms the boundary of the region R shown in the figure below. The region R is the area enclosed by the union of a 2 × 3 rectangle and a semi-circle of radius 1. The line integral evaluates to

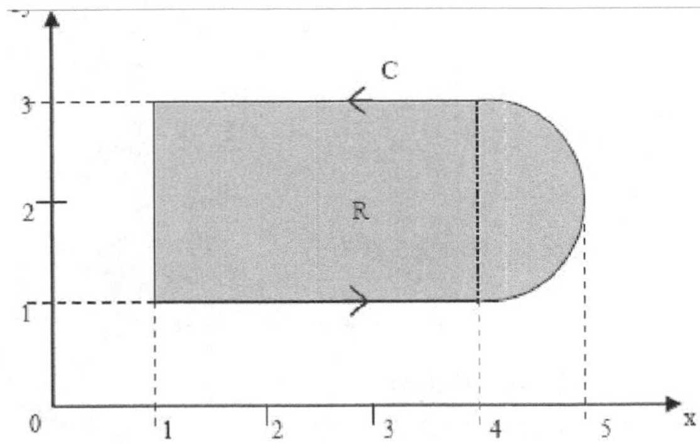

(A) $16 + 2\pi$
(B) $6 + \pi/2$
(C) $12 + \pi$
(D) $8 + \pi$

9. Consider the homogeneous oridinary differential equation $x^2\frac{d^2y}{dx^2} - 3x\frac{dy}{dx} + 3y = 0,\ x > 0$ with y(x) as a general solution. Given that y(1) = 1 and y(2) = 14 the value of y(1.5), (rounded off of to two decimal places), is ______.

SET-5

1. For the equation $\frac{dy}{dx} + 7x^2y = 0$, if y(0) = 3/7, then the value of y(1) is

(a) $\frac{3}{7}e^{-7/3}$ (b) $\frac{3}{7}e^{-3/7}$

(c) $\frac{7}{3}e^{-3/7}$ (d) $\frac{7}{3}e^{-7/3}$

2. Consider the matrix :

$$P = \begin{bmatrix} 1 & 1 & 0 \\ 0 & 1 & 1 \\ 0 & 0 & 1 \end{bmatrix}$$

The number of distinct eigenvalues of P is :

(a) 3 (b) 0

(c) 1 (d) 2

3. The position vector $\overline{OP}$ of point P(20, 10) is rotated anti-clockwise in X-Y plane by an angle θ = 30° a such that point P occupies position Q, as shown in the figure. The coordinates (x, y) of Q are

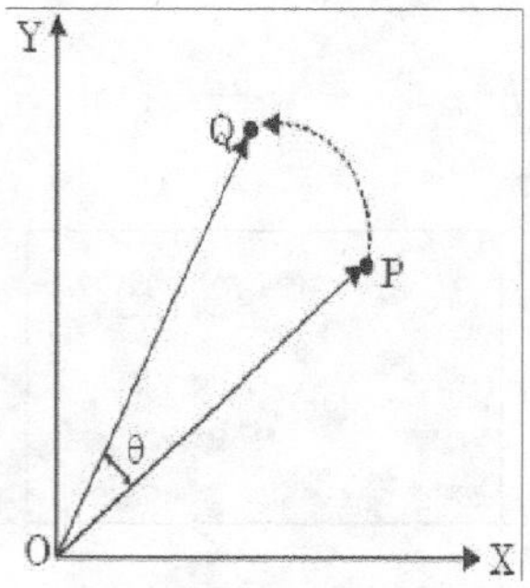

(a) (12.32, 18.66)

(b) (18.66, 12.32)

(c) (22.32 ,8.26)

(d) (13.40, 22.32)

4. Evaluation of $\int_2^4 x^3 dx$ using a 2-equal-segment trapezoidal rule gives value of _______.

5. The variable x takes a value between 0 and 10 with uniform probability distribution. The variable y takes a value between 0 and 20 with uniform probability distribution. The probability of the sum of variables (x + y) being greater than 20 is

(a) 0.33

(b) 0.25

(c) 0

(d) 0.50

6. A harmonic function is analytic if it satisfies the Laplace equation. If $u(x, y) = 2x^2 - 2y^2 + 4xy$ is a harmonic function, then its conjugate harmonic function v (x, y) is

(a) $4y^2 - 4xy$ + constant

(b) $4xy - 2x^2 + 2y^2$ + constant

(c) $2x^2 - 2y^2 + xy$ + constant

(d) $-4xy + 2y^2 - 2x^2$ + constant

7. The set of equations

$x + y + z = 1$

$a - ay + 3z = 5$

$5x - 3y + az = 6$

has infinite solution, if a =

(a) –4

(b) 3

(c) 4

(d) –3

8. The value of the following definite integral is _____ (round off to there decimal places).

$$\int_1^e (x\ln x)\,dx$$

ANSWERS AND EXPLANATIONS

GENERAL APTITUDE

SET-1

ANSWERS

1. (*c*)	2. (*a*)	3. (*b*)	4. (*d*)	5. (*b*)
6. (*c*)	7. (*d*)	8. (*a*)	9. (*a*)	10. (*b*)

EXPLANATIONS

3. Given E = 10, J = 20, O = 30 and T = 40

P + E + S + T ?

A	B	C	D	E	F	G	H	I	J	K	L	M
1	2	3	4	5	6	7	8	9	10	11	12	13
N	O	P	Q	R	S	T	U	V	W	X	Y	Z
14	15	16	17	18	19	20	21	22	23	24	25	26

If $E = 10\,[\because 2 \times 5]$; $J = 20[10 \times 2]$; $O = 30[15 \times 2]$; $T = 40[20 \times 2]$; then

$P = 32[16 \times 2]$; $S = 38[19 \times 2]$; $T = 40[20 \times 2]$

$\therefore P + E + S + T = 32 + 10 + 38 + 40 = 120.$

5.

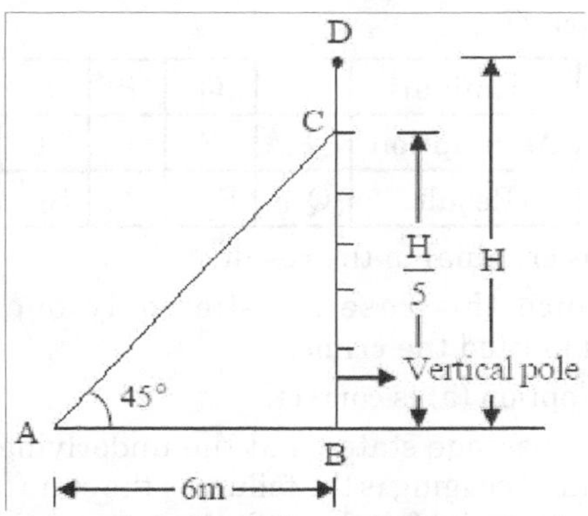

H denotes height of the pole from the bottom.

In ΔABC;

$$\tan 45^\circ = \frac{H/5}{6}$$

$$\Rightarrow 1 = \frac{H/5}{6} \qquad [\because \tan 45^\circ = 1]$$

$$\Rightarrow H/5 = 6 \Rightarrow H = 30 \text{ m.}$$

7. Given,

Play only cricket → 105

Play only hockey → 70

Play only football → 50

Play both cricket and hockey → 25

Play both hockey and football → 15

Play both cricket and football → 30

Total number of people in a sports academy = 300

Number of people who can play only one sport = 105 + 70 + 50 = 225

Number of people who can play at least two sports = Total – (people who can play only one sport) = 300 – 225 = 75

$$\therefore \text{ Percentage} = \left[\frac{75}{300} \times 100\right]\% = 25\%$$

8. Given,

P is the brother of S,

Q is the wife of P,

R and T are the children of the siblings P and S respectively.

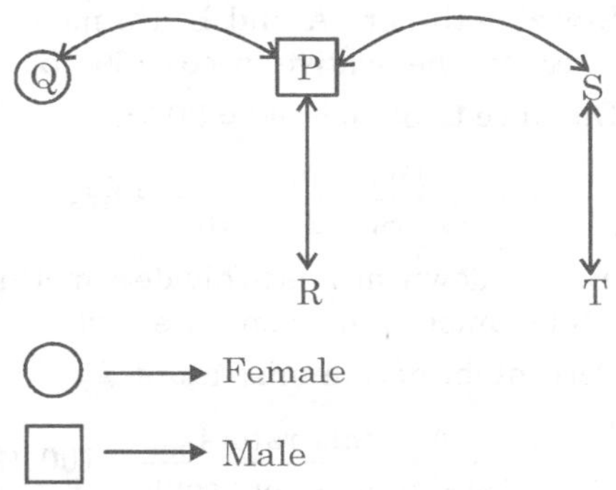

∴ T is son/daughters of S.

∴ S is the aunt of T is false.

10. Let, Side of smaller square = x cm

Side of larger square = (x + 5) cm

Area of small square = x^2

Area of larger square = $(x + 5)^2$

Given,

$(x + 5)^2 = 4x^2$ [∵ Area of the large square is four times the area of the smaller square]

$$\Rightarrow x^2 + 25 + 10x = 4x^2$$

$$\Rightarrow 3x^2 - 10x - 25 = 0$$

$$\Rightarrow x = 5;\ -\frac{5}{3} \rightarrow \text{which is not possible.}$$

∴ Side if larger square 10 cm.

SET-2

ANSWERS

1. (*b*)	2. (*c*)	3. (*b*)	4. (*a*)	5. (*c*)
6. (*b*)	7. (*d*)	8. (*c*)	9. (*a*)	10. (*c*)

EXPLANATIONS

1. Let share of each student be Rs. x

Now, cost of gift = 10 × x

According to question

$x = 8(x + 150)$

$x = 600$

∴ cost of gift = 10 × 600 = (6000)

2. Court is a place where a judge works.

Similarly, school is a place where a teacher works.

3. Fulcrum is any thing that plays a central or an essential role in an activity, event, or situation.

Hence among the given options, 'revolves' is the most appropriate.

4. Speed of 1st car = 50 km/hr

Speed of 2nd car = 60 km/hr

Since, both cars A and B are moving in same direction, the relative speed = (60 – 50) = 10 km/hr

Distance to be covered = 20 km

$\therefore \text{ Time} = \frac{\text{Distance}}{\text{Speed}} = \frac{20}{10} = 2 \text{ hrs}$

5. 'Breaks down' means to divide something such as a total amount into separate parts.

6. Percentage of Administrators

$= \left[\frac{\text{Administrators}}{\text{Total number of people}} \times 100\right]\%$

$= \left[\frac{50}{160} \times 100\right]\% = 31.25\%$

7. According to question, following venn diagram is formed

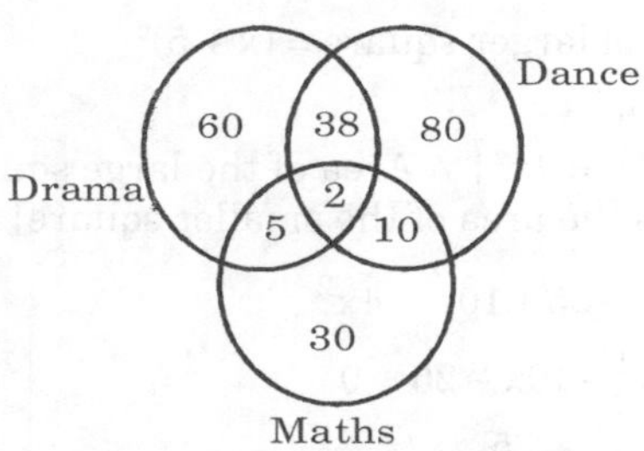

Total number of students = 60 + 80 + 30 + 38 + 5 + 10 + 2 = 225

25% = 225

If 25% = 225

$\therefore \quad 100\% = \frac{225}{25} \times 100 = 900$

9. Suffix C as criminal, NC as not criminal.

Case I :

Criminals	P	Q	R	S
Assumption	T	F	F	F
Result	Q_C	S_{NC}	R_C	S_C

Q and R are criminal in the result is impossible because only one person committed the crime.

Case II :

Criminals	P	Q	R	S
Assumption	F	F	T	F
Result	Q_{NC}	S_{NC}	R_{NC}	S_C

S_{NC} and S_C in the result which is contradiction. [S committed crime and same time not committed crime which is contradiction]

Case III :

Criminals	P	Q	R	S
Assumption	F	T	F	F
Result	Q_{NC}	S_C	R_C	S_C

S and R are criminal in the result is impossible because only one person committed the crime.

Case IV :

Criminals	P	Q	R	S
Assumption	F	F	F	T
Result	Q_{NC}	S_{NC}	R_C	S_{NC}

R is criminal in the result.

Hence this case satisfies only one person committed the crime.

So option (a) is correct.

10. The passage states that the underlying disease behind begging is the failure of the state to protect citizens who fall through the social security net. Only option (b) can be inferred from this.

SET-3

ANSWERS

1. (c)	2. (*d*)	3. (*a*)	4. (*b*)	5. (*c*)
6. (*b*)	7. (*c*)	8. (*b*)	9. (*a*)	10. (*d*)

EXPLANATIONS

2. $7^3 = 343$

$11^3 = 1331$

$17^3 = 4913$

Hence the given series is cube of prime number

$\therefore \quad (13)^3 = 2197$

4. Time taken by working to gether

$$= \left(\frac{4 \times 2}{4+2}\right) \text{hr}$$

$$= \frac{8}{6}\text{hr} = \left(\frac{8}{6} \times 60\right) \text{min} = 80 \text{ min}$$

6. Given two sets

X = {1, 2, 3} and Y = {2, 3, 4}

set of all possible fraction is given by Z

$$\therefore \quad Z = \left\{\frac{1}{2}, \frac{1}{3}, \frac{1}{4}, \frac{2}{2}, \frac{2}{3}, \frac{2}{4}, \frac{3}{2}, \frac{3}{3}, \frac{3}{4}\right\}$$

set of all possible fractions; where the numerators belong to set X and the denominators belong to set Y.

∴ minimum value in Z is $\frac{1}{4}$ and maximum value in Z is $\frac{3}{2}$

Now according to question

∴ Required product = $\frac{1}{4} \times \frac{3}{2} = \frac{3}{8}$

7. From the given information,

Mita > Ganga > Rekha & Lakshmi > Sana

8. Integers between 100 and 1000 all of the whose digits are even = 100; since

from 100 to 199 → No integers

from 200 to 300 → 25 integers

Similarly; 400 to 500 → 25 integers; 600 to 700 → 25 integers; 800 to 900 → 25 integers

∴ Total number of integers between 100 to 1000 all of whase digite are even

= 25 + 25 + 25 + 25 = 100

10. Given, ratio of number of boys to girls who participated in the exam are 4 : 3

Let, total no. of students who participated in the examination be 7x;

Given, total pass percentage = 80%

∴ Total number of students passed in the examination = $\left(7x \times \frac{80}{100}\right) = 5.6x$

and the percentage of girls who passed = 90%
i.e., number of girls who passed

$$= \left(3x \times \frac{90}{100}\right) = 2.7x$$

∴ Number of boys who passed in the exam

∴ 5.6x – 2.7x = 2.9x

∴ Required percent = $\left(\frac{2.9x}{4x} \times 100\right)\% = 72.50\%$

SET-4

ANSWERS

1. (*c*)	2. (*b*)	3. (*c*)	4. (*b*)	5. (*a*)
6. (*c*)	7. (*c*)	8. (*a*)	9. (*b*)	10. (*d*)

EXPLANATIONS

1. ∴ The number of different orders in which P, Q and T arranged = 3! × 2 × 1 = 6

3. Time taken = $\left(\frac{4 \times 2}{4+2}\right)$ hr $= \frac{4}{3}$ hours

7. Given, P and Q started working from 8 A.M. for a client

Total budget = USD 3000

P worked exactly 7 hours and took 30 min break in between.

⇒ 'P' worked number of hours = 6.5 hours

'Q' worked exactly 8 hours and took 20 min break in between.

⇒ 'Q' worked number of hours ≈ 7.67 hours

Client amount paid by for both P and Q

= USD 200/hr

∴ Total USD Paid = (6.5 × 200 + 7.67 × 200)

= (1300 + 1534) = 2834

Remaining amount left

= 3000 – 2834 ≈ 166 (approx)

9. According to the given data;

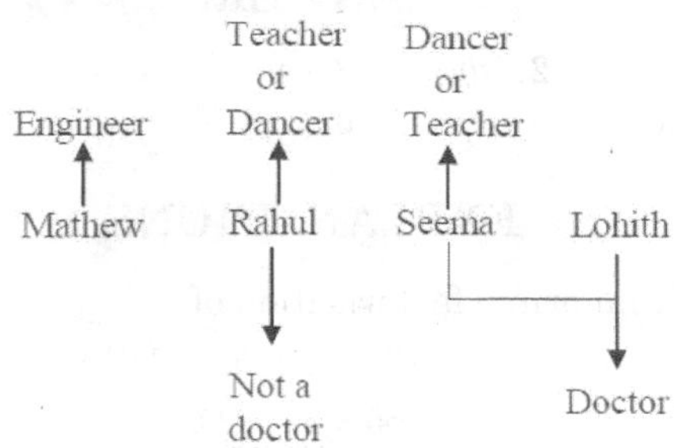

∴ Mathew must be an Engineer.

10. Let us assume, that population = 100 [2001 – 2011]

From panel (a);	From panel (b & c)
Percentage of male literates in 2001 = 50%	Number of males in 2001 = 60
Percentage of female literates in 2001 = 40%	Number of females in 2001 = 40
Percentage of male literates in 2011 = 60%	Number of males in 2011 = 50
Percentage of female literates in 2011 = 60%	Number of females in 2011 = 50

$\therefore$ Number of male literates in 2001

$$= 60 \times \frac{50}{100} = 30$$

Number of male literates in 2011

$$= 50 \times \frac{60}{100} = 30$$

Number of female literates in 2001

$$= 40 \times \frac{40}{100} = 16$$

Total number of literates in 2001

$$= 30 + 16 = 46$$

Number of female literates in 2011

$$= 50 \times \frac{60}{100} = 30$$

Total number of literates in 2011

$$= 30 + 30$$
$$= 60$$

$\therefore$ Percentage increase in the total number of literates from 2001–2011

$$= \left[\frac{60-46}{46} \times 100\right]\%$$

$$= \left[\frac{14}{46} \times 100\right]\%$$

$$= 30.43\%$$

SET-5

ANSWERS

1. (*c*)	2. (*d*)	3. (*a*)	4. (*c*)	5. (*d*)
6. (*b*)	7. (*c*)	8. (*d*)	9. (*a*)	10. (*b*)

EXPLANATIONS

3. From prime factorisation of

$$165 = 3 \times 5 \times 11$$
$$165 = 15 \times 11$$

So the two numbers whose sum is 26 and multiplication is 165 is 15 and 11

Now, difference = 15 – 11 = 4

5. Since, we know that

$$360° = 12 \text{ hours}$$

$$\therefore \quad 10 = \frac{12}{360} \text{ hours}$$

$$\therefore \quad 225° = \frac{12 \times 225}{360} \text{ hours}$$

$$= 7.5 \text{ hours}$$

6. Let the first part = x

So second part = 100,000x

According to question,

$$[0.1x + 0.12(100,000 - x)] - [0.12x + 0.1(1000,000 - x)] = 120$$

$$\Rightarrow 0.2x - 0.24x + 1200 - 1000 = 120$$

$$\Rightarrow \quad -0.04x = -1880$$

$$x = 47000$$

$$\frac{x}{100,000 - x} = \frac{47000}{53000}$$

$$= \frac{47}{53}$$

7. If prisoner make the statement "I will be shot" then only judge have no option and prisoner will be set to free.

8.

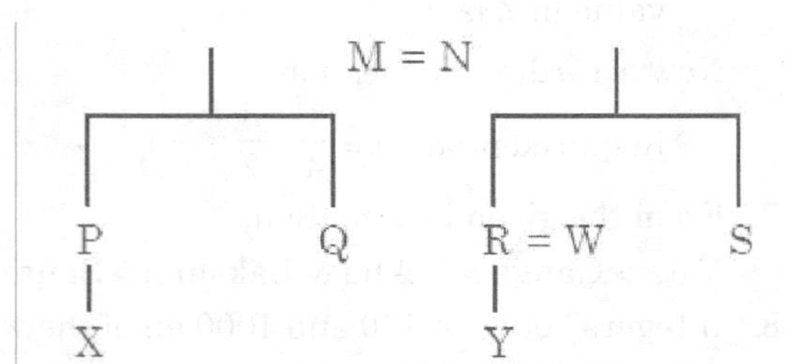

From the above family chart of 'M' and 'N', we can conclude 'W' is the wife of P.

10. 2010 2016

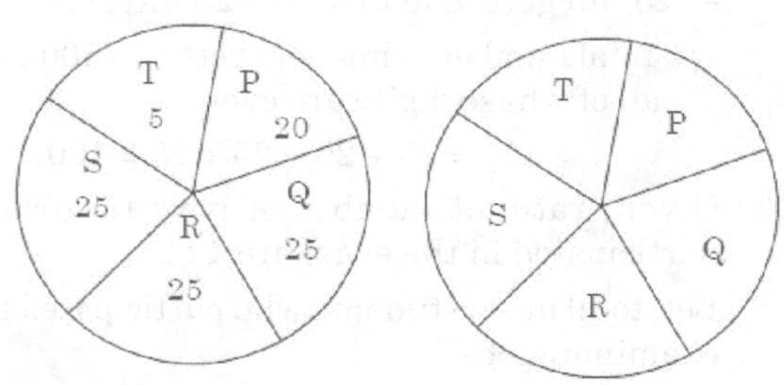

$$P = 20\% = 120\% \xrightarrow{\uparrow 15\%} 1.15 \times 600$$

$$P = 25\% = 150 \xrightarrow{\uparrow 0\%} 150$$

$$R = 25\% = 150 \xrightarrow{\uparrow 0\%} 150$$

$$S = 25\% = 150 \xrightarrow{\uparrow 40\%} 210$$

$$T = 5\% = 30 \longrightarrow ?$$

Now, employees at skill level T in 2016,

$$= 690 - (120 + 150 + 150 + 210)$$
$$= 60$$

ENGINEERING MATHEMATICS

SET-1

ANSWERS

1. (*d*) 2. (*d*) 3. (*b*) 4. (6) 5. (*a*)
6. (*a*)

EXPLANATIONS

1. The Taylor's series expansion for f(x + h) is

$$f(x+h) = f(x) + hf'(x) + \frac{h^2}{2!}f''(x) + \frac{h^3}{3!} + f'''(x) + \ldots\infty$$

2. $\lim_{x\to 0}\frac{\sin 4x}{\sin 2x} \quad \left(\frac{0}{0}\right)$

$= \lim_{x\to 0}\frac{4.\cos 4x}{2.\cos 2x}$ [∵ from L'hospital rule]

$= \frac{4}{2} = 2$

$\lim_{x\to 0}\frac{\tan x}{x} \quad \left(\frac{0}{0}\right)$

$= \lim_{x\to 0}\frac{\sec^2 x}{1} = 1$ [∵ from L'hospital rule]

3. Let $y = x\sqrt{x} = x^{\frac{1}{x}} \Rightarrow \ell n\, y = \ell n\, x^{\frac{1}{x}} = \frac{\ell n\, x}{x}$

$$\Rightarrow y = e^{\frac{\ell n x}{x}}$$

∴ If $\frac{\ell n x}{x}$ is maximum then $y = \frac{\ell n x}{x}$ is maximum for same 'x'

If $\frac{\ell n x}{x}$ is minimum then $y = \frac{\ell n x}{x}$ is minimum for same 'x'

∴ Consider the function $f(x) = \frac{\ell n x}{x}$

$$\Rightarrow f'(x) = 0 \Rightarrow \frac{x\left(\frac{1}{x}\right) - \ell n\, x}{x^2} = 0$$

$$\Rightarrow \frac{1-\ell n\, x}{x^2} = 0$$

$\Rightarrow x = e \to$ stationary point

$$f''(x) = \frac{x^2\left[0 - \frac{1}{x}\right] - [1-\ell n\, x]2x}{x^4}$$

$$\Rightarrow f''(x) = \frac{-x - 2x + 2x\,\ell n\, x}{x^4} = \frac{-3x + 2x\ell n\, x}{x^4}$$

$$\therefore f''(e) = \frac{-3e + 2e(1)}{e^4} = \frac{-e}{e^4} = -e^{-3} < 0$$

∴ x = e is a point of maximum

$\therefore y = e^{\frac{\ell n x}{x}} \Rightarrow y = \sqrt[x]{x}$ has global maximum at x = e.

4. Given D.E

$$x^2\frac{d^2y}{dx^2} - 2x\frac{dy}{dx} + 2y = 0 \;\; \& \; y(1) = 0,\; y(2) = 2.$$

→ (1) cauchy – Euler linear D.E.

$\Rightarrow \left[x^2D^2 - 2xD + 2\right]y = 0$

Let $xD = \theta$; $x^2D^2 = \theta(\theta - 1)$; where $\theta = \frac{d}{dz}$ & $x = e^z$

$[\theta(\theta - 1) - 2\theta + 2]y = 0$

$\Rightarrow \left[\theta^2 - 3\theta + 2\right]y = 0$

The A.E is $\theta^2 - 3\theta + 2 = 0$

$\Rightarrow (\theta - 2)(\theta - 1) = 0$

$\Rightarrow \theta = 1, 2 \to$ real & distinct

⇒ The solution of eq^n(1) is

$y = C_1e^z + C_2e^{2z}$

$\Rightarrow y = C_1x + C_2x^2 \left[\because x = e^z\right]$ ...(2)

Given ∴ y = 0 at x = 1

(2) $\Rightarrow 0 = C_1 + C_2 \Rightarrow C_1 = -C_2$ & y = 2 at x = 2

(2) $\Rightarrow 2 = 2C_1 + 4C_2 \Rightarrow 2 = -2C_2 + 4C_2$

$C_2 = 1$ and $C_1 = -1$

From (2) ; the solution is

$y = -x + x^2 \Rightarrow y(3) = -3 + 3^2 \Rightarrow y(3) = 6.$

5. $\frac{\partial v}{\partial t} = \beta\frac{\partial^2 V}{\partial x^2}$

$$\frac{\partial v}{\partial t} = \frac{V_i^{(n+1)} V_i^{(n)}}{\Delta t}$$

(Using forward time finite difference approximation)

Also; $f^{11}(x) = \frac{\partial^2 f}{\partial x^2} = \frac{f(x+h) - 2f(x) + f(x-h)}{h^2}$

(Using centered space finite difference approximation)

$$\Rightarrow \frac{\partial^2 V}{\partial x^2} = \frac{V_{i+1}^{(n)} - 2V_i^{(n)} + V_{i-1}^{(n)}}{(\Delta x)^2}$$

$\therefore \frac{\partial V}{\partial t} = \beta\frac{\partial^2 v}{\partial x^2}$ can be represented as

$$\frac{V_i^{(n+1)} - V_i^{(n)}}{\Delta t} = \beta\left[\frac{V_{i+1}^{(n)} - 2V_i^{(n)} + V_{i-1}^{(n)}}{(\Delta x)^2}\right]$$

6. Given, $x = \psi \ell n\phi$ & $y = \phi \ell n \psi$

$$\Rightarrow \psi = \frac{x}{\ell n\phi};$$

$$\Rightarrow \psi = \frac{x}{\ell n\left[\frac{y}{\ell n\psi}\right]} \left(\because y = \phi \ell n \psi \Rightarrow \phi = \frac{y}{\ell n\psi}\right)$$

$$\Rightarrow \psi = \frac{x}{\ell n[y] - \ell n[\ell n\psi]}$$

$$\left(\because \ell n\left[\frac{A}{B}\right] = \ell n(A) - \ell n(B)\right)$$

$\therefore$ Differentiating ψ partially w.r.t 'x', we have

$$\frac{\partial\psi}{\partial x} = \frac{\left[(\ell n(y) - \ell n[\ell n\psi])(1) - x\left[0 - \frac{1}{\ell n\psi}\frac{1}{\psi}\frac{\partial\psi}{\partial x}\right]\right]}{[\ell n(y) - \ell n(\ell n\psi)]^2};$$

since y is constant

$$= \frac{\frac{x}{\psi} + x\left[\frac{1}{\ell n\psi}\frac{1}{\psi}\frac{\partial\psi}{\partial x}\right]}{\left(\frac{x}{\psi}\right)2};$$

since $\ell ny - \ell n(\ell n\psi) = \ell n\phi = \left[\frac{x}{\psi}\right]$

$$\Rightarrow \frac{\partial\psi}{\partial x} = \frac{1 + \frac{1}{\ell n\psi}\frac{\partial\psi}{\partial x}}{\frac{x}{\psi}}$$

$$\Rightarrow \frac{x}{\psi}\frac{\partial\psi}{\partial x} = 1 + \frac{1}{\ell n\psi}\frac{\partial\psi}{\partial x}$$

$$\Rightarrow \frac{x}{\psi}\frac{\partial\psi}{\partial x} - \frac{1}{\ell n\psi}\frac{\partial\psi}{\partial x} = 1$$

$$\Rightarrow \frac{\partial\psi}{\partial x}\left[\frac{x}{\psi} - \frac{1}{\ell n\psi}\right] = 1$$

$$\Rightarrow \frac{\partial\psi}{\partial x}\left[\frac{x\ell n\psi - \psi}{\psi\ell n\psi}\right] = 1$$

$$\Rightarrow \frac{\partial\psi}{\partial x} = \frac{\psi\ell n\psi}{\psi\ell n\psi - \psi} = \frac{\ell n\psi}{\frac{x}{\psi}\ell n\psi - 1} = \frac{\ell n\psi}{\ell n\phi\ell n\psi - 1};$$

Since $\frac{x}{\psi} = \ell n\phi \Rightarrow \frac{\partial\psi}{\partial x} = \frac{\ell n\psi}{\ell n\phi\ell n\psi - 1}$

SET-2

ANSWERS

1. (*d*)	2. (*a*)	3. (*d*)	4. (*b*)	5. (*c*)
6. 0.8	7. 12			

EXPLANATIONS

1. $\lim_{x\to 3}\frac{x^4 - 81}{2x^2 - 5x - 3}\left(\frac{0}{0}\text{form}\right)$

Now applying L' Hospital rule we get

$$= \lim_{x\to 3}\frac{4x^3}{4x - 5} = \frac{108}{7}$$

3. $A = \{(x, X), x \in X \text{ and } X \subseteq U\}$

The number of k element subsets of a set U with n elements = ${}^nC_k$.

The number of possible ordered pairs (x, X) where $x \in X$ is $k \cdot {}^nC_k$ $1 \le k \le n$ 1 to n.

So total number of ordered pairs in A is given

$$= |A| = \sum_{k=1}^{n} k \cdot {}^nC_k = \sum_{k=1}^{n} k \cdot \binom{n}{k}$$

So II is correct.

But by the combinational identity

$$\sum_{k=1}^{n} k \cdot \binom{n}{k} = n \cdot 2^{n-1}$$

So I is also correct.

So both I and II are correct.

4. In a complete graph we can traverse the n vertices in any order and return to the starting vertex and form a Hamiltonian cycle. The number of such cycles required is n!

While for is circular rotations ignored. Since for example K_4 with vertices {1, 2, 3, 4}, the cycle 1-2-3-4 is same as 2-3-4-1 is same as 3-4-1-2 etc. we now get only (n – 1)! distinct Hamiltonian cycles. Further, the cycle 1-2-3-4 and 1-4-3-2 are also same (clockwise and anticlockwise).

$\therefore$ we get $\frac{(n-1)!}{2}$ distinct Hamiltonian cycles

5. $R_1 : \forall\ a, b \in G, a R_1 b$ if and only if $\exists g \in G$ such that $a = g^{-1} bg$

Reflexive: $a = g^{-1}$

ag can be satisfied by putting g = e, identity "e" always exists in a group.

So reflexive

Symmetric: aRb $\Rightarrow$ a = g^{-1} bg for some g

$\Rightarrow b = gag^{-1} = (g^{-1})^{-1} ag^{-1}$

g^{-1} always exists for every g $\in$ G.

So symmetric

Transitive: aRb and bRc $\Rightarrow a = g_1^{-1} bg_1$ and $b = g_2^{-1} cg_2$ for some $g_1 g_2 \in G$.

Now $a = g_1^{-1} g_2^{-1} cg_2 g_1 = (g_2 g_1)^{-1} cg_2 g_1$

$g_1 \in G$ and $g_2 \in G \Rightarrow g_2 g_1 \in G$ since group is closed so aRb and aRb $\Rightarrow$ aRc hence transitive

Clearly R_1 is equivalence relation.

R_2 is not equivalence it need not even be reflexive, since $aR_2 a \Rightarrow a = a^{-1}$ $\forall$ a which not be true in a group.

R_1 is equivalence relation is the correct answer.

6. Polynomial $3x^2 + 6xY + 3Y + 6$ has only real roots

$\therefore\ b^2 - 4ax \geq 0$

$(6Y)^2 - 4(3)(3Y + 6) \geq 0$

$Y^2 - Y + 2 \geq 0$

$Y \in (-\infty, -1] \cap [2, \infty)$

$\Rightarrow Y \in [2, 6)$

Now y is uniformly distributed in (1, 6)

Probability distribution function

$$f(Y) = \frac{1}{5} \qquad 1 < y < 6$$

$$\therefore\ p(2 \leq y < 6) = \int_2^6 f(Y)dy$$

$$\frac{1}{5}[Y]_2^6 = \frac{1}{5}(6-2) = \frac{4}{5} = 0.8$$

7. Product of eigen values is same as the determinant of a matrix.

$$\begin{bmatrix} 1 & 2 & 2^2 & 2^3 \\ 1 & 3 & 3^2 & 3^3 \\ 1 & 4 & 4^2 & 4^3 \\ 1 & 5 & 5^2 & 5^3 \end{bmatrix} = \begin{bmatrix} 1 & 2 & 2^2 & 2^3 \\ 0 & 3-2 & 3^2-2^2 & 3^3-2^3 \\ 0 & 4-2 & 4^2-2^2 & 4^3-2^3 \\ 0 & 5-2 & 5^2-2^2 & 5^3-2^3 \end{bmatrix}$$

$$= (3-2)(4-2)(5-2)\begin{bmatrix} 1 & 5 & 19 \\ 1 & 6 & 28 \\ 1 & 7 & 39 \end{bmatrix}$$

$$= 1.2.3\begin{bmatrix} 1 & 5 & 19 \\ 0 & 1 & 9 \\ 0 & 2 & 20 \end{bmatrix}$$

$= 1 \times 2 \times 3 \times 2 = 12$

So the product of eigen value is 12.

SET-3

ANSWERS

1. (139)	2. (*d*)	3. (*a*)	4. (*a*)	5. (3)
6. (45)	7. (*c*)	8. (*d*)	9. (*d*)	

EXPLANATIONS

1. Given,

$$f = 2x^3 + 3y^2 + 4z$$

$$\nabla f = \left(\frac{\partial \hat{i}}{\partial x} + \frac{\partial \hat{j}}{\partial y} + \frac{\partial \hat{k}}{\partial z}\right)(2x^3 + 3y^2 + 4z)$$

$$\nabla f = \hat{i}[6x^2] + \hat{j}[6y] + \hat{k}[4]$$

Now curl of $(\nabla f) = 0$

$\therefore\ \nabla f$ is irrotational

$$\therefore\ \int_C \text{grad} f.dr = \int 6x^2dx + 6ydy + 4dz$$

$$= \int_C d(2x^3 + 3y^2 + 4z)$$

$$= \int_{(-3,-3,2)}^{(2,-3,2)} d(2x^3 + 3y^2 + 4z) + \int_{(2,-3,2)}^{(2,6,2)} d(2x^3 + 3y^2 + 4z) + \int_{(2,6,2)}^{(2,6,-1)} d(2x^3 + 3y^2 + 4z)$$

$$\therefore\ \int_C \text{grad}\, f.dr = \int_{(-3,-3,2)}^{(2,6,-1)} d(2x^3 + 3y^2 + 4z)$$

$$= \left[2x^3 + 3y^2 + 4z\right]_{(-3,-3,2)}^{(2,6,-1)}$$

$$= 120 - (-19) = (120 + 19) = 139$$

2. Given, $H(s) = \dfrac{s+3}{s^2 + 2s + 1}$ for $t \geq 0$

Taking inverse laplace transform,

$$L^{-1}[H(s)] = L^{-1}\left[\frac{s+3}{s^2+2s+1}\right] = L^{-1}\left[\frac{s+3}{(s+1)^2}\right]$$

$$= L^{-1}\left[\frac{s+1+2}{(s+1)^2}\right] = L^{-1}\left[\frac{1}{s+1} + \frac{2}{(s+1)^2}\right]$$

$$= L^{-1}\left[\frac{1}{s+1}\right] + 2L^{-1}\left[\frac{1}{(s+1)^2}\right]$$

$$= e^{-t}(1) + 2e^{-t}t \qquad \left[\because L^{-1}\left[\frac{1}{s^2}\right] = t\right]$$

$$\therefore\ L^{-1}[H(s)] = 2te^{-t} + e^{-t}$$

3. Given partial differential equation.

$$\frac{\partial^2 u}{\partial t^2} - C^2\left(\frac{\partial^2 u}{\partial x^2} + \frac{\partial^2 u}{\partial y^2}\right) = 0,$$

$$\therefore \quad \frac{\partial^2 u}{\partial t^2} = C^2\left(\frac{\partial^2 u}{\partial x^2} + \frac{\partial^2 u}{\partial y^2}\right)$$

It represent two dimensional wave equation.

4. Given,

M is a 2 × 2 matrix with eigen values 4 and 9.

$\therefore$ i.e., $\lambda = 4, 9$ [Where 'λ' represents eigen values of M]

If λ is an eigen value of the matrix M then λ^2 is an eigen value of the matrix M^2.

$\therefore$ $\lambda^2 = 4^2, 9^2 = 16, 81$ are eigen values of matrix M^2.

5. Given,

$$\begin{bmatrix} 0 & 1 & 1 \\ 1 & 0 & 1 \\ 1 & 1 & 0 \end{bmatrix}$$

Applying $R_3 \to R_3 - R_2$

$$\begin{bmatrix} 0 & 1 & 1 \\ 1 & 0 & 1 \\ 0 & 1 & -1 \end{bmatrix}$$

Applying $R_3 \to R_3 - R_1$

$$\begin{bmatrix} 0 & 1 & 1 \\ 1 & 0 & 1 \\ 0 & 0 & -2 \end{bmatrix}$$

Applying $R_2 \leftrightarrow R_3$;

$$\begin{bmatrix} 1 & 0 & 1 \\ 0 & 1 & 1 \\ 0 & 0 & -2 \end{bmatrix}$$

The above matrix is in the form of echelon matrix.

$\therefore$ $\rho(M)$ = Number of non-zeros rows in echelon matrix.

$\Rightarrow \rho(M) = 3$

6. Given, $A = 2xi + 3y\hat{j} + 4z\hat{k}$ and $u = x^2 + y^2 + z^2$

$\text{div}(uA) = \nabla.(uA)$

$$\text{Now div}(uA) = u(\nabla.A) + \nabla u.A \qquad \text{...(i)}$$

$$\text{Now } \nabla.A = \frac{\partial}{\partial x}(2x) + \frac{\partial}{\partial y}(3y) + \frac{\partial}{\partial z}(4z)$$

$$= 2 + 3 + 4 = 9$$

$$\therefore \quad \nabla.A = 9 \qquad \text{...(ii)}$$

$$\text{Now } \nabla u = i\frac{\partial u}{\partial x} + j\frac{\partial u}{\partial y} + k\frac{\partial u}{\partial z}; \text{ where } u = x^2 + y^2 + z^2$$

$$= i[2x] + j[2y] + k[2z] = 2[xi + yj + zk]$$

$$\therefore \nabla u.A = [(2x)i + (2y)j + (2z)k].[(2x)i + (3y)j + (4z)k]$$

$$\Rightarrow \nabla u.A = 4x^2 + 6y^2 + 8z^2 \qquad \text{...(iii)}$$

From eq. (i), (ii) and (iii), we get

$$\text{div}(uA) = (x^2 + y^2 + z^2)9 + (4x^2 + 6y^2 + 8z^2)$$

$$\text{div}(uA/_{(1,1,1)} = (1 + 1 + 1)9 + (4 + 6 + 8)$$

$$= 27 + 18 = 45.$$

7. Let $F(z) = \dfrac{z^3 + z^2 + 8}{z + 2}$

$\therefore$ Singular point of F(z) is at $z = -2$; which lies inside $C : |Z| = 5$

Now using Cauchy's integral formula,

$$\oint_C F(z)dz = \oint_C \frac{z^3 + z^2 + 8}{z + 2}dz = \oint_C \frac{z^3 + z^2 + 8}{z - (-2)}dz$$

$$\left[\text{By cauchy's formula; } \oint_C \frac{f(z)}{z - z_0}dz = 2\pi j f(z_0)\right]$$

$$= 2\pi j\left[z^3 + z^2 + 8\right]_{z=-2}$$

$$\Rightarrow \oint_C \frac{z^3 + z^2 + 8}{z + 2}dz = 2\pi j(-8 + 4 + 8) = 8\pi j$$

8. As per the quient its waveform is drawn below.

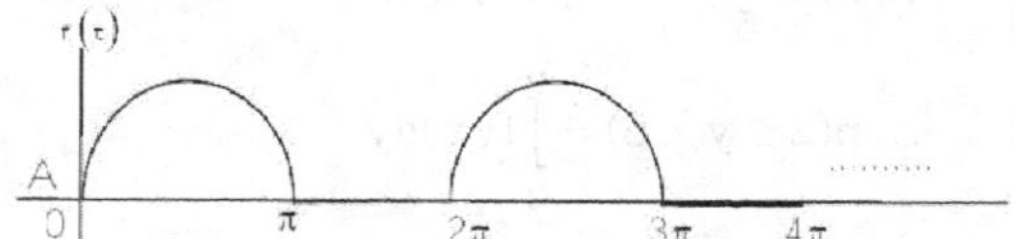

$$f(t) = f_o(t)1 + fe(t) = \left[\frac{f(t) - f(-t)}{2}\right] + \left[\frac{f(t) + f(-t)}{2}\right]$$

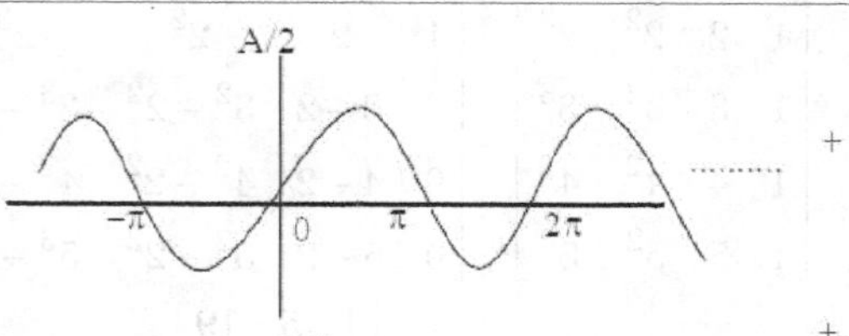

+

f(t)

+

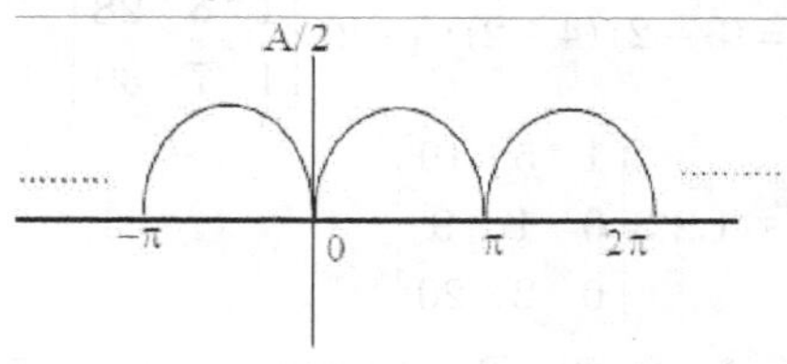

$f_e(t)$ $= f(t)$

$$f_0(t) = \left[\frac{A}{2}\sin\omega_0 t\right] + \sum_{n=1}^{N} a_n \cos\omega_0 t.$$

From this $b_1 = \frac{A}{2}$

9. Let $M_{2\times2} = \begin{bmatrix} a_{11} & a_{12} \\ a_{21} & a_{22} \end{bmatrix}$, and

$v_1 = \begin{bmatrix} a_{11} \\ a_{21} \end{bmatrix}, v_2 = \begin{bmatrix} a_{12} \\ a_{22} \end{bmatrix}$

$\therefore\ M^{-1} = \frac{1}{a_{11}a_{22} - a_{21}a_{12}}\begin{bmatrix} a_{22} & -a_{12} \\ -a_{21} & a_{11} \end{bmatrix},$

where $u_1^T = \frac{1}{a_{11}a_{22} - a_{21}a_{12}}[a_{22} \quad -a_{12}],$

$u_1^T = \frac{1}{a_{11}a_{22} - a_{21}a_{12}}[-a_{21} \quad a_{11}]$

$\therefore u_1^T v_1 = \left[\frac{a_{22}}{a_{11}a_{22} - a_{21}a_{12}} \quad \frac{-a_{12}}{a_{11}a_{22} - a_{21}a_{12}}\right]\begin{bmatrix} a_{11} \\ a_{21} \end{bmatrix} = 1$

$u_2^T v_2 = \left[\frac{-a_{21}}{a_{11}a_{22} - a_{21}a_{12}} \quad \frac{a_{11}}{a_{11}a_{22} - a_{21}a_{12}}\right]\begin{bmatrix} a_{12} \\ a_{22} \end{bmatrix} = 1$

∴ Statement-I is true.

$u_1^T v_2 = \left[\frac{a_{22}}{a_{11}a_{22} - a_{21}a_{12}} \quad \frac{-a_{12}}{a_{11}a_{22} - a_{21}a_{12}}\right]\begin{bmatrix} a_{12} \\ a_{22} \end{bmatrix} = 0$

$u_2^T v_1 = \left[\frac{-a_{21}}{a_{11}a_{22} - a_{21}a_{12}} \quad \frac{a_{11}}{a_{11}a_{22} - a_{21}a_{12}}\right]\begin{bmatrix} a_{11} \\ a21 \end{bmatrix} = 0$

∴ Statement-II is true. So, both Statements are true.

SET-4

ANSWERS

1. (B)	2. (2)	3. (3)	4. (A)	5. (11)
6. (0.37)	7. (0)	8. (C)	9. (5.25)	

EXPLANATIONS

1. $\ell n(z)$ is not analytic at z = 0;

and $e^{\frac{1}{z}}$ is also not analytic at z = 0 and

$\frac{1}{1-z}$ is also not analytic at z = 0

But cos (z) is analytic over the entire complex plane,

Since $\cos z = \cos(x + iy) = \cos x \cos(iy) - \sin x \sin(iy)$

$= \cos x \cos h\, y - i \sin x \sin h\, y;$

Now comparing with u + iv we get

Where $u(x, y) = \cos x \cos h\, y;$

$v(x,y) = -\sin x \sin h\, y$

$u_x = -\sin x.\cosh y \quad | \quad v_x = -\cos x.\sinh y$

$u_y = \cos x \sinh y \quad | \quad v_y = -\sin x.\cosh y$

2.

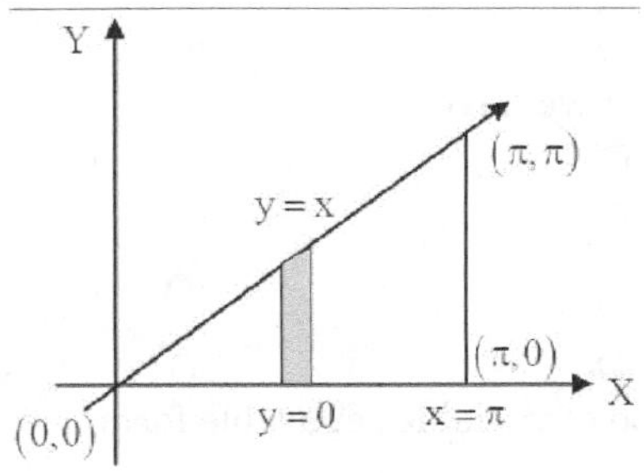

$\int_0^\pi \int_0^\pi \frac{\sin x}{x} dx\, dy$

Given limits are x = y → x = π & y = 0 → y = π.

For change of order of integration.

Consider a strip parallel to Y-axis. Then

Limits of y are : y = 0 → y = x and

Limits of x : x = 0 → x = π

$$\therefore \int_0^\pi\int_0^\pi \frac{\sin x}{x} dx\, dy = \int_{x=0}^{\pi}\left[\int_{y=0}^{\pi} \frac{\sin x}{x}.dy\right] dx$$

$$= \int_{x=0}^{\pi} \frac{\sin x}{x}[y]_0^x . dx$$

$$= \int_{x=0}^{\pi} \sin x.dx$$

$$= [-\cos x]_0^\pi$$

$$= -[-1 - 1] = 2$$

$$\Rightarrow \int_0^\pi\int_y^\pi \frac{\sin y}{x} dx\, dy = 2$$

3. Given,

$$\text{Matrix } A = \begin{bmatrix} 2 & 2 & 3 & 3 \\ 0 & 1 & 1 & 1 \\ 0 & 0 & 3 & 3 \\ 0 & 0 & 0 & 2 \end{bmatrix}$$

→ Upper triangular matrix

∴ Eigen values of A are λ = 2, 1, 3, 2 [diagonal elements]

∴ The number of distinct eigen values of the matrix = 3. i.e. 1, 2, 3.

4. Given differential equation is

$$\frac{dy}{dx} = -\left(\frac{x}{y}\right)^n$$

$$\therefore \quad \frac{dy}{dx} = -\frac{x^n}{y^n} \qquad ...(1)$$

For n = –1; we have

$$\frac{dy}{dx} = -\frac{x^{-1}}{y^{-1}}$$

$$\Rightarrow \frac{dy}{dx} = -\frac{y}{x}$$

By method of variable separable form.

$$\Rightarrow \frac{1}{y}dy = \frac{-1}{x}dx$$

Integrative on both sides we get,

$$\ell ny = -\ell nx + \ell nc \Rightarrow \ell n(xy) = \ell nc$$

$$\Rightarrow xy = c^2$$

Which represents family of hyperbolas.

For n = 1; we have

$$\frac{dy}{dx} = -\frac{x}{y} \qquad [\because (1)]$$

By method of variable separable

$$\Rightarrow ydy = -xdx$$

Integrating on bothsides get

$$\frac{y^2}{2} = -\frac{x^2}{2} + c \Rightarrow \frac{x^2}{2} + \frac{y^2}{2} = c$$

$$\Rightarrow x^2 + y^2 = c$$

$\Rightarrow x^2 + y^2 = (\sqrt{c})^2$ which represents family of circles.

5. Given, E[2x + y] = 0

and E[x + 2y] = 33

$$\Rightarrow \quad 2E(x) + E(y) = 0 \qquad ...(i)$$

$$\Rightarrow \quad E(x) + 2E(y) = 33 \qquad ...(ii)$$

Solving eq. (i) and (ii) we get

$$E(y) = 22 \text{ and } E(x) = -11$$

$$\therefore \quad E(x) + E(y) = -11 + 22 = 11$$

$$\Rightarrow \quad E(x) + E(y) = 11$$

6. Given,

Cumulative distribution function,

$$F_z(x) = \begin{cases} 1 - e^{-x} & \text{if } x \ge 0 \\ 0 & \text{if } x < 0 \end{cases}$$

→ probability density function.

$$\Rightarrow F_z'(x) = f_z(x) = \begin{cases} e^{-x}; & x \ge 0 \\ 0; & x < 0 \end{cases}$$

Where Z be an exponential R.V with mean '1'

Using conditional probability

$$\therefore \Pr\left[\frac{z>2}{z>1}\right] = \frac{\Pr[z>2 \cap z>1]}{\Pr[z>1]}$$

$$= \frac{\Pr[z>2]}{\Pr[z>1]} = \frac{\int_2^\infty f(x)dx}{\int_1^\infty f(x)dx} = \frac{\int_0^\infty e^{-x}dx}{\int_1^\infty e^{-x}dx}$$

$$\Rightarrow \Pr\left[\frac{z>2}{z>1}\right] = \frac{\left[-e^{-x}\right]_2^\infty}{\left[-e^{-x}\right]_1^\infty}$$

$$= \frac{\left[0+e^{-2}\right]}{\left[0+e^{-1}\right]}$$

$$= \frac{e^{-2}}{e^{-1}} = \frac{1}{e} \approx 0.37$$

7. $$\frac{1}{2\pi j}\oint\left(z+\frac{1}{z}\right)^2 dz = \frac{1}{2\pi j}\oint\frac{(z^2+1)^2}{z^2}dz \qquad ...(i)$$

∴ Singular point z = 0; which lies inside unit circle |z| = 1

∴ Using Cauchy's generalization integration formula, we get.

$$\oint_c \frac{f(z)}{(z-z_0)^n}dz = 2\pi j\frac{\left[f^{n-1}(z_0)\right]}{(n-1)!}$$

$$\Rightarrow \therefore \oint_C \frac{(z^2+1)^2}{(z-0)^2}dz = 2\pi j\frac{\left[\left(z^2+1\right)^2\right]'}{1!}\Big/ z = 0;$$

$$= 2\pi j[2(z^2+1)(2z)]_{z=0}$$

$$= 2\pi j[0] = 0.$$

From eq. (i) we get,

$$\frac{1}{2\pi j}\oint\left(z+\frac{1}{z}\right)^2 dz = \frac{1}{2\pi j}\oint\frac{(z^2+1)^2}{(z-0)^2} = \frac{1}{2\pi j}(0) = 0$$

8.

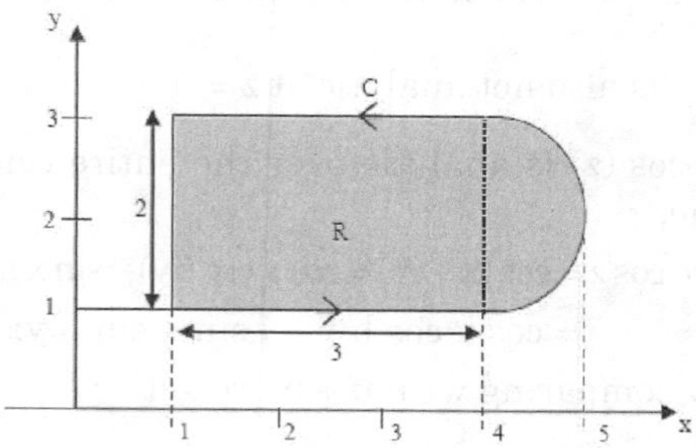

$\int_C (xdy - ydx) = \int_C (-y\,dx + x\,dy)$; where C is a

Closed figure formed by rectangle and a semi circle having radius 1.

To countert line integral to area integral.

Now Using Green's theorem, we have

$$\oint_C \left(-\underset{M}{\underset{\downarrow}{y}}\,dx + \underset{N}{\underset{\downarrow}{y}}\,dy \right) = \int_R (1+1)\,dx\,dy$$

$$= 2\iint_R dx\,dy \rightarrow \text{Area of the region R.}$$

$$= 2[\text{Area of rectangle + Area of semi-circle}]$$

$$= 2\left[(3\times 2) + \frac{1}{2}\pi(1^2)\right]$$

$$= 2\left[6 + \frac{\pi}{2}\right] = (12+\pi)$$

9. Given differential equatin,

$$x^2\frac{d^2y}{dx^2} - 3x\frac{dy}{dx} + 3y = 0, \quad \text{...(i)}$$

$x > 0$ & $y(1) = 1$ and $y(2) = 14$

(Cauchy-Euler Linear D. E.)

Equation (i) can be written as

$[x^2D^2 - 3xD + 3]y = 0$

Let $xD = \theta$; $x^2D^2 = \theta(\theta - 1)$;

where $\theta = \frac{d}{dz}$ and $x = e^z$

$\therefore [\theta(\theta-1) - 3\theta + 3]y = 0 \Rightarrow [\theta^2 - 4\theta + 3]y = 0$

Consider the Auxillary equation as

$\theta^2 - 4\theta + 3 = 0 \Rightarrow (\theta - 3)(\theta - 1) = 0$

$\Rightarrow \theta = 1, 3 \rightarrow$ (Roots are real and distinct)

$\therefore$ The solution is,

$y = C_1e^{1.z} + C_2e^{3.z} \Rightarrow y = C_1x + C_2x^3 [\because x = e^z]$ ...(ii)

Given y = 1at x = 1

$\therefore C_1 + C_2$ ...(iii)

and y = 14 at x = 2

$\therefore 14 = 2C_1 + 8C_2$

$\therefore C_1 + 4C_2 = 7$ ...(iv)

Solving eq. (iii) and (iv) we get

$\boxed{C_2 = 2 \text{ and } C_1 = 1}$

Now from eq. (ii),

$y = (-1)x + (2)x^3$

$\therefore y(1.5) = (-1)(1.5) + 2(1.5)^3$

$\therefore y(1.5) = 5.25$

SET-5

ANSWERS

1. (*a*)	2. (*c*)	3. (*a*)	4. (63)	5. (*b*)
6. (*b*)	7. (*c*)	8. (2.097)		

EXPLANATIONS

1. From question, we have

$$\frac{dy}{dx} + 7x^2y = 0,\ y(0) = \frac{3}{7},\ y(1) = ?$$

$$\frac{dy}{dx} = -7x^2 y$$

$$\frac{dy}{y} = -7x^2\,dx$$

$$ln\,y = -\frac{7x^3}{3} + C$$

$$y = e^{-\frac{7x^3}{3}+C}$$

$$y(0) = e^C = \frac{3}{7}$$

$$C = ln\left(\frac{3}{7}\right)$$

$$ln\,y = -\frac{7x^3}{3} + \ln\left(\frac{3}{7}\right)$$

$$ln\,y - ln\left(\frac{3}{7}\right) = -\frac{7x^3}{3}$$

$$ln\left(\frac{7y}{3}\right) = -\frac{7x^3}{3}$$

$$y = \frac{3}{7}e^{-\frac{7x^3}{3}}$$

$$\therefore \quad y(1) = \frac{3}{7}e^{-\frac{7}{3}}$$

2. $$|P-\lambda I| = \begin{vmatrix} 1-\lambda & 1 & 0 \\ 0 & 1-\lambda & 1 \\ 0 & 0 & 1-\lambda \end{vmatrix}$$

$$(1-\lambda)[(1-\lambda)(1-\lambda)-0] = 0$$

Eigen values are,

$\lambda = 1,1,1$

So, No. of distinct eigen value = 1

3. Let Co-ordinates of Q are (x', y') and rotation about x-y plane anti clockwise.

$$\begin{bmatrix} x' \\ y' \end{bmatrix} = \begin{bmatrix} \cos\theta & -\sin\theta \\ \sin\theta & \cos\theta \end{bmatrix}\begin{bmatrix} x \\ y \end{bmatrix}$$

$$\begin{bmatrix} x' \\ y' \end{bmatrix} = \begin{bmatrix} \cos 30 & -\sin 30 \\ \sin 30 & \cos 30 \end{bmatrix}\begin{bmatrix} 20 \\ 10 \end{bmatrix}$$

$$= \begin{bmatrix} 0.866 & -0.5 \\ 0.5 & 0.866 \end{bmatrix}\begin{bmatrix} 20 \\ 10 \end{bmatrix}$$

$$\begin{bmatrix} x' \\ y' \end{bmatrix} = \begin{bmatrix} 0.866\times 20 - 0.5\times 10 \\ 0.5\times 20 + 0.866\times 10 \end{bmatrix}$$

$$= \begin{bmatrix} 12.32 \\ 18.66 \end{bmatrix}$$

$$(x', y') = (12.32, 18.66)$$

4. Trapezoidal Rule, h = 1

$$\int y\,dx = \frac{h}{2}\left[(y_0+y_1)+2(y_1+y_2+.....)\right]$$

$$= \frac{1}{2}\left[(2^3+4^3)+2(3^3)\right]$$

$$= \frac{1}{2}[8+64+54] = 63$$

5. 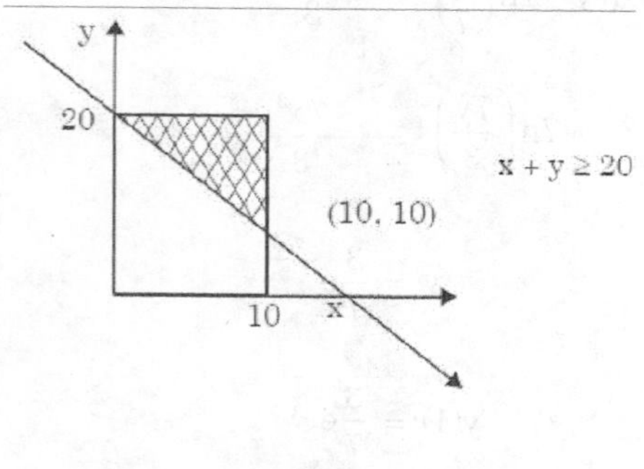

Required Probability (x + y > 20)

$$= \frac{\text{Shaded Area}}{\text{Total Area}}$$

$$= \frac{\frac{1}{2}\times 10\times 10}{20\times 10} = \frac{50}{200} = \frac{1}{4} = 0.25$$

6. $u(x, y) = 2x^2 - 2y^2 + 4xy$

As harmonic function is analytic there fore,

$u_x = V_y$

$u_y = -V_x$

$u_x = 4x + 4y$

$u_y = -4y + 4x$

$$\frac{\partial v}{\partial y} = V_y = 4x + 4y$$

$$V = 4xy + 2y^2 + f(x)$$

Now $$\frac{\partial v}{\partial y} = 4y + f'(x) = 4y - 4x$$

$f'(x) = -4x$

$f(x) = -2x^2 + C$

So, conjugate harmonic function v(x, y) is,

$V = 2y^2 - 2x^2 + 4xy + C$

7. Given set of equations are,

$x + y + z = 1$, $a - ay + 32 = 5$ and $5x - 3y + az = 6$

Now $\rho(A) = \rho(AB) <$ no. of variables

$|A| = 0$

$$\begin{vmatrix} 1 & 1 & 1 \\ a & -a & 3 \\ 5 & -3 & a \end{vmatrix} = 0$$

$(-a^2 + 9) - (a^2 - 15) + (-3a + 5a) = 0$

$9 - a^2 + 15 - a^2 + 2a = 0$

$2a^2 - 2a - 24 = 0$

$a^2 - a - 12 = 0$

$a^2 - 4a + 3a - 12 = 0$

$(a = 4, -3)$

$\rho(AB) <$ (No. of variable)

$$\begin{vmatrix} 1 & 1 & 1 \\ -a & 3 & 5 \\ -3 & a & 6 \end{vmatrix} = 0$$

$1(18-5a)-1(-6a+15)+1(-a^2+9)=0$

$18-5a-15+6a+9-a^2=0$

$a^2-a-12=0$

$a = 3, 4$ or $a = 4$

$$\begin{vmatrix} 1 & 1 & 1 \\ a & -a & 5 \\ 5 & -3 & 6 \end{vmatrix} = 0$$

$(-6a+15)-(6a-25)+(-3a+5a)=0$

$-6a+15-6a+25+2a=0$

$10a = 40$

$a = 4$

if $a = 4$

$|AB| = 0$

$$\begin{vmatrix} 1 & 1 & 1 \\ a & 3 & 5 \\ 5 & a & 6 \end{vmatrix} = 0$$

$(18-5a)-(6a-25)+(a^2-15)=0$

$a^2-11a+28=0$

$a = 7, 4$ or $\boxed{a=4}$

8. $$I = \int_1^e x\, ln\, x dx$$

$$= \frac{ln\, x^2}{2} - \int\left[\frac{1}{x}\times\frac{x^2}{2}\right]dx$$

$$= \left.\frac{x^2 ln\, x}{2} - \frac{x^2}{4}\right|_1^e$$

$$= \left(\frac{e^2}{2}-\frac{e^2}{4}\right)+\frac{1}{4} = \frac{e^2+1}{4} = 2.097$$

SOLVED PAPER – 2020

GENERAL APTITUDE

SET-1

Q 1 – Q 5 carry one mark each

1. It is a common criticism that most of the academicians live in their _____., so, they are not aware of the real life challenges.

(a) homes (b) ivory towers
(c) glass palaces (d) big flats

2. His hunger for reading is insatiable. He reads indiscriminately. He is most certainly a/an _____ reader.

(a) all-round (b) precocious
(c) voracious (d) wise

3. Select the word that fits the analogy:

Fuse : Fusion :: Use: _______.

(a) Usage (b) User
(c) Uses (d) Usion

4. If 0, 1, 2, ..., 7, 8, 9 are coded as O, P, Q ....., V, W, X, then 45 will be coded as _______.

(a) TS (b) ST
(c) SS (d) SU

5. The sum of two positive numbers is 100. After subtracting 5 from each number, the product of the resulting numbers is 0. One of the original numbers is _______ .

(a) 80 (b) 85
(c) 90 (d) 95

Q 6 – Q 10 carry two marks each

6. The American psychologist Howard Gardener expounds that human intelligence can be subcategorised into multiple kinds, in such a way that individuals differ with respect to their relative competence in each kind. Based on this theory, modern educationists insist on prescribing multi-dimensional curriculum and evaluation parameters that enable development and assessment of multiple intelligences.

Which of the following statements can be inferred from the given text?

(a) Howard Gardner insists that the teaching curriculum and evaluation needs to be multi-dimensional

(b) Howard Gardner wants to develop and assess the theory of multiple intelligences

(c) Modern educationists want to develop and assess the theory of multiple intelligences.

(d) Modern educationists insist that the teaching curriculum and evaluation needs to be multidimensional

7. Five friends P, Q, R, S and T went camping. At night, they had to sleep in a row inside the tent. P, Q and T refused to sleep next to R since he snored loudly. P and S wanted to avoid Q as he usually hugged people in sleep.

Assuming everyone was satisfied with the sleeping arrangements. What is the order in which they slept?

(a) RSPTQ
(b) SPRTQ
(c) QRSPT
(d) QTSPR

8. Insert seven numbers between 2 and 34, such that the resulting sequence including 2 and 34 is an arithmatic progression. The sum of these inserted seven numbers is _______.

(a) 120 (b) 124
(c) 126 (d) 130

9. The unit's place in 26591749^{110016} is _______.

(a) 1 (b) 3
(c) 6 (d) 9

10. The total expenditure of a family, on different activities in a month, is shown in the pie-chart. The extra money spent on education as compared to transport (in percent) is _______ .

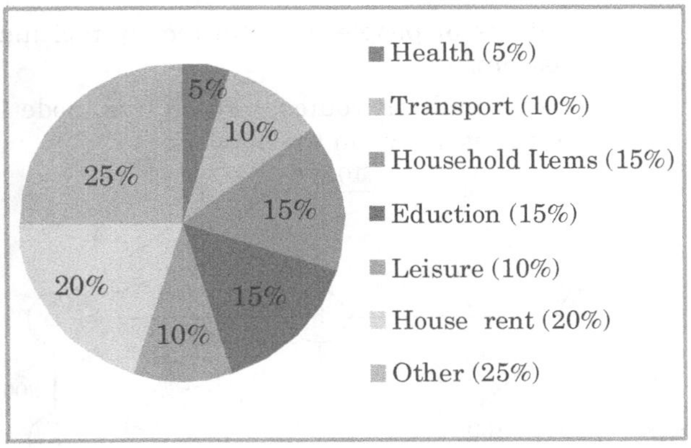

(a) 5 (b) 33.3
(c) 50 (d) 100

SET-2

Q. 1 to Q. 5 : Carry One Mark Each.

1. Raman is confident of speaking English ________ six months as he has been practising regularly ________the last three weeks.

(a) during, for (b) for, since
(c) for, in (d) within, for

2. His knowledge of the subject was excellent but his classroom performance was ______.

(a) extremely poor (b) good
(c) desirable (d) praiseworthy

3. Select the word that fits the analogy:
Cook : Cook :: Fly : ________

(a) Flyer (b) Flying
(c) Flew (d) Flighter

4. The dawn of the 21st century witnessed the melting glaciers oscillating between giving too much and too little to billions of people who depend on them for fresh water. The UN climate report estimates that without deep cuts to man-made emissions, at least 30% of the northern hemisphere's surface permafrost could melt by the end of the century. Given this situation of imminent global exodus of billions of people displaced by rising seas, nation-states need to rethink their carbon footprint for political concerns, if not for environmental ones.

Which one of the following statements can be inferred from the given passage?

(a) Nation-states do not have environmental concerns.
(b) Nation-states are responsible for providing fresh water to billions of people.
(c) Billions of people are responsible for man-made emissions.
(d) Billions of people are affected by melting glaciers.

5. There are multiple routes to reach from node 1 to node 2, as shown in the network.

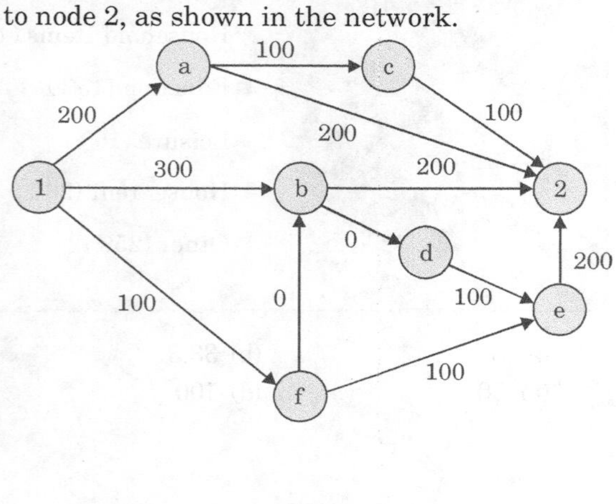

The cost of travel on an edge between two nodes is given in rupees. Nodes 'a', 'b', 'c', 'd', 'e' and 'f' are toll booths. The toll price at toll booths marked 'a' and 'e' is Rs. 200 and is Rs. 100 for the other toll booths. Which is the cheapest route from node 1 to node 2?

(a) 1-a-c-2 (b) 1-f-b-2
(c) 1-b-2 (d) 1-f-e-2

Q. 6 to Q. 10 : Carry Two Marks Each.

6. Goods and Services Tax (GST) is an indirect tax introduced in India in 2017 that is imposed on the supply of goods and services, and it subsumes all indirect taxes except few. It is a destination-based tax imposed on goods and services used, and it is not imposed at the point of origin from where goods come. GST also has a few components specific to state governments, central government and Union Territories (UTs).

Which one of the following statements can be inferred from the given passage?

(a) GST is imposed on the production of goods and services.
(b) GST includes all indirect taxes.
(c) GST does not have a component specific to UT.
(d) GST is imposed at the point of usage of goods and services.

7. If P = 3, R = 27, T = 243, then Q + S = ______.

(a) 40 (b) 80
(c) 90 (d) 110

8. The figure below shows an annular ring with outer and inner radii as b and a, respectively. The annular space has been painted in the form of blue colour circles touching the outer and inner periphery of annular space. If maximum n number of circles can be painted, then the unpainted area available in annular space is _____.

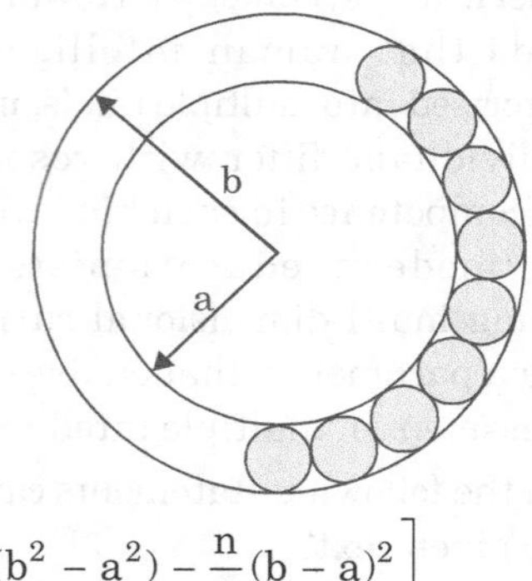

(a) $\pi\left[(b^2 - a^2) - \frac{n}{4}(b - a)^2\right]$

(b) $\pi\,[(b^2 - a^2) - n\,(b - a)^2]$

(c) $\pi\left[(b^2 - a^2) + \frac{n}{4}(b - a)^2\right]$

(d) $\pi[(b^2 - a^2) + n(b - a)^2]$

9. Two straight lines are drawn perpendicular to each other in X-Y plane. If α and β are the acute angles the straight lines make with the X-axis, then $\alpha + \beta$ is ________.

(a) 60° (b) 90°

(c) 120° (d) 180°

10. The total revenue of a company during 2014-2018 is shown in the bar graph. If the total expenditure of the company in each year is 500 million rupees, then the aggregate profit or loss (in percentage) on the total expenditure of the company during 2014-2018 is ________.

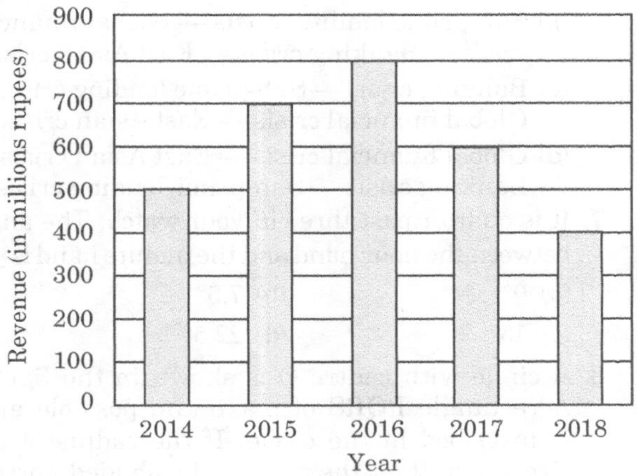

(a) 16.67% profit (b) 16.67% loss

(c) 20% profit (d) 20% loss

SET-3

Q 1 – Q 5 Carry one mark each.

1. This book, including all its chapters, _____ interesting. The students as well as instructor ____ in agreement about it.

(a) is, was (b) are, are

(c) is, are (d) were, was

2. People were prohibited _____ their vehicles near the entrance of the main administrative building

(a) to park (b) from parking

(c) parking (d) to have parked

3. Select the word that fits the analogy

Do : Undo :: Trust: ______.

(a) Entrust (b) Intruct

(c) Distrust (d) Untrust

4. Stock markets _____ at the news of the coup

(a) poised

(b) plunged

(c) plugged

(d) probed

5. If P, Q, R, S are four individuals, how many teams of size exceeding one can be formed, with Q as a member?

(a) 5 (b) 6

(c) 7 (d) 8

Q 6 – Q 10 Carry two marks each.

6. Non-performing assets (NPAs) of a bank in India is defined as an asset, which remains unpaid by a borrower for a certain period of time in terms of interest. Principal, or both. Reserve Bank of India (RBI) has changed the definition of NPA thrice during 1993 - 2004 in terms of holding period of loans. The holding period was reduced by one quarter each time. In 1993, the holding period was four quarters (360 days).

Based on the above paragraph, the holding period of loans in 2004 after the third revision was _____ days.

(a) 45 (b) 90

(c) 135 (d) 180

7. Select the next element of the series Z, WV, RQP, ______.

(a) LKJI (b) JIHG

(c) KJIH (d) NMLK

8. In four-digit integer numbers from 1001 to 9999 the digit group "37" (in the same sequence) appears ______ times.

(a) 270 (b) 279

(c) 280 (d) 299

9. Given a semicircle with 'O' as the centre, as shown in the figure, the ratio $\frac{\overline{AC} + \overline{CB}}{\overline{AB}}$ is _____.

Where $\overline{AC}$, $\overline{CB}$ and $\overline{AB}$ are chords.

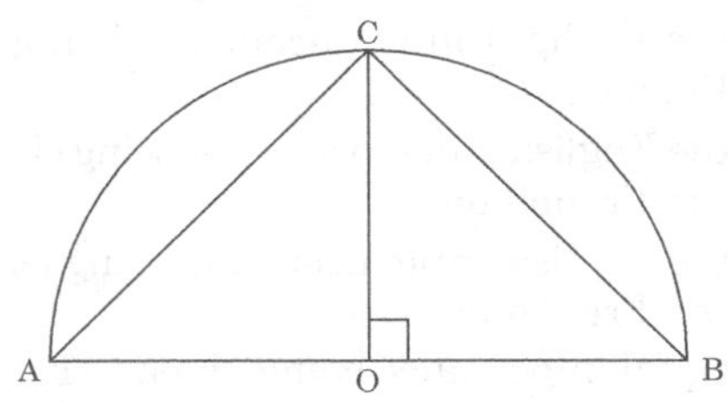

(a) $\sqrt{2}$ (b) $\sqrt{3}$

(c) 2 (d) 3

10. The revenue and expenditure of four different companies P, Q, R and S in 2015 are shown in the figure. If the revenue of company Q in 2015 was 20% more than that in 2014, and company Q had earned a profit of 10% on expenditure in 2014, then its expenditure (in million rupees) in 2014 was ______.

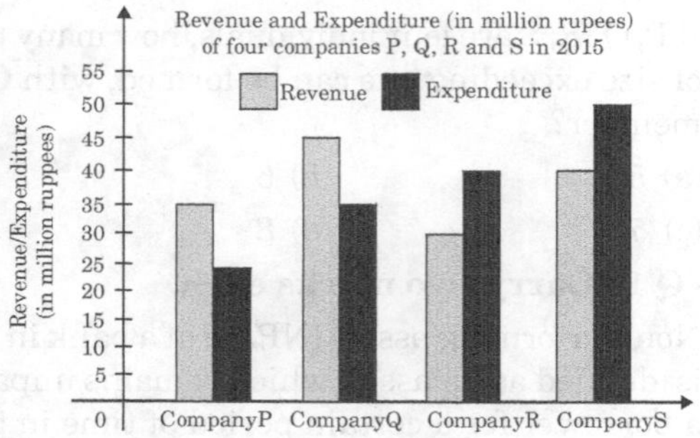

(a) 32.7 (b) 33.7
(c) 34.1 (d) 35.1

SET-4

Q. 1 to Q. 5 : Carry One Mark Each.

1. The untimely loss of life is a cause of serious global concern as thousands of people get killed ____ accidents every year while many other die ____ diseases like cardio vascular disease, cancer, etc.

(a) in, of (b) from, of
(c) during, from (d) from, from

2. He was not only accused of theft ___ of conspiracy.

(a) rather (b) but also
(c) but even (d) rather than

3. Select the word that fits the analogy:

Explicit : Implicit :: Express : _____

(a) Impress (b) Repress
(c) Compress (d) Suppress

4. The Canadian constitution requires that equal importance be given to English and French. Last year, Air Canada lost a lawsuit, and had to pay a six- figure fine to a French-speaking couple after they filed complaints about formal in-flight announcements in English lasting 15 seconds, as opposed to informal 5 second messages in French.

The French-speaking couple were upset at ____.

(a) the in-flight announcements being made in English.
(b) the English announcements being clearer than the French ones.
(c) the English announcements being longer than the French ones.
(d) equal importance being given to English and French.

5. A superadditive function f(.) satisfies the following property

$$f(x_1 + x_2) \geq f(x_1) + f(x_2)$$

Which of the following functions is a superadditive function for x > 1 ?

(a) e^x (b) $\sqrt{x}$
(c) $\frac{1}{x}$ (d) e^{-x}

Q. 6 to Q. 10 : Carry Two Marks Each.

6. The global financial crisis in 2008 is considered to be the most serious world-wide financial crisis, which started with the sub-prime lending crisis in USA in 2007. The sub-prime lending crisis led to the banking crisis in 2008 with the collapse of Lehman Brothers in 2008. The sub-prime lending refers to the provision of loans to those borrowers who may have difficulties in repaying loans, and it arises because of excess liquidity following the East Asian crisis. Which one of the following sequences shows the correct precedence as per the given passage?

(a) East Asian crisis — Sub-prime lending crisis — Banking crisis — Global financial crisis
(b) Sub-prime lending crisis — Global financial crisis — Banking crisis — East Asian crisis
(c) Banking crisis — Sub-prime lending crisis — Global financial crisis — East Asian crisis
(d) Global financial crisis — East Asian crisis — Banking crisis — Sub-prime lending crisis

7. It is quarter past three in your watch. The angle between the hour hand and the minute hand is ___

(a) 0° (b) 7.5°
(c) 15° (d) 22.5°

8. A circle with centre O is shown in the figure. A rectangle PQRS of maximum possible area is inscribed in the circle. If the radius of the circle is a, then the area of the shaded portion is _____.

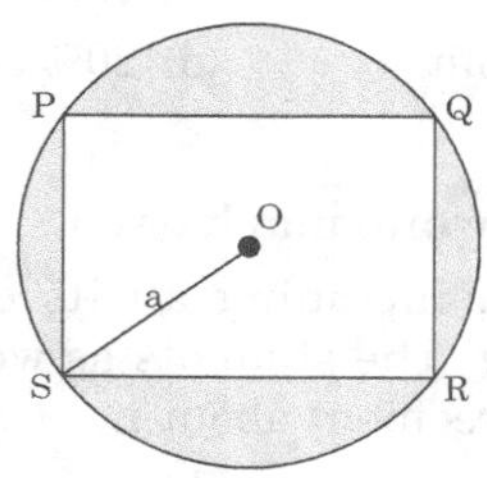

(a) $\pi a^2 - a^2$
(b) $\pi a^2 - \sqrt{2}a^2$
(c) $\pi a^2 - 2a^2$
(d) $\pi a^2 - 3a^2$

9. a, b, c are real numbers. The quadratic equation $ax^2 - bx + c = 0$ has equal roots, which is β, then

(a) $\beta = b/a$ (b) $\beta^2 = ac$
(c) $\beta^3 = bc/(2a^2)$ (d) $b^2 \neq 4ac$

10. The following figure shows the data of students enrolled in 5 years (2014 to 2018) for two schools P and Q. During this period, the ratio of the average number of the students enrolled in school P to the average of the difference of the number of students enrolled in schools P and Q is ____

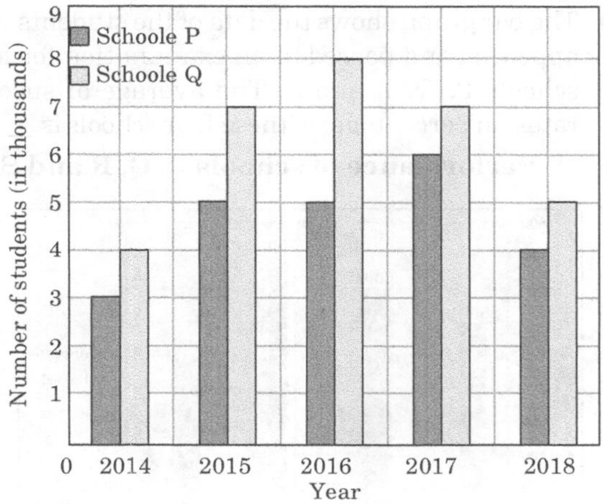

(a) 8 : 23 (b) 23 : 8

(c) 23 : 31 (d) 31 : 23

SET-5

Q. 1 to Q. 5 : Carry One Mark Each.

1. He is known for his unscrupulous ways. He always sheds ________ tears to deceive people.

(a) fox's

(b) crocodile's

(c) crocodile

(d) fox

2. Jofra Archer, the England fast bowler, is _______ than accurate.

(a) more fast (b) faster

(c) less fast (d) more faster

3. Select the word that fits the analogy:

Build : Building : : Grow :

(a) Grown

(b) Grew

(c) Growth

(d) Growed

4. I do not think you know the case well enough to have opinions. Having said that, I agree with your other point.

What does the phrase "having said that" mean in the given text?

(a) as opposed to what I have said

(b) despite what I have said

(c) in addition to what I have said

(d) contrary to what I have said

5. Define [x] as the greatest integer less than or equal to x, for each $x \in (-\infty, \infty)$. If y = [x], then area under y for $x \in [1, 4]$ is ________ .

(a) 1 (b) 3

(c) 4 (d) 6

Q. 6 to Q. 10 : Carry Two Marks Each.

6. Crowd funding deals with mobilisation of funds for a project from a large number of people, who would be willing to invest smaller amounts through web-based platforms in the project.

Based on the above paragraph, which of the following is correct about crowd funding?

(a) Funds raised through unwilling contributions on web-based platforms.

(b) Funds raised through large contributions on web-based platforms.

(c) Funds raised through coerced contributions on web-based platforms.

(d) Funds raised through voluntary contributions on web-based platforms.

7. P, Q, R and S are to be uniquely coded using α and β. If P is coded as $\alpha\alpha$ and Q as $\alpha\beta$, then R and S, respectively, can be coded as

(a) $\beta\alpha$ and $\alpha\beta$ (b) $\beta\beta$ and $\alpha\alpha$

(c) $\alpha\beta$ and $\beta\beta$ (d) $\beta\alpha$ and $\beta\beta$

8. The sum of the first n terms in the sequence 8, 88, 888, 8888, ... is ________ .

(a) $\frac{81}{80}(10^n - 1) + \frac{9}{8}n$

(b) $\frac{81}{80}(10^n - 1) - \frac{9}{8}n$

(c) $\frac{80}{81}(10^n - 1) + \frac{8}{9}n$

(d) $\frac{80}{81}(10^n - 1) - \frac{8}{9}n$

9. Select the graph that schematically represents BOTH $y = x^m$ and $y = x^{1/m}$ properly in the interval $0 \le x \le 1$, for integer values of m, where m > 1.

(a)

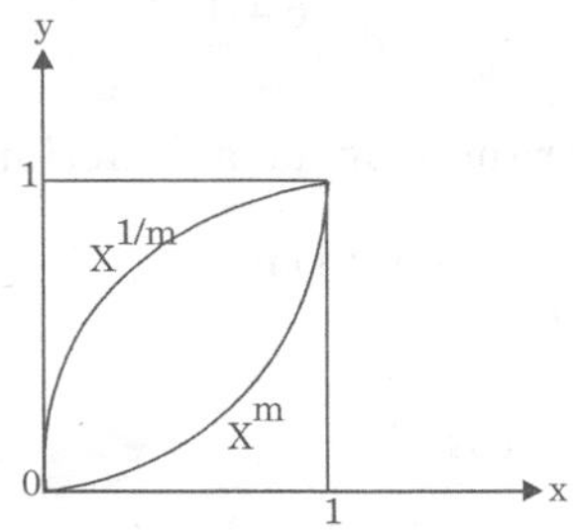

(b)

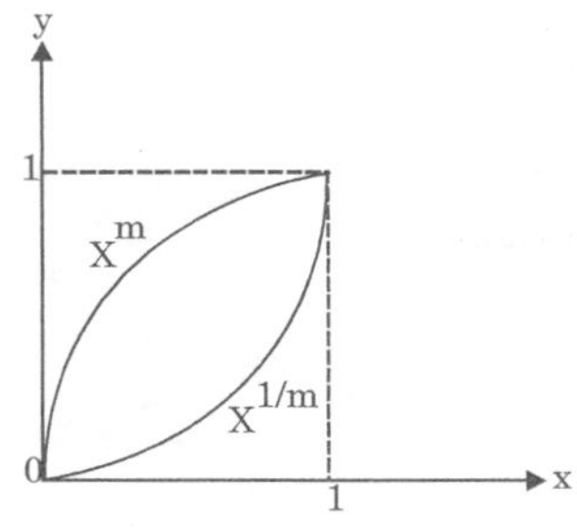

(c)

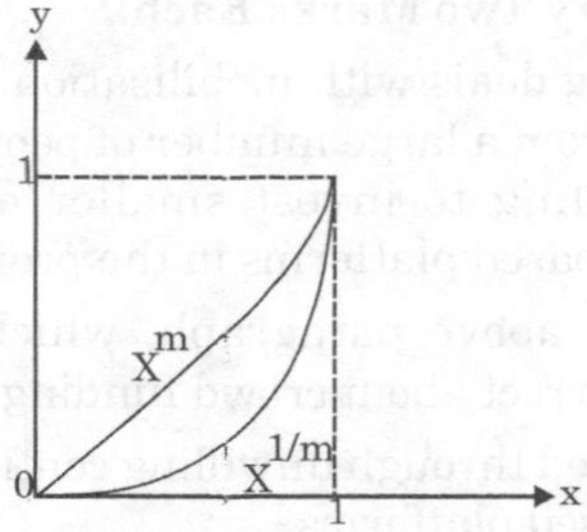

(d)

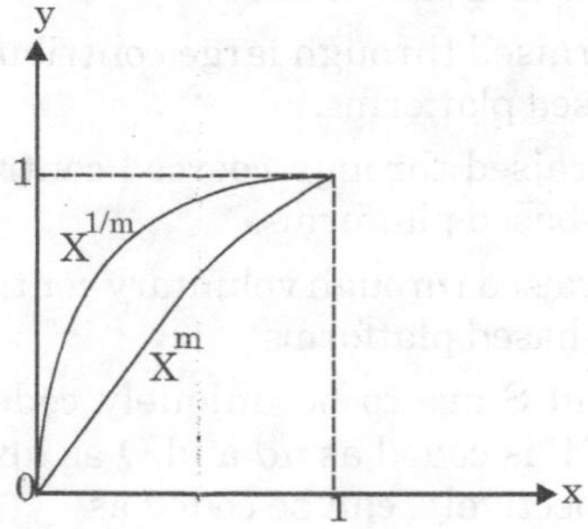

10. The bar graph shows the data of the students who appeared and passed in an examination for four schools P, Q, R and S. The average of success rates (in percentage) of these four schools is _____.

Performance of schools P, Q, R and S

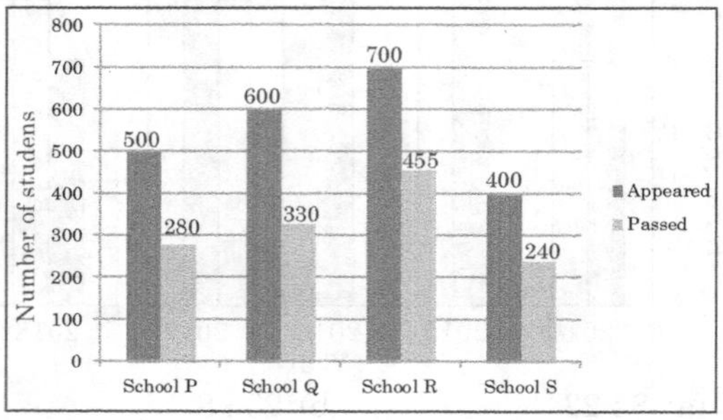

(a) 58.5% (b) 58.8%

(c) 59.0% (d) 59.3%

ENGINEERING MATHEMATICS

February 1 (Forenoon Session)

1. Multiplication of real valued square matrices of same dimension is

(a) associative

(b) commutative

(c) always positive definite

(d) not always possible to compute

2. The value of $\lim_{x\to 1}\left(\dfrac{1-e^{-c(1-x)}}{1-xe^{-c(1-x)}}\right)$ is

(a) c (b) c + 1

(c) $\dfrac{c}{c+1}$ (d) $\dfrac{c+1}{c}$

3. The Laplace transform of a function f(t) is $L(f) = \dfrac{1}{(s^2+\omega^2)}$. Then, f(t) is

(a) $f(t) = \dfrac{1}{\omega^2}(1 - \cos\omega t)$

(b) $f(t) = \dfrac{1}{\omega}\cos\omega t$

(c) $f(t) = \dfrac{1}{\omega}\sin\omega t$

(d) $f(t) = \dfrac{1}{\omega^2}(1 - \sin\omega t)$

4. Which of the following function f(z), of the complex variable z, is NOT analytic at all the points of the complex plane?

(a) $f(z) = z^2$

(b) $f(z) = e^z$

(c) $f(z) = \sin z$

(d) $f(z) = \log z$

5. For three vectors $\vec{A} = 2\hat{j} - 3\hat{k}$, $\vec{B} = -2\hat{i} + \hat{k}$, and $\vec{C} = 3\hat{i} - \hat{j}$ where $\hat{i}, \hat{j}$ and $\hat{k}$ are unit vectors along the axes of a right-handed rectangular Cartesian coordinate system, the value of $(\vec{A}\cdot(\vec{B}\times\vec{C}) + 6)$ is ________ .

6. A company is hiring to fill four managerial vacancies. The candidates are five men and three women. If every candidate is equally likely to be chosen then the probability that at least one woman will be selected is ________ (round off to 2 decimal places).

7. The evaluation of the definite integral $\int_{-1}^{1.4} x|x|\,dx$ by using Simpson's 1/3rd (one third) rule with step size h = 0.6 yields.

(a) 0.914

(b) 1.248

(c) 0.581

(d) 0.592

8. A vector field is defined as

$$\vec{f}(x, y, z) = \frac{x}{\left[x^2 + y^2 + z^2\right]^{\frac{3}{2}}}\hat{i} + \frac{y}{\left[x^2 + y^2 + z^2\right]^{\frac{3}{2}}}\hat{j} + \frac{z}{\left[x^2 + y^2 + z^2\right]^{\frac{3}{2}}}\hat{k}$$

where, $\hat{i}, \hat{j}, \hat{k}$ are unit vectors along the axes of a right handed rectangular– Cartesian coordinate system. The surface integral $\iint \vec{f} \cdot d\vec{S}$ (where $d\vec{S}$ is an elemental surface area vector) evaluated over the inner and outer surfaces of a spherical shell formed by two concentric spheres with origin as the center, and internal and external radii of 1 and 2, respectively is

(a) 0
(b) 2π
(c) 4π
(d) 8π

9. Consider two exponentially distributed random variables X and Y, both having a mean of 0.50. Let Z = X + Y and r be the corrrelation coefficient between X and Y. If the variance of Z equals 0, then the value of r is ________ (round off to 2 decimal places)

10. An analytic function of a complex variable

$z = x + iy\left(i = \sqrt{-1}\right)$ is defined as

$f(z) = x^2 - y^2 + i\psi(x, y)$

where $\psi(x, y)$ is a real function. The value of the imaginary part of f(z) at z = (1 + i) is ________ (round off to 2 decimal places)

February 1 (Afternoon Session)

11. The sum of two normally distributed random variables X and Y is

(a) always normally distributed
(b) normally distributed, only if X and Y are independent
(c) normally distributed, only if X and Y have the same standard deviation
(d) normally distributed, only if X and Y have the same mean

12. A matrix P is decomposed into its symmetric part S and skew symmetric part V. If

$$S = \begin{pmatrix} -4 & 4 & 2 \\ 4 & 3 & 7/2 \\ 2 & 7/2 & 2 \end{pmatrix}, V = \begin{pmatrix} 0 & -2 & 3 \\ 2 & 0 & 7/2 \\ -3 & -7/2 & 0 \end{pmatrix}$$

then matrix P is

(a) $\begin{pmatrix} -4 & 6 & -1 \\ 2 & 3 & 0 \\ 5 & 7 & 2 \end{pmatrix}$

(b) $\begin{pmatrix} -4 & 2 & 5 \\ 6 & 3 & 7 \\ -1 & 0 & 2 \end{pmatrix}$

(c) $\begin{pmatrix} 4 & -6 & 1 \\ -2 & -3 & 0 \\ -5 & -7 & -2 \end{pmatrix}$

(d) $\begin{pmatrix} -2 & 9/2 & -1 \\ -1 & 81/4 & 11 \\ -2 & 45/2 & 73/4 \end{pmatrix}$

13. Let $I = \int_{x=0}^{1}\int_{y=0}^{x^2} xy^2 dy\, dx$. Then, I may also be expressed as

(a) $\int_{y=0}^{1}\int_{x=0}^{\sqrt{y}} xy^2 dx\, dy$ (b) $\int_{y=0}^{1}\int_{x=\sqrt{y}}^{1} yx^2 dx\, dy$

(c) $\int_{y=0}^{1}\int_{x=\sqrt{y}}^{1} xy^2 dx\, dy$ (d) $\int_{y=0}^{1}\int_{x=0}^{\sqrt{y}} yx^2 dy\, dx$

14. The solution of

$$\frac{d^2y}{dt^2} - y = 1,$$

which additionally satisfies $y\Big|_{t=0} = \frac{dy}{dt}\Big|_{t=0} = 0$ in the Laplace s-domain is

(a) $\frac{1}{s(s+1)(s-1)}$

(b) $\frac{1}{s(s+1)}$

(c) $\frac{1}{s(s-1)}$

(d) $\frac{1}{s-1}$

15. Let I be a 100 dimensional identity matrix and E be the set of its distinct (no value appears more than once in E) real eigenvalues. The number of elements in E is ________.

16. The directional derivative of f (x, y, z) = xyz at point (–1, 1, 3) in the direction of vector $\hat{i} - 2\hat{j} + 2\hat{k}$ is

(a) $3\hat{i} - 3\hat{j} - \hat{k}$ (b) $-\frac{7}{3}$

(c) $\frac{7}{3}$ (d) 7

17. The function f(z) of complex variable z = x + iy, where $i = \sqrt{-1}$, is given as $f(z) = (x^3 - 3xy^2) + i\,v(x, y)$. For this function to be analytic, v(x, y) should be

(a) $(3xy^2 - y^3)$ + constant

(b) $(3x^2y^2 - y^3)$+ constant

(c) $(x^3 - 3x^2y)$ + constant

(d) $(3x^2y - y^3)$ + constant

18. For the integral $\int_0^{\pi/2} (8 + 4\cos x)\,dx$, the absolute percentage error in numerical evaluation with the Trapezoidal rule, using only the end points, is ________ (round off to one decimal place).

19. A fair coin is tossed 20 times. The probability that 'head' will appear exactly 4 times in the first ten tosses, and 'tail' will appear exactly 4 times in the next ten tosses is ________ (round off to 3 decimal places).

February 2 (Forenoon Session)

20. Which one of the following methods requires specifying an initial interval containing the root (i.e., bracketing) to obtain the solution of f(x) = 0, where f(x) is a continuous non-linear algebraic function?

(a) Newton-Raphson method

(b) regula falsi method

(c) secant method

(d) fixed point iteration method

21. Consider the hyperbolic functions in Group-1 and their definitions in Group-2.

Group-1	Group-2
P tanh x	I $\frac{e^x + e^{-x}}{e^x - e^{-x}}$
Q coth x	II $\frac{2}{e^x + e^{-x}}$
R sech x	III $\frac{2}{e^x - e^{-x}}$
S cosech x	IV $\frac{e^x - e^{-x}}{e^x + e^{-x}}$

The correct combination is

(a) P-IV, Q-I, R-III, S-II

(b) P-II, Q-III, R-I, S-IV

(c) P-IV, Q-I, R-II, S-III

(d) P-I, Q-II, R-IV, S-III

22. Consider the following continuously differentiable function

$\mathbf{v}(x, y, z) = 3x^2 y\,\mathbf{i} + 8y^2 z\,\mathbf{j} + 5\,xyz\,\mathbf{k}$

where **i**, **j** and **k** represent the respective unit vectors along the x, y and z directions in the Cartesian coordinate system. The curl of this function is

(a) $-3x^2\,\mathbf{i} - 8y^2\,\mathbf{j} + 5z\,(x + y)\,\mathbf{k}$

(b) $6xy\,\mathbf{i} - 16\,yz\,\mathbf{j} + 5\,xy\,\mathbf{k}$

(c) $(5xz - 8y^2)\,\mathbf{i} - 5\,yz\,\mathbf{j} - 3x^2\,\mathbf{k}$

(d) y(11x + 16z)

23. Consider the following unit step function.

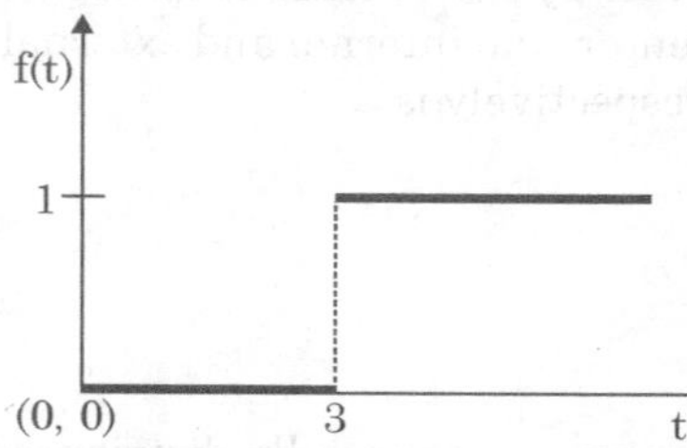

The Laplace transform of this function is

(a) $\frac{e^{-3s}}{S}$ (b) $\frac{e^{-3s}}{S^2}$

(c) $\frac{e^{-3s}}{3S}$ (d) $\frac{e^{-6s}}{S}$

24. Sum of the eigenvalues of the matrix $\begin{bmatrix} 2 & 4 & 6 \\ 3 & 5 & 9 \\ 12 & 1 & 7 \end{bmatrix}$ is ________ (round off to nearest integer).

25. In a box, there are 5 green balls and 10 blue balls, a person picks 6 balls randomly. The probability that the person has picked 4 green balls and 2 blue balls is

(a) $\frac{42}{1001}$ (b) $\frac{45}{1001}$

(c) $\frac{240}{1001}$ (d) $\frac{420}{1001}$

26. The maximum value of the function $f(x) = -\frac{5}{3}x^3 + 10x^2 - 15x + 16$ in the interval (0.5, 3.5) is

(a) 0 (b) 8

(c) 16 (d) 48

27. Given $\frac{dy}{dx} = y - 20$, and $y|_{x=0} = 40$, the value of y at x = 2 is ________ (round off to nearest integer).

28. Consider the following dataset.

x	1	3	5	15	25
f(x)	6	8	10	12	5

The value of the integral $\int_1^{25} f(x)\,dx$ using Simpson's 1/3rd rule is ________ (round off to 1 decimal place).

February 2 (Afternoon Session)

29. For f(x) = |x|, with $\frac{df}{dx}$ denoting the derivative, the mean value theorem is not applicable because

(a) f(x) is not continuous at x = 0

(b) f(x) = 0 at x = 0

(c) $\frac{df}{dx}$ is not defined at x = 0

(d) $\frac{df}{dx} = 0$ at x = 0

30. For the function $f(x) = \frac{e^{-\lambda}}{\sigma\sqrt{2\pi}}$, where $\lambda = \frac{1}{2\sigma^2}(x-\mu)^2$, and σ and μ are constants, the maximum occurs at

(a) $x = \sigma$ (b) $x = \sigma\sqrt{2\pi}$

(c) $x = 2\sigma^2$ (d) $x = \mu$

31. $y = Ae^{mx} + Be^{-mx}$, where A, B and m are constants, is a solution of

(a) $\frac{d^2y}{dx^2} - m^2y = 0$ (b) $A\frac{d^2y}{dx^2} + m^2y = 0$

(c) $B\frac{d^2y}{dx^2} + Ay = 0$ (d) $\frac{d^2y}{dx^2} + my = m^2$

32. Given $A = \begin{pmatrix} \sin\theta & \tan\theta \\ 0 & \cos\theta \end{pmatrix}$, the sum of squares of eigen values of A is

(a) $\tan^2\theta$ (b) 1

(c) $\sin^2\theta$ (d) $\cos^2\theta$

33. The similarity solution for the diffusion equation, $\frac{\partial u}{\partial t} = \alpha\frac{\partial^2 u}{\partial x^2}$ is $u(x, t) = u(\eta)$, where similarity variable, $\eta = \frac{x}{\sqrt{\alpha t}}$. If $u(x,0) = e^{-x^2}$ the ratio, $\frac{u(0,1)}{u(0,4)} =$ ________ (*round off to one decimal place*).

34. The equation $x\frac{dx}{dy} + y = c$, where c is a constant, represents a family of

(a) exponential curves

(b) parabolas

(c) circles

(d) hyperbolas

35. In the equation $AX = B$, $A = \begin{bmatrix} \frac{1}{\sqrt{2}} & 0 & \frac{1}{\sqrt{2}} \\ 0 & 1 & 0 \\ \frac{1}{\sqrt{2}} & 0 & \frac{-1}{\sqrt{2}} \end{bmatrix}$, $X = \begin{bmatrix} x \\ y \\ z \end{bmatrix}$

$B = \begin{bmatrix} 0 \\ 1 \\ -\sqrt{2} \end{bmatrix}$, where A is an orthogonal matrix, the sum of the unknowns, $x + y + z =$ ______ (*round off to one decimal place*).

36. If $\int_0^1 (x^2 - 2x + 1)dx$ is evaluated numerically using trapezoidal rule with four intervals, the difference between the numerically evaluated value and the analytical value of the integral is equal to ________ (*round off to three decimal places*).

February 8 (Forenoon Session)

37. $ax^3 + bx^2 + cx + d$ is a polynomial on real x over real coefficients a, b, c, d, wherein $a \neq 0$. Which of the following statements is true?

(a) d can be chosen to ensure that x = 0 is a root for any given set a, b, c

(b) No choice of coefficients can make all roots identical

(c) a, b, c, d can be chosen to ensure that all roots are complex

(d) c alone cannot ensure that all roots are real

38. Which of the following is true for all possible non-zero choices of integers m, n; m ≠ n, or all possible non-zero choices of real numbers p, q; p ≠ q as applicable?

(a) $\frac{1}{\pi}\int_0^{\pi} \sin m\theta \sin n\theta d\theta = 0$

(b) $\frac{1}{2\pi}\int_{-\pi/2}^{\pi/2} \sin p\theta \sin q\theta d\theta = 0$

(c) $\frac{1}{2\pi}\int_{-\pi}^{\pi} \sin p\theta \cos q\theta d\theta = 0$

(d) $\lim_{\alpha\to\infty} \frac{1}{2\alpha}\int_{-\alpha}^{\alpha} \sin p\theta \sin q\theta d\theta = 0$

39. Which of the following statements is true about the two-sided Laplace transform?

(a) It exists for every signal that may or may not have a Fourier Transform

(b) It has no poles for any bounded signal that is non-zero only inside a finite time interval.

(c) The number of finite poles and finite zeroes is must be equal.

(d) If a signal can be expressed as a weighted sum of shifted one sided exponentials, then its Laplace transform will have no poles.

40. Consider a signal $x(n) = \left(\frac{1}{2}\right)^n 1(n)$, where $1(n) = 0$ if $n < 0$ and $1(n) = 1$ if $n \geq 0$. The z-transform of $x(n - k)$, $k > 0$ is $\dfrac{z^{-k}}{1 - \frac{1}{2}z^{-1}}$ with region of convergence being

(a) $|z| < 2$ (b) $|z| > 2$

(c) $|z| < \frac{1}{2}$ (d) $|z| > \frac{1}{2}$

41. The value of the following complex integral, with 'C' representing the unit circle centered at origin in the counter clockwise sense, is : $\int_c \dfrac{Z^2 + 1}{Z^2 - 2Z} dZ$

(a) $8\pi i$

(b) $-8\pi i$

(c) $-\pi i$

(d) πi

42. Consider the initial value problem below. The value of y at x = ln 2, (rounded off to 3 decimal places) is _____ $\dfrac{dy}{dx} = 2x - y,\ y(0) = 1$

43. For real number x and y with $y = 3x^2 + 3x + 1$, the maximum and minimum value of y for $x \in [-2, 0]$ are respectively

(a) 7 and $\frac{1}{4}$ (b) 7 and 1

(c) –2 and $-\frac{1}{2}$ (d) 1 and $\frac{1}{4}$

44. The vector function expressed by

$F = a_x(5y - K_1z) + a_y(3z + K_2x) + a_z(K_3y - 4x)$.

Represents a conservative field, where a_x, a_y, a_z are unit vectors along x, y and z directions respectively. The values of constant K_1, K_2, K_3 are given by

(a) $K_1 = 3, K_2 = 3, K_3 = 7$,

(b) $K_1 = 3, K_2 = 8, K_3 = 5$

(c) $K_1 = 4, K_2 = 5, K_3 = 3$

(d) $K_1 = 0, K_2 = 0, K_3 = 0$

45. The number of purely real elements in a lower triangular respresentation of the given 3 × 3 matrix, obtained through the given decomposition is _____.

$$\begin{bmatrix} 2 & 3 & 3 \\ 3 & 2 & 1 \\ 3 & 1 & 7 \end{bmatrix} = \begin{bmatrix} a_{11} & 0 & 0 \\ a_{12} & a_{22} & 0 \\ a_{13} & a_{23} & a_{33} \end{bmatrix} \begin{bmatrix} a_{11} & 0 & 0 \\ a_{12} & a_{22} & 0 \\ a_{13} & a_{23} & a_{33} \end{bmatrix}^T$$

(a) 5 (b) 6

(c) 8 (d) 9

46. Let a_x and a_y be unit vectors along x and y directions, respectively. A vector function is given by $F = a_xy - a_yx$.

The line integral of the above function $\int_c F.d\ell$ along the curve C, which follows the parabola $y = x^2$ as shown below is ____ (rounded off to 2 decimal places).

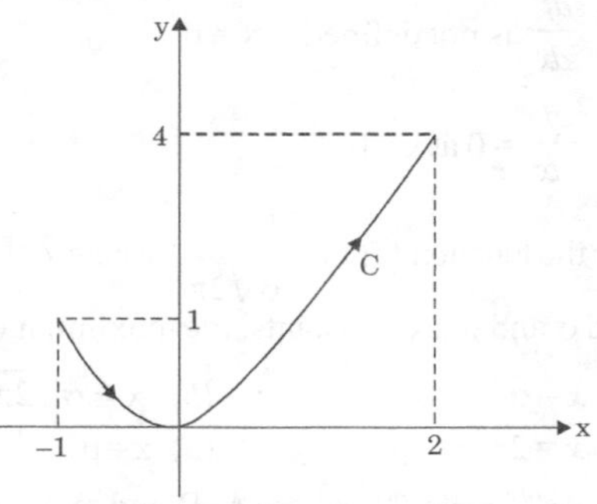

February 8 (Afternoon Session)

47. Consider the functions:

I. e^{-x}

II. $x^2 - \sin x$

III. $\sqrt{x^3 + 1}$

Which of the above functions is/are increasing everywhere in [0, 1]?

(a) III only (b) II only

(c) II and III only (d) I and III only

48. Let R be the set of all binary relations on the set {1,2,3}. Suppose a relation is chosen from R at random. The probability that the chosen relation is reflexive(round off to 3 decimal places) is _____.

49. Let G be a group of 35 elements. Then the largest possible size of a subgroup of G other than G itself is ________.

50. Let G = (V, E) be a weighted undirected graph and let T be a Minimum Spanning Tree (MST) of G maintained using adjacency lists. Suppose a new weighted edge $(u, v) \in V \times V$ is added to G. The worst case time complexity of determining if T is still an MST of the resultant graph is

(a) $\Theta(|E| + |V|)$ (b) $\Theta(|E| |V|)$

(c) $\Theta(|E| \log |V|)$ (d) $\Theta(|V|)$

51. Let G = (V, E) be a directed, weighted graph with weight function $w : E \to R$. For some function $f : V \to R$, for each edge $(u, v) \in E$, define $w'(u, v)$ as $w(u, v) + f(u) - f(v)$. Which one of the options completes the following sentence so that it is TRUE?

"The shortest paths in G under w are shortest paths under w′ too, ________".

(a) for every $f : V \to R$

(b) if and only if $\forall u \in V$, f (u) is positive

(c) if and only if $\forall u \in V$, f (u) is negative

(d) if and only if f (u) is the distance from s to u in the graph obtained by adding a new vertex s to G and edges of zero weight from s to every vertex of G

52. Consider a graph G = (V, E), where $V = \{v_1, v_2, \ldots, v_{100}\}$, $E = \{(v_i, v_j) \mid 1 \le i < j \le 100\}$, and weight of the edge (v_i, v_j) is $|i - j|$. The weight of minimum spanning tree of G is ______ .

53. Graph G is obtained by adding vertex s to $K_{3,4}$ and making s adjacent to every vertex of $K_{3,4}$. The minimum number of colours required to edge-colour G is ________.

February 9 (Forenoon Session)

54. In the following partial differential equation, θ is a function of t and z, and D and K are functions of θ

$$D(\theta)\frac{\partial^2\theta}{\partial z^2} + \frac{\partial K(\theta)}{\partial z} - \frac{\partial\theta}{\partial t} = 0$$

The above equation is

(a) a second order linear equation

(b) a second degree linear equation

(c) a second order non-linear equation

(d) a second degree non-linear equation

55. The value of $\lim_{x \to \infty} \frac{x^2 - 5x + 4}{4x^2 + 2x}$ is

(a) 0 (b) $\frac{1}{4}$

(c) $\frac{1}{2}$ (d) 1

56. The true value of $ln(2)$ is 0.69. If the value of ln (2) is obtained by linear interpolation between ln(1) and ln(6), the percentage of absolute error (round off to the nearest integer), is

(a) 35 (b) 48

(c) 69 (d) 84

57. The area of an ellipse represented by an equation $\frac{x^2}{a^2} + \frac{y^2}{b^2} = 1$ is

(a) $\frac{\pi ab}{4}$ (b) $\frac{\pi ab}{2}$

(c) πab (d) $\frac{4\pi ab}{3}$

58. For the Ordinary Differential Equation

$\frac{d^2x}{dt^2} - 5\frac{dx}{dt} + 6x = 0$, with initial conditions

$x(0) = 0$ and $\frac{dx}{dt}(0) = 10$, the solution is

(a) $-5e^{2t} + 6e^{3t}$

(b) $5e^{2t} + 6e^{3t}$

(c) $-10e^{2t} + 10e^{3t}$

(d) $10e^{2t} + 10e^{3t}$

59. A continuous function f(x) is defined. If the third derivative at x_i is to be computed by using the fourth order central finite-divided-difference scheme (with step length = h), the correct formula is

(a) $f'''(x_i) = \frac{-f(x_{i+3}) - 8f(x_{i+2}) - 13f(x_{i+1}) + 13f(x_{i-1}) - 8f(x_{i-2}) + f(x_{i-3})}{8h^3}$

(b) $f'''(x_i) = \frac{f(x_{i+3}) - 8f(x_{i+2}) - 13f(x_{i+1}) + 13f(x_{i-1}) + 8f(x_{i-2}) + f(x_{i-3})}{8h^3}$

(c) $f'''(x_i) = \frac{-f(x_{i+3}) - 8f(x_{i+2}) - 13f(x_{i-1}) + 13f(x_{i-1}) + 8f(x_{i-2}) - f(x_{i-3})}{8h^3}$

(d) $f'''(x_i) = \frac{f(x_{i+3}) - 8f(x_{i+2}) - 13f(x_{i+1}) + 13f(x_{i-1}) + 8f(x_{i-2}) - f(x_{i-3})}{8h^3}$

60. If C represents a line segment between (0, 0, 0) and (1, 1, 1) in Cartesian coordinate system, the value (expressed as integer) of the line integral

$$\int_C [(y + z)dx + (x + z)dy + (x + y)dx]$$

is ________.

61. Consider the system of equations

$$\begin{vmatrix} 1 & 3 & 2 \\ 2 & 2 & -3 \\ 4 & 4 & -6 \\ 2 & 5 & 2 \end{vmatrix} \begin{vmatrix} x_1 \\ x_2 \\ x_3 \end{vmatrix} = \begin{vmatrix} 1 \\ 1 \\ 2 \\ 1 \end{vmatrix}$$

The value of x_3 (round off to the nearest integer), is _______.

February 9 (Afternoon Session)

62. The ordinary differential equation

$$\frac{d^2u}{dx^2} - 2x^2u + \sin x = 0 \text{ is}$$

(a) linear and homogeneous

(b) linear and non homogeneous

(c) nonlinear and homogeneous

(d) non linear and non homogeneous

63. The value of $\lim_{x\to\infty} \frac{\sqrt{9x^2+2020}}{x+7}$ is

(a) $\frac{7}{9}$ (b) 1

(c) 3 (d) indeterminable

64. The integral

$$\int_0^1 (5x^3 + 4x^2 + 3x + 2)\,dx$$

is estimated numerically using three alternative methods namely the rectangular, trapezoidal and Simpson's rules with a common step size. In this context, which one of the following statements is **TRUE**?

(a) Simpson's rule as well as rectangular rule of estimation will give NON-zero error.

(b) Simpson's rule, rectangular rule as well as trapezoidal rule of estimation will give NON-zero error.

(c) Only the rectangular rule of estimation will give zero error.

(d) Only Simpson's rule of estimation will give zero error.

65. The following partial differential equation is defined for u : u (x, y)

$$\frac{\partial u}{\partial y} = \frac{\partial^2 u}{\partial x^2};\quad y \geq 0;\ x_1 \leq x \leq x_2$$

The set of auxiliary conditions necessary to solve the equation uniquely, is

(a) three initial conditions

(b) three boundary conditions

(c) two initial condition and one boundary condition

(d) one initial condition and two boundary conditions

66. A fair (unbiased) coin is tossed 15 times. The probability of getting exactly 8 Heads (round off to three decimal places), is _____.

67. An ordinary differential equation is given below

$$6\frac{d^2y}{dx^2} + \frac{dy}{dx} - y = 0$$

The general solution of the above equation (with constants C_1 and C_2), is

(a) $y(x) = C_1 e^{-\frac{x}{3}} + C_2 e^{\frac{x}{2}}$

(b) $y(x) = C_1 e^{\frac{x}{3}} + C_2 e^{-\frac{x}{2}}$

(c) $y(x) = C_1 x e^{-\frac{x}{3}} + C_2 e^{\frac{x}{2}}$

(d) $y(x) = C_1 e^{\frac{x}{3}} + C_2 x e^{\frac{x}{2}}$

68. A 4 × 4 matrix [P] is given below

$$[P] = \begin{bmatrix} 0 & 1 & 3 & 0 \\ -2 & 3 & 0 & 4 \\ 0 & 0 & 6 & 1 \\ 0 & 0 & 1 & 6 \end{bmatrix}$$

The eigen values of [P] are

(a) 0, 3, 6, 6 (b) 1, 2, 3, 4

(c) 3, 4, 5, 7 (d) 1, 2, 5, 7

69. The Fourier series to represent $x - x^2$ for $-\pi \leq x \leq \pi$ is given by

$$x - x^2 = \frac{a_o}{2} + \sum_{n=1}^{\infty} a_n \cos nx + \sum_{n=1}^{\infty} b_n \sin nx$$

The value of a_o (round off to two decimal places), is _____ .

ANSWERS AND EXPLANATIONS

GENERAL APTITUDE

SET-1

ANSWERS

1. (b)	**2.** (c)	**3.** (a)	**4.** (b)	**5.** (d)
6. (d)	**7.** (a)	**8.** (c)	**9.** (a)	**10.** (c)

EXPLANATIONS

1. Ivory towers means lack of concern with or interest in practical matters.
2. Voracious means hungry for knowledge.
3. Verb and noun analogy.
4. O – 0, P – 1, Q – 2, R – 4, T – 5, U – 6, V – 7, W – 8, X – 9

   So, ST = 45
5. **As Given that :**

   a + b = 100

   and (a – 5) (b – 5) = 0

   So, either a = 5 or b = 5

   Since 5 is not in any of the option.

   If a = 5, then b = 95 or if b = 5, then a = 95

   Hence of the original number is 95.
6. The last two lines support the answer.
7. P, Q & T refused to sleep next to R

   So option (c, b, d) eliminated

   Only option (a) is satisfies the given conditions in the paragraph.
8. 2, –, –, –, –, –, –, –, 34

   $$t_1 = a = 2 \qquad ta = a + (n-1)d = 34$$
   $$= 2 + 8d = 34$$
   $$d = 4$$

   So, 2, 6, 10, 14, 18, 22, 26, 30, 34

   Sum of inserted 7 AP numbers

   $$= \frac{n}{2}(I^{st} \text{ terms} + \text{last term}) = \frac{n}{2}[a + (a + 6d)]$$

   $$= \frac{7}{2}(6 + 30) = 18 \times 7 = 126$$
9. Units place in $26591749^{110016} = 9^{110016}$

   $= 9^{even} = 1 \qquad (\because 9^2 = 81)$

   Hence, answer will be 1.
10. Money spent on Education $= \frac{15}{100} \times 100$

    = 15%

    Money spent on transport $= \frac{10}{100} \times 100$

    = 10%

    So, required % of extra money spent on education compared to transport :

    $$= \frac{15 - 10}{10} \times 100$$

    = 50%

SET-2

ANSWERS

1. (d)	**2.** (a)	**3.** (a)	**4.** (d)
5. (b)	**6.** (d)	**7.** (c)	**8.** (a)
9. (b)	**10.** (c)		

EXPLANATIONS

1. "Within" Preposition that is used to express something that occurs inside a particular period of time.

   Here, because Sentance is in 'present perfect continuous tense' "For is"used when we talk about a period of time.
2. Here, we have to note that the confunction used is 'but'. 'But' is used for introducing an idea which contrasts with the statement that has been already said.
3. (i) Flyer : Relation is Verb : Noun

   (ii) One who cooks is a cook, and one who flies any aircraft is a flyer.
5. • Cost of 1-a-C-2 : 200 + 200 + 100 + 100 + 100 = 700
   • Cost of 1-f-e-2 : 100 + 100 + 100 + 200 + 200 = 700
   • Cost of 1-b-2 : 300 + 100 + 200 = 600
   • Cost of 1-f-b-2 : 100 + 100 + 100 +200 = 500

   Hence, 1-f-b-2 is having minimum cost.
6. According to given passage GST is imposed on the usage or supply of goods and services hence, option (d) is correct.
7. As given that

   $P = 3^1, Q = 9 = 3^2, R = 27 = 3^3, S = 81 = 3^4,$

   $T = 243 = 3^5$

   Hence, Q + S = 9 + 81 = 90.

8. Unpainted area = Area of annular ring – Area of n blue color circles

Area of annular ring = $\pi b^2 - \pi a^2$

i.e. $\pi(b^2 - a^2)$

Area of 1 blue circle = $\pi\left(\frac{b-a}{2}\right)^2$

Hence, area of n blue circles = $n\pi\left(\frac{b-a}{2}\right)^2$

$\Rightarrow$ Unpainted area = $\pi\left[(b^2 - a^2) - \frac{\pi}{4}(b-a)^2\right]$

9. Since, α and β are acute angles hence it will form the interior angles of straight lines $\alpha + \beta = 180° - 90° = 90°$

10. As given that,

Total expenditure = 2500 million

Total revenue = 3000 million

Total profit = 3000 – 2500 = 500

So, profit % = $\frac{500}{2500} \times 100 = 20\%$

SET-3

ANSWERS

1. (c)	2. (b)	3. (c)	4. (b)	5. (c)
6. (b)	7. (c)	8. (b)	9. (a)	10. (c)

EXPLANATIONS

1. Subject + Verb + agreement

↓ is ↓ are

2. Prohibited from parking means "to officially refuse allow something"

3. Do : Undo : Trust : Distrust

The obvious analogical relationship is do and undo are antonyms just as trust and distrust are antonyms.

4. Plunged means plummet especially (prices, temperature, etc.) to decrease suddenly and quickly.

5. When team size is more than 1 then

Case 1 : When team size = 2, then Q is fixed, only 41 member to be selected among P, R & S

$^3C_1 = 3$ ways

Case 2 : When team size = 3, then Q is fixed, only 2 members to be selected among P, R & S

$^3C_2 = 3$ ways

Case 3 : When team size = 4, then Q is fixed, only 3 members to be selected among P, R & S

$^3C_3 = 1$ ways

Case 4 : When team size ≥ 5; That Q Not possible.

So, total possible ways = 3 + 3 + 1 = 7 ways

Hence, possible combinations are :

PQ, RQ, SQ, PRQ, PSQ, RSQ, PQRS.

6. Based on the given paragraph the holding period of loans was reduced by one quater in 2004. Therefore after the third revision equal to one quarter days of the year.

$= \frac{1}{4} \times 360 = 90$ days.

7. 26 23,22 18,17,16 11,10,9,8

Z, WV, RQP, KJIH

↓ ↓ ↓ ↓

YX UTS ONML GFED

8. Four digit integer numbers from 1001 to 9999, with 37 as sequence appears as :

(1) 3 7 – –

$10 \times 10 = 100$ integers

(2) – 3 7 –

$9 \times 10 = 90$ Integers

(3) – – 3 7

9×10 Integers

This question asking how many times 37 appears not how many number.

So, total ways = 100 + 90 + 90 = 280 ways

9. AB = Diameter = 2r

OC = radius = r = AO = OB

consider is Δ AOC

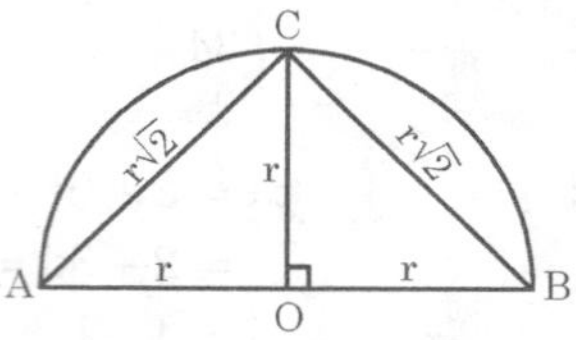

using pythagerous theorem

$AC^2 = r^2 + r^2 \Rightarrow AC^2 = 2r$

$\Rightarrow AC = r\sqrt{2}$

similarly Δ COB $\Rightarrow$ BC = $r\sqrt{2}$

so, the ratio : $\frac{\overline{AC} + \overline{CB}}{\overline{AB}} = \frac{r\sqrt{2} + r\sqrt{2}}{2r} = \sqrt{2}$

10. Revenue of company Q in 2015 = 45 revenue of company Q in 2015 was 20% more than that in 2014.

Let, Revenue in 2014 = x,

$R_{2015} = 120\% (R_x)$

$45 = \frac{120}{100} x$

x = 37.5 in million rupees.

Revenue of company Q in 2014 = 37.5

10% profit in 2014 for company Q.

⇒ Revenue in $(R_{2014})x$ = 110% Expenditure of company in 2014.

$37.5 = \frac{110}{100} \times$ Expenditure Q_{2014}

Expenditure of company Q in 2014 = $\frac{375}{11}$ = 34.1

(Note : profit & loss is calculated on Expenditure)

SET-4

ANSWERS

1. (a)	**2.** (b)	**3.** (b)	**4.** (c)	**5.** (a)
6. (a)	**7.** (b)	**8.** (c)	**9.** (c)	**10.** (b)

EXPLANATIONS

5. x^m is super additive function, if x > 1.
example: $2^3 > 2^1 + 2^2$
$f(x_1 + x_2) > f(x_1) + f(x_2)$ Satisfying
Hence, e^x is super additive fucntion.

7. As given that, Time: 03 : 15
So, h = 3, m = 15

Then, the angle between hands $(\theta) = \left|30h - \frac{11m}{2}\right|$

Hence, $\theta = \left|30 \times 3 - \frac{165}{2}\right| = 75°$

8. From given figure:
Area of shaded portion
= Area of circle – Area of rectangle
As we know that,
Area of the circle = πa^2 And
The maximum possible, area of rectangle inscribed in the circle, when it becomes square.

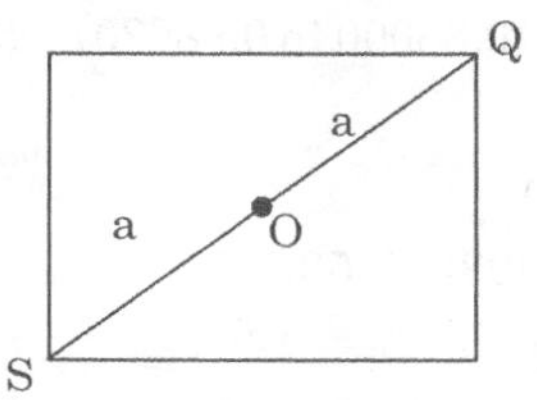

Diagonal = 2a

Area of square = $\frac{1}{2}$ (diagonal)^2 = $2a^2$

So, required shaded Area.
= Area of circle – Area of square = $\pi a^2 - 2a^2$

9. As given that, the quadratic equation:
$ax^2 - bx + c = 0$
According to question it has equal roots (β).
So, sum of roots = $\beta + \beta = \frac{b}{a} \Rightarrow 2\beta = \frac{b}{a}$ ...(1)
And, product of roots = $\beta * \beta = \frac{c}{a} \Rightarrow \beta^2 = \frac{c}{a}$ ...(2)
By multiplying (1) & (2) equation:
$\Rightarrow (2\beta)(\beta^2) = (b/a)(c/a)$
$(2b^3) = bc/a^2$
$b^3 = bc/(2a^2)$

10. Average number of students in school (P) in 5 years
$= \frac{3000 + 5000 + 5000 + 6000 + 4000}{5} = 4600$
Average number of students in school (Q) in 5 years
$= \frac{4000 + 7000 + 8000 + 7000 + 5000}{5} = 6200$
The difference of the number of students enrolled in school P and Q = (6200 – 4600) = 1600
The ratio of the average number of the student enrolled in school P to the average of the diffrence of the number of student enrolled in school P and Q then,
Required ratio $= \frac{4600}{1600} = \frac{23}{8}$ = 23 : 8

SET-5

ANSWERS

1. (c)	**2.** (a)	**3.** (c)	**4.** (b)
5. (d)	**6.** (d)	**7.** (d)	**8.** (d)
9. (a)	**10.** (c)		

EXPLANATIONS

1. Crocodile tears: A false, insincere display of emotion.

2. Jofra Archar, the England fast bowler, is more fast than accurate.

3. Build : Building (Noun)
so, Grow : Growth (Noun)

4. having said that : despite what one just said.

5. Graph of y = [x] for x ∈ [1, 4]

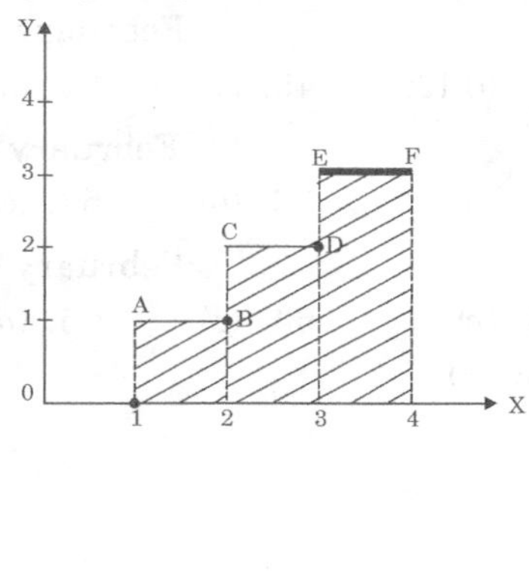

The total area, $A = (1 \times 1) + (2 \times 1) + (3 \times 1)$

$= 1 + 2 + 3 = 6$

6. The correct statement about crowd funding is to fund raised through voluntary contributions on web-based platforms.

7. P, Q, R, S

αα αβ $\boxed{\beta\alpha}$ $\boxed{\beta\beta}$

8. $S = 8 + 88 + 888 + \ldots n$

$$= \frac{8}{9}(9 + 99 + 999 + \ldots.. n)$$

$$= \frac{8}{9}\left[(10^1 - 1) + (100^2 - 1) + \ldots + (10^n - 1)\right]$$

$$= \frac{8}{9}\left[(10^1 + 100^2 + \ldots 100^n) - n\right]$$

$$= \frac{8}{9}\left[\frac{10 \times (10^n - 1)}{10 - 1}\right] - \frac{8}{9}n$$

$$= \frac{80}{81}(10^n - 1) - \frac{8}{9}n$$

9. Graph of $y = x^m$ and $y = x^{1/m}$ for $m > 1$

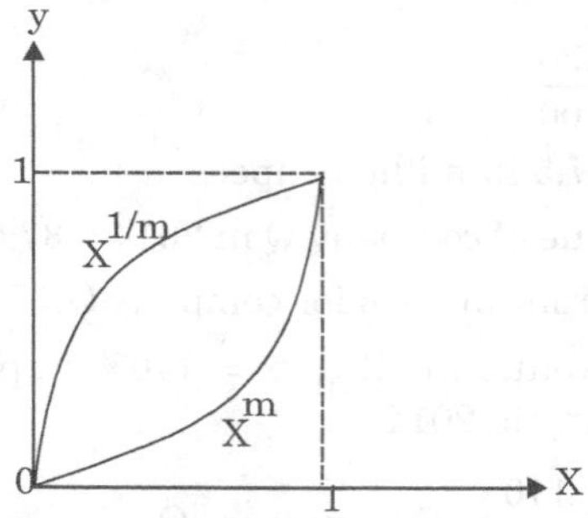

Correct option is (a).

10. $\text{Success rate} = \dfrac{\text{Number of candidtes who passed the exam}}{\text{Number of appeared candidates}}$

So, Average success rate,

$$= \frac{\frac{280}{500} + \frac{330}{600} + \frac{455}{700} + \frac{240}{400}}{4} = 0.59 \text{ or } 59\%$$

Correct option is (c).

ENGINEERING MATHEMATICS

ANSWERS

February 1 (Forenoon Session)

1. (a) 2. (c) 3. (c) 4. (d) 5. (6 to 6) 6. (0.90 to 0.95) 7. (d) 8. (a) 9. (–1.00 to –0.98) 10. (1.99 to 2.01)

February 1 (Afternoon Session)

11. (a) 12. (b) 13. (c) 14. (a) 15. (1 to 1) 16. (c) 17. (d) 18. (5.1 to 5.3) 19. (0.041 to 0.043)

February 2 (Forenoon Session)

20. (b) 21. (c) 22. (c) 23. (a) 24. (14 to 14) 25. (b) 26. (c) 27. (162 to 170) 28. (241.4 to 242.4)

February 2 (Afternoon Session)

29. (c) 30. (d) 31. (a) 32. (b) 33. (1.9 to 2.1) 34. (c) 35. (0.9 to 1.1) 36. (0.010 to 0.012)

February 8 (Forenoon Session)

37. (a) 38. (d) 39. (b) 40. (d) 41. (c) 42. (0.88000 to 0.88630) 43. (a) 44. (c) 45. (c) 46. (–3.05 to –2.95)

February 8 (Afternoon Session)

47. (a) 48. (0.125) 49. (7) 50. (d) 51. (a) 52. (99) 53. (7)

February 9 (Forenoon Session)

54. (c) 55. (b) 56. (b) 57. (c) 58. (c) 59. (a) 60. (3 to 3) 61. (3 to 3)

February 9 (Afternoon Session)

62. (b) 63. (c) 64. (d) 65. (d) 66. (0.190 to 0.200) 67. (b) 68. (d) 69. (–6.61 to – 6.55)

EXPLANATIONS

February 1 (Forenoon Session)

1. If A, B and C are real valued square matrices of same dimension,

   (1) $AB \neq BA$

   i.e. multiplication of matrices are not commutative.

   (2) $A(BC) = (AB)C$

   i.e. multiplication of matrices are associative.

   so, the correct option is (a)

2. $\lim_{x\to 1}\left(\frac{1 - e^{-c(1-x)}}{1 - xe^{-c(1-x)}}\right)$

   At x = 1, $\frac{1 - e^{-c(1-x)}}{1 - xe^{-c(1-x)}} = \frac{0}{0}$

   Apply L-Hospital rule,

   $$\lim_{x\to 1}\left[\frac{-ce^{-c(1-x)}}{0 - e^{-c(1-x)} - xe^{-c(1-x)} \times c}\right]$$

   At x = 1,

   $$\Rightarrow \lim_{x\to 1}\left[\frac{-ce^{-c(1-x)}}{0 - e^{-c(1-x)} - xce^{-c(1-x)}}\right] = \frac{-c \times 1}{-1 - c} = \frac{c}{1 + c}$$

3. $L(\sin \omega t) = \frac{\omega}{s^2 + \omega^2}$ (Formula to be used)

   Given, $L(f) = \frac{1}{s^2 + \omega^2} = \frac{1}{\omega} \times \frac{\omega}{s^2 + \omega^2}$

   $L(f) = \frac{1}{\omega} L(\sin \omega t)$

   $\boxed{f(t) = \frac{1}{\omega}\sin \omega t}$

4. **Analytic function:** A function f(z) is said to be analytic in a given region R, if it is continuous, single-valued and has definite derivartive at every point inside the region R.

   f(z) = log z is differentiable and analytic everywhere except at origin.

5. Given: $\vec{A} - = 2\hat{j} - 3\hat{k}$

   $\vec{B} = -2\hat{i} + \hat{k}$

   and $\vec{C} = 3\hat{i} - \hat{j}$

   $(\vec{B} \times \vec{C}) = (-2\hat{i} + \hat{k}) \times (3\hat{i} - \hat{j})$

   $= -2\hat{i} + 3\hat{i} + 2\hat{i} \times \hat{j} + \hat{k} \times 3\hat{i} - \hat{k} \times \hat{j}$

   $[\because \hat{i} \times \hat{i} = 0, \hat{i} \times \hat{j} = \hat{k}, \hat{k} \times \hat{i} = \hat{j} \text{ and } \hat{k} \times \hat{j} = -\hat{i}]$

   so, $(\vec{B} \times \vec{C}) = 0 + 2\hat{k} + 3\hat{j} + \hat{i}$
   $= (2\hat{k} + 3\hat{j} + \hat{i})$

   Now, $\vec{A} \cdot (\vec{B} \times \vec{C}) = (2\hat{j} - 3\hat{k}) \cdot (2\hat{k} + 3\hat{j} + \hat{i})$

   $= 4\hat{j} \cdot \hat{k} + 6\hat{j} \cdot \hat{j} + 2\hat{j} \cdot \hat{i} - 6\hat{k} \cdot \hat{k} - 9\hat{k} \cdot \hat{j} - 3\hat{k} \cdot \hat{i}$

   $\because [\hat{j} \cdot \hat{k} = 0, \hat{k} \cdot \hat{k} = 1, \hat{j} \cdot \hat{j} = 1, \hat{j} \cdot \hat{i} = 0, \hat{k} \cdot \hat{j} = 0,$ and $\hat{k} \cdot \hat{i} = 0]$

   So $(\vec{A} \cdot (\vec{B} \times \vec{C}) = 0 + 6 + 0 - 6 - 0 - 0$

   The value of $(\vec{A} \cdot (\vec{B} \times \vec{C}) + 6)$ is 0 + 6 = 6

6. Number of Men, M = 5

   Number of Women N = 3

   We have to hire four candidates.

   The required probability (i.e. the Probability that at least one woman will be selected).

   $$P = \frac{^5C_3 \times {^3C_1}}{^8C_4} + \frac{^5C_2 + {^3C_2}}{^8C_4} + \frac{^5C_1 \times {^3C_3}}{^8C_4}$$

   $$= \frac{3}{7} + \frac{3}{7} + \frac{1}{14} = \frac{13}{14} = 0.93$$

7. $\int_{-1}^{1.4} x|x|\,dx$

   $f(x) = -x^2 \quad -1 \le x \le 0$

   $f(x) = x^2 \quad 0 \le x < 14$

x	–1	–0.4	0.2	0.8	1.4
f(x)	–1	–0.16	0.04	0.64	1.96
	y_0	y_1	y_2	y_3	y_4

   By simpson's 1/3rd rule

   $$\int_{-1}^{1} x|x|\,dx = \frac{h}{3}[(y_0 + y_4) + 2(y_2) + 4(y_1 + y_2)]$$

   $$= \frac{0.6}{3}[(-1 + 1.96) + 2 \times 0.04 + 4(-0.16 + 0.64)]$$

   = 0.592

8. $\vec{f}(x, y, z) = \frac{x}{(x^2 + y^2 + z^2)^{3/2}}\hat{i} + \frac{y}{(x^2 + y^2 + z^2)^{3/2}}\hat{j} + \frac{z}{[x^2 + y^2 + z^2]^{3/2}}\hat{k}$

   By gauss-divergence theorem

   $$\iint_s \bar{F}\hat{n}ds = \iiint_v \vec{\nabla} \cdot \vec{F}\,dV \quad \text{...(i)}$$

So, $\vec{\nabla}\cdot\vec{F} = \frac{\partial}{\partial x}\left(\frac{x}{(x^2+y^2+z^2)^{\frac{3}{2}}}\right) + \frac{\partial}{\partial y}\left(\frac{y}{(x^2+y^2+z^2)^{\frac{3}{2}}}\right) + \frac{\partial}{\partial z}\left(\frac{z}{(x^2+y^2+z^2)^{\frac{3}{2}}}\right)$

$$\vec{\nabla}\cdot\vec{F} = \frac{1}{(x^2+y^2+z^2)^{\frac{3}{2}}} - \frac{\frac{3}{2}\times x\times 2x}{(x^2+y^2+z^2)^{\frac{5}{2}}}$$

$$+\frac{1}{(x^2+y^2+z^2)^{\frac{3}{2}}} - \frac{\frac{3}{2}\times y\times 2y}{(x^2+y^2+z^2)^{\frac{5}{2}}}$$

$$+\frac{1}{(x^2+y^2+z^2)^{\frac{3}{2}}} - \frac{\frac{3}{2}\times z\times 2z}{(x^2+y^2+z^2)^{\frac{5}{2}}} = 0$$

So by equation (i),

$$\iint_S \vec{F}\hat{n}6dS = \iiint_V \vec{\nabla}\cdot\vec{F}\,dV = 0$$

9. Mean of exponential distribution $= \frac{1}{\lambda}$

Standard deviation of exponential

distribution $= \frac{1}{\lambda}$

For x, $\lambda_1 = \frac{1}{0.5} = 2$

Standard deviation, $\sigma_x = 0.5$

Variance, $V(X) = (\sigma_x)^2 = 0.25$

For Y, $\lambda_2 = \frac{1}{0.5} = 2$

Standard deviation, $\sigma_Y = 0.5$

Variance $V(Y) = (\sigma_Y)^2 = 0.25$

$\because V(X+Y) = V(X) + V(Y) + 2COV(X, Y)$

Given : $V(X+Y) = 0$

$\Rightarrow \quad 0 = 0.25 + 0.25 + 2\,COV(X, Y)$

$\Rightarrow COV(X, Y) = -0.25$

Corelation factor,

$$r = \frac{COV(X,Y)}{\sigma_X\sigma_Y}$$

$$= \frac{-0.25}{0.5\times 0.5} = -1$$

10.
$$f(z) = u + iv$$
$$u = x^2 - y^2$$
$$v = \psi(x, y) \quad ...(i)$$

f(z) is an analytic function,

$$\frac{\partial u}{\partial x} = \frac{\partial v}{\partial y}$$

$$\frac{\partial v}{\partial x} = -\frac{\partial u}{\partial y}$$

so, $$dv = \frac{\partial v}{\partial x}dx + \frac{\partial v}{\partial y}dy$$

$$dv = -\frac{\partial u}{\partial y}dx + \frac{\partial u}{\partial x}dy$$

$$dv = 2y\,dx + 2x\,dy$$

$$\int dv = 2\int d(xy)$$

$$v = 2xy \quad ...(ii)$$

Given, $z = 1 + i$

So, $x = 1$

$y = 1$

From equation (1) and (2)

So; $v = \psi(x, y)_{(1,1)} = 2$

February 1 (Afternoon Session)

11. The sum of two normally distributed random variables X and Y is normally distributed only if X and Y are independent or jointly normal.

12. When the matrix P is decomposed into its symmetric part.

$$S = \begin{bmatrix} -4 & 4 & 2 \\ 4 & 3 & 7/2 \\ 2 & 7/2 & 2 \end{bmatrix}$$

$$V = \begin{bmatrix} 0 & -2 & 3 \\ 2 & 0 & 7/2 \\ -3 & -7/2 & 0 \end{bmatrix}$$

For matrix P its symmetric part is

$$S = \frac{P + P^T}{2}$$

For matrix P its skew symmetric part is

$$V = \frac{P - P^T}{2}$$

$$P = S + V$$

$$= \begin{bmatrix} -4 & 4 & 2 \\ 4 & 3 & 7/2 \\ 2 & 7/2 & 2 \end{bmatrix} + \begin{bmatrix} 0 & -2 & 3 \\ 2 & 0 & 7/2 \\ -3 & -7/2 & 0 \end{bmatrix}$$

$$= \begin{bmatrix} -4 & 2 & 5 \\ 6 & 3 & 7 \\ -1 & 0 & 2 \end{bmatrix}$$

13. Given that

Then by changing the order of integration

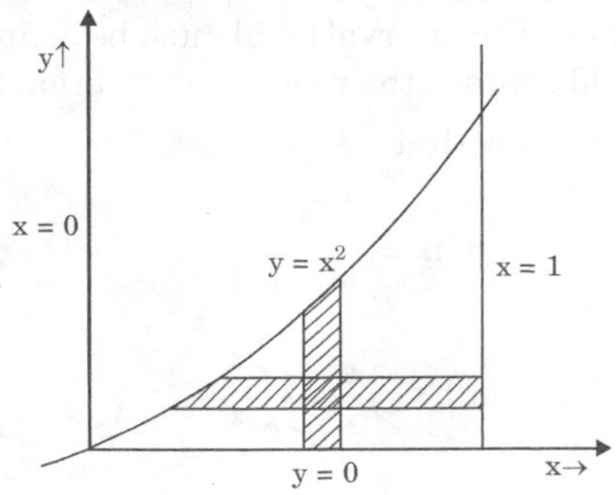

Limit $x = 0$ to $x = 1$

$y = 0$ to $y = x^2$

By changing the order of the integration the strip change from vertical to horizontal.

The new limit, $y = 0$ to $y = 1$

$$x = \sqrt{y} \text{ to } x = 1$$

So the integration

$$I = \int_{y=0}^{1}\int_{x=\sqrt{y}}^{1} xy^2 dx\, dy$$

14. Given $\frac{d^2y}{dt^2} - y = 1$

$$y\Big|_{t=0} \frac{dy}{dt}\Big|_{t=0} = 0$$

$$\frac{d^2y}{dt^2} - y = 1$$

$$y''(t) - y(t) = 1$$

Taking Laplace both side

$Ly''(t) - Ly(t) = L(1)$

$$S^2y(s) - sy(0) - y'(0) - y(s) = \frac{1}{s}$$

Given $y(0) = 0$, $y'(0) = 0$

Then $s^2y(s) - y(s) = \frac{1}{s}$

$$y(s)(s^2 - 1) = \frac{1}{s}$$

$$y(s) = \frac{1}{s(s-1)(s+1)}$$

15.
$$I = \begin{bmatrix} 1 & 0 & 0 & --- & 0 \\ 0 & 1 & 0 & --- & 0 \\ 0 & 0 & 1 & --- & 0 \\ - & - & - & --- & 0 \\ - & - & - & --- & 0 \\ - & - & - & --- & 1 \\ - & - & - & --- & 1 \end{bmatrix}_{100\times 100}$$

Since, diagonal elements are eigen values of identify matrix.

So, E has only 1 element i.e., 1.

16. Directional Derivative = $(\Delta f)_\phi \cdot \hat{a}$

$$f(x) = xyz$$

$$\vec{a} = \hat{i} - 2\hat{j} + 2\hat{k}$$

$$\hat{a} = \frac{\vec{a}}{|a|} = \frac{\hat{i} - 2\hat{j} + 2\hat{k}}{\sqrt{1^2 + (2)^2 + 2^2}}$$

$$= \frac{1}{3}(\hat{i} - 2\hat{j} + 2\hat{k})$$

$$\text{Gradient of (f)} = (\nabla f)_\phi = \left(\frac{\partial}{\partial x}\hat{i} + \frac{\partial}{\partial y}\hat{j} + \frac{\partial}{\partial z}\hat{k}\right)f$$

$$= yz\hat{i} + xz\hat{j} + xy\hat{k}$$

Directional derivative

$$= \left(yz\hat{i} + xz\hat{j} + xy\hat{k}\right)\cdot\frac{1}{3}\left(\hat{i} - 2\hat{j} + 2\hat{k}\right)$$

$$DD = \frac{1}{3}[yz - 2xz + 2xy]$$

DD at $(-1, 1, 3)$ is,

$$DD = \frac{1}{3}[1 \times 3 - 2 \times (-1) \times 3 + 2 \times (-1) \times 1] = \frac{7}{3}$$

17. Given: $u = x^3 - 3xy^2$

For function to be analytic it should follow Cauchy-Riemann's equations

$$u_x = v_y \text{ and } u_y = -v_x \quad \text{... (2)}$$

Now, ∵ $u_x = v_y$

$$u_x = \frac{\partial u}{\partial x} = 3x^2 - 3y^2$$

$$= v_y = \frac{\partial v}{\partial y}$$

$$\Rightarrow \quad v = \int (3x^2 - 3y^2)dy + f(x) + C_1$$

$$\Rightarrow \quad v = 3x^2y - \frac{3y^3}{3} + f(x) + C_1 \quad \text{...(3)}$$

$$\Rightarrow \quad \frac{\partial v}{\partial x} = 6xy - 0 + f'(x)$$

$$\Rightarrow \quad v_x = 6xy + f'(x) \quad \text{...(4)}$$

Now, $u_y = \frac{\partial u}{\partial y} = 0 - 6xy$ ...(5)

From (2)

$$u_y = -v_x$$

$$-6xy = -v_x \text{ [From (5)]}$$

$$-6xy = -(6xy + f'(x)) \quad \text{[From (4)]}$$

$$\Rightarrow \quad 6xy = 6xy + f'(x)$$

$\Rightarrow \quad f'(x) = 0$

$\Rightarrow \quad f(x) = \text{Constant} \quad ...(6)$

So, from (3)

$$v = 3x^2y - y^3 + \text{Constant} + C_1$$

[From (6)]

$$v = 3x^2y - y^3 + C$$

[Let C = Constant + C_1]

18. Conventional Method (I)

$$I = \int_0^{\pi/2} (8 + 4\cos x)\cdot dx$$

$$= (8x + 4\sin x)_0^{\pi/2}$$

$$= 4\pi + 4 = 16.566$$

Numerical Method using trapezoidal rule,

$$II = \frac{h}{2}(y_0 + y_n)$$

where, $h = \frac{\pi}{2} - 0 = \frac{\pi}{2}$

$$y_0 = 8 + 4 \times \cos(0) = 12$$

$$y_n = 8 + 4 \times \cos\left(\frac{\pi}{2}\right) = 8$$

$\Rightarrow \quad II = \frac{\pi}{4}(12 + 8) = 15.708$

$\therefore \quad \% \text{ error} = \frac{II - I}{I} \times 100$

$$= \left(\frac{II}{I} - 1\right) \times 100$$

$$= \left(\frac{15.708}{16.566} - 1\right) \times 100$$

$$= -5.18$$

$$= 5.18 \text{ (absolute \%)}$$

19. Fair coin is tossed 20 times,

Probability (Head exactly appear 4 times in the first 10 tosses)

$$P(X) = {}^{10}C_4 \left(\frac{1}{2}\right)^4 \left(\frac{1}{2}\right)^6 = 0.205$$

Probability (Tail exactly appears 4 times in the next 10 tosses)

$$P(Y) = {}^{10}C_4 \left(\frac{1}{2}\right)^4 \left(\frac{1}{2}\right)^6 = 0.205$$

Total probability, P(T) = P(X) × P(Y)

$$= 0.205 \times 0.205$$

$$= 0.042$$

February 2 (Forenoon Session)

20. Initial interval needs to be specified in Regular Fahi method. The interval [a, b] must be defined which should contain the root in the interval [a, b].

21. As we know that

$$\tan hx = \frac{e^x - e^{-x}}{e^x + e^{-x}}$$

$$\cot hx = \frac{e^x + e^{-x}}{e^x - e^{-x}}$$

$$\sec hx = \frac{2}{e^x + e^{-x}}$$

$$\text{cosec } hx = \frac{2}{e^x - e^{-x}}$$

Hence, option (b) is correct.

22. As given that,

$$V(x, y, z) = 3x^2y\hat{i} + 8y^2z\hat{j} + 5xyz\hat{k}$$

So, curl of the function

$$\nabla \times F = \left(\frac{\partial F_z}{\partial y} - \frac{\partial F_y}{\partial z}\right)\hat{i} + \left(\frac{\partial F_x}{\partial z} - \frac{\partial F_z}{\partial x}\right)\hat{j} + \left(\frac{\partial F_y}{\partial x} - \frac{\partial F_x}{\partial y}\right)\hat{k}$$

where, $F_x = 3x^2y$, $F_y = 8y^2z$

$F_z = 5xyz$

Now, putting the values

$$\nabla \times F = (5xz - 8y^2)i - 5yzj - 3x^2k$$

23. As given that

The delay in this case is 3 seconds.

Time lag = 3 seconds

As we know that

$$f(t) = 1 \times u(t - 3)$$

$$\text{Laplace transform} = \frac{e^{-\tau ds}}{s} = \frac{e^{-3s}}{s}$$

For unit step function with time lag of 3 seconds.

24. $A = \begin{bmatrix} 2 & 4 & 6 \\ 3 & 5 & 9 \\ 12 & 1 & 7 \end{bmatrix} \Rightarrow \begin{bmatrix} 2 & - & - \\ - & 5 & - \\ - & - & 7 \end{bmatrix}$

Sum of eigen values of the matrix

= the sum of diagonal elements

= trace of a matrix

= 2 + 5 + 7 = 14.

25. According to the question :

5 green balls, 10 blue balls, 4 green balls and 2 blue balls can be picked in $\left(5_{C_4} \times 10_{C_2}\right)$ ways

Total number of ways of picking 6 balls = 15_{C_6}

$\therefore$ Probability (P) = $\dfrac{5_{C_4} \times 10_{C_2}}{15_{C_6}} = \dfrac{5 \times 45}{5005} = \dfrac{45}{1001}$

26. As given funcition

$$f(x) = -\frac{5}{3}x^3 + 10x^2 - 15x + 16$$

By differentiating w.r.t 'x'

$$\frac{df(x)}{dx} = -5x^2 + 20x - 15$$

$$\frac{df(x)}{dx} = 0 \text{ (for maxima or minima)}$$

$\Rightarrow$ $-5x^2 - 20x + 15 = 0$

$(x - 3)(x - 1) = 0$

$$\frac{d^2f(x)}{dx^2} = -10x + 20$$

For x = 3, $\dfrac{d^2f(x)}{dx^2} = -10 < 0$ [Maxima]

$\therefore$ Maximum value of f(x) is for x = 3

$$f(3) = -45 + 90 - 45 + 16 = 16$$

27. As given that,

$$\frac{dy}{dx} = y - 20$$

$$y|_{x=0} = 40$$

By using variable separable method

$$\int \frac{dy}{y-20} = \int dx$$

$\ln(y - 20) = x + c$

$y - 20 = e^{x + c}$

By putting x = 0, y = 40

$\log(40 - 20) = 0 + c$

$\ln 20 = c$

$\ln(y - 20) = x + \ln 20$ $\quad\because x = 2$

$\ln(y - 20) = 2 + \ln 20$

$$\ln\left(\frac{y-20}{20}\right) = 2$$

$\therefore$ $y = 167.78 = 168$

28. $\displaystyle\int_{x_o}^{x_n} y\,dx = \frac{h}{3}[(y_o + y_n) + 4(y_1 + y_3 + ...) + 2(y_2 + y_4 + ...)]$

$$= \left\{\frac{2}{3}[(6+10)+4\times(8)]\right\} + \left\{\frac{10}{3}[(10+5)+4\times(12)]\right\}$$

$= 48 + 210 = 258$

February 2 (Afternoon Session)

29.

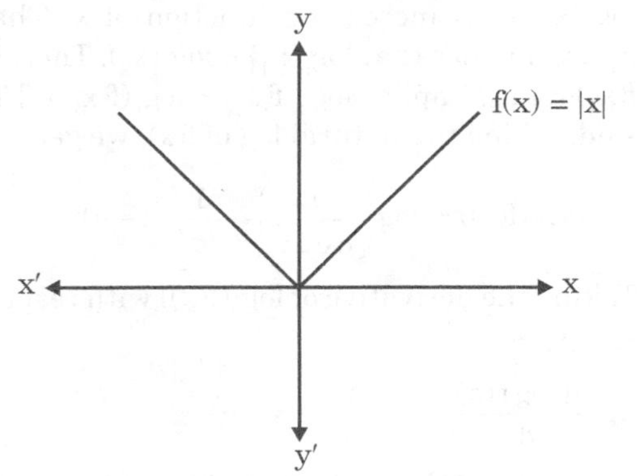

$$f(x) = |x|$$

$$|x| = \begin{cases} x, & x \geq 0 \\ -x, & x < 0 \end{cases}$$

$$f'(0) = \lim_{x\to 0} \frac{f(x) - f(0)}{x - 0}$$

$$= \lim_{x\to 0} \frac{|x| - |0|}{x - 0}$$

$$= \lim_{x\to 0} \frac{|x|}{0}$$

Check : $f(x) = |x|$

$$\lim_{x\to 0^+} \frac{|x|}{x} = \lim_{x\to 0^+} \frac{x}{x}$$

$$= \lim_{x\to 0^+}(1) = 1 \qquad ...(1)$$

$$\lim_{x\to 0^-} \frac{|x|}{x} = \lim_{x\to 0^-} \frac{-x}{x}$$

$$= \lim_{x\to 0^-}(-1) = -1 \qquad ...(2)$$

① ≠ ②

$$f'(x) = \begin{cases} 1 \text{ if } x > 0 \\ -1 \text{ if } x < 0 \end{cases}$$

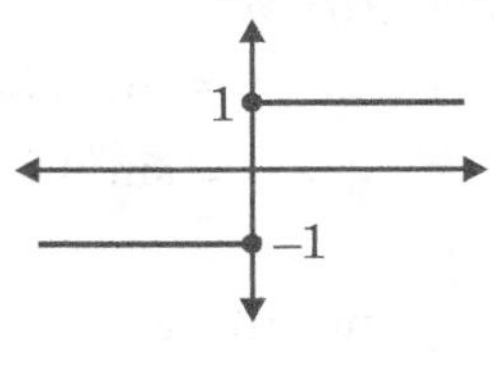

$\dfrac{\partial f}{\partial x}$ is not defined at x = 0 because derivative does not exist at the point

30. Differentiate f(x) with respect to x and solve for x to find the maximum. Because our f(x) contains the natural exponential function, however, it is easier to take the derivative of the natural log of f(x) with respect to x and solve for x to find the maximum. [The maximum of f(x) is the same as the maximum of the natural log of f(x), because $\log_e(x)$ is an increasing function of x. That is $x_1 < x_2$ implies that $\log_e(x_1) < \log_e(x_2)$. Therefore, $f(x_1) < f(x_2)$ implies $\log e(f(x_1)) < \log_e(f(x_2))$).] That said, taking the natural log of f(x), we get :

$$\text{loge}(f(x)) = \log\left(\frac{1}{\sigma\sqrt{2\pi}}\right) - \frac{1}{2\sigma^2}(x-\mu)^2$$

Taking the derivative of $\log_e(f(x))$ with respect to x, we get :

$$\frac{d \log f(x)}{dx} = -\frac{1}{2\sigma^2}.2(x-\mu)$$

Now, setting the derivative of log e(f(x)) to 0 :

$$\frac{d \log f(x)}{dx} = -\frac{1}{2\sigma^2}.2(x-\mu) \text{ skr } 0$$

and solving for x, we get that x = μ. Taking the second derivative of $\log_e(f(x))$ with respect to x, we get :

$$\frac{d^2 \log f(x)}{dx^2} = \frac{1}{\sigma^2}$$

Because the second derivative of $\log_e(f(x))$ is negative (for all x, in fact), the point x = μ is deemed a local maximum.

31.
$$y = Ae^{mx} + Be^{-mx}$$
$$y' = mAe^{mx} + (-m)Be^{-mx}$$
$$y'' = m^2Ae^{mx} + (-m \times -m)Be^{-mx}$$
$$y'' = m^2(Ae^{mx} + Be^{-mx})$$
$$y'' = m^2y$$
$$y'' - m^2y = 0$$
$$\frac{d^2y}{dx^2} - m^2y = 0$$

32.
$$A = \begin{pmatrix} \sin\theta & \tan\theta \\ 0 & \cos\theta \end{pmatrix}$$

In case of diagonal matrix, diagonal elements are eigen values (roots)

$$\lambda_1 = \sin\theta,\ \lambda_2 = \cos\theta$$

$$\therefore \quad \lambda_1^2 + \lambda_2^2 = \sin^2\theta + \cos^2\theta = 1$$

33. Given,

$$u(x, 0) = e^{-x2}$$
$$\eta = \frac{x}{\sqrt{\alpha t}}$$
$$u(x, t) = u(\eta)$$

Now, $u(0, t) = e^0 = 1$

Therefore, $\frac{u(0,1)}{u(0,4)} = \frac{1}{\sqrt{\alpha}} \times \frac{\sqrt{\alpha 4}}{1} = 2$

34. $x\frac{dx}{dy} + y = c$

$\Rightarrow xdx + ydy = cy$

Integrating we get,

$$\Rightarrow \frac{x^2}{2} + \frac{y^2}{2} = cy + k$$
$$\Rightarrow x^2 + y^2 - 2cy + 2k$$
$$\Rightarrow x^2 + y^2 - 2cy = 2k$$
$$\Rightarrow x^2 + y^2 - 2cy + c^2 = 2k + c^2$$
$$\Rightarrow x^2 + (y-c)^2 = 2k + c^2 \ldots\ldots\ldots\ldots(i)$$

Thus, euation 1 represent the equation of circle.

35.
$$A = \begin{bmatrix} \frac{1}{\sqrt{2}} & 0 & \frac{1}{\sqrt{2}} \\ 0 & 1 & 0 \\ \frac{1}{\sqrt{2}} & 0 & \frac{-1}{\sqrt{2}} \end{bmatrix},\ X = \begin{bmatrix} x \\ y \\ z \end{bmatrix}$$

$$B = \begin{bmatrix} 0 \\ 1 \\ -\sqrt{2} \end{bmatrix}$$

Now, $AX = B$

$$\begin{bmatrix} \frac{1}{\sqrt{2}} & 0 & \frac{1}{\sqrt{2}} \\ 0 & 1 & 0 \\ \frac{1}{\sqrt{2}} & 0 & \frac{-1}{\sqrt{2}} \end{bmatrix}\begin{bmatrix} x \\ y \\ z \end{bmatrix} = \begin{bmatrix} 0 \\ 1 \\ -\sqrt{2} \end{bmatrix}$$

$$\begin{bmatrix} \frac{x}{\sqrt{2}} + \frac{z}{\sqrt{2}} \\ y \\ \frac{x}{\sqrt{2}} - \frac{z}{\sqrt{2}} \end{bmatrix} = \begin{bmatrix} 0 \\ 1 \\ -\sqrt{2} \end{bmatrix}$$

$$\frac{x}{\sqrt{2}} + \frac{z}{\sqrt{2}} = 0 \quad \ldots(i)$$
$$y = 1 \quad \ldots(ii)$$
$$\frac{x}{\sqrt{2}} - \frac{z}{\sqrt{2}} = -\sqrt{2} \quad \ldots(iii)$$

From equation (i) and (ii), we get

x = – 1 and by then putting x = – 1 in equation (i), we get

$$z = 1$$

$$\therefore \quad x + y + z = -1 + 1 + 1 = 1$$

36. $\int_0^1 (x^2 - 2x + 1)dx$

As given to evaluate with four intervals

$$f(x) = x^2 - 2x + 1$$

x	0	0.25	0.5	0.75	1
$f(x)$	1	0.562	0.25	0.062	0

Trapezoidal rule in given by

$$\int_a^b f(x)dx = \frac{b-a}{2n}\left[f(x_o)+f(x_n)+2[f(x_1) + f(x_2)]+......+f(x_{n-1})\right]$$

For given Interval

$$\int_a^b f(x)dx = \frac{0.25}{2}\left[f(0)+f(1)+2\left[f(0.25)+f(0.5) + f(0.75)\right]\right]$$

$$= 0.3435$$

For analytical value

$$\int_0^1 (x^2 - 2x + 1)dx = \int_0^1 x^2dx - 2\int_0^1 xdx + \int_0^1 dx$$

$$= \left|\frac{x^3}{3}\right|_0^1 - 2\left|\frac{x^2}{2}\right|_0^1 + |x|_0^1$$

$$= \frac{1}{3} - 1 + 1 = \frac{1}{3}$$

$$\text{Difference} = \left|0.3435 - \frac{1}{3}\right| = 0.0102$$

February 8 (Forenoon Session)

37. As given that

$ax^3 + bx^2 + cx + d = 0; a \neq 0$

When, d = 0 and at least one of a, b, c ≠ 0 then x = 0 is a root. 'C' is true.

38. $\frac{1}{2\pi}\int_{-\pi}^{\pi} \sin(p\theta)\cos(q\theta)d\theta$, the integral is an odd function its value is zero.

39. It has no poles for any bounded signal i.e., non-zero only inside a finite time interval.

40. $x(n) = \left(\frac{1}{2}\right)^n u(n);$

ROC of x(n) : $|z| > \frac{1}{2}$

$$\Rightarrow x(z) = \frac{z}{z - \frac{1}{2}} = \frac{1}{1 - \frac{1}{2}z^{-1}}; \ |z| > \frac{1}{2}$$

$x(n - k) \leftrightarrow z^{-k}. x(z)$

$$\Rightarrow x(n-k) \leftrightarrow z^{-k}\frac{1}{\left(1 - \frac{1}{2}z^{-1}\right)} = \frac{z^{-k}}{\left(1 - \frac{1}{2}z^{-1}\right)},$$

ROC of x(n – k) : $|z| > \frac{1}{2}$

$|z| > \frac{1}{2}$

Time shifting property will not effect ROC

41. $\int_C \frac{z^2+1}{(z^2 - 2z)}dz$ (where C is |Z| = 1)

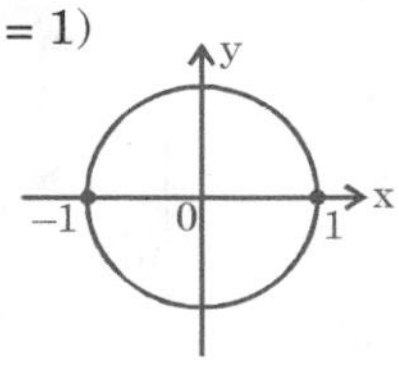

$$\int_C \frac{z^2+1}{z(z-2)}dz$$

Z = 0 is inside

it is rotating in anti-clockwise direction

$$\Rightarrow 2\pi i \lim_{z\to 0} z\left(\frac{z^2+1}{z(z-2)}\right) = 2\pi i\left(\frac{-1}{2}\right) = -\pi i$$

42. $\frac{dy}{dx} = 2x - y;$

$$\frac{dy}{dx} + y = 2x$$

$$I.F = e^{\int 1.dx} = e^x$$

$\therefore ye^x = 2\int xe^x\,dx = 2e^x(x-1) + C$

$y(0) = 1 \Rightarrow 1 = -2 + C$

$\therefore C = 3$

$y = 2x - 2 + 3e^{-x}$

At, x = ln (2)

$y = 2(\log 2) - 2 + 3e^{\ln 2}$

$= 1.386 - 2 + \frac{3}{2}$

$\therefore y = 0.886$

43. As given that

$f(x) = y\ 3x^2 + 3x + 1, x \in (-2, 0)$

$f'(x) = \frac{\delta y}{\delta x} = 0$

$\Rightarrow (6x + 3) = 0$

$$\frac{\delta^2 y}{\delta x^2} = 6$$

$$x = \frac{-1}{2} \in (-2, 0)$$

$f''(x) = 6 > 0$

$$\text{Min}\begin{bmatrix} f(-2) = 7 \\ f(0) = 1 \\ f\left(-\frac{1}{2}\right) = \frac{3}{4} - \frac{3}{2} + 1 = \frac{1}{4} \end{bmatrix} = \frac{1}{4}$$

So, maximum value = 7 & minimum value = 1/4

44. As given that,

$$\vec{F} = \hat{a}_x(5y - k_1z) + \hat{a}_y(3z + k_2x) + \hat{a}_z(k_3y - 4x)$$

$\bar{F} = (5y - k_1z)\hat{i} + (3z + k_2x)\hat{j} + (k_3y - 4x)\hat{k}$

As $\vec{F}$ is conservative field (or) irrotational and hence $\nabla \times \bar{F} = 0$ (or) A null vector.

$$\begin{vmatrix} \hat{a}_x & \hat{a}_y & \hat{a}_z \\ \frac{\partial}{\partial x} & \frac{\partial}{\partial y} & \frac{\partial}{\partial z} \\ (5y - k_1z) & (3z + k_2x) & (k_3y - 4x) \end{vmatrix}$$

$$= 0.\hat{a}_x + 0.\hat{a}_y + 0.\hat{a}_z = 0$$

$$k_3 - 3 = 0$$

$$\Rightarrow \quad k_3 = 3$$

$$-4 + k_1 = 0$$

$$\Rightarrow \quad k_1 = 4$$

$$k_2 - 5 = 0$$

$$\Rightarrow \quad k_2 = 5$$

Hence, $k_1 = 4$, $k_2 = 5$ & $k_3 = 3$

45. $\begin{pmatrix} 2 & 3 & 3 \\ 3 & 2 & 1 \\ 3 & 1 & 7 \end{pmatrix} = \begin{pmatrix} a & 0 & 0 \\ b & c & 0 \\ d & e & f \end{pmatrix}\begin{pmatrix} a & 0 & 0 \\ b & c & 0 \\ d & e & f \end{pmatrix}^T$

$$= \begin{pmatrix} a & 0 & 0 \\ b & c & 0 \\ d & e & f \end{pmatrix}\begin{pmatrix} a & b & d \\ 0 & c & e \\ 0 & 0 & f \end{pmatrix}$$

$$\begin{pmatrix} 2 & 3 & 3 \\ 3 & 2 & 1 \\ 3 & 1 & 7 \end{pmatrix} = \begin{pmatrix} a^2 & ab & ad \\ ab & b^2 + c^2 & bd + ce \\ ad & bd + ce & d^2 + c^2 + f^2 \end{pmatrix}$$

Comparing both sides, we get $a^2 = 2$, $ab = 3$, $b^2 + c^2 = 2$, $bd + ce = 1$ and $d^2 + e^2 + f^2 = 7$

Now by solving all of the above equation we get

$a = \pm\sqrt{2}$, is purely real

$b = \pm\frac{3}{\sqrt{2}}$, is also real

$d = \pm\frac{3}{\sqrt{2}}$, is also real

$c = \pm\sqrt{\frac{5}{2}}i$, is imaginary

$e = \pm\frac{7}{\sqrt{10}}i$, is imaginary

$f = \pm\sqrt{\frac{37}{5}}$, is real

The no. of purely real elements is 4.

46. Given, $\bar{F} = y\hat{a}_x - x\hat{a}_y$

and given curve, $y = x^2$

Now, $\int_C F.dr = \int_{(-1)}^{(2)} (y.dx - xdy)$

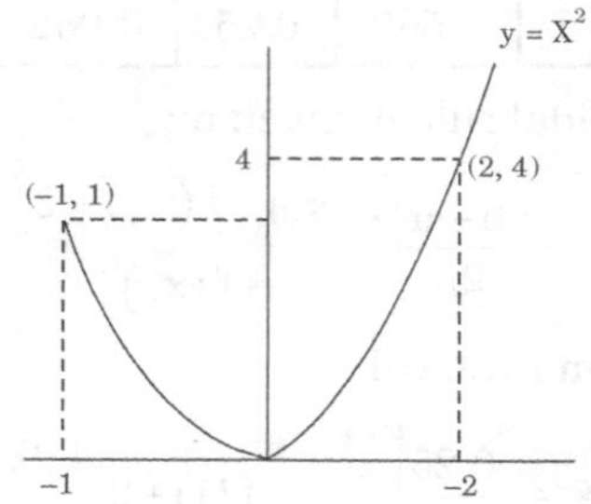

$$\int_C \bar{F}.d\bar{r} = \int_{x=-1}^{2} x^2.dx - x(2x)dx = \int_{-1}^{2} -x^2.dx$$

$$= \left[\frac{-x^3}{3}\right]_{-1}^{2} \Rightarrow \frac{-1}{3}(8 + 1) = -3$$

February 8 (Afternoon Session)

47. If, $f(x) = \sqrt{x^3 + 1}$

$$f'(x) = \frac{1}{\sqrt{x^3 + 1}} \cdot 3x^2 \geq 0$$

Here, for all points in [0, 1]

$\therefore$ Therefore, it is increasing function.

48. As given that,

$A = \{1, 2, 3\}$

$n = |A| = 3$

Number of relations on $A = 2^{n^2} = 2^{3^2} = 29$

Number of reflexive relations on

$A = 2^{n^2 - n} = 2^{3^2 - 3} = 2^6$

$$P[\text{reflexive}] = \frac{2^{n^2 - n}}{2^{n^2}} = \frac{2^6}{2^9} = \frac{1}{8} = 0.125$$

49. The order of a subgroup always divides the order of groups.

Size of group = O(G) = 35

Let H be subgroup of G

$\therefore$ O(H) | O(G)

Possible orders of H are 1,5,7,35

So, size of largest possible proper subgroup of G other than G itself would be = 7.

51. As given that:

$$w(u, v) = (u, v) \text{ edge weight}$$

$$w(u,v) = w(u, v) + \underbrace{f(u) - f(v)}_{0}$$

$$= w(u, v)$$

Hence, option (a) correct.

52. Consider small instance of 4-vertices

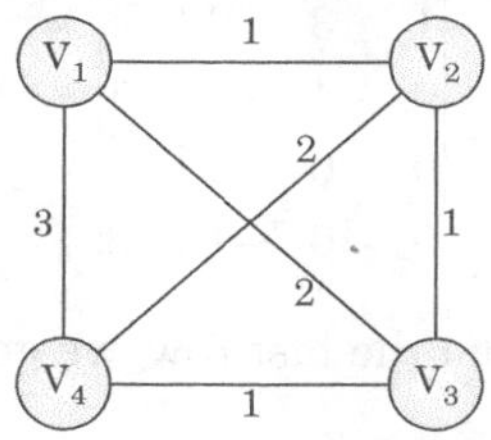

Clearly, 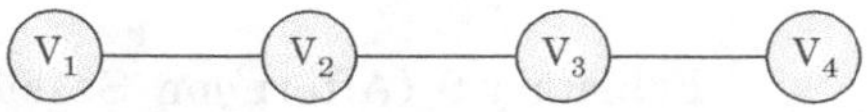

would be a MST with cost 3.

The number of edges of the MST is $|V| - 1$, in case of the given question $|V| = 100$, now the cost for this is $= |V| - 1 = 99$.

Hence, for 100 vertices, 99 will the weight of minumum spanning tree.

53.

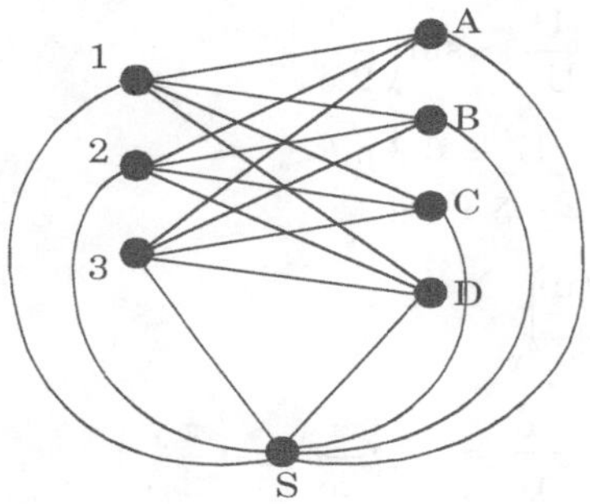

Therefore, 4 colors (for the k3, 4) + 3 new colors from vertex (s) to vertices 1, 2, 3.

Hence, Minimum number of edge-colors required = 3 + 4 = 7.

February 9 (Forenoon Session)

54. As given that, the partial differential equation :

$$D(\theta)\frac{\delta^2\theta}{\delta z^2} + \frac{\delta k(\theta)}{\delta z} - \frac{\delta\theta}{\delta t} = 0$$

The 1st term of given partial differential equation contains product of dependent variable with its derivative, So it is non-linear and also we have 2nd order derivative, Hence, its order is two i.e., 2nd order non-linear equation.

55. $\lim\limits_{x\to\infty}\frac{x^2 - 5x + 4}{4x^2 + 2x}$

By L-hospital rule :

$$= \lim_{x\to\infty}\frac{x^2\left(1 - \frac{5}{x} + \frac{4}{x^2}\right)}{x^2\left(4 + \frac{2}{x}\right)}$$

$$= \frac{1 - 0 + 0}{4 + 0} = \frac{1}{4}$$

56. The linear interpolation for two data points (1, ln(1), (6, ln (6))

$$\frac{y - \ln(1)}{\ln(6) - \ln(1)} = \frac{x - 1}{5}$$

$$y = \left(\frac{x - 1}{5}\right)(\ln(6) - \ln(1))$$

$$y = \left(\frac{x - 1}{5}\right)\ln(6)$$

$$y = \frac{\ln(6)}{5} = 0.358 = A \text{ (approximate value)}$$

$T = \ln(2) = 0.69$ (exact value)

Absolute error $= |(T - A)|$

$= |0.69 - 0.358| = 0.332$

$$\% \text{ error} = \frac{T - A}{T} \times 100$$

$$= \frac{0.332}{0.69} \times 100 = 48.11\% = 48 \text{ (approx)}$$

57. Area of the ellipse $\frac{x^2}{a^2} + \frac{y^2}{b^2} = 1$

$$\text{Area} = \iint (1)dy\,dx$$

$$= \int_{x=-a}^{a}\int_{y=-\frac{b}{a}\sqrt{-x^2+a^2}}^{+\frac{b}{a}\sqrt{-x^2+a^2}}$$

$$= 4\int_{x=0}^{a}\int_{y=0}^{+\frac{b}{a}\sqrt{a^2-x^2}} (1)dydx = \pi ab$$

58. As given that, the ordinary differential equation :

$$\frac{d^2x}{dt^2} - 5\frac{dx}{dt} + 6x = 0$$

The auxiliary equation is

$D^2 - 5D + 6 = 0$

Roots are D = 2, 3

The general solution is

$$x = C_1e^{2t} + C_2\,e^{3t} \quad ...(1)$$

$$\frac{dx}{dt} = 2C_1e^{2t} + 3C_2e^{3t} \quad ...(2)$$

Now, by using initial condition;

$x(0) = 0$, (1)

$\Rightarrow 0 = C_1 + C_2$

$\Rightarrow C_2 = -C_1$

$\Rightarrow \frac{dx(0)}{dt} = 10$, (2)

$\Rightarrow 10 = 2C_1 + 3C_2$

$\Rightarrow 10 = 2C_1 - 3C_1 = -C_1$

$\Rightarrow C_1 = -10$ & $C_2 = 10$

∴ The particular solution is

$x = -10\,e^{2t} + 10\,e^{3t}$

Note: Among four options, only option (d) satisfies $x(0) = 0$. So option (d) is correct.

59. The six point central difference formula for third derivative is

$$\left.\frac{\delta^3 u}{\delta x^3}\right|_{x_i} = \left[\frac{-u_{i+3} + 8u_{i+2} - 13u_{i+1} + 13u_{i-1} - 8u_{i-2} + u_{i-3}}{8\Delta h^3}\right]$$

$$f''(x_i) = \left[\frac{[-f(x_{i-3}) - 8f(x_{i-2}) + 13f(x_{i-1}) - 13f(x_{i+1}) + 8f(x_{i+2}) - f(x_{i+3})]}{8h^3}\right]$$

60. $I = \int_c ((y+z)dx + (x+z)dy + (x+y)dz)$

$= \int_c [ydx + zdx + xdy + zdy + xdz + ydz]$

$= \int_c [(ydx + xdy) + (zdx + xdz) + (zdy + ydz)]$

$= \int_{(0,0,0)}^{(1,1,1)} [d(xy) + d(xz) + d(yz)]$

$= [xy + xz + yz]_{(0,0,0)}^{(1,1,1)} = (1+1+1) - (0+0+0) = 3$

61. The augmented matrix is

$$[A\,|\,B] = \left|\begin{array}{ccc|c} 1 & 3 & 2 & 1 \\ 2 & 2 & -3 & 1 \\ 4 & 4 & -6 & 2 \\ 2 & 5 & 2 & 1 \end{array}\right|$$

By converting into an echelon form :

$R_2 \to R_2 - 2R_1$

$R_3 \to R_3 - 4R_1$

$R_4 \to R_4 - 2R_1$

$$\text{So, } [A\,|\,B] = \left|\begin{array}{ccc|c} 1 & 3 & 2 & 1 \\ 0 & -4 & -7 & -1 \\ 0 & -8 & -14 & -2 \\ 0 & -1 & -2 & -1 \end{array}\right|$$

Now, $R_3 \to R_3 - 2R_2$

$R_4 \to 4R_4 - R_2$

$$\text{So, } [A\,|\,B] = \left|\begin{array}{ccc|c} 1 & 3 & 2 & 1 \\ 0 & -4 & -7 & -1 \\ 0 & 0 & 0 & 0 \\ 0 & 0 & -1 & -3 \end{array}\right|$$

Hence, From the last row, we get

$-x_3 = -3 \Rightarrow x_3 = 3$

February 9 (Afternoon Session)

62. Hydrologic channel routing is done by Muskingum method.

63. $\frac{d^2u}{dx^2} - 2x^2u + \sin x = 0$

It is a linear non-homogeneous equation

64. Given : $\frac{U}{U_\infty} = \sin\left(\frac{\pi}{2}\frac{y}{\delta}\right)$

$U_\infty = 0.3$ m/s

$\delta = 1$ m

To find: $\left(\frac{du}{dy}\right)_{y=0}$

$$\frac{du}{dy} = U_\infty \cdot \frac{\pi}{2\delta} \cdot \cos\left(\frac{\pi}{2}\frac{y}{\delta}\right)$$

$$\cos\left(\frac{\pi}{2}\frac{y}{\delta}\right) = 1$$

$$\Rightarrow \left(\frac{du}{dy}\right)_{y=0} = \frac{0.3 \times \pi}{2 \times 1} = 0.471\ s^{-1}$$

65. As per clause 5.4.2 of IS 456 : 2000, the pH value of water should not be less than 6.

66. The integrand is

$f(x) = 5x^3 + 4x^2 + 3x + 2$

It is a polynomial of degree 3

Simpson's rule gives the exact value of the integral

67. $CO_3^{-2} = 3.04 \times 10^{-4}$ mole/L

$= 3.04 \times 10^{-4} \times 60 \times 10^3$

$= 18.24$ mg/L

$HCO_3^- = 2 \times 10^{-3}$ mole/L

$= 2 \times 10^{-3} \times 61 \times 10^3$

$= 122$ mg/L

$$\therefore \text{Total Alkalinity} = [CO_3^{-2}] \times \frac{50}{30} + [HCO_3] \times \frac{50}{61}$$

$$\therefore \quad TA = 18.24 \times \frac{50}{30} + 122 \times \frac{50}{61}$$

$= 30.4 + 100$

$= 130.4$ mg/L as $CaCO_3$

68.

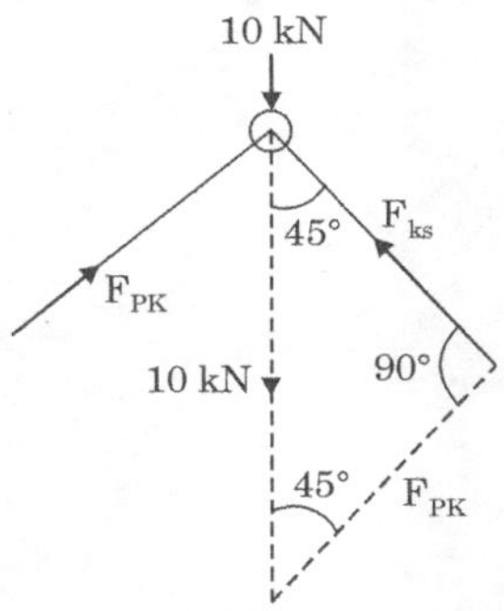

By applying Lami's Theorem

$$\frac{F_{KS}}{\sin 45} = \frac{10}{\sin 90}$$

$$F_{KS} = \frac{10}{\sqrt{2}} \text{ kN(compression)}$$

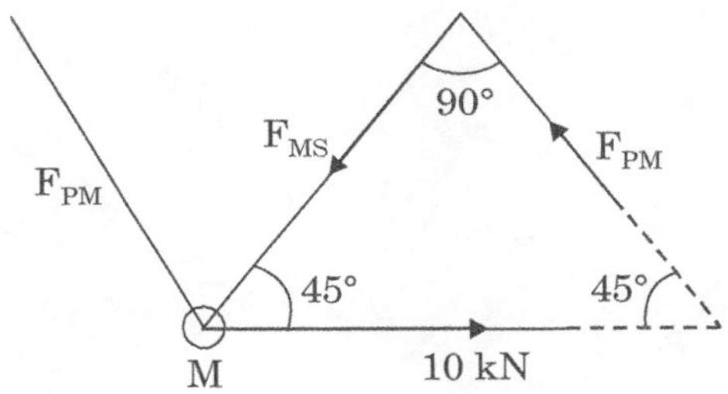

By applying Lami's theorem

$$\frac{F_{MS}}{\sin 45} = \frac{10}{\sin 90}$$

$$F_{MS} = \frac{10}{\sqrt{2}} \text{(compression)}$$

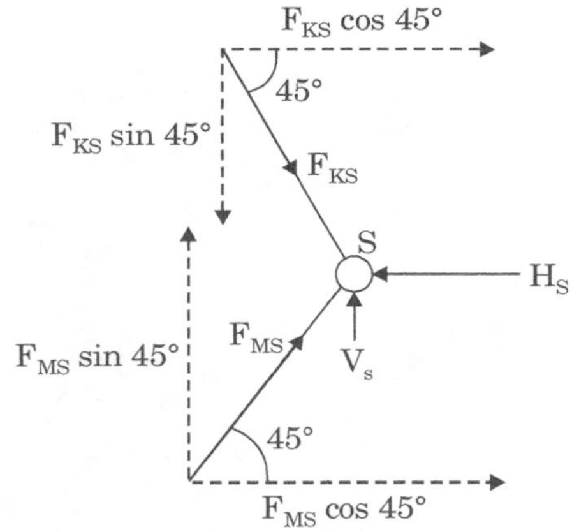

$V_s = 0$

The magnitude of the horizontal component of reaction at S

$$H_S = F_{KS} \cos 45° + F_{MS} \cos 45°$$

$$= \frac{10}{\sqrt{2}} \times \frac{1}{\sqrt{2}} + \frac{10}{\sqrt{2}} \times \frac{1}{\sqrt{2}} = 10 \text{ kN}$$

69. Design Shear capacity of fillet weld (P_{dw}) = P

$$P_{dw} = P = \frac{L_w + t_t + f_u}{\sqrt{3} \times \gamma_{mw}}$$

$L_w = 2 \times 200 + 120 = 520$ mm

$t_t = K \times S = 0.7 \times 6 = 4.2$ mm

$\gamma_{mw} = 1.25$ (for workshop welding)

$= 1.50$ (for site welding)

$$P = P_{dw} = \frac{520 \times 4.2 \times 410}{\sqrt{3} \times 1.25}$$

= 413586.02 N

= 413.586 kN

■■

SOLVED PAPER – 2021

GENERAL APTITUDE

SET 1

Q. 1 to Q. 5 : Carry One Mark Each.

1. Getting to the top is ________ than staying on top.

(a) more easy (b) much easy

(c) easiest (d) easier

2. 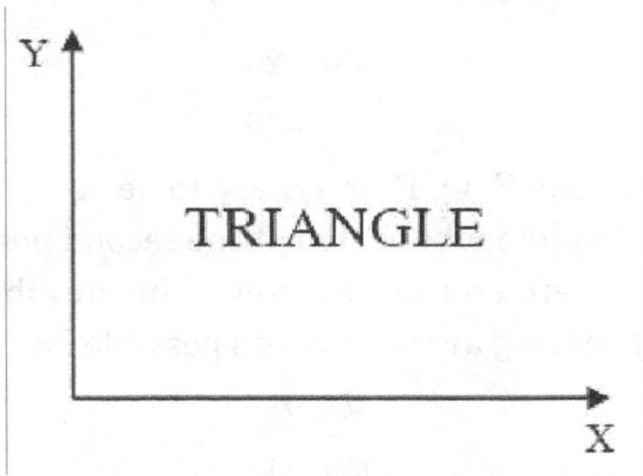

The mirror image of the above text about the x-axis is

(a)  (b) 

(c)  (d) 

3. In a company, 35% of the employees drink coffee, 40% of the employees drink tea and 10% of the employees drink both tea and coffee. What % of employees drink neither tea nor coffee?

(a) 15 (b) 25

(c) 35 (d) 40

4. $\oplus$ and $\odot$ are two operators on numbers p and q such that

$$p \oplus q = \frac{p^2 + q^2}{pq} \text{ and } p \odot q = \frac{p^2}{q};$$

If $x \oplus y = 2 \odot 2$, then $x =$

(a) $\frac{y}{2}$ (b) y

(c) $\frac{3y}{2}$ (d) 2y

5. Four persons P, Q, R and S are to be seated in a row, all facing the same direction, but not necessarily in the same order. P and R cannot sit adjacent to each other. S should be seated to the right of Q. The number of distinct seating arrangements possible is:

(a) 2 (b) 4

(c) 6 (d) 8

Q. 6 to Q. 10 : Carry Two Marks Each.

6. Statement: Either P marries Q or X marries Y

Among the options below, the logical NEGATION of the above statement is:

(a) P does not marry Q and X marries Y.

(b) Neither P marries Q nor X marries Y.

(c) X does not marry Y and P marries Q.

(d) P marries Q and X marries Y.

7. Consider two rectangular sheets, Sheet M and Sheet N of dimensions 6 cm x 4 cm each.

Folding operation 1: The sheet is folded into half by joining the short edges of the current shape.

Folding operation 2: The sheet is folded into half by joining the long edges of the current shape.

Folding operation 1 is carried out on Sheet M three times.

Folding operation 2 is carried out on Sheet N three times.

The ratio of perimeters of the final folded shape of Sheet N to the final folded shape of Sheet M is ________.

(a) 13 : 7 (b) 3 : 2

(c) 7 : 5 (d) 5 : 13

8. 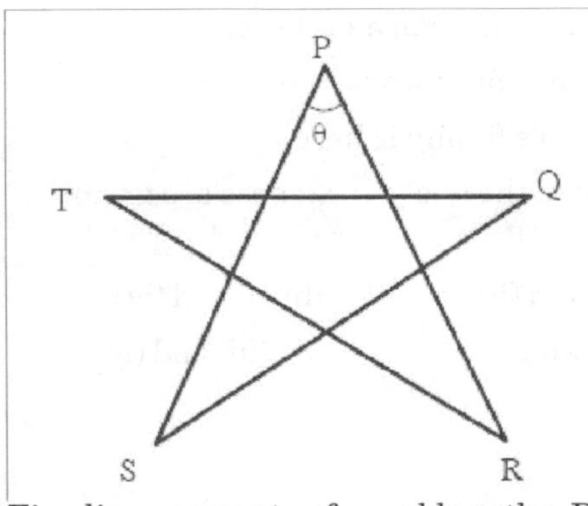

Five line segments of equal lengths, PR, PS, QS, QT and RT are used to form a star as shown in the figure above.

The value of θ, in degrees, is

(a) 36 (b) 45

(c) 72 (d) 108

9. A function, λ, is defined by

$$\lambda(p, q) = \begin{cases} (p-q)^2, & \text{if } p \geq q, \\ p+q, & \text{if } p < q. \end{cases}$$

The value of the expression

$\frac{\lambda(-(-3+2),(-2+3))}{(-(-2+1))}$ is :

(a) –1 (b) 0

(c) $\frac{16}{3}$ (d) 16

10. Humans have the ability to construct worlds entirely in their minds, which don't exist in the physical world. So far as we know, no other species possesses this ability. This skill is so important that we have different words to refer to its different flavors, such as imagination, invention and innovation.

Based on the above passage, which one of the following is TRUE?

(a) No species possess the ability to construct worlds in their minds.

(b) The terms imagination, invention and innovation refer to unrelated skills.

(c) We do not know of any species other than humans who possess the ability to construct mental worlds.

(d) Imagination, invention and innovation are unrelated to the ability to construct mental worlds.

SET 2

Q. 1 – Q. 5 carry one mark each

1. (i) Arun and Aparna are here.

(ii) Arun and Aparna is here.

(iii) Arun's families is here.

(iv) Arun's family is here.

Which of the above sentences are grammatically CORRECT?

(a) (i) and (ii) (b) (i) and (iv)

(c) (ii) and (iv) (d) (iii) and (iv)

2.

The mirror image of the above text about the x-axis is

(a) 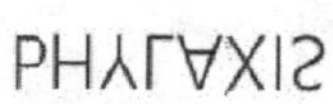 (b) 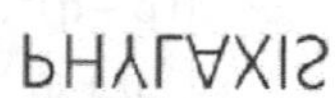

(c)  (d) 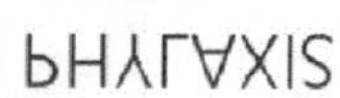

3. Two identical cube shaped dice each with faces numbered 1 to 6 are rolled simultaneously. The probability that an even number is rolled out on each dice is:

(a) 1/36 (b) 1/12

(c) 1/8 (d) 1/4

4. $\oplus$ and $\odot$ are two operators on numbers p and q such that $p \odot q = p - q$, and $p \oplus q = p \times q$

Then, $(9 \odot (6 \oplus 7)) \odot (7 \oplus (6 \odot 5)) =$

(a) 40 (b) –26

(c) –33 (d) –40

5. Four persons P, Q, R and S are to be seated in a row. R should not be seated at the second position from the left end of the row. The number of distinct seating arrangements possible is:

(a) 6 (b) 9

(c) 18 (d) 24

Q. 6 – Q. 10 carry two marks each

6. On a planar field, you travelled 3 units East from a point O. Next you travelled 4 units South to arrive at point P. Then you travelled from P in the North-East direction such that you arrive at a point that is 6 units East of point O. Next, you travelled in the North-West direction, so that you arrive at point Q that is 8 units North of point P. The distance of point Q to point O, in the same units, should be _____

(a) 3 (b) 4

(c) 5 (d) 6

7. The author said, "Musicians rehearse before their concerts. Actors rehearse their roles before the opening of a new play. On the other hand, I find it strange that many public speakers think they can just walk on to the stage and start speaking. In my opinion, it is no less important for public speakers to rehearse their talks."

Based on the above passage, which one of the following is TRUE?

(a) The author is of the opinion that rehearsing is important for musicians, actors and public speakers.

(b) The author is of the opinion that rehearsing is less important for public speakers than for musicians and actors.

(c) The author is of the opinion that rehearsing is more important only for musicians than public speakers.

(d) The author is of the opinion that rehearsal is more important for actors than musicians.

8. 1. Some football players play cricket.

2. All cricket players play hockey.

Among the options given below, the statement that logically follows from the two statements 1 and 2 above, is:

(a) No football player plays hockey.

(b) Some football players play hockey.

(c) All football players play hockey.

(d) All hockey players play football.

9.

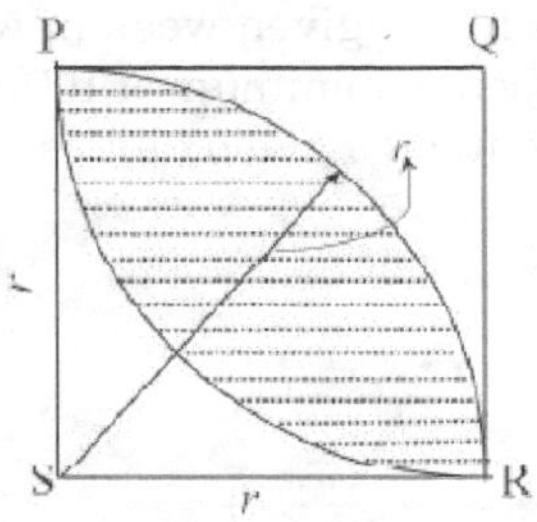

In the figure shown above, PQRS is a square. The shaded portion is formed by the intersection of sectors of circles with radius equal to the side of the square and centers at S and Q.

The probability that any point picked randomly within the square falls in the shaded area is ______

(a) $4-\frac{\pi}{2}$ (b) 1/2

(c) $\frac{\pi}{2}-1$ (d) $\frac{\pi}{4}$

10. In an equilateral triangle PQR, side PQ is divided into four equal parts, side QR is divided into six equal parts and side PR is divided into eight equal parts. The length of each subdivided part in cm is an integer.

The minimum area of the triangle PQR possible, in cm^2, is

(a) 18 (b) 24

(c) $48\sqrt{3}$ (d) $144\sqrt{3}$

SET 3

Q.1 to Q.5: Multiple Choice Questions (MCQ), carry ONE mark each.

1. The current population of a city is 11,02,500. If it has been increasing at the rate of 5% per annum, what was its population 2 years ago?

(a) 9,92,500 (b) 9,95,006

(c) 10,00,000 (d) 12,51,506

2. p and q are positive integers and $\frac{p}{q}+\frac{q}{p}=3$, then,

$\frac{p^2}{q^2}+\frac{q^2}{p^2}=$

(a) 3 (b) 7

(c) 9 (d) 11

3.

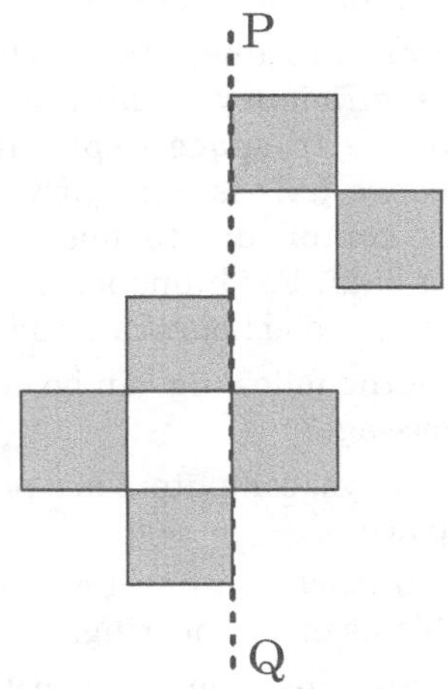

The least number of squares that must be added so that the line P-Q becomes the line of symmetry is ______

(a) 4 (b) 3

(c) 6 (d) 7

4. *Nostalgia* is to *anticipation* as _____ is to ______

Which one of the following options maintains a similar logical relation in the above sentence?

(a) Present, past (b) Future, past

(c) Past, future (d) Future, present

5. Consider the following sentences:

(i) I woke up from sleep.

(ii) I woked up from sleep.

(iii) I was woken up from sleep.

(iv) I was wokened up from sleep.

Which of the above sentences are grammatically CORRECT?

(a) (i) and (ii) (b) (i) and (iii)

(c) (ii) and (iii) (d) (i) and (iv)

Q.6 to Q.10: Multiple Choice Question (MCQ), carry TWO marks each.

6. Given below are two statements and two conclusions.

Statement 1 : All purple are green.

Statement 2 : All black are green.

Conclusion I : Some black are purple.

Conclusion II : No black is purple.

Based on the above statements and conclusions, which one of the following options is logically CORRECT?

(a) Only conclusion I is correct.

(b) Only conclusion II is correct.

(c) Either conclusion I or II is correct.

(d) Both conclusion I and II are correct.

7. Computers are ubiquitous. They are used to improve efficiency in almost all fields from agriculture to space exploration. Artificial intelligence (AI) is currently a hot topic. AI enables computers to learn, given enough training data. For humans, sitting in front of a computer for long hours can lead to health issues.

Which of the following can be deduced from the above passage?

(i) Nowadays, computers are present in almost all places.

(ii) Computers cannot be used for solving problems in engineering.

(iii) For humans, there are both positive and negative effects of using computers.

(iv) Artificial intelligence can be done without data.

(a) (ii) and (iii)

(b) (ii) and (iv)

(c) (i), (iii) and (iv)

(d) (i) and (iii)

8. Consider a square sheet of side 1 unit. In the first step, it is cut along the main diagonal to get two triangles. In the next step, one of the cut triangles is revolved about its short edge to form a solid cone. The volume of the resulting cone, in cubic units, is ________

(a) $\frac{\pi}{3}$

(b) $\frac{2\pi}{3}$

(c) $\frac{3\pi}{2}$

(d) 3π

9.

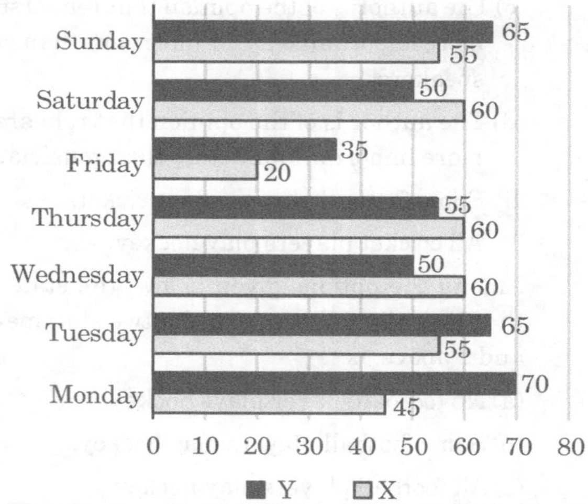

The number of minutes spent by two students, X and Y, exercising every day in a given week are shown in the bar chart above.

The number of days in the given week in which one of the students spent a minimum of 10% more than the other student, on a given day, is

(a) 4 (b) 5

(c) 6 (d) 7

10.

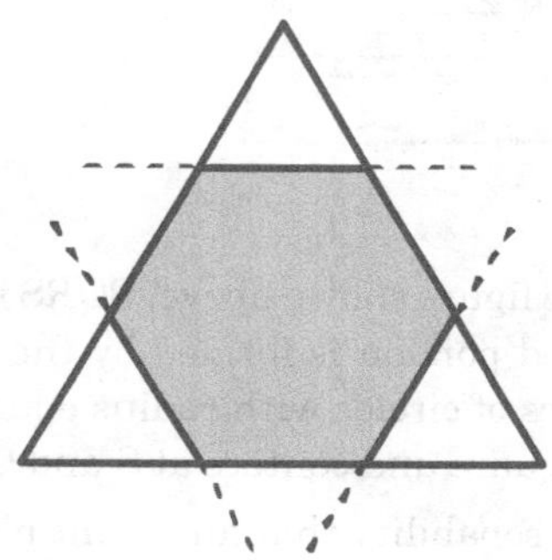

Corners are cut from an equilateral triangle to produce a regular convex hexagon as shown in the figure above.

The ratio of the area of the regular convex hexagon to the area of the original equilateral triangle is

(a) 2 : 3 (b) 3 : 4

(c) 4 : 5 (d) 5 : 6

SET 4

Q.1 to Q.5: Multiple Choice Questions (MCQ), carry ONE mark each.

1. The people ________ were at the demonstration were from all sections of society.

(a) whose

(b) which

(c) who

(d) whom

2. 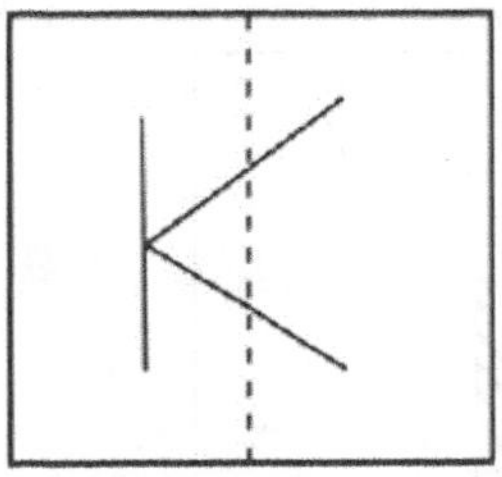

A transpart square sheet shown above is folded along the dotted line. The folded sheet will look like______.

(a) 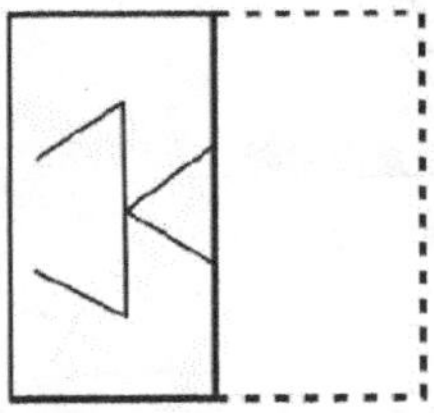

(b) 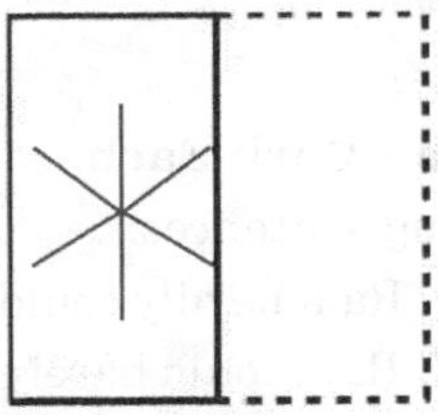

(c) 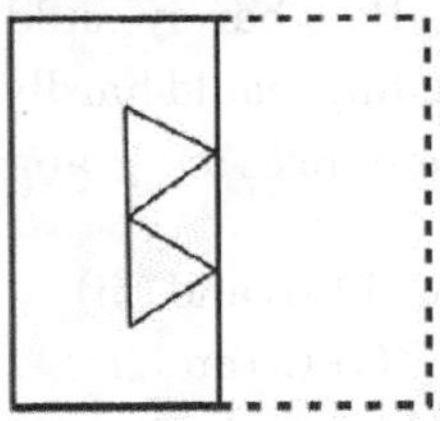

(d) 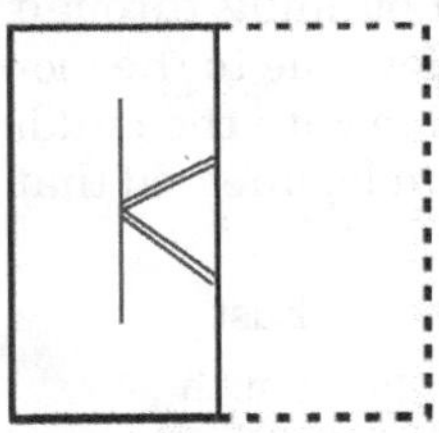

3. For a regular polygon having 10 sides, the interior angle between the sides of the polygon, in degrees, is :

(a) 396 (b) 324

(c) 216 (d) 144

4. Which one of the following numbers is exactly divisible by $(11^{13} + 1)$?

(a) $11^{26} + 1$ (b) $11^{33} + 1$

(c) $11^{39} - 1$ (d) $11^{52} - 1$

5. *Oasis* is to *sand* as *island* is to ________

Which one of the following options maintains a similar logical relation in the above sentence?

(a) Stone (b) Land

(c) Water (d) Mountain

Q. 6 to Q. 10 : Multiple Choice Question (MCQ), carry Two marks each.

6. The importance of sleep is often overlooked by students when they are preparing for exams. Research has consistently shown that sleep deprivation greatly reduces the ability to recall the material learnt. Hence, cutting down on sleep to study longer hours can be counter productive.

Which one of the following statements is the CORRECT inference from the above passage?

(a) Sleeping well alone is enough to prepare for an exam. Studying has lesser benefit.

(b) Students are efficient and are not wrong in thinking that sleep is a waste of time.

(c) If a student is extremely well prepared for an exam, he needs little or no sleep.

(d) To do well in an exam, adequate sleep must be part of the preparation.

7. 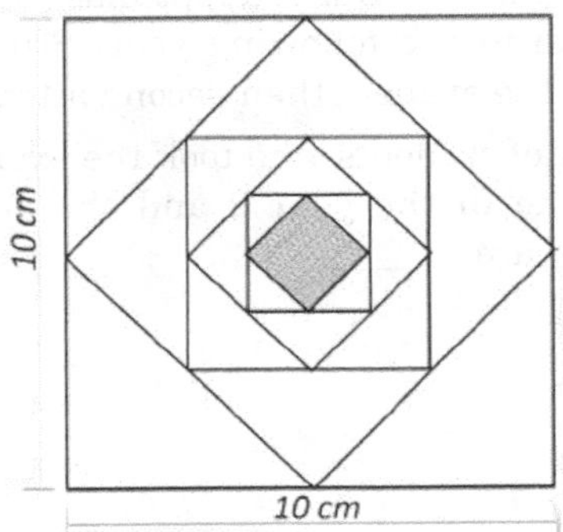

In the figure shown above, each inside square is formed by joining the midpoints of the sides of the next larger square. The area of the smallest square (shaded) as shown, in cm^2 is :

(a) 12.50

(b) 6.25

(c) 3.125

(d) 1.5625

8. Let X be a continuous random variable denoting the temperature measured. The range of temperature is [0, 100] degree Celsius and let the probability density function of X be $f(x) = 0.01$ for $0 \le X \le 100$.

The mean of X is ___________

(a) 2.5 (b) 5.0

(c) 25.0 (d) 50.0

9. 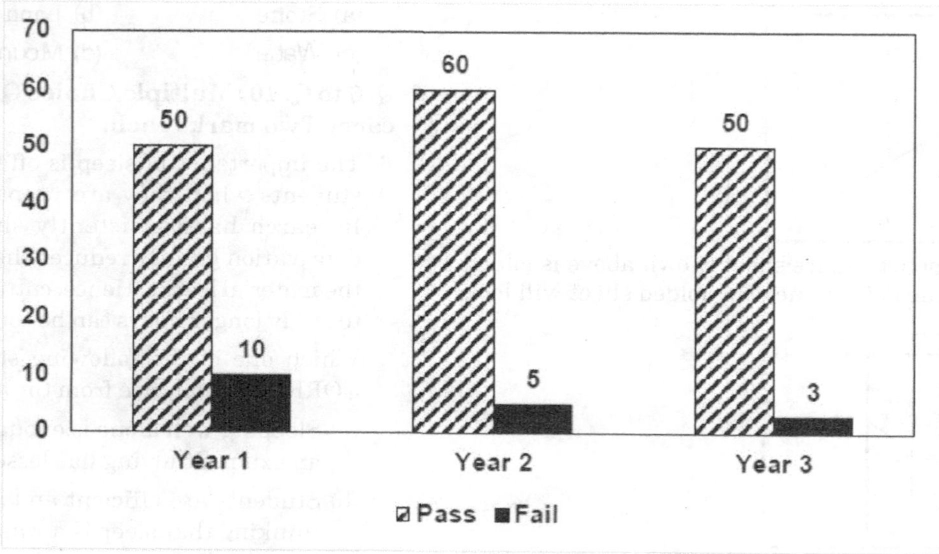

The number of students passing or failing in an exam for a particular subject are presented in the bar chart above. Students who pass the exam cannot appear for the exam again. Students who fail the exam in the first attempt must appear for the exam in the following year. Students always pass the exam in their second attempt.

The number of students who took the exam for the first time in the year 2 and the year 3 respectively, are ________

(a) 65 and 53

(b) 60 and 50

(c) 55 and 53

(d) 55 and 48

10. Seven cars P, Q, R, S, T, U and V are parked in a row not necessarily in that order. The cars T and U should be parked next to each other. The cars S and V also should be parked next to each other, whereas P and Q cannot be parked next to each other. Q and S must be parked next to each other. R is parked to the immediate right of V. T is parked to the left of U.

Based on the above statements, the only INCORRECT option given below is :

(a) There are two cars parked in between Q and V.

(b) Q and R are not parked together.

(c) V is the only car parked in between S and R.

(d) Car P is parked at the extreme end.

SET 5

Q. 1 to Q. 5 : Carry One Mark Each.

1. Consider the following sentences:

(i) After his surgery, Raja hardly could walk.

(ii) After his surgery, Raja could barely walk.

(iii) After his surgery, Raja barely could walk.

(iv) After his surgery, Raja could hardly walk.

Which of the above sentences are grammatically CORRECT?

(a) (i) and (ii) (b) (i) and (iii)

(c) (iii) and (iv) (d) (ii) and (iv)

2. Ms. X came out of a building through its front door to find her shadow due to the morning sun falling to her right side with the building to her back. From this, it can be inferred that building is facing ____________

(a) North (b) East

(c) West (d) South

3. In the above figure, O is the center of the circle and, M and N lie on the circle.

The area of the right triangle MON is 50 cm^2.

What is the area of the circle in cm^2 ?

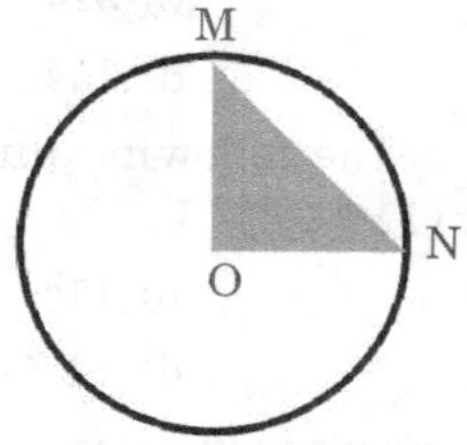

(a) 2π (b) 50π

(c) 75π (d) 100π

4. If $\begin{cases} "\oplus" \text{ means} "-", \\ "\otimes" \text{ means} "\div", \\ "\Delta" \text{ means} "+", \\ "\nabla" \text{ means} "\times", \end{cases}$

then, the value of the expression $\Delta\, 2 \oplus 3\, \Delta\, ((4 \otimes 2) \nabla 4) =$

(a) –1 (b) –0.5

(c) 6 (d) 7

5. "The increased consumption of leafy vegetables in the recent months is a clear indication that the people in the state have begun to lead a healthy lifestyle"

Which of the following can be logically inferred from the information presented in the above statement?

(a) The people in the state did not consume leafy vegetables earlier.

(b) Consumption of leafy vegetables may not be the only indicator of healthy lifestyle.

(c) Leading a healthy lifestyle is related to a diet with leafy vegetables.

(d) The people in the state have increased awareness of health hazards causing by consumption of junk foods.

Q. 6 to Q. 10: Multiple Choice Question (MCQ), Carry Two Marks Each

6. Oxpeckers and rhinos manifest a symbiotic relationship in the wild. The oxpeckers warn the rhinos about approaching poachers, thus possibly saving the lives of the rhinos. Oxpeckers also feed on the parasitic ticks found on rhinos.

In the symbiotic relationship described above, the primary benefits for oxpeckers and rhinos respectively are,

(a) Oxpeckers get a food source, rhinos have no benefit.

(b) Oxpeckers save their habitat from poachers while the rhinos have no benefit.

(c) Oxpeckers get a food source, rhinos may be saved from the poachers.

(d) Oxpeckers save the lives of poachers, rhinos save their own lives.

7.

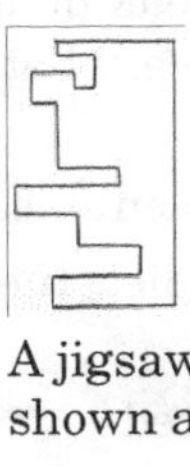

A jigsaw puzzle has 2 pieces. One of the pieces is shown above. Which one of the given options for the missing piece when assembled will form a rectangle? The piece can be moved, rotated or flipped to assemble with the above piece.

(a) 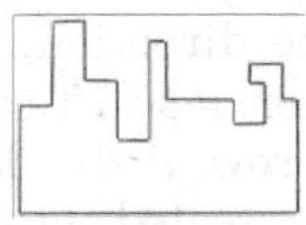 (b) 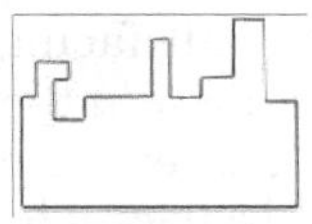

(c) 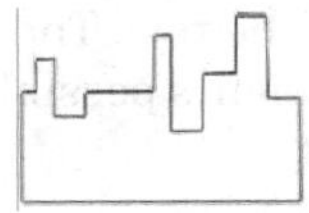 (d) 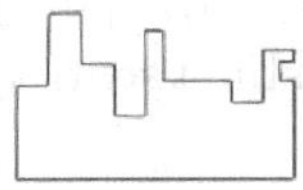

8. The number of hens, ducks and goats in farm P are 65, 91 and 169, respectively. The total number of hens, ducks and goats in a nearby farm Q is 416. The ratio of hens:ducks:goats in farm Q is 5:14:13. All the hens, ducks and goats are sent from farm Q to farm P.

The new ratio of hens:ducks:goats in farm P is

(a) 5:7:13 (b) 5:14:13

(c) 10:21:26 (d) 21:10:26

9.

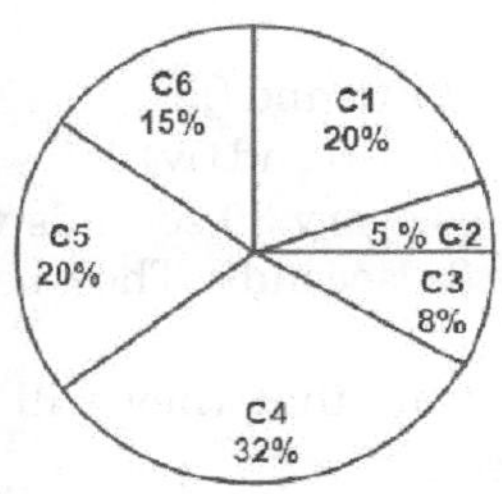

Company	Ratio
C1	3:2
C2	1:4
C3	5:3
C4	2:3
C5	9:1
C6	3:4

The distribution of employees at the rank of executives, across different companies C1, C2, ..., C6 is presented in the chart given above. The ratio of executives with a management degree to those without a management degree in each of these companies is provided in the table above. The total number of executives across all companies is 10,000.

The total number of management degree holders among the executives in companies C2 and C5 together is ______________.

(a) 225 (b) 600

(c) 1900 (d) 2500

10. Five persons P, Q, R, S and T are sitting in a row not necessarily in the same order. Q and R are separated by one person, and S should not be seated adjacent to Q.

The number of distinct seating arrangements possible is:

(a) 4 (b) 8

(c) 10 (d) 16

SET 6

Q. 1 to Q. 5 : Carry One Mark Each.

1. Five persons P, Q, R, S and T are to be seated in a row, all facing the same direction, but not necessarily in the same order. P and T cannot be seated at either end of the row. P should not be seated adjacent to S. R is to be seated at the second position from the left end of the row. The number of distinct seating arrangements possible is:

(*a*) 2 (*b*) 3
(*c*) 4 (*d*) 5

2. Consider the following sentences:

(i) The number of candidates who appear for the GATE examination is staggering.
(ii) A number of candidates from my class are appearing for the GATE examination.
(iii) The number of candidates who appear for the GATE examination are staggering.
(iv) A number of candidates from my class is appearing for the GATE examination.

Which of the above sentences are grammatically CORRECT?

(*a*) (i) and (ii) (*b*) (i) and (iii)
(*c*) (ii) and (iii) (*d*) (ii) and (iv)

3. A digital watch X beeps every 30 seconds while watch Y beeps every 32 seconds. They beeped together at 10 AM.

The immediate next time that they will beep together is _________

(*a*) 10.08 AM (*b*) 10.42 AM
(*c*) 11.00 AM (*d*) 10.00 PM

4. If $\oplus \div \odot = 2; \oplus \div \Delta = 3; \odot + \Delta = 5; \Delta \times \otimes = 10,$

Then, the value of $(\otimes - \oplus)^2$, is:

(*a*) 0 (*b*) 1
(*c*) 4 (*d*) 16

5. The front door of Mr. X's house faces East. Mr. X leaves the house, walking 50 m straight from the back door that is situated directly opposite to the front door. He then turns to his right, walks for another 50 m and stops. The direction of the point Mr. X is now located at with respect to the starting point is _________

(*a*) South-East (*b*) North-East
(*c*) West (*d*) North-West

Q. 6 to Q. 10: Multiple Choice Questions (MCQ), Carry Two Marks Each

6. Given below are two statements 1 and 2, and two conclusions I and II.

Statement 1: All entrepreneurs are wealthy.

Statement 2: All wealthy are risk seekers.

Conclusion I: All risk seekers are wealthy.

Conclusion II: Only some entrepreneurs are risk seekers.

Based on the above statements and conclusions, which one of the following options is CORRECT?

(*a*) Only conclusion I is correct
(*b*) Only conclusion II is correct
(*c*) Neither conclusion I nor II is correct
(*d*) Both conclusions I and II are correct

7. A box contains 15 blue balls and 45 black balls. If 2 balls are selected randomly, without replacement, the probability of an outcome in which the first selected is a blue ball and the second selected is a black ball, is _________

(*a*) $\frac{3}{16}$ (*b*) $\frac{45}{236}$
(*c*) $\frac{1}{4}$ (*d*) $\frac{3}{4}$

8.

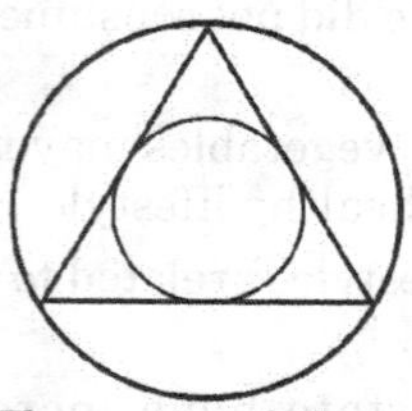

The ratio of the area of the inscribed circle to the area of the circumscribed circle of an equilateral triangle is _________

(*a*) $\frac{1}{8}$ (*b*) $\frac{1}{6}$
(*c*) $\frac{1}{4}$ (*d*) $\frac{1}{2}$

9. Consider a square sheet of side 1 unit. The sheet is first folded along the main diagonal. This is followed by a fold along its line of symmetry. The resulting folded shape is again folded along its line of symmetry. The area of each face of the final folded shape, in square units, equal to _____

(*a*) $\frac{1}{4}$ (*b*) $\frac{1}{8}$
(*c*) $\frac{1}{16}$ (*d*) $\frac{1}{32}$

10. The world is going through the worst pandemic in the past hundred years. The air travel industry is facing a crisis, as the resulting quarantine requirement for travelers led to weak demand. In relation to the first sentence above, what does the second sentence do?

(*a*) Restates an idea from the first sentence.
(*b*) Second sentence entirely contradicts the first sentence.
(*c*) The two statements are unrelated.
(*d*) States an effect of the first sentence.

SET 7

Q. 1 to Q. 5 : Carry One Mark Each.

1. The ratio of boys to girls in a class is 7 to 3.

Among the options below, an acceptable value for the total number of students in the class is :

(a) 21 (b) 37

(c) 50 (d) 73

2. A polygon is convex if, for every pair of points, P and Q belonging to the polygon, the line segment PQ lies completely inside or on the polygon.

Which one of the following is NOT a convex polygon ?

(a) 

(b) 

(c) 

(d) 

3. Consider the following sentences :

(i) Everybody in the class is prepared for the exam.

(ii) Babu invited Danish to his home because he enjoys playing chess.

Which of the following is the CORRECT observation about the above two sentences ?

(a) (i) is grammatically correct and (ii) is unambiguous

(b) (i) is grammatically incorrect and (ii) is unambiguous

(c) (i) is grammatically correct and (ii) is ambiguous

(d) (i) is grammatically incorrect and (ii) is ambiguous

4. 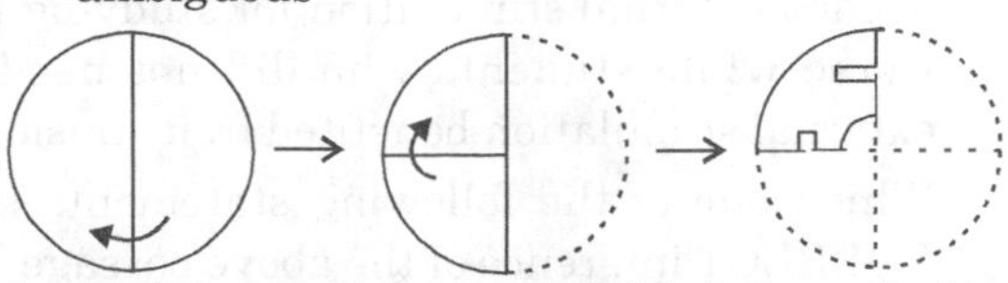

A circular sheet of paper is folded along the lines in the directions shown. The paper, after being punched in the final folded state as shown and unfolded in the reverse order of folding, will look like ____.

(a) 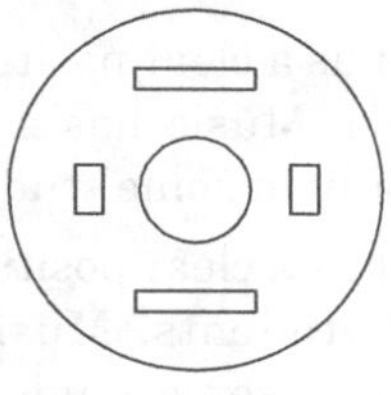

(b) 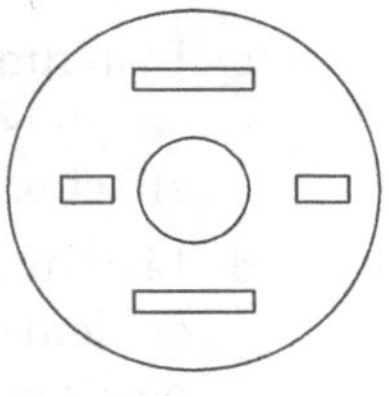

(c) 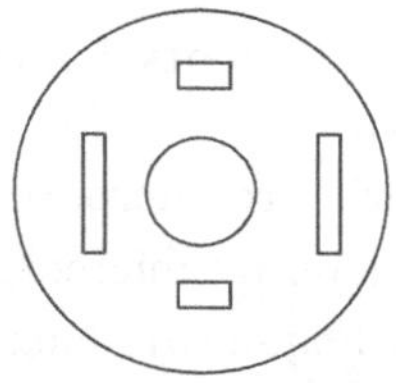

(d) 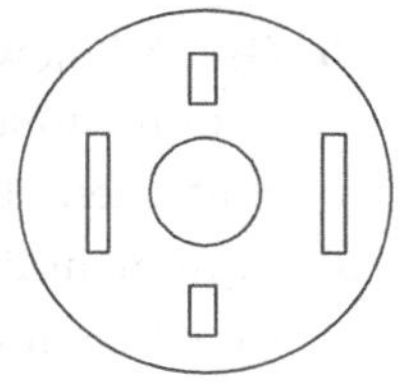

5. ____ is to *surgery* as *writer* is to ______.

Which one of the following options maintains a similar logical relation in the above sentence ?

(a) Plan, outline (b) Hospital, library

(c) Doctor, book (d) Medicine, grammar

Q. 6 to Q. 10 : Carry Two Marks Each

6. We have 2 rectangular sheets of paper, M and N, of dimensions 6 cm × 1 cm each. Sheet M is rolled to form an open cylinder by bringing the short edges of the sheet together. Sheet N is cut into equal square patches an assembled to form the largest possible closed cube. Assuming the ends of the cylinder are closed, the ratio of the volume of the cylinder to that of the cube is ______.

(a) $\frac{\pi}{2}$ (b) $\frac{3}{\pi}$

(c) $\frac{9}{\pi}$ (d) 3π

7.

Items	Cost (₹)	Profit %	Marked Price (₹)
P	5,400	--	5,860
Q	---	25	10,000

Details of prices of two items P and Q are presented in the above table. The ratio of cost of item P to cost of item Q is 3:4. Discount is calculated as the difference between the marked price and the selling price. The profit percentage is calculated as the ratio of the difference between selling price and cost, to the cost

$$\left(\text{Profit } \% = \frac{\text{Selling price} - \text{Cost}}{\text{Cost}} \times 100\right).$$

The discount on item Q, as a percentage of its marked price, is _____.

(a) 25 (b) 12.5

(c) 10 (d) 5

8. There are five bags each containing identical sets of ten distinct chocolates. One chocolate is picked form each bag.

The probability that at least two chocolates are identical is ________.

(a) 0.3024 (b) 0.4235

(c) 0.6976 (d) 0.8125

9. Given below are two statements 1 and 2, and two conclusions I and II.

Statement 1 : All bacteria are microorganisms.

Statement 2 : All pathogens are microorganisms.

Conclusion I : Some pathogens are bacteria.

Conclusion II : All pathogens are not bacteria.

Based on the above statements and conclusions, which one of the following option is logically CORRECT ?

(a) Only conclusion I is correct

(b) Only conclusion II is correct

(c) Either conclusion I or II is correct.

(d) Neither conclusion I nor II is correct.

10. Some people suggest anti-obesity measures (AOM) such as displaying calorie information in restaurant menus. Such measures sidestep addressing the core problems that cause obesity: poverty and income inequality.

Which one of the following statements summarizes the passage ?

(a) The proposed AOM addresses the core problems that cause obesity.

(b) If obesity reduces, poverty will naturally reduce, sine obesity causes poverty.

(c) AOM are addressing the core problems and are likely to succeed.

(d) AOM are addressing the problem superficially.

SET 8

Q. 1 to Q. 5 : Carry One Mark Each.

1. Gauri said that she can play the keyboard ____ her sister.

(a) as well as (b) as better as

(c) as nicest as (d) as worse as

2.

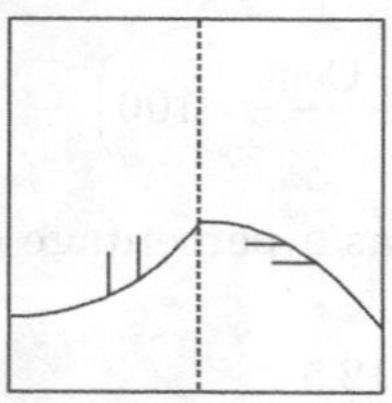

As transparent square sheet shown above is folded along the dotted line. The folded sheet will look like ____.

(a) 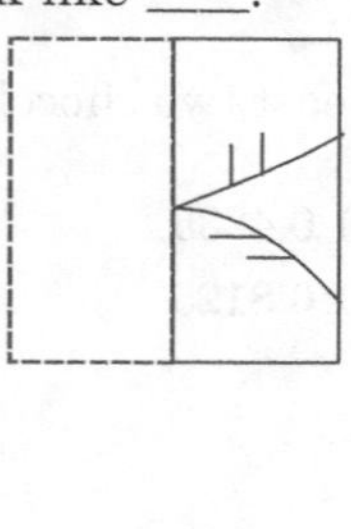 (b) 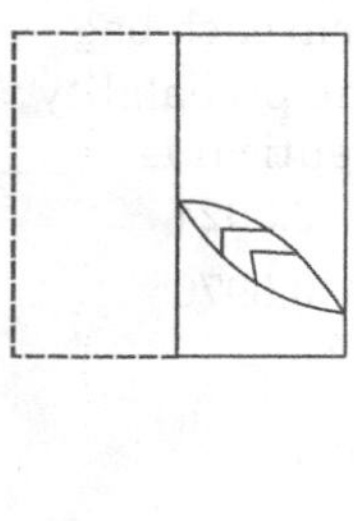

(c) 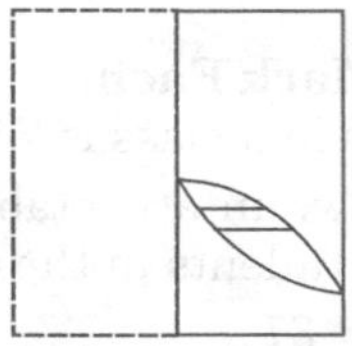 (d) 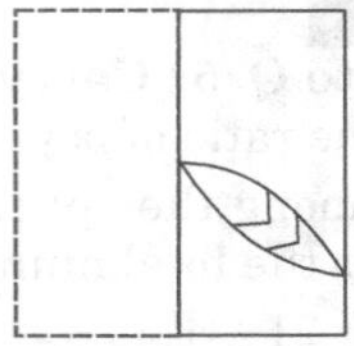

3. If θ is the angle, in degrees, between the longest diagonal of the cube and any one of the edges of the cube, then, cos θ =

(a) $\frac{1}{2}$ (b) $\frac{1}{\sqrt{3}}$

(c) $\frac{1}{\sqrt{2}}$ (d) $\frac{\sqrt{3}}{2}$

4. If $\left(x-\frac{1}{2}\right)^2-\left(x-\frac{3}{2}\right)^2 = x+2$, then the value of x is :

(a) 2 (b) 4

(c) 6 (d) 8

5. Pen : Write :: Knife : ____.

Which one of the following options maintains a similar logical relation in the above ?

(a) Vegetables

(b) Sharp

(c) Cut

(d) Blunt

Q. 6 to Q. 10 : Carry Two Marks Each.

6. Listening to music during exercise improve exercise performance and reduces discomfort. Scientists researched whether listening to music while studying can help students learn better and the results were inconclusive. Students who needed external stimulation for studying fared worse while students who did not need any external stimulation benefited from music.

Which one of the following statements is the CORRECT inference of the above passage ?

(a) Listening to music has no effect on learning and a positive effect on physical exercise.

(b) Listening to music has a clear positive effect both on physical exercise and on learning.

(c) Listening to music has a clear positive effect on physical exercise. Music has a positive effect on learning only in some students.

(d) Listening to music has a clear positive effect on learning in all students. Music has a positive effect only in some students who exercise.

7. 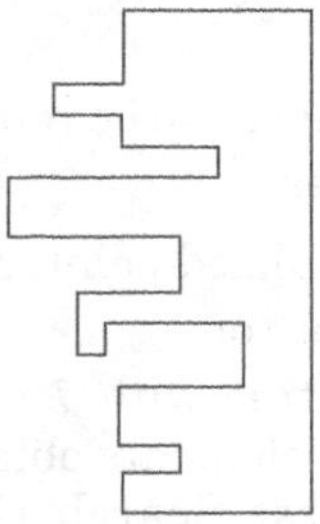

A jigsaw puzzle has 2 pieces. One of the pieces is shown above. Which one of the given options for the missing piece when assembled will form a rectangle ? The piece can be moved, rotated or flipped to assemble with the above piece.

(a) 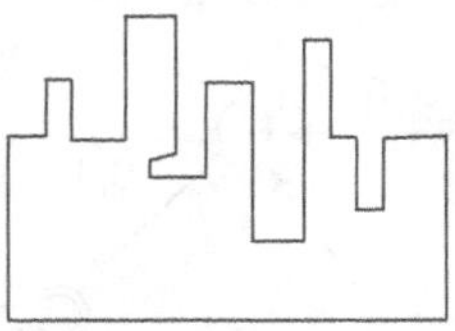

(b) 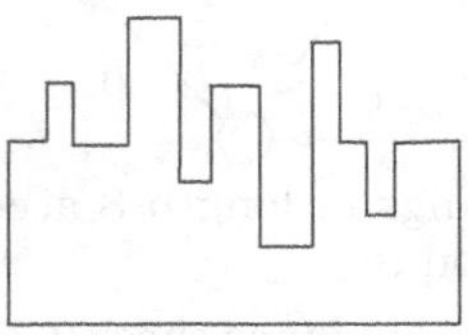

(c) 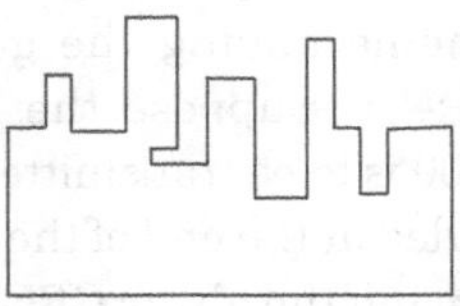

(d) 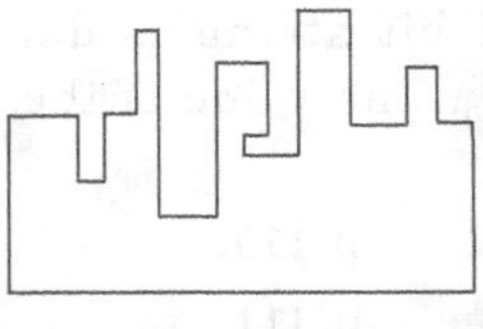

8. The number of students in three classes is in the ratio 3 : 13 : 6. If 18 students are added to each class, the ratio changes to 15 : 35 : 21.

The total number of students in all the three classes in the beginning was :

(a) 22 (b) 66

(c) 88 (d) 110

9. 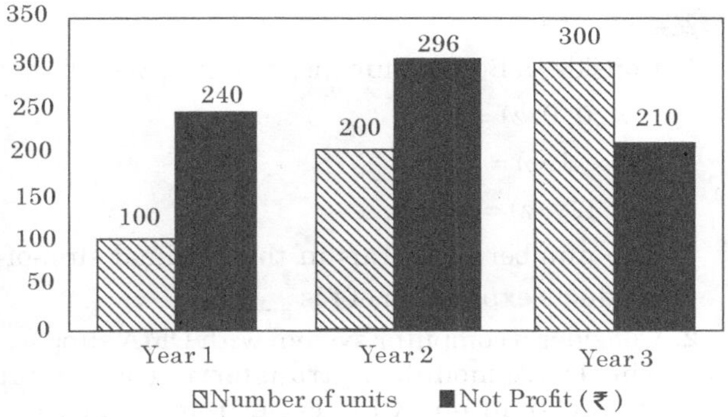

The number of units of a product sold in three different years and the respective net profits are presented in the figure above. The cost/unit in Year 3 was ´1, which was half the cost/unit in Year 2. The cost/unit in Year 3 was one-third of the cost/unit in Year 1. Taxes were paid on the selling price at 10%, 13% and 15% respectively for the three years. Net profit is calculated as the difference between the selling price and the sum of cost and taxes paid in that year.

The ratio of the selling price in Year 2 to the selling price in Year 3 is ____.

(a) 4 : 3 (b) 1 : 1

(c) 3 : 4 (d) 1 : 2

10. Six students P, Q, R, S, T and U, with distinct heights, compare their heights and make the following observations.

Observation I: S is taller than R.

Observation II: Q is the shortest of all.

Observation III: U is taller than only one student.

Observation IV: T is taller than S but is not the tallest.

The number of students that are taller than R is the same as the number of students shorter than ____.

(a) T

(b) R

(c) S

(d) P

ENGINEERING MATHEMATICS

SET-1

1. Consider a Boolean function f(w, x, y, z) such that

   f (w, 0, 0, z) = 1

   f(1, x, 1, z) = x + z

   f(w, 1, y, z) = wz + y

   The number of literals in the minimal sum-of-products expression of f is ____.

2. Consider a computer system with DMA support. The DMA module is transferring one 8-bit character in one CPU cycle from a device to memory through cycle stealing at regular intervals. Consider a 2 MHz processor. If 0.5% processor cycles are used for DMA, the data transfer rate of the device is ____ bits per second.

3. If x and y are two decimal digits and $(0.1101)_2 = (0.8\,xy5)_{10}$, the decimal value of x + y is ____.

4. Consider the following ANSI C program :

```
#include <stdio.h>
#include <stdlib.h>
struct Node{
        int value;
        struct Node *next;};
int main () {
    struct Node *boxE, *head, *boxN; int index
    = 0;
    boxE = head = (struct Node *) malloc (sizeof
    (struct Node)) ;
    head->value = index ;
    for (index = 1; index <= 3; index++) {
        boxN = (struct Node *) malloc (sizeof
        (struct Node)) ;
        boxE–>next = boxN;
        boxN–>value = index;
        boxE = box; N; }
    for (index = 0; index <= 3; index++) {
        printf ("Value at index %d is %d\n", index,
        head–> value) ;
        head = head–>next ;
        printf ("Value at index %d is %d\n", index
        + 1, head–>value) ; } }
```

   Which one of the statements below is correct about the program ?

   (a) Upon execution, the program creates a linked-list of five nodes.

   (b) Upon execution, the program goes into an infinite loop.

   (c) It has a missing return which will be reported as an error by the compiler.

   (d) It dereferences an uninitialized pointer that may result in a run-time error.

5. Consider a complete binary tree with 7 nodes. Let A denote the set of first 3 elements obtained by performing Breadth-First Search (BFS) starting from the root. Let B denote the set of first 3 elements obtained by performing Depth-First Search (DFS) starting from the root.

   The Value of |A – B| is ____.

6. Consider the following deterministic finite automaton (DFA).

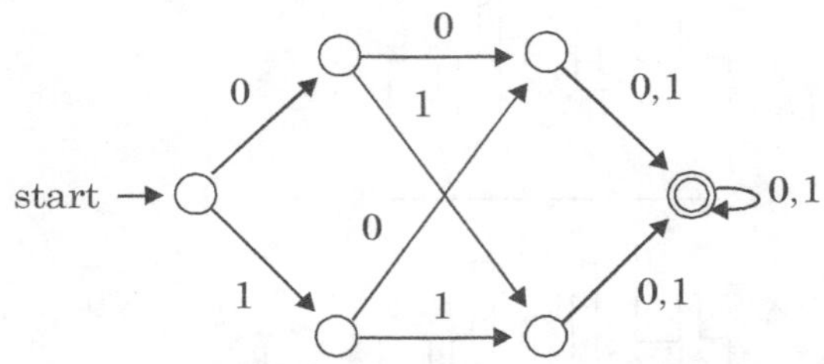

   The number of strings of length 8 accepted by the above automaton is ____.

7. Consider the cyclic redundancy check (CRC) based error detecting scheme having the generator polynomial $X^3 + X + 1$. Suppose the message $m_4m_3m_2m_1m_0$ = 11000 is to be transmitted. Check bits $c_2c_1c_0$ are appended at the end of the message by the transmitter using the above CRC scheme. The transmitted bit string is denoted by $m_4m_3m_2m_1m_0c_2c_1c_0$. The value of the checkbit sequence $c_2c_1c_0$ is

   (a) 101 (b) 110

   (c) 100 (d) 111

SET-2

1. Consider the following representation of a number in IEEE 754 single-precision floating point format with a bias of 127.

   S : 1 E : 10000001 F : 11110000000000000000000

   Here S, E and F denote the sign, exponent and fraction components of the floating point representation.

   The decimal value corresponding to the above representation (rounded to 2 decimal places) is ________ .

2. Three processes arrive at time zero with CPU bursts of 16, 20 and 10 milliseconds. If the scheduler has prior knowledge about the length

of the CPU bursts, the minimum achievable average waiting time for these three processes in a non-preemptive scheduler (rounded to nearest integer) is __________ milliseconds.

3. Consider a computer system with a byte-addressable primary memory of size 2^{32} bytes. Assume the computer system has a direct-mapped cache of size 32 KB (1 KB = 2^{10} bytes), and each cache block is of size 64 bytes.

The size of the tag field is ________ bits.

4. Consider the relation R(P, Q, S, T, X, Y, Z, W) with the following functional dependencies.

$PQ \rightarrow X; P \rightarrow YX; Q \rightarrow Y; Y \rightarrow ZW$

Consider the decomposition of the relation R in to the constituent relations according to the following two decomposition schemes.

D_1 : R = [(P, Q, S, T); (P, T, X); (Q, Y); (Y, Z, W)]

D_2 : R = [(P, Q, S); (T, X); (Q, Y); (Y, Z, W)]

Which one of the following options is correct ?

(a) D_1 is a lossless decomposition, but D_2 is a lossy decomposition.

(b) D_1 is a lossy decomposition, but D_2 is a lossless decomposition.

(c) Both D_1 and D_2 are lossless decompositions.

(d) Both D_1 and D_2 are lossy decompositions.

5. In the context of operating systems, which of the following statements is/are correct with respect to paging ?

(a) Paging helps solve the issue of external fragmentation.

(b) Page size has no impact on internal fragmentation.

(c) Paging incurs memory overheads.

(d) Multi-level paging is necessary to support pages of different sizes.

6. Consider a linear list based directory implementation in a file system. Each directory is a list of nodes, where each node contains the file name along with the file metadata, such as the list of pointers to the data blocks. Consider a given directory foo.

Which of the following operations will necessarily require a full scan of foo for successful completion ?

(a) Creation of a new file in foo

(b) Deletion of an existing file from foo

(c) Renaming of an existing file in foo

(d) Opening of an existing file in foo

7. Consider the following C code segment :

```
a = b + c;
e = a + 1;
d = b + c;
f = d + 1;
g = e + f;
```

In a compiler, this code segment is represented internally as a directed acyclic graph (DAG). The number of nodes in the DAG is ________ .

SET-3

1. If the vectors (1. 0, –1. 0, 2. 0), (7. 0, 3. 0, x) and (2. 0, 3. 0, 1. 0) in $\mathbb{R}^3$ are linearly dependent, the value of x is __________

2. A real 2 × 2 non-singular matrix A with repeated eigenvalue is given as

$$A = \begin{bmatrix} x & -3.0 \\ 3.0 & 4.0 \end{bmatrix}$$

where x is a real positive number. The value of x (*rounded off to one decimal place*) is __________

3. The vector function $F(r) = -x\hat{i} + y\hat{j}$ is defined over a circular arc C shown in the figure.

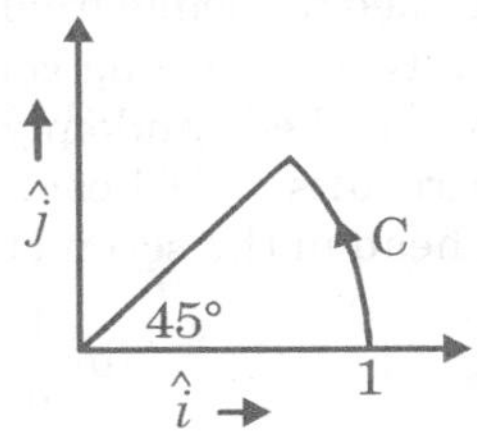

The line integral of $\int_C F(r).\,dr$ is

(a) $\frac{1}{2}$ (b) $\frac{1}{4}$

(c) $\frac{1}{6}$ (d) $\frac{1}{3}$

4. Consider the differential equation given below.

$$\frac{dy}{dx} + \frac{x}{1-x^2}y = x\sqrt{y}$$

The integrating factor of the differential equation is

(a) $(1-x^2)^{-\frac{3}{4}}$ (b) $(1-x^2)^{-\frac{1}{4}}$

(c) $(1-x^2)^{-\frac{3}{2}}$ (d) $(1-x^2)^{-\frac{1}{2}}$

5. Consider the vector field $F = a_x(4y - c_1 z) + a_y(4x + 2z) + a_z(2y + z)$ in a rectangular coordinate system (x, y, z) with unit vectors a_x, a_y, and a_z. If the field F is irrotational (conservative), then the constant c_1 (in integer) is _______

6. Consider the integral

$$\oint_C \frac{\sin(x)}{x^2(x^2+4)}dx$$

where C is a counter-clockwise oriented circle defined as $|x - i| = 2$. The value of the integral is

(a) $-\frac{\pi}{8}\sin(2i)$ (b) $\frac{\pi}{8}\sin(2i)$

(c) $-\frac{\pi}{4}\sin(2i)$ (d) $\frac{\pi}{4}\sin(2i)$

7. Two continuous random variables X and Y are related as

Y = 2X + 3

Let σ_X^2 and σ_Y^2 denote the variances of X and Y, respectively. The variances are related as

(a) $\sigma_Y^2 = 2\sigma_X^2$ (b) $\sigma_Y^2 = 4\sigma_X^2$

(c) $\sigma_Y^2 = 5\sigma_X^2$ (d) $\sigma_Y^2 = 25\sigma_X^2$

8. A box contains the following three coins.

I. A fair coin with head on one face and tail on the other face.

II. A coin with heads on both the faces.

III. A coin with tails on both the faces.

A coin is picked randomly from the box and tossed. Out of the two remaining coins in the box, one coin is then picked randomly and tossed. If the first toss results in a head, the probability of getting a head in the second toss is

(a) $\frac{2}{5}$ (b) $\frac{1}{3}$

(c) $\frac{1}{2}$ (d) $\frac{2}{3}$

SET-4

1. Consider a vector p in 2-dimensional space. Let its direction (counter-clockwise angle with the positive x-axis) be θ. Let p be an eigenvector of a 2 × 2 matrix A with corresponding eigenvalue λ, λ > 0. If we denote the magnitude of a vector v by $||v||$, identify the VALID statement regarding p', where $p' = Ap$.

(a) Direction of $p' = \lambda\theta$, $||p'|| = ||p||$

(b) Direction of $p' = \theta$, $||p'|| = \lambda||p||$

(c) Direction of $p' = \lambda\theta$, $||p'|| = \lambda||p||$

(d) Direction of $p' = \theta$, $||p'|| = ||p||$

2. The value of $\lim_{x\to 0}\left(\frac{1-\cos x}{x^2}\right)$ is

(a) $\frac{1}{4}$ (b) $\frac{1}{3}$

(c) $\frac{1}{2}$ (d) 1

3. The Dirac-delta function $(\delta(t - t_0))$ for $t, t_0 \in \mathbb{R}$, has the following property

$$\int_a^b \phi(t)\delta(t-t_0)\,dt = \begin{cases}\phi(t_0) & a < t_0 < b\\ 0 & \text{otherwise}\end{cases}$$

The Laplace transform of the Dirac-delta function $\delta(t - a)$ for $a > 0$;

$L(\delta(t - a)) = F(s)$ is

(a) 0 (b) ∞

(c) e^{sa} (d) e^{-sa}

4. Let $f(x) = x^2 - 2x + 2$ be a continuous function defined on $x \in [1,3]$. The point x at which the tangent of f(x) becomes parallel to the straight line joining f(1) and f(3) is

(a) 0 (b) 1

(c) 2 (d) 3

5. If $y(x)$ satisfies the differential equation

$$(\sin x)\frac{dy}{dx} + y\cos x = 1,$$

subject to the condition $y(\pi/2) = \pi/2$, then $y(\pi/6)$ is

(a) 0 (b) $\frac{\pi}{6}$

(c) $\frac{\pi}{3}$ (d) $\frac{\pi}{2}$

6. Consider a binomial random variable X. If $X_1, X_2, \ldots, X_n$ are independent and identically distributed samples from the distribution of X with sum $Y = \sum_{i=1}^{n} X_i$, then the distribution of Y as $n \to \infty$ can be approximated as

(a) Exponential (b) Bernoulli

(c) Binomial (d) Normal

7. The ordinary differential equation $\frac{dy}{dt} = -\pi y$ subject to an initial condition $y(0) = 1$ is solved numerically using the following scheme:

$$\frac{y(t_{n+1}) - y(t_n)}{h} = -\pi y(t_n)$$

where h is the time step, $t_n = nh$, and $n = 0,1,2,\ldots$. This numerical scheme is stable for all values of h in the interval__________.

(a) $0 < h < \frac{2}{\pi}$ (b) $0 < h < 1$

(c) $0 < h < \frac{\pi}{2}$ (d) for all $h > 0$

8. Let C represent the unit circle centered at origin in the complex plane, and complex variable, $z = x + iy$. The value of the contour integral $\int_C \frac{\cosh 3z}{2z} dz$ (where integration is taken counter clockwise) is

(a) 0 (b) 2

(c) πi (d) $2\pi i$

SET-5

1. Consider an $n \times n$ matrix A and a non-zero $n \times l$ vector p. Their product $Ap = \alpha^2 p$, where $\alpha \in R$ and $\alpha \not\in \{-1,0,1\}$. Based on the given information, the eigen value of A^2 is:

(a) α (b) α^2

(c) $\sqrt{\alpha}$ (d) α^4

2. Let the superscript T represent the transpose operation. Consider the function $f(X) = \frac{1}{2}x^T Qx - r^T x$, where x and r are $n \times 1$ vectors and Q is a symmetric $n \times n$ matrix. The stationary point of f(x) is

(a) $Q^T r$ (b) $Q^{-1} r$

(c) $\frac{r}{r^T r}$ (d) r

3. If the Laplace transform of a function f(t) is given by $\frac{s+3}{(s+1)(s+2)}$, then f(0) is

(a) 0 (b) $\frac{1}{2}$

(c) 1 (d) $\frac{3}{2}$

4. The value of $\int_0^{\frac{\pi}{2}} \int_0^{\cos\theta} r \sin\theta \, dr \, d\theta$ is

(a) 0 (b) $\frac{1}{6}$

(c) $\frac{4}{3}$ (d) π

5. Consider the following differential equation

$$(1+y)\frac{dy}{dx} = y$$

The solution of the equation that satisfies the condition y(1) = 1 is

(a) $2ye^y = e^x + e$ (b) $y^2 e^y = e^x$

(c) $ye^y = e^x$ (d) $(1+y)e^y = 2e^x$

6. The mean and variance, respectively, of a binomial distribution for n independent trials with the probability of success as p, are

(a) $\sqrt{np}$, $np(1 - 2p)$ (b) $\sqrt{np}, \sqrt{np(1-p)}$

(c) np, np (d) $np, np(1-p)$

7. Value of $\int_4^{5.2} \ln x \, dx$ using Simpson's one-third rule with interval size 0.3 is

(a) 1.83 (b) 1.60

(c) 1.51 (d) 1.06

8. Find the positive real root of $x^3 - x - 3 = 0$ using Newton-Raphson method. If the starting guess (x_0) is 2, the numerical value of the root after two iterations (x_2) is__________ *(round off to two decimal places).*

9. Value of $(1+i)^8$, where $i = \sqrt{-1}$, is equal to

(a) 4 (b) 16

(c) $4i$ (d) $16i$

SET-6

1. Let p and q be real numbers such that $p^2 + q^2 = 1$. The eigen values of the matrix $\begin{bmatrix} p & q \\ q & -p \end{bmatrix}$ are

(a) 1 and 1 (b) 1 and – 1

(c) j and –j (d) pq and – pq

2. Let $p(z) = z^3 + (1+j)z^2 + (2+j)z + 3$, where z is a complex number. Which one of the following is true?

(a) conjugate {p(z)} = p (conjugate {z}) for all z

(b) The sum of the roots of p(z) = 0 is a real number

(c) The complex roots of the equation p(z) = 0 come in conjugate pairs

(d) All the roots cannot be real

3. Let f(x) be a real-valued function such that $f'(x_0) = 0$ for some $x_0 \in (0, 1)$, and $f''(x) > 0$ for all $x \in (0, 1)$. Then f(x) has

(a) no local minimum in (0, 1)

(b) one local maximum in (0, 1)

(c) exactly one local minimum in (0, 1)

(d) two distinct local minima in (0, 1)

4. Suppose the circles $x^2 + y^2 = 1$ and $(x-1)^2 + (y-1)^2 = r^2$ intersect each other orthogonally at the point (u, v). Then u + v = _________.

5. In the open interval (0, 1), the polynomial $p(x) = x^4 - 4x^3 + 2$ has

(a) two real roots (b) one real root

(c) three real roots (d) no real roots

6. Suppose the probability that a coin toss shows "head" is p, where 0 < p < 1. The coin is tossed repeatedly until the first "head" appears. The expected number of tosses required is

(a) p/(1 – p) (b) (1 – p)/p

(c) 1/p (d) $1/p^2$

7. Let (–1 – j), (3 – j), (3 + j) and (– 1 + j) be the vertices of a rectangle C in the complex plane. Assuming that C is traversed in counter-clockwise direction, the value of the contour integral

$\oint_C \frac{dz}{z^2(z-4)}$ is

(a) $j\pi/2$ (b) 0
(c) $-j\pi/8$ (d) $j\pi/16$

8. The causal signal with z-transform $z^2(z-a)^{-2}$ is (u[n] is the unit step signal)

(a) $a^{2n}u[n]$ (b) $(n+1)a^n u[n]$
(c) $n^{-1}a^n u[n]$ (d) $n^2 a^n u[n]$

9. Let f(t) be an even function, i.e., f(–t) = f(t) for all t. Let the Fourier transform of f(t) be defined as

$F(\omega) = \int_{-\infty}^{\infty} f(t)e^{-j\omega t}dt$. Suppose $\frac{dF(\omega)}{d\omega} = -\omega F(\omega)$ for all ω, and F(0) = 1. Then

(a) f(0) < 1 (b) f(0) > 1
(c) f(0) = 1 (d) f(0) = 0

10. Let A be a 10 × 10 matrix such that A^5 is a null matrix, and let I be the 10 × 10 identity matrix. The determinant of A + I is ________.

11. Consider a continuous-time signal x(t) defined by x(t) = 0 for |t| > 1, and x(t) = 1 – |t| for |t| ≤ 1. Let the Fourier transform of x(t) be defined as

$X(\omega) = \int_{-\infty}^{\infty} x(t)e^{-j\omega t}dt$. The maximum magnitude of X(ω) is _____.

SET-7

1. The rank of matrix $\begin{bmatrix} 1 & 2 & 2 & 3 \\ 3 & 4 & 2 & 5 \\ 5 & 6 & 2 & 7 \\ 7 & 8 & 2 & 9 \end{bmatrix}$ is

(a) 1 (b) 2
(c) 3 (d) 4

2. If $P = \begin{bmatrix} 1 & 2 \\ 3 & 4 \end{bmatrix}$ and $Q = \begin{bmatrix} 0 & 1 \\ 1 & 0 \end{bmatrix}$ then Q^TP^T is

(a) $\begin{bmatrix} 1 & 2 \\ 3 & 4 \end{bmatrix}$ (b) $\begin{bmatrix} 1 & 3 \\ 2 & 4 \end{bmatrix}$

(c) $\begin{bmatrix} 2 & 1 \\ 4 & 3 \end{bmatrix}$ (d) $\begin{bmatrix} 2 & 4 \\ 1 & 3 \end{bmatrix}$

3. The shape of the cumulative distribution function of Gaussian distribution is

(a) Horizontal line
(b) Straight line at 45 degree angle
(c) Bell-shaped
(d) S-shaped

4. Consider the limit:

$\lim_{x \to 1}\left(\frac{1}{\ln x} - \frac{1}{x-1}\right)$

The limit (*correct up to one decimal place*) is______

5. The volume determined from ∭v 8 xyz dV for V = [2, 3] × [1, 2] × [0,1] will be (*in integer*)______

6. The solution of the second-order differential equation $\frac{d^2y}{dx^2} + 2\frac{dy}{dx} + y = 0$ with boundary conditions y(0) = 1 and y(1) = 3 is

(a) $e^{-x} + (3e-1)xe^{-x}$
(b) $e^{-x} - (3e-1)xe^{-x}$
(c) $e^{-x} + \left[3e\sin\left(\frac{\pi x}{2}\right) - 1\right]xe^{-x}$
(d) $e^{-x} - \left[3e\sin\left(\frac{\pi x}{2}\right) - 1\right]xe^{-x}$

7. The value of $\int_0^1 e^x dx$ using the trapezoidal rule with four equal subintervals is

(a) 1.718 (b) 1.727
(c) 2.192 (d) 2.718

8. The values of abscissa (x) and ordinate (y) of a curve are as follows:

X	y
2.0	5.00
2.5	7.25
3.0	10.00
3.5	13.25
4.0	17.00

By Simpson's 1/3rd rule, the area under the curve (*round off to two decimal places*) is______

SET-8

1. The value of $\lim_{x \to \infty} \frac{x\ln(x)}{1+x^2}$ is

(*a*) 0 (*b*) 1.0
(*c*) 0.5 (*d*) ∞

2. The rank of the matrix $\begin{bmatrix} 5 & 0 & -5 & 0 \\ 0 & 2 & 0 & 1 \\ -5 & 0 & 5 & 0 \\ 0 & 1 & 0 & 2 \end{bmatrix}$ is

(*a*) 1 (*b*) 2

(*c*) 3 (*d*) 4

3. The unit normal vector to the surface $X^2 + Y^2 + Z^2 - 48 = 0$ at the point (4, 4, 4) is

(*a*) $\frac{1}{\sqrt{2}}, \frac{1}{\sqrt{2}}, \frac{1}{\sqrt{2}}$ (*b*) $\frac{1}{\sqrt{3}}, \frac{1}{\sqrt{3}}, \frac{1}{\sqrt{3}}$

(*c*) $\frac{2}{\sqrt{2}}, \frac{2}{\sqrt{2}}, \frac{2}{\sqrt{2}}$ (*d*) $\frac{1}{\sqrt{5}}, \frac{1}{\sqrt{5}}, \frac{1}{\sqrt{5}}$

4. If A is a square matrix then orthogonality property mandates

(*a*) $AA^T = I$

(*b*) $AA^T = 0$

(*c*) $AA^T = A^{-1}$

(*d*) $AA^T = A^2$

5. The value (round off to one decimal place) of $\int_{-1}^{1} xe^{|x|}dx$ is ________

6. If k is a constant, the general solution of $\frac{dy}{dx} - \frac{y}{x} = 1$ will be in the form of

(*a*) $y = x \ln(kx)$ (*b*) $y = k \ln(kx)$

(*c*) $y = x \ln(x)$ (*d*) $y = xk \ln(k)$

7. The smallest eigenvalue and the corresponding eigenvector of the matrix $\begin{bmatrix} 2 & -2 \\ -1 & 6 \end{bmatrix}$, respectively, are

(*a*) 1.55 and $\begin{Bmatrix} 2.00 \\ 0.45 \end{Bmatrix}$ (*b*) 2.00 and $\begin{Bmatrix} 1.00 \\ 1.00 \end{Bmatrix}$

(*c*) 1.55 and $\begin{Bmatrix} -2.55 \\ -0.45 \end{Bmatrix}$ (*d*) 1.55 and $\begin{Bmatrix} 2.00 \\ -0.45 \end{Bmatrix}$

8. A function is defined in Cartesian coordinate system as $f(x, y) = xe^y$. The value of the directional derivative of the function (in integer) at the point (2, 0) along the direction of the straight line segment from point (2, 0) to point $\left(\frac{1}{2}, 2\right)$ is ________

9. Numerically integrate, $f(x) = 10x - 20x^2$ from lower limit a = 0 to upper limit b = 0.5. Use Trapezoidal rule with five equal subdivisions. The value (in units, round off to two decimal places) obtained is ____________

ANSWERS AND EXPLANATIONS

GENERAL APTITUDE

SET-1

ANSWERS

1. (d)	**2.** (b)	**3.** (c)	**4.** (b)	**5.** (c)
6. (b)	**7.** (a)	**8.** (a)	**9.** (b)	**10.** (c)

EXPLANATIONS

SET 1

1. When the comparison is between two things we use the second degree of the adjective the degree from of easy are : easy $\rightarrow$ easier $\rightarrow$ easiest.

Hence, comparative degree is easier.

2. The mirror image of TRIANGLE about the X-axis is

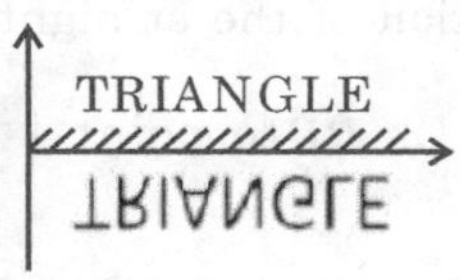

$\because$ Top to Bottom will interchange (or) symmetrical about X-axis.

3.

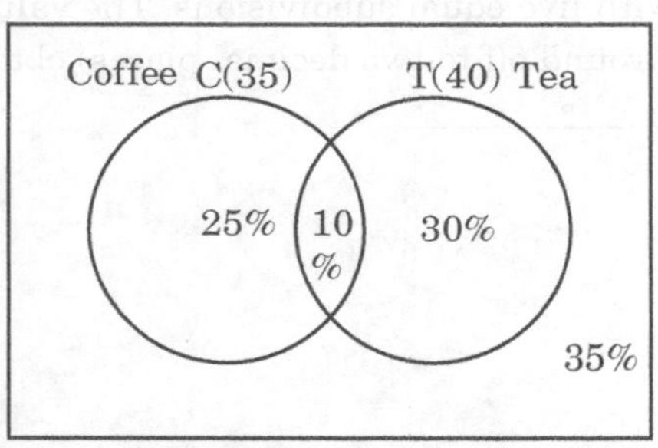

C $\rightarrow$ Coffee $\rightarrow$ Percentage of employees drink Coffee = 35%

T $\rightarrow$ Tea $\rightarrow$ Percentage of employees drink Tea = 40%

$\rightarrow$ Percentage of employees drink both Tea and Coffee = 10%

$\rightarrow$ Percentage of employees drink neither Tea and Coffee = 100% – n(TUE)

= 100% – (25% + 10% + 30%)

= 35%

4. As given that,

$$\left[\because p \oplus q = \frac{p^2+q^2}{pq} \text{ and } p \odot q = \frac{p^2}{q}\right]$$

Now, L.H.S $\Rightarrow$ $x \oplus y = \dfrac{x^2+y^2}{x.y}$

R.H.S $\Rightarrow$ $x \oplus y = 2 \odot 2 = \dfrac{2^2}{2}$

So, L.H.S = R.H.S

$$\Rightarrow \frac{x^2+y^2}{xy} = \frac{2^2}{2}$$

$\Rightarrow x^2 + y^2 = 2xy \Rightarrow x^2 + y^2 - 2xy = 0$

$\Rightarrow (x-y)^2 = 0 \quad \Rightarrow x - y = 0 \quad \Rightarrow \boxed{x = y}$

5. According to the given information,

Possible way are

R Q S P

P Q S R

R Q P S

P Q R S

Q R S P

Q P S R

$\Rightarrow$ Total number of ways = 2 + 2 + 2 = 6

$\therefore$ The number of distinct seating arrangements possible is 6.

6. As Given statement:

Either P marries Q or X marries Y.

The logical negation of the Either P marries Q or X marries Y

"above statement is Neither P marries Q nor X marries Y."

Hence option (b) is correct.

7. Shortest side of folding:

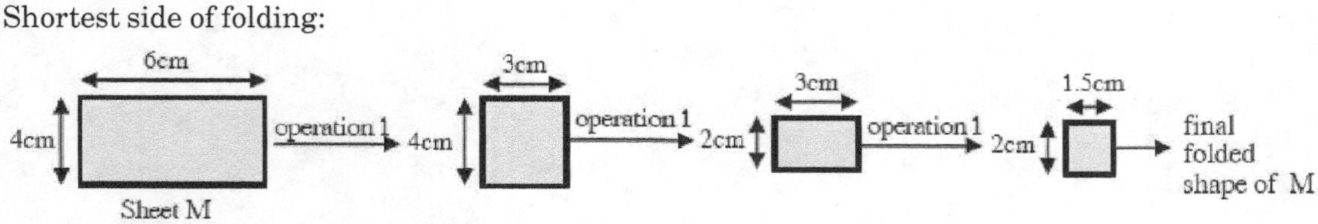

So, perimeter of final folded shape of M = 2 (2 + 1.5) = 7

Longest side in folded:

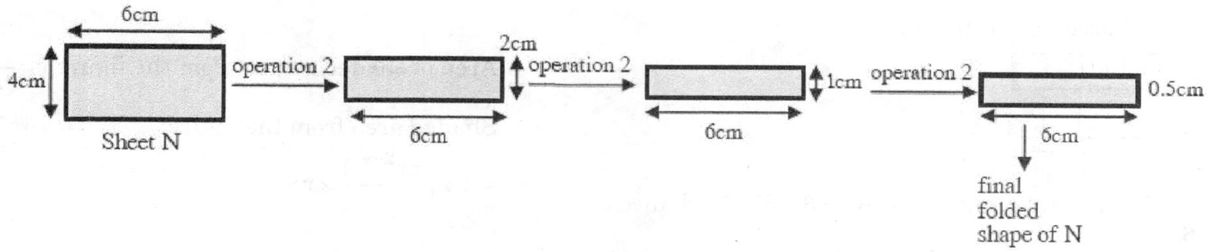

$\therefore$ Perimeter of final folded shape of N = 2(6 + 0.5) = 13

$\therefore$ Required ratio is (N:M) = 13:7

8.

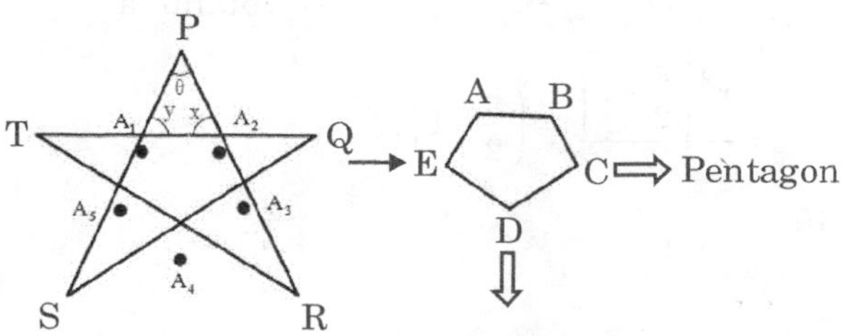

PR = PS = QS = QT = RT (Given)

Then ABCDE will be regular pentagon.

So, Sum of internal angles formed at pentagon = (5 – 2) × 180

$\Rightarrow$ Each internal angle = $\frac{3\times180}{5}$ = 108°

So, $\angle\theta$ = ?; $\angle x$ = 180° – 108° = 72°; $\angle y$ = 180 – 108 = 72

$\therefore \angle x + \angle y + \theta = 180$

$\Rightarrow \theta = 180 - 72 - 72 = 36°$

9. According to the question

$$\frac{\lambda(-(-3+2),(-2+3)}{(-(-2+1)} = \frac{\lambda(-1(-1),1)}{1}$$

So, 1st definition will be applicable as p = q

$[\therefore \lambda(p, q) = (p - q)^2 \text{ if } p = q]$

So, $\lambda(1, 1) = (1 - 1)^2 = 0$

10. Only humans have the ability to construct worlds entirely in their minds and no other species has this.

Hence, answer is option (c) which reflects the information given in the passage.

SET-2

ANSWERS

1. (*b*)	**2.** (*b*)	**3.** (*d*)	**4.** (*d*)	**5.** (*c*)
6. (*c*)	**7.** (*a*)	**8.** (*b*)	**9.** (*c*)	**10.** (*d*)

EXPLANATIONS

1. When two or more singular nouns joined by 'and' the verb should be plural.

2.

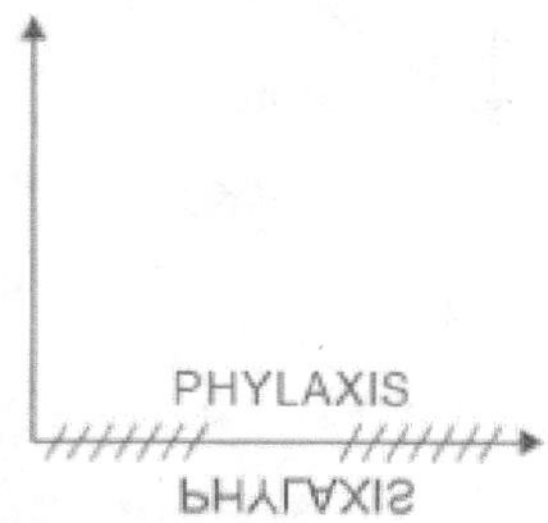

With respect 'x' – axis.

3. Let, $E_1 \rightarrow$ occurrence of even number on first dice

$E_2 \rightarrow$ occurrence of even number on second dice

Probability of even number on a first dice is:

$$P(E_1) = \frac{3}{6} = \frac{1}{2}$$

And probability of even number on a second dice is:

$$P(E_2) = \frac{3}{6} = \frac{1}{2}$$

So, the probability that an even number is rolled out on each dice is:

$$P(E_1 \cap E_2) = P(E_1) \times P(E_2)$$

$$= \frac{1}{2} \times \frac{1}{2} = \frac{1}{4}$$

4. As we know that

$p \odot q = p - q$

$p \oplus q = p \times q$

$\Rightarrow 9 \odot (6 \oplus 7) \odot (7 \oplus (6 \odot 5))$

= 9 – (6 × 7) – (7 × (6 – 5))

= 9 – 42 – (7 × 1)

= –40

5. Possible arrangents:

$\because$ 'R' should not be seated at second position from the left end:

(–, R, –, –)

So, possible arrangments:

| P | | | | = 3! |

| | Q | | | = 3! |

| | S | | | = 3! |

Hence, total number of ways = $3 \times 3! = 18$ chances.

6.

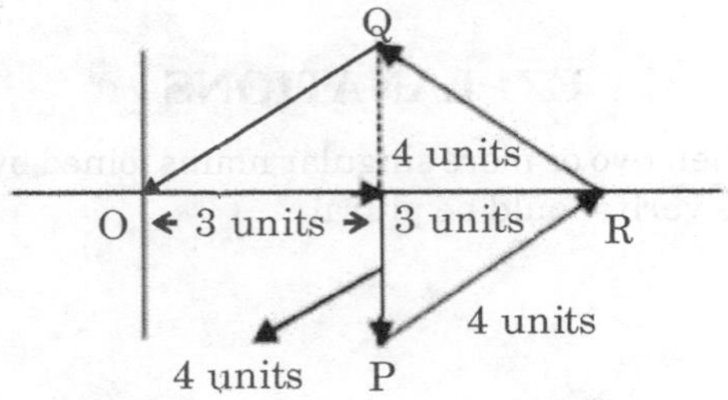

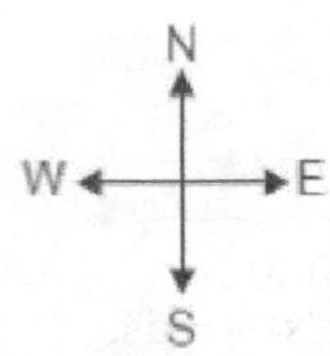

$\therefore \quad OQ = \sqrt{(3)^2 + (4)^2}$
$= 5$ units

7. The author states musicians rechearse, actors rehearse and like wise he states public speakers also should rechearse and this is stated in option (a).

8. (i) Some foot ball player play cricket:

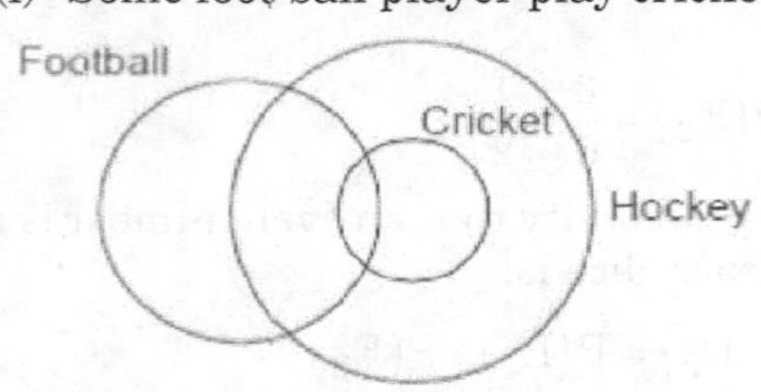

(ii) All, cricket players play hockey:

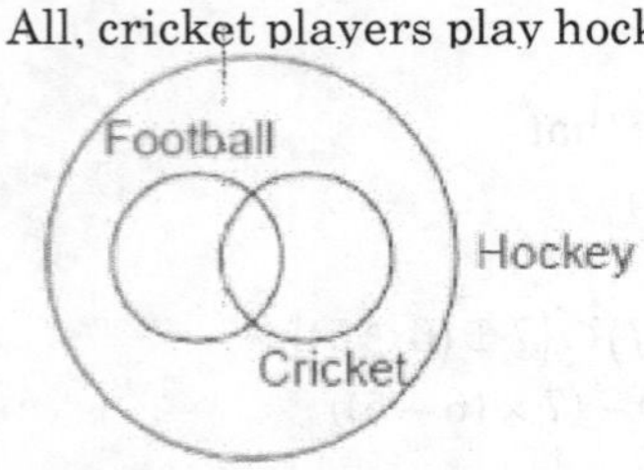

Hence, some football players play hockey is satisfied.

9. According to question:

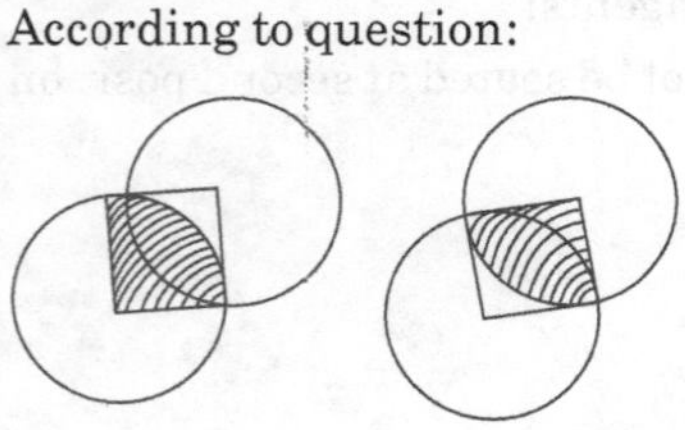

Area of shaded region from the figure is $\frac{\pi r^2}{4}$.

Shaded area from the figure:

$$= 2 \times \left[\frac{\pi r^2}{4} - \frac{1}{2} \times r \times r\right]$$

$$= r^2 \left[\frac{\pi}{2} - 1\right]$$

Total area = r^2

As we know that,

$$\therefore \quad \text{Required probability} = \left[\frac{\text{Shaded area}}{\text{Total area}}\right]$$

$$= \frac{r^2 \left[\frac{\pi}{2} - 1\right]}{r^2} = \left(\frac{\pi}{2} - 1\right)$$

10.

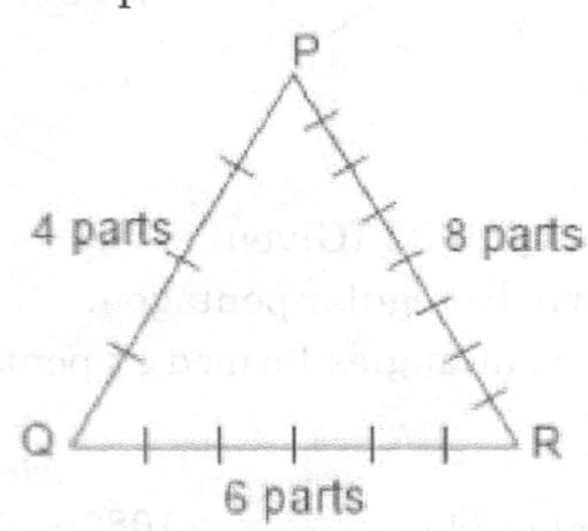

Let the dimension of equilateral $\Delta PQR = a$ units

As per question,

$\left(\frac{a}{4}, \frac{a}{6}, \frac{a}{8}\right)$ is integer,

Minimum area of triangle

(A) = LCM (4, 6, 8) = 24 units

$$\therefore \quad \text{Area of } \Delta PQR = \frac{\sqrt{3}}{4} \times a^2$$

$$= \left[\frac{\sqrt{3}}{4} \times 24 \times 24\right] = 144\sqrt{3} \text{ sq. units}$$

SET-3

ANSWERS

1. (c)	2. (b)	3. (c)	4. (c)	5. (b)
6. (c)	7. (d)	8. (a)	9. (c)	10. (a)

EXPLANATIONS

1. As given that,

Current population of a city = 11, 02, 500

Increasing rate = 5%/Annum

The population 2 years ago = ?

Let us Consider, the population 2 years ago = x
Using compound interest formula, we have

$$A = P\left(1+\frac{R}{100}\right)^n$$

$$\Rightarrow \quad 11{,}02{,}500 = x\left(1+\frac{5}{100}\right)^2$$

$$\Rightarrow \quad 11{,}02{,}500 \times \frac{100\times 100}{105\times 105} = x$$

$$\Rightarrow \quad \boxed{x = 10{,}00{,}000}$$

Hence, the population 2 years ago = 10,00,000.

2. As Given that, $\frac{p}{q}+\frac{q}{p} = 3$

By Squaring on both sides;

$$\left(\frac{p}{q}+\frac{q}{p}\right)^2 = 9$$

$$\Rightarrow \frac{p^2}{q^2}+\frac{q^2}{p^2}+2 = 9$$

Hence, $\boxed{\frac{p^2}{q^2}+\frac{q^2}{p^2} = 7}$

3.

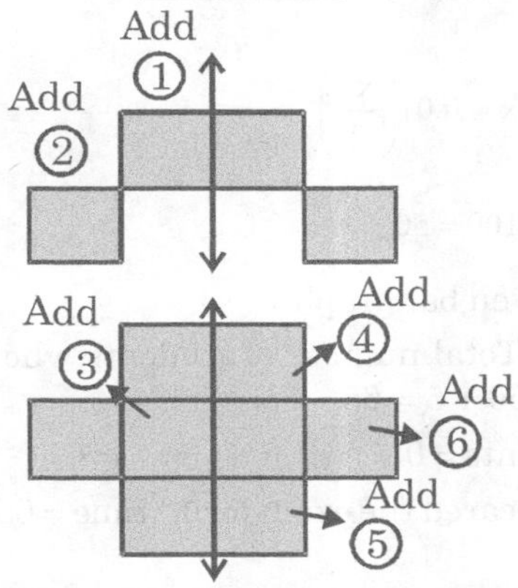

Hence, the least number of squares added = 6.

4. Nostalgia (means excessively sentimental yearning for return to or of some past period) is to anticipation. (means visualization of a future event or state)

5. Wake pastense is woke and third form is woken so, the verb forms are:

V1	V2	V3
Wake	Woke	Woken

6.

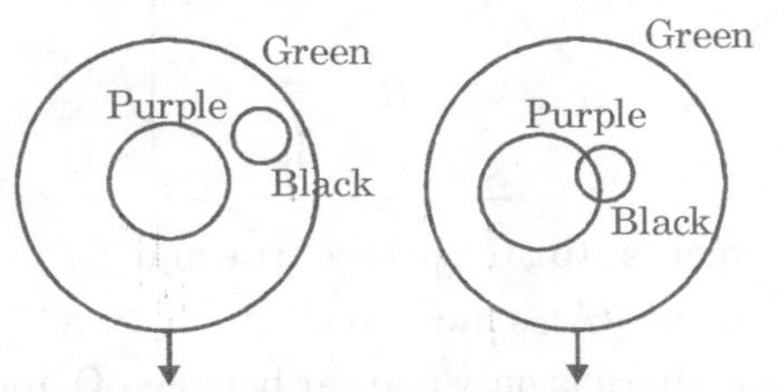

No black is purple Some black are purple

7. The passage deduces that computers are present every where and they have both positive and negative effects on humans.

8. Square:

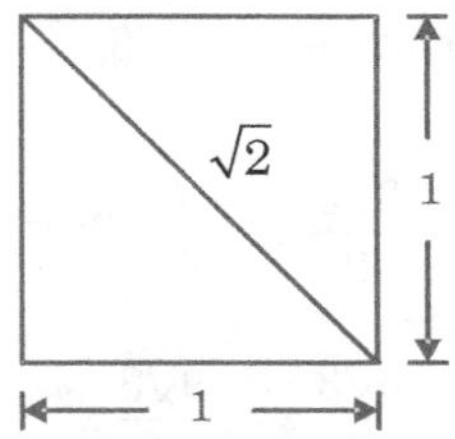

Solid core formed by revolving tringle about its short edge.

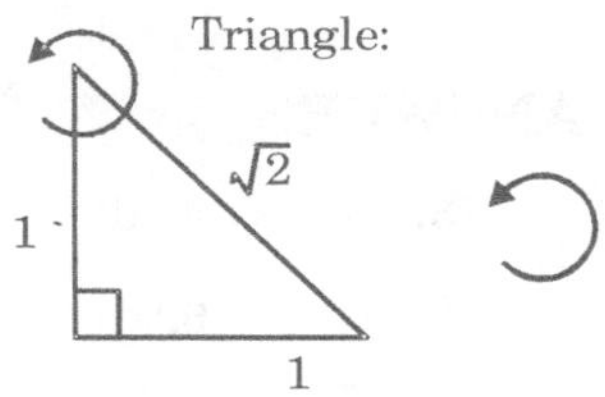

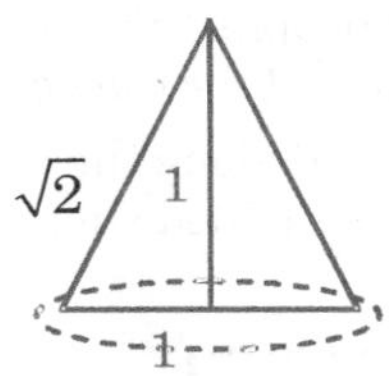

Radius of base circle(r) = 1 unit

Height of cone(h) = 1 unit

Volume of cone = $\frac{\pi}{3}\times r^2h = \frac{\pi}{3}\times 1\times 1 = \frac{\pi}{3}$ unit^3

9. From the bar graph, it is clear that except Thursday, total 6 days one of the students spent a minimum of 10% more than the other student.

10. In regular hexagon, each interior angle = 120°,
So, if we cut corners of an equilateral triangle, then the removed triangles are also equilateral triangles.

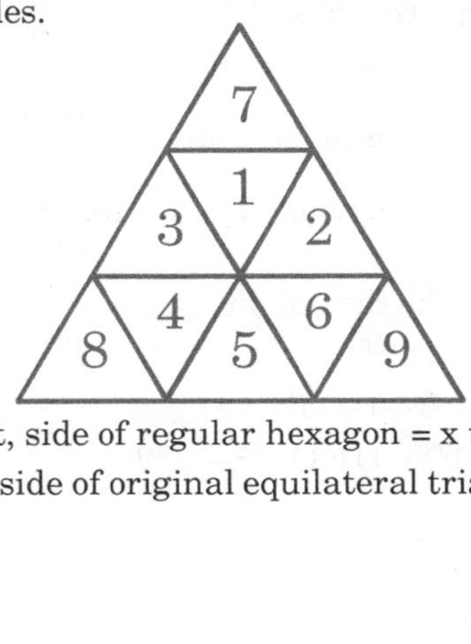

∴ Let, side of regular hexagon = x units
Then, side of original equilateral triangle = 3x

$\therefore$ Area of regular hexagon $= \frac{3\sqrt{3}}{2}x^2$

Area of equilateral triangle $= \frac{\sqrt{3}}{4}(3x)^2$

$\therefore$ Required ratio $= \frac{3\sqrt{3}}{2}x^2 : \frac{\sqrt{3}}{4}(3x)^2$

$= \frac{3\sqrt{3}}{2}x^2 : \frac{9\sqrt{3}}{4}x^2$

$= 1 : \frac{3}{2} = 2 : 3$

SET-4

ANSWERS

1. (c)	**2.** (c)	**3.** (d)	**4.** (d)	**5.** (c)
6. (d)	**7.** (c)	**8.** (d)	**9.** (d)	**10.** (a)

EXPLANATIONS

1. Here the subject of the verb is 'who' because we can not use 'Which' for people, 'Who' is the only word which acts as a subject for the verb.
2. If the square sheet is folded along the dotted line by upon rotation, then the folded sheet will look like as follows.

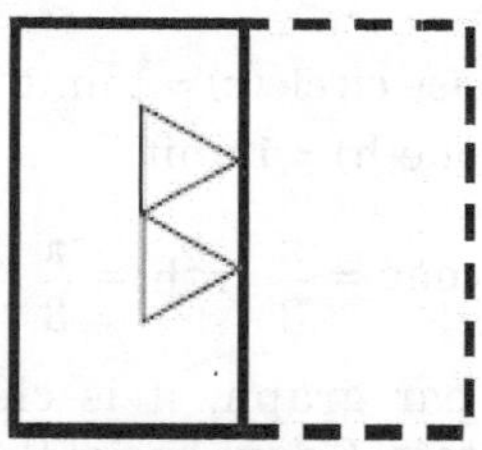

3. As given that,

n = 10

As we know that

$\therefore$ The sum of interior angles for a regular polygon = (n – 2) × 180°

= (10 –2) × 180 = 8 × 180° = 1440

So, Each interior side angle $= \frac{8\times180°}{10} = 144°$.

4. $(a^n - b^n)$ is divisible by (a + b), only when 'n' is even

So, we take the option (d).

$11^{52} - 1 = (11^{26})^2 - 1^2$

$= (11^{26} + 1)(11^{26} - 1)$

$= (11^{26} + 1)[(11^{13})^2 - (1^2)]$

$= (11^{26} + 1)[(11^{13}) - 1](11^{13} + 1)$

$\therefore$ $11^{13} + 1$ is divisor of $11^{52} - 1$.

5. Oasis is a water pool amidst sand just as island is a piece of land amidst water.
6. Deprivation of sleep during exams will have negative effect on exams. So, considerable sleep is necessary to well in exams.
7. According to the question.

$\therefore$ Total number of squares = 6

Now, by using Pythagoras theorem, in each square, we can find side length.

$\therefore$ Side of smallest square becomes $= \frac{5}{\sqrt{8}}$

So, Area of smallest square $= \frac{5}{\sqrt{8}} \times \frac{5}{\sqrt{8}} = \frac{25}{8} = 3.125$

8. As we know that,

$E(X) = \text{Mean of } X = \int_{-\infty}^{\infty} x.f_X(x)dx$

$E[x] = \int_0^{100} x.f(x)dx \ (\because 0 \le x \le 100)$

$= \int_0^{100} x.(0.01)dx = 0.01\left(\frac{x^2}{2}\right)_0^{100}$

$= \frac{1}{200} \times 100 \times 100 = 50$

9. From the given bar graph:

In year 2 : Total number of students who took the exam = 60 + 5 = 65

Failed students = 0

But who appeared the exam for 1st time = 65 – 10 = 55.

In year 3 : Total number of students who took the exam = 50 + 3 = 53

Failed students = 5

But who took the exam for 1st exam = 53 –5 = 48

Hence option (d) is correct.

10. According to the given information, possible arrange:

T	U	P	Q	S	V	R
U	Q	S	V	R	P	I
I	U	O	S	V	R	P

Statements : (b), (c), (d) are true and

Statements (A) is incorrect.

Hence, there is only one car between Q and V.

SET-5

ANSWERS

1. (d)	**2.** (d)	**3.** (d)	**4.** (d)	**5.** (c)
6. (c)	**7.** (a)	**8.** (c)	**9.** (c)	**10.** (d)

EXPLANATIONS

1. When you say you could hardly do some thing you are emphasing that it is very difficult for you to do it.

$\because$ Could + Adverb + Verb (I).

Hence, option (d) is the correct answer.

2. According to the questions:

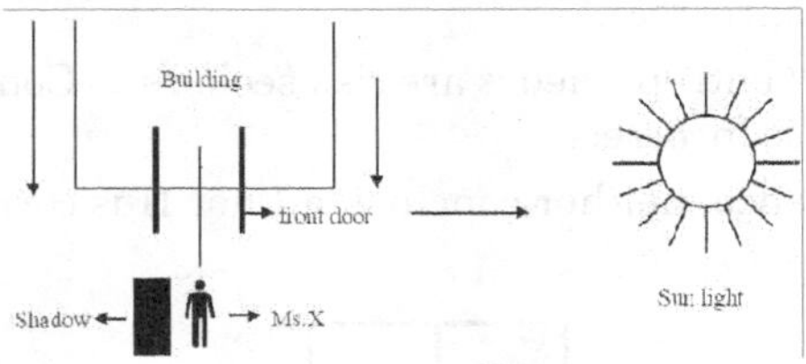

$\therefore$ From the above figure, it is clear that building is facing towards South.

Hence, option (d) is the correct answer.

3. Given that,

Area of right triangle MON = 50

According to given figure:

$\Rightarrow \frac{1}{2} \times ON \times OM = 50$

$[\because ON = OM = \text{radius of circle} = r(\text{say})]$

$\Rightarrow \frac{1}{2} \times r \times r = 50$

$\Rightarrow r^2 = 100$

$\therefore$ Area of circle with radius $r = \pi r^2$

$= \pi(100) = 100\pi$.

Hence option (d) is the correct answer.

4. As given expression =

$\Delta 2 \oplus 3\Delta((4 \otimes 2) \nabla 4)$

$= +2 - 3 + ((4 \div 2) \times 4)$

$= 2 - 3 + (2 \times 4) = 2 - 3 + 8 = 7$

Hence option (d) is the correct answer.

5. One can clearly infer that eating leafy vegetables undoubtely lead a healthy style.

7. From the options, the figure in option (b) is the missing piece. If the figure is rotated clockwise then it will form a rectangular with the given piece in the picture.

8. For farm P:

Number of hens = 65

Number of ducks = 91

Number of goats = 169

For farm Q:

Number of hens $= 416 \times \frac{5}{32} = 65$

Number of ducks $= 416 \times \frac{14}{32} = 182$

Number of goats $= 416 \times \frac{13}{32} = 169$

$\therefore$ Total number of hens = 65 + 65 = 130

Total number of ducks = 91 + 182 = 273

Total number of goats = 169 + 169 = 338

$\therefore$ The new ratio of hens: ducks: goats = 130: 273: 338 = 10: 21:26.

9. As given that,

Total number of executives, across all companies = 10,000

$\therefore$ Number of executives in $C_2 = 10{,}000 \times \frac{5}{100} = 500$

Number of executives in $C_5 = 10{,}000 \times \frac{20}{100} = 2000$

$\therefore$ Number of management degree holders among the executives in $C_2 = 500 \times \frac{1}{5} = 100$

And, Number of management degree holders among the executives in $C_5 = 2000 \times \frac{9}{10} = 1800$

So, the total number of management degree holders among the executives in companies C_5 and C_5 together = 100 + 1800 = 1900.

Hence, option (c) is the correct answer.

10. As given that

Total block is (Q_R)

Total number of ways,

$\frac{1}{Q}\frac{2}{-}\times\frac{1}{R}\times\frac{2}{S}\times\frac{1}{-} \rightarrow 4$ choices

(OR)

$\frac{2}{-}\times\frac{-}{Q}\times\frac{1}{R}\times\frac{S}{-} \rightarrow 2$ choices

$\frac{S\,2}{-\,-}\times\frac{-}{Q}\times\frac{1}{R} \rightarrow 2$ choices

(OR)

$\frac{-}{R}\frac{2}{-}\times\frac{-}{Q}\times\frac{1}{S} \rightarrow 2$ choices

(OR)

$\frac{S}{\ } \times \frac{R}{\ } \times \frac{2}{\ } \times \frac{Q}{\ } \times \frac{1}{\ } \rightarrow$ 2 choices

(OR)

$\frac{2}{S} \times \frac{2}{\ } \times \frac{\ }{R} \times \frac{1}{\ } \times \frac{\ }{Q} \rightarrow$ 4 choices

Hence, total number of distinct seating arrangement possible = 4 + 2 + 2 + 2 + 2 + 4 = 16.

SET-6

ANSWERS

1. (*b*)	**2.** (*a*)	**3.** (*a*)	**4.** (*b*)	**5.** (*d*)
6. (*c*)	**7.** (*b*)	**8.** (*c*)	**9.** (*b*)	**10.** (*d*)

EXPLANATIONS

1. Positive arrangements are

<u>S R P T Q</u> (OR) <u>Q R P T Q</u> (OR) <u>S R T P Q</u>

Total number of combinations possible is 3.

2. (i) The first sentence informs us about the number with is singular.

(ii) A number of means several, some.

The second sentence takes about the students which is plural word.

3. As given that,

Watch x beeps every 30 sec

Watch y beeps every 32 sec

L.C.M of 30, 32 = 480 seconds = 8 minutes

They beeps together every 8 min

∴ i.e., 10 AM, 10.08 AM; 10.16 AM, 10.24 AM, ...

Hence option (a) is the correct answer.

4. (i) $\oplus \div \odot \Rightarrow \frac{\oplus}{\odot} = 2$

(ii) $\oplus \div \Delta = 3 \Rightarrow \frac{\oplus}{\Delta} = 3$

(iii) $\odot + \Delta = 5 \Rightarrow \frac{\oplus}{2} + \frac{\oplus}{3} = 5$, $\oplus = 6$ and $\Delta = \frac{6}{3} = 2$.

(iv) $\Delta \times \otimes = 10 \Rightarrow 2 \times \otimes = 10\ \otimes = 5$

So, $(\otimes - \oplus)^2 = (5-6)^2 = 1$

5. From the figure, it is clear that, the direction of X is North West

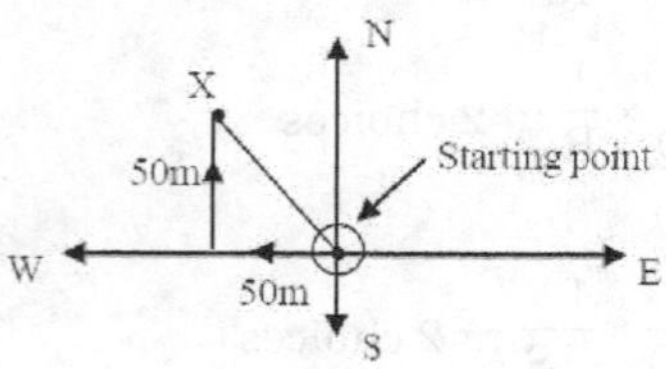

6. E → Entrepreneurs

W → Wealthy

R → Risk seekers

Statement-1

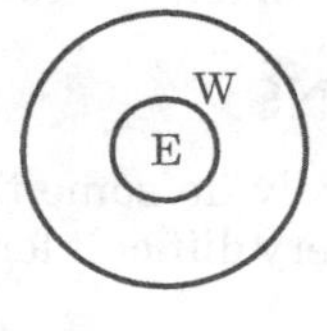

All entrepreneurs are wealthy

Statement-2

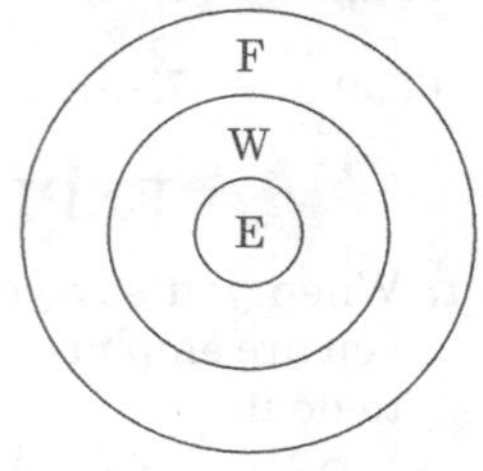

All wealthy are risk seekers

∴ All risk sectors are not wealthy ⇒ Conclusion I is incorrect.

All entrepreneurs are risk seekers ⇒ Conclusion II is incorrect

Hence, neither conclusion I nor II is correct.

7. Box

Total balls =

15 Blue	45 Black

If the two balls are drawn without replacement the required probability of getting first ball blue and second ball black will be

∴ Required probability = $\frac{15}{60} \times \frac{45}{59} = \frac{45}{236}$.

8. ∴ From the figure,

$\sin 30° = \frac{r}{R} \Rightarrow \frac{1}{2} = \frac{r}{R} \Rightarrow R = 2r$

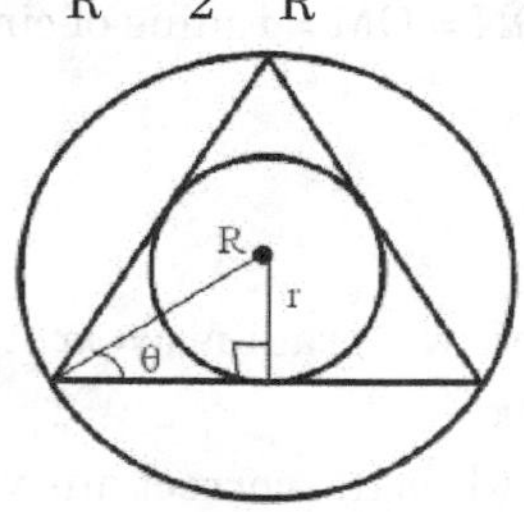

r → radius of inscribed circle

R → radius of circumscribed circle

∴ Area of inscribed circle = πr^2

Area of circumscribed circle = πR^2

$= \pi(4r^2)$

∴ Required ratio = $\pi r^2 : 4\pi r^2$

$= 1 : 4 = \frac{1}{4}$

9. ∴ Area of each face of the final folded shape. (Right angle triangle)

$$= \frac{1}{2} \times \frac{1}{2} \times \frac{1}{2}\left(\frac{1}{2} \times \text{Box} \times \text{weight}\right)$$

$$= \frac{1}{8} \text{ square units}$$

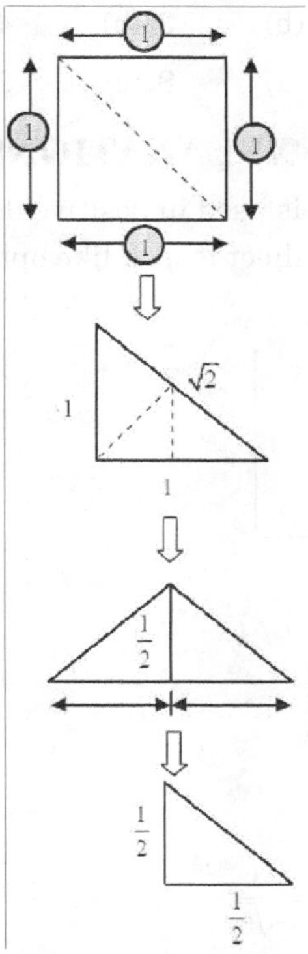

Hence, option (b) is the correct answer.

10. Due to worst pandemic the air travel industry is facing a crisis is nothing but restarting the idea from first sentence.

SET-7

ANSWERS

1. (c)	**2.** (a)	**3.** (c)	**4.** (a)	**5.** (c)
6. (c)	**7.** (c)	**8.** (c)	**9.** (d)	**10.** (d)

EXPLANATIONS

1. Let us consider the number of boys = B
And total number of students = S
So, we have

$$\frac{B}{(S-B)} = \frac{7}{3}$$

$$B = \frac{7S}{10}$$

Thus, S must be multiple of 10. Hence, answer is 50.

2. If we choose two points P, Q such that P is the top-left corner and Q is the bottom left corner then the line joining P, Q will not lie on the polygon. Hence, option (a) is the correct answer.

3. Every body is prepared means they are ready for the exam, the second sentence does not clearly state whether babu will play with Danish or not.

Babu loves playing chess the does not means danish knows how to play chess. So statement

(i) is grammatically correct and

(ii) is ambiguous.

Statement (ii) is ambiguous because we do not know who enjoys playing chees, Babu or Danish.

So, statement (i) is grammatically correct.

Hence, option (c) is the correct answer.

4. Unforhed in reverse order of folding, it's look like option (a)

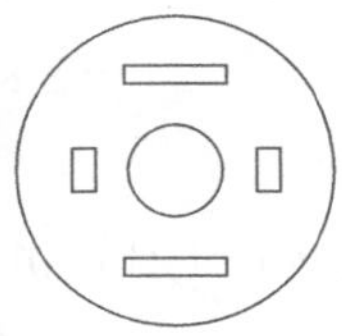

5. A doctor performs surgery just as a writer writes a book. It is the relation of person and skill.

6.

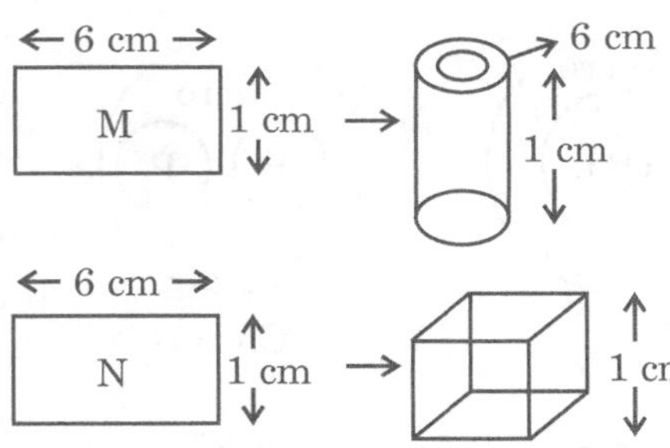

The circumference of lower circle $2\pi r = 6$

$$\Rightarrow r = \frac{3}{\pi}$$

Volume of a cylinder $= \pi r^2 h = \pi \times \frac{3}{\pi} \times \frac{3}{\pi} \times 1 = \frac{9}{\pi}$

Volume of a cube = $(1)^3$

So, request ratio

$$= \frac{\left(\frac{9}{\pi}\right)}{1} = \frac{9}{\pi}$$

Hence, option (c) is the correct answer.

7. As given that

$CP_p = 5400$

$MP_p = 5860$

Cost of $Q(CP_q) = \left(\frac{4}{3}\right) * CP_p = 7200$

Let, selling price (S.P) of Q = S

$$25 = \left(SP_q - 7200\right) * \frac{100}{7200}$$

$SP_q = 9000$

Discount on $Q = MP_q - SP_q = 10000 - 9000 = 1000$

$$\text{Discount\%} = \frac{1000}{10000} \times 100 = 10\%$$

Discount on SPq i.e. 1000 is 10% of marked price (MP_q) of Q i.e. 10, 000.

So, answer is 10%.

Hence, option (c) is the correct answer.

8. P (exactly 1 =) Probability that all the picked chocolates are different:

$$= \frac{(10*9*8*7*6)}{10^5} = 0.3024$$

Probability that at least two of the picked chocolates are identical:

$\Rightarrow 1 -$ (Probability that all the picked chocolates are different)

$\Rightarrow 1 - 0.3024 = 0.6976$

9.

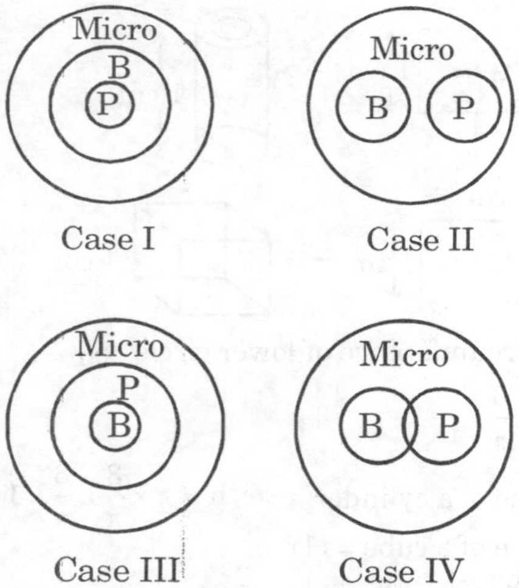

So, none of the two conclusions will satisfy all the 4 case. Hence, option (d) is the correct answer.

10. Superficially is the deciding key word which means apparently or seemingly.

As AOM are not addressing the core problems, they are superficial.

Superficial means shallow, cursory mean lacking in depth or solidity. Superficial implies a concern only with surface aspects or obvious features. A superficial analysis of the problem shallow is more generally derogatory in implying lack of depth in knowledge, reasoning, emotions, or character. Hence, option (d) is the correct answer.

SET-8

ANSWERS

1. (a)	**2.** (b)	**3.** (b)	**4.** (b)	**5.** (c)
6. (c)	**7.** (a)	**8.** (c)	**9.** (a)	**10.** (c)

EXPLANATIONS

1. 'As well as' is used in positive degree.

2. The folded sheet is look like option (b).

3.

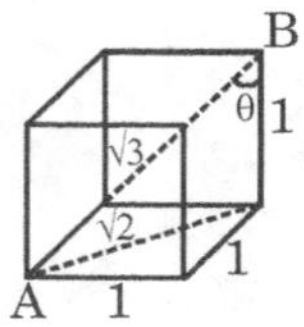

$$\cos\theta = \frac{BX}{AB} = \frac{1}{\sqrt{3}}$$

4. As we know that

$(a - b)^2 = a^2 + b^2 - 2ab$

As given expression

$$\left(x - \frac{1}{2}\right)^2 - \left(x - \frac{3}{2}\right)^2 = (x + 2)$$

So,

$$\Rightarrow \left[x^2 + \left(\frac{1}{2}\right)^2 - 2(x)\left(\frac{1}{2}\right)\right] - \left[x^2 + \left(\frac{9}{4}\right) - 2(x).\left(\frac{3}{2}\right)\right]$$

$= [x + 2]$

$$\Rightarrow \left[x^2 + \frac{1}{4} - x - x^2 - \frac{9}{4} + 3x\right] = [x + 2]$$

$\Rightarrow [-2 + 2x] = [x + 2]$

$x = 4$

5. Pen is to write just as knife is to cut.

6. Listening to music improves exercise performance and reduces discomfort. Music as positive effort only on few students during learning.

7. When the piece can be moved, rotated or flipped to assemble with the given piece is option (a)

8. The total number of students in three classes in the beginning.

Let us consider:

$3a + 13a + 6a = 22a = n$

after addition of 18 student total number of students.

$15b + 35b + 21b = 71b$

$3a + 18 + 13a + 18 + 6a + 18 = 22a + 54$

$22a + 54 = 71b$

$n + 54 = 71b$

$n = 71b - 54$

Let $(b = 2)$

So, $n = 71 \times 2 - 54$

$= 88$

9. Let us consider, the selling price in year '2' be 'x'.

$\therefore$ Net profit = Selling price – (Cost + Taxes)

$296 = x - (400 + 0.13x)$

$296 = 0.87x - 400$

$$x = \frac{696}{0.87} = 800$$

So, selling price in year '2' = Rs.800

In year 3, total cost = 300×1 = Rs.300

Taxes paid = 0.15% of selling price.

Now, let us consider, the selling price in year 3 be 'y'

$\therefore 210 = y - (300 + 0.15y)$

$\Rightarrow 210 = (0.85y - 300)$

$$y = \frac{510}{0.85} = 600$$

$\therefore$ Selling price in year '3' = 600

$$\text{So, } \frac{x}{y} = \frac{800}{600} = \frac{4}{3}.$$

10. I: $S > R$

II: $Q < P, R, S, T, U$

From III & IV observation

$P > T > S > R > U > Q$

The number of students that are taller than 'R' is same as the number of students shorter than S.

ENGINEERING MATHEMATICS

SET-1

ANSWERS

1. (6 to 6) **2.** (80000 to 80000) **3.** (3 to 3)

4. (d) **5.** (1 to 1) **6.** (256 to 256)

7. (c)

EXPLANATIONS

1. As given that for a function f (w, x, y, z)

f (w, 0, 0, z) = 1

⇒ i.e., when x = 0, y = 0, f = 1

f (w, x, 1, z) = x + z

⇒ when x = 1, f = wz + y

So, we need to obtain number of literal in minimum SOP of f.

x	x	y	z	x = 0, y = 0, f = 1	w = 1, y = 1, f = x + z	x = 1 f = wz + y	Overall f
0	0	0	0	1			1
0	0	0	1				
0	0	1	0				
0	0	1	1			0	0
0	1	0	1			0	1
0	1	1	0			1	1
0	1	1	1			1	1
0	1	1	1			0	1
1	0	0	0	1		0	1
1	0	0	1	1		0	1
1	0	1	0		0	0	0
1	0	1	1		1	0	1
1	1	0	0			0	0
1	1	0	1			1	1
1	1	1	0		1	1	1
1	1	1	1		1	1	1

Now information for w, x, y, z = 00010 and 0011, so f is taken don't care for unit combination f (w, x, yz)

= Σm (0, 1, 6, 7, 8, 9, 11, 13, 14, 15) + d(2, 3)

By using k-map

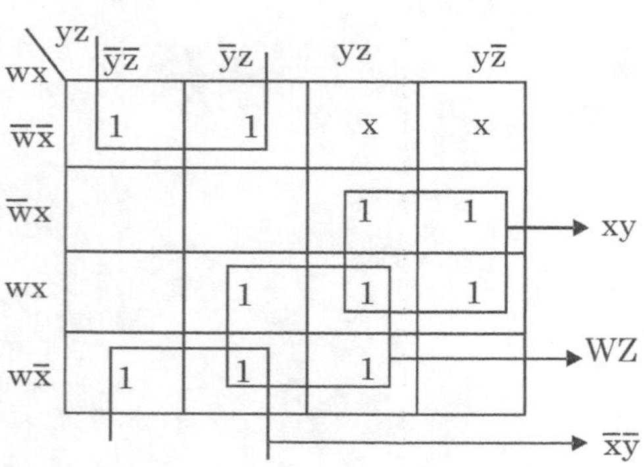

So $f = \bar{x}\bar{y} + xy + wz$ (minimized SOP expression)

ear term of f containing 2 lit eral, so total 6 literal in expression of f.

2. In One sec number type of data transfer is cycle steel i.e. one 8 bit characters one request of cycle generated by the CPU is 2×10^6 Since 0.5% cycles are used By DMA The number of cycle used by the DMA is

$$= 2 \times 10^6 \times \frac{0.5}{100}$$

$$= 2 \times 10^6 \times \frac{1}{200}$$

$= 10^4$

In 1 cycle 8 bits transferred

$\therefore$ 10^4 cycles ------- $10^4 \times 8$

= 80000 bits

3. As given that $(0.1101)_2 = (0.8\,xy5)_{10}$, we need to obtain decimal value of x + y.

$(0.1101)_2 = (1 \times 2^{-1}) + (1 \times 2^{-2}) + (0 \times 2^{-3}) + (1 \times 2^{-4})$

$= \frac{1}{2} + \frac{1}{4} + 0 + \frac{1}{16}$

$= \frac{13}{16}$

$= 0.8125$

$\Rightarrow (0.110)_2 = (0.8125)_0 = (0.8\,xyz)_{10}$

Now, By comparison x = 1, y = 2.

Hence x + y = 3.

4. It will result in run time error because it dereferences an uninitialized pointer in the loop when index value is 3.

Hence, option (d) is correct answer.

5. According to the question consider a complete binary tree with 7 nodes:

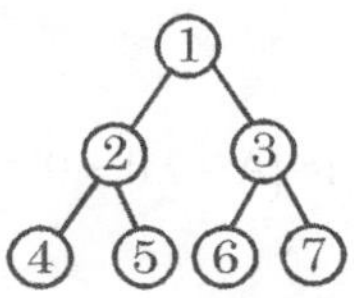

If the given CBT has root node 1, its children 2,3, and children of 2 are 4,5 then BFS will visit from root {1,2,3}, A = 3 as first three nodes, and DFS will visit from root {1,2,4}, B = 4 so, cardinality of the set |A – B| = 1

6. The regular expression for the given DFA is

= (0(0 + 1) (0 + 1) +1(0 + 1) (0 + 1)) (0 +1)*

It can be written as:

= (0 + 1) (0 + 1) (0 + 1) (0 + 1)*

Which can be defined as all string over {0, 1} of length greater than 3

As per the given condition in the string length it should be exactly 8.

We can write the expression for length 8 string as:

$\Rightarrow$ (0 + 1)(0 + 1)(0 + 1)(0 + 1)(0 + 1)(0 + 1)(0 + 1) (0 + 1) = $[0 + 1]^8$

So, all possible combination

For length 8, all 2^8 = 256 string will be accepted.

7. By doing the simple CRC based calculation where the general or polynomial is 1011

$x^3 + x + 1 = 1011$

```
1011 ) 11000000
       1011
       ----
        1110
        1011
        ----
         1010
         1011
         ----
          100
```

SET-2

ANSWERS

1. (-7.75 to -7.75) **2.** (12 to 12) **3.** (17 to 17)

4. (a) **5.** (a; c) **6.** (a; c)

7. (6 to 6)

EXPLANATIONS

1. As given that IEEE 754 format with bias 127

S = 1 $\Rightarrow$ Number is negative

E =10000001 $\Rightarrow 2^7 + 2^6 = 128 + 1 = 129$

F =1111

Biased exponent: 129

Exponent: Biased exponent – 127 = 129 – 127 = 2

Now, Binary number:

$= -1.1111 \times 2^2$

$= -(111.11)_2$

$= -[(1 \times 2^2) + (1 \times 2^1) + (1 \times 2^0) + (1 \times 2^{-1}) + (1 \times 2^{-2}]$

$= [4 + 2 + 1 + 0.5 + 0.25] = -7.75$

2. Minimum waiting time will be given by SJF algo. So, execution order will be P3,P1, P2. Waiting times of P1,P2,P3 respectively will be 10, 26,0.

Gantt chart:

CPU	P_3	P_1	P_2

0 10 26 46

Waiting time $(P_1) = 10$

Waiting time $(P_2) = 26$

Waiting time $(P_3) = 0$

So, Avg waiting time $= \frac{36}{3} = 12$

Hence, average waiting time = 12 ms.

3. As given the MA = 32 KB

Cache block size is 64 bits, so block offset = 6 bits

Number of cache lines $= \frac{32 \text{ KB}}{64 \text{ B}} = 512$ lines

So, bits for line number = 9

CA = (9 + 6) = 15 bits

Hence, tag bits = (MA – CA) = (32 – 15) = 17 bits

4. Decomposition D1: R_3 (Q,Y) and R_4 (Y,W,Z) have common attribute Y which is key in R_4 (Y,W,Z), so we can merge them in lossless manner. Now (Q,Y,W,Z) and R_1(P,Q,S,T) have Q in common which is key in (Q,Y,W,Z). Then (Q,Y,W,Z,P,S,T) and R_2(P,T,X) have P,T in common which is key in R_2(P,T,X)

So, D1 is lossless.

Decompasition D_2:

R_1 (PQS), R_2 (TX), R_3(QY), R_4 (YZW)

(R_1 join R_3 join R_4) $\cap R_2 = \phi$

So, D2 is lossy because (T,X) has nothing common in above relations.

Hence, option (a) is the correct answer.

5. Paging: does not suffer from external fragmentation and help to solve the issue of external fragmentation. And Paging suffers from internal fragmentation.

Page tables are memory overheads,

So, option (c) is also correct.

Hence, option (a) &)c) are correct answer.

6. While creating new file we have to check whether same name file exist or not , so we always need full scan for this

Renaming a file:

While renaming a file we have to check whether file with same name as newly named file exist or not , so we always need full scan for this.

Deletion, Opening of an existing file:

For deleting file, file can be present at any node (best case if present as first node, worst if present at last). So, we don't always need to full scan list. Similarly, for Opening a file. So, option (a) & (c) would require to traverse the directly foofully. The other two operation does not required.

7. According to the question:

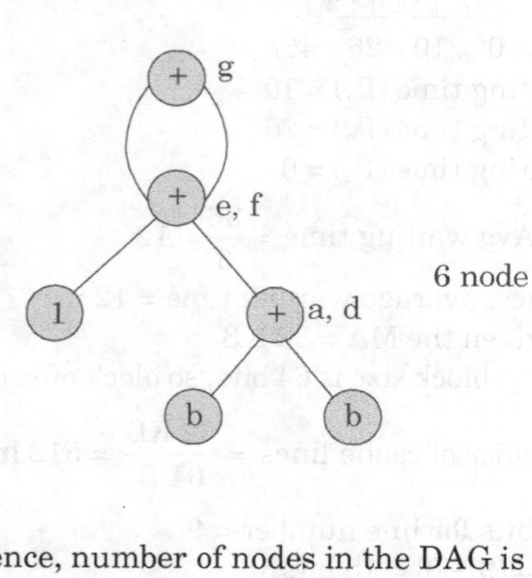

Hence, number of nodes in the DAG is 6 nodes.

SET-3

ANSWERS

1. (8.0 to 8.0)		2. (10.0 to 10.0)		3. (a)
4. (b)	5. (0 to 0)	6. (a)	7. (b)	8. (b)

EXPLANATIONS

1. As given that, $X_1 = (1.0, -1.0, 2.0)$, $X_2 = (7.0, 3.0, x)$ and $X_3 = (2.0, 3.0, 1.0)$ be three linearly dependent vectors. Consider these vectors as rows in a matrix A. For linearly dependent vectors, we have

$$|A| = 0$$

$$\Rightarrow \begin{vmatrix} 1 & -1 & 2 \\ 7 & 3 & x \\ 2 & 3 & 1 \end{vmatrix} = 0$$

$$\Rightarrow 1 \times (3 - 3x) + 1 \times (7 - 2x) + 2 \times (21 - 6) = 0$$

$$\Rightarrow -5x + 10 + 30 = 0$$

$$\Rightarrow x = 8$$

2. As we know that

The characteristic equation is $|A - \lambda I| = 0$

$$\begin{bmatrix} x-\lambda & -3 \\ 3 & 4-\lambda \end{bmatrix} = 0$$

$\Rightarrow (x - \lambda)(4 - \lambda) + 9 = 0$

$\Rightarrow \lambda^2 + (-4 - x)\lambda + (4_x + 9) = 0$, a quadratic equatin in λ

For repeated eigen values, we have

$B^2 - 4AC = 0$, where $A = 1$, $B = -4 - x$, $C = 4x + 9$

$(-4 - x)^2 - 4(1)(4x + 9) = 0$

$16 + x^2 + 8x - 16x - 36 = 0$

$x^2 - 8x - 20 = 0$

$(x - 10)(x + 2) = 0$

$\Rightarrow x = 10$ or -2

$\because$ x is positive real number.

$\therefore\ x = 10$

3. Parametric equations of the curve C (circular arc) $x = \cos t$; $y = \sin t$ where t is parameter, and radius = 1

$$\int_C F(r)\,dr = \int(-x\hat{i} + y\hat{j})(dx\hat{i} + dy\hat{j})$$

$$\therefore \int_C F(r).dr = \int_C (-xdx + ydy)$$

$$= \int_C \{(-\cos t)(-\sin t\, dt) + (\sin t)(\cos t\, dt)\}$$

where $(t \to 0 \text{ to } \frac{\pi}{4})$

$$= \int_0^{\frac{\pi}{4}} 2\sin t \cos t \, dt$$

$$= \int_0^{\frac{\pi}{4}} \sin 2t \, dt$$

$$= \left(\frac{-\cos 2t}{2}\right)_0^{\frac{\pi}{4}}$$

$$= \frac{-1}{2}(0-1) = \frac{1}{2}$$

Hence option (a) is correct answer.

4. As given differential equation $x(t) \times x(\frac{t}{2})$

$$\frac{dy}{dx} + \underbrace{\left(\frac{x}{1-x^2}\right)}_{P} y = \underbrace{(x)}_{Q} y^{1/2} \quad ...(i)$$

is a Bernoulli's equation in y.

Dividing both sides by $y^{1/2}$

$$y^{-1/2}\frac{dy}{dx} + \left(\frac{x}{1-x^2}\right)y^{1/2} = x \quad ...(ii)$$

Let $y^{1/2} = t$

$$\Rightarrow \frac{1}{2}y^{-\frac{1}{2}}\frac{dy}{dx} = \frac{dt}{dx}$$

$\therefore$ (2) becomes $2\frac{dt}{dx} + \left(\frac{x}{1+x^2}\right)t = x$

$$\Rightarrow \frac{dt}{dx} + \left(\frac{x}{2(1+x^2)}\right)t = \frac{x}{2} \quad ...(iii)$$

is a linear equation in t.

$$\therefore \quad I.F = e^{\int \frac{x}{2(1-x^2)}dx}$$

$$= e^{-\frac{1}{4}\int \frac{-2x}{1-x^2}dx}$$

$$= e^{\ell n\left((1-x^2)^{-\frac{1}{4}}\right)}$$

$$\Rightarrow \boxed{I \cdot F = (1-x^2)^{-\frac{1}{4}}}$$

Hence, option (b) is correct answer.

5. For irrotational field, we have curl $(\nabla \times \vec{F}) = 0$

$$\Rightarrow \begin{vmatrix} \hat{i}_x & \hat{j}_y & \hat{k}_z \\ \frac{\partial}{\partial x} & \frac{\partial}{\partial y} & \frac{\partial}{\partial z} \\ 4y - c_1 z & 4x + 2z & 2y + z \end{vmatrix} = 0$$

$$\Rightarrow \hat{i}_x(2-z) - \hat{j}_y(0+c_1) + \hat{k}_z(4-4) = 0$$

$$\Rightarrow c_1 = 0$$

6. C(curve) is $|x - i \Rightarrow = 2$.

$x^2(x^2 + 4) = 0$ gives $x = 0, 2i$, $2i$ are the singular points.

Only $x = 0$, $2i$ lies inside C.

$$\text{Let } f(x) = \frac{\sin x}{x^2(x^2+4)} = \left[\frac{\sin x}{4x^2}\right] - \left[\frac{\sin x}{4(x^2+4)}\right]$$

$$\therefore \oint_C \frac{\sin x}{x^2(x^2+4)}dx = \frac{1}{4}\oint_C \frac{\sin x}{(x-0)^2}dx - \frac{1}{4}\oint_C \frac{\left(\frac{\sin x}{x+2i}\right)}{x-2i}dx$$

$$\int_C f(z)\,dz = 2\pi i\begin{bmatrix} \text{Res} & +\text{Res} \\ x=0 & x=2i \end{bmatrix}$$

$$= \frac{1}{4} \times \frac{2\pi i}{1!}\left[\frac{d}{dx}(\sin x)\right]_{x=0} - \frac{1}{4} \times 2\pi i\left[\frac{\sin x}{x+2i}\right]_{x=2i}$$

($\because$ Using Cauchy's integral formula)

$$= 2\pi i\left[\frac{1}{4} - \frac{\sin 2i}{16i}\right]$$

$$= \frac{\pi i}{2} - \frac{\pi}{8}\sin(2i)$$

Hence, no option matched.

7. As Given that, $Y = 2X + 3$

$$Var(Y) = E(Y^2) - (E(Y))^2$$
$$= E[(2X+3)^2] - [E(2X+3)]^2$$
$$= 4[E(X^2) - (E(X))^2]$$

$$\sigma_Y^2 = 4\sigma_X^2$$

Hence, option (b) is correct answer.

8. As givne that,

A box contains the three coins such as,(i) HT (ii) HH (iii) TT.

So, probability of head in first toss $P(H_1) = \frac{1}{2}$

As we know that $P[H_2/H_1] = P[H_2 \cap H_1]/P(H_1)$

So, $P(H_2 \cap H_1) = \left(\frac{1}{12} + \frac{1}{12}\right) = \frac{1}{6}$

Therefore, $P\left(\frac{H_2}{H_1}\right) = \frac{1}{6} \times \frac{2}{1} = \frac{1}{3}$

Hence, option (b) is correct answer.

SET-4

ANSWERS

1. (b)	**2.** (c)	**3.** (d)	**4.** (c)	**5.** (c)
6. (d)	**7.** (a)	**8.** (c)		

EXPLANATIONS

1. As we know that

$Ap = \lambda p$,

where λ is eigen values and p is the corresponding eigen vector of 2 × 2 A

$\Rightarrow \|p'\| = |\lambda p'\| = \lambda\ |p\|$

Since p is 2-dimensional vector

$\Rightarrow \lambda$ p and p' are also 2-dimensional vectors. Also p and λp has same direction vectors of p' = θ

Hence, option (b) is the correct answer.

2. Let, $\ell = \lim_{x\to 0}\left(\frac{1-\cos x}{x^2}\right)$ is $\frac{0}{0}$ form

Then, $\ell = \lim_{x\to 0}\frac{\sin x}{2x}$ (using L – Hospital's rule)

$\therefore \ell = \frac{1}{2}\lim_{x\to 0}\frac{\sin x}{x} = \frac{1}{2}\times 1 = \frac{1}{2}$

Hence, option (c) is correct answer.

3. As we know that, Dirac Delta function

$$\delta(t-a) = \lim_{\in\to 0}\begin{bmatrix} 0, & -\infty < t < a \\ 1/\in, & -a \le t \le a+\in \\ 0, & a+\in < t < \in \end{bmatrix}$$

$L\{\delta(t-a)\} = e^{-as} \to$ direct result

$L\delta(t) = 1$

4. Let $f(x) = x^2 - 2x + 2$

and [a, b] = [1, 3]

than $f'((x) = (2x - 2)$

Since the tangent of f(x) is parallel to the straight line joining f(1) and f(3), which is geometrical interpretation of Lagrange's mean value theorem.

$\therefore f'(x) = \frac{f(b)-f(a)}{b-a}$, where a = 1, b = 3

$\Rightarrow 2x - 2 = \frac{5-1}{2} = 2$

$\Rightarrow x = 2 \in [1, 3]$

$\therefore x = 2$

Hence, option (c) is the correct answer.

5. As given differential equation,

$$\sin(x)\frac{dy}{dx} + y\cos x = 1$$

Differential equation can be written as

$$\frac{dy}{dx} + (\cot x)\, y = \text{cosec}\, x \quad \text{.... (i)}$$

$\therefore$ I.F = $e^{\int \cot x\, dx} = e^{\log(\sin x)} = \sin x$

$\therefore$ solution is $y \times (\sin x) = \int \text{cosec}\, x \times \sin x\, dx + c$

$= \int 1 dx + c$

$\Rightarrow y \sin x = x + c$ ...(ii)

Using $y\left(\frac{\pi}{2}\right) = \frac{\pi}{2}$, (ii) gives

$\frac{\pi}{2}\times 1 = \frac{\pi}{2} + c \Rightarrow c = 0$

$\therefore y \sin x = x$ ....(iii)

at $x = \frac{\pi}{6}$, (iii) gives $y \times \frac{1}{2} = \frac{\pi}{6}$

$\Rightarrow y = \frac{\pi}{3}$ is the value of $y\left(\frac{\pi}{6}\right)$

Hence, option (c) is the correct answer.

6. Since sum of independent Binomial random variables is also a Binomial random variable and Normal distribution is limiting case of Binomial distribution, if n is indefinitely large i.e., $n \to \infty$

$\therefore y = \sum_{i=1}^{n} X_i = X_1 + X_2 + ... + X_n$

And distribution of y as $n \to \infty$ can be approximated by Normal.

Hence option (d) is the correct answer.

7. As we know that,

The Euler's iterative formula (forward) is

$y_{n+1} = y_n + hf(t_n, y_n)$

$= y_n + h \times (-\pi\, y_n)$ ...(i)

$\Rightarrow \frac{y(t_{n+1}) - y(t_n)}{h} = -\pi y(t_n)$

Now (i) can be expressed as $y_{n+1} = (1 - h\pi)\, y_n$

For stable, we have $|1 - h\pi| < 1$

$\Rightarrow -1 < 1 - h\pi < 1$

$\Rightarrow 2 < -h\pi < 0$

$\Rightarrow 0 < h < \frac{2}{\pi}$ $(\because |x| < \ell \Leftrightarrow -\ell < x < \ell)$

Hence, option (a) is the correct answer.

8. Let, $I = \oint \frac{\cosh 3z}{2z}$, where

C : $|z| = 1$ and $2z = 0$ gives $z = 0$ is a singular point lies inside C.

$$I = \int_c \frac{\cosh 3z}{2z} dz = \int_C \frac{\left(\frac{\cosh 3z}{2}\right)}{z-0} dz$$

$$I = 2\pi i \left(\frac{\cosh 3z}{2}\right)_{at\, z=0}$$

(By using Cauchys integral formula)

$$I = 2\pi i \frac{1}{2} = \pi i$$

SET-5

ANSWERS

1. (*d*)	**2.** (*b*)	**3.** (*c*)	**4.** (*b*)	**5.** (*c*)
6. (*d*)	**7.** (*a*)	**8.** 1.66 to 1.68	**9.** (*b*)	

EXPLANATIONS

1. As given that,

$Ap = \alpha^2 p$ ...(i)

Pre-multiply both sides by A, we get

$A(AP) = A(\alpha^2 p)$

$\Rightarrow (AA)p = \alpha^2 (Ap)$

$\Rightarrow A^2 p = \alpha^2(\alpha^2 p) = \alpha^4 p$ $(\because$ From (i))

Hence, α^4 is the eigenvalue of A^2

2. As we know that

If x is $(n \times 1)$ vector = $\begin{bmatrix} x_1 \\ x_2 \\ \vdots \\ x_n \end{bmatrix}$

So, x^T is $(1 \times n)$ vector = $[x_1, x_2, \ldots x_n]$

As given that,

Take n = 2 and Let $Q = \begin{bmatrix} a & c \\ c & b \end{bmatrix}$ be a symmetric (2×2) matrix, $x = \begin{bmatrix} x_1 \\ x_2 \end{bmatrix}; r = \begin{bmatrix} r_1 \\ r_2 \end{bmatrix}$

Then $f(x) = \frac{1}{2}\{x, x_2\} \begin{bmatrix} a & c \\ c & b \end{bmatrix} \begin{bmatrix} x_1 \\ x_2 \end{bmatrix} - [r_1 \quad r_2] \begin{bmatrix} x_1 \\ x_2 \end{bmatrix}$

$$= \frac{1}{2}\left[ax_1^2 + bx_2^2 + 2cx_1x_2\right] - r_1x_1 - r_2x_2$$

Let $f(x) = \phi(x_1, x_2) = \frac{ax_1^2}{2} + \frac{bx_2^2}{2} + cx_1x_2 - r_1x_1 - r_2x_2$

For stationary points, we have = $\frac{\partial \phi}{\partial x_1} = 0$ and $\frac{\partial \phi}{\partial x_2} = 0$

$\Rightarrow ax_1 + cx_2 - r_1 = 0$ ....(i)

And $bx_2 + cx_1 - r_2 = 0$ ....(ii)

Equation (i) and (ii) in matrix form is

$$\begin{bmatrix} a & c \\ c & b \end{bmatrix} \begin{bmatrix} x_1 \\ x_2 \end{bmatrix} = \begin{bmatrix} r_1 \\ r_2 \end{bmatrix}$$

$\Rightarrow Q.x = r$

Multiply by Q^{-1}, we get

Hence, $x = Q^{-1}r$, option (b) is the stationary point.

3. As given that,

$$F(s) = L\{f(t)\} = \frac{s+3}{(s+1)(s+2)}$$

$\therefore f(0) = \lim_{s\to\infty} sF(s)$ (Using initial value theorem)

$$= \lim_{s\to\infty} \frac{s(s+3)}{s(s+1)(s+2)}$$

$$= \lim_{s\to\infty} \frac{\left(1+\frac{3}{s}\right)}{\left(1+\frac{1}{s}\right)\left(1+\frac{2}{s}\right)} = \frac{1}{1\times 1} = 1$$

Hence, option (c) is correct answer.

4. As given double integral

$$\Rightarrow \int_{\theta=0}^{\pi/2}\left[\int_{r=0}^{\cos\theta} r\sin\theta\, dr\right] d\theta$$

$$= \int_{\theta=0}^{\pi/2} \sin\theta \left(\frac{r^2}{2}\right)_0^{\cos\theta} d\theta = \frac{1}{2}\int_0^{\pi/2} \sin\theta\cos^2\theta\, d\theta$$

$$= \frac{1}{2}\int_0^1 t^2 dt$$

[$\because$ By putting $\cos\theta = t$, $\sin\theta\, d\theta = dt$ & $\theta = \frac{\pi}{2} \Rightarrow t = 1$]

$$= \frac{1}{2}\left(\frac{t^3}{3}\right)_0^1 = \frac{1}{6}$$

Hence, option (b) is the correct answer.

5. As given that,

The differential equation

$\Rightarrow \left(\frac{1+y}{y}\right) dy = dx$, which is VSF

Integrating, we get

$$\int\left(\frac{1}{y}+1\right) dy = \int dx + C$$

$\Rightarrow \ell ny + y = x + C$ ...(1)

Using $y(1) = 1$, (1) gives $0 + 1 = 1 + C$

$\Rightarrow C = 0$

$\therefore \ell ny + y = x$

$\Rightarrow y = e^{x-y}$ or $ye^y = e^x$

Hence option (c) is the correct answer.

6. Mean and variance of a binomial distribution are np and npq respectively, where q = 1 – p i.e.., np and np(1 – p)

Hence, option (d) is correct answer.

7. As given that,

$f(x) = fnx, h = 0.3; a = 4, b = 5.2$

$y_0 = f(4) = \ln 4 = 1.386$

$y_0 = f(a + h) = \ln(4.3) = 1.458$

$y_2 = f(a + 2h) = l.n(4.6) = 1.526$

$y_3 = f(a + 3h) = \ln(4.9) = 1.589$

$y_4 = f(a + 4h) = \ln(5.2) = 1.648$

∴ By Simpson's one-third rule,

$$\int_4^{5.2} \ln x dx = \frac{h}{3}\left[y_0 + y_4 + 4 \times (y_1 + y_3) + 2(y_2)\right]$$

$$= \frac{0.3}{3}[(1.386 + 1.648) + 4(1.458 + 1.589) + 2\pi(1.526)]$$

$$= 1.827 \approx 1.83$$

Hence, option (a) is the correct answer.

8. As given that,

$f(x) = x^3 - x - 3$

$\Rightarrow f'(x) = 3x^2 - 1$ and $x_0 = 2$

As we know that,

Newton-Raphson method :

$$x_1 = x_0 - \frac{f(x_0)}{f'(x_0)} = 2 - \frac{(3)}{11} = 1.73$$

$$x_2 = x_1 - \frac{f(x_1)}{f'(x_1)} = 1.73 - \frac{(0.48)}{7.98} \approx 1.67$$

Hence the numerical value of the root after two iterations (x_2) is 1.67.

9. As we know that,

$(1 + i)^2 = 1^2 + i^2 + 2i = 2i$

$(1 + i)^4 = 2^2 i^2 = -4$

$\therefore (1 + i)^8 = (-4)^2 = 16.$

Hence, option (b) is the correct answer.

SET-6

ANSWERS

1. (b)	**2.** (d)	**3.** (c)	**4.** (1 to 1)	**5.** (b)
6. (c)	**7.** (c)	**8.** (b)	**9.** (a)	
10. (1 to 1)	**11.** (1 to 1)			

EXPLANATIONS

1. Let $|A| = \begin{bmatrix} p & q \\ q & -p \end{bmatrix}$

(Hint : $\lambda^2 - \lambda \times$ (trace) + (determinant of matrix))

Characteristic equation of the matrix is

$\lambda^2 - \lambda \times (-p + p) + (-p^2 - q^2) = 0$

$\lambda^2 - (P^2 + q^2) = 0$

$\Rightarrow \lambda^2 - 1 = 0$

$\Rightarrow \lambda = -1, 1$ (since $p^2 + q^2 = 1$)

2. As given that,

$P(z) = z^3 + (1 + j)z^2 + (2 + j)z + 3$

Let z = j be a complex number

$\Rightarrow \bar{z} = -j$

$p(z) = j^3 + (1 + j)j^2 + (2 + j)j + 3$

$= -j - 1 - j + 2j - 1 + 3 = 1$

$\Rightarrow \overline{p(z)} = 1$

And $p(\bar{z}) = p(-j) = (-j)^3 + (1 + j)(-j)^2 + (2 + j)(-j) + 3$

$= j - 1 - j - 2j + 1 + 3 = 3 - 2j$

$\Rightarrow \overline{p(z)} \neq p(\bar{z})$

So, option (a) is not true.

Now, take example:

(i) Let $p(z) = z^3 - jz^2 + z - j$ has a 3 complex roots $z = -j, j, j$

(ii) Let $P(z) = z^3 - z^2 + z - 1$ has 2 complex roots – j, j and one real root

∴ From the above two examples, sum of the roots contain imaginary term so all the roots cannot be real.

∴ Hence, option (d) is the correct answer.

3. Since $f'(x_0) = 0 \Rightarrow x_0$ is called stationary point and $x_0 \in (0, 1)$

So, $f''(x_0) = 0$ and $f(x) > 0$ for all $x \in (0, 1)$

$\Rightarrow f''(x_0) > 0$ for all $x \in (0, 1)$

Hence, f(x) has exactly one local minimum in at x_0 in (0, 1)

4. Letus consider $f(x, y) = x^2 + y^2 = 1$ ...(i)

and $g(x, y) = (x - 1)^2 + (y - 1)^2 = r^2$ ...(ii)

As we know that

$$m_1 = \frac{dy}{dx} = \frac{-\frac{\partial f}{\partial x}}{\frac{\partial f}{\partial y}}$$

$m_1 = \frac{-2x}{2y} = -\frac{x}{y}$ this is the slope of curve (1).

We also know that,

$$m_2 = \frac{dy}{dx} = \frac{-\frac{\partial g}{\partial x}}{\frac{\partial g}{\partial y}}$$

$$m_2 = \frac{-2(x-1)}{2(y-1)} = \frac{-(x-1)}{y-1}$$

This is the slope of curve (2)

Now we also know that,

So, for orthogonal curves, product of slopes is – 1

$m_1 \times m_2 = -1$

$\Rightarrow \frac{x}{y} \times \frac{(x-1)}{(y-1)} = -1$

$\Rightarrow x^2 - x = -y^2 + y$

$\Rightarrow x^2 + y^2 = x + y$

$\Rightarrow x + y = 1 \qquad (\because x^2 + y^2 = 1)$

Since (i) and (ii) intersect at (u, v)

$\therefore \boxed{u + v = 1}$

5. From intermediate value theorem:

$p(0) = 2 > 0$ and $p(1) = 1 - 4 + 2 = -1 < 0$

So, exactly one real root lies between 0 and 1

Signs of p(x) : 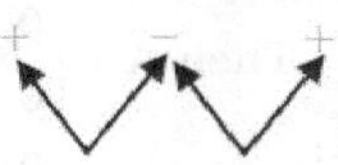

$\Rightarrow$ two sign changes in p(x)

Signs of p(–x) : 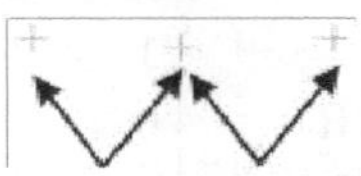

So, no sign changes in p(–x)

Thus, p(x) has at most two positive roots and no negative root. We know that complex roots occur in pairs and p(x) is a 4th degree polynomial also,

Now, $p(3) = 81 - 108 + 2 < 0$ and

$p(4) = 256 - 256 + 2 > 0$

$\Rightarrow$ root lies between 3 and 4

Hence, p(x) has one real root in (0, 1)

6. Let us consider $p = P_r$ (head)

$q = P_r$ (tail) $= 1 - p$

Let X be a random variable denote number of tosses till to get one head then X = 1, 2, 3, ... is a discrete variable

So,

X = x	P(x)
1	p
2	qp
3	qqp
4	qqqp
⋮	⋮
∞	∞

$$\therefore \quad E(x) = \sum_{x=1}^{\infty} x.P(x)$$

$= 1 \times p + 2 \times qp + 3 \times q^2p + q^3p + ... \infty$

$= p \times [1 + 2q + 3q^2 + ...]$

$= p \times (1 - q)^{-2}$

$= p \times p^{-2} = \frac{1}{p}$

Hence, option (c) is the correct answer.

7. Let us consider A(–1, –1) = –1 – j; B(3, –1); D(3, 1) and E(–1, 1) be the vertices of rectangle.

$z^2(z - 4) = 0$

$\Rightarrow z = 0, 4$

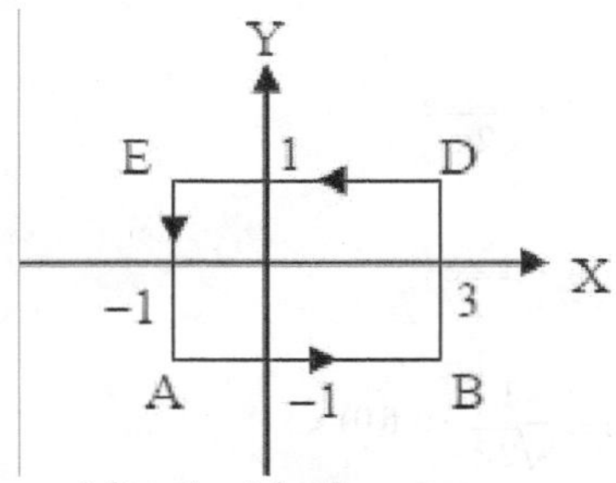

Only z = 0 lies inside the curve:

$$\therefore \oint_C \frac{dz}{z^2(z-4)} = \oint_C \frac{\left(\frac{1}{z-4}\right)}{(z-0)^2} dz$$

$$= \frac{2\pi i}{1!}\left(\frac{d}{dz}\left(\frac{1}{z-4}\right)\right)_{\text{at } z=0}$$

Now by using Cauchy's integral formula

$$= 2\pi i x\left(\frac{-1}{(z-4)^2}\right)_{\text{at } z=0} = \frac{-\pi i}{8} \text{ or } \frac{-j\pi}{8}$$

8. As given that $X(z) = \frac{z^2}{(z-a)^2} = \frac{1}{(1-az^{-1})^2}$

We know that, $\frac{1}{(1-az^{-1})^2} \leftrightarrow (n+1)a^n u(n)$

(standard z-transform)

9. As given that

$f(-t) = f(t) = f(t)$, At

$\frac{dF(\omega)}{d\omega} = -\omega F(\omega), \forall \omega$

$F(0) = 1$

$\frac{dF(\omega)}{F(\omega)} = -\omega d\omega$

$\Rightarrow \ell n F(\omega) = \frac{-\omega^2}{2} + C$

$\Rightarrow\ F(\omega) = e^{-\frac{\omega^2}{2}+C}$

$\Rightarrow\ F(0) = e^{0+C}$

$\Rightarrow\ 1 = e^{C}$

$\Rightarrow\ C = 0$

So $F(\omega) = e^{-\frac{\omega^2}{2}}$

As we know that,

$$e^{-at^2} \leftrightarrow \sqrt{\frac{\pi}{a}}\, e^{-\frac{\omega^2}{4a}}$$

If $a = \frac{1}{2}$ then $e^{-\frac{t_2}{2}} \leftrightarrow \sqrt{2\pi}\, e^{-\frac{\omega^2}{2}}$

$$e^{-\frac{\omega^2}{2}} \leftrightarrow \frac{1}{\sqrt{2\pi}} e^{-t^2/2}$$

So, $f(t) = \frac{1}{\sqrt{2\pi}} e^{-\frac{t^2}{2}}$

$f(0) = \frac{1}{\sqrt{2\pi}}$ so $f(0) < 1$

10. As given that

$A^5 = 0$ where,

A is Nilpotent matrix.

⇒ eigen values of A are all '0'

⇒ eigen values of the matrix A + I are all equal to 0 + 1 = 1

∴ Determinant of (A + I) is product of eigen values of (A + I)

= 1 × 1 × 1 .... × 1 (10 times)

= 1 (∵ By using properties of eigen values)

11. As given that,

Signal : CTFT

$$x(t) = \begin{cases} 1-|t|; & |t| \le 1 \\ 0; & |t| > 1 \end{cases}$$

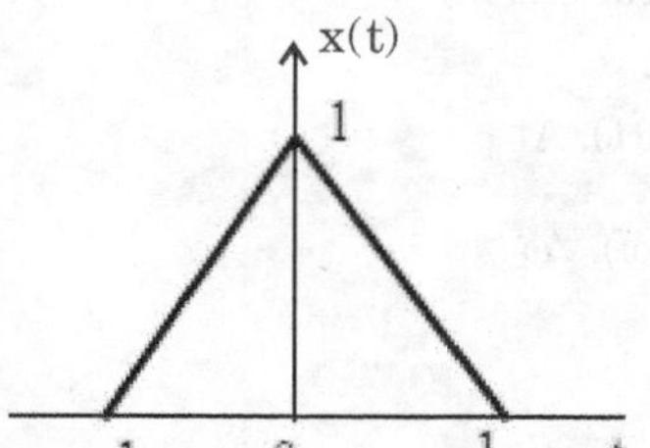

$X(\omega) = (\text{Area of triangle})\ X\ Sa^2\left(\omega\frac{T}{2}\right)$

$= \left(\frac{1}{2}\times 2\times 1\right) Sa^2\left(\frac{\omega}{2}\right) = Sa^2\left(\frac{\omega}{2}\right)\quad [\because T = 1]$

$\Rightarrow X(0) = Sa^2(0)$

$\Rightarrow X(0) = 1$

Hence, the maximum magnitude of x(W) is '1'.

SET-7

ANSWERS

1. (b)	**2.** (d)	**3.** (d)	**4.** (0.5 to 0.5)
5. (15 to 15)		**6.** (a)	**7.** (b)
8. (20.00 to 21.00)			

EXPLANATIONS

1. Let $A = \begin{bmatrix} 1 & 2 & 2 & 3 \\ 3 & 4 & 2 & 5 \\ 5 & 6 & 2 & 7 \\ 7 & 8 & 2 & 9 \end{bmatrix}$

Convert given matrix A into part row echelon form using element row transfer.

$R_2 \to R_2 - 3R_1$
$R_3 \to R_3 - 5R_1$ Then $A \sim \begin{bmatrix} 1 & 2 & 2 & 3 \\ 0 & -2 & -4 & -4 \\ 0 & -4 & -8 & -8 \\ 0 & -6 & -12 & -12 \end{bmatrix}$
$R_4 \to R_4 - 7R_1$

$R_3 \to R_3 - 2R_2$
$R_4 \to R_4 - 3R_2$ $A \sim \begin{bmatrix} 1 & 2 & 2 & 3 \\ 0 & -2 & -4 & -4 \\ 0 & 0 & 0 & 0 \\ 0 & 0 & 0 & 0 \end{bmatrix}$ is row

Echelon form

Hence, rank (A) is the number of non zero rows = 2

2. As we know that,

$(PQ)^T = Q^TP^T$

Where $Q^TP^T = (PQ)^T$

$= (QP)^T$

$Q^T = \begin{bmatrix} 0 & 1 \\ 1 & 0 \end{bmatrix};\ P^T = \begin{bmatrix} 1 & 3 \\ 2 & 4 \end{bmatrix}$

$Q^TP^T = \left[\begin{bmatrix} 0 & 1 \\ 1 & 0 \end{bmatrix}\begin{bmatrix} 1 & 3 \\ 2 & 4 \end{bmatrix}\right]^T = \begin{bmatrix} 2 & 4 \\ 1 & 3 \end{bmatrix}$

3. As we know that,

The Gaussian distribution also known as normal distribution is a bell-shaped curve. So, the cummulative gaussion distribution curve is S-shaped curve.

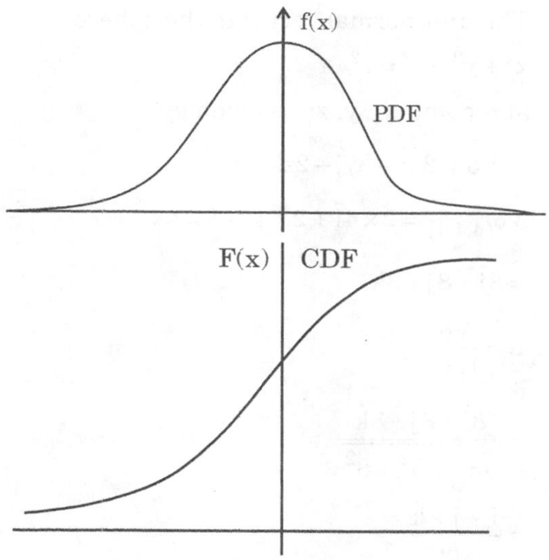

So, P.D.F: $f(x) = \frac{1}{\sigma\sqrt{2\pi}} e^{-(x-\mu)^2/(2\sigma^2)}$

C.D.F: $f(x) = \frac{1}{2}\left[1 + \text{erf}\left[\frac{(x-\mu)}{\sigma\sqrt{2}}\right]\right]$

4. Let, $l = \lim_{x\to1}\left(\frac{1}{\ell nx} - \frac{1}{x-1}\right) \to (\infty - \infty \text{ form})$

Then,

$$l = \lim_{x\to1}\left(\frac{x-1-\ell nx}{(x-1)\ell nx}\right) \to \left(\frac{0}{0}\text{form}\right)$$

$$l = \lim_{x\to1}\left(\frac{0+\frac{1}{x^2}}{(x-1)\left(-\frac{1}{x^2}\right)+\frac{1}{x}+\frac{1}{x}}\right) = \frac{+1}{0+1+1} = \frac{1}{2} = +0.5$$

$\boxed{\therefore l = +0.5}$

5. x limits : 2 to 3 ; y limits : 1 to 2 and z limits: 0 to 1

$$\therefore \text{Volume} = 8\int_{x=2}^{3}\int_{y=1}^{2}\int_{z=0}^{1} xyz\,dx\,dy\,dz$$

$$= 8\left(\int_{x=2}^{3} xdx\right)\left(\int_{y=1}^{2} ydy\right)\left(\int_{z=0}^{1} zdz\right)$$

$$= 8\left(\frac{x^2}{2}\right)_2^3 \times \left(\frac{y^2}{2}\right)_1^2 \times \left(\frac{z^2}{2}\right)_0^1$$

$$V = 8\left[\frac{9-4}{2}\right]\left[\frac{4-1}{2}\right]\left[\frac{1-0}{2}\right]$$

$= 5 \times 3 \times 1 = 15$

6. As given D.E is

$$\frac{d^2y}{dx^2} + \frac{2dy}{dx} + y = 0$$

A.E of given D.E is $m^2 + 2m + 1 = 0$

$\Rightarrow (m + 1)(m + 1) = 0$

$\Rightarrow m = -1, -1$

(two equal real roots), since R.H.S is 0

$\therefore$ D.E is homogenous equation

$\therefore$ Solution is y = C.F

$\Rightarrow y = (c_1 + c_2x)e^{-x}$ ...(1)

Using boundary condition y(0) =1 (i.e., y = 1, x = 0) and y (1) = 3 (i.e., y = 3, x = 1), equation(1) gives

$1 = c_1$ & $3 = (c_1 + c_2)e^{-1}$

$\Rightarrow 3e = 1 + c_2$

$\Rightarrow c_2 = 3e - 1$

Now, particular solution:

$\therefore y = (1 + (3e - 1)x)e^{-x}$

7. As given that, $f(x) = e^x$, a = 0, b = 1, n = 4

So,

$$h = \frac{b-a}{n} = \frac{1-0}{4} = 0.25$$

$y_0 = f(a) = e^0 = 1$; $y_1 = f(a + h) = e^{0.25} = 1.284$

$y_2 = f(a + 2h) = e^{0.5} = 1.648$; $y_3 = f(a + 3h) = e^{0.75} = 2.117$

$y_4(a + 4h) = e^1 = 2.718$

$\therefore$ By Trapezoidal rule, Numerical integration is:

$$\int_0^1 e^x dx = \frac{h}{2}[(y_0 + y_4) + 2(y_1 + y_2 + y_3)]$$

$$= \frac{1}{8}[(1 + 2.718) + 2(1.284 + 1.648 + 2.117)]$$

≈ 1.727

8. As given that,

$h = 0.5$, $y_0 = 5$, $y_1 = 7.25$; $y_2 = 10$, $y_3 = 13.25$, $y_4 = 17$

The formula of Simpson's $\frac{1}{3}$rd rule,

So, numerical integration of the area $= \int_2^4 ydx$

$$= \frac{h}{3}[(y_0 + y_4) + 4(y_1 + y_3) + 2(y_2)]$$

$$= \frac{0.5}{3}[(5 + 17) + 4(7.25 + 13.25) + 2(10)]$$

$$= \frac{0.5}{3}[22 + 82 + 20] = \frac{124}{6} = 20.66$$

SET-8

ANSWERS

1. (*a*) 2. (*c*) 3. (*b*) 4. (*a*)
5. (0.0 to 0.0) 6. (*a*) 7. (*a*)
8. (*1 to 1*) 9. (0.38 to 0.42)

EXPLANATIONS

1. $\lim_{x\to\infty}\left[\frac{x\times \ln x}{1+x^2}\right]=\left[\frac{\infty}{\infty}\text{form}\right]$ (Indeterminate form)

$$=\lim_{x\to\infty}\left[\frac{1\times\ln x+x\times\left(\frac{1}{x}\right)}{0+2x}\right]=\lim_{x\to\infty}\left[\frac{1\times \ln x}{2x}\right]$$

$$=\left[\text{Again}\left(\frac{\infty}{\infty}\right)\text{form}\right]$$

Apply 'L' Hospital rule again, we get,

$$=\lim_{x\to\infty}\left[\frac{0+\left(\frac{1}{x}\right)}{2}\right]$$

$$=\lim_{x\to\infty}\left(\frac{1}{2x}\right)=\left(\frac{1}{2\times\infty}\right)=0$$

$$\therefore \lim_{x\to\infty}\frac{x\ln(x)}{(1+x^2)}=0$$

2. As given that,

$$A=\begin{bmatrix}5&0&-5&0\\0&2&0&1\\-5&0&5&0\\0&1&0&2\end{bmatrix}$$

$$\sim\begin{bmatrix}5&0&-5&0\\0&2&0&1\\0&0&0&0\\0&1&0&2\end{bmatrix}(R_3\to R_1+R_3)$$

$$\sim\begin{bmatrix}5&0&-5&0\\0&2&0&1\\0&0&0&0\\0&0&0&1.5\end{bmatrix}\left(R_4\to R_4-\frac{R_2}{2}\right)$$

which is the form of Echelon matrix.

So, the rank of matrix is the number of non-zero

Hence, the rank of the matrix, i.e. $\rho(A) = 3$.

3. The unit normal vector to the sphere

$x^2 + y^2 + z^2 = r^2$

at a point (x_1, y, z_1) is given by

$\because \nabla\phi = 2x\hat{i}+2y\hat{j}+2z\hat{k}$

$\nabla\phi_{(4,4,4)} = 2\times 4\hat{i}+2\times 4\hat{j}+2\times 4\hat{k}$

$= 8\hat{i}+8\hat{j}+8\hat{k}$

$$\hat{n}=\frac{\nabla\phi}{|\nabla\phi|}$$

$$=\frac{8\hat{i}+8\hat{j}+8\hat{k}}{\sqrt{8^2+8^2+8^2}}$$

$$=\frac{\hat{i}+\hat{j}+\hat{k}}{\sqrt{3}}$$

Hence, option (b) is correct.

4. A square matrix A is called an orthogonal matrix if the product of matrix A with its transpose matrix

A^T is an identity matrix, i.e.,

$AA^T = A^TA = I.$

or $A^T = A^{-1}$

5. As given expression

$$\int_{-1}^{1} xe^{|x|}dx=\sum_{-1}^{0} xe^{-x}dx+\int_{0}^{1} xe^{x}dx$$

$$=\left[x\left(-e^{-x}\right)-\int\left(-e^{-x}\right)dx\right]_{-1}^{0}+\left[x\times e^{x}-\int e^{x}dx\right]_{0}^{1}$$

$$=\left[-e^{-x}(x+1)\right]_{-1}^{0}+\left[e^{x}(x-1)\right]_{0}^{1}$$

$$=\left[\begin{matrix}-e^{-0}(0+1)-\\(-e\times(-1+1))\end{matrix}\right]+\left[\begin{matrix}e(1-1)\\-e^{0}(0-1)\end{matrix}\right]$$

Here, $f(x) = xe^{|x|}$

$f(-x) = -xe^{|-x|} = -xe^{|x|} = -f(x)$

So, the given function is odd.

$$\therefore \int_{-1}^{1} f(x)dx=\int xe^{|x|}dx=0$$

6. As given that,

$$\frac{dy}{dx}-\frac{y}{x}=1$$

Compaire D. E is in the form of $\frac{dy}{dx}+Py=Q$

$$\Rightarrow \frac{dy}{dx}+\underbrace{\left(\frac{-1}{x}\right)}_{P}y=\underbrace{1}_{Q} \qquad \text{...(i)}$$

∴ Integrating factor (I.F.)

$$= e^{\int P dx} = e^{\int -\frac{1}{x} dx}$$

$$= e^{-1 \times \ln x}$$

$$= x^{-1} = \left(\frac{1}{x}\right)$$

Solution of the D. E is given by:

$$y \times (\text{I.F.}) = \int Q(\text{I.F.})dx + C$$

$$\Rightarrow y \times \left(\frac{1}{x}\right) = \int 1 \times \frac{1}{x} dx + C$$

$$\Rightarrow y \times \left(\frac{1}{x}\right) = \ln x + \ln k = \ln(kx)$$

$$\Rightarrow y = x\ln(kx)$$

7. As we know that,

The question the characteristic equation

$$|A - \lambda| = 0 \Rightarrow \begin{vmatrix} 2-\lambda & -2 \\ -1 & 6-\lambda \end{vmatrix} = 0$$

$$\Rightarrow (2-\lambda)(6-\lambda) - 2 = 0$$

$$\Rightarrow 12 - 2\lambda - 6\lambda + \lambda^2 - 2 = 0$$

$$\lambda^2 - 8\lambda + 10 = 0$$

$$\lambda = 4 \pm \sqrt{6}$$

Smallest eigen value = $4 - \sqrt{6} = 1.55$

$$\begin{bmatrix} 2-\lambda & -2 \\ -1 & 6-\lambda \end{bmatrix}\begin{bmatrix} x_1 \\ x_2 \end{bmatrix} = \begin{bmatrix} 0 \\ 0 \end{bmatrix}$$

$$\begin{bmatrix} 2-(4-\sqrt{6}) & -2 \\ -1 & 6-(4-\sqrt{6}) \end{bmatrix}\begin{bmatrix} x_1 \\ x_2 \end{bmatrix} = \begin{bmatrix} 0 \\ 0 \end{bmatrix}$$

$$\begin{bmatrix} -2+\sqrt{6} & -2 \\ -1 & 2+\sqrt{6} \end{bmatrix}\begin{bmatrix} x_1 \\ x_2 \end{bmatrix} = \begin{bmatrix} 0 \\ 0 \end{bmatrix}$$

$$-2x_1 + \sqrt{6}x_1 - 2x_2 = 0$$

$$\left(-2+\sqrt{6}\right)x_1 = 2x_2$$

$$x_1 = \frac{2x_2}{-2+\sqrt{6}}$$

$$\frac{x_1}{2} = \frac{x_2}{-2+\sqrt{6}}$$

$$\frac{x_1}{2} = \frac{x_2}{0.45}$$

Hence, $\begin{bmatrix} x_1 \\ x_2 \end{bmatrix} = \begin{bmatrix} 2 \\ 0.45 \end{bmatrix}$

8. As given that,

$f(x, y) = xe^y$

$\nabla\phi = \nabla f$

$$\nabla\phi = \frac{\partial f}{\partial x}\hat{i} + \frac{\partial f}{\partial y}\hat{j} + \frac{\partial f}{\partial z}\hat{k}$$

$$\nabla\phi = e^y\hat{i} + xe^y\hat{j} + 0\hat{k}$$

$$\nabla\phi_{(2,0)} = \hat{i} + 2\hat{j}$$

$$\vec{a} = (x_2 - x_1)\hat{i} + (y_2 - y_1)\hat{j}$$

$$= -\frac{3}{2}\hat{i} + 2\hat{j}$$

$$\hat{n} = \frac{\vec{a}}{|\vec{a}|} = \frac{-\frac{3}{2}\hat{i} + 2\hat{j}}{\sqrt{\left(-\frac{3}{2}\right)^2 + (2)^2}}$$

$$= \frac{2}{5}\left(-\frac{3}{2}\hat{i} + 2\hat{i}\right)$$

$$= -\frac{3\hat{i}}{5} + \frac{4}{5}\hat{j}$$

$$DD = \nabla f . \hat{n}$$

$$= -\frac{3}{5} + 2 \times \frac{4}{5} = 1$$

9. As given that,

$a = 0, b = 0.5, n = 5$

As we know that,

$$h = \left(\frac{b-a}{n}\right)$$

$$= \frac{0.5-0}{5} = 0.1$$

$x_0 = 0, y_0 = 0$

$x_1 = 0.1, y_1 = 0.8$

$x_2 = 0.2, y_2 = 1.2$

$x_3 = 0.3, y_3 = 1.2$

$x_4 = 0.4, y_4 = 0.8$

$x_5 = 0.5, y_5 = 0$

So, Numerical intergration

$$= \frac{h}{2}[(y_0 + y_5) + 2(y_1 + y_2 + y_3 + y_4)]$$

$$= \frac{0.1}{2}[(0 + 0) + 2(0.8 + 1.2 + 1.2 + 0.8)]$$

$$= \frac{0.1 \times 8}{2} = 0.4$$